T0189810

Lecture Notes in Computer Science   10210

Commenced Publication in 1973
Founding and Former Series Editors:
Gerhard Goos, Juris Hartmanis, and Jan van Leeuwen

More information about this series at http://www.springer.com/series/7410

Jean-Sébastien Coron · Jesper Buus Nielsen (Eds.)

Advances in Cryptology – EUROCRYPT 2017

36th Annual International Conference on the Theory
and Applications of Cryptographic Techniques
Paris, France, April 30 – May 4, 2017
Proceedings, Part I

 Springer

Editors
Jean-Sébastien Coron
University of Luxembourg
Luxembourg
Luxembourg

Jesper Buus Nielsen
Aarhus University
Aarhus
Denmark

ISSN 0302-9743 ISSN 1611-3349 (electronic)
Lecture Notes in Computer Science
ISBN 978-3-319-56619-1 ISBN 978-3-319-56620-7 (eBook)
DOI 10.1007/978-3-319-56620-7

Library of Congress Control Number: 2017936355

LNCS Sublibrary: SL4 – Security and Cryptology

Printed on acid-free paper

This Springer imprint is published by Springer Nature
The registered company is Springer International Publishing AG
The registered company address is: Gewerbestrasse 11, 6330 Cham, Switzerland

Preface

Eurocrypt 2017, the 36th annual International Conference on the Theory and Applications of Cryptographic Techniques, was held in Paris, France, from April 30 to May 4, 2017. The conference was sponsored by the International Association for Cryptologic Research (IACR). Michel Abdalla (ENS, France) was responsible for the local organization. He was supported by a local organizing team consisting of David Pointcheval (ENS, France), Emmanuel Prouff (Morpho, France), Fabrice Benhamouda (ENS, France), Pierre-Alain Dupoint (ENS, France), and Tancrède Lepoint (SRI International). We are indebted to them for their support and smooth collaboration.

The conference program followed the now established parallel track system where the works of the authors were presented in two concurrently running tracks. Only the invited talks spanned over both tracks.

We received a total of 264 submissions. Each submission was anonymized for the reviewing process and was assigned to at least three of the 56 Program Committee members. Submissions co-authored by committee members were assigned to at least four members. Committee members were allowed to submit at most one paper, or two if both were co-authored. The reviewing process included a first-round notification followed by a rebuttal for papers that made it to the second round. After extensive deliberations the Program Committee accepted 67 papers. The revised versions of these papers are included in these three-volume proceedings, organized topically within their respective track.

The committee decided to give the Best Paper Award to the paper "Scrypt Is Maximally Memory-Hard" by Joël Alwen, Binyi Chen, Krzysztof Pietrzak, Leonid Reyzin, and Stefano Tessaro. The two runners-up to the award, "Computation of a 768-bit Prime Field Discrete Logarithm," by Thorsten Kleinjung, Claus Diem, Arjen K. Lenstra, Christine Priplata, and Colin Stahlke, and "Short Stickelberger Class Relations and Application to Ideal-SVP," by Ronald Cramer, Léo Ducas, and Benjamin Wesolowski, received honorable mentions. All three papers received invitations for the *Journal of Cryptology*.

The program also included invited talks by Gilles Barthe, titled "Automated Proof for Cryptography," and by Nigel Smart, titled "Living Between the Ideal and Real Worlds."

We would like to thank all the authors who submitted papers. We know that the Program Committee's decisions, especially rejections of very good papers that did not find a slot in the sparse number of accepted papers, can be very disappointing. We sincerely hope that your works eventually get the attention they deserve.

We are also indebted to the Program Committee members and all external reviewers for their voluntary work, especially since the newly established and unified page limits and the increasing number of submissions induce quite a workload. It has been an honor to work with everyone. The committee's work was tremendously simplified by Shai Halevi's submission software and his support, including running the service on IACR servers.

Finally, we thank everyone else —speakers, session chairs, and rump session chairs — for their contribution to the program of Eurocrypt 2017. We would also like to thank Thales, NXP, Huawei, Microsoft Research, Rambus, ANSSI, IBM, Orange, Safran, Oberthur Technologies, CryptoExperts, and CEA Tech for their generous support.

May 2017   Jean-Sébastien Coron
                                                              Jesper Buus Nielsen

Eurocrypt 2017

The 36th Annual International Conference on the Theory and Applications of Cryptographic Techniques

Sponsored by *the International Association for Cryptologic Research*

30 April – 4 May 2017
Paris, France

General Chair

Michel Abdalla · ENS, France

Program Co-chairs

Jean-Sébastien Coron · University of Luxembourg
Jesper Buus Nielsen · Aarhus University, Denmark

Program Committee

Gilad Asharov · Cornell Tech, USA
Nuttapong Attrapadung · AIST, Japan
Fabrice Benhamouda · ENS, France and IBM, USA
Nir Bitansky · MIT, USA
Andrey Bogdanov · Technical University of Denmark
Alexandra Boldyreva · Georgia Institute of Technology, USA
Chris Brzuska · Technische Universität Hamburg, Germany
Melissa Chase · Microsoft, USA
Itai Dinur · Ben-Gurion University, Israel
Léo Ducas · CWI, Amsterdam, The Netherlands
Stefan Dziembowski · University of Warsaw, Poland
Nicolas Gama · Inpher, Switzerland and University of Versailles, France
Pierrick Gaudry · CNRS, France
Peter Gaži · IST Austria, Austria
Niv Gilboa · Ben-Gurion University, Israel
Robert Granger · EPFL, Switzerland
Nathan Keller · Bar Ilan University, Israel
Aggelos Kiayias · University of Edinburgh, UK
Eike Kiltz · Ruhr-Universität Bochum, Germany

Vladimir Kolesnikov   Bell Labs, USA
Ranjit Kumaresan   MIT, USA
Eyal Kushilevitz   Technion, Israel
Gregor Leander   Ruhr-University Bochum, Germany
Tancrède Lepoint   SRI International, USA
Benoît Libert   ENS de Lyon, France
San Ling   Nanyang Technological University, Singapore
Anna Lysyanskaya   Brown University, USA
Tal Malkin   Columbia University, USA
Willi Meier   FHNW, Switzerland
Florian Mendel   Graz University of Technology, Austria
Bart Mennink   K.U. Leuven, Belgium
Ilya Mironov   Google, USA
María Naya-Plasencia   Inria, France
Ivica Nikolić   Nanyang Technological University, Singapore
Miyako Ohkubo   NICT, Japan
Rafail Ostrovsky   UCLA, USA
Omkant Pandey   Stony Brook University, USA
Omer Paneth   Boston University, USA
Chris Peikert   University of Michigan, USA
Thomas Peters   UCL, Belgium
Krzysztof Pietrzak   IST Austria, Austria
Emmanuel Prouff   Morpho, France
Leonid Reyzin   Boston University, USA
Louis Salvail   University of Montreal, Canada
Yu Sasaki   NTT Secure Platform Laboratories, Japan
Abhi Shelat   University of Virginia, USA
Elaine Shi   Cornell University, USA
Martijn Stam   University of Bristol, UK
Damien Stehlé   ENS de Lyon, France
John P. Steinberger   Tsinghua University, China
Ingrid Verbauwhede   K.U. Leuven, Belgium
Brent Waters   University of Texas, USA
Daniel Wichs   Northeastern University, USA
Mark Zhandry   Princeton University, USA

Additional Reviewers

Michel Abdalla   Martin Albrecht   Daniel Apon
Masayuki Abe   Ghada Almashaqbeh   Benny Applebaum
Aysajan Abidin   Jacob Alperin-Sheriff   Christian Badertscher
Hamza Abusalah   Joël Alwen   Saikrishna
Divesh Aggarwal   Abdelrahaman Aly   Badrinarayanan
Shashank Agrawal   Elena Andreeva   Shi Bai
Navid Alamati   Yoshinori Aono   Josep Balasch

Foteini Baldimtsi
Marshall Ball
Valentina Banciu
Subhadeep Banik
Razvan Barbulescu
Guy Barwell
Carsten Baum
Anja Becker
Christof Beierle
Amos Beimel
Sonia Belaïd
Shalev Ben-David
Iddo Bentov
Jean-François Biasse
Begul Bilgin
Olivier Blazy
Xavier Bonnetain
Joppe Bos
Christina Boura
Florian Bourse
Luis Brandao
Dan Brownstein
Chris Campbell
Ran Canetti
Anne Canteaut
Angelo De Caro
Ignacio Cascudo
David Cash
Wouter Castryck
Hubert Chan
Nishanth Chandran
Jie Chen
Yilei Chen
Nathan Chenette
Mahdi Cheraghchi
Alessandro Chiesa
Ilaria Chillotti
Sherman S.M. Chow
Kai-Min Chung
Michele Ciampi
Ran Cohen
Craig Costello
Alain Couvreur
Claude Crépeau
Edouard Cuvelier
Guillaume Dabosville

Ivan Damgård
Jean Paul Degabriele
Akshay Degwekar
David Derler
Apoorvaa Deshpande
Julien Devigne
Christoph Dobraunig
Frédéric Dupuis
Nico Döttling
Maria Eichlseder
Keita Emura
Xiong Fan
Pooya Farshim
Sebastian Faust
Omar Fawzi
Dario Fiore
Ben Fisch
Benjamin A. Fisch
Nils Fleischhacker
Georg Fuchsbauer
Eiichiro Fujisaki
Steven Galbraith
Chaya Ganesh
Juan Garay
Sumegha Garg
Romain Gay
Ran Gelles
Mariya Georgieva
Benedikt Gierlichs
Oliver W. Gnilke
Faruk Göloğlu
Sergey Gorbunov
Dov Gordon
Rishab Goyal
Hannes Gross
Vincent Grosso
Jens Groth
Daniel Gruss
Jian Guo
Siyao Guo
Qian Guo
Benoît Gérard
Felix Günther
Britta Hale
Carmit Hazay
Felix Heuer

Shoichi Hirose
Viet Tung Hoang
Justin Holmgren
Fumitaka Hoshino
Pavel Hubáček
Ilia Iliashenko
Laurent Imbert
Takanori Isobe
Tetsu Iwata
Malika Izabachene
Kimmo Jarvinen
Eliane Jaulmes
Dimitar Jetchev
Daniel Jost
Marc Joye
Herve Kalachi
Seny Kamara
Chethan Kamath
Angshuman Karmakar
Pierre Karpman
Nikolaos Karvelas
Marcel Keller
Elena Kirshanova
Fuyuki Kitagawa
Susumu Kiyoshima
Thorsten Kleinjung
Lars Knudsen
Konrad Kohbrok
Markulf Kohlweiss
Ilan Komargodski
Venkata Koppula
Thomas Korak
Lucas Kowalczyk
Thorsten Kranz
Fabien Laguillaumie
Kim Laine
Virginie Lallemand
Adeline Langlois
Hyung Tae Lee
Jooyoung Lee
Kwangsu Lee
Troy Lee
Kevin Lewi
Huijia (Rachel) Lin
Jiao Lin
Wei-Kai Lin

Feng-Hao Liu
Atul Luykx
Vadim Lyubashevsky
Xiongfeng Ma
Houssem Maghrebi
Mohammad Mahmoody
Daniel Malinowski
Alex Malozemoff
Antonio Marcedone
Daniel P. Martin
Daniel Masny
Takahiro Matsuda
Christian Matt
Alexander May
Sogol Mazaheri
Peihan Miao
Kazuhiko Minematsu
Ameer Mohammed
Tal Moran
Fabrice Mouhartem
Pratyay Mukherjee
Elke De Mulder
Pierrick Méaux
Michael Naehrig
Yusuke Naito
Kashif Nawaz
Kartik Nayak
Khoa Nguyen
Ryo Nishimaki
Olya Ohrimenko
Elisabeth Oswald
Ayoub Otmani
Giorgos Panagiotakos
Alain Passelègue
Kenneth G. Paterson
Serdar Pehlivanoglou
Alice Pellet–Mary
Pino Persiano
Cécile Pierrot
Rafaël Del Pino
Bertram Poettering
David Pointcheval
Antigoni Polychroniadou

Romain Poussier
Thomas Prest
Erick Purwanto
Carla Rafols
Ananth Raghunathan
Srinivasan Raghuraman
Sebastian Ramacher
Somindu Ramanna
Francesco Regazzoni
Ling Ren
Oscar Reparaz
Silas Richelson
Thomas Ricosset
Thomas Ristenpart
Florentin Rochet
Mike Rosulek
Yannis Rouselakis
Sujoy Sinha Roy
Michal Rybár
Carla Ràfols
Robert Schilling
Jacob Schuldt
Nicolas Sendrier
Yannick Seurin
Ido Shahaf
Sina Shiehian
Siang Meng Sim
Dave Singelee
Luisa Siniscalchi
Daniel Slamanig
Benjamin Smith
Akshayaram Srinivasan
François-Xavier Standaert
Ron Steinfeld
Noah
    Stephens-Davidowitz
Katerina Stouka
Koutarou Suzuki
Alan Szepieniec
Björn Tackmann
Stefano Tessaro
Adrian Thillard
Emmanuel Thomé

Mehdi Tibouchi
Elmar Tischhauser
Yosuke Todo
Ni Trieu
Roberto Trifiletti
Yiannis Tselekounis
Furkan Turan
Thomas Unterluggauer
Margarita Vald
Prashant Vasudevan
Philip Vejre
Srinivas Vivek Venkatesh
Daniele Venturi
Frederik Vercauteren
Ivan Visconti
Vanessa Vitse
Damian Vizár
Petros Wallden
Michael Walter
Lei Wang
Huaxiong Wang
Mor Weiss
Weiqiang Wen
Mario Werner
Benjamin Wesolowski
Carolyn Whitnall
Friedrich Wiemer
David Wu
Keita Xagawa
Sophia Yakoubov
Shota Yamada
Takashi Yamakawa
Avishay Yanay
Kan Yasuda
Eylon Yogev
Kazuki Yoneyama
Henry Yuen
Thomas Zacharias
Karol Zebrowski
Rina Zeitoun
Bingsheng Zhang
Ryan Zhou
Dionysis Zindros

Advances in Computer-Aided Cryptography
(Invited Talk)

Gilles Barthe

IMDEA Software Institute, Madrid, Spain

Designing, analyzing and implementing correct, secure and efficient cryptography are challenging tasks. Computer-aided cryptography is a young field of research which aims to provide rigorous tools that ease these tasks. Computer-aided cryptography leverages advances in the broad area of formal methods, concerned with the development of safe and correct high-assurance systems, and in particular program verification. For security proofs, computer-aided cryptography exploits connections between reductionist arguments in provable security and a program verification method for verifying probabilistic couplings. To date, computer-aided cryptography has been used for checking reductionistic security of primitives and protocols, for analyzing the strength of implementations against side channels and physical attacks, and for synthesizing new algorithms that achieve different trade-offs between efficiency and security. The talk will present recent developments in computer-aided cryptography and reflect on some of the challenges, benefits and opportunities in computer-aided cryptography.

Contents – Part I

Functional Encryption I

Elliptic Curves

Contents – Part II

Blockchain

Blockchain

Contents – Part III

Lattice Attacks and Constructions I

Lattice Attacks and Constructions I

Revisiting Lattice Attacks on Overstretched NTRU Parameters

Paul Kirchner[1,2] and Pierre-Alain Fouque[2,3(✉)]

[1] École Normale Supérieure, Paris, France
pkirchner@clipper.ens.fr
[2] IRISA, Rennes, France
[3] Université de Rennes 1 & Institut Universitaire de France, Paris, France
pierre-alain.fouque@univ-rennes1.fr

Abstract. In 2016, Albrecht, Bai and Ducas and independently Cheon, Jeong and Lee presented very similar attacks to break the NTRU cryptosystem with larger modulus than in the NTRUEncrypt standard. They allow to recover the secret key given the public key of Fully Homomorphic Encryption schemes based on NTRU ideas. Hopefully, these attacks do not endanger the security of the NTRUEncrypt, but shed new light on the hardness of the NTRU problem. The idea consists in decreasing the dimension of the NTRU lattice using the multiplication matrix by the norm (resp. trace) of the public key in some subfield instead of the public key itself. Since the dimension of the subfield is smaller, so is the dimension of the lattice and better lattice reduction algorithms perform.

In this paper, we first propose a new variant of the subfield attacks that outperforms both of these attacks in practice. It allows to break several concrete instances of YASHE, a NTRU-based FHE scheme, but it is not as efficient as the hybrid method on smaller concrete parameters of NTRUEncrypt. Instead of using the norm and trace, the multiplication by the public key in a subring allows to break smaller parameters and we show that in $\mathbb{Q}(\zeta_{2^n})$, the time complexity is polynomial for $q = 2^{\Omega(\sqrt{n \log \log n})}$. Then, we revisit the lattice reduction part of the hybrid attack of Howgrave-Graham and analyze the success probability of this attack using a new technical tool proposed by Pataki and Tural. We show that, under some heuristics, this attack is more efficient than the subfield attack and works in any ring for large q, such as the NTRU Prime ring. We insist that the improvement on the analysis applies even for relatively small modulus; although if the secret is sparse, it may not be the fastest attack. We also derive a tight estimation of security for (Ring-) LWE and NTRU assumptions and perform many practical experiments.

1 Introduction

NTRU has been introduced by Hoffstein, Pipher and Silverman since 1996 in [26] and has since resisted many attacks [13,21,22,27]. NTRU is one of the most attractive lattice-based cryptosystems since it is very efficient, and many

© International Association for Cryptologic Research 2017
J.-S. Coron and J.B. Nielsen (Eds.): EUROCRYPT 2017, Part I, LNCS 10210, pp. 3–26, 2017.
DOI: 10.1007/978-3-319-56620-7_1

Ring-LWE cryptosystems have a NTRU variant. Ducas, Lyubashevsky and Prest propose an Identity Based Encryption scheme based on NTRU [20] (albeit with a much larger standard deviation), López-Alt, Tromer and Vaikuntanathan describe a Fully Homomorphic Encryption scheme [32], which is improved in a scheme called YASHE [6,31], and Ducas *et al.* propose a very fast signature scheme called BLISS [19].

Currently, the most efficient and heuristic attack on NTRU has been given by Kirchner and Fouque in [29] which has subexponential-time complexity in $2^{(n/2+o(n))/\log\log q}$, but the $o(n)$ is too large to lead to attack for given parameters. To date, the most efficient attack on practical NTRU parameters is the so-called hybrid attack described by Howgrave-Graham in [27].

The key recovery problem of NTRU is the following problem: given a public key $\mathbf{h} = \mathbf{f}/\mathbf{g}$ in some polynomial ring $R_q = \mathbb{Z}_q[X]/(X^n - 1)$ for n prime, q a small integer and the euclidean norms of $\mathbf{f}, \mathbf{g}$ are small, recover $\mathbf{f}$ and $\mathbf{g}$ or a small multiple of them. In NTRUEncrypt, $\mathbf{f}$ and $\mathbf{g}$ are two sparse polynomials of degrees strictly smaller than n and coefficients $\{-1, 0, 1\}$. It is easy to see that the public key cannot be uniformly distributed in the whole ring, since the entropy is too small. In [42], Stehlé and Steinfeld, show that if $\mathbf{f}$ and $\mathbf{g}$ are generated using a Gaussian distribution of standard deviation $\sigma \approx q^{1/2}$, then the distribution of the public key is statistically indistinguishable from the uniform distribution.

State-of-the-Art Lattice Algorithm on NTRU. In [13], Coppersmith and Shamir show that the $(2n)$-dimensional lattice L^{cs} generated by the columns of the matrix

$$\begin{pmatrix} q\mathbf{I}_n & \mathbf{M}_{\mathbf{h}}^{R_q} \\ \mathbf{0} & \mathbf{I}_n \end{pmatrix},$$

where $\mathbf{M}_{\mathbf{h}}^{R_q}$ denotes the multiplication by the public key $\mathbf{h}$ in the ring R_q, contains the vector $(\mathbf{f}, \bar{\mathbf{g}})$. It is easy to show that for $\bar{\mathbf{g}} = \mathbf{g}(1/X)$ in R_q, we have $\mathbf{h} \cdot \bar{\mathbf{g}} = \mathbf{f}$. By reducing this lattice, it is possible to find $(\mathbf{f}, \bar{\mathbf{g}})$ which is short if $(\mathbf{f}, \mathbf{g})$ is. Finally, Coppersmith and Shamir show that for cryptographic purposes, it is sufficient to recover a small solution, maybe not the smallest one to decrypt.

In 2001, May showed in [33] how to exploit that the shifts of the target vector, i.e. $\mathbf{x}^i \cdot \mathbf{f}$ in R_q are also contained in the L^{cs} lattice. Consequently, we only have to recover one of the n shifts of the target vector and the smallest vector is not unique. The idea of May consists in constructing a lattice that contains as a short vector only one of the shift and such that the gap between the first and second minima of the lattice will be higher. This gap is an important parameter when running lattice reduction algorithm. If we take into account that the vector of the secret key contains $\{0, \pm1\}$-coefficients, there is a unique long run of 0-coefficients. For one of the n shifts, we can assume that this run is for instance in the first r coefficients and if we multiply the $(n+1)$th to $(n+r)$th columns of L^{cs} matrix by a suitable large constant, only this shift will be a solution for the new lattice. He also introduces the *projection* technique to reduce the dimension of L^{cs} from $2n$ to $(1 + \alpha)n$ for $0 < \alpha \leq 1$ by removing the last columns of the matrix $\mathbf{M}_{\mathbf{h}}^{\mathcal{O}_{\mathbb{K}}}$ or of the last rows of the original matrix. The main idea is that it suffices that among the n equations corresponding to $\mathbf{h} \cdot \bar{\mathbf{g}} = \mathbf{f}$, some of them will not be fulfilled. Experimentally, since there is no other small vector except the n shifts, then there will be no other vector with small entries in these coefficients and we will recover the target.

In [27], Howgrave-Graham makes various experiments on NTRU lattice and proposes a mix between lattice reduction and a combinatorial technique, known as Odlyzko's meet-in-the-middle attack on NTRU. The first phase of the algorithm starts by reducing the original matrix corresponding to L^{cs} and we can see that lattice algorithms first reduce the column vectors in the middle of the matrix. This process that treats the columns in a symmetric manner between $[n - r, n + r]$ is also used in [21] in the symplectic reduction. Consequently, it is more efficient to begin by reducing a small dimensional matrix in the center of the original Coppersmith-Shamir matrix and then another combinatorial technique can take into account the small coefficients in the short vector by guessing some part of the secret key. In the following, we will speak of the *middle* technique.

More recently, in [1,12], Cheon, Jeong and Lee at ANTS 2016 and Albrecht, Bai and Ducas at CRYPTO 2016, described a new attack on NTRU-like cryptosystems. An attack based on similar ideas was proposed by Jonsson, Nguyen and Stern in [23, Sect. 6]. It uses the fact that for cyclotomic number fields, there exist subfields that allow to reduce the dimension of the lattice. The *subfield* attack recovers the norm of the secret key in these subfields, which are smaller than in the classical NTRU lattice. In the maximal real subfield $\mathbb{K}^+$ of a power of two cyclotomic field $\mathbb{K}$ for instance, the norm can be written as $N_{\mathbb{K}/\mathbb{K}^+}(\mathbf{f}) = \mathbf{f}\bar{\mathbf{f}}$ which is small if $\mathbf{f}$ is small and $N_{\mathbb{K}/\mathbb{K}^+}(\mathbf{f})$ is of dimension half. The lattice L^{norm} is generated by the columns of the matrix of dimension n:

$$\begin{pmatrix} q\mathbf{I}_{n/2} & \mathbf{M}_{\mathbf{h}\bar{\mathbf{h}}}^{\mathcal{O}_{\mathbb{K}^+}} \\ \mathbf{0} & \mathbf{I}_{n/2} \end{pmatrix}.$$

The vector $(N_{\mathbb{K}/\mathbb{K}^+}(\mathbf{f}), N_{\mathbb{K}/\mathbb{K}^+}(\mathbf{g}))$ is small in L^{norm}. By the Gaussian heuristic, the expected length of the shortest vector in the lattice L^{norm} is $\sqrt{qn/(2\pi e)}$, and the norm of $\mathbf{f}$ depends on the density of non-zero coefficients is of size around n. For standard NTRU parameters and when n is greater than q, lattice reduction algorithms will not recover the secret key. However, if q is large as in the case of FHE cryptosystems to allow a large number of multiplication steps before boostraping, then this attack can be interesting. We have not been able to apply it for other cryptosystems, for instance on IBE and signature schemes [20].

The drawback of this technique is that q has to be very large compared to n. We estimate asymptotically $q = 2^{\Omega(\sqrt{n \log \log n})}$ for a polynomial time complexity.

Our Results. In this paper, we revisit the lattice attacks on NTRU by considering the subfield idea, the projection of May and the middle lattice of Howgrave-Graham in the context of large modulus.

1. We first propose a new *subfield* attack and give, contrary to [1,12], a precise analysis by considering the *projection* technique for power of two cyclotomic

fields. We show that using the multiplication matrix by the public key in a subring (which has the same size as the subfield), leads to more efficient attacks. In particular, we were able to attack concrete parameters proposed in YASHE based on overstretched NTRU [6,7,10,14–16,30,31], meaning that we can recover a decryption key for smaller modulus q, compared to the previous approaches [1,12]. The previous attacks use the norm over the subfield in [1] or the trace in [12]. It would also be possible for instance to use all the coefficients of the characteristic polynomial. Our attack using the subring is better than the two previous ones since in the same configuration, we can choose exactly the size of the subfield as the number of coordinates (remove some rows or *project* the lattice) in Sect. 3.

2. Secondly, we analysis lattice reduction algorithm on the full L^{cs} lattice using a nice lemma due to Petaki and Tural [38] on the volume of sublattices with high rank (Sect. 4). We show that reducing this lattice allows us to achieve similar performances as in the projection and subfield attacks. We do not rely in our analysis on the Hermite factor (or approximate factor). This is the first time that the high number of small lattice vectors (shifts) are used to improve the analysis of the attack against NTRU. May used it to run lattice reduction on smaller dimensional lattices. The high dimensional low volume sublattice (formed by the shift vectors) makes the approximate-SVP problem for NTRU lattices substantially easier to solve by lattice reduction than generic lattices of the same dimension when the modulus is sufficiently large. This result is true in any ring and can be applied for instance on NTRUPrime with large q. In practice, we run experiment using the *middle* technique in order to use small dimension lattices.

3. We make experiments (Sect. 5) to understand the behaviour of lattice reduction algorithm and derive precise predictions when this attack will work (Sect. 6). We show that also experimentally the subfield attack is not more efficient than the *middle* technique on the original matrix. Consequently, we mount this attack to break FHE with NTRU and overstretched NTRU Prime scheme. Experimental computations show that if we are able to reduce this matrix, we recover a basis consisting of n small vectors, which are rotated version of the secret key. Finally, we provide a tight asymptotical security estimate of NTRU and LWE schemes in order to give exact predictions for these attacks by considering the Dual-BKZ [37].

We want to stress that the subfield attack we propose is not needed to break the schemes. We first discovered our subfield attack and the experiments shown in Fig. 1 have been obtained using it. The experiments on NTRUPrime with overstretched parameters (Fig. 2) have been achieved by reducing the middle lattice in the standard lattice. We experimentally recovered the same results for Fig. 1 using the middle lattice later and we conclude that the subfield attack is not needed to improve results on NTRU, but it could be useful to attack multilinear maps [1,12].

2 Preliminaries

Algebraic Number Field. An *algebraic number field* (or simply number field) K is a finite (algebraic) field extension of the field of rational numbers $\mathbb{Q}$. An *algebraic number* $\zeta \in \mathbb{C}$ is a root of a polynomial $f(x) \in \mathbb{Q}[x]$ and is called an *algebraic integer* if $f(x)$ is a monic (leading coefficient is 1), polynomial in $\mathbb{Z}[x]$. The *minimal polynomial* of ζ is a monic polynomial $f(x) \in \mathbb{Q}[x]$ of least positive degree such that $f(\zeta) = 0$ and the minimal polynomial of an algebraic integer is in $\mathbb{Z}[x]$. The set of all algebraic integers form a ring: the sum and product of two algebraic integers is an algebraic integer. The *ring of integers of a number field* $K = \mathbb{Q}[\zeta]$, obtained by adjoining ζ to $\mathbb{Q}$, is the ring $\mathcal{O}_K = \{\mathbf{x} \in K : \mathbf{x} \text{ is an algebraic integer}\}$. Let $f(x)$ be the minimal polynomial of ζ of degree n, then as $f(\zeta) = 0$, there is an isomorphism between $\mathbb{Q}[x] \bmod f(x)$ and K, defined by $x \mapsto \zeta$ and K can be seen as an n-dimensional vector space over $\mathbb{Q}$ with *power basis* $\{1, \zeta, \ldots, \zeta^{n-1}\}$. The *conjugates* of ζ are defined as all the roots of its minimal polynomial.

A number field $K = \mathbb{Q}[\zeta]$ of degree n has exactly n field homomorphisms $\sigma_i : K \hookrightarrow \mathbb{C}$ that fix every element of $\mathbb{Q}$ and they map ζ to each of its conjugates. An embedding whose image lies in $\mathbb{R}$ (real root of $f(x)$) is called a *real embedding*; otherwise it is called a *complex embedding*. Since complex root of $f(x)$ come in pairs, so do complex embeddings. The number of real ones is denoted s_1 and the number of pairs of complex ones s_2, so we get $n = s_1 + 2s_2$. By convention, we let $\{\sigma_j\}_{j \in [s_1]}$ be the real embedding and order the complex embeddings so that $\sigma_{s_1+s_2+j} = \overline{\sigma_{s_1+j}}$ for $j \in [s_2]$. The *canonical embedding* $\sigma : K \to \mathbb{R}^{s_1} \times \mathbb{C}^{2s_2}$ is defined by

$$\sigma(\mathbf{x}) = (\sigma_1(\mathbf{x}), \ldots, \sigma_n(\mathbf{x})).$$

The canonical embedding σ is a field homomorphism from K to $\mathbb{R}^{s_1} \times \mathbb{C}^{2s_2}$, where multiplication and addition in $\mathbb{R}^{s_1} \times \mathbb{C}^{2s_2}$ are component-wise. The discriminant Δ_K of K is the determinant of the matrix $(\sigma_i(\alpha_j))_{i,j}$, where (α_j) is a set of n elements of K.

For elements $H \subseteq \mathbb{R}^{s_1} \times \mathbb{C}^{2s_2} \subset \mathbb{C}^n$ where

$$H = \{(x_1, \ldots, x_n) \in \mathbb{R}^{s_1} \times \mathbb{C}^{2s_2} : x_{s_1+s_2+j} = \overline{x_{s_1+j}}, \forall j \in [s_2]\},$$

we can identify elements of K to their canonical embeddings in H and speak of the geometric canonical norms on K as $\|\mathbf{x}\|$ as $\|\sigma(\mathbf{x})\|_2 = (\sum_{i \in [n]} |\sigma_i(\mathbf{x})|^2)^{1/2}$. The field norm of an element $\mathbf{a} \in K$ is defined as $\mathsf{N}_{K/\mathbb{Q}}(\mathbf{a}) = \prod_{i \in [n]} \sigma_i(\mathbf{a})$. Note that the norm of an algebraic integer is in $\mathbb{Z}$ as the constant coefficient of the minimal polynomial. Let L a subfield of K, the relative norm of $\mathsf{N}_{K/L}(\mathbf{a}) = \prod_{\sigma_i \in \mathrm{Gal}(K/L)} \sigma_i(\mathbf{a})$, where $\mathrm{Gal}(K/L)$ contains the elements that fix L. The trace of $\mathbf{a} \in K$ is defined $\mathsf{Tr}_{K/\mathbb{Q}}(\mathbf{a}) = \sum_{i \in [n]} \sigma_i(\mathbf{a})$ and is the trace of the endomorphism $\mathbf{y} \mapsto \mathbf{a}\mathbf{y}$ and of its matrix representation.

Let K a number field of dimension n, which has a subfield L of dimension $m \mid n$. For simplicity, we assume that K is a Galois extension of $\mathbb{Q}$, with Galois group G; and G' is the subgroup of G fixing L. It is a standard fact that $|G'| = n/m$.

Notice that elements of the Galois group permute or conjugate the coordinates in $\mathbb{R}^r \times \mathbb{C}^s$, and therefore the norm is invariant by elements of G:

$$\forall \sigma \in G, \|\sigma(\mathbf{x})\| = \|\mathbf{x}\|.$$

We call $\mathsf{N}_{K/L} : K \to L$ the relative norm, with $\mathsf{N}_{K/L}(\mathbf{a})$ the determinant of the L-linear endomorphism $\mathbf{x} \mapsto \mathbf{ax}$. It is known that we have:

$$\mathsf{N}_{K/L}(\mathbf{a}) = \prod_{\sigma \in H} \sigma(\mathbf{a}).$$

We can bound the norm using the inegality of arithmetic and geometric means:

$$|\mathsf{N}_{K/\mathbb{Q}}(\mathbf{a})| \leq \left(\frac{\|\mathbf{a}\|}{\sqrt{n}} \right)^n.$$

The operator norm for the euclidean norm is denoted $\| \cdot \|_{op}$ and is defined as $\|\mathbf{a}\|_{op} = \sup_{\mathbf{x} \in \mathbb{K}^*} \|\mathbf{ax}\|/\|\mathbf{x}\|$. Remark that it is simply the maximum of the norm of the coordinates in $\mathbb{R}^r \times \mathbb{C}^s$. Also, it is sub-multiplicative and $\|\mathbf{x}\| \leq \sqrt{n}\|\mathbf{x}\|_{op}$.

Let $\mathcal{O}$ be an *order* of K, that is $\mathcal{O} \subset K$ and $\mathcal{O}$ is a commutative group which is isomorphic as an abelian group to $\mathbb{Z}^n$. We define $\mathcal{O}_L$ as $\mathcal{O} \cap L$, and is an order of L. We denote by $\mathrm{Vol}(\mathcal{L})$ the volume of the lattice $\mathcal{L}$, which is the square root of the determinant of the Gram matrix corresponding to any basis of $\mathcal{L}$. We define Δ to be the square of the volume of $\mathcal{O}$, and likewise for Δ_L with respect to $\mathcal{O}_L$.

We define

$$\mathbf{M}_{\mathbf{a}}^{\mathcal{L}} : \begin{array}{ccc} \mathcal{L} & \longrightarrow & \mathcal{O} \\ \mathbf{x} & \longmapsto & \mathbf{ax} \end{array}$$

for any lattice $\mathcal{L} \subset \mathcal{O}$ and $\mathbf{a} \in \mathcal{O}$; and we also denote $\mathbf{M}_{\mathbf{a}}^{\mathcal{L}}$ the corresponding matrix for some basis of $\mathcal{L}$.

Cyclotomic Field. In the case of cyclotomic field defined by $\Phi_{\mathfrak{f}}(x) = \prod_{k \in \mathbb{Z}_{\mathfrak{f}}^*} (x - \zeta_{\mathfrak{f}}^k)$, where $\zeta_{\mathfrak{f}} = e^{2i\pi/\mathfrak{f}} \in \mathbb{C}$, a primitive $\mathfrak{f}$-root of unity. Thus, $\Phi_{\mathfrak{f}}(x)$ has degree $n = \varphi(\mathfrak{f})$, is monic and irreducible over $\mathbb{Q}$ and its the minimal polynomial of the algebraic integer $\zeta_{\mathfrak{f}}$. The $\mathfrak{f}$th *cyclotomic field* is $\mathbb{Q}[\zeta_{\mathfrak{f}}]$ and its ring of integers is $\mathbb{Z}[\zeta_{\mathfrak{f}}]$, also called the *cyclotomic ring*. In this case, there are $2s_2 = n = \varphi(\mathfrak{f})$ complex canonical embeddings (no real ones), defined by $\sigma_i(\zeta_{\mathfrak{f}}) = \zeta_{\mathfrak{f}}^i$ for $i \in \mathbb{Z}_{\mathfrak{f}}^*$. For an element $\mathbf{x} = \zeta^j \in K$ in the power basis of K, all the embeddings of $\mathbf{x}$ have magnitude 1, and hence $\|\mathbf{x}\|_2^{can} = \sqrt{n}$ and $\|\mathbf{x}\|_{\infty}^{can} = 1$ as well as the coefficient embedding. The discriminant of the $\mathfrak{f}$th cyclotomic field of degree $n = \varphi(\mathfrak{f})$ is $\Delta_K \leq n^n$.

In the cyclotomic case, we can define the maximal real subfield $K^+ = \mathbb{Q}[\zeta_{\mathfrak{f}} + \zeta_{\mathfrak{f}}^{-1}]$, which only contains real numbers. It has index 2 in K and its degree is $n/2$. The rings of integers $\mathcal{O}_{K^+}$ of K^+ is simply $\mathbb{Z}[\zeta_{\mathfrak{f}} + \zeta_{\mathfrak{f}}^{-1}]$. The embeddings σ_1, σ_{-1} both fix every elements in $\mathbb{K}^+$ and the relative norm $\mathsf{N}_{K/K^+}(\mathbf{a}) = \sigma_1(\mathbf{a}) \cdot \sigma_{-1}$

$(\mathbf{a}) = \mathbf{a} \cdot \bar{\mathbf{a}}$. If we represent $\mathbf{a}$ as a polynomial $\mathbf{a}(\mathbf{x}) = \sum_{i=0}^{n-1} \mathbf{a}_i \mathbf{x}^i \in \mathbb{Q}[\mathbf{x}]/\Phi_{\mathfrak{f}}(\mathbf{x})$, then $\bar{\mathbf{a}}(\mathbf{x}) = \mathbf{a}(1/\mathbf{x}) = a_0 - \sum_{i=1}^{n-1} a_i \mathbf{x}^i$.

Ideals in the Ring of Integers. The ring of integers $\mathcal{O}_K$ of a number field K of degree n is a free $\mathbb{Z}$-module of rank n, i.e. the set of all $\mathbb{Z}$-linear combinations of some *integral basis* $\{\mathbf{b}_1, \ldots, \mathbf{b}_n\} \subset \mathcal{O}_K$. It is also a $\mathbb{Q}$-basis for K. In the case of cyclotomic field, the power basis $\{1, \zeta_{\mathfrak{f}}, \ldots, \zeta_{\mathfrak{f}}^{n-1}\}$ is an integral basis of the cyclotomic ring $\mathbb{Z}[\zeta_{\mathfrak{f}}]$ which is isomorphic to $\mathbb{Z}^n$ with $n = \varphi(\mathfrak{f})$.

It is well known that

$$\mathrm{Vol}(\mathbb{Z}[\zeta_{\mathfrak{f}}])^2 = \frac{\mathfrak{f}^{\phi(\mathfrak{f})}}{\prod_{p|\mathfrak{f}} p^{\phi(\mathfrak{f})/(p-1)}}.$$

In particular, if $\mathfrak{f}$ is a power of two, $\mathrm{Vol}(\mathbb{Z}[\zeta_{\mathfrak{f}}]) = (\mathfrak{f}/2)^{\mathfrak{f}/4}$. In this case, we also have that $(\zeta_{\mathfrak{f}}^i)_{i=0}^{\mathfrak{f}/2-1}$ is an orthogonal basis for the norm $\|\cdot\|$.

Lattices. Let $\mathbf{B} = \{\mathbf{b}_1, \ldots, \mathbf{b}_n\}$ be a basis of a lattice $\mathcal{L}$. Given $\mathbf{B}$, the LLL algorithm outputs a vector $\mathbf{v} \in \mathcal{L}$ satisfying $\|\mathbf{v}\|_2 \leq 2^{n/2} \cdot \det(\mathcal{L})^{1/n}$ in polynomial time in the size of its input.

Theorem 1 *(Minkowski)*. *For any lattice $\mathcal{L}$ of dimension n, there exists $\mathbf{x} \in \mathcal{L} \setminus \{\mathbf{0}\}$ with $\|\mathbf{x}\| \leq \sqrt{n}\mathrm{Vol}(\mathcal{L})^{1/n}$.*

We give a theorem for estimating the running time of lattice based algorithms:

Theorem 2. *Given a lattice $\mathcal{L}$ of dimension n, we can find a non-zero vector in $\mathcal{L}$ of norm less than $\beta^{n/\beta}\mathrm{Vol}(\mathcal{L})^{1/n}$ in deterministic time smaller than $2^{O(\beta)}$ times the size of the description of $\mathcal{L}$, for any $\beta < n/2$. With b_i^* the Gram-Schmidt norms of the output basis, we have $b_i^*/b_j^* \leq \beta^{O((j-i)/\beta+\log\beta)}$. Furthermore, the maximum of the Gram-Schmidt norms of the output basis is at most the maximum of the Gram-Schmidt norms of the input basis.*

Proof. Combine the semi-block Korkin-Zolotarev reduction [40] and the efficient deterministic shortest vector algorithm [36] with block size $\Theta(\beta)$ for the first point. Schnorr's algorithm combines the use of LLL reduction on a (possibly) linearly dependent basis, which is known to not increase the maximum of the Gram-Schmidt norms, and the insertion of a vector in position i whose projected norm is less than b_i^*. Also, the b_i^* decrease by a factor of at most $\beta^{O(\log\beta)}$ in a block, and the first Gram-Schmidt norms of blocks decrease by a factor of at most $\beta^{O(\beta)}$. $\square$

Lattice Analysis. We also use the GSA assumption [41], which states that the Gram-Schmidt norms output by a lattice reduction follow a geometric sequence. If we draw the the curve with the log of the Gram-Schmidt norms, we see a line with slope $\log\beta/\beta$ is the case of BKW (it is not accurate for the last ones than follows a parabola instead). Usually, we use the fact that the minimum of the Gram-Schmidt norms has to be smaller than the norm of the smallest vector in order to find it and so, the slope has to be close to horizontal, which implies that β is large.

In our analysis, we will a result of Pataki and Tural [38] in order to take into account that in NTRU lattice, all the shifts form a sublattice with small volume. They proved that the volume of the sublattice generated by r vectors is larger than the product of the r smallest Gram-Schmidt norms.

Lemma 1 [38]. *Let $\mathcal{L} \subseteq \mathbb{R}^n$ be a full-rank lattice and $r \geq 1$. Then for any basis $(\mathbf{b}_1, \ldots, \mathbf{b}_n)$ of $\mathcal{L}$, and any r-dimensional sublattice $\mathcal{L}'$ of $\mathcal{L}$, we have*

$$\det(\mathcal{L}') \geq \min_{1 \leq t_1 < \cdots < t_r \leq n} \prod_{1 \leq i \leq r} b_{t_i}^*.$$

Distribution on Ideal Lattices. The discrete Gaussian distribution over a lattice $\mathcal{L}$ is noted $D_{\mathcal{L},s}$, where the probability of sampling $\mathbf{x} \in \mathcal{L}$ is proportional to $\exp(-\pi\|\mathbf{x}\|^2/s^2)$. The continuous Gaussian distribution over K is noted D_s, and its density in $\mathbf{x}$ is proportional to $\exp(-\pi\|\mathbf{x}\|^2/s^2)$. We define

$$\rho_s(E) = \sum_{\mathbf{x} \in E} \exp(-\pi\|\mathbf{x}\|^2/s^2).$$

We will denote by $\mathbb{E}[X]$ the expectation of a random variable X.

We now recall two results from [35] and Banaszczyk's lemma [4] about discrete gaussian sampling over a lattice.

Lemma 2. *Given a lattice $\mathcal{L} \subset \mathbb{R}^n$, for any s and $\mathbf{c} \in \mathbb{R}^n$, we have*

$$\rho_s(\mathcal{L} + \mathbf{c}) \leq \rho_s(\mathcal{L}).$$

Lemma 3. *For a lattice $\mathcal{L}$, any $t \geq 1$, the probability that $\mathbf{x}$ sampled according to $D_{\Lambda,s}$ verifies $\|\mathbf{x}\| > st\sqrt{\frac{n}{2\pi}}$ is at most*

$$\exp\left(-n(t-1)^2/2\right).$$

We now prove a standard bound on ideal lattices, which indicates that they do not have very short vectors:

Lemma 4. *Let $M \subset (K \otimes \mathbb{R})^d$ be an $\mathcal{O}$ module of rank 1. Then, for any $0 \neq \mathbf{v} \in M$, we have $\mathrm{Vol}(M) \leq \sqrt{\Delta}\|\mathbf{v}/\sqrt{n}\|^n$.*

Proof. Since we can build a K-linear isometry from $\mathbb{R} \otimes M$ to $K \otimes \mathbb{R}$, we can assume $d = 1$. Then,

$$\mathrm{Vol}(M) \leq \mathrm{Vol}(\mathbf{v}\mathcal{O}) = \mathsf{N}_{K/\mathbb{Q}}(\mathbf{v})\sqrt{\Delta} \leq \|\mathbf{v}/\sqrt{n}\|^n\sqrt{\Delta}.$$

$\square$

3 Projection of a Subring Attack

In this section, we propose a new subfield attack, that we call subring, since we use the multiplication by the original public key $\mathbf{h}$, which is an element of the n-dimensional ring R_q, in a subring for instance the maximal real ring of integers $\mathbb{Z}[X + 1/X]$ of dimension $n/2$, or in a smaller subring. First, we first show that small vectors in this lattice are linked to the norms and in the case of the maximal real ring, the short vector is $(\mathbf{f}\bar{\mathbf{g}}, \mathbf{g}\bar{\mathbf{g}})$. For some parameters, we also show that the norm is not the smallest element: this explains some experiments in [1]. Then, we show that in the case of power of two cyclotomic fields, if we project the matrix represented the subring lattice on the last d rows and columns, we can precisely analyze the running time of the algorithm. Moreover, removing some rows allows to reach optimal parameters for our subring attack, which is not possible in other subfield attacks.

3.1 Description of the Basic Subring Attack

We show that in our subring attack, the lattice vector we are looking for is short. We first make sure that $\mathcal{O}$ is stable by all elements of H. This can be done by computing the Hermite normal form of the concatenation of the basis of $\sigma(\mathcal{O})$ for all $\sigma \in H$. We may then call $\mathcal{O}$ the order generated by this matrix. The attack consists in finding short vectors of the lattice generated by

$$\mathbf{A} = \begin{pmatrix} q\mathbf{I}_n & \mathbf{M}_{\mathbf{h}}^{\mathcal{O}_{\mathrm{L}}} \\ \mathbf{0} & \mathbf{I}_m \end{pmatrix}$$

by using lattice reduction. We recall that $\mathbf{h}$ is the public key, so that a basis of this lattice can be built. We want to show that $\begin{pmatrix} \mathbf{f}\mathsf{N}_{K/L}(\mathbf{g})/\mathbf{g} \\ \mathsf{N}_{K/L}(\mathbf{g}) \end{pmatrix}$ is a short vector of this lattice.

The quadratic form we reduce is actually the one induced by $\| \cdot \|$, i.e. $\|(\mathbf{x}, \mathbf{y})\|^2 = \|\mathbf{x}\|^2 + \|\mathbf{y}\|^2$, on this lattice.

Lemma 5. *For any* $\mathbf{g} \in \mathcal{O}$*, we have*

$$\mathsf{N}_{K/L}(\mathbf{g}) \in \mathbf{g}\mathcal{O} \cap \mathcal{O}_{\mathrm{L}}.$$

Proof. We have

$$\mathsf{N}_{K/L}(\mathbf{g}) = \mathbf{g} \prod_{\sigma \in H - \{1\}} \sigma(\mathbf{g})$$

so that $\mathsf{N}_{K/L}(\mathbf{g}) \in \mathbf{g}\mathcal{O}$. By definition of $\mathsf{N}_{K/L}$, we have $\mathsf{N}_{K/L}(\mathbf{g}) \in \mathrm{L}$. Therefore, $\mathsf{N}_{K/L}(\mathbf{g}) \in \mathbf{g}\mathcal{O} \cap \mathcal{O}_{\mathrm{L}}$. □

Using Banaszczyk's lemma, we will now show that integers sampled from a discrete Gaussian distribution behaves in a way similar to a continuous Gaussian distribution.

Lemma 6. *Let* $\mathbf{x}$ *be sampled according to* $D_{\mathcal{O},s}$. *Then, the probability that*

$$\|\mathbf{x}\|_{op} \geq s\sqrt{2\ln(2n/\epsilon)/\pi}$$

is at most ϵ.

Proof. Let $\mathbf{u}$ be a unit vector, i.e. $\|\mathbf{u}\| = 1$. Then,

$$\rho_s(\mathcal{O})\mathbb{E}[\exp(2\pi t\langle \mathbf{x}, \mathbf{u}\rangle/s^2)] = \sum_{\mathbf{x}\in\mathcal{O}} \exp(-\pi(\langle \mathbf{x}, \mathbf{x}\rangle - 2\langle \mathbf{x}, t\mathbf{u}\rangle)/s^2)$$

$$= \exp(\pi t^2/s^2)\sum_{\mathbf{x}\in\mathcal{O}} \exp(-\pi\|\mathbf{x} - t\mathbf{u}\|^2/s^2)$$

$$= \exp(\pi t^2/s^2)\rho_s(\mathcal{O} - t\mathbf{u}).$$

We deduce with the previous lemma

$$\mathbb{E}[\exp(2\pi t\langle \mathbf{x}, \mathbf{u}\rangle/s^2)] \leq \exp(\pi t^2/s^2).$$

Using Markov's inequality and the union bound with $-\mathbf{u}$, we have that the probability of $|\langle \mathbf{x}, \mathbf{u}\rangle| \geq t$ is at most $2\exp(-\pi t^2/s^2)$.

We now use $t = s\sqrt{\ln(2n/\epsilon)/\pi}$, so that the probability of any real or imaginary part of a coordinate of $\mathbf{x}$ in $\mathbb{R}^r\mathbb{C}^s$ is larger than

$$s\sqrt{\ln(2n/\epsilon)/\pi}$$

is at most ϵ.   □

Theorem 3. *Let* $\mathbf{f}$ *be sampled according to* $D_{\mathcal{O},\sigma}$, $\mathbf{g}$ *according to* $D_{\mathcal{O},s}$ *and set* $\mathbf{h} = \mathbf{f}/\mathbf{g}$. *Assume* $\mathbf{h}$ *is well defined, except with probability at most* $\epsilon/3$. *Then, there exists* $\mathbf{x} \neq \mathbf{0}$ *where* $\mathbf{x}$ *is an integer vector, such that*

$$\|\mathbf{Ax}\| \leq \sqrt{n(1 + \sigma^2/s^2)}\left(s\sqrt{2\ln(6n/\epsilon)/\pi}\right)^{n/m}$$

except with probability at most ϵ.

Proof. With probability at least $1 - \epsilon$, we have

$$\|\mathbf{f}\|_{op} \leq \sigma\sqrt{2\ln(6n/\epsilon)/\pi}$$

and

$$\|\mathbf{g}\|_{op} \leq s\sqrt{2\ln(6n/\epsilon)/\pi}.$$

In this case, we consider $\mathbf{y}$ such that $\mathbf{h}\mathsf{N}_{K/L}(\mathbf{g}) + q\mathbf{y} = \mathbf{f}\mathsf{N}_{K/L}(\mathbf{g})/\mathbf{g}$ and consider

$$\mathbf{x} = \begin{pmatrix} \mathbf{y} \\ \mathsf{N}_{K/L}(\mathbf{g}) \end{pmatrix}.$$

Using the multiplicativity of operator norms, we have

$$\|N_{K/L}(\mathbf{g})\|_{op} \leq \left(s\sqrt{2\ln(6n/\epsilon)/\pi} \right)^{|H|}$$

and

$$\|\mathbf{f}N_{K/L}(\mathbf{g})/\mathbf{g}\|_{op} \leq \sigma/s \left(s\sqrt{2\ln(6n/\epsilon)/\pi} \right)^{|H|}.$$

Finally,

$$\|\mathbf{A}\mathbf{x}\|^2 = \|\mathbf{f}N_{K/L}(\mathbf{g})/\mathbf{g}\|^2 + \|N_{K/L}(\mathbf{g})\|^2 \leq n \left(\|\mathbf{f}N_{K/L}(\mathbf{g})/\mathbf{g}\|_{op}^2 + \|N_{K/L}(\mathbf{g})\|_{op}^2 \right).$$

□

We now try to get rid of the factor $\Theta(\ln(6n/\epsilon))^{n/2m}$ which is significant when s is small and n/m is large. To do so, we *heuristically* assume that $D_{\mathcal{O},\sigma}$ has properties similar to a *continuous* Gaussian here.

Theorem 4. *Let* $\mathbf{f}$ *be sampled according to* D_s *and* $E \subset G$. *Then, except with probability at most* ϵ *and under heuristics, we have:*

$$\left\| \prod_{\sigma \in E} \sigma(\mathbf{f}) \right\|_{op} \leq \Theta(s)^{|E|} \exp\left(\Theta(\sqrt{|E|\log(n/\epsilon)}) \right)$$

under the condition $|E| = \Omega\left(\log(n/\epsilon) \log^2(\log(n/\epsilon)) \right)$.

Proof. Let X be a random variable over $\mathbb{R}^+$, with a probability density function proportional to $\exp(-\pi x^2/s^2)$; and $Y = \sqrt{X_0^2 + X_1^2}$ where X_0 and X_1 are independent copies of X.

We have $\mathbb{E}[\log(X)] = \log(s) + \Theta(1)$ and $\mathrm{Var}[\log(X)] = \Theta(1)$ and $\log(X) < \log(s) + \Theta(\log(\log(n/\epsilon)))$ except with probability $\epsilon/(2n^2)$, due to standard bounds on Gaussian tails. Also, the same is true for Y.

We can now use the one-sided version of Bernstein's inequality [8, Theorem 3]: for Z the average of $|E|$ independent copies of $\log(X)$ or $\log(Y)$, we have:

$$\Pr[Z > t + \log(s)] \leq \epsilon/(2n) + \exp\left(-\frac{|E|t^2}{2(\Theta(1) + \Theta(\log(\log(n/\epsilon)))t/3)} \right).$$

We then choose some $t = \Theta(\sqrt{\log(n/\epsilon)/|E|})$, so that with our lower bound on $|E|$, this probability is at most ϵ/n.

The result follows from the union bound over the coordinates in the canonical embedding of $\prod_{\sigma \in E} \sigma(\mathbf{f})$.   □

For some parameters, the norm may not be the shortest element, as demonstrated by the following theorem.

Theorem 5. *There exists an element* $\mathbf{v} \in \mathbf{g}\mathcal{O} \cap \mathcal{O}_\mathbb{L}$ *with*

$$0 < \|\mathbf{v}\| \le \sqrt{m}\Delta^{1/(2n)}\sigma^{n/m}$$

with probability $1 - 2^{-\Omega(n)}$.

Proof. We use Banaszczyk's lemma with $t = 2$, so that $\|\mathbf{g}\| \le \sigma\sqrt{2n/\pi}$ except with probability $\exp(-n/2)$. Then, the determinant of $\mathbf{v} \in \mathbf{g}\mathcal{O} \cap \mathcal{O}_\mathbb{L}$ is smaller than the determinant of $\mathsf{N}_{K/L}(\mathbf{g})\mathcal{O}_\mathbb{L}$, which is $\mathsf{N}_{K/\mathbb{Q}}(\mathbf{g})\sqrt{\Delta_\mathbb{L}}$. But we have $\mathsf{N}_{K/\mathbb{Q}}(\mathbf{g}) \le \left(\frac{\|\mathbf{g}\|}{\sqrt{n}}\right)^n$ and $\Delta_\mathbb{L} \le \Delta^{m/n}$ so we conclude with Minkowski's theorem. $\square$

This implies that for most parameters, the norm of the shortest non-zero vector is around $O(\sigma)^{n/m}$. Since this is smaller than the previous value as soon as n/m is a bit large, it explains why [1] found vectors shorter than the solution.

3.2 Asymptotic Analysis for Power of Two Cyclotomic Fields

We set here $\mathbb{K} = \mathbb{Q}[X]/(X^n + 1) \simeq \mathbb{Q}[\zeta_{2n}]$ for n a power of two, and $\mathcal{O} = \mathbb{Z}[X]/(X^n + 1) \simeq \mathbb{Z}[\zeta_{2n}]$ which is popular in cryptosystems. For some $r \mid n$ (any such r works), we select $\mathbb{L} = \mathbb{Q}[X^r]$ so that $\mathcal{O}_\mathbb{L} = \mathbb{Z}[X^r]$ and $|H| = r$, so that m, the dimension of $\mathbb{L}$ is n/r. Since the $\mathbf{X}^i$ forms an orthogonal basis, we have that the coordinates of $\mathbf{f}$ and $\mathbf{g}$ are independent discrete Gaussians of parameter $s/\sqrt{n}$. Also, we can directly reduce the lattice generated by $\mathbf{A}$ with the canonical quadratic form.

We restrict our study to power of two cyclotomic fields because $\mathcal{O}$ has a known orthogonal basis, so that we can derive a closed-form expression of the results. In more complicated cases, it is clear that we can deduce the result using a polynomial time algorithm.

For the rest of this section, we assume that when the previous algorithm is used on our orthogonal projection of $\mathbf{A}\mathbb{Z}^{n+m}$, and finds a vector shorter than $\sqrt{k}\mathrm{Vol}(\mathcal{L})^{1/k}$ (which is about the size of the shortest vector of a random lattice if the lattice dimension is k), then it must be a short multiple of the key. This assumption is backed by all experiments in the literature, including ours, and can be justified by the fact that decisional problems over lattices are usually as hard as their search counterpart (see [34] for example).

We also assume the size of the input is in $n^{O(1)}$, which is the usual case.

Theorem 6. *Let* $nB^2 = \|\mathbf{f}\mathsf{N}_{K/L}(\mathbf{g})/\mathbf{g}\|^2 + \|\mathsf{N}_{K/L}(\mathbf{g})\|^2$. *Assume* $\frac{\log(qB)}{\log(q/B)} \le r$. *Then, for*

$$\frac{\beta}{\log \beta} = \frac{2m \log q}{\log(q/B)^2}$$

we can find a non-zero element $\mathbf{A}\mathbf{x}$ *such that* $\|\mathbf{A}\mathbf{x}\|^2 = O(nB^2)$ *in time* $2^{O(\beta+\log n)}$.

Proof. We extract the last $d \approx m \frac{\log(q^2)}{\log(q/B)} \leq n + m$ rows and columns of

$$\mathbf{A} = \begin{pmatrix} q\mathbf{I} & \mathbf{M}_\mathbf{h}^{\mathcal{O}_L} \\ \mathbf{0} & \mathbf{I} \end{pmatrix}$$

and call the generated lattice $\mathcal{L}$. Note that it is the lattice generated by $\mathbf{A}$ projected orthogonally to the first columns, so that it contains a non-zero vector $\mathbf{y}$ such that $\|\mathbf{y}\|^2 \leq nB^2$. Then, we can compute the needed β by

$$\frac{1}{d} \log \left(\frac{\sqrt{n} \mathrm{Vol}(\mathcal{L})^{1/d}}{\sqrt{n}B} \right) = \frac{d-m}{d^2} \log(q) - \frac{1}{d} \log(B)$$

$$\approx \frac{\log(q/B)}{m \log(q^2)} \left(\frac{\log(qB) \log(q/B) \log(q)}{\log(q/B) \log(q^2)} - \log(B) \right)$$

$$= \frac{\log(q/B)}{m \log(q^2)} \left(\frac{\log(qB)}{2} - \log(B) \right)$$

$$= \frac{\log^2(q/B)}{2m \log(q)}.$$

The previous theorem indicates we can recover a short vector $\mathbf{z} \neq \mathbf{0}$ in $\mathcal{L}$ with $\|\mathbf{z}\| \leq nB^2$ in time $2^{\Theta(\beta + \log n)}$, and our assumption implies it is in fact a short vector in $\mathbf{A}\mathbb{Z}^{n+m}$. □

Notice that for $B \leq q$, a necessary condition for the problem to be solvable, we have $d \geq 2m$. It implies that the optimal dimension d cannot be reached by previous algorithms.

Theorem 7. *Let* $\mathbf{f}$ *and* $\mathbf{g}$ *be sampled according to* $D_{\mathcal{O},\sigma}$, *and* $\mathbf{h} = \mathbf{f}/\mathbf{g} \bmod q$ *which is well defined with probability at least* $1 - \epsilon$. *Assume* $\sigma = n^{\Omega(1)}$ *and* $\sigma < q^{1/4}$. *Then, we can recover a non-zero multiple of* $(\mathbf{f}, \mathbf{g})$ *of norm at most* $\sqrt{q}$ *in time*

$$\exp \left(O \left(\max \left(\log n, \frac{n \log \sigma}{\log^2 q} \log \left(\frac{n \log \sigma}{\log^2 q} \right) \right) \right) \right)$$

with a probability of failure of at most $\epsilon + 2^{-n}$.
   This is polynomial time for

$$\log \sigma = O \left(\frac{\log^2 q \log n}{n \log \log n} \right).$$

Proof. We choose $m = \Theta(\max(1, \frac{n \log \sigma}{\log q})) \leq n$ so that we can set $B = \sqrt{q}$, except with probability ϵ. The corresponding β is given by

$$\frac{\beta}{\log \beta} = \frac{2m \log q}{\log(q/B)^2} = \Theta(m/\log(q)) = \Theta \left(\frac{n \log \sigma}{\log^2 q} \right).$$

□

If we use $\log \sigma = \Theta(\log n)$ as in many applications, we are in polynomial time when

$$q = 2^{\Omega(\sqrt{n \log \log n})}.$$

If $\sigma = \Theta(\sqrt{n})$, the best generic algorithm runs in time $2^{\Theta(n/\log \log q)}$, which is slower for any $q \geq n^{\Theta(\sqrt{\log \log n})}$.

3.3 Comparison with Other Subfield Attacks

We consider the lattice generated by $\begin{pmatrix} q\mathbf{I}_n & \mathbf{M}_\mathbf{h}^{\mathcal{O}_\mathbb{L}} \\ \mathbf{0} & \mathbf{I}_{n/r} \end{pmatrix}$ while Albrecht *et al.* for

instance consider $\begin{pmatrix} q\mathbf{I}_{n/r} & \mathbf{M}_{N_{K/L}(\mathbf{h})}^{\mathcal{O}_\mathbb{L}} \\ \mathbf{0} & \mathbf{I}_{n/r} \end{pmatrix}$, where $\mathbf{M}_\mathbf{h}^{\mathcal{O}_\mathbb{L}}$ represents the multiplication

by the element $\mathbf{h}$ in the subring $\mathcal{O}_\mathbb{L}$ of $\mathbb{K}$. Our lattice is of dimension $n + n/r$, which is *larger* than Albrecht *et al.* attack, but smaller than the $2n$ original lattice. Since the running time of lattice reduction algorithms depends on the dimension of the matrix, we may think that our variant is less efficient than the subfield attack. First of all, in order to improve the running time, we will show that we can work in a projected lattice and not on the full $(n + n/r, n + n/r)$-matrix by considering the matrix forms using the last d rows and columns. The idea of working in this lattice is that the second important parameter is the *approximation factor*. This parameter depends on the size of the Gram-Schmidt coefficients. If we use the logarithm of their size, these coefficients draw a decreasing line of slope correlated with the approximation factor, so that the smaller the approximation factor be, the more horizontal the line will be. However, if we have only a $(2n/r)$-dimensional matrix, as in the subfield attack, *the determinant is too small to produce large Gram-Schmidt norms*. This problem is bypassed with our approach since we can choose the number of coordinates and the size of the subfield. Using this attack, we were able to break in practice proposed parameters by YASHE and in other papers, which were not the case in Albrecht *et al.*. We also show a tight estimation of the parameters broken by lattice reduction, and in particular that working in the original field works well. Experiments were conducted in an extensive way, and over much larger parameters.

4   Analysis of Lattice Reduction on NTRU Lattices

We now show how to predict when this attack will work, and compare our theoretical analysis with experiments. While Albrecht *et al.* compare the subfield attack to the attack on the full dimension lattice, we will show that, the classical attack, used in Howgrave-Graham work on the hybrid attack, performs a lattice reduction on the matrix centered in the original Coppersmith-Shamir lattice. This gives a better result and we show that considering subfield is not helpful. Consequently, this attack can also be mounted on NTRU prime with overstretched parameters and works well.

4.1   Analysis of the Simple Method

Here, we consider the lattice reduction algorithm described in Theorem 2 applied to the full Coppersmith-Shamir matrix. We show that using Pataki-Tural lemma and the above heuristics, we can actually achieve the same efficiency regardless of the presence of a subfield, as long as we know an orthogonal basis of $\mathcal{O}$.

The analysis hinges on the fact that the difficulty for lattice reduction to find a vector in a sublattice of low volume depends on the rank of the sublattice. Previous analysis relied on its special case where the rank is one, so that the volume is the length of the generator. In this case, we can prove using the GSA and the determinant of the lattice, that $\beta/\log\beta = O(n/\log(q/\sigma^2))$. In the following, using the Pataki-Tural lemma, we show that we can achieve the same efficiency as in the case of subfield, directly on the Coppersmith-Shamir lattice, i.e. $\beta/\log\beta = O(n\log(\sigma)/\log^2(q))$.

The following theorem, identical to [1, Theorem 2], indicates that short vectors are multiples of the secret key.

Theorem 8. *Let* $\mathbf{f}, \mathbf{g} \in \mathcal{O}$ *with* $\mathbf{g}$ *invertible modulo* q *and* $\mathbf{f}$ *coprime to* $\mathbf{g}$. *Then, any vector shorter than* $\frac{nq}{\|(\mathbf{f},\mathbf{g})\|}$ *in*

$$\begin{pmatrix} q\mathbf{I}_n & \mathbf{M}^{\mathcal{O}}_{\mathbf{f}/\mathbf{g}} \\ 0 & \mathbf{I}_n \end{pmatrix} \mathcal{O}^2 \; is \; in \; \begin{pmatrix} \mathbf{f} \\ \mathbf{g} \end{pmatrix} \mathcal{O}.$$

Proof. By coprimality, there exists $\mathbf{F}, \mathbf{G}$ such that $\mathbf{f}\mathbf{G} - \mathbf{g}\mathbf{F} = q$. Then,

$$\begin{pmatrix} \mathbf{f} & \mathbf{F} \\ \mathbf{g} & \mathbf{G} \end{pmatrix}$$

generates the same lattice. We let $\Lambda = \begin{pmatrix} \mathbf{f} \\ \mathbf{g} \end{pmatrix}\mathcal{O} \subset (\mathbb{R}\otimes\mathbb{K})^2$ and Λ^* the projection of $\begin{pmatrix} \mathbf{F} \\ \mathbf{G} \end{pmatrix}\mathcal{O}$ orthogonally to Λ. We have $\mathrm{Vol}(\Lambda)\mathrm{Vol}(\Lambda^*) = q^n\Delta$. Finally, let $0 \neq \mathbf{x} \in \Lambda^*$. Using twice Lemma 4, we have

$$\|\mathbf{x}/\sqrt{n}\|^n \geq \frac{q^n\Delta}{\sqrt{\Delta}\mathrm{Vol}(\Lambda^*)} = (\frac{q\sqrt{n}}{\|(\mathbf{f},\mathbf{g})\|})^n.$$

$\square$

In the following, we show that the Pataki-Tural lemma allows us to have a lattice reduction algorithm with β around $\tilde{\Theta}(n\log\sigma/\log^2 q)$, which is close to Theorem 7 in the case of subfield.

Theorem 9. *Let* $\mathbf{f}, \mathbf{g}$ *sampled according* $D_{\mathcal{O},s}$ *such that* $\mathbf{g}$ *is invertible with probability* $1-\epsilon$, *and an orthogonal basis of* $\mathcal{O}$ *is known. Reducing the lattice generated by* $\begin{pmatrix} q\mathbf{I}_n & \mathbf{M}^{\mathcal{O}}_{\mathbf{h}} \\ 0 & \mathbf{I}_n \end{pmatrix}$ *using the algorithm of Theorem 2, assuming the minimum of the Gram-Schmidt norms does not decrease, with*

$$\beta = \Theta\left(\frac{n \log \sigma}{\log^2 q} \log\left(\frac{n \log \sigma}{\log^2 q}\right)\right),$$

we recover at least $n/2$ vectors of a basis of $\mathbf{f}\mathcal{O}$ and $\mathbf{g}\mathcal{O}$, if $\Delta^{1/2n}\sigma = q^{O(1)}$ and

$$\log q = \Omega\left(\log^2\left(\frac{n \log \sigma}{\log q}\right)\right),$$

with probability $1 - \epsilon - 2^{-\Omega(n)}$.

Proof. Before calling the lattice reduction algorithm, the Gram-Schmidt norms are $q\Delta^{1/2n}$ for the first n vectors, and $\Delta^{1/2n}$ for the next n vectors. The lattice contains $\begin{pmatrix}\mathbf{f}\\\mathbf{g}\end{pmatrix}\mathcal{O}$ so that the lattice spanned has a volume of $\sigma^n\sqrt{\Delta}$ except with probability $2^{-\Omega(n)}$, thanks to Lemma 3.

We consider now the $2n$-dimensional basis outputted by the reduction algorithm (Theorem 2), and call b_i^* 'small' when it is amongst the n smallest Gram-Schmidt norms, and 'large' otherwise. Let $\ell = O\left(\frac{n \log \sigma}{\log q}\right) \leq n$. We consider two cases, depending whether there is a small b_i^* that has a large value or not. We show that either case is is impossible, which will complete the argument by contradiction. Assume first that there is an $i \leq n$, such that

$$b_i^* \geq \frac{\sqrt{n}q}{\sigma} \geq q^{1/4}\Delta^{1/2n}.$$

Suppose then again by contradiction, that there is a $b_j^* \geq q^{1/4}\Delta^{1/2n}$ which is small (Case 1). Consequently by Theorem 2,

$$b_k^* \geq q^{1/4}\Delta^{1/2n}\beta^{-O(\ell/\beta + \log \beta)} \geq q^{1/4}\Delta^{1/2n}\beta^{-O(\ell/\beta)}$$

for all the ℓ first $k \geq i$ such that b_k^* is small (we use here the assumption that the minimum of the Gram-Schmidt norms does not decrease). Hence, the product of the n smallest b_i^* is at least $\Delta^{1/2}q^{\ell/4}\beta^{-O(\ell^2/\beta)}$ by lower bounding the last $(n-\ell)$ ones by $\Delta^{1/2n}$. We deduce that for small enough constants, this is impossible using the Pataki-Tural lemma: otherwise we get a contradiction with the fact that this product should be smaller than the smallest volume of a sublattice of dimension n, $\sqrt{\Delta}\sigma^n$.

Suppose now every small b_j^* satisfies $b_j^* < q^{(1/4)}\Delta^{1/(2n)}$ (Case 2). Let $j \geq i$ be the smallest such that b_j^* is small. Then, we have

$$b_k^* \leq q^{1/4}\Delta^{1/2n}\beta^{O(\ell/\beta + \log \beta)}$$

for all the last ℓ indices $k \leq j$ such that b_k^* is large. Thus, the product of the large Gram-Schmidt norms is at most $\Delta^{1/2}q^{n-\ell/4}\beta^{O(\ell^2/\beta)}$, as all b_k^*'s remain $\leq q\Delta^{1/(2n)}$, by Theorem 2. Since the determinant is preserved during the running of the algorithm, the product of the small Gram-Schmidt norms is at least $\Delta^{1/2}q^{\ell/4}\beta^{-O(\ell^2/\beta)}$, which is impossible by using again the Pataki-Tural lemma.

To sum up, we have proved that for all the n first b_i^*, we have $b_i^* < \sqrt{n}q/\sigma$ and so, $b_i < nq/\sigma$ and using Lemma 8, we can show that all the first $n/2$ vectors are in $\begin{pmatrix} \mathbf{f} \\ \mathbf{g} \end{pmatrix}\mathcal{O}$. $\qquad\qquad\qquad\qquad\qquad\qquad\qquad\qquad\qquad\qquad$ □

4.2 Generalization and the Middle Technique

As we can see from the formula, considering a subfield is not helpful since the quantity $n \log \sigma$ is essentially constant; unless we have reasons to believe there are huge factors of $\mathbf{g}\mathcal{O}$ which are in the subfield. Even worse, it actually decreases the efficiency when $\sigma \geq \sqrt{q}$ because the value of ℓ is forced to a suboptimal. We also observe that the significant reduction in the dimension due to the use of subfields, allowing to break instances of high dimension is also present here: indeed, we can project orthogonally to the first $2n - \ell$ vectors the next ℓ vectors so that we reduce a lattice of dimension ℓ instead of $2n$.

Also, when we choose to work with $\mathcal{O} = \mathbb{Z}[X]/(X^n - X - 1)$ as in NTRU Prime [5], where we can use $(X^i)_{i=0}^{n-1}$ as an orthogonal basis due to the choice of the error distribution made by the authors (the coordinates are almost independent and uniform over $\{-1, 0, 1\}$), the same result applies.

We stress that while our theorem does not prove much - assuming the maximum of the Gram-Schmidt norms decreases is wrong, except for LLL - experiments indicate that either the middle part of the lattice behaves as a 'random' lattice as it is evaluated in [25], or the first n vectors are a basis of $\begin{pmatrix} \mathbf{f} \\ \mathbf{g} \end{pmatrix}\mathcal{O}$.

Furthermore, the phase transition between the two possible outputs is almost given by the impossibility of the first case. As lattice reduction algorithms are well understood (see [11,24]), it is thus easy to compute the actual β.

5 Implementation

Heuristically, we have that for reduced random lattices, the sequence b_i^* is (mostly) decreasing and therefore the relevant quantity is $\prod_{i=n-r}^{n-1} b_i^*$. It means that when the b_i^*s decrease geometrically and $\det(\mathcal{L}')^{1/r}$ is about the length of the shortest vector, we need $b_{\lfloor n-r/2 \rfloor}^*$ to be larger than the shortest vector instead of the b_{n-1}^* given by a standard analysis. We now remark that for $r = 1$, this is pessimistic. Indeed, for a "random" short vector, we expect the projection to reduce its length by a $\simeq \sqrt{n}$ factor. In our case, we can expect the projection to reduce the length by a $\simeq \sqrt{n/(n-r)}$ factor.

For our predictions, we assumed that the determinant of the quadratic form

$$\mathbf{x} \mapsto \mathbf{f} \mathsf{N}_{K/L}(\mathbf{g})/\mathbf{g}\mathbf{x}\overline{\mathbf{f}\mathsf{N}_{K/L}(\mathbf{g})/\mathbf{g}\mathbf{x}} + \mathsf{N}_{K/L}(\mathbf{g})\mathbf{x}\overline{\mathsf{N}_{K/L}(\mathbf{g})\mathbf{x}},$$

which corresponds to the $\det(\mathbf{U}^t\mathbf{G}\mathbf{U})$ above, is about the square of the norm over $\mathbb{Z}$ of $\mathbf{g}$. This quantity can be evaluated in quasi-linear time when we work within a cyclotomic field with smooth conductor by repeatedly computing the norm over a subfield, instead of the generic quadratic algorithm, or its variants such as in [2, Sect. 5.2]. We observe a very good agreement between the experiments and the prediction, while considering only the fact that the lattice has a short vector would lead a much higher bound. Also, while $N_{K/L}(\mathbf{g})$ has a predicted size of $n^{r/2}\exp(\sqrt{r\log(n/r)})$ with $\sigma = \sqrt{n}$, we expect LLL to find a multiple of size $n^{r/2}\exp(n/r)$ (possibly smaller) but none of these quantities are actually relevant for determining whether or not LLL will recover a short element.

Finally, we may have $(N_{K/L}(\mathbf{g}))/((\mathbf{g})\cap \mathcal{O}_L)$ which is non-trivial. However, if it is an ideal of norm κ, we have that κ^2 divides the norm over $\mathbb{Z}$ of $\mathbf{g}$, which is exceedingly unlikely for even small values of $\kappa^{r/n}$.

Our predictions indicate all proposed parameters of [6, Table 1] are broken by LLL. We broke the first three using fplll and about three weeks of computation. The last three where broken in a few days over a 16-core processor (Intel Xeon E5-2650).

The parameters proposed for schemes using similar overstretched NTRU assumption, such as in homomorphic encryption [7,10,14–16,18,30,31] or in private information retrieval [17], are also broken in practical time using LLL. For example, we recovered a decryption key of the FHE described in [15] in only 10 h. For comparison, they evaluated AES in 29 h: that means that we can more efficiently than the FHE evalation, recover the secret, perform the AES evaluation, and then re-encrypt the result! A decryption key was recovered for [18] in 4 h. Other instanciations such as [9,28] are harder, but within range of practical cryptanalysis, using BKZ with moderate block-size [11].

6   Explicit Complexity

We now turn towards the problem of deriving the first order of the asymptotical complexity of *heuristic* algorithms. Before the dual BKZ algorithm [37], simple derivations (as in [29, Appendix B]) could only be done using the Geometric Series Assumption, since the heuristic Gram-Schmidt norms outputted by the BKZ algorithm have a fairly complicated nature (see [24]), making an exact derivation quite cumbersome if not intractable. We are only interested in the part of the Gram-Schmidt norms known to be geometrically decreasing, which simplifies the computations[1].

We emphasize that we are only using standard heuristics, checked in practice, and *tight* at the first order. We compute the necessary block-size β to solve the problems and assume $\log\beta \approx \log n$. More precisely, if $\log\beta = (1 + o(1))\log n$, then the exponent in the running time is within $1 + o(1)$ of its actual value.

For more information on the dual BKZ algorithm and dual lattices, see [37]. We denote by dual BKZ algorithm their Algorithm 1 followed by a forward (i.e. primal) round, so that it attempts to *minimize* the *first* Gram-Schmidt norm (as the previous algorithms), rather than *maximizing* the *last* Gram-Schmidt norm.

---

[1] We remark that the last Gram-Schmidt norms have no constraints in the original algorithm. However, we can always assume they are HKZ-reduced, so that their logarithms are a parabola.

log n	log q	log r	Success	Method	Coordinates used	Origin
11	165	4	Yes	[1]	128	-
11	115	4	Yes	Ours	510	-
11	114	4	No	Ours	630	-
11	95	3	Yes	[1]	256	-
11	81	3	Yes	Ours	600	-
11	80	3	No	Ours	600	-
11	79	3	No	Ours	860	YASHE[6]
11	70	2	Yes	Ours	600	-
11	69	2	No	Ours	600	-
12	190	4	Yes	[1]	256	-
12	157	4	Yes	Ours	430	YASHE[6]
12	144	4	Yes	Ours	850	-
12	143	4	No	Ours	850	-
13	383	4	Yes	Ours	512	[18]
13	312	5	Yes	Ours	470	YASHE[6]
14	622	5	Yes	Ours	470	YASHE[6]
15	1271	5	Yes	Ours	512	[15]
15	1243	6	Yes	Ours	660	YASHE[6]
16	2485	7	Yes	Ours	820	YASHE[6]

log n	Prediction	log r
11	116	4
11	82	3
11	71	2
12	146	4
12	105	1
13	271	5
13	155	1
14	525	6
14	228	1
15	1045	7
15	335	1
16	2121	8
16	491	1

Fig. 1. Experiments with LLL for solving the NTRU problem in the ring $\mathbb{Z}[X]/(q, X^n + 1)$, where the coefficients of the polynomials are uniform in $\{-1, 0, 1\}$. The lattice dimension used is equal to the number of coordinates used added to n/r. The values of [1] are the smallest moduli for which their algorithm works, up to one, one and five bits. The prediction is the minimum $\log q$ an LLL reduction can solve assuming we use all the (necessary) coordinates.

log n	log q	ℓ	Success
11	72	1116	Yes
11	70	1200	Yes
11	69	1200	No
12	118	1024	Yes
12	117	1024	No
12	105	1700	Yes
12	104	1700	No
13	230	1024	Yes
14	450	1024	Yes
15	930	1024	Yes

log n	ℓ	Prediction
11	1033	71
12	1472	106
13	2275	156
14	3357	230
15	5127	337
16	7124	477

Fig. 2. Experiments with LLL for solving the NTRU problem in the ring $\mathbb{Z}[X]/(q, X^p - X - 1)$, where the coefficients of the polynomials are uniform in $\{-1, 0, 1\}$ and p is the smallest prime larger than n. The lattice dimension used is ℓ. The prediction is the minimum $\log q$ an LLL reduction can solve.

We remark that all uses of NTRU for "standard" cryptography (key-exchange, signature and IBE) are instantiated with a modulus below n^2, so that the lattice reduction algorithms are *not* affected by the property.

6.1 Security of Learning with Errors

The following heuristic analysis applies for NTRU, but also for any LWE problem with dimension n and exactly $2n$ samples[2], or Ring-LWE with two samples. The primal algorithm searches for a short vector in a lattice.

As usual, we build the lattice

$$\mathbf{A} = \begin{pmatrix} q\mathbf{I}_n & \mathbf{M}_h^{\mathcal{O}_L} \\ \mathbf{0} & \mathbf{I}_m \end{pmatrix}$$

and apply the dual BKZ algorithm on its dual. We assume it did not find the key, and suppose the projection of $(\mathbf{f}, \mathbf{g})$ orthogonally to the first $2n - 1$ vector has a norm of $\sigma/\sqrt{n}$. Then, the last Gram-Schmidt norm must be smaller than $\sigma/\sqrt{n}$ and we compute the smallest block-size β such that it is not the case. Hopefully, this means that applying the algorithm with a block-size β will find the key.

Once the dual BKZ algorithm has converged, the $2n - \beta$ first Gram-Schmidt norms are decreasing with a rate of $\approx \beta^{-1/\beta}$ and the $2n - \beta$th norm is about $\sqrt{\beta}V^{1/\beta}$ where V is the product of the last β norms. We deduce that the volume of the dual lattice is

$$q^{-n} = \left(\frac{\sigma}{\sqrt{n}}\right)^{-2n} \beta^{-(2n-\beta)^2/2\beta-n} = \left(\frac{\sigma}{\sqrt{n}}\right)^{-2n} \beta^{-2n^2/\beta}$$

so with $q = n^a$, $\sigma = n^b$ and $\beta = nc$ we have

$$-a \approx 1 - 2b - 2/c$$

and we deduce $c = 2/(a + 1 - 2b)$.

Another possibility is to apply the dual BKZ algorithm on the basis. If it reduces the last $m + n$ vectors, then the $m + n - \beta$th Gram-Schmidt norm cannot be smaller than the size of the key, σ. Now, if $m = n$ this norm is $\sqrt{q}\beta^{n/\beta-(2n-\beta)/\beta}$, and we deduce $a/2 - 1/c + 1 = b$ or $c = 2/(a + 2 - 2b)$ which happens when $c \geq 2/a$ (iff $b \geq 1$). Else, we take m maximum so that $q^{m/(m+n)}\beta^{(m+n)/2\beta} = q$ or $m = n(\sqrt{2ca} - 1)$ which gives $q\beta^{-(m+n-\beta)/\beta} = \sigma$ or $a - (\sqrt{2ca} - 1 + 1 - c)/c = b$ and hence $c = 2a/(a + 1 - b)^2$ when $b \leq 1$.

The dual algorithm searches for $2^{o(n)}$ short vectors in the dual lattice, so that the inner product with a gaussian of standard deviation σ can be distinguished. Applying the dual BKZ algorithm on the dual lattice gives a vector of norm $\beta^{n/\beta}q^{-m/(n+m)} = \sigma/n$. The norm is minimized for $m = \sqrt{2ac} - 1$ or $m = n$, which gives $c = 2a/(a + 1 - b)^2$ when $b < 1$, and $2/(a + 2 - 2b)$ else.

In all cases, the best complexity is given by $c = \max(2a/(a + 1 - b)^2, 2/(a + 2 - 2b))$ (and when the number of samples is unlimited, this is $2a/(a + 1 - b)^2$).

---

[2] Beware that an element sampled in the ring with standard deviation σ has coordinates of size only $\sigma/\sqrt{n}$.

6.2 Security of NTRU

Here, the analysis is specific to NTRU. We apply the dual BKZ algorithm to the same lattice, and compute the β such that the product of the n last Gram-Schmidt norms is equal to σ^n. Note that it is equivalent to having the product of the n first Gram-Schmidt norms equal to q/σ^n.

We first compute m such that the dual BKZ algorithm changes only the $2m$ middle norms. This is given by:

$$q = \sqrt{q}\beta^{m/\beta}$$

so that $m \approx a\beta/2$. For $a \geq 2$, we have $\beta \leq m$ so that, assuming $m \leq n$, the product of the m first norms is $q^m\beta^{-m^2/2\beta}$. Hence, we need $\beta^{m^2/2\beta} = \sigma^n$. We deduce

$$a^2c^2/8c = b$$

so that $c = 8b/a^2$.

When $m > n$, the first vector is of norm only $\sqrt{q}\beta^{n/\beta}$, so that for $c \leq 1$, we must have

$$q^{n/2}\beta^{n^2/2\beta - n^2/\beta} = \sigma^n$$

so that $a/2 - 1/2c \approx b$ and $c = 1/(a - 2b)$. For this formula to be correct, we need $8b/a^2a/2 \geq 1$, or $4b \geq a$.

We can show that this is better than the algorithms against Ring-LWE when $b = 1/2$ ($\approx$ binary errors) when $a \geq (4 + \sqrt[3]{262 - 6\sqrt{129}} + \sqrt[3]{262 + 6\sqrt{129}})/6 \approx 2.783$. When $b \geq 1$ which is the proven case, it is better for all $a > 4$ and $b < a/2 - 1$.

We again remark that going to a subfield, so that nb is constant, does not improve the complexity.

7 Conclusion

We conclude that the shortest vector problem over module lattices seems strictly easier than the bounded distance decoding. Since the practical cost of transforming a NTRU-based cryptosystem into a Ring-LWE-based cryptosystem is usually small, especially for key-exchange (e.g. [3]), we recommend to dismiss the former, in particular since it is known to be weaker (see [39, Sect. 4.4.4]). One important difference between NTRU and Ring-LWE instances is the fact that in NTRU lattices, there exist many short vectors. This has been used by May and Silverman in [33] and in our case, the determinant of the sublattice generated by these short vectors is an important parameter to predict the behaviour of our algorithm.

We remark that the only proven way to use NTRU is to use $\sigma \approx \sqrt{n^3q}$ [42]. We showed here that attacks are more efficient against NTRU than on a Ring-LWE lattice until $\sigma \approx n^{-1}\sqrt{q}$, which suggests their result is essentially optimal. Furthermore, the property we use is present until $\sigma \approx \sqrt{nq}$, i.e. until the public key $\mathbf{h}$ is (heuristically) indistinguishable from uniform.

Our results show that the root approximation factor is a poor indicator in the NTRU case: indeed, we reached 1.0059 using a mere LLL. We suggest to switch the complexity measure to the maximum dimension used in shortest vector routines (i.e. the block size of the lattice reduction algorithm) of a successful attack. While there are less problems with LWE-based cryptosystems, the root approximation factor has also several shortcomings which are corrected by this modification. Indeed, highly reduced basis do not obey to the Geometric Series Assumption, so that the root approximation factor also depends on the dimension of the lattice. Even when the dimension is much larger than the block-size, converting the factor into a block-size - which is essentially inverting the function $\beta \mapsto \left(\frac{(\beta/2)!}{\pi^{\beta/2}}\right)^{1/\beta^2}$ - is very cumbersome. Finally, the complexity of shortest vector algorithms is more naturally expressed as a function of the dimension than the asymptotical root approximation factor they can achieve.

Acknowledgments. We also would like to thank the reviewers and particularly Damien Stehlé for his help and his advices for the final version. We would like to thank the Crypto Team at ENS for providing us computational ressources to perform our experimentations.

References

1. Albrecht, M., Bai, S., Ducas, L.: A subfield lattice attack on overstretched NTRU assumptions. In: Robshaw, M., Katz, J. (eds.) CRYPTO 2016. LNCS, vol. 9814, pp. 153–178. Springer, Heidelberg (2016). doi:10.1007/978-3-662-53018-4_6
2. Albrecht, M.R., Cocis, C., Laguillaumie, F., Langlois, A.: Implementing candidate graded encoding schemes from ideal lattices. In: Iwata, T., Cheon, J.H. (eds.) ASIACRYPT 2015. LNCS, vol. 9453, pp. 752–775. Springer, Heidelberg (2015). doi:10.1007/978-3-662-48800-3_31
3. Alkim, E., Ducas, L., Pöppelmann, T., Schwabe, P.: Post-quantum key exchange - a new hope. Cryptology ePrint Archive, Report 2015/1092 (2015). http://eprint.iacr.org/2015/1092
4. Banaszczyk, W.: New bounds in some transference theorems in the geometry of numbers. Math. Ann. **296**(1), 625–635 (1993)
5. Bernstein, D.J., Chuengsatiansup, C., Lange, T., van Vredendaal, C.: NTRU prime (2016). http://eprint.iacr.org/
6. Bos, J.W., Lauter, K., Loftus, J., Naehrig, M.: Improved security for a ring-based fully homomorphic encryption scheme. In: Stam, M. (ed.) IMACC 2013. LNCS, vol. 8308, pp. 45–64. Springer, Heidelberg (2013). doi:10.1007/978-3-642-45239-0_4
7. Bos, J.W., Lauter, K., Naehrig, M.: Private predictive analysis on encrypted medical data. J. Biomed. Inform. **50**, 234–243 (2014)
8. Boucheron, S., Lugosi, G., Bousquet, O.: Concentration inequalities. In: Bousquet, O., von Luxburg, U., Rätsch, G. (eds.) ML -2003. LNCS (LNAI), vol. 3176, pp. 208–240. Springer, Heidelberg (2004). doi:10.1007/978-3-540-28650-9_9
9. Çetin, G.S., Dai, W., Doröz, Y., Sunar, B.: Homomorphic autocomplete. Cryptology ePrint Archive, Report 2015/1194 (2015). http://eprint.iacr.org/2015/1194
10. Çetin, G.S., Doröz, Y., Sunar, B., Savaş, E.: Depth optimized efficient homomorphic sorting. In: Lauter, K., Rodríguez-Henríquez, F. (eds.) LATINCRYPT 2015. LNCS, vol. 9230, pp. 61–80. Springer, Cham (2015). doi:10.1007/978-3-319-22174-8_4

11. Chen, Y., Nguyen, P.Q.: BKZ 2.0: better lattice security estimates. In: Lee, D.H., Wang, X. (eds.) ASIACRYPT 2011. LNCS, vol. 7073, pp. 1–20. Springer, Heidelberg (2011). doi:10.1007/978-3-642-25385-0_1

12. Cheon, J.H., Jeong, J., Lee, C.: An algorithm for NTRU problems and cryptanalysis of the GGH multilinear map without an encoding of zero. Cryptology ePrint Archive, Report 2016/139 (2016). http://eprint.iacr.org/

13. Coppersmith, D., Shamir, A.: Lattice attacks on NTRU. In: Fumy, W. (ed.) EURO-CRYPT 1997. LNCS, vol. 1233, pp. 52–61. Springer, Heidelberg (1997). doi:10.1007/3-540-69053-0_5

14. Dai, W., Doröz, Y., Sunar, B.: Accelerating SWHE based PIRs using GPUs. In: Brenner, M., Christin, N., Johnson, B., Rohloff, K. (eds.) FC 2015. LNCS, vol. 8976, pp. 160–171. Springer, Heidelberg (2015). doi:10.1007/978-3-662-48051-9_12

15. Doröz, Y., Yin, H., Sunar, B.: Homomorphic AES evaluation using the modified LTV scheme. Des. Codes Cryptogr., 1–26 (2015)

16. Doröz, Y., Shahverdi, A., Eisenbarth, T., Sunar, B.: Toward practical homomorphic evaluation of block ciphers using prince. In: Böhme, R., Brenner, M., Moore, T., Smith, M. (eds.) FC 2014. LNCS, vol. 8438, pp. 208–220. Springer, Heidelberg (2014). doi:10.1007/978-3-662-44774-1_17

17. Doröz, Y., Sunar, B., Hammouri, G.: Bandwidth efficient PIR from NTRU. In: Böhme, R., Brenner, M., Moore, T., Smith, M. (eds.) FC 2014. LNCS, vol. 8438, pp. 195–207. Springer, Heidelberg (2014). doi:10.1007/978-3-662-44774-1_16

18. Dowlin, N., Gilad-Bachrach, R., Laine, K., Lauter, K., Naehrig, M., Wernsing, J.: Cryptonets: applying neural networks to encrypted data with high throughput and accuracy (2015)

19. Ducas, L., Durmus, A., Lepoint, T., Lyubashevsky, V.: Lattice signatures and bimodal Gaussians. In: Canetti, R., Garay, J.A. (eds.) CRYPTO 2013. LNCS, vol. 8042, pp. 40–56. Springer, Heidelberg (2013). doi:10.1007/978-3-642-40041-4_3

20. Ducas, L., Lyubashevsky, V., Prest, T.: Efficient identity-based encryption over NTRU lattices. In: Sarkar, P., Iwata, T. (eds.) ASIACRYPT 2014. LNCS, vol. 8874, pp. 22–41. Springer, Heidelberg (2014). doi:10.1007/978-3-662-45608-8_2

21. Gama, N., Howgrave-Graham, N., Nguyen, P.Q.: Symplectic lattice reduction and NTRU. In: Vaudenay, S. (ed.) EUROCRYPT 2006. LNCS, vol. 4004, pp. 233–253. Springer, Heidelberg (2006). doi:10.1007/11761679_15

22. Gentry, C.: Key recovery and message attacks on NTRU-composite. In: Pfitzmann, B. (ed.) EUROCRYPT 2001. LNCS, vol. 2045, pp. 182–194. Springer, Heidelberg (2001). doi:10.1007/3-540-44987-6_12

23. Gentry, C., Szydlo, M.: Cryptanalysis of the revised NTRU signature scheme. In: Knudsen, L.R. (ed.) EUROCRYPT 2002. LNCS, vol. 2332, pp. 299–320. Springer, Heidelberg (2002). doi:10.1007/3-540-46035-7_20

24. Hanrot, G., Pujol, X., Stehlé, D.: Analyzing blockwise lattice algorithms using dynamical systems. In: Rogaway, P. (ed.) CRYPTO 2011. LNCS, vol. 6841, pp. 447–464. Springer, Heidelberg (2011). doi:10.1007/978-3-642-22792-9_25

25. Hoffstein, J., Pipher, J., Schanck, J.M., Silverman, J.H., Whyte, W., Zhang, Z.: Choosing parameters for NTRUEncrypt. In: Handschuh, H. (ed.) CT-RSA 2017. LNCS, vol. 10159, pp. 3–18. Springer, Cham (2017). doi:10.1007/978-3-319-52153-4_1

26. Hoffstein, J., Pipher, J., Silverman, J.H.: NTRU: a ring-based public key cryptosystem. In: Buhler, J.P. (ed.) ANTS 1998. LNCS, vol. 1423, pp. 267–288. Springer, Heidelberg (1998). doi:10.1007/BFb0054868

27. Howgrave-Graham, N.: A hybrid lattice-reduction and meet-in-the-middle attack against NTRU. In: Menezes, A. (ed.) CRYPTO 2007. LNCS, vol. 4622, pp. 150–169. Springer, Heidelberg (2007). doi:10.1007/978-3-540-74143-5_9

28. Kim, M., Lauter, K.: Private genome analysis through homomorphic encryption. BMC Med. Inf. Decis. Mak. **15**(Suppl 5), S3 (2015)

29. Kirchner, P., Fouque, P.-A.: An improved BKW algorithm for LWE with applications to cryptography and lattices. In: Gennaro, R., Robshaw, M. (eds.) CRYPTO 2015. LNCS, vol. 9215, pp. 43–62. Springer, Heidelberg (2015). doi:10.1007/978-3-662-47989-6_3

30. Lauter, K., López-Alt, A., Naehrig, M.: Private computation on encrypted genomic data. In: Aranha, D.F., Menezes, A. (eds.) LATINCRYPT 2014. LNCS, vol. 8895, pp. 3–27. Springer, Cham (2015). doi:10.1007/978-3-319-16295-9_1

31. Lepoint, T., Naehrig, M.: A comparison of the homomorphic encryption schemes FV and YASHE. In: Pointcheval, D., Vergnaud, D. (eds.) AFRICACRYPT 2014. LNCS, vol. 8469, pp. 318–335. Springer, Cham (2014). doi:10.1007/978-3-319-06734-6_20

32. López-Alt, A., Tromer, E., Vaikuntanathan, V.: On-the-fly multiparty computation on the cloud via multikey fully homomorphic encryption. In: Karloff, H.J., Pitassi, T. (eds.) 44th Annual ACM Symposium on Theory of Computing, 19–22 May, pp. 1219–1234. ACM Press, New York (2012)

33. May, A., Silverman, J.H.: Dimension reduction methods for convolution modular lattices. In: Silverman, J.H. (ed.) CaLC 2001. LNCS, vol. 2146, pp. 110–125. Springer, Heidelberg (2001). doi:10.1007/3-540-44670-2_10

34. Micciancio, D., Goldwasser, S.: Complexity of Lattice Problems: A Cryptographic Perspective, vol. 671. Springer Science & Business Media, Berlin (2012)

35. Micciancio, D., Regev, O.: Worst-case to average-case reductions based on Gaussian measures. SIAM J. Comput. **37**(1), 267–302 (2007)

36. Micciancio, D., Voulgaris, P.: Faster exponential time algorithms for the shortest vector problem. In: Charika, M. (ed.) 21st Annual ACM-SIAM Symposium on Discrete Algorithms, 17–19 January, pp. 1468–1480. ACM-SIAM, Austin (2010)

37. Micciancio, D., Walter, M.: Practical, predictable lattice basis reduction. In: Fischlin, M., Coron, J.-S. (eds.) EUROCRYPT 2016. LNCS, vol. 9665, pp. 820–849. Springer, Heidelberg (2016). doi:10.1007/978-3-662-49890-3_31

38. Pataki, G., Tural, M.: On sublattice determinants in reduced bases. arXiv preprint arXiv:0804.4014 (2008)

39. Peikert, C.: A decade of lattice cryptography. Cryptology ePrint Archive, Report 2015/939 (2015). http://eprint.iacr.org/2015/939

40. Schnorr, C.-P.: A hierarchy of polynomial time lattice basis reduction algorithms. Theor. Comput. Sci. **53**(2), 201–224 (1987)

41. Schnorr, C.P.: Lattice reduction by random sampling and birthday methods. In: Alt, H., Habib, M. (eds.) STACS 2003. LNCS, vol. 2607, pp. 145–156. Springer, Heidelberg (2003). doi:10.1007/3-540-36494-3_14

42. Stehlé, D., Steinfeld, R.: Making NTRU as secure as worst-case problems over ideal lattices. In: Paterson, K.G. (ed.) EUROCRYPT 2011. LNCS, vol. 6632, pp. 27–47. Springer, Heidelberg (2011). doi:10.1007/978-3-642-20465-4_4

Short Generators Without Quantum Computers: The Case of Multiquadratics

Jens Bauch[1](✉), Daniel J. Bernstein[2,3](✉), Henry de Valence[2](✉),
Tanja Lange[2](✉), and Christine van Vredendaal[2](✉)

[1] Department of Mathematics, Simon Fraser University,
8888 University Dr, Burnaby, BC V5A 1S6, Canada
jbauch@sfu.ca

[2] Department of Mathematics and Computer Science,
Technische Universiteit Eindhoven, P.O. Box 513,
5600 MB Eindhoven, The Netherlands
djb@cr.yp.to, hdevalence@hdevalence.ca, tanja@hyperelliptic.org,
c.v.vredendaal@tue.nl

[3] Department of Computer Science, University of Illinois at Chicago,
Chicago, IL 60607-7045, USA

Abstract. Finding a short element g of a number field, given the ideal generated by g, is a classic problem in computational algebraic number theory. Solving this problem recovers the private key in cryptosystems introduced by Gentry, Smart–Vercauteren, Gentry–Halevi, Garg–Gentry–Halevi, et al. Work over the last few years has shown that for some number fields this problem has a surprisingly low *post-quantum* security level. This paper shows, and experimentally verifies, that for some number fields this problem has a surprisingly low *pre-quantum* security level.

Keywords: Public-key encryption · Lattice-based cryptography · Ideal lattices · Soliloquy · Gentry · Smart–Vercauteren · Units · Multiquadratic fields

1 Introduction

Gentry's breakthrough ideal-lattice-based homomorphic encryption system at STOC 2009 [29] was shown several years later to be breakable by a fast quantum

This work was supported by the Netherlands Organisation for Scientific Research (NWO) under grants 613.001.011 and 639.073.005; by the Commission of the European Communities through the Horizon 2020 program under project number 645622 (PQCRYPTO), project number 643161 (ECRYPT-NET), and project number 645421 (ECRYPT-CSA); and by the U.S. National Science Foundation under grant 1314919. Calculations were carried out on the Saber cluster of the Cryptographic Implementations group at Technische Universiteit Eindhoven, and the Saber2 cluster at the University of Illinois at Chicago. Permanent ID of this document: 62a8ffc6a6765bdc6d92754e8ae99f1d. Date: 2017.02.16.

J.-S. Coron and J.B. Nielsen (Eds.): EUROCRYPT 2017, Part I, LNCS 10210, pp. 27–59, 2017.
DOI: 10.1007/978-3-319-56620-7_2

algorithm *if* the underlying number field[1] is chosen as a cyclotomic field (with "small h^+", a condition very frequently satisfied). Cyclotomic fields were considered in Gentry's paper ("As an example, $f(x) = x^n \pm 1$"), in a faster cryptosystem from Smart–Vercauteren [38], and in an even faster cryptosystem from Gentry–Halevi [31]. Cyclotomic fields were used in all of the experiments reported in [31,38]. Cyclotomic fields are also used much more broadly in the literature on lattice-based cryptography, although many cryptosystems are stated for more general number fields.

The secret key in the systems of Gentry, Smart–Vercauteren, and Gentry–Halevi is a short element g of the ring of integers $\mathcal{O}$ of the number field. The public key is the ideal $g\mathcal{O}$ generated by g. The attack has two stages:

- Find some generator of $g\mathcal{O}$, using an algorithm of Biasse and Song [10], building upon a unit-group algorithm of Eisenträger, Hallgren, Kitaev, and Song [24]. This is the stage that uses quantum computation. The best known pre-quantum attacks reuse ideas from NFS, the number-field sieve for integer factorization, and take time exponential in $N^{c+o(1)}$ for a real number c with $0 < c < 1$ where N is the field degree. If N is chosen as an appropriate power of the target security level then the pre-quantum attacks take time exponential in the target security level, but the Biasse–Song attack takes time polynomial in the target security level.
- Reduce this generator to a short generator, using an algorithm introduced by Campbell, Groves, and Shepherd [17, p. 4]: "A simple generating set for the cyclotomic units is of course known. The image of $\mathcal{O}^\times$ under the logarithm map forms a lattice. The determinant of this lattice turns out to be much bigger than the typical log-length of a private key α [i.e., g], so it is easy to recover the causally short private key given *any* generator of $\alpha\mathcal{O}$ e.g. via the LLL lattice reduction algorithm."[2] This is the stage that relies on the field being cyclotomic.

A quantum algorithm for the first stage was stated in [17] before [10], but the effectiveness of this algorithm was disputed by Biasse and Song (see [9]) and was not defended by the authors of [17]. The algorithm in [17, p. 4] quoted above for the second stage does not rely on quantum computers, and its effectiveness is easily checked by experiment.

It is natural to ask whether quantum computers play an essential role in this polynomial-time attack. It is also natural to ask whether the problem of finding g given $g\mathcal{O}$ is weak for *all* number fields, or whether there is something that makes cyclotomic fields particularly weak.

---

[1] We assume some familiarity with algebraic number theory, although we also review some background as appropriate.

[2] Beware that the analysis in [17, p. 4] is incomplete: the analysis correctly states that the secret key is short, but fails to state that the textbook basis for the cyclotomic units is a very good basis; LLL would not be able to find the secret key starting from a bad basis. A detailed analysis of the basis appeared in a followup paper [22] by Cramer, Ducas, Peikert, and Regev.

1.1 Why Focus on the Problem of Finding g Given $g\mathcal{O}$?

There are many other lattice-based cryptosystems that are not broken by the Biasse–Song–Campbell–Groves–Shepherd attack. For example, the attack does not break a more complicated homomorphic encryption system introduced in Gentry's thesis [28,30]; it does not break the classic NTRU system [32]; and it does not break the BCNS [12] and New Hope [3] systems. But the simple problem of finding g given $g\mathcal{O}$ remains of interest for several reasons.

First, given the tremendous interest in Gentry's breakthrough paper, the scientific record should make clear whether Gentry's original cryptosystem is completely broken, or is merely broken for some special number fields.

Second, despite burgeoning interest in post-quantum cryptography, most cryptographic systems today are chosen for their pre-quantum security levels. Fast quantum attacks have certainly not eliminated the interest in RSA and ECC, and also do not end the security analysis of Gentry's system.

Third, the problem of finding a generator of a principal ideal has a long history of being considered hard—even if the ideal actually has a short generator, and even if the output is allowed to be a long generator. There is a list of five "main computational tasks of algebraic number theory" in [19, p. 214], and the problem of finding a generator is the fifth on the list. Smart and Vercauteren describe their key-recovery problem as an "instance of a classical and well studied problem in algorithmic number theory", point to the Buchmann–Maurer–Möller cryptosystem [13] a decade earlier relying on the hardness of this problem, and summarize various slow solutions.

Fourth, this problem has been reused in various attempts to build secure multilinear maps, starting with the Garg–Gentry–Halevi construction [27]. We do not mean to overstate the security or applicability of multilinear maps (see, e.g., [18,21]), but there is a clear pattern of this problem appearing in the design of advanced cryptosystems. Future designers need to understand whether this problem should simply be discarded, or whether it can be a plausible foundation for security.

Fifth, even when cryptosystems rely on more complicated problems, it is natural for cryptanalysts to begin by studying the security of simpler problems. Successful attacks on complicated problems are usually outgrowths of successful attacks on simpler problems. As explained in Appendix B (in the full version of this paper), the Biasse–Song–Campbell–Groves–Shepherd attack has already been reused to attack a more complicated problem.

1.2 Contributions of This Paper

We introduce a *pre-quantum* algorithm that, for a large class of number fields, computes a short g given $g\mathcal{O}$. Plausible heuristic assumptions imply that, for a wide range of number fields in this class, this algorithm (1) has success probability converging rapidly to 100% as the field degree increases and (2) takes time *quasipolynomial* in the field degree.

One advantage of building pre-quantum algorithms is that the algorithms can be tested experimentally. We have implemented our algorithm within the Sage computer-algebra system; the resulting measurements are consistent with our analysis of the performance of the algorithm.

The number fields that we target are *multiquadratics*, such as the degree-256 number field $\mathbb{Q}(\sqrt{2}, \sqrt{3}, \sqrt{5}, \sqrt{7}, \sqrt{11}, \sqrt{13}, \sqrt{17}, \sqrt{19})$, or more generally any $\mathbb{Q}(\sqrt{d_1}, \sqrt{d_2}, \ldots, \sqrt{d_n})$. Sometimes we impose extra constraints for the sake of simplicity: for example, in a few steps we require $d_1, \ldots, d_n$ to be coprime and squarefree, and in several steps we require them to be positive.

A preliminary step in the attack (see Sect. 5.1) is to compute a full-rank subgroup of "the unit group of" the number field (which by convention in algebraic number theory means the unit group of the ring of integers of the field): namely, the subgroup generated by the units of all real quadratic subfields. We dub this subgroup the set of "multiquadratic units" by analogy to the standard terminology "cyclotomic units", with the caveat that "multiquadratic units" (like "cyclotomic units") are not guaranteed to be *all* units.

The degree-256 example above has exactly 255 real quadratic subfields

$$\mathbb{Q}(\sqrt{2}), \mathbb{Q}(\sqrt{3}), \mathbb{Q}(\sqrt{6}), \ldots, \mathbb{Q}(\sqrt{2 \cdot 3 \cdot 5 \cdot 7 \cdot 11 \cdot 13 \cdot 17 \cdot 19}).$$

Each of these has a unit group quickly computable by standard techniques. For example, the units of $\mathbb{Q}(\sqrt{2})$ are $\pm(1 + \sqrt{2})^{\mathbb{Z}}$, and the units of the last field are $\pm(6915878018249487671 9 + 22205900901368228\sqrt{2 \cdot 3 \cdot 5 \cdot 7 \cdot 11 \cdot 13 \cdot 17 \cdot 19})^{\mathbb{Z}}$.

This preliminary step generally becomes slower as $d_1, \ldots, d_n$ grow, but it takes time quasipolynomial in the field degree N, assuming that $d_1, \ldots, d_n$ are quasipolynomial in N.

In the next step (the rest of Sect. 5) we go far beyond the multiquadratic units: we quickly compute the *entire* unit group of the multiquadratic field. This is important because the gap between the multiquadratic units and all units would interfere, potentially quite heavily, with the success probability of our algorithm, the same way that a "large h^+" (a large gap between cyclotomic units and all units) would interfere with the success probability of the cyclotomic attacks. Note that computing the unit group is another of the five "main computational tasks of algebraic number theory" listed in [19]. There is an earlier algorithm by Wada [42] to compute the unit group of a multiquadratic field, but that algorithm takes exponential time.

We then go even further (Sect. 6), quickly computing a generator of the input ideal. The generator algorithm uses techniques similar to, but not the same as, the unit-group algorithm. The unit-group computation starts from unit groups computed recursively in three subfields, while the generator computation starts from generators computed recursively in those subfields *and* from the unit group of the top field.

There is a very easy way to extract *short* generators when $d_1, \ldots, d_n$ are large enough, between roughly N and any quasipolynomial bound. This condition is satisfied by a wide range of fields of each degree.

We do more work to extend the applicability of our attack to allow smaller $d_1, \ldots, d_n$, using LLL to shorten units and indirectly generators. Analysis of this extension is difficult, but experiments suggest that the success probability converges to 1 even when $d_1, \ldots, d_n$ are chosen to be as small as the first n primes starting from n^2.

There are many obvious opportunities for precomputation in our algorithm, and in particular the unit group can be reused for attacking many targets $g\mathcal{O}$ in the same field. We separately measure the cost of computing the unit group and the cost of subsequently finding a generator.

1.3   Why Focus on Multiquadratics?

Automorphisms and subfields play critical roles in several strategies to attack discrete logarithms. These strategies complicate security analysis, and in many cases they have turned into successful attacks. For example, small-characteristic multiplicative-group discrete logarithms are broken in quasipolynomial time; there are ongoing disputes regarding a strategy to attack small-characteristic ECC; and very recently pairing-based cryptography has suffered a significant drop in security level, because of new optimizations in attacks exploiting subfields of the target field. See, e.g., [5, 26, 33].

Do automorphisms and subfields also damage the security of lattice-based cryptography? We chose multiquadratics as an interesting test case because they have a huge number of subfields, presumably amplifying and clarifying any impact that subfields might have upon security.

A degree-2^n multiquadratic field is Galois: i.e., it has 2^n automorphisms, the maximum possible for a degree-2^n field. The Galois group, the group of automorphisms, is isomorphic to $(\mathbb{Z}/2)^n$. The number of subfields of the field is the number of subgroups of $(\mathbb{Z}/2)^n$, i.e., the number of subspaces of an n-dimensional vector space over $\mathbb{F}_2$. The number of k-dimensional subspaces is the 2-binomial coefficient

$$\binom{n}{k}_2 = \frac{(2^n - 1)(2^{n-1} - 1) \cdots (2^1 - 1)}{(2^k - 1)(2^{k-1} - 1) \cdots (2^1 - 1)(2^{n-k} - 1)(2^{n-k-1} - 1) \cdots (2^1 - 1)},$$

which is approximately $2^{n^2/4}$ for $k \approx n/2$. This turns out to be overkill from the perspective of our attack: as illustrated in Figs. 5.1 and 5.2, the number of subfields we use ends up essentially linear in 2^n.

2   Multiquadratic Fields

A **multiquadratic field** is, by definition, a field that can be written in the form $\mathbb{Q}(\sqrt{r_1}, \ldots, \sqrt{r_m})$ where $(r_1, \ldots, r_m)$ is a finite sequence of rational numbers. The notation $\mathbb{Q}(\sqrt{r_1}, \ldots, \sqrt{r_m})$ means the smallest subfield of $\mathbb{C}$, the field of complex numbers, that contains $\sqrt{r_1}, \ldots, \sqrt{r_m}$.

When we write $\sqrt{r}$ for a nonnegative real number r, we mean specifically the nonnegative square root of r. When we write $\sqrt{r}$ for a negative real number r, we mean specifically $i\sqrt{-r}$, where i is the standard square root of -1 in

$\mathbb{C}$; for example, $\sqrt{-2}$ means $i\sqrt{2}$. These choices do not affect the definition of $\mathbb{Q}(\sqrt{r_1}, \ldots, \sqrt{r_m})$, but many other calculations rely on each $\sqrt{r}$ having a definite value.

See full version of paper on multiquad.cr.yp.to for proofs.

Theorem 2.1. *Let n be a nonnegative integer. Let $d_1, \ldots, d_n$ be integers such that, for each nonempty subset $J \subseteq \{1, \ldots, n\}$, the product $\prod_{j \in J} d_j$ is not a square. Then the 2^n complex numbers $\prod_{j \in J} \sqrt{d_j}$ for all subsets $J \subseteq \{1, \ldots, n\}$ form a basis for the multiquadratic field $\mathbb{Q}(\sqrt{d_1}, \ldots, \sqrt{d_n})$ as a $\mathbb{Q}$-vector space. Furthermore, for each $j \in \{1, \ldots, n\}$ there is a unique field automorphism of $\mathbb{Q}(\sqrt{d_1}, \ldots, \sqrt{d_n})$ that preserves $\sqrt{d_1}, \ldots, \sqrt{d_n}$ except for mapping $\sqrt{d_j}$ to $-\sqrt{d_j}$.*

Consequently $\mathbb{Q}(\sqrt{d_1}, \ldots, \sqrt{d_n})$ is a degree-2^n number field.

Theorem 2.2. *Every multiquadratic field can be expressed in the form of Theorem 2.1 with each d_j squarefree.*

3   Fast Arithmetic in Multiquadratic Fields

See full version of paper on multiquad.cr.yp.to.

4   Recognizing Squares

This section explains how to recognize squares in a multiquadratic field $L = \mathbb{Q}(\sqrt{d_1}, \ldots, \sqrt{d_n})$. The method does not merely check whether a single element $u \in L$ is a square: given nonzero $u_1, \ldots, u_r \in L$, the method rapidly identifies the set of exponent vectors $(e_1, \ldots, e_r) \in \mathbb{Z}^r$ such that $u_1^{e_1} \cdots u_r^{e_r}$ is a square.

The method here was introduced by Adleman [2] as a speedup to NFS. The idea is to apply a group homomorphism χ from $L^\times$ to $\{-1, 1\}$, or more generally from T to $\{-1, 1\}$, where T is a subgroup of $L^\times$ containing $u_1, \ldots, u_r$. Then χ reveals a linear constraint, hopefully nontrivial, on $(e_1, \ldots, e_r)$ modulo 2. Combining enough constraints reveals the space of $(e_1, \ldots, e_r) \bmod 2$.

One choice of χ is the sign of a real embedding of L, but this is a limited collection of χ (and empty if L is complex). Adleman suggested instead taking χ as a quadratic character defined by a prime ideal. There is an inexhaustible supply of prime ideals, and thus of these quadratic characters.

Section 3.6 (in the full version of this paper) used this idea for $L = \mathbb{Q}$, but only for small r (namely $r = n$), where one can afford to try 2^r primes. This section handles arbitrary multiquadratics and allows much larger r.

4.1   Computing Quadratic Characters

Let q be an odd prime number modulo which all the d_i are nonzero squares. For each i, let s_i be a square root of d_i modulo q. The map $\mathbb{Z}[x_1, \ldots, x_n] \to \mathbb{F}_q$ defined by $x_i \mapsto s_i$ and reducing coefficients modulo q induces a homomorphism

$\mathbb{Z}[x_1, \ldots, x_n]/(x_1^2 - d_1, \ldots, x_n^2 - d_n) \to \mathbb{F}_q$, or equivalently a homomorphism $\varphi : \mathbb{Z}[\sqrt{d_1}, \ldots, \sqrt{d_n}] \to \mathbb{F}_q$.

Let $\mathfrak{P}$ be the kernel of φ. Then $\mathfrak{P}$ is a degree-1 prime ideal of $\mathbb{Z}[\sqrt{d_1}, \ldots, \sqrt{d_n}]$ above q, i.e., a prime ideal of prime norm q. Write $\mathcal{O}_L$ for the ring of integers of L; then $\mathfrak{P}$ extends to a unique degree-1 prime ideal of $\mathcal{O}_L$. The map φ extends to the set R_φ of all $u \in L$ having nonnegative valuation at this prime ideal. For each $u \in R_\varphi$ define $\chi(u) \in \{-1, 0, 1\}$ as the Legendre symbol of $\varphi(u) \in \mathbb{F}_q$. Then $\chi(uu') = \chi(u)\chi(u')$, since $\varphi(uu') = \varphi(u)\varphi(u')$ and the Legendre symbol is multiplicative. In particular, $\chi(u^2) \in \{0, 1\}$.

More explicitly: Given a polynomial $u \in \mathbb{Z}[x_1, \ldots, x_n]/(x_1^2 - d_1, \ldots, x_n^2 - d_n)$ represented as coefficients of 1, x_1, x_2, $x_1 x_2$, etc., first take all coefficients modulo q to obtain $u \bmod q \in \mathbb{F}_q[x_1, \ldots, x_n]/(x_1^2 - d_1, \ldots, x_n^2 - d_n)$. Then substitute $x_n \mapsto s_n$: i.e., write $u \bmod q$ as $u_0 + u_1 x_n$, where $u_0, u_1 \in \mathbb{F}_q[x_1, \ldots, x_{n-1}]/(x_1^2 - d_1, \ldots, x_{n-1}^2 - d_{n-1})$, and compute $u_0 + u_1 s_n$. Inside this result substitute $x_{n-1} \mapsto s_{n-1}$ similarly, and so on through $x_1 \mapsto s_1$, obtaining $\varphi(u) \in \mathbb{F}_q$. Finally compute the Legendre symbol modulo q to obtain $\chi(u)$.

As in Sect. 3 (in the full version of this paper), assume that each coefficient of u has at most B bits, and choose q (using the GoodPrime function from Sect. 3.2) to have $n^{O(1)}$ bits. Then the entire computation of $\chi(u)$ takes time essentially NB, mostly to reduce coefficients modulo q. The substitutions $x_j \mapsto s_j$ involve a total of $O(N)$ operations in $\mathbb{F}_q$, and the final Legendre-symbol computation takes negligible time.

More generally, any element of L is represented as u/h for a positive integer denominator h. Assume that q is coprime to h; this is true with overwhelming probability when q is chosen randomly. (It is also guaranteed to be true for any $u/h \in \mathcal{O}_L$ represented in lowest terms, since q is coprime to $2d_1 \cdots d_n$.) Then $\varphi(u/h)$ is simply $\varphi(u)/h$, and computing the Legendre symbol produces $\chi(u/h)$.

4.2 Recognizing Squares Using Many Quadratic Characters

Let $\chi_1, \ldots, \chi_m$ be quadratic characters. Define T as the subset of L on which all χ_i are defined and nonzero. Then T is a subgroup of $L^\times$, the intersection of the unit groups of the rings R_φ defined above. Define a group homomorphism $X : T \to (\mathbb{Z}/2)^m$ as $u \mapsto (\log_{-1} \chi_1, \ldots, \log_{-1} \chi_m)$.

Given nonzero $u_1, \ldots, u_r \in L$, choose m somewhat larger than r, and then choose $\chi_1, \ldots, \chi_m$ randomly using GoodPrime. Almost certainly $u_1, \ldots, u_r \in T$; if any $\chi(u_j)$ turns out to be undefined or zero, simply switch to another prime.

Define U as the subgroup of T generated by $u_1, \ldots, u_r$. If a product $\pi = u_1^{e_1} \cdots u_r^{e_r}$ is a square in L then its square root is in T so $X(\pi) = 0$, i.e., $e_1 X(u_1) + \cdots + e_r X(u_r) = 0$. Conversely, if $X(\pi) = 0$ and m is somewhat larger than r then almost certainly π is a square in L, as we now explain.

The group $U/(U \cap L^2)$ is an $\mathbb{F}_2$-vector space of dimension at most r, so its dual group $\mathrm{Hom}(U/(U \cap L^2), \mathbb{Z}/2)$ is also an $\mathbb{F}_2$-vector space of dimension at most r. As in [16, Sect. 8], we heuristically model $\log_{-1} \chi_1, \ldots, \log_{-1} \chi_m$ as independent uniform random elements of this dual; then they span the dual with probability at least $1 - 1/2^{m-r}$ by [16, Lemma 8.2]. If they do span the dual, then any $\pi \in U$ with $X(\pi) = 0$ must have $\pi \in U \cap L^2$.

The main argument for this heuristic is the fact that, asymptotically, prime ideals are uniformly distributed across the dual. Restricting to degree-1 prime ideals does not affect this heuristic: prime ideals are counted by norm, so asymptotically 100% of all prime ideals have degree 1. Beware that taking more than one prime ideal over a single prime number q would not justify the same heuristic.

Computing $X(u_1), \ldots, X(u_r)$ involves $mr \approx r^2$ quadratic-character computations, each taking time essentially NB. We do better by using remainder trees to merge the reductions of B-bit coefficients mod q across all r choices of q; this reduces the total time from essentially $r^2 NB$ to essentially $rN(r + B)$.

We write EnoughCharacters($L, (v_1, \ldots, v_s)$) for a list of m randomly chosen characters that are defined and nonzero on $v_1, \ldots, v_s$. In higher-level algorithms in this paper, the group $\langle v_1, \ldots, v_s \rangle$ can always be expressed as $\langle u_1, \ldots, u_r \rangle$ with $r \leq N + 1$, and we choose m as $N + 64$, although asymptotically one should replace 64 by, e.g., $\sqrt{N}$. The total time to compute $X(u_1), \ldots, X(u_R)$ is essentially $N^2(N + B)$. The same heuristic states that these characters have probability at most $1/2^{63}$ (or asymptotically at most $1/2^{\sqrt{N}-1}$) of viewing some non-square $u_1^{e_1} \cdots u_r^{e_r}$ as a square. Our experiments have not encountered any failing square-root computations.

5   Computing Units

This section presents a fast algorithm to compute the unit group $\mathcal{O}_L^\times$ of a multiquadratic field L. For simplicity we assume that L is real, i.e., that $L \subseteq \mathbb{R}$. Note that a multiquadratic field is real if and only if it is totally real, i.e., every complex embedding $L \to \mathbb{C}$ has image in $\mathbb{R}$. For $L = \mathbb{Q}(\sqrt{d_1}, \ldots, \sqrt{d_n})$ this is equivalent to saying that each d_j is nonnegative.

Like Wada [42], we recursively compute unit groups for three subfields K_σ, K_τ, $K_{\sigma\tau}$, and then use the equation $u^2 = N_{L:K_\sigma}(u) N_{L:K_\tau}(u) / \sigma(N_{L:K_{\sigma\tau}}(u))$ to glue these groups together into a group U between $\mathcal{O}_L^\times$ and $(\mathcal{O}_L^\times)^2$. At this point Wada resorts to brute-force search to identify the squares in U, generalizing an approach taken by Kubota in [34] for degree-4 multiquadratics ("biquadratics"). We reduce exponential time to polynomial time by using quadratic characters as explained in Sect. 4.

5.1   Fundamental Units of Quadratic Fields

A **quadratic field** is, by definition, a degree-2 multiquadratic field; i.e., a field of the form $\mathbb{Q}(\sqrt{d})$, where d is a non-square integer.

Fix a positive non-square integer d. Then $L = \mathbb{Q}(\sqrt{d})$ is a real quadratic field, and the unit group $\mathcal{O}_L^\times$ is

$$\{\ldots, -\varepsilon^2, -\varepsilon, -1, -\varepsilon^{-1}, -\varepsilon^{-2}, \ldots, \varepsilon^{-2}, \varepsilon^{-1}, 1, \varepsilon, \varepsilon^2, \ldots\}$$

for a unique $\varepsilon \in \mathcal{O}_L^\times$ with $\varepsilon > 1$. This ε, the smallest element of $\mathcal{O}_L^\times$ larger than 1, is the **normalized fundamental unit** of $\mathcal{O}_L$. For example, the normalized fundamental unit is $1 + \sqrt{2}$ for $d = 2$; $2 + \sqrt{3}$ for $d = 3$; and $(1 + \sqrt{5})/2$ for $d = 5$.

Sometimes the literature says "fundamental unit" instead of "normalized fundamental unit", but sometimes it defines all of $\varepsilon, -\varepsilon, 1/\varepsilon, -1/\varepsilon$ as "fundamental units". The phrase "normalized fundamental unit" is unambiguous.

The size of the normalized fundamental unit ε is conventionally measured by the **regulator** $R = \ln(\varepsilon)$. A theorem by Hua states that $R < \sqrt{d}(\ln(4d) + 2)$, and experiments suggest that R is typically $d^{1/2+o(1)}$, although it is often much smaller. Write ε as $a + b\sqrt{d}$ with $a, b \in \mathbb{Q}$; then both $2a$ and $2b\sqrt{d}$ are very close to $\exp(R)$, and there are standard algorithms that compute a, b in time essentially R, i.e., at most essentially $d^{1/2}$. See generally [36, 43].

For our time analysis we assume that d is quasipolynomial in N, i.e., $\log d \in (\log N)^{O(1)}$. Then the time to compute ε is also quasipolynomial in N.

Take, for example, $d = d_1 \cdots d_n$, where $d_1, \ldots, d_n$ are the first n primes, and write $N = 2^n$. The product of primes $\leq y$ is approximately $\exp(y)$, so $\ln d \approx n \ln n = (\log_2 N) \ln \log_2 N$. As a larger example, if $d_1, \ldots, d_n$ are primes between N^3 and N^4, and again $d = d_1 \cdots d_n$, then $\log_2 d$ is between $3n^2$ and $4n^2$, i.e., between $3(\log_2 N)^2$ and $4(\log_2 N)^2$. In both of these examples, d is quasipolynomial in N.

Subexponential Algorithms. There are much faster algorithms that compute ε as a product of powers of smaller elements of L. There is a deterministic algorithm that provably takes time essentially $R^{1/2}$, i.e., at most essentially $d^{1/4}$; see [11]. Heuristic algorithms take subexponential time $\exp((\ln(d))^{1/2+o(1)})$, and thus time polynomial in N if $\ln(d) \in O((\log N)^{2-\epsilon})$; see [1, 15, 19, 40]. Quantum algorithms are even faster, as mentioned in the introduction, but in this paper we focus on pre-quantum algorithms.

This representation of units is compatible with computing products, quotients, quadratic characters (see Sect. 4), and automorphisms, but we also need to be able to compute square roots. One possibility here is to generalize from "product of powers" to any algebraic algorithm, i.e., any chain of additions, subtractions, multiplications, and divisions. This seems adequate for our square-root algorithm in Sect. 3.7 (in the full version of this paper): for example, h_0 inside Algorithm 3.3 can be expressed as the chain $(h + \sigma(h))/2$ for an appropriate automorphism σ, and the base case involves square roots of small integers that can be computed explicitly. However, it is not clear whether our recursive algorithms produce chains of polynomial size. We do not explore this possibility further.

5.2   Units in Multiquadratic Fields

Let $d_1, \ldots, d_n$ be integers satisfying the conditions of Theorem 2.1. Assume further that $d_1, \ldots, d_n$ are positive. Then $L = \mathbb{Q}(\sqrt{d_1}, \ldots, \sqrt{d_n})$ is a real multiquadratic field.

This field has $N - 1 = 2^n - 1$ quadratic subfields, all of which are real. Each quadratic subfield is constructed as follows: take one of the $N - 1$ nonempty subsets $J \subseteq \{1, \ldots, n\}$; define $d_J = \prod_{j \in J} d_j$; the subfield is $\mathbb{Q}(\sqrt{d_J})$. We write the normalized fundamental units of these $N - 1$ quadratic subfields as $\varepsilon_1, \ldots, \varepsilon_{N-1}$.

The set of **multiquadratic units** of L is the subgroup $\langle -1, \varepsilon_1, \ldots, \varepsilon_{N-1} \rangle$ of $\mathcal{O}_L^\times$; equivalently, the subgroup of $\mathcal{O}_L^\times$ generated by -1 and all units of rings of integers of quadratic subfields of L. (The "-1 and" can be suppressed except for $L = \mathbb{Q}$.) A unit in $\mathcal{O}_L$ is not necessarily a multiquadratic unit, but Theorem 5.2 states that its Nth power must be a multiquadratic unit.

The group $\mathcal{O}_L^\times$ is isomorphic to $(\mathbb{Z}/2) \times \mathbb{Z}^{N-1}$ by Dirichlet's unit theorem. For $N \geq 2$ this isomorphism takes the Nth powers to $\{0\} \times (N\mathbb{Z})^{N-1}$, a subgroup having index $2^{1+n(N-1)}$. The index of the multiquadratic units in $\mathcal{O}_L^\times$ is therefore a divisor of $2^{1+n(N-1)}$. One corollary is that $\varepsilon_1, \ldots, \varepsilon_{N-1}$ are multiplicatively independent: if $\prod \varepsilon_j^{a_j} = 1$, where each $a_j \in \mathbb{Z}$, then each $a_j = 0$.

Lemma 5.1. *Let L be a real multiquadratic field and let σ, τ be distinct non-identity automorphisms of L. Define $\sigma\tau = \sigma \circ \tau$. For $\ell \in \{\sigma, \tau, \sigma\tau\}$ let K_ℓ be the subfield of L fixed by ℓ. Define $U = \mathcal{O}_{K_\sigma}^\times \cdot \mathcal{O}_{K_\tau}^\times \cdot \sigma(\mathcal{O}_{K_{\sigma\tau}}^\times)$. Then*

$$(\mathcal{O}_L^\times)^2 \leq U \leq \mathcal{O}_L^\times.$$

Proof. $\mathcal{O}_{K_\sigma}^\times$, $\mathcal{O}_{K_\tau}^\times$, and $\mathcal{O}_{K_{\sigma\tau}}^\times$ are subgroups of $\mathcal{O}_L^\times$. The automorphism σ preserves $\mathcal{O}_L^\times$, so $\sigma(\mathcal{O}_{K_{\sigma\tau}}^\times)$ is a subgroup of $\mathcal{O}_L^\times$. Hence U is a subgroup of $\mathcal{O}_L^\times$.

For the first inclusion, let $u \in \mathcal{O}_L^\times$. Then $N_{L:K_\ell}(u) \in \mathcal{O}_{K_\ell}^\times$ for $\ell \in \{\sigma, \tau, \sigma\tau\}$. Each non-identity automorphism of L has order 2, so in particular each $\ell \in \{\sigma, \tau, \sigma\tau\}$ has order 2 (if $\sigma\tau$ is the identity then $\sigma = \sigma\sigma\tau = \tau$, contradiction), so $N_{L:K_\ell}(u) = u \cdot \ell(u)$. We thus have

$$\frac{N_{L:K_\sigma}(u) N_{L:K_\tau}(u)}{\sigma(N_{L:K_{\sigma\tau}}(u))} = \frac{u \cdot \sigma(u) \cdot u \cdot \tau(u)}{\sigma(u \cdot \sigma\tau(u))} = u^2.$$

Hence $u^2 = N_{L:K_\sigma}(u) N_{L:K_\tau}(u) \sigma(N_{L:K_{\sigma\tau}}(u^{-1})) \in U$. This is true for each $u \in \mathcal{O}_L^\times$, so $(\mathcal{O}_L^\times)^2$ is a subgroup of U. $\qquad\square$

Theorem 5.2. *Let L be a real multiquadratic field of degree N. Let Q be the group of multiquadratic units of L. Then $\mathcal{O}_L^\times = Q$ if $N = 1$, and $(\mathcal{O}_L^\times)^{N/2} \leq Q$ if $N \geq 2$. In both cases $(\mathcal{O}_L^\times)^N \leq Q$.*

Proof. Induct on N. If $N = 1$ then $L = \mathbb{Q}$ so $\mathcal{O}_L^\times = \langle -1 \rangle = Q$. If $N = 2$ then L is a real quadratic field so $\mathcal{O}_L^\times = \langle -1, \varepsilon_1 \rangle = Q$ where ε_1 is the normalized fundamental unit of L.

Assume from now on that $N \geq 4$. By Theorem 2.2, L can be expressed as $\mathbb{Q}(\sqrt{d_1}, \ldots, \sqrt{d_n})$ where $d_1, \ldots, d_n$ are positive integers meeting the conditions of Theorem 2.1 and $N = 2^n$.

Define σ as the automorphism of L that preserves $\sqrt{d_1}, \ldots, \sqrt{d_n}$ except for negating $\sqrt{d_n}$. The field K_σ fixed by σ is $\mathbb{Q}(\sqrt{d_1}, \ldots, \sqrt{d_{n-1}})$, a real multiquadratic field of degree $N/2$. Write Q_σ for the group of multiquadratic units of K_σ. By the inductive hypothesis, $(\mathcal{O}_{K_\sigma}^\times)^{N/4} \leq Q_\sigma \leq Q$.

Define τ as the automorphism of L that preserves $\sqrt{d_1}, \ldots, \sqrt{d_n}$ except for negating $\sqrt{d_{n-1}}$. Then the field K_τ fixed by τ is $\mathbb{Q}(\sqrt{d_1}, \ldots, \sqrt{d_{n-2}}, \sqrt{d_n})$, and the field $K_{\sigma\tau}$ fixed by $\sigma\tau$ is $\mathbb{Q}(\sqrt{d_1}, \ldots, \sqrt{d_{n-2}}, \sqrt{d_{n-1}d_n})$. Both of these are real multiquadratic fields of degree $N/2$, so $(\mathcal{O}_{K_\tau}^\times)^{N/4} \leq Q$ and $(\mathcal{O}_{K_{\sigma\tau}}^\times)^{N/4} \leq Q$. The automorphism σ preserves Q, so $\sigma(\mathcal{O}_{K_{\sigma\tau}}^\times)^{N/4} \leq Q$.

By Lemma 5.1, $(\mathcal{O}_L^\times)^2 \leq \mathcal{O}_{K_\sigma}^\times \cdot \mathcal{O}_{K_\tau}^\times \cdot \sigma(\mathcal{O}_{K_{\sigma\tau}}^\times)$. Simply take $(N/4)$th powers: $(\mathcal{O}_L^\times)^{N/2} \leq (\mathcal{O}_{K_\sigma}^\times)^{N/4} \cdot (\mathcal{O}_{K_\tau}^\times)^{N/4} \cdot \sigma(\mathcal{O}_{K_{\sigma\tau}}^\times)^{N/4} \leq Q$. $\qquad\square$

5.3 Representing Units: Logarithms and Approximate Logarithms

Sections 5.4 and 5.5 will use Lemma 5.1, quadratic characters, and square-root computations to obtain a list of generators for $\mathcal{O}_L^\times$. However, this is usually far from a minimal-size list of generators. Given this list of generators we would like to produce a **basis** for $\mathcal{O}_L^\times$. This means a list of $N-1$ elements $u_1, \ldots, u_{N-1} \in \mathcal{O}_L^\times$ such that each element of $\mathcal{O}_L^\times$ can be written uniquely as $\zeta u_1^{e_1} \cdots u_{N-1}^{e_{N-1}}$ where ζ is a root of unity; i.e., as $\pm u_1^{e_1} \cdots u_{N-1}^{e_{N-1}}$. In other words, it is a list of independent generators of $\mathcal{O}_L^\times/\{\pm 1\}$.

A basis $u_1, \ldots, u_{N-1}$ for $\mathcal{O}_L^\times$ is traditionally viewed as a lattice basis in the usual sense: specifically, as the basis $\mathrm{Log}\, u_1, \ldots, \mathrm{Log}\, u_{N-1}$ for the lattice $\mathrm{Log}\, \mathcal{O}_L^\times$, where Log is Dirichlet's logarithm map. However, this view complicates the computation of a basis. We instead view a basis $u_1, \ldots, u_{N-1}$ for $\mathcal{O}_L^\times$ as a basis $\mathrm{ApproxLog}\, u_1, \ldots, \mathrm{ApproxLog}\, u_{N-1}$ for the lattice $\mathrm{ApproxLog}\, \mathcal{O}_L^\times$, where $\mathrm{ApproxLog}$ is an "approximate logarithm map". We define our approximate logarithm map here, explain why it is useful, and explain how we use the approximate logarithm map in our representation of units. In Sect. 5.5 we use $\mathrm{ApproxLog}$ to reduce a list of generators to a basis.

Dirichlet's Logarithm Map. Let $\sigma_1, \sigma_2, \ldots, \sigma_N$ be (in some order) the embeddings of L into $\mathbb{C}$, i.e., the ring homomorphisms $L \to \mathbb{C}$. Since L is Galois, these are exactly the automorphisms of L. **Dirichlet's logarithm map** $\mathrm{Log} : L^\times \to \mathbb{R}^N$ is defined as follows:

$$\mathrm{Log}(u) = (\ln|\sigma_1(u)|, \ln|\sigma_2(u)|, \ldots, \ln|\sigma_N(u)|).$$

This map has several important properties. It is a group homomorphism from the multiplicative group $L^\times$ to the additive group $\mathbb{R}^N$. The kernel of Log restricted to $\mathcal{O}_L^\times$ is the cyclic group of roots of unity in L, namely $\{1, -1\}$. The image $\mathrm{Log}(\mathcal{O}_L^\times)$ forms a lattice of rank $N-1$, called the *log-unit lattice*.

Given units $u_1, \ldots, u_b$ generating $\mathcal{O}_L^\times$, one can compute $\mathrm{Log}(u_1), \ldots, \mathrm{Log}(u_b)$ in $\mathbb{R}^N$, and then reduce these images to linearly independent vectors in $\mathbb{R}^N$ by a chain of additions and subtractions, obtaining a basis for the log-unit lattice. Applying the corresponding chain of multiplications and divisions to the original units produces a basis for $\mathcal{O}_L^\times$.

However, elements of $\mathbb{R}$ are conventionally represented as nearby rational numbers. "Computing" $\mathrm{Log}(u_1), \ldots, \mathrm{Log}(u_b)$ thus means computing nearby vectors of rational numbers. The group generated by these vectors usually has rank larger than $N - 1$: instead of producing $N - 1$ linearly independent vectors and $b - (N - 1)$ zero vectors, reduction can produce as many as b linearly independent vectors.

One can compute approximate linear dependencies by paying careful attention to floating-point errors. An alternative is to use p-adic techniques as in [7]. Another alternative is to represent logarithms in a way that allows all of the necessary real operations to be carried out without error: for example, one can verify that Log u > Log v by using interval arithmetic in sufficiently high precision, and one can verify that Log u = Log v by checking that u/v is a root of unity.

Approximate Logarithms. We instead sidestep these issues by introducing an approximate logarithm function ApproxLog as a replacement for the logarithm function Log. This new function is a group homomorphism from $\mathcal{O}_L^\times$ to $\mathbb{R}^N$. Its image is a lattice of rank $N - 1$, which we call the *approximate unit lattice*. Its kernel is the group of roots of unity in L. The advantage of ApproxLog over Log is that all the entries of ApproxLog(u) are rationals, allowing exact linear algebra.

To define ApproxLog, we first choose N linearly independent vectors

$$(1, 1, \ldots, 1), \text{ApproxLog}(\varepsilon_1), \ldots, \text{ApproxLog}(\varepsilon_{N-1}) \in \mathbb{Q}^N,$$

where $\varepsilon_1, \ldots, \varepsilon_{N-1}$ are the normalized fundamental units of the quadratic subfields of L as before; $(1, 1, \ldots, 1)$ is included here to simplify other computations. We then extend the definition by linearity to the group $\langle -1, \varepsilon_1, \ldots, \varepsilon_{N-1} \rangle$ of multiquadratic units: if

$$u = \pm \prod_{j=1}^{N-1} \varepsilon_j^{e_j}$$

then we define ApproxLog(u) as $\sum_j e_j$ ApproxLog(ε_j). Finally, we further extend the definition by linearity to all of $\mathcal{O}_L^\times$: if $u \in \mathcal{O}_L^\times$ then u^N is a multiquadratic unit by Theorem 5.2, and we define ApproxLog(u) as ApproxLog(u^N)/N. It is easy to check that ApproxLog is a well-defined group homomorphism.

For example, one can take ApproxLog(ε_1) = $(1, 0, \ldots, 0, 0)$, ApproxLog (ε_2) = $(0, 1, \ldots, 0, 0)$, and so on through ApproxLog(ε_{N-1}) = $(0, 0, \ldots, 1, 0)$. Then ApproxLog(u) = $(e_1/N, e_2/N, \ldots, e_{N-1}/N, 0)$ if $u^N = \pm \varepsilon_1^{e_1} \varepsilon_2^{e_2} \cdots \varepsilon_{N-1}^{e_{N-1}}$. In other words, write each unit modulo ± 1 as a product of powers of $\varepsilon_1, \ldots, \varepsilon_{N-1}$; ApproxLog is then the exponent vector.

We actually define ApproxLog to be numerically much closer to Log. We choose a precision parameter β, and we choose each entry of ApproxLog(ε_j) to be a multiple of $2^{-\beta}$ within $2^{-\beta}$ of the corresponding entry of Log(ε_j). Specifically, we build ApproxLog(ε_j) as follows:

- Compute the regulator $R = \ln(\varepsilon_j)$ to slightly more than $\beta + \log_2 R$ bits of precision.
- Round the resulting approximation to a (nonzero) multiple R' of $2^{-\beta}$.

- Build a vector with R' at the $N/2$ positions i for which $\sigma_i(\varepsilon_j) = \varepsilon_j$, and with $-R'$ at the remaining $N/2$ positions i.

The resulting vectors $\text{ApproxLog}(\varepsilon_1), \ldots, \text{ApproxLog}(\varepsilon_{N-1})$ are orthogonal to each other and to $(1, 1, \ldots, 1)$.

How Units are Represented. Each unit in Algorithms 5.1 and 5.2 is implicitly represented as a pair consisting of (1) the usual representation of an element of L and (2) the vector $\text{ApproxLog}(u)$. After the initial computation of $\ln(\varepsilon_j)$ for each j, all subsequent units are created as products (or quotients) of previous units, with sums (or differences) of the ApproxLog vectors; or square roots of previous units, with the ApproxLog vectors multiplied by $1/2$. This approach ensures that we do not have to compute $\ln|\sigma(u)|$ for the subsequent units u.

As mentioned in Sect. 5.1, we assume that each quadratic field $\mathbb{Q}(\sqrt{d})$ has $\log d \in (\log N)^{O(1)} = n^{O(1)}$, so $\log R \in n^{O(1)}$. We also take $\beta \in n^{O(1)}$, so each entry of $\text{ApproxLog}(\varepsilon_j)$ has $n^{O(1)}$ bits. One can deduce an $n^{O(1)}$ bound on the number of bits in any entry of any ApproxLog vector used in our algorithms, so adding two such vectors takes time $n^{O(1)}N$, i.e., essentially N.

For comparison, recall that multiplication takes time essentially NB, where B is the maximum number of bits in any *coefficient* of the field elements being multiplied. For normalized fundamental units, this number of bits is essentially R, i.e., quasipolynomial in N, rather than $\log R$, i.e., polynomial in n.

5.4 Pinpointing Squares of Units Inside Subgroups of the Unit Group

Algorithm 5.1, UnitsGivenSubgroup, is given generators $u_1, \ldots, u_b$ of any group U with $(\mathcal{O}_L^\times)^2 \leq U \leq \mathcal{O}_L^\times$. It outputs generators of $\mathcal{O}_L^\times/\{\pm 1\}$.

The algorithm begins by building enough characters $\chi_1, \ldots, \chi_m$ that are defined and nonzero on U. Recall from Sect. 4.2 that m is chosen to be slightly larger than N.

For each $u \in U$ define $X(u)$ as the vector $(\log_{-1}(\chi_1(u)), \ldots, \log_{-1}(\chi_m(u))) \in (\mathbb{Z}/2)^m$. If $u \in (\mathcal{O}_L^\times)^2$ then $X(u) = 0$. Conversely, if $u \in U$ and $X(u) = 0$ then (heuristically, with overwhelming probability) $u = v^2$ for some $v \in L$; this v must be a unit, so $u \in (\mathcal{O}_L^\times)^2$.

The algorithm assembles the rows $X(u_1), \ldots, X(u_b)$ into a matrix M; computes a basis S for the left kernel of M; lifts each element $(S_{i1}, \ldots, S_{ib})$ of this basis to a vector of integers, each entry 0 or 1; and computes $s_i = u_1^{S_{i1}} \cdots u_b^{S_{ib}}$. By definition $X(s_i) = S_{i1}X(u_1) + \cdots + S_{ib}X(u_b) = 0$, so $s_i \in (\mathcal{O}_L^\times)^2$. The algorithm computes a square root v_i of each s_i, and it outputs $u_1, \ldots, u_b, v_1, v_2, \ldots$.

To see that $-1, u_1, \ldots, u_b, v_1, v_2, \ldots$ generate $\mathcal{O}_L^\times$, consider any $u \in \mathcal{O}_L^\times$. By definition $u^2 \in (\mathcal{O}_L^\times)^2$, so $u^2 \in U$, so $u^2 = u_1^{e_1} \cdots u_b^{e_b}$ for some $e_1, \ldots, e_b \in \mathbb{Z}$. Furthermore $X(u^2) = 0$ so $e_1 X(u_1) + \cdots + e_b X(u_b) = 0$; i.e., the vector $(e_1 \bmod 2, \ldots, e_b \bmod 2)$ in $(\mathbb{Z}/2)^b$ is in the left kernel of M. By definition S is a basis for this left kernel, so $(e_1 \bmod 2, \ldots, e_b \bmod 2)$ is a linear combination of the rows of S modulo 2; i.e., $(e_1, \ldots, e_b)$ is some $(2f_1, \ldots, 2f_b)$ plus a linear

---

Algorithm 5.1. UnitsGivenSubgroup($L, (u_1, \ldots, u_b)$)

Input: A real multiquadratic field L; elements $u_1, \ldots, u_b$ of $\mathcal{O}_L^\times$ such that $(\mathcal{O}_L^\times)^2 \subseteq \langle u_1, \ldots, u_b \rangle$.
Result: Generators for $\mathcal{O}_L^\times / \{\pm 1\}$.

1   $\chi_1, \ldots, \chi_m \leftarrow$ EnoughCharacters($L, (u_1, \ldots, u_b)$)
2   $M \leftarrow [\log_{-1}(\chi_k(u_j))]_{1 \le j \le b, 1 \le k \le m}$
3   $S \leftarrow$ BASIS(LEFTKERNEL(M))
4   **for** $i = 1, \ldots, \#S$ **do**
5     $s_i \leftarrow \prod_j u_j^{S_{ij}}$, interpreting exponents in $\mathbb{Z}/2$ as $\{0, 1\}$ in $\mathbb{Z}$
6     $v_i \leftarrow \sqrt{s_i}$

7   **return** $u_1, \ldots, u_b, v_1, \ldots, v_{\#S}$

---

combination of the rows of S; i.e., u^2 is $u_1^{2f_1} \cdots u_b^{2f_b}$ times a product of powers of s_i; i.e., u is $\pm u_1^{f_1} \cdots u_b^{f_b}$ times a product of powers of v_i.

Complexity Analysis and Improvements. Assume that the inputs $u_1, \ldots, u_b$ have at most B bits in each coefficient. Each of the products $s_1, s_2, \ldots$ is a product of at most b inputs, and thus has, at worst, essentially bB bits in each coefficient.

Computing the character matrix M takes time essentially $bN(b + B)$; see Sect. 4.2. Computing S takes $O(N^3)$ operations by Gaussian elimination over $\mathbb{F}_2$; one can obtain a better asymptotic exponent here using fast matrix multiplication, but this is not a bottleneck in any case. Computing one product s_i takes time essentially bNB with a product tree, and computing its square root v_i takes time essentially $bN^{\log_2 3}B$. There are at most b values of i.

Our application of this algorithm has $b \in \Theta(N)$. The costs are essentially $N^3 + N^2 B$ for characters, N^3 for kernel computation, $N^3 B$ for products, and $N^{2 + \log_2 3}B$ for square roots.

These bounds are too pessimistic, for three reasons. First, experiments show that products often have far fewer factors, and are thus smaller and faster to compute. Second, one can enforce a limit upon output size by integrating the algorithm with lattice-basis reduction (see Sect. 5.5), computing products and square roots only after reduction. Third, we actually use the technique of Sect. 3.5 (in the full version of this paper) to compute products of powers.

5.5 A Complete Algorithm to Compute the Unit Group

Algorithm 5.2 computes a basis for $\mathcal{O}_L^\times$, given a real multiquadratic field L.

As usual write N for the degree of L. There is no difficulty if $N = 1$. For $N = 2$, the algorithm calls standard subroutines cited in Sect. 5.1. For $N \ge 4$, the algorithm calls itself recursively on three subfields of degree $N/2$; merges the results into generators for a subgroup $U \le \mathcal{O}_L^\times$ such that $(\mathcal{O}_L^\times)^2 \le U$; calls UnitsGivenSubgroup to find generators for $\mathcal{O}_L^\times$; and then uses lattice-basis reduction to find a basis for $\mathcal{O}_L^\times$. A side effect of lattice-basis reduction is that the basis is short, although it is not guaranteed to be minimal.

Algorithm 5.2. Units(L)

Input: A real multiquadratic field L. As a side input, a parameter $H > 0$.
Result: Independent generators of $\mathcal{O}_L^\times/\{\pm 1\}$.

1 **if** $[L : \mathbb{Q}] = 1$ **then**
2 $\quad$ | $\quad$ **return** ()

3 **if** $[L : \mathbb{Q}] = 2$ **then**
4 $\quad$ | $\quad$ **return** the normalized fundamental unit of L

5 $\sigma, \tau \leftarrow$ distinct non-identity automorphisms of L
6 **for** $\ell \in \{\sigma, \tau, \sigma\tau\}$ **do**
7 $\quad$ | $\quad G_\ell \leftarrow$ Units(fixed field of ℓ)

8 $G \leftarrow -1, G_\sigma, G_\tau, \sigma(G_{\sigma\tau})$
9 $(u_1, \ldots, u_b) \leftarrow$ UnitsGivenSubgroup(L, G)

10 $A \leftarrow \begin{pmatrix} 1 & 0 & \ldots & 0 & H \cdot \mathrm{ApproxLog}(u_1) \\ 0 & 1 & \ldots & 0 & H \cdot \mathrm{ApproxLog}(u_2) \\ \vdots & \vdots & \ddots & \vdots & \vdots \\ 0 & 0 & \ldots & 1 & H \cdot \mathrm{ApproxLog}(u_b) \end{pmatrix}$

11 $A' \leftarrow$ LLL(A), putting shortest vectors first
12 **for** $i = 1, \ldots, N - 1$ *where* $N = [L : \mathbb{Q}]$ **do**
13 $\quad$ | $\quad w_i \leftarrow \prod_{1 \leq j \leq b} u_j^{A'_{b-(N-1)+i,j}}$

14 **return** $w_1, \ldots, w_{N-1}$

The Subgroup and the Generators. Lemma 5.1 defines $U = \mathcal{O}_{K_\sigma}^\times \cdot \mathcal{O}_{K_\tau}^\times \cdot \sigma(\mathcal{O}_{K_{\sigma\tau}}^\times)$ where σ, τ are distinct non-identity automorphisms of L.

The three subfields used in the algorithm are K_σ, K_τ, and $K_{\sigma\tau}$. The recursive calls produce lists of generators for $\mathcal{O}_{K_\sigma}^\times/\{\pm 1\}$, $\mathcal{O}_{K_\tau}^\times/\{\pm 1\}$, and $\mathcal{O}_{K_{\sigma\tau}}^\times/\{\pm 1\}$ respectively. The algorithm builds a list G that contains each element of the first list; each element of the second list; σ applied to each element of the third list; and -1. Then G generates U. As a speedup, we sort G to remove duplicates.

We cache the output of Units(L) for subsequent reuse (without saying so explicitly in Algorithm 5.2). For example, if $L = \mathbb{Q}(\sqrt{2}, \sqrt{3}, \sqrt{5})$, then the three subfields might be $\mathbb{Q}(\sqrt{2}, \sqrt{3})$, $\mathbb{Q}(\sqrt{2}, \sqrt{5})$, and $\mathbb{Q}(\sqrt{2}, \sqrt{15})$, and the next level of recursion involves $\mathbb{Q}(\sqrt{2})$ three times. We perform the Units($\mathbb{Q}(\sqrt{2})$) computation once and then simply reuse the results the next two times.

The overall impact of caching depends on how σ and τ are chosen (which is also not specified in Algorithm 5.2). We use the following specific strategy. As usual write L as $\mathbb{Q}(\sqrt{d_1}, \ldots, \sqrt{d_n})$, where $d_1, \ldots, d_n$ are integers meeting the conditions of Theorem 2.1. Assume that $0 < d_1 < \cdots < d_n$. Choose σ and τ such that $K_\sigma = \mathbb{Q}(\sqrt{d_1}, \sqrt{d_2}, \ldots, \sqrt{d_{n-1}})$ and $K_\tau = \mathbb{Q}(\sqrt{d_1}, \sqrt{d_2}, \ldots, \sqrt{d_{n-2}}, \sqrt{d_n})$. We depict the resulting set of subfields in Figs. 5.1 and 5.2. Notice that, in Figs. 5.1 and 5.2, the leftmost field in each horizontal layer is a subfield used by all fields in the horizontal layer above it.

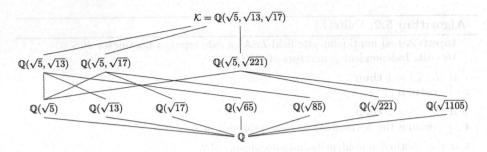

Fig. 5.1. How to pick subfields for the recursive algorithm for multiquadratic fields of degree 8.

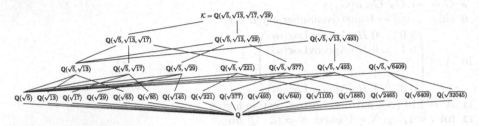

Fig. 5.2. How to pick subfields for the recursive algorithm for multiquadratic fields of degree 16.

With this strategy, the recursion reaches exactly $2^{n-\ell+1} - 1$ subfields of degree 2^ℓ, namely the subfields of the form $\mathbb{Q}(\sqrt{d_1}, \ldots, \sqrt{d_{\ell-1}}, \sqrt{D})$ where D is a product of a nonempty subset of $\{d_\ell, \ldots, d_n\}$. With a less disciplined strategy, randomly picking 3 subfields of degree $N/2$ at each step, we would instead end up with nearly $3^{n-\ell}$ subfields of degree 2^ℓ. "Nearly" accounts for accidental collisions and for the limited number of subfields of low degree.

Finding Short Bases Given Generators. Applying Pohst's modified LLL algorithm [37] to the vectors ApproxLog(u_1), ..., ApproxLog(u_b) would find $b - (N - 1)$ zero vectors and $N - 1$ independent short combinations of the input vectors. The algorithm is easily extended to produce an invertible $b \times b$ transformation matrix T that maps the input vectors to the output vectors. (The algorithm in [37] already finds the part of T corresponding to the zero outputs.) We could simply use the entries of any such T as exponents of u_j in our algorithm. It is important to realize, however, that there are many possible choices of T (except in the extreme case $b = N - 1$), and the resulting computations are often much slower than necessary. For example, if $u_3 = u_1 u_2$, then an output u_1/u_2 might instead be computed as $u_1^{1001} u_2^{999}/u_3^{1000}$.

We instead apply LLL to the matrix A shown in Algorithm 5.2. This has three effects. First, if H is chosen sufficiently large, then the right side of A is reduced to $b - (N - 1)$ zero vectors and $N - 1$ independent short combinations of the vectors $H \cdot$ ApproxLog(u_1), ..., $H \cdot$ ApproxLog(u_b). (We check that there are exactly $b - (N - 1)$ zero vectors.) Second, the left side of A keeps track of the transformation matrix that is used. Third, this transformation matrix is automatically reduced: short coefficients are found for the $b - (N - 1)$ zero vectors, and these coefficients are used to reduce the coefficients for the $N - 1$ independent vectors.

An upper bound on LLL cost can be computed as follows. LLL in dimension N, applied to integer vectors where each vector has $O(B)$ bits, uses $O(N^4 B)$ arithmetic operations on integers with $O(NB)$ bits; see [35, Proposition 1.26]. The total time is bounded by essentially $N^5 B^2$. To bound B one can bound each $H \cdot \mathrm{ApproxLog}(\cdots)$. To bound H one can observe that the transformation matrix has, at worst, essentially N bits in each coefficient (see, e.g., [39]), while the required precision of ApproxLog is essentially 1, so it suffices to take essentially N bits in H. The total time is, at worst, essentially N^7.

Our experiments show much better LLL performance for these inputs. We observe LLL actually using very few iterations; evidently the input vectors are already very close to being reduced. It seems plausible to conjecture that the entries of the resulting transformation matrix have at most $n^{O(1)}$ bits, and that it suffices to take H with $n^{O(1)}$ bits, producing B bounded by $n^{O(1)}$. The total time might be as small as essentially N^3, depending on how many iterations there are.

6 Finding Generators of Ideals

This section presents the main contribution of this paper: a fast pre-quantum algorithm to compute a nonzero g in a multiquadratic ring, given the ideal generated by g. For simplicity we focus on the real case, as in Sect. 5. The algorithm takes quasipolynomial time under reasonable heuristic assumptions if $d_1, \ldots, d_n$ are quasipolynomial.

The algorithm reuses the equation $g^2 = N_{L:K_\sigma}(g) N_{L:K_\tau}(g) / \sigma(N_{L:K_{\sigma\tau}}(g))$ that was used for unit-group computation in Sect. 5. To compute $N_{L:K}(g)$, the algorithm computes the corresponding norm of the input ideal, and then calls the same algorithm recursively.

The main algebraic difficulty here is that there are many generators of the same ideal: one can multiply g by any unit, such as -1 or $1 + \sqrt{2}$, to obtain another generator. What the algorithm actually produces is some ug where u is a unit. This means that the recursion produces unit multiples of $N_{L:K_\sigma}(g)$ etc., and thus produces some vg^2 rather than g^2. The extra unit v might not be a square, so we cannot simply compute the square root of vg^2. Instead we again use the techniques of Sect. 4, together with the unit group computed in Sect. 5, to find a unit u such that $u(vg^2)$ is a square, and we then compute the square root.

6.1 Representing Ideals and Computing Norms of Ideals

Let L be a real multiquadratic field of degree $N = 2^n$. Let $\mathcal{R}$ be an order inside L, such as $\mathbb{Z}[\sqrt{d_1}, \ldots, \sqrt{d_n}]$ inside $\mathbb{Q}(\sqrt{d_1}, \ldots, \sqrt{d_n})$. Our algorithm does not require $\mathcal{R}$ to be the ring of integers $\mathcal{O}_L$, although its output allows arbitrary units from the ring of integers; i.e., if the input is a principal ideal $\mathcal{I}$ of $\mathcal{R}$ then the output is some $g \in \mathcal{O}_L$ such that $g\mathcal{O}_L = \mathcal{I}\mathcal{O}_L$. Equivalently, one can (with or without having computed $\mathcal{O}_L$) view $\mathcal{I}$ as representing the ideal $\mathcal{I}\mathcal{O}_L$ of $\mathcal{O}_L$.

We consider three representations of an ideal $\mathcal{I}$ of $\mathcal{R}$:

- One standard representation is as a $\mathbb{Z}$-basis $\omega_1, \omega_2, \ldots, \omega_N \in \mathcal{R}$, i.e., a basis of $\mathcal{I}$ as a lattice.
- A more compact standard representation is the "two-element representation" (α_1, α_2) representing $\mathcal{I} = \alpha_1\mathcal{R} + \alpha_2\mathcal{R}$, typically with $\alpha_1 \in \mathbb{Z}$. If $\mathcal{R} \neq \mathcal{O}_L$ then $\mathcal{I}$ might not have a two-element representation, but failure to convert $\mathcal{I}$ to a two-element representation reveals a larger order.
- Our target cryptosystem in Appendix A uses another representation that works for many, but certainly not all, ideals of $\mathcal{R} = \mathbb{Z}[\sqrt{d_1}, \ldots, \sqrt{d_n}]$: namely, $(q, s_1, \ldots, s_n) \in \mathbb{Z}^{n+1}$, where each s_j is a nonzero square root of d_j modulo q and where q is odd, representing $\mathcal{I} = q\mathcal{R} + (\sqrt{d_1} - s_1)\mathcal{R} + \cdots + (\sqrt{d_n} - s_n)\mathcal{R}$.

Our algorithm works with any representation that allows basic ideal operations, such as ideal norms, which we discuss next. Performance depends on the choice of representation.

Let σ be a nontrivial automorphism of L, and let K be its fixed field; then K is a subfield of L with $[L : K] = 2$. Assume that $\sigma(\mathcal{R}) = \mathcal{R}$, and let $\mathcal{S}$ be the order $K \cap \mathcal{R}$ inside K. For example, if $\mathcal{R} = \mathbb{Z}[\sqrt{d_1}, \ldots, \sqrt{d_n}]$ and σ preserves $\sqrt{d_1}, \ldots, \sqrt{d_{n-1}}$ while negating $\sqrt{d_n}$, then $\mathcal{S} = \mathbb{Z}[\sqrt{d_1}, \ldots, \sqrt{d_{n-1}}]$; if $\mathcal{R} = \mathbb{Z}[\sqrt{d_1}, \ldots, \sqrt{d_n}]$ and σ preserves $\sqrt{d_1}, \ldots, \sqrt{d_{n-2}}$ while negating $\sqrt{d_{n-1}}, \sqrt{d_n}$, then $\mathcal{S} = \mathbb{Z}[\sqrt{d_1}, \ldots, \sqrt{d_{n-2}}, \sqrt{d_{n-1}d_n}]$.

The relative norm $N_{L:K}(\mathcal{I})$ is, by definition, $\mathcal{I}\sigma(\mathcal{I}) \cap K$, which is the same as $\mathcal{I}\sigma(\mathcal{I}) \cap \mathcal{S}$. This is an ideal of $\mathcal{S}$. It has two important properties: it is not difficult to compute; and if $\mathcal{I} = g\mathcal{R}$ then $N_{L:K}(\mathcal{I}) = N_{L:K}(g)\mathcal{S}$. See, e.g., [20].

Given a $\mathbb{Z}$-basis of $\mathcal{I}$, one can compute a $\mathbb{Z}$-basis of $N_{L:K}\mathcal{I}$ by computing $\{\omega_i \cdot \sigma(\omega_j) \mid 1 \leq i \leq j \leq N\}$, transforming this into a Hermite-Normal-Form (HNF) basis for $\mathcal{I}\sigma(\mathcal{I})$, and intersecting with $\mathcal{S}$. A faster approach appears in [6]: compute a two-element representation of $\mathcal{I}$; multiply the two elements by a $\mathbb{Z}$-basis for $\sigma(\mathcal{I})$; convert to HNF form; and intersect with $\mathcal{S}$, obtaining a $\mathbb{Z}$-basis for $N_{L:K}\mathcal{I}$. This takes total time essentially N^5B.

The $(q, s_1, \ldots, s_n)$ representation allows much faster norms, and is used in our software. The norm to $\mathbb{Z}[\sqrt{d_1}, \ldots, \sqrt{d_{n-1}}]$ is simply $(q, s_1, \ldots, s_{n-1})$, and the norm to $\mathbb{Z}[\sqrt{d_1}, \ldots, \sqrt{d_{n-2}}, \sqrt{d_{n-1}d_n}]$ is simply $(q, s_1, \ldots, s_{n-2}, s_{n-1}s_n)$.

6.2 Computing a Generator of $\mathcal{I}$ from a Generator of $\mathcal{I}^2$

Assume now that we have a nonzero principal ideal $\mathcal{I} \subseteq \mathcal{O}_L$, and a generator h for $\mathcal{I}^2$. To find a generator g for $\mathcal{I}$, it is sufficient to find a square generator for $\mathcal{I}^2$ and take its square root. To this end we seek a unit $u \in \mathcal{O}_L^\times$ such that $uh = g^2$ for some g. Applying the map X from Sect. 4.2 to this equation, we obtain

$$X(uh) = X(g^2) = 2X(g) = 0.$$

Therefore $X(u) = X(h)$.

Algorithm 6.1. IdealSqrt(L, h)

Input: A real multiquadratic field L; an element h of $\mathcal{O}_L^\times \cdot (L^\times)^2$.
Result: Some $g \in L^\times$ such that $h/g^2 \in \mathcal{O}_L^\times$.

1 $u_1, \ldots, u_{N-1} \leftarrow$ Units(L)
2 $u_0 \leftarrow -1$
3 $\chi_1, \ldots, \chi_m \leftarrow$ EnoughCharacters($L, (u_0, \ldots, u_{N-1}, h)$)
4 $M \leftarrow [\log_{-1} \chi_j(u_i)]_{0 \leq i \leq N-1, 1 \leq j \leq m}$
5 $V \leftarrow [\log_{-1} \chi_j(h)]_{1 \leq j \leq m}$
6 $[e_0, \ldots, e_{N-1}] \leftarrow$ SOLVELEFT(M, V)
7 $u \leftarrow \prod_j u_j^{e_j}$, interpreting exponents in $\mathbb{Z}/2$ as $\{0,1\}$ in $\mathbb{Z}$
8 $g \leftarrow \sqrt{uh}$
9 **return** g

We start by computing $X(h)$ from h. We then compute a basis $u_1, \ldots, u_{N-1}$ for $\mathcal{O}_L^\times$, and we define $u_0 = -1$, so $u_0, u_1, \ldots, u_{N-1}$ generate $\mathcal{O}_L^\times$. We then solve the matrix equation

$$[e_0, e_1, \ldots, e_{N-1}] \begin{bmatrix} X(u_0) \\ X(u_1) \\ \vdots \\ X(u_{N-1}) \end{bmatrix} = X(h)$$

for $[e_0, e_1, \ldots, e_{N-1}] \in (\mathbb{Z}/2)^N$ and set $u = \prod_j u_j^{e_j}$. Then uh is (almost certainly) a square, so its square root g is a generator of $\mathcal{I}$. This algorithm is summarized in Algorithm 6.1.

The subroutine SOLVELEFT(M, V) solves the matrix equation $eM = V$ for the vector e. One can save time by precomputing the inverse of an invertible full-rank submatrix of M, and using only the corresponding characters.

Note that for this computation to work we need a basis of the full unit group. If we instead use units $v_1, \ldots, v_{N-1}$ generating, e.g., the group $U = (\mathcal{O}_L^\times)^2$, and if $h = vg^2$ for some $v \in \mathcal{O}_L^\times - U$, then uh cannot be a square for any $u \in U$: if it were then h would be a square (since every $u \in U$ is a square), so v would be a square, so v would be in U, contradiction.

There are several steps in this algorithm beyond the unit-group precomputation. Characters for $u_0, \ldots, u_{N-1}$ take time essentially $N^3 + N^2 B$ and can also be precomputed. Characters for h take time essentially $N^2 + NB$. Linear algebra mod 2 takes time essentially N^3, or better with fast matrix multiplication; most of this can be precomputed, leaving time essentially N^2 to multiply a precomputed inverse by $X(h)$. The product of powers takes time essentially $N^2 B$, and the square root takes time essentially $N^{1+\log_2 3} B$, although these bounds are too pessimistic for the reasons mentioned in Sect. 5.4.

6.3   Shortening

Algorithm 6.2, ShortenGen, finds a bounded-size generator g of a nonzero principal ideal $\mathcal{I} \subseteq \mathcal{O}_L$, given any generator h of $\mathcal{I}$. See Sect. 8 for analysis of the success probability of this algorithm at finding the short generators used in a cryptosystem.

Recall the log-unit lattice $\mathrm{Log}(\mathcal{O}_L^\times)$ defined in Sect. 5.3. The algorithm finds a lattice point $\mathrm{Log}\, u$ close to $\mathrm{Log}\, h$, and then computes $g = h/u$.

In more detail, the algorithm works as follows. Start with a basis $u_1, \ldots, u_{N-1}$ for $\mathcal{O}_L^\times$. Compute $\mathrm{Log}\, h$, and write $\mathrm{Log}\, h$ as a linear combination of the vectors $\mathrm{Log}(u_1), \ldots, \mathrm{Log}(u_{N-1}), (1, 1, \ldots, 1)$; recall that $(1, 1, \ldots, 1)$ is orthogonal to each $\mathrm{Log}(u_j)$. Round the coefficients in this combination to integers $(e_1, \ldots, e_N)$. Compute $u = u_1^{e_1} \cdots u_{N-1}^{e_{N-1}}$ and $g = h/u$.

The point here is that $\mathrm{Log}\, h$ is close to $e_1 \mathrm{Log}(u_1) + \cdots + e_{N-1} \mathrm{Log}(u_{N-1}) + e_N(1, 1, \ldots, 1)$, and thus to $\mathrm{Log}\, u + e_N(1, 1, \ldots, 1)$. The gap $\mathrm{Log}\, g = \mathrm{Log}\, h - \mathrm{Log}\, u$ is between -0.5 and 0.5 in each of the $\mathrm{Log}(u_j)$ directions, plus some irrelevant amount in the $(1, 1, \ldots, 1)$ direction.

Normally the goal is to find a generator that is known in advance to be short. If the logarithm of this target generator is between -0.5 and 0.5 in each of the $\mathrm{Log}(u_j)$ directions then this algorithm will find this generator (modulo ± 1). See Sect. 8 for further analysis of this event.

Approximations. The algorithm actually computes $\mathrm{Log}\, h$ only approximately, and uses ApproxLog u_j instead of $\mathrm{Log}\, u_j$, at the expense of marginally adjusting the 0.5 bounds mentioned above.

Assume that h has integer coefficients with at most B bits. (We discard the denominator in any case: it affects only the irrelevant coefficient of $(1, 1, \ldots, 1)$.) Then $|\sigma_j(h)| \leq 2^B \prod_i (1 + \sqrt{|d_i|})$, so $\ln |\sigma_j(h)| \leq B \ln 2 + \sum_i \ln(1 + \sqrt{|d_i|})$. By assumption each d_i is quasipolynomial in N, so $\ln |\sigma_j(h)| \leq B \ln 2 + n^{O(1)}$.

To put a *lower* bound on $\ln |\sigma_j(h)|$, consider the product of the other conjugates of h. Each coefficient of this product is between -2^C and 2^C where C is bounded by essentially NB. Dividing this product by the absolute norm of h, a nonzero integer, again produces coefficients between -2^C and 2^C, but also produces exactly $1/\sigma_j(h)$. Hence $\ln |1/\sigma_j(h)| \leq C \ln 2 + n^{O(1)}$.

In short, $\ln |\sigma_j(h)|$ is between essentially $-NB$ and B, so an approximation to $\ln |\sigma_j(h)|$ within $2^{-\beta}$ uses roughly $\beta + \log(NB)$ bits. We use interval arithmetic with increasing precision to ensure that we are computing $\mathrm{Log}\, h$ accurately; the worst-case precision is essentially NB. Presumably it would save time here to augment our representation of ideal generators to include approximate logarithms, the same way that we augment our representation of units, but we have not implemented this yet.

Other Reduction Approaches. Finding a lattice point close to a vector, with a promised bound on the distance, is called the *Bounded-Distance Decoding Problem* (BDD). There are many BDD algorithms in the literature more sophisticated than simple rounding: for example, Babai's nearest-plane algorithm [4]. See generally [25].

Algorithm 6.2. ShortenGen(L, h)

Input: A real multiquadratic field L, and a nonzero element $h \in L$. As a side input, a positive integer parameter β.
Result: A short $g \in L$ with $g/h \in \mathcal{O}_L^\times$.

1 $u_1, \ldots, u_{N-1} \leftarrow \text{Units}(L)$

2 $M \leftarrow \begin{pmatrix} \text{ApproxLog}(u_1) \\ \vdots \\ \text{ApproxLog}(u_{N-1}) \\ 1 \quad 1 \quad \ldots \quad 1 \quad 1 \end{pmatrix}$

3 $v \leftarrow$ approximation to $\text{Log}(h)$ within $2^{-\beta}$ in each coordinate
4 $e \leftarrow \lfloor -vM^{-1} \rceil$
5 $g \leftarrow hu_1^{e_1} \cdots u_{N-1}^{e_{N-1}}$
6 **return** g

Algorithm 6.3. QPIP($Q, \mathcal{I}$)

Input: Real quadratic field Q and a principal ideal $\mathcal{I}$ of an order inside Q
Result: A short generator g for $\mathcal{I}\mathcal{O}_Q$
1 $h \leftarrow \text{FindQGen}(Q, \mathcal{I})$
2 $g \leftarrow \text{ShortenGen}(Q, h)$
3 **return** g

Our experiments show that, unsurprisingly, failures in rounding are triggered most frequently by the shortest vectors in our lattice bases. One cheap way to eliminate these failures is to enumerate small combinations of the shortest vectors.

6.4 Finding Generators of Ideals for Quadratics

We now have all the ingredients for the attack algorithm. It will work in a recursive manner and in this subsection we will treat the base case.

Recall from Sect. 5.1 that there are standard algorithms to compute the normalized fundamental unit ε of a real quadratic field $\mathbb{Q}(\sqrt{d})$ in time essentially $R = \ln(\varepsilon)$, which is quasipolynomial under our assumptions. There is, similarly, a standard algorithm to compute a generator of a principal ideal of $\mathcal{O}_{\mathbb{Q}(\sqrt{d})}$ in time essentially $R + B$, where B is the number of bits in the coefficients used in the ideal. We call this algorithm FindQGen.

There are also algorithms that replace R by something subexponential in d; see [8,14,41]. As in Sect. 5.1, these algorithms avoid large coefficients by working with products of powers of smaller field elements, raising other performance questions in our context.

Algorithm 6.3, QPIP, first calls FindQGen to find a generator h, and then calls ShortenGen from Sect. 6.3 to find a short generator g. For quadratics this is guaranteed to find a generator with a minimum-size logarithm, up to the limits of the approximations used in computing logarithms.

Algorithm 6.4. MQPIP($L, \mathcal{I}$)

Input: Real multiquadratic field L and a principal ideal $\mathcal{I}$ of an order inside L
Result: A short generator g for $\mathcal{I}\mathcal{O}_L$

1   **if** $[L : \mathbb{Q}] = 1$ **then**
2   ⌊ **return** the smallest positive integer in $\mathcal{I}$

3   **if** $[L : \mathbb{Q}] = 2$ **then**
4   ⌊ **return** QPIP($L, \mathcal{I}$)

5   $\sigma, \tau \leftarrow$ distinct non-identity automorphisms of L
6   **for** $\ell \in \{\sigma, \tau, \sigma\tau\}$ **do**
7 | $K_\ell \leftarrow$ fixed field of ℓ
8 | $\mathcal{I}_\ell \leftarrow N_{L:K_\ell}(\mathcal{I})$
9   ⌊ $g_\ell \leftarrow$ MQPIP($K_\ell, \mathcal{I}_\ell$)

10   $h \leftarrow g_\sigma g_\tau / \sigma(g_{\sigma\tau})$
11   $g' \leftarrow$ IdealSqrt(L, h)
12   $g \leftarrow$ ShortenGen(L, g')
13   **return** g

6.5   Finding Generators of Ideals for Multiquadratics

Algorithm 6.4 recursively finds generators of principal ideals of orders in real multiquadratic fields. The algorithm works as follows.

Assume, as usual, that $d_1, \ldots, d_n$ are positive integers meeting the conditions of Theorem 2.1. Let L be the real multiquadratic field $\mathbb{Q}(\sqrt{d_1}, \ldots, \sqrt{d_n})$ of degree $N = 2^n$. Let $\mathcal{I}$ be a principal ideal of an order inside L, for which we want to find a generator.

If $N = 1$ then there is no difficulty. If $N = 2$, we find the generator with the QPIP routine of the previous section. Assume from now on that $N \geq 4$.

As in Sect. 5.5, choose distinct non-identity automorphisms σ, τ of L, and let $K_\sigma, K_\tau, K_{\sigma\tau}$ be the fields fixed by $\sigma, \tau, \sigma\tau$ respectively. These are fields of degree $N/2$.

For each $\ell \in \{\sigma, \tau, \sigma\tau\}$, compute $\mathcal{I}_\ell = N_{L:K_\ell}(\mathcal{I})$ as explained in Sect. 6.1, and call MQPIP($K_\ell, \mathcal{I}_\ell$) recursively to compute a generator g_ℓ for each $\mathcal{I}_\ell \mathcal{O}_{K_\ell}$. Notice that if g is a generator of $\mathcal{I}\mathcal{O}_L$, then $g\ell(g)$ generates $\mathcal{I}_\ell \mathcal{O}_{K_\ell}$, so $g_\ell = u_\ell g\ell(g)$ for some $u_\ell \in \mathcal{O}_{K_\ell}^\times$. Therefore

$$\frac{g_\sigma g_\tau}{\sigma(g_{\sigma\tau})} = \frac{u_\sigma g\sigma(g) u_\tau g\tau(g)}{\sigma(u_{\sigma\tau} g\sigma\tau(g))} = g^2 u_\sigma u_\tau \sigma(u_{\sigma\tau}^{-1}),$$

so that $h = g_\sigma g_\tau / \sigma(g_{\sigma\tau})$ is a generator of $\mathcal{I}^2 \mathcal{O}_L$. Now use IdealSqrt to find a generator of $\mathcal{I}\mathcal{O}_L$, and ShortenGen to find a bounded-size generator.

Table 6.1 summarizes the scalability of the subroutines inside MQPIP. Many of the costs are in precomputations that we share across many ideals $\mathcal{I}$, and these costs involve larger powers of N than the per-ideal costs. On the other hand, the per-ideal costs can dominate when the ideals have enough bits B per coefficient.

Table 6.1. Complexities of subroutines at the top and bottom levels of recursion of MQPIP. Logarithmic factors are suppressed. B is assumed to be at least as large as regulators. "UGS" means UnitsGivenSubgroup; "IS" means IdealSqrt; "SG" means ShortenGen. "Precomp" means that the results of the computation can be reused for many inputs $\mathcal{I}$.

Precomp?	Subroutine	Cost
Yes	Units for all quadratic fields	NB
Yes	Characters of units (in UGS, IS)	$N^3 + N^2 B$
Yes	Linear algebra (in UGS, IS)	N^3 without fast matrix multiplication
Yes	Basis reduction (in Units)	N^7; experimentally closer to N^3
Yes	Products (in UGS, Units)	$N^3 B$
Yes	Square roots (in UGS)	$N^{2+\log_2 3} B$
No	Generators for all quadratic fields	NB
No	Characters for h (in IS)	$N^2 + NB$
No	Linear algebra for h (in IS)	N^2
No	Products (in IS, SG, MQPIP)	$N^2 B$
No	Square roots (in IS)	$N^{1+\log_2 3} B$

7 Timings

This section reports experiments on the timings of our software for our algorithms: specifically, the number of seconds used for various operations in the Sage [23] computer-algebra system on a single core of a 4 GHz AMD FX-8350 CPU.

7.1 Basic Subroutine Timings

Table 7.1 shows the time taken for multiplication, squaring, etc., rounded to the nearest 0.0001 s: e.g., 0.0627 s to multiply two elements of a degree-256 multiquadratic ring, each element having random 1000-bit coefficients. The table is consistent with the analysis earlier in the paper: e.g., doubling the degree approximately doubles the cost of multiplication, and approximately triples the cost of square roots.

We have, for comparison, also explored the performance of multiquadratics using Sage's tower-field functions, Sage's absolute-number-field functions (using the polynomial F defined in Appendix A), and Sage's ring constructors. The underlying polynomial-arithmetic code inside Sage is written in C, avoiding Python overhead, but suffers from poor algorithm scalability. Sage's construction of degree-2 relative extensions (in towers of number fields or in towers of rings) uses Karatsuba arithmetic, losing a factor of 3 for each extension, with no obvious way to enable FFTs. Working with one variable modulo F produces good scalability for multiplication but makes norms difficult. Division is very slow in any case: for example, it takes 0.2 s, 2.8 s, and 93 s in degrees 32, 64,

Table 7.1. Observed time for basic operations in $\mathbb{Z}[\sqrt{d_1}, \ldots, \sqrt{d_n}]$, with $d_1 = 2$, $d_2 = 3$, $d_3 = 5$, etc., and $\lambda = 64$. The "mult" column is the time to compute $h = fg$ where f, g have each coefficient chosen randomly between -2^{1000} and $2^{1000} - 1$. The "square" column is the time to compute f^2. The "relnorm" column is the time to compute $f\sigma(f)$ where σ is any of the automorphisms in Theorem 2.1. The "absnorm" column is the time to compute $N_{\mathbb{Q}(\sqrt{d_1}, \ldots, \sqrt{d_n}):\mathbb{Q}} f$. The "div" column is the time to divide $h = fg$ by g, recovering f. The "sqrt" column is the time to recover $\pm f$ from f^2. Each timing is the median of 21 measurements.

n	2^n	mult	square	relnorm	absnorm	div	sqrt
3	8	0.0084	0.0062	0.0080	0.0515	0.0140	0.3547
4	16	0.0100	0.0075	0.0088	0.1119	0.0153	1.0819
5	32	0.0132	0.0101	0.0106	0.2364	0.0176	3.3507
6	64	0.0209	0.0163	0.0145	0.5013	0.0231	10.2689
7	128	0.0347	0.0275	0.0221	1.0199	0.0341	31.2408
8	256	0.0627	0.0501	0.0367	2.1024	0.0573	93.9827

and 128 respectively using the tower-field representation, and it takes 0.15 s, 1.16 s, and 11.3 s in degrees 32, 64, and 128 respectively using the single-variable representation, while we use under 0.06 s in degree 256.

7.2   Timings to Compute the Unit Group and Generators

The difference in scalability is much more striking for unit-group computation, as shown in Table 7.2. Our algorithm uses 2.34 s for degree 16, 7.80 s for degree

Table 7.2. Observed time to compute (once) the unit group of $\mathbb{Z}[\sqrt{d_1}, \ldots, \sqrt{d_n}]$, with $d_1 = 2$, $d_2 = 3$, $d_3 = 5$, etc.; and to break the cryptosystem presented in Appendix A. The "tower" column is the time used by Sage's tower-field unit-group functions (with `proof=False`); for $n = 6$ these functions ran out of memory after approximately 710000 s. The "absolute" column is the time used by Sage's absolute-field unit-group functions (also with `proof=False`), starting from the polynomial F defined in Appendix A. The "new" column is the time used by this paper's unit-group algorithm. The "attack" column is the time to find a generator of the public key, after the unit group is precomputed. In "new2" and "attack2" the same timings are given for the field with the first n consecutive primes after n. In "new3" and "attack3" the same timings are given for the field with the first n consecutive primes after n^2.

n	2^n	tower	absolute	new	new2	new3	attack	attack2	attack3
3	8	0.05	0.03	0.63	0.65	0.66	0.10	0.11	0.11
4	16	0.51	0.24	2.34	2.21	2.18	0.27	0.35	0.36
5	32	7.24	4.80	7.80	7.71	8.22	0.96	1.36	1.47
6	64	>700000	>700000	26.62	28.08	81.78	4.36	6.68	7.48
7	128			146.60	192.19	2332.79	26.14	37.23	42.30
8	256			942.36	2364.18	65932	181.26	239.05	239.90

32, 26.62 s for degree 64, 146.60 s for degree 128, etc., slowing down by a factor considerably below 2^5 for each doubling in the degree. Sage's internal C library uses 4.8 s for degree 32, but we did not see it successfully compute a unit group for degree 64.

Table 7.2 also shows that our short-generator algorithm has similar scaling to our unit-group algorithm, as one would expect from the structure of the algorithms. As inputs we used public keys from a Gentry-style multiquadratic cryptosystem;[3] see Appendix A. The number of bits per coefficient in this cryptosystem grows almost linearly with 2^n, illustrating another dimension of scalability of our algorithm. See Sect. 8 for analysis of the success probability of the algorithm as an attack against the cryptosystem.

8 Key-Recovery Probabilities

In this section we analyze the success probability of our algorithm recovering the secret key g in a Gentry-style multiquadratic cryptosystem.

The specific system that we target is the system defined in Appendix A (in the full version of this paper), the same system used for timings in Sect. 7.2. The secret key g in this cryptosystem is $g_0 + g_1\sqrt{d_1} + g_2\sqrt{d_2} + g_3\sqrt{d_1}\sqrt{d_2} + \cdots + g_{N-1}\sqrt{d_1}\cdots\sqrt{d_n}$, where $g_0, g_1, g_2, \ldots$ are independent random integers chosen from intervals $[-G, G], [-G/\sqrt{d_1}, G/\sqrt{d_1}], [-G/\sqrt{d_2}, G/\sqrt{d_2}], \ldots$. The distribution within each interval is uniform, except for various arithmetic requirements (e.g., g must have odd norm) that do not appear to have any impact on the performance of our attack.

Section 8.1 presents heuristics for the expected size of $\text{Log } g$ on the basis $\text{Log } \varepsilon_1, \ldots, \text{Log } \varepsilon_{N-1}$ for the logarithms of multiquadratic units, a sublattice of the log-unit lattice. Section 8.2 presents experimental data confirming these heuristics. Section 8.3 presents experimental data regarding the size of $\text{Log } g$ on the basis that we compute for the full log-unit lattice. Section 8.4 presents an easier-to-analyze way to find g when $\text{Log } \varepsilon_1, \ldots, \text{Log } \varepsilon_{N-1}$ are large enough.

8.1 MQ Unit Lattice: Heuristics for Log g

Write U_L for the group of multiquadratic units in L. Recall that U_L is defined as the group $\langle -1, \varepsilon_1, \ldots, \varepsilon_{N-1} \rangle$, where $\varepsilon_1, \ldots, \varepsilon_{N-1}$ are the normalized fundamental units of the $N-1$ quadratic subfields $\mathbb{Q}(\sqrt{D_1}), \ldots, \mathbb{Q}(\sqrt{D_{N-1}})$.

The logarithms $\text{Log } \varepsilon_1, \ldots, \text{Log } \varepsilon_{N-1}$ form a basis for the **MQ unit lattice** $\text{Log } U_L$. This is an orthogonal basis: for example, for $\mathbb{Q}(\sqrt{2}, \sqrt{3})$, the basis vectors are $(x, -x, x, -x)$, $(y, y, -y, -y)$, and $(z, -z, -z, z)$ with $x = \log(1 + \sqrt{2})$,

---

[3] The dimensions we used in these experiments are below the $N = 8192$ recommended by Smart and Vercauteren for 2^{100} security against standard lattice-basis-reduction attacks, specifically BKZ. However, the Smart–Vercauteren analysis shows that BKZ scales quite poorly as N increases; see Appendix A. Our attack should still be feasible for $N = 8192$, and a back-of-the-envelope calculation suggests that $N \approx 2^{20}$ is required for 2^{100} security against our attack.

$y = \log(2 + \sqrt{3})$, and $z = \log(5 + 2\sqrt{6})$. The general pattern (as in Sect. 5.3) is that Log ε_j is a vector with $R_j = \ln \varepsilon_j$ at $N/2$ positions and $-R_j$ at the other $N/2$ positions, specifically with R_j at position i if and only if $\sigma_i(\varepsilon_j) = \varepsilon_j$.

One consequence of orthogonality is that rounding on this basis is a perfect solution to the closest-vector problem for the MQ unit lattice. If 0 is the closest lattice point to Log g, and u is any multiquadratic unit, then rounding Log gu produces Log u. One can decode beyond the closest-vector problem by enumerating some combinations of basis vectors, preferably the shortest basis vectors, but for simplicity we skip this option.

Write c_j for the coefficient of Log g on the jth basis vector Log ε_j; note that if each c_j is strictly between -0.5 and 0.5 then 0 is the closest lattice point to Log g. Another consequence of orthogonality is that c_j is simply the dot product of Log g with Log ε_j divided by the squared length of Log ε_j; i.e., the dot product of Log g with a pattern of $N/2$ copies of R_j and $N/2$ copies of $-R_j$, divided by NR_j^2; i.e., $Y/(NR_j)$, where Y is the dot product of Log g with a pattern of $N/2$ copies of 1 and $N/2$ copies of -1.

We heuristically model g_0 as a uniform random real number from the interval $[-G, G]$; g_1 as a uniform random real number from $[-G/\sqrt{d_1}, G/\sqrt{d_1}]$; etc. In this model, each conjugate $\sigma_i(g)$ is a sum of N independent uniform random real numbers from $[-G, G]$. For large N, the distribution of this sum is close to a Gaussian distribution with mean 0 and variance $G^2 N/3$; i.e., the distribution of $(G\sqrt{N/3})\mathcal{N}$, where $\mathcal{N}$ is a normally distributed random variable with mean 0 and variance 1. The distribution of $\ln|\sigma_i(g)|$ is thus close to the distribution of $\ln(G\sqrt{N/3}) + \ln|\mathcal{N}|$.

Recall that Log(g) is the vector of $\ln|\sigma_i(g)|$ over all i, so Y is $\ln|\sigma_1(g)| - \ln|\sigma_2(g)| + \cdots$ modulo an irrelevant permutation of indices. The mean of $\ln|\sigma_1(g)|$ is close to the mean of $\ln(G\sqrt{N/3}) + \ln|\mathcal{N}|$, while the mean of $-\ln|\sigma_2(g)|$ is close to the mean of $-\ln(G\sqrt{N/3}) - \ln|\mathcal{N}|$, etc., so the mean of Y is close to 0. (For comparison, the mean of the sum of entries of Log(g) is close to $N\ln(G\sqrt{N/3}) + Nc$. Here c is a universal constant, the average of $\ln|\mathcal{N}|$.)

To analyze the variance of Y, we heuristically model $\sigma_1(g), \ldots, \sigma_N(g)$ as independent. Then the variance of Y is the variance of $\ln|\sigma_1(g)|$ plus the variance of $-\ln|\sigma_2(g)|$ etc. Each term is close to the variance of $\ln|\mathcal{N}|$, a universal constant V, so the variance of Y is close to VN. The deviation of Y is thus close to $\sqrt{VN}$, and the deviation of $c_j = Y/(NR_j)$ is close to $\sqrt{V}/(\sqrt{N}R_j) \approx 1.11072/(\sqrt{N}R_j)$.

To summarize, this model predicts that the coefficient of Log g on the jth basis vector Log ε_j has average approximately 0 and deviation approximately $1.11072/(\sqrt{N}R_j)$, where $R_j = \ln \varepsilon_j$. Recall that R_j typically grows as $D_j^{1/2+o(1)}$.

8.2   MQ Unit Lattice: Experiments for Log g

The experiments in Fig. 8.1 confirm the prediction of Sect. 8.1. For each n, we took possibilities for n consecutive primes $d_1, \ldots, d_n$ below 100. For each corresponding multiquadratic field, there are $N - 1$ blue dots. For each D in $\{d_1, d_2, d_1 d_2, \ldots, d_1 d_2 \cdots d_n\}$, one of these $N-1$ dots is at horizontal position D.

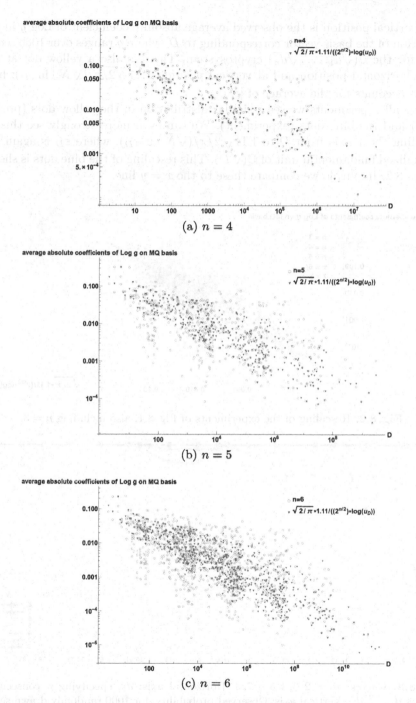

Fig. 8.1. Blue dots: For $n = 4, 5, 6$, the observed average absolute coefficient of $\mathrm{Log}(g)$ in the direction of the basis vector corresponding to $\mathbb{Q}(\sqrt{D})$. Yellow dots: Predicted values. (Color figure online)

The vertical position is the observed average absolute coefficient of Log g in the direction of the basis vector corresponding to D, where g ranges over 1000 secret keys for the $\mathbb{Q}(\sqrt{d_1}, \ldots, \sqrt{d_n})$ cryptosystem. There is also a yellow dot at the same horizontal position and at vertical position $1.11\sqrt{2/\pi}/(\sqrt{N} \cdot \ln \varepsilon_D)$; here $\sqrt{2/\pi}$ accounts for the average of $|\mathcal{N}|$.

For all experiments we see a similar distribution in the yellow dots (predictions) and the blue dots (experiment). We can even more strongly see this by rescaling the x-axis from D to $1.11\sqrt{2/\pi}/(\sqrt{N} \cdot \ln \varepsilon_D)$, where ε_D is again the normalized fundamental unit of $\mathbb{Q}(\sqrt{D})$. This rescaling of the blue dots is shown in Fig. 8.2. In purple we compare these to the $x = y$ line.

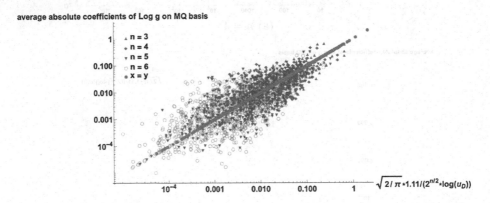

Fig. 8.2. Rescaling of the experiments of Fig. 8.1, also including $n = 3$.

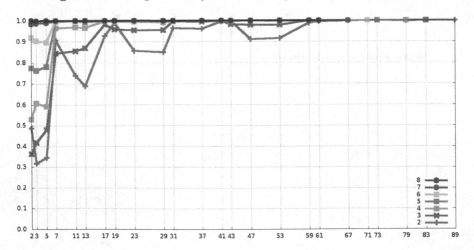

Fig. 8.3. Curves: $n = 2, 3, 4, 5, 6, 7, 8$. Horizontal axis: d_1, specifying n consecutive primes $d_1, \ldots, d_n$. Vertical axis: Observed probability, for 1000 randomly drawn secret keys g in the cryptosystem, that Log g is successfully rounded to 0 in the MQ unit lattice.

After exploring these geometric aspects of the MQ unit lattice, we ran experiments on the success probability of rounding in the lattice. Figure 8.3 shows how often $\mathrm{Log}(g)$ is rounded to 0 (by simple rounding without enumeration) in our basis for the MQ unit lattice.

This graph shows a significant probability of failure if d_1 and n are both small. Fields that contain the particularly short unit $(1 + \sqrt{5})/2$ seem to be the worst case, as one would expect from our heuristics. However, even in this case, failures disappear as n increases. The success probability seems to be uniformly bounded away from 0, seems to be above 90% for all fields with $d_1 \geq 7$ and $n \geq 4$, and seems to be above 90% for all fields with $n \geq 7$.

8.3   Full Unit Lattice: Experiments for Log g

Analyzing the full unit lattice is difficult, so we proceed directly to experiments. We first numerically compare the MQ unit lattice basis to the full unit lattice basis. The results of this are shown in Table 8.1. The index of $\mathrm{Log}(U_L)$ in $\mathrm{Log}(\mathcal{O}_L^\times)$ seems to grow as roughly $N^{0.3N}$.

Table 8.1. Experimental comparison of the MQ unit lattice $\mathrm{Log}(U_L)$, with basis formed by logarithms of the fundamental units of the quadratic subfields, and the full unit lattice $\mathrm{Log}(\mathcal{O}_L^\times)$, with basis produced by Algorithm 5.2. For each dimension 2^n, U_L and $\mathcal{O}_L^\times$ were computed for 1130 (except for $n = 8$: first 832 that have finished) random multiquadratic fields $L = \mathbb{Q}(\sqrt{d_1}, \ldots, \sqrt{d_n})$, with d_i primes bounded by $2n^2$. First row shows the average over these fields of $\log_2 \|u^*\|$, where $\|u^*\|$ is the length of the smallest Gram–Schmidt vector of the basis for U_L. Second row shows the same for $\mathcal{O}_L^\times$. Third row shows the average of $\log_2$ of the index of $\mathrm{Log}(U_L)$ in $\mathrm{Log}(\mathcal{O}_L^\times)$.

n	3	4	5	6	7	8
Average $\log_2 \|u^*\|$ for U_L	1.095	1.762	2.345	2.899	3.487	4.040
Average $\log_2 \|u^*\|$ for $\mathcal{O}_L^\times$	0.964	1.642	2.223	2.797	3.386	3.926
Average $\log_2(\#(\mathcal{O}_L^\times/U_L))$	5.711	17.462	44.095	108.133	253.722	580.099

Table 8.2. Observed attack success probabilities for various multiquadratic fields. By definition $L_j = \mathbb{Q}(\sqrt{d_1}, \ldots, \sqrt{d_n})$, with d_i the first n consecutive primes larger than or equal to j; and $p_{\mathrm{suc}}(L_j)$ is the fraction of keys out of at least 1000 trials (except 100 for $n = 8, j = 1$) that were successfully recovered without any enumeration by our attack on field L_j. Table covers $j \in \{1, n, n^2\}$.

n	3	4	5	6	7	8
$p_{\mathrm{suc}}(L_1)$	0.112	0.130	0.145	0.084	0.003	0.00
$p_{\mathrm{suc}}(L_n)$	0.204	0.497	0.649	0.897	0.783	0.348
$p_{\mathrm{suc}}(L_{n^2})$	0.782	0.980	0.999	1.000	1.000	1.000

In Table 8.2 we see the total success probability of the attack, with public keys provided as inputs, and with each successful output verified to match the corresponding secret key times ± 1.

We see that as the size and the number of the primes grow, the success probability increases, as was the case for the MQ unit basis. Specifically for the first n primes after n^2 the success probability seems to rapidly converge towards 1, as was mentioned in Sect. 1.

8.4   Full Unit Lattice: An Alternative Strategy

The following alternative method of computing g is easier to analyze asymptotically, because it does not require understanding the effectiveness of reduction in the full unit lattice. It does require $d_1, \ldots, d_n$ to be large enough compared to N, say larger than $N^{1.03}$, and it will obviously fail for many smaller d_i where our experiments succeed, but it still covers a wide range of real multiquadratic number fields.

The point of requiring $d_1, \ldots, d_n$ to be larger than $N^{1.03}$ is that, for sufficiently large N and most such choices of $d_1, \ldots, d_n$, the n corresponding regulators $\log \varepsilon$ are heuristically expected to be larger than $N^{0.51}$, and the remaining regulators for $d_1 d_2$ etc. are heuristically expected to be even larger. The coefficients of $\mathrm{Log}\, g$ on the MQ unit basis are then predicted to have deviation at most $1.11072/N^{1.01}$; see Sect. 8.1. We will return to this in a moment.

Compute, by our algorithm, some generator gu of the public key $\mathcal{I}$. From Theorem 5.2 we know that u^N is an MQ unit. Compute $N\mathrm{Log}\, gu$ and round in the MQ unit lattice. The coefficients of $N\mathrm{Log}\, g$ on the MQ unit basis are predicted to have deviation at most $1.11072/N^{0.01}$, so for sufficiently large N these coefficients have negligible probability of reaching 0.5 in absolute value. Rounding thus produces $\mathrm{Log}(u^N)$ with high probability, revealing $\mathrm{Log}(g^N)$ and thus $\pm g^N$. Use a quadratic character to deduce g^N, compute the square root $\pm g^{N/2}$, use a quadratic character to deduce $g^{N/2}$, and so on through $\pm g$.

One can further extend the range of applicability of this strategy by finding a smaller exponent e such that u^e is always an MQ unit. Theorem 5.2 says $N/2$ for $N \geq 2$. By computing the MQ units for a particular field one immediately sees the minimum value of e for that field; our computations suggest that $N/2$ is usually far from optimal.

A   A Multiquadratic Cryptosystem

See full version of paper on multiquad.cr.yp.to.

B   Recent Progress in Attacking Ideal-SVP

See full version of paper on multiquad.cr.yp.to.

References

1. Abel, C.S.: Ein Algorithmus zur Berechnung der Klassenzahl und des Regulators reell-quadratischer Ordnungen. Ph.D. thesis, Universität des Saarlandes, Saarbrücken, Germany (1994)
2. Adleman, L.M.: Factoring numbers using singular integers. In: STOC 1991, pp. 64–71 (1991)
3. Alkim, E., Ducas, L., Pöppelmann, T., Schwabe, P.: Post-quantum key exchange - a new hope. USENIX Security 2016, pp. 327–343 (2016)
4. Babai, L.: On Lovász' lattice reduction and the nearest lattice point problem. Combinatorica 6(1), 1–13 (1986)
5. Barbulescu, R., Gaudry, P., Joux, A., Thomé, E.: A heuristic quasi-polynomial algorithm for discrete logarithm in finite fields of small characteristic. In: Nguyen, P.Q., Oswald, E. (eds.) EUROCRYPT 2014. LNCS, vol. 8441, pp. 1–16. Springer, Heidelberg (2014). doi:10.1007/978-3-642-55220-5_1
6. Belabas, K.: Topics in computational algebraic number theory. J. de Théorie des Nombres de Bordeaux 16(1), 19–63 (2004)
7. Biasse, J.-F., Fieker, C.: Improved techniques for computing the ideal class group and a system of fundamental units in number fields. In: ANTS-IX. Open Book Series, vol. 1, pp. 113–133. Mathematical Sciences Publishers (2012)
8. Biasse, J.-F., Jacobson Jr., M.J., Silvester, A.K.: Security estimates for quadratic field based cryptosystems. In: Steinfeld, R., Hawkes, P. (eds.) ACISP 2010. LNCS, vol. 6168, pp. 233–247. Springer, Heidelberg (2010). doi:10.1007/978-3-642-14081-5_15
9. Biasse, J.-F., Song, F.: On the quantum attacks against schemes relying on the hardness of finding a short generator of an ideal in $\mathbb{Q}(\zeta_{p^n})$ (2015). http://cacr.uwaterloo.ca/techreports/2015/cacr2015-12.pdf
10. Biasse, J.-F., Song, F.: Efficient quantum algorithms for computing class groups and solving the principal ideal problem in arbitrary degree number fields. In: SODA 2016, pp. 893–902 (2016)
11. Biehl, I., Buchmann, J.: Algorithms for quadratic orders. In: Mathematics of Computation 1943–1993: A Half-century of Computational Mathematics, pp. 425–451. AMS (1994)
12. Bos, J.W., Costello, C., Naehrig, M., Stebila, D.: Post-quantum key exchange for the TLS protocol from the ring learning with errors problem. In: IEEE S&P 2015, pp. 553–570 (2015)
13. Buchmann, J., Maurer, M., Möller, B.: Cryptography based on number fields with large regulator. J. de Théorie des Nombres de Bordeaux 12(2), 293–307 (2000)
14. Buchmann, J., Vollmer, U.: Binary Quadratic Forms: An Algorithmic Approach. Algorithms and Computation in Mathematics. Springer, Heidelberg (2007)
15. Buchmann, J.A.: A subexponential algorithm for the determination of class groups and regulators of algebraic number fields. In: Séminaire de Théorie des Nombres, Paris 1988–1989, pp. 27–41 (1990)
16. Buhler, J.P., Lenstra Jr., H.W., Pomerance, C.: Factoring integers with the number field sieve. In: Lenstra, A.K., Lenstra, H.W. (eds.) Algorithms and Computation in Mathematics. Springer, Heidelberg (2007)
17. Campbell, P., Groves, M., Shepherd, D.: Soliloquy: a cautionary tale (2014). http://docbox.etsi.org/Workshop/2014/201410_CRYPTO/S07_Systems_and_Attacks/S07_Groves_Annex.pdf

18. Cheon, J.H., Han, K., Lee, C., Ryu, H., Stehlé, D.: Cryptanalysis of the multilinear map over the integers. In: Oswald, E., Fischlin, M. (eds.) EUROCRYPT 2015. LNCS, vol. 9056, pp. 3–12. Springer, Heidelberg (2015). doi:10.1007/978-3-662-46800-5_1

19. Cohen, H.: A Course in Computational Algebraic Number Theory. Springer, Heidelberg (1993)

20. Cohen, H.: Advanced Topics in Computational Number Theory. Springer, New York (1999)

21. Coron, J.-S., Lepoint, T., Tibouchi, M.: New multilinear maps over the integers. In: Gennaro, R., Robshaw, M. (eds.) CRYPTO 2015. LNCS, vol. 9215, pp. 267–286. Springer, Heidelberg (2015). doi:10.1007/978-3-662-47989-6_13

22. Cramer, R., Ducas, L., Peikert, C., Regev, O.: Recovering short generators of principal ideals in cyclotomic rings. In: Fischlin, M., Coron, J.-S. (eds.) EUROCRYPT 2016. LNCS, vol. 9666, pp. 559–585. Springer, Heidelberg (2016). doi:10.1007/978-3-662-49896-5_20

23. The Sage Developers: SageMath, the Sage Mathematics Software System (Version 7.5.1) (2017). http://www.sagemath.org

24. Eisenträger, K., Hallgren, S., Lauter, K.: Weak instances of PLWE. In: Joux, A., Youssef, A. (eds.) SAC 2014. LNCS, vol. 8781, pp. 183–194. Springer, Cham (2014). doi:10.1007/978-3-319-13051-4_11

25. Galbraith, S.D.: Mathematics of Public Key Cryptography. Cambridge University Press, Cambridge (2012)

26. Galbraith, S.D., Gaudry, P.: Recent progress on the elliptic curve discrete logarithm problem. Des. Codes Cryptogr. **78**(1), 51–72 (2016)

27. Garg, S., Gentry, C., Halevi, S.: Candidate multilinear maps from ideal lattices. In: Johansson, T., Nguyen, P.Q. (eds.) EUROCRYPT 2013. LNCS, vol. 7881, pp. 1–17. Springer, Heidelberg (2013). doi:10.1007/978-3-642-38348-9_1

28. Gentry, C.: A fully homomorphic encryption scheme. Ph.D. thesis, Stanford University (2009). https://crypto.stanford.edu/craig

29. C. Gentry.: Fully homomorphic encryption using ideal lattices. In: STOC 2009, pp. 169–178 (2009)

30. Gentry, C.: Toward basing fully homomorphic encryption on worst-case hardness. In: Rabin, T. (ed.) CRYPTO 2010. LNCS, vol. 6223, pp. 116–137. Springer, Heidelberg (2010). doi:10.1007/978-3-642-14623-7_7

31. Gentry, C., Halevi, S.: Implementing Gentry's fully-homomorphic encryption scheme. In: Paterson, K.G. (ed.) EUROCRYPT 2011. LNCS, vol. 6632, pp. 129–148. Springer, Heidelberg (2011). doi:10.1007/978-3-642-20465-4_9

32. Hoffstein, J., Pipher, J., Silverman, J.H.: NTRU: a ring-based public key cryptosystem. In: Buhler, J.P. (ed.) ANTS 1998. LNCS, vol. 1423, pp. 267–288. Springer, Heidelberg (1998). doi:10.1007/BFb0054868

33. Kim, T., Barbulescu, R.: Extended tower number field sieve: a new complexity for the medium prime case. In: Robshaw, M., Katz, J. (eds.) CRYPTO 2016. LNCS, vol. 9814, pp. 543–571. Springer, Heidelberg (2016). doi:10.1007/978-3-662-53018-4_20

34. Kubota, T.: Über den bizyklischen biquadratischen Zahlkörper. Nagoya Math. J. **10**, 65–85 (1956)

35. Lenstra, A.K., Lenstra, H.W., Lovász, L.: Factoring polynomials with rational coefficients. Math. Ann. **261**, 515–534 (1982)

36. Lenstra, H.W.: Solving the Pell equation. Notices Amer. Math. Soc. **49**, 182–192 (2002)

37. Pohst, M.: A modification of the LLL reduction algorithm. J. Symb. Comput. **4**, 123–127 (1987)
38. Smart, N.P., Vercauteren, F.: Fully homomorphic encryption with relatively small key and ciphertext sizes. In: Nguyen, P.Q., Pointcheval, D. (eds.) PKC 2010. LNCS, vol. 6056, pp. 420–443. Springer, Heidelberg (2010). doi:10.1007/978-3-642-13013-7_25
39. van der Kallen, W.: Complexity of an extended lattice reduction algorithm (1998). http://www.staff.science.uu.nl/~kalle101/complexity.pdf
40. Vollmer, U.: Asymptotically fast discrete logarithms in quadratic number fields. In: Bosma, W. (ed.) ANTS 2000. LNCS, vol. 1838, pp. 581–594. Springer, Heidelberg (2000). doi:10.1007/10722028_39
41. Vollmer, U.: Rigorously analyzed algorithms for the discrete logarithm problem in quadratic number fields. Ph.D. thesis, Technische Universität, Darmstadt (2004)
42. Wada, H.: On the class number and the unit group of certain algebraic number fields. J. Fac. Sci. Univ. Tokyo Sect. I **13**(13), 201–209 (1966)
43. Williams, H.C.: Solving the Pell equation. In: Number theory for the millennium III, pp. 397–435. A K Peters (2002)

Computing Generator in Cyclotomic Integer Rings

A Subfield Algorithm for the Principal Ideal Problem in $L_{|\Delta_\mathbb{K}|}\left(\frac{1}{2}\right)$ and Application to the Cryptanalysis of a FHE Scheme

Jean-François Biasse[1($\boxtimes$)], Thomas Espitau[2], Pierre-Alain Fouque[3,4], Alexandre Gélin[2($\boxtimes$)], and Paul Kirchner[5]

[1] Department of Mathematics and Statistics, University of South Florida, Tampa, USA
biasse@usf.edu

[2] Sorbonne Universités, UPMC Paris 6, UMR 7606, LIP6, Paris, France
{thomas.espitau,alexandre.gelin}@lip6.fr

[3] Institut Universitaire de France, Paris, France
pierre-alain.fouque@univ-rennes1.fr

[4] Université de Rennes 1, Rennes, France

[5] École Normale Supérieure, Paris, France
paul.kirchner@ens.fr

Abstract. The Principal Ideal Problem (resp. Short Principal Ideal Problem), shorten as PIP (resp. SPIP), consists in finding a generator (resp. short generator) of a principal ideal in the ring of integers of a number field. Several lattice-based cryptosystems rely on the presumed hardness of these two problems. In practice, most of them do not use an arbitrary number field but a power-of-two cyclotomic field. The Smart and Vercauteren fully homomorphic encryption scheme and the multilinear map of Garg, Gentry, and Halevi epitomize this common restriction. Recently, Cramer, Ducas, Peikert, and Regev showed that solving the SPIP in such cyclotomic rings boiled down to solving the PIP. In this paper, we present a heuristic algorithm that solves the PIP in prime-power cyclotomic fields in subexponential time $L_{|\Delta_\mathbb{K}|}(1/2)$, where $\Delta_\mathbb{K}$ denotes the discriminant of the number field. This is achieved by descending to its totally real subfield. The implementation of our algorithm allows to recover in practice the secret key of the Smart and Vercauteren scheme, for the smallest proposed parameters (in dimension 256).

1 Introduction

Hard Problem in Lattices. Lattice-based problems appear to be among the most attractive alternatives to the integer factorization and discrete logarithm problems due to their conjectured resistance to quantum computations. Fortunately, all cryptographic primitives can be instantiated on the hardness of

© International Association for Cryptologic Research 2017
J.-S. Coron and J.B. Nielsen (Eds.): EUROCRYPT 2017, Part I, LNCS 10210, pp. 60–88, 2017.
DOI: 10.1007/978-3-319-56620-7_3

solving lattice problems, such as signature, basic encryption, Identity Based Encryption (IBE) as well as Fully Homomorphic Encryption (FHE) [21]. Not all these schemes rely on the same lattice-based problem. For instance, the NTRU cryptosystem [24], which is one of the most efficient encryption scheme related to lattices, is based on the Shortest Vector Problem (SVP). Besides, the authors of NTRU were the first to consider specific kinds of lattices, namely those related to polynomial rings. This idea was followed by the definition of another lattice-based problem that is the topic of a large body of works [31–34,44]: the Ring Learning With Error Problem (RLWE). Cryptosystems based on RLWE present both an efficient key size reduction and improved performance (for instance decryption, encryption and signature are faster than with arbitrary lattices). Yet, RLWE belongs to the specific family of *ideal-lattice* problems, which stem from algebraic number theory. This raises a potential drawback, since those lattices carry more structure than classical lattices, as they are derived from ideals in integer rings of number fields.

SPIP and PIP. Another presumably hard problem related to these ideals is called the Short Principal Ideal Problem (SPIP). It consists in finding a short[1] generator of an ideal, assuming it is principal. For instance, recovering the secret key from the public key in the Smart and Vercauteren FHE scheme [43] and in the Garg, Gentry, and Halevi multilinear map scheme [20], consists in solving an instance of the SPIP. This problem turns out to hinge on two distinct phases: on the one hand finding an arbitrary generator — known as the Principal Ideal Problem (PIP) — and on the other hand reducing such a generator to a short one. The problem of finding a generator of a principal ideal, which is the aim of this article, and even testing the principality of an ideal, are difficult problems in algorithmic number theory, as precised in [15, Chap. 4] and [45, Sect. 7].

From SPIP to PIP in Cyclotomic Fields. Recently, Cramer, Ducas, Peikert, and Regev [17] showed how to recover a small generator of a principal ideal in a prime-power cyclotomic field from an arbitrary generator in polynomial time. This work was based on an observation of Campbell, Groves, and Shepherd [12] who first proposed an efficient algorithm for reduction, essentially by decoding the log-unit lattice. The correctness of this approach was corroborated by Schank in an independent replication study [39].

Studying SPIP and PIP in this very specific class of number fields is motivated by the concrete instantiations of the various schemes. Again the Smart and Vercauteren FHE scheme [43] and the Garg, Gentry, and Halevi Multilinear Map scheme [20] exemplify this restriction to cyclotomic fields.

Prior Work on the PIP. Solving the PIP essentially requires the computation of the ideal class group $Cl(\mathbb{K})$ of the number field $\mathbb{K}$ where the ideals are defined. This approach is described in [15, Algorithm 6.5.10] (see [5, Algorithm 7] for a

---

[1] Short means that we have a norm. In our case, it is derived from the canonical embedding of the number field into a Euclidean space.

description in line with the approach of this paper). The first subexponential algorithm for computing $Cl(\mathbb{K})$ was due to Hafner and McCurley [23]. It applies to imaginary quadratic fields, and it was later generalized by Buchmann [11] to classes of number fields of fixed degree. In [8], Biasse and Fieker presented an algorithm for computing $Cl(\mathbb{K})$ in subexponential time in arbitrary classes of number fields. Combined with [5, Algorithm 7], this yielded a subexponential time algorithm for solving the PIP in arbitrary classes of number fields. In a prime-power cyclotomic field of degree N, the Biasse-Fieker algorithm solves the PIP in time $L_{|\Delta_{\mathbb{K}}|}(2/3 + \varepsilon) \left(\approx 2^{N^{2/3+o(1)}} \right)$, for $\varepsilon > 0$ arbitrarily small. Biasse also described[2] in [6] an $L_{|\Delta_{\mathbb{K}}|}(1/2 + \varepsilon)$-algorithm that computes $Cl(\mathbb{K})$ and solves the PIP in fields of the form $\mathbb{Q}(\zeta_{p^k})$. Note that the PIP is also the subject of research on quantum algorithms for its resolution. Recently, Biasse and Song [9] described a quantum polynomial time algorithm for the PIP in classes of number fields of arbitrary degree.

Our Results. The main contribution of this paper is an algorithm for computing the class group $Cl(\mathbb{K}^+)$ and solving the PIP in $\mathbb{K}^+$ in time $L_{|\Delta_{\mathbb{K}}|}(1/2) \left(\approx 2^{N^{1/2+o(1)}} \right)$ where $\mathbb{K}^+$ is the maximal real subfield of prime-power cyclotomic field $\mathbb{K}$ and N denotes its degree. Thanks to the Gentry-Szydlo algorithm, our algorithm also provides a solution to the PIP in $\mathbb{K}$ with the same $L_{|\Delta_{\mathbb{K}}|}(1/2)$-complexity.

In addition to this theoretical study, we implement an attack against a FHE scheme that relies on the hardness of finding a small generator of ideals in those fields. We were able to recover in practice a generator in the field $\mathbb{Q}(\zeta_{512})$. Such parameters were proposed by Smart and Vercauteren as toy parameters in [43]. The most challenging part of the computation was to efficiently implement the Gentry-Szydlo algorithm [22]. We used the version of Gentry-Szydlo described by Kirchner in [26]. We also implemented an algorithm for descending to the subfield $\mathbb{K}^+$ from $\mathbb{K}$ and for collecting relations between generators of $Cl(\mathbb{K}^+)$.

Organization of the Paper. In Sect. 2, we recall mathematical results for lattices and algebraic number theory that we use in the rest of the paper. Then, Sect. 3 presents the principal ideal problem (PIP) and the cryptosystems based on this problem such as the Smart-Vercauteren fully homomorphic encryption scheme. Next, we describe the different steps of the algorithm to solve PIP in Sect. 4. Finally, Sect. 5 gives information about our experimentations.

2   Mathematical Background

We recall briefly here basic facts on lattices and algebraic number theory. A more detailed introduction is provided in the Appendix A.

---

[2] There was a small mistake in the original description which was corrected in a subsequent version.

General Notations. For dealing with complexities, we introduce the L-notation, that is classical when presenting index calculus algorithms with subexponential complexity. Given two constants a and c with $a \in [0,1]$ and $c \geq 0$, we denote by:

$$L_{|\Delta_{\mathbb{K}}|}(a,c) = e^{(c+o(1))(\log|\Delta_{\mathbb{K}}|)^a (\log\log|\Delta_{\mathbb{K}}|)^{1-a}},$$

where $o(1)$ tends to 0 as $|\Delta_{\mathbb{K}}|$, the discriminant of the number field, tends to infinity. We also encounter the notation $L_{|\Delta_{\mathbb{K}}|}(a)$ when specifying c is superfluous, that is considering quantities in $L_{|\Delta_{\mathbb{K}}|}(a, O(1))$.

2.1 Lattices

Lattices are defined as additive discrete subgroups of $\mathbb{R}^n$, i.e. the integer span $L(\mathbf{b}_1, \ldots, \mathbf{b}_d) = \bigoplus_{i=1}^{d} \mathbb{Z}\mathbf{b}_i$ of a linearly independent family of vectors $\mathbf{b}_1, \ldots, \mathbf{b}_d$ in $\mathbb{R}^n$. Such a family is called a *basis* of the lattice, and is not unique. Nevertheless, all the bases of a given lattice have the same number of elements, d, which is called the *dimension* of the lattice. Among the infinite number of different bases of an n-dimensional lattice with $n \geq 2$, some have interesting properties, such as having reasonably small vectors and low orthogonality defect — that means that they are almost orthogonal.

The problem of finding such good bases is the aim of *lattice reduction*. There are in short two kinds of reduction algorithms: *approximation* algorithms on the one hand, like the celebrated LLL algorithm and its blockwise variants such as BKZ and DBKZ [35], and *exact* algorithms on the other hand, such as enumeration or sieving, that are exponential in time and space. In high dimension, only approximation algorithms — which run in polynomial time in the dimension[3] — can be used to find relatively short vectors, but usually not the shortest ones.

The DBKZ Algorithm and Cheon's Determinant Trick. In this part, we recall the complexity of DBKZ algorithm, introduced by Micciancio and Walter in [35], its approximation factor, and a trick due to Cheon and Lee [14] that improves this factor for integer lattices with small determinant.

Theorem 1 (Bounds for DBKZ output). *The smallest vector output by DBKZ algorithm with block-size β has a norm bounded by:*

$$\beta^{\frac{n-1}{2(\beta-1)}} \cdot \mathrm{Vol}\,(\mathcal{L})^{\frac{1}{n}}.$$

The algorithm runs in time $Poly(n, size(\mathbf{B})) \cdot (3/2 + o(1))^{\beta/2}$, where $\mathbf{B}$ is the input basis and $(3/2 + o(1))^{\beta/2}$ stands for the cost of solving the Shortest Vector Problem in dimension β, using sieving techniques (see [3]).

Proof. This is a direct application of [35, Theorem 1], where the Hermite constant γ_β is upper bounded by β.

---

[3] BKZ and DBKZ are exponential in the block size.

In a note [14] of 2015, Cheon and Lee suggest to convert the basis of an integer lattice having small determinant, to its Hermite normal form (HNF) before reducing it, for instance with the DBKZ algorithm. This algorithm seems to be folklore. In particular, Biasse uses a similar strategy in the context of class group computations in [5, Sect. 3.3]. This note gives a detailed analysis and we refer to this method as *Cheon's trick*. We develop here this idea and derive corresponding bounds. For completeness purpose, the definition of HNF is recalled in Appendix A.1. More precisely, we have the following lemma.

Lemma 1. *Given* $\mathbf{B} = [\mathbf{b}_1, \ldots, \mathbf{b}_n]$ *a basis in HNF of a n-dimensional lattice $\mathcal{L}$, we have for any* $1 \leq i < n$:

$$\mathrm{Vol}\left([\mathbf{b}_1, \ldots, \mathbf{b}_i]\right) \leq \mathrm{Vol}\left([\mathbf{b}_1, \ldots, \mathbf{b}_{i+1}]\right).$$

In particular, for any sublattice $\mathcal{L}'$ generated by the m first vectors of $\mathbf{B}$, we have $\mathrm{Vol}\left(\mathcal{L}'\right) \leq \mathrm{Vol}\left(\mathcal{L}\right)$.

Remark that both the n-th root of the determinant and an exponential factor of n appear in the bound of Theorem 1. Hence we can perform the DBKZ reduction on a sublattice only generated by the first m columns of the HNF in order to minimize this upper bound, as a trade-off between these quantities.

Explicitly we fix $m = \left\lfloor \sqrt{\frac{2\beta}{\log \beta} \log(\mathrm{Vol}\left(\mathcal{L}\right))} \right\rceil$ and run the algorithm of Fig. 1 on the basis $\mathbf{B} = (\mathbf{b}_1, \ldots, \mathbf{b}_n)$:

---

1. Compute the HNF $(\mathbf{b}_1', \ldots, \mathbf{b}_n')$ of $\mathbf{B}$.
2. Run DBKZ with block-size β on $(\mathbf{b}_1', \ldots, \mathbf{b}_m')$ with $m = \left\lfloor \sqrt{\frac{2\beta}{\log \beta} \log(\mathrm{Vol}\left(\mathcal{L}\right))} \right\rceil$.
3. Return the first vector of the output of DBKZ.

Fig. 1. Approx-SVP algorithm with HNF+DBKZ with block-size β.

Theorem 2. *For any n-dimensional integer lattice $\mathcal{L}$ such that* $\mathrm{Vol}\left(\mathcal{L}\right) \leq \beta^{\frac{n^2}{2\beta}}$, *the output $\mathbf{v}$ of the previous Approx-SVP algorithm satisfies:*

$$\|\mathbf{v}\| \leq \beta^{(1+o(1))}\sqrt{2\log_\beta(\mathrm{Vol}(\mathcal{L})/\beta}.$$

This algorithm takes time $\mathrm{Poly}(n, \mathrm{size}(\mathbf{B}))(3/2 + o(1))^{\beta/2}$.

Proof. The condition on the covolume of $\mathcal{L}$ ensures that $m \leq n$.
Then, by Theorem 1 and Lemma 1 we have

$$
\begin{aligned}
\|\mathbf{v}\| &\leq & \beta^{\frac{m}{2\beta}} \cdot \mathrm{Vol}\left(\mathcal{L}'\right)^{\frac{1}{m}} \\
&\leq & \beta^{\frac{m}{2\beta}} \cdot \mathrm{Vol}\left(\mathcal{L}\right)^{\frac{1}{m}} \\
&\leq & \beta^{\sqrt{2\log_\beta(\mathrm{Vol}(\mathcal{L}))/\beta}},
\end{aligned}
$$

which yields the announced result.

2.2 Number Fields

Let $\mathbb{K} = \mathbb{Q}(\alpha)$ be a number field of degree N, then there exists a monic irreducible degree-N polynomial $P \in \mathbb{Z}[X]$ such that $\mathbb{K} \simeq \mathbb{Q}[X]/(P)$. Denoting by $(\alpha_1, \ldots, \alpha_N) \in \mathbb{C}^N$ its distinct complex roots, each embedding (field homomorphism) $\sigma_i : \mathbb{K} \to \mathbb{C}$ is the evaluation of $\mathbf{a} \in \mathbb{K}$, viewed as a polynomial modulo P, at the root α_i, i.e. $\sigma_i : \mathbf{a} \mapsto \mathbf{a}(\alpha_i)$. Let r_1 be the number of real roots and r_2 be the number of pairs of complex roots ($N = r_1 + 2r_2$), we have $\mathbb{K} \otimes \mathbb{R} \simeq \mathbb{R}^{r_1} \times \mathbb{C}^{r_2}$. We define the norm $\| \cdot \|$ over $\mathbb{K}$ as the canonical Euclidean norm of $\sigma(\mathbf{x}) \in \mathbb{R}^{r_1} \times \mathbb{C}^{r_2}$ where $\sigma(\mathbf{x}) = (\sigma_1(\mathbf{x}), \ldots, \sigma_{r_1+r_2}(\mathbf{x})) \in \mathbb{R}^{r_1} \times \mathbb{C}^{r_2}$, where $\sigma_1, \ldots, \sigma_{r_1}$ are the real embeddings of $\mathbb{K}$ and $\sigma_{r_1+1}, \ldots, \sigma_N$ are the complex embeddings of $\mathbb{K}$, each σ_{r_1+j} being paired with its complex conjugate $\sigma_{r_1+r_2+j}$. The number field $\mathbb{K}$ is viewed as a Euclidean $\mathbb{Q}$-vector space endowed with the inner product $\langle \mathbf{a}, \mathbf{b} \rangle = \sum_\sigma \sigma(\mathbf{a}) \bar{\sigma}(\mathbf{b})$ where σ ranges over all the $r_1 + 2r_2$ embeddings $\mathbb{K} \to \mathbb{C}$. This defines the euclidean norm denoted $\| \cdot \|$. The algebraic norm on $\mathbb{K}$ is defined as $\mathcal{N}_{\mathbb{K}/\mathbb{Q}}(\mathbf{v}) = \prod_{i=1}^{N} \sigma_i(\mathbf{v})$.

Coefficient Embedding and Ideal Lattices. Let α be one of the roots α_i (it may differ from the initial α if this one is not an algebraic integer). Considering the natural isomorphism between $\mathbb{Z}[\alpha] \subset \mathcal{O}_\mathbb{K}$ and $\mathbb{Z}[X]/(P)$ gives rise to an embedding of $\mathbb{Z}[\alpha]$ trough the coefficients of associated polynomials. More precisely, we have the following sequence of abelian groups

$$
\begin{array}{ccccc}
\mathbb{Z}^N & \xrightarrow{\iota} & \mathbb{Z}[X] & \xrightarrow{\pi} & \mathbb{Z}[X]/(P) \simeq \mathbb{Z}[\alpha] \\
(c_0, \cdots, c_{N-1}) & \longmapsto & \sum_{0 \le i < N} c_i X^i & \longmapsto & \sum_{0 \le i < N} c_i \alpha^i,
\end{array}
$$

defining the announced embedding by coefficients as $\mathcal{C} = \iota^{-1} \circ \pi^{-1}$. Such an embedding provides a norm in the field, namely: $\|\mathbf{a}\|_\mathcal{C} = \|\mathcal{C}(\mathbf{a})\|_2$.

Let us state a basic result on the link between field norm and polynomial representation:

Lemma 2. *For algebraic integers defined as polynomials in α, namely $\mathbf{a} = T(\alpha)$ for $T \in \mathbb{Z}[X]$, we can bound the norm by*

$$
|\mathcal{N}_{\mathbb{K}/\mathbb{Q}}(\mathbf{a})| \le (N+1)^{m/2}(m+1)^{N/2} H(T)^N H(P)^m,
$$

where $m = \deg T$, $N = \deg P$ and $H(P)$ is the absolute maximum of the coefficients of P.

Proof. Remark first that the norm of this element corresponds to the resultant of the polynomials T and P [15, Proposition 4.3.4]. Then we apply the bounds of [10, Theorem 7] for the resultant of two polynomials and conclude.

As a result, we can directly relate the norm of the embedding with the field norm:

Corollary 1. *For any $\mathbf{a} \in \mathbb{Z}[\alpha]$: $|\mathcal{N}_{\mathbb{K}/\mathbb{Q}}(\mathbf{a})|^{\frac{1}{N}} \le (N+1) \cdot H(P) \cdot \|\mathbf{a}\|_\mathcal{C}$.*

Canonical Embedding and Ideals. A remarkable property of the canonical embedding is the way it represents the ring of integers and more generally every integral ideal. Indeed, the embedding $\sigma(\mathfrak{a})$ of any integral ideal $\mathfrak{a}$ is a Euclidean lattice. In particular, for the ring of integers, we have that $\sigma(\mathcal{O}_\mathbb{K})$ is a lattice. Its (co)volume is called the *discriminant* $\Delta_\mathbb{K}$ of the field $\mathbb{K}$. Therefore, one can compute the discriminant as a determinant: for $(\mathbf{b}_1, \ldots, \mathbf{b}_N)$ an integral basis of $\mathcal{O}_\mathbb{K}$, we have

$$
\Delta_\mathbb{K} = \left| \det \begin{pmatrix} \sigma_1(\mathbf{b}_1) & \sigma_1(\mathbf{b}_2) & \cdots & \sigma_1(\mathbf{b}_N) \\ \sigma_2(\mathbf{b}_1) & \ddots & & \vdots \\ \vdots & & \ddots & \vdots \\ \sigma_N(\mathbf{b}_1) & \cdots & \cdots & \sigma_N(\mathbf{b}_N) \end{pmatrix} \right)^2 .
$$

Loosely speaking, the discriminant is a size measure of the integer ring. That is why we use it to express the complexity when we work with number fields or rings of integers. Moreover, it acts as a proportionality coefficient between the norm of an ideal and the covolume of its embedding:

Lemma 3. *For any integral ideal $\mathfrak{a}$ of $\mathbb{K}$, we have $\sigma(\mathfrak{a})$ is a lattice of $\mathbb{R}^N$ and*

$$
\mathrm{Vol}\,(\sigma(\mathfrak{a})) = \sqrt{|\Delta_\mathbb{K}|}\mathcal{N}\,(\mathfrak{a}),
$$

where $\mathrm{Vol}\,(\mathcal{L})$ is the covolume of the lattice $\mathcal{L}$.

Smoothness of Ideals. To evaluate the probability of smoothness of ideals, we need to assume the same unproven heuristic as in [5,8], directly derived from what has been proved for integers by Canfield, Erdős and Pomerance [13]. Let $\mathcal{P}(x, y)$ be the probability that a principal ideal of $\mathcal{O}_\mathbb{K}$ of norm bounded by x is a power-product of prime ideals of norm bounded by y. Then, we have

Heuristic 1 [5, Heuristic 1]. *We assume that under the Generalized Riemann Hypothesis (GRH), the probability $\mathcal{P}(x, y)$ satisfies*

$$
\mathcal{P}(x, y) \geq e^{-u \log u(1 + o(1))} \quad \textit{for } u = \frac{\log x}{\log y}.
$$

Heuristic 1 was put in perspective with Scourfield's work [40] by Biasse and Fieker [8, Sect. 3.1]. In the number field setting, the previous heuristic admits a neat rewriting in terms of the handy L-notation:

Corollary 2 [5, Corollary 2.1]. *Let $x = \lfloor \log L_{|\Delta_\mathbb{K}|}\,(a, c) \rfloor$ and the smoothness bound $y = \lceil \log L_{|\Delta_\mathbb{K}|}\,(b, c') \rceil$. Then assuming Heuristic 1, the probability $\mathcal{P}(x, y)$ that an ideal of $\mathcal{O}_\mathbb{K}$ of norm bounded by x is a power-product of prime ideals of norm bounded by y satisfies*

$$
\mathcal{P}(x, y) \geq L_{|\Delta_\mathbb{K}|} \left(a - b, \frac{-c}{c'}(a - b) \right).
$$

A similar assertion for smoothness of ideals was proved by Seysen [41] in 1985 for the quadratic case, but for arbitrary degree, it remains conjectural, even under GRH. This is one of the reasons why the complexity of the number field sieve (NFS) [29] is still a heuristic estimation.

2.3   Cyclotomic Fields and Cyclotomic Integers

We denote by Φ_m the *m-th cyclotomic polynomial*, that is the unique irreducible polynomial in $\mathbb{Q}[X]$ dividing $X^m - 1$ that is not a divisor of any of the $X^k - 1$ for $k < m$. Its roots are thus the m-th primitive roots of the unity. Therefore, cyclotomic polynomials can be written in closed form as:

$$\Phi_m = \prod_{\substack{1 \leq k \leq m \\ \gcd(k,m)=1}} \left(X - e^{2i\pi \frac{k}{m}} \right).$$

The m-th cyclotomic field $\mathbb{Q}(\zeta_m)$ is obtained by adjoining a primitive m-th root ζ_m of unity to the rational numbers. As such, $\mathbb{Q}(\zeta_m)$ is isomorphic to the splitting field $\mathbb{Q}[X]/(\Phi_m)$. Its degree over $\mathbb{Q}$ is $\deg(\Phi_m)$, that is $\varphi(m)$, where φ is the Euler totient function. In this specific class of number fields, the ring of integer is precisely $\mathbb{Z}[X]/(\Phi_m) \cong \mathbb{Z}[\zeta_m]$ (see [46, Theorem 2.6] for a proof of this statement).

The canonical embedding can also be easily presented since the embeddings are the linear functions sending ζ_m to ζ_m^j, for $j \in (\mathbb{Z}/m\mathbb{Z})^*$. Since the roots come in conjugate pairs ($\zeta_m^j = -\zeta_m^{m-j}$ for all j), we can write down the Log-embedding by indexing over the quotient $G = (\mathbb{Z}/m\mathbb{Z})^*/\{-1,1\}$:

$$\mathrm{Log}(x): \quad \mathbb{K} \quad \longrightarrow \mathbb{R}^{\varphi(m)/2}$$
$$P \mod \Phi_m \mapsto \left(\log |P(\zeta_m^j)| \right)_{j \in G}.$$

The discriminant of $\mathbb{Q}(\zeta_m)$ has a closed form expression [46, Proposition 2.7]:

$$\Delta_{\mathbb{Q}(\zeta_m)} = (-1)^{\varphi(m)/2} \frac{m^{\varphi(m)}}{\prod_{p|m} p^{\varphi(m)/(p-1)}},$$

where the product in the denominator is over primes p dividing m.

Example 1. For a prime-power cyclotomic field, we get $\left| \Delta_{\mathbb{Q}(\zeta_{p^k})} \right| = p^{(kp-k-1)p^{k-1}}$. In particular, when $p = 2$, $\left| \Delta_{\mathbb{Q}(\zeta_{2^{n+1}})} \right| = 2^{n2^n}$.

For power-of-two cyclotomic fields, we then have $L_{|\Delta_{\mathbb{K}}|}(\alpha) = 2^{O(N^\alpha \log(N))}$. Thus, writing the complexity as $L_{|\Delta_{\mathbb{K}}|}(\alpha)$ or $2^{O(N^\alpha \log(N))}$ is equivalent. We choose to use the L-notation, since it eases the exposition of the complexities presented in this paper.

2.4   Cyclotomic Units

Giving the complete description of the units of a generic number field is a computationally hard problem of algorithmic number theory. However it is possible to describe a subgroup of finite index of the unit group, called the *cyclotomic units*. This subgroup contains all the units that are products of numbers[4] of the form $\zeta_m^i - 1$ for any $1 \leq i \leq m$. More precisely we have

Lemma 4 (Lemma 8.1 of [46]). *Let m be a prime power, then the group C of cyclotomic units is generated by $\pm\zeta_m$ and $(\mathbf{b}_i)_{1 \leq i \leq m}$, where*

$$\mathbf{b}_i = \frac{\zeta_m^i - 1}{\zeta_m - 1}.$$

The index of the subgroup of cyclotomic units in the group of units is $h^+(m)$, the class number of the totally real subfield of $\mathbb{Q}(\zeta_m)$ (see for instance [46]). In the case of power-of-two m, a well supported conjecture clarifies the value of h^+.

Heuristic 2 (Weber's class number problem). *We assume that for power-of-two cyclotomic fields, the class number of its totally real subfield is 1.*

Thus, under Weber's heuristic, the cyclotomic units and the units coincide in the power-of-two cyclotomic fields.

3   Principal-Ideal Problem and Cryptography

Among all the FHE schemes proposed in the last decade, the security of a couple of them directly relies on the ability to find relatively short generators in principal ideals. This is the case of the proposal of Smart and Vercauteren [43], which is a simplified version of the original scheme of Gentry [21]. Other schemes based on the same security assumptions include the Soliloquy scheme of Campbell, Groves and Shepherd [12] and the candidates for multilinear maps [20, 28]. More formally, the underlying — presumably hard — problem is the following one, already known as SPIP (Short Principal Ideal Problem) or SG-PIP (Short Generator-Principal Ideal Problem): given some $\mathbb{Z}$-basis of a principal ideal with a promise that it possesses a "short" generator $\mathbf{g}$ for the Euclidean norm, find this generator or at least a short enough generator of this ideal.

The strategy to address this problem roughly splits in two main steps:

1. Given the $\mathbb{Z}$-basis of the ideal, find a generator, not necessarily short, that is $\mathbf{g}' = \mathbf{g} \cdot \mathbf{u}$ for a unit $\mathbf{u}$.
2. From $\mathbf{g}'$, find a short generator of the ideal.

Recently, several results have allowed to deal with the second step. Indeed, Campbell, Groves and Shepherd [12] claimed in 2014 an — although unproven — efficient solution for power-of-two cyclotomic fields, confirmed by experiments

---

[4] One should notice that if m is a prime power, $\zeta_m^i - 1$ is not a unit, but $\mathbf{b}_i$ is.

conducted by Schank [39] in 2015. Eventually, the proof was provided by Cramer, Ducas, Peikert, and Regev [17] in 2015. Throughout this paper, we focus on the resolution of the first step, known as PIP (Principal Ideal Problem). Nonetheless, for completeness, we present briefly the reduction from SPIP to PIP in Sect. 4.4.

As a direct illustration of the resolution of this problem, we present an attack on the scheme that Smart and Vercauteren present in [43], which leads to a *full key recovery*. This attack is our key thread through the exposition of the algorithm. Before going any further in the details of the attack, we recall in Fig. 2 the key generation process in the case of power-of-two cyclotomic fields. This instantiation is the one chosen by the authors for presenting their implementation results.

---

1. Fix the security parameter $N = 2^n$.
2. Let $F(X) = X^N + 1$ be the polynomial defining the cyclotomic field $\mathbb{K} = \mathbb{Q}(\zeta_{2N})$.
3. Set $G(X) = 1 + 2 \cdot S(X)$ for $S(X)$ of degree $N - 1$ with coefficients absolutely bounded by $2^{\sqrt{N}}$, such that the norm $\mathcal{N}(\langle G(\zeta_{2N}) \rangle)$ is prime.
4. Set $\mathbf{g} = G(\zeta_{2N}) \in \mathcal{O}_\mathbb{K}$.
5. Return $(\mathsf{sk} = \mathbf{g}, \mathsf{pk} = \mathrm{HNF}(\langle \mathbf{g} \rangle))$.

Fig. 2. Key generation of the scheme [43].

Remark 1. The public key can be any $\mathbb{Z}$-basis of the ideal generated by $\mathbf{g}$, or even a two-elements representation of this ideal. Precisely, [43] provides the public key as a pair of elements that generates the lattice. This is always possible, see [15, Sect. 4.7.2]. We make the choice of the Hermite Normal Form representation[5].

As our attack consists in a full secret key recovery, realized directly from the public key, we do not mention here the encryption and decryption procedures. Even though this work tackles more on the principal ideal problem than on this reduction, we emphasize the fact that the output of this reduction to a short generator can be any one of the $\mathbf{g} \cdot \zeta_{2N}^i$, having same Euclidean norm for any $1 \leq i \leq 2N$. Nonetheless, this does not represent an issue, since all of these keys are equivalent with regard to the decryption procedure. In addition, in this precise construction of the Smart and Vercauteren FHE scheme, the only odd coefficient of $G(X)$ is the last one, so that we may recover the exact generator $\mathbf{g}$ readily.

The whole complexity of our attack is subexponential, in $L_{|\Delta_\mathbb{K}|}(1/2)$. This beats the previous state-of-the-art in $L_{|\Delta_\mathbb{K}|}(2/3)$, derived from the combined work of [8,17].

---

[5] The definition of the HNF is recalled for completeness in Appendix A.

4 Solving the PIP or How to Perform a Full Key Recovery?

We recall that our ultimate goal is to perform a full key recovery given only the public elements. As mentioned in [43], this problem is obviously much more difficult than recovering a plain-text from a cipher-text which is based on the bounded distance decoding problem and the security level is set according to this latter problem. We first give an overview of the whole strategy and then get an in-depth view of each part. But before going any further into the details of the attack, let us fix the notations and recurrent objects we are going to use. The number field where the PIP is defined is $\mathbb{Q}(\zeta_{2N})$, for $N = 2^n$, defined by the polynomial $X^N + 1$, in the same fashion as in Sect. 2.3. For the sake of notation simplicity, ζ_{2N} is simply denoted by ζ. Though we focus on power-of-two cyclotomic fields, all our results can be easily generalized to arbitrary prime-power cyclotomic fields. Our starting point is the public key, that is, a somewhat "bad" basis of the principal ideal $\mathcal{I} = \langle \mathbf{g} \rangle$, generated by the secret key $\mathbf{g}$.

Before any other operations, the dimension of the ideals involved is shrunk by half by reducing the problem to an equivalent one in the *totally real subfield* $\mathbb{Q}(\zeta + \zeta^{-1})$. This is not mandatory (see [6]), but it eases the computation. This part of the algorithm is a straightforward consequence of the Gentry-Szydlo algorithm introduced in [22]. The problem is now reduced to the research of a generator of an ideal $\mathcal{I}^+$ in the totally real subfield. Then, the strategy appears to be recursive reductions of ideals, until we eventually reach a B-smooth ideal $\mathcal{I}^s$, for a fixed bound $B > 0$ and an algebraic integer $\mathbf{h}$ such that $\langle \mathbf{h} \rangle = \mathcal{I}^+ \cdot \mathcal{I}^s$. This is the q-*descent phase*.

We are now interested in finding a generator of $\mathcal{I}^s$. We use a strategy based on *class group computation*. It consists in finding a generating set of all the relations between generators of the class group, and then rewrite the input ideal with respect to these generators. Then we can recover a generator $\mathbf{h_0}$ of $\mathcal{I}^s$ by solving a linear system of equations. It then permits to derive the generator of the ideal $\mathcal{I}^+$: $\mathbf{h} \cdot \mathbf{h_0}^{-1}$. A generator of the public-key ideal is then obtained by lifting it from the totally real subfield to the initial number field $\mathbb{Q}(\zeta)$. It suffices to multiply the current generator by another integer obtained during the computation. Now the PIP is solved, it only remains a final step to recover the secret key: perform the reduction from this generator to a short one, using the method of [17].

Consequently, the full algorithm can be split in four main steps, which are, in a nutshell:

1. Perform a reduction from the cyclotomic field to its totally real subfield, allowing to work in smaller dimension.
2. Then a q-descent makes the size of involved ideals decrease.
3. Collect relations and run linear algebra to construct small ideals and a generator.
4. Eventually run the derivation of the small generator from a bigger one.

Let us now get into the details of all these parts.

4.1   Step 1: Reduction to the Totally Real Subfield

Starting with the public key, we get a $\mathbb{Z}$-basis $(\mathbf{b}_1, \ldots, \mathbf{b}_N)$ of an ideal $\mathcal{I}$ belonging to the cyclotomic field $\mathbb{Q}(\zeta)$ of dimension[6] N. The larger the dimension is, the harder it is to handle and even only represent such objects. However, it is possible to halve the dimension. The main part of this step relies on the so-called *Gentry-Szydlo* (GS) algorithm, first described in [22] as an attack on the NTRU scheme and later revised and generalized by Lenstra and Silverberg in [30].

This original algorithm takes as input a $\mathbb{Z}$-basis of an ideal $\mathcal{I}$ in the ring $\mathbb{Z}[X]/(X^N + 1)$ — with the promise to be principal — and the algebraic integer $\mathbf{u} \cdot \bar{\mathbf{u}}$, for $\mathbf{u}$ a generator of $\mathcal{I}$. Here, $\bar{\mathbf{u}}$ denotes the conjugate of $\mathbf{u}$ for the automorphism defined by $\zeta \mapsto \zeta^{-1}$. It then recovers in polynomial time the element $\mathbf{u}$. In our case, we can not perform the recovery of the generator $\mathbf{g}$, secret key of the scheme, since *a priori* we do not have access to any kind of information about the product $\mathbf{g} \cdot \bar{\mathbf{g}}$.

To overcome this difficulty, we introduce another integer $\mathbf{u} = \mathcal{N}(\mathbf{g}) \, \mathbf{g} \bar{\mathbf{g}}^{-1}$, as described by Garg, Gentry, and Halevi in [20, Sect. 7.8.1]. One should notice that the norm factor is only there to avoid introduction of denominators in the definition of $\mathbf{u}$. Although $\mathbf{u}$ is still unknown at this point, thanks to the $\mathbb{Z}$-basis of $\langle \mathbf{g} \rangle$ we can construct a $\mathbb{Z}$-basis of $\langle \mathbf{u} \rangle$ and deriving the product $\mathbf{u} \cdot \bar{\mathbf{u}}$ which simply corresponds to $\mathcal{N}(\mathbf{g})^2$.

Hence, we get access to $\mathbf{u}$ in polynomial time using GS. From this element $\mathbf{u}$, we directly reconstruct $\mathbf{g} \bar{\mathbf{g}}^{-1}$ and using the basis of $\mathcal{I}$, we then introduce the family of vectors

$$\mathbf{c}_i = \mathbf{b}_i \left(1 + \frac{\bar{\mathbf{g}}}{\mathbf{g}} \right),$$

providing a basis of the ideal $\mathcal{I}^+$ generated by $\mathbf{g} + \bar{\mathbf{g}}$. The reader should notice that this ideal belongs to the totally real subfield $\mathbb{Q}(\zeta + \zeta^{-1})$, of index 2 in $\mathbb{Q}(\zeta)$. From now on, we denote by $\mathcal{O}_{\mathbb{K}}^+$ the ring of integers of $\mathbb{Q}(\zeta + \zeta^{-1})$, corresponding to $\mathcal{O}_{\mathbb{K}} \cap \mathbb{Q}(\zeta + \zeta^{-1})$.

Let us suppose briefly that we know the generator $\mathbf{g} + \bar{\mathbf{g}}$ of $\mathcal{I}^+$. Then it would be sufficient to multiply it by $\frac{1}{1 + \mathbf{g} \bar{\mathbf{g}}^{-1}}$ to recover the secret key $\mathbf{g}$. Hence, we have reduced the problem of finding a generator of the idea $\mathcal{I}$ belonging to the cyclotomic field of dimension N to the one of finding a generator of ideal $\mathcal{I}^+$ that belongs to the totally real subfield, whose dimension is $\frac{N}{2}$. For a more detailed presentation of this technique, see [20, Theorem 8].

Note that even though the generator is known up to a unit — i.e. $(\mathbf{g} + \bar{\mathbf{g}}) \cdot \mathbf{v}$ for $\mathbf{v} \in \mathcal{U}_{\mathbb{Q}(\zeta)}$ — the generator of $\mathcal{I}$ recovered is $\mathbf{g} \cdot \mathbf{v}$. This suffices, thanks to the last reduction part, to recover a short generator.

One could wonder if working in a real field has some relevant matter with the upcoming parts of the attack. The answer is up to our knowledge negative and we are only interested in the halving of dimension. For the asymptotic complexity, this initial reduction is somehow not meaningful since it only gives a speedup

---

[6] The smallest security parameters of the Smart and Vercauteren scheme is $N = 256$.

of a constant factor in the exponent. But in practice, it allows to double the dimension of the tractable cases, implying tackling security parameters twice bigger!

4.2   Step 2: q-descent Phase

Let us momentarily set aside the algebraic integer obtained in the previous phase and only focus on the ideal $\mathcal{I}^+$. By construction, it is principal and generated by $\mathbf{g} + \bar{\mathbf{g}}$. From now on, all the computations are performed in the totally real subfield of dimension $\frac{N}{2}$, and from then on N becomes $\frac{N}{2}$.

The goal of this phase is to find an integer $\mathbf{h}$ and a B-smooth principal ideal $\mathcal{I}^s$, such that $\langle \mathbf{h} \rangle = \mathcal{I}^+ \cdot \mathcal{I}^s$, for a certain bound $B > 0$. These objects are discovered recursively, by generating at each step ideals of norm smaller and smaller. This descent strategy derives from discrete logarithm computations (see [1,25]) and has been adapted to number fields of large degree by Biasse [5, Sect. 3.2]. Since we want a global complexity in $L_{|\Delta_{\mathbb{K}}|}(1/2)$, the smoothness bound B is chosen[7] in $L_{|\Delta_{\mathbb{K}}|}(1/2)$. In order to bootstrap this q-descent, we first need to find an ideal that splits in the class group as a product of multiple prime ideals of controlled norm, that is in our case, upper bounded by $L_{|\Delta_{\mathbb{K}}|}(1)$.

Initial Round: Classical DBKZ Reduction. As announced, we aim to construct efficiently a $L_{|\Delta_{\mathbb{K}}|}(1)$-smooth principal ideal from $\mathcal{I}^+$. Formally, we want to prove the following:

Theorem 3. *Let $\mathbb{K}$ be a number field. Assuming Heuristic 1, from any ideal $\mathfrak{a} \subset \mathcal{O}_{\mathbb{K}}$, it is possible to generate in expected time $L_{|\Delta_{\mathbb{K}}|}(1/2)$ an integral ideal $\mathfrak{b}$ that is $L_{|\Delta_{\mathbb{K}}|}(1)$-smooth and an integer $\mathbf{v}$ such that:*

$$\langle \mathbf{v} \rangle = \mathfrak{a} \cdot \mathfrak{b}.$$

The difficulty of this preliminary part is that a priori the norm of the input ideal $\mathfrak{a}$ can be large. We thus want to construct at first an ideal whose norm is bounded independently from $\mathcal{N}(\mathfrak{a})$ in the same ideal class as $\mathfrak{a}$. We proceed by ideal-lattice reduction, as Biasse did in [5, Sect. 2.2]. Through the canonical embedding, any integral ideal $\mathfrak{a}$ can be viewed as a Euclidean lattice. As usual when dealing with lattice reduction, we are interested in small vectors, or equivalently here, integers with small Euclidean norm. Let us first study the guarantees that a classical DBKZ-reduction offers on the embedding of $\mathfrak{a}$.

Lemma 5. *Let $\mathbb{K}$ be a number field of degree N, $\beta \in \{1, \dots, N\}$, and $\mathfrak{a}$ be an ideal of $\mathcal{O}_{\mathbb{K}}$. Then it is possible to find a short element $\mathbf{v} \in \mathfrak{a}$ in time $Poly(N, \log \mathcal{N}(\mathfrak{a}))(3/2 + o(1))^{\beta/2}$, that satisfies:*

$$\|\mathbf{v}\| \leq \beta^{\frac{N}{2\beta}} \cdot |\Delta_{\mathbb{K}}|^{\frac{1}{2N}} \cdot \mathcal{N}(\mathfrak{a})^{\frac{1}{N}}$$

where $\|.\|$ denotes the Euclidean norm.

---

[7] Justification of this choice appears explicitly when we study the complexity of the q-descent in the algorithm.

Proof. This is only a direct application of Theorem 1 and Lemma 3. Indeed, let $\mathbf{v}$ be the short vector output by DBKZ applied to the lattice of the embedding of $\mathfrak{a}$. It has determinant $\mathcal{N}(\mathfrak{a})\sqrt{|\Delta_{\mathbb{K}}|}$, yielding the announced upper bound.

Since the ideal $\mathfrak{a}$ contains $\langle \mathbf{v} \rangle$, there exists a unique integral ideal $\mathfrak{b}$ satisfying $\langle \mathbf{v} \rangle = \mathfrak{a} \cdot \mathfrak{b}$. From the guarantees on $\|\mathbf{v}\|$, we can bound the norm of this new ideal $\mathfrak{b}$.

Corollary 3. *With the same notations as in Lemma 5, we have*

$$\mathcal{N}(\mathfrak{b}) \leq \beta^{\frac{N^2}{2\beta}} \cdot \sqrt{|\Delta_{\mathbb{K}}|}.$$

Proof. From Lemma 5, we have

$$\|\mathbf{v}\| \leq \beta^{\frac{N}{2\beta}} \cdot |\Delta_{\mathbb{K}}|^{\frac{1}{2N}} \cdot \mathcal{N}(\mathfrak{a})^{\frac{1}{N}}.$$

Thus, its field norm is below the N-th power of this bound — the N^N term is negligible here — and so:

$$\mathcal{N}(\langle \mathbf{v} \rangle) \leq \beta^{\frac{N^2}{2\beta}} \cdot \sqrt{|\Delta_{\mathbb{K}}|} \cdot \mathcal{N}(\mathfrak{a}).$$

As a consequence, since $\langle \mathbf{v} \rangle = \mathfrak{a} \cdot \mathfrak{b}$, we have by the multiplicative property of the norm $\mathcal{N}(\mathfrak{b}) \leq \beta^{\frac{N^2}{2\beta}} \cdot \sqrt{|\Delta_{\mathbb{K}}|}$.

Remark 2. Because $\mathbb{K}$ is a cyclotomic field, we may choose a block-size β in $\log L_{|\Delta_{\mathbb{K}}|}(1/2)$ since $\log L_{|\Delta_{\mathbb{K}}|}(1/2) = N^{1/2+o(1)} \leq N$. Then Corollary 3 generates in time $L_{|\Delta_{\mathbb{K}}|}(1/2)$ an integral ideal of norm bounded by $L_{|\Delta_{\mathbb{K}}|}(3/2)$.

This last result allows us to find an ideal of norm bounded independently from $\mathcal{N}(\mathfrak{a})$. We then want this new ideal to split in the class group as a product of multiple prime ideals of controlled norms. Thanks to Corollary 2, the probability of an integral ideal $\mathfrak{b}$ of norm bounded by $L_{|\Delta_{\mathbb{K}}|}(3/2)$ to be $L_{|\Delta_{\mathbb{K}}|}(1)$-smooth is greater than $L_{|\Delta_{\mathbb{K}}|}(1/2)^{-1}$. In addition, using ECM for testing smoothness keeps the complexity in $L_{|\Delta_{\mathbb{K}}|}(1/2)$. The analysis of this part is left for Sect. 4.5. Therefore, repeating the last construction $L_{|\Delta_{\mathbb{K}}|}(1/2)$ times on randomized independent inputs eventually yields a $L_{|\Delta_{\mathbb{K}}|}(1)$-smooth ideal. The simplest strategy to perform this randomization of the input ideal is to compose it with some factors of norm less than $B = L_{|\Delta_{\mathbb{K}}|}(1/2)$. Formally, we denote by $\mathcal{B} = \{\mathfrak{p}_1, \ldots, \mathfrak{p}_{|\mathcal{B}|}\}$ the set of all prime ideals of norm upper bounded by $L_{|\Delta_{\mathbb{K}}|}(1/2)$. Let $k, A > 0$ be fixed integers. We choose $\mathfrak{p}_{j_1}, \ldots, \mathfrak{p}_{j_k}$ prime ideals of norm $L_{|\Delta_{\mathbb{K}}|}(1/2)$. Then for any k-uple $(e_1, \ldots, e_k) \in \{1, \ldots, A\}^k$, we have

$$\mathcal{N}\left(\mathfrak{a} \cdot \prod_{i=1}^{k} \mathfrak{p}_{j_i}^{e_i}\right) \leq \mathcal{N}(\mathfrak{a}) \cdot \prod_{i=1}^{k} \mathcal{N}(\mathfrak{p}_{j_i})^{e_i} \leq \mathcal{N}(\mathfrak{a}) \cdot L_{|\Delta_{\mathbb{K}}|}(1/2)^{k \cdot A} = \mathcal{N}(\mathfrak{a}) \cdot L_{|\Delta_{\mathbb{K}}|}(1/2).$$

We know from the Landau prime ideal theorem [27] that in every number field $\mathbb{K}$, the number of prime ideals of norm bounded by X, denoted by $\pi_{\mathbb{K}}(X)$, satisfies

$$\pi_{\mathbb{K}}(X) \sim \frac{X}{\log X}. \tag{1}$$

Thus, the randomization can be done by choosing uniformly at random the tuple $(e_1, \ldots, e_k)$ and k prime ideals in $\mathcal{B}$. Since $|\mathcal{B}| = L_{|\Delta_K|}(1/2)$, set of possible samples is large enough for our purposes.

Other ways to perform the randomization may be by randomizing directly the lattice reduction algorithm or by enumerating points of the lattice of norm close to the norm guarantee and change the basis vectors by freshly enumerated ones. The latter would be useful in practice as it reduces the number of reductions.

This last remark concludes the proof of Theorem 3. The full outline of this bootstrap section is given in Fig. 3.

---

1. CurrentIdeal $\leftarrow \mathfrak{a}$.
2. **While** CurrentIdeal is not $L_{|\Delta_K|}(1)$-smooth **do**:
3.   Choose $\mathfrak{p}_{j_1}, \cdots, \mathfrak{p}_{j_k}$ uniformly at random in $\mathcal{B}$.
4. $\mathfrak{c} \leftarrow \mathfrak{a} \cdot \prod \mathfrak{p}_{j_i}^{e_i}$ for random $e_i \in \{1, \ldots, A\}$.
5.   Generate $\mathfrak{b}$ from $\mathfrak{c}$ as in Lemma 3.
6.   CurrentIdeal $\leftarrow \mathfrak{b}$.
7. **End while**
8. **Return** CurrentIdeal.

Fig. 3. First reduction to a $L_{|\Delta_K|}(1)$-smooth ideal.

Interlude: Reduction with Cheon's Trick. In the proof of Theorem 3, we use the *classical*-DBKZ reduction in order to find a short element in the embedding of the considered ideal. We could not use directly Cheon's trick here since the norm of the ideal $\mathcal{I}^+$ — and so the determinant of its coefficient embedding — is potentially large. Nonetheless, the norm of prime ideals appearing in the factorization are by construction bounded, hence a natural question is to look at the guarantees offered when applying the sub-cited trick. The systematic treatment of this question is the aim of Theorem 4.

Theorem 4. *Let $\mathfrak{a}$ be an integral ideal of norm below $L_{|\Delta_K|}(\alpha)$, for $\frac{1}{2} \leq \alpha \leq 1$. Then, in expected time $L_{|\Delta_K|}(1/2)$, it is possible to construct an algebraic integer $\mathbf{v}$ and an $L_{|\Delta_K|}((2\alpha + 1)/4)$-smooth ideal $\mathfrak{b}$ such that:*

$$\langle \mathbf{v} \rangle = \mathfrak{a} \cdot \mathfrak{b}.$$

Proof. The core of the proof is somehow similar to the proof of Theorem 3 as it heavily relies on lattice reduction and randomization techniques. Nonetheless, the major difference is on the embedding with respect to which the reduction is performed. In Theorem 3, the canonical embedding is used, whereas we use here the coefficient embedding $\mathcal{C}$. It avoids the apparition of a power of the discriminant in the field norm of the output of DBKZ. Nonetheless, remark that since we work in the totally real subfield, we cannot use a naive coefficients embedding of this subfield. In order to benefit from the nice shape of the defining polynomial $X^N + 1$ of the cyclotomic field, we use instead a fold-in-two strategy: the embedding of $\mathcal{O}_{\mathbb{K}}^+$ is defined as the coefficient embedding $\mathcal{C}^+$ for the $\mathbb{Z}$-base $(\zeta^i + \zeta^{-i})_i$. Let us denote by $\|.\|_{\mathcal{C}^+}$ the induced norm. Hence, for any $v \in \mathcal{O}_{\mathbb{K}}^+$:

$$\|\mathbf{v}\|_{\mathcal{C}} = \sqrt{2}\|\mathbf{v}\|_{\mathcal{C}^+}.$$

Let $\mathcal{L} = \mathcal{C}^+(\mathfrak{a})$ be the embedding of $\mathfrak{a}$. Its covolume is by definition its index in $\mathbb{Z}^n$, that is the index of $\mathfrak{a}$ as a $\mathbb{Z}$-module in $\mathcal{O}_{\mathbb{K}}^+$, which is $\mathcal{N}(\mathfrak{a})$. Then, with the same block-size $\beta = \log L_{|\Delta_{\mathbb{K}}|}(1/2) = \mathcal{O}(\sqrt{N}\log(N))$, we have

$$\text{Vol}(\mathcal{L}) \leq L_{|\Delta_{\mathbb{K}}|}(\alpha) = 2^{\mathcal{O}(N^\alpha \log(N))} \leq \beta^{\frac{N^2}{2\beta}}.$$

Using the Approx-SVP algorithm of Theorem 2 yields in time $L_{|\Delta_{\mathbb{K}}|}(1/2)$ an integer $\mathbf{v}$ satisfying:

$$\|\mathbf{v}\|_{\mathcal{C}^+} \leq \beta^{(1+o(1))}\sqrt{2^{\frac{\log_\beta(\det(\mathcal{L}))}{\beta}}} \leq \beta^{(1+o(1))}\sqrt{4^{\frac{N^\alpha}{\sqrt{N}\log N}}} = L_{|\Delta_{\mathbb{K}}|}(\alpha/2 - 1/4).$$

Using Corollary 1 to fall back on the field norm induces:

$$\mathcal{N}_{\mathbb{K}/\mathbb{Q}}(\mathbf{v}) \leq (\sqrt{2}(N+1))^N \cdot \|\mathbf{v}\|_{\mathcal{C}}^N = L_{|\Delta_{\mathbb{K}}|}(1) \cdot L_{|\Delta_{\mathbb{K}}|}(\alpha/2 + 3/4).$$

Since $\alpha \geq 1/2$, we then have $\mathcal{N}(\langle \mathbf{v} \rangle) = \mathcal{N}_{\mathbb{K}/\mathbb{Q}}(\mathbf{v}) \leq L_{|\Delta_{\mathbb{K}}|}(\alpha/2 + 3/4)$.

Because the ideal $\mathfrak{a}$ contains $\langle \mathbf{v} \rangle$, there exists a unique ideal $\mathfrak{b}$, satisfying $\langle \mathbf{v} \rangle = \mathfrak{a} \cdot \mathfrak{b}$. We get that $\mathcal{N}(\mathfrak{b}) \leq L_{|\Delta_{\mathbb{K}}|}(\alpha/2 + 3/4)$ from the multiplicative property of the norm and $\mathcal{N}(\mathfrak{a}) = L_{|\Delta_{\mathbb{K}}|}(1) \leq L_{|\Delta_{\mathbb{K}}|}(\alpha/2 + 3/4)$. Under Heuristic 1, this ideal is $L_{|\Delta_{\mathbb{K}}|}(\alpha/2 + 1/4)$-smooth with probability $L_{|\Delta_{\mathbb{K}}|}(1/2)$. Eventually performing the randomization-and-repeat technique as in the initial round, this reduction in the coefficient embedding yields the desired couple $(\mathbf{v}, \mathfrak{b})$ in expected time $L_{|\Delta_{\mathbb{K}}|}(1/2)$.

Descending to B-smoothness. After the first round, we end up with an $L_{|\Delta_{\mathbb{K}}|}(1)$-smooth ideal, denoted by $\mathcal{I}^{(0)}$, and an algebraic integer $\mathbf{h}^{(0)}$ satisfying

$$\langle \mathbf{h}^{(0)} \rangle = \mathcal{I}^+ \cdot \mathcal{I}^{(0)},$$

with $\mathcal{I}^+$ the ideal of the totally real subfield obtained after phase Sect. 4.1. The factorization of $\mathcal{I}^{(0)}$ gives

$$\mathcal{I}^{(0)} = \prod_j \mathcal{I}_j^{(0)},$$

where the $\mathcal{I}_j^{(0)}$ are integral prime ideals of norm upper bounded by $L_{|\Delta_{\mathbb{K}}|}(1)$. Taking the norms of the ideals involved in this equality ensures that the number of terms in this product is $O(n_{\mathcal{I}})$, with $n_{\mathcal{I}} = \frac{\log|\Delta_{\mathbb{K}}|}{\log\log|\Delta_{\mathbb{K}}|} = O(N)$. Then applying Theorem 4 on each small ideal $\mathcal{I}_j^{(0)}$ gives rise to ideals $\mathcal{I}_j^{(1)}$ in expected time $L_{|\Delta_{\mathbb{K}}|}(1/2)$ that are $L_{|\Delta_{\mathbb{K}}|}\left(\frac{2\times 1 + 1}{4}\right) = L_{|\Delta_{\mathbb{K}}|}(3/4)$-smooth and integers $\mathbf{h}_j^{(1)}$ such that for every j,

$$\langle \mathbf{h}_j^{(1)} \rangle = \mathcal{I}_j^{(0)} \cdot \mathcal{I}_j^{(1)}.$$

For each factor $\mathcal{I}_j^{(1)}$, let us write its prime decomposition:

$$\mathcal{I}_j^{(1)} = \prod_k \mathcal{I}_{j,k}^{(1)}.$$

Once again, the number of terms appearing is $O(n_\mathcal{I})$. Because we have the inequality $\mathcal{N}\left(\mathcal{I}_{j,k}^{(1)}\right) \le L_{|\Delta_\mathbb{K}|}(3/4)$, then performing the same procedure on each ideal $\mathcal{I}_{j,k}^{(1)}$ now yields $L_{|\Delta_\mathbb{K}|}(5/8)$-smooth ideals $\mathcal{I}_{j,k}^{(2)}$ and integers $\mathbf{h}_{j,k}^{(2)}$ such that

$$\langle \mathbf{h}_{j,k}^{(2)} \rangle = \mathcal{I}_{j,k}^{(1)} \cdot \mathcal{I}_{j,k}^{(2)},$$

once again in expected time $L_{|\Delta_\mathbb{K}|}(1/2)$. Remark that this smoothness bound in $L_{|\Delta_\mathbb{K}|}(5/8)$ is obtained as $L_{|\Delta_\mathbb{K}|}\left(\frac{2\times 3/4+1}{4}\right)$, as exposed in Theorem 4. This reasoning naturally leads to a recursive strategy for reduction. At step k, we want to reduce an ideal $\mathcal{I}_{a_1,\ldots,a_{k-1}}^{(k-1)}$ which is $L_{|\Delta_\mathbb{K}|}\left(1/2 + 1/2^{k+1}\right)$-smooth. As before, we have a decomposition — in $O(n_\mathcal{I})$ terms — in smaller ideals:

$$\mathcal{I}_{a_1,\ldots,a_{k-1}}^{(k-1)} = \prod_j \mathcal{I}_{a_1,\ldots,a_{k-1},j}^{(k-1)}.$$

Using Theorem 4 on each factor $\mathcal{I}_{a_1,\ldots,a_{k-1},j}^{(k-1)}$ which have norm bounded by $L_{|\Delta_\mathbb{K}|}\left(1/2 + 1/2^{k+1}\right)$ leads to $L_{|\Delta_\mathbb{K}|}\left(1/2 + 1/2^{k+2}\right)$-smooth ideals $\mathcal{I}_{a_1,\ldots,a_{k-1},j}^{(k)}$ and algebraic integers $\mathbf{h}_{a_1,\ldots,a_{k-1},j}^{(k)}$ such that

$$\langle \mathbf{h}_{a_1,\ldots,a_{k-1},j}^{(k)} \rangle = \mathcal{I}_{a_1,\ldots,a_{k-1},j}^{(k-1)} \cdot \mathcal{I}_{a_1,\ldots,a_{k-1},j}^{(k)},$$

since $\frac{2\times(1/2+1/2^{k+1})+1}{4} = 1/2 + 1/2^{k+2}$.

As a consequence, one can generate $L_{|\Delta_\mathbb{K}|}(1/2 + 1/\log N)$-smooth ideals with the previous method in at most $\lceil \log_2(\log N) \rceil$ recursive steps. At this point only $(n_\mathcal{I})^{\lceil \log_2(\log N) \rceil}$ ideals and algebraic integers appear since at each step this number is multiplied by a factor $O(n_\mathcal{I})$. As deriving one couple integer/ideal is done in expected time $L_{|\Delta_\mathbb{K}|}(1/2)$, the whole complexity remains in $L_{|\Delta_\mathbb{K}|}(1/2)$. However, as $|\Delta_\mathbb{K}| = N^N$, a quick calculation entails that

$$\log L_{|\Delta_\mathbb{K}|}\left(\frac{1}{2} + \frac{1}{\log(N)}\right) = O(N^{\frac{1}{2} + \frac{1}{\log N}} \log(N))$$

$$= O(N^{\frac{1}{2}} \log(N)) \cdot N^{\frac{1}{\log N}}.$$

Since the last factor is $e = \exp(1)$, we obtain that

$$\log L_{|\Delta_\mathbb{K}|}\left(\frac{1}{2} + \frac{1}{\log(N)}\right) = \log L_{|\Delta_\mathbb{K}|}\left(\frac{1}{2}\right),$$

so that after at most $\lceil \log_2(\log N) \rceil$ steps, we have ideals that are $L_{|\Delta_\mathbb{K}|}(1/2)$-smooth.

At the end of this final round, we may express the input ideal as the product of ideals for which we know a generator and others that have by construction norms bounded by $L_{|\Delta_{\mathbb{K}}|}(1/2)$. Let us denote $\mathcal{K}$ the final step. For avoiding to carry inverse ideals, we may assume without loss of generality[8] that $\mathcal{K}$ is even. Explicitly we have

$$\langle \mathbf{h}^{(0)} \rangle = \mathcal{I}^+ \cdot \mathcal{I}^{(0)}$$

$$= \mathcal{I}^+ \cdot \prod_{a_1} \mathcal{I}^{(0)}_{a_1}$$

$$= \mathcal{I}^+ \cdot \left\langle \frac{\prod_{a_1} \mathbf{h}^{(1)}_{a_1} \prod_{a_1,a_2,a_3} \mathbf{h}^{(3)}_{a_1,a_2,a_3}}{\prod_{a_1,a_2} \mathbf{h}^{(2)}_{a_1,a_2}} \right\rangle \cdot \prod_{a_1,a_2,a_3} \mathcal{I}^{(3)}_{a_1,a_2,a_3}$$

$$= \mathcal{I}^+ \cdot \left\langle \prod_{a_1,\ldots,a_{\mathcal{K}+1}} \frac{\prod_{t\in 2\mathbb{Z}+1} \mathbf{h}^{(t)}_{a_1,\ldots,a_t}}{\prod_{s\in 2\mathbb{Z}} \mathbf{h}^{(s)}_{a_1,\ldots,a_s}} \right\rangle \cdot \underbrace{\prod_{a_1,\ldots,a_{\mathcal{K}+1}} \mathcal{I}^{(\mathcal{K})}_{a_1,\ldots,a_{\mathcal{K}+1}}}_{=\mathcal{I}^s}.$$

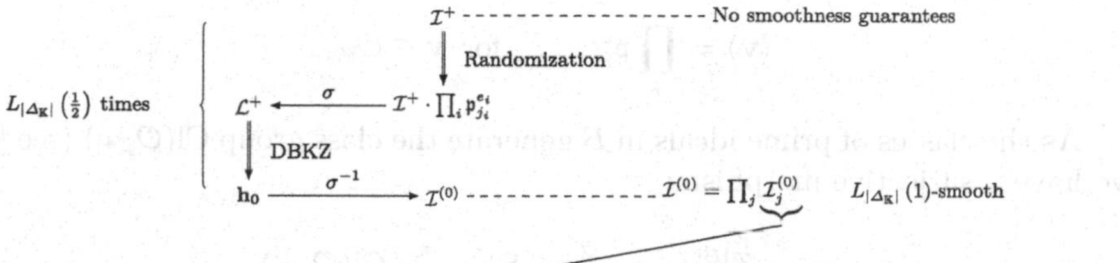

Fig. 4. The q-descent algorithm.

---

[8] We can always run an additional step in the q-descent without changing the whole complexity.

In this last expression, the indices are chosen such that $1 \leq t \leq \mathcal{K}$ and $2 \leq s \leq \mathcal{K}$. We also recall that all the quantities involved here belong to the totally real subfield $\mathbb{Q}(\zeta + \zeta^{-1})$.

By construction, $\mathcal{I}^s$ is $L_{|\Delta_{\mathbb{K}}|}(1/2)$-smooth and we directly get $\mathbf{h} \in \mathcal{O}_{\mathbb{K}}^+$ such that $\langle \mathbf{h} \rangle = \mathcal{I}^+ \cdot \mathcal{I}^s$. The full outline of this descent phase is sketched in Fig. 4.

Remark that the number of terms, which is at most $O(N)^{\mathcal{K}}$ is in $L_{|\Delta_{\mathbb{K}}|}(o(1))$, is negligible in the final complexity estimate.

4.3   Step 3: Case of $L_{|\Delta_{\mathbb{K}}|}(1/2)$-smooth Ideals

At this point, we have reduced the search for a generator of a principal ideal of large norm to the search for a generator of a principal ideal $\mathcal{I}^s$ which is $L_{|\Delta_{\mathbb{K}}|}(1/2)$-smooth. If we can find a generator of $\mathcal{I}^s$ in time $L_{|\Delta_{\mathbb{K}}|}(1/2)$, from the previous steps we directly recover the generator of $\mathcal{I}^+$, and so the generator of $\mathcal{I}$, that is the secret key. To tackle this final problem, we follow the approach relying on class group computation (see [15, Algorithm 6.5.10] or [5, Algorithm 7]): we consider the previously introduced set $\mathcal{B}$ of prime ideals of norm below $B > 0$ where $B \in L_{|\Delta_{\mathbb{K}}|}(1/2)$ and look for relations of the shape

$$\langle \mathbf{v} \rangle = \prod_i \mathfrak{p}_i^{e_i}, \qquad \text{for } \mathbf{v} \in \mathcal{O}_{\mathbb{K}^+}.$$

As the classes of prime ideals in $\mathcal{B}$ generate the class group $\mathrm{Cl}(\mathcal{O}_{\mathbb{K}^+})$ (see [2]), we have a surjective morphism:

$$\begin{array}{ccccc}
\mathbb{Z}^{|\mathcal{B}|} & \xrightarrow{\phi} & \mathcal{S}_{\mathcal{I}} & \xrightarrow{\pi} & \mathrm{Cl}(\mathcal{O}_{\mathbb{K}^+}) \\
(e_1, \cdots, e_{|\mathcal{B}|}) & \longmapsto & \prod_i \mathfrak{p}_i^{e_i} & \longmapsto & \prod_i [\mathfrak{p}_i]^{e_i}.
\end{array}$$

Formally, a relation is an element of $\mathrm{Ker}\,(\pi \circ \phi)$, which is a full-rank sublattice of $\mathbb{Z}^{|\mathcal{B}|}$. Following the subexponential approach of [8,11,23], we need to find at least $|\mathcal{B}| \in L_{|\Delta_{\mathbb{K}}|}(1/2)$ linearly independent relations to generate this lattice. The relation collection is performed in a similar way as [4]: due to the good shape of the defining polynomial $X^N + 1$, the algebraic integers whose representation as polynomials in ζ have small coefficients also have small norms.

Let us fix an integer $0 < A \leq L_{|\Delta_{\mathbb{K}}|}(0) = \log |\Delta_{\mathbb{K}}|$. Then for any integers $(v_0, \ldots, v_{\frac{N}{2}-1}) \in \{-A, \ldots, A\}^{\frac{N}{2}}$, we define the element $\mathbf{v} = v_0 + \sum_{i \geq 1} v_i(\zeta^i + \zeta^{-i})$. The norm of this element in $\mathbb{K}^+$ is upper bounded by $L_{|\Delta_{\mathbb{K}}|}(1)$. Indeed, it corresponds to the square root of its norm in $\mathbb{K}$, which is below $N^N \cdot A^N = L_{|\Delta_{\mathbb{K}}|}(1)$ by Lemma 2. Then under Heuristic 1, the element $\mathbf{v}$ generates an ideal $\langle \mathbf{v} \rangle$ that is $L_{|\Delta_{\mathbb{K}}|}(1/2)$-smooth with probability $L_{|\Delta_{\mathbb{K}}|}(1/2)^{-1}$. This means that we need to draw on average $L_{|\Delta_{\mathbb{K}}|}(1/2)$ independent algebraic integers to find one relation.

To bound the run time of the algorithm, we need to assume that the relation we collect by this method are independent. This is a commonly used heuristic in the analysis of index calculus algorithms for computing $\mathrm{Cl}(\mathbb{K})$.

Heuristic 3 [4, Heuristic 2]. *There exists Q negligible with respect to $|\mathcal{B}|$ such that collecting $Q \cdot |\mathcal{B}|$ relations suffices to generate the whole lattice of relations.*

Thanks to Eq. (1), we know that $\mathcal{B}$ contains about $L_{|\Delta_{\mathbb{K}}|}(1/2)$ elements. Therefore, $L_{|\Delta_{\mathbb{K}}|}(1/2)$ relations are needed thanks to Heuristic 3, implying that $L_{|\Delta_{\mathbb{K}}|}(1/2)^2 = L_{|\Delta_{\mathbb{K}}|}(1/2)$ independently drawn algebraic integers suffice to generate the whole lattice of relations. Of course, the set of integers arising from the previous construction is large enough to allow such repeated sampling, because its size is $L_{|\Delta_{\mathbb{K}}|}(1)$. We store the relations in a $|\mathcal{B}| \times Q|\mathcal{B}|$ matrix M, as well as the corresponding algebraic integers in a vector G.

$$
\left.
\begin{array}{c}
M
\begin{pmatrix}
e_{1,1} & \cdots & e_{1,i} & \cdots & e_{1,Q|\mathcal{B}|} \\
e_{2,1} & \cdots & e_{2,i} & \cdots & e_{2,Q|\mathcal{B}|} \\
\vdots & & \vdots & & \vdots \\
e_{|\mathcal{B}|,1} & \cdots & e_{|\mathcal{B}|,i} & \cdots & e_{|\mathcal{B}|,Q|\mathcal{B}|}
\end{pmatrix} \\
\hline
G \begin{pmatrix} \mathbf{v}_1 & \cdots & \mathbf{v}_i & \cdots & \mathbf{v}_{|\mathcal{B}|} \end{pmatrix}
\end{array}
\right\}
\quad
\forall i, (\mathbf{v}_i) = \prod_{j=1}^{|\mathcal{B}|} \mathfrak{p}_i^{e_{j,i}}.
$$

The $L_{|\Delta_{\mathbb{K}}|}(1/2)$-smooth ideal $\mathcal{I}^s$ splits over the set $\mathcal{B}$, so that there exists a vector Y of $\mathbb{Z}^{|\mathcal{B}|}$ containing the exponents of the factorization

$$
\mathcal{I}^s = \prod_i \mathfrak{p}_i^{Y_i}.
$$

As the relations stored in M generate the lattice of all elements of this form, the vector Y necessarily belongs to it. Hence solving the equation $MX = Y$ yields a vector $X \in \mathbb{Z}^{Q|\mathcal{B}|}$ from which we can recover a generator of the ideal since:

$$
\prod_i \mathfrak{p}_i^{Y_i} = \langle \mathbf{v}_1^{X_1} \cdots \mathbf{v}_{Q|\mathcal{B}|}^{X_{Q|\mathcal{B}|}} \rangle. \tag{2}
$$

By construction, $\mathcal{N}(\mathcal{I}^s) \leq L_{|\Delta_{\mathbb{K}}|}(\mathcal{K}/2 + 1/2)$ so that the coefficients of Y are below $L_{|\Delta_{\mathbb{K}}|}(0)$. Since solving such a linear system with Dixon's p-adic method [18] can be done in time $Poly(d, \log \|M\|)$ where d is the dimension of the matrix and $\|M\| = \max |M_{i,j}|$ the maximum of its coefficients, we are able to recover X with a complexity in $L_{|\Delta_{\mathbb{K}}|}(1/2)$.

4.4   Final Step: Reduction to a Short Generator

As mentioned in Sect. 3, this part of the algorithm is a result of Cramer, Ducas, Peikert, and Regev [17]. They state that recovering a short generator from an arbitrary one can be solved in polynomial time in any prime-power cyclotomic ring. For completeness purposes, we give here a brief overview of this reduction.

As a liminary observation, note that for those fields, a set of fundamental units is given for free, whereas their computation in arbitrary number fields is computationally hard. A second remark is that we get the promise that there exists a small generator of the considered ideal. Then, instead of solving a general *closest vector problem* (CVP), we solve an instance of *bounded-distance decoding* problem (BDD). The key argument is based on a precise study of the geometry of the log-unit lattice of prime-power cyclotomic fields (see Appendix A.3 for basic recalls about this lattice). Finally, their geometric properties make possible to solve BDD in this lattice in polynomial time, instead of exponential time as for generic instances.

Theorem 5 [17, Theorem 4.1]. *Let D be a distribution over $\mathbb{Q}(\zeta)$ with the property that for any tuple of vectors $\mathbf{v}_1, \ldots, \mathbf{v}_{N/2-1} \in \mathbb{R}^{N/2-1}$ of Euclidean norm 1 that are orthogonal to the all-1 vector $\mathbf{1}$, the probability that the inequation $|(\mathrm{Log}(\mathbf{g}), \mathbf{v}_i)| < c\sqrt{2N} \cdot \log(2N)^{-3/2}$ holds for all i is at least some $\alpha > 0$, where $\mathbf{g}$ is chosen from D and c is a universal constant. Then there is an efficient algorithm that, given $\mathbf{g}' = \mathbf{g} \cdot \mathbf{u}$, where $\mathbf{g}$ chosen from D and $\mathbf{u} \in C$ is a cyclotomic unit, outputs an element of the form $\zeta^j \cdot \mathbf{g}$ with probability at least α.*

The reader might argue that, in order to use this theorem on the output of our algorithm, we should ensure that we recover a generator up to a *cyclotomic unit* and not up to an arbitrary unit. In the specific case of power-of-two cyclotomic fields, we can rely on Weber's Heuristic 2 to ensure this constraint. In case $h^+(N) > 1$, two solutions are given in [17]. The first one is to directly compute the group of units, which is hopefully determined by the kernel of the matrix M arising in the third stage[9]. One can then enumerate the $h^+(N)$ classes of the group of units modulo the subgroup of cyclotomic units. Another possibility is to generate a list of ideals, sampled according to the same distribution as the input ideal, with a known generator. Then, we run the PIP algorithm on these ideals, and deduce the cosets of the group of units modulo the subgroup of cyclotomic units, which are likely to be output.

The whole key recovery, combining our PIP algorithm and the aforementioned reduction is outlined in Fig. 5.

---

1. Compute a generator $\mathbf{g_0}$ of $\mathcal{I}$ with Gentry-Szydlo, q-descent and relation collection.
2. Let $\mathbf{B}$ be the basis defined by the $\mathrm{Log}(\mathbf{b}_i)$ for $\mathbf{b}_i = \frac{\zeta_m^i - 1}{\zeta_m - 1}$.
3. Set $t = \mathrm{Log}(\mathbf{g_0}) + \mathrm{Log}(\mathcal{O}_{\mathbb{K}})$.
4. Return Babai's rounding of $\mathbf{B} \lfloor (\mathbf{B}^\vee)^t \cdot t \rceil$.

---

where $\lfloor \cdot \rceil$ denotes the rounding function: $\lfloor c \rceil = \lfloor c + \frac{1}{2} \rfloor$.

Fig. 5. Recovery of the secret key by PIP + [17].

---

[9] Another possibility is to use the saturation method which might run in polynomial time [7].

4.5 Complexity Analysis

The whole runtime of our attack is $L_{|\Delta_\mathbb{K}|}(1/2)$, that is about $2^{N^{1/2+o(1)}}$ operations. We have already mentioned the complexity of most parts of our algorithm. However, we provide a brief summary in this paragraph to ensure the entirety of our result.

For the reduction algorithms, DBKZ and Cheon's trick, the block-size is always in $\log L_{|\Delta_\mathbb{K}|}(1/2)$ so that the complexity is $L_{|\Delta_\mathbb{K}|}(1/2)$. Our choice for the smoothness bound $B = L_{|\Delta_\mathbb{K}|}(1/2)$ ensures that the step of relation collection together with the linear system solution are derived in time $L_{|\Delta_\mathbb{K}|}(1/2)$.

In addition, from the work of [20], we get that the first part of the algorithm, corresponding to the reduction to the totally real subfield, is performed in polynomial time.

The last part, which corresponds to the generation of a small generator from an arbitrary one, runs in polynomial time with respect to the input $(\mathbf{B}, t)$ of Babai's round-off algorithm (see Step 4 of the algorithm in Fig. 5), thanks to the results of [17]. However, $t = \text{Log}(\mathbf{g_0}) + \text{Log}(\mathcal{O}_\mathbb{K})$ is of subexponential size at this stage. Indeed, according to Eq. (2),

$$\text{Log}(\mathbf{g_0}) = X_1 \text{Log}(\mathbf{v}_1) + \cdots + X_{Q|\mathcal{B}|}\text{Log}(\mathbf{v}_{Q|\mathcal{B}|}),$$

where each $\mathbf{v}_i$ is of polynomial size while, by Hadamard's bound, the X_i satisfy $X_i \leq Q|\mathcal{B}|^{Q|\mathcal{B}|/2}\|M\|^{Q|\mathcal{B}|-1}\max_j\|Y_j\|$. Therefore, the bit size of the X_i are in $L_{|\Delta_\mathbb{K}|}(1/2)$, and the fixed point approximations of $\text{Log}(\mathbf{v}_i)$ must be taken at precision $b \in L_{|\Delta_\mathbb{K}|}(1/2)$ to ensure the accuracy of the value of $\text{Log}(\mathbf{g_0})$ (and therefore t). Babai's round-off computation $\mathbf{B}\lfloor(\mathbf{B}^\vee)^t \cdot t\rceil$ has an asymptotic cost in $L_{|\Delta_\mathbb{K}|}(1/2)$ and returns $e_1, \ldots, e_r$ where the e_i have bit size in $L_{|\Delta_\mathbb{K}|}(1/2)$ and where

$$\mathbf{g'} = \mathbf{g_0} \cdot \mathbf{b}_1^{e_1} \cdots \mathbf{b}_r^{e_r} = \left(\mathbf{v}_1^{X_1} \cdots \mathbf{v}_{Q|\mathcal{B}|}^{X_{Q|\mathcal{B}|}}\right) \cdot (\mathbf{b}_1^{e_1} \cdots \mathbf{b}_r^{e_r}),$$

is a short generator of the input ideal. This product cannot be evaluated directly since the intermediate terms may have exponential size, but it may be performed modulo distinct prime ideals $\mathfrak{p}_1, \ldots, \mathfrak{p}_k$ such that $\mathcal{N}\left(\prod \mathfrak{p}_i\right) > \mathcal{N}(\mathbf{g'})$ and then reconstructed by the Chinese Remainder Theorem. The complexity of this process is in $L_{|\Delta_\mathbb{K}|}(1/2)$.

We highlight now two points whose complexity were eluded in the exposition of the algorithm:

– *Arithmetic of ideals.* All the operations made on ideals are classical, with complexities polynomial in the dimension and in the size of the entries (see for instance [15, Chap. 4]), which is way below the bound of $L_{|\Delta_\mathbb{K}|}(1/2)$.
– *Smoothness tests.* The strategy is to deal with the norms of ideals, that are integers. The largest norm arising in the computations is in $L_{|\Delta_\mathbb{K}|}(3/2)$ and appears after the initial DBKZ reduction. Testing $L_{|\Delta_\mathbb{K}|}(1)$-smoothness for an integer of this size is easier than completely factorizing it, even if both methods share the same asymptotic complexity in $L_{|\Delta_\mathbb{K}|}(1/2)^{10}$. Hence all the smoothness tests performed have complexity dominated by $L_{|\Delta_\mathbb{K}|}(1/2)$.

---

[10] Factorizing an integer N is done in $L_N(1/3)$.

As a consequence the global complexity is given by the first and last steps of the q-descent, that is in $L_{|\Delta_\mathbb{K}|}(1/2)$.

Remark 3. This algorithm has a complexity in $L_{|\Delta_\mathbb{K}|}(1/2)$ in the discriminant, that represents the size of the number field involved. However, it is important to figure out that the parameters of the keys have $N^{3/2}$ bits. Therefore we present an algorithm that is "sort of" $L(1/3)$ in the size of the inputs.

5   Implementation Results

In addition to the theoretical improvement, our algorithm permits in practice to break concrete cryptosystems. Our discussion is based on the scheme presented by Smart and Vercauteren at PKC 2010. In [43, Sect. 7], security estimations are given for parameters $N = 2^n$ for $8 \leq n \leq 11$ since they are unable to generate keys for larger parameters. Our implementation allows us to recover the secret key from the public key for $N = 2^8 = 256$ in less than a day. The code runs with **PARI-GP** [38], with an external call to **fplll** [19], and all the computations are performed on an Intel(R) Xeon(R) CPU E3-1275 v3 @ 3.50 GHz with 32 GB of memory. Indeed the Gentry-Szydlo algorithm requires large storage.

We perform the key generation as recalled in Fig. 2. We then obtain a generator for the ideal as a polynomial in $\zeta = \zeta_{512}$, of degree 255 and coefficients absolutely bounded by $2^{\sqrt{256}} + 1 = 65537$. That corresponds to ideals whose norm has about 4800 bits in average, that is below the bound 6145 from Lemma 2, but above the size given in [43] (4096). As for every timing arising in this section, we have derived a set of 10 keys, and the given time is the average one. Thus, deriving a secret key takes on average 30 s. We test 1381 algebraic integers for finding 10 having prime norm. Then the public key is derived from the secret key in about 96 s.

While, in theory, the first reduction to the totally real subfield seems to be of limited interest, it is clearly the main part of the practical results: indeed, it reduces in our example the size of the matrices involved from 256×256 to 128×128. As we know that lattice-reduction is getting worse while the dimension grows, this part is the key point of the algorithm. Our code essentially corresponds to the Gentry-Szydlo algorithm together with the trick explained in Sect. 4.1, in order to output the element $\mathbf{u}$ and a basis of the ideal $\mathcal{I}^+$ generated by $\mathbf{g} + \bar{\mathbf{g}}$. This part of the algorithm has the largest runtime, about 20 h, and requires 24Go of memory.

At this point, we put aside $\mathbf{u}$ and only consider the ideal $\mathcal{I}^+$. Our goal is to recover one generator of this ideal, and a multiplication with $\frac{1}{1+\mathbf{u}}$ is going to lead to the generator of the input ideal. The method we have presented is to reduce step by step the norm of the ideals involved by performing lattice reductions. However, we observe that for the cases we run, the first reduction suffices: the short vector we find corresponds to the generator. We make use of the BKZ algorithm implemented in **fplll** [19], with block-size 24 to begin. It gives a correct generator with probability higher than 0.75 and runs in less than 10 minutes. If the output is not correct, we increase the block-size to 30. This always works and requires between 2 and 4 h.

In addition to the good behavior of this reduction, the generator we exhibit is already small, by construction. More precisely, it corresponds to $\mathbf{g} + \bar{\mathbf{g}}$, up to a factor that is a power of ζ. Hence, we recover $\mathbf{g} \cdot \zeta^i$ thanks to $\mathbf{u}$ and the decoding algorithm analyzed in [17] is unnecessary for our concern. The key recovery is already completed after these two first steps. We still implement this part together with a method for recovering the actual private key (up to sign). Indeed, because all its coefficients are even except the constant one, it is easy to identify the power of ζ that appears as a factor during the computation.

Additional Work. To illustrate the practical performances of our method, we look at one of the main other steps of the algorithm: namely the relation collection between generators of $\mathrm{Cl}(\mathbb{K}^+)$. Thanks to the good behavior of BKZ, the relation collection is not necessary for the attack in $\mathbb{Q}(\zeta_{512})$, but it is an important part of the computation in higher dimension.

We fix our factor base as all the prime ideals in the totally-real field that lie above a prime number p that is below the bound $c \left(\log |\Delta_\mathbb{K}|\right)^2$, for a parameter $c \in \{0.1, 0.2, 0.3\}$. We give in Table 1 the values, together with the size of the factor base and the time required for building it in MAGMA [16]. The computations are performed on a laptop with Intel(R) Core(TM) i7-4710MQ CPU @ 2.50 GHz and 8Go of RAM for this part.

Naturally, this choice of bound would not be sufficient for the descent described in Fig. 4, because it is polynomial and not subexponential. However, it provides a relation matrix for the computation of the class group. Reaching a subexponential bound seems unlikely in that way, that supports the fact that our implementation results are consequences of the small dimension obtained by the Gentry-Szydlo algorithm.

Table 1. Construction of differently parametrized factor bases.

c	Bound	# primes	# factor base	Time (sec)
0.1	201516	149	18945	1240
0.2	403033	274	35073	2320
0.3	604549	385	49281	3320

The relation collection is performed using algebraic integers of the shape

$$\sum_{i=1}^{5} \zeta^{a_i} + \zeta^{-a_i} = \sum_{i=1}^{5} \zeta^{a_i} - \zeta^{256-a_i},$$

for a_i chosen at random in $\{1, \ldots, 255\}$. This is inspired from the work of Miller [37]. We use C++ code with NTL Library [42] for finding a set of integers with different norms that suffice for generating the full lattice of relations (see Sect. 4.3). The size of these sets depends on the bound we have chosen and on the relations picked, so that the timings may vary. Our results are provided in Table 2. Once we know these integers, we use Magma for building the entire matrix of relations. In particular, we make use of the automorphisms on the field for deriving 128 relations from each integer — this is the reason we use integers of different norms. Eventually, the matrices we get are full-rank.

Table 2. Relation collection for the different parameters.

c	# relations	Time (hours)	
		Relation collection	Matrix construction
0.1	1500	8.6	1.7
0.2	3400	13.8	4.9
0.3	6300	23.9	10.7

We also run our code for the algorithm described in [17] on inputs constructed as a secret key multiplied by a random non-zero vector of the log-unit lattice (because in the full attack described previously, we only have the null vector). This runs in 150 s.

To conclude, for the parameter $N = 2^8$, the time of the key recovery is below 24 h, and the main part of the computation comes from the reduction to the totally real subfield. Hence, one may wonder if this step is mandatory, and the answer is yes, because the surprisingly good practical behavior of the BKZ reduction is a conjoint consequence of the dimension of lattices involved on the one hand — the regime for such medium dimension allows better practical output bounds than the theoretical worst case — and the specificity of the geometry of the considered ideals induced by the abnormally small norm of its generator.

Acknowledgments. We would like to sincerely thank Claus Fieker for his comments about our implementation. These discussions were very rewarding. We would also like to thank the anonymous reviewers for their insightful comments and Joseph de Vilmarest for his help with the implementation of Gentry-Szydlo algorithm. This work has been supported in part by the European Union's H2020 Programme under grant agreement number ICT-644209.

A Mathematical Background Recalls

A.1 Hermite Normal Form

Definition 1. A $m \times n$ matrix $\mathbf{B}$ with integer entries has a (unique) Hermite Normal Form (HNF) $\mathbf{H}$ such that there exists a square unimodular matrix $\mathbf{U}$ satisfying $\mathbf{H} = \mathbf{B}\mathbf{U}$ and

1. $\mathbf{H}$ is lower triangular, $h_{i,j} = 0$ for $i < j$, and any columns of zeros are located on the right.

2. *The leading coefficient (the first nonzero entry from the top, also called the pivot) of a nonzero column is always strictly below the leading coefficient of the column before it and is positive.*
3. *The elements to the right of pivots are zero and elements to the left are non-negative and strictly smaller than the pivot.*

The computation of the HNF can be done efficiently in $O(n^\theta \mathsf{M}(n \log M))$ time and $O(n^2 \log M)$ space, where n^θ is the arithmetic complexity of the multiplication of two $n \times n$ matrices and $\mathsf{M}(b) = O(b)$ the complexity of the multiplication of two b-bit integers (see [36] for more details).

A.2   Ring of Integers, Integer Ideals

Integers of a Number Field. An element γ of $\mathbb{K}$ is said to be *integral* if its minimal polynomial has integer coefficients and is monic. The *ring of integers* of $\mathbb{K}$ is the ring of all integral elements contained in $\mathbb{K}$, and is denoted by $\mathcal{O}_{\mathbb{K}}$. Noticeably, the norm of any integer of the number field is an integer.

For α a primitive element of $\mathbb{K}$, we have $\mathbb{Z}[\alpha] \subset \mathcal{O}_{\mathbb{K}}$, but $\mathbb{Z}[\alpha]$ can be strictly included in $\mathcal{O}_{\mathbb{K}}$. Yet, as a finite-rank sub-module of the field $\mathbb{K}$, there exists a finite family $(b_i)_{i \in i}$ such that $\mathcal{O}_{\mathbb{K}} \cong \bigoplus_{i \in I} \mathbb{Z} \cdot b_i$. Such a family is called an *integral basis* of the number field.

Ideals and Norms. An additive subgroup $\mathfrak{a}$ of $\mathcal{O}_{\mathbb{K}}$ such that for every $\mathbf{x} \in \mathfrak{a}$, the coset $\mathbf{x} \cdot \mathcal{O}_{\mathbb{K}} = \{\mathbf{x} \cdot \mathbf{a} | \mathbf{a} \in \mathcal{O}_{\mathbb{K}}\}$ lies in $\mathfrak{a}$, is called an *integral ideal* of the number field. One can generalize the notion of norm of an element in the number field to integral ideals: let define the norm[11] $\mathcal{N}$ as the integer valued map:

$$\mathfrak{a} \mapsto [\mathcal{O}_{\mathbb{K}} : \mathfrak{a}] = |\mathcal{O}_{\mathbb{K}}/\mathfrak{a}|.$$

The ideal norm is multiplicative: for any ideals $\mathfrak{a}, \mathfrak{b}$, $\mathcal{N}(\mathfrak{a} \cdot \mathfrak{b}) = \mathcal{N}(\mathfrak{a}) \cdot \mathcal{N}(\mathfrak{b})$. Moreover this norm is closely linked to the norm of integers in the sense that for every $\mathbf{a} \in \mathcal{O}_{\mathbb{K}}$, $\mathcal{N}(\langle \mathbf{a} \rangle) = |\mathcal{N}_{\mathbb{K}/\mathbb{Q}}(\mathbf{a})|$, where $\langle \mathbf{a} \rangle$ denotes the principal ideal generated by $\mathbf{a}$: $\langle \mathbf{a} \rangle = \{\mathbf{a} \cdot \mathbf{x} | \mathbf{x} \in \mathcal{O}_{\mathbb{K}}\}$.

The norm of an ideal $\mathfrak{a}$ can be used to give an upper bound on the norm of the smallest nonzero element it contains: there always exists a nonzero $\mathbf{a} \in \mathfrak{a}$ for which:

$$0 < |\mathcal{N}_{\mathbb{K}/\mathbb{Q}}(\mathbf{a})| \leq \left(\frac{2}{\pi}\right)^{r_2} \sqrt{|\Delta_{\mathbb{K}}|} \mathcal{N}(\mathfrak{a}),$$

where $\Delta_{\mathbb{K}}$ is the discriminant of $\mathbb{K}$ and r_2 is the number of pairs of complex embeddings, defined as previously.

A.3   Dirichlet Unit Theorem

Unit Group of a Number Field. Let $\mathbb{K}$ be a number field. The *unit group* $\mathcal{U}_{\mathbb{K}}$ of $\mathbb{K}$ is the group of all integers in $\mathcal{O}_{\mathbb{K}}$ whose inverse also lies in $\mathcal{O}_{\mathbb{K}}$. The unit group has a simple geometric characterization in term of norm:

---

[11] We define here the *absolute norm* of an ideal.

Lemma 6. *An element* $\mathbf{a} \in \mathcal{O}_{\mathbb{K}}$ *is a unit if and only if* $\mathcal{N}_{\mathbb{K}/\mathbb{Q}}(\mathbf{a}) = 1$.

Log-Unit Lattice. Let $N = [\mathbb{K} : \mathbb{Q}]$ be the degree of the number field, written as $n = r_1 + 2r_2$, where r_1 and r_2 are defined respectively as the number of real embeddings and the number of pairs of complex embeddings. Define the map Log by

$$
\begin{cases}
\mathbb{K} \longrightarrow \mathbb{R}^{r_1+r_2} \\
\mathbf{x} \longmapsto \left(\log |\sigma_1(\mathbf{x})|, \dots, \log |\sigma_{r_1}(\mathbf{x})|, 2\log |\sigma_{r_1+1}(\mathbf{x})|, \dots, 2\log |\sigma_{r_1+r_2}(\mathbf{x})| \right)
\end{cases}
$$

The image of the kernel of Log by the canonical embedding σ lies in the intersection between the embedding $\sigma(\mathcal{O}_{\mathbb{K}})$ and the set of points of coordinates lower than 1. Since the embedding of $\mathcal{O}_{\mathbb{K}}$ is discrete, we deduce that $\sigma(\text{Ker Log})$ and so Ker Log are discrete.

Moreover, the image $\text{Log}(\mathcal{U}_{\mathbb{K}})$ lies in the hyperplane of equation $\sum x_i = 0$. A careful analysis of this image shows that it is in fact a full-rank lattice of this hyperplane. It called the *log-unit* lattice associated to $\mathbb{K}$. These remarks on the map Log lead then to the complete description of the structure of $\mathcal{U}_{\mathbb{K}}$.

Theorem 6 (Dirichlet's Unit Theorem). *Let* $\mathbb{K}$ *be a number field of degree* $N = r_1 + 2r_2$ *with* r_1 *and* r_2 *the number of real and pairs of complex embeddings. Then, the unit group of* $\mathbb{K}$ *is a direct product of a discrete cyclic group with a free abelian group of rank* $r = r_1 + r_2 - 1$.

References

1. Adleman, L.M., DeMarrais, J.: A Subexponential algorithm for discrete logarithms over all finite fields. In: Stinson, D.R. (ed.) CRYPTO 1993. LNCS, vol. 773, pp. 147–158. Springer, Heidelberg (1994). doi:10.1007/3-540-48329-2_13
2. Bach, E.: Explicit bounds for primality testing and related problems. Math. Comput. **55**, 355–380 (1990)
3. Becker, A., Ducas, L., Gama, N., Laarhoven, T.: New directions in nearest neighbor searching with applications to lattice sieving. In: Proceedings of the Twenty-Seventh Annual ACM-SIAM Symposium on Discrete Algorithms, SODA 2016, pp. 10–24 (2016)
4. Biasse, J.F.: An L(1/3) algorithm for ideal class group and regulator computation in certain number fields. Math. Comput. **83**, 2005–2031 (2014)
5. Biasse, J.F.: Subexponential time relations in the class group of large degree number fields. Adv. Math. Commun. **8**(4), 407–425 (2014)
6. Biasse, J.F.: A fast algorithm for finding a short generator of a principal ideal of $\mathbb{Q}(\zeta_{2^n})$. arXiv:1503.03107v1 (2015)
7. Biasse, J.F., Fieker, C.: Improved techniques for computing the ideal class group and a system of fundamental units in number fields. In: Proceedings of the 10th Algorithmic Number Theory Symposium (ANTS X) 2012, vol. 1, pp. 113–133 (2012)
8. Biasse, J.F., Fieker, C.: Subexponential class group and unit group computation in large degree number fields. LMS J. Comput. Math. **17**, 385–403 (2014)

9. Biasse, J.F., Song, F.: Efficient quantum algorithms for computing class groups and solving the principal ideal problem in arbitrary degree number fields. In: Proceedings of the Twenty-Seventh Annual ACM-SIAM Symposium on Discrete Algorithms, SODA 2016, pp. 893–902 (2016)
10. Bistritz, Y., Lifshitz, A.: Bounds for resultants of univariate and bivariate polynomials. Linear Algebra Appl. **432**, 1995–2005 (2010)
11. Buchmann, J.: A subexponential algorithm for the determination of class groups and regulators of algebraic number fields. Séminaire de Théorie des Nombres, Paris 1988–1989, pp. 27–41 (1990)
12. Campbell, P., Groves, M., Shepherd, D.: SOLILOQUY: a cautionary tale. In: ETSI 2nd Quantum-Safe Crypto Workshop (2014). http://docbox.etsi.org/workshop/2014/201410_CRYPTO/S07_Systems_and_Attacks/S07_Groves.pdf
13. Canfield, E.R., Erdős, P., Pomerance, C.: On a problem of Oppenheim concerning 'factorisatio numerorum'. J. Number Theory **17**, 1–28 (1983)
14. Cheon, J.H., Lee, C.: Approximate algorithms on lattices with small determinant. Cryptology ePrint Archive, Report 2015/461 (2015). http://eprint.iacr.org/2015/461
15. Cohen, H.: A Course in Computational Algebraic Number Theory. Graduate Texts in Mathematics, vol. 138. Springer, New York (1993)
16. Computational Algebra Group, University of Sydney: MAGMA, version 2.22.2 (2016). http://magma.maths.usyd.edu.au/magma/
17. Cramer, R., Ducas, L., Peikert, C., Regev, O.: Recovering short generators of principal ideals in cyclotomic rings. In: Fischlin, M., Coron, J.-S. (eds.) EUROCRYPT 2016. LNCS, vol. 9666, pp. 559–585. Springer, Heidelberg (2016). doi:10.1007/978-3-662-49896-5_20
18. Dixon, J.D.: Exact solution of linear equations using p-adic expansions. Numer. Math. **40**, 137–141 (1982)
19. The FPLLL development team: fplll, version 5.0 (2016). https://github.com/fplll/fplll
20. Garg, S., Gentry, C., Halevi, S.: Candidate multilinear maps from ideal lattices. In: Johansson, T., Nguyen, P.Q. (eds.) EUROCRYPT 2013. LNCS, vol. 7881, pp. 1–17. Springer, Heidelberg (2013). doi:10.1007/978-3-642-38348-9_1
21. Gentry, C.: Fully homomorphic encryption using ideal lattices. In: Proceedings of the 41st Annual ACM Symposium on Theory of Computing, STOC 2009, pp. 169–178 (2009)
22. Gentry, C., Szydlo, M.: Cryptanalysis of the revised NTRU signature scheme. In: Knudsen, L.R. (ed.) EUROCRYPT 2002. LNCS, vol. 2332, pp. 299–320. Springer, Heidelberg (2002). doi:10.1007/3-540-46035-7_20
23. Hafner, J.L., McCurley, K.S.: A rigorous subexponential algorithm for computation of class groups. J. Am. Math. Soc. **2**, 839–850 (1989)
24. Hoffstein, J., Pipher, J., Silverman, J.H.: NTRU: a ring-based public key cryptosystem. In: Buhler, J.P. (ed.) ANTS 1998. LNCS, vol. 1423, pp. 267–288. Springer, Heidelberg (1998). doi:10.1007/BFb0054868
25. Joux, A., Lercier, R., Smart, N., Vercauteren, F.: The number field sieve in the medium prime case. In: Dwork, C. (ed.) CRYPTO 2006. LNCS, vol. 4117, pp. 326–344. Springer, Heidelberg (2006). doi:10.1007/11818175_19
26. Kirchner, P.: Algorithms on ideal over complex multiplication order. Cryptology ePrint Archive, Report 2016/220 (2016). http://eprint.iacr.org/2016/220
27. Landau, E.: Neuer beweis des primzahlsatzes und beweis des primidealsatzes. Math. Ann. **56**, 645–670 (1903)

28. Langlois, A., Stehlé, D., Steinfeld, R.: GGHLite: more efficient multilinear maps from ideal lattices. In: Nguyen, P.Q., Oswald, E. (eds.) EUROCRYPT 2014. LNCS, vol. 8441, pp. 239–256. Springer, Heidelberg (2014). doi:10.1007/978-3-642-55220-5_14
29. Lenstra, A.K., Lenstra Jr., H.W., Manasse, M.S., Pollard, J.M.: The number field sieve. In: Proceedings of the 22nd Annual ACM Symposium on Theory of Computing, pp. 564–572 (1990)
30. Lenstra, H.W., Silverberg, A.: Revisiting the Gentry-Szydlo algorithm. In: Garay, J.A., Gennaro, R. (eds.) CRYPTO 2014. LNCS, vol. 8616, pp. 280–296. Springer, Heidelberg (2014). doi:10.1007/978-3-662-44371-2_16
31. Lyubashevsky, V., Micciancio, D., Peikert, C., Rosen, A.: SWIFFT: a modest proposal for FFT hashing. In: Fast Software Encryption, FSE 2008, pp. 54–72 (2008)
32. Lyubashevsky, V., Peikert, C., Regev, O.: On ideal lattices and learning with errors over rings. J. ACM **60**(6), 1–23 (2013)
33. Lyubashevsky, V., Peikert, C., Regev, O.: A toolkit for ring-LWE cryptography. In: Johansson, T., Nguyen, P.Q. (eds.) EUROCRYPT 2013. LNCS, vol. 7881, pp. 35–54. Springer, Heidelberg (2013). doi:10.1007/978-3-642-38348-9_3
34. Micciancio, D.: Generalized compact knapsacks, cyclic lattices, and efficient one-way functions from worst-case complexity assumptions. In: Proceedings of the 43rd Symposium on Foundations of Computer Science, FOCS 2002, pp. 356–365 (2002)
35. Micciancio, D., Walter, M.: Practical, predictable lattice basis reduction. In: Fischlin, M., Coron, J.-S. (eds.) EUROCRYPT 2016. LNCS, vol. 9665, pp. 820–849. Springer, Heidelberg (2016). doi:10.1007/978-3-662-49890-3_31
36. Micciancio, D., Warinschi, B.: A linear space algorithm for computing the Hermite normal form. In: Proceedings of the 2001 International Symposium on Symbolic and Algebraic Computation, ISSAC 2001, pp. 231–236 (2001)
37. Miller, J.C.: Class numbers of totally real fields and applications to the Weber class number problem. Acta Arith. **164**, 381–397 (2014)
38. The PARI Group, Bordeaux: PARI/GP, version 2.7.6 (2016). http://pari.math.u-bordeaux.fr/
39. Schank, J.: LOGCVP, Pari implementation of CVP in $\log \mathbb{Z}[\zeta_{2^n}]^*$ (2015). https://github.com/jschanck-si/logcvp
40. Scourfield, E.: On ideals free of large prime factors. J. Théorie Nombres Bordx. **16**(3), 733–772 (2004)
41. Seysen, M.: A probabilistic factorization algorithm with quadratic forms of negative discriminant. Math. Comput. **84**, 757–780 (1987)
42. Shoup, V.: NTL: A Library for doing Number Theory, version 9.11.0 (2016). http://www.shoup.net/ntl/
43. Smart, N.P., Vercauteren, F.: Fully homomorphic encryption with relatively small key and ciphertext sizes. In: Nguyen, P.Q., Pointcheval, D. (eds.) PKC 2010. LNCS, vol. 6056, pp. 420–443. Springer, Heidelberg (2010). doi:10.1007/978-3-642-13013-7_25
44. Stehlé, D., Steinfeld, R., Tanaka, K., Xagawa, K.: Efficient public key encryption based on ideal lattices. In: Matsui, M. (ed.) ASIACRYPT 2009. LNCS, vol. 5912, pp. 617–635. Springer, Heidelberg (2009). doi:10.1007/978-3-642-10366-7_36
45. Thiel, C.: On the complexity of some problems in algorithmic algebraic number theory. Ph.D. thesis, Universität des Saarlandes (1995). https://www.cdc.informatik.tu-darmstadt.de/reports/reports/Christoph_Thiel.diss.pdf
46. Washington, L.C.: Introduction to Cyclotomic Fields. Graduate Texts in Mathematics, vol. 83, 2nd edn. Springer, New York (1997)

Obfuscation and Functional Encryption

Robust Transforming Combiners from Indistinguishability Obfuscation to Functional Encryption

Prabhanjan Ananth[1], Aayush Jain[1(✉)], and Amit Sahai[2]

[1] Center for Encrypted Functionalities, Computer Science Department,
UCLA, Los Angeles, USA
prabhanjan@cs.ucla.edu, aayushjain1728@gmail.com
[2] University of California Los Angeles and Center for Encrypted Functionalities,
Los Angeles, USA
sahai@cs.ucla.edu

Abstract. Indistinguishability Obfuscation (iO) has enabled an incredible number of new and exciting applications. However, our understanding of how to actually build secure iO remains in its infancy. While many candidate constructions have been published, some have been broken, and it is unclear which of the remaining candidates are secure.

This work deals with the following basic question: *Can we hedge our bets when it comes to iO candidates?* In other words, if we have a collection of iO candidates, and we only know that at least one of them is secure, can we still make use of these candidates?

This topic was recently studied by Ananth, Jain, Naor, Sahai, and Yogev [CRYPTO 2016], who showed how to construct a robust iO combiner: Specifically, they showed that given the situation above, we can construct a single iO scheme that is secure as long as (1) at least one candidate iO scheme is a subexponentially secure iO, and (2) either the subexponential DDH or LWE assumptions hold.

In this work, we make three contributions:

- **(Better robust iO combiners.)** First, we work to improve the assumptions needed to obtain the same result as Ananth et al.: namely we show how to replace the DDH/LWE assumption with the assumption that subexponentially secure one-way functions exist.

P. Ananth—This work was partially supported by grant #360584 from the Simons Foundation and the grants listed under Amit Sahai.
A. Jain—This work was supported by the grants listed under Amit Sahai.
A. Sahai—Research supported in part from a DARPA/ARL SAFEWARE award, NSF Frontier Award 1413955, NSF grants 1619348, 1228984, 1136174, and 1065276, BSF grant 2012378, a Xerox Faculty Research Award, a Google Faculty Research Award, an equipment grant from Intel, and an Okawa Foundation Research Grant. This material is based upon work supported by the Defense Advanced Research Projects Agency through the ARL under Contract W911NF-15-C-0205. The views expressed are those of the authors and do not reflect the official policy or position of the Department of Defense, the National Science Foundation, or the U.S. Government.

© International Association for Cryptologic Research 2017
J.-S. Coron and J.B. Nielsen (Eds.): EUROCRYPT 2017, Part I, LNCS 10210, pp. 91–121, 2017.
DOI: 10.1007/978-3-319-56620-7_4

- **(Transforming Combiners from iO to FE and NIKE.)** Second, we consider a broader question: what if we start with several iO candidates where only one works, but we don't care about achieving iO itself, rather we want to achieve concrete applications of iO? In this case, we are able to work with the *minimal* assumption of just polynomially secure one-way functions, and where the working iO candidate only achieves polynomial security. We call such combiners *transforming combiners*. More generally, a transforming combiner from primitive A to primitive B is one that takes as input many candidates of primitive A, out of which we are guaranteed that at least one is secure and outputs a secure candidate of primitive B. We can correspondingly define robust transforming combiners. We present transforming combiners from indistinguishability obfuscation to *functional encryption* and *non-interactive multiparty key exchange (NIKE)*.
- **(Correctness Amplification for iO from polynomial security and one-way functions.)** Finally, along the way, we obtain a result of independent interest: Recently, Bitansky and Vaikuntanathan [TCC 2016] showed how to amplify the correctness of an iO scheme, but they needed subexponential security for the iO scheme and also require subexponentially secure DDH or LWE. We show how to achieve the same correctness amplification result, but requiring only polynomial security from the iO scheme, and assuming only polynomially secure one-way functions.

1 Introduction

Indistinguishability Obfuscation (iO), first defined by [4], has been a major revelation to cryptography. The discovery of the punctured programming technique by Sahai and Waters [46] has led to several interesting applications of indistinguishability obfuscation. A very incomplete list of such results includes functional encryption [2,24,47], the feasibility of succinct randomized encodings [7,13,39], time lock puzzles [8], software watermarking [16], instantiating random oracles [34] and hardness of Nash equilibrium [10,26].

On the construction side, however, iO is still at a nascent stage. The first candidate was proposed by Garg et al. [24] from multilinear maps [19,23,29]. Since then there have many proposals of iO [3,29,48]. All these constructions are based on multilinear maps. The constructions of multilinear maps have come under scrutiny after several successful cryptanalytic attacks [14,15,17,18,35,44] were mounted against them. In fact, there have also been direct attacks on some of the iO candidates as well [17,44]. However, there are (fortunately) still many candidates that have survived all known cryptanalytic attacks. We refer the reader to Appendix A in [1] for a partial list of these candidates[1]. In light of this, its imperative to revisit the applications of iO and hope to weaken the trust we place on any specific known iO candidate to construct these applications.

In other words, *can we hedge our bets when it comes to iO candidates?*

---

[1] Several recent candidates such as [25,42,43] have not been included in this list. There are currently no attacks known on these candidates as well.

If we're wrong about some candidates that seem promising right now, but not others, then can we still give explicit constructions that achieve the amazing applications of iO?

Robust iO Combiners. Recently, Ananth et al. [1] considered the closely related problem of constructing an iO scheme starting from many iO candidates, such that the final iO scheme is guaranteed to be secure as long as *even one* of the iO candidates is secure. In fact, they only assume that the secure candidate satisfies correctness, and in particular, the insecure candidates could also be incorrect. This notion is termed as a *robust iO combiners* (also studied by Fischlin et al. [22] in a relaxed setting where multiple underlying iO candidates must be secure) and are useful in constructing *universal iO* [32] [2]. The work of [1] constructs robust iO combiners assuming the existence of a *sub-exponentially secure* iO scheme and *sub-exponentially secure DDH/ LWE*. As a consequence of this result, we can construct the above applications by combining all known iO candidates as long as one of the candidates is sub-exponentially secure.

While the work of [1] is a major advance, it leaves open two very natural questions, that we study in this work. The first question is: do we really need to assume DDH or LWE? In other words:

1. *What assumption suffices to construct a robust iO combiner?*
   In particular, are (sub-exponentially secure) one-way functions sufficient?

The second, broader, question is: if we care about constructing *applications* of iO, can we do better in terms of assumptions? In particular, recent work [27] has shown that functional encryption – itself an application of iO – can be directly used to construct several applications of iO. Let us then define an *transforming combiner* as an object that takes several iO candidates, with the promise that at least one of them is only polynomially secure, and outputs an explicit secure functional encryption scheme. Then, let us consider the following question, which truly addresses a *minimal* assumption:

2. *Assuming only polynomially secure one-way functions, can we construct a transforming combiner from iO to functional encryption?*

Note that since the existence of iO does not even imply that **P≠NP**, while functional encryption implies one-way functions, the above question lays out a minimal assumption for constructing a transforming combiner from iO to FE.

1.1 Our Contribution

We address questions 1 and 2 in this work. We show,

Theorem 1 (Transforming Combiners). *Given many iO candidates out of which at least one of them is correct and secure and additionally assuming one-way functions, we can construct a compact functional encryption scheme.*

---

[2] A scheme Π is said to be a universal secure iO scheme if the following holds: if there exists a secure iO scheme (whose explicit description is unknown) then Π is a secure iO scheme.

As a corollary, we can construct an explicit functional encryption scheme assuming the *existence* of iO and one-way functions. In other words, we show that it suffices that iO exists (rather than relying on a constructive proof of it) to construct an explicit functional encryption scheme.

Corollary 1 (Informal). *Assuming polynomially secure iO and one-way functions exists, we can construct an explicit compact functional encryption scheme. In particular, the construction of functional encryption does not rely on an explicit description of the iO scheme.*

Combining this result with the works of [2,11] who show how to construct iO from sub-exponentially secure compact FE, we obtain the following result.

Theorem 2 (Informal). *There exists a robust iO combiner assuming sub-exponentially secure one-way functions as long as one of the underlying iO candidates is sub-exponentially secure.*

This improves upon the result of Ananth et al. [1] who achieve the same result assuming sub-exponentially secure DDH or LWE.

Explicit NIKE from several iO candidates: Recent works of Garg and Srinivasan [28], Li and Micciancio [41], show how to achieve collusion resistant functional encryption from compact functional encryption and Garg et al. [27] show how to build multi-party non interactive key exchange (NIKE) from collusion resistant functional encryption. When combined with these results, our work shows how to obtain an explicit NIKE protocol when given any one-way function, and many iO candidates with the guarantee that only one of the candidates is secure.

New Correctness Amplification Theorem for iO. En route to achieving this result, we demonstrate a new correctness amplification theorem for iO. In particular, we show how to obtain almost-correct iO starting from polynomially secure approximately-correct iO[3] and one-way functions. Prior to our work, [12] showed how to achieved a correctness amplification theorem starting from *sub-exponentially secure* iO and *sub-exponentially secure* DDH/ LWE.

Theorem 3 (Informal). *There is a transformation from a polynomially secure approximately-correct iO to polynomially secure almost-correct iO assuming one-way functions.*

2   Technical Overview

The goal of our work is to construct a compact functional encryption scheme starting many iO candidates out of which one of them is secure. Let us start with the more ambitious goal of building a robust compact FE combiner. If

---

[3] An iO scheme is ε-approximately correct if every obfuscated circuit agrees with the original circuit on ε fraction of the inputs.

we have such a combiner, then we achieve our goal since the ith compact FE candidate used in the combiner can be built from the ith iO candidate using prior works [24].

To build a compact FE combiner, we view this problem via the lens of secure multi-party computation: we view every compact FE candidate as corresponding to a party in the MPC protocol; insecure candidates correspond to adversaries. Ananth et al. [1] took the same viewpoint when building an iO combiner and in particular, used *non-interactive* MPC techniques that relied on DDH/ LWE to solve this problem. Our goal is however to base our combiner only on one-way functions and to achieve that, we start with an interactive MPC protocol.

A first attempt is the following: Let $\Pi_1, \ldots, \Pi_n$ be the n compact FE candidates. We start with an interactive MPC protocol for parties $P_1, \ldots, P_n$.

- To encrypt a message x, we secret share x into n additive shares. Each of these shares are encrypted using candidates $\Pi_1, \ldots, \Pi_n$.
- To generate a functional key for function f, we generate a functional key for the following function g_i using FE candidate Π_i: this function g_i takes as input message m and executes the next message function of Π_i to obtain message m'. If m' has to be sent to Π_j then it encrypts m' under the public key of Π_j and outputs the ciphertext.

The decryption algorithm proceeds as in the evaluation of the multi-party secure computation protocol. Since one of the candidates is secure, say $\mathbf{i}^{th}$ candidate, the hope is that the $\mathbf{i}^{th}$ ciphertext hides the $\mathbf{i}^{th}$ share of x and thus security of FE is guaranteed.

However, implementing the above high level idea faces the following obstacles.

Statelessness: While a party participating in a MPC protocol is stateful, the functional key is not. Hence, the next message function as part of the functional key expects to receive the previous state as input. Its not clear how to ensure without sharing state information with all the other candidates.

Oblivious Transfer: Recall that our goal was to base the combiner only on one-way functions. However, MPC requires oblivious transfer and from Impagliazzo and Rudich's result [36] we have strong evidence to believe that oblivious transfer cannot be based on one-way functions. Given this, it is unclear how to directly use MPC to achieve our goal.

Randomized Functions: The functional key in the above solution encrypts a message with respect to another candidate. Since encryption is a probabilistic process, we need to devise a mechanism to generate randomness for encrypting the ciphertext.

Correctness Amplification: A recent elegant work of Bitansky and Vaikuntanathan [12] study correctness amplification techniques in the context of indistinguishability obfuscation and functional encryption. Their correctness amplification theorems assume DDH/ LWE to achieve this result. Indeed, this work was also employed by Ananth et al. to construct an iO combiner. We need a different mechanism to handle the correctness issue if our goal is to base our construction on one-way functions.

TACKLING ISSUES: We propose the following ideas to tackle the above issues.

Use 2-ary FE instead of compact FE: The first idea is to replace compact FE candidates with 2-ary FE[4] candidates. We can build each of 2-ary FE candidates starting from iO candidates. The advantage of using 2-ary FE is two fold:

1. It helps in addressing the issue of statelessness. The functional keys, of say i^{th} candidate, are now associated with 2-ary functions, where the first input of the function takes as input the previous state and the other input takes as input the message from another candidate. The output of this function is the updated state encrypted under the public key of the ith candidate and encryption of message under public key of jth candidate, where j^{th} candidate is supposed to receive this message. This way, the state corresponding to the ith candidate is never revealed to any other candidate.
2. It also helps in addressing the issue of randomized functions. The first input to the function could also contain a PRF key. This key will be used to generate the randomness required to encrypt messages with respect to public keys of other candidates.

Getting Rid of OT: To deal with this issue, we use the idea of pre-processing OTs that is extensively used in the MPC literature [5,6,20,37][5]. We pre-compute polynomially many OTs [5] ahead of time. Once we have pre-computed OTs, we can construct an information theoretically secure MPC protocol that is secure upto $n-1$ corruptions, where n is the number of parties. Note that we can only achieve semi-honest security in this setting, achieving malicious security would require that the pre-processing phase outputs exponentially many bits [37].

Next, we consider whether to perform the OT pre-computation as part of the key generation or the encryption algorithm. Depending on where we perform the pre-computation phase, we are faced with the following issues:

1. *Reusability*: In a secure MPC protocol, the pre-computed OTs are used only in one execution of the MPC protocol. So, if we perform the OT pre-computation as part of the key generation algorithm, then the pre-computed OTs need to be reused across different ciphertexts. In this case, no security is guaranteed.
2. *Compactness*: In the current secure MPC with pre-processing solutions, it turns out that the number of OTs to be pre-computed depends on the size of the circuit implementing the MPC functionality. So if we implement the

---

[4] A 2-ary FE scheme is a functional encryption corresponding to 2-ary functions. A functional key of 2-ary function f decrypts two ciphertexts CT_1 (of message x) and CT_2 (of message y) to obtain $f(x,y)$.

[5] The key difference is that in prior works, the pre-processing phase is generally independent of the inputs and in our case, it is input dependent. We require that this pre-processing phase is compatible with any MPC functionality that will be defined after the pre-processing phase.

OT pre-computation as part of the encryption algorithm, we need to make sure that the encryption complexity is independent of the number of pre-processed OTs.

We perform the OT pre-computation as part of the encryption algorithm. Hence, we have to deal with the compactness issue stated above. To resolve this, we "compress" the OTs using PRF keys. That is, to generate OTs between two parties P_i and P_j, we use a PRF key K_{ij}. The next problem is under which public key do we encrypt K_{ij}. Encrypting this under either i^{th} candidate or j^{th} candidate could compromise the key completely. The guarantee we want is that as long as one of the two candidates is honest, this key is not compromised. To solve this problem, we employ a 1-out-2 combiner of 2-ary FE – given two candidates, 1-out-2 combiner is secure as long as one of them is secure. This can be achieved by computing an "onion" of two FE candidates. We refer the reader to the technical section for more details.

Correctness Amplification: [12] showed how to transform ε-approximately correct iO into an almost correct iO scheme. They do this in two steps: (i) the first step is the self reducibility step, where they transform approximately correct iO scheme into one, where the iO scheme is correct on every input with probability close to ε, (ii) then they apply BPP amplification techniques to get almost correct iO. Their self reducibility step involves using a type of secure function evaluation scheme and they show how to construct this based on DDH and LWE. We instead show how to achieve the self reducibility step using a single key private key functional encryption scheme. The main idea is as follows: to obfuscate a circuit C, we generate a functional key of C and then obfuscate the FE decryption algorithm with the functional key hardwired inside it. Additionally, we give out the master secret key in the clear along with this obfuscated circuit. To evaluate on an input x, first encrypt this using the master secret key and feed this ciphertext to the obfuscated circuit, which evaluates the decryption algorithm to produce the output. This approach leads to the following issues: (i) firstly, revealing the output of the FE decryption could affect the correctness of iO: for instance, the obfuscated circuit could output $\perp$ for all inputs on which the FE decryption outputs 1, (ii) since the evaluator has the master secret key, he could feed in maliciously generated FE ciphertexts into the obfuscated circuit.

We solve (i) by using by masking the output of the circuit. Here, the mask is supplied as input to the obfuscated circuit. We solve (ii) by using NIZKs with pre-processing, a tool used by Ananth et al. to construct witness encryption combiners. This primitive can be based on one-way functions.

OUR SOLUTION IN A NUTSHELL: Summarizing, we take the following approach to build compact FE starting from many iO candidates out of which at least one of them is correct and secure.

1. First check if the candidates are approximately correct. If not, discard the candidates.
2. Apply the new correctness amplification mechanism on all the remaining iO candidates.

3. Construct n 2-ary FE candidates from the n iO candidates obtained from the previous step.
4. Then using an onion-based approach, obtain a 2-ary FE combiner that only combines two candidates. This will lead to $N = n^2 - n$ candidates.
5. Construct a compact FE scheme starting from the above N 2-ary FE candidates and an n-party MPC protocol with OT preprocessing phase. Essentially every $(i, j)^{th}$ 2-ary FE candidate implements a channel between i^{th} and j^{th} party.

We expand on the above high level approach in the relevant technical sections.

3   Preliminaries

Let λ be the security parameter. For a distribution $\mathcal{D}$ we denote by $x \xleftarrow{\$} $ an element chosen from $\mathcal{D}$ uniformly at random. We denote that $\{\mathcal{D}_{1,\lambda}\} \approx_{c,\mu} \{\mathcal{D}_{2,\lambda}\}$, if for every PPT distinguisher $\mathcal{A}$, $\left| \Pr\left[\mathcal{A}(1^\lambda, x \xleftarrow{\$} \mathcal{D}_{1,\lambda}) = 1\right] - \Pr\left[\mathcal{A}(1^\lambda, x \xleftarrow{\$} \mathcal{D}_{2,\lambda}) = 1\right] \right| \leq \mu(\lambda)$ where μ is a negligible function. For a language L associated with a relation R with denote by $(x, w) \in R$ an instance $x \in L$ with a valid witness w. For an integer $n \in \mathbb{N}$ we denote by $[n]$ the set $\{1, \ldots, n\}$. By negl we denote a negligible function. We assume that the reader is familiar with the concepts of one-way functions, pseudorandom functions, functional encryption, NIWI, statistically binding commitments and in particular sub-exponential security of these primitives. We say that the one-way function is sub-exponentially secure if no polynomial time adversary inverts a random image with a probability greater than inverse sub-exponential in the length of the input. We refer the reader to full version for the definitions of these primitives.

Important Notation. We introduce some notation that will be useful throughout this work. Consider an algorithm A. We define the *time function* of A to be T if the runtime of $A(x) \leq T(|x|)$. We are only interested in time functions which satisfy the property that $T(\text{poly}(n)) = |\text{poly}(T(n))|$. In this section, we describe NIZK with Pre-Processing.

3.1   NIZK with Pre-Processing

We consider a specific type of zero knowledge proof system where the messages exchanged is independent of the input instance till the last round. We call this zero knowledge proof system with pre-processing. The pre-processing algorithm essentially simulates the interaction between the prover and the verifier till the last round and outputs views of the prover and the verifier.

Definition 1. *Let L be a language with relation R. A scheme* PZK = (PZK.Pre, PZK.Prove, PZK.Verify) *of PPT algorithms is a zero knowledge proof system with pre-processing,* PZK, *between a verifier and a prover if they satisfy the following* *properties. Let $(\sigma_V, \sigma_P) \leftarrow$ PZK.Pre(1^λ) be a preprocessing stage where the prover and the verifier interact. Then:*

1. **Completeness**: *for every $(x, w) \in R$ we have that:*

$$\Pr[\text{PZK.Verify}(\sigma_V, x, \pi) = 1 \ : \ \pi \leftarrow \text{PZK.Prove}(\sigma_P, x, w)] = 1.$$

   where the probability is over the internal randomness of all the PZK *algorithms.*

2. **Soundness**: *for every $x \notin L$ we have that:*

$$\Pr[\exists \pi : \text{PZK.Verify}(\sigma_V, x, \pi) = 1] < 2^{-n}$$

   where the probability is only over PZK.Pre.

3. **Zero-Knowledge**: *there exists a PPT algorithm S such that for any x, w where $V(x, w) = 1$ there exists a negligible function μ such that it holds that:*

$$\{\sigma_V, \text{PZK.Prove}(\sigma_P, x, w)\} \approx_{c, \mu} \{S(x)\}$$

We say that PZK *is sub-exponentially secure if $\mu(\lambda) = O(2^{-\lambda^c})$ for a constant $c > 0$.*

Such schemes were studied in [21,40] where they proposed constructions based on one-way functions. Sub-exponentially secure PZK can be built from sub-exponentially secure one-way functions.

4 Definitions: IO Combiner

We recall the definition of IO combiners from [1]. Suppose we have many indistinguishability obfuscation (IO) schemes, also referred to as *IO candidates*. We are additionally guaranteed that one of the candidates is secure. No guarantee is placed on the rest of the candidates and they could all be potentially broken. Indistinguishability obfuscation combiners provides a mechanism of combining all these candidates into a single monolithic IO scheme *that is secure*. We emphasize that the only guarantee we are provided is that one of the candidates is secure and in particular, it is unknown exactly which of the candidates is secure.

We formally define IO combiners next. We start by providing the syntax of an obfuscation scheme. We then present the definitions of an IO candidate and a secure IO candidate. To construct IO combiner, we need to also consider functional encryption candidates. Once we give these definitions, we present our construction in Sect. 5.2.

Syntax of Obfuscation Scheme. An obfuscation scheme associated to a class of circuits $\mathcal{C} = \{\mathcal{C}_\lambda\}_{\lambda \in \mathbb{N}}$ with input space $\mathcal{X}_\lambda$ and output space $\mathcal{Y}_\lambda$ consists of two PPT algorithms (Obf, Eval) defined below.

- **Obfuscate**, $\overline{C} \leftarrow$ Obf$(1^\lambda, C)$: It takes as input security parameter λ, a circuit $C \in \mathcal{C}_\lambda$ and outputs an obfuscation of C, $\overline{C}$.

- **Evaluation,** $y \leftarrow \mathsf{Eval}\left(\overline{C}, x\right)$: This is usually a deterministic algorithm. But sometimes we will treat it as a randomized algorithm. It takes as input an obfuscation $\overline{C}$, input $x \in \mathcal{X}_\lambda$ and outputs $y \in \mathcal{Y}_\lambda$.

Throughout this work, we will only be concerned with *uniform* Obf algorithms. That is, Obf and Eval are represented as Turing machines (or equivalently uniform circuits).

We require that each candidate satisfy the following property called polynomial slowdown.

Definition 2 (Polynomial Slowdown). *An obfuscation scheme $\Pi = (\mathsf{Obf}, \mathsf{Eval})$ is an IO candidate for a class of circuits $\mathcal{C} = \{\mathcal{C}_\lambda\}_{\lambda \in \mathbb{N}}$, with every $C \in \mathcal{C}_\lambda$ has size $\mathrm{poly}(\lambda)$, if it satisfies the following property:*

Polynomial Slowdown: *For every $C \in \mathcal{C}_\lambda$, we have the running time of Obf on input $(1^\lambda, C)$ to be $\mathrm{poly}(|C|, \lambda)$. Similarly, we have the running time of Eval on input $(\overline{C}, x)$ for $x \in \mathcal{X}_\lambda$ is $\mathrm{poly}(|\overline{C}|, \lambda)$.*

We now define various notions of correctness.

Definition 3 (Almost/Perfect Correct IO candidate). *An obfuscation scheme $\Pi = (\mathsf{Obf}, \mathsf{Eval})$ is an almost correct IO candidate for a class of circuits $\mathcal{C} = \{\mathcal{C}_\lambda\}_{\lambda \in \mathbb{N}}$, with every $C \in \mathcal{C}_\lambda$ has size $\mathrm{poly}(\lambda)$, if it satisfies the following property:*

- **Almost Correctness***: For every $C : \mathcal{X}_\lambda \to \mathcal{Y}_\lambda \in \mathcal{C}_\lambda, x \in \mathcal{X}_\lambda$ it holds that:*

$$\Pr\left[\forall x \in \mathcal{X}_\lambda, \mathsf{Eval}\left(\mathsf{Obf}(1^\lambda, C), x\right) = C(x)\right] \geq 1 - \mathsf{negl},$$

over the random coins of Obf. The candidate is called a correct IO candidate if this probability is 1.

Definition 4 (α−worst-case Correctness). *An obfuscation scheme $\Pi = (\mathsf{Obf}, \mathsf{Eval})$ is α−worst-case correct IO candidate for a class of circuits $\mathcal{C} = \{\mathcal{C}_\lambda\}_{\lambda \in \mathbb{N}}$, with every $C \in \mathcal{C}_\lambda$ has size $\mathrm{poly}(\lambda)$, if it satisfies the following property:*

- α−**worst-case Correctness***: For every $C : \mathcal{X}_\lambda \to \{0, 1\} \in \mathcal{C}_\lambda, x \in \mathcal{X}_\lambda$ it holds that:*
$$\Pr\left[\mathsf{Eval}\left(\mathsf{Obf}(1^\lambda, C), x\right) = C(x)\right] \geq \alpha,$$

over the random coins of Obf and Eval. The candidate is correct if this probability is 1.

Remark 1. Given any α−worst case correct IO candidate where $\alpha > 1/2 + 1/\mathrm{poly}(\lambda)$, as observed by [12] we can gen an almost correct IO candidate while retaining security via BPP amplification.

$\epsilon - Secure\ IO\ candidate.$ If any IO candidate additionally satisfies the following (informal) security property then we define it to be a *secure* IO candidate: for every pair of circuits C_0 and C_1 that are equivalent[6] we have obfuscations of C_0 and C_1 to be indistinguishable by any PPT adversary.

Definition 5 (ϵ-Secure IO candidate). *An obfuscation scheme* $\Pi =$ (Obf, Eval) *for a class of circuits* $C = \{C_\lambda\}_{\lambda \in \mathbb{N}}$ *is a ϵ-secure IO candidate if it satisfies the following conditions:*

- **Security.** *For every PPT adversary* $\mathcal{A}$, *for every sufficiently large* $\lambda \in \mathbb{N}$, *for every* $C_0, C_1 \in C_\lambda$ *with* $C_0(x) = C_1(x)$ *for every* $x \in \mathcal{X}_\lambda$ *and* $|C_0| = |C_1|$, *we have:*

$$\left| \Pr\left[0 \leftarrow \mathcal{A}\Big(\mathsf{Obf}(1^\lambda, C_0), C_0, C_1\Big)\right] - \Pr\left[0 \leftarrow \mathcal{A}\Big(\mathsf{Obf}(1^\lambda, C_1), C_0, C_1\Big)\right] \right| \leq \epsilon(\lambda)$$

Remark 2. We say that Π is a secure IO candidate if it is a ϵ-secure IO candidate with $\epsilon(\lambda) = \mathsf{negl}(\lambda)$, for some negligible function negl.

We remarked earlier that the identity function is an IO candidate. However, note that the identity function is *not* a secure IO candidate. Whenever we refer an IO candidate we will specify the correctness and the security notion it satisfies. For example [4, 24, 33] are examples of negl-secure correct IO candidate. In particular, an IO candidate need not necessarily have any security/correctness property associated with it.

We have the necessary ingredients to define an IO combiner.

4.1   Definition of IO Combiner

We present the formal definition of IO combiner below. First, we provide the syntax of the IO combiner. Later we present the properties associated with an IO combiner.

There are two PPT algorithms associated with an IO combiner, namely, CombObf and CombEval. Procedure CombObf takes as input circuit C along with the description of multiple correct IO candidates[7] and outputs an obfuscation of C. Procedure CombEval takes as input the obfuscated circuit, input x, the description of the candidates and outputs the evaluation of the obfuscated circuit on input x.

Syntax of IO Combiner. We define an IO combiner $\Pi_{\mathsf{comb}} =$ (CombObf, CombEval) for a class of circuits $C = \{C_\lambda\}_{\lambda \in \mathbb{N}}$.

- **Combiner of Obfuscate algorithms,** $\overline{C} \leftarrow \mathsf{CombObf}(1^\lambda, C, \Pi_1, \ldots, \Pi_n)$: It takes as input security parameter λ, a circuit $C \in C$, description of correct IO candidates $\{\Pi_i\}_{i \in [n]}$ and outputs an obfuscated circuit $\overline{C}$.

---

[6] Two circuits C_0 and C_1 are equivalent if they (a) have the same size, (b) have the same input domain and, (c) for every x in the input domain, $C_0(x) = C_1(x)$.

[7] The description of an IO candidate includes the description of the obfuscation and the evaluation algorithms.

- **Combiner of Evaluation algorithms,** $y \leftarrow \mathsf{CombEval}(\overline{C}, x, \Pi_1, \ldots, \Pi_n)$: It takes as input obfuscated circuit $\overline{C}$, input x, descriptions of IO candidates $\{\Pi_i\}_{i \in [n]}$ and outputs y.

We define the properties associated to any IO combiner. There are three main properties – correctness, polynomial slowdown, and security. The correctness and the polynomial slowdown properties are defined on the same lines as the corresponding properties of the IO candidates.

The intuitive security notion of IO combiner says the following: suppose one of the candidates is a secure IO candidate then the output of obfuscator (CombObf) of the IO combiner on C_0 is computationally indistinguishable from the output of the obfuscator on C_1, where C_0 and C_1 are equivalent circuits.

Definition 6 (((ϵ', ϵ)-secure IO combiner). *Consider a circuit class* $\mathcal{C} = \{\mathcal{C}_\lambda\}_{\lambda \in \mathbb{N}}$. *We say that* $\Pi_{\mathsf{comb}} = (\mathsf{CombObf}, \mathsf{CombEval})$ *is a* (ϵ', ϵ)-secure IO **combiner** *if the following conditions are satisfied: Let* $\Pi_1, \ldots, \Pi_n$ *be* n *correct IO candidates for P/poly, and* ϵ *is a function of* ϵ'.

- **Correctness.** *Let* $C \in \mathcal{C}_{\lambda \in \mathbb{N}}$ *and* $x \in \mathcal{X}_\lambda$. *Consider the following process: (a)* $\overline{C} \leftarrow \mathsf{CombObf}(1^\lambda, C, \Pi_1, \ldots, \Pi_n)$, *(b)* $y \leftarrow \mathsf{CombEval}(\overline{C}, x, \Pi_1, \ldots, \Pi_n)$. *Then with overwhelming probability over randomness of* $\mathsf{CombObf}$, $\Pr[y = C(x)] \geq 1$, *where the probability is over* $x \xleftarrow{\$} \mathcal{X}_\lambda$.
- **Polynomial Slowdown.** *For every* $C : \mathcal{X}_\lambda \to \mathcal{Y}_\lambda \in \mathcal{C}_\lambda$, *we have the running time of* $\mathsf{CombObf}$ *on input* $(1^\lambda, C, \Pi_1, \ldots, \Pi_n)$ *to be at most* $\mathsf{poly}(|C| + n + \lambda)$. *Similarly, we have the running time of* $\mathsf{CombEval}$ *on input* $(\overline{C}, x, \Pi_1, \ldots, \Pi_n)$ *to be at most* $\mathsf{poly}(|\overline{C}| + n + \lambda)$.
- **Security.** *Let* Π_i *be* ϵ-secure correct IO candidate for some $i \in [n]$. For every PPT adversary $\mathcal{A}$, for every sufficiently large $\lambda \in \mathbb{N}$, for every $C_0, C_1 \in \mathcal{C}_\lambda$ with $C_0(x) = C_1(x)$ for every $x \in \mathcal{X}_\lambda$ and $|C_0| = |C_1|$, we have:*

$$\left| \Pr\left[0 \leftarrow \mathcal{A}\left(\overline{C_0}, C_0, C_1, \Pi_1, \ldots, \Pi_n\right) \right] - \Pr\left[0 \leftarrow \mathcal{A}\left(\overline{C_1}, C_0, C_1, \Pi_1, \ldots, \Pi_n\right) \right] \right|$$
$$\leq \epsilon'(\lambda),$$

where $\overline{C_b} \leftarrow \mathsf{CombObf}(1^\lambda, C_b, \Pi_1, \ldots, \Pi_n)$ *for* $b \in \{0, 1\}$.

Some remarks are in order.

Remark 3. We say that Π_{comb} is an IO combiner if it is a (ϵ', ϵ)-secure IO combiner, where, (c) $\epsilon' = \mathsf{negl}'$ and, (d) $\epsilon = \mathsf{negl}$ with negl and negl' being negligible functions.

Remark 4. We alternatively call the IO combiner defined in Definition 6 to be a 1-out-n IO combiner. In our construction we make use of 1-out-2 IO combiner. This can be instantiated using a folklore "onion combiner" in which to obfuscate any given circuit one uses both the obfuscation algorithms to obfuscate the circuit one after the other in a nested fashion.

Remark 5. We also define robust combiner, where the syntax is the same as above except that security and correctness properties hold even if there is only one input candidate that is secure and correct. No restriction about correctness and security is placed on other candidates.

As seen in [1], a robust combiner for arbitrary many candidates imply universal obfuscation as defined below.

Definition 7 $((T, \epsilon)$-Universal Obfuscation). *We say that a pair of Turing machines $\Pi_{univ} = (\Pi_{univ}.\text{Obf}, \Pi_{univ}.\text{Eval})$ is a **universal obfuscation**, parameterized by T and ϵ, if there exists a correct ϵ-secure indistinguishability obfuscator for $P/poly$ with time function T then Π_{univ} is an indistinguishability obfuscator for $P/poly$ with time function $\text{poly}(T)$.*

4.2   Definition of 2-ary Functional Encryption Candidate

We now define 2-ary (public-key) functional encryption candidates, also referred to as MIFE *candidates*). We start by providing the syntax of a MIFE scheme.

Syntax of 2-ary Functional Encryption Scheme. A MIFE scheme associated to a class of circuits $\mathcal{C} = \{\mathcal{C}_\lambda\}_{\lambda \in \mathbb{N}}$ consists of four polynomial time algorithms (Setup, Enc, KeyGen, Dec) defined below. Let $\mathcal{X}_\lambda$ be the message space of the scheme and $\mathcal{Y}_\lambda$ be the space of outputs for the scheme (same as the output space of $\mathcal{C}_\lambda$).

- **Setup,** $(\text{EK}_1, \text{EK}_2, \text{MSK}) \leftarrow \text{Setup}(1^\lambda)$: It is a randomized algorithm takes as input security parameter λ and outputs a keys $(\text{EK}_1, \text{EK}_2, \text{MSK})$. Here EK_1 and EK_2 are encryption keys for indices 1 and 2 and MSK is the master secret key.
- **Encryption,** $\text{CT} \leftarrow \text{Enc}(\text{EK}_i, m)$: It is a randomized algorithm takes the encryption key EK_i for any index $i \in [2]$ and a message $m \in \mathcal{X}_\lambda$ and outputs an encryption of m (encrypted under EK_i).
- **Key Generation,** $sk_C \leftarrow \text{KeyGen}(\text{MSK}, C)$: This is a randomized algorithm that takes as input the master secret key MSK and a 2-input circuit $C \in \mathcal{C}_\lambda$ and outputs a function key sk_C.
- **Decryption,** $y \leftarrow \text{Dec}(sk_C, \text{CT}_1, \text{CT}_2)$: This is a deterministic algorithm that takes as input the function secret key sk_C and a ciphertexts CT_1 and CT_2 (encrypted under EK_1 and EK_2 respectively). Then it outputs a value $y \in \mathcal{Y}_\lambda$.

Throughout this work, we will only be concerned with *uniform* algorithms. That is, (Setup, Enc, KeyGen, Dec) are represented as Turing machines (or equivalently uniform circuits).

We define the notion of an MIFE candidate below. The following definition of multi-input functional encryption scheme incorporates only the correctness and compactness properties of a multi-input functional encryption scheme [31]. In particular, an MIFE candidate need not necessarily have any security property associated with it. Formally,

Definition 8 (Correct MIFE candidate). *A multi-input functional encryption scheme* MIFE = (Setup, Enc, KeyGen, Dec) *is a correct MIFE candidate for a class of circuits* $C = \{C_\lambda\}_{\lambda \in \mathbb{N}}$, *with every* $C \in C_\lambda$ *has size* $\mathrm{poly}(\lambda)$, *if it satisfies the following properties:*

- **Correctness:** *For every* $C : X_\lambda \times X_\lambda \to \{0,1\} \in C_\lambda, m_1, m_2 \in X_\lambda$ *it holds that:*

$$
\Pr \left[
\begin{array}{c}
(\mathsf{EK}_1, \mathsf{EK}_2, \mathsf{MSK}) \leftarrow \mathsf{Setup}(1^\lambda) \\
\mathsf{CT}_i \leftarrow \mathsf{Enc}(\mathsf{EK}_i, m_i) \ i \in [2] \\
sk_C \leftarrow \mathsf{KeyGen}(\mathsf{MSK}, C) \\
C(m_1, m_2) \leftarrow \mathsf{Dec}(sk_C, \mathsf{CT}_1, \mathsf{CT}_2)
\end{array}
\right] \geq 1 - \mathsf{negl}(\lambda),
$$

  where negl *is a negligible function and the probability is taken over the coins of the setup only.*
- **Compactness:** *Let* $(\mathsf{EK}_1, \mathsf{EK}_2, \mathsf{MSK}) \leftarrow \mathsf{Setup}(1^\lambda)$, *for every* $m \in X_\lambda$ *and* $i \in [2]$, $\mathsf{CT} \leftarrow \mathsf{Enc}(\mathsf{EK}_i\, m)$. *We require that* $|\mathsf{CT}| < poly(|m|, \lambda)$.

A scheme is an MIFE candidate if it only satisfies the correctness and compactness property.

Selective Security. We recall indistinguishability-based selective security for MIFE. This security notion is modeled as a game between a challenger C and an adversary A where the adversary can request for functional keys and ciphertexts from C. Specifically, A can submit 2-ary function queries f and respond with the corresponding functional keys. It submits message queries of the form (m_1^0, m_2^0) and (m_1^1, m_2^1) and receive encryptions of messages m_i^b for $i \in [2]$, and for some random bit $b \in \{0,1\}$. The adversary A wins the game if she can guess b with probability significantly more than $1/2$ if the following properties are satisfied:

- $f(m_1^0, \cdot)$ is functionally equivalent to $f(m_1^1, \cdot)$.
- $f(\cdot, m_2^0)$ is functionally equivalent to $f(\cdot, m_2^1)$
- $f(m_1^0, m_2^0) = f(m_1^1, m_2^1)$

Formal definition is presented next.

$\epsilon - Secure\ MIFE\ candidate.$ If any MIFE candidate additionally satisfies the following (informal) security property then we define it to be a *secure* MIFE candidate:

Definition 9 (ϵ-Secure MIFE candidate). *A scheme* MIFE *for a class of circuits* $C = \{C_\lambda\}_{\lambda \in \mathbb{N}}$ *and message space* X_λ *is a ϵ-secure FE candidate if it satisfies the following conditions:*

- **MIFE** *is a correct and compact MIFE candidate with respect to* C,

– **Security.** *For every PPT adversary $\mathcal{A}$, for every sufficiently large $\lambda \in \mathbb{N}$, we have:*

$$\left| \Pr\left[0 \leftarrow \mathsf{Expt}_{\mathcal{A}}^{\mathsf{MIFE}}\left(1^{\lambda}, 0\right) \right] - \Pr\left[0 \leftarrow \mathsf{Expt}_{\mathcal{A}}^{\mathsf{MIFE}}\left(1^{\lambda}, 1\right) \right] \right| \leq \epsilon(\lambda)$$

where the probability is taken over coins of all algorithms. For each $b \in B$ and $\lambda \in \mathbb{N}$, the experiment $\mathsf{Expt}_{\mathcal{A}}^{\mathsf{MIFE}}(1^{\lambda}, b)$ is defined below:

1. **Challenge message queries:** $\mathcal{A}$ *outputs* (m_1^0, m_2^0) *and* (m_1^0, m_2^0) *where each* $m_j^i \in \mathcal{X}_{\lambda}$

2. *The challenger computes* $\mathsf{Setup}(1^{\lambda}) \to (\mathsf{EK}_1, \mathsf{EK}_2, \mathsf{MSK})$. *It then computes* $\mathsf{CT}_1 \leftarrow \mathsf{Enc}(\mathsf{EK}_1, m_1^b)$ *and* $\mathsf{CT}_2 \leftarrow \mathsf{Enc}(\mathsf{EK}_1, m_2^b)$. *Challenger hands* $\mathsf{CT}_1, \mathsf{CT}_2$ *to the adversary.*

3. $\mathcal{A}$ *submits functions* f_i *to the challenger satisfying the constraint given below.*
   - $f_i(m_1^0, \cdot)$ *is functionally equivalent to* $f_i(m_1^1, \cdot)$.
   - $f_i(\cdot, m_2^0)$ *is functionally equivalent to* $f_i(\cdot, m_2^1)$
   - $f_i(m_1^0, m_2^0) = f_i(m_1^1, m_2^1)$
   For every i, the adversary gets $\mathsf{sk}_{f_i} \leftarrow \mathsf{KeyGen}(\mathsf{MSK}, f_i)$.

4. *Adversary submits the guess b'. The output of the game is b'.*

Remark 6. We say that MIFE is a secure MIFE candidate if it is a ϵ-secure FE candidate with $\epsilon(\lambda) = \mathsf{negl}(\lambda)$, for some negligible function negl.

5 Construction of IO Combiner

In this section we describe our construction for IO combiner. We first define an MPC framework that will be used in our construction.

5.1 MPC Framework

We consider an MPC framework in the pre-processing model described below. Intuitively the input is pre-processed and split amongst n deterministic parties which are also given some correlated randomness. Then, they run a protocol together to compute $f(x)$ for any function f of the input x. The syntax consists of the following algorithms:

– $\mathsf{Preproc}(1^{\lambda}, n, x) \to (x_1, \mathsf{corr}_1, .., x_n, \mathsf{corr}_n)$: This algorithm takes as input $x \in \mathcal{X}_{\lambda}$, the number of parties computing the protocol n, and the security parameter λ. It outputs strings x_i, corr_i for $i \in [n]$. Each $\mathsf{corr}_i = \mathsf{corr}_i(r)$ is represented both a function and a value depending on the context. $(x_1, .., x_n)$ forms a secret sharing of x.

– $\mathsf{Eval}(\mathsf{Party}_1(x_1, \mathsf{corr}_1), .., \mathsf{Party}_n(x_n, \mathsf{corr}_n), f) \to f(x)$: The evaluate algorithm is a protocol run by n parties with Party_i having input x_i, corr_i. Each Party_i is deterministic. The algorithm also takes as input the function $f \in \mathcal{C}_{\lambda}$ of size bounded by $\mathsf{poly}(\lambda)$ and it outputs $f(x)$.

We now list the notations used for the protocol.

1. The number of rounds in the protocol is given by a polynomial $t_f(\lambda, n, |x|)$.
2. For every $i \in [n]$, $\mathsf{corr}_i = \{\mathsf{corr}_{i,j}\}_{j \neq i}$. Let $len_f = len_f(\lambda, n)$ denote a polynomial. Then, for each $i, j \in [n]$ such that $i \neq j$, $\mathsf{corr}_{i,j}$ and $\mathsf{corr}_{j,i}$ are generated as follows. Sample $r_{i,j} \xleftarrow{\$} \{0,1\}^{len_f}$ then compute $\mathsf{corr}_{i,j} = \mathsf{corr}_{i,j}(r_{i,j})$ and $\mathsf{corr}_{j,i} = \mathsf{corr}_{j,i}(r_{i,j})$.
3. There exists an efficiently computable function ϕ_f that takes as input a round number $k \in [t_f]$ and outputs $\phi_f(k) = (i, j)$. Here, (i, j) represents that the sender of the message at k^{th} round is Party_i and the recipient is Party_j.
4. The efficiently computable next message function for every round $k \in [t_f]$, M_k does the following. Let $\phi_f(k) = (i, j)$. Then, M_k takes as input $(x_i, y_1, .., y_{k-1}, \mathsf{corr}_{i,j})$ and outputs the next message as y_k.

Correctness : We require the following correctness property to be satisfied by the protocol. For every $n, \lambda \in \mathbb{N}$, $x \in \mathcal{X}_\lambda$, $f \in \mathcal{C}_\lambda$ it holds that:

$$Pr\left[\begin{array}{c} (x_1, \mathsf{corr}_1, .., x_n, \mathsf{corr}_n) \leftarrow \mathsf{Preproc}(1^\lambda, n, x) \\ \mathsf{Eval}(\mathsf{Party}_1(x_1, \mathsf{corr}_1), ..., \mathsf{Party}_n(x_n, \mathsf{corr}_n), f) \rightarrow f(x) \end{array} \right] = 1,$$

Here the probability is taken over coins of the algorithm $\mathsf{Preproc}$.

Security Requirement. We require the security against static corruption of $n - 1$ semi-honest parties. Informally the security requirement is the following. There exists a polynomial time algorithm that takes as input $f(x)$ and inputs of $n - 1$ corrupt parties $\{(\mathsf{corr}_i, x_i)\}_{i \neq i^*}$ and simulates the outgoing messages of Party_{i^*}. Formally, consider a PPT adversary $\mathcal{A}$. Let the associated PPT simulator be *Sim*. We define the security experiment below.

$\underline{\mathsf{Expt}_{real, \mathcal{A}}(1^\lambda)}$

- $\mathcal{A}$ on input 1^λ outputs n, the circuit f and input x along with the index of the honest party, $i^* \in [n]$.
- Secret share x into $(x_1, .., x_n)$.
- Part of the pre-processing step is performed by the adversary. For every $i > j$ such that $i^* \neq i$ and $i^* \neq j$, $\mathcal{A}$ samples $r_{i,j} \xleftarrow{\$} \{0,1\}^{len_f}$. Then, it computes, $\mathsf{corr}_{i,j} = \mathsf{corr}_{i,j}(r_{i,j})$ and $\mathsf{corr}_{j,i} = \mathsf{corr}_{j,i}(r_{i,j})$.
- Sample $r_j \xleftarrow{\$} \{0,1\}^{len_f}$ for $j \neq i^*$. Then compute $\mathsf{corr}_{i^*,j} = \mathsf{corr}_{i^*,j}(r_j)$ and $\mathsf{corr}_{j,i^*} = \mathsf{corr}_{j,i^*}(r_j)$. We denote $\mathsf{corr}_i = \{\mathsf{corr}_{i,j}\}_{j \neq i}$. This completes the pre-processing step.
- Let $y_1, .., y_{t_f}$ be the messages computed by the parties in the protocol computing $f(x)$. Output $(\{x_i, \mathsf{corr}_i\}_{i \neq i^*}, y_1, .., y_{t_f})$. In MPC literature $(\{x_i, \mathsf{corr}_i\}_{i \neq i^*}, y_1, .., y_{t_f})$ is referred to the view of the adversary in this experiment. We refer this as $\mathsf{view}_{\mathsf{Expt}_{real, \mathcal{A}}}$.

$\underline{\mathsf{Expt}_{ideal, \mathcal{A}}(1^\lambda)}$

- $\mathcal{A}$ on input 1^λ outputs n, the circuit f and input x along with the index of the honest party, $i^* \in [n]$.

- Secret share x into $(x_1, .., x_n)$.
- Part of the pre-processing step is performed by the adversary. For every $i > j$ such that $i^* \neq i$ and $i^* \neq j$, $\mathcal{A}$ samples $r_{i,j} \xleftarrow{\$} \{0,1\}^{len_f}$. Then, it computes, $\mathsf{corr}_{i,j} = \mathsf{corr}_{i,j}(r_{i,j})$ and $\mathsf{corr}_{j,i} = \mathsf{corr}_{j,i}(r_{i,j})$.
- Sample $r_j \xleftarrow{\$} \{0,1\}^{len_f}$ for $j \neq i^*$. Then compute $\mathsf{corr}_{i^*,j} = \mathsf{corr}_{i^*,j}(r_j)$ and $\mathsf{corr}_{j,i^*} = \mathsf{corr}_{j,i^*}(r_j)$. This completes the pre-processing step.
- Compute $Sim(1^\lambda, 1^{|f|}, f(x), \{x_i, \mathsf{corr}_i\}_{i \neq i^*})$. Output the result. We refer this as $\mathsf{view}_{\mathsf{Expt}_{ideal,\mathcal{A}}}$.

We require that the output of both the above experiments is computationally indistinguishable from each other. That is,

Definition 10 (Security). *Consider a PPT adversary $\mathcal{A}$ and let the associated PPT simulator be Sim. For every PPT distinguisher $\mathcal{D}$, for sufficiently large security parameter λ, it holds that:*

$$\left| \Pr\left[1 \leftarrow \mathcal{D}\left(\mathsf{Expt}_{real,\mathcal{A}}(1^\lambda) \right) \right] - \Pr\left[1 \leftarrow \mathcal{D}(\mathsf{Expt}_{ideal,\mathcal{A}}(1^\lambda)) \right] \right| \leq \mathsf{negl}(\lambda),$$

where negl *is some negligible function.*

Instantiation of MPC Framework: We show how to instantiate this MPC framework. We use a 1-out-of-n (i.e., $n-1$ of them are insecure) information theoretically secure MPC protocol secure against passive adversaries [30,38] in the OT hybrid model. We then replace the OT oracle by preprocessing all the OTs [5] before the execution of the protocol begins. Note that every OT pair is associated exactly with a pair of parties.

5.2 Construction Roadmap

In this section, we describe the roadmap of our construction. We start with n IO candidates, $\Pi_1, .., \Pi_n$ and construct $n^2 - n$ IO candidates $\Pi_{i,j}$ where $i \neq j$. $\Pi_{i,j}$ is constructed by using an onion obfuscation combiner (one in which each obfuscation candidate is run sequentially on the circuit). Each candidate $\Pi_{i,j}$ is now used to construct a 2-ary public-key multi-input functional encryption scheme $\mathsf{FE}_{i,j}$ candidates using [9,31] (this step uses the existence of one-way function). This is because [31] uses an existence of a public-key encryption, statistically binding commitments and statistically sound non-interactive witness-indistinguishable proofs. All these primitives can be constructed using IO and one-way functions as shown in works such as [9,46]. These primitives maintain binding/soundness as long as the underlying candidate is correct.

Any candidate $\mathsf{FE}_{i,j}$ is secure as long as either Π_i or Π_j is secure. This follows from the security of onion obfuscation combiner. We describe below how to construct a compact functional encryption FE from these multi-input functional encryption candidates and MPC framework in Sect. 5.3. Finally, using [2,11] and relying on complexity leveraging we construct a secure IO candidate Π_{comb} from FE. Below is a flowchart describing the roadmap.

5.3   Constructing Compact FE from $n^2 - n$ FE Candidates

Consider the circuit class $\mathcal{C}$. We now present our construction for a compact functional encryption scheme FE for $\mathcal{C}$ starting from compact multi-input functional encryption candidates $\mathsf{FE}_{i,j}$ for $\mathcal{C}$. Let Γ be a secure MPC protocol described in Sect. 5.1. Let λ be the security parameter and F denote a pseudorandom function (PRF) where $F : \{0,1\}^{\lambda} \times \{0,1\}^{*} \to \{0,1\}^{len(\lambda)}$ where len is some large enough polynomial. Finally let Com be a statistically binding commitment scheme.

$\mathsf{FE.Setup}(1^{\lambda})$ Informally, the setup algorithm samples encryption and master secret keys for candidates $\mathsf{FE}_{i,j}$ such that $i \neq j$ and $i,j \in [n]$. These candidates act as a channel between candidate i and j. It also samples $\mathsf{NIWI}_{i,j}$ prover strings for these candidates to prove consistency of the messages computed during the protocol.

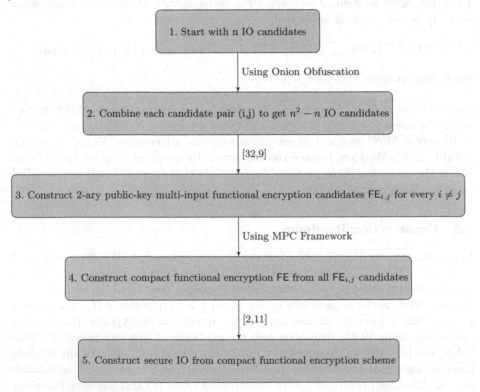

1. **Setting up MIFE candidates:**
   - For every $i,j \in [n]$ and $i \neq j$ run $\mathsf{FE}_{i,j}.\mathsf{Setup}(1^{\lambda}) \to (\mathsf{EK}_{i,j,1}, \mathsf{EK}_{i,j,2}, \mathsf{MSK}_{i,j})$
2. **Sample NIWI prover strings**
   - Run $\mathsf{NIWI}_{i,j}.\mathsf{Setup} \to \sigma_{i,j}$ for $i,j \in [n]$ and $i \neq j$. Recall, $\mathsf{NIWI}_{i,j}$ is a non-interactive statistically sound witness-indistinguishable proof scheme (in the CRS model) constructed using IO candidate $\Pi_{i,j}$ and any one-way

function as done in [9][8]. This proof scheme remains sound if the underlying obfuscation candidate is correct/almost correct. The proof retains witness indistinguishability if the candidate is additionally secure.

- Output $\mathsf{MPK} = \{\mathsf{EK}_{i,j,1}, \mathsf{EK}_{i,j,2}, \sigma_{i,j}\}_{i,j \in [n], i \neq j}$ and $\mathsf{MSK} = \{\mathsf{MSK}\}_{i,j \in [n], i \neq j}$.

$\mathsf{FE.Enc}(\mathsf{MPK}, m)$. Informally, the encryption algorithm takes the message m and runs preprocessing to get $(m_1, \mathsf{corr}'_1, ..., m_n, \mathsf{corr}'_n)$. It discards corr'_i (which is allowed by our MPC framework). Then it samples PRF keys $K_{i,j}$ for $i \neq j$, which are used to generate randomness for next message function (via computing corr_i for every decryption). It also commits these message shares m_i and PRF keys, which are used to compute proofs about messages of the MPC protocol. Finally, these shares and PRF keys are encrypted using an appropriate FE candidate.

1. **MPC Preprocessing**
   - Run $\mathsf{Preproc}(1^\lambda, n, m) \to (m_1, \mathsf{corr}'_1, .., m_n, \mathsf{corr}'_n)$. Compute commitments $Z_{\mathsf{in},i} = \mathsf{Com}(m_i)$ for all $i \in [n]$. Let $r_{\mathsf{in},i}$ be the corresponding randomness.
2. **Sample and commit PRF keys**
   - Sample PRF keys $K_{i,j}$ for $i, j \in [n]$ and $i \neq j$ with the constraint that $K_{i,j} = K_{j,i}$. Compute $Z_{i,j} = \mathsf{Com}(K_{i,j})$ for $i, j \in [n]$ and $i \neq j$. Let $r_{i,j}$ be the corresponding randomness.
   - Sample PRF keys $K'_{i,j}$ for $i, j \in [n]$ such that $i \neq j$.
3. **Compute encryptions**
   - For every $i, j \in [n]$ and $i \neq j$ compute $\mathsf{CT}_{i,j} = \mathsf{FE}_{i,j}.\mathsf{Enc}(\mathsf{EK}_{i,j,1}, m_i, K_{i,j}, K'_{i,j}, \{Z_{\mathsf{in},k}, Z_{k,j}\}_{k,j \in [n], k \neq j}, r_{i,j}, r_{\mathsf{in},i}, \bot)$. Here $\bot$ is a slot of size $\mathsf{poly}(\lambda)$, which is described later.
   - Output $\mathsf{CT} = \{\mathsf{CT}_{i,j}\}_{i \neq j}$

$\mathsf{FE.KeyGen}(\mathsf{MSK}, C)$. Let t_C denote the number of rounds for the MPC protocol Γ for computing the circuit C. Let len_{msg} denote the maximum length of any message sent in the protocol while computing C on input. Informally, this algorithm generates FE keys for the circuits implementing next message function (used to compute $C(m)$) for every round $k \in [t_C]$.

1. **Computing commitments**
   - Compute $Z_{\mathsf{out},i} \leftarrow \mathsf{Com}(\bot^{len_{msg}})$ for $i \in [t_C]$.
2. **Compute secret-key encryptions**
   - Let E by a secret-key encryption scheme. Run $\mathsf{E.Setup}(1^\lambda) \to sk$. For every $i \in [t_C]$, compute $c_i = \mathsf{E.Enc}(sk, 0)$. These encryptions encrypt messages of sufficient length (described later).
3. **Generate keys**
   - Sample a random $\mathsf{tag} \in \{0, 1\}^\lambda$.
   - For every round $k \in [t_C]$, let $\phi(k) = (i', j')$, generate a key $sk_{C,k} \leftarrow \mathsf{FE}_{i',j'}.\mathsf{KeyGen}(\mathsf{MSK}_{i',j'}, G_k)$ where G_k is described in Fig. 1. Output $\{sk_{C,k}\}_{k \in [t_C]}$.

---

[8] We note that we could have also used NIZKs with pre-processing based on one-way functions. The construction becomes a little complicated with that.

$$G_k\left[\{\sigma_{i,j}\}_{i,j\in[n].i\neq j}, c_k, \{Z_{\mathsf{out},i}\}_{i\in[t_C]}, \mathsf{tag}\right]$$

Hardwired values: NIWI verifier strings $\{\sigma_i\}_{i\in[n]}$, Encryption c_k, Commitments $\{Z_{\mathsf{out},i}\}_{i\in[t_C]}$, a random tag

Inputs 1: $m_{i'}, K_{i',j'}, K'_{i',j'}, \{Z_{\mathsf{in},i}, Z_{i,j}\}_{i,j\in[n],i\neq j}, r_{i',j'}, r_{\mathsf{in},i'}, s$

Inputs 2: $x_1, \pi_1, ..., x_{k-1}, \pi_{k-1}$

- Initialise counter $= 0$. For every counter $\leq k-1$ do the following:
- **Check to see if** $(x_1, .., x_{k-1})$ **are according to the MPC protocol**
    - $\phi(\mathsf{counter}) = (i, j)$. Check that $\mathsf{NIWI}_{i,j}.\mathsf{Verify}(\sigma_{i,j}, x_1, .., x_{\mathsf{counter}}, \pi_{\mathsf{counter}}) = 1$ for the language L_i described below.
    - Output $\perp$ if any check fails.
- **Trapdoor condition, output hardwiring** . If $s \neq \perp$ output $\mathsf{E.Dec}(s, c_k)$.
- **Compute the next message and the proof of consistency**
    - Let $\phi(k) = (i', j')$. Compute $\mathsf{corr}_{i',j'} = \mathsf{corr}_{i',j'}(F(K_{i',j'}, \mathsf{tag}))$. Compute $x_k = M_k(m_{i'}, x_1, .., x_{k-1}, \mathsf{corr}_{i',j'})$.
    - Compute a $\mathsf{NIWI}_{i',j'}$ proof π_k using the reference string $\sigma_{i',j'}$ and randomness $F(K'_{i',j'}, x_1, .., x_{k-1}, x_k)$ that the statement $(x_1, .., x_k) \in L_{i'}$ (language defined below). This proof is computed using the witness as $m_{i'}, K_{i',j'}, r_{i',j'}, r_{\mathsf{in},i'}, 0$ where $\phi(k) = (i', j')$.
- Output (x_k, π_k)

Language L_i: An NP language L_i is defined by the relation as defined below:

Instance: $x_1, .., x_k$ where $k \in [t_C]$

Witness: $m_i, K_{i,j}, r_{i,j}, r_{\mathsf{in},i}, r_k$

Hardwired: $\{Z_{i,j}\}_{j\neq i}, \mathsf{tag}, Z_{\mathsf{in},i}, Z_{\mathsf{out},k}$

- Check $\phi(k) = (i, j)$ for some j. If not, output 0.
- Let $\mathsf{corr}_{i,j} = \mathsf{corr}_{i,j}(F(K_{i,j}, \mathsf{tag}))$. Check that either, $x_k = M_k(m_i, x_1, .., x_{k-1}, \mathsf{corr}_{i,j})$, $Z_{i,j} = \mathsf{Com}(K_{i,j}; r_{i,j})$ and $Z_{\mathsf{in},i} = \mathsf{Com}(m_i; r_{\mathsf{in},i})$. Or,
- $Z_{\mathsf{out},k} = \mathsf{Com}(y_k; r_k)$, where $y_k \neq \perp$ and $y_k = x_k$. If the check passes output 1 otherwise 0.

Fig. 1. Circuit G_k

$\mathsf{FE.Dec}(sk_C, \mathsf{CT})$w

1. **Evaluating the MPC protocol for circuit C**
    - Let $\phi(1) = (i_1, j_1)$. Compute $\mathsf{CT}_1 = \mathsf{FE}_{i_1,j_1}.\mathsf{Enc}(\mathsf{EK}_{i_1,j_1,2}, \perp, \perp)$. Set $(x_1, \pi_1) = \mathsf{FE}_{i,j}.\mathsf{Dec}(sk_{C,1}, \mathsf{CT}_{i_1,j_1}, \mathsf{CT}_1)$.
    - For every round $k \in [t_C]$, compute x_k, π_k iteratively from $x_1, \pi_1, .., x_{k-1}, \pi_{k-1}$ as described below.
        a Compute $\phi(k) = (i, j)$. Then, compute $\mathsf{CT}_k = \mathsf{FE}_{i,j}.\mathsf{Enc}(\mathsf{EK}_{i,j,2}, x_1, \pi_1, .., x_{k-1}, \pi_{k-1})$.
        b Run $(x_k, \pi_k) \leftarrow \mathsf{FE}_{i,j}.\mathsf{Dec}(sk_{C,k}, \mathsf{CT}_{i,j}, \mathsf{CT}_k)$
    - Output x_{t_C}

Correctness: If the underlying MPC protocol is correct and the multi-input functional encryption candidates $FE_{i,j}$ are correct then one can inspect that our scheme satisfies correctness.

Compactness. Compactness is discussed next. The cipher-text encrypting any message m, consists of $FE_{i,j}$ encryptions $CT_{i,j}$ for any $i, j \in [n]$ such that $i \neq j$. Each $CT_{i,j}$ encrypts $m_i, K_{i,j}, K'_{i,j}, \{Z_{in,k}, Z_{k,j}\}_{k,j \in [n], k \neq j}, r_{i,j}, r_{in,i}, \bot$. Note that m_i is of the same length of the message where as $K_{i,j}, K'_{i,j}$ are just the PRF keys that are of length λ. $\{Z_{in,k}, Z_{k,j}\}_{k,j \in [n], k \neq j}$ are commitments of m_i and the PRF keys respectively while $r_{i,j}$ and $r_{in,i}$ is the randomness used for the commitments $Z_{i,j}$ and $Z_{in,i}$. $\bot$ is a slot of size $\text{poly}(\lambda)$ (which is the length of the decryption key for scheme E). All these strings are of a fixed polynomial size (polynomial in $n, \lambda, |m|$). If the underlying scheme $FE_{i,j}$ is compact, the scheme FE is also compact. We give a brief sketch of proof here. We refer the reader to our full version for a detailed proof.

Theorem 4. *Consider the circuit class $\mathcal{C} = P/poly$. Assuming Γ is a secure MPC protocol for $\mathcal{C}$ according to the framework described in Sect. 5.1 and one-way functions exist, then scheme FE is a secure functional encryption scheme as long as there is $i^* \in [n]$ such that Π_{i^*} is a secure candidate.*

Proof (Sketch). We now sketch the security proof of this theorem. Assume Π_{i^*} is a secure IO candidate. This implies $FE_{i^*,j}$ and FE_{j,i^*} is secure for any $j \neq i^*$. We use this crucially in our proofs. We employ the standard hybrid argument to prove the theorem. In the first hybrid (Hyb_1), the message M_b is encrypted honestly with $b \xleftarrow{\$} \{0,1\}$. In the final hybrid ($\text{Hyb}_9$), the ciphertext contains no information about b. At this point, the probability of guessing the bit b is exactly $1/2$. By arguing indistinguishability of every consecutive intermediate hybrids, we show that the probability of guessing b in the first hybrid is negligibly close to $1/2$ (or the advantage is 0), which proves the theorem.

The first hybrid corresponds to the regular FE security game. Then we switch to a hybrid where the secret-key encryption cipher-text c_i in the function keys for all rounds $i \in [t_C]$ are hard-wired as encryptions of the output of the MIFE decryption in those rounds. This can be done, because the cipher-text and the function key fixes these outputs (as a function of PRF keys, e.t.c). Then, we change the commitments $Z_{out,k}$ to commitments of message output in round k (for k such that the i^* is the receiving or sending party in that round). This security holds due to the security of the commitment. In the next hybrid, we rely on the security of the scheme $FE_{i^*,j}$ and FE_{j,i^*} by generating encryptions that does not contain the PRF keys and the openings of the commitments but only contain the secret key for the encryption scheme E. Now we invoke the security of the PRF to generate proofs π_k hard-wired in c_k for any round k (for k such that the i^* is the receiving or sending party in that round) randomly. Next, we rely on the security of $NIWI_{i^*,j}$ and $NIWI_{j,i^*}$ to use the opening of $Z_{out,k}$ (for k such that the i^* is the receiving or sending party in that round) to generate the proofs. Now relying on the security of commitment scheme, we make the commitments $Z_{i*,j}, Z_{j,i*}$ and $Z_{in,i*}$ to commit to $\perp$. Then we use the security of the PRF to generate $\mathsf{corr}_{i*,j}$ and $\mathsf{corr}_{j,i*}$ (used for generating outputs x_n) randomly. Finally, we invoke the security of the MPC framework (by using the simulator) to make the game independent of b.

5.4   Summing Up: Combiner Construction

We now give the combiner construction:

- $\mathsf{CombObf}(1^\lambda, C, \Pi_1, .., \Pi_n)$: Use $\Pi_1, .., \Pi_n$ and any one-way function to construct a compact functional encryption FE as in Sect. 5.2. Use [2,11] to construct an obfuscator Π_{comb}. Output $\overline{C} \leftarrow \Pi_{comb}(1^\lambda, C)$.
- $\mathsf{CombEval}(\overline{C}, x)$: Output $\Pi_{comb}.\mathsf{Eval}(\overline{C}, x)$.

Correctness of the scheme is straight-forward to see because of the correctness of FE as shown in Sect. 5.2. The security follows from the sub-exponential security of construction in Sect. 5.2. The construction in Sect. 5.2 is sub-exponentially secure as long as the underlying primitives are sub-exponentially secure.

We now state the theorem.

Theorem 5. *Assuming sub-exponentially secure one-way functions, the construction described above is a* $(\mathsf{negl}, 2^{-\lambda^c})-secure$ *IO combiner for P/poly where* $c > 0$ *is a constant and* negl *is some negligible function.*

6   From Combiner to Robust Combiner

The combiner described in Sect. 5.2, is not robust. It guarantees no security/correctness if the underlying candidates are not correct. A robust combiner provides security/correctness as long as there exists one candidate Π_{i*} such that it is secure and correct. There is no other restriction placed on the other set of candidates. A robust combiner for arbitrary many candidates imply universal obfuscation [1].

In this section we describe how to construct a robust combiner. The idea is the following.

- We correct the candidates (upto overwhelming probability) before feeding it as input to the combiner.
- First, we leverage the fact that secure candidate is correct. We transform each candidate so that all candidates are $(1 - 1/\lambda)-$worst case correct while maintaining security of the secure candidate.
- Then using [12] we convert a worst-case correct candidate to an almost correct candidate.

In the discussion below, we assume $\mathcal{C}$ consists polynomial size circuits with one bit output. One can construct obfuscator for circuits with multiple output bits from obfuscator with one output bit. For simplicity let us assume that $\mathcal{C}_\lambda$ consists of circuits with input length $p(\lambda)$ for some polynomial p.

6.1 Generalised Secure Function Evaluation

The starting point to get a worst-case correct IO candidate is a variant of "Secure Function Evaluation" (SFE) scheme as considered in [12]. They use SFE to achieve worst-case correctness by obfuscating evaluation function of SFE for the desired circuit C. To evaluate on input x, the evaluator first encodes x according to the SFE scheme and feeds it as an input to the obfuscated program. Then, it finally decodes the result as the output of the obfuscated program. Worst case correctness is guaranteed because using the information hard-wired in the obfuscated program its hard to distinguish an encoding of any input x_1 from that of x_2.

We essentially use the same idea except that we consider a variant of SFE with a setup algorithm (which produces secret parameters), and the evaluation function for the circuit C is not public. It requires some helper information to perform evaluation on the input encodings.

We consider a generalised variant of secure function evaluation [12] with the following properties. Let $\mathcal{C}_\lambda$ be the allowed set of circuits. Let $\mathcal{X}_\lambda$ and $\mathcal{Y}_\lambda$ denote the ensemble of inputs and outputs. A secure function evaluation scheme consists of the following algorithms:

– Setup(1^λ): On Input 1^λ, the setup algorithm outputs secret parameters SP.
– CEncode(SP, C): The randomized circuit encoding algorithm on input a circuit $C \in \mathcal{C}_\lambda$ and SP outputs another $\tilde{C} \in \mathcal{C}_\lambda$.
– InpEncode(SP, x): The randomized input encoding algorithm on input $x \in \mathcal{X}_\lambda$ and SP outputs $(\tilde{x}, z) \in \mathcal{X}_\lambda \times \mathcal{Z}_\lambda$.
– Decode(y, z): Let $y = \tilde{C}(\tilde{x})$. The deterministic decoding algorithm takes as input y and z to recover $C(x) \in \mathcal{Y}_\lambda$.

We require the following properties:

Input Secrecy: For any $x_1, x_2 \in \mathcal{X}_\lambda$, any circuit $C \in \mathcal{C}_\lambda$ and SP $\leftarrow$ Setup(1^λ), it holds that:

$$\{\mathsf{CEncode}(\mathsf{SP}, C), \mathsf{InpEncode}(\mathsf{SP}, x_1)\} \approx_c$$
$$\{\mathsf{CEncode}(\mathsf{SP}, C), \mathsf{InpEncode}(\mathsf{SP}, x_2)\}$$

Correctness: For any circuit $C \in \mathcal{C}_\lambda$ and any input x, it holds that:

$$Pr[\mathsf{Decode}(\tilde{C}(\tilde{x}), z) = C(x)] = 1$$

where SP $\leftarrow$ Setup(1^λ), $\tilde{C} \leftarrow$ CEncode(SP, C), $(\tilde{x}, z) \leftarrow$ InpEncode(SP, x) and the probability is taken over coins of all the algorithms.

Functionality: For any equivalent circuits C_0, C_1, SP $\leftarrow$ Setup(1^λ), $\tilde{C}_0 \leftarrow$ CEncode(SP, C_0) and $\tilde{C}_1 \leftarrow$ CEncode(SP, C_1), it holds that $\tilde{C}_0$ is equivalent to $\tilde{C}_1$ with overwhelming probability over the coins of setup and the circuit encoding algorithm. This captures the behaviour of the circuit encodings when evaluated on maliciously generated input encodings.

6.2   Modified Obfuscation Candidate

In this section we achieve the following. Given any candidate Π, we transform it to a candidate Π' such that the following holds:

- If Π is both secure and correct, then so is Π'.
- Otherwise Π' is guaranteed to be $(1 - 1/\lambda)-$worst-case correct.

In either case, we can amplify its correctness to get an almost correct IO candidate, which can be used by our combiner construction. Given any IO candidate Π we now describe a modified IO candidate Π'. For simplicity let us assume that $\mathcal{C}_\lambda$ consists of circuits with one bit output and input space $\mathcal{X}_\lambda$ corresponds to the set $\{0, 1\}^{p(\lambda)}$ for some polynomial p. Let SFE be a secure function evaluation scheme as described in Sect. 6.1 for $\mathcal{C}_\lambda$ with $\mathcal{Z}_\lambda = \mathcal{Y}_\lambda = \{0, 1\}$.

- **Obfuscate:** On input the security parameter 1^λ and $C \in \mathcal{C}_\lambda$, first run SP $\leftarrow$ SFE.Setup(1^λ), compute $\tilde{C} \leftarrow$ CEncode(SP, C). We now define an algorithm Obf$_{int,\Pi}$ that takes as input $\tilde{C}$ and 1^λ and does the following:
    - Compute $\overline{C} \leftarrow \Pi.\mathsf{Obf}(1^\lambda, \tilde{C})$.
    - Then sample randomly $x_1, .., x_{\lambda^2} \in \{0, 1\}^{p(\lambda)}$. Compute $(\tilde{x}_i, z_i) \leftarrow$ InpEncode(SP, x_i). Check that $\Pi.\mathsf{Eval}(\overline{C}, \tilde{x}_i) = \tilde{C}(\tilde{x}_i)$ for all $i \in [\lambda^2]$.
    - If the check passes output $\overline{C}$, otherwise output $\tilde{C}$ [9].
  Output of the obfuscate algorithm is $(\mathsf{SP}, \mathsf{Obf}_{int,\Pi}(\tilde{C}))$.
- **Evaluate:** On input $(\mathsf{SP}, \overline{C})$ and an input x, first compute $(\tilde{x}, z) \leftarrow$ InpEncode(SP, x). Then compute $\tilde{y} \leftarrow \Pi.\mathsf{Eval}(\overline{C}, \tilde{x})$ or $\tilde{y} \leftarrow \overline{C}(\tilde{x})$ depending on the case if $\overline{C} = \tilde{C}$ or not. We define as an intermediate evaluate algorithm, i.e. $\tilde{y} = \mathsf{Eval}_{int,\Pi}(\overline{C}, \tilde{x})$.
  Output $y = \mathsf{SFE.Decode}(\tilde{y}, z)$.

Few claims are in order:

Theorem 6. *Assuming* SFE *is a secure function evaluation scheme as described in Sect. 6.1, if Π is a secure and correct candidate, then so is, Π'.*

Proof. We deal with this one by one. First we argue security. Note that when Π is correct, the check at λ^2 random points passes. In this case the obfuscation algorithm always outputs $(\mathsf{SP}, \Pi.\mathsf{Obf}(1^\lambda, \tilde{C}_b))$ where $\tilde{C}_b \leftarrow \mathsf{SFE.CEncode}(\mathsf{SP}, C_b)$ for $b \in \{0, 1\}$ and SP $\leftarrow$ SFE.Setup(1^λ). Since, $\tilde{C}_0$ is equivalent to $\tilde{C}_1$ due to functionality property of the SFE scheme, the security holds due to the security of Π.

The correctness holds due to the correctness of SFE and Π.

Theorem 7. *Assuming* SFE *is a secure function evaluation scheme as described in Sect. 6.1, if Π is an IO candidate, then Π' is $(1 - 2/\lambda)$-worst case correct IO candidate.*

---

[9] This step ensures circuit-specific correctness. Note that any correct candidate will always pass the step. Any candidate that is not correct with high enough probability will not pass the check. In this case, the algorithm outputs the circuit in the clear.

Proof. The check step in the obfuscate algorithm ensures the following: Using Chernoff bound it follows that, with overwhelming probability, for any circuit C,

$$Pr[\Pi'.\mathsf{Eval}(\mathsf{SP}, \overline{C}, x) = C(x) | (\mathsf{SP}, \overline{C}) \leftarrow \Pi'.\mathsf{Obf}(1^\lambda, C), x \xleftarrow{\$} \mathcal{U}_{p(\lambda)}] \geq (1 - 1/\lambda) \tag{1}$$

We now prove that for any x_1, x_2 it holds that, with overwhelming probability over coins of obfuscate algorithm, for any circuit C,

$$\begin{aligned} |Pr[\Pi'.\mathsf{Eval}(\mathsf{SP}, \overline{C}, x_1) = C(x_1) | (\mathsf{SP}, \overline{C}) \leftarrow \Pi'.\mathsf{Obf}(1^\lambda, C)] - \\ Pr[\Pi'.\mathsf{Eval}(\mathsf{SP}, \overline{C}, x_2) = C(x_2) | (\mathsf{SP}, \overline{C}) \leftarrow \Pi'.\mathsf{Obf}(1^\lambda, C)]| \leq \mathsf{negl}(\lambda) \end{aligned} \tag{2}$$

This is because for any input x,

$$\begin{aligned} Pr[\Pi'.\mathsf{Eval}(\mathsf{SP}, \overline{C}, x) = C(x) | (\mathsf{SP}, \overline{C}) \leftarrow \Pi'.\mathsf{Obf}(1^\lambda, C)] = \\ Pr[\mathsf{Eval}_{int,\Pi}(\overline{C}, \tilde{x}) = \tilde{C}(\tilde{x}) | \mathsf{SP} \leftarrow \mathsf{Setup}(1^\lambda), \tilde{C} \leftarrow \mathsf{CEncode}(\mathsf{SP}, C), \\ (\tilde{x}, z) \leftarrow \mathsf{InpEncode}(\mathsf{SP}, x), \overline{C} \leftarrow \mathsf{Obf}_{int,\Pi}(1^\lambda, \tilde{C})] \end{aligned} \tag{3}$$

Note that due to the input secrecy property of the SFE scheme we have that,

$$\begin{aligned} |Pr[\mathsf{Eval}_{int,\Pi}(\overline{C}, \tilde{x}_1) = \tilde{C}(\tilde{x}_1) | \mathsf{SP} \leftarrow \mathsf{Setup}(1^\lambda), \tilde{C} \leftarrow \mathsf{CEncode}(\mathsf{SP}, C), \\ (\tilde{x}_1, z_1) \leftarrow \mathsf{InpEncode}(\mathsf{SP}, x_1), \overline{C} \leftarrow \mathsf{Obf}_{int,\Pi}(1^\lambda, \tilde{C})] - \\ Pr[\mathsf{Eval}_{int,\Pi}(\overline{C}, \tilde{x}_2) = \tilde{C}(\tilde{x}_2) | \mathsf{SP} \leftarrow \mathsf{Setup}(1^\lambda), \tilde{C} \leftarrow \mathsf{CEncode}(\mathsf{SP}, C), \\ (\tilde{x}_2, z_2) \leftarrow \mathsf{InpEncode}(\mathsf{SP}, x_2), \overline{C} \leftarrow \mathsf{Obf}_{int,\Pi}(1^\lambda, \tilde{C})]| < \mathsf{negl}(\lambda) \end{aligned} \tag{4}$$

for a negligible function negl. Otherwise we can build a reduction $\mathcal{R}$ that given any circuit-encoding, input encoding pair $\tilde{C}, \tilde{x}_b$ decides if $b = 0$ or $b = 1$ with a non-negligible probability. The reduction just computes $\overline{C} \leftarrow \mathsf{Obf}_{int,\Pi}(\tilde{C})$ and checks if $\mathsf{Eval}_{int,\Pi}(\overline{C}, \tilde{x}_b) = \tilde{C}(\tilde{x}_b)$.

Using the pigeon-hole principle and Eq. 1, for any $C \in \mathcal{C}_\lambda$ there exists x^* such that,

$$|Pr[\Pi'.\mathsf{Eval}(\mathsf{SP}, \overline{C}, x^*) = C(x^*) | (\mathsf{SP}, \overline{C}) \leftarrow \Pi'.\mathsf{Obf}(1^\lambda, C)] \geq (1 - 1/\lambda) \tag{5}$$

Now substituting $x_1 = x$ and $x_2 = x^*$ in Eq. 4 and then plugging into Eq. 3 gives us,

$$\begin{aligned} |Pr[\Pi'.\mathsf{Eval}(\mathsf{SP}, \overline{C}, x) = C(x) | (\mathsf{SP}, \overline{C}) \leftarrow \Pi'.\mathsf{Obf}(1^\lambda, C)] - \\ Pr[\Pi'.\mathsf{Eval}(\mathsf{SP}, \overline{C}, x^*) = C(x^*) | (\mathsf{SP}, \overline{C}) \leftarrow \Pi'.\mathsf{Obf}(1^\lambda, C)]| \leq \mathsf{negl}(\lambda) \end{aligned} \tag{6}$$

Substituting result of Eq. 5 gives us the desired result. That is, For any circuit C and input x, it holds that,

$$\begin{aligned} |Pr[\Pi'.\mathsf{Eval}(\mathsf{SP}, \overline{C}, x) = C(x) | (\mathsf{SP}, \overline{C}) \leftarrow \Pi'.\mathsf{Obf}(1^\lambda, C)] \geq (1 - 1/\lambda) - \mathsf{negl}(\lambda) \\ > (1 - 2/\lambda) \end{aligned} \tag{7}$$

This proves the result.

6.3   Instantiation of SFE

To instantiate SFE as described in Sect. 6.1, we use any single-key functional encryption scheme. To compute the circuit encoding for any circuit C we compute a function key for a circuit that has hard-wired a function key sk_{H_C} for a circuit H_C that takes as input (x, b) and outputs $C(x) \oplus b$. This circuit uses the hard-wired function key to decrypt the input. To encode the input x, we just compute an FE encryption of (x, b). But this does not suffice, because then for any equivalent circuits C_0 and C_1, the circuit encodings are not equivalent. Hence, we use a one-time zero-knowledge proof system to prove that the ciphertext is consistent. The details follow next. Our decoding/evaluation operation is not randomized in contrast to [12], hence this allows us to directly argue security with polynomial loss, instead of going input by input.

Theorem 8. *Assuming (non-compact) public-key functional encryption scheme for a single function key query exists, there exists an* SFE *according to the definition in Sect. 6.1.*

Proof. Let FE denote any public-key functional encryption scheme for a single function key. Let PZK denote a non-interactive zero knowledge proof system with pre-processing. We now describe the scheme.

- Setup(1^λ): The setup takes as input the security parameter 1^λ. It first runs (MPK, MSK) $\leftarrow$ FE.Setup(1^λ) and PZK.Pre(1^λ) $\rightarrow (\sigma_P, \sigma_V)$. Output SP = (MPK, MSK, σ_P, σ_V).
- CEncode(SP, C): The algorithm on input (SP, C) does the following.
    - Compute $sk_{H_C} \leftarrow$ FE.KeyGen(MSK, H_C). H_C represents a circuit that on input (x, b) outputs $C(x) \oplus b$.
    - Let H be the circuit described in Fig. 2. Output H.

$H[\sigma_V, \text{MPK}, sk_{H_C}]$

Hardwired values: NIZK Verifier sting σ_V, Functional encryption key MPK, Function secret key sk_{H_C}
Inputs: FE ciphertext CT, Proof π
- First run PZK.Verify(σ_V, CT, π). The verification is with respect to the NP language L defined below. If it fails output $\perp$.
- Otherwise, output FE.Dec(sk_{H_C}, CT)

Language L: An NP language L is defined by the relation as defined below:
Instance: A functional encryption key MPK and an FE ciphertext CT
Witness: x, b, r
- CT = FE.Enc(MPK, $x, b; r$) where $b \in \{0, 1\}$.

Fig. 2. Circuit H

- InpEncode(SP, x): On input an SP and an input x do the following:
  - Sample a random bit b and compute CT $=$ FE.Enc(MPK, $x, b; r_1$).
  - Compute the NIZK with pre-processing proof π proving that CT $\in L$ using the witness (x, b, r_1).
  - Output $((\mathsf{CT}, \pi), b) = (\tilde{x}, b)$
- Decode(y, b): Output $y \oplus b$.

We now discuss the properties:

Correctness. It is straightforward to see correctness as it follows from the completeness of the proof system and correctness of the functional encryption scheme.

Functionality: Let H_b denote an circuit encoding of circuit C_b for $b \in \{0, 1\}$ where C_0 and C_1 are equivalent circuits. The circuit takes as input (CT, π) where π is a proof that CT is an encryption of (x, b') for some x and a bit b'. Then it verifies the proof and decrypts the cipher-text using a function key that computes $C_b(x) \oplus b'$. Since, the proof system is statistically sound and FE scheme is correct this property is satisfied with overwhelming probability over the coins of the setup.

Input Secrecy: We want to show that for any circuit C and inputs x_0, x_1:

$$\{(H, \tilde{x}_0) | H \leftarrow \mathsf{CEncode}(\mathsf{SP}, C), \mathsf{SP} \leftarrow \mathsf{Setup}(1^\lambda), (\tilde{x}_0, z) \leftarrow \mathsf{InpEncode}(\mathsf{SP}, x_0)\} \approx_c$$

$$\{(H, \tilde{x}_1) | H \leftarrow \mathsf{CEncode}(\mathsf{SP}, C), \mathsf{SP} \leftarrow \mathsf{Setup}(1^\lambda), (\tilde{x}_1, z) \leftarrow \mathsf{InpEncode}(\mathsf{SP}, x_1)\}$$

We claim this in a number of hybrids. The first one corresponds to the actual game where $\tilde{x}_0$ is given while the last one corresponds to the case of $\tilde{x}_1$. We also show that the hybrids are indistinguishable.

Hyb_0: This hybrid corresponds to the following experiment for C, x_0. To run setup we run (MPK, MSK) $\leftarrow$ FE.Setup(1^λ). Then we sample a random bit b and compute CT $\leftarrow$ FE.Enc(MPK, x_0, b). We sample $(\sigma_P, \sigma_V) \leftarrow$ PZK.Pre(1^λ) and compute a proof π using PZK prover string σ_P and a witness of CT $\in L$. We output SP $=$ (MPK, MSK, σ_P, σ_V). This SP is used to encode C by computing a functional encryption key for circuit H_C first (sk_{H_C}). Call this circuit H (this circuit depends upon, σ_V and sk_{H_C}).

Hyb_1: This hybrid is the same as the previous one except that we generate π, σ_V differently. We run the simulator of PZK system and compute $(\sigma_V, \pi) \leftarrow Sim(\mathsf{CT})$. Hyb_0 is indistinguishable to Hyb_1 due to the zero-knowledge security of PZK proof.

Hyb_2: This hybrid is the same as the previous one except that we generate CT differently. CT $=$ FE.Enc(MPK, $x_1, b \oplus C(x_0) \oplus C(x_1)$). Hyb_1 is indistinguishable to Hyb_2 due to the security of FE proof.

Hyb$_3$: This hybrid is the same as the previous one except that (σ_P, σ_V) are generated honestly. Hyb$_2$ is indistinguishable to Hyb$_3$ due to the zero-knowledge security of PZK.

Hyb$_4$: This hybrid is the same as the previous one except that CT is generated as FE.Enc(MPK, x_1, b). this corresponds to the experiment for input x_1. Hyb$_3$ is identical to Hyb$_4$ as b is a random bit.

Corollary 2. *Assuming public-key encryption exists [45], there exists an SFE scheme satisfying requirements described in Sect. 6.1.*

Remark 7. We note that such a scheme can be instantiated from one-way functions alone. The idea is to use a secret-key functional encryption for single function query along with a statistically binding commitment scheme. The public parameters now include a commitment of a master secret key which is used to proof consistency of the cipher-text. Since the end result of constructing public key functional encryption from IO candidates itself imply PKE, we do not describe this construction.

6.4   Robust Combiner: Construction

We now describe our robust combiner. On input the candidates $\Pi_1, .., \Pi_n$, we transform them using the SFE scheme as done in Sect. 6.2 so that they are $(1 - 1/\lambda)$−worst-case correct. Then using majority trick as in [12], we convert them to almost correct. Plugging it to the construction in Sect. 5.2, gives us the desired result. Finally, we also state our theorem about universal obfuscation.

Theorem 9. *Assuming sub-exponentially secure one-way functions, there exists a* (poly, ϵ)-*Universal Obfuscation with* $\epsilon = O(2^{-\lambda^c})$ *for any constant $c > 0$ and any polynomial* poly.

References

1. Ananth, P., Jain, A., Naor, M., Sahai, A., Yogev, E.: Universal constructions and robust combiners for indistinguishability obfuscation and witness encryption. In: Robshaw, M., Katz, J. (eds.) CRYPTO 2016. LNCS, vol. 9815, pp. 491–520. Springer, Heidelberg (2016). doi:10.1007/978-3-662-53008-5_17
2. Ananth, P., Jain, A.: Indistinguishability obfuscation from compact functional encryption. In: Gennaro, R., Robshaw, M. (eds.) CRYPTO 2015. LNCS, vol. 9215, pp. 308–326. Springer, Heidelberg (2015). doi:10.1007/978-3-662-47989-6_15
3. Barak, B., Garg, S., Kalai, Y.T., Paneth, O., Sahai, A.: Protecting obfuscation against algebraic attacks. In: Nguyen, P.Q., Oswald, E. (eds.) EUROCRYPT 2014. LNCS, vol. 8441, pp. 221–238. Springer, Heidelberg (2014). doi:10.1007/978-3-642-55220-5_13
4. Barak, B., Goldreich, O., Impagliazzo, R., Rudich, S., Sahai, A., Vadhan, S., Yang, K.: On the (Im)possibility of obfuscating programs. In: Kilian, J. (ed.) CRYPTO 2001. LNCS, vol. 2139, pp. 1–18. Springer, Heidelberg (2001). doi:10.1007/3-540-44647-8_1

5. Beaver, D.: Precomputing oblivious transfer. In: Coppersmith, D. (ed.) CRYPTO 1995. LNCS, vol. 963, pp. 97–109. Springer, Heidelberg (1995). doi:10.1007/3-540-44750-4_8

6. Beimel, A., Ishai, Y., Kumaresan, R., Kushilevitz, E.: On the cryptographic complexity of the worst functions. In: Lindell, Y. (ed.) TCC 2014. LNCS, vol. 8349, pp. 317–342. Springer, Heidelberg (2014). doi:10.1007/978-3-642-54242-8_14

7. Bitansky, N., Garg, S., Lin, H., Pass, R., Telang, S.: Succinct randomized encodings and their applications. In: STOC (2015)

8. Bitansky, N., Goldwasser, S., Jain, A., Paneth, O., Vaikuntanathan, V., Waters, B.: Time-lock puzzles from randomized encodings. In: ITCS (2016)

9. Bitansky, N., Paneth, O.: ZAPs and non-interactive witness indistinguishability from indistinguishability obfuscation. In: Dodis, Y., Nielsen, J.B. (eds.) TCC 2015. LNCS, vol. 9015, pp. 401–427. Springer, Heidelberg (2015). doi:10.1007/978-3-662-46497-7_16

10. Bitansky, N., Paneth, O., Rosen, A.: On the cryptographic hardness of finding a Nash equilibrium. In: FOCS (2015)

11. Bitansky, N., Vaikuntanathan, V.: Indistinguishability obfuscation from functional encryption. In: FOCS (2015)

12. Bitansky, N., Vaikuntanathan, V.: Indistinguishability obfuscation: from approximate to exact. In: Kushilevitz, E., Malkin, T. (eds.) TCC 2016. LNCS, vol. 9562, pp. 67–95. Springer, Heidelberg (2016). doi:10.1007/978-3-662-49096-9_4

13. Canetti, R., Holmgren, J., Jain, A., Vaikuntanathan, V.: Indistinguishability obfuscation of iterated circuits and RAM programs. In: STOC (2015)

14. Cheon, J.H., Fouque, P.-A., Lee, C., Minaud, B., Ryu, H.: Cryptanalysis of the new CLT multilinear map over the integers. In: Fischlin, M., Coron, J.-S. (eds.) EUROCRYPT 2016. LNCS, vol. 9665, pp. 509–536. Springer, Heidelberg (2016). doi:10.1007/978-3-662-49890-3_20

15. Cheon, J.H., Han, K., Lee, C., Ryu, H., Stehlé, D.: Cryptanalysis of the multilinear map over the integers. In: Oswald, E., Fischlin, M. (eds.) EUROCRYPT 2015. LNCS, vol. 9056, pp. 3–12. Springer, Heidelberg (2015). doi:10.1007/978-3-662-46800-5_1

16. Cohen, A., Holmgren, J., Nishimaki, R., Vaikuntanathan, V., Wichs, D.: Watermarking cryptographic capabilities. In: STOC (2016)

17. Coron, J.-S., et al.: Zeroizing without low-level zeroes: new MMAP attacks and their limitations. In: Gennaro, R., Robshaw, M. (eds.) CRYPTO 2015. LNCS, vol. 9215, pp. 247–266. Springer, Heidelberg (2015). doi:10.1007/978-3-662-47989-6_12

18. Coron, J.-S., Lee, M.S., Lepoint, T., Tibouchi, M.: Cryptanalysis of GGH15 multilinear maps. In: Robshaw, M., Katz, J. (eds.) CRYPTO 2016. LNCS, vol. 9815, pp. 607–628. Springer, Heidelberg (2016). doi:10.1007/978-3-662-53008-5_21

19. Coron, J.-S., Lepoint, T., Tibouchi, M.: Practical multilinear maps over the integers. In: Canetti, R., Garay, J.A. (eds.) CRYPTO 2013. LNCS, vol. 8042, pp. 476–493. Springer, Heidelberg (2013). doi:10.1007/978-3-642-40041-4_26

20. Damgård, I., Zakarias, S.: Constant-overhead secure computation of boolean circuits using preprocessing. In: Sahai, A. (ed.) TCC 2013. LNCS, vol. 7785, pp. 621–641. Springer, Heidelberg (2013). doi:10.1007/978-3-642-36594-2_35

21. De Santis, A., Micali, S., Persiano, G.: Non-interactive zero-knowledge with preprocessing. In: Goldwasser, S. (ed.) CRYPTO 1988. LNCS, vol. 403, pp. 269–282. Springer, New York (1990). doi:10.1007/0-387-34799-2_21

22. Fischlin, M., Herzberg, A., Noon, H.B., Shulman, H.: Obfuscation combiners. Cryptology ePrint Archive, Report 2016/289 (2016). http://eprint.iacr.org/

23. Garg, S., Gentry, C., Halevi, S.: Candidate multilinear maps from ideal lattices. In: Johansson, T., Nguyen, P.Q. (eds.) EUROCRYPT 2013. LNCS, vol. 7881, pp. 1–17. Springer, Heidelberg (2013). doi:10.1007/978-3-642-38348-9_1
24. Garg, S., Gentry, C., Halevi, S., Raykova, M., Sahai, A., Waters, B.: Candidate indistinguishability obfuscation and functional encryption for all circuits. In: FOCS (2013)
25. Garg, S., Miles, E., Mukherjee, P., Sahai, A., Srinivasan, A., Zhandry, M.: Secure obfuscation in a weak multilinear map model. In: Hirt, M., Smith, A. (eds.) TCC 2016. LNCS, vol. 9986, pp. 241–268. Springer, Heidelberg (2016). doi:10.1007/978-3-662-53644-5_10
26. Garg, S., Pandey, O., Srinivasan, A.: Revisiting the cryptographic hardness of finding a nash equilibrium. In: Robshaw, M., Katz, J. (eds.) CRYPTO 2016. LNCS, vol. 9815, pp. 579–604. Springer, Heidelberg (2016). doi:10.1007/978-3-662-53008-5_20
27. Garg, S., Pandey, O., Srinivasan, A., Zhandry, M.: Breaking the sub-exponential barrier in obfustopia. IACR Cryptology ePrint Archive 2016 (2016)
28. Garg, S., Srinivasan, A.: Single-Key to multi-key functional encryption with polynomial loss. In: Hirt, M., Smith, A. (eds.) TCC 2016. LNCS, vol. 9986, pp. 419–442. Springer, Heidelberg (2016). doi:10.1007/978-3-662-53644-5_16
29. Gentry, C., Gorbunov, S., Halevi, S.: Graph-induced multilinear maps from lattices. In: Dodis, Y., Nielsen, J.B. (eds.) TCC 2015. LNCS, vol. 9015, pp. 498–527. Springer, Heidelberg (2015). doi:10.1007/978-3-662-46497-7_20
30. Goldreich, O., Micali, S., Wigderson, A.: How to play any mental game or a completeness theorem for protocols with honest majority. In: STOC (1987)
31. Goldwasser, S., Gordon, S.D., Goyal, V., Jain, A., Katz, J., Liu, F.-H., Sahai, A., Shi, E., Zhou, H.-S.: Multi-input functional encryption. In: Nguyen, P.Q., Oswald, E. (eds.) EUROCRYPT 2014. LNCS, vol. 8441, pp. 578–602. Springer, Heidelberg (2014). doi:10.1007/978-3-642-55220-5_32
32. Goldwasser, S., Tauman Kalai, Y.: Cryptographic assumptions: a position paper. In: Kushilevitz, E., Malkin, T. (eds.) TCC 2016. LNCS, vol. 9562, pp. 505–522. Springer, Heidelberg (2016). doi:10.1007/978-3-662-49096-9_21
33. Goldwasser, S., Rothblum, G.N.: On best-possible obfuscation. In: Vadhan, S.P. (ed.) TCC 2007. LNCS, vol. 4392, pp. 194–213. Springer, Heidelberg (2007). doi:10.1007/978-3-540-70936-7_11
34. Hohenberger, S., Sahai, A., Waters, B.: Replacing a random oracle: full domain hash from indistinguishability obfuscation. In: Nguyen, P.Q., Oswald, E. (eds.) EUROCRYPT 2014. LNCS, vol. 8441, pp. 201–220. Springer, Heidelberg (2014). doi:10.1007/978-3-642-55220-5_12
35. Hu, Y., Jia, H.: Cryptanalysis of GGH map. In: Fischlin, M., Coron, J.-S. (eds.) EUROCRYPT 2016. LNCS, vol. 9665, pp. 537–565. Springer, Heidelberg (2016). doi:10.1007/978-3-662-49890-3_21
36. Impagliazzo, R., Rudich, S.: Limits on the provable consequences of one-way permutations. In: STOC (1989)
37. Ishai, Y., Kushilevitz, E., Meldgaard, S., Orlandi, C., Paskin-Cherniavsky, A.: On the power of correlated randomness in secure computation. In: Sahai, A. (ed.) TCC 2013. LNCS, vol. 7785, pp. 600–620. Springer, Heidelberg (2013). doi:10.1007/978-3-642-36594-2_34
38. Ishai, Y., Prabhakaran, M., Sahai, A.: Founding cryptography on oblivious transfer - efficiently. In: CRYPTO (2008)
39. Koppula, V., Lewko, A.B., Waters, B.: Indistinguishability obfuscation for turing machines with unbounded memory. In: STOC (2015)

40. Lapidot, D., Shamir, A.: Publicly verifiable non-interactive zero-knowledge proofs. In: Menezes, A.J., Vanstone, S.A. (eds.) CRYPTO 1990. LNCS, vol. 537, pp. 353–365. Springer, Heidelberg (1991). doi:10.1007/3-540-38424-3_26

41. Li, B., Micciancio, D.: Compactness vs collusion resistance in functional encryption. In: Hirt, M., Smith, A. (eds.) TCC 2016. LNCS, vol. 9986, pp. 443–468. Springer, Heidelberg (2016). doi:10.1007/978-3-662-53644-5_17

42. Lin, H.: Indistinguishability obfuscation from constant-degree graded encoding schemes. In: Fischlin, M., Coron, J.-S. (eds.) EUROCRYPT 2016. LNCS, vol. 9665, pp. 28–57. Springer, Heidelberg (2016). doi:10.1007/978-3-662-49890-3_2

43. Lin, H., Vaikunthanathan, V.: Indistinguishability obfuscation from DDH-like assumptions on constant-degree graded encodings. In: FOCS (2016)

44. Miles, E., Sahai, A., Zhandry, M.: Annihilation attacks for multilinear maps: Cryptanalysis of indistinguishability obfuscation over GGH13. Cryptology ePrint Archive, Report 2016/147 (2016). http://eprint.iacr.org/

45. Sahai, A., Seyalioglu, H.: Worry-free encryption: functional encryption with public keys. In: Proceedings of the 17th ACM Conference on Computer and Communications Security, pp. 463–472. ACM (2010)

46. Sahai, A., Waters, B.: How to use indistinguishability obfuscation: deniable encryption, and more. In: STOC (2014)

47. Waters, B.: A punctured programming approach to adaptively secure functional encryption. Cryptology ePrint Archive, Report 2014/588 (2014)

48. Zimmerman, J.: How to obfuscate programs directly. In: Oswald, E., Fischlin, M. (eds.) EUROCRYPT 2015. LNCS, vol. 9057, pp. 439–467. Springer, Heidelberg (2015). doi:10.1007/978-3-662-46803-6_15

From Minicrypt to Obfustopia via Private-Key Functional Encryption

Ilan Komargodski[1](✉) and Gil Segev[2]

[1] Weizmann Institute of Science, 76100 Rehovot, Israel
ilan.komargodski@weizmann.ac.il
[2] Hebrew University of Jerusalem, 91904 Jerusalem, Israel
segev@cs.huji.ac.il

Abstract. Private-key functional encryption enables fine-grained access to symmetrically-encrypted data. Although private-key functional encryption (supporting an unbounded number of keys and ciphertexts) seems significantly weaker than its public-key variant, its known realizations all rely on public-key functional encryption. At the same time, however, up until recently it was not known to imply any public-key primitive, demonstrating our poor understanding of this extremely-useful primitive.

Recently, Bitansky et al. [TCC '16B] showed that sub-exponentially-secure private-key function encryption bridges from nearly-exponential security in Minicrypt to slightly super-polynomial security in Cryptomania, and from sub-exponential security in Cryptomania to Obfustopia. Specifically, given any sub-exponentially-secure private-key functional encryption scheme and a nearly-exponentially-secure one-way function, they constructed a public-key encryption scheme with slightly super-polynomial security. Assuming, in addition, a sub-exponentially-secure public-key encryption scheme, they then constructed an indistinguishability obfuscator.

We show that quasi-polynomially-secure private-key functional encryption bridges from sub-exponential security in Minicrypt all the way to Cryptomania. First, given any quasi-polynomially-secure private-key functional encryption scheme, we construct an indistinguishability obfuscator for circuits with inputs of poly-logarithmic length. Then, we observe that such an obfuscator can be used to instantiate many natural applications of indistinguishability obfuscation. Specifically, relying on sub-exponentially-secure one-way functions, we show that quasi-polynomially-secure private-key functional encryption implies not just public-key encryption but leads all the way to public-key functional

I. Komargodski—Supported by a Levzion fellowship and by a grant from the Israel Science Foundation.

G. Segev—Supported by the European Union's 7th Framework Program (FP7) via a Marie Curie Career Integration Grant, by the European Union's Horizon 2020 Framework Program (H2020) via an ERC Grant (Grant No. 714253), by the Israel Science Foundation (Grant No. 483/13), by the Israeli Centers of Research Excellence (I-CORE) Program (Center No. 4/11), by the US-Israel Binational Science Foundation (Grant No. 2014632), and by a Google Faculty Research Award.

J.-S. Coron and J.B. Nielsen (Eds.): EUROCRYPT 2017, Part I, LNCS 10210, pp. 122–151, 2017.
DOI: 10.1007/978-3-319-56620-7_5

encryption for circuits with inputs of poly-logarithmic length. Moreover, relying on sub-exponentially-secure injective one-way functions, we show that quasi-polynomially-secure private-key functional encryption implies a hard-on-average distribution over instances of a PPAD-complete problem.

Underlying our constructions is a new transformation from single-input functional encryption to multi-input functional encryption in the private-key setting. The previously known such transformation [Brakerski et al., EUROCRYPT '16] required a sub-exponentially-secure single-input scheme, and obtained a scheme supporting only a slightly super-constant number of inputs. Our transformation both relaxes the underlying assumption and supports more inputs: Given any quasi-polynomially-secure single-input scheme, we obtain a scheme supporting a poly-logarithmic number of inputs.

1   Introduction

Functional encryption [16,49,51] allows tremendous flexibility when accessing encrypted data: Such encryption schemes support restricted decryption keys that allow users to learn specific functions of the encrypted data without leaking any additional information. We focus on the most general setting where the functional encryption schemes support an unbounded number of functional keys in the public-key setting, and an unbounded number of functional keys and ciphertexts in the private-key setting. In the public-key setting, it has been shown that functional encryption is essentially equivalent to indistinguishability obfuscation [6,7,12,33,54], and thus it currently seems somewhat challenging to base its security on standard cryptographic assumptions (especially given the various attacks on obfuscation schemes and their underlying building blocks [21,25 29,40,47,48] – see [5, Appendix A] for a summary of these attacks).

Luckily, when examining the various applications of functional encryption (see, for example, the survey by Boneh et al. [17]), it turns out that *private-key* functional encryption suffices in many interesting scenarios.[1] However, although private-key functional encryption may seem significantly weaker than its public-key variant, constructions of private-key functional encryption schemes are currently known based only on public-key functional encryption.[2]

Minicrypt, Cryptomania, or Obfustopia? For obtaining a better understanding of private-key functional encryption, we must be able to position it correctly within the hierarchy of cryptographic primitives. Up until recently,

---

[1] As a concrete (yet quite general) example, consider a user who stores her data on a remote server: The user uses the master secret key both for encrypting her data, and for generating functional keys that will enable the server to offer her various useful services.

[2] This is not true in various restricted cases, for example, when the functional encryption scheme has to support an a-priori bounded number of functional keys or ciphertexts [39]. However, as mentioned, we focus on schemes that support an unbounded number of functional keys and ciphertexts.

private-key functional encryption was not known to imply any cryptographic primitives other than those that are essentially equivalent to one-way functions (i.e., Minicrypt primitives [42]). Moreover, Asharov and Segev [8] proved that as long as a private-key functional encryption scheme is invoked in a black-box manner, it cannot be used as a building block to construct any public-key primitive (i.e., Cryptomania primitives [42]).[3] This initial evidence hinted that private-key functional encryption may belong to Minicrypt, and thus may be constructed based on extremely well-studied cryptographic assumptions.

Recently, Bitansky et al. [10] showed that private-key functional encryption is more powerful than suggested by the above initial evidence. They proved that any sub-exponentially-secure private-key functional encryption scheme and any (nearly) exponentially-secure one-way function can be used to construct a public-key encryption scheme.[4] Although their underlying building blocks are at least sub-exponentially secure, the resulting public-key scheme is only slightly super-polynomially secure. In addition, Bitansky et al. proved that any sub-exponentially-secure private-key functional encryption scheme and any sub-exponentially-secure public-key encryption scheme can be used to construct a full-fledged indistinguishability obfuscator. Overall, their work shows that sub-exponentially-secure private-key functional encryption bridges from nearly-exponential security in Minicrypt to slightly super-polynomial security in Cryptomania, and from sub-exponential security in Cryptomania to Obfustopia (see Fig. 1).

1.1   Our Contributions

We show that quasi-polynomially-secure private-key functional encryption bridges from sub-exponential security in Minicrypt all the way to Cryptomania. First, given any *quasi-polynomially*-secure private-key functional encryption scheme, we construct a (quasi-polynomially-secure) indistinguishability obfuscator for circuits with inputs of poly-logarithmic length and sub-polynomial size. We prove the following theorem:

Theorem 1.1 (Informal). *Assuming a quasi-polynomially-secure private-key functional encryption scheme for polynomial-size circuits, there exists an indistinguishability obfuscator for the class of circuits of size $2^{(\log \lambda)^{\epsilon}}$ with inputs of length $(\log \lambda)^{1+\delta}$ bits, for some positive constants ϵ and δ.*

Underlying our obfuscator is a new transformation from single-input functional encryption to multi-input functional encryption in the private-key setting. The previously known such transformation of Brakerski et al. [22] required a

---

[3] This holds even if the construction is allowed to generate functional keys (in a non-black-box manner) for any circuit that invokes one-way functions in a black-box manner.

[4] Bitansky et al. overcome the black-box barrier introduced by Asharov and Segev [8] by relying on the non-black-box construction of a private-key multi-input functional encryption scheme of Brakerski et al. [22].

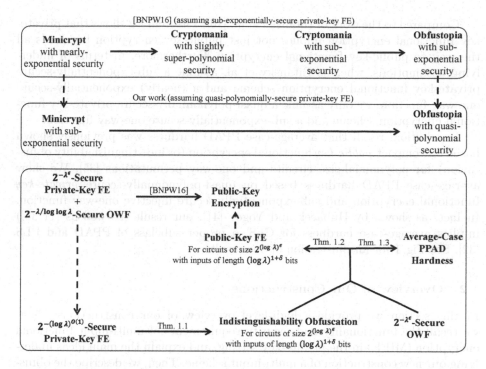

Fig. 1. An illustration of our results (dashed arrows correspond to trivial implications).

sub-exponentially-secure single-input scheme, and obtained a multi-input scheme supporting only a slightly super-constant number of inputs. Our transformation both relaxes the underlying assumption and supports more inputs: Given any quasi-polynomially-secure single-input scheme, we obtain a multi-input scheme supporting a poly-logarithmic number of inputs.

We demonstrate the wide applicability of our obfuscator by observing that it can be used to instantiate many natural applications of (full-fledged) indistinguishability obfuscation for polynomial-size circuits. We exemplify this observation by constructing a public-key functional encryption scheme (based on [54]), and a hard-on-average distribution of instances of a PPAD-complete problem (based on [11]).

Theorem 1.2 (Informal). *Assuming a quasi-polynomially-secure private-key functional encryption scheme for polynomial-size circuits, and a sub-exponentially-secure one-way function, there exists a public-key functional encryption scheme for the class of circuits of size $2^{(\log \lambda)^\epsilon}$ with inputs of length $(\log \lambda)^{1+\delta}$ bits, for some positive constants ϵ and δ.*

Theorem 1.3 (Informal). *Assuming a quasi-polynomially-secure private-key functional encryption scheme for polynomial-size circuits, and a sub-exponentially-secure injective one-way function, there exists a hard-on-average distribution over instances of a PPAD-complete problem.*

Compared to the work of Bitansky at el. [10], Theorem 1.2 shows that private-key functional encryption implies not just public-key encryption but leads all the way to public-key functional encryption. Furthermore, in terms of underlying assumptions, whereas Bitansky et al. assume a sub-exponentially-secure private-key functional encryption scheme and a (nearly) exponentially-secure one-way function, we only assume a quasi-polynomially-secure private-key functional encryption scheme and a sub-exponentially-secure one-way function.

In addition, recall that average-case PPAD hardness was previously shown based on compact *public-key* functional encryption (or indistinguishability obfuscation) for polynomial-size circuits and one-way permutations [35]. We show average-case PPAD hardness based on quasi-polynomially-secure *private-key* functional encryption and sub-exponentially-secure injective one-way function. In fact, as shown by Hubáček and Yogev [41], our result (as well as [11,35]) implies average-case hardness for CLS, a proper subclass of PPAD and PLS [32]. See Fig. 1 for an illustration of our results.

1.2 Overview of Our Constructions

In this section we provide a high-level overview of our constructions. First, we recall the functionality and security requirements of multi-input functional encryption (MIFE) in the private-key setting, and explain the main ideas underlying our new construction of a multi-input scheme. Then, we describe the obfuscator we obtain from our multi-input scheme, and briefly discuss its applications to public-key functional encryption and to average-case PPAD hardness.

Multi-input Functional Encryption. In a private-key t-input functional encryption scheme [37], the master secret key msk of the scheme is used for encrypting any message x_i to the ith coordinate, and for generating functional keys for t-input functions. A functional key sk_f corresponding to a function f enables to compute $f(x_1, \ldots, x_t)$ given $\mathsf{Enc}(x_1, 1), \ldots, \mathsf{Enc}(x_t, t)$. Building upon the previous notions of security for private-key multi-input functional encryption schemes [13,37], we consider a strengthened notion of security that combines both message privacy and function privacy (as in [2,23] for single-input schemes and as in [6,22] for multi-input schemes), to which we refer as *full security*. Specifically, we consider adversaries that are given access to "left-or-right" key-generation and encryption oracles.[5] These oracles operate in one out of two modes corresponding to a randomly-chosen bit b. The key-generation oracle receives as input pairs of the form (f_0, f_1) and outputs a functional key for the function f_b. The encryption oracle receives as input triples of the form (x^0, x^1, i), and outputs an encryption of the message x^b with respect to coordinate i.

---

[5] In this work we focus on selectively-secure schemes, where an adversary first submits all of its encryption queries, and can then adaptively interact with the key-generation oracle (see Definition 2.7). This notion of security suffices for the applications we consider in this paper.

We require that no efficient adversary can guess the bit b with probability noticeably higher than $1/2$, as long as for each such $t + 1$ queries (f_0, f_1), $(x_1^0, x_1^1), \ldots, (x_t^0, x_t^1)$ it holds that $f_0(x_1^0, \ldots, x_t^0) = f_1(x_1^1, \ldots, x_t^1)$.

The BKS Approach. Given any private-key single-input functional encryption scheme for all polynomial-size circuits, Brakerski et al. [22] constructed a $t(\lambda)$-input scheme for all circuits of size $s(\lambda) = 2^{(\log \lambda)^\epsilon}$, where $t(\lambda) = \delta \cdot \log \log \lambda$ for some fixed positive constants ϵ and δ, and $\lambda \in \mathbb{N}$ is the security parameter.

Their transformation is based on extending the number of inputs the scheme supports one by one. That is, for any $t \geq 1$, given a t-input scheme they construct a $(t + 1)$-input scheme. Relying on the function privacy of the underlying scheme, Brakerski et al. observed that ciphertexts for one of the coordinates can be treated as a functional key for a function that has the value of the input hardwired. In terms of functionality, this idea enabled them to support $t + 1$ inputs using a scheme that supports t inputs. The transformation is implemented such that every step of it incurs a *polynomial* blowup in the size of the ciphertexts and functional keys.[6] Thus, applying this transformation t times, the size of a functional key for a function of size s is roughly $(s \cdot \lambda)^{O(1)^t}$. Therefore, Brakerski et al. could only apply their transformation $t(\lambda) = \delta \cdot \log \log \lambda$ times, and this required assuming that their underlying single-input scheme is sub-exponentially secure, and that $s(\lambda) = 2^{(\log \lambda)^\epsilon}$.

Our Construction. We present a new transformation that constructs a $2t$-inputs scheme directly from any t-input scheme. Our transformation shares the same polynomial efficiency loss as in [22], so applying the transformation t times makes a functional key be of size $(s \cdot \lambda)^{O(1)^t}$. But now, since each transformation doubles the number of inputs, applying the transformation t times gets us all the way to a scheme that supports $2^t = (\log \lambda)^\delta$ inputs, as required. We further observe, by a careful security analysis, that for the resulting scheme to be secure it suffices that the initial scheme is only *quasi-polynomially secure* (and the resulting scheme can be made quasi-polynomially secure as well).

Doubling the Number of Inputs via Dynamic Key Encapsulation. As opposed to the approach of [22] (and the similar idea of [6]), it is much less clear how to combine the ciphertexts and functional keys of a t-input scheme to satisfy the required functionality (and security) of a $2t$-input scheme.

Our high-level idea is as follows. Given a $2t$-input function f, we will generate a functional key for a function f^* that gets t inputs each of which is composed of two inputs: $f^*(x_1 \parallel x_{1+t}, \ldots, x_t \parallel x_{2t}) = f(x_1, \ldots, x_{2t})$. We will encrypt each input such that it is possible to compute an encryption of each pair $(x_\ell, x_{\ell+t})$, and evaluate the function in two steps. First, we concatenate each such pair to get an encryption of $x_\ell \parallel x_{\ell+t}$. Then, given such t ciphertexts, we will apply

---

[6] A similar strategy was also employed by Ananth and Jain [6], that showed how to use any t-input private-key scheme to get a private-key $(t + 1)$-input scheme under the additional assumption that a *public-key* functional encryption scheme exists. Their construction, however, did not incur the polynomial blowup and could be applied all the way to get a scheme that supports a polynomial number of inputs.

a functional key that corresponds to f^*. By the correctness of the underlying primitives, the output must be correct. There are three main issues that we have to overcome: (1) We need to be able to generate the encryption of $x_\ell \parallel x_{\ell+t}$, (2) we need to make sure all of these ciphertexts are with respect to the same master secret key and that the functional key for f^* is also generated with respect to the same key, and (3) we need to prove the security of the resulting scheme. We now describe our solution.

The master secret key for our scheme is a master secret key for a t-input scheme msk and a PRF key K. We split the $2t$ input coordinates into two parts: (1) the first t coordinates $1, \ldots, t$ which we call the "master coordinates" and (2) the last t coordinates $1+t, \ldots, 2t$ which we call the "slave coordinates". Our main idea is to let each combination of the master coordinates implicitly define a master secret "encapsulation" key $\mathsf{msk}_{x_1 \ldots, x_t}$ for a t-input scheme. Details follow.

To encrypt a message x_ℓ with respect to a master coordinate $1 \leq \ell \leq t$, we encrypt x_ℓ with respect to coordinate ℓ under the key msk. To encrypt a message $x_{\ell+t}$ with respect to a slave coordinate $1 \leq \ell \leq t$, we generate a functional key for a t-input function $\mathsf{AGG}_{x_{\ell+t}, K}$ under the key msk. To generate a functional key for a $2t$-input function f, we generate a functional key for a t-input function $\mathsf{Gen}_{f,K}$ under msk. Both $\mathsf{AGG}_{x_{\ell+t}, K}$ and $\mathsf{Gen}_{f,K}$ first compute a pseudorandom master secret key $\mathsf{msk}_{x_1 \ldots x_t}$ using randomness generated via the PRF key K on input $x_1 \ldots x_t$. Then, $\mathsf{AGG}_{x_{\ell+t}, K}$ computes an encryption of $(x_\ell \parallel x_{\ell+t})$ to coordinate ℓ under this master secret key, and $\mathsf{Gen}_{f,K}$ computes a functional key for f^* (described above) under this master secret key (see Fig. 2).

$\mathsf{Gen}_{f,K}(x_1, x_2, \ldots, x_t):$

1. $\mathsf{msk}_{x_1 \ldots x_t} = \mathsf{Setup}(\mathsf{PRF}(K, x_1 \ldots x_t)).$
2. Output $\mathsf{KG}(\mathsf{msk}_{x_1 \ldots x_t}, f^*).$

$\mathsf{AGG}_{x_{\ell+t}, K}(x_1, x_2, \ldots, x_t):$

1. $\mathsf{msk}_{x_1 \ldots x_t} = \mathsf{Setup}(\mathsf{PRF}(K, x_1 \ldots x_t)).$
2. Output $\mathsf{Enc}(\mathsf{msk}_{x_1 \ldots x_t}, (x_\ell \parallel x_{\ell+t}), \ell).$

Fig. 2. The t-input functions $\mathsf{Gen}_{f,K}$ and $\mathsf{AGG}_{x_{\ell+t}, K}$.

It is straightforward to verify that the above scheme indeed provides the required functionality of a $2t$-input scheme. Indeed, given t ciphertexts corresponding to the master coordinates $\mathsf{ct}_{x_1}, \ldots, \mathsf{ct}_{x_t}$, t ciphertexts corresponding to the slave coordinates $\mathsf{ct}_{x_{1+t}}, \ldots, \mathsf{ct}_{x_{2t}}$, and a functional key sk_f for a $2t$-input function f, we first combine $\mathsf{ct}_{x_1}, \ldots, \mathsf{ct}_{x_t}$ with each $\mathsf{ct}_{x_{\ell+t}}$ to get $\mathsf{ct}_{x_\ell \parallel x_{\ell+t}}$, which is an encryption of $x_\ell \parallel x_{\ell+t}$ under $\mathsf{msk}_{x_1 \ldots x_t}$. Then, we combine $\mathsf{ct}_{x_1}, \ldots, \mathsf{ct}_{x_t}$ with sk_f to get a functional key sk_{f^*} for f^* under the same $\mathsf{msk}_{x_1 \ldots x_t}$. Finally, we combine $\mathsf{ct}_{x_1 \parallel x_{1+t}}, \ldots, \mathsf{ct}_{x_t \parallel x_{2t}}$ with sk_{f^*} to get $f^*(x_1 \parallel x_{1+t}, \ldots, x_t \parallel x_{2t}) = f(x_1, \ldots, x_{2t})$, as required.

The security proof is done by a sequence of hybrid experiments, where we "attack" each possible sequence of master coordinates separately, namely, we handle each $\mathsf{msk}_{x_1 \ldots x_t}$ separately so that it will not be explicitly needed. A typical approach for such a security proof is to embed all possible encryptions and key-generation queries under $\mathsf{msk}_{x_1\ldots x_t}$ in the ciphertexts that are generated under msk. Handling the key-generation queries using $\mathsf{msk}_{x_1\ldots x_t}$ is rather standard: whenever a key-generation query is requested we compute the corresponding functional key under $\mathsf{msk}_{x_1\ldots x_t}$ and embed it into the functional key. Handling encryption queries under $\mathsf{msk}_{x_1\ldots x_t}$ is significantly more challenging since for every $x_1\ldots x_t$ sequence, there are many possible ciphertexts $x_{\ell+t}$ of slave coordinates that will be paired with it to get the encryption of $x_\ell \,\|\, x_{\ell+t}$. It might seem as if there is not enough space to embed all these possible ciphertexts, but we observe that we can embed each ciphertext $\mathsf{ct}_{x_\ell\|x_{\ell+t}}$ in the ciphertext corresponding to $x_{\ell+t}$ (for each such $x_{\ell+t}$). This way, $\mathsf{msk}_{x_1\ldots x_t}$ is not explicitly needed in the scheme and we can use the security of the underlying t-input scheme. In total, the number of hybrids is roughly T^t, where T is an upper bound on the running time of the adversary. Thus, since t is roughly logarithmic in the security parameter, we have to start with a quasi-polynomially-secure scheme.

From MIFE to Obfuscation. Goldwasser et al. [37] observed that multi-input functional encryption is tightly related to indistinguishability obfuscation [9,33]. Specifically, a multi-input scheme that supports a polynomial number of inputs (i.e., $t(\lambda) = \mathsf{poly}(\lambda)$) readily implies an indistinguishability obfuscator (and vice-versa). We use a more fine-grained relationship (as observed by [10]) that is useful when $t(\lambda)$ is small compared to λ: A multi-input scheme that supports all circuits of size $s(\lambda)$ and $t(\lambda)$ inputs implies an indistinguishability obfuscator for all circuits of size $s(\lambda)$ that have at most $t(\lambda) \cdot \log \lambda$ input bits.

This transformation works as follows. An obfuscation of a function f of circuit-size at most $s(\lambda)$ that has at most $t(\lambda) \cdot \log \lambda$ bits as input, is composed of $t(\lambda) \cdot \lambda$ ciphertexts and one functional key. We think of f as a function f^* that gets $t(\lambda)$ inputs each of which is of length $\log \lambda$ bits. The obfuscation now consists of a functional key for the circuit f^*, denoted by $\mathsf{sk}_f = \mathsf{KG}(f^*)$, and a ciphertext $\mathsf{ct}_{x,i} = \mathsf{Enc}(x,i)$ for every $(x,i) \in \{0,1\}^{\log \lambda} \times [t(\lambda)]$. To evaluate C at a point $x = (x_1 \ldots x_{t(\lambda)}) \in (\{0,1\}^{\log \lambda})^{t(\lambda)}$ one has to compute and output $\mathsf{Dec}(\mathsf{sk}_f, \mathsf{ct}_{x_1,1}, \ldots, \mathsf{ct}_{x_{t(\lambda)},t(\lambda)}) = f(x)$. Correctness and security of the obfuscator follow directly from the correctness and security of the multi-input scheme.

Given the relationship described above and given our multi-input scheme that supports circuits of size at most $s(\lambda) = 2^{(\log \lambda)^\epsilon}$ that have $t(\lambda) = (\log \lambda)^\delta$ inputs for some fixed positive constants ϵ and δ, we obtain Theorem 1.1.

Applications of our Obfuscator. One of the main conceptual contributions of this work is the observation that an indistinguishability obfuscator as described above (that supports circuits with a poly-logarithmic number of input bits) is in fact sufficient for many of the applications of indistinguishability obfuscation for all polynomial-size circuits. We exemplify this observation by showing how to adapt the construction of Waters [54] of a public-key functional encryption scheme and the construction of Bitansky et al. [11] of a hard-on-average distribution for PPAD, to our obfuscator. Such an adaptation is quite delicate and involves a careful choice of the additional primitives that are involved in the construction. In a very high level, since the obfuscator supports only a poly-logarithmic number of inputs, a primitive that has to be secure when applied on (part of) the input (say a one-way function), must be sub-exponentially secure. We believe that this observation may find additional applications beyond the scope of our work.

Using the Multi-input Scheme of [22]. Using the multi-input scheme of [22], one can get that sub-exponentially-secure private-key functional encryption implies indistinguishability obfuscation for inputs of length slightly super-logarithmic. However, using such an obfuscator as a building block seems to inherently require to additionally assume nearly-exponentially-secure primitives and the resulting primitives are (at most) slightly super-polynomially-secure.

Our approach, on the other hand, requires quasi-polynomially-secure private-key functional encryption. In addition, our additional primitives are only sub-exponentially-secure and the resulting primitives are quasi-polynomially secure.

1.3   Additional Related Work

Constructions of FE Schemes. Private-key *single-input* functional encryption schemes that are sufficient for our applications are known to exist based on a variety of assumptions, including indistinguishability obfuscation [33,54], differing-input obfuscation [3,19], and multilinear maps [34]. Restricted functional encryption schemes that support either a bounded number of functional keys or a bounded number of ciphertexts can be based on the Learning with Errors (LWE) assumption (where the length of ciphertexts grows with the number of functional-key queries and with a bound on the depth of allowed functions) [38], and even based on pseudorandom generators computable by small-depth circuits (where the length of ciphertexts grows with the number of functional-key queries and with an upper bound on the circuit size of the functions) [39].

In the work of Bitansky et al. [10, Proposition 1.2 & Footnote 1] it has been shown that, assuming weak PRFs in NC^1, any public-key encryption scheme can be used to transform a private-key functional encryption scheme into a public-key functional encryption scheme (which can be used to get PPAD-hardness [35]). This gives a better reduction than ours in terms of security loss, but requires a public-key primitive to begin with.

Constructions of MIFE Schemes. There are several constructions of private-key multi-input functional encryption schemes. Mostly related to our work is the construction of Brakerski et al. [22] which we significantly improve (see Sect. 1.2 for more details). Other constructions [6,13,37] are incomparable as they either rely on stronger assumptions or could be proven secure only in an idealized generic model. Goldwasser et al. [37] constructed a multi-input scheme that supports a polynomial number of inputs assuming indistinguishability obfuscation for all polynomial-size circuits. Ananth and Jain [6] constructed a multi-input functional encryption scheme that supports a polynomial number of inputs assuming any sub-exponentially-secure (single-input) *public-key* functional encryption scheme. Boneh et al. [13] constructed a multi-input scheme that supports a polynomial number of inputs based on multilinear maps, and was proven secure in the idealized generic multilinear map model.

Proof Techniques. Parts of our proof rely on two useful techniques from the functional encryption literature: key encapsulation (also known as "hybrid encryption") and function privacy.

Key encapsulation is an extremely useful approach in the design of encryption schemes, both for improved efficiency and for improved security. Specifically, key encapsulation typically means that instead of encrypting a message m under a fixed key sk, one can instead sample a random key k, encrypt m under k and then encrypt k under sk. The usefulness of this technique in the context of functional encryption was demonstrated by [4,22]. Our constructions incorporate key encapsulation techniques, and exhibit additional strengths of this technique in the context of functional encryption schemes. Specifically, as discussed in Sect. 1.2, we use key encapsulation techniques for our *dynamic key-generation* technique, a crucial ingredient in our constructions and proofs of security.

The security guarantees of functional encryption typically focus on *message privacy* that ensures that a ciphertext does not reveal any unnecessary information on the plaintext. In various cases, however, it is also useful to consider *function privacy* [2,14,15,23,53], asking that a functional key sk_f does not reveal any unnecessary information on the function f. Brakerski and Segev [23] (and the follow-up of Ananth and Jain [6]) showed that any private-key (multi-input) functional encryption scheme can be generically transformed into one that satisfies both message privacy and function privacy. Function privacy was found useful as a building block in the construction of several functional encryption schemes [4,22,46]. In particular, functional encryption allows to successfully apply proof techniques "borrowed" from the indistinguishability obfuscation literature (including, for example, a variant of the punctured programming approach of Sahai and Waters [52]).

1.4   Paper Organization

The remainder of this paper is organized as follows. In Sect. 2 we provide an overview of the notation, definitions, and tools underlying our constructions. In Sect. 3 we present our construction of a private-key multi-input functional encryption scheme based on any single-input scheme. In Sect. 4 we present our construction of an indistinguishability obfuscator for circuits with inputs of polylogarithmic length, and its applications to public-key functional encryption and average-case PPAD hardness.

2   Preliminaries

In this section we present the notation and basic definitions that are used in this work. For a distribution X we denote by $x \leftarrow X$ the process of sampling a value x from the distribution X. Similarly, for a set $\mathcal{X}$ we denote by $x \leftarrow \mathcal{X}$ the process of sampling a value x from the uniform distribution over $\mathcal{X}$. For a randomized function f and an input $x \in \mathcal{X}$, we denote by $y \leftarrow f(x)$ the process of sampling a value y from the distribution $f(x)$. For an integer $n \in \mathbb{N}$ we denote by $[n]$ the set $\{1, \ldots, n\}$.

Throughout the paper, we denote by λ the security parameter. A function $\mathsf{neg} : \mathbb{N} \to \mathbb{R}^+$ is *negligible* if for every constant $c > 0$ there exists an integer N_c such that $\mathsf{neg}(\lambda) < \lambda^{-c}$ for all $\lambda > N_c$. Two sequences of random variables $X = \{X_\lambda\}_{\lambda \in \mathbb{N}}$ and $Y = \{Y_\lambda\}_{\lambda \in \mathbb{N}}$ are *computationally indistinguishable* if for any probabilistic polynomial-time algorithm $\mathcal{A}$ there exists a negligible function $\mathsf{neg}(\cdot)$ such that $\big| \Pr[\mathcal{A}(1^\lambda, X_\lambda) = 1] - \Pr[\mathcal{A}(1^\lambda, Y_\lambda) = 1] \big| \leq \mathsf{neg}(\lambda)$ for all sufficiently large $\lambda \in \mathbb{N}$.

2.1   One-Way Functions and Pseudorandom Generators

We rely on the standard (parameterized) notions of one-way functions and pseudorandom generators.

Definition 2.1 (One-way function). *An efficiently computable function $f \colon \{0,1\}^* \to \{0,1\}^*$ is (t,μ)-one-way if for every probabilistic algorithm $\mathcal{A}$ that runs in time $t = t(\lambda)$ it holds that*

$$\mathsf{Adv}^{\mathsf{OWF}}_{f,\mathcal{A}}(\lambda) \overset{\text{def}}{=} \Pr_{x \leftarrow \{0,1\}^\lambda}[\mathcal{A}(1^\lambda, f(x)) \in f^{-1}(f(x))] \leq \mu(\lambda),$$

for all sufficiently large $\lambda \in \mathbb{N}$, where the probability is taken over the choice of $x \in \{0,1\}^\lambda$ and over the internal randomness of $\mathcal{A}$.

Whenever $t = t(\lambda)$ is a super-polynomial function and $\mu = \mu(\lambda)$ is a negligible function, we will often omit t and μ and simply call the function *one-way*. In case $t(\lambda) = 1/\mu(\lambda) = 2^{\lambda^\epsilon}$, for some constant $0 < \epsilon < 1$, we will say that f is sub-exponentially one-way.

Definition 2.2 (Pseudorandom generator). *Let $\ell(\cdot)$ be a function. An efficiently computable function $\mathsf{PRG} \colon \{0,1\}^{\ell(\lambda)} \to \{0,1\}^{2\ell(\lambda)}$ is a (t,μ)-secure pseudorandom generator if for every probabilistic algorithm $\mathcal{A}$ that runs in time $t = t(\lambda)$ it holds that*

$$\mathsf{Adv}^{\mathsf{PRG}}_{f,\mathcal{A}} = \left| \Pr_{x \leftarrow \{0,1\}^{\ell(\lambda)}}[\mathcal{A}(1^\lambda, \mathsf{PRG}(x)) = 1] - \Pr_{r \leftarrow \{0,1\}^{2\ell(\lambda)}}[\mathcal{A}(1^\lambda, r) = 1] \right| \leq \mu(\lambda)$$

for all sufficiently large $\lambda \in \mathbb{N}$.

Whenever $t = t(\lambda)$ is a super-polynomial function and $\mu = \mu(\lambda)$ is a negligible function, we will often omit t and μ and simply call the function a *pseudorandom generator*. In case $t(\lambda) = 1/\mu(\lambda) = 2^{\lambda^\epsilon}$, for some constant $0 < \epsilon < 1$, we will say that PRG is sub-exponentially secure.

2.2   Pseudorandom Functions

Let $\{\mathcal{K}_\lambda, \mathcal{X}_\lambda, \mathcal{Y}_\lambda\}_{\lambda \in \mathbb{N}}$ be a sequence of sets and let $\mathsf{PRF} = (\mathsf{PRF.Gen}, \mathsf{PRF.Eval})$ be a function family with the following syntax:

- $\mathsf{PRF.Gen}$ is a probabilistic polynomial-time algorithm that takes as input the unary representation of the security parameter λ, and outputs a key $K \in \mathcal{K}_\lambda$.
- $\mathsf{PRF.Eval}$ is a deterministic polynomial-time algorithm that takes as input a key $K \in \mathcal{K}_\lambda$ and a value $x \in \mathcal{X}_\lambda$, and outputs a value $y \in \mathcal{Y}_\lambda$.

The sets $\mathcal{K}_\lambda$, $\mathcal{X}_\lambda$, and $\mathcal{Y}_\lambda$ are referred to as the *key space*, *domain*, and *range* of the function family, respectively. For easy of notation we may denote by $\mathsf{PRF.Eval}_K(\cdot)$ or $\mathsf{PRF}_K(\cdot)$ the function $\mathsf{PRF.Eval}(K, \cdot)$ for $K \in \mathcal{K}_\lambda$. The following is the standard definition of a pseudorandom function family.

Definition 2.3 (Pseudorandomness). *A function family* $\mathsf{PRF} = (\mathsf{PRF.Gen}, \mathsf{PRF.Eval})$ *is* (t, μ)-*secure pseudorandom if for every probabilistic algorithm* $\mathcal{A}$ *that runs in time* $t(\lambda)$, *it holds that*

$$\mathsf{Adv}_{\mathsf{PRF}, \mathcal{A}}(\lambda) \overset{\mathsf{def}}{=}$$
$$\left| \Pr_{K \leftarrow \mathsf{PRF.Gen}(1^\lambda)} \left[\mathcal{A}^{\mathsf{PRF.Eval}_K(\cdot)}(1^\lambda) = 1 \right] - \Pr_{f \leftarrow F_\lambda} \left[\mathcal{A}^{f(\cdot)}(1^\lambda) = 1 \right] \right| \leq \mu(\lambda),$$

for all sufficiently large $\lambda \in \mathbb{N}$, *where* F_λ *is the set of all functions that map* $\mathcal{X}_\lambda$ *into* $\mathcal{Y}_\lambda$.

In addition to the standard notion of a pseudorandom function family, we rely on the seemingly stronger (yet existentially equivalent) notion of a *puncturable* pseudorandom function family [18, 20, 43, 52]. In terms of syntax, this notion asks for an additional probabilistic polynomial-time algorithm, $\mathsf{PRF.Punc}$, that takes as input a key $K \in \mathcal{K}_\lambda$ and a set $S \subseteq \mathcal{X}_\lambda$ and outputs a "punctured" key K_S. The properties required by such a puncturing algorithm are captured by the following definition.

Definition 2.4 (Puncturable PRF). *A* (t, μ)-*secure pseudorandom function family* $\mathsf{PRF} = (\mathsf{PRF.Gen}, \mathsf{PRF.Eval})$ *is* puncturable *if there exists a probabilistic polynomial-time algorithm* $\mathsf{PRF.Punc}$ *such that the following properties are satisfied:*

1. **Functionality:** *For all sufficiently large* $\lambda \in \mathbb{N}$, *for every set* $S \subseteq \mathcal{X}_\lambda$, *and for every* $x \in \mathcal{X}_\lambda \setminus S$ *it holds that*

$$\Pr_{\substack{K \leftarrow \mathsf{PRF.Gen}(1^\lambda); \\ K_S \leftarrow \mathsf{PRF.Punc}(K,S)}} [\mathsf{PRF.Eval}_K(x) = \mathsf{PRF.Eval}_{K_S}(x)] = 1.$$

2. **Pseudorandomness at punctured points:** *Let* $\mathcal{A} = (\mathcal{A}_1, \mathcal{A}_2)$ *be any probabilistic algorithm that runs in time at most* $t(\lambda)$ *such that* $\mathcal{A}_1(1^\lambda)$ *outputs a*

set $S \subseteq \mathcal{X}_\lambda$, a value $x \in S$, and state information state. Then, for any such $\mathcal{A}$ it holds that

$$\mathsf{Adv}_{\mathsf{PRF},\mathcal{A}}(\lambda) \overset{\text{def}}{=}$$
$$|\Pr[\mathcal{A}_2(K_S, \mathsf{PRF.Eval}_K(x), \mathsf{state}) = 1] - \Pr[\mathcal{A}_2(K_S, y, \mathsf{state}) = 1]| \leq \mu(\lambda)$$

for all sufficiently large $\lambda \in \mathbb{N}$, where $(S, x, \mathsf{state}) \leftarrow \mathcal{A}_1(1^\lambda)$, $K \leftarrow \mathsf{PRF.Gen}$ (1^λ), $K_S = \mathsf{PRF.Punc}(K, S)$, and $y \leftarrow \mathcal{Y}_\lambda$.

For our constructions we rely on pseudorandom functions that need to be punctured only at one point (i.e., in both parts of Definition 2.4 it holds that $S = \{x\}$ for some $x \in \mathcal{X}_\lambda$). As observed by [18,20,43,52] the GGM construction [36] of PRFs from any one-way function can be easily altered to yield such a puncturable pseudorandom function family.

2.3   Private-Key Multi-Input Functional Encryption

In this section we define the functionality and security of private-key t-input functional encryption. For $i \in [t]$ let $\mathcal{X}_i = \{(\mathcal{X}_i)_\lambda\}_{\lambda \in \mathbb{N}}$ be an ensemble of finite sets, and let $\mathcal{F} = \{\mathcal{F}_\lambda\}_{\lambda \in \mathbb{N}}$ be an ensemble of finite t-ary function families. For each $\lambda \in \mathbb{N}$, each function $f \in \mathcal{F}_\lambda$ takes as input t strings, $x_1 \in (\mathcal{X}_1)_\lambda, \ldots, x_t \in (\mathcal{X}_t)_\lambda$, and outputs a value $f(x_1, \ldots, x_t) \in \mathcal{Z}_\lambda$.

A private-key t-input functional encryption scheme Π for $\mathcal{F}$ consists of four probabilistic polynomial time algorithm Setup, Enc, KG and Dec, described as follows. The setup algorithm $\mathsf{Setup}(1^\lambda)$ takes as input the security parameter λ, and outputs a master secret key msk. The encryption algorithm $\mathsf{Enc}(\mathsf{msk}, m, \ell)$ takes as input a master secret key msk, a message m, and an index $\ell \in [t]$, where $m \in (\mathcal{X}_\ell)_\lambda$, and outputs a ciphertext ct_ℓ. The key-generation algorithm $\mathsf{KG}(\mathsf{msk}, f)$ takes as input a master secret key msk and a function $f \in \mathcal{F}_\lambda$, and outputs a functional key sk_f. The (deterministic) decryption algorithm Dec takes as input a functional key sk_f and t ciphertexts, $\mathsf{ct}_1, \ldots, \mathsf{ct}_t$, and outputs a string $z \in \mathcal{Z}_\lambda \cup \{\bot\}$.

Definition 2.5 (Correctness). A private-key t-input functional encryption scheme $\Pi = (\mathsf{Setup}, \mathsf{Enc}, \mathsf{KG}, \mathsf{Dec})$ for $\mathcal{F}$ is correct if there exists a negligible function $\mathsf{neg}(\cdot)$ such that for every $\lambda \in \mathbb{N}$, for every $f \in \mathcal{F}_\lambda$, and for every $(x_1, \ldots, x_t) \in (\mathcal{X}_1)_\lambda \times \cdots \times (\mathcal{X}_t)_\lambda$, it holds that

$$\Pr\left[\mathsf{Dec}(\mathsf{sk}_f, \mathsf{Enc}(\mathsf{msk}, x_1, 1), \ldots, \mathsf{Enc}(\mathsf{msk}, x_t, t)) = f(x_1, \ldots, x_t)\right] \geq 1 - \mathsf{neg}(\lambda),$$

where $\mathsf{msk} \leftarrow \mathsf{Setup}(1^\lambda)$, $\mathsf{sk}_f \leftarrow \mathsf{KG}(\mathsf{msk}, f)$, and the probability is taken over the internal randomness of Setup, Enc and KG.

In terms of security, we rely on the private-key variant of the standard indistinguishability-based notion that considers both message privacy and function privacy [2,22,23]. Intuitively, we say that a t-input scheme is secure if for any two t-tuples of messages $(x_1^0, \ldots, x_t^0)$ and $(x_1^1, \ldots, x_t^1)$ that are encrypted with respect to indices $\ell = 1$ through $\ell = t$, and for every pair of functions (f_0, f_1), the triplets $(\mathsf{sk}_{f_0}, \mathsf{Enc}(\mathsf{msk}, x_1^0, 1), \ldots, \mathsf{Enc}(\mathsf{msk}, x_t^0, t))$ and $(\mathsf{sk}_{f_1}, \mathsf{Enc}(\mathsf{msk}, x_1^1, 1), \ldots,$ $\mathsf{Enc}(\mathsf{msk}, x_t^1, t))$ are computationally indistinguishable as long as $f_0(x_1^0, \ldots, x_t^0) = f_1(x_1^1, \ldots, x_t^1)$ (note that this captures both message privacy and function privacy). The formal notions of security build upon this intuition and capture the fact that an adversary may in fact hold many functional keys and ciphertexts, and may combine them in an arbitrary manner. We formalize our notions of security using left-or-right key-generation and encryption oracles. Specifically, for each $b \in \{0, 1\}$ and $\ell \in \{1, \ldots, t\}$ we let the left-or-right key-generation and encryption oracles be $\mathsf{KG}_b(\mathsf{msk}, f_0, f_1) \overset{\mathsf{def}}{=} \mathsf{KG}(\mathsf{msk}, f_b)$ and $\mathsf{Enc}_b(\mathsf{msk}, (m_0, m_1), \ell) \overset{\mathsf{def}}{=} \mathsf{Enc}(\mathsf{msk}, m_b, \ell)$. Before formalizing our notions of security we define the notion of a *valid t-input adversary*. Then, we define *selective security*.

Definition 2.6 (Valid adversary). *A probabilistic polynomial-time algorithm $\mathcal{A}$ is called* valid *if for all private-key t-input functional encryption schemes $\Pi = (\mathsf{Setup}, \mathsf{KG}, \mathsf{Enc}, \mathsf{Dec})$ over a message space $\mathcal{X}_1 \times \cdots \times \mathcal{X}_t = \{(\mathcal{X}_1)_\lambda\}_{\lambda \in \mathbb{N}} \times \cdots \times \{(\mathcal{X}_t)_\lambda\}_{\lambda \in \mathbb{N}}$ and a function space $\mathcal{F} = \{\mathcal{F}_\lambda\}_{\lambda \in \mathbb{N}}$, for all $\lambda \in \mathbb{N}$ and $b \in \{0, 1\}$, and for all $(f_0, f_1) \in \mathcal{F}_\lambda$ and $((x_i^0, x_i^1), i) \in \mathcal{X}_i \times \mathcal{X}_i \times [t]$ with which $\mathcal{A}$ queries the left-or-right key-generation and encryption oracles, respectively, it holds that $f_0(x_1^0, \ldots, x_t^0) = f_1(x_1^1, \ldots, x_t^1)$.*

Definition 2.7 (Selective security). *Let $t = t(\lambda)$, $T = T(\lambda)$, $Q_{\mathsf{key}} = Q_{\mathsf{key}}(\lambda)$, $Q_{\mathsf{enc}} = Q_{\mathsf{enc}}(\lambda)$ and $\mu = \mu(\lambda)$ be functions of the security parameter $\lambda \in \mathbb{N}$. A private-key t-input functional encryption scheme $\Pi = (\mathsf{Setup}, \mathsf{KG}, \mathsf{Enc}, \mathsf{Dec})$ over a message space $\mathcal{X}_1 \times \cdots \times \mathcal{X}_t = \{(\mathcal{X}_1)_\lambda\}_{\lambda \in \mathbb{N}} \times \cdots \times \{(\mathcal{X}_t)_\lambda\}_{\lambda \in \mathbb{N}}$ and a function space $\mathcal{F} = \{\mathcal{F}_\lambda\}_{\lambda \in \mathbb{N}}$ is $(T, Q_{\mathsf{key}}, Q_{\mathsf{enc}}, \mu)$-selectively-secure if for any valid adversary $\mathcal{A}$ that on input 1^λ runs in time $T(\lambda)$ and issues at most $Q_{\mathsf{key}}(\lambda)$ key-generation queries and at most $Q_{\mathsf{enc}}(\lambda)$ encryption queries for each index $i \in [t]$, it holds that*

$$\mathsf{Adv}_{\Pi, \mathcal{F}, \mathcal{A}}^{\mathsf{selFE}_t} \overset{\mathsf{def}}{=} \left| \Pr\left[\mathsf{Exp}_{\Pi, \mathcal{F}, \mathcal{A}}^{\mathsf{selFE}_t}(\lambda) = 1 \right] - \frac{1}{2} \right| \leq \mu(\lambda),$$

for all sufficiently large $\lambda \in \mathbb{N}$, where the random variable $\mathsf{Exp}_{\Pi, \mathcal{F}, \mathcal{A}}^{\mathsf{selFE}_t}(\lambda)$ is defined via the following experiment:

1. *$(\vec{x}_1, \ldots, \vec{x}_t, \mathsf{state}) \leftarrow \mathcal{A}_1(1^\lambda)$, where $\vec{x}_i = ((x_{i,1}^0, x_{i,1}^1), \ldots, (x_{i,T}^0, x_{i,T}^1))$ for $i \in [t]$.*
2. *$\mathsf{msk} \leftarrow \mathsf{Setup}(1^\lambda)$, $b \leftarrow \{0, 1\}$.*
3. *$\mathsf{ct}_{i,j} \leftarrow \mathsf{Enc}(\mathsf{msk}, x_{i,j}^b, 1)$ for $i \in [t]$ and $j \in [T]$.*
4. *$b' \leftarrow \mathcal{A}_2^{\mathsf{KG}_b(\mathsf{msk}, \cdot, \cdot)}\left(1^\lambda, \{\mathsf{ct}_{i,j}\}_{i \in [t], j \in [T]}, \mathsf{state}\right).$*
5. *If $b' = b$ then output 1, and otherwise output 0.*

Known Constructions for $t = 1$. Private-key *single-input* functional encryption schemes that satisfy the above notion of full security and support circuits of any a-priori bounded polynomial size are known to exist based on a variety of assumptions.

Ananth et al. [4] gave a generic transformation from selective security to full security. Moreover, Brakerski and Segev [23] showed how to transform any message-private functional encryption scheme into a functional encryption scheme which is fully secure, and the resulting scheme inherits the security guarantees of the original one. Therefore, based on [4,23], given any selectively-secure message-private functional encryption scheme we can generically obtain a fully secure scheme. This implies that schemes that are fully secure for *any* number of encryption and key-generation queries can be based on indistinguishability obfuscation [33,54], differing-input obfuscation [3,19], and multilinear maps [34]. In addition, schemes that are fully secure for a bounded number of key-generation queries Q_{key} can be based on the Learning with Errors (LWE) assumption (where the length of ciphertexts grows with Q_{key} and with a bound on the depth of allowed functions) [38], and even based on pseudorandom generators computable by small-depth circuits (where the length of ciphertexts grows with Q_{key} and with an upper bound on the circuit size of the functions) [39].

Known Constructions for $t > 1$. Private-key multi-input functional encryption schemes are much less understood than single-input ones. Goldwasser et al. [37] gave the first construction of a selectively-secure multi-input functional encryption scheme for a polynomial number of inputs relying on indistinguishability obfuscation and one-way functions [9,33,44]. Following the work of Goldwasser et al., a *fully-secure* private-key multi-input functional encryption scheme for a polynomial number of inputs based was constructed based on multilinear maps [13]. Later, Ananth, Jain, and Sahai, and Bitasnky and Vaikuntanathan [6,7,12] showed a selectively-secure multi-input functional encryption scheme for a polynomial number of inputs based on any sub-exponentially secure single-input *public-key* functional encryption scheme. Brakerski et al. [22] showed that a *fully-secure* single-input *private-key* scheme implies a *fully-secure* multi-input scheme for any *constant* number of inputs. Furthermore, Brakerski et al. observed that their construction can be used to get a fully-secure t-input scheme for $t = O(\log \log \lambda)$ inputs, where λ is the security parameter, if the underlying single-input scheme is sub-exponentially secure.

2.4   Public-Key Functional Encryption

In this section we define the functionality and security of public-key (single-input) functional encryption. Let $\mathcal{X} = \{\mathcal{X}_\lambda\}_{\lambda \in \mathbb{N}}$ be an ensemble of finite sets, and let $\mathcal{F} = \{\mathcal{F}_\lambda\}_{\lambda \in \mathbb{N}}$ be an ensemble of finite function families. For each $\lambda \in \mathbb{N}$, each function $f \in \mathcal{F}_\lambda$ takes as input a string, $x \in \mathcal{X}_\lambda$, and outputs a value $f(x) \in \mathcal{Z}_\lambda$.

A public-key functional encryption scheme Π for $\mathcal{F}$ consists of four probabilistic polynomial time algorithm Setup, Enc, KG and Dec, described as follows. The setup algorithm Setup(1^λ) takes as input the security parameter λ, and outputs a master secret key msk and a master public key mpk. The encryption algorithm Enc(mpk, m) takes as input a master public key mpk and a message $m \in \mathcal{X}_\lambda$, and outputs a ciphertext ct. The key-generation algorithm KG(msk, f)

takes as input a master secret key msk and a function $f \in \mathcal{F}_\lambda$, and outputs a functional key sk_f. The (deterministic) decryption algorithm Dec takes as input a functional key sk_f and t ciphertexts, $\mathsf{ct}_1, \ldots, \mathsf{ct}_t$, and outputs a string $z \in \mathcal{Z}_\lambda \cup \{\bot\}$.

Definition 2.8 (Correctness). *A public-key functional encryption scheme $\Pi = (\mathsf{Setup}, \mathsf{Enc}, \mathsf{KG}, \mathsf{Dec})$ for $\mathcal{F}$ is correct if there exists a negligible function $\mathsf{neg}(\cdot)$ such that for every $\lambda \in \mathbb{N}$, for every $f \in \mathcal{F}_\lambda$, and for every $x \in \mathcal{X}_\lambda$, it holds that*

$$\Pr\left[\mathsf{Dec}(\mathsf{sk}_f, \mathsf{Enc}(\mathsf{mpk}, x)) = f(x)\right] \geq 1 - \mathsf{neg}(\lambda),$$

where $(\mathsf{msk}, \mathsf{mpk}) \leftarrow \mathsf{Setup}(1^\lambda)$, $\mathsf{sk}_f \leftarrow \mathsf{KG}(\mathsf{msk}, f)$, and the probability is taken over the internal randomness of Setup, Enc and KG.

In terms of security, we rely on the public-key variant of the existing indistinguishability-based notions for message privacy.[7] Intuitively, we say that a scheme is secure if the encryption of any pair of messages $\mathsf{Enc}(\mathsf{mpk}, m_0)$ and $\mathsf{Enc}(\mathsf{mpk}, m_1)$ cannot be distinguished as long as for any function f for which a functional key is queries, it holds that $f(m_0) = f(m_1)$. The formal notions of security build upon this intuition and capture the fact that an adversary may in fact hold many functional keys and ciphertexts, and may combine them in an arbitrary manner. We formalize our notions of security using left-or-right key-generation (similarly to the private-key setting). Specifically, for each $b \in \{0, 1\}$ we let the left-or-right key-generation and encryption oracles be $\mathsf{KG}_b(\mathsf{msk}, f_0, f_1) \overset{\text{def}}{=} \mathsf{KG}(\mathsf{msk}, f_b)$ and $\mathsf{Enc}_b(\mathsf{msk}, (m_0, m_1)) \overset{\text{def}}{=} \mathsf{Enc}(\mathsf{msk}, m_b)$, respectively. Before formalizing our notions of security we define the notion of a *valid adversary*. Then, we define *selective security*.[8]

Definition 2.9 (Valid adversary). *A probabilistic polynomial-time algorithm $\mathcal{A}$ is called valid if for all public-key functional encryption schemes $\Pi = (\mathsf{Setup}, \mathsf{KG}, \mathsf{Enc}, \mathsf{Dec})$ over a message space $\mathcal{X} = \{\mathcal{X}_\lambda\}_{\lambda \in \mathbb{N}}$ and a function space $\mathcal{F} = \{\mathcal{F}_\lambda\}_{\lambda \in \mathbb{N}}$, for all $\lambda \in \mathbb{N}$ and $b \in \{0, 1\}$, and for all $f \in \mathcal{F}_\lambda$ and $((x^0, x^1) \in (\mathcal{X})^2$ with which $\mathcal{A}$ queries the left-or-right encryption oracle, it holds that $f(x^0) = f(x^1)$.*

Definition 2.10 (Selective security). *Let $t = t(\lambda)$, $T = T(\lambda)$, $Q_\mathsf{key} = Q_\mathsf{key}(\lambda)$ and $\mu = \mu(\lambda)$ be functions of the security parameter $\lambda \in \mathbb{N}$. A public-key functional encryption scheme $\Pi = (\mathsf{Setup}, \mathsf{KG}, \mathsf{Enc}, \mathsf{Dec})$ over a message space $\mathcal{X} = \{\mathcal{X}_\lambda\}_{\lambda \in \mathbb{N}}$ and a function space $\mathcal{F} = \{\mathcal{F}_\lambda\}_{\lambda \in \mathbb{N}}$ is (T, Q_key, μ)-selectively secure if for any valid adversary $\mathcal{A}$ that on input 1^λ runs in time $T(\lambda)$ and issues at most $Q_\mathsf{key}(\lambda)$ key-generation queries, it holds that*

$$\mathsf{Adv}_{\Pi, \mathcal{F}, \mathcal{A}}^{\mathsf{sel\text{-}pkFE}} \overset{\text{def}}{=} \left| \Pr\left[\mathsf{Exp}_{\Pi, \mathcal{F}, \mathcal{A}}^{\mathsf{sel\text{-}pkFE}}(\lambda) = 1\right] - \frac{1}{2} \right| \leq \mu(\lambda),$$

---

[7] We note that the notion of *function privacy* is very different from the one in the private-key setting, and in particular, natural definitions already imply obfuscation.

[8] We focus on selective securiy and do not define full security since there is a generic transfomation [4].

for all sufficiently large $\lambda \in \mathbb{N}$, *where the random variable* $\mathsf{Exp}_{\Pi,\mathcal{F},\mathcal{A}}^{\mathsf{sel-pkFE}}(\lambda)$ *is defined via the following experiment:*

1. $(x^0, x^1, \mathsf{state}) \leftarrow \mathcal{A}_1(1^\lambda)$.
2. $(\mathsf{msk}, \mathsf{mpk}) \leftarrow \mathsf{Setup}(1^\lambda)$, $b \leftarrow \{0, 1\}$.
3. $b' \leftarrow \mathcal{A}_2^{\mathsf{KG}_b(\mathsf{msk},\cdot,\cdot)}(1^\lambda, \mathsf{Enc}(\mathsf{mpk}, x^b), \mathsf{state})$.
4. *If* $b' = b$ *then output 1, and otherwise output 0.*

2.5 Indistinguishability Obfuscation

We consider the standard notion of indistinguishability obfuscation [9,33]. We say that two circuits, C_0 and C_1 are *functionally equivalent*, and denote it by $C_0 \equiv C_1$, if for every x it holds that $C_0(x) = C_1(x)$.

Definition 2.11 (Indistinguishability obfuscation). *Let* $\mathcal{C} = \{\mathcal{C}_n\}_{n \in \mathbb{N}}$ *be a class of polynomial-size circuits operating on inputs of length* n. *An efficient algorithm* $i\mathcal{O}$ *is called a* (t, μ) *-indistinguishability obfuscator for the class* $\mathcal{C}$ *if it takes as input a security parameter* λ *and a circuit in* $\mathcal{C}$ *and outputs a new circuit so that following properties are satisfied:*

1. **Functionality:** *For any input length* $n \in \mathbb{N}$, *any* $\lambda \in \mathbb{N}$, *and any* $C \in \mathcal{C}_n$ *it holds that*
$$\Pr\left[C \equiv i\mathcal{O}(1^\lambda, C)\right] = 1,$$
   where the probability is taken over the internal randomness of $i\mathcal{O}$.

2. **Indistinguishability:** *For any probabilistic adversary* $\mathcal{A} = (\mathcal{A}_1, \mathcal{A}_2)$ *that runs in time* $t = t(\lambda)$, *it holds that*
$$\mathsf{Adv}_{i\mathcal{O},\mathcal{C},\mathcal{A}}^{i\mathcal{O}} \overset{\mathsf{def}}{=} \left|\Pr\left[\mathsf{Exp}_{i\mathcal{O},\mathcal{C},\mathcal{A}}^{i\mathcal{O}}(\lambda) = 1\right] - \frac{1}{2}\right| \leq \mu(\lambda),$$

   for all sufficiently large $\lambda \in \mathbb{N}$, *where the random variable* $\mathsf{Exp}_{i\mathcal{O},\mathcal{C},\mathcal{A}}^{i\mathcal{O}}(\lambda)$ *is defined via the following experiment:*
   (a) $(C_0, C_1, \mathsf{state}) \leftarrow \mathcal{A}_1(1^\lambda)$ *such that* $C_0, C_1 \in \mathcal{C}$ *and* $C_0 \equiv C_1$.
   (b) $\widehat{C} \leftarrow i\mathcal{O}(C_b)$, $b \leftarrow \{0, 1\}$.
   (c) $b' \leftarrow \mathcal{A}_2\left(1^\lambda, \widehat{C}, \mathsf{state}\right)$.
   (d) *If* $b' = b$ *then output 1, and otherwise output 0.*

3 Private-Key MIFE for a Poly-Logarithmic Number of Inputs

In this section we present our construction of a private-key multi-input functional encryption scheme. The main technical tool underlying our approach is a transformation from a t-input scheme to a $2t$-input scheme which is described in Sect. 3.1. Then, in Sects. 3.2 and 3.3 we show that by iteratively applying our transformation $O(\log \log \lambda)$ times, and by carefully controlling the security loss and the efficiency loss by adjusting the security parameter appropriately, we obtain a t-input scheme, where $t = (\log \lambda)^\delta$ for some constant $0 < \delta < 1$ (recall that $\lambda \in \mathbb{N}$ denotes the security parameter).

3.1   From t Inputs to $2t$ Inputs

Let $\mathcal{F} = \{\mathcal{F}_\lambda\}_{\lambda \in \mathbb{N}}$ be a family of $2t$-input functionalities, where for every $\lambda \in \mathbb{N}$ the set $\mathcal{F}_\lambda$ consists of functions of the form $f \colon (\mathcal{X}_1)_\lambda \times \cdots \times (\mathcal{X}_{2t})_\lambda \to \mathcal{Z}_\lambda$. Our construction relies on the following building blocks:

1. A private-key t-input functional encryption scheme $\mathsf{FE}_t = (\mathsf{FE}_t.\mathsf{S}, \mathsf{FE}_t.\mathsf{KG}, \mathsf{FE}_t.\mathsf{E}, \mathsf{FE}_t.\mathsf{D})$.
2. A puncturable pseudorandom function family $\mathsf{PRF} = (\mathsf{PRF}.\mathsf{Gen}, \mathsf{PRF}.\mathsf{Eval})$.

Our scheme $\mathsf{FE}_{2t} = (\mathsf{FE}_{2t}.\mathsf{S}, \mathsf{FE}_{2t}.\mathsf{KG}, \mathsf{FE}_{2t}.\mathsf{E}, \mathsf{FE}_{2t}.\mathsf{D})$ is defined as follows.

- **The setup algorithm.** On input the security parameter 1^λ the setup algorithm $\mathsf{FE}_{2t}.\mathsf{S}$ samples a master secret key for a t-input scheme $\mathsf{msk}_{\mathsf{in}} \leftarrow \mathsf{FE}_t.\mathsf{S}(1^\lambda)$, and a PRF key $K^{\mathsf{msk}} \leftarrow \mathsf{PRF}.\mathsf{Gen}(1^\lambda)$, and outputs $\mathsf{msk} = (\mathsf{msk}_{\mathsf{in}}, K^{\mathsf{msk}})$.
- **The key-generation algorithm.** On input the master secret key msk and a function $f \in \mathcal{F}_\lambda$, the key-generation algorithm $\mathsf{FE}_{2t}.\mathsf{KG}$ samples a PRF key $K^{\mathsf{key}} \leftarrow \mathsf{PRF}.\mathsf{Gen}(1^\lambda)$ and outputs $\mathsf{sk}_f \leftarrow \mathsf{FE}_t.\mathsf{KG}(\mathsf{msk}_{\mathsf{in}}, \mathsf{Gen}_{f,\perp,K^{\mathsf{msk}},K^{\mathsf{key}},\perp})$, where $\mathsf{Gen}_{f,\perp,K^{\mathsf{msk}},K^{\mathsf{key}},\perp}$ is the t-input function that is defined in Fig. 3.

$\mathbf{Gen}_{f^0, f^1, K^{\mathsf{msk}}, K^{\mathsf{key}}, w}$

$((x_1^0, x_1^1, \tau_1, \mathsf{c}_1, \mathsf{thr}_1, \ldots, \mathsf{thr}_t), (x_2^0, x_2^1, \tau_2, \mathsf{c}_2), \ldots, (x_t^0, x_t^1, \tau_t, \mathsf{c}_t))$:

1. For $i = 1, \ldots, t$ do:
   (a) If $\mathsf{c}_i < \mathsf{thr}_i$, then set $f = f_1$ and exit loop.
   (b) If $\mathsf{c}_i > \mathsf{thr}_i$, then set $f = f_0$ and exit loop.
   (c) If $\mathsf{c}_i = \mathsf{thr}_i$ and $i < t$, continue to next iteration (with $i = i + 1$).
   (d) If $\mathsf{c}_i = \mathsf{thr}_i$ and $i = t$, then output w and HALT.
2. Compute $r_1 = \mathsf{PRF}.\mathsf{Eval}(K^{\mathsf{msk}}, \tau_1 \ldots \tau_t)$.
3. Compute $r_2 = \mathsf{PRF}.\mathsf{Eval}(K^{\mathsf{key}}, \tau_1 \ldots \tau_t)$.
4. Compute $\mathsf{msk}_{\tau_1, \ldots, \tau_t} = \mathsf{FE}_t.\mathsf{S}(1^\lambda, r_1)$.
5. Output $\mathsf{FE}_t.\mathsf{KG}(\mathsf{msk}_{\tau_1, \ldots, \tau_t}, C_f; r_2)$.

$C_f((x_1, x_{t+1}), \ldots, (x_t, x_{2t}))$:

1. Output $f(x_1, \ldots, x_{2t})$.

Fig. 3. The t-input functions $\mathsf{Gen}_{f^0, f^1, K^{\mathsf{msk}}, K^{\mathsf{key}}, w}$ and C_f.

- **The encryption algorithm.** On input the master secret key msk, a message x and an index $\ell \in [2t]$, the encryption algorithm $\mathsf{FE}_{2t}.\mathsf{E}$ distinguished between the following three cases:
   - If $\ell = 1$, it samples a random string $\tau \in \{0,1\}^\lambda$, and then outputs ct_ℓ defined as follows:

$$\mathsf{ct}_\ell \leftarrow \mathsf{FE}_t.\mathsf{E}(\mathsf{msk}_{\mathsf{in}}, (x, \perp, \tau, 1, \underbrace{1, \ldots, 1}_{t \text{ slots}}, 0), \ell).$$

- If $1 < \ell \leq t$, it samples a random string $\tau \in \{0,1\}^\lambda$, and then outputs ct_ℓ defined as follows:

$$\mathsf{ct}_\ell \leftarrow \mathsf{FE}_t.\mathsf{E}(\mathsf{msk}_{\mathsf{in}}, (x, \bot, \tau, 1), \ell).$$

- If $t < \ell \leq 2t$, it samples a PRF key $K^{\mathsf{enc}} \leftarrow \mathsf{PRF}.\mathsf{Gen}(1^\lambda)$ and outputs sk_ℓ defined as follows:

$$\mathsf{sk}_\ell \leftarrow \mathsf{FE}_t.\mathsf{KG}(\mathsf{msk}_{\mathsf{in}}, \mathsf{AGG}_{x, \bot, \ell, K^{\mathsf{msk}}, K^{\mathsf{enc}}, \bot}),$$

where $\mathsf{AGG}_{x, \bot, \ell, K^{\mathsf{msk}}, K^{\mathsf{enc}}, \bot}$ is the t-input function that is defined in Fig. 4.

$\mathsf{AGG}_{x^0_{\ell+t}, x^1_{\ell+t}, \ell+t, K^{\mathsf{msk}}, K^{\mathsf{enc}}, v}$

$((x^0_1, x^1_1, \tau_1, \mathsf{c}_1, \mathsf{thr}_1, \ldots, \mathsf{thr}_t), (x^0_2, x^1_2, \tau_2, \mathsf{c}_2), \ldots, (x^0_t, x^1_t, \tau_t, \mathsf{c}_t))$:

1. For $i = 1, \ldots, t$ do:
   (a) If $\mathsf{c}_i < \mathsf{thr}_i$, then set $x_i = x^1_i$ for all $i \in [t]$ and exit loop.
   (b) If $\mathsf{c}_i > \mathsf{thr}_i$, then set $x_i = x^0_i$ for all $i \in [t]$ and exit loop.
   (c) If $\mathsf{c}_i = \mathsf{thr}_i$ and $i < t$, continue to next iteration (with $i = i + 1$).
   (d) If $\mathsf{c}_i = \mathsf{thr}_i$ and $i = t$, output v and HALT.
2. Compute $r_1 = \mathsf{PRF}.\mathsf{Eval}(K^{\mathsf{msk}}, \tau_1 \ldots \tau_t)$.
3. Compute $r_2 = \mathsf{PRF}.\mathsf{Eval}(K^{\mathsf{enc}}, \tau_1 \ldots \tau_t)$.
4. Compute $\mathsf{msk}_{\tau_1, \ldots, \tau_t} = \mathsf{FE}_t.\mathsf{S}(1^\lambda, r_1)$
5. Output $\mathsf{FE}_t.\mathsf{E}(\mathsf{msk}_{\tau_1, \ldots, \tau_t}, (x_\ell, x_{\ell+t}), \ell; r_2)$.

Fig. 4. The t-input function $\mathsf{AGG}_{x^0_{\ell+t}, x^1_{\ell+t}, \ell+t, K^{\mathsf{msk}}, K^{\mathsf{enc}}, v}$.

- **The decryption algorithm.** On input a functional key sk_f and ciphertexts $\mathsf{ct}_1, \ldots, \mathsf{ct}_t, \mathsf{sk}_{t+1}, \ldots, \mathsf{sk}_{2t}$, the decryption algorithm $\mathsf{FE}_t.\mathsf{D}$ computes

$$\forall i \in \{t+1, \ldots, 2t\}: \mathsf{ct}'_i = \mathsf{FE}_t.\mathsf{D}(\mathsf{sk}_i, \mathsf{ct}_1, \ldots, \mathsf{ct}_t)$$
$$\mathsf{sk}' = \mathsf{FE}_t.\mathsf{D}(\mathsf{sk}_f, \mathsf{ct}_1, \ldots, \mathsf{ct}_t),$$

and outputs $\mathsf{FE}_t.\mathsf{D}(\mathsf{sk}', \mathsf{ct}'_{t+1}, \ldots, \mathsf{ct}'_{2t})$.

Correctness. For any $\lambda \in \mathbb{N}$, $f \in \mathcal{F}_\lambda$ and $(x_1, \ldots, x_{2t}) \in (\mathcal{X}_1)_\lambda \times \cdots \times (\mathcal{X}_{2t})_\lambda$, let sk_f denote a functional key for f and let $\mathsf{ct}_1, \ldots, \mathsf{ct}_t, \mathsf{sk}_{t+1}, \ldots, \mathsf{sk}_{2t}$ denote encryptions of $x_1, \ldots, x_{2t}$. Then, for every $i \in \{1, \ldots, t\}$, it holds that

$$\mathsf{ct}'_{i+t} = \mathsf{FE}_t.\mathsf{D}(\mathsf{sk}_{i+t}, \mathsf{ct}_1, \ldots, \mathsf{ct}_t)$$
$$= \mathsf{AGG}_{x_{i+t}, \bot, i+t, K^{\mathsf{msk}}, K^{\mathsf{enc}}_{i+t}, \bot}((x_1, \bot, \tau_1, 1, 1, \ldots, 1, 0), (x_2, \bot, \tau_2, 1), \ldots,$$
$$(x_t, \bot, \tau_t, 1))$$
$$= \mathsf{FE}_t.\mathsf{E}(\mathsf{msk}_{\tau_1, \ldots, \tau_t}, (x_i, x_{i+t}), i; \mathsf{PRF}.\mathsf{Eval}(K^{\mathsf{enc}}_{i+t}, \tau_1 \ldots \tau_t))$$

and

$$\mathsf{sk}' = \mathsf{FE}_t.\mathsf{D}(\mathsf{sk}_f, \mathsf{ct}_1, \ldots, \mathsf{ct}_t)$$

$$= \mathsf{Gen}_{f, \perp, K^{\mathsf{msk}}, K^{\mathsf{key}}_f, \perp}((x_1, \perp, \tau_1, 1, 1, \ldots, 1, 0), (x_2, \perp, \tau_2, 1), \ldots, (x_t, \perp, \tau_t, 1))$$

$$= \mathsf{FE}_t.\mathsf{KG}(\mathsf{msk}_{\tau_1, \ldots, \tau_t}, C_f; \mathsf{PRF}.\mathsf{Eval}(K^{\mathsf{key}}_f, \tau_1 \ldots \tau_t))$$

where $\mathsf{msk}_{\tau_1, \ldots, \tau_t} = \mathsf{FE}_t.\mathsf{S}(1^\lambda, \mathsf{PRF}.\mathsf{Eval}(K^{\mathsf{msk}}, \tau_1 \ldots \tau_t))$. Therefore,

$$\mathsf{FE}_t.\mathsf{D}(\mathsf{sk}', \mathsf{ct}'_{t+1}, \ldots, \mathsf{ct}'_{2t}) = C_f((x_1, x_{t+1}), \ldots, (x_t, x_{2t})) = f(x_1, \ldots, x_{2t}).$$

Security. The following theorem captures the security our transformation. The proof can be found in the full version [45].

Theorem 3.1. *Let $t = t(\lambda)$, $T = T(\lambda)$, $Q_{\mathsf{key}} = Q_{\mathsf{key}}(\lambda)$, $Q_{\mathsf{enc}} = Q_{\mathsf{enc}}(\lambda)$ and $\mu = \mu(\lambda)$ be functions of the security parameter $\lambda \in \mathbb{N}$, and assume that FE_t is a $(T, Q_{\mathsf{key}}, Q_{\mathsf{enc}}, \mu)$-selectively-secure t-input functional encryption scheme and that PRF is a (T, μ)-secure puncturable pseudorandom function family. Then, FE_{2t} is $(T', Q'_{\mathsf{key}}, Q'_{\mathsf{enc}}, \mu')$-selectively-secure, where*

- *$T'(\lambda) = T(\lambda) - Q_{\mathsf{key}}(\lambda) \cdot \mathsf{poly}(\lambda)$, for some fixed polynomial $\mathsf{poly}(\cdot)$.*
- *$Q'_{\mathsf{key}}(\lambda) = Q_{\mathsf{key}}(\lambda) - t(\lambda) \cdot Q_{\mathsf{enc}}(\lambda)$.*
- *$Q'_{\mathsf{enc}}(\lambda) = Q_{\mathsf{enc}}(\lambda)$.*
- *$\mu'(\lambda) = 8t(\lambda) \cdot (Q_{\mathsf{enc}}(\lambda))^{t(\lambda)+1} \cdot Q_{\mathsf{key}}(\lambda) \cdot \mu(\lambda)$.*

3.2 Efficiency Analysis

In this section we analyze the overhead incurred by our transformation. Specifically, for a message space $\mathcal{X}_1 \times \cdots \times \mathcal{X}_{2t}$ and a function space $\mathcal{F}$ that consists of $2t$-input functions, we instantiate our scheme (by applying our transformation $\log t$ times) and analyze the size of a master secret key, the size of a functional-key, the size of a ciphertext and the time it takes to evaluate a functional-key with $2t$ ciphertexts.

Let $\lambda \in \mathbb{N}$ be a security parameter with which we instantiate the $2t$-input scheme, let us assume that $\mathcal{F}$ consists of functions of size at most $s = s(\lambda)$ and that each $\mathcal{X}_i$ consists of messages of size at most $m = m(\lambda)$. Assuming that $\log t \leq \mathsf{poly}(\lambda)$ (to simplify notation), we show that there exists a fixed constant $\mathsf{c} \in \mathbb{N}$ such that:

- the setup procedure takes time λ^{c},
- the key-generation procedure takes time $(s \cdot \lambda)^{t^{\log \mathsf{c}}}$,
- the encryption procedure takes time $(m \cdot \lambda)^{t^{\log \mathsf{c}}}$, and
- the decryption procedure takes time $t^{\log t} \cdot \lambda^{\mathsf{c}}$.

In Sect. 3.3 we will choose s, m, t and λ to satisfy Lemma 3.2.

For a circuit A that receives inputs of lengths $x_1 \ldots, x_m$, we denote by $\mathsf{Time}(A, x_1, \ldots, x_m)$ the size of the circuit when applied to inputs of length

$\sum_{i=1}^{m} x_i$. For a function family $\mathcal{F}$, we denote by $\mathsf{Size}(\mathcal{F})$ the maximal size of the circuit that implements a function from $\mathcal{F}$.

We analyze the overhead incurred by our transformation

The Setup Procedure. The setup procedure of FE_{2t} is composed of sampling a key for a scheme FE_t and generating a PRF key. Iterating this, we see that a master secret key in our final scheme consists of a single master secret key for a single-input scheme and $\log t$ additional PRF keys. Namely,

$$\mathsf{Time}(\mathsf{FE}_{2t}.\mathsf{S}, 1^\lambda) = \mathsf{Time}(\mathsf{FE}_t.\mathsf{S}, 1^\lambda) + p_1(\lambda),$$

where p_1 is a fixed polynomial that depends on the key-generation time of the PRF, and thus

$$\mathsf{Time}(\mathsf{FE}_{2t}.\mathsf{S}, \lambda) = \mathsf{Time}(\mathsf{FE}_1.\mathsf{S}, \lambda) + \log t \cdot p_1(\lambda).$$

The Key-Generation Procedure. The key-generation procedure of FE_{2t} depends on the complexity of the key-generation procedure of the FE_t scheme. Let $\mathcal{F}^{2t}$ be the function family that is supported by the scheme FE_{2t}.

$$\mathsf{Time}(\mathsf{FE}_{2t}.\mathsf{KG}, \lambda, \mathsf{Size}(\mathsf{FE}_{2t}.\mathsf{S}, \lambda), \mathsf{Size}(\mathcal{F}^{2t})) =$$
$$\mathsf{Time}(\mathsf{FE}_t.\mathsf{KG}, \lambda, 2\mathsf{Size}(\mathcal{F}^{2t}), \mathsf{Time}(\mathsf{FE}_t.\mathsf{S}, \lambda), \mathsf{Time}(\mathsf{FE}_t.\mathsf{KG}, \mathsf{Size}(\mathcal{F}^{2t})), p_2(\lambda)))$$
$$+ p_3(\lambda),$$

where p_2 subsumes the size of the embedded PRF keys and the complexity of the simple operations that are done in Gen, and p_3 subsumes the running time of the generation of the PRF key K^{key}.

The dominant part in the above equation is that the time it takes to generate a key with respect to FE_{2t} for a function whose size is $\mathsf{Size}(\mathcal{F}^{2t})$ depends on the circuit size of key-generation in the scheme FE_t for a function whose size is $\mathsf{Time}(\mathsf{FE}_t.\mathsf{KG}, \mathsf{Size}(\mathcal{F}^{2t}))$ (namely, it is a function that outputs a functional key for a function whose size is $\mathsf{Size}(\mathcal{F}^{2t})$). Thus, applying this equation recursively, we get that for large enough $c \in \mathbb{N}$ (that depends on the exponents of p_2 and p_3), it holds that

$$\mathsf{Time}(\mathsf{FE}_{2t}.\mathsf{KG}, \lambda, \mathsf{Time}(\mathsf{FE}_{2t}.\mathsf{S}, \lambda), \mathsf{Size}(\mathcal{F}^{2t})) \leq$$
$$(\mathsf{Size}(\mathcal{F}^{2t}) \cdot \lambda)^{c^{\log t}} = (\mathsf{Size}(\mathcal{F}^{2t}) \cdot \lambda)^{t^{\log c}}.$$

The Encryption Procedure. The encryption procedure of FE_{2t} depends on the complexity of encryption and key-generation of the FE_t scheme. Let m be the length of a message to encrypt. For $\ell \leq t$, the complexity is at most

$$\mathsf{Time}(\mathsf{FE}_{2t}.\mathsf{E}, \lambda, \mathsf{Size}(\mathsf{FE}_{2t}.\mathsf{S}, \lambda), m) \leq \mathsf{Time}(\mathsf{FE}_t.\mathsf{E}, \lambda, 2m, (t+2)\lambda).$$

For $t + 1 \leq \ell \leq 2t$, the complexity of encryption is

$$\mathsf{Time}(\mathsf{FE}_{2t}.\mathsf{E}, \lambda, \mathsf{Size}(\mathsf{FE}_{2t}.\mathsf{S}, \lambda), m) \leq$$
$$\mathsf{Time}(\mathsf{FE}_t.\mathsf{KG}, \lambda, \mathsf{Time}(\mathsf{FE}_t.\mathsf{S}, \lambda), \mathsf{Time}(\mathsf{FE}_t.\mathsf{E}, 2m), p_4(\lambda)),$$

where p_4 subsumes the running time of the key-generation procedure of the PRF and the various other simple operations made by AGG.

The dominant part is that an encryption of a message with respect to the scheme FE_{2t} requires generating a key with respect to the scheme FE_t for a function whose size is $\mathsf{Time}(\mathsf{FE}_t.\mathsf{E}, 2m)$. Thus, similarly to the analysis of the key-generation procedure, we get that for some fixed $c \in \mathbb{N}$ (that depends on the exponents of p_4 and the time it takes to encrypt a message with respect to FE_1), we get that

$$\mathsf{Time}(\mathsf{FE}_{2t}.\mathsf{E}, \lambda, \mathsf{Size}(\mathsf{FE}_{2t}.\mathsf{S}, \lambda), m) \leq (m \cdot \lambda)^{t^{\log c}}.$$

The Decryption Procedure. Decryption in the scheme FE_{2t} requires $t + 2$ decryption operations with respect to the scheme FE_t. Let $\mathsf{ct}(t)$ and $\mathsf{sk}(t)$ be the length of a ciphertext and a key in the scheme FE_t, respectively. We get that

$$\mathsf{Time}(\mathsf{FE}_{2t}.\mathsf{D}, \mathsf{sk}(t), 2t \cdot \mathsf{ct}(t)) =$$
$$(t + 2) \cdot \mathsf{Time}(\mathsf{FE}_t.\mathsf{D}, \mathsf{sk}(t), t \cdot \mathsf{ct}(t)) \leq (t + 2)^{\log t} \cdot p_5(\lambda),$$

where p_5 is a polynomial that subsumes the complexity of decryption in FE_1.

3.3 Iteratively Applying Our Transformation

In this section we show that by iteratively applying our transformation $O(\log \log \lambda)$ times we obtain a t-input scheme, where $t = (\log \lambda)^\delta$ for some constant $0 < \delta < 1$. We prove the following two theorems:

Lemma 3.2. *Let $T = T(\lambda)$, $Q_{\mathsf{key}} = Q_{\mathsf{key}}(\lambda)$, $Q_{\mathsf{enc}} = Q_{\mathsf{enc}}(\lambda)$ and $\mu = \mu(\lambda)$ be functions of the security parameter $\lambda \in \mathbb{N}$ and let $\epsilon \in (0, 1)$. Assume any $(T, Q_{\mathsf{key}}, Q_{\mathsf{enc}}, \mu)$-selectively-secure single-input private-key functional encryption scheme with the following properties:*

1. *it supports circuits and messages of size $\mathsf{poly}(2^{(\log \lambda)^{2\epsilon}})$ and*
2. *the size of a ciphertext and a functional key is bounded by $\mathsf{poly}(2^{(\log \lambda)^{2\epsilon}})$,*

then for some constant $\delta \in (0, 1)$, there exists a $\left(T', Q'_{\mathsf{key}}, Q'_{\mathsf{enc}}, \mu'\right)$-selectively-secure $(\log \lambda)^\delta$-input private-key functional encryption scheme with the following properties:

1. *it supports circuits and messages of size $\mathsf{poly}(2^{(\log \lambda)^\epsilon})$,*
2. *$T'(\lambda) \geq T(\lambda) - (\log \log \lambda) \cdot p(\lambda)$,*
3. *$Q'_{\mathsf{key}}(\lambda) \geq Q_{\mathsf{key}}(\lambda) - (2 \log \lambda) \cdot Q_{\mathsf{enc}}(\lambda)$,*
4. *$Q'_{\mathsf{enc}}(\lambda) = Q_{\mathsf{enc}}(\lambda)$, and*
5. *$\mu'(\lambda) \leq 2^{(3 \log \log \lambda)^2} \cdot (Q_{\mathsf{enc}}(\lambda))^{2(\log \lambda)^\delta + 2} \cdot (Q_{\mathsf{key}}(\lambda))^{\log \log \lambda} \cdot \mu(\lambda)$.*

Proof. Let FE_1 be a $(T, Q_{\mathsf{key}}, Q_{\mathsf{enc}}, \mu)$-selectively-secure single-input scheme with the properties from the statement.

Let us analyze the complexity of the t-input scheme where $t(\lambda) = (\log \lambda)^\delta$, where $\delta > 0$ is some fixed constant that we fix later. In terms of complexity, using the properties of the single-input scheme and our efficiency analysis from Sect. 3.2, we have that *setup* takes a polynomial time in λ, *key-generation* for a function of size s takes time at most $(s \cdot \lambda)^{t^{\log c}}$ and *encryption* of a message of length m takes time $(m \cdot \lambda)^{t^{\log c}}$ for some large enough constant $c > 1$ (recall that c is an upper bound on the exponents of the running time of key generation and encryption procedures of the underlying single-input scheme). Plugging in $\delta = 2\epsilon/(3 \log c)$, $t = (\log \lambda)^\delta$ and $s, m \leq 2^{c' \cdot (\log \lambda)^\epsilon}$ for any $c' \in \mathbb{N}$, we get that key-generation and encryption take time at most $2^{c' \cdot (\log \lambda)^{2\epsilon/3} \cdot (\log \lambda)^\epsilon} = 2^{c' \cdot (\log \lambda)^{5\epsilon/3}}$. Notice that for large enough λ, decryption of such a key-message pair takes time at most $\mathsf{poly}(2^{(\log \lambda)^{5\epsilon/3}}) \cdot (t+2)^{\log t} \leq 2^{(\log \lambda)^{2\epsilon}}$.

In terms of security, by Theorem 3.1, we have that if FE_t is $(T^{(t)}, Q_{\mathsf{key}}^{(t)}, Q_{\mathsf{enc}}^{(t)}, \mu^{(t)})$-selectively-secure and PRF is a $(T^{(t)}, \mu^{(t)})$-secure puncturable pseudorandom function family, then FE_{2t} is $(T^{(2t)}, Q_{\mathsf{key}}^{(2t)}, Q_{\mathsf{enc}}^{(2t)}, \mu^{(2t)})$-selectively-secure, where

1. $T^{(2t)}(\lambda) = T^{(t)}(\lambda) - p(\lambda)$,
2. $Q_{\mathsf{key}}^{(2t)}(\lambda) = Q_{\mathsf{key}}^{(t)}(\lambda) - t \cdot Q_{\mathsf{enc}}^{(t)}$,
3. $Q_{\mathsf{enc}}^{(2t)}(\lambda) = Q_{\mathsf{enc}}^{(t)}(\lambda)$, and
4. $\mu^{(2t)}(\lambda) = 2^{(3 \log \log \lambda)^2} \cdot (Q_{\mathsf{enc}}(\lambda))^{2(\log \lambda)^\delta + 2} \cdot (Q_{\mathsf{key}}(\lambda))^{\log \log \lambda} \cdot \mu(\lambda)$.

Iterating these recursive equations, using the fact that $Q_{\mathsf{key}}^{(2t)} \leq Q_{\mathsf{key}}^{(t)}$, and plugging in our initial scheme parameters, we get that

$$Q'_{\mathsf{enc}}(\lambda) = Q_{\mathsf{enc}}^{(1)}(\lambda) = Q_{\mathsf{enc}}(\lambda)$$

$$\begin{aligned} Q'_{\mathsf{key}}(\lambda) &= Q_{\mathsf{key}}^{(t)}(\lambda) - t(\lambda) \cdot Q_{\mathsf{enc}}(\lambda) \\ &\geq Q_{\mathsf{key}}(\lambda) - 2t(\lambda) \cdot Q_{\mathsf{enc}}(\lambda) \\ &\geq Q_{\mathsf{key}}(\lambda) - (2 \log(\lambda)) \cdot Q_{\mathsf{enc}}(\lambda) \end{aligned}$$

$$\begin{aligned} T'(\lambda) &\geq T(\lambda) - \log t(\lambda) \cdot p(\lambda) \\ &\geq T(\lambda) - (\log \log \lambda) \cdot p(\lambda) \end{aligned}$$

$$\begin{aligned} \mu'(\lambda) &\leq (8t(\lambda))^{\log t(\lambda)} \cdot (Q_{\mathsf{enc}}(\lambda))^{2t(\lambda)+2} \cdot (Q_{\mathsf{key}}(\lambda))^{\log t(\lambda)} \cdot \mu(\lambda) \\ &\leq 2^{(3 \log t(\lambda))^2} \cdot (Q_{\mathsf{enc}}(\lambda))^{2t(\lambda)+2} \cdot (Q_{\mathsf{key}}(\lambda))^{\log t(\lambda)} \cdot \mu(\lambda) \\ &\leq 2^{(3 \log \log \lambda)^2} \cdot (Q_{\mathsf{enc}}(\lambda))^{2(\log \lambda)^\delta + 2} \cdot (Q_{\mathsf{key}}(\lambda))^{\log \log \lambda} \cdot \mu(\lambda) \end{aligned}$$

∎

Claim 3.3. *Let $\lambda \in \mathbb{N}$ be a security parameter and fix any constant $\epsilon \in (0,1)$. Assuming any $(2^{2 \cdot (\log \lambda)^{1/\epsilon}}, 2^{2 \cdot (\log \lambda)^{1/\epsilon}}, 2^{(\log \lambda)^{1/\epsilon}}, 2^{-(\log \lambda)^{1.5/\epsilon}})$-selectively-secure single-input private-key functional encryption scheme supporting polynomial-size circuits, there exists a $(2^{2 \cdot (\log \lambda)^2}, 2^{2 \cdot (\log \lambda)^2}, 2^{(\log \lambda)^2}, 2^{-(\log \lambda)^3})$-selectively-secure single-input private-key functional encryption scheme with the following properties*

1. it supports circuits and messages of size $\mathsf{poly}(2^{(\log \lambda)^{2\epsilon}})$ and
2. the size of a ciphertext and a functional key is bounded by $\mathsf{poly}(2^{(\log \lambda)^{2\epsilon}})$.

Proof. We instantiate the given scheme with security parameter $\tilde{\lambda} = 2^{(\log \lambda)^{2\epsilon}}$. The resulting scheme is $\left(2^{2 \cdot (\log \lambda)^2}, 2^{2 \cdot (\log \lambda)^2}, 2^{(\log \lambda)^2}, 2^{-(\log \lambda)^3}\right)$-selectively-secure and for a circuit (resp., message) of size $\tilde{\lambda}$, the size of a functional key (resp., ciphertext) is bounded by $\mathsf{poly}(\tilde{\lambda})$. ∎

Combining Theorem 3.3 and Lemma 3.2 we get the following theorem.

Theorem 3.4. *Let $\lambda \in \mathbb{N}$ be a security parameter and fix any constant $\epsilon \in (0,1)$. Assuming any $(2^{2 \cdot (\log \lambda)^{1/\epsilon}}, 2^{1 \cdot (\log \lambda)^{2/\epsilon}}, 2^{(\log \lambda)^{1/\epsilon}}, 2^{-(\log \lambda)^{1.5/\epsilon}})$-selectively-secure single-input private-key functional encryption scheme supporting polynomial-size circuits, then for some $\delta \in (0,1)$, there exists a $(2^{(\log \lambda)^2}, 2^{(\log \lambda)^2}, 2^{(\log \lambda)^2}, 2^{-(\log \lambda)^2})$-selectively-secure $(\log \lambda)^\delta$-input private-key functional encryption scheme supporting circuits of size $2^{(\log \lambda)^\epsilon}$.*

Proof. Assuming any $\left(2^{2 \cdot (\log \lambda)^{1/\epsilon}}, 2^{2 \cdot (\log \lambda)^{1/\epsilon}}, 2^{(\log \lambda)^{1/\epsilon}}, 2^{-(\log \lambda)^{1.5/\epsilon}}\right)$-selectively-secure single-input private-key functional encryption scheme supporting polynomial-size circuits. By 3.3, it implies a $\left(2^{2 \cdot (\log \lambda)^2}, 2^{2 \cdot (\log \lambda)^2}, 2^{(\log \lambda)^2}, 2^{-(\log \lambda)^3}\right)$-selectively-secure single-input private-key functional encryption scheme with the following properties:

1. it supports circuits and messages of size $\mathsf{poly}(2^{(\log \lambda)^{2\epsilon}})$ and
2. the size of a ciphertext and a functional key is bounded by $\mathsf{poly}(2^{(\log \lambda)^{2\epsilon}})$.

Using Lemma 3.2, we get that for some constant $\delta \in (0,1)$, there exists a $\left(T', Q'_{\mathsf{key}}, Q'_{\mathsf{enc}}, \mu'\right)$-selectively-secure $(\log \lambda)^\delta$-input private-key functional encryption scheme with the following properties:

1. it supports circuits and messages of size at most $\mathsf{poly}(2^{(\log \lambda)^{\epsilon/2}})$,
2. $T'(\lambda) \geq 2^{2 \cdot (\log \lambda)^2} - (\log \log \lambda) \cdot p(\lambda) \geq 2^{(\log \lambda)^2}$,
3. $Q'_{\mathsf{key}}(\lambda) \geq 2^{2 \cdot (\log \lambda)^2} - (2 \log \lambda) \cdot 2^{(\log \lambda)^2} \geq 2^{(\log \lambda)^2}$,
4. $Q'_{\mathsf{enc}}(\lambda) = 2^{(\log \lambda)^2}$, and
5. $\mu'(\lambda) \leq 2^{(3 \log \log \lambda)^2} \cdot (2^{(\log \lambda)^2})^{2^{(\log \lambda)^\delta}+2} \cdot (2^{(\log \lambda)^2})^{\log \log \lambda} \cdot 2^{-(\log \lambda)^3} \leq 2^{-(\log \lambda)^2}$. ∎

4   Applications of Our Construction

In this section we present our construction of an indistinguishability obfuscator for circuits with inputs of poly-logarithmic length, and its applications to public-key functional encryption and average-case PPAD hardness.

4.1   Obfuscation for Circuits with Poly-Logarithmic Input Length

We show that any selectively-secure t-input private-key functional encryption scheme that supports circuits of size s can be used to construct an indistinguishability obfuscator that supports circuits of size s that have at most $t \cdot \log \lambda$ inputs, where $\lambda \in \mathbb{N}$ is the security parameter. This is similar to the proof of Goldwasser et al. [37] that showed that private-key multi-input functional encryption for a polynomial number of inputs implies indistinguishability obfuscation (and a follow-up refinement of Bitansky et al. [10]).

We consider the following restricted class of circuits:

Definition 4.1. *Let $\lambda \in \mathbb{N}$ and let $s(\cdot)$ and $t'(\cdot)$ be functions. Let $\mathcal{C}_\lambda^{s,t'}$ denoet the class of all circuits of size at most $s(\lambda)$ that get as input $t'(\lambda)$ bits.*

Lemma 4.2. *Let $t = t(\lambda)$, $s = s(\lambda)$, $T = T(\lambda)$, $Q_{\mathsf{key}} = Q_{\mathsf{key}}(\lambda)$, $Q_{\mathsf{enc}} = Q_{\mathsf{enc}}(\lambda)$ and $\mu = \mu(\lambda)$ be functions of the security parameter $\lambda \in \mathbb{N}$, and assume a $(T, Q_{\mathsf{key}}, Q_{\mathsf{enc}}, \mu)$-selectively-secure t-input private-key functional encryption scheme for functions of size at most s, where $Q_{\mathsf{key}}(\lambda) \geq 1$ and $Q_{\mathsf{enc}}(\lambda) \geq \lambda$. Then, there exists a $(T(\lambda) - \lambda \cdot t(\lambda) \cdot p(\lambda), \mu(\lambda))$-secure indistinguishability obfuscator for the circuit class $\mathcal{C}_\lambda^{s,t'}$, where $p(\cdot)$ is some fixed polynomial and $t'(\lambda) = t(\lambda) \cdot \log \lambda$.*

Proof. Let FE_t be a t-input scheme as in the statement of the lemma. We construct an obfuscator for circuits of size at most $s(\lambda)$ that receive $t(\lambda) \cdot \log \lambda$ bits as input. On input a circuit $C \in \mathcal{C}_\lambda^{s,t'}$, the obfuscator works as follows:

1. Sample a master secret key $\mathsf{msk} \leftarrow \mathsf{FE}_t.\mathsf{S}(1^\lambda)$.
2. Compute $\mathsf{ct}_{i,j} = \mathsf{FE}_t.\mathsf{E}(\mathsf{msk}, i, j)$ for every $i \in \{0,1\}^{\log \lambda}$ and $j \in [t(\lambda)]$.
3. Compute $\mathsf{sk}_C = \mathsf{FE}_t.\mathsf{KG}(\mathsf{msk}, C)$
4. Output $\widehat{C} = \{\mathsf{sk}_C\} \cup \{\mathsf{ct}_{i,j}\}_{i \in \{0,1\}^{\log \lambda}, j \in [t(\lambda)]}$.

Evaluation of an obfuscated circuit $\widehat{C}$ on an input $x \in (\{0,1\}^{\log \lambda})^t$, where we view x as $x = x_1 \ldots x_t$ and $x_i \in \{0,1\}^{\log \lambda}$, is done by outputting the result of a single execution of the decryption procedure of the t-input scheme $\mathsf{FE}_t.\mathsf{D}(\mathsf{sk}_C, \mathsf{ct}_{x_1,1}, \ldots, \mathsf{ct}_{x_t,t})$. Notice that the description size of the obfuscated circuit is upper bounded by some fixed polynomial in λ.

For security, notice that a single functional key is generated and it is for a circuit of size at most $s(\lambda)$. Moreover, the number of ciphertexts is bounded by λ ciphertexts per coordinate. Thus, following [37], one can show that an adversary that can break the security of the above obfuscator can be used to break the security of the FE_t scheme with the same success probability (it can even break FE_t that satisfies a weaker security notion in which the functional keys are also fixed ahead of time, before seeing any ciphertext). ∎

Applying Lemma 4.2 with the t-input scheme from Theorem 3.4 we obtain the following corollary.

Corollary 4.3. *Let* $\lambda \in \mathbb{N}$ *be a security parameter and fix any constant* $\epsilon \in (0,1)$. *Assume a* $(2^{2(\log \lambda)^{1/\epsilon}}, 2^{2(\log \lambda)^{1/\epsilon}}, 2^{(\log \lambda)^{1/\epsilon}}, 2^{-(\log \lambda)^{1.5/\epsilon}})$-selectively-secure single-input private-key functional encryption scheme for all functions of polynomial size. Then, for some constant* $\delta \in (0,1)$, *there exists a* $(2^{(\log \lambda)^2}, 2^{-(\log \lambda)^2})$-secure indistinguishability obfuscator for the circuit class* $\mathcal{C}_\lambda^{2^{O((\log \lambda)^\epsilon)}, (\log \lambda)^{1+\delta}}$.

4.2 Public-Key Functional Encryption

In this section we present a construction of a public-key functional encryption scheme based on our multi-input private-key scheme.

Theorem 4.4. *Let* $\lambda \in \mathbb{N}$ *be a security parameter and fix any* $\epsilon \in (0,1)$. *There exists a constant* $\delta > 0$ *for which the following holds. Assume a* $(2^{2(\log \lambda)^{1/\epsilon}}, 2^{2(\log \lambda)^{1/\epsilon}}, 2^{(\log \lambda)^{1/\epsilon}}, 2^{-(\log \lambda)^{1.5/\epsilon}})$-selectively-secure single-input private-key functional encryption scheme for all functions of polynomial size, and that* $(2^{2\lambda^{\epsilon'}}, 2^{-2\lambda^{\epsilon'}})$-secure one-way functions exist for* $\epsilon' > 1/(1+\delta)$. *Then, for some constant* $\zeta > 1$, *there exists a* $(2^{(\log \lambda)^\zeta}, 2^{(\log \lambda)^\zeta}, 2^{-(\log \lambda)^\zeta})$-selectively-secure public-key encryption scheme for the circuit class* $\mathcal{C}_\lambda^{2^{O((\log \lambda)^\epsilon)}, (\log \lambda)^{1+\delta}}$.

Our construction is essentially the construction of Waters [54], who showed how to construct a public-key functional encryption scheme for the set of all polynomial-size circuits assuming indistinguishability obfuscation for all polynomial-size circuits. We make a more careful analysis of his scheme and show that for a specific range of parameters, it suffices to use the obfuscator we have obtained in Corollary 4.3. The proof of Theorem 4.4 can be found in the full version [45].

4.3 Average-Case PPAD Hardness

We present a construction of a hard-on-average distribution of Sink-of-Verifiable-Line (SVL) instances assuming any quasi-polynomially-secure private-key (single-input) functional encryption scheme and sub-exponentially-secure one-way function. Following the work of Abbot et al. [1] and Bitansky et al. [11], this shows that the complexity class PPAD [24,30,31,50] contains complete problems that are hard on average (we refer the reader to [11] for more details). In what follows we first recall the SVL problem, and then state our hardness result. The proof can be found in the full version [45].

Definition 4.5 (Sink-of-Verifiable-Line). *An SVL instance* $(\mathsf{S}, \mathsf{V}, x_s, T)$ *consists of a source* $x_s \in \{0,1\}^\lambda$, *a target index* $T \in [2^\lambda]$, *and a pair of circuits* $\mathsf{S}: \{0,1\}^\lambda \to \{0,1\}^\lambda$ *and* $\mathsf{V}: \{0,1\}^\lambda \times [T] \to \{0,1\}$, *such that for* $(x,i) \in \{0,1\}^\lambda \times [T]$, *it holds that* $\mathsf{V}(x,i) = 1$ *if and only if* $x = x_i = \mathsf{S}^{i-1}(x_s)$, *where* $x_1 = x_s$. *A string* $w \in \{0,1\}^\lambda$ *is a valid witness if and only if* $\mathsf{V}(w,T) = 1$.

Theorem 4.6. *Let $\lambda \in \mathbb{N}$ be a security parameter and fix any constant $\epsilon \in (0,1)$. Assume a $(2^{2(\log \lambda)^{1/\epsilon}}, 2^{2(\log \lambda)^{1/\epsilon}}, 2^{(\log \lambda)^{1/\epsilon}}, 2^{-(\log \lambda)^{1.5/\epsilon}})$-selectively-secure single-input private-key functional encryption scheme for all functions of polynomial size, and that $(2^{\lambda^{2\epsilon'}}, 2^{-\lambda^{2\epsilon'}})$-secure injective one-way functions exist for some large enough constant $\epsilon' \in (0,1)$. Then, there exists a distribution with an associated efficient sampling procedure that generates instances of sink-of-verifiable-line which are hard to solve for any polynomial-time algorithm.*

Acknowledgments. We thank Zvika Brakerski and the anonymous referees for many valuable comments. The first author thanks his advisor Moni Naor for his support and guidance.

References

1. Abbot, T., Kane, D., Valiant, P.: On algorithms for Nash equilibria (2004). http://web.mit.edu/tabbott/Public/final.pdf
2. Agrawal, S., Agrawal, S., Badrinarayanan, S., Kumarasubramanian, A., Prabhakaran, M., Sahai, A.: Function private functional encryption and property preserving encryption: new definitions and positive results. Cryptology ePrint Archive, Report 2013/744 (2013)
3. Ananth, P., Boneh, D., Garg, S., Sahai, A., Zhandry, M.: Differing-inputs obfuscation and applications. Cryptology ePrint Archive, Report 2013/689 (2013)
4. Ananth, P., Brakerski, Z., Segev, G., Vaikuntanathan, V.: From selective to adaptive security in functional encryption. In: Gennaro, R., Robshaw, M. (eds.) CRYPTO 2015. LNCS, vol. 9216, pp. 657–677. Springer, Heidelberg (2015). doi:10.1007/978-3-662-48000-7_32
5. Ananth, P., Jain, A., Naor, M., Sahai, A., Yogev, E.: Universal constructions and robust combiners for indistinguishability obfuscation and witness encryption. In: Robshaw, M., Katz, J. (eds.) CRYPTO 2016. LNCS, vol. 9815, pp. 491–520. Springer, Heidelberg (2016). doi:10.1007/978-3-662-53008-5_17
6. Ananth, P., Jain, A.: Indistinguishability obfuscation from compact functional encryption. In: Gennaro, R., Robshaw, M. (eds.) CRYPTO 2015. LNCS, vol. 9215, pp. 308–326. Springer, Heidelberg (2015). doi:10.1007/978-3-662-47989-6_15
7. Ananth, P., Jain, A., Sahai, A.: Achieving compactness generically: indistinguishability obfuscation from non-compact functional encryption. Cryptology ePrint Archive, Report 2015/730 (2015)
8. Asharov, G., Segev, G.: Limits on the power of indistinguishability obfuscation and functional encryption. In: Proceedings of the 56th Annual IEEE Symposium on Foundations of Computer Science, pp. 191–209 (2015)
9. Barak, B., Goldreich, O., Impagliazzo, R., Rudich, S., Sahai, A., Vadhan, S.P., Yang, K.: On the (im)possibility of obfuscating programs. J. ACM **59**(2), 6 (2012)
10. Bitansky, N., Nishimaki, R., Passelègue, A., Wichs, D.: From cryptomania to obfustopia through secret-key functional encryption. In: Hirt, M., Smith, A. (eds.) TCC 2016. LNCS, vol. 9986, pp. 391–418. Springer, Heidelberg (2016). doi:10.1007/978-3-662-53644-5_15
11. Bitansky, N., Paneth, O., Rosen, A.: On the cryptographic hardness of finding a Nash equilibrium. In: Proceedings of the 56th Annual IEEE Symposium on Foundations of Computer Science, pp. 1480–1498 (2015)

12. Bitansky, N., Vaikuntanathan, V.: Indistinguishability obfuscation from functional encryption. In: Proceedings of the 56th Annual IEEE Symposium on Foundations of Computer Science, pp. 171–190 (2015)
13. Boneh, D., Lewi, K., Raykova, M., Sahai, A., Zhandry, M., Zimmerman, J.: Semantically secure order-revealing encryption: multi-input functional encryption without obfuscation. In: Oswald, E., Fischlin, M. (eds.) EUROCRYPT 2015. LNCS, vol. 9057, pp. 563–594. Springer, Heidelberg (2015). doi:10.1007/978-3-662-46803-6_19
14. Boneh, D., Raghunathan, A., Segev, G.: Function-private identity-based encryption: hiding the function in functional encryption. In: Canetti, R., Garay, J.A. (eds.) CRYPTO 2013. LNCS, vol. 8043, pp. 461–478. Springer, Heidelberg (2013). doi:10.1007/978-3-642-40084-1_26
15. Boneh, D., Raghunathan, A., Segev, G.: Function-private subspace-membership encryption and its applications. In: Sako, K., Sarkar, P. (eds.) ASIACRYPT 2013. LNCS, vol. 8269, pp. 255–275. Springer, Heidelberg (2013). doi:10.1007/978-3-642-42033-7_14
16. Boneh, D., Sahai, A., Waters, B.: Functional encryption: definitions and challenges. In: Ishai, Y. (ed.) TCC 2011. LNCS, vol. 6597, pp. 253–273. Springer, Heidelberg (2011). doi:10.1007/978-3-642-19571-6_16
17. Boneh, D., Sahai, A., Waters, B.: Functional encryption: a new vision for public-key cryptography. Commun. ACM 55(11), 56–64 (2012)
18. Boneh, D., Waters, B.: Constrained pseudorandom functions and their applications. In: Sako, K., Sarkar, P. (eds.) ASIACRYPT 2013. LNCS, vol. 8270, pp. 280–300. Springer, Heidelberg (2013). doi:10.1007/978-3-642-42045-0_15
19. Boyle, E., Chung, K.-M., Pass, R.: On extractability obfuscation. In: Lindell, Y. (ed.) TCC 2014. LNCS, vol. 8349, pp. 52–73. Springer, Heidelberg (2014). doi:10.1007/978-3-642-54242-8_3
20. Boyle, E., Goldwasser, S., Ivan, I.: Functional signatures and pseudorandom functions. In: Krawczyk, H. (ed.) PKC 2014. LNCS, vol. 8383, pp. 501–519. Springer, Heidelberg (2014). doi:10.1007/978-3-642-54631-0_29
21. Brakerski, Z., Gentry, C., Halevi, S., Lepoint, T., Sahai, A., Tibouchi, M.: Cryptanalysis of the quadratic zero-testing of GGH. Cryptology ePrint Archive, Report 2015/845 (2015)
22. Brakerski, Z., Komargodski, I., Segev, G.: Multi-input functional encryption in the private-key setting: stronger security from weaker assumptions. In: Fischlin, M., Coron, J.-S. (eds.) EUROCRYPT 2016. LNCS, vol. 9666, pp. 852–880. Springer, Heidelberg (2016). doi:10.1007/978-3-662-49896-5_30
23. Brakerski, Z., Segev, G.: Function-private functional encryption in the private-key setting. In: Dodis, Y., Nielsen, J.B. (eds.) TCC 2015. LNCS, vol. 9015, pp. 306–324. Springer, Heidelberg (2015). doi:10.1007/978-3-662-46497-7_12
24. Chen, X., Deng, X., Teng, S.: Settling the complexity of computing two-player Nash equilibria. J. ACM 56(3) (2009). http://doi.acm.org/10.1145/1516512.1516516
25. Cheon, J.H., Fouque, P.A., Lee, C., Minaud, B., Ryu, H.: Cryptanalysis of the new CLT multilinear map over the integers. Cryptology ePrint Archive, Report 2016/135 (2016)
26. Cheon, J.H., Han, K., Lee, C., Ryu, H., Stehlé, D.: Cryptanalysis of the multilinear map over the integers. In: Oswald, E., Fischlin, M. (eds.) EUROCRYPT 2015. LNCS, vol. 9056, pp. 3–12. Springer, Heidelberg (2015). doi:10.1007/978-3-662-46800-5_1
27. Cheon, J.H., Jeong, J., Lee, C.: An algorithm for NTRU problems and cryptanalysis of the GGH multilinear map without an encoding of zero. Cryptology ePrint Archive, Report 2016/139 (2016)

28. Cheon, J.H., Lee, C., Ryu, H.: Cryptanalysis of the new CLT multilinear maps. Cryptology ePrint Archive, Report 2015/934 (2015)
29. Coron, J.-S., et al.: Zeroizing without low-level zeroes: new MMAP attacks and their limitations. In: Gennaro, R., Robshaw, M. (eds.) CRYPTO 2015. LNCS, vol. 9215, pp. 247–266. Springer, Heidelberg (2015). doi:10.1007/978-3-662-47989-6_12
30. Daskalakis, C., Goldberg, P.W., Papadimitriou, C.H.: The complexity of computing a Nash equilibrium. Commun. ACM **52**(2), 89–97 (2009)
31. Daskalakis, C., Goldberg, P.W., Papadimitriou, C.H.: The complexity of computing a Nash equilibrium. SIAM J. Comput. **39**(1), 195–259 (2009)
32. Daskalakis, C., Papadimitriou, C.H.: Continuous local search. In: Proceedings of the 22nd Annual ACM-SIAM Symposium on Discrete Algorithms, pp. 790–804 (2011)
33. Garg, S., Gentry, C., Halevi, S., Raykova, M., Sahai, A., Waters, B.: Candidate indistinguishability obfuscation and functional encryption for all circuits. In: Proceedings of the 54th Annual IEEE Symposium on Foundations of Computer Science, pp. 40–49 (2013)
34. Garg, S., Gentry, C., Halevi, S., Zhandry, M.: Functional encryption without obfuscation. In: Kushilevitz, E., Malkin, T. (eds.) TCC 2016. LNCS, vol. 9563, pp. 480–511. Springer, Heidelberg (2016). doi:10.1007/978-3-662-49099-0_18
35. Garg, S., Pandey, O., Srinivasan, A.: Revisiting the cryptographic hardness of finding a nash equilibrium. In: Robshaw, M., Katz, J. (eds.) CRYPTO 2016. LNCS, vol. 9815, pp. 579–604. Springer, Heidelberg (2016). doi:10.1007/978-3-662-53008-5_20
36. Goldreich, O., Goldwasser, S., Micali, S.: How to construct random functions. J. ACM **33**(4), 792–807 (1986)
37. Goldwasser, S., et al.: Multi-input functional encryption. In: Nguyen, P.Q., Oswald, E. (eds.) EUROCRYPT 2014. LNCS, vol. 8441, pp. 578–602. Springer, Heidelberg (2014). doi:10.1007/978-3-642-55220-5_32
38. Goldwasser, S., Kalai, Y., Popa, R.A., Vaikuntanathan, V., Zeldovich, N.: Reusable garbled circuits and succinct functional encryption. In: Proceedings of the 45th Annual ACM Symposium on Theory of Computing, pp. 555–564 (2013)
39. Gorbunov, S., Vaikuntanathan, V., Wee, H.: Functional encryption with bounded collusions via multi-party computation. In: Safavi-Naini, R., Canetti, R. (eds.) CRYPTO 2012. LNCS, vol. 7417, pp. 162–179. Springer, Heidelberg (2012). doi:10.1007/978-3-642-32009-5_11
40. Hu, Y., Jia, H.: Cryptanalysis of GGH map. Cryptology ePrint Archive, Report 2015/301 (2015)
41. Hubáček, P., Yogev, E.: Hardness of continuous local search: query complexity and cryptographic lower bounds. In: Proceedings of the 28th Annual ACM-SIAM Symposium on Discrete Algorithms, SODA, pp. 1352–1371 (2017)
42. Impagliazzo, R.: A personal view of average-case complexity. In: Proceedings of the 10th Annual Structure in Complexity Theory Conference, pp. 134–147 (1995)
43. Kiayias, A., Papadopoulos, S., Triandopoulos, N., Zacharias, T.: Delegatable pseudorandom functions and applications. In: Proceedings of the 20th Annual ACM Conference on Computer and Communications Security, pp. 669–684 (2013)
44. Komargodski, I., Moran, T., Naor, M., Pass, R., Rosen, A., Yogev, E.: One-way functions and (im)perfect obfuscation. In: Proceedings of the 55th Annual IEEE Symposium on Foundations of Computer Science, pp. 374–383 (2014)
45. Komargodski, I., Segev, G.: From Minicrypt to Obfustopia via private-key functional encryption. Cryptology ePrint Archive, Report 2017/080

46. Komargodski, I., Segev, G., Yogev, E.: Functional encryption for randomized functionalities in the private-key setting from minimal assumptions. In: Dodis, Y., Nielsen, J.B. (eds.) TCC 2015. LNCS, vol. 9015, pp. 352–377. Springer, Heidelberg (2015). doi:10.1007/978-3-662-46497-7_14
47. Miles, E., Sahai, A., Zhandry, M.: Annihilation attacks for multilinear maps: cryptanalysis of indistinguishability obfuscation over GGH13. Cryptology ePrint Archive, Report 2016/147 (2016)
48. Minaud, B., Fouque, P.A.: Cryptanalysis of the new multilinear map over the integers. Cryptology ePrint Archive, Report 2015/941 (2015)
49. O'Neill, A.: Definitional issues in functional encryption. Cryptology ePrint Archive, Report 2010/556 (2010)
50. Papadimitriou, C.H.: On the complexity of the parity argument and other inefficient proofs of existence. J. Comput. Syst. Sci. 48(3), 498–532 (1994)
51. Sahai, A., Waters, B.: Slides on functional encryption (2008). http://www.cs.utexas.edu/bwaters/presentations/files/functional.ppt
52. Sahai, A., Waters, B.: How to use indistinguishability obfuscation: deniable encryption, and more. In: Proceedings of the 46th Annual ACM Symposium on Theory of Computing, pp. 475–484 (2014)
53. Shen, E., Shi, E., Waters, B.: Predicate privacy in encryption systems. In: Reingold, O. (ed.) TCC 2009. LNCS, vol. 5444, pp. 457–473. Springer, Heidelberg (2009). doi:10.1007/978-3-642-00457-5_27
54. Waters, B.: A punctured programming approach to adaptively secure functional encryption. In: Gennaro, R., Robshaw, M. (eds.) CRYPTO 2015. LNCS, vol. 9216, pp. 678–697. Springer, Heidelberg (2015). doi:10.1007/978-3-662-48000-7_33

Projective Arithmetic Functional Encryption and Indistinguishability Obfuscation from Degree-5 Multilinear Maps

Prabhanjan Ananth[✉] and Amit Sahai

Center for Encrypted Functionalities and Department of Computer Science,
UCLA, Los Angeles, USA
{prabhanjan,sahai}@cs.ucla.edu

Abstract. In this work, we propose a variant of functional encryption called *projective arithmetic functional encryption* (PAFE). Roughly speaking, our notion is like functional encryption for arithmetic circuits, but where secret keys only yield partially decrypted values. These partially decrypted values can be linearly combined with known coefficients and the result can be tested to see if it is a small value.

We give a *degree-preserving* construction of PAFE from multilinear maps. That is, we show how to achieve PAFE for arithmetic circuits of degree d using only degree-d multilinear maps. Our construction is based on an assumption over such multilinear maps, that we justify in a generic model. We then turn to applying our notion of PAFE to one of the most pressing open problems in the foundations of cryptography: building secure indistinguishability obfuscation ($i\mathcal{O}$) from simpler building blocks.

$i\mathcal{O}$ **from degree-5 multilinear maps.** Recently, the works of Lin [Eurocrypt 2016] and Lin-Vaikuntanathan [FOCS 2016] showed how to build $i\mathcal{O}$ from constant-degree multilinear maps. However, no explicit constant

The full version of this paper is available in [AS16].

P. Ananth—Work done in part while visiting the Simons Institute for Theoretical Computer Science, supported by the Simons Foundation and by the DIMACS/Simons Collaboration in Cryptography through NSF grant #CNS-1523467. This work was partially supported by grant #360584 from the Simons Foundation and the grants listed under Amit Sahai.

A. Sahai—Work done in part while visiting the Simons Institute for Theoretical Computer Science, supported by the Simons Foundation and by the DIMACS/Simons Collaboration in Cryptography through NSF grant #CNS-1523467. Research supported in part from a DARPA/ARL SAFEWARE award, NSF Frontier Award 1413955, NSF grants 1619348, 1228984, 1136174, and 1065276, BSF grant 2012378, a Xerox Faculty Research Award, a Google Faculty Research Award, an equipment grant from Intel, and an Okawa Foundation Research Grant. This material is based upon work supported by the Defense Advanced Research Projects Agency through the ARL under Contract W911NF-15-C-0205. The views expressed are those of the authors and do not reflect the official policy or position of the Department of Defense, the National Science Foundation, or the U.S. Government.

J.-S. Coron and J.B. Nielsen (Eds.): EUROCRYPT 2017, Part I, LNCS 10210, pp. 152–181, 2017.
DOI: 10.1007/978-3-319-56620-7_6

was given in these works, and an analysis of these published works shows that the degree requirement would be in excess of 30. The ultimate "dream" goal of this line of work would be to reduce the degree requirement all the way to 2, allowing for the use of well-studied bilinear maps, or barring that, to a low constant that may be supportable by alternative secure low-degree multilinear map candidates. We make substantial progress toward this goal by showing how to leverage PAFE for degree-5 arithmetic circuits to achieve $i\mathcal{O}$, thus yielding the first $i\mathcal{O}$ construction from degree-5 multilinear maps.

1   Introduction

Functional encryption (FE), introduced by Sahai and Waters [SW05, SW08], allows for the creation of secret keys sk_f corresponding to functions f, such that when such a secret key sk_f is applied to an encryption of x, decryption yields $f(x)$ but, intuitively speaking, nothing more is revealed about x. In this work, we will focus on the secret-key variant of FE where knowledge of the master secret key is needed to perform encryption. Functional encryption has proven to be remarkably versatile: it captures as special cases efficient applications like attribute-based encryption for formulas [GPSW06, BSW07] and predicate encryption for inner products [KSW08] from bilinear maps. At the same time, the general notion of functional encryption implies remarkably powerful primitives, including most notably indistinguishability obfuscation ($i\mathcal{O}$) [AJ15, BV15, AJS15, BNPW16].

In this work, we continue the study of functional encryption notions, constructions, and implications. As a byproduct of our study, we tackle the one of the most pressing open problems in theoretical cryptography: building secure $i\mathcal{O}$ from simpler building blocks. In particular, we give the first construction of $i\mathcal{O}$ using only degree-5 multilinear maps.

FE in the Arithmetic Context. For a number of cryptographic objects that deal with general computations, *arithmetic* circuits have been considered in addition to boolean circuits. The primary motivation for this arises when we wish to apply these objects to cryptographic computations, since many cryptographic computations can be better expressed as arithmetic circuits rather than boolean circuits. For example, zero-knowledge proofs [GMR89] for arithmetic circuits (e.g. [GS08] in the bilinear setting) have been influential because they allow for the construction of zero-knowledge protocols whose structure and complexity more closely match the structure and complexity of algebraic cryptographic algorithms.

In a similar spirit, we study general FE in the context where secret keys should correspond to arithmetic circuits. Notably however, our motivation will not (primarily) be efficiency, but rather achieving new feasibility results, as we will elaborate below.

Previous work has studied FE for arithmetic circuits in two special cases: The work of Boneh et al. [BNS13, BGG+14] studied attribute-based encryption for arithmetic circuits from the LWE assumption. (Our work will diverge technically from this.) Another line of work started with the work of Katz, Sahai, and Waters [KSW08], studying FE where secret keys corresponded to arithmetic inner product computations, using bilinear groups as the underlying cryptographic tool. There has been several followup papers on FE for inner products [ABCP15, AAB+15, BJK15, ABCP16, DDM16, LV16] with various security notions and correctness properties. An issue that will be important to us, and that arises already in the context of inner products, concerns the *correctness* property of the FE scheme. Ideally, a secret key for an arithmetic circuit C, when applied to an encryption of x, should allow the decryptor to learn $C(x)$. However, FE constructions typically store values "in the exponent," and thus the difficulty of discrete logarithms in bilinear groups implies that if $C(x)$ is superpolynomial, it will be difficult to recover. This issue has been dealt with in the past either by requiring that decryption only reveals whether $C(x) = 0$, as in [KSW08], or by requiring that decryption only reveals $C(x)$ if $C(x)$ is polynomially bounded, such as in the works of Abdalla et al. and others [ABCP15, BJK15, ABCP16, DDM16]. We will diverge from past work when dealing with this issue, in order to provide greater flexibility, and in so doing, we introduce our notion of *projective*[1] *arithmetic FE*.

1.1   Our Contributions

Projective Arithmetic FE (PAFE). In projective arithmetic FE, like in FE, encrypting a value x yields a ciphertext c. Also like in (arithmetic) FE, in PAFE each secret key sk_C is associated with an *arithmetic* circuit[2] C. However, unlike in FE, in PAFE when the secret key sk_C is applied to the ciphertext c, it does not directly yield the decrypted value $C(x)$, but rather this yields a partial decryption p_C. We call this process *projective decryption*. We envision a party holding a collection of secret keys $\{sk_C\}_C$ would apply projective decryption using these secret keys to the ciphertext c to obtain a collection of partial decryptions $\{p_C\}_C$. Finally, this party can choose any collection of small coefficients $\{\alpha_C\}_C$ arbitrarily, and then call a different efficient *recovery* algorithm which is given all the partial decryptions $\{p_C\}_C$ and coefficients $\{\alpha_C\}_C$. The recovery algorithm then outputs a bit that indicates whether $\sum_C \alpha_C C(x) = 0$

---

[1] We call our notion *projective* FE because, roughly speaking, a user holding a collection of keys $\{sk_C\}_C$ for several arithmetic circuits C can only learn information about various *linear projections* $\sum_C \alpha_C C(x)$ for known small coefficients $\{\alpha_C\}_C$. We discuss this in more detail below. Our name is also loosely inspired by the notion of projective hash functions, introduced by Cramer and Shoup [CS02], where keys (called projective keys) only allow one to evaluate the hash function on inputs x in some NP language, but not on all strings. In our setting, as well, our keys are similarly only "partially functional" in that they only allow the user to learn information about various linear projections, and they do not in general reveal the full information that should be learned by obtaining all $C(x)$ values. However, to the best of our knowledge, only this loose relationship exists between projective hash functions and our notion of projective FE.

[2] We only are interested in arithmetic circuits of fan-in 2.

or not. (More generally, we can allow the user to recover the value of $\sum_C \alpha_C C(x)$ as long as it is bounded by a polynomial.)

Thus, projective arithmetic FE can be seen as relaxing the correctness guarantee that would be provided by the standard notion of FE when applied to arithmetic circuits over fields of superpolynomial size (which is not known to be achievable). Of course, if decryption actually allowed a user to learn $\{C(x)\}_C$ for several arithmetic circuits C, then the user would be able to compute $\sum_C \alpha_C C(x)$ for any set of small coefficients $\{\alpha_C\}_C$ of her choice. Note that our notion is more permissive than *only* revealing whether $C(x) = 0$, as in the original work for FE for inner products [KSW08], or only revealing $C(x)$ if it is polynomially bounded, such as in other works on FE for inner products [ABCP15, BJK15, ABCP16, DDM16]. With regard to security, our notion will, intuitively speaking, only require indistinguishability of encryptions of x from encryptions of y, if $C(x) = C(y)$ for all secret keys sk_C obtained by the adversary. However, for our application of PAFE to iO, we require a stronger notion of security that we call *semi-functional security*. We give an intuitive explanation of this notion in the technical overview.

Degree-Preserving Construction of PAFE from Multilinear Maps. The first main technical contribution of our work is a construction of (secret-key) PAFE for degree-d arithmetic circuits, from degree-d *asymmetric* multilinear maps[3]. Furthermore, it suffices that the groups over which the multilinear maps are defined are prime order. Our construction is based on an explicit pair of assumptions over such multilinear maps, that we can justify in the standard generic multilinear model.

Theorem 1 (Informal). *There exists secret-key PAFE for degree-d arithmetic circuits from degree-d prime order asymmetric multilinear maps under Assumptions #1 and #2 (see Sect. 4.1).*

Our assumptions do not require any low-level encodings of 0 to be given to the adversary, and we thus believe them to be instantiable using existing candidate multilinear maps. Indeed, because of some pseudorandomness properties of our construction and generic proof of security, we believe that our assumptions can be proven secure in the Weak MMap model considered in the works of Miles et al. and Garg et al. [MSZ16, GMM+16], which would give further evidence of its instantiability. Because we want to posit instantiable assumptions, we do not formulate a succinct version of our assumption together with a reduction of security as was done in the works of Gentry et al. or Lin and Vaikuntanathan [GLSW15, LV16], because unfortunately no existing candidate multilinear map construction is known to securely support such reductions, and indeed the assumptions of [GLSW15, LV16] are broken when instantiated with existing candidates. We stress that, like in the recent work of [Lin16, LV16], if

---

[3] Roughly speaking, asymmetric multilinear maps disallows pairing of elements from the same group structure.

the degree d is constant, then our pair of assumptions would only involve a constant-degree multilinear map.

Our construction can be seen as a generalizing FE for inner products (degree 2 functions) from bilinear maps, to higher degrees in a degree preserving manner. Thus, our construction can be applied to cryptographic computations that are naturally represented as arithmetic functions of low degree, but not as inner products. In more detail, we introduce the notion of slotted encodings that has the same flavor of multilinear maps defined over composite order groups. We then show how to emulate slotted encodings using prime-order multilinear maps. However, this emulation strategy only works in the case of constant degree. We hope that this technique will be useful to transform constructions based on constant degree composite order multilinear maps (for example [Lin16]) to constructions based on constant degree prime order multilinear maps.

iO from Degree-5 Multilinear Maps. Our motivation for building PAFE for arithmetic circuits in a degree-preserving manner is to achieve new feasibility results for iO from low-degree multilinear maps. The concept of iO was first defined by Barak et al. [BGI+01]. Informally speaking, iO converts a program (represented by a boolean circuit) into a "pseudo-canonical form." That is, for any two equivalent programs P_0, P_1 of the same size, we require that $iO(P_0)$ is computationally indistinguishable from $iO(P_1)$. The first candidate construction of iO was given by Garg et al. [GGH+13b], and especially since the introduction of punctured programming techniques of Sahai and Waters [SW14], iO has found numerous applications, with numerous papers published since 2013 that use iO to accomplish cryptographic tasks that were not known to be feasible before (see, e.g., [GGH+13b, SW14, GGHR14, HSW14, GGG+14, BPR15, BP15, CHN+16, BGJ+16]). However, it is still not known how to build iO from standard cryptographic assumptions. Given the enormous applicability of iO to a wide variety of cryptographic problems, one of the most pressing open problems in the foundations of cryptography is to find ways to construct iO from simpler building blocks. Indeed, while there have been dozens of papers published showing how to use iO to accomplish amazing things, only a handful of papers have explored simpler building blocks that suffice for constructing iO.

One line of work toward this objective is by Lin [Lin16] and Lin and Vaikuntanathan [LV16], who showed how to build iO from constant-degree multilinear maps. Unfortunately, no explicit constant was given in these works, and an analysis of these published works shows that the degree requirement would be in excess of 100. The ultimate "dream" goal of this line of work would be to reduce the degree requirement all the way to 2, allowing for the use of well-studied bilinear maps, or barring that, to a low constant that may be supportable by alternative secure low-degree multilinear map candidates.

We make substantial progress toward this goal by showing how to achieve iO starting from PAFE. Specifically, we first construct ε-sublinear secret key functional encryption for NC^1 circuits, with constant $\varepsilon < 1$, starting from PAFE[4] for

---

[4] We additionally require that PAFE has encryption complexity to be multiplicative overhead in the message size. Our construction of PAFE satisfies this property.

degree-d arithmetic circuits and a specific type of degree d-randomizing polynomials [IK00, AIK06][5]. We require that the randomizing polynomials satisfy some
additional properties such as the encoding polynomials should be homogenous,
the randomness complexity[6] is ε-sub-linear in the circuit size and the decoding
algorithm should be executed as a sequence of linear functions. We call a scheme
that satisfies these additional properties as homogenous randomizing polynomials with ε-sub-linear randomness complexity. As we will see later, we can achieve
ε-sub-linear randomness complexity property for free by employing an appropri-
ate pseudorandom generator of $\frac{1}{\varepsilon'}$-stretch, where constant $\varepsilon' > 1$ is related to ε.
Hence, we only care about constructing homogenous randomizing polynomials
(without sublinear property) and we provide an information theoretic construction achieving the same.

Once we construct ε-sublinear secret key functional encryption, we can then
invoke the result of [BNPW16] and additionally assume learning with errors to
obtain iO. For this transformation, we are required to assume that the underlying
FE scheme and learning with errors is sub-exponentially secure. Thus,

Theorem 2 (Informal). *We construct an indistinguishability obfuscation
scheme for P/poly assuming the following: for some constant d,*

1. *Sub-exponentially secure PAFE scheme for degree d arithmetic circuits with
   multiplicative overhead in encryption complexity. From Theorem 1, this can
   be based on sub-exponentially secure Assumptions #1 and #2 (Sect. 4.1).*
2. *Sub-exponentially secure degree d homogenous randomizing polynomials with
   ε-sub-linear randomness complexity. This can be based on sub-exponentially
   secure pseudorandom generators of stretch $\frac{1}{\varepsilon'}$, where constant $\varepsilon' > 1$ is related
   to ε.*
3. *Sub-exponentially secure learning with errors.*

Instantiation: We show how to leverage PAFE for degree-5 arithmetic circuits
to achieve iO, thus yielding the first iO construction from degree-5 multilinear
maps. The crucial step in this transformation is to first construct homogenous
randomizing polynomials with sub-linear randomness complexity of degree 15.
We first identify that the work of [AIK06] satisfies the required properties of
a degree-3 homogenous randomizing polynomials scheme. To achieve sublinear
randomness complexity, we assume an explicit degree-2 pseudo-random generator (PRGs) achieving super-linear stretch in the boolean setting, and a related
explicit degree-3 PRG achieving super-quadratic stretch in the arithmetic setting. In particular we use a boolean PRG of stretch 1.49 and an algebraic PRG
of stretch 2.49 [OW14] (see also [AL16]). We then observe that for a special
class of circuits $\mathcal{C}$, the degree of the above polynomials can be reduced to 5 if
we additionally allow for pre-processing of randomness. Also, we show how to

---

[5] The degree of a randomizing polynomial is defined to be the maximum degree of the
   polynomials computing the encoding function.
[6] Randomness complexity in this context refers to the size of the random string used
   in the encoding algorithm.

remove the algebraic PRG part in the construction of randomizing polynomials for C.

As alluded to above, the fact that our PAFE can directly deal with an arithmetic PRG in a degree-preserving manner is critical to allowing us to achieve i$\mathcal{O}$ with just degree-5 mutlilinear maps.

Theorem 3 (Informal). *We construct an indistinguishability obfuscation scheme for P/poly assuming the following: for some constant d,*

1. *Sub-exponentially secure PAFE scheme for degree 5 arithmetic circuits with multiplicative overhead in encryption complexity. From Theorem 1, this can be based on sub-exponentially secure Assumptions #1 and #2 (Sect. 4.1).*
2. *Sub-exponentially secure degree 5 homogenous randomizing polynomials for C with ε-sub-linear randomness complexity. This can be based on sub-exponentially secure boolean PRG of stretch 1.01.*
3. *Sub-exponentially secure learning with errors.*

Concurrent Work(s). In a concurrent work, Lin obtains a new IO construction with a security reduction to (1) L-linear maps with the subexponential symmetric external Diffie-Hellman (SXDH) assumption, (2) subexponentially secure locality-L PRG, and (3) subexponential LWE. When using a locality 5 PRG, 5-linear maps with the SXDH assumption suffice. The L-linear maps consist of L source groups $G_1, \cdots, G_L$, whose elements $g_1^{a_1}, \cdots, g_L^{a_L}$ can be "paired" together to yield an element in a target group $g_T^{a_1 \cdots a_L}$. The SXDH assumption on such multilinear maps is a natural generalization of the SXDH assumption on bilinear maps: It postulates that the DDH assumption holds in every source group G_d, that is, elements g_d^a, g_d^b, g_d^{ab} are indistinguishable from g_d^a, g_d^b, g_d^r, for random a, b and r.

To obtain IO, she first constructs collusion-resistant FE schemes for computing degree-L polynomials from L-linear maps, and then bootstraps such FE schemes to IO for P, assuming subexponentially secure locality-L PRG and LWE.

A corollary of our degree-preserving PAFE construction is a construction of FE for degree-2 polynomials from bilinear maps. Concurrently, two works [BCF16, Gay16] achieved the same result based on concrete assumptions on bilinear maps.

We now give a technical overview of our approach.

1.2 Technical Overview

We give an informal description of the algorithms of projective arithmetic functional encryption (PAFE). We focus on secret-key setting in this work.

- **Setup**: It outputs secret key MSK.
- **Key Generation**: On input an arithmetic circuit C and master secret key, it produces a functional key sk_C.
- **Encryption**: On input message x, it outputs a ciphertext CT.

- **Projective Decryption**: On input a functional key sk_C and ciphertext CT, it produces a partial decrypted value ι.
- **Recover**: On input many partial decrypted values $\{\iota_i\}$ and a linear function (specified as co-efficients), it outputs the result of applying the linear function on the values contained in $\{\iota_i\}$.

We first show how to achieve iO starting from secret-key PAFE. Later, we show how to obtain PAFE for degree D polynomials starting from degree D multilinear maps.

iO from Secret-Key PAFE: We start with the goal of constructing a sub-linear secret-key FE scheme for NC^1 (from which we can obtain iO [BNPW16]) starting from PAFE for constant degree arithmetic circuits. Our goal is to minimize the degree of arithmetic circuits that suffices us to achieve sub-linear FE.

We start with the standard tool of randomizing polynomials to implement NC^1 using a constant degree arithmetic circuit. We use randomizing polynomials with a special decoder: the decoder is a sequence of linear functions chosen adaptively[7]. At a high level the construction proceeds as follows: let the randomizing polynomial of circuit C, input x and randomness r be of the form $p_1(x; r), \ldots, p_N(x; r)$. The sub-linear FE functional key corresponding to a circuit C are a collection of PAFE keys for $p_1, \ldots, p_N$. The encryption of x w.r.t sublinear FE scheme is a PAFE encryption of (x, r). To obtain $C(x)$, first execute the projective decryption algorithm on key of p_i and ciphertext of (x, r) to obtain partial decrypted values corresponding to $p_i(x, r)$. Now, execute the recover algorithm on input a linear function and the above partial decrypted values, where the linear function is chosen by the decoder of the randomizing polynomials scheme. Depending on the output of the recover algorithm, the decoder picks a new linear function. This process is repeated until we finally recover the output of the circuit C.

Before we justify why this scheme is secure, we remark as to why this scheme satisfies the sub-linear efficiency property. In order to achieve sub-linear efficiency, we require that $|r| = |C|^{1-\varepsilon}$ for some $\varepsilon > 0$. Thus, we require randomizing polynomials with sub-linear randomness complexity. We remark later how to achieve this.

The next goal is to argue security: prior works either employ function privacy properties [BS15] or Trojan techniques [CIJ+13, ABSV15] to make the above approach work. However, going through these routes is going to increase the degree of arithmetic circuits required to achieve sub-linear FE. Instead, we start with a PAFE scheme with a stronger security guarantee called *semi-functional* security. This notion is inspired by the dual system methodology introduced by Waters [Wat09] in different context and later employed by several other works

---

[7] That is, choice of every linear function could depend on the output of the previously chosen linear functions on the encoding of computation.

(see for example, [LOS+10, GGHZ14]). Associated with this notion, there are two types of objects:

- *Semi-Functional Keys:* A semi-functional key is associated with an arithmetic circuit C and a hardwired value v.
- *Semi-Functional Ciphertexts:* A semi-functional ciphertext is generated just using the master secret key.

We define how honestly generated keys, honestly generated ciphertexts and semi-functional keys, semi-functional ciphertexts are required to behave with each other in Table 1. Honestly generated key or ciphertext refers to generation of key or ciphertext according to the description of the scheme.

Table 1. We consider four possibilities of decryption: (a) honestly generated keys correctly decrypts honestly generated ciphertexts (from correctness property), (b) semi-functional keys also correctly decrypts honestly generated ciphertexts, (c) there is no correctness guarantee on the decryption of honestly generated keys on semi-functional ciphertexts, (d) Finally, the decryption of semi-functional keys on semi-functional ciphertexts yields the hardwired value associated with the key.

	Honestly generated keys	Semi-functional keys
Honestly generated ciphertexts	Honest decryption	Honest decryption
Semi-functional ciphertexts	Not defined	Output hardwired value

A PAFE scheme is said to satisfy semi-functional security if both the following definitions are satisfied:

- *Indistinguishability of Semi-functional keys*: It should be hard to distinguish an honestly generated functional key of C from a semi-functional key of C associated with any hardwired value v.
- *Indistinguishability of Semi-functional Ciphertexts*: It should be hard to distinguish an honestly generated ciphertext of x from a semi-functional ciphertext if every functional key of C issued is a semi-functional key associated with hardwired value $C(x)$.

Once we have a secret key PAFE scheme that satisfies semi-functional security then we can prove the security as follows: we consider a simple case when the adversary only submits one message query (x_0, x_1).

- We first turn the functional key associated with an arithmetic circuit C into a semi-functional key with the hardwired value $C(x_0)$.
- Once all the functional keys are semi-functional, we can now switch the ciphertext of x_0 to semi-functional ciphertext.

- Since $C(x_0) = C(x_1)$, we can switch back the semi-functional keys to be honestly generated functional keys.
- Finally, we switch back the ciphertext from semi-functional to honestly generated ciphertext of x_1.

If the adversary requests multiple message queries, then the above process is to be repeated one message query at a time.

Choice of Randomizing Polynomials with Sub-linear Randomness: The next question is what randomizing polynomials do we choose to instantiate the above approach. As we will see later, if we choose randomizing polynomials with sub-linear randomness complexity of degree $\mathbf{D}$ then it suffices build PAFE from degree $\mathbf{D}$ multilinear maps. Also, we will require the polynomials to be homogenous.

Hence, our goal is to choose a homogenous randomizing polynomials with minimal degree and also satisfying (i) linear decodability and (ii) sub-linear randomness complexity properties. We achieve this in the following steps:

1. First, build randomizing polynomials with minimal degree. We start with [AIK06] for NC^1, where the polynomials are of degree 3. In spirit, this is essentially information theoretic Yao with the wire keys being elements over $\mathbb{F}_p$ and every wire key is associated with a random mask (which is represented as a bit) that helps in figuring out which of the four entries to be decoded for the next gate.
2. The above scheme already satisfies linear decodability property. This is because the decryption of every garbled gate is a linear operation. The linear function chosen to decrypt one garbled gate now depends on the linear functions chosen to decrypt its children gates.
3. Next, we tackle sub-linear randomness complexity: we generate the wire keys and the random masks as the output of a PRG. The total length of all the wire keys is roughly square the size of the NC^1 circuit. This is because, the size of the wire keys at the bottom most (input) layer are proportional to the size of the circuit. We use an algebraic PRG of stretch $(2 + \varepsilon)$ to generate the wire keys and we use a boolean PRG to generate the random masks. The degree of the algebraic PRG over $\mathbb{F}_p$ is 3 while the degree of the boolean PRG represented over $\mathbb{F}_p$ is 5. When the above PRGs are plugged into the randomizing polynomials construction from the above step, we get the degree of the polynomials to be 15.
4. Finally, we show how to make the above randomizing polynomials homogenous. This is done using a standard homogenization argument: add dummy variables to the polynomials such that the degree of all the terms in the polynomials are the same. While evaluating these polynomials, set all these dummy variables to 1. This retains the functionality and at the same time ensures homogeneity.

We can now use the above randomizing polynomials scheme to instantiate the above approach. After partial decryption, we get partial decrypted values associated with $\{p_i(x; r)\}$. Now, since the decoding is composed of many linear functions, we can execute the Recover algorithm (multiple times) to recover the output.

Reducing the Degree: We can apply some preprocessing to reduce the degree of the above polynomials further. We remark how to reduce the degree to 5. Later, in the technical sections, we explore alternate ways of reducing the degree, as well.

Suppose we intend to construct sublinear FE for a specific class of circuits $\mathcal{C}$. In this case, we are required to construct randomizing polynomials only for $\mathcal{C} \in NC^1$.

We define $\mathcal{C}$ as follows: every circuit $C \in \mathcal{C}$ of output length $\mathbf{N}$ is of the form $C = (C_1, \ldots, C_\mathbf{N})$, where (i) C_i outputs the ith output bit of C, (ii) $|C_i| = $ poly(λ) for a fixed polynomial poly, (iii) Depth of C_i is $c \cdot \log(\lambda)$, where c is a constant independent of $|C|$ and, (iv) C_i for every $i \in [\mathbf{N}]$ has the same topology–what is different, however, are the constants associated with the wires. We show later that it suffices to build sublinear FE for $\mathcal{C}$ to obtain iO. We now focus on obtain randomizing polynomials for $\mathcal{C}$.

We start with the randomizing polynomials scheme that we described above. Recall that it involved generating a garbled table for every gate in the circuit C. Moreover, the randomness to generate this garbled table is derived from an algebraic and a boolean PRG. We make the following useful changes: let $C = (C_1, \ldots, C_\mathbf{N})$ such that C_i outputs the ith output bit of C. Let $w_1^i, \ldots, w_{nw}^i$ be the set of wires in C_i and $G_1^i, \ldots, G_{ng}^i$ be the set of gates in C_i.

- We invoke nw number of instantiations of boolean PRGs $\mathsf{bPRG}_1^w, \ldots, \mathsf{bPRG}_{nw}^w$ and $\mathsf{bPRG}_1^r, \ldots, \mathsf{bPRG}_{nw}^r$. All these PRGs have the same structure (i.e., same predicates is used) and have degree 5 over arbitrary field (with slightly super-linear stretch $1+\varepsilon$). Pseudorandom generator bPRG_j^w is used to generate wire keys for wires $w_j^1, \ldots, w_j^\mathbf{N}$. Recall that earlier we were using an algebraic PRG of quadratic stretch. This is because the size of wire keys was proportional to exponential in depth, which could potentially be linear in the size of the circuit. However, since we are considering the specific circuit class $\mathcal{C}$, the depth of every circuit is $c \log(\lambda)$. And thus the size of the wire keys is independent of the security parameter. This is turn allows us to use just a PRG of superlinear stretch $1+\varepsilon$. Finally, bPRG_j^r is used to generate random masks for the wires $w_j^1, \ldots, w_j^\mathbf{N}$.
- We now consider the [AIK06] randomizing polynomials associated with circuit C. As before, we substitute the variables associated with wire keys and random masks with the polynomials associated with the appropriate PRGs. The formal variables in the PRG polynomials are associated with the seed.
- The result of the above process is the encoding of C consisting of polynomials $p_1, \ldots, p_N$ with variables associated with the seeds of PRGs. Note that the degree of these polynomials is still 15.

- We then observe that there are polynomials $q_1, \ldots, q_T$ in seed variables such that $p_1, \ldots, p_N$ can be rewritten in terms of $q_1, \ldots, q_T$ and moreover, the degree of p_i in the new variables $\{q_i\}$ is 5. The advantage of doing this is that the polynomials $\{q_i\}$ can be evaluated during the encryption phase[8]. The only thing we need to be wary of is the fact that T could be as big as $|C|$. If this is the case then the encryption complexity would be at least linear in $|C|$, which violate the sublinearity of the FE scheme. We show how to carefully pick $q_1, \ldots, q_T$ such that T is sub-linear in $|C|$ and the above properties hold. We refer the reader to the technical sections for more details.

The only missing piece here is to show that sublinear FE for this special class of circuits $\mathcal{C}$ with sub-exponential security loss implies iO. To show this, it suffices to show that sublinear FE for $\mathcal{C}$ implies sublinear FE for all circuits. Consider the transformation from FE for NC^1 to FE for all circuits by [ABSV15] – the same transformation also works for single-key sublinear secret key FE. We consider a variant of their transformation. In this transformation, a sublinear FE key for circuit C' is generated by constructing a circuit C that has hardwired into it C' and value v. Circuit C takes as input x, PRF key K and mode b. If $b = 0$ it outputs a Yao's garbled circuit of (C, x) computed w.r.t randomness derived from K. If $b = 1$ it outputs the value v. We can re-write C as being composed of sub-circuits $C_1, \ldots, C_N$ such that each of C_i is in NC^1, $|C_i| = \text{poly}(\lambda)$ and depth of C_i is $c \cdot \log(\lambda)$ for a fixed polynomial poly and fixed constant c. Intuitively, C_i, has hardwired into it gate G_i of C' and i^{th} block of v. It computes a garbled table corresponding to G_i if $b = 0$, otherwise it outputs the ith block of v.

Constructing PAFE: We now focus on building PAFE from multilinear maps. The first attempt to encrypt the input $x = (x_1, \ldots, x_{\ell_{\text{inp}}})$ would be to just encode every x_i separately. Now, during evaluation of circuits $C_1, \ldots, C_N$ on these encodings will yield a top level encoding of $C_i(x)$. This homomorphic evaluation would correspond to projective decryption operation. The recover algorithm would just compute a linear function on all the top level encodings of $C_i(x)$ and using zero test parameters, recover the answer if the output of the linear function is 0.

However, we cannot allow the adversary to evaluate recover outputs for circuits C_i of his choice. We should ensure that he recovers outputs only for circuits corresponding to which he has been issued functional keys. The main challenge in designing a functional key for C is to guarantee authenticity – how do we ensure that if the adversary, given a functional key corresponding to C, can only evaluate C on these inputs? To ensure this, we introduce a parallel branch of computation: we instead encode (x_i, α_i) where $\{\alpha_i\}$ are random elements determined during the setup. Then as part of the functional key associated with C, we give out an encoding of $C(\{\alpha_i\})$ at the top level that will allow us to cancel the α_i part after computing C on encodings of $\{(x_i, \alpha_i)\}$ and in the end, just get an encoding of $C(x)$. However, to implement this, we need to make sure that

---

[8] This idea is similar in spirit to the recent work of Bitansky et al. [BLP16], who introduced degree reduction techniques in a different context.

the computation of C on $\{x_i\}$ and $\{\alpha_i\}$ are done separately even though x_i and α_i are encoded together.

The work of [Zim15, AB15] used the above idea in the context of designing iO. As we will discuss below, we extend their techniques in several ways, to deal with the problem of mixing ciphertext components and achieving the semi-functional security properties we need from our PAFE scheme. However, before we discuss these difficulties, we note that the work of [Zim15, AB15] implement parallel branches by using composite order multilinear maps. Composite order multilinear maps allow for jointly encoding for a vector of elements such that addition and multiplication operations can be homomorphically performed on every component of the vector separately.

However, one of the primary motivations for this line of work on building constructions for $i\mathcal{O}$ from low-degree multilinear maps is to enable the use of future candidate low-degree multilinear maps, where achieving composite order may not be possible. Indeed, current instantiations of composite order multlinear maps [CLT13] have poorly understood security properties, and have been subject to efficient cryptanalytic attacks in some settings (see, e.g., [CHL+15, CGH+15]). Thus, instead of relying on composite order multilinear maps, we do the following: we introduce a primitive called a slotted encoding scheme, that allows for the same functionality as offered by composite order multilinear maps. This then helps us in implementing the idea of [Zim15, AB15] using a slotted encoding scheme. We later show how to realize a constant degree slotted encoding scheme using prime order multilinear maps. We define slotted encodings next.

Slotted Encoding: A slotted encoding scheme, parameterized by L (number of slots), has the following algorithms: (i) Setup: this generates the secret parameters, (ii) Encode: it takes as input $(a_1, \ldots, a_L)$ and outputs an encoding of it, (iii) Arithmetic operations: it takes two encodings of $(a_1, \ldots, a_L)$ and $(b_1, \ldots, b_L)$ and performs arithmetic operations on every component separately. For instance, addition of encoding of $(a_1, \ldots, a_L)$ and $(b_1, \ldots, b_L)$ would lead to encoding of $(a_1 + b_1, \ldots, a_L + b_L)$, (iv) Zero Testing: It outputs success if the encoding of $(a_1, \ldots, a_L)$ is such that $a_i = 0$ for every i.

In this work, we will be interested in asymmetric slotted encodings, where the slotted encodings is associated with a tree T such that every encoding is associated with a node in T and two encodings can be paired only if their associated nodes are siblings. The degree of slotted encodings is defined to be the maximum degree of polynomials the scheme lets us evaluate.

Constant Degree Slotted Encoding from Prime Order MMaps: We start with the simple case when degree of slotted encodings is 2 (the bilinear case). The idea of dual vector spaces were introduced by [OT08] and further developed as relevant to us by [OT09, BJK15] to address this problem for bilinear maps. In this framework, there is an algorithm that generates $2n$ vectors $(\mu_1, \ldots, \mu_n), (\nu_1, \ldots, \nu_n)$ of dimension n such that: (i) inner product, $\langle \mu_i, \nu_i \rangle = 1$ and, (ii) inner product, $\langle \mu_i, \nu_j \rangle = 0$ when $i \neq j$. Using this, we can encode

$(a_1, \ldots, a_n)$ associated with some node u in the tree as follows: encode every element of the vector $a_1\mu_1 + \cdots + a_n\mu_n$. The encoding of $(b_1, \ldots, b_n)$ associated with a node v, which is a sibling of u, will be encodings of the vector $b_1\nu_1 + \cdots + b_n\nu_n$. Now, computing inner product of both these encodings will lead to an encoding of $a_1 \cdot b_1 + \cdots + a_n \cdot b_n$.

This idea doesn't suffice for degree 3. So our idea is to work modularly, and consider multiple layers of vectors. The encoding of $(a_1, \ldots, a_n)$ under node u will be encodings of the vector $(a_1\mu_1 \otimes \mu_1' + \cdots + a_n\mu_n \otimes \mu_n')^9$, where $\{\mu_i'\}$ is a basis of a vector space associated with the parent of node u. Now, when this is combined with encoding of $b_1\nu_1 + \cdots + b_n\nu_n$, computed under node v, we get encoding of $(a_1 b_1 \mu_1' + \cdots a_n b_n \mu_n')$. Using this we can then continue for one more level.

To generalize this for higher degrees we require tensoring of multiple vectors (potentially as many as the depth of the tree). This means that the size of the encodings at the lower levels is exponential in the depth and thus, we can only handle constant depth trees. Implementing our tensoring idea for multiple levels is fairly technical, and we refer the reader to the relevant technical section for more details.

PAFE from Slotted Encodings: Using slotted encodings, we make a next attempt in constructing PAFE:

- To encrypt $x = (x_1, \ldots, x_{\ell_{\text{inp}}})$, we compute a slotted encoding of (x_i, α_i), where α_i are sampled uniformly at random during the setup phase.
- A functional key of C consists of a slotted encoding of $(0, C(\{\alpha_i\}))$ at the top level.

The partial decryption first homomorphically evaluates C on slotted encodings of (x_i, α_i) to get a slotted encoding of $(C(\{x_i\}), C(\{\alpha_i\}))$. The second slot can be 'canceled' using top level encoding of $(0, C(\{\alpha_i\}))$ to get an encoding of $(C(\{x_i\}), 0)$. The hope is that if the evaluator uses a different circuit C' then the second slot will not get canceled and hence, he would be unable to get a zero encoding.

However, choosing a different C' is not the only thing an adversary can do. He could also mix encodings from different ciphertexts and try to compute C on it – the above approach does not prevent such attacks. In order to handle this, we need to ensure that the evaluation of ciphertexts can never be mixed. In order to solve this problem, we use a mask γ that be independently sampled for every ciphertext. Every encoding will now be associated with this mask. Implementing this idea will crucially make use of the fact that the polynomial computed by the arithmetic circuit is a homogenous polynomial.

Yet another problem arises is in the security proof: for example, to design semi-functional keys, we need to hardwire a value in the functional key. In order to enable this, we introduce a third slot. With this new modification, we put

---

[9] Here, $\mu_i \otimes \mu_j$ denotes the tensoring of μ_i and μ_j.

forward a template of our construction. Our actual construction involves more details which we skip to keep this section informal.

- To encrypt $x = (x_1, \ldots, x_{\ell_{\text{inp}}})$, we compute a slotted encoding of $(x_i, \alpha_i, 0)$, where α_i are sampled uniformly at random during the setup phase. Additionally, you give out encoding of $(0, S, 0)$ at one level lower than the top level, where S is also picked at random in the setup phase.
- A functional key of C consists of a slotted encoding of $(0, C(\{\alpha_i\}) \cdot S^{-1}, 0)$ at the top level.

The decryption proceeds as before, except that the encodings of $(0, C(\{\alpha_i\}) \cdot S^{-1}, 0)$ and $(0, S, 0)$ are paired together before we proceed.

Note that in both the ciphertext and the functional key, the third slot is not used at all. The third slot helps in the security proof. To see how we describe the semi-functional parameters at a high level as follows:

- *Semi-functional Ciphertexts*: To encrypt $x = (x_1, \ldots, x_{\ell_{\text{inp}}})$, we compute a slotted encoding of $(0, \alpha_i, 0)$, where α_i is computed as before. Additionally, you give out encoding of $(0, S, 1)$ at one level lower than the top level, where S is also picked at random in the setup phase. Note that the third slot now contains 1 which signals that it is activated.
- *Semi-functional Keys*: A functional key of C consists of a slotted encoding of $(0, C(\{\alpha_i\}), v)$ at the one level lower than top level, where v is the hardwired value associated with the semi-functional key.

During the decryption of semi-functional key with honestly generated ciphertext, the third slot will not be used since it will be deactivated in the ciphertext. So the decryption proceeds normally. However, during the decryption of semi-functional key with semi-functional ciphertexts, the third slot is used since the third slot is activated in the ciphertext. We argue the security of our construction in the ideal multilinear map model.

Comparison with [LV16]. We now compare our work with the recent exciting work of [LV16], in order to illustrate some differences that allow us to achieve lower degree. The work of [LV16] first defines FE for NC^0 with a non-trivial efficiency property and give a new bootstrapping theorem[10] to achieve compact FE. They then show how to achieve FE for NC^0 from constant degree multilinear maps[11]. Interestingly, they use arithmetic randomizing polynomials within their construction of FE for NC^0 – this will be important as we note below.

---

[10] Their bootstrapping theorem also works if we start with FE for constant degree polynomials over $\mathbb{F}_2$.

[11] Note that, in particular, the security of their scheme reduces to a succinct assumption called the multilinear joint SXDH assumption. As we noted earlier, unfortunately this assumption is not known to be instantiable with existing multilinear map candidates. However, one can posit a different assumption that directly assumes their FE for NC^0 scheme to be secure, and we do not know of any attacks on that (non-succinct) assumption.

In contrast, we do not build FE for NC^0, but rather show how to proceed directly from projective arithmetic FE for degree-5 arithmetic circuits to iO (without additional use of multilinear maps). Furthermore, our construction of PAFE is *degree preserving*, so to achieve PAFE for degree-5 arithmetic circuits, we only need degree-5 multilinear maps. In contrast, in [LV16], to build FE for NC^0, their work has to "pay" in degree not only based on the depth of the NC^0 circuit that underlies each secret key, but also for the arithmetic randomizing polynomial that they apply to the NC^0 circuit. This adds a significant overhead in the constant degree their multilinear map must support. Our approach is simpler, as our randomizing polynomials are only used in the path from PAFE to iO, which does not use multilinear maps in any additional way. There are, of course, many other technical differences between our work and [LV16], as well. Another conceptual idea that we introduce, and that is different from [LV16], is the notion of slotted encodings, an abstraction of composite order multilinear maps, and our method for emulating slotted encodings using prime order multilinear maps without increasing the degree.

Organization. We define the notion of projective arithmetic functional encryption and present a degree-preserving construction of PAFE from slotted encodings. In the full version, we show how to combine PAFE and (a stronger notion of) randomizing polynomials to obtain secret key functional encryption that can then bootstrapped to obtain iO.

2   Projective Arithmetic Functional Encryption

Throughout this paper we will use standard cryptographic notation and concepts; for details, refer to the full version. In this section, we introduce the notion of projective arithmetic functional encryption scheme. There are two main differences from a (standard) functional encryption scheme:

– Functional keys are associated with arithmetic circuits.
– The projective decryption algorithm only outputs partial decrypted values. There is a recover algorithm that computes on the partial decrypted values and produces an output.

2.1   Definition

We can consider either a public key projective arithmetic FE scheme or a secret key projective arithmetic secret key FE scheme. In this work, we define and construct a secret key projective arithmetic FE scheme.

A secret-key projective arithmetic functional encryption (FE) scheme PAFE over field $\mathbb{F}_p$ is associated with a message space $\mathcal{X} = \{\mathcal{X}_\lambda\}_{\lambda \in \mathbb{N}}$ and a arithmetic circuit class $\mathcal{C} = \{\mathcal{C}_\lambda\}_{\lambda \in \mathbb{N}}$ over $\mathbb{F}_p$. Here, $\mathcal{X}$ comprises of strings with every symbol in the string belongs to $\mathbb{F}_p$.

PAFE comprises of a tuple (Setup, KeyGen, Enc, ProjectDec) of PPT algorithms with the following properties:

- Setup(1^λ): The setup algorithm takes as input the unary representation of the security parameter, and outputs a secret key MSK.
- KeyGen(MSK, C): The key-generation algorithm takes as input the secret key MSK and a arithmetic circuit $C \in \mathcal{C}_\lambda$, over $\mathbb{F}_\mathbf{p}$, and outputs a functional key sk_C.
- Enc(MSK, x): The encryption algorithm takes as input the secret key MSK and a message $x \in \mathcal{X}_\lambda$, and outputs a ciphertext CT.
- ProjectDec(sk_C, CT): The projective decryption algorithm takes as input a functional key sk_C and a ciphertext CT, and outputs a partial decrypted value ι.
- Recover($c_1, \iota_1, \ldots, c_{\ell_\mathfrak{f}}, \iota_{\ell_\mathfrak{f}}$): The recover algorithm takes as input co-efficients $c_1, \ldots, c_{\ell_\mathfrak{f}} \in \mathbb{F}_\mathbf{p}$, partial decrypted values $\iota_1, \ldots, \iota_{\ell_\mathfrak{f}}$ and outputs out.

We first define the correctness property and later, define the security property.

B-Correctness. The correctness is parameterized by a set $B \subseteq \mathbb{F}_\mathbf{p}$. We emphasize that B is a set of polynomial size, i.e., $|B| = \mathrm{poly}(\lambda)$. Consider an honestly generated ciphertext CT of input x. Consider honestly generated keys $sk_{C_1}, \ldots, sk_{C_{\ell_\mathfrak{f}}}$. Denote the corresponding decrypted values to be $\iota_1, \ldots, \iota_{\ell_\mathfrak{f}}$. If it holds that $\sum_{i=1}^{\ell_\mathfrak{f}} c_i \cdot C_i(x) = \mathsf{out}^* \in B$ then we require that Recover($c_1, \iota_1, \ldots, c_{\ell_\mathfrak{f}}, \iota_{\ell_\mathfrak{f}}$), where $c_i \in \mathbb{F}_\mathbf{p}$, always outputs out^*.

Remark 1. Our construction only supports the case when $B = \{0\}$ when implemented by multilinear maps that only allows for zero testing at the final level. However, if encodings of 1 are given out at the top level, then B can be defined to be the set $\{0, \ldots, \mathrm{poly}(\lambda)\}$, where poly is a fixed polynomial.

Remark 2 ((B,B')-Correctness). We can also consider a property that we call (B, B')-correctness. It is the same as B-correctness except that the co-efficients c_i input to the above evaluation algorithm has to be in the set $B' \subseteq \mathbb{F}_\mathbf{p}$.

Remark 3 (Alternate Notation of Evaluation). Instead of feeding coefficients to the evaluation algorithm, we can directly feed in the description of the linear function. That is, if $\mathsf{out}^* \leftarrow$ Recover($\mathfrak{f}, (\iota_1, \ldots, \iota_{\ell_{\ell_\mathfrak{f}}})$) with $\mathfrak{f}$ being a linear function then we require that $\mathfrak{f}(C_1(x), \ldots, C_{\ell_{\ell_\mathfrak{f}}}) = \mathsf{out}^*$, where ι_i is obtained by decrypting a functional key of C_i with x.

2.2 Semi-Functional Security

We introduce a notion of semi-functional security associated with projective arithmetic FE. We refer the reader to the technical overview for an informal intuition behind the notion of semi-functional security.

We define the following two auxiliary algorithms.

Semi-Functional Key Generation, $\mathsf{sfKG}(\mathsf{MSK}, C, \theta)$: On input master secret key MSK, arithmetic circuit C, value θ, it outputs a semi-functional key sk_C.

Semi-Functional Encryption, $\mathsf{sfEnc}(\mathsf{MSK}, 1^{\ell_{\mathsf{inp}}})$: On input master secret key MSK and ℓ_{inp}, it outputs a semi-functional ciphertext CT.

We now introduce two security properties. We start with the first property, namely indistinguishability of semi-functional keys.

This property states that it should be hard for an efficient adversary to distinguish a semi-functional key associated with circuit C and value v from an honestly generated key associated with C. Additionally, the adversary can request for other semi-functional keys or honestly generated keys. The ciphertexts will be honestly generated.

Definition 1 (Indistinguishability of Semi-Functional Keys). *Consider a projective arithmetic functional encryption scheme* PAFE = (Setup, KeyGen, Enc, ProjectDec, Recover). *We say that* PAFE *satisfies* **indistinguishability of semi-functional keys** *with respect to* sfKG *if for any PPT adversary* $\mathcal{A}$ *there exists a negligible function* $\mathsf{negl}(\cdot)$ *such that*

$$\mathsf{Advtge}_{\mathcal{A}}^{\mathsf{PAFE}}(\lambda) = \left| \Pr[\mathsf{Expt}_{\mathcal{A}}^{\mathsf{PAFE}}(\lambda, 0) = 1] - \Pr[\mathsf{Expt}_{\mathcal{A}}^{\mathsf{PAFE}}(\lambda, 1) = 1] \right| \leq \mathsf{negl}(\lambda),$$

for all sufficiently large $\lambda \in \mathbb{N}$, *where for each* $b \in \{0, 1\}$ *and* $\lambda \in \mathbb{N}$ *the experiment* $\mathsf{Expt}_{\mathcal{A}}^{\mathsf{PAFE}}(1^{\lambda}, b)$, *modeled as a game between the adversary* $\mathcal{A}$ *and a challenger, is defined as follows:*

1. **Setup phase:** *The challenger samples* MSK $\leftarrow$ Setup(1^{λ}).
2. **Message queries:** *On input* 1^{λ} *the adversary submits* $(x_1, \ldots, x_{\ell_{\mathbf{x}}})$ *for some polynomial* $\ell_{\mathbf{x}} = \ell_{\mathbf{x}}(\lambda)$.
3. **Function queries:** *The adversary also submits arithmetic circuit queries to the challenger. There are three tuples the adversary submits:*
   - *This comprises of circuits and values associated with every circuit;* $(C_1^0, \theta_1, \ldots, C_{\ell_{\mathbf{f}}}^0, \theta_{\ell_{\mathbf{f}}})$. *Here,* $\theta_j \in \mathbb{F}_{\mathbf{p}}$.
   - *This comprises of just circuits;* $(C_1^1, \ldots, C_{\ell_{\mathbf{f}}'}^1)$.
   - *This corresponds to a challenge circuit pair query* (C^*, θ^*)
4. **Challenger's response:** *The challenger replies with* $(\mathsf{CT}_1, \ldots, \mathsf{CT}_{\ell_{\mathbf{x}}})$, *where* $\mathsf{CT}_i \leftarrow \mathsf{Enc}(\mathsf{MSK}, x_i)$ *for every* $i \in [\ell_{\mathbf{x}}]$. *It also sends the following functional keys: for every* $j \in [\ell_{\mathbf{f}}]$,
   - $sk_{C_j^0} \leftarrow \mathsf{sfKG}(\mathsf{MSK}, C_j^0, \theta_j)$.
   - $sk_{C_j^1} \leftarrow \mathsf{KeyGen}(\mathsf{MSK}, C_j^1)$.
   - *If* $b = 0$, *generate* $sk_{C^*} \leftarrow \mathsf{sfKG}(\mathsf{MSK}, C^*, \theta^*)$. *Otherwise generate* $sk_{C^*} \leftarrow \mathsf{KeyGen}(\mathsf{MSK}, C^*)$.
5. **Output phase:** *The adversary outputs a bit* b' *which is defined as the output of the experiment.*

The second property is indistinguishability of semi-functional ciphertexts. This property states that it should be hard for an efficient adversary to distinguish honestly generated ciphertext of x from a semi-functional ciphertext. In this experiment, it is required that the adversary only gets semi-functional keys associated with circuits C_i and value v_i such that $v_i = C_i(x)$.

Definition 2 (Indistinguishability of Semi-Functional Ciphertexts).
Consider a projective arithmetic functional encryption scheme PAFE $=$ (Setup,
KeyGen, Enc, ProjectDec, Recover). *We say that* PAFE *satisfies* **indistinguishability of semi-functional ciphertexts** *with respect to* sfEnc *if for any PPT
adversary* $\mathcal{A}$ *there exists a negligible function* negl($\cdot$) *such that*

$$\mathsf{Advtge}_{\mathcal{A}}^{\mathsf{PAFE}}(\lambda) = \left| \Pr[\mathsf{Expt}_{\mathcal{A}}^{\mathsf{PAFE}}(\lambda, 0) = 1] - \Pr[\mathsf{Expt}_{\mathcal{A}}^{\mathsf{PAFE}}(\lambda, 1) = 1] \right| \leq \mathsf{negl}(\lambda),$$

for all sufficiently large $\lambda \in \mathbb{N}$, *where for each* $b \in \{0, 1\}$ *and* $\lambda \in \mathbb{N}$ *the experiment* $\mathsf{Expt}_{\mathcal{A}}^{\mathsf{PAFE}}(1^\lambda, b)$, *modeled as a game between the adversary* $\mathcal{A}$ *and a challenger, is defined as follows:*

1. **Setup phase:** *The challenger samples* MSK $\leftarrow$ Setup(1^λ).
2. **Message queries:** *On input* 1^λ *the adversary submits* $(x_1, \ldots, x_{\ell_\mathbf{x}})$ *for some polynomial* $\ell_\mathbf{x} = \ell_\mathbf{x}(\lambda)$ *and it also sends the challenge query* x^*.
3. **Function queries:** *The adversary also submits arithmetic circuit queries to the challenger. The query is of the form* $(C_1, \theta_1, \ldots, C_{\ell_\mathbf{f}}, \theta_{\ell_\mathbf{f}})$. *It should hold that* $\theta_j = C_j(x^*)$ *for every* $j \in [\ell_\mathbf{f}]$. *If it does not hold, the experiment is aborted.*
4. **Challenger's response:** *The challenger replies with* $(\mathsf{CT}_1, \ldots, \mathsf{CT}_{\ell_\mathbf{x}})$, *where* $\mathsf{CT}_i \leftarrow \mathsf{Enc}(\mathsf{MSK}, x_i)$ *for every* $i \in [\ell_\mathbf{x}]$. *It sends* $\mathsf{CT}^* \leftarrow \mathsf{Enc}(\mathsf{MSK}, x^*)$ *only if* $b = 0$, *otherwise it sends* $\mathsf{CT}^* \leftarrow \mathsf{sfEnc}\left(\mathsf{MSK}, 1^{|x^*|}\right)$. *Finally, it sends the following functional keys: for every* $j \in [\ell_\mathbf{f}]$, *compute* $sk_{C_j} \leftarrow \mathsf{sfKG}(\mathsf{MSK}, C_j, \theta_j)$.
5. **Output phase:** *The adversary outputs a bit* b' *which is defined as the output of the experiment.*

Remark 4. One can also define a stronger property where instead of submitting one challenge message x^*, the challenger submits a challenge message pair (x_0^*, x_1^*) and the requirement that for every circuit C_j query, $C_j(x_0^*) = C_j(x_1^*)$. The reduction, in response, encrypts x_b^* where b is the challenge bit. It can be seen that this stronger security property is implied by the above property.

We now define semi-functional security property.

Definition 3. *We say that a projective arithmetic FE scheme, over* $\mathbb{F}_\mathbf{p}$, *is said to be* **semi-functionally secure** *if it satisfies both (i) indistinguishability of semi-functional keys property and, (ii) indistinguishability of semi-functional ciphertexts property.*

2.3   Other Notions

We also consider the following two notions of projective arithmetic FE.

Constant Degree Projective Arithmetic FE. In this work, we are interested in projective arithmetic FE for circuits that compute constant degree arithmetic circuits. In particular, we consider constant degree arithmetic circuits over arbitrary field $\mathbb{F}_\mathbf{p}$.

Multiplicative Overhead in Encryption Complexity. We say that a projective arithmetic FE scheme, over field $\mathbb{F}_\mathbf{p}$, satisfies multiplicative overhead in encryption complexity property if the complexity of encrypting x is $|x| \cdot \text{poly}(\lambda, \log(\mathbf{p}))$. That is,

Definition 4 (Multiplicative Overhead in Encryption Complexity). *Consider a projective arithmetic FE scheme* PAFE = (Setup, KeyGen, Enc, ProjectDec), *over field* $\mathbb{F}_\mathbf{p}$. *We say that* PAFE *satisfies multiplicative overhead in encryption complexity if* $|\text{Enc}(\text{MSK}, x)| = |x| \cdot \text{poly}(\lambda, \log(\mathbf{p}))$, *where* MSK *is the secret key generated during setup.*

Circuits versus Polynomials. Often in this manuscript, we interchangeably use arithmetic circuits over $\mathbb{F}_\mathbf{p}$ with polynomials computed over $\mathbb{F}_\mathbf{p}$. If there is a polynomial p over $\mathbb{F}_\mathbf{p}$ having $\text{poly}(\lambda)$ number of terms then there is a $\text{poly}'(\lambda)$-sized arithmetic circuit over $\mathbb{F}_\mathbf{p}$, where poly and poly' are polynomials. However, the reverse in general need not be true: if there is a $\text{poly}'(\lambda)$-sized arithmetic circuit over $\mathbb{F}_\mathbf{p}$ then the associated polynomial could have exponentially many terms. For example: $(x_1 + x_2) \cdots (x_{2n-1} + x_{2n})$ has a succinct circuit representation but when expanded as a polynomial has exponential number of terms.

In this work, we are only interested in arithmetic circuits which can be expressed as polynomials efficiently. In particular, we consider arithmetic circuits of constant fan-in and constant depth.

3 Slotted Encodings

We define the notion of slotted encodings: this concept can be thought of as abstraction of composite order multilinear maps. It allows for jointly encoding a vector of elements. Given the encodings of two vectors, using the addition and multiplication operations it is possible to either homomorphically add the vectors component-wise or multiply them component-wise.

To define this primitive, we first define the notion of structured asymmetric multilinear maps in Sect. 3.1. We show in Sect. 3.2 how to instantiate this form of structured asymmetric multilinear maps using current known instantiations of multilinear maps. Once we have armed ourselves with the definition of structured multilinear maps, we define the notion of slotted encodings (a special type of structured multilinear maps) in Sect. 3.3. In the full version, we show how to realize slotted encodings using structured asymmetric multilinear maps for the constant degree[12] case.

3.1 Structured (Asymmetric) Multilinear Maps

We define the notion of structured asymmetric multilinear maps. It is associated with a binary tree T. Every node is associated with a group structure and

---

[12] As we see later, this corresponds to the scenario where the structured multilinear maps is associated with constant number of bilinear maps.

additionally, every non leaf node is associated with a *noisy* bilinear map. Every element in this group structure has multiple noisy representations as in the case of recent multilinear map candidates [GGH13a, CLT13, GGH15].

Suppose nodes u and v are children of node w in tree T. And let the respective associated groups be $\mathbf{G}_u, \mathbf{G}_v$ and $\mathbf{G}_w$ respectively. Let e_{uv} be the bilinear map associated with node w. Then $e_{uv} : \mathbf{G}_u \times \mathbf{G}_v \to \mathbf{G}_w$.

Before we define structured multilinear maps we first put forward some notation about trees and also define some structural properties that will be useful later.

NOTATION ABOUT TREES: Consider a tree $T = (V, E)$, where V denotes the set of vertices and E denotes the set of edges. We are only interested in binary trees (every node has only two children) in this work.

1. We define the function $\mathbf{lc} : [V] \to \{0, 1\}$ such that $\mathbf{lc}(u) = 0$ if u is the left child of its parent, else $\mathbf{lc}(u) = 1$ if u is the right child of its parent.
2. We define $\mathbf{par} : [V] \to [V]$ such that $\mathbf{par}(u) = v$ if v is the parent of u.
3. $\mathbf{rt}(T) = w$ if the root of T is w.

Definition of Structured Multilinear Maps. A structured multilinear maps is defined by the tuple $\mathsf{SMMap} = (T = (V, E), \{\mathbf{G}_u\}_{u \in V})$ and associated with ring R, where:

- $T = (V, E)$ is a tree.
- $\mathbf{G}_u$ is a group structure associated with node $u \in V$. The order of the group is N.

The encoding of elements and operations performed on them are specified by the following algorithms:

- **Secret Key Generation,** $\mathsf{Gen}(1^\lambda)$: It outputs secret key $\mathbf{sk}$ and zero test parameters ztpp.
- **Encoding,** $\mathsf{Encode}(\mathbf{sk}, a, u \in V)$: In addition to secret key $\mathbf{sk}$, it takes as input $a \in R$ and a node $u \in V$. It outputs an encoding $[a]_{\mathbf{u}}$.
- **Add,** $[a]_{\mathbf{u}} + [b]_{\mathbf{u}} = [a + b]_{\mathbf{u}}$. Note that only elements corresponding to the same node in the tree can be added.
- **Multiply,** $[a]_{\mathbf{u}} \circ [b]_{\mathbf{v}} = [a \cdot b]_{\mathbf{w}}$. Here, w is the parent of u and v, i.e., $w = \mathbf{par}(u)$ and $w = \mathbf{par}(v)$.
- **Zero Test,** $\mathsf{ZeroTest}(\mathsf{ztpp}, [a]_{\mathfrak{r}})$: On input zero test parameters ztpp and an encoding $[a]_{\mathfrak{r}}$ at level $\mathfrak{r}$, where $\mathfrak{r} = \mathbf{rt}(T)$, output 0 if and only if $a = 0$.

We define degree of structured multilinear maps.

Definition 5 (Degree of SMMAP). *Consider a structured multilinear maps scheme given by* $\mathsf{SMMap} = (T = (V, E), \{\mathbf{G}_u\}_{u \in V})$. *The* **degree** *of* SMMap *is defined recursively as follows.*
We assign degree to every node in the tree as follows:

- *Degree of every leaf node u is 1.*

– *Consider a non leaf node w. Let u and v be its children. The degree of w is the sum of degree of u and degree of v.*

The degree of SMMap *is defined to be the degree of the root node.*

Remark 5. If we restrict ourselves to only binary trees (which is the case in our work) and if d is the depth of the binary tree T then the degree of SMMap, associated with $(T, \{\mathbf{G}_u\}_{u \in V})$ is 2^d.

Useful Notation: We employ the following notation that will be helpful later. Suppose $[v_1]_{\mathbf{i}}, \ldots, [v_m]_{\mathbf{i}}$ be a vector of encodings and let $\mathbf{v} = (v_1, \ldots, v_m) \in \mathbb{Z}_N^m$. Then, $[\mathbf{v}]_{\mathbf{i}}^m$ denotes $([v_1]_{\mathbf{i}}, \ldots, [v_m]_{\mathbf{i}})$. If the dimension of the vector is clear, we just drop m from the subscript and write $[\mathbf{v}]_{\mathbf{i}}$.

3.2   Instantiations of Structured Multilinear Maps

We can instantiate structured multilinear maps using the 'asymmetric' version of existing multilinear map candidates [GGH13a, CLT13]. For example, in asymmetric GGH, every encoding is associated with set S. Two encodings associated with the same set can be added. If there are two encodings associated with sets S_1 and S_2 respectively, then they can be paired if and only if $S_1 \cap S_2 = \emptyset$. The encoding at the final level is associated with the universe set, that is the union of all the sets.

To construct a structure multilinear map associated with $(T = (V, E), \phi)$, we can start with a universal set $U = \{1, \ldots, |V'|\}$, where $V' \subseteq V$ is the set of leaves in T. That is, there are as many elements as the number of leaves in V. We then design a bijection $\psi : U \to [V']$. An encoding is encoded at a leaf node u under the set $S_u = \{\psi^{-1}(u)\}$. For a non leaf node w, the encoding is performed under the set $S_w = S_u \cup S_v$, where u and v are the children of w.

3.3   Definition

A L-slotted encoding SEnc is a type of structured multilinear maps SMMap $= (T = (V, E), \{\mathbf{G}_u\}_{u \in V})$ associated with ring R and is additionally parameterized by L. It consists of the following algorithms:

– **Secret Key Generation,** Gen(1^λ): It outputs secret key **sk** and zero test parameters ztpp.
– **Encoding,** Encode(**sk**, $a_1, \ldots, a_L, u \in V$): In addition to secret key **sk**, it takes as input $a_1, \ldots, a_L \in R$ and a node $u \in V$. If u is not the root node, it outputs an encoding $[a_1| \cdots |a_L]_{\mathbf{u}}$. If u is indeed the root node, it outputs an encoding $\left[\sum_{i=1}^L a_i\right]_{\mathbf{u}}$.
– **Add,** $[a_1| \cdots |a_L]_{\mathbf{u}} + [b_1| \cdots |b_L]_{\mathbf{u}} = [a_1 + b_1| \cdots |a_L + b_L]_{\mathbf{u}}$. Note that only elements corresponding to the same node in the tree can be added. Further, the elements in the vector are added component-wise.

– **Multiply:** Suppose $w = \mathbf{par}(u)$ and $w = \mathbf{par}(v)$.

$$[a_1|\cdots|a_L]_{\mathbf{u}} \circ [b_1|\cdots|b_L]_{\mathbf{v}} = \begin{cases} [a_1b_1|\cdots|a_Lb_L]_{\mathbf{w}} & \text{if } \mathbf{rt}(T) \neq w \\[2mm] \left[\displaystyle\sum_{i=1}^{L} a_ib_i\right]_{\mathbf{w}} & \text{otherwise} \end{cases}$$

The elements in the vectors are multiplied component-wise.

– **Zero Test,** ZeroTest(ztpp, $[a]_{\mathfrak{r}}$): On input zero test parameters ztpp and an encoding $[a]_{\mathfrak{r}}$ at level $\mathfrak{r}$, where $\mathfrak{r} = \mathbf{rt}(T)$, output 0 if and only if $a = 0$.

Remark 6. The degree of slotted encodings can be defined along the same lines as the degree of structured multilinear maps.

3.4 Evaluation of Polynomials on Slotted Encodings

We consider the homomorphic evaluation of (T, ϕ)-respecting polynomials on slotted encodings. We first define evaluation of (T, ϕ)-respecting monomials on slotted encodings and then using this notion define evaluation of $(T, \overrightarrow{\phi})$-respecting polynomials on slotted encodings.

HomEval $(t, \mathsf{SMMap}, \{E_{1,u}\}_{u \in V}, \ldots, \{E_{n,u}\}_{u \in V})$: The input to this algorithm is (T, ϕ)-respecting monomial $t \in \mathbb{F}_{\mathbf{p}}[y_1, \ldots, y_n]$, slotted encoding scheme $\mathsf{SMMap} = (T = (V, E), \{\mathbf{G}_u\}_{u \in V})$ and slotted encodings $E_{i,u}$, for every $i \in [n]$ and every $u \in V$, encoded under $\mathbf{G}_u$.

The evaluation proceeds recursively as follows: for every non leaf node $u \in V$, set $\widetilde{E_u} = E_{\phi(u),u}$. Consider the case when u is a non-leaf node and let v and w be the children of u. Compute encoding associated with node u as $\widetilde{E_u} = \widetilde{E_v} \circ \widetilde{E_w}$. Let $\mathbf{rt}$ be the root of T. Output the encoding $\widetilde{E_{\mathbf{rt}}}$ associated with $\mathbf{rt}$.

HomEval $(p, \mathsf{SMMap}, \{E_{1,u}\}_{u \in V}, \ldots, \{E_{n,u}\}_{u \in V})$: The input to this algorithm is $(T, \overrightarrow{\phi})$-respecting polynomial $p \in \mathbb{F}_{\mathbf{p}}[y_1, \ldots, y_n]$, slotted encoding scheme $\mathsf{SMMap} = (T = (V, E), \{\mathbf{G}_u\}_{u \in V})$ and slotted encodings $E_{i,u}$, for every $i \in [n]$ and every $u \in V$, encoded under $\mathbf{G}_u$.

Let $p = \sum_{i=1}^{n} c_i t_i$, for $c_i \in \mathbb{F}_{\mathbf{p}}$ and t_i is a (T, ϕ_i)-respecting monomial for every $i \in [n]$. The evaluation proceeds as follows: for every $i \in [n]$, execute $\widetilde{E_{\mathbf{rt}}}^{(i)} \leftarrow \mathsf{HomEval}(t_i, \mathsf{SMMap}, \{E_{1,u}\}_{u \in V}, \ldots, \{E_{n,u}\}_{u \in V})$. Compute $E_{\mathbf{rt}} = \sum_{i=1}^{n} c_i \widetilde{E_{\mathbf{rt}}}^{(i)}$. Output the encoding $E_{\mathbf{rt}}$.

Remark 7. Based on the current implementation of multilinear maps, given an encoding of an element $a \in \mathbb{F}_{\mathbf{p}}$, we don't know how to securely obtain encoding of $c \cdot a$ for some scalar $c \in \mathbb{F}_{\mathbf{p}}$ of our choice. But instead, we can still obtain encoding of $c \cdot a$, when c is small (for instance, polynomial in security parameter). This can achieved by adding encoding of a, c number of times.

4   Projective Arithmetic FE from Slotted Encodings

We show how to construct projective arithmetic FE starting from the notion of slotted encodings defined in Sect. 3.3.

Consider a L-slotted encoding scheme SEnc, defined with respect to structured multilinear maps SMMap $= (T = (V, E), \{\mathbf{G}_u\}_{u \in V})$ and is parameterized by L. We construct a multi-key secret key projective arithmetic functional encryption scheme PAFE for a function class $\mathcal{C} = \{\mathcal{C}_\lambda\}_{\lambda \in \mathbb{N}}$ as follows. Here, $\mathcal{C}_\lambda$ consists of functions with input length λ and output length poly(λ).

Setup(1^λ): On input security parameter λ,

- It executes the secret key generation algorithm of the slotted encoding scheme to obtain $\mathbf{sk} \leftarrow \mathsf{Gen}(1^\lambda)$.
- Sample values $\alpha_{i,u} \in \mathbb{F}_\mathbf{p}$ for every $i \in [\ell_{\mathsf{inp}}], u \in V$ at random. We define ℓ_{inp} later. Denote $\overrightarrow{\alpha} = (\alpha_{i,u})_{i \in [\ell_{\mathsf{inp}}], u \in V}$.
- Sample a random value $S \in \mathbb{F}_\mathbf{p}$.

It outputs MSK $= (\mathbf{sk}, \overrightarrow{\alpha}, S)$.

KeyGen(MSK, p): It takes as input master secret key MSK and a T-respecting polynomial $p \in \mathbb{F}_\mathbf{p}[y_1, \dots, y_{\ell_{\mathsf{inp}}}]$ associated with an arithmetic circuit C, where T is the same tree associated with the structured multilinear maps. Since p is T-respecting, we have the following: There exists $\phi = (\phi_1, \dots, \phi_K)$ with $\phi_i : [V] \rightarrow [\ell_{\mathsf{inp}}]$ such that:

- $p = \sum_{j=1}^K c_i t_i$, where $c_i \in \mathbb{F}_\mathbf{p}$.
- t_i is a (T, ϕ_i)-respecting monomial in ℓ_{inp} variables.

Let δ_i be obtained by first assigning $\alpha_{\phi_i(u), u}$ to every leaf node u and then evaluating T^{13}. That is, δ_i is the value obtained at the root of T. Assign $\Delta = \sum_{i=1}^K c_i \cdot \delta_i$.

Let $\mathbf{rt}$ be the root of T and let $\mathfrak{u}$ be its left child and $\mathfrak{v}$ be its right child. Compute $\mathsf{E}^C = \mathsf{Encode}(\mathbf{sk}, (0, \Delta \cdot S, p(0; 0)), \mathfrak{u})$ for every $i \in [n]$. Output $sk_C = (C, \mathsf{E}^C)$.

Enc(MSK, x): It takes as input master secret key MSK and input $x \in \{0, 1\}^{\ell_x}$. Let $\mathsf{inp} = x$ and $\ell_{\mathsf{inp}} = |x|$.

It also samples an element $\gamma \in \mathbb{F}_\mathbf{p}$ at random. For every $i \in [\ell_{\mathsf{inp}}], u \in V$ and u is a leaf node, encode the tuple $(\mathsf{inp}_i, \gamma \cdot \alpha_{i,u}, 0)$ with inp_i denoting the ith bit of inp, as follows: $\mathsf{E}_{i,u}^{\mathsf{inp}} = \mathsf{Encode}(\mathsf{MSK}, (\mathsf{inp}_i, \gamma \cdot \alpha_{i,u}, 0), u)$. Also encode γ^D under group $\mathbf{G}_\mathfrak{v}$, where $\mathfrak{v}$ is the right child of $\mathbf{rt}$: $\mathsf{E}_\gamma = \mathsf{Encode}(\mathsf{MSK}, (0, \gamma^D \cdot S^{-1}, 0), \mathfrak{v})$. Recall that D is the degree of homogeneity of RP.

Output the ciphertext $\mathsf{CT} = ((\mathsf{E}_{i,u})_{i \in [\mathsf{inp}], u \in V}, \mathsf{E}_\gamma)$.

ProjectDec(sk_C, CT): It takes as input functional key sk_C and ciphertext CT. It parses sk_C as (C, E^C) and CT as $((\mathsf{E}_{i,u})_{i \in [\mathsf{inp}], u \in V}, \mathsf{E}_\gamma)$. It executes the following:

---

[13] Note that every non leaf node is treated as a multiplication gate.

- Compute $\mathsf{out}_1 = \mathsf{HomEval}(p, \mathsf{SMMap}, (\mathsf{E}_{i,u})_{i \in [\mathsf{inp}], u \in V})$.
- Compute $\mathsf{out}_2 = \mathsf{E}^C \circ \mathsf{E}_\gamma$.

Output the partial decrypted value $\iota = \mathsf{out}_1 - \mathsf{out}_2$.

Recover$(c_1, \iota_1, \ldots, c_{\ell_f}, \iota_{\ell_f})$: On input co-efficients $c_i \in \mathbb{F}_\mathbf{p}$, partial decrypted values ι_i, it first computes:

$$temp = c_1 \iota_1 + \cdots + c_{\ell_f} \iota_{\ell_f}$$

The addition carried out above corresponds to the addition associated with the slotted encodings scheme. Now, perform $\mathsf{ZeroTest}(\mathsf{ztpp}, temp)$ and output the result. Note that the output is either in $\{0, \ldots, B\}$ or its $\perp$.

(B, B')-*Correctness.* From the correctness of $\mathsf{HomEval}$ and slotted encodings, it follows that out_1 is an encoding of $(p(x), \gamma^D \cdot p(\{\alpha_{i,u}\}), 0)$. Further, out_2 is an encoding of $(0, \gamma^D \cdot p(\{\alpha_{i,u}\}), 0)$. Thus, the partial decrypted value $\mathsf{out}_1 - \mathsf{out}_2$ is an encoding of $(p(x), 0, 0)$.

With this observation, we remark that for many polynomials $p_1, \ldots, p_N$, the decryption of functional key of p_i on encryption of x yields as partial decrypted values, encodings of $(p_i(x), 0, 0)$. Thus, sum of all encodings of $(c_i \cdot p_i(x), 0, 0)$, where $c_i \in B'$ and $B' = \{0, \ldots, \mathsf{poly}(\lambda)\}$, yields a successful zero test query if and only if $\sum_{i=1}^N c_i p_i(x) = 0$.

We remark that if ztpp just contains parameters to test whether a top level encoding is zero or not, then the above construction only supports $B = \{0\}$. If it additionally contains encoding of 1, then we can set $B = \mathsf{poly}(\lambda)$.

Encryption Complexity: Multiplicative Overhead. We calculate the encryption complexity as follows.

$$|\mathsf{Enc}(\mathsf{MSK}, x)| = |x| \cdot (\text{Number of groups in } \mathsf{SMMap}) \cdot \mathsf{poly}(\lambda)$$

Thus, the above scheme satisfies the multiplicative overhead property.

4.1 Proof of Security

SEMI-FUNCTIONAL ALGORITHMS: We describe the semi-functional encryption and the key generation algorithms. We start with the semi-functional key generation algorithm.

sfKG$(\mathsf{MSK}, p, \theta)$: Parse MSK as $(\mathbf{sk}, \overrightarrow{\alpha}, S)$. In addition, it takes as input a (T, ϕ)-respecting polynomial p and value θ to be hardwired in the third slot. Let $p = \sum_{j=1}^K c_j t_j$, where t_j is a (T, ϕ_j)-respecting monomial in ℓ_{inp} variables. Let δ_j be obtained by first assigning $\alpha_{\phi_j(u), u}$ to every leaf node u and then evaluating T. That is, δ_j is the value obtained at the root of T. Assign $\Delta = \sum_{j=1}^K c_{i,j} \cdot \delta_j$.

Let $\mathbf{rt}$ be the root of T and let u be its left child and v be its right child. Compute $\mathsf{E}^p = \mathsf{Encode}(\mathbf{sk}, (0, \Delta \cdot S, p(0; 0) - \theta), \mathsf{u})$ for every $i \in [n]$. Output $sk_C = (p, \mathsf{E}^p)$.

We now describe the semi-functional encryption algorithm.

$\mathsf{sfEnc}(\mathsf{MSK}, 1^{\ell_{\mathsf{inp}}})$: Parse MSK as $(\mathbf{sk}, \overrightarrow{\alpha})$. It samples an element $\gamma \in \mathbb{F}_\mathbf{p}$ at random. *

For every $i \in [\ell_{\mathsf{inp}}]$, $u \in V$ and u is a leaf node, encode the tuple $(0, \gamma \cdot \alpha_{i,u}, 0)$ as follows: $\mathsf{E}^{\mathsf{inp}}_{i,u} = \mathsf{Encode}(\mathsf{MSK}, (0, \gamma \cdot \alpha_{i,u}, 0), u)$. Also encode γ^D under group $\mathbf{G}_\mathfrak{v}$, where $\mathfrak{v}$ is the right child of $\mathbf{rt}$: $\mathsf{E}_\gamma = \mathsf{Encode}(\mathsf{MSK}, (0, \gamma^D, 1), \mathfrak{v})$. Recall that D is the degree of homogeneity of RP.

Output the ciphertext $\mathsf{CT} = \left((\mathsf{E}_{i,u})_{i \in [\mathsf{inp}], u \in V}, \mathsf{E}_\gamma \right)$.

We now prove the indistinguishability of semi-functional ciphertexts and indistinguishability of functional keys properties. Before that we state the assumptions on the slotted encodings upon which we prove the security of our scheme.

Assumptions. We define the following two assumptions.

Assumption #1: For all (i) inputs $\mathbf{x} = (x_1, \ldots, x_\mu) \in \{0,1\}^{\mu \cdot \ell_x}$, (ii) polynomials $p \in \mathbb{F}_\mathbf{p}[y_1, \ldots, y_n]$, $\mathbf{q} = (q_1, \ldots, q_N) \in \mathbb{F}_\mathbf{p}[y_1, \ldots, y_n]^N$ be (T, ϕ)-respecting polynomials, (iii) subset $I \subseteq [n]$ and finally, (iv) values $\theta \in \mathbb{F}_\mathbf{p}$, $\Theta = (\theta_i)_{i \in I} \in \mathbb{F}_\mathbf{p}^{|I|}$ and for every sufficiently large $\lambda \in \mathbb{N}$, the following holds:

$$\{ \mathsf{KeyGen}(\mathsf{MSK}, p), \ aux[\mathbf{x}, \mathbf{q}, I, \Theta] \} \cong_c \{ \mathsf{sfKG}(\mathsf{MSK}, p, \theta), \ aux[\mathbf{x}, \mathbf{q}, I, \Theta] \}$$

– $\mathsf{MSK} \leftarrow \mathsf{Setup}(1^\lambda)$
– $aux[\mathbf{x}, \mathbf{q}, I, \Theta] = (\mathsf{CT}_1, \ldots, \mathsf{CT}_\mu, sk_1, \ldots, sk_N)$ *consists of two components:*
  1. *For every* $i \in [n]$, *compute* $\mathsf{CT}_i \leftarrow \mathsf{Enc}(\mathsf{MSK}, x_i)$.
  2. *For every* $i \in [N]$ *and* $i \in I$, *compute* $sk_i \leftarrow \mathsf{sfKG}(\mathsf{MSK}, q_i, \theta_i)$. *Else if* $i \notin I$, *compute* $sk_i \leftarrow \mathsf{KeyGen}(\mathsf{MSK}, q_i)$.

Assumption #2: For all (i) inputs $x^* \in \{0,1\}^{\ell_x}$, $\mathbf{x} = (x_1, \ldots, x_\mu) \in \{0,1\}^{\mu \cdot \ell_x}$, (ii) polynomials $\mathbf{q} = (q_1, \ldots, q_N) \in \mathbb{F}_\mathbf{p}[y_1, \ldots, y_n]^N$ be (T, ϕ)-respecting polynomials and finally, (iii) values $\Theta = (\theta_i)_{i \in [N]}$ and for every sufficiently large $\lambda \in \mathbb{N}$, the following holds:

$$\{ \mathsf{sfEnc}(\mathsf{MSK}, 1^{\mathsf{inp}}), \ aux[\mathbf{x}, \mathbf{q}, \Theta] \} \cong_c \{ \mathsf{Enc}(\mathsf{MSK}, x^*), \ aux[\mathbf{x}, \mathbf{q}, \Theta] \}$$

– $\mathsf{MSK} \leftarrow \mathsf{Setup}(1^\lambda)$
– $aux[\mathbf{x}, \mathbf{q}, \Theta] = (\mathsf{CT}_1, \ldots, \mathsf{CT}_\mu, sk_1, \ldots, sk_N)$ *is computed in the following way:*
  1. *For every* $i \in [n]$, *compute* $\mathsf{CT}_i \leftarrow \mathsf{Enc}(\mathsf{MSK}, x_i)$.
  2. *For every* $i \in [N]$ $\theta_i = q_i(x^*)$.
  3. *For every* $i \in [N]$, *compute* $sk_i \leftarrow \mathsf{sfKG}(\mathsf{MSK}, q_i, \theta_i)$.

The following two theorems directly follow from the above two assumptions.

Theorem 4. *The scheme* PAFE *satisfies indistinguishability of semi-functional keys under Assumption #1.*

Theorem 5. *The scheme* PAFE *satisfies indistinguishability of semi-functional ciphertexts under Assumption #2.*

From the above two theorems, we have the following theorem.

Theorem 6. *The* PAFE *satisfies semi-functional security under Assumptions #1 and #2.*

References

[AAB+15] Agrawal, S., Agrawal, S., Badrinarayanan, S., Kumarasubramanian, A., Prabhakaran, M., Sahai, A.: On the practical security of inner product functional encryption. In: Katz, J. (ed.) PKC 2015. LNCS, vol. 9020, pp. 777–798. Springer, Heidelberg (2015). doi:10.1007/978-3-662-46447-2_35

[AB15] Applebaum, B., Brakerski, Z.: Obfuscating circuits via composite-order graded encoding. In: Dodis, Y., Nielsen, J.B. (eds.) TCC 2015. LNCS, vol. 9015, pp. 528–556. Springer, Heidelberg (2015). doi:10.1007/978-3-662-46497-7_21

[ABCP15] Abdalla, M., Bourse, F., Caro, A., Pointcheval, D.: Simple functional encryption schemes for inner products. In: Katz, J. (ed.) PKC 2015. LNCS, vol. 9020, pp. 733–751. Springer, Heidelberg (2015). doi:10.1007/978-3-662-46447-2_33

[ABCP16] Abdalla, M., Bourse, F., De Caro, A., Pointcheval, D.: Better security for functional encryption for inner product evaluations. IACR Cryptology ePrint Archive 2016:11 (2016)

[ABSV15] Ananth, P., Brakerski, Z., Segev, G., Vaikuntanathan, V.: From selective to adaptive security in functional encryption. In: Gennaro, R., Robshaw, M. (eds.) CRYPTO 2015. LNCS, vol. 9216, pp. 657–677. Springer, Heidelberg (2015). doi:10.1007/978-3-662-48000-7_32

[AIK06] Applebaum, B., Ishai, Y., Kushilevitz, E.: Computationally private randomizing polynomials and their applications. Comput. Compl. 15(2), 115–162 (2006)

[AJ15] Ananth, P., Jain, A.: Indistinguishability obfuscation from compact functional encryption. In: Gennaro, R., Robshaw, M. (eds.) CRYPTO 2015. LNCS, vol. 9215, pp. 308–326. Springer, Heidelberg (2015). doi:10.1007/978-3-662-47989-6_15

[AJS15] Ananth, P., Jain, A., Sahai, A.: Achieving compactness generically: Indistinguishability obfuscation from non-compact functional encryption. IACR Cryptology ePrint Archive 2015:730 (2015)

[AL16] Applebaum, B., Lovett, S.: Algebraic attacks against random local functions and their countermeasures. In: STOC, pp. 1087–1100 (2016)

[AS16] Ananth, P., Sahai, A.: Projective arithmetic functional encryption and indistinguishability obfuscation from degree-5 multilinear maps. Cryptology ePrint Archive, Report 2016/1097 (2016). http://eprint.iacr.org/2016/1097

[BCF16] Elisabetta, C., Baltico, Z., Catalano, D., Fiore, D.: Practical functional encryption for bilinear forms. Cryptology ePrint Archive, Report 2016/1104 (2016). http://eprint.iacr.org/2016/1104

[BGG+14] Boneh, D., Gentry, C., Gorbunov, S., Halevi, S., Nikolaenko, V., Segev, G., Vaikuntanathan, V., Vinayagamurthy, D.: Fully key-homomorphic encryption, arithmetic circuit ABE and Compact garbled circuits. In: Nguyen, P.Q., Oswald, E. (eds.) EUROCRYPT 2014. LNCS, vol. 8441, pp. 533–556. Springer, Heidelberg (2014). doi:10.1007/978-3-642-55220-5_30

[BGI+01] Barak, B., Goldreich, O., Impagliazzo, R., Rudich, S., Sahai, A., Vadhan, S., Yang, K.: On the (im)possibility of obfuscating programs. In: Kilian, J. (ed.) CRYPTO 2001. LNCS, vol. 2139, pp. 1–18. Springer, Heidelberg (2001). doi:10.1007/3-540-44647-8_1

[BGJ+16] Bitansky, N., Goldwasser, S., Jain, A., Paneth, O., Vaikuntanathan, V., Waters, B.: Time-lock puzzles from randomized encodings. In: ITCS 2016

[BJK15] Bishop, A., Jain, A., Kowalczyk, L.: Function-hiding inner product encryption. In: Iwata, T., Cheon, J.H. (eds.) ASIACRYPT 2015. LNCS, vol. 9452, pp. 470–491. Springer, Heidelberg (2015). doi:10.1007/978-3-662-48797-6_20

[BLP16] Bitansky, N., Lin, H., Paneth, O.: On removing graded encodings from functional encryption. Cryptology ePrint Archive, Report 2016/962, 2016 http://eprint.iacr.org/2016/962

[BNPW16] Bitansky, N., Nishimaki, R., Passelègue, A., Wichs, D.: From cryptomania to obfustopia through secret-key functional encryption. Cryptology ePrint Archive, Report 2016/558 (2016). http://eprint.iacr.org/2016/558

[BNS13] Boneh, D., Nikolaenko, V., Segev, G.: Attribute-based encryption for arithmetic circuits. IACR Cryptology ePrint Archive 2013:669 (2013)

[BP15] Bitansky, N., Paneth, O.: ZAPs and non-interactive witness indistinguishability from indistinguishability obfuscation. In: Dodis, Y., Nielsen, J.B. (eds.) TCC 2015. LNCS, vol. 9015, pp. 401–427. Springer, Heidelberg (2015). doi:10.1007/978-3-662-46497-7_16

[BPR15] Bitansky, N., Paneth, O., Rosen, A.: On the cryptographic hardness of finding a Nash equilibrium. In: FOCS (2015)

[BS15] Brakerski, Z., Segev, G.: Function-private functional encryption in the private-key setting. In: Dodis, Y., Nielsen, J.B. (eds.) TCC 2015. LNCS, vol. 9015, pp. 306–324. Springer, Heidelberg (2015). doi:10.1007/978-3-662-46497-7_12

[BSW07] Bethencourt, J., Sahai, A., Waters, B.: Ciphertext-policy attribute-based encryption. In: IEEE (S&P 2007), pp. 321–334 (2007)

[BV15] Bitansky, N., Vaikuntanathan, V.: Indistinguishability obfuscation from functional encryption. In: FOCS, IEEE (2015)

[CGH+15] Coron, J.-S., Gentry, C., Halevi, S., Lepoint, T., Maji, H.K., Miles, E., Raykova, M., Sahai, A., Tibouchi, M.: Zeroizing without low-level zeroes: new MMAP attacks and their limitations. In: Gennaro, R., Robshaw, M. (eds.) CRYPTO 2015. LNCS, vol. 9215, pp. 247–266. Springer, Heidelberg (2015). doi:10.1007/978-3-662-47989-6_12

[CHL+15] Cheon, J.H., Han, K., Lee, C., Ryu, H., Stehlé, D.: Cryptanalysis of the multilinear map over the integers. In: Oswald, E., Fischlin, M. (eds.) EUROCRYPT 2015. LNCS, vol. 9056, pp. 3–12. Springer, Heidelberg (2015). doi:10.1007/978-3-662-46800-5_1

[CHN+16] Cohen, A., Holmgren, J., Nishimaki, R., Vaikuntanathan, V., Wichs, D.: Watermarking cryptographic capabilities. In: STOC (2016)

[CIJ+13] Caro, A., Iovino, V., Jain, A., O'Neill, A., Paneth, O., Persiano, G.: On the achievability of simulation-based security for functional encryption. In: Canetti, R., Garay, J.A. (eds.) CRYPTO 2013. LNCS, vol. 8043, pp. 519–535. Springer, Heidelberg (2013). doi:10.1007/978-3-642-40084-1_29

[CLT13] Coron, J.-S., Lepoint, T., Tibouchi, M.: Practical multilinear maps over the integers. In: Canetti, R., Garay, J.A. (eds.) CRYPTO 2013. LNCS, vol. 8042, pp. 476–493. Springer, Heidelberg (2013). doi:10.1007/978-3-642-40041-4_26

[CS02] Cramer, R., Shoup, V.: Universal hash proofs and a paradigm for adaptive chosen ciphertext secure public-key encryption. In: Knudsen, L.R. (ed.) EUROCRYPT 2002. LNCS, vol. 2332, pp. 45–64. Springer, Heidelberg (2002). doi:10.1007/3-540-46035-7_4

[DDM16] Datta, P., Dutta, R., Mukhopadhyay, S.: Functional encryption for inner product with full function privacy. In: Cheng, C.-M., Chung, K.-M., Persiano, G., Yang, B.-Y. (eds.) PKC 2016. LNCS, vol. 9614, pp. 164–195. Springer, Heidelberg (2016). doi:10.1007/978-3-662-49384-7_7

[Gay16] Gay, R.: Functional encryption for quadratic functions, and applications to predicate encryption. Cryptology ePrint Archive, Report 2016/1106 (2016). http://eprint.iacr.org/2016/1106

[GGG+14] Goldwasser, S., Gordon, S.D., Goyal, V., Jain, A., Katz, J., Liu, F.-H., Sahai, A., Shi, E., Zhou, H.-S.: Multi-input functional encryption. In: Nguyen, P.Q., Oswald, E. (eds.) EUROCRYPT 2014. LNCS, vol. 8441, pp. 578–602. Springer, Heidelberg (2014). doi:10.1007/978-3-642-55220-5_32

[GGH13a] Garg, S., Gentry, C., Halevi, S.: Candidate multilinear maps from ideal lattices. In: Johansson, T., Nguyen, P.Q. (eds.) EUROCRYPT 2013. LNCS, vol. 7881, pp. 1–17. Springer, Heidelberg (2013). doi:10.1007/978-3-642-38348-9_1

[GGH+13b] Garg, S., Gentry, C., Halevi, S., Raykova, M., Sahai, A., Waters, B.: Candidate indistinguishability obfuscation and functional encryption for all circuits. In: FOCS, pp. 40–49 (2013)

[GGH15] Gentry, C., Gorbunov, S., Halevi, S.: Graph-induced multilinear maps from lattices. In: Dodis, Y., Nielsen, J.B. (eds.) TCC 2015. LNCS, vol. 9015, pp. 498–527. Springer, Heidelberg (2015). doi:10.1007/978-3-662-46497-7_20

[GGHR14] Garg, S., Gentry, C., Halevi, S., Raykova, M.: Two-round secure MPC from indistinguishability obfuscation. In: Lindell, Y. (ed.) TCC 2014. LNCS, vol. 8349, pp. 74–94. Springer, Heidelberg (2014). doi:10.1007/978-3-642-54242-8_4

[GGHZ14] Garg, S., Gentry, C., Halevi, S., Zhandry, M.: Fully secure attribute based encryption from multilinear maps. IACR Cryptology ePrint Archive 2014:622 (2014)

[GLSW15] Gentry, C., Lewko, A.B., Sahai, A., Waters, B.: Indistinguishability obfuscation from the multilinear subgroup elimination assumption. In: FOCS, pp. 151–170 (2015)

[GMM+16] Garg, S., Miles, E., Mukherjee, P., Sahai, A., Srinivasan, A., Zhandry, M.: Secure obfuscation in a weak multilinear map model. Cryptology ePrint Archive, Report 2016/817 (2016). http://eprint.iacr.org/2016/817

[GMR89] Goldwasser, S., Micali, S., Rackoff, C.: The knowledge complexity of interactive proof systems. SIAM J. Comput. 18(1), 186–208 (1989)

[GPSW06] Goyal, V., Pandey, O., Sahai, A., Waters, B.: Attribute-based encryption for fine-grained access control of encrypted data. In: ACM CCS (2006)

[GS08] Groth, J., Sahai, A.: Efficient non-interactive proof systems for bilinear groups. In: Smart, N. (ed.) EUROCRYPT 2008. LNCS, vol. 4965, pp. 415–432. Springer, Heidelberg (2008). doi:10.1007/978-3-540-78967-3_24

[HSW14] Hohenberger, S., Sahai, A., Waters, B.: Replacing a random oracle: full domain hash from indistinguishability obfuscation. In: Nguyen, P.Q., Oswald, E. (eds.) EUROCRYPT 2014. LNCS, vol. 8441, pp. 201–220. Springer, Heidelberg (2014). doi:10.1007/978-3-642-55220-5_12

[IK00] Ishai, Y., Kushilevitz, E.: Randomizing polynomials: a new representation with applications to round-efficient secure computation. In: FOCS, pp. 294–304 (2000)

[KSW08] Katz, J., Sahai, A., Waters, B.: Predicate encryption supporting disjunctions, polynomial equations, and inner products. In: Smart, N. (ed.) EUROCRYPT 2008. LNCS, vol. 4965, pp. 146–162. Springer, Heidelberg (2008). doi:10.1007/978-3-540-78967-3_9

[Lin16] Lin, H.: Indistinguishability obfuscation from constant-degree graded encoding schemes. In: Fischlin, M., Coron, J.-S. (eds.) EUROCRYPT 2016. LNCS, vol. 9665, pp. 28–57. Springer, Heidelberg (2016). doi:10.1007/978-3-662-49890-3_2

[LOS+10] Lewko, A., Okamoto, T., Sahai, A., Takashima, K., Waters, B.: Fully secure functional encryption: attribute-based encryption and (hierarchical) inner product encryption. In: Gilbert, H. (ed.) EUROCRYPT 2010. LNCS, vol. 6110, pp. 62–91. Springer, Heidelberg (2010). doi:10.1007/978-3-642-13190-5_4

[LV16] Lin, H., Vaikuntanathan, V.: Indistinguishability obfuscation from DDII-like assumptions on constant-degree graded encodings. In FOCS (2016)

[MSZ16] Miles, E., Sahai, A., Zhandry, M.: Annihilation attacks for multilinear maps: cryptanalysis of indistinguishability obfuscation over GGH13. In: Robshaw, M., Katz, J. (eds.) CRYPTO 2016. LNCS, vol. 9815, pp. 629–658. Springer, Heidelberg (2016). doi:10.1007/978-3-662-53008-5_22

[OT08] Okamoto, T., Takashima, K.: Homomorphic encryption and signatures from vector decomposition. In: Galbraith, S.D., Paterson, K.G. (eds.) Pairing 2008. LNCS, vol. 5209, pp. 57–74. Springer, Heidelberg (2008). doi:10.1007/978-3-540-85538-5_4

[OT09] Okamoto, T., Takashima, K.: Hierarchical predicate encryption for inner-products. In: Matsui, M. (ed.) ASIACRYPT 2009. LNCS, vol. 5912, pp. 214–231. Springer, Heidelberg (2009). doi:10.1007/978-3-642-10366-7_13

[OW14] O'Donnell, R., Witmer, D.: Goldreich's PRG: evidence for near-optimal polynomial stretch. In: CCC, pp. 1–12 (2014)

[SW05] Sahai, A., Waters, B.: Fuzzy identity-based encryption. In: Cramer, R. (ed.) EUROCRYPT 2005. LNCS, vol. 3494, pp. 457–473. Springer, Heidelberg (2005). doi:10.1007/11426639_27

[SW08] Sahai, A., Waters, B.: Slides on functional encryption. Powerpoint presentation (2008)

[SW14] Sahai, A., Waters, B.: How to use indistinguishability obfuscation: deniable encryption, and more. In STOC, pp. 475–484 (2014)

[Wat09] Waters, B.: Dual system encryption: realizing fully secure IBE and HIBE under simple assumptions. In: Halevi, S. (ed.) CRYPTO 2009. LNCS, vol. 5677, pp. 619–636. Springer, Heidelberg (2009). doi:10.1007/978-3-642-03356-8_36

[Zim15] Zimmerman, J.: How to obfuscate programs directly. In: Oswald, E., Fischlin, M. (eds.) EUROCRYPT 2015. LNCS, vol. 9057, pp. 439–467. Springer, Heidelberg (2015). doi:10.1007/978-3-662-46803-6_15

Discrete Logarithm

Computation of a 768-Bit Prime Field Discrete Logarithm

Thorsten Kleinjung[1,2]([✉]), Claus Diem[2], Arjen K. Lenstra[1],
Christine Priplata[2], and Colin Stahlke[2]

[1] EPFL IC LACAL, Station 14, CH-1015 Lausanne, Switzerland
768b.dlp@gmail.com
[2] Mathematisches Institut, Universität Leipzig, D-04009 Leipzig, Germany

Abstract. This paper reports on the number field sieve computation of
a 768-bit prime field discrete logarithm, describes the different parameter
optimizations and resulting algorithmic changes compared to the factorization of a 768-bit RSA modulus, and briefly discusses the cryptologic
relevance of the result.

Keywords: Discrete logarithm · DSA · ElGamal · Number field sieve

1 Introduction

Let $p = [2^{766}\pi] + 62762$, which is the smallest 768-bit prime number larger than
$2^{766}\pi$ for which $\frac{p-1}{2}$ is prime too[1]. Let $g = 11$, which is a generator of the
multiplicative group $\mathbf{F}_p^\times$ of the prime field $\mathbf{F}_p$. On June 16, 2016, we finished the
computation of the discrete logarithm of $t = [2^{766}e]$ with respect to g. We found
that the smallest non-negative integer x for which $g^x \equiv t \bmod p$ equals

325923617918270562238615985978623709128341338833721058543950813
521768156295090163834803063792023717563811735244229923404165874 8
471079911977497864301995972638266781162575370644813703762423329
7831296215671274794172806874952314633488 12.

By itself, this is a useless result. What is interesting is how we found it, that
we did so with much less effort than we expected, and what the result implies
for cryptographic security that relies on the difficulty of larger similar problems.
These issues are discussed in this paper.

The result was obtained using the number field sieve (NFS, [13, 28]). It
required the equivalent of about 5300 core years on a single core of a 2.2 GHz
Xeon E5-2660 processor, mostly harvested during the period May to December,
2015, on clusters at the authors' universities. On average each additional discrete logarithm requires two core days. This result is a record for computing

---

[1] Here $[x]$ denotes the classical entier function, the largest integer less than or equal
to x.

© International Association for Cryptologic Research 2017
J.-S. Coron and J.B. Nielsen (Eds.): EUROCRYPT 2017, Part I, LNCS 10210, pp. 185–201, 2017.
DOI: 10.1007/978-3-319-56620-7_7

prime field discrete logarithms. It closes the gap between record calculations for general purpose integer factoring and computing arbitrary prime field discrete logarithms, with the 768-bit integer factorization record [21] dating back to 2009. Although our effort was substantial, we spent a fraction of what we originally expected. The purpose of this paper is to describe how this was achieved.

Records of this sort are helpful to get an impression of the security offered by cryptographic systems that are used in practice. The 768-bit number field sieve factorization from [21], for instance, required about 1700 core years. Because factoring a single 1024-bit RSA modulus [34] using the number field sieve is about three orders of magnitude more work (cf. end of Sect. 2), an educated guess follows for the worst-case effort to break a 1024-bit RSA key. Interpretation of the resulting estimate is another matter. Depending on one's perception, applications, incentives, taste, ..., it may boost or undermine one's confidence in the security of 1024-bit RSA moduli.

The ratio is similar between the difficulties of computing 768-bit and 1024-bit prime field discrete logarithms (cf. Sect. 2). It follows that even the nonchalant users of 1024-bit RSA, ElGamal [11], or DSA [36] have no reason to be nervous anytime soon if their concern is an "academic attack" such as the one presented here (cf. [6]). They have to be a bit more concerned, however, than suggested by [2, Sect. 4.1]. Also, we explicitly illustrate in Sect. 3 that continued usage of 1024-bit prime field ElGamal or DSA keys is much riskier than it is for 1024-bit RSA (all are still commonly used), because once a successful attack has been conducted against a single well-chosen prime field all users of that prime field [27, Sect. 4] may be affected at little additional effort [2].

As shown in Sect. 5 our result gives a good indication for the difficulty of computing discrete logarithms in multiplicative groups of other 768-bit prime fields as well. One such group, the so-called First Oakley Default Group, is of some historical interest as it was one of the groups supported by the Internet Key Exchange standard from 1998 [15], a standard that has been obsolete since 2005 [16]. In some cryptographic applications, however, one may prefer to use a generator of a relatively small prime order subgroup of $\mathbf{F}_p^\times$ that is chosen in such a way that comparable efforts would be required by Pollard's rho in the subgroup and by the number field sieve in $\mathbf{F}_p^\times$. Our choice of p assures that no (published) shortcut can be taken for our discrete logarithm computation. It also represents the most difficult case for the number field sieve, in particular for its linear algebra step. It follows from the numbers presented below that, for a discrete logarithm computation, our choice is overall more difficult than a subgroup order that may sometimes be preferred for cryptographic applications. With independently optimized parameters the two efforts are however of the same order of magnitude (cf. Sect. 4).

Two simple methods can be used to give an a priori estimate of the effort to solve our 768-bit prime field discrete logarithm problem. The first is direct extrapolation (cf. Sect. 2): given that solving a 596-bit prime field discrete logarithm problem took 130 core years (cf. [7]), extrapolation suggests that our 768-bit problem should be doable in about thirty thousand core years. For the second method we observe that the number field sieve for factoring or for prime field discrete logarithms is essentially the same algorithm. When applied to 768-bit composites or 768-bit prime fields and when using comparable number fields, they deal with similar probabilities and numbers of comparable sizes, with the sole exception occurring in the linear algebra step: although in both cases the matrix is very sparse and all non-zero entries are (absolutely) very small, when factoring linear algebra is done in a matrix modulo two, but for discrete logarithm problems the matrix elements are taken modulo the group order (a 767-bit integer in our case). An opposite effect, however, is caused by the fact that, with proper care, the number fields will *not* be comparable because modulo large primes polynomial selection methods can be used that do not work modulo large composites.

It follows that the numbers reported in [21] can be used to derive an *upper bound* for the 768-bit prime field discrete logarithm effort, simply by using a 767-fold increase (cf. Sect. 3) of the linear algebra effort from [21] while leaving the other steps unchanged. With [21, Sect. 2.4] we find that fifty thousand core years should suffice for our problem. If we would switch to a 768-bit prime that allows a much smaller but cryptographically still interesting subgroup this rough overall estimate would be reduced by a factor of about five.

Thirty or fifty thousand core years would be a waste of resources for a calculation of this sort, and the more doable small subgroup alternative would be of insufficient interest; independent of our estimates, a very similar figure was derived in [2, Sect. 4.1]. All these estimates, however, overlook several points. Direct extrapolation of the 596-bit effort turned out to be meaningless due to software improvements and because the limited size did not allow an optimization that applies to our case. But more importantly, the very different nature and size of the moduli used in, respectively, the polynomial selection and linear algebra steps imply a radical shift in the trade-off between the steps of the number field sieve, which in turn leads to very different parameter and algorithmic choices compared to what is done for factoring. We are not aware of a satisfactory theoretical analysis of this different trade-off and the resulting parameter selection, or of a reliable way to predict the practical implication for the relative hardness of integer factoring and prime field discrete logarithm problems. It is clear, however, that the issue is more subtle than recognized in the literature, such as [26,31] and, more recently, [2, Sect. 4.1].

As described in Sect. 3, adapting the parameter choices and algorithms to the case at hand – and guided by multiple experiments – it was found that it should be possible to reduce the fifty thousand core years estimate by almost an order of magnitude. This led to the conclusion that actually doing the full calculation would be a worthwhile undertaking: in the first place because it shows that for our current range of interest k-bit factoring and computing k-bit prime field discrete logarithms require a comparable effort; and in the second place, and possibly more interesting, because it required more than just a casual application of known methods.

The previous 596-bit and current 768-bit prime field discrete logarithm records should not be confused with extension field discrete logarithm records. Due to recent developments, we now have much better methods than the number field sieve to compute discrete logarithms in small characteristic extension fields. As a consequence, those fields are no longer relevant for basic cryptographic applications such as DSA. Indeed, recent extension field records imply that impractically large extension fields would have to be used to get an appreciable level of security: for instance, computing discrete logarithms in the multiplicative group of the 9234-bit field $\mathbf{F}_{2^2 \cdot 3^5 \cdot 19}$ took less than fifty core years [14], and in the 3796-bit group $\mathbf{F}^\times_{3^{5 \cdot 479}}$ the problem was dealt with in less than a single core year [18]. On the other hand, the current characteristic two *prime extension degree* record involved the much smaller finite field $\mathbf{F}_{2^{1279}}$ and took between three and four core years [20]: the advantage of the new methods over the number field sieve strongly depends on properties of the extension degree, but for favorable degrees the advantage is much bigger than the advantage for the number field sieve when factoring special numbers (such as Mersenne or Fermat numbers) compared to general ones (such as RSA moduli).

While the correctness of the outcome of our calculation can simply be verified, independent validation of the other claims made in this paper requires access to suitable source code and data. We have established a long-standing tradition of open collaborations [30] with other leading researchers in this field (see [21,22] and the references therein) which applies to anything relevant for the present project as well.

The paper is organized as follows. Section 2 presents the background for the rest of the paper. Section 3 describes the impact of the parameter selection on the way one of the main steps of the number field sieve is best implemented for the problem solved here and lists all relevant details of our new record calculation. Section 4 gives more details about the trade-off between the main steps of the number field sieve, and presents estimates for the effort required to solve a discrete logarithm problem in a small subgroup. In Sect. 5 it is shown that our choice of $p = \lceil 2^{766}\pi \rceil + 62762$ is not more or less favorable than other primes of the same size.

2   Algorithm Overview

Descriptions of the number field sieve are available in the literature, ranging from the high level narrative [33] to the somewhat simplified and fully detailed versions in [29] and [28], respectively.

Index Calculus Method [1,24,25]. Let $\mathbf{F}_p$ be a finite field of cardinality p, identified with $\{0, 1, \ldots, p-1\}$ in the usual manner, and let g generate its multiplicative group $\mathbf{F}^\times_p$. To compute discrete logarithms with respect to g, an index calculus method fixes a so-called *factor base* $B \subset \mathbf{F}^\times_p$, collects more than $\#B$ multiplicative *relations* between the elements of $B \cup \{g\}$, and uses linear algebra modulo the order of g to determine for all elements of B their discrete logarithm with respect to g. Given this information, the discrete logarithm of any $h \in \mathbf{F}^\times_p$ is then found by finding a multiplicative relationship between h and the elements of $B \cup \{g\}$.

Doing more or less the same modulo a composite N (as opposed to modulo p) and using linear algebra modulo two (as opposed to modulo the order of g) an integer solution to $x^2 \equiv y^2 \bmod N$ may be found, and thus a chance to factor N by computing $\gcd(N, x - y)$. This explains the similarity in the algorithms for factoring and computing discrete logarithms as well as the difference between the matrices for factoring and discrete logarithms that was pointed out in the introduction. The effect of the prime p versus the composite N, as also mentioned in the introduction, is touched upon below and in Sect. 3.

Different index calculus methods vary mostly in the way the multiplicative relations are found. This affects the way B is chosen. For prime p for instance, relations may be collected by considering g^e for random integers e and keeping those that factor over B. With B the set of primes up to some bound b one would thus be collecting b-smooth g^e-values. Faster methods increase the smoothness probabilities by generating smaller values in $\{1, 2, \ldots, p - 1\}$; select the values in an arithmetic progression so that *sieving* can be used to faster recognize smooth values; allow in relations a few *large primes* between b and a large prime bound b_ℓ; or they manage to combine those speedups. Dan Gordon [13] was the first to show how for prime p the ideas from the number field sieve for integer factorization [28] can be included as well. Many other variants have been proposed since then; the most accurate reference for the one used here is [35].

Relations in the Number Field Sieve. A property of the number field sieve that sets it apart from the earlier index calculus methods is that for a relation two distinct numbers must be smooth (with both numbers asymptotically significantly smaller than the values considered before). Let f and g be two coprime irreducible polynomials in $\mathbf{Z}[X]$ of degrees d_f and d_g, respectively, chosen in such a way that they have a root m in common modulo p (see Sect. 3 for how this may be done). A relation corresponds to a coprime pair of integers (a, b) with $b \geq 0$ such that the two integers $\mathcal{N}_f(a, b) = b^{d_f} f(\frac{a}{b})$ and $\mathcal{N}_g(a, b) = b^{d_g} g(\frac{a}{b})$ are smooth with respect to appropriately chosen bounds.

This is, very briefly, explained as follows. The integer $\mathcal{N}_f(a, b)$ is essentially (except for the leading coefficient of f) the norm of $a - \alpha_f b \in \mathbf{Z}[\alpha_f] \subset \mathbf{Q}(\alpha_f)$, where α_f denotes a zero of f and $\mathbf{Q}(\alpha_f)$ is the algebraic number field $\mathbf{Q}[X]/(f(X))$. The smoothness of $\mathcal{N}_f(a, b)$ then implies a factorization into small prime ideals in $\mathbf{Q}(\alpha_f)$ of the ideal $(a - \alpha_f b)$ (cf. [8]). Noting that mapping α_f to the common root m results in a ring homomorphism φ_f from $\mathbf{Z}[\alpha_f]$ to $\mathbf{F}_p$, and defining α_g and φ_g in a similar manner for g, a relation (a, b) thus corresponds to factorizations of the ideals $(a - \alpha_f b)$ and $(a - \alpha_g b)$ that map, via φ_f and φ_g, respectively, to the same element $a - bm \in \mathbf{F}_p$.

The 768-bit factorization from [21] used degrees $d_f = 6$ and $d_g = 1$; consequently, the labels *algebraic* or *rational* were used to distinguish values and computations related to f or g, respectively. These intuitive labels can no longer be used here, because the primality of our modulus (p) offers flexibility in the polynomial selection that is not available for composite moduli and that resulted, as could be expected, in the "better" choices $d_f = 3$ and $d_g = 4$ for the present paper. Though the f- and g-related parts here are thus both algebraic, it will be seen that the g-part is easier to deal with, and thus, to some extent, corresponds to the rational side in [21]. Because f and g have rather different properties, different considerations come into play when selecting the factor bases for $\mathcal{N}_f(a, b)$ and $\mathcal{N}_g(a, b)$. The single factor base B is therefore replaced by two distinct factor bases, denoted by B_f and B_g. For the present purposes it may be assumed that B_f and B_g consist of the primes bounded by b_f and b_g. We make no distinction between f and g for the large prime bound (which is thus still denoted by b_ℓ, with $\#B_\ell$ denoting the number of primes bounded by b_ℓ), but may allow different numbers of large primes in $\mathcal{N}_f(a, b)$ and $\mathcal{N}_g(a, b)$, denoted by n_f and n_g.

Finding Relations in the Number Field Sieve. As each relation requires two numbers being smooth, collecting relations is a two-stage process: in the first stage pairs (a, b) for which $\mathcal{N}_f(a, b)$ is smooth are located; in the second stage, from the pairs found those for which $\mathcal{N}_g(a, b)$ is smooth as well are selected. Thus, the first stage treats the numbers that are least likely to be smooth, thereby minimizing the number of pairs to be considered for the second stage: switching the roles of f and g would have led to more pairs to be treated in the second stage. Depending on the factor base sizes, various methods may be used to find relations.

The search for relations is typically limited to a (large) rectangular region S of the lattice $\mathbf{Z}^2$. For the first stage (and numbers in the current range of interest) index-q sublattices L_q of $\mathbf{Z}^2$ are identified such that q divides $\mathcal{N}_f(a, b)$ for all pairs $(a, b) \in L_q$ and for primes q close to and often somewhat larger than b_f (these primes are referred to as *special q primes*). The process described below is repeated for different special q primes until, after removal of unavoidable duplicates, enough relations have been collected.

Given a special q prime, *lattice sieving* is conducted over a rectangular region S_q (which roughly approximates $L_q \cap S$), to locate (a, b) pairs for which $\mathcal{N}_f(a, b)$ is b_f-smooth (except for q and at most n_f large primes $\leq b_\ell$). The number of "surviving" pairs thus found is denoted by y_f. If y_f is large (and the pairs are not spread too widely as for instance in [9,22]), it is best to again use lattice sieving in S_q to collect from those y_f pairs the y_g pairs that are actually relations, i.e., for which $\mathcal{N}_g(a, b)$ is b_g-smooth as well (again with at most n_g large primes $\leq b_\ell$). This is the regular approach to the second stage, and was used in [21]. But there are circumstances where the second stage is best done in another manner: in [22], for instance, factorization trees (cf. [12, Sect. 4] and [4]) were used. This is also the approach taken here, as further described in Sect. 3.

Effort Required by the Number Field Sieve. With natural logarithms, let

$$E(x, c) = \exp\left(\left(\left(\tfrac{64}{9}\right)^{\frac{1}{3}} + c\right)(\log x)^{\frac{1}{3}}(\log\log x)^{\frac{2}{3}}\right);$$

this slight variation on a well-known and more common notation allows us to focus on what is of greatest interest in the present context. The current best heuristic expected effort to compute a discrete logarithm in $\mathbf{F}_p^\times$ using the number field sieve is $E(p, o(1))$, asymptotically for $p \to \infty$ [13]. This is the same as the effort $E(N, o(1))$ (for $N \to \infty$) to factor a composite N using the number field sieve [8]. This "same" should, however, be taken with a grain of salt because the $o(1)$ hides different functions for the two cases.

Optimal Factor Base Sizes. The smoothness probabilities, the number of relations to be collected, and the dimension of the matrix handled by the linear algebra all increase with the smoothness parameters b_f, b_g, b_ℓ, n_f and n_g. The resulting trade-off leads to optimal factor base sizes $\#B_f$ and $\#B_g$, namely $E(p, o(1))^{\frac{1}{2}}$ for discrete logarithms and $E(N, o(1))^{\frac{1}{2}}$ for factoring. As noted above, even if $[\log p] = [\log N]$, in a given model of computation the optimal values for both factoring and discrete logarithm computation may be very different because the two $o(1)$-functions behave quite differently. Moreover, in practice the situation is further complicated because of the software and hardware actually used. Thus, naively using factor base sizes that worked well for a factoring problem for a similarly sized prime field discrete logarithm problem, as done in the introduction and despite the "correction" attempted there, will at best result in a rough upper bound. Section 3 discusses this issue in more detail.

Remark on Using $E(x, c)$ in Practice. The uncertain function hiding in the $o(1)$ makes it challenging to use $E(p, o(1))$ to give an absolute estimate for the effort to solve a discrete logarithm problem in $\mathbf{F}_p^\times$. It turns out, however, that a somewhat pessimistic indication can be obtained for the relative effort for $\mathbf{F}_{\bar{p}}^\times$ compared to $\mathbf{F}_p^\times$, for $\bar{p}$ not much bigger than p (say, $\bar{p} \leq p^{\frac{4}{3}}$), by dropping the $o(1)$. Obviously, this assumes similar software that suffers no ill side-effects nor profits from new optimizations when moving to the larger $\bar{p}$. The same works for factoring.

As an example, the three orders of magnitude difference between the efforts of factoring 768-bit and 1024-bit moduli, as mentioned in the introduction, follows from $\frac{E(2^{1024}, 0)}{E(2^{768}, 0)} \approx 1200$; the jump from 130 core years for a 596-bit prime field discrete logarithm problem to about thirty thousand core years for 768 bits follows from $\frac{E(2^{768}, 0)}{E(2^{596}, 0)} \approx 275$ – an extrapolation that failed to be useful because of the reasons mentioned in the introduction.

3   Computational Details

This section provides some background on our parameter choices. For comparison, we also provide the parameters that were used for the 768-bit factoring effort from [21].

Polynomial Selection. To get an initial impression of the feasibility of the calculation an extensive search was conducted using the method from [17]. First all integer polynomials g of degree four with coefficients absolutely bounded by 165 (and noting that $g(X)$, $g(-X)$, and $X^4 g(\frac{1}{X})$ and thus $X^4 g(\frac{-1}{X})$ are equivalent) were inspected, by using, for all the roots of g modulo p, lattice reduction to find a corresponding degree three integer polynomial f, and measuring the overall quality of all resulting pairs (f, g) (as usual with respect to their small modular roots and size properties). For the second search, with bound 330, roots and size properties of g were first considered and only for the most promising candidates the roots of g modulo p were calculated and, if any, the polynomial f was derived. The best pair was found during the first search:

$$f(X) = 3708634038864161411505055239195276772319326181841000095924X^3$$
$$- 19379813128330387785656174698293955440652559380159203096799X^2$$
$$- 2175832936269478997875774411283330276175410950047347364159X$$
$$+ 2772607304003495228904226184734981485287061150033379351509,$$
$$g(X) = 140X^4 + 34X^3 + 86X^2 + 5X - 55.$$

Because it requires root finding modulo p, the above search does not work to find polynomials for the number field sieve for integer factorization. There one is limited to more restrictive methods that cannot be expected to result in polynomials of comparable "quality", with respect to the metric used in this context: indeed, the above pair is noticeably better than the degree $(6, 1)$ pair used for the slightly smaller 768-bit modulus factored in [21]. A more quantitative statement requires a more careful analysis than we are ready to provide here. No significant amount of time was spent on searching for pairs (f, g) of other degrees than $d_f = 3$ and $d_g = 4$.

Parameter Selection Background. The two main steps of the number field sieve after the polynomial selection, relation collection and linear algebra, are of a very different nature. Relation collection is long-term but low-maintenance: core years are easily harvested on any number of otherwise idle independent cores on any number of clusters that one can get access to, progress will be steady, and the process requires almost no human interaction. The results can easily be checked for correctness (cf. [22, Sect. 6]) and results that are lost or forgotten are easily replaced by others. Compared to this almost "happy-go-lucky" relation collection process, the linear algebra is tedious and cumbersome, despite the elegance of the block Wiedemann method used for it [10,37]. It involves careful orchestration of a (modest number of) substeps each of which requires as many tightly coupled cores as needed to store the data (easily on the order of hundreds of GB), frequent checkpointing, and a central step that is even more memory-demanding but otherwise fortunately relatively swift. Overall, based on past experience core years are collected at about half the rate compared to relation collection.

For both main steps the required effort is well understood:

- Given relation collection software and any choice of smoothness parameters a small number of experiments suffices to get an accurate indication for the effort required to collect any specified number of relations (it follows from the description in Sect. 5 how this may be done).
- Similarly, given block Wiedemann software and any matrix dimension, weight, and modulus-size, the overall linear algebra effort can be reliably estimated based on the effort required for a few matrix × vector multiplications on the processor network of one's choice.

However, the relations as collected are never directly used for the linear algebra, because doing so would be hugely inefficient. Instead, a linear algebra preprocessing step is applied to the relations in order to reduce the dimension of the matrix while keeping its weight under control, thereby (substantially) reducing the linear algebra effort. This preprocessing becomes more effective as more relations are available (cf. Sect. 4) but the precise behavior of both dimension and weight depends on how (large) primes in a relation can be matched with the same primes in other relations and is thus uncertain. In practice one collects relations while occasionally doing a preprocessing attempt, and stops when the resulting linear algebra effort is within the targeted range. When to stop is a judgment call as more often than not the additional effort invested in relation collection is more than the expected linear algebra savings: it thus serves more to reduce the linear algebra headaches than to reduce the overall effort. As an example, for the current 768-bit factoring record about twice the strictly necessary relation collection effort was spent to make the linear algebra more manageable, an extra effort that was commented on as being "well spent" (cf. [21, Introduction]). These "negative returns" are further illustrated in Sect. 4.

Based on consistent past behavior of the preprocessing and given specific smoothness parameters, it can be roughly estimated how many relations have to be collected for a targeted matrix dimension and weight. Given the uncertainty alluded to above, this estimate can only be a guess, though it is a mildly educated one. With the known behavior of the software, an overall effort estimate assuming those specific smoothness parameters can be derived. Repeating this for different smoothness parameters, the "best" – despite a lack of clear optimization criteria – overall effort then follows.

Parameter Selection. Our starting point was that on current equipment the linear algebra effort for the 768-bit modulus factored in [21] would be about 75 core years. Given the similarity of the algorithms and sizes, and using the same smoothness parameters as in [21], the overall effort to solve our discrete logarithm problem can be estimated as $1500 + 767 \cdot 75 = 59025$ core years; due to the small entries of the matrix the linear algebra effort only depends linearly on the size of the group order. The fifty thousand core years estimate mentioned in the introduction then follows from the expected favorable comparison of polynomials found using the method from [17] compared to the method used in [21]; we refer to the papers involved for an explanation of this effect.

All that is clear at this point is that attempts to lower this estimate must focus on lowering the linear algebra effort; thus the smoothness parameters must be reduced, but by how much and what the overall effect is going to be is unclear. Because the block Wiedemann effort is roughly proportional to the product of the matrix dimension and weight, reducing the matrix dimension by a factor of c while keeping the same average row-weight, cuts the linear algebra effort by a factor of c^2. Thus, given any targeted linear algebra effort a reduction factor c for the dimension follows. Assuming that, for our problem, a thousand core years would be acceptable for the linear algebra effort, a dimension reduction factor of about 7.6 follows, because $\frac{767 \cdot 75}{7.6^2} \approx 1000$. Compared to the parameters used in [21], such a drastic reduction requires severely cutting the smoothness parameters and the number of special q primes one may use. This in turn entails a more than proportional increase in the search space and thus a substantial increase in the overall relation collection effort compared to the 1500 core years spent in [21]. A priori, however, and as argued above, the effect of any of the changes that one may wish to consider cannot be accurately predicted.

While conducting experiments with a 640-bit example to better understand the increase in the relation collection effort depending on various possible combinations of smaller smoothness parameters and larger search spaces, we observed a mildly beneficial side-effect which – once observed – is obvious, but which was unanticipated: in the notation of Sect. 2, if B_f decreases, the number y_f of b_f-smooth norms $\mathcal{N}_f(a, b)$ becomes smaller too, at a given point to an extent that it becomes more efficient to replace sieving for the second search stage (as used in [21]) by factorization trees. For the 640-bit example the effect was still small, i.e., y_f was still relatively large. But for our 768-bit prime the impact soon turned out to be considerable, almost halving the (inflated, compared to [21]) relation collection effort.

The resulting "best" parameters that we settled for are listed in Table 1 (though for some special q primes larger factor bases were used), along with the parameters used in [21] for the 768-bit number field sieve factorization. The clear difference is that the choices for 768-bit factoring were optimized for speed during relation collection (collecting relations until the after-preprocessing matrix dimension and weight were found to be acceptable), whereas our choices try to squeeze as many relations as possible out of every special q prime under a relatively restrictive smoothness regime. Compared to [21], $\#B_f$ is reduced by a factor of a bit more than two, the number of special q primes is reduced by a factor of more than twenty, the number of large primes per relation is cut from $4+3$ to $2+2$ with a large prime bound that is reduced by a factor of 2^4 from 2^{40} to 2^{36}, while $\#B_g$ remains unchanged and the search space is on average (forced to be) more than 2^8 times larger. As a result the number of relations per core unit of time drops by a factor of about sixteen compared to [21].

The first preprocessing attempt that resulted in the hoped-for matrix dimension and weight occurred when 1.09e10 relations had been collected, i.e., about six times fewer relations than in [21]. The resulting overall relation collection effort thus became $\frac{16}{6} \cdot 1500 = 4000$ core years. With 920 core years the linear algebra effort was close to but less than the thousand core years that we had hoped to achieve. A much smaller set of relations would in principle have sufficed too, but it would have resulted in a larger linear algebra effort; Sect. 4 below describes the trade-off in more detail.

Some of the details in Table 1 are listed for completeness; for explanations we refer to [21]. Like most of our computational number theory colleagues, we missed the fact (which apparently had not escaped numerical analysts) that the evaluation stage of the block Wiedemann method can be sped up considerably using Horner's rule [19]; it would have reduced our overall effort to approximately 5000 core years.

Table 1. Comparison of 768-bit factoring and computing 768-bit prime field discrete logarithms.

	768-bit factorization from [21]	768-bit discrete logarithm
polynomial selection	2005 and 2007	2015:02:14 − 2015:05:04
	more than 2e18 pairs (f, g) were considered, spending 40 core years (20 in 2005 and 20 in 2007)	$\|g_{max}\|\leq165$ { all 5.9e11 g-candidates / 110 core years ; $\|g_{max}\|\leq333$ { best 2e13 g-candidates / 90 core years
d_f, d_g	6, 1	3, 4
relation collection	2007:08 − 2009:04	2015:05:04 − 2015:12:13
method	lattice sieving for both f and g	lattice sieving for f (>98% of effort) factorization tree for g (<2% of effort)
smoothness bounds	≥2GB RAM { b_f=min(q,1.1e9) b_g=2e8 ; #B_f=5.6e7 #B_g=1.1e7 } <2GB RAM { b_f=4.5e8 b_g=1e8 ; #B_f=2.4e7 #B_g=5.8e6 }	{ b_f=min(q,4.4e8) b_g=2e8 ; #B_f=2.3e7 #B_g=1.1e7 }
large primes parameters	$b_\ell=2^{40}$ (#B_ℓ=4.1e10), n_f=4, n_g=3	$b_\ell=2^{36}$ (#B_ℓ=2.9e9), n_f=2, n_g=2
#S_q { bounds on q ; #q ; time per q ; (y_f, y_g) per q	2^{31} { 4.5e8 < q < 1.1e10 ; 4.8e8 ; < 100 seconds ; ((y_f large and irrelevant),134) }	2^{40} { 1.9e8 < q < 3e8 ; 5.7e6 ; ≈ 8750 seconds ; (7.0e5, 590) } 2^{39} { 3e8 < q < 6e8 ; 1.5e7 ; ≈ 4950 seconds ; (7.3e5, 480) } 2^{38} { 6e8 < q < 6.3e8 ; 1.5e6 ; ≈ 2550 seconds ; (4.6e5, 300) }
totals: number of special q primes	4.8e8	2.2e7
yield { with duplicates & unfactored	64 334 489 730	10 802 334 123[†]
unique & factored	47 705 019 942	9 060 739 382
free relations	57 223 462	19 967 617
effort	≈ 1500 core years	≈ 4000 core years
linear algebra preprocessing		
duplicate & singleton removal	2009:05	2015:08:11 − 2015:12:20
filtering	2009:06	2015:12:21 − 2015:12:25
result { dimensions	192 796 550 × 192 795 550	23 504 483 × 23 504 413
weight	27 797 115 920 bits	3 140 911 353 mostly very small entries
average non-zeros per row	144	134
effort	a few core years	a few core years
linear algebra		
block Wiedemann parameters	$m = 16 \times 64, n = 8 \times 64$	$m = 32, n = 16$
scalar products	2009:08:10 − 2009:11:03 { 8 independent sequences ; 43 core years (current cluster)	2015:12:28 − 2016:03:23 { 16 independent sequences ; 560 core years
Berlekamp–Massey	2009:11:03 { 17.3 hours on 224 of 672 cores ; 896GB RAM ; 0.5(+1 idle) core years (old cluster)	2016:04:03 − 2016:04:04 { 33.85 hours on 256 of 4096 cores ; 8TB RAM ; 1(+15 idle) core years
evaluation	2009:11:05 − 2009:12:09 { many independent jobs ; 30 core years (current cluster)	2016:04:05 − 2016:05:18 { 480 independent jobs ; about 355 core years
effort	75 core years (current cluster)	920 core years
final calculation	2009:12:07 − 2009:12:12	2016:05:18 − 2016:06:16
	20 core hours per square root attempt	200 core years to build database; 43 core hours on average (was 100) per individual logarithm, varying between 3 and 220 core hours

†: this excludes about 1e8 forgotten relations

Database. The linear algebra step resulted in the (virtual) logarithms of 24 million prime ideals. Spending less than 200 core years for additional sieving and further processing, a database was built containing the about 3e9 logarithms of all prime ideals of norms up to 2^{35}.

Individual Logarithms. Using the database and q-descent, the logarithm of the target t from the introduction was computed in approximately 115 core hours. Similar computations for $t + 1, t + 2, \ldots, t + 10$ took on average about 96 core hours per logarithm.

With improved software any individual logarithm can now be computed at an average effort of 43 core hours (and a rather large variation, cf. Table 1). Further software enhancements are easily conceivable, but this already insignificant effort underscores the point made in the introduction that once a single number field sieve based attack has been carried out successfully, other attacks against the same prime field are straightforward.

4   Trade-Off

During relation collection occasional preprocessing attempts were made, until the resulting matrix was found to be acceptable. Data about the non-final attempts were not kept, so for the purpose of the present paper part of this work was redone to be able to give an impression how the linear algebra effort decreases as more relations become available.

Table 2 summarizes the results of these "after the fact" preprocessing attempts of the sets of relations found for special q primes up to increasing bounds b_q, along with the resulting extrapolations of the block Wiedemann efforts (*not* using Horner's rule for the evaluation stage). Estimates are also listed for the effort required for a discrete logarithm problem in a 160-bit subgroup of the multiplicative group of a 768-bit prime field. For each set of relations up to five preprocessing attempts were made, and the best was selected depending on the subgroup size; this explains why for five of the 160-bit subgroup entries the matrix dimension and weight are different from those in the 767-bit subgroup entry. The last two rows show the effect of the 1e8 forgotten relations (cf. Table 1): including those and spending more time on constructing the matrix could have reduced the matrix effort by 20 (or 5) core years.

The difference between the linear algebra estimate in the last row for the matrix as actually used and the effort as reported in Table 1 is due to a lower number of nodes on which the experiment was run: for a full execution it would lower the linear algebra effort, but increase the calendar time. The effort required for polynomial selection and individual logarithms is independent of the b_q-value, and is not included in the "combined effort". The database building effort may be up to three times larger for the smallest feasible b_q-value, but is not included either.

The numbers in the "combined effort" columns of Table 2 illustrate the negative returns mentioned in Sect. 3: with more patience to deal with a larger linear algebra problem (that would have required disproportionally more calendar time), our overall effort could have been reduced from 5300 to less than 4000 core years. As in [21], the additional relation collection effort was well spent, because a large block Wiedemann job requires constant attention and any way to reduce the calendar time is welcome.

Note that for the smaller subgroup problem the overall least effort is reduced by a factor smaller than two.

Table 2. Relation collection effort, matrix dimension and weight as a result of preprocessing, estimated linear algebra effort, and the combined effort (all efforts are in core years), when using special q primes up to b_q and both for 767-bit and 160-bit subgroup orders. The overshoot factor is the ratio of the number of relations and the number of relations for the least b_q (2.8e8) for which enough relations had been found. Relations are unique and factored and include the free relations.

b_q	Relation collection			767-bit subgroup order (our problem)				160-bit subgroup order			
	Relation count	Effort	Overshoot factor	Dimension and weight	Nodes	Matrix effort	Combined effort	Dimension and weight	Nodes	Matrix effort	Combined effort
2.70e8	2.33e9	Insufficient									
2.80e8	2.58e9	1300	1.000	5.62e7 9.5e9	25	6575	7875	5.62e7 9.5e9	9	1780	3080
3.06e8	3.24e9	1625	1.255	3.27e7 6.2e9	12	2095	3720	4.00e7 4.7e9	4	500	2125
3.35e8	3.90e9	1850	1.508	2.96e7 4.5e9	9	1420	3270	2.96e7 4.5e9	4	325	2175
3.67e8	4.52e9	2100	1.751	2.62e7 4.4e9	9	1120	3220	2.76e7 3.9e9	4	270	2370
4.03e8	5.15e9	2400	1.995	2.47e7 4.2e9	9	1000	3400	2.57e7 3.8e9	4	240	2640
4.75e8	6.50e9	2975	2.516	2.36e7 3.7e9	9	870	3845	2.48e7 3.3e9	4	210	3185
5.37e8	7.74e9	3475	2.997	2.41e7 3.1e9	6	790	4265	2.41e7 3.1e9	4	190	3665
6.30e8	9.15e9	4000	3.542	2.17e7 3.6e9	6	740	4740	2.08e7 4.0e9	4	180	4180
(used)	9.08e9	4000	3.515	2.35e7 3.1e9	6	760	4760	2.35e7 3.1e9	4	185	4185

5 Other Prime Fields

To convince ourselves that our results were not due to unexpected, lucky properties of our choice of prime field, we tested ten other similarly chosen 768-bit primes and roughly compared them to our p with respect to their sieving yield. Define the following eleven transcendental or supposed-to-be transcendental numbers:

$\rho_0 = \pi$;

$\rho_1 = e$, Euler's number;

$\rho_2 = \gamma$, the Euler-Mascheroni constant;

$\rho_3 = \sqrt{2}^{\sqrt{2}}$;

$\rho_4 = \zeta(3)$, where ζ is the Riemann zeta function;

$\rho_5 = \log(\frac{1+\sqrt{5}}{2})$, the regulator of the "smallest" real quadratic number field;

$\rho_6 = \Omega_{X_0(11)}$, the real period of the "smallest" elliptic curve, namely $X_0(11)$ given by $y^2 + y = x^3 - x^2 - 10x - 20$;

$\rho_7 = \hat{h}_{X_0(37)}(P_{37})$, the canonical height of a generator $P_{37} = (0,0)$ of the "smallest" rank 1 elliptic curve, namely $X_0(37)$ given by $y^2 + y = x^3 - x$;

$\rho_8 = t_0$, the imaginary part of the first zero $\frac{1}{2} + t_0 i$ on the critical strip of ζ;

$\rho_9 = \pi^e$;

$\rho_{10} = \sum_{i=1}^{\infty} 10^{-i!}$, Liouville's constant.

For $0 \leq i \leq 10$ let $\epsilon_i = 767 - \lceil \frac{\log \rho_i}{\log 2} \rceil$ and let p_i be the least prime larger than $2^{\epsilon_i} \rho_i$ for which $\frac{p_i - 1}{2}$ is prime as well. Then $p_0 = p$. Let π_j be the number of primes in $[j \cdot 1e7, (j+1) \cdot 1e7]$; for $19 \leq j \leq 62$ these intervals cover our range of special q primes (cf. Table 1).

For each of the eleven primes p_i with $0 \leq i \leq 10$ the following calculation was carried out:

Polynomial Selection. Find the best pair (f_i, g_i) for p_i among the first 5e9 candidate polynomials for g_i. (This requires about one core year.)

Sieving Experiments. For $19 \leq j \leq 62$ find the number r_j of relations when sieving with the parameters as in Table 1 but with the polynomials f_i and g_i and the prime p_i and for the least special q prime larger than $j \cdot 1e7 + 5e6$. (This requires less than four core days, cf. Table 1.)

Overall Yield Estimate. Let $R_i = \sum_{j=19}^{62} \pi_j r_j$.

Table 3. Relative performance of $p = p_0$ compared to ten other choices.

	p_0	p_1	p_2	p_3	p_4	p_5	p_6	p_7	p_8	p_9	p_{10}
p_i/p_0	1.000	0.865	0.735	1.039	0.765	1.225	0.808	1.041	1.125	0.894	1.120
R_i/R	0.844	0.821	0.847	0.820	0.864	0.800	0.795	0.884	0.848	0.798	0.823
R_i/R_0	1.000	0.973	1.004	0.972	1.024	0.948	0.942	1.048	1.005	0.946	0.975

We also carried out the same sieving experiments for the polynomial pair (f, g) from Sect. 3 and $p = p_0$, finding an overall yield estimate $R = 1.02\mathrm{e}10$. This is less than the 1.09e10 relations reported in Table 1, because there some of the sieving jobs used a larger factor base bound than reported in Table 1, thus producing more duplicates. But it is more than R_0 (which was found to be 8.6e9), matching the expectation that (f, g) is considerably better than (f_0, g_0). Table 3 lists the relative performance of our p compared to the ten new choices: as can be seen in the final row, four of the ten perform better and six are worse, but they are all within a 6% margin from p. It also follows that the core years spent on polynomial selection for our p were well spent.

Although our tests counter suspicions about p being special, it may be argued that in practice primes used in cryptography would be chosen with high entropy [5]. Testing a few "random" primes as well might strengthen our argument. It is unclear to us, however, how such primes may be obtained in a manner that is sufficiently convincing to any suspicious reader, without input from that reader [32].

6   Conclusion

We presented the computation of a discrete logarithm in the multiplicative group of a 768-bit prime field. This is a new record in its category, beating the previous 596-bit record. We showed the beneficial effect of judicious choice of parameters and algorithms, and highlighted the differences with integer factorization. Based on our findings we may conclude that for sizes that are currently within reach of an academic effort, the hardness of factoring and computing prime field discrete logarithms is comparable, though discrete logarithms are harder. Although this was always suspected to be the case, the gap between the two problems is quite a bit smaller than we expected. Compared to the 768-bit factoring record (which required 1700 core years as opposed to our 5300 core years) we used less calendar time and a smaller collaborative and less heterogeneous effort [23]. We also conclude that the explicit 1024-bit estimates from [2, Sect. 4.1] should be redone, as they require not entirely straightforward re-optimization efforts.

Unless algorithmic improvements are proposed or new insights may be expected, pushing for actual new factoring or prime field discrete logarithm records – as opposed to studies that result in reliable estimates – is mostly a waste of energy. We are not aware of any developments based on which we could realistically expect publication of a 1024-bit record within the next, say, five years. As usual, this may change at any moment, but so far the predictions made back in 2009 (cf. [6]) have already turned out to be accurate, or remain valid. In this context it is relevant to note that the project embarked on in [3] is still ongoing.

Acknowledgements. We thank Rob Granger and the anonymous Eurocrypt 2017 reviewers for their useful comments. Part of the computation was carried out on equipment sponsored by the Swiss National Science Foundation under grant number 206021-144981.

References

1. Adleman, L.: A subexponential algorithm for the discrete logarithm problem with applications to cryptography. In: FOCS, pp. 55–60 (1979)
2. Adrian, D., Bhargavan, K., Durumeric, Z., Gaudry, P., Green, M., Halderman, J.A., Heninger, N., Springall, D., Thomé, E., Valenta, L., VanderSloot, B., Wustrow, E., Zanella-Béguelin, S., Zimmermann, P.: Imperfect forward secrecy: how Diffie-Hellman fails in practice. In: 22nd ACM Conference on Computer and Communications Security, October 2015 (2015)
3. Bailey, D.V., Baldwin, B., Batina, L., Bernstein, D.J., Birkner, P., Bos, J.W., van Damme, G., de Meulenaer, G., Fan, J., Güneysu, T., Gurkaynak, F., Kleinjung, T., Lange, T., Mentens, N., Paar, C., Regazzoni, F., Schwabe, P., Uhsadel, L.: The certicom challenges ECC2-X. Special-Purpose Hardware for Attacking Cryptographic Systems - SHARCS 2009 (2009). http://www.hyperelliptic.org/tanja/SHARCS/record2.pdf
4. Bernstein, D.J.: How to find small factors of integers, june 2002. http://cr.yp.to/papers.html
5. Bernstein, D.J., Chou, T., Chuengsatiansup, C., Hülsing, A., Lange, T., Niederhagen, R., van Vredendaal, C.: How to manipulate curve standards: a white paper for the black hat. Cryptology ePrint Archive, Report 2014/571 (2014). http://eprint.iacr.org/2014/571
6. Bos, J.W., Kaihara, M.E., Kleinjung, T., Lenstra, A.K., Montgomery, P.L.: On the security of 1024-bit RSA and 160-bit elliptic curve cryptography. Cryptology ePrint Archive, Report 2009/389 (2009). http://eprint.iacr.org/

7. Bouvier, C., Gaudry, P., Imbert, L., Hamza, J., Thomé, E.: Discrete logarithms in GF(p) - 180 digits. NMBRTHRY list, 11/6/2014
8. Buhler, J.P., Lenstra Jr., H.W., Pomerance, C.: Factoring integers with the number field sieve. In: Lenstra, A.K., Lenstra, H.W. (eds.) The development of the number field sieve. LNM, vol. 1554, pp. 50–94. Springer, Heidelberg (1993). doi:10.1007/BFb0091539
9. Coppersmith, D.: Modifications to the number field sieve. J. Cryptol. **6**(3), 169–180 (1993)
10. Coppersmith, D.: Solving homogeneous linear equations over GF(2) via block Wiedemann algorithm. Math. Comput. **62**(205), 333–350 (1994)
11. ElGamal, T.: A public key cryptosystem and a signature scheme based on discrete logarithms. In: Blakley, G.R., Chaum, D. (eds.) CRYPTO 1984. LNCS, vol. 196, pp. 10–18. Springer, Heidelberg (1985). doi:10.1007/3-540-39568-7_2
12. Franke, J., Kleinjung, T., Morain, F., Wirth, T.: Proving the primality of very large numbers with fastECPP. In: Buell, D.A. (ed.) Algorithmic Number Theory - ANTS-VI. Lecture Notes in Computer Science, vol. 3076, pp. 194–207. Springer, Heidelberg (2004)
13. Gordon, D.M.: Discrete logarithms in GF(p) using the number field sieve. SIAM J. Discret. Math. **6**, 124–138 (1993)
14. Granger, R., Kleinjung, T., Zumbrägel, J.: Discrete Logarithms in $GF(2^{9234})$. NMBRTHRY list, 31/1/2014
15. IETF. RFC 2409, November 1998. https://tools.ietf.org/html/rfc2409
16. IETF. RFC 4306, December 2005. https://tools.ietf.org/html/rfc4306
17. Joux, A., Lercier, R.: Improvements to the general number field sieve for discrete logarithms in prime fields. A comparison with the Gaussian integer method. Math. Comput. **72**(242), 953–967 (2003). (electronic)
18. Joux, A., Pierrot, C.: Improving the polynomial time precomputation of frobenius representation discrete logarithm algorithms. In: Sarkar, P., Iwata, T. (eds.) ASIACRYPT 2014. LNCS, vol. 8873, pp. 378–397. Springer, Heidelberg (2014). doi:10.1007/978-3-662-45611-8_20
19. Kaltofen, E.: Analysis of Coppersmith's block Wiedemann algorithm for the parallel solution of sparse linear systems. Math. Comput. **64**, 777–806 (1995)
20. Kleinjung, T.: Discrete logarithms in GF(2^{1279}). NMBRTHRY list, 17/10/2014
21. Kleinjung, T., Aoki, K., Franke, J., Lenstra, A.K., Thomé, E., Bos, J.W., Gaudry, P., Kruppa, A., Montgomery, P.L., Osvik, D.A., te Riele, H., Timofeev, A., Zimmermann, P.: Factorization of a 768-bit RSA modulus. In: Rabin, T. (ed.) CRYPTO 2010. LNCS, vol. 6223, pp. 333–350. Springer, Heidelberg (2010). doi:10.1007/978-3-642-14623-7_18
22. Kleinjung, T., Bos, J.W., Lenstra, A.K.: Mersenne factorization factory. In: Sarkar, P., Iwata, T. (eds.) ASIACRYPT 2014. LNCS, vol. 8873, pp. 358–377. Springer, Heidelberg (2014). doi:10.1007/978-3-662-45611-8_19
23. Kleinjung, T., Bos, J.W., Lenstra, A.K., Osvik, D.A., Aoki, K., Contini, S., Franke, J., Thomé, E., Jermini, P., Thiémard, M., Leyland, P., Montgomery, P.L., Timofeev, A., Stockinger, H.: A heterogeneous computing environment to solve the 768-bit RSA challenge. Cluster Comput. **15**, 53–68 (2012)
24. Kraitchik, M.: Théorie des nombres, Tome I. Gauthiers-Villars, Paris (1922)
25. Kraitchik, M.: Recherches sur le théorie des nombres, Tome I. Gauthiers-Villars, Paris (1924)
26. Lenstra, A.K.: Unbelievable security: matching AES security using public key systems. In: Boyd, C. (ed.) ASIACRYPT 2001. LNCS, vol. 2248, pp. 67–86. Springer, Heidelberg (2001). doi:10.1007/3-540-45682-1_5

27. Lenstra, A.K., Hughes, J.P., Augier, M., Bos, J.W., Kleinjung, T., Wachter, C.: Ron was wrong, Whit is right. Cryptology ePrint Archive, Report 2012/064 (2012). http://eprint.iacr.org/2012/064
28. Lenstra, A.K., Lenstra Jr., H.W. (eds.): The development of the number field sieve. LNM, vol. 1554. Springer, Heidelberg (1993)
29. Lenstra, A.K., Lenstra Jr., H.W., Manasse, M.S., Pollard, J.M.: The factorization of the ninth Fermat number. Math. Comput. **61**(203), 319–349 (1993)
30. Lenstra, A.K., Manasse, M.S.: Factoring by electronic mail. In: Quisquater, J.-J., Vandewalle, J. (eds.) EUROCRYPT 1989. LNCS, vol. 434, pp. 355–371. Springer, Heidelberg (1990). doi:10.1007/3-540-46885-4_35
31. Lenstra, A.K., Verheul, E.R.: Selecting cryptographic key sizes. J. Cryptol. **14**(4), 255–293 (2001)
32. Lenstra, A.K., Wesolowski, B.: A random zoo: sloth, unicorn, and trx. Cryptology ePrint Archive, Report 2015/366, 2015. http://eprint.iacr.org/2015/366, to appear in the International Journal of Applied Cryptology as Trustworthy public randomness with sloth, unicorn, and trx
33. Pomerance, C.: A tale of two sieves. Not. AMS **43**(12), 1473–1485 (1996)
34. Rivest, R.L., Shamir, A., Adleman, L.M.: A method for obtaining digital signature and public-key cryptosystems. Commun. Assoc. Comput. Mach. **21**(2), 120–126 (1978)
35. Schirokauer, O.: Virtual logarithms. J. Algorithm. **57**(2), 140–147 (2005)
36. U.S. Department of Commerce/National Institute of Standards and Technology. Digital Signature Standard (DSS). FIPS-186-4 (2013). http://nvlpubs.nist.gov/nistpubs/FIPS/NIST.FIPS.186-4.pdf
37. Wiedemann, D.: Solving sparse linear equations over finite fields. IEEE Trans. Inf. Theory **32**, 54–62 (1986)

A Kilobit Hidden SNFS Discrete Logarithm Computation

Joshua Fried[1], Pierrick Gaudry[2]([⊠]), Nadia Heninger[1], and Emmanuel Thomé[2]

[1] University of Pennsylvania, Philadelphia, USA
{josh.fried,nadiah}@cis.upenn.edu
[2] Inria, CNRS, Université de Lorraine, Nancy, France
pierrick.gaudry@loria.fr, emmanuel.thome@inria.fr

Abstract. We perform a special number field sieve discrete logarithm computation in a 1024-bit prime field. To our knowledge, this is the first kilobit-sized discrete logarithm computation ever reported for prime fields. This computation took a little over two months of calendar time on an academic cluster using the open-source CADO-NFS software.

Our chosen prime p looks random, and $p - 1$ has a 160-bit prime factor, in line with recommended parameters for the Digital Signature Algorithm. However, our p has been trapdoored in such a way that the special number field sieve can be used to compute discrete logarithms in $\mathbb{F}_p^*$, yet detecting that p has this trapdoor seems out of reach. Twenty-five years ago, there was considerable controversy around the possibility of backdoored parameters for DSA. Our computations show that trapdoored primes are entirely feasible with current computing technology. We also describe special number field sieve discrete log computations carried out for multiple conspicuously weak primes found in use in the wild.

As can be expected from a trapdoor mechanism which we say is hard to detect, our research did not reveal any trapdoored prime in wide use. The only way for a user to defend against a hypothetical trapdoor of this kind is to require verifiably random primes.

1 Introduction

In the early 1990's, NIST published draft standards for what later became the Digital Signature Algorithm (DSA) [40]. DSA is now widely used. At the time, many members of the cryptographic community voiced concerns about the proposal. Among these concerns were that the standard encouraged the use of a global common prime modulus p [45], and that a malicious party could specially craft a trapdoored prime so that signatures would be easier to forge for the trapdoor owner [31]. This latter charge was the subject of a remarkable panel at Eurocrypt 1992 [1,15]. Most of the panelists agreed that it appeared to be difficult to construct an undetectably trapdoored modulus, and that such trapdoors appeared unlikely. To protect against possible trapdoored primes, the Digital Signature Standard suggests that primes for DSA be chosen in a "verifiably random" way, with a published seed value [49]. Yet DSA primes used in the wild today are seldom published with the seed.

© International Association for Cryptologic Research 2017
J.-S. Coron and J.B. Nielsen (Eds.): EUROCRYPT 2017, Part I, LNCS 10210, pp. 202–231, 2017.
DOI: 10.1007/978-3-319-56620-7_8

Concerns about cryptographic backdoors have a long history (for instance, it has been formalized as "kleptography" in the 90's [54]) and regained prominence in recent years since the disclosure of NSA documents leaked by Edward Snowden. A set of leaked documents published in September 2013 by the NY Times [43], The Guardian [4], and ProPublica [30] describe an NSA "SIGINT Enabling Project" that included among its goals to "Influence policies, standards, and specification for commercial public key technologies". The newspaper articles describe documents making specific reference to a backdoor in the Dual EC random number generator, which had been standardized by both NIST and ANSI. NIST responded by withdrawing its recommendation for the Dual EC DRBG, writing "This algorithm includes default elliptic curve points for three elliptic curves, the provenance of which were not described. Security researchers have highlighted the importance of generating these elliptic curve points in a trustworthy way. This issue was identified during the development process, and the concern was initially addressed by including specifications for generating different points than the default values that were provided. However, recent community commentary has called into question the trustworthiness of these default elliptic curve points" [38]. There is evidence that the ability to backdoor the Dual EC algorithm has been exploited in the wild: Juniper Networks had implemented Dual EC in NetScreen VPN routers, but had used it with custom-generated parameters. In December 2015 Juniper published a security advisory [25] announcing that an attacker had made unauthorized modifications to the source code for these products to substitute a different curve point in the Dual EC implementation [10].

In this paper, we demonstrate that constructing and exploiting trapdoored primes for Diffie-Hellman and DSA is feasible for 1024-bit keys with modern academic computing resources. Current estimates for 1024-bit discrete log in general suggest that such computations are likely within range for an adversary who can afford hundreds of millions of dollars of special-purpose hardware [2]. In contrast, we were able to perform a discrete log computation on a specially trapdoored prime in two months on an academic cluster. While the Dual EC algorithm appears to have only rarely been used in practice [9], finite-field Diffie-Hellman and DSA are cornerstones of public-key cryptography. We neither show nor claim that trapdoored primes are currently in use. However, the near-universal failure of implementers to use verifiable prime generation practices means that use of weak primes would be undetectable in practice and unlikely to raise eyebrows.

The Special Number Field Sieve Trapdoor

The Number Field Sieve (NFS), still very much in its infancy at the beginning of the 1990's, was originally proposed as an integer factoring algorithm [32]. Gordon adapted the algorithm to compute discrete logarithms in prime fields [22]. Both for the integer factoring and the discrete logarithm variants, several theoretical and computational obstacles had to be overcome before the NFS was practical to use for large scale computations. For the past twenty years, the NFS has been routinely used in record computations, and the underlying algorithms have been thoroughly improved. The NFS is now a versatile algorithm which can handle an arbitrary prime p, and compute discrete logarithms in $\mathbb{F}_p^*$ in asymptotic time $L_p(1/3, (64/9)^{1/3})^{1+o(1)}$, using the usual L-notation (see Sect. 3.2).

Current computational records for the number field sieve include a 768-bit factorization of an RSA modulus, completed in December 2009 by Kleinjung et al. [27] and a 768-bit discrete log for a safe prime, completed in June 2016 by Kleinjung et al. [28].

Very early on in the development of NFS, it was observed that the algorithm was particularly efficient for inputs of a special form. Some composite integers are particularly amenable to being factored by NFS, and primes of a special form allow easier computation of discrete logarithms. This relatively rare set of inputs defines the Special Number Field Sieve (SNFS). It is straightforward to start with parameters that give a good running time for the NFS—more precisely, a pair of irreducible integer polynomials meeting certain degree and size constraints—and derive an integer to be factored, or a prime modulus for a discrete logarithm. In general, moving in the other direction, from a computational target to SNFS parameters, is known to be possible only in rare cases (e.g. the Cunningham project). The complexity of SNFS is $L_p(1/3, (32/9)^{1/3})^{1+o(1)}$, much less than its general counterpart. A 1039-bit SNFS factorization was completed in 2007 by Aoki et al. [3].

In 1992, Gordon [21] suggested several methods that are still the best known for trapdooring primes to give the best running time for the SNFS, without the trapdoored SNFS property being conspicuous[1]. Most of his analysis remains valid, but there has been significant improvement in the NFS algorithm in the past 25 years. In the early days, an NFS computation had to face issues of dealing with class groups and explicit units. That meant much less flexibility in creating the trapdoor, to the point that it was indeed difficult to conceal it. It is now well understood that these concerns were artificial and can be worked around [46], much to the benefit of the trapdoor designer. Gordon's analysis and much of the discussion of trapdoored DSA primes in 1992 focused on 512-bit primes, the suggested parameter sizes for NIST's DSS draft at the time. However, 25 years later, 1024-bit targets are of greater cryptanalytic interest.

We update the state of the art in crafting trapdoored 1024-bit primes for which practical computation of discrete logarithms is possible, and demonstrate that exploitation is practical by performing a 1024-bit SNFS discrete log computation.

We begin by reviewing in Sect. 2 the origin of primes found in multiple practical cryptographic contexts. Section 3 recalls a brief background on the Number Field Sieve. In Sect. 4, we reevaluate Gordon's work on trapdooring primes for SNFS given the modern understanding of the algorithm, and explain for a given target size which polynomial pair yields the fastest running time. This answers a practical question for the problem at hand—how to optimally select trapdoor

---

[1] In 1991, another method was suggested by Lenstra in [31], and played a role in triggering the Eurocrypt panel [15]. Gordon's trap design is more general.

parameters to simplify computations with the prime—and is also of wider interest for NFS-related computations.

We then run a full 1024-bit experiment to show that trapdoored primes are indeed a practical threat, and perform a full 1024-bit SNFS discrete log computation for our trapdoored prime. We describe our computation in Sect. 5. We show how various adaptations to the block Wiedemann algorithm are essential to minimizing its running time, compared to previous computations of the same kind. We detail the descent procedure, and the various challenges which must be overcome so as to complete individual logs in a short time. We also provide an extensive appendix giving details on the analysis of individual logarithm computation in Appendix A, as this portion of the computation is not well detailed in the literature.

Finally, we evaluate the impact of our results in Sect. 6. Our computation required roughly a factor of 10 less resources than the recent 768-bit GNFS discrete log announced by Kleinjung et al. However, we have found a number of primes amenable to non-hidden SNFS DLP computations in use in the wild. We describe additional SNFS computations we performed on these primes in Sect. 6.2.

2   Modern Security Practices for Discrete Log Cryptosystems

Verifiable Prime Generation. It is legitimate to wonder whether one should worry about trapdoored primes at all. Good cryptographic practice recommends that publicly agreed parameters must be "verifiably random". For example, Appendix A.1.1.2 of the FIPS 186 standard [40] proposes a method to generate DSA primes p and q from a random seed and a hash function, and suggests that one should publish that seed alongside with p and q. The publication of this seed is marked *optional*. Primes of this type are widely used for a variety of cryptographic primitives; for example NIST SP 800-56A specifies that finite-field parameters for key exchange should be generated using FIPS 186 [41, Sect. 5.5.1.1].

While it is true that some standardized cryptographic data includes "verifiable randomness"[2] or rigidity derived from "nothing up my sleeve" numbers, it is noteworthy that this is not always the case. For example, both France and China standardized elliptic curves for public use without providing any sort of justification for the chosen parameters [8, Sect. 3.1]. RFC 5114 [33] specifies a number of groups for use with Diffie-Hellman, and states that the parameters were drawn from NIST test data, but neither the NIST test data [39] nor RFC 5114 itself contain the seeds used to generate the finite field parameters. In a similar vein, the origin of the default 2048-bit prime in the Helios voting system used in the most recent IACR Board of Directors Election in 2015 is undocumented. Most

---

[2] This still leaves the question of whether the seed is honest, see e.g. [8,47]. We do not address this concern here.

users would have to go out of their way to generate verifiable primes: the default behavior of OpenSSL does not print out seeds when generating Diffie-Hellman or DSA parameter sets. However, some implementations do provide seeds. Java's `sun.security.provider` package specifies hard-coded 512-, 768-, and 1024-bit groups together with the FIPS 186 seeds used to generate them.

Standardized and Hard-Coded Primes. It is also legitimate to wonder whether one should be concerned about widespread reuse of primes. For modern computers, prime generation is much less computationally burdensome than in the 1990s, and any user worried about a backdoor could easily generate their own group parameters. However, even today, many applications use standardized or hard-coded primes for Diffie-Hellman and DSA. We illustrate this by several examples.

In the TLS protocol, the server specifies the group parameters that the client and server use for Diffie-Hellman key exchange. Adrian et al. [2] observed in 2015 that 37% of the Alexa Top 1 Million web sites supported a single 1024-bit group for Diffie-Hellman key exchange. The group parameters were hard-coded into Apache 2.2, without any specified seed for verification. They also observed that in May 2015, 56% of HTTPS hosts selected one of the 10 most common 1024-bit groups when negotiating ephemeral Diffie-Hellman key exchange. Among 13 million recorded TLS handshakes negotiating ephemeral Diffie-Hellman key exchange, only 68,000 distinct prime moduli were used. The TLS 1.3 draft restricts finite-field Diffie-Hellman to a set of five groups modulo safe primes ranging in size from 2048 to 8196 bits derived from the nothing-up-my-sleeve number e [19].

In the IKE protocol for IPsec, the initiator and responder negotiate a group for Diffie-Hellman key exchange from a set list of pre-defined groups; Adrian et al. observed that 66% of IKE responder hosts preferred the 1024-bit Oakley Group 2 over other choices. The Oakley groups specify a collection of primes derived from a "nothing-up-my-sleeve" number, the binary expansion of π, and have been built into standards, including IKE and SSH, for decades [42]. The additional finite-field Diffie-Hellman groups specified in RFC 5114 are widely used in practice: Internet-wide scans from September 2016 found that over 900,000 (2.25%) of TLS hosts on port 443 chose these groups [16]. Scans from February 2016 of IKE hosts on port 500 revealed that 340,000 (13%) supported the RFC 5114 finite-field Diffie-Hellman parameters [53].

RFC 4253 specifies two groups that must be supported for SSH Diffie-Hellman key exchange: Oakley Group 2 (which is referred to as SSH group 1) and Oakley Group 14, a 2048-bit prime. SSH group 1 key exchange was disabled by default in OpenSSH version 7.0, released in August 2015 [18]. Optionally, SSH clients and servers may negotiate a different group using the group exchange handshake. However, OpenSSH chooses the group negotiated during this exchange from a pre-generated list that is generally shipped with the software package. The `/etc/ssh/moduli` file on an Ubuntu 16.04 machine in our cluster contained 267 entries in size between 1535 and 8191 bits. The 40 to 50 primes at each size appear to have been generated by listing successive primes from a fixed starting point, and differ only in the least significant handful of bits. We examined data from a full IPv4 SSH scan performed in October 2015 [53] that offered Diffie-Hellman group exchange only, and found 11,658 primes in use from 10.9 million responses, many of which could be clustered into groups differing only in a handful of least significant bits.

The SSH protocol also allows servers to use long term DSA keys to authenticate themselves to clients. We conducted a scan of a random 1% portion of the IPv4 space for hosts running SSH servers on port 22 with DSA host keys in September 2016, and found that most hosts seemed to generate unique primes for their DSA public keys. The scan yielded 27,380 unique DSA host keys from 32,111 host servers, of which only 557 shared a prime with another key. DSA host key authentication was also disabled by default in OpenSSH 7.0 [18].

1024-Bit Primes in Modern Cryptographic Deployments. It is well understood that 1024-bit factorization and discrete log computations are within the range of government-level adversaries [2], but such computations are widely believed by practitioners to be *only* within the range of such adversaries, and thus that these key sizes are still safe for use in many cases. While NIST has recommended a minimum prime size of 2048 bits since 2010 [6], 1024-bit primes remain extremely common in practice. Some of this is due to implementation and compatibility issues. For example, versions of Java prior to Java 8, released in 2014, did not support Diffie-Hellman or DSA group sizes larger than 1024 bits. DNSSEC limits DSA keys to a maximum size of 1024-bit keys [29], and stated, in 2012, that with respect to RSA keys, "To date, despite huge efforts, no one has broken a regular 1024-bit key; ... it is estimated that most zones can safely use 1024-bit keys for at least the next ten years." SSL Labs SSL Pulse estimated in September 2016 that 22% of the 200,000 most popular HTTPS web sites performed a key exchange with 1024-bit strength [50].

3   The Number Field Sieve for Discrete Logarithms

3.1   The NFS Setting

We briefly recall the Number Field Sieve (NFS) algorithm for computing discrete logarithms in finite fields. This background is classical and can be found in a variety of references. NFS appeared first as an algorithm for factoring integers [32], and has been adapted to the computation of discrete logarithms over several works [22,23].

Let $\mathbb{F}_p$ be a prime field, let $\gamma \in \mathbb{F}_p^*$ be an element of prime order $q \mid p - 1$. We wish to solve discrete logarithms in $\langle \gamma \rangle$. The basic outline of the NFS-DL algorithm is as follows:

Polynomial selection. Select irreducible integer polynomials f and g, sharing a common root m modulo p. Both polynomials define number fields, which we denote by $\mathbb{Q}(\alpha)$ and $\mathbb{Q}(\beta)$.

Sieving. Find many pairs a, b such that the two integers (called *norms* – albeit improperly if f or g are not monic) $\mathrm{Res}(f(x), a - bx)$ and $\mathrm{Res}(g(x), a - bx)$ factor completely into primes below a chosen smoothness bound B.

Filtering. Form multiplicative combinations of the (a, b) pairs to reduce the number of prime ideals appearing in the corresponding ideal factorizations.

Compute maps. Compute q-adic characters (known as Schirokauer maps [46]). This yields a relation matrix which can be written as $(M\|S)$, with M the block with ideal valuations, and S the block with Schirokauer maps.

Linear algebra. Solve the linear system $(M\|S)x = 0$, which gives *virtual logarithms*.

Individual logarithm. Given a target value $z \in \langle\gamma\rangle$, derive its logarithm as a linear combination of a subset of the virtual logarithms.

3.2   Complexity Analysis

NFS complexity analysis involves the usual L-notation, defined as

$$L_p(e, c) = \exp(c(\log p)^e (\log\log p)^{1-e}). \tag{1}$$

This notation interpolates between polynomial ($e = 0$) and exponential ($e = 1$) complexities. It adapts well to working with smoothness probabilities. In this formula and elsewhere in the paper, we use the log notation for the natural logarithm. In the few places where we have formulae that involve bit sizes, we always use $\log_2$ for the logarithm in base 2.

The polynomials f and g have a crucial impact on the size of the integers $\mathrm{Res}(f(x), a - bx)$ and $\mathrm{Res}(g(x), a - bx)$, and therefore on the probability that these integers factor into primes below B (in other words, are B-*smooth*).

Table 1. Polynomial selection choices for NFS variants.

Variant	$\deg f$	$\|f\|$	$\deg g$	$\|g\|$	Complexity exponent
General NFS (base-m)	d	$p^{1/(d+1)}$	1	$p^{1/(d+1)}$	$(64/9)^{1/3}$
General NFS (Joux-Lercier)	$d' + 1$	$O(1)$	d'	$p^{1/(d'+1)}$	$(64/9)^{1/3}$
Special NFS (for example)	d	$O(1)$	1	$p^{1/(d+1)}$	$(32/9)^{1/3}$

The analysis of the NFS depends on the prime p. When no assumptions are made on p, we have the so-called *general number field sieve* (GNFS). It is possible to perform polynomial selection so that $(\deg f, \deg g)$ is $(1, d)$ (by choosing $m \approx N^{1/d}$ and writing p in "base-m", or similar), or $(d' + 1, d')$ (the Joux-Lercier method [23]). The degrees d and d' in each case are integer parameters, for which an educated guess is provided by their asymptotic optimum, namely $d = (3\log p/\log\log p)^{1/3}$, and $d' = d/2$. Both approaches lead to an overall complexity $L_p(1/3, (64/9)^{1/3})^{1+o(1)}$ for a discrete logarithm computation, as indicated in Table 1, where $\|f\|$ denotes the maximum absolute value of the coefficients of f.

In contrast, some prime numbers are such that *there exist* exceptionally small polynomial pairs (f, g) sharing a common root modulo p. This makes a considerable difference in the efficiency of the algorithm, to the point that the exponent in the complexity drops from $(64/9)^{1/3}$ to $(32/9)^{1/3}$—a difference which is also considerable in practice. In most previously considered cases, this special structure is clear from the number itself. For the SNFS factorization performed by Aoki et al. for the composite integer $2^{1039} - 1$, they chose $f(x) = 2x^6 - 1$ and $g(x) = x - 2^{173}$ [3].

4   Heidi Hides Her Polynomials

Early on in the development of NFS, Gordon [21] suggested that one could craft primes so that SNFS polynomials exist, but may not be apparent to the casual observer. Heidi, a mischievous designer for a crypto standard, would select a pair of SNFS polynomials to her liking *first*, and publish only their resultant p (if it is prime) *afterwards*. The hidden trapdoor then consists in the pair of polynomials which Heidi used to generate p, and that she can use to considerably ease the computation of discrete logarithms in $\mathbb{F}_p$.

Twenty-five years later, we reconsider the best-case scenario for Heidi: given a target size, what type of polynomial pair will give the fastest running time for a discrete logarithm computation? For the current state of the art in algorithmic development and computation power, is there a parameter setting for which the computations are simultaneously within reach, Heidi can efficiently generate a trapdoored prime, and defeat attempts at unveiling it?

4.1   Best Form for SNFS Polynomials

The Special Number Field Sieve has been mostly used in the context of integer factorization, in particular for numbers from the Cunningham Project. In that case the integers are given, and typically the only way to find SNFS polynomials for these numbers is to take one linear polynomial with large coefficients and a polynomial of larger degree, with tiny coefficients. In our situation, the construction can go the opposite way: we are free to choose first the form of the polynomials, hoping that their resultant will be a prime number. We thus have more freedom in the construction of our polynomial pair.

Let n be the number of bits of the SNFS prime p we will construct. We consider first the case where we have two polynomials f and g that are non-skewed, i.e. all the coefficients of each polynomial have roughly the same size. We denote d_f and d_g their respective degrees, and $\|f\|$ and $\|g\|$ the respective maximum absolute values of their coefficients. Since the resultant must be almost equal to p, we have

$$d_f \log_2 \|g\| + d_g \log_2 \|f\| \approx n. \tag{2}$$

Let A be a bound on the a and b integers we are going to consider during relation collection. Then the product of the norms that have to be tested for smoothness can be approximated by $\|f\| \|g\| A^{d_f + d_g}$. We will try to make its size as small as possible, so we want to minimize

$$\log_2 \|f\| + \log_2 \|g\| + (d_f + d_g) \log_2 A. \tag{3}$$

Of course, the value taken by A will also depend on the size of the norms. If this size is larger than expected, then the probability of finding a relation is too small and the sieving range corresponding to A will not allow the creation of enough relations. But assuming A is fixed is enough to compare various types of polynomial constructions: if one of them gives larger norms, then for this construction, the value of A should be larger, leading to even larger norms. In other words, the optimal value is unchanged whether we consider A fixed or let it depend on d_f and d_g.

The Best Asymmetric Construction is the Classical SNFS. We first analyze the case where d_f and d_g are distinct. Let us assume $d_f > d_g$. We first remark that subject to constraint (2), Expression (3) is minimized by taking $\|f\|$ as small as possible (i.e. $\log_2 \|f\| = 0$) and $\log_2 \|g\| = n/d_f$. This yields an optimal norm size equal to $n/d_f + (d_f + d_g) \log_2 A$. It follows that given d_f, we should choose d_g to be minimal, which leads us precisely to the classical case, the example construction listed in the third row of Table 1. The optimal d_f yields an optimal norm size equal to $2\sqrt{n \log_2 A}$.

An All-Balanced Construction. In many situations, the optimal value is obtained by balancing each quantity as much as possible. Unfortunately, this is suboptimal in our case. If $d_f = d_g$, Expression (3) becomes $n/d_f + 2d_f \log_2 A$. Choosing the best possible value for d_f, we obtain $2\sqrt{2n \log_2 A}$. This is much worse than in the classical construction and in fact, pushing the analysis to its end would lead to a GNFS complexity with a $(64/9)^{1/3}$ exponent.

More General Constructions. Unfortunately, it seems to be impossible to combine the SNFS construction with Coppersmith's multiple number field strategy [11,12,35] and obtain a complexity with an exponent smaller than $(32/9)^{1/3}$. Any linear combination of f and g will lead to a polynomial having both high degree and large coefficients, which must be avoided to achieve SNFS complexity.

In principle, one could also perform an analysis allowing skewed polynomials, where the ratio between two consecutive coefficients is roughly a constant different from 1. This general analysis would require still more parameters than the one we did, so we skip the details, since we did not find a situation where this could lead to a good asymptotic complexity.

4.2   Hiding the Special Form

The conclusion of the previous discussion is that the best form for a pair of SNFS polynomials is still the same as the one considered by Gordon more than

20 years ago. His discussion about how to hide them is still valid. We recall it here for completeness.

The goal is to find a prime p, and possibly a factor q of $p - 1$, together with an SNFS pair of polynomials, such that from the knowledge of p (and q) it is harder to guess the SNFS pair of polynomials than to run a discrete logarithm computation with the general NFS algorithm, or using Pollard Rho in the subgroup of order q.

We enumerate requirements on the construction below.

The Polynomial f Must Be Chosen Within a Large Enough Set. If f is known, then its roots modulo p can be computed. With the Extended Euclidean Algorithm, it can be efficiently checked whether one of them is equal to a small rational number modulo p. If this is the case, then the numerator and the denominator are (up to sign) the coefficients of g. Therefore, if f has been chosen among a small set, an exhaustive search over the roots modulo p of all these polynomials will reveal the hidden SNFS pair of polynomials. Thus we must choose f from a large enough set so that this exhaustive search takes at least as much time as a direct discrete logarithm computation.

The Two Coefficients of g Must Be Large. If g is a monic polynomial $g = x - g_0$, then, since $p = f(g_0)$, the most significant bits of p depend only on g_0 and the leading coefficient f_d of f. In that case, recovering the hidden SNFS polynomials reduces to an exhaustive search on the leading coefficient of f: we can use the LLL algorithm to minimize the other coefficients of f by writing a multiple of p as a sum of powers of g_0. Examining the least significant bits of p shows that having a polynomial g with a constant term equal to 1 is equally bad. More generally, having one of the coefficients of g belonging to a small set also leads to a faster exhaustive search than if both are large. In the following, we will therefore always consider linear polynomials g for which the two coefficients have similar sizes; compared to using a monic g, this has only a marginal impact on the effectiveness of the SNFS efficiency in our context.

Attempts to Unveil the Trapdoor. Heidi does not want her trapdoor to be unveiled, as she would not be able to plausibly deny wrongdoing. It is therefore highly important that Heidi convinces herself that the criteria above are sufficient for the trapdoor to be well hidden. We tried to improve on the method mentioned above that adapts to monic g. In particular, we tried to take advantage of the possibility that the leading coefficient of f might be divisible by small primes. This did not lead to a better method.

4.3   Adapting the Prime to the Hider's Needs

Algorithm to Build a DSA-like Prime. In Algorithm 1, we recall the method of Gordon to construct hidden SNFS parameters in a DSA setting. The general idea is to start from the polynomial f and the prime q, then derive a polynomial g such that q divides the resultant of f and g minus 1, and only at the end check if this resultant is a prime p. This avoids the costly factoring of $p - 1$ that would be needed to check whether there is a factor of appropriate size to play the role of q. Our version is slightly more general than Gordon's, since we allow signed coefficients for the polynomials. As a consequence, we do not ensure the sign of the resultant, so that the condition $q \mid p - 1$ can fail. This explains the additional check in Step 8. The size of the coefficients of f are also adjusted so that an exhaustive search on all the polynomials will take more or less the same time as the Pollard Rho algorithm in the subgroup of order q, namely $2^{s_q/2}$ where s_q is the bit-length of q.

In Step 6, it is implicitly assumed that 2^{s_q} is smaller than $2^{s_p/d}/\|f\|$, that is $s_q < s_p/d - s_q/2(d + 1)$. This condition will be further discussed in the next subsection. We note however that if it fails to hold by only a few bits, it is possible to run the algorithm and hope that the root r produced at Step 5 will be small enough. We can expect that r will behave like a uniformly random element modulo q, so that the probability that this event occurs can be estimated.

Input : The bit-sizes s_p and s_q for p and q; the degree d of f.
Output: HSNFS parameters f, g, p, q.

1 Pick a random irreducible polynomial f, with $\|f\| \approx 2^{s_q/2(d+1)}$;
2 Pick a random prime q of s_q bits;
3 Pick a random integer $g_0 \approx 2^{s_p/d}/\|f\|$;
4 Consider the polynomial $G_1(g_1) = \mathrm{Res}_x(f(x), g_1 x + g_0) - 1$ of degree d in g_1;
5 Pick a root r of G_1 modulo q; if none exists go back to Step 1;
6 Add a random multiple of q to r to get an integer g_1 of size $\approx 2^{s_p/d}/\|f\|$;
7 Let $p = |\mathrm{Res}_x(f(x), g_1 x + g_0)|$;
8 If p has not exactly s_p bits or if p is not prime or if q does not divide $p - 1$, then go back to Step 1;
9 Return f, g, p, q.

Algorithm 1. Gordon's hidden SNFS construction algorithm

Selecting Good f-Polynomials. In Algorithm 1, in the case of failure at Step 5 or Step 8, we could restart only at Step 2, in order to keep using the same f-polynomial for a while. More generally, the polynomial f could be given as input of the algorithm, opening the opportunity for the hider to use a polynomial f with nice algebraic properties that accelerate the NFS algorithm. The so-called Murphy-α value [37, Sect. 3.2] has a measurable influence on the probability of the norms to be smooth. A norm of s bits is expected to have a smoothness probability similar to the one of a random integer of $s + \frac{\alpha}{\log 2}$ bits. A negative α-value is therefore helping the relation collection.

Experimentally, for an irreducible polynomial of fixed degree over $\mathbb{Z}$ with coefficients uniformly distributed in an interval, the α-value follows a centered normal law with standard deviation around 0.94 (measured empirically for degree 6). From this, it is possible to estimate the expected minimum α-value after trying N polynomials: we get $\alpha_{\min} \sim -0.94\sqrt{2 \log N}$.

In a set of 2^{80} candidates for the f-polynomial, we can therefore expect to find one with an α-value around -10. But it is *a priori* very hard to find this polynomial, and if it were easy, then it would not be a good idea for the hider to choose it, because then it would not be hidden anymore. A compromise is for the hider to try a small proportion of the candidates and keep the one with the best α. Since checking the α-value of a polynomial is not really faster than checking its roots modulo p, the attacker gains no advantage by knowing that f has a smaller value than average. For instance, after trying 2^{20} polynomials, one can expect to find an f that has an α-value of -5 which gives a nice speed-up for the NFS without compromising the hidden property.

Apart from the α-value, another well-known feature of polynomials that influences the smoothness properties is the number of real roots: more real roots translates into finding relations more easily. We did not take this into account in our proof of concept experiments, but this could certainly also be used as a criterion to select f.

4.4   Size Considerations

Algorithm 1 does not work if the size s_q of the subgroup order q is too large compared to the size of the coefficients of the g-polynomials that are optimal for the size of p. The condition is

$$s_q < s_p/d - s_q/2(d+1),$$

where d is the optimal degree of the f-polynomial for running SNFS on a prime of s_p bits. We can plug in the asymptotic formula for d in terms of s_p: it is proportional to $(s_p/\log(s_p))^{1/3}$, leading to a condition of the form

$$s_q < c(s_p)^{2/3}(\log(s_p))^{1/3} = \log(L_p(2/3, c)),$$

for a constant c. Now, s_q will be chosen so that the running time of Pollard Rho in the subgroup of order q matches the running time of the NFS algorithm modulo p. The former grows like $2^{s_q/2}$, while the latter grows like $L_p(1/3, c') \approx 2^{s_p^{1/3}}$. Therefore, asymptotically, it makes sense to have s_q close to proportional to $s_p^{1/3}$, and the condition for Algorithm 1 to work is easily satisfied.

Back in 1992, when Gordon studied the question, the complexity analysis of the Number Field Sieve was not as well understood, and the available computing power was far less than today. At that time, $s_q = 160$ and $s_p = 512$ were the proposed parameter sizes for DSA, leading to difficulties satisfying the condition of Algorithm 1 unless a suboptimal d was chosen. Nowadays, popular DSA parameters are $s_p = 1024$ and $s_q = 160$, leaving much room for the condition to hold, and it is possible to choose $d = 6$, which is optimal for our NFS implementation. Therefore, the relevant parameters for today are beyond the point where Gordon's algorithm would need to be run with suboptimal parameters.

4.5   Reassessing the Hiding Problem

The prudent conclusions of cryptographers in the 1990's was that it might be difficult to put a useful and hard to detect trapdoor in a DSA prime. For example in [15], Lenstra concludes that "this kind of trap can be detected", based on the trap design from [31]. It is true that whether for Lenstra's trap method in [31], or Gordon's trap in Algorithm 1, f had to be chosen within a too-small set given the state of the art with NFS back in 1992. This stance is also found in reference books from the time, such as the *Handbook of Applied Cryptography* by Menezes, van Oorschot, and Vanstone [36, note Sect. 8.4] which remain influential today.

This is no longer true. It is now clearly possible to hide an SNFS pair of polynomials for a DSA prime p of 1024 bits with a 160-bit subgroup. It remains to show that this SNFS computation is indeed feasible, even with moderate academic computing resources.

5   Computation of a 1024-Bit SNFS DLP

In addition to the computational details, we describe the algorithmic improvements and parameter choices that played a key role in the computation (Table 2).

Table 2. Our 1024-bit hidden SNFS discrete log computation took around two months of calendar time to complete. We used a variety of resources for sieving, so the total number of cores in use varied over time.

	sieving	linear algebra			individual log
		sequence	generator	solution	
cores	≈3000	2056	576	2056	500–352
CPU time (1 core)	240 years	123 years	13 years	9 years	10 days
calendar time	1 month	1 month			80 minutes

5.1   Selecting a Target

We ran Algorithm 1 to find a hidden SNFS prime p of 1024 bits such that $\mathbb{F}_p$ has a subgroup of prime order q of 160 bits. For these parameters, a polynomial f of degree $d = 6$ is the most appropriate. After a small search among the polynomials with (signed) coefficients of up to 11 bits, we selected

$$f = 1155\,x^6 + 1090\,x^5 + 440\,x^4 + 531\,x^3 - 348\,x^2 - 223\,x - 1385,$$

for which the α-value is about -5.0. The set of all polynomials with this degree that satisfy the coefficient bound is a bit larger than 2^{80}, which is the expected cost of Pollard Rho modulo q. We note that the cost of testing a polynomial f (a root finding modulo p and the rational reconstruction of these roots) is much higher than one step of Pollard Rho (one multiplication modulo p), so this is a conservative setting.

We then ran the rest of Algorithm 1 exactly as it is described. The resulting public parameters are

$p = 1633239872404436791014020700930491550309894398069175191735800707915692277289328503584988628543993514237336976605348001944927248287213149802482594503587920692359918265889442004406870941366695063490936917689024405553414932372965552524247379422702221515929837629813600812082006124038089463610239236157651252180491$

$q = 1120320311183071261988433674300182306029096710473,$

and the trapdoor polynomial pair is

$$f = 1155\,x^6 + 1090\,x^5 + 440\,x^4 + 531\,x^3 - 348\,x^2 - 223\,x - 1385$$
$$g = 5671623128181204324899915687856269867712018292374 08\,x$$
$$-663612177378148694314176730818181556491705934826717.$$

This computation took 12 core-hours, mostly spent in selecting a polynomial f with a good α-value. No effort was made to optimize this step.

5.2   Choosing Parameters for the Sieving Step

The sieving step (also known as relation collection) consists of finding many (a, b)-pairs such that the two norms $\mathrm{Res}(f(x), a - bx)$ and $\mathrm{Res}(g(x), a - bx)$ are simultaneously smooth.

We use the special-q sieving strategy, where we concentrate the search in the positions where we know in advance that one of the two norms will be divisible by a large prime: the special q. For the general number field sieve, it is always the case that one norm is much larger than the other, so it makes sense to choose the special q on the corresponding side. In our case, the norms have almost the same size (about 200 bits each), so there is no obvious choice. Therefore, we decided to sieve with special q's on both sides. As a consequence, the largest special q that we had to consider were 1 or 2 bits smaller than if we had allowed special q's to be only on one side; the norms were accordingly a bit smaller.

The general strategy used for a given special q is classical: among a vast quantity of candidates, we mark those that are divisible by primes up to a given *sieving bound* using a sieve à la Eratosthenes; then the most promising candidates are further scrutinized using the ECM method, trying to extract primes up to the smoothness bound B. The criterion for selecting those promising candidates is best expressed as the number of times the smoothness bound is allowed for the remaining part of the norms once the sieved primes have been removed. This is usually referred to as the *number of large primes* allowed on a given side.

For the 1024-bit computation, we used the following parameters. On the rational side, we sieved all the prime special q in the 150M–1.50G range (that is, with $1.5 \cdot 10^8 < q < 1.5 \cdot 10^9$), and on the algebraic side, we sieved special-q prime ideals in the range 150M–1.56G. The difference between the two is due to the fact that we used two different clusters for this step, and when we stopped sieving, one was slightly ahead of the other.

For each special q, the number of (a, b)-pairs considered was about 2^{31}. This number includes the pairs where both a and b are even, but almost no time is spent on those, since they cannot yield a valid relation.

All primes on both sides up to 150M were extracted using sieving, and the remaining primes up to the smoothness bound $B = 2^{31}$ were extracted using ECM. On the side where the special q was placed, we allowed 2 large primes, while 3 large primes were allowed on the other side.

This relation collection step can be parallelized almost infinitely with no overhead since each special q is handled completely separately from the others. We used a variety of computational resources for the sieving, and in general took advantage of hyperthreading and in addition oversubscribed our virtual cores with multiple threads. Aggregating reported CPU time for virtual cores over all of the machine types we used, we spent $5.08 \cdot 10^9$ CPU seconds, or 161 CPU years sieving the rational side, and $5.03 \cdot 10^9$ CPU seconds, or 159 CPU years sieving the algebraic side. In order to obtain a more systematic estimate of the CPU effort dedicated to sieving without these confounding factors, we ran sampling experiments on a machine with 2 Intel Xeon E5-2650 processors running at 2.00 GHz with 16 physical cores in total. From these samples, we estimate that sieving would have taken 15 years on this machine, or 240 core-years. We spent about one month of calendar time on sieving.

The total number of collected relations was 520M relations: 274M from the rational side and 246M from the algebraic side. Among them, 249M were unique, involving 201M distinct prime ideals. After filtering these relations, we obtained a matrix with 28M rows and columns, with 200 non-zero entries per row on average.

Before entering the linear algebra step, we calculated the dense block of "Schirokauer maps", which are q-adic characters introduced by Schirokauer in [46]. These consist, for each matrix row, in 3 full-size integers modulo q (the number 3 is here the unit rank of the number field defined by our polynomial f).

5.3   Linear Algebra

The linear algebra problem to be solved can be viewed in several ways. One is to consider a square matrix of size $N \times N$, whose left-hand side M of size $N \times (N-r)$ is the matrix produced by the filtering task, while the right block S of size $N \times r$ is made of dense Schirokauer maps. Recent work [24] has coined the term "nearly sparse" for such matrices. We seek a non-trivial element of the right nullspace of the square matrix $(M\|S)$.[3] This approach has the drawback that an iterative linear algebra algorithm based on the matrix $(M\|S)$ is hampered by the weight of the block S, which contributes to each matrix-times-vector product.

---

[3] The integer factorization case, in contrast, has $q = 2$, and requires an element of the *left* nullspace. The latter fact allows for a two-stage algorithm selecting first many solutions to $xM = 0$, which can then be recombined to satisfy $x(M\|S) = 0$. No such approach works for the right nullspace.

Shirokauer Maps Serve as Initialization Vectors. An alternative method, originally proposed by Coppersmith [13, Sect. 8] alongside the introduction of the block Wiedemann algorithm, is to use this algorithm to constructively write a zero element in the sum of the column spaces of M and S. In this case, the iterative algorithm is run on the square matrix $M_0 = (M\|0)$, which is of considerably lower weight than $(M\|S)$. More precisely, the block Wiedemann algorithm, with two *blocking factors* which are integers m and n, achieves this by proceeding through the following steps. The blocking factor n is chosen so that $n \geq r$, and we let $D(t)$ be the diagonal $n \times n$ matrix with coefficients t (r times) and 1 ($n - r$ times).

Initialization. Pick blocks of *projection vectors* $x \in \mathbb{F}_q^{N \times m}$ and *starting vectors* $y \in \mathbb{F}_q^{N \times n}$. The block x is typically chosen of very low weight, while we set $y = (S\|R)$, with R a random block in $\mathbb{F}_q^{N \times (n-r)}$.

Sequence. Compute the sequence of matrices $a_i = {}^t x M_0^i y$, for $0 \leq i < L$, with $L = \lceil N/m \rceil + \lceil N/n \rceil + \lceil m/n + n/m \rceil$.

Linear generator. Let $A(t) = \sum_i a_i t^i$. Let $A'(t) = A(t)D(t)$ div t. Compute an $n \times n$ matrix of polynomials $F(t)$ such that $A'(t)F(t)$ is a matrix of polynomials of degree less than $\deg F$, plus terms of degree above L (see [13, 52]). We typically have $\deg F \approx N/n$.

Solution. Consider one column of degree d of $F(t)$. Write the corresponding column of $D(t)F(t)$ as $ct^{d+1} + f_0 t^d + \cdots + f_d$ with $c, f_i \in \mathbb{F}_q^{n \times 1}$. With high probability, we have $c \neq 0$ and $w = (S\|0)c + M_0 \sum_{i \geq 0} M_0^i y f_i = 0$. Rewrite that as $Mu + Sv = 0$, where u and v are:

$$u = \text{first } N - r \text{ coefficients of } \sum_{i \geq 0} M_0^i y f_i,$$

$$v = \text{first } r \text{ coefficients of } c$$

This readily provides a solution to the problem.

Solving the Linear System with $(1 + o(1))N$ SpMVs. The most expensive steps above are the *sequence* and *solution* steps. The dominating operation is the sparse matrix-times-vector operation (SpMV), which multiplies M_0 by a column vector in $\mathbb{F}_q^{N \times 1}$. It is easy to see that the sequence step can be run as n independent computations, each requiring L SpMV operations (therefore $nL = (1 + n/m)N$ in total): matrices a_i are computed piecewise, column by column. Once all these computations are completed, the fragments of the matrices a_i need to be collated to a single place in order to run the linear generator step.

It is tempting to regard the solution step in a directly similar way (as was done, e.g., in the context of the recent 768-bit DLP computation [28]). However, as was pointed out very early on by Kaltofen [26, Step C3, p. 785, and corollary to Theorem 7] yet seems to have been overlooked since, one should proceed differently. Assume that some of the vectors $M_0^i y$ from the sequence step have been kept as regular checkpoints (an obvious choice is $M_0^{Kj} y$ for some well chosen checkpoint period K). For an arbitrary j, we compute $\sum_{i=0}^{i=K-1} M_0^i M_0^{Kj} y f_{Kj+i}$

with a Horner evaluation scheme which costs K SpMV operations only. These expressions together add up to u, and can be computed independently (using as many independent tasks as K allows). This adds up to $\deg F \approx N/n$ SpMV operations.

In total, this evaluation strategy yields a cost of $(1 + n/m + 1/n)N$ SpMV operations (see [26, Theorem 7]), which can be freely parallelized n-fold for the sequence step, and possibly much more for the solution step. It is important to note that as blocking factors m and n grow with $m \gg n$, this brings the total cost close to N SpMV operations, a count which to our knowledge beats all other exact sparse linear algebra algorithms. The only limiting point to that is the linear generator step, whose cost depends roughly linearly on $(m + n)$. Thanks to the use of asymptotically fast algorithms [7,20,52], this step takes comparatively little time.

Linear Algebra for 1024-Bit SNFS DLP. The matrix M_0 had 28 million rows and columns, and 200 non-zero coefficients per row on average. We used the linear algebra code in CADO-NFS [51]. We chose blocking factors $m = 24$, $n = 12$. Consequently, a total of 44 million SpMV operations were needed. We ran these in two computing facilities in the respective research labs of the authors, with a roughly even split. For the sequence step, each of the 12 independent computations used between 4 and 8 nodes, each with up to 44 physical cores. The nodes we used were interconnected with various fabrics, including Mellanox 56 Gbps Infiniband FDR and 40 Gbps Cisco UCS Interconnects. The total time for the sequence step was about 123 core-years. The linear generator step was run on 36 nodes, and cost 13 core-years. The solution step was split in $48 = \frac{\deg F = 2400000}{K = 50000}$ independent tasks. Each used a fairly small number of nodes (typically one or two), which allowed us to minimize the communication cost induced by the parallelization. Despite this, each iteration was 33% more expensive than the ones for the sequence step, because of the extra cost of the term $M_0^{Kj} y f_{Kj+i}$ which is to be added at each step. The total time for the solution step was 9 core-years, which brings the total linear algebra cost for this computation below 150 core-years. In total we spent about one month of calendar time on linear algebra. Table 4 in Appendix B gives more details of the iteration times for the different machine architectures present in our clusters.

After this step and propagating the knowledge to relations that were eliminated during the filtering step we obtained the logarithm of 198M elements, or 94.5% of the prime ideals less than 2^{31}.

5.4   Individual Logarithms

In our scenario where a malicious party has generated a trapdoored prime with a goal of breaking many discrete logarithms in the corresponding group, it is interesting to give details on the individual logarithm step, which is often just quickly mentioned as an "easy" step (with the notable exception of [2] where it is at the heart of the attack).

From now on, we denote by z the element of the group G for which we want the discrete logarithm (modulo q). The database of discrete logarithms computed thus far is with respect to an a priori unknown generator. In order to obtain the logarithm of z with respect to a generator specified by the protocol being attacked, it is typical that two individual logarithm queries are necessary. This aspect will not be discussed further.

The individual logarithm step can itself be decomposed in two sub-steps:

Initialization. Find an exponent e such that $z' = z^e \equiv u/v \mod p$, where u and v are B_{init}-smooth numbers of size about half of the size of p. Note that B_{init} has to be much larger than B to get a reasonable smoothness probability.

Descent. For every factor of u or v that is larger than the smoothness bound, treat it as a special-q to rewrite its discrete logarithm in terms of smaller elements, and continue recursively until it is rewritten in terms of elements of known logarithm.

We emphasize that the Initialization step does not use the polynomials selected for the NFS computation, and therefore, it does not take advantage of the SNFS nature of the prime p. For the Descent step, on the other hand, this makes heavy use of the polynomials, and here knowing the hidden polynomials corresponding to p helps significantly.

Asymptotic Complexities. In terms of asymptotic complexity, the Initialization step is more costly than the Descent step, and in a theoretical analysis, the bound B_{init} is chosen in order to minimize the expected time of the Initialization step only. The early-abort analysis of Barbulescu [5, Chap. 4] gives $B_i = L_p(2/3, 0.811)$, for a running time of $L_p(1/3, 1.232)$.

For the Descent step, the complexity analysis can be found in two different flavours in the literature: either we use polynomials of degree 1 (the polynomial $a - bx$ corresponding to an (a, b)-pair) like in [48], or polynomials of possibly higher degrees, depending on where we are in the descent tree similarly to [17].

Using higher degree polynomials, we get a complexity of $L_p(1/3, 0.763)$, where the last steps of the descent are the most costly. Sticking with polynomials of degree 1, the first steps become more difficult and the complexity is $L_p(1/3, 1.117)$. Both are lower than the complexity of the Initialization step.

We give further details on the complexity analysis of individual logarithms in Appendix A.

Practical Approach. This theoretical behaviour gives only a vague indication of the situation for our practical setting. For the initialization step, we follow the general idea of Joux-Lercier [23]. For a random integer e, we compute $z' \equiv z^e \mod p$, and take two consecutive values in the extended Euclidean algorithm to compute u_0, v_0, u_1, v_1 of size about $\sqrt{p}$ such that $z' \equiv \frac{u_0}{v_0} \equiv \frac{u_1}{v_1} \mod p$. We then look for two integers a and b such that $u = au_0 + bu_1$ and $v = av_0 + bv_1$ are both smooth. Since this is an unlikely event, and testing for smoothness is very costly, we do a 3-step filtering strategy.

First, we sieve on the set of small (a, b)-pairs, to detect pairs for which the corresponding u and v are divisible by many small primes. For this, we re-use the sieving code that helps collecting relations. After this first step, we keep only (a, b)-pairs for which the remaining unfactored part is less than a given threshold on each side.

Then, many ECM factorizations are run on each remaining cofactor; this is tuned so that we expect most of the prime factors up to a given bit size to be extracted. After this step, we again keep only the candidates for which the remaining unfactored parts are smaller than another threshold. The cofactors of the surviving pairs are then fully factored using MPQS.

At each stage, if a prime factor larger than the smoothness bound B is found, we naturally abort the computation.

This practical strategy keeps the general spirit of filters used in the theoretical analysis of Barbulescu which relies only on ECM, but we found that combined with sieving and MPQS, it is much faster.

Parameters for the 1024-Bit SNFS Computation. For the 1024-bit computation, we used a bound $B_{init} = 135$ bits for this Initialization step. After applying the Joux-Lercier trick, we first sieved to extract primes up to 2^{31}, and we kept candidates for which the unfactored parts are both less than 365 bits. Then we used GMP-ECM with 600 curves and $B_1 = 500,000$, hoping to remove most of the prime factors of 100 bits and less. After this second step, we kept only the candidates for which the unfactored parts are both less than 260 bits.

For the Descent step, we used only polynomials of degree 1. Polynomials of degree 2 do not seem to yield smaller norms even for the largest 135-bit primes to be descended. The depth of the recursion was rarely more than 7. A typical example of a sequence of degrees encountered while following the tree from the top to a leaf is

$$135 \rightarrow 90 \rightarrow 65 \rightarrow 42 \rightarrow 33 \rightarrow 31,$$

but this is of course different if the ideals along the path are on the rational or the algebraic side.

Both the Initialization and the Descent steps can be heavily parallelized. The expected CPU-time for computing an individual logarithm is a few days on a typical core, distributed more or less equally between the two steps. Using parallelization, we managed to get a result in 80 mins of wall-clock time: the initialization took around 20 min parallelized across 500 cores, and the descent took 60 min parallelized across 352 cores. (We used 44 cores for each large special q to be descended.)

As an example, we computed the discrete logarithm of

$$z = \lceil \pi 10^{307} \rceil = 3141592653 \cdots 7245871,$$

taking 2 as a generator. More precisely, we are talking about their images in the subfield of order q obtained by raising 2 and z to the power $(p-1)/q$. We obtained:

$$\log z / \log 2 \equiv 40955910136077466935944380848964570408223951 3256 \mod q,$$

which can be easily checked to be the correct answer.

6 Discussion

6.1 Comparison with GNFS DLP for Various Sizes

The recently reported GNFS 768-bit discrete logarithm computation [28] took about 5000 core-years. It is tempting to directly compare this number to the 400 core-years that we spent in our experiments. As a rule of thumb, one would expect the 768-bit GNFS to be about a factor of 10 more difficult than a 1024-bit SNFS computation, and this appears to hold in the numbers we report. However, we note that first, the software used in both experiments are different (the CADO-NFS sieving implementation is slower than Kleinjung's), and second, the GNFS-768 computation was done with a safe prime, while we used a DSA prime, thus saving a factor of 6 in the linear algebra running time.

It is possible to get another hint for the comparison by considering the typical sizes of the norms in both contexts. For GNFS-768, they appear to be roughly 20 bits larger (in total) than for SNFS-1024. Taking all correcting factors into account, like the α-values, the (un-)balance of the norms, and the special-q, this translates to roughly a factor of 8 in the smoothness probability, thus more or less confirming the ratio of running times observed in practice.

Asymptotically, the difference of complexities between GNFS and SNFS (namely the factor $2^{1/3}$ in the exponent) means that we would expect to obtain similar running times when SNFS is run on an input that is twice the size of the one given to GNFS. However, key sizes relevant to current practice and practical experiments are still too small for these asymptotic bounds to be accurate.

To get concrete estimates for these smaller key sizes, we can compare the size of the norms and estimate that an 1152-bit SNFS computation would correspond to the same amount of time as a GNFS-768. For an SNFS of 2048 bits, the equivalent would be around a GNFS of 1340 bits. And finally, for an SNFS of 4096 bits, the equivalent would be around a GNFS of 2500 bits. Of course, for such large sizes these are more educated guesses than precise estimates.

6.2 Non-hidden SNFS Primes in Real Use

We have found multiple implementations using non-hidden SNFS primes in the real world. 150 hosts used the 512-bit prime $2^{512} - 38117$ for export-grade Diffie-Hellman key exchange in a full IPv4 HTTPS scan performed by Adrian et al. [2] in March 2015. Performing the full NFS discrete log computation for this prime took about 215 min on 1288 cores, with 8 min spent on the sieving stage, 145 min spent on linear algebra, and the remaining time spent filtering relations and reconstructing logarithms. In September 2016, 134 hosts were observed still using this prime.

We also found 170 hosts using the 1024-bit prime $2^{1024} - 1093337$ for non-export TLS Diffie-Hellman key exchange in scans performed by Adrian et al. In September 2016, 106 hosts were still using this prime. We estimate that performing a SNFS-DL computation for this prime would require about 3 times the amount of effort for the sieving step as the 1024-bit SNFS computation that we performed. This difference is mostly due to the α-value of the f-polynomial that can not easily be made small. The linear algebra step will suffer at the very least a 7-fold slowdown. Indeed, since this prime is safe, the linear algebra must be performed modulo $(p-1)/2$, which is more expensive than the 160-bit linear algebra we used for a DSA prime in our computation. Furthermore, since the smoothness probabilities are worse, we expect also the matrix to be a bit larger, and the linear algebra step cost to grow accordingly.

The LibTomCrypt library [14], which is widely distributed and provides public domain implementations of a number of cryptographic algorithms, includes several hard-coded choices for Diffie-Hellman groups ranging in size from 768 to 4096 bits. Each of the primes has a special form amenable to the SNFS. The 768-bit strength (actually a 784-bit prime) is $2^{784} - 2^{28} + 1027679$. We performed a SNFS discrete log for this prime. On around a thousand cores, we spent 10 calendar days sieving and 13 calendar days on linear algebra. The justification for the special-form primes appears to be the diminished radix form suggested by Lim and Lee [34], which they suggest for decreasing the cost of modular reduction. We examined the TLS and SSH scan datasets collected by Adrian et al. [2] and did not find these primes in use for either protocol.

We also carried out a perfunctory search for poorly hidden SNFS primes among public key datasets, based on the rather straightforward strategy in Sect. 4.2, hoping for monic g, and f such that $2 \le d \le 9$ and $|f_d| \le 1024$. We carried out this search for the 11,658 distinct SSH group exchange primes, 68,126 distinct TLS ephemeral Diffie-Hellman primes, and 2,038,232 distinct El Gamal and DSA primes from a dump of the PGP public key database. This search rediscovered the special-form TLS primes described above, but did not find any other poorly hidden primes susceptible to SNFS. We cannot rule out the existence of trapdoored primes using this method, but if hidden SNFS primes are in use the designers must have followed Gordon's advice.

6.3 Lessons

It is well known among the cryptographic community that 1024-bit primes are insufficient for cryptosystems based on the hardness of discrete logarithms. Such primes should have been removed from use years ago. NIST recommended transitioning away from 1024-bit key sizes for DSA, RSA, and Diffie-Hellman in 2010 [6]. Unfortunately, such key sizes remain in wide use in practice. Our results are yet another reminder of the risk, and we show this dramatically in the case of primes which lack verifiable randomness. The discrete logarithm computation for our backdoored prime was only feasible because of the 1024-bit size.

The asymptotic running time estimates suggest that a SNFS-based trapdoor for a 2048-bit key would be roughly equivalent to a GNFS computation for a 1340-bit key. We estimate that such a computation is about 16 million times harder than the 1024-bit computation that we performed, or about $6.4 \cdot 10^9$ core-years. Such a computation is likely still out of range of even the most sophisticated adversaries in the near future, but is well below the security guarantees that a 2048-bit key should provide. Since 2048-bit keys are likely to remain in wide usage for many years, standardized primes should be published together with their seeds.

In the 1990s, key sizes of interest were largely limited to 512 or 1024 bits, for which a SNFS computation was already known to be feasible in the near future. Both from this perspective, and from our more modern one, dismissing the risk of trapdoored primes in real usage appears to have been a mistake, as the apparent difficulties encountered by the trapdoor designer in 1992 turn out to be easily circumvented. A more conservative design decision for FIPS 186 would have required mandatory seed publication instead of making it optional. As a result, there are opaque, standardized 1024-bit and 2048-bit primes in wide use today that cannot be properly verified.

Acknowledgements. We are grateful to Paul Zimmermann for numerous discussions all along this work. Rafi Rubin performed invaluable system administration for the University of Pennsylvania cluster. Shaanan Cohney and Luke Valenta contributed to sieving for the 784-bit SNFS-DL computation. Part of the experiments presented in this paper were carried out using the Grid'5000 testbed, supported by a scientific interest group hosted by Inria and including CNRS, RENATER and several Universities as well as other organizations. We are grateful to Cisco for donating the Cisco UCS hardware that makes up most of the University of Pennsylvania cluster. Ian Goldberg donated time on the CrySP RIPPLE Facility at the University of Waterloo and Daniel J. Bernstein donated time on the Saber cluster at TU Eindhoven for the 784-bit SNFS-DL computation. This work was supported by the U.S. National Science foundation under grants CNS-1513671, CNS-1505799, and CNS-1408734, and a gift from Cisco.

A   Complexity Analysis of Individual Logarithms

The complexity analysis of individual logarithms is not well detailed in the literature, in particular in the SNFS case. For convenience we summarize the results in this appendix. As usual in the NFS context, the claimed complexities are not rigorously proven and rely on heuristics.

The notation is the same as in the main body of the paper: p is a prime and we have to compute the discrete logarithm of an element z in a prime order subgroup of $\mathbb{F}_p^*$. We are given f and g a pair of polynomials that have been used for an NFS computation so that the (virtual) logarithms of all the ideals of norm less than a bound B should have been pre-computed. The bound B, the degrees of f and g, and the sizes of their coefficients depend on the General vs Special NFS variant.

We recall the classical corollary of the Canfield-Erdős-Pomerance theorem that expresses smoothness probabilities in terms of the L-notation:

Theorem 1. *Let a, b, u, v be real numbers such that $a > b > 0$ and $u, v > 0$. As $x \to \infty$, the proportion of integers below $L_x(a, u)$ that are $L_x(b, v)$-smooth is*

$$L_x\left(a - b, \frac{u}{v}(a - b) + o(1)\right)^{-1}.$$

A.1 Initialization of the Descent

This step consists of first "smoothing" z in order to bootstrap the subsequent descent step. We choose a random integer e, compute the element $z' \equiv z^e$ mod p, and test it for smoothness. Many elements are tested until one is found to be B_{init}-smooth. The best known algorithm for smoothness testing is the ECM algorithm: it extracts (with high probability) all primes up to a bound K in time $L_K(1/2, \sqrt{2} + o(1))$. The dependence in the size of the integer from which we extract primes is polynomial, so we omit it: in our context this type of factor ends up being hidden in the $o(1)$ in the exponents.

From this estimate, one can derive that if we want to allow a running time in $L_p(1/3, \cdot)$, then B_{init} can only be as large as $L_p(2/3, \cdot)$; otherwise, testing the smoothness would be too costly. At the same time, the probability that z' is B_{init}-smooth drives the number of attempts and puts additional constraints. It is remarkable that it also imposes a smoothness bound B_{init} in $L_p(2/3, \cdot)$ to get an $L_p(1/3, \cdot)$ number of attempts. Following Commeine-Semaev [11], if we set $B_{\text{init}} = L_p(2/3, c)$, one can show that the expected running time for the basic algorithm for the initialization step is in $L_p(1/3, \frac{1}{3c} + 2\sqrt{c/3} + o(1))$, which is minimal for $c = 1/\sqrt[3]{3}$, yielding a complexity of $L_p(1/3, \sqrt[3]{3} + o(1)) \approx L_p(1/3, 1.442)$.

Inspired by the early abort strategy that Pomerance [44] had developed in the context of the quadratic sieve, Barbulescu [5] has shown that this complexity can be reduced. The idea is to start the smoothness test with a bound smaller than the target B_{init} smoothness bound: this allows one to extract the smallest factors. Then, we make a decision based on the size of the remaining unfactored part: if it is too large, the probability that this will yield a B_{init}-smooth number is too small and we start again with another random exponent e. In other words, instead of testing immediately for B_{init}-smoothness, we first run a filter, with cheaper ECM parameters, that allows us to select promising candidates for which the full test is run. Analyzing this technique and optimizing all the involved parameters is not a simple task; according to [5], we obtain a final complexity of $\approx L_p(1/3, 1.296)$ for a smoothness bound $B_{\text{init}} = L_p(1/3, 0.771)$.

This is not the end of the story: instead of just one filter, one can add more. The analysis becomes even more involved, but this improves again on the complexity. Numerical experiments indicate that performance does not improve beyond 6 filters, and for 6 filters, the final complexity given in [5] is summarized in the following fact:

Fact. The initialization step of the descent can be done in time $L_p(1/3, 1.232)$ with a smoothness bound $B_{init} = L_p(1/3, 0.811)$.

Finally, we mention that writing $z' \equiv \frac{u}{v} \bmod p$ for u and v that are about half the size of p, and testing them for smoothness does not change the asymptotic complexities, but it yields a huge practical improvement, especially when combined with sieving as in Joux-Lercier [23].

On the other hand, when neither f nor g are linear polynomials, the smoothing test has to be done in one of the number fields, and then, in this context, using half-size elements is necessary to get the appropriate complexity; we refer to [5, Sect. 8.4.3] for details about this.

A.2 Descent Step

After the initialization step, the discrete logarithm of z can be expressed in terms of the virtual logarithms of a few ideals of degree 1 in one of the number fields associated to f or g. Those whose norm is less than the smoothness bound B that was used in the sieving and linear algebra steps are assumed to be already known. Since $B = L_p(1/3, \cdot)$ while $B_{init} = L_p(2/3, \cdot)$, we expect to have a handful of prime ideals whose logarithms are not known. These are the ones that will be subject to this descent step. We do the analysis for one ideal of maximal size B_{init}; since there are only polynomially many of them, doing all of them will contribute only to the $o(1)$ in the final exponent.

Let $\mathfrak{q}$ be an ideal of norm $q = L_p(\alpha, c)$, where $\alpha \in [\frac{1}{3}, \frac{2}{3}]$. We consider the lattice of polynomials $\varphi(x) = a_0 + a_1 x + \cdots + a_{k-1} x^{k-1}$ that, after being mapped to a principal ideal in the number field where $\mathfrak{q}$ belongs, become divisible by $\mathfrak{q}$. For $k = 2$, this would correspond to the (a, b)-pairs corresponding to $\mathfrak{q}$ seen as a special-q, but we allow larger degrees. Since we are going to allow a search that takes a time T in $L_p(1/3, \cdot)$ for handling $\mathfrak{q}$, the a_i's can be bounded by $(qT)^{1/k} = L_p(\alpha, c/k) L_p(1/3, \cdot)$.

Let us analyze first the case where $\alpha > 1/3$ so that the second factor can be neglected. The product of the norms is given by

$$\operatorname{Res}(f(x), \varphi(x)) \operatorname{Res}(g(x), \varphi(x)) \approx \|\varphi\|^{\deg f + \deg g} \|f\|^{k-1} \|g\|^{k-1}$$
$$\approx L_p(\alpha, c/k(\deg f + \deg g))(\|f\| \|g\|)^{k-1}.$$

Let us write $\deg f + \deg g = \delta(\log p / \log \log p)^{1/3}$, so that we can cover all the variants. Then $\|f\| \|g\|$ is $L_p(2/3, 2/\delta)$ in the case of GNFS and $L_p(2/3, 1/\delta)$ in the case of SNFS. Finally, the product of the norms is

$$L_p\left(\alpha + 1/3, \frac{c\delta}{k}\right) L_p\left(2/3, \left\{{\scriptstyle 1 \text{ for SNFS} \atop \scriptstyle 2 \text{ for GNFS}}\right\} \frac{k-1}{\delta}\right).$$

Here there are two strategies: we can fix $k = 2$, so that the second factor does not contribute to the complexity, or we can let k grow in order to balance the two factors.

Descending with Higher Degree Polynomials. The best value for k is proportional to $(\log p/\log\log p)^{\alpha/2-1/6}$ (we deliberately omit to analyze the proportionality ratio). In that case, the product of the norms takes the form

$$L_p(\alpha/2 + 1/2, \cdot),$$

so that, since we allow a time $L_p(1/3, \cdot)$, we can expect to find an element that is $L_p(\alpha/2 + 1/6, \cdot)$-smooth. The smoothness test implies multiplying the cost by $L_p(\alpha/4 + 1/12, \cdot)$, which is bounded by $L_p(1/4, \cdot)$ since $\alpha \le 2/3$, and therefore does not contribute to the final complexity. As a consequence, as long as α is more than $\frac{1}{3}$, it is possible to descend a $\mathfrak{q}$ whose norm is in $L_p(\alpha, \cdot)$ in prime ideals of norms at most $L_p(\alpha/2 + 1/2, \cdot)$, in time bounded by $L_p(1/3, \cdot)$. We can choose the exponent constant smaller than the other steps of the descent so that these first steps become negligible. This is true whether we are dealing with GNFS or SNFS.

As we get close to $\alpha = \frac{1}{3}$, the value of k tends to a constant. We postpone the corresponding analysis.

Descending with Degree-1 Polynomials. In the case where we force $k = 2$, the product of the norms is dominated by the first factor and we get $L_p(\alpha + \frac{1}{3}, c\delta/2)$. Let us try to descend $\mathfrak{q}$ in prime ideals of norms slightly smaller than the norm q of $\mathfrak{q}$, namely we target $L_p(\alpha, c\lambda)$, for some value λ that we hope to be strictly less than 1. The probability of the product of the norms being q^λ-smooth is then in $L_p(\frac{1}{3}, \frac{\delta}{6\lambda} + o(1))^{-1}$. The cost of smoothness testing with ECM is in $L_p(\frac{\alpha}{2}, \cdot)$, which is negligible as soon as $\alpha < 2/3$. Hence, the cost of the descent with degree-1 polynomials is dominated by the case $\alpha = 2/3$, which we will now focus on. In this limiting case, the cost of ECM is $L_{L_p(2/3, c\lambda)}(1/2, \sqrt{2} + o(1)) = L_p(1/3, 2\sqrt{c\lambda/3} + o(1))$, so that the time to descend $\mathfrak{q}$ in prime ideals of norms bounded by q^λ is in $L_p(1/3, \frac{\delta}{6\lambda} + 2\sqrt{c\lambda/3} + o(1))$. This is minimized for $\lambda = \sqrt[3]{\delta^2/12c}$ and yields a running time of $L_p(1/3, (3c\delta/2)^{1/3} + o(1))$. In the case of GNFS, we have $\delta = 3^{1/3}$, while it is $\delta = (3/2)^{1/3}$ for SNFS. We fix λ so that we minimize the time when dealing with the largest $\mathfrak{q}$ coming out from the initialization step, namely for $q = L_p(2/3, 0.811)$; this value $c = 0.811$ gives $\lambda = 0.598$ in the case of GNFS, and $\lambda = 0.513$ in the case of SNFS. Both are less than 1, which means that the descent process indeed descends. Finally, we obtain the following:

Fact. If we use degree-1 polynomials, the cost of the first stages of the descent is $L_p(1/3, 1.206)$ for GNFS and $L_p(1/3, 1.117)$ for SNFS.

Last Steps of the Descent. We now deal with the final steps of the descent where $\mathfrak{q} = L_p(1/3, c)$, with c larger than the constant involved in the smoothness bound $B = L_p(1/3, \cdot)$, which depends on whether we are in the GNFS or SNFS case. In this setting, there is no gain in considering $k > 2$, so we keep $k = 2$. The factor that was neglected when evaluating the size of the a_i's is no longer negligible, so we start again, and assume that we are going to spend time

$T = L_p(1/3, \tau + o(1))$. This propagates into the formulae and gives a bound $L_p(1/3, (\tau + c)/2)$ for the a_i's, which in turn gives

$$L_p(2/3, (\tau + c)\delta/2)\|f\|\,\|g\|$$

for the product of the norms. Let us denote $B = L_p(1/3, \beta)$ the smoothness bound used for sieving and linear algebra, and write $c = \beta + \varepsilon$, where $\varepsilon > 0$. We omit the details, but it can be checked that if we allow time $L_p(1/3, \beta)$, we can descend $\mathfrak{q}$ in prime ideals of norms at most $L_p(1/3, \beta + \frac{\varepsilon}{4})$. This analysis is valid both for GNFS and SNFS, even though the values of β and δ are different for these two cases. This is no surprise that this is the cost of finding one relation in the sieving step, since when $\mathfrak{q}$ is just above the smoothness bound, descending involves essentially the same procedure as what we do during sieving with special-$\mathfrak{q}$ that are marginally smaller. We obtain therefore:

Fact. The cost of the last stages of the descent is $L_p(1/3, 0.961)$ for GNFS and $L_p(1/3, 0.763)$ for SNFS.

In this analysis, we have not studied the transition between the two modes where we decrease the value α or the value c when descending an ideal of size $L_p(\alpha, c)$. This technicality is dealt with in [17] in the context of the Function Field Sieve, but it applies *mutatis mutandis* to our NFS situation.

In the following table, we summarize the exponent constants in the $L_p(1/3, \cdot)$ complexities of the various steps of the descent, for GNFS and SNFS, allowing or not sieving with higher degree polynomials:

	Initialization step	Descent step		Small q
		Large q		
		Sieving deg = 1	Sieving higher deg	
GNFS	1.232	1.206	$o(1)$	0.961
SNFS	1.232	1.117	$o(1)$	0.763

B   Block Wiedemann Algorithm Timings

Table 3. We ran both sieving and linear algebra on various clusters of different configurations. For the CPU clock speed, we give both nominal and turbo speeds.

Location	Nodes	CPU type	Clock speed	Cores	RAM	Interconnect
UPenn	20	2 × Xeon E5-2699v4	2.2–2.8 GHz	44	512 GB	eth40g
	8	2 × Xeon E5-2680v3	2.5–2.9 GHz	24	512 GB	eth40g
	6	2 × Xeon E5-2699v3	2.3–2.8 GHz	36	128 GB	eth10g
Nancy	48	2 × Xeon E5-2650v1	2.0–2.4 GHz	16	64 GB	ib56g

Table 4. Timings for the Block Wiedemann algorithm as run on the various clusters for the 1024-bit SNFS Discrete Log computation. Table 3 gives details on the node configurations.

CPU type	Interconnect	Nodes/job	Seconds per iteration		
			Sequence	Solution	Communication
Xeon E5-2699v4	eth40g	1		2.42	0.12
		4	0.41		0.17
		8	0.19		0.17
		12	0.13		0.14
		16	0.10	0.21	0.13
Xeon E5-2680v3	eth40g	2		2.24	0.30
		8	0.35		0.15
Xeon E5-2699v3	eth10g	6	0.36		0.33
Xeon E5-2650v1	ib56g	2		3.7	0.19
		8	0.60		0.10

References

1. (author redacted): Eurocrypt '92 reviewed. Cryptolog, March 1994. https://www.nsa.gov/news-features/declassified-documents/cryptologs/
2. Adrian, D., Bhargavan, K., Durumeric, Z., Gaudry, P., Green, M., Halderman, J.A., Heninger, N., Springall, D., Thomé, E., Valenta, L., VanderSloot, B., Wustrow, E., Béguelin, S.Z., Zimmermann, P.: Imperfect forward secrecy: how Diffie-Hellman fails in practice. In: Ray, I., Li, N., Kruegel, C. (eds.) ACM CCS 2015: 22nd Conference on Computer and Communications Security, Denver, CO, USA, 12–16 October 2015, pp. 5–17. ACM Press (2015)
3. Aoki, K., Franke, J., Kleinjung, T., Lenstra, A.K., Osvik, D.A.: A kilobit special number field sieve factorization. In: Kurosawa, K. (ed.) ASIACRYPT 2007. LNCS, vol. 4833, pp. 1–12. Springer, Heidelberg (2007). doi:10.1007/978-3-540-76900-2_1
4. Ball, J., Borger, J., Greenwald, G.: Revealed: how US and UK spy agencies defeat internet privacy and security. The Guardian, 5 September 2013. https://www.theguardian.com/world/2013/sep/05/nsa-gchq-encryption-codes-security
5. Barbulescu, R.: Algorithmes de logarithmes discrets dans les corps finis. Ph.D. thesis, Université de Lorraine, France (2013)
6. Barker, E., Roginsky, A.: Transitions: recommendation for transitioning the use of cryptographic algorithms and key lengths. Technical report, National Institute of Standards and Technology (2011). http://nvlpubs.nist.gov/nistpubs/Legacy/SP/nistspecialpublication800-131a.pdf
7. Beckerman, B., Labahn, G.: A uniform approach for the fast computation of matrix-type Padé approximants. SIAM J. Matrix Anal. Appl. **15**(3), 804–823 (1994)
8. Bernstein, D.J., Chou, T., Chuengsatiansup, C., Hülsing, A., Lambooij, E., Lange, T., Niederhagen, R., van Vredendaal, C.: How to manipulate curve standards: a white paper for the black hat http://bada55.cr.yp.to. In: Chen, L., Matsuo, S. (eds.) SSR 2015. LNCS, vol. 9497, pp. 109–139. Springer, Heidelberg (2015). doi:10.1007/978-3-319-27152-1_6

9. Checkoway, S., Fredrikson, M., Niederhagen, R., Everspaugh, A., Green, M., Lange, T., Ristenpart, T., Bernstein, D.J., Maskiewicz, J., Shacham, H.: On the practical exploitability of Dual EC in TLS implementations. In: Fu, K. (ed.) Proceedings of USENIX Security 2014, pp. 319–335. USENIX, August 2014

10. Checkoway, S., Maskiewicz, J., Garman, C., Fried, J., Cohney, S., Green, M., Heninger, N., Weinmann, R.-P., Rescorla, E., Shacham, H.: A systematic analysis of the juniper dual EC incident. In: Weippl, E.R., Katzenbeisser, S., Kruegel, C., Myers, A.C., Halevi, S. (eds.) ACM CCS 2016: 23rd Conference on Computer and Communications Security, Vienna, Austria, 24–28 October 2016, pp. 468–479. ACM Press (2016)

11. Commeine, A., Semaev, I.: An algorithm to solve the discrete logarithm problem with the number field sieve. In: Yung, M., Dodis, Y., Kiayias, A., Malkin, T. (eds.) PKC 2006. LNCS, vol. 3958, pp. 174–190. Springer, Heidelberg (2006). doi:10. 1007/11745853_12

12. Coppersmith, D.: Modifications to the number field sieve. J. Cryptol. 6(3), 169–180 (1993)

13. Coppersmith, D.: Solving linear equations over GF(2) via block Wiedemann algorithm. Math. Comp. 62(205), 333–350 (1994)

14. Denis, T.S.: LibTomCrypt. http://www.libtom.net/

15. Desmedt, Y., Landrock, P., Lenstra, A.K., McCurley, K.S., Odlyzko, A.M., Rueppel, R.A., Smid, M.E.: The Eurocrypt'92 controversial issue trapdoor primes and moduli. In: Rueppel, R.A. (ed.) EUROCRYPT 1992. LNCS, vol. 658, pp. 194–199. Springer, Heidelberg (1993). doi:10.1007/3-540-47555-9_17

16. Durumeric, Z., Adrian, D., Mirian, A., Bailey, M., Halderman, J.A.: A search engine backed by internet-wide scanning. In: Ray, I., Li, N., Kruegel, C. (eds.) ACM CCS 2015: 22nd Conference on Computer and Communications Security, Denver, CO, USA, 12–16 October 2015, pp. 542–553. ACM Press (2015)

17. Enge, A., Gaudry, P., Thomé, E.: An $L(1/3)$ discrete logarithm algorithm for low degree curves. J. Cryptol. 24(1), 24–41 (2011)

18. Friedl, M., Provos, N., de Raadt, T., Steves, K., Miller, D., Tucker, D., McIntyre, J., Rice, T., Lindstrom, D.: Announce: OpenSSH 7.0 released, August 2015. http://www.openssh.com/txt/release-7.0

19. Gillmor, D.K.: Negotiated FFDHE for TLS, August 2016. https://datatracker.ietf.org/doc/rfc7919/

20. Giorgi, P., Lebreton, R.: Online order basis algorithm and its impact on the block Wiedemann algorithm. In: ISSAC 2014, pp. 202–209. ACM (2014)

21. Gordon, D.M.: Designing and detecting trapdoors for discrete log cryptosystems. In: Brickell, E.F. (ed.) CRYPTO 1992. LNCS, vol. 740, pp. 66–75. Springer, Heidelberg (1993). doi:10.1007/3-540-48071-4_5

22. Gordon, D.M.: Discrete logarithms in GF(p) using the number field sieve. SIAM J. Discret. Math. 6(1), 124–138 (1993)

23. Joux, A., Lercier, R.: Improvements to the general number field sieve for discrete logarithms in prime fields. A comparison with the gaussian integer method. Math. Comp. 72(242), 953–967 (2003)

24. Joux, A., Pierrot, C.: Nearly sparse linear algebra and application to discrete logarithms computations. In: Canteaut, A., Effinger, G., Huczynska, S., Panario, D., Storme, L. (eds.) Contemporary Developments in Finite Fields and Applications, pp. 119–144. World Scientific Publishing Company, Singapore (2016)

25. Juniper Networks: 2015-12 Out of Cycle Security Bulletin: ScreenOS: Multiple Security issues with ScreenOS (CVE-2015-7755, CVE-2015-7756), December 2015

26. Kaltofen, E.: Analysis of Coppersmith's block Wiedemann algorithm for the parallel solution of sparse linear systems. Math. Comp. **64**(210), 777–806 (1995)
27. Kleinjung, T., et al.: Factorization of a 768-bit RSA modulus. In: Rabin, T. (ed.) CRYPTO 2010. LNCS, vol. 6223, pp. 333–350. Springer, Heidelberg (2010). doi:10. 1007/978-3-642-14623-7_18
28. Kleinjung, T., Diem, C., Lenstra, A.K., Priplata, C., Stahlke, C.: Discrete logarithms in GF(p) - 768 bits. E-mail on the NMBRTHRY mailing list, 16 June 2016
29. Kolkman, O.M., Mekking, W.M., Gieben, R.M.: DNSSEC Operational Practices, Version 2. RFC 6781, Internet Society, December 2012
30. Larson, J., Perlroth, N., Shane, S.: Revealed: the NSA's secret campaign to crack, undermine internet security. ProPublica, 5 September 2013. https://www.pro publica.org/article/the-nsas-secret-campaign-to-crack-undermine-internet-encryp tion
31. Lenstra, A.K.: Constructing trapdoor primes for the proposed DSS. Technical report (1991). https://infoscience.epfl.ch/record/164559
32. Lenstra, A.K., Lenstra Jr., H.W. (eds.): The Development of the Number Field Sieve. LNM, vol. 1554. Springer, Heidelberg (1993)
33. Lepinski, M., Kent, S.: Additional Diffie-Hellman groups for use with IETF standards (2010). http://ietf.org/rfc/rfc5114.txt
34. Lim, C.H., Lee, P.J.: Generating efficient primes for discrete log cryptosystems (2006). http://citeseerx.ist.psu.edu/viewdoc/summary?doi=10.1.1.43.8261
35. Matyukhin, D.V.: On asymptotic complexity of computing discrete logarithms over $GF(p)$. Discret. Math. Appl. **13**(1), 27–50 (2003)
36. Menezes, A.J., van Oorschot, P.C., Vanstone, S.A.: Handbook of Applied Cryptography. CRC Press, Boca Raton (1997)
37. Murphy, B.A.: Polynomial selection for the number field sieve integer factorisation algorithm. Ph.D. thesis, Australian National University (1999)
38. National Institute of Standards and Technology: Supplemental ITL bulletin for september 2013:NIST opens draft special publication 800–90A, recommendation for random number generation using deterministic random bit generators, for review and comment. http://csrc.nist.gov/publications/nistbul/itlbul2013_09_ supplemental.pdf
39. National Institute of Standards and Technology: Examples for NIST 800–56A (2006). http://csrc.nist.gov/groups/ST/toolkit/documents/KS_FFC_Prime.pdf
40. National Institute of Standards and Technology: Digital signature standard (DSS, FIPS-186-4). Fourth revision (2013)
41. National Institute of Standards and Technology: Recommendation for pair-wise key establishment schemes using discrete logarithm cryptography, SP 800–56A, Second revision (2013)
42. Orman, H.: The Oakley key determination protocol. RFC 2412, November 1998
43. Perlroth, N., Larson, J., Shane, S.: N.S.A. able to foil basic safeguards of privacy on Web. The New York Times, 5 September 2013. http://www.nytimes.com/2013/ 09/06/us/nsa-foils-much-internet-encryption.html
44. Pomerance, C.: Analysis and comparison of some integer factoring algorithms. In: Lenstra Jr., H.W., Tijdeman, R. (eds.) Computational Methods in Number Theory Mathematical Center Tracts, vol. 154, pp. 89–140. Mathematisch Centrum, Amsterdam (1982)
45. Rivest, R., Hellman, M., Anderson, J.C., Lyons, J.W.: Responses to NIST's proposal. CACM **35**(7), 41–54 (1992)

46. Schirokauer, O.: Discrete logarithms and local units. Philos. Trans. Roy. Soc. Lond. Ser. A **345**(1676), 409–423 (1993)
47. Scott, M.: Re: NIST announces set of elliptic curves. `sci.crypt` newsgroup posting dated 1999/06/17. https://groups.google.com/forum/message/raw?msg=sci. crypt/mFMukSsORmI/FpbHDQ6hM_MJ
48. Semaev, I.A.: Special prime numbers and discrete logs in finite prime fields. Math. Comput. **71**(237), 363–377 (2002)
49. Smid, M.E., Branstad, D.K.: Response to comments on the NIST proposed digital signature standard. In: Brickell, E.F. (ed.) CRYPTO 1992. LNCS, vol. 740, pp. 76–88. Springer, Heidelberg (1993). doi:10.1007/3-540-48071-4_6
50. SSL Labs: SSL pulse. https://www.trustworthyinternet.org/ssl-pulse/
51. The CADO-NFS Development Team: CADO-NFS, an implementation of the number field sieve algorithm. Development version (prior to release 2.3) (2016). http:// cado-nfs.gforge.inria.fr/
52. Thomé, E.: Subquadratic computation of vector generating polynomials and improvement of the block Wiedemann algorithm. J. Symb. Comput. **33**(5), 757– 775 (2002)
53. Valenta, L., Adrian, D., Sanso, A., Cohney, S., Fried, J., Hastings, M., Halderman, J.A., Heninger, N.: The most dangerous groups in the world: exploiting DSA groups for Diffie-Hellman (2016)
54. Young, A., Yung, M.: The dark side of "Black-Box" cryptography or: should we trust Capstone? In: Koblitz, N. (ed.) CRYPTO 1996. LNCS, vol. 1109, pp. 89–103. Springer, Heidelberg (1996). doi:10.1007/3-540-68697-5_8

Multiparty Computation I

Improved Private Set Intersection Against Malicious Adversaries

Peter Rindal$^{(\boxtimes)}$ and Mike Rosulek

Oregon State University, Corvallis, USA
{rindalp,rosulekm}@eecs.oregonstate.edu

Abstract. Private set intersection (PSI) refers to a special case of secure two-party computation in which the parties each have a set of items and compute the intersection of these sets without revealing any additional information. In this paper we present improvements to practical PSI providing security in the presence of *malicious* adversaries.

Our starting point is the protocol of Dong, Chen & Wen (CCS 2013) that is based on Bloom filters. We identify a bug in their malicious-secure variant and show how to fix it using a cut-and-choose approach that has low overhead while simultaneously avoiding one the main computational bottleneck in their original protocol. We also point out some subtleties that arise when using Bloom filters in malicious-secure cryptographic protocols.

We have implemented our PSI protocols and report on its performance. Our improvements reduce the cost of Dong et al.'s protocol by a factor of $14 - 110\times$ on a single thread. When compared to the previous fastest protocol of De Cristofaro et al., we improve the running time by $8 - 24\times$. For instance, our protocol has an online time of 14 s and an overall time of 2.1 min to securely compute the intersection of two sets of 1 million items each.

1 Introduction

Private set intersection (PSI) is a cryptographic primitive that allows two parties holding sets X and Y, respectively, to learn the intersection $X \cap Y$ while not revealing any additional information about X and Y.

PSI has a wide range of applications: contact discovery [19], secret handshakes [12], measuring advertisement conversion rates, and securely sharing security incident information [22], to name a few.

There has been a great deal of recent progress in efficient PSI protocols that are secure against *semi-honest* adversaries, who are assumed to follow the protocol. The current state of the art has culminated in extremely fast PSI protocols. The fastest one, due to Kolesnikov et al. [16], can securely compute the intersection of two sets, each with 2^{20} items, in less than 4 s.

P. Rindal—Partially supported by NSF awards 1149647, 1617197 and a Google Research Award. The first author is also supported by an ARCS foundation fellowship.

© International Association for Cryptologic Research 2017
J.-S. Coron and J.B. Nielsen (Eds.): EUROCRYPT 2017, Part I, LNCS 10210, pp. 235–259, 2017.
DOI: 10.1007/978-3-319-56620-7_9

Looking more closely, the most efficient semi-honest protocols are those that are based on **oblivious transfer (OT) extension.** Oblivious transfer is a fundamental cryptographic primitive (see Fig. 1). While in general OT requires expensive public-key computations, the idea of OT extension [3,13] allows the parties to efficiently realize any number of *effective* OTs by using only a small number (e.g., 128) of *base OTs* plus some much more efficient symmetric-key computations. Using OT extension, oblivious transfers become extremely inexpensive in practice. Pinkas et al. [23] compared many paradigms for PSI and found the ones based on OTs are much more efficient than those based on algebraic & public-key techniques.

Our Contributions. In many settings, security against semi-honest adversaries is insufficient. *Our goal in this paper is to translate the recent success in semi-honest PSI to the setting of **malicious security.*** Following the discussion above, this means focusing on PSI techniques based on oblivious transfers. Indeed, recent protocols for OT extension against *malicious* adversaries [1,15] are almost as efficient as (only a few percent more expensive than) OT extension for semi-honest adversaries.

Our starting point is the protocol paradigm of Dong et al. [8] (hereafter denoted DCW) that is based on OTs and Bloom filter encodings. We describe their approach in more detail in Sect. 3. In their work they describe one of the few malicious-secure PSI protocols based primarily on OTs rather than algebraic public-key techniques. We present the following improvements and additions to their protocol:

1. Most importantly, we show that their protocol has a subtle security flaw, which allows a malicious sender to induce inconsistent outputs for the receiver. We present a fix for this flaw, using a very lightweight cut-and-choose technique.
2. We present a full simulation-based security proof for the Bloom-filter-based PSI paradigm. In doing so, we identify a subtle but important aspect about using Bloom filters in a protocol meant to provide security in the presence of malicious adversaries. Namely, the simulator must be able to extract all items stored in an adversarially constructed Bloom filter. We argue that this capability is an *inherently* non-standard model assumption, in the sense that it seems to require the Bloom filter hash functions to be modeled as (non-programmable) random oracles. Details are in Sect. 5.1.
3. We implement both the original DCW protocol and our improved version. We find that the major bottleneck in the original DCW protocol is not in the cryptographic operations, but actually in a polynomial interpolation computation. The absence of polynomial interpolation in our new protocol (along with our other improvements) decreases the running time by a factor of over $8 - 75\times$.

1.1 Related Work

As mentioned above, our work builds heavily on the protocol paradigm of Dong et al. [8] that uses Bloom filters and OTs. We discuss this protocol in great detail in Sect. 3. We identify a significant bug in that result, which was independently discovered by Lambæk [17] (along with other problems not relevant to our work).

Several other paradigms for PSI have been proposed. Currently the fastest protocols in the *semi-honest* setting are those in a sequence of works initiated by Pinkas et al. [16,22,23] that rely heavily on oblivious transfers. Adapting these protocols to the malicious setting is highly non-trivial, and we were unsuccessful in doing so. However, Lambæk [17] observes that the protocols can easily be made secure against a malicious *receiver* (but not also against a malicious sender).

Here we list other protocol paradigms that allow for malicious security when possible. The earliest technique for PSI is the elegant Diffie-Hellman-based protocol of [12]. Protocols in this paradigm achieving security against malicious adversaries include the one of De Cristofaro et al. [7]. We provide a performance analysis comparing their protocol to ours.

Freedman et al. [9] describe a PSI paradigm based on oblivious polynomial evaluation, which was extended to the malicious setting in [6].

Huang et al. [11] explored using general-purpose 2PC techniques (e.g., garbled circuits) for PSI. Several improvements to this paradigm were suggested in [22]. Malicious security can be achieved in this paradigm in a generic way, using any cut-and-choose approach, e.g., [18].

Kamara et al. [14] presented PSI protocols that take advantage of a semi-trusted server to achieve extremely high performance. Our work focuses on the more traditional setting with just 2 parties.

2 Preliminaries

We use κ to denote a computational security parameter (e.g., $\kappa = 128$ in our implementations), and λ to denote a statistical security parameter (e.g., $\lambda = 40$ in our implementations). We use $[n]$ to denote the set $\{1, \ldots, n\}$.

2.1 Efficient Oblivious Transfer

Our protocol makes use of 1-out-of-2 oblivious transfer (OT). The ideal functionality is described in Fig. 1. We require a large number of such OTs, secure against malicious adversaries. These can be obtained efficiently via OT extension [3]. The idea is to perform a fixed number (e.g., 128) of "base OTs", and from this correlated randomness derive a large number of effective OTs using only symmetric-key primitives.

The most efficient OT extension protocols providing malicious security are those of [2,15,21], which are based on the semi-honest secure paradigm of [13].

Parameters: ℓ is the length of the OT strings.

- Oninput($m_0, m_1) \in (\{0,1\}^\ell)^2$ from the sender and $b \in \{0,1\}$ from the receiver, give output m_b to the receiver.

Fig. 1. Ideal functionality for 1-out-of-2 OT

2.2 Private Set Intersection

In Fig. 2 we give the ideal functionality that specifies the goal of private set intersection. We point out several facts of interest. (1) The functionality gives output only to Bob. (2) The functionality allows corrupt parties to provide larger input sets than the honest parties. This reflects that our protocol is unable to strictly enforce the size of an adversary's set to be the same as that of the honest party. We elaborate when discussing the security of the protocol.

We define security of a PSI protocol using the standard paradigm of 2PC. In particular, our protocol is secure in the *universal composability (UC)* framework of Canetti [4]. Security is defined using the real/ideal, simulation-based paradigm that considers two interactions:

- In the **real interaction**, a malicious adversary $\mathcal{A}$ attacks an honest party who is running the protocol π. The honest party's inputs are chosen by an *environment* $\mathcal{Z}$; the honest party also sends its final protocol output to $\mathcal{Z}$. The environment also interacts arbitrarily with the adversary. Our protocols are in a *hybrid* world, in which the protocol participants have access to an ideal random-OT functionality (Fig. 1). We define REAL$[\pi, \mathcal{Z}, \mathcal{A}]$ to be the (random variable) output of $\mathcal{Z}$ in this interaction.
- In the **ideal interaction**, a malicious adversary $\mathcal{S}$ and an honest party simply interact with the ideal functionality $\mathcal{F}$ (in our case, the ideal PSI protocol of Fig. 2). The honest party simply forwards its input from the environment to $\mathcal{F}$ and its output from $\mathcal{F}$ to the environment. We define IDEAL$[\mathcal{F}, \mathcal{Z}, \mathcal{S}]$ to be the output of $\mathcal{Z}$ in this interaction.

We say that a protocol π **UC-securely realizes** functionality $\mathcal{F}$ if: for all PPT adversaries $\mathcal{A}$, there exists a PPT simulator $\mathcal{S}$, such that for all PPT environments $\mathcal{Z}$:

Parameters: σ is the bit-length of the parties' items. n is the size of the honest parties' sets. $n' > n$ is the allowed size of the corrupt party's set.

- On input $Y \subseteq \{0,1\}^\sigma$ from Bob, ensure that $|Y| \leq n$ if Bob is honest, and that $|Y| \leq n'$ if Bob is corrupt. Give output BOB-INPUT to Alice.
- Thereafter, on input $X \subseteq \{0,1\}^\sigma$ from Alice, likewise ensure that $|X| \leq n$ if Alice is honest, and that $|X| \leq n'$ if Alice is corrupt. Give output $X \cap Y$ to Bob.

Fig. 2. Ideal functionality for private set intersection (with one-sided output)

$$\text{REAL}[\pi, \mathcal{Z}, \mathcal{A}] \approx \text{IDEAL}[\mathcal{F}, \mathcal{Z}, \mathcal{S}]$$

where "$\approx$" denotes computational indistinguishability.

Our protocol uses a (non-programmable) random oracle. In Sect. 5.4 we discuss technicalities that arise when modeling such global objects in the UC framework.

2.3   Bloom Filters

A **Bloom filter (BF)** is an N-bit array B associated with k random functions $h_1, \ldots, h_k : \{0,1\}^* \rightarrow [N]$. To store an item x in the Bloom filter, one sets $B[h_i(x)] = 1$ for all i. To check the presence of an item x in the Bloom filter, one simply checks whether $B[h_i(x)] = 1$ for all i. Any item stored in the Bloom filter will therefore be detected when queried; however, *false positives* are possible.

3   The DCW Protocol Paradigm

The PSI protocol of Dong et al. [8] (hereafter DCW) is based on representing the parties' input sets as Bloom filters (BFs). We describe the details of their protocol in this section.

If B and B' are BFs for two sets S and S', using the same parameters (including the same random functions), then it is true that $B \wedge B'$ (bit-wise AND) is a BF for $S \cap S'$. However, one cannot construct a PSI protocol simply by computing a bit-wise AND of Bloom filters. The reason is that $B \wedge B'$ leaks more about S and S' than their intersection $S \cap S'$. For example, consider the case where $S \cap S' = \emptyset$. Then the most natural Bloom filter for $S \cap S'$ is an all-zeroes string, and yet $B \wedge B'$ may contain a few 1 s with noticeable probability. The location of these 1 s depends on the items in S and S', and hence cannot be simulated just by knowing that $S \cap S' = \emptyset$.

DCW proposed a variant Bloom filter that they call a **garbled Bloom filter** (GBF). In a GBF G meant to store m-bit strings, each $G[i]$ is itself an m-bit string rather than a single bit. Then an item x is stored in G by ensuring that $x = \bigoplus_i G[h_i(x)]$. That is, the positions indexed by hashing x should store additive secret shares of x. All other positions in G are chosen uniformly.

The **semi-honest** PSI protocol of DCW uses GBFs in the following way. The two parties agree on Bloom filter parameters. Alice prepares a GBF G representing her input set. The receiver Bob prepares a standard BF B representing his input set. For each position i in the Bloom filters, the parties use oblivious transfer so that Bob can learn $G[i]$ (a string) iff $B[i] = 1$. These are exactly the positions of G that Bob needs to probe in order to determine which of his inputs is stored in G. Hence Bob can learn the intersection. DCW prove that this protocol is secure. That is, they show that Bob's view $\{G[i] \mid B[i] = 1\}$ can be simulated given only the intersection of Alice and Bob's sets.

DCW also describe a **malicious-secure** variant of their GBF-based protocol. The main challenge is that nothing in the semi-honest protocol prevents a malicious Bob from learning *all* of Alice's GBF G. This would reveal Alice's entire input, which can only be simulated in the ideal world by Bob sending the entire universe $\{0,1\}^\sigma$ as input. Since in general the universe is exponentially large, this behavior is unsimulatable and hence constitutes an attack.

To prevent this, DCW propose to use 1-out-of-2 OTs in the following way. Bob can choose to either pick up a position $G[i]$ in Alice's GBF (if Bob has a 1 in $B[i]$) or else learn a value s_i (if Bob has a 0 in $B[i]$). The values s_i are an $N/2$-out-of-N secret sharing of some secret s^* which is used to encrypt all of the $G[i]$ values. Hence, Alice's inputs to the ith OT are $(s_i, \mathsf{Enc}(s^*, G[i]))$, where Enc is a suitable encryption scheme. Intuitively, if Bob tries to obtain too many positions of Alice's GBF (more than half), then he cannot recover the key s^* used to decrypt them.

As long as $N > 2k|Y|$ (where Y is Bob's input set), an honest Bob is guaranteed to have at least half of his BF bits set to zero. Hence, he can reconstruct s^* from the s_i shares, decrypt the $G[i]$ values, and probe these GBF positions to learn the intersection. We describe the protocol formally in Fig. 3.

Parameters: X is Alice's input, Y is Bob's input. N is the required Bloom filter size; We assume the parties have agreed on common BF parameters.

1. Alice chooses a random key $s^* \in \{0,1\}^\kappa$ and generates an $N/2$-out-of-N secret sharing $(s_1, \ldots, s_N)$.
2. Alice generates a GBF G encoding her inputs X. Bob generates a standard BF B encoding his inputs Y.
3. For $i \in [N]$, the parties invoke an instance of 1-out-of-2 OT, where Alice gives inputs $(s_i, c_i = \mathsf{Enc}(s^*, G[i]))$ and Bob uses choice bit $B[i]$.
4. Bob reconstructs s^* from the set of shares $\{s_i \mid B[i] = 0\}$ he obtained in the previous step. Then he uses s^* to decrypt the ciphertexts $\{c_i \mid B[i] = 1\}$, obtaining $\{G[i] \mid B[i] = 1\}$. Finally, he outputs $\{y \in Y \mid y = \bigoplus_i G[h_i(y)]\}$.

Fig. 3. The malicious-secure protocol of DCW [8].

3.1   Insecurity of the DCW Protocol

Unfortunately, the malicious-secure variant of DCW is not secure[1]! We now describe an a attack on their protocol, which was independently & concurrently discovered by Lambæk [17]. A corrupt Alice will generate s_i values that are *not* a valid $N/2$-out-of-N secret sharing. DCW do not specify Bob's behavior when obtaining invalid shares. However, we argue that no matter what Bob's behavior is (e.g., to abort in this case), Alice can violate the security requirement.

As a concrete attack, let Alice honestly generate shares s_i of s^*, but then change the value of s_1 in any way. She otherwise runs the protocol as instructed.

---

[1] We contacted the authors of [8], who confirmed that our attack violates malicious security.

If the first bit of Bob's Bloom filter is 1, then this deviation from the protocol is invisible to him, and Alice's behavior is indistinguishable from honest behavior. Otherwise, Bob will pick up s_1 which is not a valid share. If Bob aborts in this case, then his abort probability depends on whether his first BF bit is 1. The effect of this attack on Bob's output cannot be simulated in the ideal PSI functionality, so it represents a violation of security.

Even if we modify Bob's behavior to gracefully handle some limited number of invalid shares, there must be some threshold of invalid shares above which Bob (information theoretically) cannot recover the secret s^*. Whether or not Bob recovers s^* therefore depends on *individual bits* of his Bloom filter. And whether we make Bob abort or do something else (like output $\emptyset$) in the case of invalid shares, the result cannot be simulated in the ideal world. Lambæk [17] points out further attacks, in which Alice can cleverly craft shares and encryptions of GBF values to cause her effective input to depend on Bob's inputs (hence violating input independence).

4 Our Protocol

The spirit of DCW's malicious protocol is to restrict the adversary from setting too many 1 s in its Bloom filter, thereby learning too many positions in Alice's GBF. In this section, we show how to achieve the spirit of the DCW protocol using a lightweight cut-and-choose approach.

The high-level idea is to generate slightly more 1-out-of-2 OTs than the number of BF bits needed. Bob is supposed to use a limited number of 1 s for his choice bits. To check this, Alice picks a small random fraction of the OTs and asks Bob to prove that an appropriate number of them used choice bit 0. If Alice uses *random* strings as her choice-bit-0 messages, then Bob can prove his choice bit by simply reporting this string.[2] If Bob cannot prove that he used sufficiently many 0 s as choice bits, then Alice aborts. Otherwise, Alice has high certainty that the unopened OTs contain a limited number of choice bits 1.

After this cut-and-choose, Bob can choose a permutation that reorders the unopened OTs into his desired BF. In other words, if $c_1, \ldots, c_N$ are Bob's choice bits in the unopened OTs, Bob sends a random π such that $c_{\pi(1)}, \ldots, c_{\pi(N)}$ are the bits of his desired BF. Then Alice can send her GBF, masked by the choice-bit-1 OT messages permuted in this way.

We discuss the required parameters for the cut-and-choose below. However, we remark that the overhead is minimal. It increases the number of required OTs by only 1–10%.

4.1 Additional Optimizations

Starting from the basic outline just described, we also include several important optimizations. The complete protocol is described formally in Fig. 4.

---

[2] This *committing* property of an OT choice bit was pointed out by Rivest [24].

Parameters: X is Alice's input, Y is Bob's input. N_{bf} is the required Bloom filter size; k is the number of Bloom filter hash functions; N_{ot} is the number of OTs to generate. H is modeled as a random oracle with output length κ. The choice of these parameters, as well as others $\alpha, p_{chk}, N_{maxones}$, is described in Section 5.2.

1. **[setup]** The parties perform a secure coin-tossing subprotocol to choose (seeds for) random Bloom filter hash functions $h_1, \ldots, h_k : \{0,1\}^* \to [N_{bf}]$.
2. **[random OTs]** Bob chooses a random string $b = b_1 \ldots b_{N_{ot}}$ with an α fraction of 1s. Parties perform N_{ot} OTs of random messages (of length κ), with Alice choosing random strings $m_{i,0}, m_{i,1}$ in the ith instance. Bob uses choice bit b_i and learns $m_i^* = m_{i,b_i}$.
3. **[cut-and-choose challenge]** Alice chooses a set $C \subseteq [N_{ot}]$ by choosing each index with independent probability p_{chk}. She sends C to Bob. Bob aborts if $|C| > N_{ot} - N_{bf}$.
4. **[cut-and-choose response]** Bob computes the set $R = \{i \in C \mid b_i = 0\}$ and sends R to Alice. To prove that he used choice bit 0 in the OTs indexed by R, Bob computes $r^* = \bigoplus_{i \in R} m_i^*$ and sends it to Alice. Alice aborts if $|C| - |R| > N_{maxones}$ or if $r^* \neq \bigoplus_{i \in R} m_{i,0}$.
5. **[permute unopened OTs]** Bob generates a Bloom filter BF containing his items Y. He chooses a random injective function $\pi : [N_{bf}] \to ([N_{ot}] \setminus C)$ such that $BF[i] = b_{\pi(i)}$, and sends π to Alice.
6. **[randomized GBF]** For each item x in Alice's input set, she computes a summary value

$$K_x = H\left(x \,\middle\|\, \bigoplus_{i \in h_*(x)} m_{\pi(i),1} \right),$$

where $h_*(x) \stackrel{\text{def}}{=} \{h_i(x) : i \in [k]\}$. She sends a random permutation of $K = \{K_x \mid x \in X\}$.
7. **[output]** Bob outputs $\{y \in Y \mid H(y \,\|\, \bigoplus_{i \in h_*(y)} m_{\pi(i)}^*) \in K\}$.

Fig. 4. Malicious-secure PSI protocol based on garbled Bloom filters.

Random GBF. In their treatment of the *semi-honest* DCW protocol, Pinkas et al. [23] suggested an optimization that eliminates the need for Alice to send her entire masked GBF. Suppose the parties use 1-out-of-2 OT of *random* messages (i.e., the sender Alice does not choose the OT messages; instead, they are chosen randomly by the protocol/ideal functionality). In this case, the concrete cost of OT extension is greatly reduced (cf. [1]). Rather than generating a GBF of her inputs, Alice generates an array G where $G[i]$ is the random OT message in the ith OT corresponding to bit 1 (an honest Bob learns $G[i]$ iff the ith bit of his Bloom filter is 1).

Rather than arranging for $\bigoplus_i G[h_i(x)] = x$, as in a garbled BF, the idea is to let the G-values be random and have Alice directly send to Bob a **summary value** $K_x = \bigoplus_i G[h_i(x)]$ for each of her elements x. For each item y in Bob's input set, he can likewise compute K_y since he learned the values of G corresponding to 1s in his Bloom filter. Bob can check to see whether K_y is in the list of strings sent by Alice. For items x not stored in Bob's Bloom filter, the value K_x is random from his point of view.

Pinkas et al. show that this optimization significantly reduces the cost, since most OT extension protocols require less communication for OT of random messages. In particular, Alice's main communication now depends on the number of items in her set rather than the size of the GBF encoding her set. Although the optimization was suggested for the semi-honest variant of DCW, we point out that it also applies to the malicious variant of DCW and to our cut-and-choose protocol.

In the malicious-secure DCW protocol, the idea is to prevent Bob from seeing GBF entries unless he has enough shares to recover the key s^*. To achieve the same effect with a random-GBF, we let the choice-bit-1 OT messages be random (choice-bit-0 messages still need to be chosen messages: secret shares of s^*). These choice-bit-1 OT messages define a random GBF G for Alice. Then instead of sending a summary value $\bigoplus_i G[h_i(x)]$ for each x, Alice sends $[\bigoplus_i G[h_i(x)]] \oplus F(s^*, x)$, where F is a pseudorandom function. If Bob does not use choice-bit-0 enough, he does not learn s^* and all of these messages from Alice are pseudorandom.

In our protocol, we can let both OT messages be random, which significantly reduces the concrete overhead. The choice-bit-0 messages are used when Bob proves his choice bit in the cut-and-choose step. The choice-bit-1 messages are used as a random GBF G, and Alice sends summary values just as in the semi-honest variant.

We also point out that Pinkas et al. and DCW overlook a subtlety in how the summary values and the GBF should be constructed. Pinkas et al. specify the summary value as $\bigoplus_i G[h_i(x)]$ where h_i are the BF hash functions. Suppose that there is a collision involving two BF hash functions under the same x — that is, $h_i(x) = h_{i'}(x)$. Note that since the range of the BF hash functions is polynomial in size ($[N_{\mathsf{bf}}]$), such a collision is indeed possible with noticeable probability. When such a collision happens, the term $G[h_i(x)] = G[h_{i'}(x)]$ can cancel itself out from the XOR summation and the summary value will not depend on this term. The DCW protocol also has an analogous issue.[3] If the $G[h_i(x)]$ term was the only term unknown to the Bob, then the collision allows him to guess the summary value for an item x that he does not have. We fix this by computing the summary value using an XOR expression that eliminates the problem of colliding terms:

$$\bigoplus_{j \in h_*(x)} G[j], \qquad \text{where } h_*(x) \overset{\text{def}}{=} \{h_i(x) : i \in [k]\}.$$

Note that in the event of a collision among BF hash functions, we get $|h_*(x)| < k$.

---

[3] Additionally, if one strictly follows the DCW pseudocode then correctness may be violated in the event of a collision $h_i(x) = h_{i'}(x)$. If $h_i(x)$ is the first "free" GBF location then $G[h_i(x)]$ gets set to a value and then erroneously overwritten later.

Finally, for technical reasons, it turns out to be convenient in our protocol to define the summary value of x to be $H(x \| \bigoplus_{j \in h_*(x)} G[j])$ where H is a (non-programmable) random oracle.[4]

Hash Only "On Demand." In OT-extension for random messages, the parties compute the protocol outputs by taking a hash of certain values derived from the base OTs. Apart from the base OTs (whose cost is constant), these hashes account for essentially all the cryptographic operations in our protocol. We therefore modify our implementation of OT extension so that these hashes are not performed until the values are needed. In our protocol, only a small number (e.g., 1%) of the choice-bit-0 OT messages are ever used (for the cut-and-choose check), and only about half of the choice-bit-1 OT messages are needed by the sender (only the positions that would be 1 in a BF for the sender's input). Hence, the reduction in cost for the receiver is roughly 50%, and the reduction for the sender is roughly 75%. A similar optimization was also suggested by Pinkas et al. [23], since the choice-bit 0 messages are not used at all in the semi-honest protocol.

Aggregating Proofs-of-Choice-Bits. Finally, we can reduce the communication cost of the cut-and-choose step. Recall that Bob must prove that he used choice bit 0 in a sufficient number of OTs. For the ith OT, Bob can simply send $m_{i,0}$, the random output he received from the ith OT. To prove he used choice bit 0 for an entire *set I* of indices, Bob can simply send the single value $\bigoplus_{i \in I} m_{i,0}$, rather than sending each term individually.

Optimization for Programmable Random Oracles. The formal description of our protocol is one that is secure in the *non-programmable* random oracle model. However, the protocol can be significantly optimized by assuming a programmable random oracle. The observation is that Alice's OT input strings are always chosen randomly. Modern OT extension protocols natively give OT of random strings and achieve OT of chosen strings by sending extra correction data (cf. [1]). If the application allows the OT extension protocol itself to determine the sender's strings, then this additional communication can be eliminated. In practice, this reduces communication cost for OTs by a factor of 2.

We can model OT of random strings by modifying the ideal functionality of Fig. 1 to choose m_0, m_1 randomly itself. The OT extension protocol of [21] securely realizes this functionality in the presence of malicious adversaries, in the programmable random oracle model. We point out that even in the semi-honest model it is not known how to efficiently realize OT of strings randomly *chosen by the functionality*, without assuming a programmable random oracle.

---

[4] In practice H is instantiated with a SHA-family hash function. The XOR expression and x itself are each 128 bits, so both fit in a single SHA block.

5 Security

5.1 BF Extraction

The analysis in DCW argues for malicious security in a property-based manner, but does not use a standard simulation-based notion of security. This turns out to mask a non-trivial subtlety about how one can prove security about Bloom-filter-based protocols.

One important role of a simulator is to extract a corrupt party's input. Consider the case of simulating the effect of a corrupt Bob. In the OT-hybrid model the simulator sees Bob's OT choice bits as well as the permutation π that he sends in 5. Hence, the simulator can easily extract Bob's "effective" Bloom filter. However, the simulator actually needs to extract the receiver's *input set* that corresponds to that Bloom filter, so that it can send the set itself to the ideal functionality.

In short, the simulator must *invert* the Bloom filter. While invertible Bloom filters do exist [10], they require storing a significant amount of data beyond that of a standard Bloom filter. Yet this PSI protocol only allows the simulator to extract the receiver's OT choice bits, which corresponds to a *plain* Bloom filter. Besides that, in our setting we must invert a Bloom filter that may not have been honestly generated.

Our protocol achieves extraction by modeling the Bloom filter hash functions as (non-programmable) random oracles. The simulator must *observe* the adversary's queries to the Bloom filter hash functions.[5] Let Q be the set of queries made by the adversary to any such hash function. This set has polynomial size, so the simulator can probe the extracted Bloom filter to test each $q \in Q$ for membership. The simulator can take the appropriate subset of Q as the adversary's extracted input set. More details are given in the security proof below.

Simulation/extraction of a corrupt Alice is also facilitated by observing her oracle queries. Recall that the *summary value* of x is (supposed to be) $H(x \|$ $\bigoplus_{j \in h_*(x)} m_{\pi(j),1})$. Since H is a non-programmable random oracle, the simulator can obtain candidate x values from her calls to H.

More details about malicious Bloom filter extraction are given in the security proof in Sect. 5.3.

Necessity of Random Oracles. We show that random oracles are necessary, when using plain Bloom filters for a PSI protocol.

Lemma 1. *There is **no** PSI protocol that simultaneously satisfies the following conditions:*

- *The protocol is UC secure against malicious adversaries in the standard model.*
- *When Bob is corrupted in a semi-honest manner, the view of the simulator can be sampled given only on a Bloom filter representation of Bob's input.*
- *The parameters of the Bloom filter depend only on the number of items in the parties' sets, and in particular not on the bitlength of those items.*

---

[5] The simulator does not, however, require the ability to *program* the random oracle.

In our protocol, the simulator's indeed gets to see the receiver's OT choice bits, which correspond to a plain Bloom filter encoding of their input set. However, the simulator also gets to observe the receiver's random oracle queries, and hence the statement of the lemma does not apply.

The restriction about the Bloom filter parameters is natural. One important benefit of Bloom filters is that they do not depend on the bit-length of the items being stored.

Proof. Consider an environment that chooses a random set $S \subseteq \{0,1\}^\ell$ of size n, and gives it as input to both parties (ℓ will be chosen later). An adversary corrupts Bob but runs semi-honestly on input S as instructed. The environment outputs 1 if the output of the protocol is S (note that it does not matter if only one party receives output). In this real execution, the environment outputs 1 with overwhelming probability due to the correctness of the protocol.

We will show that if the protocol satisfies all three conditions in the lemma statement, then the environment will output 0 with constant probability in the ideal execution, and hence the protocol will be insecure.

Suppose the simulator for a corrupt Bob sees only a Bloom filter representation of Bob's inputs. Let N be the total length of the Bloom filter representation (the Bloom filter array itself as well as the description of hash functions). Set the length of the input items $\ell > 2N$. Now the simulator's view can be sampled given only N bits of information about S, whereas S contains randomly chosen items of length $\ell > 2N$. The simulator must extract a value S' and send it on behalf of Bob to the ideal functionality. With constant probability this S' will fail to include some item of S (it will likely not include any of them). Then since the honest party gave input S, the output of the functionality will be $S \cap S' \neq S$, and the environment outputs zero.

5.2   Cut-and-Choose Parameters

The protocol mentions various parameters:

N_{ot}: the number of OTs
N_{bf}: the number of Bloom filter bits
k: the number of Bloom filter hash functions
α: the fraction of 1s among Bob's choice bits
p_{chk}: the fraction of OTs to check
$N_{maxones}$: the maximum number of 1 choice bits allowed to pass the cut-and-choose.

As before, we let κ denote the computational security parameter and λ denote the statistical security parameter.

We require the parameters to be chosen subject to the following constraints:

– *The cut-and-choose restricts Bob to few 1s.* Let N_1 denote the number of OTs that remain after the cut and choose, in which Bob used choice bit 1. In the security proof we argue that the difficulty of finding an element stored in the

Bloom filter *after the fact* is $(N_1/N)^k$ (i.e., one must find a value which all k random Bloom filter hash functions map to a 1 in the BF).

Let $\mathcal{B}$ denote the "bad event" that no more than N_{maxones} of the checked OTs used choice bit one (so Bob can pass the cut-and-choose), and yet $(N_1/N_{\text{bf}})^k \geq 2^{-\kappa}$. We require $\Pr[\mathcal{B}] \leq 2^{-\lambda}$.

As mentioned above, the spirit of the protocol is to restrict a corrupt receiver from setting too many 1 s in its (plain) Bloom filter. DCW suggest to restrict the receiver to 50% 1s, but do not explore how the fraction of 1 s affects security (except to point out that 100% 1s is problematic). Our analysis pinpoints precisely how the fraction of 1 s affects security.

- *The cut-and-choose leaves enough OTs unopened for the Bloom filter.* That is, when choosing from among N_{ot} items, each with independent p_{chk} probability, the probability that less than N_{bf} remain unchosen is at most $2^{-\lambda}$.
- *The honest Bob has enough one choice bits after the cut and choose.* When inserting n items into the bloom filter, at most nk bits will be set to one. We therefore require that no fewer than this remain after the cut and choose.

Our main technique is to apply the Chernoff bound to the probability that Bob has too many 1 s after the cut and choose. Let $m_h^1 = \alpha N_{\text{ot}}$ (resp. $m_h^0 = (1 - \alpha)N_{\text{ot}}$) be the number of 1 s (resp. 0 s) Bob is supposed to select in the OT extension. Then in expectation, there should be $m_h^1 p_{\text{chk}}$ ones in the cut and choose open set, where each OT message is opened with independent probability p_{chk}. Let ϕ denote the number of ones in the open set. Then applying the Chernoff bound we obtain,

$$\Pr[\phi \geq (1 + \delta)m_h^1 p_{\text{chk}}] \leq e^{-\frac{\delta^2}{2+\delta}m_h^1 p_{\text{chk}}} \leq 2^{-\lambda}$$

where the last step bounds this probability to be negligible in the statistical security parameter λ. Solving for δ results in,

$$\delta \leq \frac{\lambda + \sqrt{\lambda^2 + 8\lambda m_h^1 p_{\text{chk}}}}{2m_h^1 p_{\text{chk}}}.$$

Therefore an honest Bob should have no more than $N_{\text{maxones}} = (1 + \delta)m_h^1 p_{\text{chk}}$ 1s revealed in the cut and choose, except with negligible probability. To ensure there are at least nk ones[6] remaining to construct the bloom filter, set $m_h^1 = nk + N_{\text{maxones}}$. Similarly, there must be at least N_{bf} unopened OTs which defines the total number of OTs to be $N_{\text{ot}} = N_{\text{bf}} + (1 + \delta^*)N_{\text{ot}}p_{\text{chk}}$ where δ^* is analogous to δ except with respect to the total number of OTs opened in the cut and choose.

A malicious Bob can instead select $m_a^1 \geq m_h^1$ ones in the OT extension. In addition to Bob possibly setting more 1 s in the BF, such a strategy will increase the probability of the cut and choose revealing more than N_{maxones} 1s. A Chernoff bound can then be applied to the probability of seeing a δ' factor fewer 1s than

---

[6] nk ones is an upper bound on the number of ones required. A tighter analysis could be obtained if collisions were accounted for.

expected. Bounding this to be negligible in the statistical security parameter λ, we obtain,

$$\Pr[\phi \leq (1 - \delta')p_{\mathsf{chk}}m_a^1] \leq e^{-\frac{\delta'^2}{2}p_{\mathsf{chk}}m_a^1} \leq 2^{-\lambda}.$$

Solving for δ' then yields $\delta' \leq \sqrt{\frac{2\lambda}{p_{\mathsf{chk}}m_a^1}}$. By setting N_{maxones} equal to $(1 - \delta')p_{\mathsf{chk}}m_a^1$ we can solve for m_a^1 such that the intersection of these two distribution is negligible. Therefore the maximum number of 1 s remaining is $N_1 = (1 - p_{\mathsf{chk}})m_a^1 + \sqrt{2\lambda p_{\mathsf{chk}}m_a^1}$.

For a given p_{chk}, n, k, the above analysis allows us to bound the maximum advantage a malicious Bob can have. In particularly, a honest Bob will have at least nk 1 s and enough 0 s to construct the bloom filter while a malicious Bob can set no more than N_1/N_{bf} fraction of bits in the bloom filter to 1. Modeling the bloom filter hash function as random functions, the probability that all k index the boom filter one bits is $(N_1/N_{\mathsf{bf}})^k$. Setting this to be negligible in the computational security parameter κ we can solve for N_{bf} given N_1 and k. The overall cost is therefore $\frac{N_{\mathsf{bf}}}{(1-p_{\mathsf{chk}})}$. By iterating over values of k and p_{chk} we obtain set of parameters shown in Fig. 5.

n	p_{chk}	k	N_{ot}	N_{bf}	α	N_{maxones}
2^8	0.099	94	99,372	88,627	0.274	3,182
2^{12}	0.053	94	1,187,141	1,121,959	0.344	22,958
2^{16}	0.024	91	16,992,857	16,579,297	0.360	150,181
2^{20}	0.010	90	260,252,093	257,635,123	0.366	962,092

Fig. 5. Optimal Bloom filter cut and choose parameters for set size n to achieve statistical security $\lambda = 40$ and computational security $\kappa = 128$. N_{ot} denotes the total number of OTs used. N_{bf} denotes the bit count of the bloom filer. α is the faction of ones which should be generated. N_{maxones} is the maximum number of ones in the cut and choose to pass.

5.3   Security Proof

Theorem 2. *The protocol in Fig. 4 is a UC-secure protocol for PSI in the random-OT-hybrid model, when H and the Bloom filter hash functions are non-programmable random oracles, and the other protocol parameters are chosen as described above.*

Proof. We first discuss the case of a corrupt receiver Bob, which is the more difficult case since we must not only extract Bob's input but simulate the output. The simulator behaves as follows:

The simulator plays the role of an honest Alice and ideal functionalities in steps 1 through 5, but also extracts all of Bob's choice bits b for the OTs. Let N_1 be the number of OTs with choice bit 1 that remain after the cut and choose. The simulator artificially aborts if Bob succeeds at the cut and choose and yet $(N_1/N_{\mathsf{bf}})^k \geq 2^{-\kappa}$. From the choice of parameters, this event happens with probability only $2^{-\lambda}$.

After receiving Bob's permutation π in step 5, the simulator computes Bob's effective Bloom filter $BF[i] = b_{\pi(i)}$. Let Q be the set of queries made by Bob to *any* of the Bloom filter hash functions (random oracles). The simulator computes $\tilde{Y} = \{q \in Q \mid \forall i : BF[h_i(q)] = 1\}$ as Bob's effective input, and sends $\tilde{Y}$ to the ideal functionality. The simulator receives $Z = X \cap \tilde{Y}$ as output, as well as $|X|$. For $z \in Z$, the simulator generates $K_z = H(z \parallel \bigoplus_{j \in h_*(z)} m_{\pi(j),1})$. The simulator sends a random permutation of K_z along with $|X| - |Z|$ random strings to simulate Alice's message in step 6.

To show the soundness of this simulation, we proceed in the following sequence of hybrids:

1. The first hybrid is the real world interaction. Here, an honest Alice also queries the random oracles on her actual inputs $x \in X$. For simplicity later on, assume that Alice queries her random oracle as late as possible (in step 6 only).
2. In the next hybrid, we artifically abort in the event that $(N_1/N_{bf})^k \geq 2^{-\kappa}$. As described above, our choice of parameters ensures that this abort happens with probability at most $2^{-\lambda}$, so the hybrids are indistinguishable.
   In this hybrid, we also observe Bob's OT choice bits. Then in step 5 of the protocol, we compute Q, BF, and $\tilde{Y}$ as in the simulator description above.
3. We next consider a sequence of hybrids, one for each item x of Alice such that $x \in X \setminus \tilde{Y}$. In each hybrid, we replace the summary value $K_x = H(x \parallel \bigoplus_{j \in h_*(x)} m_{\pi(j),1})$ with a uniformly random value.
   There are two cases for $x \in X \setminus \tilde{Y}$:
   – Bob queried some h_i on x before step 5: If this happened but x was not included in $\tilde{Y}$, then x is *not* represented in Bob's effective Bloom filter BF. There must be an i such that Bob did not learn $m_{\pi(h_i(x)),1}$.
   – Bob did *not* query any h_i on x: Then the value of $h_i(x)$ is random for all i. The probability that x is present in BF is the probability that $BF[h_i(x)] = 1$ for all i, which is $(N_1/N_{bf})^k$ since Bob's effective Bloom filter has N_1 ones. Recall that the interaction is already conditioned on the event that $(N_1/N_{bf})^k < 2^{-\kappa}$. Hence it is with overwhelming probability that Bob did not learn $m_{\pi(h_i(x)),1}$ for some i.
   In either case, there is an i such that Bob did not learn $m_{\pi(h_i(x)),1}$, so that value is random from Bob's view. Then the corresponding sum $\bigoplus_{j \in h_*(x)} m_{\pi(j),1}$ is uniform in Bob's view.[7] It is only with negligible probability that Bob makes the oracle query $K_x = H(x \parallel \bigoplus_{j \in h_*(x)} m_{\pi(j),1})$. Hence K_x is pseudorandom and the hybrids are indistinguishable.

In the final hybrid, the simulation does not need to know X, it only needs to know $X \cap \tilde{Y}$. In particular, the values $\{K_x \mid x \in X \setminus \tilde{Y}\}$ are now being

---

[7] This is part of the proof that breaks down if we compute a summary value using $\bigoplus_i m_{\pi(h_i(x)),1}$ instead of $\bigoplus_{j \in h_*(x)} m_{\pi(j),1}$. In the first expression, it may be that $h_{i'}(x) = h_i(x)$ for some $i' \neq i$ so that the randomizing term $m_{\pi(h_i(x)),1}$ cancels out in the sum.

simulated as random strings. The interaction therefore describes the behavior of our simulator interacting with corrupt Bob.

Now consider a corrupt Alice. The simulation is as follows:

The simulator plays the role of an honest Bob and ideal functionalities in steps 1 through 4. As such, the simulator knows Alice's OT messages $m_{i,b}$ for all i, b, and can compute the correct r^* value in step 4. The simulator sends a completely random permutation π in step 5.

In step 6, the simulator obtains a set K as Alice's protocol message. Recall that each call made to random oracle H has the form $q\|s$. The simulator computes $Q = \{q \mid \exists s : \text{Alice queried } H \text{ on } q\|s\}$. The simulator computes $\tilde{X} = \{q \in Q \mid H(q \parallel \bigoplus_{j\in h_*(q)} m_{\pi(j),1}) \in K\}$ and sends $\tilde{X}$ to the ideal functionality as Alice's effective input. Recall Alice receives no output.

It is straight-forward to see that Bob's protocol messages in steps 4 & 5 are distributed independently of his input.

Recall that Bob outputs $\{y \in Y \mid H(y \parallel \bigoplus_{j\in h_*(y)} m^*_{\pi(j)}) \in K\}$ in the last step of the protocol. In the ideal world (interacting with our simulator), Bob's output from the functionality is $\tilde{X} \cap Y = \{y \in Y \mid y \in \tilde{X}\}$. We will show that the two conditions are the same except with negligible probability. This will complete the proof.

We consider two cases:

- If $y \in \tilde{X}$, then $H(y \parallel \bigoplus_{j\in h_*(y)} m^*_{\pi(j)}) = H(y \parallel \bigoplus_{j\in h_*(y)} m_{\pi(j),1}) \in K$ by definition.
- If $y \notin \tilde{X}$, then Alice never queried the oracle $H(y\|\cdot)$ before fixing K, hence $H(y \parallel \bigoplus_{j\in h_*(y)} m^*_{\pi(j)})$ is a fresh oracle query, distributed independently of K. The output of this query appears in K with probability $|K|/2^\kappa$.

Taking a union bound over $y \in Y$, we have that, except with probability $|K||Y|/2^\kappa$,

$$H(y \parallel \bigoplus_{j\in h_*(y)} m^*_{\pi(j)}) \in K \iff y \in \tilde{X}$$

Hence Bob's ideal and real outputs coincide.

Size of the Adversary's Input Set. When Alice is corrupt, the simulator extracts a set $\tilde{X}$. Unless the adversary has found a collision under random oracle H (which is negligibly likely), we have that $|\tilde{X}| \le |K|$. Thus the protocol enforces a straightforward upper bound on the size of a corrupt Alice's input.

The same is not true for a corrupt Bob. The protocol enforces an upper bound only on the size on Bob's *effective Bloom filter* and a bound on the number of 1 s in that BF. We now translate these bounds to derive a bound on the size of the set extracted by the simulator. Note that the ideal functionality for PSI (Fig. 2) explicitly allows corrupt parties to provide larger input sets than honest parties.

First, observe that only queries made by the adversary before step 5 of the protocol are relevant. Queries made by the adversary *after* do not affect the simulator's extraction. As in the proof, let Q be the set of queries made by Bob before step 5. Bob is able to construct a BF with at most N_1 ones, and causing the simulator to extract items $\tilde{Y} \subseteq Q$, only if:

$$\left| \bigcup_{y \in \tilde{Y}; i \in [k]} h_i(y) \right| \leq N_1.$$

Then by a union bound over all Bloom filters with N_1 bits set to 1, and all $\tilde{Y} \subseteq Q$ of size $|\tilde{Y}| = n'$, we have:

$$\Pr \left[\begin{array}{c} \text{simulator extracts} \\ \text{some set of size } n' \end{array} \right] \leq \binom{|Q|}{n'} \binom{N_{bf}}{N_1} \left(\frac{N_1}{N_{bf}} \right)^{kn'}.$$

The security proof already conditions on the event that $(N_1/N_{bf})^k \leq 2^{-\kappa}$, so we get:

$$\Pr \left[\begin{array}{c} \text{simulator extracts} \\ \text{some set of size } n' \end{array} \right] \leq \binom{|Q|}{n'} \binom{N_{bf}}{N_1} 2^{-\kappa n'}$$

$$\leq \left(|Q|^{n'} \right) \left(2^{N_{bf}} \right) 2^{-\kappa n'}$$

To make the probability less than $2^{-\kappa}$ it therefore suffices to have $n' = (\kappa + N_{bf})/(\kappa - \log |Q|)$.

In our instantiations, we always have $N_{bf} \leq 3\kappa n$, where n denotes the *intended* size of the parties' sets. Even in the pessimistic case that the adversary makes $|Q| = 2^{\kappa/2}$ queries to the Bloom filter hash functions, we have $n' \approx 6n$. Hence, the adversary is highly unlikely to produce a Bloom filter containing 6 times the intended number of items. We emphasize that this is a very loose bound, but show it just to demonstrate that the simulator indeed extracts from the adversary a modestly sized effective input set.

5.4 Non-Programmable Random Oracles in the UC Model

Our protocol makes significant use of a non-programmable random oracle. In the standard UC framework [4], the random oracle must be treated as *local* to each execution for technical reasons. The UC framework does not deal with global objects like a single random oracle that is used by many protocols/instances. Hence, as currently written, our proof implies security when instantiated with a highly local random oracle.

Canetti et al. [5] proposed a way to model global random oracles in the UC framework (we refer to their model as UC-gRO). One of the main challenges is that (in the plain UC model) the simulator can observe the adversary's oracle queries, but an adversary can ask the environment to query the oracle on its behalf, hidden from the simulator. In the UC model, every functionality and party in the UC model is associated with a *session id* (sid) for the protocol instance in which it participates. The idea behind UC-gRO is as follows:

- There is a functionality gRO that implements an ideal random oracle. Furthermore, this functionality is **global** in the sense that all parties and all functionalities can query it.
- Every oracle query in the system must be prefixed with some sid.
- There is no enforcement that oracle queries are made with the "correct" sid. Rather, if a party queries gRO with a sid that does not match its own, that query is marked as **illegitimate** by gRO.
- A functionality can ask gRO for all of the illegitimate queries made using that functionality's sid.

Our protocol and proof can be modified in the following ways to provide security in the UC-gRO model:

1. In the protocol, all queries to relevant random oracles (Bloom filter functions h_i and outer hash function H) are prefixed with the sid of this instance.
2. The ideal PSI functionality is augmented in a standard way of UC-gRO: When the adversary/simulator gives the functionality a special command `illegitimate`, the functionality requests the list of illegitimate queries from gRO and forwards them to the adversary/simulator.
3. In the proof, whenever the simulator is described as obtaining a list of the adversary's oracle queries, this is done by observing the adversary's queries and also obtaining the illegitimate queries via the new mechanism.

With these modifications, our proof demonstrates security in the UC-gRO model.

6   Performance Evaluation

We implemented our protocol in addition to the protocols of DCW [8] outlined in Sect. 3 and that of DKT [7]. In this section we report on their performance and analyze potential trade offs.

6.1   Implementation and Test Platform

In the offline phase, our protocol consists of performing 128 base OTs using the protocol of [20]. We extend these base OTs to N_{ot} OTs using an optimized implementation of the Keller et al. [15] OT extension protocol. Our implementation uses the programmable-random-oracle optimization for OT of random strings, described in Sect. 4.1. In the multi-threaded case, the OT extension and Base OTs are performed in parallel. Subsequently, the cut and choose seed is published which determines the set of OT messages to be opened. Then one or more threads reports the choice bits used for the corresponding OT and the XOR sum of the messages. The sender validates the reported value and proceeds to the online phase.

The online phase begins with both parties inserting items into a plaintext bloom filter using one or more threads. As described in Sect. 5.1, the BF hash functions should be modeled as (non-programmable) random oracles. We use

SHA1 as a random oracle but then expand it to a suitable length via a fast PRG (AES in counter mode) to obtain:[8]

$$h_1(x)\|h_2(x)\| \cdots \|h_k(x) = \mathsf{PRG}(\mathsf{SHA1}(x)).$$

Hence we use just one (slow) call to SHA to compute all BF hash functions for a single element, which significantly reduces the time for generating Bloom filters. Upon the computing the plaintext bloom filter, the receiver selects a random permutation mapping the random OT choice bits to the desired bloom filter. The permutation is published and the sender responds with the random garbled bloom filter masks which correspond to their inputs. Finally, the receiver performs a plaintext intersection of the masks and outputs the corresponding values.

We evaluated the prototype on a single server with simulated network latency and bandwidth. The server has 2 36-cores Intel(R) Xeon(R) CPU E5-2699 v3 @ 2.30 GHz and 256 GB of RAM (e.i. 36 cores & 128 GB per party). We executed our prototype in two network settings: a LAN configuration with both parties in the same network with 0.2 ms round-trip latency, 1 Gbps; and a WAN configuration with a simulated 95 ms round-trip latency, 60 Mbps. All experiments we performed with a computational security parameter of $\kappa = 128$ and statistical security parameter $\lambda = 40$. The times reported are an average over 10 trials. The variance of the trials was between 0.1%–5.0% in the LAN setting and 0.5%–10% in the WAN setting with a trend of smaller variance as n becomes larger. The CPUs used in the trials had AES-NI instruction set for fast AES computations.

6.2 Parameters

We demonstrate the scalability of our implementation by evaluating a range of set sizes $n \in \{2^8, 2^{12}, 2^{16}, 2^{20}\}$ for strings of length $\sigma = 128$. In all of our tests, we use system parameters specified in Fig. 5. The parameters are computed using the analysis specified in Sect. 5.2. Most importantly they satisfy that except with probability negligible in the computation security parameter κ, a receiver after step 5 of Fig. 4 will not find an x not previously queried which is contained in the garbled bloom filter.

The parameters are additionally optimized to reduce the overall cost of the protocol. In particular, the total number of OTs $N_{ot} = N_{bf}/(1 - p_{chk})$ is minimized. This value is derived by iterating over all the region of $80 \le k \le 100$ hash functions and cut-and-choose probabilities $0.001 \le p_{chk} \le 0.1$. For a given value of n, k, p_{chk}, the maximum number of ones N_1 which a possibly malicious receiver can have after the cut and choose is defined as shown in Sect. 5.2. This in turn determines the minimum value of N_{bf} such that $(N_{bf}/N_1)^{-k} \le 2^{-\kappa}$ and therefore the overall cost N_{ot}. We note that for κ other than 128, a different range for the number of hash functions should be considered.

---

[8] Note that if we model SHA1 as having its queries observable to the simulator, then this property is inherited also when expanding the SHA1 output with a PRG.

6.3   Comparison to Other Protocols

For comparison, we implemented two other protocol paradigms, which we describe here:

DCW Protocol. Our first point of comparison is to the protocol of Dong et al. [8], on which ours is based. The protocol is described in Sect. 3. While their protocol has issues with its security, our goal here is to illustrate that our protocol also has significantly better performance.

In [8], the authors implement only their semi-honest protocol variant, not the malicious one. An aspect of the malicious DCW protocol that is easy to overlook is its reliance on an $N/2$-out-of-N secret sharing scheme. When implementing the protocol, it becomes immediately clear that such a secret-sharing scheme is a major computational bottleneck.

Recall that the sender generates shares from such a secret sharing scheme, and the receiver reconstructs such shares. In this protocol, the required N is the number of bits in the Bloom filter. As a concrete example, for PSI of sets of size 2^{20}, the Bloom filter in the DCW protocol has roughly 2^{28} bits. Using Shamir secret sharing, the sender must evaluate a random polynomial of degree $\sim 2^{27}$ on $\sim 2^{28}$ points. The sender must interpolate such a polynomial on $\sim 2^{27}$ points to recover the secret. Note that the polynomial will be over $GF(2^{128})$, since the protocol secret-shares an (AES) encryption key.

We chose not to develop a full implementation of the malicious DCW protocol. Rather, we fully implemented the [garbled] Bloom filter encoding steps and the OTs. We then **simulated** the secret-sharing and reconstruction steps in the following way. We calculated the number of field multiplications that would be required to evaluate a polynomial of the suitable degree by the Fast Fourier Transform (FFT) method, and simply had each party perform the appropriate number of field multiplications in $GF(2^{128})$. The field was instantiated using the NTL library with all available optimizations enabled. Our simulation significantly underestimates the cost of secret sharing in the DCW protocol, since: (1) it doesn't account for the cost associated with virtual memory accesses when computing on such a large polynomial; and (2) evaluating/interpolating the polynomial via FFT reflects a *best-case scenario*, when the points of evaluation are roots of unity. In the protocol, the receiver Bob in particular does not have full control over which points of the polynomial he will learn.

Despite this optimistic simulation of the secret-sharing step, its cost is substantial, accounting for 97% of the execution time. In particular, when comparing our protocol to the DCW protocol, the main difference in the online phase is the secret sharing reconstruction which accounts for a 113× increase in the online running time for $n = 2^{16}$.

We simulated two variants of the DCW malicious-secure protocol. One variant reflects the DCW protocol as written, using OTs of chosen messages. The other variant includes the "random GBF" optimization inspired by [23] and described in Sect. 4. In this variant, one of the two OT messages is set randomly by the protocol itself, and not chosen by the sender. This reduces the online communication cost of the OTs by roughly half. However, it surprisingly has a slight negative effect on total time. The reason is that during the online phase Alice has more than enough time to construct and send a plain GBF while Bob performs the more time intensive secret-share reconstruction step. For $n = 2^{16}$, the garbled bloom filter takes less than 5% of the secret share reconstruction time to be sent. When using a randomized GBF, Alice sends summary values to Bob, which he must compare to his own summary values. Note that there is a summary value for each item in a party's set (e.g., 2^{20}), so these comparisons involve lookups in some non-trivial data structure. This extra computational effort is part of the the critical path since the Bob has to do it. In summary, the "random GBF" optimization does reduce the required communication, however it also increases the critical path of the protocol due to the secret-share reconstruction hiding the effects of this communication savings and the small additional overhead of performing n lookups.

DH-Based PSI Protocols. Another paradigm for PSI uses public-key techniques and is based on Diffie-Hellman-type assumptions in cyclic groups. The most relevant protocol in this paradigm that achieves malicious security is that of De Cristofaro et al. [7] which we refer to as DKT. While protocols in this paradigm have extremely low communication complexity, they involve a large number of computationally expensive public-key operations (exponentiations). Another potential advantage of the DKT protocol over schemes based on Bloom filters is that the receiver can be restricted to a set size of exactly n items. This is contrasted with our protocol where the receiver can have a set size of $n' \approx 6n$.

We fully implemented the [7] PSI protocol both in the single and multi threaded setting. In this protocol, the parties perform $5n$ exponentiations and $2n$ related zero knowledge proofs of discrete log equality. Following the suggestions in [7], we instantiate the zero knowledge proofs in the RO model with the Fiat-Shamir transform applied to a sigma protocol. The resulting PSI protocol has in total $12n$ exponentiations along with several other less expensive group operations. The implementation is built on the Miracl elliptic curve library using Curve 25519 achieving 128 bit computational security. The implementation also takes advantage of the Comb method to perform a precomputation to increase the speed of exponentiations (point multiplication). Additionally, all operations are performed in a streaming manner allowing for the greatest amount of work to be performed concurrently by the parties.

6.4 Results

The running time of our implementation is shown in Fig. 7. We make the distinction of reporting the running times for both the total time and online phase when applicable. The offline phase contains all operations which are independent of the input sets. For the bloom filter based protocols the offline phase consists of performing the OT extension and the cut and choose. Out of these operations, the most time-consuming is the OT extension. For instance, with $n = 2^{20}$ we require 260 million OTs which requires 124 s; the cut and choose takes only 3 s.

For the smaller set size of $n = 2^{12}$, the OT extension required 461 ms and the cut and choose completed in 419 ms. The relative increase in the cut and choose running time is primarily due to the need to open a larger portion of the OTs when n is smaller.

The online phase consists of the receiver first computing their bloom filter. For set size $n = 2^{20}$, computing the bloom filter takes 6.4 s. The permutation mapping the receiver's OTs to the bloom filter then computed in less than a second and sent. Upon receiving the permutation, the sender computes their PSI summary values and sends them to the receiver. This process when $n = 2^{20}$ takes roughly 6 s. The receiver then outputs the intersection in less than a second.

As expected, our optimized protocol achieves the fastest running times compared to the other malicious secure constructions. When evaluating our implementation with a set size of $n = 2^8$ on a single thread in the LAN setting, we obtain an online running time of 3 ms and an overall time of 0.2 s. The next fastest is that of DH-based DKT protocol which required 1.7 s, an 8.5× slowdown compared to our protocol. For the larger set size of $n = 2^{12}$, our overall running time is 0.9 s with an online phase of just 40 ms. The DKT protocol is again the next fastest requiring 25× longer resulting in a total running time of 22.6 s. The DCW protocol from which ours is derived incurs more than a 60× overhead. For the largest set size performed of $n = 2^{20}$, our protocol achieves an online phase of 14 s and an overall time of 127 s. The DKT protocol overall running time was more than 95 min, a 47× overhead compared to our running time. The DCW protocol took prohibitively long to run but is expected to take more than 100× longer than our optimized protocol.

When evaluating our protocol in the WAN setting with 95 ms round trip latency our protocol again achieves the fastest running times. For the small set size of $n = 2^8$, the protocol takes an overall running time of 0.95 s with the online phase taking 0.1 s. DKT was the next fastest protocol requiring a total time of 1.7 s, an almost 2× slowdown. Both variants of the DCW protocol experience a more significant slowdown of roughly 4×. When increasing the set size, our protocol experiences an even greater relative speedup. For $n = 2^{16}$, our protocol takes 56 s, with 11 of the seconds consisting of the online phase. Comparatively, DKT takes 393 s resulting in our protocol being more than 7× faster. The DCW protocols are even slower requiring more than 19 min, a 20× slowdown. This is primarily due to the need to perform the expensive secret-sharing operations and send more data.

In addition to faster serial performance, our protocol also benefits from easily being parallelized, unlike much of the DCW online phase. Figure 6 shows the running times of our protocol and that of DKT when parallelized using p threads per party in the LAN setting. With $p = 4$ we obtain a speedup of 2.3× for set size $n = 2^{16}$ and 2× speedup for $n = 2^{20}$. However, the DKT protocol benefits from being trivially parallelizable. As such, they enjoy a nearly one-to-one speedup when more threads are used. This combined with the extremely small communication overhead of the DKT protocol could potentially allow their protocol to outperform ours when the network is quite slow and the parties have many threads available.

Setting	Protocol	Set size n							
		2^8		2^{12}		2^{16}		2^{20}	
		Total	Online	Total	Online	Total	Online	Total	Online
LAN	*DCW (Fig. 3)	3.0	(1.4)	58.5	(27.8)	1,134	(532)	-	
	*DCW + RGBF	2.9	(1.4)	58.4	(27.6)	1,145	(542)	-	
	DKT	1.7		22.6		358		3,050	
	Ours (Fig 4)	**0.2**	(0.003)	**0.9**	(0.04)	**9.7**	(0.7)	**127**	(14)
WAN	*DCW (Fig. 3)	4.2	(1.8)	61.3	(28.8)	1,185	(532)	-	
	*DCW + RGBF	4.0	(1.6)	60.6	(28.6)	1,189	(530)	-	
	DKT	1.7		23.1		393		5,721	
	Ours (Fig 4)	**0.95**	(0.1)	**4.6**	(0.8)	**56**	(11)	**935**	(175)

Fig. 6. Total running time in seconds for the DKT and our protocol when 4, 16, and 64 threads per party are used. The evaluations were performed in the LAN setting with a 0.2 ms round trip time.

Threads	Protocol	Set size n			
		2^8	2^{12}	2^{16}	2^{20}
4	DKT	0.79	6.75	98.1	1,558
	Ours (Fig 4)	0.17	0.63	4.3	66
16	DKT	0.36	2.56	31.0	461
	Ours (Fig 4)	0.17	0.46	3.8	51
64	DKT	0.17	1.30	20.1	309
	Ours (Fig 4)	0.17	0.30	2.3	37

Fig. 7. Total time in seconds, with online time in parentheses, for PSI of two sets of size n with elements of 128 bits. The LAN (resp. WAN) setting has 0.2 ms (resp. 95 ms) round trip time latency. As noted in Sect. 6.3, when the protocol is marked with an asterisk, we report an optimistic underestimate of the running time. Missing times (-) took >5 h.

In Fig. 8 we report the empirical and asymptotic communication costs of the protocols. Out of the bloom filter based protocols, ours consumes significantly less bandwidth. For $n = 2^8$, only 1.9 MB communication was required with most

	set size n				asymptotic	
	2^8	2^{12}	2^{16}	2^{20}	Offline	Online
DCW (Fig. 3)	3.2	50.7	810	-	$2n\kappa^2$	$4n\kappa^2$
DCW + RGBF	2.4	33.9	541	-	$2n\kappa^2$	$2n\kappa^2 + n\kappa$
DKT	0.05	0.8	14	213	0	$6n\phi + 6\phi + n\kappa$
Ours (Fig 4)	1.9	23	324	4,970	$2n\kappa^2$	$2n\kappa \log_2(2n\kappa) + n\kappa$

Fig. 8. The empirical and asymptotic communication cost for sets of size n reported in megabytes, and bits respectively. $\phi = 283$ is the size of the elliptic curve elements. Missing entries had prohibitively long running times and are estimated to be greater than 8,500 MB.

of that cost in the offline phase. Then computing the intersection for $n = 2^{16}$, our protocol uses 324 MB of communication, approximately 5 KB per item. The largest amount of communication occurs during the OT extension and involves the sending of a roughly $2n\kappa^2$-bit matrix. The cut and choose contributes minimally to the communication and consists of np_{chk} choice bits and the xor of the corresponding OT messages. In the online phase, the sending of the permutation consisting of $N_{bf} \log_2(N_{ot}) \approx 2n\kappa \log(2n\kappa)$ bits that dominates the communication.

References

1. Asharov, G., Lindell, Y., Schneider, T., Zohner, M.: More efficient oblivious transfer and extensions for faster secure computation. In: Sadeghi et al. [25], pp. 535–548
2. Asharov, G., Lindell, Y., Schneider, T., Zohner, M.: More efficient oblivious transfer extensions with security for malicious adversaries. In: Oswald, E., Fischlin, M. (eds.) EUROCRYPT 2015. LNCS, vol. 9056, pp. 673–701. Springer, Heidelberg (2015). doi:10.1007/978-3-662-46800-5_26
3. Beaver, D.: Correlated pseudorandomness and the complexity of private computations. In: 28th ACM STOC, pp. 479–488. ACM Press, May 1996
4. Canetti, R.: Universally composable security: a new paradigm for cryptographic protocols. In: 42nd FOCS, pp. 136–145. IEEE Computer Society Press, October 2001
5. Canetti, R., Jain, A., Scafuro, A.: Practical UC security with a global random oracle. In: Ahn, G.-J., Yung, M., Li, N. (eds.) ACM CCS 14, pp. 597–608. ACM Press, New York (2014)
6. Dachman-Soled, D., Malkin, T., Raykova, M., Yung, M.: Efficient robust private set intersection. In: Abdalla, M., Pointcheval, D., Fouque, P.-A., Vergnaud, D. (eds.) ACNS 2009. LNCS, vol. 5536, pp. 125–142. Springer, Heidelberg (2009). doi:10.1007/978-3-642-01957-9_8
7. De Cristofaro, E., Kim, J., Tsudik, G.: Linear-complexity private set intersection protocols secure in malicious model. In: Abe, M. (ed.) ASIACRYPT 2010. LNCS, vol. 6477, pp. 213–231. Springer, Heidelberg (2010). doi:10.1007/978-3-642-17373-8_13
8. Dong, C., Chen, L., Wen, Z.: When private set intersection meets big data: an efficient and scalable protocol. In: Sadeghi et al. [25], pp. 789–800
9. Freedman, M.J., Nissim, K., Pinkas, B.: Efficient private matching and set intersection. In: Cachin, C., Camenisch, J.L. (eds.) EUROCRYPT 2004. LNCS, vol. 3027, pp. 1–19. Springer, Heidelberg (2004). doi:10.1007/978-3-540-24676-3_1
10. Goodrich, M.T., Mitzenmacher, M.: Invertible bloom lookup tables. In: 49th Annual Allerton Conference on Communication, Control, and Computing, Allerton 2011, Allerton Park & Retreat Center, Monticello, IL, USA, 28–30 September 2011, pp. 792–799. IEEE (2011)
11. Huang, Y., Evans, D., Katz, J.: Private set intersection: are garbled circuits better than custom protocols? In: NDSS 2012. The Internet Society, Feburary 2012
12. Huberman, B.A., Franklin, M.K., Hogg, T.: Enhancing privacy and trust in electronic communities. In: EC, pp. 78–86 (1999)
13. Ishai, Y., Kilian, J., Nissim, K., Petrank, E.: Extending oblivious transfers efficiently. In: Boneh, D. (ed.) CRYPTO 2003. LNCS, vol. 2729, pp. 145–161. Springer, Heidelberg (2003). doi:10.1007/978-3-540-45146-4_9

14. Kamara, S., Mohassel, P., Raykova, M., Sadeghian, S.: Scaling private set intersection to billion-element sets. In: Christin, N., Safavi-Naini, R. (eds.) FC 2014. LNCS, vol. 8437, pp. 195–215. Springer, Heidelberg (2014). doi:10.1007/978-3-662-45472-5_13

15. Keller, M., Orsini, E., Scholl, P.: Actively secure OT extension with optimal overhead. In: Gennaro, R., Robshaw, M. (eds.) CRYPTO 2015. LNCS, vol. 9215, pp. 724–741. Springer, Heidelberg (2015). doi:10.1007/978-3-662-47989-6_35

16. Kolesnikov, V., Kumaresan, R., Rosulek, M., Trieu, N.: Efficient batched oblivious PRF with applications to private set intersection. In: Weippl, E.R., Katzenbeisser, S., Kruegel, C., Myers, A.C., Halevi, S. (eds.) ACM CCS 16, pp. 818–829. ACM Press, New York (2016)

17. Lambæk, M.: Breaking and fixing private set intersection protocols. Master's thesis, Aarhus University (2016). https://eprint.iacr.org/2016/665

18. Lindell, Y.: Fast cut-and-choose based protocols for malicious and covert adversaries. In: Canetti, R., Garay, J.A. (eds.) CRYPTO 2013. LNCS, vol. 8043, pp. 1–17. Springer, Heidelberg (2013). doi:10.1007/978-3-642-40084-1_1

19. Marlinspike, M.: The difficulty of private contact discovery (2014). Blog post https://whispersystems.org/blog/contact-discovery

20. Naor, M., Pinkas, B.: Efficient oblivious transfer protocols. In: Kosaraju, S.R. (ed.) 12th SODA, pp. 448–457. ACM-SIAM, New York (2001)

21. Orrù, M., Orsini, E., Scholl, P.: Actively secure 1-out-of-N OT extension with application to private set intersection. In: Handschuh, H. (ed.) CT-RSA 2017. LNCS, vol. 10159, pp. 381–396. Springer, Cham (2017). doi:10.1007/978-3-319-52153-4_22

22. Pinkas, B., Schneider, T., Segev, G., Zohner, M.: Phasing: private set intersection using permutation-based hashing. In: Jung, J., Holz, T. (eds.) 24th USENIX Security Symposium, USENIX Security 15, pp. 515–530. USENIX Association, Berkeley (2015)

23. Pinkas, B., Schneider, T., Zohner, M.: Faster private set intersection based on OT extension. In: Fu, K., Jung, J. (eds.) 23rd USENIX Security Symposium, USENIX Security 14, pp. 797–812. USENIX Association, Berkeley (2014)

24. Rivest, R.L.: Unconditionally secure commitment and oblivious transfer schemes using private channels and a trusted initializer (1999, unpublished manuscript). http://people.csail.mit.edu/rivest/Rivest-commitment.pdf

25. Sadeghi, A.-R., Gligor, V.D., Yung, M. (eds.): ACM CCS 13. ACM Press, New York (2013)

Formal Abstractions for Attested Execution Secure Processors

Rafael Pass[1], Elaine Shi[2], and Florian Tramèr[3]([✉])

[1] Cornell Tech, New York, USA
[2] Cornell University, Ithaca, USA
[3] Stanford University, Stanford, USA
tramer@stanford.edu

Abstract. Realistic secure processors, including those built for academic and commercial purposes, commonly realize an "attested execution" abstraction. Despite being the *de facto* standard for modern secure processors, the "attested execution" abstraction has not received adequate formal treatment. We provide formal abstractions for "attested execution" secure processors and rigorously explore its expressive power. Our explorations show both the expected and the surprising.

On one hand, we show that just like the common belief, attested execution is extremely powerful, and allows one to realize powerful cryptographic abstractions such as stateful obfuscation whose existence is otherwise impossible even when assuming virtual blackbox obfuscation and *stateless* hardware tokens. On the other hand, we show that surprisingly, realizing *composable* two-party computation with attested execution processors is not as straightforward as one might anticipate. Specifically, only when both parties are equipped with a secure processor can we realize composable two-party computation. If one of the parties does not have a secure processor, we show that composable two-party computation is impossible. In practice, however, it would be desirable to allow multiple legacy clients (without secure processors) to leverage a server's secure processor to perform a multi-party computation task. We show how to introduce minimal additional setup assumptions to enable this. Finally, we show that *fair* multi-party computation for general functionalities is impossible if secure processors do not have trusted clocks. When secure processors have trusted clocks, we can realize fair two-party computation if both parties are equipped with a secure processor; but if only one party has a secure processor (with a trusted clock), then fairness is still impossible for general functionalities.

1 Introduction

The science of cybersecurity is founded atop one fundamental guiding principle, that is, to minimize a system's Trusted Computing Base (TCB) [69]. Since it is notoriously difficult to have "perfect" software in practice especially in the presence of legacy systems, the architecture community have advocated a new paradigm to bootstrap a system's security from trusted hardware (henceforth

© International Association for Cryptologic Research 2017
J.-S. Coron and J.B. Nielsen (Eds.): EUROCRYPT 2017, Part I, LNCS 10210, pp. 260–289, 2017.
DOI: 10.1007/978-3-319-56620-7_10

also referred to as secure processors). Roughly speaking, secure processors aim to reduce a sensitive application's trusted computing base to only the processor itself (possibly in conjunction with a minimal software TCB such as a secure hypervisor). In particular, besides itself, a sensitive application (e.g., a banking application) should not have to trust any other software stack (including the operating system, drivers, and other applications) to maintain the confidentiality and/or integrity of mission-critical data (e.g., passwords or credit card numbers). Security is retained even if the software stack can be compromised (as long as the sensitive application itself is intact). Besides a software adversary, some secure processors make it a goal to defend against physical attackers as well. In particular, even if the adversary (e.g., a rogue employee of a cloud service provider or a system administrator) has physical access to the computing platform and may be able to snoop or tamper with memory or system buses, he should not be able to harvest secret information or corrupt a program's execution.

Trusted hardware is commonly believed to provide a very powerful abstraction for building secure systems. Potential applications are numerous, ranging from cloud computing [11,28,50,61,62], mobile security [60], web security, to cryptocurrencies [73]. In the past three decades, numerous secure processors have been proposed and demonstrated by both academia and industry [6,22, 27,31,32,48,49,51,66,72]; and several have been commercialized, including the well-known Trusted Platform Modules (TPMs) [1], Arm's TrustZone [5,7], and others. Notably, Intel's recent release of its new x86 security extensions called SGX [6,26,51] has stirred wide-spread interest to build new, bullet-proof systems that leverage emerging trusted hardware offerings.

1.1 Attested Execution Secure Processors

Although there have been numerous proposals for the design of trusted hardware, and these designs vary vastly in terms of architectural choices, instruction sets, implementation details, cryptographic suites, as well as adversarial models they promise to defend against — amazingly, it appears that somehow most of these processors have converged on providing a common abstraction, henceforth referred to as the *attested execution* abstraction [1,6,27,51,66,68]. Roughly speaking, an attested execution abstraction enables the following:

– A platform equipped with an attested execution processor can send a program and inputs henceforth denoted (prog, inp) to its local secure processor. The secure processor will execute the program over the inputs, and compute outp := prog(inp). The secure processor will then sign the tuple (prog, outp) with a secret signing key to obtain a digital signature σ — in practice, a hash function is applied prior to the signing. Particularly, this signature σ is commonly referred to as an "attestation", and therefore this entire execution is referred to as an "attested execution".
– The execution of the aforementioned program is conducted in a sandboxed environment (henceforth referred to as an *enclave*), in the sense that a software adversary and/or a physical adversary cannot tamper with the execution,

or inspect data that lives inside the enclave. This is important for realizing privacy-preserving applications. For example, a remote client who knows the secure processor's public key can establish a secure channel with a secure processor residing on a remote server S. The client can then send encrypted and authenticated data (and/or program) to the secure processor — while the messages are passed through the intermediary S, S cannot eavesdrop on the contents, nor can it tamper with the communication.

- Finally, various secure processors make different concrete choices in terms of how they realize such secure sandboxing mechanisms as mentioned above — and the choices are closely related to the adversarial capabilities that the secure processor seeks to protect against. For example, roughly speaking, Intel's SGX technology [6,51] defends against a restricted software adversary that does not measure timing or other possible side channels, and does not observe the page-swap behavior of the enclave application (e.g., the enclave application uses small memory or is by design data-oblivious); it also defends against a restricted physical attacker capable of tapping memory, but not capable of tapping the addresses on the memory bus or measuring side-channel information such as timing and power consumption.

We refer the reader to Shi et al. [64] for a general-purpose introduction of trusted hardware, and for a comprehensive comparison of the different choices made by various secure processors.

The fact that the architecture community has converged on the "attested execution" abstraction is intriguing. How exactly this has become the *de facto* abstraction is beyond the scope of this paper, but it is helpful to observe that the attested execution abstraction is cost-effective in practice in the following senses:

- *General-purpose:* The attested execution abstraction supports the computation of general-purpose, user-defined programs inside the secure enclave, and therefore can enable a broad range of applications;
- *Reusability:* It allows a single trusted hardware token to be reused by multiple applications, and by everyone in the world — interestingly, it turns out such reusability actually gives rise to many of the technicalities that will be discussed later in the paper;
- *Integrity and privacy:* It offers both integrity and privacy guarantees. In particular, although the platform P that is equipped with the trusted hardware serves an intermediary in every interaction with the trusted hardware, privacy guarantees can be bootstrapped by having remote users establish a secure channel with the secure processor.

In the remainder of the paper, whenever we use the term "secure processors" or "trusted hardware", unless otherwise noted we specifically mean attested execution secure processors.

1.2   Why Formal Abstractions for Secure Processors?

Although attested execution has been accepted by the community as a *de facto* standard, to the best of our knowledge, no one has explored the following fundamental questions:

1. Precisely and formally, what is the attested execution abstraction?
2. What can attested execution express and what can it not express?

If we can formally and precisely articulate the answers to these questions, the benefits can be wide-spread. It can help both the producer as well as the consumer of trusted hardware, in at least the following ways:

– *Understand whether variations in abstraction lead to differences in expressive power.* First, various secure processors may provide similar but subtly different abstractions — do these differences matter to the expressive power of the trusted hardware? If we wish to add a specific feature to a secure processor (say, timing), will this feature increase its expressive power?
– *Enable formally correct use of trusted hardware.* Numerous works have demonstrated how to use trusted hardware to build a variety of secure systems [11,12,23,28,50,55,59,61–63]. Unfortunately, since it is not even clear what precise abstraction the trusted hardware offers, the methodology adopted by most existing works ranges from heuristic security to semi-formal reasoning.
  Moreover, most known secure processors expose cryptography-related instructions (e.g., involving hash chains or digital signatures [1,6,26,51]), and this confounds the programming of trusted hardware — in particular, the programmer essentially has to design cryptographic protocols to make use of trusted hardware. It is clear that user-friendly higher-level programming abstractions that hide away the cryptographic details will be highly desirable, and may well be the key to the democratization of trusted hardware programming (and in fact, to security engineering in general) — and yet without precisely articulating the formal abstraction trusted hardware offers, it would clearly be impossible to build formally correct higher-level programming abstractions atop.
– *Towards formally secure trusted hardware.* Finally, understanding what is a "good" abstraction for trusted hardware can provide useful feedback to the designers and manufacturers of trusted hardware. The holy grail would be to design and implement a formally secure processor. Understanding what cryptography-level formal abstraction to realize is a necessary first step towards this longer-term goal — but to realize this goal would obviously require additional, complementary techniques and machinery, e.g., those developed in the formal methods community [31,57,58,72], that can potentially allow us to ensure that the actual secure processor implementation meets the specification.

1.3   Summary of Our Contributions

To the best of our knowledge, we are the first to investigate cryptographically sound and composable formal abstractions for realistic, attested execution secure processors. Our findings demonstrate both the "expected" and the (perhaps) "surprising".

The Expected and the Surprising. On one hand, we show that attested execution processors are indeed extremely powerful as one might have expected, and allow us to realize primitives that otherwise would have been impossible even when assuming stateless hardware tokens or virtual blackbox secure cryptographic obfuscation.

On the other hand, our investigation unveils subtle technical details that could have been easily overlooked absent an effort at formal modeling, and we draw several conclusions that might have come off as surprising initially (but of course, natural in hindsight). For example,

- We show that universally composable two-party computation is impossible if a single party does not have such a secure processor (and the other party does);

This was initially surprising to us, since we commonly think of an attested execution processor as offering an "omnipotent" trusted third party that can compute general-purpose, user-defined programs. When such a trusted third party exists, it would appear that any function can be evaluated securely and non-interactively, hiding both the program and data. One way to interpret our findings is that such intuitions are technically imprecise and dangerous to presume — while attested execution processors indeed come close to offering such a "trusted third party" ideal abstraction, there are aspects that are "imperfect" about this ideal abstraction that should not be overlooked, and a rigorous approach is necessary towards formally correct usage of trusted hardware.

Additional Results for Multi-party Computation. We additionally show the following results:

- Universally composable two-party computation is indeed possible when both parties are equipped with an attested execution processor. We give an explicit construction and show that there are several interesting technicalities in its design and proof (which we shall comment on soon). Dealing with these technicalities also demonstrates how a *provably* secure protocol candidate would differ in important details from the most natural protocol candidates [41,55,62] practitioners would have adopted (which are not known to have provable composable security). This confirms the importance of formal modeling and provable security.
- Despite the infeasibility of multi-party computation when even a single party does not have a secure processor, in practice it would nonetheless be desirable to build multi-party applications where multiple possibly legacy clients outsource data and computation to a single cloud server equipped with a secure processor.

We show how to introduce minimal global setup assumptions — more specifically, by adopting a global augmented common reference string [18] (henceforth denoted $\mathcal{G}_{acrs}$) — to circumvent this impossibility. Although the theoretical feasibility of general UC-secure MPC is known with $\mathcal{G}_{acrs}$ even without secure processors [18], existing constructions involve cryptographic computation that is (at least) linear in the runtime of the program to be securely evaluated. By contrast, we are specifically interested in *practical* constructions that involve only $O(1)$ amount of cryptographic computations, and instead perform all program-dependent computations inside the secure processor (and not cryptographically).

Techniques. Several interesting technicalities arise in our constructions. First, composition-style proofs typically require that a simulator intercepts and modifies communication to and from the adversary (and the environment), such that the adversary cannot distinguish whether it is talking to the simulator or the real-world honest parties and secure processors. Since the simulator does not know honest parties' inputs (beyond what is leaked by the computation output), due to the indistinguishability, one can conclude that the adversary cannot have knowledge of honest parties inputs either.

– *Equivocation.* Our simulator's ability to perform such simulation is hampered by the fact that the secure processors sign attestations for messages coming out — since the simulator does not possess the secret signing key, it cannot modify these messages and must directly forward them to the adversary. To get around this issue would require new techniques for performing *equivocation*, a technicality that arises in standard protocol composition proofs. To achieve equivocation, we propose new techniques that place special backdoors inside the enclave program. Such backdoors must be carefully crafted such that they give the simulator more power without giving the real-world adversary additional power. In this way, we get the best of both worlds: (1) honest parties' security will not be harmed in the real-world execution; and (2) the simulator in the proof can "program" the enclave application to sign any output of its choice, provided that it can demonstrate the correct trapdoors. This technique is repeatedly used in different forms in almost all of our protocols.
– *Extraction.* Extraction is another technical issue that commonly arises in protocol composition proofs. The most interesting manifestation of this technical issue is in our protocol that realizes multi-party computation in the presence of a global common reference string ($\mathcal{G}_{acrs}$) and a single secure processor (see Sect. 2.5). Here again, we leverage the idea of planting special backdoors in the enclave program to allow for such extraction. Specifically, when provided with the correct identity key of a party, the enclave program will leak the party's inputs to the caller. Honest parties' security cannot be harmed by this backdoor, since no one ever learns honest parties' identity keys in the real world, not even the honest parties themselves. In the simulation, however, the simulator learns the corrupt parties' identity keys, and therefore it can extract corrupt parties' inputs.

Trusted Clocks and Fairness. Finally, we formally demonstrate how differences in abstraction can lead to differences in expressive power. In particular, many secure processors provide a trusted clock, and we explore the expressive power of such a trusted clock in the context of *fair* 2-party computation. It is well-known that in the standard setting fairness is impossible in 2-party computation for general functionalities [25]. However, several recent works have shown that the impossibility for general functionalities does not imply impossibility for every functionality — interestingly, there exist a broad class of functionalities that can be fairly computed in the plain setting [8,38,39]. We demonstrate several interesting findings in the context of attested execution processors:

- First, even a single attested execution processor already allows us to compute more functionalities fairly than in the plain setting. Specifically, we show that fair two-party coin flipping, which is impossible in the plain setting, is possible if only one party is equipped with an attested execution processor.
- Unfortunately, we show that a single attested execution processor is insufficient for fairly computing general 2-party functionalities;
- On the bright side, we prove that if both parties are equipped with an attested execution processor, it is indeed possible to securely compute any function fairly.

Variant Models and Additional Results. Besides the trusted clock, we also explore variations in abstraction and their implications — for example, we compare non-anonymous attestation and anonymous attestation since various processors seem to make different choices regarding this.

We also explore an interesting model called "transparent enclaves" [70], where secret data inside the enclave can leak to the adversary due to possible side-channel attacks on known secure processors, and we show how to realize interesting tasks such as UC-secure commitments and zero-knowledge proofs in this weaker model — here again our protocols must deal with interesting technicalities related to extraction and equivocation.

1.4 Non-goals and Frequently Asked Questions

Trusted hardware has been investigated by multiple communities from different angles, ranging from how to architect secure processors [6,22,27,31,32,48, 49,51,66,72], how to apply them in applications [11,12,23,28,50,55,59,61–63], side-channels and other attacks [36,46,47,67,71,74] and protection against such attacks [32,49,72,74]. Despite the extensive literature, cryptographically sound formal abstractions appear to be an important missing piece, and this work aims to make an initial step forward towards this direction. In light of the extensive literature, however, several natural but frequently asked questions arise regarding the precise scope of this paper, and we address such questions below.

First, although we base our modeling upon what realistic secure processors aim to provide, it is not our intention to claim that any existing secure processors provably realize our abstraction. We stress that to make any claim of this nature (that a secure processor correctly realizes any formal specification) is an area of active research in the formal methods and programming language communities [31,57,58,72], and thus still a challenging open question — let alone the fact that some commercial secure processor designs are closed-source.

Second, a frequently asked question is what adversarial models our formal abstraction defends against. The answer to such a question is processor-specific, and thus outside the scope of our paper — we leave it to the secure processor itself to articulate the precise adversarial capabilities it protects against. The formal models and security theorems in this paper hold assuming that the adversary is indeed confined to the capabilities assumed by the specific secure processor. As mentioned earlier, some processors defend only against software adversaries [27]; others additionally defend against physical attackers [32–34,49]; others defend against a restricted class of software and/or physical attackers that do not exploit certain side channels [1,6,48,51,66]. We refer the reader to a comprehensive systematization of knowledge paper by Shi et al. [64] for a taxonomy and comparison of various secure processors.

Finally, it is also not our goal to propose new techniques that defend against side-channel attacks, or suggest how to better architect secure processors — these questions are being explored in an orthogonal but complementary line of research [27,31–34,49,72,74].

2 Technical Roadmap

2.1 Formal Modeling

Modeling Choices. To enable cryptographically sound reasoning, we adopt the universal composition (UC) paradigm in our modeling [17,18,21]. At a high level, the UC framework allows us to abstract complex cryptographic systems as simple ideal functionalities, such that protocol composition can be modularized. The UC framework also provides what is commonly referred to as "concurrent composition" and "environmental friendliness": in essence, a protocol π proven secure in the UC framework can run in any environment such that (1) any other programs or protocols executing possibly simultaneously will not affect the security of the protocol π, and (2) protocol π will not inject undesirable side effects (besides those declared explicitly in the ideal abstraction) that would affect other programs and protocols in the system.

More intuitively, if a system involving cryptography UC-realizes some ideal functionality, henceforth, a programmer can simply program the system pretending that he is making remote procedural calls to a trusted third party without having to understand the concrete cryptography implementations. We refer the reader to the full version of this work [56] for a more detailed overview of the UC framework in our context. Before we proceed, we stress the importance of cryptographically sound reasoning: by contrast, earlier works in the formal methods community would make assumptions that cryptographic primitives such as encryption and signatures realize the "most natural" ideal box without formal justification — and such approaches have been shown to be flawed when the ideal box is actually instantiated with cryptography [2–4,9,13,20,43,44,53,54].

Roadmap for Formal Modeling. We first describe an ideal functionality $\mathcal{G}_{att}$ that captures the core abstraction that a broad class of attested execution processors intend to provide. We are well aware that various attested execution processors make different design choices — most of them are implementation-level details that do not reflect at the abstraction level, but a few choices do matter at the abstraction level — such as whether the secure processor provides a trusted clock and whether it implements anonymous or non-anonymous attestation.

In light of such differences, we first describe a basic, anonymous attestation abstraction called $\mathcal{G}_{att}$ that lies at the core of off-the-shelf secure processors such as Intel SGX [6,51]. We explore the expressive power of this basic abstraction in the context of stateful obfuscation and multi-party computation. Later in the paper, we explore variants of the abstraction such as non-anonymous attestation and trusted clocks. Therefore, in summary our results aim to be broadly applicable to a wide class of secure processor designs.

$$\mathcal{G}_{att}[\Sigma, \mathsf{reg}]$$

// *initialization:*
On initialize: $(\mathsf{mpk}, \mathsf{msk}) := \Sigma.\mathsf{KeyGen}(1^\lambda)$, $T = \emptyset$

// *public query interface:*
On receive* **getpk**() from some $\mathcal{P}$: send mpk to $\mathcal{P}$

Enclave operations

// *local interface — install an enclave:*
On receive* **install**(idx, prog) from some $\mathcal{P} \in \mathsf{reg}$:
  if $\mathcal{P}$ is honest, assert $idx = sid$
  generate nonce $eid \in \{0,1\}^\lambda$, store $T[eid, \mathcal{P}] := (idx, \mathsf{prog}, 0)$, send eid to $\mathcal{P}$

// *local interface — resume an enclave:*
On receive* **resume**(eid, inp) from some $\mathcal{P} \in \mathsf{reg}$:
  let $(idx, \mathsf{prog}, \mathsf{mem}) := T[eid, \mathcal{P}]$, abort if not found
  let $(\mathsf{outp}, \mathsf{mem}) := \mathsf{prog}(\mathsf{inp}, \mathsf{mem})$, update $T[eid, \mathcal{P}] := (idx, \mathsf{prog}, \mathsf{mem})$
  let $\sigma := \Sigma.\mathsf{Sig}_{\mathsf{msk}}(idx, eid, \mathsf{prog}, \mathsf{outp})$, and send (outp, σ) to $\mathcal{P}$

Fig. 1. A global functionality modeling an SGX-like secure processor. Blue (and starred*) activation points denote *reentrant* activation points. Green activation points are executed at most once. The enclave program prog may be probabilistic and this is important for privacy-preserving applications. Enclave program outputs are included in an anonymous attestation σ. For honest parties, the functionality verifies that installed enclaves are parametrized by the session id *sid* of the current protocol instance. (Color figure online)

The $\mathcal{G}_{att}$ Abstraction. We first describe a basic $\mathcal{G}_{att}$ abstraction capturing the essence of SGX-like secure processors that provide anonymous attestation (see Fig. 1). Here we briefly review the $\mathcal{G}_{att}$ abstraction and explain the technicalities that arise in the formal modeling. More detailed discussions can be found in the full version [56, Sect. 3].

1. **Registry.** First, $\mathcal{G}_{att}$ is parametrized with a registry reg that is meant to capture all the platforms that are equipped with an attested execution processor. For simplicity, we consider a static registry reg in this paper.
2. **Stateful enclave operations.** A platform $\mathcal{P}$ that is in the registry reg may invoke enclave operations, including
   - install: installing a new enclave with a program prog, henceforth referred to as the enclave program. Upon installation, $\mathcal{G}_{att}$ simply generates a fresh enclave identifier eid and returns the eid. This enclave identifier may now be used to uniquely identify the enclave instance.
   - resume: resuming the execution of an existing enclave with inputs inp. Upon a resume call, $\mathcal{G}_{att}$ executes the prog over the inputs inp, and obtains an output outp. $\mathcal{G}_{att}$ would then sign the prog together with outp as well as additional metadata, and return both outp and the resulting attestation.

   Each installed enclave can be resumed multiple times, and we stress that the enclave operations store state across multiple resume invocations. This stateful property will later turn out to be important for several of our applications.
3. **Anonymous attestation.** Secure processors such as SGX rely on group signatures and other anonymous credential techniques [15,16] to offer "anonymous attestation". Roughly speaking, anonymous attestation allows a user to verify that the attestation is produced by some attested execution processor, without identifying which one. To capture such anonymous attestation, our $\mathcal{G}_{att}$ functionality has a manufacturer public key and secret key pair denoted (mpk, msk), and is parametrized by a signature scheme Σ. When an enclave resume operation is invoked, $\mathcal{G}_{att}$ signs any output to be attested with msk using the signature scheme Σ. Roughly speaking, if a group signature scheme is adopted as in SGX, one can think of Σ as the group signature scheme parametrized with the "canonical" signing key. $\mathcal{G}_{att}$ provides the manufacturer public key mpk to any party upon query — this models the fact that there exists a secure key distribution channel to distribute mpk. In this way, any party can verify an anonymous attestation signed by $\mathcal{G}_{att}$.

Globally Shared Functionality. Our $\mathcal{G}_{att}$ functionality essentially captures all attested execution processors in the world. Further, we stress that $\mathcal{G}_{att}$ is globally shared by all users, all applications, and all protocols. In particular, rather than generating a different (mpk, msk) pair for each different protocol instance, the same (mpk, msk) pair is globally shared.

More technically, we capture such sharing across protocols using the Universal Composition with Global Setup (GUC) paradigm [18]. As we show later, such global sharing of cryptographic keys becomes a source of "imperfectness" — in particular, due to the sharing of (mpk, msk), attestations signed by msk from one protocol instance (i.e., or application) may now carry meaning in a completely unrelated protocol instance, thus introducing potentially undesirable side effects that breaks composition.

Additional Discussions and Clarifications. More detailed discussions of our modeling choices, and importantly, clarifications on how the environment $\mathcal{Z}$ interacts with $\mathcal{G}_{att}$ are deferred to our technical report [56, Sect. 3].

Throughout this paper, we assume that parties interact with each other over *secure* channels. It is possible to realize (UC-secure) secure channels from authenticated channels through key exchange. Whenever applicable, our results are stated for the case of *static* corruption.

2.2 Power of Attested Execution: Stateful Obfuscation

We show that the attested execution abstraction is indeed extremely powerful as one would have expected. In particular, we show that attested execution processors allow us to realize a new abstraction which we call "stateful obfuscation".

Theorem 1 (Informal). *Assume that secure key exchange protocols exist. There is a $\mathcal{G}_{att}$-hybrid protocol that realizes non-interactive stateful obfuscation, which is not possible in plain settings, even when assuming stateless hardware tokens or virtual-blackbox secure cryptographic obfuscation.*

Stateful obfuscation allows an (honest) client to obfuscate a program and send it to a server, such that the server can evaluate the obfuscated program on multiple inputs, while the obfuscated program keeps (secret) internal state across multiple invocations. We consider a simulation secure notion of stateful obfuscation, where the server should learn only as much information as if it were interacting with a stateful oracle (implementing the obfuscated program) that answers the server's queries. For example, stateful obfuscation can be a useful primitive in the following application scenario: imagine that a client (e.g., a hospital) outsources a sensitive database (corresponding to the program we wish to obfuscate) to a cloud server equipped with trusted hardware. Now, an analyst may send statistical queries to the server and obtain *differentially private* answers. Since each query consumes some privacy budget, we wish to guarantee that after the budget is depleted, any additional query to the database would return $\bot$. We formally show how to realize stateful obfuscation from attested execution processors. Further, as mentioned, we prove that stateful obfuscation is not possible in the plain setting, even when assuming the existence of stateless hardware tokens or assuming virtual-blackbox secure obfuscation.

2.3 Impossibility of Composable 2-Party Computation with a Single Secure Processor

One natural question to ask is whether we can realize universally composable (i.e., UC-secure) multi-party computation, which is known to be impossible in the plain setting without any setup assumptions — but feasible in the presence of a common reference string [17,19], i.e., a public random string that is generated in a trustworthy manner freshly and independently for each protocol instance. On the surface, $\mathcal{G}_{att}$ seems to provide a much more powerful functionality than a common reference string, and thus it is natural to expect that it will enable UC-secure multi-party computation. However, upon closer examination, we find that perhaps somewhat surprisingly, such intuition is subtly incorrect, as captured in the following informal theorem.

Theorem 2 (Informal). *If at least one party is not equipped with an attested execution processor, it is impossible to realize UC-secure multi-party computation absent additional setup assumptions (even when all others are equipped with an attested execution processor).*

Here the subtle technicalities arise exactly from the fact that $\mathcal{G}_{att}$ is a global functionality shared across all users, applications, and protocol instances. This creates a *non-deniability* issue that is well-known to the cryptography community. Since the manufacturer signature key (mpk, msk) is globally shared, attestations produced in one protocol instance can carry side effects into another. Thus, most natural protocol candidates that send attestations to other parties will allow an adversary to implicate an honest party of having participated in a protocol, by demonstrating the attestation to a third party. Further, such non-deniability exists even when the secure processor signs *anonymous* attestations: since if not all parties have a secure processor, the adversary can at least prove that *some* honest party that is in $\mathcal{G}_{att}$'s registry has participated in the protocol, even if he cannot prove which one. Intuitively, the non-deniability goes away if all parties are equipped with a secure processor — note that this necessarily means that the adversary himself must have a secure processor too. Since the attestation is anonymous, the adversary will fail to prove whether the attestation is produced by an honest party or he simply asked his own local processor to sign the attestation. This essentially allows the honest party to deny participation in a protocol.

Impossibility of Extraction. We formalize the above intuition, and show that not only natural protocol candidates that send attestations around suffer from non-deniability, in fact, it is impossible to realize UC-secure multi-party computation if not all parties have secure processors. The impossibility is analogous to the impossibility of UC-secure commitments in the plain setting absent a common reference string [19]. Consider when the real-world committer $\mathcal{C}$ is corrupt and the receiver is honest. In this case, during the simulation proof, when the real-world $\mathcal{C}$ outputs a commitment, the ideal-world simulator Sim must capture the corresponding transcripts and extract the value v committed, and send v to the commitment ideal functionality $\mathcal{F}_{com}$. However, if the ideal-world simulator Sim can perform such extraction, the real-world receiver must be able too (since Sim does not have extra power than the real-world receiver) — and this violates the requirement that the commitment must be hiding. As Canetti and Fischlin show [19], a common reference string allows us to circumvent this impossibility by giving the simulator more power. Since a common reference string (CRS) is a *local* functionality, during the simulation, the simulator can program the CRS and embed a trapdoor — this trapdoor will allow the simulator to perform extraction. Since the real-world receiver does not possess such a trapdoor, the protocol still retains confidentiality against a real-world receiver.

Indeed, if our $\mathcal{G}_{att}$ functionality were also local, our simulator Sim could have programmed $\mathcal{G}_{att}$ in a similar manner and extraction would have been easy. In practice, however, a local $\mathcal{G}_{att}$ function would mean that a fresh key manufacturer pair (mpk, msk) must be generated for each protocol instance (i.e., even for multiple applications of the same user). Thus, a local $\mathcal{G}_{att}$ clearly fails to capture the reusability of real-world secure processors, and this justifies why we model attested execution processors as a globally shared functionality.

Unfortunately, when $\mathcal{G}_{att}$ is global, it turns out that the same impossibility of extraction from the plain setting would carry over when the committer $\mathcal{C}$ is corrupt and only the receiver has a secure processor. In this case, the simulator Sim would also have to extract the input committed from transcripts emitted from $\mathcal{C}$. However, if the simulator Sim can perform such extraction, so can the real-world receiver — note that in this case the real-world receiver is actually more powerful than Sim, since the real-world receiver, who is in the registry, is capable of meaningfully invoking $\mathcal{G}_{att}$, while the simulator Sim cannot!

It is easy to observe that this impossibility result no longer holds when the corrupt committer has a secure processor — in this case, the protocol can require that the committer $\mathcal{C}$ send its input to $\mathcal{G}_{att}$. Since the simulator captures all transcripts going in and coming out of $\mathcal{C}$, it can extract the input trivially. Indeed, we show that not only commitment, but also general 2-party computation is possible when both parties have a secure processor.

2.4   Composable 2-Party Computation when both have Secure Processors

Theorem 3 (Informal). *Assume that secure key exchange protocols exist. Then there exists an $\mathcal{G}_{att}$-hybrid protocol that UC-realizes $\mathcal{F}_{2pc}$. Further, in this protocol, all program-dependent evaluation is performed inside the enclave and not cryptographically.*

We give an explicit protocol in Fig. 2 (for concreteness, we use Diffie-Hellman key exchanges in our protocols, although the same approach extends to any secure key-exchange). The protocol is efficient in the sense that it performs only $O(1)$ (program-independent) cryptographic computations; and all program-dependent computation is performed inside the enclave. We now explain the protocol briefly.

- First, the two parties' secure processors perform a key exchange and establish a secret key sk for an authenticated encryption scheme.
- Then, each party's enclave encrypts the party's input with sk. The party then sends the resulting authenticated ciphertext ct to the other.
- Now each enclave decrypts ct and perform evaluation, and each party can query its local enclave to obtain the output.

$$\text{prog}_{2\text{pc}}[f, \mathcal{P}_0, \mathcal{P}_1, b]$$

On input ("keyex"): $y \xleftarrow{\$} \mathbb{Z}_p$, return g^y

On input ("send", g^x, inp_b):

    assert that "keyex" has been called

    $\text{sk} := (g^x)^y$, $\text{ct} := \text{AE.Enc}_{\text{sk}}(\text{inp}_b)$, return ct

On input ("compute", ct, v):

    assert that "send" has been called and ct not seen

    $\text{inp}_{1-b} := \text{AE.Dec}_{\text{sk}}(\text{ct})$, assert that decryption succeeds

    if $v \neq \bot$, return v; else return $\text{outp} := f(\text{inp}_0, \text{inp}_1)$

$$\mathbf{Prot}_{2\text{pc}}[sid, f, \mathcal{P}_0, \mathcal{P}_1, b]$$

On input inp_b from $\mathcal{Z}$:

    $eid := \mathcal{G}_{\text{att}}.\text{install}(sid, \text{prog}_{2\text{pc}}[f, \mathcal{P}_0, \mathcal{P}_1, b])$

    henceforth denote $\mathcal{G}_{\text{att}}.\text{resume}(\cdot) := \mathcal{G}_{\text{att}}.\text{resume}(eid, \cdot)$

    $(g^y, \sigma) := \mathcal{G}_{\text{att}}.\text{resume}(\text{"keyex"})$

    send (eid, g^y, σ) to $\mathcal{P}_{1-b}$, await (eid', g^x, σ')

    assert $\Sigma.\text{Ver}_{\text{mpk}}((sid, eid', \text{prog}_{2\text{pc}}[f, \mathcal{P}_0, \mathcal{P}_1, 1-b], g^x), \sigma')$

    $(\text{ct}, \_) := \mathcal{G}_{\text{att}}.\text{resume}(\text{"send"}, g^x, \text{inp}_b)$, send ct to $\mathcal{P}_{1-b}$, await ct$'$

    $(\text{outp}, \_) := \mathcal{G}_{\text{att}}.\text{resume}(\text{"compute"}, \text{ct}', \bot)$, output outp

Fig. 2. Composable 2-party computation: both parties have secure processors. AE denotes authenticated encryption. All ITIs' activation points are *non-reentrant*. When an activation point is invoked for more than once, the ITI simply outputs $\bot$. Although not explicitly noted, if $\mathcal{G}_{\text{att}}$ ever outputs $\bot$ upon a query, the protocol aborts outputting $\bot$. The group parameters (g, p) are hardcoded into $\text{prog}_{2\text{pc}}$.

- Most of the protocol is quite natural, but one technique is necessary for equivocation. Specifically, the enclave program's "compute" entry point has a backdoor denoted v. If $v = \bot$, $\mathcal{G}_{\text{att}}$ will sign the true evaluation result and return the attested result. On the other hand, if $v \neq \bot$, the enclave will simply sign and output v itself. In the real-world execution, an honest party will always supply $v = \bot$ as input to the enclave program's "compute" entry point. However, as we explain later, the simulator will leverage this backdoor v to perform equivocation and program the output.

We now explain some interesting technicalities that arise in the proof for the above protocol.

- *Extraction.* First, extraction is made possible since each party sends their input directly to its local enclave. If a party is corrupt, this interaction will be captured by the simulator who can then extract the corrupt party's input;
- *Equivocate.* We now explain how the backdoor v in the enclave program allows for equivocation in the proof. Recall that initially, the simulator does not know the honest party's input. To simulate the honest party's message for the adversary (which contains an attestation from the enclave), the simulator

must send a dummy input to $\mathcal{G}_{att}$ on behalf of the honest party to obtain the attestation. When the simulator manages to extract the corrupt party's input, it will send the input to the ideal functionality $\mathcal{F}_{2pc}$ and obtain the outcome of the computation denoted outp*. Now when the corrupt party queries its local enclave for the output, the simulator must get $\mathcal{G}_{att}$ to sign the correct outp* (commonly referred to as *equivocation*). To achieve this, the simulator will make use of the aforementioned backdoor v: instead of sending $(ct, \perp)$ to $\mathcal{G}_{att}$ as in the real-world protocol, the simulator sends $(ct, outp^*)$ to $\mathcal{G}_{att}$, such that $\mathcal{G}_{att}$ will sign outp*.

- *A note on anonymous attestation.* It is interesting to note how our protocol relies on the attestation being *anonymous* for security. Specifically, in the proof, the simulator needs to simulate the honest party's messages for the adversary $\mathcal{A}$. To do so, the simulator will simulate the honest party's enclave on its own (i.e., the adversary's) secure processor — and such simulation is possible because the attestations returned by $\mathcal{G}_{att}$ are anonymous. Had the attestation not been anonymous (e.g., binding to the party's identifier), the simulator would not be able to simulate the honest party's enclave (see our full version [56, Sect. 8.4] for more discussions).

2.5   Circumventing the Impossibility with Minimal Global Setup

In practice, it would obviously be desirable if we could allow composable multi-party computation in the presence of a single attested execution processor. As a desirable use case, imagine multiple clients (e.g., hospitals), each with sensitive data (e.g., medical records), that wish to perform some computation (e.g., data mining for clinical research) over their joint data. Moreover, they wish to outsource the data and computation to an untrusted third-party cloud provider. Specifically, the clients may not have secure processors, but as long as the cloud server does, we wish to allow outsourced secure multi-party computation.

We now demonstrate how to introduce a minimal global setup assumption to circumvent this impossibility. Specifically, we will leverage a global Augmented Common Reference String (ACRS) [18], henceforth denoted $\mathcal{G}_{acrs}$. Although the feasibility of UC-secure multi-party computation is known with $\mathcal{G}_{acrs}$ even absent secure processors [18], existing protocols involve cryptographic computations that are (at least) linear in the runtime of the program. Our goal is to demonstrate a *practical* protocol that performs any program-dependent computation inside the secure enclave, and performs only $O(1)$ cryptographic computation.

Theorem 4 (Informal). *Assume that secure key exchange protocols exist. Then, there exists a $(\mathcal{G}_{acrs}, \mathcal{G}_{att})$-hybrid protocol that UC-realizes $\mathcal{F}_{mpc}$ and makes use of only a single secure processor. Further, this protocol performs all program-dependent computations inside the secure processor's enclave (and not cryptographically).*

Minimal Global Setup $\mathcal{G}_{acrs}$. To understand this result, we first explain the minimal global setup $\mathcal{G}_{acrs}$. First, $\mathcal{G}_{acrs}$ provides a global common reference string. Second, $\mathcal{G}_{acrs}$ also allows each (corrupt) party $\mathcal{P}$ to query an *identity key* for itself. This identity key is computed by signing the party's identifier $\mathcal{P}$ using a global master secret key. Note that such a global setup is *minimal* since *honest parties should never have to query for their identity keys*. The identity key is simply a backdoor provided to corrupt parties. Although at first sight, it might seem counter-intuitive to provide a backdoor to the adversary, note that this backdoor is also provided to our simulator — and this increases the power of the simulator allowing us to circumvent the aforementioned impossibility of extraction, and design protocols where honest parties can deny participation.

MPC with a Single Secure Processor and $\mathcal{G}_{acrs}$. We consider a setting with a server that is equipped with a secure processor, and multiple clients that do not have a secure processor.

Let us first focus on the (more interesting) case when the server and a subset of the clients are corrupt. The key question is how to get around the impossibility of extraction with the help of $\mathcal{G}_{acrs}$ — more specifically, how does the simulator extract the corrupt clients' inputs? Our idea is the following — for the readers' convenience, we skip ahead and present the detailed protocol in Fig. 3 as we explain the technicalities. We defer formal notations and proofs to [56, Sect. 6].

- First, we parametrize the enclave program with the global common reference string $\mathcal{G}_{acrs}$.mpk.
- Second, we add a backdoor in the enclave program, such that the enclave program will return the secret key for $\mathcal{P}_i$'s secure channel with the enclave, if the caller provides the correct identity key for $\mathcal{P}_i$. In this way, the simulator can be a man-in-the-middle for all corrupt parties' secure channels with the enclave, and extract their inputs. We note that honest parties' security will not be harmed by this backdoor, since honest parties will never even query $\mathcal{G}_{acrs}$ for their identity keys, and thus their identity keys should never leak. However, in the simulation, the simulator will query $\mathcal{G}_{acrs}$ for all corrupt parties' identity keys, which will allow the simulator to extract corrupt parties' inputs by querying this backdoor in the enclave program.
- Third, we introduce yet another backdoor in the enclave program that allows the caller to program any party's output, provided that the caller can demonstrate that party's identity key. Again, in the real world, this backdoor should not harm honest parties' security because honest parties' identity keys never get leaked. Now in the simulation, the simulator will query $\mathcal{G}_{acrs}$ for all corrupt parties' identity keys which will give the simulator the power to query the corrupt parties' outputs. Such "programmability" is necessary, because when the simulator obtains the outcome outp from $\mathcal{F}_{mpc}$, it must somehow obtain the enclave's attestation on outp — however, since the simulator does not know honest parties' inputs, he cannot have provided honest parties' inputs to the enclave. Therefore, there must be a special execution path such that the simulator can obtain a signature on outp from the enclave.

Now, let us turn our attention to the case when the server is honest, but a subset of the clients are corrupt. In this case, our concern is how to achieve

$$\text{prog}_{\text{mpc}}[f, \mathcal{G}_{\text{acrs}}.\text{mpk}, \mathcal{S}, \mathcal{P}_1, \ldots, \mathcal{P}_n]$$

On input ("init"): for $i \in [n]$: $(\text{pk}_i, \text{sk}_i) \leftarrow \text{PKE.Gen}(1^\lambda)$; return $\{\text{pk}_1, \ldots, \text{pk}_n\}$

On input ("input", $\{\text{ct}_i\}_{i \in [n]}$):
  for $i \in [n]$: $(\text{inp}_i, k_i) := \text{PKE.Dec}_{\text{sk}_i}(\text{ct}_i)$; return $\Omega := \{\text{ct}_i\}_{i \in [n]}$

On input ("extract", $\{\text{idk}_i\}_{i \in [n]}$):
  for $i \in [n]$: if $\text{check}(\mathcal{G}_{\text{acrs}}.\text{mpk}, \mathcal{P}_i, \text{idk}) = 1$, $v_i := \text{sk}_i$, else $v_i := \bot$; return $\{v_i\}_{i \in [n]}$

On input ("program", $\{\text{idk}_i, u_i\}_{i \in [n]}$):
  for $i \in [n]$: if $\text{check}(\mathcal{G}_{\text{acrs}}.\text{mpk}, \mathcal{P}_i, \text{idk}) = 1$, $\text{outp}_i := u_i$

On input ("proceed", $\{\text{ct}'_i\}_{i \in [n]}$):
  for $i \in [n]$: assert $\text{AE.Dec}_{k_i}(\text{ct}'_i) = \text{"ok"}$
  $\text{outp}^* := f(\text{inp}_1, \ldots, \text{inp}_n)$, return "done"

On input* ("output", $\mathcal{P}_i$):
  assert outp^* has been stored
  if outp_i has been stored, $\text{ct} := \text{Enc}_{k_i}(\text{outp}_i)$, else $\text{ct} := \text{Enc}_{k_i}(\text{outp}^*)$
  return ct

$$\textbf{Prot}_{\text{mpc}}[sid, f, \mathcal{G}_{\text{acrs}}.\text{mpk}, \mathcal{S}, \mathcal{P}_1, \ldots, \mathcal{P}_n]$$

Server $\mathcal{S}$:

  let $eid := \mathcal{G}_{\text{att}}.\text{install}(sid, \text{prog}_{\text{mpc}}[f, \mathcal{G}_{\text{acrs}}.\text{mpk}, \mathcal{S}, \mathcal{P}_1, \ldots, \mathcal{P}_n])$
  henceforth let $\mathcal{G}_{\text{att}}.\text{resume}(\cdot) := \mathcal{G}_{\text{att}}.\text{resume}(eid, \cdot)$
  let $(\{\text{pk}_i\}_{i \in [n]}, \sigma) := \mathcal{G}_{\text{att}}.\text{resume}(\text{"init"})$, send $(eid, \psi(\mathcal{P}_i, \{\text{pk}_i\}_{i \in [n]}, \sigma))$ to each $\mathcal{P}_i$

  for each $\mathcal{P}_i$: await ("input", ct_i) from $\mathcal{P}_i$
  $(\Omega, \sigma) := \mathcal{G}_{\text{att}}.\text{resume}(\text{"input"}, \{\text{ct}_i\}_{i \in [n]})$, send $\psi(\mathcal{P}_i, \Omega, \sigma)$ to each $\mathcal{P}_i$

  for each $\mathcal{P}_i$: await ("proceed", ct'_i) from $\mathcal{P}_i$
  $\mathcal{G}_{\text{att}}.\text{resume}(\text{"proceed"}, \{\text{ct}'_i\}_{i \in [n]})$
  for each $\mathcal{P}_i$: $(\text{ct}_i, \sigma_i) := \mathcal{G}_{\text{att}}.\text{resume}(\text{"output"}, \mathcal{P}_i)$, send ct_i to $\mathcal{P}_i$

Remote Party $\mathcal{P}_i$: On input inp from $\mathcal{Z}$:

  await (eid, ψ) from $\mathcal{S}$
  // Henceforth for $\tilde{\psi} := (\text{msg}, C, \pi)$,
  // let $\text{Ver}(\tilde{\psi}) := \text{Ver}(\text{crs}, (sid, eid, C, \text{mpk}, \mathcal{G}_{\text{acrs}}.\text{mpk}, \mathcal{P}_i, \text{msg}), \pi)$
  assert $\text{Ver}(\psi)$, parse $\psi := (\{\text{pk}_i\}_{i \in [n]}, \text{-}, \text{-})$
  $k \leftarrow \{0,1\}^\lambda$, $\text{ct} = \text{PKE.Enc}_{\text{pk}}(\text{inp}, k)$ where $\text{pk} := \text{pk}_i$
  send ("input", ct) to $\mathcal{S}$, await ψ from $\mathcal{S}$, assert $\text{Ver}(\psi)$, parse $\psi := (\Omega, \text{-}, \text{-})$
  assert $\Omega[i] = \text{ct}$, send eid to all parties, wait for all parties to ack the same eid
  let $\text{ct}' := \text{AE.Enc}_k(\text{"ok"})$, send ("proceed", ct') to $\mathcal{S}$, await ct, assert ct not seen
  $\text{outp} := \text{Dec}_k(\text{ct})$, assert ct decryption successful, return outp

Fig. 3. Composable multi-party computation with a single secure processor. $\psi(\mathcal{P}, \text{msg}, \sigma)$ outputs a tuple (msg, C, π), where π is a witness-indistinguishable proof that the ciphertext C either encrypts a valid attestation σ on msg, or encrypts $\mathcal{P}$'s identity key. PKE and AE denote public-key encryption and authenticated encryption respectively. The notation send denotes messages sent over a secure channel.

deniability for the server — specifically, an honest server should be able to deny participation in a protocol. If the honest server sends an attestation in the clear to the (possibly corrupt) clients, we cannot hope to obtain such deniability, because a corrupt client can then prove to others that *some* honest party in $\mathcal{G}_{att}$'s registry must have participated, although it might not be able to prove which one since the attestation is anonymous. To achieve deniability, our idea is the following:

- Instead of directly sending an attestation on a message msg, the server will produce a witness indistinguishable proof that either he knows an attestation on msg, or he knows the recipient's identity key. Note that in the real world protocol, the server always provide the attestation as the witness when producing the witness indistinguishable proof.
- However, in the simulation when the server is honest but a subset of the clients are corrupt, the simulator is unable to query any enclave since none of the corrupt clients have a secure processor. However, the simulator can query $\mathcal{G}_{acrs}$ and obtain all corrupt parties' identity keys. In this way, the simulator can use these identity keys as an alternative witness to construct the witness indistinguishable proofs — and the witness indistinguishability property ensures that the adversary (and the environment) cannot distinguish which witness was provided in constructing the proof.

Implementing $\mathcal{G}_{acrs}$. In practice, the $\mathcal{G}_{acrs}$ functionality can be implemented by having a trusted third party (which may be the trusted hardware manufacturer) that generates the reference string and hands out the appropriate secret keys [18].

It is instructive to consider why $\mathcal{G}_{acrs}$ cannot be implemented from $\mathcal{G}_{att}$ itself (indeed, this would contradict our result that it is impossible to obtain composable MPC in the presence of a single attested execution processor, with no further setup assumptions). Informally, the reason this does not work is that unless all parties have access to $\mathcal{G}_{att}$ (which is the case we consider), then if only the party that does not have access to $\mathcal{G}_{att}$ is corrupted, the view of the adversary cannot be simulated—in particular, the attested generation of the CRS cannot be simulated (since the adversary does not have access to $\mathcal{G}_{att}$) and as such serves as evidence that some honest party participated in an execution (i.e., we have a "deniability attack").

2.6   Fairness

It is well-known that fairness is in general impossible in secure two-party computation in the plain model (even under weaker security definitions that do not necessarily aim for concurrent composition). Intuitively, the party that obtains the output first can simply abort from the protocol thus preventing the other party from learning the outcome. Cleve [25] formalized this intuition and demonstrated an impossibility result for fair 2-party coin tossing, which in turns suggests the impossibility of fairness in general 2-party computation. Interestingly, a sequence of recent works show that although fairness is impossible in general, there are a class of non-trivial functions that can indeed be computed fairly [8,38,39].

Since real-world secure processors such as Intel's SGX offer a "trusted clock" abstraction, we explore whether and how such trusted clocks can help in attaining fairness. It is not hard to see that Cleve's lower bound still applies, and fairness is still impossible when our attested execution processors do not have trusted clocks. We show how having trusted clocks in secure processors can help with fairness.

First, we show that fairness is indeed possible in general 2-party computation, when both parties have secure processors with trusted clocks. Specifically, we consider a clock-adjusted notion of fairness which we refer to as Δ-fairness. Intuitively, Δ-fairness stipulates that if the corrupt party receives output by some round r, then the honest party must receive output by round $\Delta(r)$, where Δ is a polynomial function.

Theorem 5 (Informal). *Assume that secure key exchange protocols exist, and that both parties have an attested execution processor with trusted clocks, then there exists a protocol that UC-realizes $\mathcal{F}_{2pc}$ with Δ-fairness where $\Delta(r) = 2r$.*

In other words, if the corrupt party learns the outcome by round r, the honest party is guaranteed to learn the outcome by round $2r$. Our protocol is a *tit-for-tat* style protocol that involves the two parties' enclaves negotiating with each other as to when to release the output to its owner. At a high level, the protocol works as follows:

- First, each party sends their respective input to its local secure processor.
- The two secure processors then perform a key exchange to establish a secret key k for an authenticated encryption scheme. Now the two enclave exchange the parties' inputs over a secure channel, at which point both enclaves can compute the output.
- However, at this point, the two enclaves still withhold the outcome from their respective owners, and the initial timeout value $\delta := 2^\lambda$ is set to exponentially large in λ. In other words, each enclave promises to release the outcome to its owner in round δ.
- At this moment, the *tit-for-tat* protocol starts. In each turn, each secure enclave sends an acknowledgment to the other over a secure channel. Upon receiving the other enclave's acknowledgment, the receiving enclave would now halve the δ value, i.e., set $\delta := \frac{\delta}{2}$. In other words, the enclave promises to release the outcome to its owner by half of the original timeout.
- If both parties are honest, then after λ turns, their respective enclaves disclose the outputs to each party.
- If one party is corrupt, then if he learns the outcome by round r, clearly the other party will learn the outcome by round $2r$.

To have provably security in the UC model, technicalities similar to our earlier 2-party computation protocol (the case when both parties have a secure processor) exist. More specifically, both parties have to send inputs to their local enclave to allow extraction in the simulation. Moreover, the enclave program needs to leave a second input (that is not used in the real-world protocol) such that the simulator can program the output for the corrupt party after learning the output from $\mathcal{F}_{2pc}$.

It is also worth noting that our protocol borrows ideas from gradual release-style protocols [14,30,35]. However, in comparison, known gradual release-style protocols rely on non-standard assumptions which are not necessary in our protocol when a clock-aware $\mathcal{G}_{att}$ is available.

We next consider whether a single secure processor enabled with trusted clock can help with fairness. We show two results: first, fairness is in impossible for generic functionalities when only one party has a clock-aware secure processor; and second, a single clock-aware secure processor allows us to fairly compute a broader class of functions than the plain setting.

Theorem 6 (Informal). *Assume that one-way functions exist, then, fair 2-party computation is impossible for general functionalities when only one party has a clock-aware secure processor (even when assuming the existence of $\mathcal{G}_{acrs}$).*

First, to prove the general fairness impossibility in the presence of a single secure processor, we consider a specific contract signing functionality $\mathcal{F}_{contract}$ in which two parties, each with a secret signing key, exchange signatures over a canonical message, say 0 (see our full version [56, Sect. 7] for a formal definition). In the plain model, there exists a (folklore) fairness impossibility proof for this functionality — and it helps to understand this proof first before presenting ours. Imprecisely speaking, if one party, say $\mathcal{P}_0$, aborts prior to sending the last protocol message, and $\mathcal{P}_0$ is able to output a correct signature over the message, then $\mathcal{P}_1$ must be able to output the correct signature as well by fairness. As a result, we can remove protocol messages one by one, and show that if the previous protocol Π_i fairly realizes $\mathcal{F}_{contract}$, then Π_{i-1} (that is, the protocol Π_i with the last message removed) must fairly realize $\mathcal{F}_{contract}$ as well. Eventually, we will arrive at the empty protocol, and conclude that the empty protocol fairly realizes $\mathcal{F}_{contract}$ as well which clearly is impossible if the signature scheme is secure. Although the intuition is simple, it turns out that the formal proof is somewhat subtle — for example, clearly the proof should not work had this been some other functionality that is not contract signing, since we know that there exist certain functions that can be computed fairly in the plain model [8,38,39]. We formalize this folklore proof and also give an alternative proof in the full version of this work [56, Sect. 7.4].

We now discuss how we can prove impossibility when only one party has a clock-aware secure processor. The overall structure of the proof is very similar to the aforementioned folklore proof where protocol messages are removed one by one, however, as we do so, we need to carefully bound the time by which the corrupt (i.e., aborting) party learns output. Without loss of generality, let us assume that party $\mathcal{P}_0$ has a secure processor and party $\mathcal{P}_1$ does not. As we remove protocol messages one by one, in each alternate round, party $\mathcal{P}_1$ is the aborting party. Suppose party $\mathcal{P}_1$ aborts in round $r \leq g(\lambda)$ where $g(\lambda)$ is the runtime of the protocol if both parties are honest. Since $\mathcal{P}_1$ does not have a secure processor, if he can learn the result in polynomially many rounds by the honest protocol, then he must be able to learn the outcome in round r too — in particular, even if the honest protocol specifies that he waits for more rounds, he can just simulate the fast forwarding of his clock in a single round and complete the remainder of his execution. This means that as we remove protocol messages one by one, in every alternate turn, the aborting party is guaranteed to obtain output by round $g(\lambda)$ — and thus even if he aborts, the other party must receive output by round $\Delta(g(\lambda))$. Similar as before, we eventually arrive at an empty protocol which we conclude to also fairly compute $\mathcal{F}_{\text{contract}}$ (where the parties do not exchange protocol messages) which clearly is impossible if the signature scheme is secure.

We stress that the ability to reset the aborting party's runtime back to $g(\lambda)$ in every alternative round is important for the proof to work. In particular, if both parties have a clock-aware secure processor, the lower bound clearly should fail in light of our upper bound — and the reason that it fails is because the runtime of the aborting party would increase by a polynomial factor every time we remove a protocol message, and after polynomially many such removals the party's runtime would become exponential.

We also note that the above is simply the intuition, and formalizing the proof is somewhat subtle which we leave to the full version of this work [56, Sect. 7.4].

Although fairness is impossible in general with only one clock-aware secure processor, we show that even one clock-aware secure processor can help with fairness too. Specifically, it broadens the set of functions that can be computed fairly in comparison with the plain setting.

Theorem 7 (Informal). *Assume that secure key exchange protocols exist, then when only a single party has a clock-aware secure processor, there exist functions that can be computed with Δ-fairness in the $(\mathcal{G}_{att}, \mathcal{G}_{acrs})$-hybrid model, but cannot be computed fairly in the $\mathcal{G}_{acrs}$-hybrid model.*

Specifically, we show that 2-party fair coin toss, which is known to be impossible in the plain model, becomes possible when only one party has a clock-aware secure processor. Intuitively, the issue in the standard setting is that the party that obtains the output first can examine the outcome coin, and can abort if he does not like the result, say abort on 0. Although the other party can now toss another coin on his own — the first party aborting already suffices to bias the remaining party's output towards 1. We now propose a $(\mathcal{G}_{att}, \mathcal{G}_{acrs})$-hybrid protocol that realizes 2-party fair toss, assuming that $\mathcal{G}_{att}$ is clock aware and that only one party has a secure processor. The idea is the following. Let the server $\mathcal{S}$ and the client $\mathcal{C}$ be the two parties involved, and suppose that the server has a secure processor but the client does not. The server's enclave first performs key exchange and establishes a secure channel with the client. Now the server's enclave flips a random coin and sends it to the client over the secure channel in a specific round, say, round 3 (e.g., assuming that key exchange takes two rounds). At this moment, the server does not see the outcome of the coin yet. If the client does not receive this coin by the end of round 3, it will flip an independent coin on its own; otherwise it outputs the coin received. Finally, in round 4, the server will receive the outcome of the coin from its local enclave. Observe that server can decide to abort prior to sending the client the coin (over the secure channel), however, the server cannot base the decision upon the value of the coin, since he does not get to see the coin until round 4. To formalize this intuition and specifically to prove the resulting protocol secure in the UC model, again we need to rely on the help of $\mathcal{G}_{\text{acrs}}$.

2.7   Additional Results

We provide some additional interesting variations in modeling and results.

The Transparent Enclave Model. Many known secure processors are known to be vulnerable to certain side-channel attacks such as cache-timing or differential power analysis. Complete defense against such side channels remains an area of active research [31–34, 49, 72].

Recently, Tramèr et al. [70] ask the question, what kind of interesting applications can we realize assuming that such side-channels are unavoidable in secure processors? Tramèr et al. [70] then propose a new model which they call the *transparent enclave* model. The transparent enclave model is almost the same as our $\mathcal{G}_{\text{att}}$, except that the enclave program leaks all internal states to the adversary $\mathcal{A}$. Nonetheless, $\mathcal{G}_{\text{att}}$ still keeps its master signing key msk secret. In practice, this model requires us to only spend effort to protect the secure processor's attestation algorithm from side channels, and we consider the entire user-defined enclave program to be transparent to the adversary.

Tramèr et al. then show how to realize interesting security tasks such as cryptographic commitments and zero-knowledge proofs with only transparent enclaves. We note that Tramèr et al. adopt modeling techniques that inherit from an earlier manuscript version of the present paper. However, Tramèr et al. model $\mathcal{G}_{\text{att}}$ as a local functionality rather than a globally shared functionality — and this lets them circumvent several technical challenges that stem from the functionality being globally shared, and allow them to achieve universally composable protocols more trivially. As mentioned earlier, if $\mathcal{G}_{\text{att}}$ were local, in practice this would mean that a fresh (mpk, msk) pair is generated for every protocol instance — even for different applications of the same user. This clearly fails to capture the reusability of real-world secure processors.

We show how to realize UC-secure commitments assuming only transparent enclaves, denoted $\widehat{\mathcal{G}}_{\text{att}}$, when both parties have a secure processor (since otherwise the task would have been impossible as noted earlier). Although intuition is quite simple — the committer could commit the value to its local enclave, and later ask the enclave to sign the opening — it turns out that this natural protocol candidate is not known to have provable security. Our actual protocol involves non-trivial techniques to achieve equivocation when the receiver is corrupt, a technical issue that arises commonly in UC proofs.

Theorem 8 (Informal). *Assume that secure key exchange protocols exist. There is a $\widehat{\mathcal{G}}_{att}$-hybrid protocol that UC-realizes $\mathcal{F}_{com}$ where $\widehat{\mathcal{G}}_{att}$ is the transparent enclave functionality.*

Challenge in Achieving Equivocation. We note that because the committer must commit its value b to its local enclave, extraction is trivial when the committer is corrupt. The challenge is how to equivocate when the receiver is corrupt. In this case, the simulator must first simulate for the corrupt receiver a commitment-phase message which contains a valid attestation. To do so, the simulator needs to ask its enclave to sign a dummy value — note that at this moment, the simulator does not know the committed value yet. Later, during the opening phase, the simulator learns the opening from the commitment ideal functionality $\mathcal{F}_{com}$. At this moment, the simulator must simulate a valid opening-phase message. The simulator cannot achieve this through the normal execution path of the enclave program, and therefore we must provide a special backdoor for the simulator to program the enclave's attestation on the opened value. Furthermore, it is important that a real-world committer who is potentially corrupt cannot make use of this backdoor to equivocate on the opening.

Our idea is therefore the following: the committer's enclave program must accept a special value c for which the receiver knows a trapdoor x such that $\mathsf{owf}(x) = c$, where owf denotes a one-way function. Further, the committer's enclave must produce an attestation on the value c such that the receiver can be sure that the correct c has been accepted by the committer's enclave. Now, if the committer produces the correct trapdoor x, then the committer's enclave will allow it to equivocate on the opening. Note that in the real-world execution, the honest receiver should never disclose x, and therefore this backdoor does not harm the security for an honest receiver. However, in the simulation when the receiver is corrupt, the simulator can capture the receiver's communication with $\widehat{\mathcal{G}}_{att}$ and extract the trapdoor x. Thus the simulator is now able to program the enclave's opening after it learns the opening from the $\mathcal{F}_{com}$ ideal functionality.

More specifically, the full protocol works as follows:

- First, the receiver selects a random trapdoor x, and sends it to its local enclave. The local enclave computes $c := \mathsf{owf}(x)$ where owf denotes a one-way function, and returns (c, σ) where σ is an attestation for c.
- Next, the committer receives (c, σ) from the receiver. If the attestation verifies, it then sends to its enclave the bit b to be committed, along with the value c that is the outcome of the one-way function over the receiver's trapdoor x. The committer's secure processor now signs the c value received in acknowledgment, and the receiver must check this attestation to make sure that the committer did send the correct c to its own enclave.
- Next, during the opening phase, the committer can ask its local enclave to sign the opening of the committed value, and demonstrate the attestation to the receiver to convince him of the opening. Due to a technicality commonly referred to as "equivocation" that arises in UC proofs, the enclave's "open" entry point provides the following backdoor: if the caller provides a pair of

values (x, b') such that $\mathsf{owf}(x) = c$ where c was stored earlier by the enclave, then the enclave will sign b' instead of the previously committed value b.

Non-anonymous Attestation. Although most of the paper is concerned about modeling anonymous attested execution as inspired by Intel's most recent SGX [6,51] and later versions of TPM [1], some secure processors instead implement non-anonymous attestation. In non-anonymous attestation, the signature binds to the platform's identity. Typically in a real-world implementation, the manufacturer embeds a long-term signing key henceforth denoted ak in each secure processor. The manufacturer then signs a certificate for the ak using its manufacturer key msk. In formal modeling, such a certificate chain can be thought of as a signature under msk, but where the message is prefixed with the platform's identity (e.g., ak).

It is not hard to see that our $(\mathcal{G}_{\mathsf{att}}, \mathcal{G}_{\mathsf{acrs}})$-hybrid protocol that realizes multiparty computation with a single secure processor can easily be adapted to work for the case of non-anonymous attestation as well. However, we point out that our 2-party protocol when both have secure processors would not be secure if we directly replaced the signatures with non-anonymous ones. Intuitively, since in the case of non-anonymous attestation, attestations bind to the platform's identity, if such signatures are transferred in the clear to remote parties, then a corrupt party can convince others of an honest party's participation in the protocol simply by demonstrating a signature from that party. In comparison, if attestations were anonymous and secure processors are omnipresent, then this would not have been an issue since the adversary could have produced such a signature on its own by asking its local secure processor.

2.8 Related Work

Trusted Hardware Built by Architects. The architecture community have been designing and building general-purpose secure processors for several decades [6,22,27,31–34,48,49,51,66,72]. The motivation for having secure processors is to minimize the trust placed in software (including the operating system and user applications) — and this seems especially valuable since software vulnerabilities have persisted and will likely continue to persist. Several efforts have been made to commercialize trusted hardware such as TPMs [1], Arm's Trustzone [5,7], and Intel's SGX [6,51]. As mentioned earlier, many of these secure processors adopt a similar attested execution abstraction despite notable differences in architectural choices, instruction sets, threat models they defend against, etc. For example, some secure processors defend against software-only adversaries [27]; others additionally defend against physical snooping of memory buses [33,34,49]; the latest Intel SGX defends against restricted classes of software and physical attackers, particularly, those that do not exploit certain side channels such as timing, and do not observe page swaps or memory access patterns (or observe but discard such information). A comprehensive survey and comparison of various secure processors is beyond the scope of this paper, and we refer the reader to the recent work by Shi et al. [64] for a systematization of knowledge and comparative taxonomy.

Besides general-purpose secure processors, other forms of trusted hardware also have been built and commercialized, e.g., hardware cryptography accelerators.

Cryptographers' Explorations of Trusted Hardware. The fact that general-purpose secure processors being built in practice have more or less converged to such an abstraction is interesting. By contrast, the cryptography community have had a somewhat different focus, typically on the minimal abstraction needed to circumvent theoretical impossibilities rather than practical performance and cost effectiveness [24,29,37,40,45]. For example, previous works showed what minimal trusted hardware abstractions are needed to realize tasks such as simulation secure program obfuscation, functional encryption, and universally composable multiparty computation — tasks known to be impossible in the plain setting. These works do not necessarily focus on practical cost effectiveness, e.g., some constructions rely on primitives such as fully homomorphic encryption [24], others require sending one or more physical hardware tokens during the protocol [37,40,42,52], thus limiting the protocol's practicality and the hardware token's global reusability. Finally, a couple recent works [42,52] also adopt the GUC framework to model hardware tokens — however, the use of GUC in these works [42,52] is to achieve composition when an adversary can possibly pass a hardware token from one protocol instance to another; in particular, like earlier cryptographic treatments of hardware tokens [24,37,45], these works [42,52] consider the same model where the hardware tokens are passed around between parties during protocol execution, and not realistic secure processors like SGX.

Use of Trusted Hardware in Applications. Numerous works have demonstrated how to apply trusted hardware to design secure cloud systems [11, 28,50,61,62], cryptocurrency systems [73], collaborative data analytics applications [55], and others [12,23,59,63]. Due to the lack of formal abstractions for secure processors, most of these works take an approach that ranges from heuristic security to semi-formal reasoning. We hope that our work can lay the foundations for formally correctly employing secure processors in applications.

Formal Security Meets Realistic Trusted Hardware. A couple earlier works have aimed to provide formal abstractions for realistic trusted hardware [10,65], however, they either do not support cryptographically sound reasoning [65], or do not support cryptographically sound composition in general protocol design [10].

We note that our goal of having cryptographically sound formal abstractions for trusted hardware is complementary and orthogonal to the goal of providing formally correct implementations of trusted hardware [31,72]. In general, building formally verified *implementations* of trusted hardware — particularly, one that realizes the abstractions proposed in this paper — still remains a grand challenge of our community.

3 Formal Definitions, Constructions, and Proofs

In the interest of space, we present our formal definitions, constructions, and proofs in a full version of this work [56] — we refer the reader to the technical roadmap section for an intuitive explanation of the key technical insights, the technicalities that arise in proofs, and how we handle them.

Acknowledgments. We thank Elette Boyle, Kai-Min Chung, Victor Costan, Srini Devadas, Ari Juels, Andrew Miller, Dawn Song, and Fan Zhang for helpful and supportive discussions. This work is supported in part by NSF grants CNS-1217821, CNS-1314857, CNS-1514261, CNS-1544613, CNS-1561209, CNS-1601879, CNS-1617676, AFOSR Award FA9550-15-1-0262, an Office of Naval Research Young Investigator Program Award, a Microsoft Faculty Fellowship, a Packard Fellowship, a Sloan Fellowship, Google Faculty Research Awards, and a VMWare Research Award. This work was done in part while a subset of the authors were visiting the Simons Institute for the Theory of Computing, supported by the Simons Foundation and by the DIMACS/Simons Collaboration in Cryptography through NSF grant CNS-1523467. The second author would like to thank Adrian Perrig and Leendert van Doorn for many helpful discussions on trusted hardware earlier in her research.

References

1. Trusted computing group. http://www.trustedcomputinggroup.org/
2. Abadi, M., Jürjens, J.: Formal eavesdropping and its computational interpretation. In: Kobayashi, N., Pierce, B.C. (eds.) TACS 2001. LNCS, vol. 2215, pp. 82–94. Springer, Heidelberg (2001). doi:10.1007/3-540-45500-0_4
3. Abadi, M., Rogaway, P.: Reconciling two views of cryptography (the computational soundness of formal encryption). J. Cryptol. **20**(3), 395 (2007)
4. Adão, P., Bana, G., Herzog, J., Scedrov, A.: Soundness of formal encryption in the presence of key-cycles. In: Vimercati, S.C., Syverson, P., Gollmann, D. (eds.) ESORICS 2005. LNCS, vol. 3679, pp. 374–396. Springer, Heidelberg (2005). doi:10.1007/11555827_22
5. Alves, T., Felton, D.: Trustzone: integrated hardware and software security. Inf. Q. **3**(4), 18–24 (2004)
6. Anati, I., Gueron, S., Johnson, S.P., Scarlata, V.R.: Innovative technology for CPU based attestation and sealing. In: HASP (2013)
7. ARM Limited: ARM Security Technology Building a Secure System using TrustZone® Technology, April 2009. Reference no. PRD29-GENC-009492C
8. Asharov, G., Beimel, A., Makriyannis, N., Omri, E.: Complete characterization of fairness in secure two-party computation of Boolean functions. In: Dodis, Y., Nielsen, J.B. (eds.) TCC 2015. LNCS, vol. 9014, pp. 199–228. Springer, Heidelberg (2015). doi:10.1007/978-3-662-46494-6_10
9. Backes, M., Pfitzmann, B., Waidner, M.: A universally composable cryptographic library. IACR Cryptology ePrint Archive, 2003:15 (2003)
10. Barbosa, M., Portela, B., Scerri, G., Warinschi, B.: Foundations of hardware-based attested computation and application to SGX. In: IEEE European Symposium on Security and Privacy, pp. 245–260 (2016)
11. Baumann, A., Peinado, M., Hunt, G.: Shielding applications from an untrusted cloud with haven. In: OSDI (2014)

12. Berger, S., Cáceres, R., Goldman, K.A., Perez, R., Sailer, R., van Doorn, L.: vTPM: virtualizing the trusted platform module. In: USENIX Security (2006)
13. Bohl, F., Unruh, D.: Symbolic universal composability. In: IEEE Computer Security Foundations Symposium, pp. 257–271 (2013)
14. Boneh, D., Naor, M.: Timed commitments. In: Bellare, M. (ed.) CRYPTO 2000. LNCS, vol. 1880, pp. 236–254. Springer, Heidelberg (2000). doi:10.1007/3-540-44598-6_15
15. Brickell, E., Camenisch, J., Chen, L.: Direct anonymous attestation. In: CCS (2004)
16. Brickell, E., Li, J.: Enhanced privacy id from bilinear pairing. IACR Cryptology ePrint Archive, 2009:95 (2009)
17. Canetti, R.: Universally composable security: a new paradigm for cryptographic protocols. In: FOCS (2001)
18. Canetti, R., Dodis, Y., Pass, R., Walfish, S.: Universally composable security with global setup. In: Vadhan, S.P. (ed.) TCC 2007. LNCS, vol. 4392, pp. 61–85. Springer, Heidelberg (2007). doi:10.1007/978-3-540-70936-7_4
19. Canetti, R., Fischlin, M.: Universally composable commitments. In: Kilian, J. (ed.) CRYPTO 2001. LNCS, vol. 2139, pp. 19–40. Springer, Heidelberg (2001). doi:10.1007/3-540-44647-8_2
20. Canetti, R., Herzog, J.: Universally composable symbolic security analysis. J. Cryptol. 24(1), 83–147 (2011)
21. Canetti, R., Rabin, T.: Universal composition with joint state. In: Boneh, D. (ed.) CRYPTO 2003. LNCS, vol. 2729, pp. 265–281. Springer, Heidelberg (2003). doi:10.1007/978-3-540-45146-4_16
22. Champagne, D., Lee, R.B.: Scalable architectural support for trusted software. In: HPCA (2010)
23. Chen, C., Raj, H., Saroiu, S., Wolman, A.: cTPM: a cloud TPM for cross-device trusted applications. In: NSDI (2014)
24. Chung, K.-M., Katz, J., Zhou, H.-S.: Functional encryption from (small) hardware tokens. In: Sako, K., Sarkar, P. (eds.) ASIACRYPT 2013. LNCS, vol. 8270, pp. 120–139. Springer, Heidelberg (2013). doi:10.1007/978-3-642-42045-0_7
25. Cleve, R.: Limits on the security of coin flips when half the processors are faulty. In: STOC 1986, pp. 364–369 (1986)
26. Costan, V., Devadas, S.: Intel SGX explained. Manuscript (2015)
27. Costan, V., Lebedev, I., Devadas, S.: Sanctum: minimal hardware extensions for strong software isolation. In: USENIX Security (2016)
28. Dinh, T.T.A., Saxena, P., Chang, E.-C., Ooi, B.C., Zhang, C.: M2R: enabling stronger privacy in MapReduce computation. In: USENIX Security (2015)
29. Döttling, N., Mie, T., Müller-Quade, J., Nilges, T.: Basing obfuscation on simple tamper-proof hardware assumptions. IACR Cryptology ePrint Archive 2011:675 (2011)
30. Even, S., Goldreich, O., Lempel, A.: A randomized protocol for signing contracts. Commun. ACM 28(6), 637–647 (1985)
31. Ferraiuolo, A., Wang, Y., Rui, X., Zhang, D., Myers, A., Edward, G.S.: Full-processor timing channel protection with applications to secure hardware compartments (2015)
32. Fletcher, C.W., van Dijk, M., Devadas, S.: A secure processor architecture for encrypted computation on untrusted programs. In: STC (2012)
33. Fletcher, C.W., Ren, L., Kwon, A., van Dijk, M., Stefanov, E., Devadas, S.: RAW Path ORAM: a low-latency, low-area hardware ORAM controller with integrity verification. IACR Cryptology ePrint Archive 2014:431 (2014)

34. Fletcher, C.W., Ren, L., Xiangyao, Y., van Dijk, M., Khan, O., Devadas, S.: Suppressing the oblivious RAM timing channel while making information leakage and program efficiency trade-offs. In: HPCA, pp. 213–224 (2014)
35. Garay, J., MacKenzie, P., Prabhakaran, M., Yang, K.: Resource fairness and composability of cryptographic protocols. In: Halevi, S., Rabin, T. (eds.) TCC 2006. LNCS, vol. 3876, pp. 404–428. Springer, Heidelberg (2006). doi:10.1007/11681878_21
36. Genkin, D., Pachmanov, L., Pipman, I., Shamir, A., Tromer, E.: Physical key extraction attacks on PCs. Commun. ACM **59**(6), 70–79 (2016)
37. Goldwasser, S., Kalai, Y.T., Rothblum, G.N.: One-time programs. In: Wagner, D. (ed.) CRYPTO 2008. LNCS, vol. 5157, pp. 39–56. Springer, Heidelberg (2008). doi:10.1007/978-3-540-85174-5_3
38. Dov Gordon, S., Katz, J.: Complete fairness in multi-party computation without an honest majority. In: Reingold, O. (ed.) TCC 2009. LNCS, vol. 5444, pp. 19–35. Springer, Heidelberg (2009). doi:10.1007/978-3-642-00457-5_2
39. Dov Gordon, S., Hazay, C., Katz, J., Lindell, Y.: Complete fairness in secure two-party computation. J. ACM **58**(6), 24:1–24:37 (2011)
40. Goyal, V., Ishai, Y., Sahai, A., Venkatesan, R., Wadia, A.: Founding cryptography on tamper-proof hardware tokens. In: Micciancio, D. (ed.) TCC 2010. LNCS, vol. 5978, pp. 308–326. Springer, Heidelberg (2010). doi:10.1007/978-3-642-11799-2_19
41. Gupta, D., Mood, B., Feigenbaum, J., Butler, K., Traynor, P.: Using intel software guard extensions for efficient two-party secure function evaluation. In: Clark, J., Meiklejohn, S., Ryan, P.Y.A., Wallach, D., Brenner, M., Rohloff, K. (eds.) FC 2016. LNCS, vol. 9604, pp. 302–318. Springer, Heidelberg (2016). doi:10.1007/978-3-662-53357-4_20
42. Hazay, C., Polychroniadou, A., Venkitasubramaniam, M.: Composable security in the tamper-proof hardware model under minimal complexity. In: Hirt, M., Smith, A. (eds.) TCC 2016. LNCS, vol. 9985, pp. 367–399. Springer, Heidelberg (2016). doi:10.1007/978-3-662-53641-4_15
43. Horvitz, O., Gligor, V.: Weak key authenticity and the computational completeness of formal encryption. In: Boneh, D. (ed.) CRYPTO 2003. LNCS, vol. 2729, pp. 530–547. Springer, Heidelberg (2003). doi:10.1007/978-3-540-45146-4_31
44. Janvier, R., Lakhnech, Y., Mazaré, L.: Completing the picture: soundness of formal encryption in the presence of active adversaries. In: Sagiv, M. (ed.) ESOP 2005. LNCS, vol. 3444, pp. 172–185. Springer, Heidelberg (2005). doi:10.1007/978-3-540-31987-0_13
45. Katz, J.: Universally composable multi-party computation using tamper-proof hardware. In: Naor, M. (ed.) EUROCRYPT 2007. LNCS, vol. 4515, pp. 115–128. Springer, Heidelberg (2007). doi:10.1007/978-3-540-72540-4_7
46. Kauer, B.: TPM reset attack. http://www.cs.dartmouth.edu/~pkilab/sparks/
47. Kocher, P., Jaffe, J., Jun, B.: Differential power analysis. In: Wiener, M. (ed.) CRYPTO 1999. LNCS, vol. 1666, pp. 388–397. Springer, Heidelberg (1999). doi:10.1007/3-540-48405-1_25
48. Lie, D., Thekkath, C., Mitchell, M., Lincoln, P., Boneh, D., Mitchell, J., Horowitz, M.: Architectural support for copy and tamper resistant software. ACM SIGPLAN Not. **35**(11), 168–177 (2000)
49. Maas, M., Love, E., Stefanov, E., Tiwari, M., Shi, E., Asanovic, K., Kubiatowicz, J., Song, D.: Phantom: practical oblivious computation in a secure processor. In: CCS (2013)

50. Martignoni, L., Poosankam, P., Zaharia, M., Han, J., McCamant, S., Song, D., Paxson, V., Perrig, A., Shenker, S., Stoica, I.: Cloud terminal: secure access to sensitive applications from untrusted systems. In: USENIX ATC (2012)
51. McKeen, F., Alexandrovich, I., Berenzon, A., Rozas, C.V., Shafi, H., Shanbhogue, V., Savagaonkar, U.R.: Innovative instructions and software model for isolated execution. In: HASP 2013: 10 (2013)
52. Mechler, J., Mller-Quade, J., Nilges, T.: Universally composable (non-interactive) two-party computation from untrusted reusable hardware tokens. Cryptology ePrint Archive, Report 2016/615 (2016). http://eprint.iacr.org/2016/615
53. Micciancio, D., Warinschi, B.: Completeness theorems for the Abadi-Rogaway language of encrypted expressions. J. Comput. Secur. **12**(1), 99–129 (2004)
54. Micciancio, D., Warinschi, B.: Soundness of formal encryption in the presence of active adversaries. In: Naor, M. (ed.) TCC 2004. LNCS, vol. 2951, pp. 133–151. Springer, Heidelberg (2004). doi:10.1007/978-3-540-24638-1_8
55. Ohrimenko, O., Schuster, F., Fournet, C., Mehta, A., Nowozin, S., Vaswani, K., Costa, M.: Oblivious multi-party machine learning on trusted processors. In: USENIX Security, August 2016
56. Pass, R., Shi, E., Tramèr, F.: Formal abstractions for attested execution secure processors. IACR Cryptology ePrint Archive 2016:1027 (2016)
57. Petcher, A., Morrisett, G.: The foundational cryptography framework. In: Focardi, R., Myers, A. (eds.) POST 2015. LNCS, vol. 9036, pp. 53–72. Springer, Heidelberg (2015). doi:10.1007/978-3-662-46666-7_4
58. Petcher, A., Morrisett, G.: A mechanized proof of security for searchable symmetric encryption. In: CSF (2015)
59. Sailer, R., Zhang, X., Jaeger, T., van Doorn, L.: Design and implementation of a TCG-based integrity measurement architecture. In: USENIX Security (2004)
60. Santos, N., Raj, H., Saroiu, S., Wolman, A.: Using arm trustzone to build a trusted language runtime for mobile applications. SIGARCH Comput. Archit. News **42**(1), 67–80 (2014)
61. Santos, N., Rodrigues, R., Gummadi, K.P., Saroiu, S.: Policy-sealed data: a new abstraction for building trusted cloud services. In: USENIX Security, pp. 175–188 (2012)
62. Schuster, F., Costa, M., Fournet, C., Gkantsidis, C., Peinado, M., Mainar-Ruiz, G., Russinovich, M.: VC3: trustworthy data analytics in the cloud. In: IEEE S&P (2015)
63. Shi, E., Perrig, A., Van Doorn, L.: BIND: a fine-grained attestation service for secure distributed systems. In: IEEE S&P (2005)
64. Shi, E., Zhang, F., Pass, R., Devadas, S., Song, D., Liu, C.: Systematization of knowledge: trusted hardware: life, the composable universe, and everything. Manuscript (2015)
65. Smith, S.W., Austel, V.: Trusting trusted hardware: towards a formal model for programmable secure coprocessors. In: Proceedings of the 3rd Conference on USENIX Workshop on Electronic Commerce, WOEC 1998, vol. 3 (1998)
66. Suh, G.E., Clarke, D., Gassend, B., Van Dijk, M., Devadas, S.: AEGIS: architecture for tamper-evident and tamper-resistant processing. In: Proceedings of the 17th Annual International Conference on Supercomputing, pp. 160–171. ACM (2003)
67. Szczys, M.: TPM crytography cracked. http://hackaday.com/2010/02/09/tpm-crytography-cracked/
68. Lie, D., Thekkath, C., Mitchell, M., Lincoln, P., Boneh, D., Mitchell, J., Horowitz, M.: Architectural support for copy and tamper resistant software. SIGOPS Oper. Syst. Rev. **34**(5), 168–177 (2000)

69. Thompson, K.: Reflections on trusting trust. Commun. ACM **27**(8), 761–763 (1984)
70. Tramèr, F., Zhang, F., Lin, H., Hubaux, J.-P., Juels, A., Shi, E.: Sealed-glass proofs: using transparent enclaves to prove and sell knowledge. In: IEEE European Symposium on Security and Privacy (2017)
71. Yuanzhong, X., Cui, W., Peinado, M.: Controlled-channel attacks: deterministic side channels for untrusted operating systems. In: IEEE S&P (2015)
72. Zhang, D., Yao Wang, G., Suh, E., Myers, A.C.: A hardware design language for timing-sensitive information-flow security. In: ASPLOS (2015)
73. Zhang, F., Cecchetti, E., Croman, K., Juels, A., Shi, E.: Town crier: an authenticated data feed for smart contracts. In: ACM CCS (2016)
74. Zhuang, X., Zhang, T., Pande, S.: Hide: an infrastructure for efficiently protecting information leakage on the address bus. SIGARCH Comput. Archit. News **32**(5), 72–84 (2004)

Lattice Attacks and Constructions II

Lattice Attacks and Constructions II

One-Shot Verifiable Encryption from Lattices

Vadim Lyubashevsky[✉] and Gregory Neven

IBM Research, Zurich, Switzerland
{vad,nev}@zurich.ibm.com

Abstract. Verifiable encryption allows one to prove properties about encrypted data and is an important building block in the design of cryptographic protocols, e.g., group signatures, key escrow, fair exchange protocols, etc. Existing lattice-based verifiable encryption schemes, and even just proofs of knowledge of the encrypted data, require parallel composition of proofs to reduce the soundness error, resulting in proof sizes that are only truly practical when amortized over a large number of ciphertexts.

In this paper, we present a new construction of a verifiable encryption scheme, based on the hardness of the Ring-LWE problem in the random-oracle model, for short solutions to linear equations over polynomial rings. Our scheme is "one-shot", in the sense that a single instance of the proof already has negligible soundness error, yielding compact proofs even for individual ciphertexts. Whereas verifiable encryption usually guarantees that decryption can recover a witness for the original language, we relax this requirement to decrypt a witness of a related but extended language. This relaxation is sufficient for many applications and we illustrate this with example usages of our scheme in key escrow and verifiably encrypted signatures.

One of the interesting aspects of our construction is that the decryption algorithm is probabilistic and uses the proof as input (rather than using only the ciphertext). The decryption time for honestly-generated ciphertexts only depends on the security parameter, while the *expected* running time for decrypting an adversarially-generated ciphertext is directly related to the number of random-oracle queries of the adversary who created it. This property suffices in most practical scenarios, especially in situations where the ciphertext proof is part of an interactive protocol, where the decryptor is substantially more powerful than the adversary, or where adversaries can be otherwise discouraged to submit malformed ciphertexts.

1 Introduction

Lattice cryptography has matured to the point where the general belief is that any primitive that can be constructed from any other assumption can also be constructed based on a lattice assumption. The main question that remains is how efficient (in a practical, rather than asymptotic, sense) one can make the lattice-based constructions. A primitive that has been getting a lot of recent attention is a "proof of plaintext knowledge."

© International Association for Cryptologic Research 2017
J.-S. Coron and J.B. Nielsen (Eds.): EUROCRYPT 2017, Part I, LNCS 10210, pp. 293–323, 2017.
DOI: 10.1007/978-3-319-56620-7_11

In a proof of plaintext knowledge, a prover who has a message μ produces a ciphertext $t = \text{Enc}(\mu)$ and a zero-knowledge proof of knowledge π showing that he knows the value of $\text{Dec}(t)$. Proving knowledge of the value of $\text{Dec}(t)$ is usually the same as proving that t is a correctly formed ciphertext along with proving the knowledge of μ that was used to construct it.

By itself, a proof of plaintext knowledge is not particularly useful, and it is almost always used as a part of a primitive known as a verifiable encryption scheme. In such a scheme, there is a relation R_L and a language

$$L = \{x \; : \; \exists w \text{ s.t. } R_L(x, w) = 1\}.$$

Thus the value w is a witness to the fact that x is in the language L. The relation R_L and the element x are public, while the prover possesses the secret witness w. He then produces an encryption $t = \text{Enc}(w)$ as well as a zero-knowledge proof of knowledge π of the value $w = \text{Dec}(t)$ and that w satisfies $R_L(x, w) = 1$. Verifiable encryption can therefore also be seen as an extractable non-interactive zero-knowledge proof. It is a building block for many primitives, e.g.,

- group signatures [CvH91], where a group manager hands distinct signing keys to all users, using which they can anonymously sign messages. A trusted opener is able to trace back a signature to the identity of the signer. A common construction [CL06] is to let users verifiably encrypt their identity under the opener's public key together with a proof that they know a signature by the group manager on the same identity.
- key escrow protocols [YY98,PS00], where users encrypt their decryption key under the public key of a trusted escrow authority. Using verifiable encryption, communication partners or network providers can check that the ciphertext indeed encrypts the user's decryption key, and not some bogus data.
- optimistic fair exchange protocols [ASW00,BDM98], where two parties can fairly exchange secrets by, in a first step, proving that they encrypted their respective secrets under the public key of a trusted authority, who can later be called upon to recover the secret in case one of the parties aborts the protocol early;
- verifiable secret sharing, where one dealer sends verifiably encrypted shares of a secret to a set of parties, and proves to an external third party that the ciphertexts contain actual shares of the secret.

1.1 Proofs of Plaintext Knowledge from Lattices – Prior Work

If one uses a lattice-based encryption scheme based on LWE or Ring-LWE, then the encryption of a message m satisfies the linear relation $\mathbf{A} \begin{bmatrix} \mathbf{r} \\ m \end{bmatrix} = \mathbf{t} \bmod q$. There are several known techniques to prove that the ciphertext $\mathbf{t}$ is well-formed and one knows the message m. One technique is adapting Stern's protocol based on permutations [Ste93] to lattice-based schemes [LNSW13]. This approach is unfortunately very impractical due to the fact that each round of the protocol has soundness error $2/3$ (and therefore needs to be repeated 192 times to achieve

128-bit security). Furthermore, if proving relations where the secret has coefficients of size k, the size of the proof increases by a factor of $\log k$ [LNSW13]. This makes schemes using Stern's protocol unsuitable for most practical applications.

Another approach is to use the "Fiat-Shamir with Aborts" zero-knowledge proof technique from [Lyu09, Lyu12] with 0/1 challenges. This also has the problem of having soundness error $1/2$ and needing to be repeated 128 times for 128-bit security. This approach, however, is more algebraic than Stern's proof of knowledge and it was shown to admit several improvements. If one uses a Ring-LWE encryption scheme, then it was shown in [BCK+14] how the soundness error can be reduced to $1/(2n)$, where n is the dimension of the ring being used (typically 1024 or 2048). For 128-bit security, one then only needs to run the protocol around a dozen times. A scenario where the Fiat-Shamir with Aborts technique leads to truly practical protocols is when one wants to simultaneously do many proofs of plaintext knowledge. If one then considers the amortized cost of the proof of knowledge, then the number of iterations is only a small constant (approaches 2 as the number of instances increases) per proof [DPSZ12, BDLN16, CD16].

Despite having received considerable attention in the literature, there seems to be no satisfactory solution for the most natural scenario where the prover has a single instance to prove and would like to do it in "one shot"—that is, without repeating a protocol to amplify soundness. It is therefore conceivable that lattice-based encryption schemes are not compatible with efficient proofs of plaintext knowledge, which would make all the applications much less efficient than their number theoretic counterparts.

1.2 Proofs of Plaintext Knowledge – Our Results

In this work, we introduce a very efficient "one-shot" protocol for proving plaintext knowledge. The caveat is that the running time of the decryption algorithm depends on the running time of the prover. More precisely, our decryption algorithm is *randomized* in that it tries to decrypt ciphertexts that are "close" to the one provided by the prover. And we show that the *expected* number of decryption tries our decryptor needs is within a small factor (essentially 1) of the number of *random oracle queries* that the prover makes while constructing the proof of knowledge π. If q is the number of queries made by the prover, then Markov's inequality implies that the probability that the decryptor will need more than $\alpha \cdot q$ decryption tries is less than $1/\alpha$. If the prover is honest, though, then the decryptor will succeed from the first try.

While tying the decryption time to the adversary's running time is unusual, this should be acceptable in many scenarios. Apart from creating out-of-band incentives such as fines to prevent cheating, there are also technical ways to limit the power of the adversary. If the protocol in which the proof of knowledge is being used is interactive, then the verifier can send the prover a fresh salt during every interaction that has to be included in the cryptographic hash function (modeled as a random oracle) and require that the prover performs the proof within a certain small amount of time. Thus the adversary will have a limited time-frame during which he can make queries to the random oracle (because each new salt in essence creates a new random oracle). The decryption algorithm, on the other hand, is almost always off-line and is therefore allowed more time. In non-interactive settings, the prover can be required to use a salt from a public "randomness beacon" (such as one provided by NIST) at the time the proof was created.

In our scheme, the verification algorithm uses one random oracle query and the decryption algorithm uses none. Thus another simple way of preventing an adversary from using too many random oracle queries during encryption would be to artificially make the computational complexity of computing the hash function high (e.g. by iterating SHA-256 some number of times to produce one output). This has the effect of significantly slowing down a cheating prover, while keeping the decryption time exactly the same.

We also show that, if one wishes, one can upper-bound the running time of the decryptor by making the protocol "k-shot" rather than one shot. In this scenario, the length of the proof of knowledge would go up by a factor k, but one could bound the decryption algorithm's running time to $k \cdot 2^{\lambda/k}$ for λ-bit security.

1.3   Verifiable Encryption Schemes – Our Results

We build upon our proof of plaintext knowledge to construct a verifiable encryption scheme that is adapted to be used as a building block for lattice constructions. The relations that are most common in lattice cryptography are those of the form

$$\mathbf{Bm} = \mathbf{u} \bmod p \qquad (1)$$

where $\mathbf{B}$ is a matrix over some ring, $\mathbf{m}$ is a vector with small coefficients, and $\mathbf{u}$ is the product of $\mathbf{Bm}$ modulo p. For example, in (Ring)-LWE encryption $\mathbf{B}, \mathbf{u}$ is the public key and $\mathbf{m}$ is the secret key. In full domain hash signatures, $\mathbf{B}$ is the public key, $\mathbf{m}$ is the signature, and $\mathbf{u} = \mathsf{H}(\mu)$ is derived from the message μ. Giving a verifiable encryption scheme for such relations is a main building block for many of the protocols listed in the introduction.

While verifiable encryption would normally guarantee that decrypting a valid ciphertext yields a witness satisfying (1), our construction relaxes this guarantee to only yield a witness $(\overline{\mathbf{m}}, \overline{c})$ with small coefficients satisfying

$$\mathbf{B}\overline{\mathbf{m}} = \overline{c}\mathbf{u} \bmod p. \qquad (2)$$

This relaxation actually turns out to be sufficient for many applications of verifiable encryption. Lattice schemes can often be slightly augmented to allow for relations of the form (2) to be "useful" whenever those of the form (1) are. We will see this in the two examples provided in Sect. 6.

Notice also how it appears as if the decryption and the proof of knowledge are disjoint. Indeed, the proof of knowledge π may prove the existence of some witness $(\overline{\mathbf{m}}, \overline{c})$, whereas the decryption algorithm may obtain a completely different witness $(\overline{\mathbf{m}}', \overline{c}')$. But in addition to still being sufficient for many applications, there is also a connection between the two tuples that is actually crucial to our construction – we have that

$$\overline{\mathbf{m}}/\overline{c} = \overline{\mathbf{m}}'/\overline{c}' \bmod p.$$

While this property is not needed in many applications, the presence of this relationship may be useful when constructing group signatures or other primitives where it is important that the decryption recovers some specific attribute of the prover rather than just a witness to a relation.

1.4 Paper Organization

We present the Ring-LWE encryption scheme and the non-interactive "Fiat-Shamir with Aborts" zero-knowledge proof protocol in Sects. 2.5 and 2.6. Slight variations of these two primitives are used throughout our constructions.

We then present the definitions of our relaxed version of verifiable encryption in Sect. 3.1, and describe all the elements of the scheme in Sect. 3.2. In Sect. 7, we give some example instantiations for the proofs of plaintext knowledge and verifiable encryption schemes. The proof of plaintext knowledge scheme requires 9 KB for the ciphertext and 9 KB for the proof. This is quite efficient since this ciphertext is only around 4 times larger than a regular Ring-LWE ciphertext.

The efficiency of the verifiable encryption scheme is mostly affected by the size of the modulus p and the witness $\mathbf{m}$ in the relation. The larger these values, the larger the proofs and ciphertexts will be. The sample instantiations in Sect. 7 are meant to support the two sample applications in Sect. 6, where we describe how our relaxed verifiable encryption scheme can be used to build key escrow schemes and verifiably encrypted signatures. In Sects. 4 and 5 we describe two variants of our schemes, the former trading longer ciphertexts for bounded decryption time, the latter adding simulatability under chosen-ciphertext to the scheme.[1]

2 Preliminaries

For a set S, we write $a \xleftarrow{\$} S$ to mean that a is chosen uniformly at random from S. If D is a distribution, then $a \xleftarrow{\$} D$ signifies that a is randomly chosen according to the distribution D. The assignment operator $a \leftarrow b$ signifies that a gets assigned the value b. We will also sometimes write column vectors of the form $\begin{bmatrix} a_1 \\ \ldots \\ a_k \end{bmatrix}$ as $[\, a_1 \,;\, \ldots \,;\, a_k \,]$.

---

[1] Unlike Camenisch and Shoup [CS03], we cannot use standard indistinguishability security notions, because our decryption algorithm needs the proof to be included in the ciphertext.

2.1   The Ring $\mathbb{Z}[\mathrm{x}]/(\mathrm{x}^n + 1)$.

Consider the ring $R = \mathbb{Z}[\mathrm{x}]/(\mathrm{x}^n + 1)$ and $R_q = \mathbb{Z}_q[\mathrm{x}]/(\mathrm{x}^n + 1)$ where n is a power of 2 integer and q is some prime. The elements of the latter ring are polynomials of degree at most $n - 1$ with coefficients between $-(q - 1)/2$ and $(q - 1)/2$ (for the ring R, there is no restriction on the sizes of coefficients). All definitions that follow apply both to R and R_q. We will denote elements of $\mathbb{Z}$ and of R by lower-case letters, elements of vectors in R^k by bold lower-case letters, and of matrices in $R^{k \times l}$ by bold upper-case letters.

We will define the ℓ_1, ℓ_2, and ℓ_∞ lengths of an element $\mathrm{a} = \sum\limits_{i=0}^{n-1} a_i \mathrm{x}^i \in R$ as

$$\|\mathrm{a}\|_1 = \sum_{i=0}^{n-1} |a_i|, \quad \|\mathrm{a}\| = \sqrt{\sum_{i=0}^{n-1} a_i^2} \quad \text{and} \quad \|\mathrm{a}\|_\infty = \max_i(|a_i|)$$

respectively.[2] For k-dimensional vectors $\mathbf{a} = [\ a_1\ |\ \dots\ |\ a_k\ ] \in R^k$, we write $\|\mathbf{a}\|_1 = \|a_1\|_1 + \dots + \|a_k\|_1$, $\|\mathbf{a}\| = \sqrt{\|a_1\|^2 + \dots + \|a_k\|^2}$ and $\|\mathbf{a}\|_\infty = \max_i \|a_i\|_\infty$. We will denote by S_i (respectively S_i^k), the set of elements of R (resp. of R^k) whose ℓ_∞ length is at most i.

It is not hard to check that for any two polynomials $\mathrm{a}, \mathrm{b} \in R$, we have $\|\mathrm{ab}\|_\infty \le \|\mathrm{a}\|_1 \cdot \|\mathrm{b}\|_\infty$ and $\|\mathrm{ab}\|_\infty \le \|\mathrm{a}\| \cdot \|\mathrm{b}\|$. Similarly for $\mathbf{a}, \mathbf{b} \in R^k$, we have the same inequalities on the ℓ_∞ norms of their inner products: that is, $\|\mathbf{a} \cdot \mathbf{b}\|_\infty \le \|\mathbf{a}\|_1 \cdot \|\mathbf{b}\|_\infty$ and $\|\mathbf{a} \cdot \mathbf{b}\|_\infty \le \|\mathbf{a}\| \cdot \|\mathbf{b}\|$.

2.2   Special Properties of $\mathbb{Z}_q[\mathrm{x}]/(\mathrm{x}^n + 1)$

The algebraic properties of the ring $R_q = Z_q[\mathrm{x}]/(\mathrm{x}^n + 1)$, where n is a power of 2, depend on the prime q. For efficiency, one often takes $q = 1 \bmod (2n)$, which results in the polynomial $\mathrm{x}^n + 1$ splitting into n linear factors modulo q. Operations within the ring can then be performed extremely efficiently using the number theory transform. On the other hand, one sometimes wants the ring to be "almost a field". In particular, it is sometimes desirable for the ring to have many invertible elements. While there do not exist q that will make R_q a field, using $q = 3 \bmod 8$ (as first suggested in [SSTX09]) has the effect that $\mathrm{x}^n + 1$ factors into two irreducible polynomials of degree $n/2$ and so the ring R_q contains $q^n - 2q^{n/2} + 1$ invertible elements. By the Chinese Remainder theorem, it is also easy to see that all elements of degree less than $n/2$ are invertible.

We have not seen this used in previous works, but it turns out that setting $q = 5 \bmod 8$ may be even more convenient. Modulo such a q, the polynomial $\mathrm{x}^n + 1$ also factors into two irreducible polynomials of degree $n/2$. And in addition to all elements of degree less than $n/2$ being invertible, one can also show that all elements (of degree up to n) with small coefficients are invertible as well. We

---

[2] We point out that since $\mathbb{Z}_q$ is a finite group, these do not correspond exactly to norms when working in R_q because we do not have $\|\alpha \cdot \mathrm{a}\| = \alpha \cdot \|\mathrm{a}\|$. The other two properties of norms (i.e. that the norm of 0 is 0 and the triangle inequality) do hold.

present the proof of this statement in Lemma 2.2. To the best of our knowledge, this lemma was first proven in an unpublished manuscript of Lyubashevsky and Xagawa.

Lemma 2.1 ([LN86], special case of Theorem 3.35, p. 88). *Let* $q = 5(\bmod 8)$ *be prime and r be an integer such that $r^2 = -1(\bmod q)$. Then for all positive integers κ, the polynomials $x^{2^\kappa} - r$ and $x^{2^\kappa} + r$ are irreducible over $\mathbb{Z}_q[x]$. And in particular, the complete factorization into irreducibles over $\mathbb{Z}_q[x]$ of the polynomial $x^{2^{\kappa+1}} + 1$ is $x^{2^{\kappa+1}} + 1 = (x^{2^\kappa} - r)(x^{2^\kappa} + r) \bmod q$.*

Lemma 2.2. *Let $R_q = \mathbb{Z}_q[x]/(x^n + 1)$ where $n > 1$ is a power of 2 and q is a prime congruent to $5(\bmod 8)$. This ring has exactly $2q^{n/2} - 1$ elements without an inverse. Moreover, every non-zero polynomial a in R_q with $\|a\|_\infty < \sqrt{q/2}$ has an inverse.*

Proof. In all that follows, the reduction modulo q will be implicit. By Lemma 2.1, $x^n + 1 = (x^{n/2} - r)(x^{n/2} + r)$ where $r^2 = -1$ and $x^{n/2} \pm r$ are irreducible. Any element $a \in R_q$ can be written as $a = a_0 + x^{n/2}a_1$, where a_0, a_1 are polynomials in $\mathbb{Z}[x]$ of degree less than $n/2$. Then the Chinese remainder decomposition of a is

$$CRT(a) = (a \bmod (x^{n/2} - r), a \bmod (x^{n/2} + r)) = (a_0 + ra_1, a_0 - ra_1).$$

If a is not invertible, it means that either $a_0 + ra_1 = 0$ or $a_0 - ra_1 = 0$. If $a_1 = 0$, then $a_0 = 0$ and a is the zero polynomial. If $a_1 \neq 0$, then some coefficient of a_0, say α_0, must be equal to $\pm r\alpha_1$, where α_1 is a non-zero coefficient of a_1. Therefore we have $\alpha_0^2 = (\pm r\alpha_1)^2 = -\alpha_1^2$. In other words, $\alpha_0^2 + \alpha_1^2 = 0$. But since we assumed that $|\alpha_0|, |\alpha_1| < \sqrt{q/2}$, this is not possible, and thus proves the second part of the lemma by contradiction. The first part of the lemma follows from the fact that CRT is a bijection and all the elements without an inverse must be 0 modulo at least one of $x^{n/2} \pm r$. □

2.3   Lattices and the Discrete Gaussian Distribution

A full-rank integer lattice Λ of dimension n is an additive subgroup of $\mathbb{Z}^n$ that is generated by some basis $\mathbf{B} = [\mathbf{b}_1 \mid \ldots \mid \mathbf{b}_n] \in \mathbb{Z}^{n \times n}$ of linearly-independent vectors. If a basis $\mathbf{B}$ is a generator for a lattice Λ, we will write $\mathcal{L}(\mathbf{B}) = \Lambda$.

For a matrix $\mathbf{A} \in \mathbb{Z}^{n \times m}$, we define

$$\mathcal{L}^\perp(\mathbf{A}) = \{\mathbf{y} \in \mathbb{Z}^m : \mathbf{Ay} = 0 \bmod q\}. \tag{3}$$

It's easy to see that $\mathcal{L}^\perp(\mathbf{A})$ is a full-rank lattice of dimension m.

For a full-rank integer lattice Λ, we define the discrete Gaussian distribution

$$D_{\Lambda,\mathbf{c},\sigma}(\mathbf{v}) = e^{\frac{-\|\mathbf{v}-\mathbf{c}\|^2}{2\sigma^2}} \bigg/ \sum_{\mathbf{w} \in \Lambda} e^{\frac{-\|\mathbf{w}-\mathbf{c}\|^2}{2\sigma^2}} \text{ for any } \mathbf{v} \in \Lambda, \text{ and 0 on all other points}$$

in space.

For the special case of $\Lambda = \mathbb{Z}^m$, we know that

$$\Pr_{\mathbf{s} \xleftarrow{\$} D_{\mathbb{Z}^m, 0, \sigma}} [\|\mathbf{s}\|_\infty > t\sigma] < 2m \cdot e^{-t^2/2},$$

which implies that for $t = 6$, the probability that any coefficient of $\mathbf{s}$ is greater than 6σ is less than $m \cdot 2^{-25}$.

2.4   Polynomial Lattices and Sampling over the Ring R

In this paper, rather than working over the ring $\mathbb{Z}_q$ (with the usual addition and multiplication operation modulo q), we will be working over the ring $R = \mathbb{Z}_q[x]/(x^n + 1)$ with the usual addition and multiplication operations modulo q and $x^n + 1$. Analogously to (3), for a vector $\mathbf{A} \in R^{1 \times m}$, a lattice $\mathcal{L}^\perp(\mathbf{A})$ can be defined as

$$\mathcal{L}^\perp(\mathbf{A}) = \{\mathbf{y} \in (\mathbb{Z}[x]/(x^n + 1))^m : \mathbf{A}\mathbf{y} = 0 \bmod q\}.$$

Note that while it is an m-dimensional lattice over $\mathbb{Z}[x]/(x^n + 1)$, it is really an nm-dimensional lattice over $\mathbb{Z}$.

If we want to generate a discrete Gaussian sample over $\mathbb{Z}[x]/(x^n + 1)$, we can simply generate it over $\mathbb{Z}^n$ and then map into $\mathbb{Z}[x]/(x^n + 1)$ using the obvious embedding of coordinates into coefficients of the polynomials. We will slightly abuse notation and write $\mathbf{y} \xleftarrow{\$} D_{R,0,\sigma}$ to mean that $\mathbf{y}$ is generated according to $D_{\mathbb{Z}^n, 0, \sigma}$ and then interpreted as an element of R. Similarly, we write $(y_1, \ldots, y_l) \xleftarrow{\$} D_{R^l, 0, \sigma}$ to mean that z is generated according to $D_{\mathbb{Z}^{ln}, 0, \sigma}$ and then gets interpreted as l polynomials y_i.

2.5   Ring-LWE Encryption Scheme

We recall the Ring-LWE encryption scheme from [LPR13]. For simplicity, we take the distribution of the secret keys and the randomness to be uniformly-random elements with ℓ_∞ norm 1. The secret keys are chosen as $s_1, s_2 \xleftarrow{\$} S_1$ and the public keys are a $\xleftarrow{\$} R_q$ and $t \leftarrow as_1 + s_2$. There is also a public parameter $p > 2$, which is a positive integer. To encrypt a message $m \in R_p$, the encryptor chooses $r, e, e' \xleftarrow{\$} S_1$ and outputs (v, w) where $v \leftarrow p(ar + e)$ and $w \leftarrow p(tr + e') + m$. The decryption procedure computes

$$w - vs_1 \bmod q \bmod p = p(rs_2 + e' - es_1) + m \bmod p = m, \qquad (4)$$

where the last equality holds in the case that $\|p(rs_2 + e' - es_1) + m\|_\infty < q/2$.

From the above equations, we see that the encryption of a plaintext m under public keys a, t is a ciphertext v, w satisfying the equation

$$\begin{bmatrix} v \\ w \end{bmatrix} = \begin{bmatrix} pa \mid p \mid 0 \mid 0 \\ pt \mid 0 \mid p \mid 1 \end{bmatrix} \begin{bmatrix} r \\ e \\ e' \\ m \end{bmatrix} \bmod q, \qquad (5)$$

Extending this, the encryption of k messages $m_1, \ldots, m_k$ under the same public key a, t satisfies the following relation:

$$
2k \left\{
\overbrace{
\begin{bmatrix}
pa & & p & & & & & \\
 & \ddots & & \ddots & & & & \\
 & & pa & & p & & & \\
pt & & & & p & & 1 & \\
 & \ddots & & & & \ddots & & \ddots \\
 & & pt & & & & p & & 1
\end{bmatrix}
}^{4k}
\begin{bmatrix}
r_1 \\
\cdots \\
r_k \\
e_1 \\
\cdots \\
e_k \\
e'_1 \\
\cdots \\
e'_k \\
m_1 \\
\cdots \\
m_k
\end{bmatrix}
=
\begin{bmatrix}
v_1 \\
\cdots \\
v_k \\
w_1 \\
\cdots \\
w_k
\end{bmatrix}
\right. \bmod q,
$$

$$(6)$$

which we will write in abbreviated form as

$$
\begin{bmatrix}
pa\mathbf{I}_k & p\mathbf{I}_k & 0^{k \times k} & 0^{k \times k} \\
pt\mathbf{I}_k & 0^{k \times k} & p\mathbf{I}_k & \mathbf{I}_k
\end{bmatrix}
\begin{bmatrix}
\mathbf{r} \\
\mathbf{e} \\
\mathbf{e'} \\
\mathbf{m}
\end{bmatrix}
=
\begin{bmatrix}
\mathbf{v} \\
\mathbf{w}
\end{bmatrix}
\bmod q,
\tag{7}
$$

where $\mathbf{I}_k$ corresponds to an identity matrix of dimension k and $0^{\ell \times k}$ corresponds to an $\ell \times k$ matrix of all zeroes. The decryption procedure is then simply the vector analogy of (4), i.e.

$$
\mathbf{m} = \mathbf{w} - \mathbf{v}s_1 \bmod q \bmod p.
$$

2.6 "Fiat-Shamir with Aborts" Proofs of Knowledge of Linear Relations

In [Lyu09, Lyu12], Lyubashevsky introduced a technique for constructing practical digital signatures (in the random oracle model) based on the hardness of lattice problems. At the heart of the construction is a zero-knowledge proof of knowledge that, given an $s \in R^k$ satisfying the relation

$$
\mathbf{As} = \mathbf{t} \bmod q,
\tag{8}
$$

proves the knowledge of low-norm $\bar{s}$ and $\bar{c}$ that satisfy

$$
\mathbf{A\bar{s}} = \bar{c}\mathbf{t} \bmod q.
$$

The idea in [Lyu09, Lyu12] was to construct a Σ-protocol with the main twist being that the prover does not always output the result. In particular, the protocols use rejection sampling to tailor the distribution so that it does not depend on the secret $\mathbf{s}$. This rejection sampling can be done by making

Algorithm 1. "Fiat-Shamir with Aborts" zero-knowledge proof of knowledge

Input: A matrix $\mathbf{A} \in R^{\ell \times k}$, vector $\mathbf{s} \in S \subset R^k$, a vector $\mathbf{t} \in R^\ell$, and a vector $\mathbf{q} \in \mathbb{Z}^\ell$ such that $\mathbf{As} = \mathbf{t} \bmod \mathbf{q}$. Challenge domain $\mathcal{C} \subset R$. Standard deviation $\sigma \in \mathbb{R}^+$ such that $\sigma \geq 11 \cdot \max_{\mathbf{s} \in S, c \in \mathcal{C}} \|\mathbf{s}c\|$. Cryptographic hash function $\mathsf{H} : \{0,1\}^* \to \mathcal{C}$.
Output: $\mathbf{z} \in R^k$ such that $\mathbf{z} \sim D_{R^k, 0, \sigma}$ and $c \in \mathcal{C}$ such that $c = \mathsf{H}(\mathbf{A}, \mathbf{t}, \mathbf{Az} - \mathbf{t}c \bmod \mathbf{q})$.

1: $\mathbf{y} \xleftarrow{\$} D_{R^k, 0, \sigma}$
2: $c \leftarrow \mathsf{H}(\mathbf{A}, \mathbf{t}, \mathbf{Ay} \bmod \mathbf{q})$
3: $\mathbf{z} \leftarrow \mathbf{s}c + \mathbf{y}$
4: with probability $\frac{D_{R^k, 0, \sigma}(\mathbf{z})}{3 \cdot D_{R^k, \mathbf{s}c, \sigma}(\mathbf{z})}$, **goto** 1
5: **if** $\|\mathbf{z}\|_\infty > 6\sigma$, **goto** 1
6: output $(c, \mathbf{z})$

Algorithm 2. "Fiat-Shamir with Aborts" Verification Algorithm

Input: A matrix $\mathbf{A} \in R^{\ell \times k}$, a vector $\mathbf{t} \in R^\ell$, a vector $\mathbf{q} \in \mathbb{Z}^\ell$, $\sigma \in \mathbb{R}^+$. A tuple $(c, \mathbf{z}) \in \mathcal{C} \times R^k$. Cryptographic hash function $\mathsf{H} : \{0,1\}^* \to \mathcal{C}$.
Output: Bits 0 or 1 corresponding to Reject/Accept.

1: **if** $\|\mathbf{z}\|_\infty > 6\sigma$, **return** 0
2: **if** $c \neq \mathsf{H}(\mathbf{A}, \mathbf{t}, \mathbf{Az} - \mathbf{t}c \bmod \mathbf{q})$, **return** 0
3: **return** 1

---

the resulting distribution uniform in a box (as in [Lyu09]), or the more efficient approach in [Lyu12] of making it a discrete Gaussian. The interactive protocol is then converted to a non-interactive one in the random-oracle model [BR93] using the Fiat-Shamir technique [FS86]. This combined technique is sometimes referred to as "Fiat-Shamir with Aborts".

A variation of the signing protocol from [Lyu12] is given in Algorithm 1. It was shown in that work that the output $\mathbf{z}$ is distributed according to $D_{R^k, 0, \sigma}$. In particular, the rejection sampling stem on line 4 has the effect that the distribution of $\mathbf{z}$ is independent of the secret vector $\mathbf{s}$ (and the challenge c). This algorithm is therefore honest-verifier zero knowledge since a simulator can simply output $\mathbf{z} \xleftarrow{\$} D_{R^k, 0, \sigma}, c \xleftarrow{\$} \mathcal{C}$ and program $c = \mathsf{H}(\mathbf{A}, \mathbf{t}, \mathbf{Az} - \mathbf{t}c \bmod \mathbf{q})$. [3] We also make the observation that one does not need to use the same modulus q for every row of the relation in (8). One can instead use a different modulus for each row, and we represent this in the protocol by a vector $\mathbf{q}$ – having different moduli is crucial to the application in this paper.

We also need simulation soundness [Sah99], meaning that an adversary cannot create proofs of incorrect statements, even after seeing simulated proofs of incorrect statements. Faust et al. [FKMV12] showed that Fiat-Shamir proofs are simulation-sound if the underlying three-move protocol is honest-verifier zero-knowledge and has "quasi-unique responses", meaning that an adversary cannot

---

[3] Because the entropy of $\mathbf{z}$ is high, there is a very low probability that the value for $\mathsf{H}(\mathbf{A}, \mathbf{t}, \mathbf{Az} - \mathbf{t}c \bmod \mathbf{q})$ was previously assigned.

create two accepting transcripts that are different only in the response value. For our proof system, this translates into finding $\mathbf{z} \neq \mathbf{z}'$ such that $\mathbf{A}\mathbf{z} = \mathbf{A}\mathbf{z}' \bmod q$ be hard. Finding such $\mathbf{z}, \mathbf{z}'$ would imply that $\mathbf{A}(\mathbf{z} - \mathbf{z}') = 0 \bmod q$ where $\mathbf{A}$ is the matrix in Eq. 7. Thus, there is either a non-zero tuple $(y_1, y_2) \in R_q$ with l_∞ norm less than 12σ such that $ay_1 + py_2 = 0 \bmod q$ or $py_1 + y_2 = 0 \bmod q$. In our applications $p > 12\sigma$ and $12\sigma p + 12\sigma < q$, which implies that the second equality is not possible. Also, for most of the parameter sets (see the table in Sect. 7), $(24\sigma)^2 < q$, and therefore a standard probabilistic argument can be used to show that for all y_1, y_2 of ℓ_∞ norm less than 12σ,

$$\Pr_{a \xleftarrow{\$} R_q} [ay_1 + py_2 = 0 \bmod q] = 2^{-\Omega(n)}.$$

Thus for almost all a, there will not be a short solution (y_1, y_2) that satisfies $ay_1 + py_2 = 0$.

If $(24\sigma)^2 > q$, then the probabilistic argument no longer applies, but then finding such (y_1, y_2) gives a solution to Ring-SIS [LM06] problem for a random a, which is a computationally hard problem when the norm of y_i is small-enough with relation to q (which it is in all applications).

3   One-Shot Verifiable Encryption for Linear Relations from Ring-LWE

We follow Camenisch and Shoup [CS03] in defining verifiable encryption as encrypting a witness for a member of a language. The class of languages that we'll be looking at will be the linear relations of short vectors in a ring. While Camenisch and Shoup defined soundness by requiring that decryption of a valid ciphertext always recovers a valid witness, we will only achieve a relaxed property that recovers a witness for a related "extended" language that includes the original language. As we will see in Sect. 6, however, this weaker property suffices for many practical applications of verifiable encryption.

3.1   Definition of Relaxed Verifiable Encryption

We relax Camenisch and Shoup's [CS03] definitions for verifiable encryption in two ways. First, as mentioned above, and analogous to relaxed knowledge extraction for proofs of knowledge [CKY09], the encryption algorithm encrypts a witness w for a member x of a language L, but soundness only guarantees that decryption of a valid ciphertext returns a witness $\bar{w}$ of an extented language $\bar{L}$ instead of L. Second, rather than looking at verifiable encryption as a combination of a standard public-key encryption scheme with an associated proof system, we consider encryption and proof as a single algorithm, producing a verifiable ciphertext that includes the proof. This generalization allows for more efficient schemes, in particular our construction that speeds up decryption using information from the proof.

Let $L \subseteq \{0,1\}^*$ be a language with witness relation R_L, i.e., $x \in L$ iff there exists a witness w such that $(x,w) \in R_L$. Let $\bar{L}$ with witness relation $R_{\bar{L}}$ be an extension of L, meaning that $L \subseteq \bar{L}$ and $R_L \subseteq R_{\bar{L}}$. For our language of linear relations over short vectors, we will consider the language L with relation

$$R_L = \{((\mathbf{B}, \mathbf{u}), (\mathbf{m}, 1)) \in (R_p^{\ell \times k} \times R_p^{\ell}) \times (R_p^k \times R_p) \ : \ \mathbf{Bm} = \mathbf{u} \bmod p \wedge \mathbf{m} \in S_\gamma^k\}$$

and the extended language $\bar{L}$ with relation

$$R_{\bar{L}} = \{((\mathbf{B}, \mathbf{u}), (\bar{\mathbf{m}}, \bar{c})) \in (R_p^{\ell \times k} \times R_p^{\ell}) \times (R_p^k \times R_p) \ : \\ \mathbf{B}\bar{\mathbf{m}} = \bar{c}\mathbf{u} \bmod p \wedge \|\bar{\mathbf{m}}\|_\infty < 6\sigma \wedge \bar{c} \in \bar{\mathcal{C}}\},$$

where $\bar{\mathcal{C}} = \{c - c' \ : \ c, c' \in \mathcal{C}\}$ for $\mathcal{C} = \{c \in R : \|c\|_\infty = 1, \|c\|_1 \leq 36\}$ and other parameters are described in Algorithm 3.

A *relaxed verifiable encryption scheme* for languages $L, \bar{L}$ is a tuple of algorithms $(\mathsf{Kg}, \mathsf{Enc}, \mathsf{V}, \mathsf{Dec})$ where the key generation algorithm $\mathsf{Kg}(1^\lambda)$ returns a public and secret key (pk, sk); the encryption algorithm $\mathsf{Enc}(pk, x, w)$ returns a verifiable ciphertext t that encrypts the witness w of language member $x \in L$; the verification algorithm $\mathsf{V}(pk, x, t)$ returns 1 or 0 indicating whether t encrypts a witness for x; the decryption algorithm $\mathsf{Dec}(sk, x, t)$ returns a witness $\bar{w}$ or a failure symbol $\perp$. We will focus on the case where the ciphertext t includes a Fiat-Shamir proof of a Σ-protocol, where the proof $\pi = (cmt, c, rsp)$ consists of a commitment cmt, a challenge $c = \mathsf{H}(pk, x, t, cmt, \ldots)$ generated through a random oracle H, and a response rsp. We require that the algorithms satisfy the following adapted properties from [CS03]:

Correctness. Correctness requires that $\mathsf{Dec}(sk, x, \mathsf{Enc}(pk, x, w)) = w$ with probability one for all $(x, w) \in R_L$ and all key pairs $(pk, sk) \xleftarrow{\$} \mathsf{Kg}(1^\lambda)$.

Completeness. For all $(x, w) \in R_L$ and all key pairs $(pk, sk) \xleftarrow{\$} \mathsf{Kg}(1^\lambda)$, $\mathsf{V}(pk, x, \mathsf{Enc}(pk, x, w)) = 1$ with probability one.

Soundness. Soundness requires that a ciphertext with a valid proof for $x \in L$ can with overwhelming probability be decrypted to a valid witness $\bar{w}$ such that $(x, \bar{w}) \in R_{\bar{L}}$, i.e., the following probability is negligible:

$$\Pr\left[b = 1 \wedge (x, \bar{w}) \notin R_{\bar{L}} \ : \ \begin{matrix} (pk, sk) \xleftarrow{\$} \mathsf{Kg}(1^\lambda), (x, t) \xleftarrow{\$} \mathsf{A}(pk, sk), \\ b \leftarrow \mathsf{V}(pk, x, t), \bar{w} \xleftarrow{\$} \mathsf{Dec}(sk, x, t) \end{matrix}\right].$$

Simulatability. There exists a simulator Sim such that no adversary A can distinguish real from simulated ciphertexts, i.e., the following advantage of A is negligible:

$$\left| \Pr\left[b' = b \ : \ \begin{matrix} b \xleftarrow{\$} \{0,1\}, (pk, sk) \xleftarrow{\$} \mathsf{Kg}(1^\lambda), (,, x, w) \xleftarrow{\$} \mathsf{A}(pk), \\ t_0 \xleftarrow{\$} \mathsf{Enc}(pk, x, w), t_1 \xleftarrow{\$} \mathsf{Sim}(pk, x), b' \xleftarrow{\$} \mathsf{A}(, t_b) \end{matrix}\right] - \frac{1}{2}\right|.$$

3.2   Construction

Given a linear relation

$$\mathbf{Bm} = \mathbf{u} \bmod p, \tag{9}$$

for a matrix $\mathbf{B} \in R_p^{\ell \times k}$, our goal is to produce a ciphertext and a proof that the decryption of this ciphertext is $(\overline{\mathbf{m}}, \overline{c})$ that satisfies the relation

$$\mathbf{B}\overline{\mathbf{m}} = \mathbf{u}\overline{c} \bmod p. \tag{10}$$

Key generation. Key pairs are generated as for the Ring-LWE encryption scheme from Sect. 2.5, i.e., by choosing $s_1, s_2 \xleftarrow{\$} S_1$ and computing a $\xleftarrow{\$} R$ and t $\leftarrow$ a$s_1 + s_2$. The public key is $pk = (a, t, p, q)$, where p is the same value as the modulus that we are proving our linear relation over. The secret key is $sk = s_1$.

Encryption and verification. The prover encrypts a witness $w = \mathbf{m}$ for language member $x = (\mathbf{B}, \mathbf{u})$ satisfying (9) with randomness $(\mathbf{r}, \mathbf{e}, \mathbf{e}') \xleftarrow{\$} S_1^{3k}$ as in (7). The prover then concatenates this with (9) to form the relation below:

$$\begin{bmatrix} pa\mathbf{I}_k & p\mathbf{I}_k & 0^{k \times k} & 0^{k \times k} \\ pt\mathbf{I}_k & 0^{k \times k} & p\mathbf{I}_k & \mathbf{I}_k \\ 0^{\ell \times k} & 0^{\ell \times k} & 0^{\ell \times k} & \mathbf{B} \end{bmatrix} \begin{bmatrix} \mathbf{r} \\ \mathbf{e} \\ \mathbf{e}' \\ \mathbf{m} \end{bmatrix} = \begin{bmatrix} \mathbf{v} \\ \mathbf{w} \\ \mathbf{u} \end{bmatrix} \begin{matrix} \bmod q \\ \bmod q \\ \bmod p \end{matrix} \tag{11}$$

As discussed, there is no practical proof of knowledge for the above relation, and so the prover instead uses the "Fiat-Shamir with Aborts" approach from Sect. 2.6 to construct a proof of knowledge π of low-weight $\overline{\mathbf{r}}, \overline{\mathbf{e}}, \overline{\mathbf{e}}', \overline{\mathbf{m}}$, and $\overline{c}$ that satisfy

$$\begin{bmatrix} pa\mathbf{I}_k & p\mathbf{I}_k & 0^{k \times k} & 0^{k \times k} \\ pt\mathbf{I}_k & 0^{k \times k} & p\mathbf{I}_k & \mathbf{I}_k \\ 0^{\ell \times k} & 0^{\ell \times k} & 0^{\ell \times k} & \mathbf{B} \end{bmatrix} \begin{bmatrix} \overline{\mathbf{r}} \\ \overline{\mathbf{e}} \\ \overline{\mathbf{e}}' \\ \overline{\mathbf{m}} \end{bmatrix} = \overline{c} \begin{bmatrix} \mathbf{v} \\ \mathbf{w} \\ \mathbf{u} \end{bmatrix} \begin{matrix} \bmod q \\ \bmod q \\ \bmod p \end{matrix} \tag{12}$$

This procedure and the corresponding verification is presented in Algorithms 3 and 4.

Decryption. The main result of our work is showing that, when given the ciphertext $t = (\mathbf{v}, \mathbf{w}, c, \mathbf{z})$, the decryptor can recover some $(\overline{\mathbf{m}}, \overline{c})$ that satisfies (10). Because the proof of knowledge $(c, \mathbf{z})$ does not imply that $\begin{bmatrix} \mathbf{v} \\ \mathbf{w} \end{bmatrix}$ is a valid Ring-LWE ciphertext, we cannot simply use the Ring-LWE decryption algorithm from (4).

Instead, the intuition is to guess a value for $\overline{c}$ and then attempt to decrypt the ciphertext $\overline{c} \begin{bmatrix} \mathbf{v} \\ \mathbf{w} \end{bmatrix} \bmod q$ in hopes of recovering $\overline{\mathbf{m}}$. The problem with this straightforward approach is that the decryption algorithm will always return something, and so one needs a way to decide whether this decryption is something

---

Algorithm 3. One-Shot Verifiable encryption $\mathsf{Enc}(pk, x, w)$

Input: Public key $pk = (a, t, p, q)$, language member $x = (\mathbf{B}, \mathbf{u})$, witness $w = \mathbf{m} \in S_\gamma^k$. Challenge domain $\mathcal{C} = \{c \in R : \|c\|_\infty = 1, \|c\|_1 \leq 36\}$. Cryptographic hash function $\mathsf{H} : \{0,1\}^* \to \mathcal{C}$. Standard deviation $\sigma = 11 \cdot \max_{c \in \mathcal{C}} \|c\|_1 \cdot \sqrt{kn(3 + \gamma)}$.

1: $\mathbf{r}, \mathbf{e}, \mathbf{e}' \overset{\$}{\leftarrow} S_1^k$

2: $\begin{bmatrix} \mathbf{v} \\ \mathbf{w} \end{bmatrix} \leftarrow \begin{bmatrix} pa\mathbf{I}_k & p\mathbf{I}_k & 0^{k \times k} & 0^{k \times k} \\ pt\mathbf{I}_k & 0^{k \times k} & p\mathbf{I}_k & \mathbf{I}_k \end{bmatrix} \begin{bmatrix} \mathbf{r} \\ \mathbf{e} \\ \mathbf{e}' \\ \mathbf{m} \end{bmatrix} \begin{array}{l} \bmod q \\ \bmod q \end{array}$

3: $\mathbf{y} \leftarrow \begin{bmatrix} \mathbf{y_r} \\ \mathbf{y_e} \\ \mathbf{y_{e'}} \\ \mathbf{y_m} \end{bmatrix} \overset{\$}{\leftarrow} D_{R^{4k}, 0, \sigma}$

4: $c \leftarrow \mathsf{H} \left(\begin{bmatrix} pa\mathbf{I}_k & p\mathbf{I}_k & 0^{k \times k} & 0^{k \times k} \\ pt\mathbf{I}_k & 0^{k \times k} & p\mathbf{I}_k & \mathbf{I}_k \\ 0^{\ell \times k} & 0^{\ell \times k} & 0^{\ell \times k} & \mathbf{B} \end{bmatrix}, \begin{bmatrix} \mathbf{v} \\ \mathbf{w} \\ \mathbf{u} \end{bmatrix}, \begin{bmatrix} pa\mathbf{I}_k & p\mathbf{I}_k & 0^{k \times k} & 0^{k \times k} \\ pt\mathbf{I}_k & 0^{k \times k} & p\mathbf{I}_k & \mathbf{I}_k \\ 0^{\ell \times k} & 0^{\ell \times k} & 0^{\ell \times k} & \mathbf{B} \end{bmatrix} \begin{bmatrix} \mathbf{y_r} \\ \mathbf{y_e} \\ \mathbf{y_{e'}} \\ \mathbf{y_m} \end{bmatrix} \begin{array}{l} \bmod q \\ \bmod q \\ \bmod p \end{array} \right)$

5: $\mathbf{s} \leftarrow \begin{bmatrix} \mathbf{r} \\ \mathbf{e} \\ \mathbf{e}' \\ \mathbf{m} \end{bmatrix} c$

6: $\mathbf{z} \leftarrow \mathbf{s} + \mathbf{y}$

7: with probability $\frac{D_{R^{4k}, 0, \sigma}(\mathbf{z})}{3 \cdot D_{R^{4k}, \mathbf{s}, \sigma}(\mathbf{z})}$, continue, else goto 3

8: if $\|\mathbf{z}\|_\infty > 6 \cdot \sigma$, goto 3

9: **return** $t = (\mathbf{v}, \mathbf{w}, c, \mathbf{z})$

Algorithm 4. One-Shot Verification $\mathsf{V}(pk, x, t)$

Input: Public key $pk = (a, t, p, q)$, language member $x = (\mathbf{B}, \mathbf{u})$, ciphertext $t = (\mathbf{v}, \mathbf{w}, c, \mathbf{z})$. Cryptographic hash function H, positive real σ as in Algorithm 3.

1: **if** $\|\mathbf{z}\|_\infty > 6 \cdot \sigma$, **return** 0

2: **if** $c \neq \mathsf{H} \left(\begin{bmatrix} pa\mathbf{I}_k & p\mathbf{I}_k & 0^{k \times k} & 0^{k \times k} \\ pt\mathbf{I}_k & 0^{k \times k} & p\mathbf{I}_k & \mathbf{I}_k \\ 0^{\ell \times k} & 0^{\ell \times k} & 0^{\ell \times k} & \mathbf{B} \end{bmatrix}, \begin{bmatrix} \mathbf{v} \\ \mathbf{w} \\ \mathbf{u} \end{bmatrix}, \begin{bmatrix} pa\mathbf{I}_k & p\mathbf{I}_k & 0^{k \times k} & 0^{k \times k} \\ pt\mathbf{I}_k & 0^{k \times k} & p\mathbf{I}_k & \mathbf{I}_k \\ 0^{\ell \times k} & 0^{\ell \times k} & 0^{\ell \times k} & \mathbf{B} \end{bmatrix} \mathbf{z} - c \begin{bmatrix} \mathbf{v} \\ \mathbf{w} \\ \mathbf{u} \end{bmatrix} \begin{array}{l} \bmod q \\ \bmod q \\ \bmod p \end{array} \right)$,

    return 0

3: **return** 1

---

valid or just garbage. In Lemma 3.1, we show that if the parameters of the Ring-LWE encryption scheme are set in a particular way, then the decryptor can test whether a particular ciphertext $\bar{c} \begin{bmatrix} \mathbf{v} \\ \mathbf{w} \end{bmatrix} \bmod q$ is "valid", and for any $\bar{c}$ and $\bar{c}'$ that lead to valid ciphertexts decrypting to $\overline{\mathbf{m}}$ and $\overline{\mathbf{m}}'$, respectively, we have the equality

$$\overline{\mathbf{m}}/\bar{c} = \overline{\mathbf{m}}'/\bar{c}' \bmod p \tag{13}$$

The implication of the above equation is that once the decryptor decrypts some pair $(\overline{\mathbf{m}}', \overline{c}')$, it is a valid solution to (10). This is because the proof of knowledge π proves knowledge of some $(\overline{\mathbf{m}}, \overline{c})$ that satisfies $\mathbf{B}\overline{\mathbf{m}} = \overline{c}\mathbf{u} \bmod p$, or equivalently $\mathbf{B}\overline{\mathbf{m}}/\overline{c} = \mathbf{u} \bmod p$. Equation (13) then implies that

$$\mathbf{B}\overline{\mathbf{m}}' = \overline{c}'\mathbf{u} \bmod p.$$

The second issue is how to find a valid $\overline{c}$. In particular, if we would like the proof of knowledge to be "one-shot", then the challenge space should be exponentially large, and so it is impractical to simply try all the possible $\overline{c}$ (of which there are actually even more than in the challenge space). We show in Lemma 3.2, however, that the decryptor can try random $\overline{c}$ (there is some relation between π and which $\overline{c}$ should be tried), and then the *expected* number of tries is essentially the number of random oracle queries that the prover makes when constructing π, where the probability is taken over the randomness of the random oracle (modeled as a random function) and the coins of the decryptor. Algorithm 5 is the decryption algorithm that guesses a random c' from $\mathcal{C}$, constructs $\overline{c} = c - c'$, where c is part of the proof π, and then checks whether $\overline{c}\begin{bmatrix}\mathbf{v}\\\mathbf{w}\end{bmatrix} \bmod q$ is a valid ciphertext (actually k valid ciphertexts because the plaintext $\mathbf{m}$ is encrypted as k independent plaintexts). If it is, then it decrypts it, and otherwise it guesses a new random c'.

Algorithm 5. One-Shot Decryption $\mathsf{Dec}(sk, x, t)$

Input: Secret key $sk = \mathbf{s}_1$, language member $x = (\mathbf{B}, \mathbf{u})$, ciphertext $t = (\mathbf{v}, \mathbf{w}, c, \mathbf{z})$, constant $C = \max\limits_{c, c' \in \mathcal{C}} \|c - c'\|_1$.

1: **if** $\mathsf{V}(pk, x, t) = 1$ **then**
2:    **loop**
3:       $c' \xleftarrow{\$} \mathcal{C}$
4:       $\overline{c} \leftarrow c - c'$
5:       $\overline{\mathbf{m}} \leftarrow (\mathbf{w} - \mathbf{v}\mathbf{s}_1)\overline{c} \bmod q$
6:       **if** $\|\overline{\mathbf{m}}\|_\infty < q/2C$ **then**
7:          $\overline{\mathbf{m}} \leftarrow \overline{\mathbf{m}} \bmod p$
8:          **return** $(\overline{\mathbf{m}}, \overline{c})$
9:       **end if**
10:   **end loop**
11: **end if**

If the prover is honest, then of course $\begin{bmatrix}\mathbf{v}\\\mathbf{w}\end{bmatrix}$ will already be a valid ciphertext, and then it's not hard to see that any $\overline{c}$ will result in a valid decryption (or the decryptor can try $\overline{c} = 1$ first). On the other hand, what Lemma 3.2 roughly implies is that if the prover can only query the random oracle a few times, then the decryptor will also expect to recover a solution to (10) within a few queries.

In Sect. 4, we propose a modification to our protocol that still retains the relation between the number of random-oracle queries the prover makes and the expected number of decryption tries that the decryptor needs, but also puts an upper-bound on the decryption time. The idea is to reduce the space of challenges in the zero-knowledge proof from, say, 2^{128} down to 2^{32} and then doing the proof in parallel 4 times. This increases the proof size by a factor of 4, but upperbounds the number of decryptions tries to $4 \cdot 2^{32}$. One can of course adjust the trade-off between the decryption-time upper bound and the size of the proof to suit the particular scenario. This is the main advantage of this parallelized protocol over the earlier idea in [BCK+14]. In the latter scheme, the size of the challenge space could not be varied – if one were working over the ring $\mathbb{Z}[x]/(x^n + 1)$, then the challenge space was exactly $2n + 1$.

3.3   Interlude: Proofs of Plaintext Knowledge

One can see proofs of plaintext knowledge as a verifiable encryption scheme without a relation, or where the relation is trivially satisfied. In our case, one could consider the scheme from the previous section with $\mathbf{B}$ and $\mathbf{u}$ being 0, or simply the row(s) containing $\mathbf{B}$ and $\mathbf{u}$ not being present in relation (11).

The soundness requirement that a valid ciphertext must decrypt to a valid witness obviously makes no sense if the relation is trivial. Instead, soundness for a proof of plaintext knowledge requires that decryption returns the same value as can be extracted from the proof of knowledge. Our randomized decryption algorithm as described in Algorithm 5 does not satisfy such a requirement, as it potentially returns a different pair $(\overline{m}, \overline{c})$ at each execution. However, because of the property that $\overline{m}/\overline{c} = \overline{m}'/\overline{c}' \bmod p$ for any $(\overline{m}, \overline{c}), (\overline{m}', \overline{c}')$ returned by the decryption algorithm, we can make the decryption deterministic by letting it return $\overline{m}/\overline{c} \bmod p$. Because this unique value can also be extracted from the proof, this turns our verifiable encryption scheme into a proof of plaintext knowledge.

3.4   Correctness and Security

Soundness. We first show the soundness property of our relaxed verifiable encryption scheme by showing that decryption of a valid ciphertext, if it finishes, yields a witness from $R_{\overline{L}}$. In Sect. 3.5, we prove that the expected running time of the decryption algorithm is proportional to the number of random-oracle queries made by the adversary who created the ciphertext.

If an adversary $\mathcal{A}$ who is trying to break the soundness of the scheme outputs a ciphertext $t = (\mathbf{v}, \mathbf{w}, c, \mathbf{z})$ that is valid for $x = (\mathbf{B}, \mathbf{u})$, then by the verification procedure described in Algorithm 4 we have that $\|\mathbf{z}\|_\infty \leq 6 \cdot \sigma$ and

$$c = \mathsf{H}\left(\mathbf{B}', \begin{bmatrix} \mathbf{v} \\ \mathbf{w} \\ \mathbf{u} \end{bmatrix}, \mathbf{B}'\mathbf{z} - c \begin{bmatrix} \mathbf{v} \\ \mathbf{w} \\ \mathbf{u} \end{bmatrix} \begin{matrix} \bmod q \\ \bmod q \\ \bmod p \end{matrix}\right) \tag{14}$$

where

$$\mathbf{B}' = \begin{bmatrix} pa\mathbf{I}_k & p\mathbf{I}_k & 0^{k\times k} & 0^{k\times k} \\ pt\mathbf{I}_k & 0^{k\times k} & p\mathbf{I}_k & \mathbf{I}_k \\ 0^{\ell\times k} & 0^{\ell\times k} & 0^{\ell\times k} & \mathbf{B} \end{bmatrix}. \tag{15}$$

Let $\mathbf{A}$ denote the first argument of the above random-oracle query and $\mathbf{y}$ the last, i.e., the above equation can be rewritten as $c = \mathsf{H}(\mathbf{A}, [\,\mathbf{v}\,;\,\mathbf{w}\,;\,\mathbf{u}\,], \mathbf{y})$.

With overwhelming probability, there exists a second challenge $c' \in \mathcal{C}\setminus\{c\}$ for which there exists a vector $\mathbf{z}'$ with $\|\mathbf{z}'\|_\infty \leq 6\cdot\sigma$ and $\mathbf{y} = \mathbf{A}\mathbf{z}' - c'[\,\mathbf{v}\,;\,\mathbf{w}\,;\,\mathbf{u}\,]$. Indeed, if c were the only such challenge, then at the moment of making the above random-oracle query, A would have had probability $1/|\mathcal{C}|$ of hitting the only challenge c for which a valid proof exists. The probability that A outputs a proof for which only one such challenge c exists is therefore at most $q_{\mathsf{H}}/|\mathcal{C}|$.

So with overwhelming probability such $c', \mathbf{z}'$ does exist, and we have that $\mathbf{y} = \mathbf{A}\mathbf{z} - c[\,\mathbf{v}\,;\,\mathbf{w}\,;\,\mathbf{u}\,] = \mathbf{A}\mathbf{z}' - c'[\,\mathbf{v}\,;\,\mathbf{w}\,;\,\mathbf{u}\,]$ with $\|\mathbf{z}\|_\infty \leq 6\cdot\sigma$ and $\|\mathbf{z}'\|_\infty \leq 6\cdot\sigma$. Hence, letting $\bar{c} = c - c'$ and $\bar{\mathbf{z}} = \mathbf{z}' - \mathbf{z} = [\,\bar{\mathbf{r}}\,;\,\bar{\mathbf{e}}\,;\,\bar{\mathbf{e}}'\,;\,\bar{\mathbf{m}}\,]$, we have that $\mathbf{A}\bar{\mathbf{z}} = \bar{c}[\,\mathbf{v}\,;\,\mathbf{w}\,;\,\mathbf{u}\,]$ with $\|\bar{\mathbf{z}}\|_\infty \leq 12\cdot\sigma$.

By choosing the scheme parameters appropriately, e.g., such that $(36p + 12)\sigma < q/2C$, one can satisfy the preconditions of the following crucial lemma that shows that for any $(\bar{\mathbf{m}}', \bar{c}')$ returned by the decryption algorithm, we have that $\bar{\mathbf{m}}'/\bar{c}' = \bar{\mathbf{m}}/\bar{c}$, and, because $\mathbf{B}\bar{\mathbf{m}} = \bar{c}\mathbf{u}$, that $\mathbf{B}\bar{\mathbf{m}}' = \bar{c}'\mathbf{u}$.

Lemma 3.1. *Let (a, t, p, q) and (s_1, s_2) be generated keys as in Sect. 3.2. If for given $v, w \in R_q$ there exist $\bar{r}, \bar{e}, \bar{e}', \bar{m}, \bar{c}$ such that*

$$\begin{bmatrix} pa & | & p & | & 0 & | & 0 \\ pt & | & 0 & | & p & | & 1 \end{bmatrix} \begin{bmatrix} \bar{r} \\ \bar{e} \\ \bar{e}' \\ \bar{m} \end{bmatrix} = \bar{c}\begin{bmatrix} v \\ w \end{bmatrix} \bmod q$$

and

$$\|p(\bar{r}s_2 + \bar{e}' - \bar{e}s_1) + \bar{m}\|_\infty < q/2C \tag{16}$$

where $C = \max\limits_{\bar{c}\in\overline{\mathcal{C}}} \|\bar{c}\|_1 = \max\limits_{c,c'\in\mathcal{C}} \|c - c'\|_1$, then

1. $\|(w - vs_1)\bar{c} \bmod q\|_\infty < q/2C$
2. *For any $\bar{c}' \in \overline{\mathcal{C}}$ for which $\|(w - vs_1)\bar{c}' \bmod q\|_\infty < q/2C$,*

$$(w - vs_1)\bar{c}' \bmod q/\bar{c}' \bmod p = \bar{m}/\bar{c} \bmod p.$$

$\square$

Proof. To prove the first part, we note that by the definition of Ring-LWE decryption,

$$(w - vs_1)\bar{c} \bmod q = p(\bar{r}s_2 + \bar{e}' - \bar{e}s_1) + \bar{m},$$

which has ℓ_∞ length less than $\frac{q}{2C}$ by the hypothesis of the lemma.

To prove the second part, we first note that

$$(\mathbf{w} - \mathbf{v}s_1)\overline{cc}' \bmod q \bmod p = (p(\bar{r}s_2 + \bar{e}' - \bar{e}s_1) + \overline{m})\overline{c}' \bmod q \bmod p$$
$$= \overline{mc}' \bmod p. \tag{17}$$

Then we can write

$$((\mathbf{w} - \mathbf{v}s_1)\overline{c}' \bmod q) \,/\, \overline{c}' \bmod p = ((\mathbf{w} - \mathbf{v}s_1)\overline{c}' \bmod q) \cdot \overline{c}/(\overline{cc}') \bmod p$$
$$= ((\mathbf{w} - \mathbf{v}s_1)\overline{cc}' \bmod q)/(\overline{cc}') \bmod p$$
$$= \overline{mc}'/(\overline{cc}') \bmod p = \overline{m}/\overline{c} \bmod p$$

The first equality is an identity, the second equality holds since $\|(\mathbf{w} - \mathbf{v}s_1)\overline{c}' \bmod q\|_\infty < \frac{q}{2C}$ and therefore multiplication by $\overline{c}$ does not cause a reduction modulo q. The third equality follows from (17). □

By checking that $\|(\mathbf{w} - \mathbf{v}s_1)\overline{c} \bmod q\|_\infty < q/2C$ in line 6 of the Dec algorithm, we ensure that the condition of the second part of Lemma 3.1 is satisfied for decryption, so that the value $(\overline{m}'(\mathbf{w} - \mathbf{v}s_1)\overline{c}' \bmod q \bmod p, \overline{c}')$ is indeed a witness for $(\mathbf{B}, \mathbf{u}) \in \overline{L}$. This proves the soundness of our scheme.

Correctness. Correctness is straightforward because a valid encryption (see (5)) satisfies the preconditions of Lemma 3.1 with $[\bar{r} ; \bar{e} ; \bar{e}' ; \overline{m}] = [r ; e ; e' ; m]$ and $\overline{c} = 1$; and it's clear that $\|p(rs_2 + e' - es_1) + m\|_\infty \leq \|p(\bar{r}s_2 + \bar{e}' - \bar{e}s_1) + \overline{m}\|_\infty$.

Completeness. Completeness follows from the completeness of the proof system of Sect. 2.6.

Simulatability. The simulator Sim creates a Ring-LWE encryption $[\mathbf{v} ; \mathbf{w}]$ of $\mathbf{m} = 1$ using the scheme of Sect. 2.5 and runs the zero-knowledge simulator for the proof system of Sect. 2.6 to create a valid-looking proof $(c, \mathbf{z})$ for $(\mathbf{B}, \mathbf{u})$. The indistinguishability from the real proof follows from the IND-CPA security of Ring-LWE encryption and the zero-knowledge property of the proof system.

3.5   Decryption Running Time

Even though the running time of the decryption algorithm is unbounded in principle, we show that its *expected* running time is proportional to the number of times that the adversary queries the random oracle. More precisely, we show that if an adversary uses q_H random-oracle queries to construct a ciphertext, then the probability that the decryption algorithm requires more than $\alpha \cdot q_H$ iterations is less than $1/\alpha$.

We prove the above information-theoretic statement for any adversary $\mathcal{A}$, and we can therefore limit the analysis to deterministic adversaries, since the coins that maximize the adversary's success can always be hardwired into its code.

Lemma 3.2. *For a given key pair* $(pk, sk) \in \mathsf{Kg}(1^\lambda)$, *consider the following experiment with an adversary* $\mathcal{A}$:

$$(x, t) \xleftarrow{\$} \mathcal{A}^{\mathsf{H}}(pk)$$
If $\mathsf{V}(pk, x, t) = 1$ *then* $\bar{w} \xleftarrow{\$} \mathsf{Dec}(sk, x, t)$.

Let $\widehat{H}$ *be the random coins of the function* H *(when it is modeled as a random function) and* $\widehat{D}$ *be the random coins of the Decryption algorithm* Dec. *Let* T *be the number of loop iterations in the execution of* Dec *(see Algorithm 5 lines 2–10) until it produces its output* $\bar{w}$. *Then there is an event* G, *such that for all algorithms* $\mathcal{A}$ *that make at most* $q_{\mathsf{H}} - 1$ *queries to* H, *all key pairs* (pk, sk), *and any positive integer* f, *it holds that*

1. $\mathrm{Exp}_{\widehat{H},\widehat{D}}[\, T \mid G\,] \leq \left(1 + \frac{1}{f}\right) \cdot q_{\mathsf{H}}.$
2. $\mathrm{Pr}_{\widehat{H},\widehat{D}}[\neg G] \leq q_{\mathsf{H}} \cdot f/|\mathcal{C}|$

By Markov's inequality and optimization over f, *this implies that for any positive* α,

$$\Pr_{\widehat{H},\widehat{D}}[\, T \geq \alpha \cdot q_{\mathsf{H}}\,] \leq \frac{1}{\alpha} + 2 \cdot \sqrt{\frac{q_{\mathsf{H}}}{\alpha \cdot |\mathcal{C}|}} + \frac{q_{\mathsf{H}}}{|\mathcal{C}|}.$$

Proof. For a given public key pk, language member $x = (\mathbf{B}, \mathbf{u})$, and valid ciphertext $t = (\mathbf{v}, \mathbf{w}, \mathbf{c}, \mathbf{z})$, let $\mathbf{A}$ and $\mathbf{y}$ be the matrix and vector in the verification algorithm (Algorithm 4) so that $c = \mathsf{H}(\mathbf{A}, [\, \mathbf{v} \; ; \; \mathbf{w} \; ; \; \mathbf{u} \,], \mathbf{y})$ and $\mathbf{y} = \mathbf{Az} - c[\, \mathbf{v} \; ; \; \mathbf{w} \; ; \; \mathbf{u} \,]$. Let $\mathcal{G}_t$ be the set of "good" challenges c' for which a valid zero-knowledge proof response $\mathbf{z}'$ exists, i.e.,

$$\mathcal{G}_t = \{c' \in \mathcal{C} \; : \; \exists\, \mathbf{z}' : \mathbf{y} = \mathbf{Az}' - c'[\, \mathbf{v} \; ; \; \mathbf{w} \; ; \; \mathbf{u} \,] \wedge \|\mathbf{z}'\|_\infty \leq 6\sigma\}.$$

Let G be the "good" event that the adversary $\mathcal{A}$ produces a ciphertext t (for the decryption algorithm) with $|\mathcal{G}_t| > f$. Let $\widehat{D}$ denote the coins of the decryption algorithm Dec and let $\widehat{H}$ denote the coins determining the random oracle H. For any ciphertext t, the probability over $\widehat{D}$ that one particular iteration of Dec decrypts successfully, i.e., hits a good challenge $c' \in \mathcal{G}_t \setminus \{c\}$, is $\frac{|\mathcal{G}_t|-1}{|\mathcal{C}|}$. We therefore have that the expected number of iterations for a ciphertext t is

$$\mathrm{Exp}_{\widehat{D}}[\, T \mid \mathcal{A}^{\mathsf{H}} \text{ outputs } t\,] = \frac{|\mathcal{C}|}{|\mathcal{G}_t| - 1}$$

and therefore, conditioned on the event G, that

$$\mathrm{Exp}_{\widehat{D}}[\, T \mid \mathcal{A}^{\mathsf{H}} \text{ outputs } t \wedge G\,] \leq \frac{|\mathcal{C}|}{f}. \tag{18}$$

Below, when we say that "$\mathcal{A}^H$ outputs t_i", we mean that $\mathcal{A}^H$ outputs a language member $x = (\mathbf{B}, \mathbf{u})$ and a ciphertext $t = (\mathbf{v}, \mathbf{w}, c, \mathbf{z})$ such that $\mathcal{A}$'s i-th random-oracle query is

$$
c = H\left(\mathbf{B}', \begin{bmatrix} \mathbf{v} \\ \mathbf{w} \\ \mathbf{u} \end{bmatrix}, \mathbf{B}'\mathbf{z} - c \begin{bmatrix} \mathbf{v} \\ \mathbf{w} \\ \mathbf{u} \end{bmatrix} \begin{matrix} \bmod q \\ \bmod q \\ \bmod p \end{matrix}\right) ,
$$

where $\mathbf{B}'$ is the matrix defined in (15). Also, for any adversary making at most $q_H - 1$ queries, there exists an adversary making at most q_H queries that include the above query; we consider the latter adversary $\mathcal{A}$ in the rest of the analysis.

Because we are conditioning on the event G, we can assume without loss of generality that $\mathcal{A}$ only makes random-oracle queries for ciphertexts t_i with $|\mathcal{G}_{t_i}| > f$. (It is easy to see that for any $\mathcal{A}$ that does not obey these rules, there exists an adversary $\mathcal{A}'$ producing the same output that does.) We now have that

$$
\underset{\widehat{H},\widehat{D}}{\text{Exp}}[\, T \mid G \,] = \sum_{i=1}^{q_H} \underset{\widehat{H}}{\Pr}[\mathcal{A}^H \text{ outputs } t_i \mid G\,] \cdot \underset{\widehat{D}}{\text{Exp}}[\, T \mid \mathcal{A}^H \text{ outputs } t_i \wedge G\,] \tag{19}
$$

The above is true because

$$
\sum_{i=1}^{q_H} \underset{\widehat{H}}{\Pr}[\mathcal{A}^H \text{ outputs } t_i \mid G\,] \cdot \underset{\widehat{D}}{\text{Exp}}[\, T \mid \mathcal{A}^H \text{ outputs } t_i \wedge G\,]
$$

$$
= \sum_{i=1}^{q_H} \underset{\widehat{H}}{\Pr}[\mathcal{A}^H \text{ outputs } t_i \mid G\,] \cdot \sum_{j \in \mathbb{Z}^+} \underset{\widehat{D}}{\Pr}[\, T = j \mid \mathcal{A}^H \text{ outputs } t_i \wedge G\,] \cdot j
$$

$$
= \sum_{j \in \mathbb{Z}^+} j \cdot \sum_{i=1}^{q_H} \underset{\widehat{H},\widehat{D}}{\Pr}[\, T = j \wedge \mathcal{A}^H \text{ outputs } t_i \mid G\,]
$$

$$
= \sum_{j \in \mathbb{Z}^+} j \cdot \underset{\widehat{D},\widehat{H}}{\Pr}[\, T = j \mid G\,]
$$

$$
= \underset{\widehat{H},\widehat{D}}{\text{Exp}}[\, T \mid G\,]
$$

For each random-oracle query that $\mathcal{A}$ makes for a ciphertext t_i (all the ciphertexts need not be distinct), the probability that $\mathcal{A}$ can output t_i (over the randomness $\widehat{H}$) is at most the probability that the output of the random-oracle query is in $\mathcal{G}_{t_i}$, because otherwise no valid response $\mathbf{z}$ exists. Thus each t_i has the probability of being output at most $|\mathcal{G}_{t_i}|/|\mathcal{C}|$, regardless of the strategy of the adversary. Plugging this and (18) into (19), we obtain

$$
\underset{\widehat{H},\widehat{D}}{\text{Exp}}[\, T \mid G\,] \leq \sum_{i=1}^{q_H} \frac{|\mathcal{G}_{t_i}|}{|\mathcal{C}|} \cdot \frac{|\mathcal{C}|}{|\mathcal{G}_{t_i}| - 1} \leq q_H \cdot \max_{i=1,\ldots,q_H} \left(\frac{|\mathcal{G}_{t_i}|}{|\mathcal{G}_{t_i}| - 1} \right) \leq \frac{q_H \cdot (f+1)}{f} .
$$

$$
\tag{20}
$$

By applying Markov's inequality, we have that for any positive β,

$$\Pr_{\widehat{H},\widehat{D}}[T \geq \beta] = \Pr_{\widehat{H},\widehat{D}}[T \geq \beta \mid G] \cdot \Pr_{\widehat{H},\widehat{D}}[G] + \Pr_{\widehat{H},\widehat{D}}[T \geq \beta \mid \neg G] \cdot \Pr_{\widehat{H},\widehat{D}}[\neg G]$$

$$\leq \frac{\operatorname{Exp}_{\widehat{H},\widehat{D}}[T \mid G]}{\beta} + \Pr_{\widehat{H},\widehat{D}}[\neg G] \tag{21}$$

It is furthermore easy to see that

$$\Pr_{\widehat{H},\widehat{D}}[\neg G] = 1 - \Pr_{\widehat{H}}[G] \leq 1 - \left(\frac{|\mathcal{C}| - f}{|\mathcal{C}|}\right)^{q_{\mathsf{H}}} \leq \frac{q_{\mathsf{H}} f}{|\mathcal{C}|}.$$

Plugging this and (20) into (21) and letting $\beta = \alpha \cdot q_{\mathsf{H}}$ yields

$$\Pr_{\widehat{H},\widehat{D}}[T \geq \alpha \cdot q_{\mathsf{H}}] \leq \frac{1}{\alpha} \cdot \left(1 + \frac{1}{f}\right) + \frac{q_{\mathsf{H}} f}{|\mathcal{C}|}.$$

To minimize this expression, we set $f = \left\lceil \sqrt{\frac{|\mathcal{C}|}{\alpha q_{\mathsf{H}}}} \right\rceil$, which gives us the claim in the Lemma. $\qquad\square$

4   Multi-shot Verifiable Encryption – Construction with a Bounded Decryption Time

If one would like to put a limit on how much computational time decryption takes in the worst case, then the idea is to reduce the size of the challenge space and repeat the proof-of-knowledge protocol from Algorithm 3 in parallel α times using the standard approach. This protocol is described in Algorithm 6. The verification procedure simply checks if all the α copies are verified correctly.

One can show that with probability approximately $1 - |\mathcal{C}|^{-\alpha}$, there will be at least one $\bar{c}$ for which $\bar{c} \begin{bmatrix} \mathbf{v} \\ \mathbf{w} \end{bmatrix}$ is a valid ciphertext satisfying Lemma 3.1, and so one simply needs to find it (and check its validity) analogously to the way $\bar{c} \begin{bmatrix} \mathbf{v} \\ \mathbf{w} \end{bmatrix}$ was found in Algorithm 5. The main difference is that the challenge space is no longer exponentially large, and so one can search through all the $\bar{c} = c^{(i)} - c'$ in time $\alpha \cdot |\mathcal{C}|$. This procedure is described in Algorithm 8. If one wants to also maintain the relationship between the number of random oracle queries of the prover to the number of decryption tries, one could "dovetail" between Algorithm 8 which systematically goes through all $c' \in \mathcal{C}$ with one that randomly guesses them at random.

The trade-off between having the decryption running time be upper-bounded by $\alpha \cdot |\mathcal{C}|$ is that the proof of knowledge is now approximately α times larger.

Algorithm 6. α-shot Verifiable encryption $\mathsf{Enc}(pk, x, w)$

Input: Public key $pk = (\mathbf{a}, \mathbf{t}, p, q)$, language member $x = (\mathbf{B}, \mathbf{u})$, witness $w = \mathbf{m} \in S_\gamma^k$. Challenge domain $\mathcal{C}$. Cryptographic hash function $\mathsf{H} : \{0,1\}^* \to \mathcal{C}^\alpha$. Standard deviation $\sigma = 11 \cdot \max_{c \in \mathcal{C}} \|c\|_1 \cdot \sqrt{\alpha k n (3 + \gamma)}$

1: $\mathbf{r}, \mathbf{e}, \mathbf{e}' \xleftarrow{\$} S_1^k$

2: $\begin{bmatrix} \mathbf{v} \\ \mathbf{w} \end{bmatrix} \leftarrow \begin{bmatrix} p\mathbf{a}\mathbf{I}_k & p\mathbf{I}_k & 0^{k \times k} & 0^{k \times k} \\ t\mathbf{I} & 0^{k \times k} & p\mathbf{I}_k & \mathbf{I}_k \end{bmatrix} \begin{bmatrix} \mathbf{r} \\ \mathbf{e} \\ \mathbf{e}' \\ \mathbf{m} \end{bmatrix} \begin{matrix} \bmod q \\ \bmod q \end{matrix}$

3: **for** $i = 1$ **to** α **do**

4: $\quad \mathbf{y}^{(i)} \leftarrow \begin{bmatrix} \mathbf{y_r}^{(i)} \\ \mathbf{y_e}^{(i)} \\ \mathbf{y_{e'}}^{(i)} \\ \mathbf{y_m}^{(i)} \end{bmatrix} \xleftarrow{\$} D_{R^{4k}, 0, \sigma}$

5: $\quad \mathbf{f}^{(i)} \leftarrow \begin{bmatrix} p\mathbf{a}\mathbf{I}_k & p\mathbf{I}_k & 0^{k \times k} & 0^{k \times k} \\ t\mathbf{I}_k & 0^{k \times k} & p\mathbf{I}_k & \mathbf{I}_k \\ 0^{\ell \times k} & 0^{\ell \times k} & 0^{\ell \times k} & \mathbf{B} \end{bmatrix} \begin{bmatrix} \mathbf{y_r}^{(i)} \\ \mathbf{y_e}^{(i)} \\ \mathbf{y_{e'}}^{(i)} \\ \mathbf{y_m}^{(i)} \end{bmatrix} \begin{matrix} \bmod q \\ \bmod q \\ \bmod p \end{matrix}$

6: **end for**

7: $(\mathbf{c}^{(1)}, \ldots, \mathbf{c}^{(\alpha)}) \leftarrow \mathsf{H}\left(\begin{bmatrix} p\mathbf{a}\mathbf{I}_k & p\mathbf{I}_k & 0^{k \times k} & 0^{k \times k} \\ t\mathbf{I}_k & 0^{k \times k} & p\mathbf{I}_k & \mathbf{I}_k \\ 0^{\ell \times k} & 0^{\ell \times k} & 0^{\ell \times k} & \mathbf{B} \end{bmatrix}, \begin{bmatrix} \mathbf{v} \\ \mathbf{w} \\ \mathbf{u} \end{bmatrix}, \mathbf{f}^{(1)}, \ldots, \mathbf{f}^{(\alpha)} \right)$

8: **for** $i = 1$ **to** ℓ **do**

9: $\quad \mathbf{s}^{(i)} \leftarrow \begin{bmatrix} \mathbf{r} \\ \mathbf{e} \\ \mathbf{e}' \\ \mathbf{m} \end{bmatrix} \mathbf{c}^{(i)}$

10: $\quad \mathbf{z}^{(i)} \leftarrow \mathbf{s}^{(i)} + \mathbf{y}^{(i)}$

11: **end for**

12: $\mathbf{s} \leftarrow \begin{bmatrix} \mathbf{s}^{(1)} \\ \ldots \\ \mathbf{s}^{(\alpha)} \end{bmatrix}, \mathbf{z} \leftarrow \begin{bmatrix} \mathbf{z}^{(1)} \\ \ldots \\ \mathbf{z}^{(\alpha)} \end{bmatrix}$

13: with probability $\dfrac{D_{R^{4k\alpha}, 0, \sigma}(\mathbf{z})}{3 \cdot D_{R^{4k\alpha}, \mathbf{s}, \sigma}(\mathbf{z})}$, continue, else goto 3

14: **if** $\|\mathbf{z}\|_\infty > 6 \cdot \sigma$, goto 3

15: **return** $t = (\mathbf{v}, \mathbf{w}, \mathbf{c}^{(1)}, \ldots, \mathbf{c}^{(\alpha)}, \mathbf{z}^{(1)}, \ldots, \mathbf{z}^{(\alpha)})$

---

5   Chosen-Ciphertext Security

Many applications require a verifiable ciphertext to hide the encrypted witness, even when the adversary has access to decryptions of other ciphertexts. As a natural analog of indistinguishability under chosen-ciphertext attack (IND-CCA) for standard public-key encryption schemes, we define chosen-ciphertext simulatability and describe a construction that satisfies it.

Algorithm 7. α-shot Verification $\mathsf{V}(pk, x, t)$

Input: Public key $pk = (a, t, p, q)$, language member $x = (\mathbf{B}, \mathbf{u})$, ciphertext $t = (\mathbf{v}, \mathbf{w}, c^{(1)}, \ldots, c^{(\alpha)}, \mathbf{z}^{(1)}, \ldots, \mathbf{z}^{(\alpha)})$. Cryptographic hash function $\mathsf{H} : \{0,1\}^* \to \mathcal{C}^\alpha$. Positive real σ as in Algorithm 6.

1: $\mathbf{z} \leftarrow \begin{bmatrix} \mathbf{z}^{(1)} \\ \ldots \\ \mathbf{z}^{(\alpha)} \end{bmatrix}$

2: **if** $\|\mathbf{z}\|_\infty > 6 \cdot \sigma$, **return** 0

3: **for** $i = 1$ **to** α **do**

4: $\quad \mathbf{f}^{(i)} \leftarrow \begin{bmatrix} pa\mathbf{I}_k & p\mathbf{I}_k & 0^{k \times k} & 0^{k \times k} \\ t\mathbf{I}_k & 0^{k \times k} & p\mathbf{I}_k & \mathbf{I}_k \\ 0^{\ell \times k} & 0^{\ell \times k} & 0^{\ell \times k} & \mathbf{B} \end{bmatrix} \mathbf{z}^{(i)} - c^{(i)} \begin{bmatrix} \mathbf{v} & \bmod q \\ \mathbf{w} & \bmod q \\ \mathbf{u} & \bmod p \end{bmatrix}$

5: **end for**

6: **if** $(c^{(1)}, \ldots, c^{(\alpha)}) \neq \mathsf{H}\left(\begin{bmatrix} pa\mathbf{I}_k & p\mathbf{I}_k & 0^{k \times k} & 0^{k \times k} \\ pt\mathbf{I}_k & 0^{k \times k} & p\mathbf{I}_k & \mathbf{I}_k \\ 0^{\ell \times k} & 0^{\ell \times k} & 0^{\ell \times k} & \mathbf{B} \end{bmatrix}, \begin{bmatrix} \mathbf{v} \\ \mathbf{w} \\ \mathbf{u} \end{bmatrix}, \mathbf{f}^{(1)}, \ldots, \mathbf{f}^{(\alpha)} \right)$, **return** 0

7: **return** 1

Algorithm 8. α-shot Decryption $\mathsf{Dec}(sk, x, t)$

Input: Secret key $sk = s_1$, language member $x = (\mathbf{B}, \mathbf{u})$, ciphertext $t = (\mathbf{v}, \mathbf{w}, c^{(1)}, \ldots, c^{(\alpha)}, \mathbf{z}^{(1)}, \ldots, \mathbf{z}^{(\alpha)})$.

1: **if** $\mathsf{V}(pk, t, \pi) = 1$ **then**

2: $\quad$ **for** $i = 1$ **to** α **do**

3: $\quad\quad$ **for all** $c' \in \mathcal{C}$ **do**

4: $\quad\quad\quad \bar{c} \leftarrow c^{(i)} - c'$

5: $\quad\quad\quad \overline{\mathbf{m}} \leftarrow (\mathbf{w} - \mathbf{v}s_1)\bar{c} \bmod q$

6: $\quad\quad\quad$ **if** $\|\overline{\mathbf{m}}\|_\infty < q/2C$ **then**

7: $\quad\quad\quad\quad \overline{\mathbf{m}} \leftarrow \overline{\mathbf{m}} \bmod p$

8: $\quad\quad\quad\quad$ **return** $(\overline{\mathbf{m}}, \bar{c})$

9: $\quad\quad\quad$ **end if**

10: $\quad\quad$ **end for**

11: $\quad$ **end for**

12: **end if**

Our construction essentially follows the Naor-Yung paradigm [NY90] where the sender encrypts the message twice under different public keys and adds a non-interactive zero-knowledge (NIZK) proof that both ciphertexts encrypt the same message. Naor and Yung only proved their approach secure under non-adaptive chosen-ciphertext (CCA1), but Sahai [Sah99] later showed that if the NIZK proof is simulation-sound, then the resulting encryption scheme is secure against adaptive chosen-ciphertext (CCA2) attacks. Faust et al. [FKMV12] showed that Fiat-Shamir proofs are simulation-sound in the random-oracle model if the underlying proof system has quasi-unique responses.

Furthermore, because the verifiable encryption scheme for a CPA-secure encryption scheme already includes a NIZK, this conversion from CPA to CCA2 security is rather cheap, increasing the size of the proof and ciphertext by factors less than 2 (see (22)).

Chosen-ciphertext simulatability. We say that a relaxed verifiable encryption scheme $(\mathsf{Kg}, \mathsf{Enc}, \mathsf{V}, \mathsf{Dec})$ is chosen-ciphertext simulatable when there exists a simulator Sim such that the following advantage is negligible for all PPT adversaries A:

$$\left| \Pr\left[b' = b : \begin{array}{c} b \xleftarrow{\$} \{0,1\}, (pk, sk) \xleftarrow{\$} \mathsf{Kg}(1^\lambda), (st, x, w) \xleftarrow{\$} \mathsf{A}(pk), \\ t_0 \xleftarrow{\$} \mathsf{Enc}(pk, x, w), t_1 \xleftarrow{\$} \mathsf{Sim}(pk, x), b' \xleftarrow{\$} \mathsf{A}^{\mathsf{Dec}(sk, \cdot, \cdot)}(st, t_b) \end{array} \right] - \frac{1}{2} \right| ,$$

where A is not allowed to query its Dec oracle on the challenge ciphertext t_b. In the random-oracle model, Sim can additionally program the random oracle.

Construction. The receiver generates a Ring-LWE key pair by choosing the secrets $s_1, s_1', s_2, s_2' \xleftarrow{\$} S_1$ and a $\xleftarrow{\$} R$, and computing $t_1 \leftarrow as_1 + s_2$ and $t_1 \leftarrow as_1' + s_2'$. The public key is $pk = (a, t_1, t_2, p, q)$, where p is modulus for proving the linear relation. The secret key is $sk = s_1$.

The sender encrypts a witness $w = \mathbf{m}$ for language member $x = (\mathbf{B}, \mathbf{u})$ by choosing randomness $(\mathbf{r}_1, \mathbf{e}_1, \mathbf{e}_1', \mathbf{r}_2, \mathbf{e}_2, \mathbf{e}_2') \xleftarrow{\$} S_1^{6k}$, computing

$$\begin{bmatrix} pa\mathbf{I}_k & p\mathbf{I}_k & 0^{k\times k} & 0^{k\times k} & 0^{k\times k} & 0^{k\times k} & 0^{k\times k} \\ pt_1\mathbf{I}_k & 0^{k\times k} & p\mathbf{I}_k & 0^{k\times k} & 0^{k\times k} & 0^{k\times k} & \mathbf{I}_k \\ 0^{k\times k} & 0^{k\times k} & 0^{k\times k} & pa\mathbf{I}_k & p\mathbf{I}_k & 0^{k\times k} & 0^{k\times k} \\ 0^{k\times k} & 0^{k\times k} & 0^{k\times k} & pt_2\mathbf{I}_k & 0^{k\times k} & p\mathbf{I}_k & \mathbf{I}_k \end{bmatrix} \begin{bmatrix} \mathbf{r}_1 \\ \mathbf{e}_1 \\ \mathbf{e}_1' \\ \mathbf{r}_2 \\ \mathbf{e}_2 \\ \mathbf{e}_2' \\ \mathbf{m} \end{bmatrix} = \begin{bmatrix} \mathbf{v}_1 \\ \mathbf{w}_1 \\ \mathbf{v}_2 \\ \mathbf{w}_2 \end{bmatrix} \begin{array}{l} \bmod q \\ \bmod q \\ \bmod q \\ \bmod q \end{array} \quad (22)$$

and concatenating a proof $(c, \mathbf{z})$ using the relaxed NIZK proof system of Sect. 2.6 for the language element:

$$\begin{bmatrix} pa\mathbf{I}_k & p\mathbf{I}_k & 0^{k\times k} & 0^{k\times k} & 0^{k\times k} & 0^{k\times k} & 0^{k\times k} \\ pt_1\mathbf{I}_k & 0^{k\times k} & p\mathbf{I}_k & 0^{k\times k} & 0^{k\times k} & 0^{k\times k} & \mathbf{I}_k \\ 0^{k\times k} & 0^{k\times k} & 0^{k\times k} & pa\mathbf{I}_k & p\mathbf{I}_k & 0^{k\times k} & 0^{k\times k} \\ 0^{k\times k} & 0^{k\times k} & 0^{k\times k} & pt_2\mathbf{I}_k & 0^{k\times k} & p\mathbf{I}_k & \mathbf{I}_k \\ 0^{\ell\times k} & 0^{\ell\times k} & 0^{\ell\times k} & 0^{\ell\times k} & 0^{\ell\times k} & 0^{\ell\times k} & \mathbf{B} \end{bmatrix} \begin{bmatrix} \mathbf{r}_1 \\ \mathbf{e}_1 \\ \mathbf{e}_1' \\ \mathbf{r}_2 \\ \mathbf{e}_2 \\ \mathbf{e}_2' \\ \mathbf{m} \end{bmatrix} = \begin{bmatrix} \mathbf{v}_1 \\ \mathbf{w}_1 \\ \mathbf{v}_2 \\ \mathbf{w}_2 \\ \mathbf{u} \end{bmatrix} \begin{array}{l} \bmod q \\ \bmod q \\ \bmod q \\ \bmod q \\ \bmod p \end{array} . \quad (23)$$

Verification of a ciphertext $(\mathbf{v}_1, \mathbf{w}_1, \mathbf{v}_2, \mathbf{w}_2, c, \mathbf{z})$ is done by verifying the zero-knowledge proof $(c, \mathbf{z})$ for the language element (23). Decryption works exactly as in Algorithm 5, using $\mathbf{w}_1$ instead of $\mathbf{w}$.

Security. Correctness, completeness, and soundness all hold under the same assumptions as the CPA-secure scheme in Sect. 3.4. The following theorem states the chosen-ciphertext simulatability of the scheme. The proof can be found in the full version.

Theorem 5.1. *If the Ring-LWE encryption scheme is IND-CPA secure and the relaxed NIZK proof system is unbounded non-interactive zero-knowledge and unbounded simulation-sound, then the above construction is chosen-ciphertext simulatable.*

6 Applications

6.1 Key Escrow for Ring-LWE Encryption

A verifiable escrow scheme for decryption keys [YY98,PS00] allows a key owner to encrypt his private decryption key under the public key of a trusted authority so that anyone can check that the ciphertext is indeed an encryption of the private key corresponding to the owner's public key, but only the trusted authority can actually recover the private key. Intuitively, the owner is giving a proof that all messages sent to his public key can also be decrypted by the trusted third party. Note that a key escrow scheme cannot prevent parties from communicating securely, because even when forced to use escrowed keys, the parties can choose to double-encrypt messages under a non-escrowed key, or apply steganography to hide the fact that they are communicating altogether. The goal, therefore, is rather to prevent "dishonest" usage of public-key infrastructures, e.g., by using it to certify non-escrowed keys.

We show how our verifiable encryption scheme can be used to verifiably escrow Ring-LWE decryption keys. While, due to our relaxation of verifiable encryption, we cannot guarantee that the trusted authority recovers the actual decryption key, we show that whatever he recovers suffices to decrypt messages encrypted under the corresponding public key.

Let the authority have a Ring-LWE public key $t = as_1 + s_2 \bmod q$ as described in Sect. 2.5. Users also have Ring-LWE encryption keys, but in R_p instead of R_q. Meaning, a secret key is a pair $(m_1, m_2) \xleftarrow{\$} S_1^2$, while the public key is $u = bm_1 + m_2 \bmod p$ for $b \xleftarrow{\$} R_p$ together with a prime $p' < p$. Encryption and decryption work as in regular Ring-LWE, i.e., the sender chooses $r, e, e' \xleftarrow{\$} S_1$ and computes

$$v = p'(br + e) \bmod p$$
$$w = p'(ur + e') + \mu \bmod p . \tag{24}$$

To decrypt, the receiver computes $\mu \leftarrow w - vs_1 \bmod p \bmod p'$.

To escrow his decryption key, the key owner creates a verifiable encryption of his secret key $\mathbf{m} = [\, m_1 \; ; \; m_2 \,]$ using our scheme from Sect. 3.2 under the authority's public t with a proof that $\mathbf{m}$ is a witness for the relation

$$[\, b \;\; 1 \,] \begin{bmatrix} m_1 \\ m_2 \end{bmatrix} = u \bmod p .$$

The soundness property of our relaxed verifiable encryption scheme guarantees that the authority can decrypt a witness $(\bar{\mathbf{m}}, \bar{c})$ such that

$$b\bar{m}_1 + \bar{m}_2 = \bar{c}u \bmod p .$$

The authority can decrypt an honestly generated ciphertext of the form (24) by computing

$$
\begin{aligned}
\bar{c}w - v\bar{m}_1 \bmod p &= \bar{c}p'(ur + e') + \bar{c}\mu - p'(br + e)\bar{m}_1 \bmod p \\
&= p'((b\bar{m}_1 + \bar{m}_2)r + \bar{c}e') + \bar{c}\mu - p'(b\bar{m}_1 r + e\bar{m}_1) \bmod p \\
&= p'(\bar{m}_2 r + \bar{c}e' - e\bar{m}_1) + \bar{c}\mu \bmod p
\end{aligned}
$$

from which μ can be recovered by reducing modulo p' and then dividing by $\bar{c}$ modulo p' (note that it is important that p' is chosen such that all differences of challenges in the challenge space are invertible), as long as the parameters are chosen such that $\|p'(\bar{m}_2 r + \bar{c}e' - e\bar{m}_1) + \bar{c}\mu\|_\infty < p/2$.

6.2   Verifiably Encrypted Signatures

Suppose two parties want to engage in a contract together and exchange signatures on the agreed contract. Neither of the parties wants to be the first to send his signature, however, fearing that the other party may not reciprocate and hold the first party liable to the conditions in the contract, without being held liable himself. Fair signature exchange protocols [ASW00, BDM98] ensure that no party can obtain a significant advantage over the other party by aborting the protocol early.

Verifiably encrypted signatures [ASW00, BDM98, BGLS03] are an important tool to build optimistic fair exchange protocols. The first party initially sends his signature encrypted under the key of a trusted adjudicator such that the other party can verify that the ciphertext indeed contains a valid signature on the agreed contract, but cannot recover the signature itself. The second party responds by sending his signature, after which the first party also sends over his signature. In case the first party refuses to send his signature in the last step, the second party can contact the adjudicator to have the encrypted signature from the first decrypted.

We show how our relaxed verifiable encryption scheme can be used to build verifiably encrypted signatures for the ring-based variant of Gentry-Peikert-Vaikuntanathan (Ring-GPV) signature scheme [GPV08] based on the hardness of the Ring-SIS or NTRU problems [SS11, DLP14]. Here, the signer's public key is a polynomial $b \in R_p$, while the secret key is a trapdoor allowing to find, for a given $u \in R_p$, short polynomials m_1, m_2 such that $bm_1 + m_2 = u$. A signature on a message μ in the usual scheme is a short vector (m_1, m_2) such that $bm_1 + m_2 = H(\mu) \bmod p$, where $H : \{0,1\}^* \to R_p$ is a random oracle. It is easy to show, however, that the scheme remains secure if one relaxes the verification algorithm to accept any tuple of short polynomials (m_1, m_2, c) such that $bm_1 + m_2 = cH(\mu) \bmod p$.

In the usual security proof, when the adversary produces a forgery, $bm_1 + m_2 = H(\mu) \bmod p$, the simulator already possesses another equality $bm'_1 + m'_2 = H(\mu) \bmod p$, and thus obtains a solution to Ring-SIS as $b(m_1 - m'_1) + (m_2 - m'_2) = 0 \bmod p$. If, on the other hand, the adversary produces a forgery $bm_1 + m_2 = cH(\mu) \bmod p$, then the simulator can obtain the equation $b(cm_1 - m'_1) + (cm_2 - m'_2) = 0 \bmod p$, which is still a (slightly longer) solution to Ring-SIS.

For this modified signature scheme that we build a verifiably encrypted signature scheme using our CCA-secure relaxed verifiable encryption scheme from Sect. 5. Namely, to encrypt an honest signature $(m_1, m_2, 1)$ under the adjudicator's public key, one encrypts the witness $\mathbf{m} = [\ m_1\ ;\ m_2\ ]$ with the encryption scheme from Sect. 5 while proving that $[\ b\ \ 1\ ]\ \mathbf{m} = H(\mu) \bmod p$. When the adjudicator decrypts the signature, it recovers $(\bar{\mathbf{m}}, \bar{c})$ such that $[\ b\ \ 1\ ]\ \bar{\mathbf{m}} = \bar{c}H(\mu) \bmod p$, which is also a valid signature on μ. Unforgeability follows from the unforgeability of the relaxed Ring-GPV scheme, while the security against extraction follows from the security of Ring-LWE encryption.

6.3 Other Applications

One of the most prominent applications of verifiable encryption is in group signatures [CvH91], where group members can sign anonymously in name of the entire group, but their anonymity can be lifted by a dedicated opening authority. A common construction paradigm [CL06, BCK+14] is to let a user's signing key consist of a signature by the group manager on the user's identity. To sign a message, the user encrypts his identity under the public key of the opener and creates a NIZK proof of knowledge of a valid signature for the encrypted identity. To recover the identity of the signer, the opener simply decrypts the ciphertext included in the signature.

Our verifiable encryption scheme could be very useful to group signatures in principle, what is missing is a *practical* signature scheme where the message $\mathbf{m}$ and the signature $\mathbf{s}$ are short vectors for which the verification equation can be expressed as a linear relation $\mathbf{B} \begin{bmatrix} \mathbf{m} \\ \mathbf{s} \end{bmatrix} = \mathbf{u}$.

7 Concrete Parameters

In this section we give some sample concrete instantiations of proofs of plaintext knowledge and verifiable encryption schemes (see Table 1). We express the

Table 1. Sample parameter sets for the verifiable encryption scheme

	I	II	III
n	1024	2048	2048
k	1	2	2
p	13	2^{15}	2^{30}
$\|\mathbf{m}\|_\infty$	1	1	2^{18}
σ	25344	50688	$\approx 2^{23.6}$
q	$\approx 2^{34}$	$\approx 2^{47}$	$\approx 2^{70}$
gamma factor	≈ 1.0046	≈ 1.0033	≈ 1.0052
proof size	9 KB	38 KB	54 KB
Ciphertext size	9 KB	48 KB	71 KB

security of each scheme in terms of the "gamma factor" from [GN08]. Values of 1.01 can be broken today, 1.007 seem to be fairly secure (conjectured at least 80-bits), and those less than 1.005 are believed to require more than 2^{128} time even for quantum computers. There have been some other attacks considered (e.g. in [ADPS16]), but those require as much memory as time, and are at this point not as useful in practice as variations of lattice-reduction attacks based on LLL (e.g. [CN11]). It is of course possible that attacks that are currently impractical can be made more practical, and at that point the concrete parameters (for all lattice-based schemes) would have to be adjusted. But the ratio between parameter sizes for verifiable encryption and regular encryption (and zero-knowledge authentication) should remain the same. One caveat would be if the algorithms for "overstretched" NTRU would become applicable to Ring-LWE. It was recently shown that when the modulus in NTRU is larger than the secrets by a sub-exponential (i.e. $2^{O(\sqrt{d})}$, where d is the lattice dimension) factor, then the NTRU problem becomes easy [ABD16, CJL16, KF16]. This is in contrast to LWE and Ring-LWE, for which efficient algorithms are only known in the case that the modulus is $2^{\omega(d)}$). If these attacks are transferred to the Ring-LWE setting, then this would have implications toward all constructions (e.g. those in this paper, most FHE schemes based on Ring-LWE, etc.) in which the secrets are significantly smaller than the modulus.

Our schemes are instantiated from the Ring-LWE cryptosystem where we take the secret and error parameters to be from the set $\{-1, 0, 1\}$. While the worst-case to average-case hardness of Ring-LWE (and LWE) was only proven with larger parameters [Reg09, LPR13], there haven't been any weaknesses found when constructing public-key encryption schemes with smaller errors. In particular, the part of the NTRU cryptosystem [HPS98] that is based on the Ring-LWE assumption has never been attacked due to having secret and error vectors having coefficients from the set $\{-1, 0, 1\}$. The most practical attacks is still to rewrite the Ring-LWE instance as a lattice problem and apply lattice reduction.

For all the parameter sets, we analyze the hardness of recovering the vector $[\mathbf{r}; \mathbf{e}; \mathbf{e}'; \mathbf{m}]$ in (11). In column I, we give parameters for a proof of plaintext knowledge where there is no message $\mathbf{m}$. The exact parameters for verifiable encryption will of course depend on the parameters of the relation in (9). In columns II and III, we give the parameters that are large enough to instantiate the two example cases in Sect. 6. All the parameters are as defined in Algorithm 3 with the value of q is taken so as to satisfy (16) in the statement of Lemma 3.1 which is required for the decryption algorithm to function correctly. We point out that in the application to key escrow, there is also an encryption in the key escrow itself. But because that encryption works over modulus p, which is smaller than q, the hardness of breaking it is at least as hard as breaking the verifiable encryption scheme.

Acknowledgements. We would like to thank the anonymous reviewers for their detailed reviews and helpful feedback. This work was supported by the SNSF ERC Transfer Grant CRETP2-166734 – FELICITY and by the European Commission's PERCY grant (agreement #321310).

References

[ABD16] Albrecht, M., Bai, S., Ducas, L.: A subfield lattice attack on overstretched NTRU assumptions. In: Robshaw, M., Katz, J. (eds.) CRYPTO 2016. LNCS, vol. 9814, pp. 153–178. Springer, Heidelberg (2016). doi:10.1007/978-3-662-53018-4_6

[ADPS16] Alkim, E., Ducas, L., Pöppelmann, T., Schwabe, P.: Post-quantum key exchange - a new hope. In: USENIX, pp. 327–343 (2016)

[ASW00] Asokan, N., Shoup, V., Waidner, M.: Optimistic fair exchange of digital signatures. IEEE J. Sel. Areas Commun. 18(4), 593–610 (2000)

[BCK+14] Benhamouda, F., Camenisch, J., Krenn, S., Lyubashevsky, V., Neven, G.: Better zero-knowledge proofs for lattice encryption and their application to group signatures. In: Sarkar, P., Iwata, T. (eds.) ASIACRYPT 2014. LNCS, vol. 8873, pp. 551–572. Springer, Heidelberg (2014). doi:10.1007/978-3-662-45611-8_29

[BDLN16] Baum, C., Damgård, I., Larsen, K.G., Nielsen, M.: How to prove knowledge of small secrets. In: Robshaw, M., Katz, J. (eds.) CRYPTO 2016. LNCS, vol. 9816, pp. 478–498. Springer, Heidelberg (2016). doi:10.1007/978-3-662-53015-3_17

[BDM98] Bao, F., Deng, R.H., Mao, W.: Efficient and practical fair exchange protocols with off-line TTP. In: IEEE Symposium on Security and Privacy, pp. 77–85 (1998)

[BGLS03] Boneh, D., Gentry, C., Lynn, B., Shacham, H.: Aggregate and verifiably encrypted signatures from bilinear maps. In: Biham, E. (ed.) EUROCRYPT 2003. LNCS, vol. 2656, pp. 416–432. Springer, Heidelberg (2003). doi:10.1007/3-540-39200-9_26

[BR93] Bellare, M., Rogaway, P.: Random oracles are practical: a paradigm for designing efficient protocols. In: CCS 1993, pp. 62–73 (1993)

[CD16] Cramer, R., Damgård, I.: Amortized complexity of zero-knowledge proofs revisited: achieving linear soundness slack. IACR Cryptology ePrint Archive, 2016:681 (2016)

[CJL16] Cheon, J.H., Jeong, J., Lee, C.: An algorithm for NTRU problems and cryptanalysis of the GGH multilinear map without an encoding of zero. IACR Cryptology ePrint Archive, 2016:139 (2016)

[CKY09] Camenisch, J., Kiayias, A., Yung, M.: On the portability of generalized Schnorr proofs. In: Joux, A. (ed.) EUROCRYPT 2009. LNCS, vol. 5479, pp. 425–442. Springer, Heidelberg (2009). doi:10.1007/978-3-642-01001-9_25

[CL06] Chase, M., Lysyanskaya, A.: On signatures of knowledge. In: Dwork, C. (ed.) CRYPTO 2006. LNCS, vol. 4117, pp. 78–96. Springer, Heidelberg (2006). doi:10.1007/11818175_5

[CN11] Chen, Y., Nguyen, P.Q.: BKZ 2.0: better lattice security estimates. In: Lee, D.H., Wang, X. (eds.) ASIACRYPT 2011. LNCS, vol. 7073, pp. 1–20. Springer, Heidelberg (2011). doi:10.1007/978-3-642-25385-0_1

[CS03] Camenisch, J., Shoup, V.: Practical verifiable encryption and decryption of discrete logarithms. In: Boneh, D. (ed.) CRYPTO 2003. LNCS, vol. 2729, pp. 126–144. Springer, Heidelberg (2003). doi:10.1007/978-3-540-45146-4_8

[CvH91] Chaum, D., Heyst, E.: Group signatures. In: Davies, D.W. (ed.) EUROCRYPT 1991. LNCS, vol. 547, pp. 257–265. Springer, Heidelberg (1991). doi:10.1007/3-540-46416-6_22

[DLP14] Ducas, L., Lyubashevsky, V., Prest, T.: Efficient identity-based encryption over NTRU lattices. In: Sarkar, P., Iwata, T. (eds.) ASIACRYPT 2014. LNCS, vol. 8874, pp. 22–41. Springer, Heidelberg (2014). doi:10.1007/978-3-662-45608-8_2

[DPSZ12] Damgård, I., Pastro, V., Smart, N., Zakarias, S.: Multiparty computation from somewhat homomorphic encryption. In: Safavi-Naini, R., Canetti, R. (eds.) CRYPTO 2012. LNCS, vol. 7417, pp. 643–662. Springer, Heidelberg (2012). doi:10.1007/978-3-642-32009-5_38

[FKMV12] Faust, S., Kohlweiss, M., Marson, G.A., Venturi, D.: On the non-malleability of the Fiat-Shamir transform. In: Galbraith, S., Nandi, M. (eds.) INDOCRYPT 2012. LNCS, vol. 7668, pp. 60–79. Springer, Heidelberg (2012). doi:10.1007/978-3-642-34931-7_5

[FS86] Fiat, A., Shamir, A.: How to prove yourself: practical solutions to identification and signature problems. In: Odlyzko, A.M. (ed.) CRYPTO 1986. LNCS, vol. 263, pp. 186–194. Springer, Heidelberg (1987). doi:10.1007/3-540-47721-7_12

[GN08] Gama, N., Nguyen, P.Q.: Predicting lattice reduction. In: Smart, N. (ed.) EUROCRYPT 2008. LNCS, vol. 4965, pp. 31–51. Springer, Heidelberg (2008). doi:10.1007/978-3-540-78967-3_3

[GPV08] Gentry, C., Peikert, C., Vaikuntanathan, V.: Trapdoors for hard lattices and new cryptographic constructions. In: STOC, pp. 197–206 (2008)

[HPS98] Hoffstein, J., Pipher, J., Silverman, J.H.: NTRU: a ring-based public key cryptosystem. In: Buhler, J.P. (ed.) ANTS 1998. LNCS, vol. 1423, pp. 267–288. Springer, Heidelberg (1998). doi:10.1007/BFb0054868

[KF16] Kirchner, P., Fouque, P.-A.: Comparison between subfield and straightforward attacks on NTRU. IACR Cryptology ePrint Archive 2016:717 (2016)

[LM06] Lyubashevsky, V., Micciancio, D.: Generalized compact knapsacks are collision resistant. In: Bugliesi, M., Preneel, B., Sassone, V., Wegener, I. (eds.) ICALP 2006. LNCS, vol. 4052, pp. 144–155. Springer, Heidelberg (2006). doi:10.1007/11787006_13

[LN86] Lidl, R., Niederreiter, H.: Introduction to Finite Fields and their Applications. Cambridge University Press, Cambridge (1986)

[LNSW13] Ling, S., Nguyen, K., Stehlé, D., Wang, H.: Improved zero-knowledge proofs of knowledge for the ISIS problem, and applications. In: Kurosawa, K., Hanaoka, G. (eds.) PKC 2013. LNCS, vol. 7778, pp. 107–124. Springer, Heidelberg (2013). doi:10.1007/978-3-642-36362-7_8

[LPR13] Lyubashevsky, V., Peikert, C., Regev, O.: On ideal lattices and learning with errors over rings. J. ACM 60(6), 43 (2013). Preliminary version appeared in EUROCRYPT 2010

[Lyu09] Lyubashevsky, V.: Fiat-shamir with aborts: applications to lattice and factoring-based signatures. In: Matsui, M. (ed.) ASIACRYPT 2009. LNCS, vol. 5912, pp. 598–616. Springer, Heidelberg (2009). doi:10.1007/978-3-642-10366-7_35

[Lyu12] Lyubashevsky, V.: Lattice signatures without trapdoors. In: Pointcheval, D., Johansson, T. (eds.) EUROCRYPT 2012. LNCS, vol. 7237, pp. 738–755. Springer, Heidelberg (2012). doi:10.1007/978-3-642-29011-4_43

[NY90] Naor, M., Yung, M.: Public-key cryptosystems provably secure against chosen ciphertext attacks. In: STOC, pp. 427–437 (1990)

[PS00] Poupard, G., Stern, J.: Fair encryption of RSA keys. In: Preneel, B. (ed.) EUROCRYPT 2000. LNCS, vol. 1807, pp. 172–189. Springer, Heidelberg (2000). doi:10.1007/3-540-45539-6_13

[Reg09] Regev, O.: On lattices, learning with errors, random linear codes, and cryptography. J. ACM **56**(6) (2009)

[Sah99] Sahai, A.: Non-malleable non-interactive zero knowledge and adaptive chosen-ciphertext security. In: FOCS, pp. 543–553 (1999)

[SS11] Stehlé, D., Steinfeld, R.: Making NTRU as secure as worst-case problems over ideal lattices. In: Paterson, K.G. (ed.) EUROCRYPT 2011. LNCS, vol. 6632, pp. 27–47. Springer, Heidelberg (2011). doi:10.1007/978-3-642-20465-4_4

[SSTX09] Stehlé, D., Steinfeld, R., Tanaka, K., Xagawa, K.: Efficient public key encryption based on ideal lattices. In: Matsui, M. (ed.) ASIACRYPT 2009. LNCS, vol. 5912, pp. 617–635. Springer, Heidelberg (2009). doi:10.1007/978-3-642-10366-7_36

[Ste93] Stern, J.: A new identification scheme based on syndrome decoding. In: Stinson, D.R. (ed.) CRYPTO 1993. LNCS, vol. 773, pp. 13–21. Springer, Heidelberg (1994). doi:10.1007/3-540-48329-2_2

[YY98] Young, A., Yung, M.: Auto-recoverable auto-certifiable cryptosystems. In: Nyberg, K. (ed.) EUROCRYPT 1998. LNCS, vol. 1403, pp. 17–31. Springer, Heidelberg (1998). doi:10.1007/BFb0054114

Short Stickelberger Class Relations
and Application to Ideal-SVP

Ronald Cramer[1,2], Léo Ducas[1(✉)], and Benjamin Wesolowski[3]

[1] Cryptology Group, CWI, Amsterdam, The Netherlands
L.Ducas@cwi.nl
[2] Mathematical Institute, Leiden University, Leiden, The Netherlands
[3] École Polytechnique Fédérale de Lausanne, EPFL IC LACAL,
Lausanne, Switzerland

Abstract. The worst-case hardness of finding short vectors in ideals of cyclotomic number fields (Ideal-SVP) is a central matter in lattice based cryptography. Assuming the worst-case hardness of Ideal-SVP allows to prove the Ring-LWE and Ring-SIS assumptions, and therefore to prove the security of numerous cryptographic schemes and protocols — including key-exchange, digital signatures, public-key encryption and fully-homomorphic encryption.

A series of recent works has shown that *Principal* Ideal-SVP is not always as hard as finding short vectors in general lattices, and some schemes were broken using quantum algorithms — the SOLILO-QUY encryption scheme, Smart-Vercauteren fully homomorphic encryption scheme from PKC 2010, and Gentry-Garg-Halevi cryptographic multilinear-maps from EUROCRYPT 2013.

Those broken schemes were using a special class of principal ideals, but these works also showed how to solve SVP for principal ideals in the *worst-case* in quantum polynomial time for an approximation factor of $\exp(\tilde{O}(\sqrt{n}))$. This exposed an unexpected hardness gap between general lattices and some structured ones, and called into question the hardness of various problems over structured lattices, such as Ideal-SVP and Ring-LWE.

In this work, we generalize the previous result to general ideals. Precisely, we show how to solve the close principal multiple problem (CPM) by exploiting the classical theorem that the class-group is annihilated by the (Galois-module action of) the so-called Stickelberger ideal. Under some plausible number-theoretical hypothesis, our approach provides a close principal multiple in quantum polynomial time. Combined with the previous results, this solves Ideal-SVP in the worst case in quantum polynomial time for an approximation factor of $\exp(\tilde{O}(\sqrt{n}))$.

Although it does not seem that the security of Ring-LWE based cryptosystems is directly affected, we contribute novel ideas to the cryptanalysis of schemes based on structured lattices. Moreover, our result shows a deepening of the gap between general lattices and structured ones.

© International Association for Cryptologic Research 2017
J.-S. Coron and J.B. Nielsen (Eds.): EUROCRYPT 2017, Part I, LNCS 10210, pp. 324–348, 2017.
DOI: 10.1007/978-3-319-56620-7_12

1   Introduction

The problem of finding the shortest vector of a Euclidean lattice (the shortest vector problem, or SVP) is a central hard problem in complexity theory. Approximated versions of this problem (approx-SVP) have become the theoretical foundation for many cryptographic constructions thanks to the average-case to worst-case reductions of Ajtai [Ajt99] — a classical reduction from approx-SVP to the Short Integer Solution (SIS) problem — and Regev [Reg05] — a quantum reduction from approx-SVP to Learning with Errors (LWE).

For efficiency reasons, it is tempting to rely on structured lattices, in particular lattices arising as ideals or modules over certain rings, the earliest example being the NTRUENCRYPT[1] proposal from Hoffstein et al. [HPS98]. Later on, variations on these foundations were also considered.

Precisely, the Ring-SIS [Mic02,LM06,PR06] and Ring-LWE [SSTX09, LPR10] problems were introduced, and shown to reduce to worst-case instances of Ideal-SVP, a specialization of SVP to ideals viewed as lattices. Both problems Ring-SIS and Ring-LWE have shown very versatile problems for building efficient cryptographic schemes upon.

The typical choices of rings for Ring-SIS, Ring-LWE and Ideal-SVP are the ring of integers of a cyclotomic number field of conductor m, that is $K = \mathbb{Q}(\omega_m)$, of degree $n = \varphi(m)$, where ω_m is a complex primitive m-th root of unity. This choice further ensures the hardness of the decisional version of Ring-LWE under the same worst-case Ideal-SVP hardness assumption [LPR10].

Attack on Principal Ideals. For some time, it seemed plausible that the *structured* versions of lattice problems should be just as hard to solve as the unstructured ones: only some (almost) linear-time advantages were known. This was challenged by a claim of Campbell et al. [CGS14]: a quantum polynomial-time attack against their schemes SOLILOQUY. The attack also applies to the fully-homomorphic encryption scheme of [SV10] and the cryptographic multilinear maps candidates [GGH13,LSS14], as they all share a common key generation procedure, describe below.

For the secret key, choose an integral element $g \in \mathcal{O}_K$ with small distortion, *i.e.* a $g \in \mathcal{O}_K$ such that

$$\frac{\max_\sigma |\sigma(g)|}{\min_\sigma |\sigma(g)|} \leq \mathrm{poly}(n) \tag{1}$$

where σ ranges over the n complex embeddings $K \mapsto \mathbb{C}$. A corresponding public key consists of the ideal $\mathfrak{I} = (g)$, described by a "bad" $\mathbb{Z}$-basis (e.g. a $\mathbb{Z}$-basis in Hermite normal form).

The attack consists of two steps, sketched in [CGS14]. First, using a quantum computer, it should be possible to solve the Principal Ideal Problem (PIP): given $\mathfrak{I} \subset \mathcal{O}_K$ find $h \in \mathcal{O}_K$ such that $\mathfrak{I} = (h)$. Second, a (classical) close-vector

---

[1] Proposal which is not supported by a worst-case hardness argument, but a variant is [SS11].

algorithm in the log-unit lattice $\text{Log}\,\mathcal{O}_K^\times$ should allow to recover the secret key[2] g from h. Both steps are claimed to be polynomial time.

While the analysis of the quantum step was unclear[3], such a result seemed plausible considering the recent breakthrough on the Hidden Subgroup Problem over $\mathbb{R}^n$ by Eisentrager et al. [EHKS14] including efficient quantum unit-group computation. And indeed Biasse and Song [BS16] generalized [EHKS14] to S-unit-group computation, allowing in particular to solve PIP [BS16, Theorem 1.3].

The claimed correctness of the short generator recovery step also raised questions: unless a particularly orthogonal basis of the log-unit lattice $\text{Log}\,\mathcal{O}_K^\times$ is known, this step should take exponential time. It was already noticed [GGH13, Full version, pp. 43] that the log-unit lattice could be efficiently decoded up to a radius of $n^{-O(\log\log n)}$ thanks to the Gentry-Szydlo algorithm [GS02], but this is far from sufficient. Yet, the claim that it can be done in polynomial time was quickly supported by convincing numerical experiments [Sch15]. And indeed, by analyzing the geometry of cyclotomic units, Cramer et al. [CDPR16, Theorem 4.1] proved that the decoding-radius given by a basis of such units is in fact much better.

A second result of Cramer et al. [CDPR16, Theorem 6.3] analyses how good of an approximation of the shortest vector is obtained in the worst-case, i.e. without condition (1). Using a variation on the algorithm of [CGS14], they prove that from any generator h of $\mathfrak{I}$, one can efficiently find a generator g of euclidean length $(N\mathfrak{I})^{1/n}\cdot\exp(\tilde{O}(\sqrt{n}))$. Combined with [BS16], this solves in quantum polynomial time the Short Vector Problem over principal ideals in the worst-case for an approximation factor $\gamma = \exp(\tilde{O}(\sqrt{n}))$.

Claim 1 ([BS16, Theorem 1.3] **Combined with** [CDPR16, Theorem 6.3]).
There exists a quantum polynomial time algorithm PRINCIPALIDEALSVP($\mathfrak{a}$), *that given an ideal of $\mathcal{O}_K$ for K a cyclotomic number field of prime power conductor, returns an generator $v\in\mathfrak{a}$ of Euclidean norm $\|v\| \leq (N\mathfrak{a})^{1/n}\cdot\exp(\tilde{O}(\sqrt{n}))$.*

In particular, v is a solution to Ideal-SVP for an approximation factor $\gamma = \|v\|/\lambda_1(\mathfrak{a}) = \exp(\tilde{O}(\sqrt{n}))$ where $\lambda_1(\mathfrak{a})$ denotes the length of the shortest vector of $\mathfrak{a}$.

It is also shown [CDPR16, Lemma 6.2] that this result is tight up to a polylog(n) factor in the exponent: the shortest generator is typically larger than the shortest element by a factor $\exp(\tilde{O}(\sqrt{n}))$.

Impact and Limitatioms of the Attack on Principal Ideals. Whereas some cryptosystems were broken by this quantum attack, the current limitations of this approach to tackle more standard problems as Ring-LWE are three-fold.

(i) First, it is restricted to principal ideals, while Ring-SIS and Ring-LWE rely on worst-case hardness of SVP over general ideals.

---

[2] Up to a root of unity.
[3] And even challenged [BS16, Sect. 6].

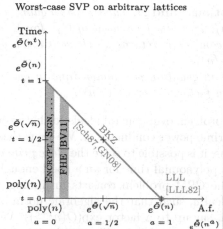

In the general case, the best known algorithms (BKZ [Sch87] and Slide [GN08]) run in time $\exp(\tilde{\Theta}(n^t))$ for an approximation factor $\exp(\tilde{\Theta}(n^a))$, where $t + a = 1$.

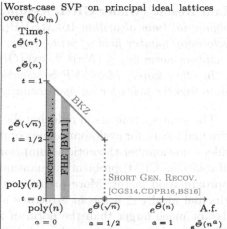

For principal ideals of cyclotomic field (of prime-power conductor), the aforementioned results give a quantum polynomial runtime (i.e., $t = 0$) for any $a \geq 1/2$.

Fig. 1. Best known (quantum) time–approximation factor tradeoffs to solve approx-SVP in arbitrary lattices (on the left) and in principal ideal lattices (on the right), in the worst case. The approximation factors of (ideal)-SVP used to build cryptography upon are typically between polynomial $\mathrm{poly}(n)$ and quasi-polynomial $\exp(\mathrm{polylog}(n))$.

(ii) Second, the approximation factor $\gamma = \exp(\tilde{O}(\sqrt{n}))$ in the worst-case is asymptotically too large to affect any actual Ring-LWE based schemes even for advanced cryptosystems such as the state of the art fully homomorphic encryption schemes (see [BV11,DM15]).

(iii) Third, Ring-LWE is known to be at least as hard as Ideal-SVP but not known to be equivalent.

But it does show an asymptotic gap between the search of mildly short vectors in general lattices and in certain structured lattices (see Fig. 1), and calls for a more thorough study of the hardness assumption over structured lattices. This work addresses the first of them.

1.1 Contributions

This work provides strong evidence that the general case of Ideal-SVP is not harder than the principal case for similar approximation factors. As a consequence, the approximation factors reachable in quantum polynomial time appear to be significantly smaller in arbitrary ideals of cyclotomic fields of prime-power conductor than known for general lattices, dropping from $\exp(\tilde{\Theta}(n))$ to $\exp(\tilde{\Theta}(\sqrt{n}))$.

Main Result (Under GRH, Assumptions 1 and 2). *There exists a quantum polynomial time algorithm* IDEALSVP($\mathfrak{a}$), *that given an ideal of* $\mathcal{O}_K$ *for* K *a cyclotomic number field of prime power conductor, returns an element* $v \in \mathfrak{a}$ *of Euclidean norm* $\|v\| \leq (N\mathfrak{a})^{1/n} \cdot \exp(\tilde{O}(\sqrt{n}))$.

In other words, Ideal-SVP is solvable in quantum polynomial time in cyclotomic number fields for an approximation factor $\gamma = \exp(\tilde{O}(\sqrt{n}))$.

The strategy consists in reducing the problem over general ideals to that over principal ideals, for cyclotomic fields of prime-power conductor m. We show that under some number-theoretic assumptions, it is possible to solve the *close principal multiple* (CPM) problem in quantum polynomial time for an a good enough approximation factor. More precisely, the CPM problem consists in finding a principal ideal $\mathfrak{c} \subset \mathfrak{a}$ for an arbitrary ideal $\mathfrak{a}$, such that the algebraic norm of $\mathfrak{c}$ is not much larger than the norm of $\mathfrak{a}$, say up to a factor $\exp(\tilde{O}(n^{1+c}))$. We will argue that one can reach $c = 1/2$, yet, any $c < 1$ will provide a better time-approximation factor tradeoff than the generic algorithms LLL and BKZ.

Our main tool to solve CPM is the classical theorem that the class-group is annihilated by the Galois-module action of the so-called Stickelberger ideal: it provides explicit class relations between an ideal and its Galois conjugates. An important fact is that this Stickelberger ideal has many short elements and that these can be explicitly constructed (see for example [Sch10]). This leads to a quantum polynomial time algorithm to solve CPM for a factor $\exp(\tilde{O}(n^{1+c}))$, where the constant c depends on how many Galois orbits of prime ideals are used to generate the (minus part of the) class group. It remains to apply the short generator recovery to $\mathfrak{c}$ to find a short vector of $\mathfrak{a}$, approximating the shortest vector by a factor $\exp(\tilde{O}(n^{\max(1/2,c)}))$.

We follow the notations of Fig. 1. If the exponent c can be made strictly smaller than 1, this gives a non-trivial result compared to generic lattice algorithms (see [Sch87, GN08]): we get $t = 0$ for any $a \geq \max(1/2, c)$, and in particular $a + t < 1$, against $a + t = 1$ for generic algorithms. If c can be made as small as $1/2$, then the asymptotic tradeoffs for Ideal-SVP are as good as the tradeoffs for Principal-Ideal-SVP.

Concluding formally on which value of c can be achieved is not straightforward, as it relies on the structure of the class group Cl_K as a $\mathbb{Z}[G]$-module (see Sect. 2.3). Based on computations of the class group structure of Schoof [Sch98] and a heuristic argument, we strongly believe it is plausible that $c = 1/2$ is reachable at least for a dense family of conductors m, if not all. This leads to the main result stated above.

1.2 Impact, Open Questions and Recommendations

To the best of our knowledge, this new result does not immediately lead to an attack on any proposed scheme, since most of them are based on Ring-LWE: obstacles (ii) and (iii) remain. Each of this obstacle leaves a crucial open cryptanalytic questions.

- The first question is whether the $\gamma = \exp(\tilde{O}(\sqrt{n}))$ approximation factors can be improved, potentially increasing the running time. One could for example consider many CPM solutions rather than just one, and hope that one of them leads to a much shorter vector.
- The second is whether an oracle for Ideal-SVP (an approx-SVP oracle for modules of rank 1) can be helpful to solve Ring-LWE, which can be summarized as an "unusually-Short Vector Problem" over a module of rank 3. Note that the natural approach of using LLL generalized to other rings as done by Napias [LLL82, Nap96] fails since only the ring of integers of a few cyclotomic fields of small conductor are Euclidean [Len75].

Despite those two serious obstacles to attack Ring-LWE based schemes by the algebraic approach developed in [CGS14, BS16, CDPR16] and in this paper, it seems a reasonable precaution to start considering weaker structured lattice assumptions, such as Module-LWE [LS15] (i.e., an "unusually-Short Vector Problem" in a module of larger rank over a smaller ring), which provides an intermediate problem between ring-LWE and general LWE.

It is also possible to consider other rings, as done in [BCLvV16]. Yet, the latter proposal surprisingly relies on the seemingly stronger NTRU assumption ("unusually-Short Vector Problem" over modules of rank 2). In the current state of affairs [KF16], there seems to be an asymptotic hardness gap between NTRU and Ring-LWE, whatever the ring[4], and down to quite small polynomial approximation factors. Should the concrete security claims of [BCLvV16] not be directly affected, the same reasonable precaution principle should favor weaker assumptions, involving modules of a larger rank.

2 Overview

2.1 Notations and Reminders

Throughout this paper, let m be a prime power, $\omega_m \in \mathbb{C}$ be a complex primitive m-th root of unity, and $K = \mathbb{Q}(\omega_m)$ be the cyclotomic number field of conductor m. It is a number field of degree $n = \varphi(m) = \Theta(m)$. Let G denote its Galois group over $\mathbb{Q}$ and $\tau \in G$ denotes the complex conjugation. We recall that the discriminant Δ_K of K asymptotically satisfies $\log |\Delta_K| = O(n \log n)$.

Ideals as Lattices. The field K is endowed with a canonical Hermitian vector space structure via its Minkowsky embedding. Concretely, its inner product is defined via the trace map $\mathrm{Tr} : K \to \mathbb{Q}$ by $\langle a, b \rangle = \mathrm{Tr}(a\tau(b))$, and the associated Euclidean norm is denoted $\| \cdot \| : a \mapsto \langle a, a \rangle = \mathrm{Tr}(a\tau(a))$.

The ring of integers of K is denoted $\mathcal{O}_K$ and in the cyclotomic case is simply given by $\mathcal{O}_K = \mathbb{Z}[\omega_m]$. Any ideal $\mathfrak{h}$ of $\mathcal{O}_K$ can be viewed as a Euclidean lattice via the above inner-product. The algebraic norm of an ideal $\mathfrak{h}$ is written $N\mathfrak{h}$.

---

[4] This actually seems to hold even without any commutative ring structure, i.e., when comparing "matrix-NTRU" to regular LWE.

The volume of $\mathfrak{h}$ as a lattice relates to its algebraic norm by $\text{Vol}(\mathfrak{h}) = \sqrt{|\Delta_K|}N\mathfrak{h}$. The length $\lambda_1(\mathfrak{h})$ of the shortest vector of $\mathfrak{h}$ is determined by its algebraic norm up to a polynomial factor:

$$\frac{1}{\text{poly}(n)} N(\mathfrak{h})^{1/n} \leq \lambda_1(\mathfrak{h}) \leq \text{poly}(n) N(\mathfrak{h})^{1/n}.$$

The right inequality is an application of Minkowsky's second theorem, whereas the left one follows from the fact that the ideal $v\mathcal{O}_K$ generated by the shortest vector v of $\mathfrak{h}$ is a multiple (a sub-ideal) of $\mathfrak{h}$, and that $\text{Vol}(v\mathcal{O}_K) \leq \|v\|^n$.

Class Group. The class group $\text{Cl}_K = \mathscr{I}_K/\mathscr{P}_K$ of K is the quotient of the (abelian) multiplicative group of fractional ideals $\mathscr{I}_K$ by the subgroup of fractional principal ideals. We denote $[\mathfrak{h}] \in \text{Cl}_K$ the class of an ideal $\mathfrak{h}$. The trivial class $[\mathcal{O}_K]$ is the class of principal ideals. Given two ideals $\mathfrak{h}$ and $\mathfrak{f}$, we write $\mathfrak{h} \sim \mathfrak{f}$ if they have the same class. The class group is written multiplicatively.

The class number $h_K = |\text{Cl}_K|$ is the order of the class group. Loosely speaking, the class group measures the lack of principality of the ring $\mathcal{O}_K$. In particular, the class group is trivial ($h_K = 1$) if and only if $\mathcal{O}_K$ is a principal ideal domain. This holds only for finitely many conductors $m \geq 1$ and, more precisely, we know that $\log h_K = \Theta(n \log m)$ [Was12, Theorem 4.20].

2.2 Overview

It has been shown [CGS14,BS16,CDPR16] (under reasonable assumptions) that given an arbitrary principal ideal $\mathfrak{a} \subset \mathcal{O}_K$, one can recover in quantum polynomial time an element $g \in \mathfrak{a}$ (in fact a generator of $\mathfrak{a}$, i.e. such that $\mathfrak{a} = g\mathcal{O}_K$) such that $\|g\| \leq (N\mathfrak{a})^{1/n} \cdot \exp(\tilde{O}(n^{1/2}))$. Our goal is to reduce the case of general ideals to the case of principal ideals.

The Close Principal Multiple Problem (CPM). To do so, a folklore approach is to search for a reasonably close multiple $\mathfrak{c} = \mathfrak{a}\mathfrak{b}$ of $\mathfrak{a}$ that is *principal*; in other words, one searches for a small integral ideal $\mathfrak{b}$ such that $\mathfrak{b} \sim \mathfrak{a}^{-1}$. If such an ideal $\mathfrak{b}$ with norm less than $\exp(\tilde{O}(n^{1+c}))$ for some constant $c > 0$ is found, this implies, by the aforementioned results, that one can find a generator g of $\mathfrak{c}$ such that

$$\|g\| \leq (N\mathfrak{c})^{1/n} \cdot \exp\left(\tilde{O}\left(n^{1/2}\right)\right)$$
$$\leq (N\mathfrak{a})^{1/n} \cdot (N\mathfrak{b})^{1/n} \cdot \exp\left(\tilde{O}\left(n^{1/2}\right)\right)$$
$$\leq (N\mathfrak{a})^{1/n} \cdot \exp\left(\tilde{O}\left(n^{\max(1/2,c)}\right)\right).$$

Because $g \in \mathfrak{c} \subset \mathfrak{a}$, one has found a short vector of $\mathfrak{a}$, larger than the shortest vector of $\mathfrak{a}$ by a sub-exponential approximation factor $\exp(\tilde{O}(n^{\max(1/2,c)}))$. This is asymptotically as good as the principal case when $c = 1/2$, and better than LLL for any $c < 1$.

CPM as a Close Vector Problem. Before searching for a solution to the CPM problem, let us discuss wether a $\exp(\tilde{O}(n^{1+c}))$-close principal multiple exists in general. A positive answer follows from the results of [JW15, Corollary 6.5][5] setting a prime factor basis $\mathfrak{B} = \{\mathfrak{p} \mid N\mathfrak{p} \leq n^{4+o(1)}\}$, for any class $C \in \mathrm{Cl}_K$, there exists a non-negative small solution $e \in \mathbb{Z}^{\mathfrak{B}}_{\geq 0}$ to the class equation $[\prod \mathfrak{p}^{e_{\mathfrak{p}}}] = C$, of ℓ_1-norm $\|e\|_1 \leq O(n^{1+o(1)})$. This proves, assuming GHR, the existence of a solution $\mathfrak{b} = \prod \mathfrak{p}^{e_{\mathfrak{p}}}$ to the CPM problem as small as $\exp(\tilde{O}(n^{1+c}))$ for $c = o(1)$.

The previous argument is based on the analysis of the expander properties of certain Caley graphs on the class group. For our purpose, existence is not enough, as we wish to efficiently find a close principal multiple. We instead write the class group using lattices. If the factor basis $\mathfrak{B}$ generates the whole class group, then one may rewrite $\mathrm{Cl}_K \simeq \mathbb{Z}^{\mathfrak{B}}/\Lambda$ where Λ is the lattice of class relations: $\Lambda = \{e \in \mathbb{Z}^{\mathfrak{B}} \mid [\prod \mathfrak{p}^{e_{\mathfrak{p}}}] = [\mathcal{O}_K]\}$. Otherly said, $\Lambda \subset \mathbb{Z}^{\mathfrak{B}}$ is the kernel of the surjection $\mu : \mathbb{Z}^{\mathfrak{B}} \twoheadrightarrow \mathrm{Cl}_K$. In fact, it will be enough to consider any full-rank sublattice $\Gamma \subset \Lambda$ of class relations, i.e. any subgroup $\Gamma \subset \Lambda$ of finite index.

The CPM problem can now be rephrased as a *close vector problem*: given a class $C = [\mathfrak{a}]^{-1} \in \mathrm{Cl}_K$, one first use the Biasse-Song quantum algorithm [BS16] to compute a representative of that class $\alpha \in \mathbb{Z}^{\mathfrak{B}}$ in base $\mathfrak{B}$ (see Proposition 2), that is an α such that $\mu(\alpha) = C$. Then one reduces this representation, by searching for a lattice vector $\beta \in \Gamma$ close to α. Note that $\mu(\alpha - \beta) = \mu(\alpha) = C$. This provides a solution[6] $\mathfrak{b} = \prod \mathfrak{p}^{\alpha_{\mathfrak{p}} - \beta_{\mathfrak{p}}}$, of norm at most $B^{\|\alpha - \beta\|_1}$, where B is a bound such that $N\mathfrak{p} \leq B$ for every $\mathfrak{p} \in \mathfrak{B}$. It is therefore sufficient to find an appropriate factor basis together with a good basis of the lattice of relations Γ to attack this problem. The condition over Γ to be of full-rank is necessary to have any guarantee on the length of the reduced representative $\alpha - \beta$.

The Stickelberger Ideal: Class Relations for Free. For this discussion, let us assume for now that the class group can be generated by a single ideal of small norm and its conjugates: $\mathfrak{B} = \{\mathfrak{p}^{\sigma} = \sigma(\mathfrak{p}) \mid \sigma \in G\}$ and $N\mathfrak{p} = \mathrm{poly}(n)$.

Stickelberger's theorem will provide *explicit class relations* between any ideal $\mathfrak{h}$ and its conjugates. More precisely, consider the group ring $\mathbb{Z}[G]$, which naturally acts on $\mathcal{O}_K$-ideals as follows:

$$\mathfrak{h}^s = \prod_{\sigma \in G} \mathfrak{h}^{s_{\sigma} \cdot \sigma} = \prod_{\sigma \in G} \sigma(\mathfrak{h})^{s_{\sigma}} \quad \text{where } s = \sum_{\sigma \in G} s_{\sigma} \cdot \sigma \in \mathbb{Z}[G].$$

Stickelberger gave an explicit construction of a $\mathbb{Z}[G]$-ideal $S \subset \mathbb{Z}[G]$ that *annihilates the class group*, i.e. $\mathfrak{h}^s \sim \mathcal{O}_K$ (i.e., $\mathfrak{h}^s$ is principal) for any ideal $\mathfrak{h} \subset \mathcal{O}_K$ and any element $s \in S$. Forgetting the multiplicative structure of $\mathbb{Z}[G]$ directly gives a lattice of class relations $\mu(S) \subset \mathbb{Z}^{\mathfrak{B}}$ by the canonical morphism of $\mathbb{Z}$-modules $\kappa : \mathbb{Z}[G] \to \mathbb{Z}^{\mathfrak{B}}$, sending σ to the canonical vector $1_{\mathfrak{p}^{\sigma}}$.

---

[5] The earlier result of [JMV09, Corollary 1.3] is not sufficient as it does not keep track of the dependence on the degree of the number fields, left hidden in the constants.

[6] One notes that this solution is not integral as desired, yet getting rid of negative exponents will be easy, at least in the relative class group Cl_K^-.

A technical issue is that the Stickelberger ideal is not of full rank in $\mathbb{Z}[G]$ as a $\mathbb{Z}$-module, so needs to be extended[7] in order to serve as the lattice of relations Γ. This can be resolved by working only with the *minus* part Cl_K^- of the class group, i.e., the relative class group of K over the maximal real subfield K^+. More formally, Cl_K^- is the kernel of the morphism $\mathrm{Cl}_K \to \mathrm{Cl}_{K^+}$ induced by the relative norm map $N_{K/K^+} : \mathfrak{h} \mapsto \mathfrak{h}\mathfrak{h}^\tau$. This subgroup $\mathrm{Cl}_K^- \subset \mathrm{Cl}_K$ is annihilated by the *augmented* Stickelberger ideal $S' = S + (1 + \tau)\,\mathbb{Z}[G]$. For this discussion, let us just assume that Cl_{K^+} is trivial, so that the whole class group $\mathrm{Cl}_K = \mathrm{Cl}_K^-$ is annihilated by the augmented Stickelberger ideal S'.

The Geometry of the Stickelberger Ideal. An important fact is that this ideal has many short elements and that these can be explicitly constructed — this remark is certainly not new, at least for prime conductors [Sch10]. Under our simplifying assumption that $\mathfrak{B} = \{\mathfrak{p}^\sigma \mid \sigma \in G\}$ generates Cl_K, and the additional assumption that the plus part of the class group Cl_{K^+} is trivial, this approach will allow to solve the close multiple problem within a norm bound

$$\exp\left(\widetilde{O}\left(n^{3/2}\right)\right).$$

Sufficient Conditions. In the result sketched above, we made two simplifying assumptions. We now sketch how those assumptions can be relaxed, and provide evidences for the relaxed assumptions. Those assumptions and their supporting evidences will be detailed in Sect. 2.3.

Triviality of Cl_{K^+}. One assumption was that the plus part Cl_{K^+} of the class group is trivial. In fact, we can rather easily handle a non-trivial plus-part as long as $h_K^+ = |\mathrm{Cl}_{K^+}| = \mathrm{poly}(n)$, using rapid-mixing properties of some Cayley graphs on Cl_{K^+}. And since h_K^+ is the class number of a totally real number field, it is actually expected to be small. This assumption is already present in [CGS14, CDPR16], and is supported by numerical evidences ([Was12, p. 420, Table 4], computed by Schoof [Sch89]), and by arguments based on the Cohen-Lenstra heuristic [BPR04].

Knowledge of a $\mathbb{Z}[G]$-*generator of* Cl_K^-. The other assumption was that we know of a factor basis of Cl_K^- of the form $\mathfrak{B} = \{\mathfrak{p}^\sigma = \sigma(\mathfrak{p}) \mid \sigma \in G\}$ for a single ideal $\mathfrak{p}$ of small norm $N\mathfrak{p} = \mathrm{poly}(n)$. Otherly said, we know of a small norm ideal $\mathfrak{p} \subseteq \mathcal{O}_K$ such that $[\mathfrak{p}]$ is a $\mathbb{Z}[G]$-generator of Cl_K^-.

This assumption can also be relaxed. We may allow a few primes and their conjugates in the factor basis. Assuming one knows a factor basis $\mathfrak{B} = \{\mathfrak{p}_i^\sigma \mid \sigma \in G, i = 1, \ldots, d\}$ composed of d Galois orbits, (with $N\mathfrak{p}_i \leq \mathrm{poly}(n)$) that generates Cl_K^-, our approach leads to solving the close principal multiple problem within a norm bound

$$\exp\left(\widetilde{O}\left(d \cdot n^{3/2}\right)\right).$$

---

[7] If a lattice is not of full rank, no close-vector algorithm can guarantee any distance bound, as any fundamental domain is unbounded.

This leads to solving approximate Ideal-SVP with a better approximation factor than pure lattice reduction for any class of conductors $m \in \mathbb{Z}$ whenever one can build a factor basis of size $d = \tilde{O}(n^a)$ for an $a < 1/2$.

Therefore, the crux of the matter is about how small of a factor basis $\mathfrak{B}$ can be built[8]. The structure of the class group Cl_K^- remains quite elusive, but it appears that it admits a very small minimum number of generators as a $\mathbb{Z}[G]$-module. Schoof [Sch98] computed that for all prime conductors $m \leq 509$, Cl_K^- is $\mathbb{Z}[G]$-cyclic (i.e., it is generated by a single element as a $\mathbb{Z}[G]$-module). This property is sufficient to argue that one can efficiently find a small generating set and reach $c = 1/2$, under the heuristic that classes of small random ideals behave similarly to uniformly random classes. Even if the minimal number of generators is not always 1 but still small, say $O(n^\epsilon)$ for some $\epsilon > 0$, this heuristic allows to reach $c = 1/2 + \epsilon$.

2.3 Assumptions

Our main result is conditionned on two assumptions concerning the asymptotic structure of the class group, sketched above and stated below. Of course, if those statement were to not hold for all prime power conductors m, our result remains meaningful if both assumptions simultaneously hold for a common infinite class of conductors, such as $\mathcal{M}_\ell = \{m = \ell^e \mid e \geq 0\}$ for a fixed prime ℓ. We also note that the second assumption can be weakened from $d = \mathrm{polylog}(n)$ to $d = n^\epsilon$ for any $\epsilon < 1/2$ to reach a non trivial approximation factor $\gamma = \exp(\tilde{O}(n^{1/2+\epsilon}))$.

The Real Class Number. The first assumption concerns the size h_K^+ of the class group of the real subfield K^+, and is already used in [CGS14,CDPR16]. For any integer m, let $h^+(m)$ be the class number of the maximal totally real subfield of the cyclotomic field of conductor m.

Assumption 1. *For prime powers m, it holds that $h^+(m) \leq \mathrm{poly}(n)$.*

The literature on h_K^+ provides strong theoretical and computational evidence that it is indeed small enough. First, the Buhler, Pomerance, Robertson [BPR04] formulate and argue in favor of the following conjecture, based on Cohen-Lenstra heuristics.

Conjecture 1 (Buhler, Pomerance, Robertson [BPR04]). *For all but finitely many pairs (ℓ, e), where ℓ is a prime and e is a positive integer, we have $h^+(\ell^{e+1}) = h^+(\ell^e)$.*

A stronger version for the case $\ell = 2$ was formulated by Weber.

---

[8] Note that, as a computational problem, this task is *non-uniform*. That is, it must be ran once for each conductor m of interest, but does not need to be re-run for each CPM instance in $\mathcal{O}_K$. A proof of existence of such a factor basis would already have a consequence in a complexity theoretic perspective. We however heuristically argue in Sect. 2.3 that a good basis can actually be found efficiently.

Conjecture 2 (Weber's Class Number Problem). *For any e, $h^+(2^e) = 1$.*

A direct consequence of Conjecture 1 is that for fixed ℓ and increasing e, $h^+(\ell^e)$ is $O(1)$, implying that Assumption 1 holds over the class $\mathcal{M}_\ell$.

But even for increasing primes ℓ, $h^+(\ell)$ itself is also small: Schoof [Sch03] computed all the values of $h^+(\ell)$ for $\ell < 10,000$ (correct under heuristics of type Cohen-Lenstra, and Miller proved in [Mil15] its correctness under GRH at least for the primes $\ell \leq 241$). According to this table, for 75.3% of the primes $\ell < 10,000$ we have $h^+(\ell) = 1$ (matching Schoof's prediction of 71.3% derived from the Cohen-Lenstra heuristics). All the non-trivial values remain very small, as $h^+(\ell) \leq \ell$ for 99.75% of the primes.

Constructing Small Factor Bases of Cl_K^-. This assumption is arguably new, and can be read as a strengthened version of a Theorem of Bach [Bac90, Theorem 4] and its generalizations from [JMV09] and [JW15, Corrolary 6.5].

Assumption 2. *There are integers $d \leq \mathrm{polylog}(n)$ and $B \leq \mathrm{poly}(n)$ such that the following holds. Choose uniformly at random d prime ideals $\mathfrak{p}_1, \ldots, \mathfrak{p}_d$ among the finitely many ideals $\mathfrak{p}$ satisfying $N\mathfrak{p} \leq B$ and $[\mathfrak{p}] \in \mathrm{Cl}_K^-$. Then, the factor basis $\mathfrak{B} = \{\mathfrak{p}_i^\sigma \mid \sigma \in G, i = 1 \ldots d\}$ generates Cl_K^- with probability at least $1/2$.*

To argue for this assumption, we prove (Proposition 1) that if Cl_K^- can be generated by r ideal classes, then $r \cdot \mathrm{polylog}(n)$ many uniformly random classes in Cl_K^- will generate it.

Proposition 1. *Let K be a cyclotomic field of conductor m, with Galois group G and relative class group Cl_K^-. Let r be the minimal number of $\mathbb{Z}[G]$-generators of Cl_K^-. Let $\alpha \geq 1$ be a parameter, and s be any integer such that*

$$s \geq r(\log_2 \log_2(h_K^-) + \alpha)$$

(note that $\log_2 \log_2(h_K^-) \sim \log_2(n)$). Let $g_1, \ldots, g_s$ be s independent uniform elements of Cl_K^-. The probability that $\{g_1, \ldots, g_s\}$ generates Cl_K^- as a $\mathbb{Z}[G]$-module is at least $\exp\left(-\frac{3}{2^\alpha}\right) = 1 - O(2^{-\alpha})$.

The proof is deferred to Appendix A.

To justify Assumption 2, we first argue that r is admittedly as small as $\mathrm{polylog}(n)$. For the case $m = 2^e$, this can be argued by just looking at the value of $h^-(2^e)$ computed up to $e = 9$ in [Was12, Table 3]. These values are square-free, so Cl_K^- is $\mathbb{Z}$-cyclic and therefore $\mathbb{Z}[G]$-cyclic; in other words, $r = 1$. The case of prime conductors was also studied by Schoof [Sch98]: he proved that Cl_K^- is $\mathbb{Z}[G]$-cyclic for every prime conductor $m \leq 509$; again, $r = 1$.

While it is unclear that this cyclicity should be the typical behavior asymptotically, it seems reasonable to assume that r remains as small as $\mathrm{polylog}(n)$, at least for a dense class of prime power conductors.

Once it is admitted that $r \leq \mathrm{polylog}(n)$, Assumption 2 simply assumes that Proposition 1 remains true when imposing that the random classes $g_1 \ldots g_s$ are chosen as the classes of random ideals of small norm, i.e. $g_i = [\mathfrak{p}_i]$ where $N\mathfrak{p}_i \leq$ poly(n). This restriction on the norms seems reasonable considering that it has been proven that prime ideals of norm poly(n) are sufficient to generate Cl_K^-, assuming GRH and Assumption 1 (see [JW15, Corrolary 6.5]).

3   Quantum Algorithms for Class Groups

Searching for a principal multiple of the ideal $\mathfrak{a}$ in $\mathcal{O}_K$ will require to perform computations in the class group in an efficient way. Classically, problems related to class group computations remain difficult, and the best known classical algorithms run in sub-exponential time (for example, see [BF14, BEF+17]). Yet, building on the recent advances on quantum algorithms for the Hidden Subgroup Problem in large dimensions [EHKS14], Biasse and Song [BS16] introduced a quantum algorithm to perform S-unit group computations. It implies class group computations, and solution to the principal ideal problem (PIP) in quantum polynomial time.

The Biasse-Song [BS16] algorithm for S-unit group computation also allows to solve the class group discrete logarithm problem: given a basis $\mathfrak{B}$ of ideals generating a subgroup of the class group Cl_K containing the class of $\mathfrak{a}$, express the class of $\mathfrak{a}$ as a product of ideals in $\mathfrak{B}$. Below, we give a formal statement and in the Appendix B, we provide a proof for completeness.[9]

Proposition 2 ([BS16]). *Let $\mathfrak{B}$ be a set of prime ideals generating a subgroup H of Cl_K. There exists a quantum algorithm $\mathrm{ClDL}_\mathfrak{B}$ which, when given as input any ideal $\mathfrak{a}$ in $\mathcal{O}_K$ such that $[\mathfrak{a}] \in H$, outputs a vector $\mathbf{y} \in \mathbb{Z}^\mathfrak{B}$ such that $\prod \mathfrak{p}^{y_\mathfrak{p}} \sim \mathfrak{a}$, and runs in polynomial time in $n = \deg(K)$, $\max_{\mathfrak{p} \in \mathfrak{B}} \log(N\mathfrak{p})$, $\log(N\mathfrak{a})$, and $|\mathfrak{B}|$.*

4   Close Multiple in the Relative Class Group

Let $K^+ = \mathbb{Q}(\omega_m + \omega_m^{-1})$ denote the maximal real subfield of K, and Cl_{K^+} the class group of K^+. The relative norm map $N_{K/K^+} : \mathrm{Cl}_K \rightarrow \mathrm{Cl}_{K^+}$ on ideal classes (which sends the class of $\mathfrak{a}$ to the class of $\mathfrak{a}\mathfrak{a}^\tau$, where τ is the complex conjugation) is a surjection, and its kernel is the relative class group Cl_K^-. In particular, it induces the isomorphism $\mathrm{Cl}_{K^+} \cong \mathrm{Cl}_K/\mathrm{Cl}_K^-$.

The core of the method to find a close principal multiple of an ideal $\mathfrak{a}$ works within the relative class group $\mathrm{Cl}_K^- \subset \mathrm{Cl}_K$. Therefore, as a first step, we need to "send" the ideal $\mathfrak{a} \in \mathrm{Cl}_K$ into this subgroup. More precisely, we want an integral ideal $\mathfrak{b}$ of small norm such that $\mathfrak{a}\mathfrak{b} \in \mathrm{Cl}_K^-$; the rest of the method then works with $\mathfrak{a}\mathfrak{b}$. Let $h_K = |\mathrm{Cl}_K|$ be the class number of K, and $h_K^- = |\mathrm{Cl}_K^-|$ its relative class number. The difficulty of this step is directly related to the index of Cl_K^- inside Cl_K, which is the real class number $h_K^+ = |\mathrm{Cl}_{K^+}|$ of K^+, and is expected to be very small.

---

[9] In fact, Proposition 2 is a corollary of [BS16, Theorem 1.1]. Even though it is not stated explicitly in that paper, it must be attributed to that paper nevertheless. Indeed, the implication is straightforward and its authors have already sketched it in public talks. Our purpose here is merely to include technical details for completeness.

4.1   Random Walks to the Relative Class Group

For any $x > 0$, consider the set $\mathcal{S}_x$ of ideals in $\mathcal{O}_K$ of prime norm at most x, and let S_x be the multiset of its image in Cl_K. Let $\mathscr{G}_x$ denote the induced Cayley (multi)graph $\mathrm{Cay}(\mathrm{Cl}_K, S_x)$. From [JW15, Corrolary 6.5] (under GRH), for any $\varepsilon > 0$ there is a constant C and a bound

$$B = O\left((n \log \Delta_K)^{2+\varepsilon}\right) = O\left((n^2 \log n)^{2+\varepsilon}\right)$$

such that any random walk in $\mathscr{G}_B$ of length at least $C \log(h_K)/\log\log(\Delta_K)$, for any starting point, lands in the subgroup Cl_K^- with probability at least $1/(2h_K^+)$.

A random walk of length $\ell = \lceil C \log(h_K)/\log\log(\Delta_K)\rceil = \tilde{O}(n)$ is a sequence $\mathfrak{p}_1, \ldots, \mathfrak{p}_\ell$ of ideals chosen independently, uniformly at random in $\mathcal{S}_B$, and their product $\mathfrak{b} = \prod \mathfrak{p}_i$ has a norm bounded by

$$N\mathfrak{b} = \prod_{i=1}^{\ell} N\mathfrak{p}_i \leq B^\ell = \exp(\mathrm{polylog}(n) \cdot \tilde{O}(\log \widetilde{h_K})) = \exp(\tilde{O}(n)),$$

If $[\mathfrak{a}]$ is the starting point of the random walk in the graph, the endpoint $[\mathfrak{ab}]$ falls in Cl_K^- with probability at least $1/(2h_K^+)$, and therefore an ideal $\mathfrak{b}$ such that $[\mathfrak{ab}] \in \mathrm{Cl}_K^-$ can be found in probabilistic polynomial time in h_K^+. Note that the PIP algorithm of Biasse and Song [BS16] allows to test the membership $[\mathfrak{ab}] \in \mathrm{Cl}_K^-$, simply by testing the principality of $N_{K/K^+}(\mathfrak{ab})$ as an ideal of $\mathcal{O}_K^+$.

The procedure is summarized as Algorithm 1, and the effiency is stated below. Under GRH and Assumption 1, this procedure runs in polynomial time.

Lemma 1 (Under GRH). *Algorithm 1 (*WALKTOCL$^-$(𝔞)*) runs in expected time $O(h_K^+) \cdot \mathrm{poly}(n, \log N\mathfrak{a})$ and is correct.*

Algorithm 1. WALKTOCL$^-$(𝔞): random walk to Cl_K^-

Require: An ideal $\mathfrak{a}$ in $\mathcal{O}_K$
Ensure: An integral ideal $\mathfrak{b}$ such that $[\mathfrak{ab}] \in \mathrm{Cl}_K^-$ and $N\mathfrak{b} \leq \exp(\tilde{O}(n))$
1: $\ell = \tilde{O}(n)$; $B = \mathrm{poly}(n)$
2: **repeat**
3:    **for all** $i = 1 \ldots \ell$ **do**
4:       Choose $\mathfrak{p}_i$ uniformly among the prime ideal of norm less than B
5:    **end for**
6:    Set $\mathfrak{b} = \prod \mathfrak{p}_i$
7: **until** $N_{K/K^+}(\mathfrak{ab})$ is principal (using the PIP algorithm of [BS16])
8: $\mathfrak{b} \leftarrow \prod_{i=1}^{d} \mathfrak{p}_i$
9: **return** $\mathfrak{b}$

---

5   Short Relations in Cl_K^- via the Stickelberger Ideal

Consider any ideal $\mathfrak{f}$ of $\mathcal{O}_K$ such that $[\mathfrak{f}] \in \mathrm{Cl}_K^-$, and its orbit under the action of the Galois group G, denoted $\mathfrak{F} = G(\mathfrak{f})$. Let R be the group ring $\mathbb{Z}[G]$. It projects to $\mathbb{Z}^{\mathfrak{F}}$, via the map sending σ to $\mathbf{1}_{\mathfrak{f}^\sigma}$.

We now show the construction of an explicit full-rank lattice of class relations in $\mathbb{Z}^{\mathfrak{F}}$ with an explicit set of *short* generators. We proceed by augmenting the *Stickelberger ideal*. This allows to reduce the representation of a given class expressed in basis $\mathfrak{F}$, as shown in Subsect. 5.3.

Recall that the Galois group G is canonically isomorphic to $(\mathbb{Z}/m\mathbb{Z})^*$ via $a \mapsto \sigma_a = \zeta_m \mapsto \zeta_m^a$. The norms $\|\cdot\|$ and $\|\cdot\|_1$ denote the usuals ℓ_2 (Euclidean) and ℓ_1 norms over $\mathbb{R}^n$, and are defined over $\mathbb{Z}[G]$ via the natural isomorphism $\mathbb{Z}[G] \cong_\mathbb{Z} \mathbb{Z}^n$.

The fractional part of a rational $x \in \mathbb{Q}$ is denoted $\{x\}$, it is defined as the unique rational in the interval $[0, 1)$ such that $\{x\} = x \mod \mathbb{Z}$; equivalently, $\{x\} = x - \lfloor x \rfloor$.

5.1   The (augmented) Stickelberger Ideal

Definition 1 (The Stickelberger ideal). *The* Stickelberger element $\theta \in \mathbb{Q}[G]$ *is defined as*

$$\theta = \sum_{a \in (\mathbb{Z}/m\mathbb{Z})^*} \left\{ \frac{a}{m} \right\} \sigma_a^{-1}.$$

The Stickelberger ideal *is defined as $S = R \cap \theta R$. We will refer to the Stickelberger lattice when S is considered as a $\mathbb{Z}$-module.*

This ideal $S \subset R$ will provide some class relations in $\mathbb{Z}^{\mathfrak{F}}$, thanks to the following theorem.

Theorem 1 (Stickelberger's theorem [Was12, Theorem 6.10]). *The Stickelberger ideal annihilates the ideal class group of K. In other words, for any ideal $\mathfrak{h}$ of $\mathcal{O}_K$ and any $s \in S$, the ideal $\mathfrak{h}^s$ is principal.*

We cannot directly use $S \subset R$ as our lattice of class relations since it does not have full rank in R as a $\mathbb{Z}$-module (precisely its $\mathbb{Z}$-rank is $n/2 + 1$ when $m \geq 2$). Indeed, if the lattice is not full rank, there can be no guarentee of how short of a represcntant will be obtained by reducing modulo the lattice. To solve this issue, we will augment the Stickelberger ideal to a full-rank ideal which still annihilates the minus part Cl_K^- of the class group.

Definition 2. *The augmented Stickelberger ideal S' is defined as*

$$S' = S + (1 + \tau)R. \tag{2}$$

We will refer to the augmented Stickelberger lattice when S' is considered as a $\mathbb{Z}$-module.

Lemma 2. *The augmented Stickelberger ideal S' annihilates Cl_K^-. In other words, for any ideal $\mathfrak{h}$ of $\mathcal{O}_K$ such that $[\mathfrak{h}] \in \mathrm{Cl}_K^-$ and any $s \in S$, the ideal $\mathfrak{h}^s$ is principal. Moreover, $S' \subset R$ has full-rank n as a $\mathbb{Z}$-module.*

Proof. For the annihilation property it suffices to show that both S and $(1+\tau)R$ annihilate Cl_K^-. By Stickelberger's theorem S annihilates Cl_K so it in particular annihilates the subgroup $\mathrm{Cl}_K^- \subset \mathrm{Cl}_K$. The ideal $(1+\tau)R$ also annihilates Cl_K^- since $\mathfrak{h}^{1+\tau} = \mathfrak{h}\bar{\mathfrak{h}} = N_{K/K^+}(\mathfrak{h})$. We conclude from the fact that Cl_K^- is exactly the kernel of the norm map $N_{K/K^+} : \mathrm{Cl}_K \to \mathrm{Cl}_K^+$.

For the rank, consider the ideal $S^- = S \cap (1-\tau)R$. A theorem from Iwasawa (originally published in [Sin80] but reformulated more conveniently in [Was12, Theorem 6.19]) states that S^- is full rank in $(1-\tau)R$. Noting that $2R \subset (1-\tau)R + (1+\tau)R$, we conclude that $S^- + (1+\tau)R$ has full rank in $2R$, and so does S'.   □

5.2   Short Generating Vectors of the Augmented Stickelberger Lattice

In the following, the elements of $(\mathbb{Z}/m\mathbb{Z})^*$ are canonically identified with the positive integers $0 < a_1 < a_2 < \cdots < a_n < m$ such that each a_i is coprime to m. The elements of G are indexed as $(\sigma_{a_1}, \ldots, \sigma_{a_n})$. Define the extra element $a_{n+1} = m + a_1$, and note that $a_2 \leq 3$ and that $a_{i+1} - a_i \leq 2$ for any i.

Lemma 3. *The Stickelberger lattice is generated by the vectors* $v_i = (a_i - \sigma_{a_i})\theta$ *for* $i \in \{2, \ldots, n+1\}$.

Proof. This is almost [Was12, Lemma 6.9]. There, S is considered as an ideal in R, whereas we need these elements to generate S as a $\mathbb{Z}$-module. Let L be the $\mathbb{Z}$-module generated by the v_i's. First, [Was12, Lemma 6.9] immediately implies that $v_i \in S$ and thereby $L \subseteq S$. Now, let $\left(\sum_{i=2}^{n+1} x_i \sigma_{a_i}\right)\theta$ be an arbitrary element of S, with $a_i \in \mathbb{Z}$. One can prove as in [Was12, Lemma 6.9] that m divides $\sum_{i=2}^{n+1} x_i a_i \in \mathbb{Z}$. Since $m = (m+1) - \sigma_{m+1}$, $m\theta$ is in L, and we deduce that $\left(\sum_{i=2}^{n+1} x_i a_i\right)\theta$ is also in L. Therefore,

$$\left(\sum_{i=2}^{n+1} x_i \sigma_{a_i}\right)\theta = \left(\sum_{i=2}^{n+1} x_i(\sigma_{a_i} - a_i)\right)\theta + \left(\sum_{i=2}^{n+1} x_i a_i\right)\theta \in L.$$

This proves that $S \subseteq L$, hence $L = S$.   □

We are now ready to construct our set of short generators for S'. Let $w_2 = v_2$ and $w_{i+1} = v_{i+1} - v_i$ for $i \in \{2, \ldots, n\}$, and let

$$W = \{w_2, \ldots, w_{n+1}\} \cup \{(1+\tau)\sigma, \sigma \in G\}.$$

Lemma 4. *The set S is a set of short generators of S'. More precisely,*

1. *W generates the augmented Stickelberger lattice S',*
2. *For any $i \in \{3 \ldots n+1\}$, $w_i = \sum_{b \in (\mathbb{Z}/m\mathbb{Z})^*} \epsilon_{i,b} \cdot \sigma_b^{-1}$, with $\epsilon_{i,j} \in \{0, 1, 2\}$,*
3. *For any $w \in W$, we have $\|w\| \leq \max(2\sqrt{n}, \sqrt{10})$.*

The second item essentially generalizes [Sch10, Proposition 9.4] from prime conductors to prime-power conductors.

Proof. We prove each item individually.

1. First note that $\{w_2, \ldots, w_{n+1}\}$ generates S: this is a direct consequence of Lemma 3 and the construction of W. By definition of $R = \mathbb{Z}[G]$, the set $\{(1+\tau)\sigma, \sigma \in G\}$ generates $(1+\tau)R$. One can conclude from the definition of $S' = S + (1+\tau)R$.

2. We follow the computation in the proof of [Was12, Lemma 6.9]:

$$v_i = (a_i - \sigma_{a_i})\theta = \sum_{b \in (\mathbb{Z}/m\mathbb{Z})^*} \left(a_i \left\{\frac{b}{m}\right\} - \left\{\frac{a_i b}{m}\right\} \right) \sigma_b^{-1}$$

$$= \sum_{b \in (\mathbb{Z}/m\mathbb{Z})^*} \left\lfloor a_i \left\{\frac{b}{m}\right\} \right\rfloor \sigma_b^{-1}$$

using the identity $x\{y\} - \{xy\} = \lfloor x\{y\} \rfloor$ for any integer x and real number y, since this difference is an integer and the term $\{xy\}$ is in the range $[0, 1)$. It remains to rewrite $w_i = \sum_{b \in (\mathbb{Z}/m\mathbb{Z})^*} \epsilon_{i,b} \sigma_b^{-1}$, where

$$\epsilon_{i,b} = \left\lfloor a_{i+1} \left\{\frac{b}{m}\right\} \right\rfloor - \left\lfloor a_i \left\{\frac{b}{m}\right\} \right\rfloor \leq a_{i+1} - a_i \leq 2.$$

3. The property follows from the previous item for any $i > 2$. For $i = 2$, we have $w_2 = v_2 = a_2 - \sigma_{a_2}$, and therefore $\|w_2\| = \sqrt{a_2^2 + 1} \leq \sqrt{3^2 + 1} = \sqrt{10}$. Finally, elements $w \in W$ of the form $(1+\tau)\sigma$ have norm $\|w\| = \sqrt{2} \leq \sqrt{10}$. □

5.3 Reducing a Class Representative in an R-cycle of Cl_K^-

We now show how to exploit the previously constructed set W of short relations to reduce class representations. More precisely, for any large $\alpha \in R$ we will find a short $\beta \in R$ such that $C^\beta = C^\alpha$, for any class $C \in \mathrm{Cl}_K^-$. We shall rely on the following close vector algorithm.

Proposition 3 (Close vector algorithm). *Let $\Gamma \subset \mathbb{R}^k$ be a lattice, and let W be a set generating Γ. There exists a (classical) polynomial time algorithm CV, that when given any $y \in \Gamma \otimes \mathbb{R}$ as input, outputs a vector $x = \mathrm{CV}(y, W) \in \Gamma$ such that $\|x - y\|_1 \leq \frac{k}{2} \cdot \max_{w \in W} \|w\|$.*

Proof. Let first $B \subset W$ be a basis of a full-rank sublattice $\Gamma' \subset \Gamma$ (this is easily built in polynomial time). Let $\tilde{B}$ denote the Gram-Schmidt orthogonalization of B. Let $g = \max_{b \in \tilde{B}} \|\tilde{b}\| \leq \max_{b \in B} \|b\| \leq \max_{w \in W} \|w\|$. Applying the Nearest Plane algorithm leads to $x \in \Gamma$ such that $x - y$ belongs to the fundamental parallelepiped $\{\tilde{B}z, z \in [-1/2, 1/2]\}$. We then have

$$\|x - y\|_2^2 \leq \frac{1}{4} \sum \|\tilde{b}_i\|^2.$$

In particular, $\|x - y\|_2 \leq \sqrt{k} \cdot g/2$ and one concludes $\|x - y\|_1 \leq kg/2$. □

Theorem 2. *Assume $n \geq 3$. There is an algorithm REDUCE, that given $\alpha \in R$, finds in polynomial time in n and $\log(\|\alpha\|)$, an element $\beta = \text{REDUCE}(\alpha) \in R$ such that $\|\beta\|_1 \leq n^{3/2}$, and $C^\alpha = C^\beta$ for any $C \in \text{Cl}_K^-$.*

Proof. Let W be the basis for the augmented Stickelberger ideal S' as in Lemma 4. From Lemma 2, it has full rank in R. So the close vector algorithm from Proposition 3 can be applied to find an element $\gamma = \text{CV}(\alpha, W) \in S'$ such that $\|\alpha - \gamma\|_1 \leq \frac{n}{2} \cdot \max_{w \in W} \|w\| \leq n^{3/2}$. Let $\beta = \alpha - \gamma$. For any $C \in \text{Cl}_K^-$, Lemma 2 implies that $C^\gamma = 0$ and therefore $C^\alpha = C^\beta$. □

6   Close Principal Multiple Within the Relative Class Group

We now show how to solve the CPM problem for ideals sitting in Cl_K^-, given a factor basis $\mathfrak{B}$ of Cl_K^-. The CPM approximation factor will depend on the size of the factor basis $\mathfrak{B}$.

Suppose the ideal $\mathfrak{a}$ is in the relative class group Cl_K^-. We are looking for an integral ideal $\mathfrak{b}$ in $\mathcal{O}_K$ of small norm such that $\mathfrak{a}\mathfrak{b}$ is principal. Let $\mathfrak{B} = \{\mathfrak{p}_i^\sigma \mid \sigma \in G, i = 1, \ldots, d\}$ be a set generating Cl_K^-, composed of d Galois orbits, such that $N\mathfrak{p}_i \leq \text{poly}(n)$ for all i. To state the algorithm and its correctness, no assumption is made on the factor basis $\mathfrak{B}$. In the final Sect. 7, we will employ Assumption 2 to provide a factor basis with $d = \text{polylog}(n)$ to this algorithm.

Algorithm 2. CLOSEPRINCIPALMULTIPLE$^-$($\mathfrak{a}, \mathfrak{B}$): close principal multiple in the relative class group

---

Require: An ideal $\mathfrak{a}$ in $\mathcal{O}_K$ such that $[\mathfrak{a}] \in \text{Cl}_K^-$, a factor basis $\mathfrak{B} = \{\mathfrak{p}_i^\sigma | i = 1 \ldots d, \sigma \in G\}$ generating Cl_K^-, such that $N\mathfrak{p}_i \leq \text{poly}(n)$ for all i.

Ensure: An (integral) ideal $\mathfrak{b}$ in $\mathcal{O}_K$ such that $\mathfrak{a}\mathfrak{b} \sim \mathcal{O}_K$ and $N\mathfrak{b} = \exp\left(\tilde{O}(dn^{3/2})\right)$

  1: $\mathbf{y} \leftarrow \text{ClDL}_{\mathfrak{B}}(\mathfrak{a})$
  2: **for** i = 1 **to** d **do**
  3:      $\alpha_i \leftarrow \sum_{\sigma \in G_i} y_{(\mathfrak{p}_i^\sigma)} \sigma \in \mathbb{Z}[G]$
  4:      $\beta_i \leftarrow \text{REDUCE}(\alpha_i)$
  5:      $(\gamma_i^+, \gamma_i^-) \leftarrow$ the pair of elements in $\mathbb{Z}[G]$ with only positive coefficients, such that $\gamma_i^+ - \gamma_i^- = -\beta_i$
  6:      $\mathfrak{b}_i \leftarrow \mathfrak{p}_i^{\gamma_i^+ + \tau \gamma_i^-}$
  7: **end for**
  8: $\mathfrak{b} \leftarrow \prod_{i=1}^d \mathfrak{b}_i$
  9: **return** $\mathfrak{b}$

Theorem 3. *Algorithm 2, CLOSEPRINCIPALMULTIPLE$^-$, runs in quantum polynomial time in $n = \deg(K)$, d and $\log(N\mathfrak{a})$, and is correct.*

Proof. Let $\mathfrak{a}, \mathfrak{B}$ be proper inputs, that is, $\mathfrak{a}$ is an ideal of $\mathcal{O}_K$ such that $[\mathfrak{a}] \in \mathrm{Cl}_K^-$, and $\mathfrak{B}$ is a factor basis $\mathfrak{B} = \{\mathfrak{p}_i^\sigma \mid i = 1 \ldots d, \sigma \in G\}$ generating Cl_K^-, such that $N\mathfrak{p}_i \leq \mathrm{poly}(n)$ for all i.

The running time follows immediately from Proposition 2 and Theorem 2. Let us now prove the correctness. We have

$$\phi(\mathbf{y}) = \prod_{\mathfrak{p} \in \mathfrak{B}} \mathfrak{p}^{y_\mathfrak{p}} = \prod_{i=1}^d \prod_{\mathfrak{p} \in \mathfrak{B}_i} \mathfrak{p}^{y_\mathfrak{p}} = \prod_{i=1}^d \prod_{\sigma \in G_i} (\mathfrak{p}_i^\sigma)^{y(\mathfrak{p}_i^\sigma)} = \prod_{i=1}^d \mathfrak{p}_i^{\alpha_i}.$$

Observe that for each i, $\mathfrak{b}_i \sim \mathfrak{p}_i^{-\beta_i}$, since $\mathfrak{p}_i^{-1} \sim \mathfrak{p}_i^\tau$. From Theorem 2, we obtain $\mathfrak{p}_i^{\alpha_i}\mathfrak{b}_i \sim \mathcal{O}_K$, which implies that $\phi(\mathbf{y})\mathfrak{b} \sim \prod_{i=1}^d \mathfrak{p}_i^{\alpha_i}\mathfrak{b}_i \sim \mathcal{O}_K$. From Proposition 2, we have $\phi(\mathbf{y}) \sim \mathfrak{a}$, and therefore $\mathfrak{a}\mathfrak{b} \sim \mathcal{O}_K$.

Now, Theorem 2 ensures that $\|\beta\|_1 \leq n^{3/2}$. So $\|\gamma_i^+\|_1 + \|\gamma_i^-\|_1$ is bounded by $n^{3/2}$ and we obtain that $N\mathfrak{b}_i \leq (N\mathfrak{p}_i)^{n^{3/2}}$. Then,

$$N\mathfrak{b} = \prod_{i=1}^d N\mathfrak{b}_i \leq \left(\max_{i=1\ldots a} N\mathfrak{p}_i\right)^{dn^{3/2}} = \exp\left(\tilde{O}(dn^{3/2})\right),$$

where the last inequality uses the fact that each $N\mathfrak{p}_i$ is polynomially bounded in n. □

7 Main Result

We now have all the ingredients to demonstrate our main result:

Main Result (Under GRH, Assumptions 1 and 2). *Assuming simultaneously the Generalized Riemann Hypothesis, Assumption 1, and Assumption 2, there exists a quantum polynomial time algorithm* IDEALSVP($\mathfrak{a}$), *that given an ideal of $\mathcal{O}_K$ for K a cyclotomic number field of prime power conductor, returns an element $v \in \mathfrak{a}$ of Euclidean norm $\|v\| \leq (N\mathfrak{a})^{1/n} \cdot \exp(\tilde{O}(\sqrt{n}))$.*

Algorithm 3. IDEALSVP($\mathfrak{a}$): finding mildly short vectors in an ideal

---

Require: An ideal $\mathfrak{a}$ in $\mathcal{O}_K$
Ensure: An element $v \in \mathfrak{a}$ of norm $\|v\| \leq (N\mathfrak{a})^{1/n} \exp(\tilde{O}(\sqrt{n}))$
1: $d = \mathrm{polylog}(n)$; $B = \mathrm{poly}(n)$
2: Set $\mathfrak{M} = \{\mathfrak{p} | N\mathfrak{p} \leq B, [\mathfrak{p}] \in \mathrm{Cl}_K^-\}$
3: Choose $\mathfrak{p}_1, \ldots, \mathfrak{p}_d$ uniformly at random in $\mathfrak{M}$
4: Set $\mathfrak{B} = \{\mathfrak{p}_i^\sigma | i \in \{1 \ldots d\}, \sigma \in G\}$
5: $\mathfrak{b}' = \mathrm{WALKToCl}^-(\mathfrak{a})$
6: $\mathfrak{b} = \mathrm{CLOSEPRINCIPALMULTIPLE}^-(\mathfrak{a}\mathfrak{b}', \mathfrak{B})$
7: $v = \mathrm{PRINCIPALIDEALSVP}(\mathfrak{a}\mathfrak{b}\mathfrak{b}')$
8: **return** v

Proof. The algorithm is given as Algorithm 3. Efficiency and correctness follow from the previous statements and assumptions:

- Step 2 is quantum polynomial time since membership in Cl_K^- can be tested by applying the Biasse-Song PIP algorithm [BS16, Theorem 1.3] to $N_{K/K^+}(\mathfrak{ab})$.
- By Assumption 2, Steps 3 and 4 produce a factor basis $\mathfrak{B}$ generating Cl_K^-. Both steps can trivially be performed in polynomial time.
- By Lemma 1, GRH and Assumption 1, Step 5 is quantum polynomial time, and produces an integral ideal $\mathfrak{b}'$ such that $N\mathfrak{b}' \le \exp(\tilde{O}(n))$ and $[\mathfrak{ab}'] \in Cl_K^-$.
- By Theorem 3, Step 6 produces (in quantum polynomial time) an integral ideal $\mathfrak{b}$ such that

$$N\mathfrak{b} \le \exp(\tilde{O}(dn^{3/2})) = \exp(\tilde{O}(n^{3/2}))$$

  and such that $\mathfrak{abb}'$ is principal.
- By Claim 1 ([CGS14,BS16,CDPR16]), Step 7 produces in quantum polynomial time a vector $v \in \mathfrak{abb}'$ of length $\|v\| \le (N\mathfrak{abb}')^{1/n} \cdot \exp(\tilde{O}(\sqrt{n}))$.

Because $\mathfrak{b}$ and $\mathfrak{b}'$ are integral, $\mathfrak{abb}' \subset \mathfrak{a}$, and $v \in \mathfrak{a}$. Finally,

$$\|v\| \le (N\mathfrak{a})^{1/n}(N\mathfrak{b})^{1/n}(N\mathfrak{b}')^{1/n} \cdot \exp(\tilde{O}(\sqrt{n}))$$
$$\le (N\mathfrak{a})^{1/n} \cdot \exp(\tilde{O}(\sqrt{n})).$$

$\square$

Acknowledgements. The authors would like to thank René Schoof for helpful and interesting discussions. We are grateful to Paul Kirchner for pointing out a mistake in the appendix of an earlier version of this paper. The second author was partly supported by a grant through a public-private partnership with NXP Semiconductors, and by a Veni Innovational Research Grant from NWO under project number 639.021.645. The third author was supported by the Swiss National Science Foundation under grant number 200021-156420.

A   Proof of Proposition 1

In this appendix, we provide the proof of Proposition 1 (restated below, used to support Assumption 2).

Proposition 1. *Let K be a cyclotomic field of conductor m, with Galois group G and relative class group Cl_K^-. Let r be the minimal number of $\mathbb{Z}[G]$-generators of Cl_K^-. Let $\alpha \ge 1$ be a parameter, and s be any integer such that*

$$s \ge r(\log_2 \log_2(h_K^-) + \alpha)$$

(note that $\log_2 \log_2(h_K^-) \sim \log_2(n)$). Let $x_1, \ldots, x_s$ be s independent uniform elements of Cl_K^-. The probability that $\{x_1, \ldots, x_s\}$ generates Cl_K^- as a $\mathbb{Z}[G]$-module is at least $\exp\left(-\frac{3}{2^\alpha}\right) = 1 - O(2^{-\alpha})$.

In other words, a set of $\Theta(r \log(n))$ random ideal classes in Cl_K^- will generate this $\mathbb{Z}[G]$-module with very good probability. Let us first prove a few lemmas.

Lemma 5. *Let R be a finite commutative local ring of cardinality ℓ^n, for some prime number ℓ. A set of s independent uniformly random elements in R generates R as an R-module with probability at least $1 - \ell^{-s}$.*

Proof. An element generates R if and only if it is invertible, meaning that it is not in the maximal ideal of R. This ideal is a fraction at most ℓ^{-1} of R, so an element does not generate R with probability at most ℓ^{-1}. Among s independent elements, the probability that none of them is a generator is at most ℓ^{-s}.

Lemma 6. *Let R be a finite commutative local ring of cardinality ℓ^n, for some prime number ℓ. Let M be a cyclic R-module. A set of s independent uniformly random elements in M generates M with probability at least $1 - \ell^{-s}$.*

Proof. Let g be a generator of M, and consider the homomorphism $\varphi : R \to M : \alpha \mapsto \alpha g$. Let $x_1, \ldots, x_s$ be s independent uniformly random element in M. For each i, let α_i be a uniformly random element of the coset $\varphi^{-1}(x_i)$. The elements α_i are independent and uniformly distributed in R, so from Lemma 5, they generate R with probability at least $1 - \ell^{-s}$. If the α_i's generate R, then the x_i's generate M, and we conclude.

Lemma 7. *Let R be a finite commutative local ring of cardinality ℓ^n, for some prime number ℓ. Let M be an R-module, and let r be the smallest number of R-generators of M. A set of s independent uniformly random elements in M generates M with probability at least $\left(1 - \ell^{-\lfloor s/r \rfloor}\right)^r$.*

Proof. Proceed by induction on r. The case $r = 1$ is Lemma 6. Suppose that for any R-module M' generated by $r - 1$ elements, and any positive s', a set of s' random elements in M' generates M' with probability at least

$$\left(1 - \ell^{-\lfloor s'/(r-1) \rfloor}\right)^{r-1}.$$

Choose s independent uniformly random elements $x_1, \ldots, x_s$ in M, and let $t = \lfloor s/r \rfloor$. Let $g_1, \ldots, g_r$ be a generating set for M. The quotient $M/(Rg_r)$ is generated by $r - 1$ elements, so the first $s - t$ random elements generate it with probability at least

$$\left(1 - \ell^{-\lfloor (s-t)/(r-1) \rfloor}\right)^{r-1} \geq \left(1 - \ell^{-\lfloor s/r \rfloor}\right)^{r-1}.$$

Now assume that these $s - t$ elements indeed generate $M/(Rg_r)$. It remains to show that adding the remaining t random elements allow to generate the full module M with probability at least $1 - \ell^{-\lfloor s/r \rfloor}$. Let $N \subset M$ be the submodule of M generated by the first $s - t$ random elements. Observe that the module M/N is generated by g_r. Indeed, let m be an arbitrary element of M. Since $M/(Rg_r)$ is generated by N, there is an $n \in N$ such that $m + Rg_r = n + Rg_r$.

This implies that there is an element $\alpha g_r \in Rg_r$ such that $m + N = \alpha g_r + N$, proving that M/N is generated by g_r. From Lemma 6, M/N is generated by the last t random elements with probability at least $1 - \ell^{-\lfloor s/r \rfloor}$. So M is generated by $x_1, \ldots, x_s$ with probability at least $\left(1 - \ell^{-\lfloor s/r \rfloor}\right)^r$.

Theorem 4. *Let R be a finite commutative ring, and M a finite R-module of cardinality m, and r be the minimal number of R-generators of M. A set of s independent uniformly random elements in M generates M with probability at least $\left(1 - 2^{-\lfloor s/r \rfloor}\right)^{\log_2 m}$.*

Proof. The ring R decomposes as an internal direct sum $\bigoplus_{i=1}^k R_i$ of finite local subrings R_i. For each i, define $e_i \in R$ the idempotent which projects to the unity of R_i and to zero in all other components of the decomposition (then, $R_i = e_i R$). In particular, we have that $M = \bigoplus_i e_i M$, and $e_i M$ may be viewed as an R_i-module.

Let $x_1, \ldots, x_s$ be s independent uniformly random elements in M. They generate M as an R-module if and only if for any i, the projections $e_i x_1, \ldots, e_i x_s$ generate M_i as an R_i-module. Let p_i be the probability that $e_i x_1, \ldots, e_i x_s$ generate M_i, and let r_i be the minimal number of generators of R_i. From Lemma 7, p_i is at least $\left(1 - 2^{-\lfloor s/r_i \rfloor}\right)^{r_i}$. We have the two bounds $r_i \leq r$ and $r_i \leq \log_2 |M_i|$, and we deduce

$$p_i \geq \left(1 - 2^{-\lfloor s/r \rfloor}\right)^{\log_2 |M_i|}.$$

Therefore $x_1, \ldots, x_s$ generate M with probability at least

$$\prod_{i=1}^k p_i = \left(1 - 2^{-\lfloor s/r \rfloor}\right)^{\sum_i \log_2 |M_i|} = \left(1 - 2^{-\lfloor s/r \rfloor}\right)^{\log_2 m},$$

concluding the proof.

Proof of Proposition 1. Note that a set of elements in Cl_K^- generate it as a $\mathbb{Z}[G]$-module if and only if they generate it as a $(\mathbb{Z}/h_K^- \mathbb{Z})[G]$-module. We deduce from Theorem 4 that $x_1, \ldots, x_s$ generate Cl_K^- with probability at least $(1 - 2^{-\lfloor s/r \rfloor})^{\log_2 h_K^-}$. For any $0 < x \leq 1/2$, we have $\ln(1-x) > -(3/2)x$. We have $2^{-\lfloor s/r \rfloor} \leq 2^{-\lfloor \alpha \rfloor} \leq 1/2$, so

$$1 \left(1 - 2^{-\lfloor s/r \rfloor}\right)^{\log_2 h_K^-} = \exp\left(\log_2 h_K^- \ln\left(1 - 2^{-\lfloor s/r \rfloor}\right)\right)$$

$$\geq \exp\left(-\frac{3}{2} \log_2(h_K^-) 2^{-\lfloor s/r \rfloor}\right).$$

With $s \geq r(\log_2 \log_2(h_K^-) + \alpha)$, we get $\lfloor s/r \rfloor \geq \log_2 \log_2(h_K^-) + \alpha - 1$ and

$$1 \left(1 - 2^{-\lfloor s/r \rfloor}\right)^{\log_2 h_K^-} \geq \exp\left(-\frac{3}{2^\alpha}\right).$$

$\square$

B Proof of Proposition 2

Given the Theorem 1.1 of [BS16] the proof of this corollary is standard, and known as the linear-algebra step of index calculus methods.

The prime factorization $\mathfrak{a} = \mathfrak{q}_1^{a_1} \ldots \mathfrak{q}_k^{a_k}$ can be obtained in polynomial time in n, $\log(\Delta_K)$ and $\log(N\mathfrak{a})$, by Shor's algorithm [Sho97, EH10]. Let $\mathfrak{C} = \mathfrak{B} \cup \{\mathfrak{q}_1 \ldots, \mathfrak{q}_k\}$, and one can assume without loss of generality that this union is disjoint. Let $r = n_1 + n_2 - 1$, where n_1 is the number of real embeddings of K, and n_2 is the number of pairs of complex embeddings. Consider the homomorphism

$$\psi : \mathbb{Z}^{\mathfrak{B}} \times \mathbb{Z}^k \longrightarrow \mathrm{Cl}_K : ((e_\mathfrak{p})_{\mathfrak{p} \in \mathfrak{B}}, (f_1, \ldots, f_k)) \longmapsto \left[\prod_{\mathfrak{p} \in \mathfrak{B}} \mathfrak{p}^{e_\mathfrak{p}} \right] \cdot \left[\prod_{i=1}^{d} \mathfrak{q}_i^{f_i} \right].$$

As described in [BS16, Sect. 4], solving the $\mathfrak{C}$-unit problem provides a generating set of size $c = r + |\mathfrak{B}| + k$ for the kernel L of ψ. From [BS16, Theorem 1.1] such a generating set $\{\mathbf{v}_i\}_{i=1}^c$ can be found by a quantum algorithm in time polynomial in n, $\max_{\mathfrak{p} \in \mathfrak{C}}\{\log(N\mathfrak{p})\}$, $\log(d_K)$ and $|\mathfrak{C}| = O(|\mathfrak{B}| + \log(N\mathfrak{a}))$. For each i, write $\mathbf{v}_i = ((w_{i,\mathfrak{p}})_{\mathfrak{p} \in \mathfrak{B}}, (v_{i,1}, \ldots, v_{i,k}))$. Since $[\mathfrak{a}] \in H$ and $\mathfrak{B}$ generates H, the system of equations $\{\sum_{j=1}^c x_j v_{j,i} = a_i\}_{i=1}^k$ has a solution $\mathbf{x} \in \mathbb{Z}^c$ which can be computed in polynomial time. We obtain

$$0 = \psi \left(\sum_{i=1}^{c} x_i \mathbf{v}_i \right) = \left[\prod_{\mathfrak{p} \in \mathfrak{B}} \mathfrak{p}^{\sum_j x_j w_{j,\mathfrak{p}}} \right] \cdot \left[\prod_{i=1}^{d} \mathfrak{q}_i^{\sum_j x_j v_{j,i}} \right] = \left[\prod_{\mathfrak{p} \in \mathfrak{B}} \mathfrak{p}^{\sum_j x_j w_{j,\mathfrak{p}}} \right] \cdot [\mathfrak{a}].$$

Then, the output of $\mathrm{ClDL}_{\mathfrak{B}}$ is $\mathbf{y} = \left(- \sum_j x_j w_{j,\mathfrak{p}} \right)_{\mathfrak{p} \in \mathfrak{B}}$. □

References

[Ajt99]   Ajtai, M.: Generating hard instances of the short basis problem. In: Wiedermann, J., Emde Boas, P., Nielsen, M. (eds.) ICALP 1999. LNCS, vol. 1644, pp. 1–9. Springer, Heidelberg (1999). doi:10.1007/3-540-48523-6_1

[Bac90]   Bach, E.: Explicit bounds for primality testing and related problems. Math. Comput. **55**(191), 355–380 (1990)

[BCLvV16] Bernstein, D.J., Chuengsatiansup, C., Lange, T., van Vredendaal, C.: NTRU prime. Cryptology ePrint Archive, Report 2016/461 (2016). http://eprint.iacr.org/2016/461

[BEF+17] Biasse, J.-F., Espitau, T., Fouque, P.-A., Gélin, A., Kirchner, P.: Computing generator in cyclotomic integer rings, a subfield algorithm for the principal ideal problem in L(1/2) and application to cryptanalysis of a FHE scheme. In: Coron, J.-S., Nielsen, J.B. (eds.) EUROCRYPT 2017. LNCS, vol. 10210, pp. 60–88. Springer, Cham (2017)

[BF14]   Biasse, J.-F., Fieker, C.: Subexponential class group and unit group computation in large degree number fields. LMS J. Comput. Math. **17**(suppl. A), 385–403 (2014)

[BPR04] Buhler, J., Pomerance, C., Robertson, L.: Heuristics for class numbers of prime-power real cyclotomic fields. In: High Primes and Misdemeanours: Lectures in Honour of the 60th Birthday of Hugh Cowie Williams, Fields Institute Communications, pp. 149–157. American Mathematical Society (2004)

[BS16] Biasse, J.-F., Song, F.: Efficient quantum algorithms for computing class groups and solving the principal ideal problem in arbitrary degree number fields. In: Proceedings of the Twenty-Seventh Annual ACM-SIAM Symposium on Discrete Algorithms, pp. 893–902. SIAM (2016)

[BV11] Brakerski, Z., Vaikuntanathan, V.: Fully homomorphic encryption from ring-LWE and security for key dependent messages. In: Rogaway, P. (ed.) CRYPTO 2011. LNCS, vol. 6841, pp. 505–524. Springer, Heidelberg (2011). doi:10.1007/978-3-642-22792-9_29

[CDPR16] Cramer, R., Ducas, L., Peikert, C., Regev, O.: Recovering short generators of principal ideals in cyclotomic rings. In: Fischlin, M., Coron, J.-S. (eds.) EUROCRYPT 2016. LNCS, vol. 9666, pp. 559–585. Springer, Heidelberg (2016). doi:10.1007/978-3-662-49896-5_20

[CGS14] Campbell, P., Groves, M., Shepherd, D.: Soliloquy: a cautionary tale. ETSI 2nd Quantum-Safe Crypto Workshop (2014). http://docbox.etsi.org/Workshop/2014/201410_CRYPTO/S07_Systems_and_Attacks/S07_Groves_Annex.pdf

[DM15] Ducas, L., Micciancio, D.: FHEW: bootstrapping homomorphic encryption in less than a second. In: Oswald, E., Fischlin, M. (eds.) EUROCRYPT 2015. LNCS, vol. 9056, pp. 617–640. Springer, Heidelberg (2015). doi:10.1007/978-3-662-46800-5_24

[EH10] Eisenträger, K., Hallgren, S.: Algorithms for ray class groups and hilbert class fields. In: Proceedings of the Twenty-first Annual ACM-SIAM Symposium on Discrete Algorithms, SODA 2010, pp. 471–483. Society for Industrial and Applied Mathematics, Philadelphia, PA, USA (2010). ISBN 978-0-898716-98-6

[EHKS14] Eisenträger, K., Hallgren, S., Kitaev, A., Song, F.: A quantum algorithm for computing the unit group of an arbitrary degree number field. In: STOC, pp. 293–302. ACM (2014)

[GGH13] Garg, S., Gentry, C., Halevi, S.: Candidate multilinear maps from ideal lattices. In: Johansson, T., Nguyen, P.Q. (eds.) EUROCRYPT 2013. LNCS, vol. 7881, pp. 1–17. Springer, Heidelberg (2013). doi:10.1007/978-3-642-38348-9_1

[GN08] Gama, N., Nguyen, P.Q.: Finding short lattice vectors within mordell's inequality. In: Proceedings of the Fortieth Annual ACM Symposium on Theory of Computing, pp. 207–216. ACM (2008)

[GS02] Gentry, C., Szydlo, M.: Cryptanalysis of the revised NTRU signature scheme. In: Knudsen, L.R. (ed.) EUROCRYPT 2002. LNCS, vol. 2332, pp. 299–320. Springer, Heidelberg (2002). doi:10.1007/3-540-46035-7_20

[HPS98] Hoffstein, J., Pipher, J., Silverman, J.H.: NTRU: a ring-based public key cryptosystem. In: Buhler, J.P. (ed.) ANTS 1998. LNCS, vol. 1423, pp. 267–288. Springer, Heidelberg (1998). doi:10.1007/BFb0054868

[JMV09] Jao, D., Miller, S.D., Venkatesan, R.: Expander graphs based on GRH with an application to elliptic curve cryptography. J. Number Theory 129(6), 1491–1504 (2009). http://dx.doi.org/10.1016/j.jnt.2008.11.006. ISSN 0022-314X

[JW15]   Jetchev, D., Wesolowski, B.: On graphs of isogenies of principally polarizable abelian surfaces and the discrete logarithm problem. CoRR, abs/1506.00522 (2015)

[KF16]   Kirchner, P., Fouque, P.-A.: Comparison between subfield and straightforward attacks on NTRU. Cryptology ePrint Archive, Report 2016/717 (2016). http://eprint.iacr.org/2016/717

[Len75]   Lenstra Jr., H.W.: Euclid's algorithm in cyclotomic fields. J. Lond. Math. Soc 10, 457–465 (1975)

[LLL82]   Lenstra, A.K., Lenstra, H.W., Lovász, L.: Factoring polynomials with rational coefficients. Math. Ann. 261(4), 515–534 (1982)

[LM06]   Lyubashevsky, V., Micciancio, D.: Generalized compact knapsacks are collision resistant. ICALP 2, 144–155 (2006)

[LPR10]   Lyubashevsky, V., Peikert, C., Regev, O.: On ideal lattices, learning with errors over rings. J. ACM 60(6), 43:1–43:35 (2013). Preliminary version in Eurocrypt 2010

[LS15]   Langlois, A., Stehlé, D.: Worst-case to average-case reductions for module lattices. Des. Codes Cryptogr. 75(3), 565–599 (2015)

[LSS14]   Langlois, A., Stehlé, D., Steinfeld, R.: GGHLite: more efficient multilinear maps from ideal lattices. In: Nguyen, P.Q., Oswald, E. (eds.) EUROCRYPT 2014. LNCS, vol. 8441, pp. 239–256. Springer, Heidelberg (2014). doi:10.1007/978-3-642-55220-5_14

[Mic02]   Micciancio, D.: Generalized compact knapsacks, cyclic lattices, and efficient one-way functions. Comput. Complex. 16(4), 365–411 (2007). Preliminary version in FOCS 2002

[Mil15]   Miller, J.C.: Real cyclotomic fields of prime conductor and their class numbers. Math. Comp. 84(295), 2459–2469 (2015)

[Nap96]   Napias, H.: A generalization of the LLL-algorithm over euclidean rings or orders. J. Théor. nombres Bordx. 8(2), 387–396 (1996)

[PR06]   Peikert, C., Rosen, A.: Efficient collision-resistant hashing from worst-case assumptions on cyclic lattices. In: Halevi, S., Rabin, T. (eds.) TCC 2006. LNCS, vol. 3876, pp. 145–166. Springer, Heidelberg (2006). doi:10.1007/11681878_8

[Reg05]   Regev, O.: On lattices, learning with errors, random linear codes, and cryptography. J. ACM 56(6), 1–40 (2009). Preliminary version in STOC 2005

[Sch87]   Schnorr, C.-P.: A hierarchy of polynomial time lattice basis reduction algorithms. Theor. Comput. Sci. 53, 201–224 (1987)

[Sch89]   Schoof, R.: The Structure of the Minus Class Groups of Abelian Number Fields. Rijksuniversiteit Utrecht, Mathematisch Instituut, Netherlands (1989)

[Sch98]   Schoof, R.: Minus class groups of the fields of the ℓ-th roots of unity. Math. Comput. Am. Math. Soc. 67(223), 1225–1245 (1998)

[Sch03]   Schoof, R.: Class numbers of real cyclotomic fields of prime conductor. Math. Comput. 72(242), 913–937 (2003)

[Sch10]   Schoof, R.: Catalan's Conjecture. Springer Science and Business Media, New York (2010)

[Sch15]   Schank, J.: LogCvp, pari implementation of CVP in $\mathrm{Log}\mathbb{Z}[\zeta_{2^n}]^*$ (2015). https://github.com/jschanck-si/logcvp

[Sho97]   Shor, P.W.: Polynomial-time algorithms for prime factorization and discrete logarithms on a quantum computer. SIAM J. Comput. 26(5), 1484–1509 (1997). doi:10.1137/S0097539795293172. ISSN 0097-5397

[Sin80] Sinnott, W.: On the Stickelberger ideal and the circular units of an abelian field. Invent. Math. **62**, 181–234 (1980)

[SS11] Stehlé, D., Steinfeld, R.: Making NTRU as secure as worst-case problems over ideal lattices. In: Paterson, K.G. (ed.) EUROCRYPT 2011. LNCS, vol. 6632, pp. 27–47. Springer, Heidelberg (2011). doi:10.1007/978-3-642-20465-4_4

[SSTX09] Stehlé, D., Steinfeld, R., Tanaka, K., Xagawa, K.: Efficient public key encryption based on ideal lattices. In: Matsui, M. (ed.) ASIACRYPT 2009. LNCS, vol. 5912, pp. 617–635. Springer, Heidelberg (2009). doi:10.1007/978-3-642-10366-7_36

[SV10] Smart, N.P., Vercauteren, F.: Fully homomorphic encryption with relatively small key and ciphertext sizes. In: Nguyen, P.Q., Pointcheval, D. (eds.) PKC 2010. LNCS, vol. 6056, pp. 420–443. Springer, Heidelberg (2010). doi:10.1007/978-3-642-13013-7_25

[Was12] Washington, L.C.: Introduction to Cyclotomic Fields, vol. 83, 2nd edn. Springer Science & Business Media, New York (2012)

Universal Composability

Concurrently Composable Security with Shielded Super-Polynomial Simulators

Brandon Broadnax[1], Nico Döttling[2], Gunnar Hartung[1], Jörn Müller-Quade[1], and Matthias Nagel[1(✉)]

[1] Karlsruhe Institute of Technology, Karlsruhe, Germany
{brandon.broadnax,gunnar.hartung,joern.mueller-quade,
matthias.nagel}@kit.edu
[2] University of California Berkeley, Berkeley, USA
nico.doettling@gmail.com

Abstract. We propose a new framework for concurrently composable security that relaxes the security notion of UC security. As in previous frameworks, our notion is based on the idea of providing the simulator with super-polynomial resources. However, in our new framework simulators are only given *restricted access* to the results computed in super-polynomial time. This is done by modeling the super-polynomial resource as a stateful oracle that may directly interact with a functionality without the simulator seeing the communication. We call these oracles "shielded oracles".

Our notion is fully compatible with the UC framework, i.e., protocols proven secure in the UC framework remain secure in our framework. Furthermore, our notion lies strictly between SPS and Angel-based security, while being closed under protocol composition.

Shielding away super-polynomial resources allows us to apply new proof techniques where we can replace super-polynomial entities by indistinguishable polynomially bounded entities. This allows us to construct secure protocols in the plain model using weaker primitives than in previous Angel-based protocols. In particular, we only use non-adaptive-CCA-secure commitments as a building block in our constructions.

As a feasibility result, we present a constant-round general MPC protocol in the plain model based on standard polynomial-time hardness assumptions that is secure in our framework. Our protocol can be made

---

B. Broadnax and M. Nagel—This work was supported by the German Federal Ministry of Education and Research within the framework of the project "Sicherheit vernetzter Infrastrukturen (SVI)" in the Competence Center for Applied Security Technology (KASTEL).

N. Döttling—This work was supported by the DAAD (German Academic Exchange Service) under the postdoctoral program (57243032) and in part supported by European Research Council Starting Grant 279447. Research supported in part from a DARPA/ARL SAFEWARE award, AFOSR Award FA9550-15-1-0274, and NSF CRII Award 1464397. The views expressed are those of the author and do not reflect the official policy or position of the Department of Defense, the National Science Foundation, or the U.S. Government.

J.-S. Coron and J.B. Nielsen (Eds.): EUROCRYPT 2017, Part I, LNCS 10210, pp. 351–381, 2017.
DOI: 10.1007/978-3-319-56620-7_13

fully black-box. As a consequence, we obtain the *first* black-box construction of a constant-round concurrently secure general MPC protocol in the plain model based on polynomial-time hardness assumptions.

1   Introduction

Cryptographic protocols typically run in a network where multiple protocols interact with each other. Some of them may even act in an adversarial manner. This makes designing protocols that are secure in such a general setting a complicated task. The universal composability (UC) framework [Can01] provides means for designing and analyzing cryptographic protocols in this concurrent setting. More specifically, it captures a security notion that implies two major properties: *general concurrent security* and *modular analysis*. The former means that a protocol remains secure even when run in an environment with multiple instances of arbitrary protocols. The latter implies that one can deduce the security of a protocol from its components. Unfortunately, there exist strong impossibility results [CF01, CKL03, Lin03, PR08, KL11] regarding the realizaility of cryptographic tasks in the UC framework: One requires trusted setup assumptions in order to design UC-secure protocols for many cryptographic tasks. UC-secure protocols have thus been constructed based on various setup assumptions [Can+02, Bar+04, Can+07, KLP07, Kat07, CPS07, LPV09, Dac+13]. However, if the trusted setup is compromised, all security guarantees are lost. In general, one would like to base the security of cryptographic protocols on as little trust as possible.

In order to drop the requirement for trusted setup, relaxed notions of security have been developed. One of the most prominent solutions is "UC security with super-polynomial time simulators" (SPS), introduced in [Pas03]. In this model, the simulator is allowed to run in *super-polynomial time*, thereby overcoming the impossibility results. Various multi-party computation protocols without trusted setup that satisfy this notion have been constructed, e.g., [Pas03, BS05, LPV09, LPV12, Gar+12, Dac+13, Ven14]. SPS security weakens the security of the UC framework because the simulator, being able to run in super-polynomial time, may now be able to carry out stronger attacks in the ideal setting. Still, this security notion is meaningful, since for many cryptographic tasks the ideal setting has an information-theoretic nature. Contrary to UC security, however, security in this model is not closed under protocol composition. As a consequence, this notion neither supports general concurrent security nor modular analysis.

"Angel-based security" [PS04] overcomes these issues. In this model, both the adversary and the simulator have access to an oracle called "(Imaginary) Angel" that provides super-polynomial resources for *specific* computational problems. Many general MPC protocols without setup have been constructed in the Angel-based framework [PS04, MMY06, CLP10, LP12, KMO14, Kiy14, Goy+15, HV16]. Like UC-security, this notion is closed under protocol composition. Furthermore, Angel-based security implies SPS security. In fact, it provides a stronger security notion since the simulator has only access to specific super- polynomial computations. [CLP10] later recast the Angel-based security model in the extended

UC (EUC) framework [Can+07] and dubbed their notion "UC with super-polynomial helpers". In contrast to the non-interactive and stateless Angels in previous works, the "helpers" in [CLP10] are highly interactive and stateful.

In this work, we take this framework a step further. In our new framework, simulators only have *restricted access* to the results computed in super-polynomial time. More specifically, we model the super-polynomial resources as stateful oracles that are "glued" to an ideal functionality. These oracles may directly interact with the functionality without the simulator observing the communication. The outputs of these oracles are therefore "shielded away" from the simulator. As with Angel-based security, our notion implies SPS security. Moreover, it can be shown that our notion is in fact strictly weaker than Angel-based security. Furthermore, our notion comes with a composition theorem guaranteeing general concurrent security. While modular analysis is not directly implied for technical reasons, using our composition theorem one can achieve modular analysis by constructing protocols with strong composition features. Protocols with these features can be "plugged" into large classes of UC-secure protocols in such a way that the composed protocol is secure in our framework. As a proof of concept, we construct a constant-round commitment scheme with such features.

In order to obtain a composable security notion, environments are "augmented" in our framework, i.e., they may invoke additional (ideal) protocols that include shielded oracles. Since the super-poly computations in these protocols are hidden away, these augmented environments have the unique property that they do not "hurt" protocols proven secure in the UC framework. Therefore, our notion is in fact fully compatible with the UC framework. Moreover, our concept of "shielding away" super-polynomial resources allows us to apply new proof techniques not applicable in previous frameworks: We are able to replace entities involving super-polynomial resources in our proofs by indistinguishable polynomially bounded entities. This allows us to construct (constant-round) protocols using weaker primitives than in previous Angel-based protocols.

1.1 Our Results

We propose a new framework that is based on the idea of granting simulators only restricted access to the results of a super-polynomial oracle. We have the following results:

- *New Composable Security Notion*: Our notion of security is closed under general composition, it implies SPS security and is strictly weaker than Angel-based security (Theorem 9, Proposition 8, Theorem 17).
- *UC-compatibility*: Protocols proven secure in the UC framework are also secure in our new framework (Theorem 12, Corollary 13).
- *Modular Composition*: As a proof of concept, we present a constant-round commitment scheme in the plain model based on OWPs that is secure in our framework and can be "plugged" into a large class of UC-secure protocols, such that the composite protocol is secure in our framework. Furthermore, this construction can be made fully black-box based on homomorphic commitment

schemes. To our best knowledge, this is the first constant-round (black-box) commitment scheme in the plain model based on a standard polynomial-time hardness assumption with such a composition feature (Theorem 21, Corollary 22, Corollary 23, Theorem 26, Corollary 30).

- *Constant-round (black-box) MPC*: We present a modular construction of a constant-round general MPC protocol in the plain model based on standard polynomial-time hardness assumptions that is secure in our framework. This protocol can be made fully black-box based on homomorphic commitment schemes. As a consequence, we obtain the first black-box construction of a constant-round concurrently secure general MPC protocol in the plain model based on polynomial-time hardness assumptions (Theorem 31).
- *Building on non-adaptive CCA-commitments*: Our constructions require weaker primitives than previous Angel-based protocols. Specifically, it suffices to use non-adaptive CCA-secure commitment schemes as a building block in our constructions instead of CCA-secure commitment schemes used previously (Theorem 21, Theorem 26).

2 Related Work

The frameworks most related to ours are SPS and Angel-based security.

SPS security, introduced by [Pas03], provides a meaningful security notion for many cryptographic tasks such as commitment schemes or oblivious transfer. However, SPS security does not come with a composition theorem. There exist many constructions (in the plain model) satisfying this notion, e.g., [Pas03, BS05, LPV09, LPV12, Gar+12, Dac+13, Ven14]. Notably, [LPV12, Gar+12] constructed (non-black-box) constant-round general MPC protocols based on standard polynomial-time hardness assumptions.

Angel-based security [PS04] implies SPS security and comes with a composition theorem. Various general MPC protocols without setup have been constructed in the Angel-based setting [PS04, MMY06, CLP10, LP12, KMO14, Kiy14, Goy+15, HV16]. Some rely on non-standard or super-polynomial time assumptions [PS04, MMY06, KMO14]. The construction in [CLP10] is the first one to rely on standard polynomial-time assumptions, but has non-constant round complexity. Later works [Goy+15, Kiy14] have improved the round-complexity, while also relying on standard assumptions. The most round-efficient construction based on standard polynomial-time assumptions is [Kiy14], which requires $\widetilde{O}(\log^2 n)$ rounds and makes only black-box use of the underlying cryptographic primitive. Some Angels in the literature, e.g., [CLP10, KMO14, Kiy14, Goy+15] come with a feature called "robustness" which guarantees that any attack mounted on a constant-round protocol using this angel can be carried out by a polytime adversary with no angels. Protocols proven secure for robust Angels can be "plugged" into UC-secure protocols, resulting in Angel-secure protocols. All known constructions for robust Angels based on standard polytime assumptions require a super-constant number of rounds. Moreover, [CLP13] construct a (super-constant-round) protocol that is secure in the Angel-based setting and additionally preserves certain security properties of other protocols running in the system. They call such protocols "environmentally friendly".

We want to note that other security notions in the concurrent setting have been proposed that are not based on the idea of simulators with super-polynomial resources. The "multiple ideal query model" [GJO10, GJ13, GGJ13, CGJ15] considers simulators that are allowed to make more than one output query per session to the ideal functionality. Another (not simulation-based) notion is "input indistinguishability" [MPR06, Gar+12] which guarantees that an adversary cannot decide which inputs have been used by the honest protocol parties. We note that this security notion is incomparable to ours.

3 Shielded Oracles

3.1 Definition of the Framework

Our model is based on the universal composability framework (UC). In this model, a protocol π carrying out a given task is defined to be secure by comparing it to an *ideal functionality* $\mathcal{F}$, which is a trusted and incorruptible party that carries out a given task in an ideally secure way. π is said to be secure if it "emulates" $\mathcal{F}$.

While the plain UC model leaves open how session identifiers and corruptions are organized, we follow the convention that both must be consistent with the hierarchical order of the protocols: The session identifier (sid) of a sub-protocol must be an extension of the session id of the calling protocol. Likewise, in order to corrupt a sub-party, an adversary must corrupt all parties that are above that sub-party in the protocol hierarchy.

We relax the UC security notion by introducing a super-polynomial time machine that may aid the simulator. This machine is modeled as a *stateful* oracle $\mathcal{O}$ that is "glued" to an the ideal functionality $\mathcal{F}$. $\mathcal{O}$ may freely interact with the simulator and $\mathcal{F}$. However, the simulator does not "see" the communication between $\mathcal{O}$ and $\mathcal{F}$. Since the output of the oracle is partially hidden from the simulator, we call $\mathcal{O}$ a *shielded oracle*.

Definition 1 (Shielded oracles). *A shielded oracle is a stateful oracle $\mathcal{O}$ that can be implemented in super-polynomial time. By convention, the outputs of a shielded oracle $\mathcal{O}$ are of the form (output-to-fnct, y) or (output-to-adv, y).*

The simulator is allowed to communicate with the functionality *only* via the shielded oracle. This way, the shielded oracle serves as an interface that carries out specific tasks the simulator could not do otherwise. The communication between the shielded oracle and the functionality is hidden away from the simulator. The actions of the shielded oracle may depend on the session identifier (sid) of the protocol session as well as the party identifiers of the corrupted parties.

Definition 2 ($\mathcal{O}$-adjoined functionalities). *Given a functionality $\mathcal{F}$ and a shielded oracle $\mathcal{O}$, define the interaction of the $\mathcal{O}$-adjoined functionality $\mathcal{F}^{\mathcal{O}}$ in an ideal protocol execution with session identifier sid as follows:*

- $\mathcal{F}^{\mathcal{O}}$ *internally runs an instance of* $\mathcal{F}$ *with session identifier* sid.
- *When receiving the first message* x *from the adversary,* $\mathcal{F}^{\mathcal{O}}$ *internally invokes* $\mathcal{O}$ *with input* (sid, x).
  All subsequent messages from the adversary are passed to $\mathcal{O}$.
- *Messages between the honest parties and* $\mathcal{F}$ *are forwarded.*
- *Corruption messages are forwarded to* $\mathcal{F}$ *and* $\mathcal{O}$.
- *When* $\mathcal{F}$ *sends a message* y *to the adversary,* $\mathcal{F}^{\mathcal{O}}$ *passes* y *to* $\mathcal{O}$.
- *The external write operations of* $\mathcal{O}$ *are treated as follows:*
  - *If* $\mathcal{O}$ *sends* (output-to-fnct, y), $\mathcal{F}^{\mathcal{O}}$ *sends* y *to* $\mathcal{F}$.
  - *If* $\mathcal{O}$ *sends* (output-to-adv, y), $\mathcal{F}^{\mathcal{O}}$ *sends* y *to the adversary.*

Let IDEAL($\mathcal{F}^{\mathcal{O}}$) be the ideal protocol with functionality $\mathcal{F}^{\mathcal{O}}$ as defined in [Can01].

In order to obtain a composable security notion, we introduce the notion of *augmented environments*. Augmented environments are UC environments that may invoke, apart form the challenge protocol, polynomially many instances of IDEAL($\mathcal{F}^{\mathcal{O}}$) for a given functionality $\mathcal{F}^{\mathcal{O}}$. The only restriction is that the session identifiers of these instances as well as the session identifier of the challenge protocol are not extensions of one another.

Augmented environments may send inputs to and receive outputs from any invoked instance of IDEAL($\mathcal{F}^{\mathcal{O}}$). In addition, augmented environments can play the role of any adversary via the adversary's interface of the functionality. In particular, augmented environments may corrupt parties sending the corresponding corruption message as input to the functionality.

In what follows we give a definition of an execution experiment with an $\mathcal{F}^{\mathcal{O}}$-augmented environment. For simplicity and due to space constraints, the description is kept informal.

Definition 3 (The $\mathcal{F}^{\mathcal{O}}$-execution experiment). *An execution of a protocol* σ *with adversary* $\mathcal{A}$ *and an* $\mathcal{F}^{\mathcal{O}}$-*augmented environment* $\mathcal{Z}$ *on input* $a \in \{0,1\}^*$ *and with security parameter* $n \in \mathbb{N}$ *is a run of a system of interactive Turing machines (ITMs) with the following restrictions:*

- *First,* $\mathcal{Z}$ *is activated on input* $a \in \{0,1\}^*$.
- *The first ITM to be invoked by* $\mathcal{Z}$ *is the adversary* $\mathcal{A}$.
- $\mathcal{Z}$ *may invoke a single instance of a challenge protocol, which is set to be* σ *by the experiment. The session identifier of* σ *is determined by* $\mathcal{Z}$ *upon invocation.*
- $\mathcal{Z}$ *may pass inputs to the adversary or the protocol parties of* σ.
- $\mathcal{Z}$ *may invoke, send inputs to and receive outputs from instances of* IDEAL($\mathcal{F}^{\mathcal{O}}$) *as long as the session identifiers of these instances as well as the session identifier of the instance of* σ *are not extensions of one another.*
- *The adversary* $\mathcal{A}$ *may send messages to protocol parties of* σ *as well as to the environment.*
- *The protocol parties of* σ *may send messages to* $\mathcal{A}$, *pass inputs to and receive outputs from subparties and give outputs to* $\mathcal{Z}$.

Denote by $\mathrm{Exec}(\sigma, \mathcal{A}, \mathcal{Z}[\mathcal{F}^{\mathcal{O}}])(n, a)$ *the output of the* $\mathcal{F}^{\mathcal{O}}$*-augmented environment* $\mathcal{Z}$ *on input* $a \in \{0, 1\}^*$ *and with security parameter* $n \in \mathbb{N}$ *when interacting with* σ *and* $\mathcal{A}$ *according to the above definition.*

Define $\mathrm{Exec}(\sigma, \mathcal{A}, \mathcal{Z}[\mathcal{F}^{\mathcal{O}}]) = \left\{ \mathrm{Exec}(\sigma, \mathcal{A}, \mathcal{Z}[\mathcal{F}^{\mathcal{O}}])(n, a) \right\}_{n \in \mathbb{N}, a \in \{0,1\}^*}$.

We will now define security in our framework in total analogy to the UC framework:

Definition 4 ($\mathcal{F}^{\mathcal{O}}$-emulation). *Let* π *and* ϕ *be protocols.* π *is said to emulate* ϕ *in the presence of* $\mathcal{F}^{\mathcal{O}}$*-augmented environments, denoted by* $\pi \geq_{\mathcal{F}^{\mathcal{O}}} \phi$*, if for any PPT adversary* $\mathcal{A}$ *there exists a PPT adversary (called "simulator")* $\mathcal{S}$ *such that for every* $\mathcal{F}^{\mathcal{O}}$*-augmented PPT environment* $\mathcal{Z}$ *it holds that*

$$\mathrm{Exec}(\pi, \mathcal{A}, \mathcal{Z}[\mathcal{F}^{\mathcal{O}}]) \stackrel{c}{\equiv} \mathrm{Exec}(\phi, \mathcal{S}, \mathcal{Z}[\mathcal{F}^{\mathcal{O}}]) \tag{1}$$

Throughout this paper, we only consider *static* corruptions.

3.2 Basic Properties and Justification

In this section, we show that that our security notion is transitive and that the dummy adversary is complete within this notion. As a justification for our notion, we show that it implies super-polynomial time simulator (SPS) security.

Definition 5 ($\mathcal{F}^{\mathcal{O}}$-emulation with respect to the dummy adversary). *The dummy adversary* $\mathcal{D}$ *is an adversary that when receiving a message* (sid, pid, m) *from the environment, sends* m *to the party with party identifier* pid *and session identifier* sid*, and that, when receiving* m *from the party with party identifier* pid *and session identifier* sid*, sends* (sid, pid, m) *to the environment.*

Let π *and* ϕ *be protocols.* π *is said to emulate* ϕ *in the presence of* $\mathcal{F}^{\mathcal{O}}$*-augmented environments with respect to the dummy adversary, if*

$$\exists \mathcal{S}_{\mathcal{D}} \; \forall \mathcal{Z} : \mathrm{Exec}(\pi, \mathcal{D}, \mathcal{Z}[\mathcal{F}^{\mathcal{O}}]) \stackrel{c}{\equiv} \mathrm{Exec}(\phi, \mathcal{S}_{\mathcal{D}}, \mathcal{Z}[\mathcal{F}^{\mathcal{O}}]). \tag{2}$$

Proposition 6 (Completeness of the dummy adversary). *Let* π *and* ϕ *be protocols. Then,* π *emulates* ϕ *in the presence of* $\mathcal{F}^{\mathcal{O}}$*-augmented environments if and only if* π *emulates* ϕ *in the presence of* $\mathcal{F}^{\mathcal{O}}$*-augmented environments with respect to the dummy adversary.*

The proof is almost exactly the same as in [Can01], and therefore only given in the full version of this work. The proof of transitivity is omitted here, too.

Proposition 7 (Transitivity). *Let* π_1, π_2, π_3 *be protocols. If* $\pi_1 \geq_{\mathcal{F}^{\mathcal{O}}} \pi_2$ *and* $\pi_2 \geq_{\mathcal{F}^{\mathcal{O}}} \pi_3$ *then it holds that* $\pi_1 \geq_{\mathcal{F}^{\mathcal{O}}} \pi_3$.

In order to justify our new notion, we prove that security with respect to $\mathcal{F}^{\mathcal{O}}$-emulation implies security with respect to SPS-emulation which we will denote by $\geq_{\mathrm{SPS}}$. See the full version for a formal definition of $\pi \geq_{\mathrm{SPS}} \phi$. The proof is straightforward: View the oracle as part of the simulator. This simulator runs in super-polynomial time, hence can be simulated by an SPS-simulator.

Proposition 8 ($\mathcal{F}^{\mathcal{O}}$-emulation implies SPS-emulation). *Let* $\mathcal{O}$ *be a shielded oracle. Assume* $\pi \geq_{\mathcal{F}^{\mathcal{O}}} \mathcal{F}^{\mathcal{O}}$*. Then it holds that* $\pi \geq_{\mathrm{SPS}} \mathcal{F}$.

3.3   Universal Composition

A central property of the UC framework is the universal composition theorem. This theorem guarantees that the security of a protocol is *closed* under protocol composition. This means that security guarantees can be given for a UC-secure protocol even if multiple other protocols interact with this protocol in a potentially adversarial manner. We prove a similar theorem in our framework. More specifically, we generalize the universal composition theorem to also include $\mathcal{F}^{\mathcal{O}}$-hybrid protocols.

Theorem 9 (Composition theorem). *Let $\mathcal{O}$ be a shielded oracle, $\mathcal{F}$ and $\mathcal{G}$ functionalities.*

1. *(Polynomial hybrid protocols) Let π, $\rho^{\mathcal{G}}$ be protocols. Assume $\pi \geq_{\mathcal{F}^{\mathcal{O}}} \mathcal{G}$. Then it holds that $\rho^{\pi} \geq_{\mathcal{F}^{\mathcal{O}}} \rho^{\mathcal{G}}$.*
2. *($\mathcal{F}^{\mathcal{O}}$-hybrid protocols) Let π be a protocol, $\rho^{\mathcal{F}^{\mathcal{O}}}$ a protocol in the $\mathcal{F}^{\mathcal{O}}$-hybrid model. Assume $\pi \geq_{\mathcal{F}^{\mathcal{O}}} \mathcal{F}^{\mathcal{O}}$. Then it holds that $\rho^{\pi} \geq_{\mathcal{F}^{\mathcal{O}}} \rho^{\mathcal{F}^{\mathcal{O}}}$.*

Proof (of the second statement). For *single instance composition* (where ρ calls only a single instance of π), treat ρ as part of the environment and use the premise that $\pi \geq_{\mathcal{F}^{\mathcal{O}}} \mathcal{F}^{\mathcal{O}}$.

For the general case iteratively apply the single instance composition theorem. In each iteration a new instance of IDEAL($\mathcal{F}^{\mathcal{O}}$) is replaced by an instance of π, and the remaining instances of π, IDEAL($\mathcal{F}^{\mathcal{O}}$) and ρ are treated as part of the augmented environment. The claim then follows using transitivity. $\square$

The universal composition theorem in the UC framework has two important implications: general concurrent security and modular analysis. The former means that a protocol remains secure even when run in an environment with multiple instances of arbitrary protocols. The latter implies that one can deduce the security of a protocol from its components.

Theorem 9 directly implies general concurrent security (with superpolynomial time simulators). However, modular analysis is not directly implied by Theorem 9. This is because the oracle $\mathcal{O}$ *may* contain all "complexity" of the protocol π, i.e., proving security of $\rho^{\mathcal{F}^{\mathcal{O}}}$ may be as complex as proving security of ρ^{π}.

Still, one can use Theorem 9 to achieve modular analysis by constructing secure protocols with strong composition features. A protocol π with such composition features allows analyzing the security of a large class of protocols $\rho^{\mathcal{F}}$ in the UC framework and achieve security in our framework when replacing $\mathcal{F}$ with π. As a proof of concept, we will show, using Theorem 9, that a large a class of protocols in the $\mathcal{F}_{\text{com}}$-hybrid model can be composed with a commitment protocol presented in this paper (Theorem 26).

The following is a useful extension of Theorem 9 for multiple oracles. The reader is referred to the full version for a proof.

Corollary 10 (Composition theorem for multiple oracles). *Let $\mathcal{O}$, $\mathcal{O}'$ be shielded oracles. Assume that $\pi \geq_{\mathcal{F}^{\mathcal{O}}} \mathcal{F}^{\mathcal{O}}$ and $\rho^{\mathcal{F}^{\mathcal{O}}} \geq_{\mathcal{F}^{\mathcal{O}},\mathcal{G}^{\mathcal{O}'}} \mathcal{G}^{\mathcal{O}'}$. Then there exists a shielded oracle $\mathcal{O}''$ such that $\rho^{\pi} \geq_{\mathcal{G}^{\mathcal{O}''}} \mathcal{G}^{\mathcal{O}''}$.*

3.4   Polynomial Simulatability

We show a unique feature of our framework: For appropriate oracles to be defined below, augmented environments do not "hurt" UC-secure protocols. This means that a protocol that was proven secure in the UC framework is secure in our framework, too. This makes our security notion fully compatible with UC security.

Definition 11 (Polynomial simulatability). *Let $\mathcal{O}$ be a shielded oracle, $\mathcal{F}$ a functionality. Say that $\mathcal{O}$ adjoined to $\mathcal{F}$ is polynomially simulatable if there exists a (PPT) functionality $\mathcal{M}$ such that for all $\mathcal{F}^{\mathcal{O}}$-augmented environments $\mathcal{Z}$ it holds that*

$$\mathcal{F}^{\mathcal{O}} \underset{\mathcal{F}^{\mathcal{O}}}{\geq} \mathcal{M} \tag{3}$$

If a functionality $\mathcal{F}^{\mathcal{O}}$ is polynomially simulatable then the super-polynomial power of the oracle $\mathcal{O}$ is totally "shielded away" from the environment. Note that in Definition 11, indistinguishability must hold for *augmented* environments not only for polynomial environments.

As a consequence, $\mathcal{F}^{\mathcal{O}}$-augmented environments can be replaced by *efficient* environments if $\mathcal{F}^{\mathcal{O}}$ is polynomially simulatable.

Theorem 12 (Reduction to polynomial time environments). *Let $\mathcal{O}$ be a shielded oracle and $\mathcal{F}$ a functionality such that $\mathcal{F}^{\mathcal{O}}$ is polynomially simulatable. Let π, ϕ be protocols that are PPT or in the $\mathcal{F}^{\mathcal{O}}$-hybrid model. It holds that*

$$\pi \underset{\mathcal{F}^{\mathcal{O}}}{\geq} \phi \iff \pi \underset{poly}{\geq} \phi \tag{4}$$

where the right-hand side means that π emulates ϕ in the presence of all $\mathcal{F}^{\mathcal{O}}$-augmented environments that never invoke an instance of IDEAL($\mathcal{F}^{\mathcal{O}}$).

Proof. Poly-emulation implies $\mathcal{F}^{\mathcal{O}}$-emulation: Replace all instances of IDEAL($\mathcal{F}^{\mathcal{O}}$) with instances of $\mathcal{M}$ using the fact that $\mathcal{F}^{\mathcal{O}}$ is polynomially simulatable. Treat all instances of $\mathcal{M}$ as part of the environment. This new environment runs in polynomial time. Substitute π by ϕ using the premise. Replace all instances of $\mathcal{M}$ with instances of IDEAL($\mathcal{F}^{\mathcal{O}}$) again. The statement follows.

The converse is trivial. $\square$

As augmented environments that never invoke instances of IDEAL($\mathcal{F}^{\mathcal{O}}$) are identical to an UC-environment, the following corollary immediately follows.

Corollary 13 (Compatibility with the UC framework). *Let $\mathcal{O}$ be a shielded oracle and $\mathcal{F}$ a functionality such that $\mathcal{F}^{\mathcal{O}}$ is polynomially simulatable. It holds that*

$$\pi \underset{\mathcal{F}^{\mathcal{O}}}{\geq} \phi \iff \pi \underset{UC}{\geq} \phi \tag{5}$$

Note that this does not contradict the classical impossibility results for the plain UC framework (cp. [CF01]): If $\pi \geq_{\mathcal{F}^{\mathcal{O}}} \mathcal{F}^{\mathcal{O}}$ for a polynomially simulatable $\mathcal{F}^{\mathcal{O}}$, then this only means that $\pi \geq_{UC} \mathcal{F}^{\mathcal{O}}$, but it does not follow that $\pi \geq_{UC} \mathcal{F}$. Although the super-polynomial power of $\mathcal{O}$ is shielded away from the outside, it is indeed necessary.

Replacing augmented environments with efficient environments will be a key property in various proofs later in this paper. In particular, it will allow us to prove the security of protocols in our framework using relatively weak primitives such as *non-adaptively*-secure-CCA commitments as opposed to CCA-secure commitments, which are commonly used in Angel-based protocols.

Next, we show that by suitably tweaking a given oracle $\mathcal{O}$ one can make $\mathcal{F}^{\mathcal{O}}$ polynomially simulatable while preserving the security relation.

Lemma 14 (Derived oracle). *Let $\mathcal{O}$ be a shielded oracle such that $\pi \geq_{\mathcal{F}^{\mathcal{O}}} \mathcal{F}^{\mathcal{O}}$. Then there exists a shielded oracle $\mathcal{O}'$ such that $\pi \geq_{\mathcal{F}^{\mathcal{O}'}} \mathcal{F}^{\mathcal{O}'}$ and additionally $\mathcal{O}'$ adjoined to $\mathcal{F}$ is polynomially simulatable.*

Proof. Since π emulates $\mathcal{F}^{\mathcal{O}}$, there exists a simulator $\mathcal{S}_D$ for the dummy adversary $\mathcal{D}$. Define the shielded oracle $\mathcal{O}'$ as follows: $\mathcal{O}'$ internally simulates $\mathcal{S}_D$ and $\mathcal{O}$, passes each message $\mathcal{S}_D$ sends to $\mathcal{F}$ to $\mathcal{O}$, sends each `output-to-fnct` output from $\mathcal{O}$ to $\mathcal{F}$ and each `output-to-adv` output from $\mathcal{O}$ to $\mathcal{S}_D$, and forwards the communication between $\mathcal{S}_D$ and the environment. By construction, for all $\mathcal{F}^{\mathcal{O}}$-augmented environments $\mathcal{Z}$ it holds that

$$\mathrm{Exec}(\pi, \mathcal{D}, \mathcal{Z}[\mathcal{F}^{\mathcal{O}}]) \overset{c}{\equiv} \mathrm{Exec}(\mathcal{F}^{\mathcal{O}}, \mathcal{S}_D, \mathcal{Z}[\mathcal{F}^{\mathcal{O}}]) \equiv \mathrm{Exec}(\mathcal{F}^{\mathcal{O}'}, \mathcal{D}, \mathcal{Z}[\mathcal{F}^{\mathcal{O}}]) \quad (6)$$

It follows from Proposition 6 that $\pi \geq_{\mathcal{F}^{\mathcal{O}}} \mathcal{F}^{\mathcal{O}'}$ and $\mathcal{F}^{\mathcal{O}'} \geq_{\mathcal{F}^{\mathcal{O}}} \pi$. Since $\mathcal{S}_D$ runs in polynomial time, $\mathcal{F}^{\mathcal{O}}$-augmented environments can simulate $\mathcal{F}^{\mathcal{O}'}$-augmented environments. Therefore, $\pi \geq_{\mathcal{F}^{\mathcal{O}'}} \mathcal{F}^{\mathcal{O}'}$ and $\mathcal{F}^{\mathcal{O}'} \geq_{\mathcal{F}^{\mathcal{O}'}} \pi$. The theorem follows by defining $\mathcal{M}$ to be the functionality that internally simulates the protocol π. □

The following corollary shows that UC-secure protocols can be used as sub-protocols in protocols proven secure in our framework, while preserving security.

Corollary 15 (Composition with UC-secure protocols). *Let π, $\rho^{\mathcal{F}}$ be protocols such that $\pi \geq_{UC} \mathcal{F}$ and $\rho^{\mathcal{F}} \geq_{\mathcal{G}^{\mathcal{O}}} \mathcal{G}^{\mathcal{O}}$. Then there exists a shielded oracle $\mathcal{O}'$ such that*

$$\rho^{\pi} \underset{\mathcal{G}^{\mathcal{O}'}}{\geq} \mathcal{G}^{\mathcal{O}'} \quad (7)$$

Proof. Since $\rho^{\mathcal{F}}$ is PPT there exists a shielded oracle $\mathcal{O}'$ such that $\mathcal{G}^{\mathcal{O}'}$ is polynomially simulatable and $\rho^{\mathcal{F}} \geq_{\mathcal{G}^{\mathcal{O}'}} \mathcal{G}^{\mathcal{O}'}$ by Lemma 14. From Corollary 13 it follows that $\pi \geq_{\mathcal{G}^{\mathcal{O}'}} \mathcal{F}$. The statement then follows from the composition theorem and the transitivity of $\mathcal{G}^{\mathcal{O}'}$-emulation. □

The last result demonstrates the compatibility of our framework with the UC framework again. While it is much more desireable to "plug" a protocol proven secure in our framework into a UC secure protocol—in order to obtain a secure protocol in the *plain model* (this will be addressed in Theorem 26 and Corollary 30)—doing it the other way around is still a convenient property. For instance, it allows one to instantiate "auxiliary" functionalities such as authenticated channels $\mathcal{F}_{\text{auth}}$ or secure channels $\mathcal{F}_{\text{SMT}}$, while preserving security.

3.5 Relation with Angel-Based Security

A natural question that arises is how our security notion compares to Angel-based security. We will prove that for a large class of Angels (which to our best knowledge includes all Angels that can be found in the literature), Angel-based security implies our security notion. However, assuming the existence of one-way functions, the converse does not hold. Thus, our notion is *strictly weaker* than Angel-based security.

In the following, we denote by $\pi \geq_{\Gamma\text{-Angel}} \phi$ if π securely realizes ϕ with respect to an angel Γ. Note that the following results also hold for "UC with super-polynomial helpers" put forward by [CLP10].

Definition 16 (Session-respecting Angel (informal)). *(See the full version for a formal treatment.) An Angel is called session-respecting if its internal state can be regarded as a vector with independent components for each session the Angel is queried for.*

Theorem 17 (Relation between angels and shielded oracles)

1. *Assume $\pi \geq_{\Gamma\text{-Angel}} \mathcal{F}$ for an imaginary Angel Γ. If Γ is session-respecting, then there exists a shielded oracle $\mathcal{O}$ such that $\pi \geq_{\mathcal{F}^{\mathcal{O}}} \mathcal{F}^{\mathcal{O}}$.*
2. *Assume the existence of one-way functions. Then there exists a protocol ρ (in the $\mathcal{F}_{\text{auth}}$-hybrid model), a functionality $\mathcal{G}$ and a shielded oracle $\mathcal{O}$ s.t. $\rho \geq_{\mathcal{G}^{\mathcal{O}}} \mathcal{G}^{\mathcal{O}}$ but no imaginary angel Γ can be found such that $\rho \geq_{\Gamma\text{-Angel}} \mathcal{G}$ holds.*

We give a proof sketch below. See the full version for a more formal treatment.

Proof (Idea of proof)

1. We consider the dummy adversary $\mathcal{D}$ only. Since $\pi \geq_{\Gamma\text{-Angel}} \mathcal{F}$ we have

$$\exists \mathcal{S}_{\mathcal{D}}^{\Gamma} \; \forall \mathcal{Z}^{\Gamma} : \text{Exec}(\pi, \mathcal{D}^{\Gamma}, \mathcal{Z}^{\Gamma}) \equiv \text{Exec}(\mathcal{F}, \mathcal{S}_{\mathcal{D}}^{\Gamma}, \mathcal{Z}^{\Gamma}) \tag{8}$$

Now, we consider the experiment with shielded oracle $\mathcal{O} = \mathcal{S}_{\mathcal{D}}^{\Gamma}$, ideal functionality $\mathcal{F}^{\mathcal{O}}$ and simulator $\mathcal{S} = \mathcal{S}_{\mathcal{D}}$. Note that the code of $\mathcal{S}_{\mathcal{D}}$ is executed twice: by $\mathcal{O}$ and by $\mathcal{S}$. As Γ is assumed to be session-respecting the operation of the Angel is split between $\mathcal{O}$, that internally runs a copy of the Angel for all queries within the challenge session, and the simulator $\mathcal{S}$, that handles all remaining queries having access to the global Angel Γ. It follows

$$\text{Exec}(\mathcal{F}, \mathcal{S}_{\mathcal{D}}^{\Gamma}, \mathcal{Z}^{\Gamma}) \equiv \text{Exec}(\mathcal{F}^{\mathcal{O}}, \mathcal{S}^{\Gamma}, \mathcal{Z}^{\Gamma}) \tag{9}$$

In order to prove $\pi \geq_{\mathcal{F}^{\mathcal{O}}} \mathcal{F}^{\mathcal{O}}$ we need to show

$$\exists \mathcal{S} \; \forall \mathcal{Z} : \mathrm{Exec}(\pi, \mathcal{D}, \mathcal{Z}[\mathcal{F}^{\mathcal{O}}]) \equiv \mathrm{Exec}(\mathcal{F}^{\mathcal{O}}, \mathcal{S}, \mathcal{Z}[\mathcal{F}^{\mathcal{O}}]) \tag{10}$$

and we claim that $\mathcal{S}$ from above suffices. Assume that (10) does not hold, i.e. there is a $\mathcal{Z}[\mathcal{F}^{\mathcal{O}}]$ that can distinguish between interacting with π and $\mathcal{D}$ or with $\mathcal{F}^{\mathcal{O}}$ and $\mathcal{S}$. Then there exists an environment $\mathcal{Z}^{\Gamma}$ that internally runs $\mathcal{F}^{\mathcal{O}}$ simulating all augmented $\mathcal{F}^{\mathcal{O}}$-sessions by means of the global Γ and thus contradicts (9).

2. Let $\tilde{\rho}$ be a commitment protocol such that $\tilde{\rho} \geq_{\mathcal{F}_{\mathrm{com}}^{\mathcal{O}}} \mathcal{F}_{\mathrm{com}}^{\mathcal{O}}$ and $\mathcal{O}$ adjoined to $\mathcal{F}_{\mathrm{com}}$ is poly-simulatable. One can find such a protocol using the Angel-based protocol in [CLP10], part 1 of this theorem and Lemma 14, assuming the existence of one-way functions. Define the protocol ρ to be identical to $\tilde{\rho}$ except for the following instruction:

Before the actual commit phase begins, the receiver chooses $a_1, \ldots, a_n$ uniformly at random (n is the security parameter) and sends $\mathrm{Commit}(a_i)$ ($i = 1, \ldots, n$) to the sender (by running the program of the honest sender in $\tilde{\rho}$ with the pid of the sender). The sender replies with $(1, \ldots, 1) \in \{0,1\}^n$. The receiver then checks if the values he received from the sender equal $(a_1, \ldots, a_n)$. If yes, the receiver outputs "11" (2-bit string). Otherwise, the protocol parties execute the protocol $\tilde{\rho}$.

By construction, it holds that $\rho \geq_{\mathcal{F}_{\mathrm{com}}^{\mathcal{O}}} \mathcal{F}_{\mathrm{com}}^{\mathcal{O}}$. This follows from the fact that every $\mathcal{F}^{\mathcal{O}}$-augmented environment can be replaced by an efficient environment (since $\mathcal{O}$ attached to $\mathcal{F}$ is polynomially simulatable) and efficient environments can guess the correct a_i only with negligible probability (otherwise $\tilde{\rho}$ would be insecure, contradicting $\tilde{\rho} \geq_{\mathcal{F}_{\mathrm{com}}^{\mathcal{O}}} \mathcal{F}_{\mathrm{com}}^{\mathcal{O}}$).

Assume for the sake of contradiction that there exists an imaginary angel Γ s.t. $\rho \geq_{\Gamma\text{-Angel}} \mathcal{F}_{\mathrm{com}}$ holds. Let the sender be corrupted. Since the adversary has access to Γ, he can run the program of the simulator. The simulator must be able to extract commitments (because $\rho \geq_{\Gamma\text{-Angel}} \mathcal{F}_{\mathrm{com}}$). This enables the adversary to extract all a_i (by relaying the commitments from the receiver each to a different internal copy of the simulator), forcing the receiver to output "11" in the real model experiment. This cannot be simulated in the ideal model experiment, however. We have thus reached a contradiction.   □

Theorem 17 raises the question if it is possible to construct secure protocols with "interesting properties" in our framework that are not (known to be) secure in the Angel-based setting. We will answer this question in the affirmative, presenting a modular construction of a general MPC protocol in the plain model that is constant-round (and black-box) and based only on standard polynomial-time hardness assumptions (Theorem 31).

We would like to briefly note that by Theorem 17 we can already conclude that we can realize every (well-formed) functionality in our framework by importing the results of [CLP10].

Proposition 18 (General MPC in the plain model). *Assume the existence of enhanced trapdoor permutations. For every (well-formed)[1] functionality $\mathcal{F}$, there exists an extraction oracle $\mathcal{O}$ and a protocol ρ (in the plain model[2]) such that*

$$\rho \underset{\mathcal{F}^{\mathcal{O}}}{\geq} \mathcal{F}^{\mathcal{O}} \tag{11}$$

4 A Constant-Round Commitment Scheme

In this section we will construct a constant-round commitment scheme that is secure in our framework. We note that we assume authenticated channels and implicitly work in the $\mathcal{F}_{\text{auth}}$-hybrid model.

Let $\langle C, R \rangle$ be a commitment scheme that we will use a building block for our bit commitment scheme Π later. We require $\langle C, R \rangle$ to be tag-based. In a tag-based commitment scheme the committer and receiver additionally use a "tag"—or identity—as part of the protocol [PR05, DDN00]. Moreover we require $\langle C, R \rangle$ to be "immediately committing" as in the following definition.

Definition 19 (Immediately committing). *A commitment scheme $\langle C, R \rangle$ is called* immediately committing *if the first message in the protocol comes from the sender and already perfectly determines the value committed to.*

The above definition implies that the commitment scheme is perfectly binding and super-polynomially extractable, i.e., given the transcript an extractor can find the unique message of the commitment by exhaustive search.

For the discussion of our commitment scheme, we settle the following notation. Let $s = ((s_{i,b})) \in \{0,1\}^{2n}$ for $i \in [n]$ and $b \in \{0,1\}$ be a $2n$-tuple of bits. For an n-bit string $I = b_1 \cdots b_n$, we define $s_I := (s_{1,b_1}, \ldots, s_{n,b_n})$. Thus I specifies a selection of n of the $s_{i,b}$, where one of these is selected from each pair $s_{i,0}, s_{i,1}$.

Construction 1. *The bit commitment scheme Π is defined as follows. Whenever the basic commitment scheme $\langle C, R \rangle$ is used, the committing party uses its pid and the sid as its tag. Let $m \in \{0,1\}$*

- Commit(m):
  - R: *Choose a random n-bit string I and commit to I using $\langle C, R \rangle$.*
  - S: *Pick n random bits $s_{i,0}$ and compute $s_{i,1} = s_{i,0} \oplus m$ for all $i \in [n]$.*
  - S and R *run $2n$ sessions of $\langle C, R \rangle$ in parallel in which S commits to the s_{i,b_i} ($i \in [n], b_i \in \{0,1\}$).*
- Unveil:
  - S: *Send all $s_{i,b_i} \in \{0,1\}$ ($i \in [n], b_i \in \{0,1\}$) to R.*
  - R: *Check if $s_{1,0} \oplus s_{1,1} = \ldots = s_{n,0} \oplus s_{n,1}$. If this holds, unveil the string I to S.*

---

[1] See [Can+02] for a definition of well-formed functionalities.

[2] A model without any trusted setup except for authenticated communication channels.

- S: *If* R *unveiled the string correctly, then unveil all* s_I.
- R: *Check if* S *unveiled correctly. If yes, let* $s'_1, \ldots, s'_n$ *be the unveiled values. Check if* $s'_i = s_{i,b_i}$ *for all* $i \in [n]$. *If so, output* $m := s_{1,0} \oplus s_{1,1}$.

The above construction is reminiscent of [DS13] who presented a compiler that transforms any ideal straight-line extractable commitment scheme into an extractable and equivocal commitment scheme.

Note that if an attacker is able to learn the index set I in the commit phase then he can easily open the commitment to an arbitrary message m' by sending "fake" shares $t_{i,b}$, such that $t_I = s_I$, and $t_{\neg I} = s_I \oplus (m', \ldots, m')$. (Here $\oplus$ is interpreted element-wise.) Hence Π is equivocal for super-polynomial machines.

We claim that this protocol securely realizes $\mathcal{F}_{\text{com}}^{\mathcal{O}}$ for a certain shielded oracle $\mathcal{O}$. We first describe $\mathcal{O}$, before we move to the theorem.

Construction 2. *We define the actions of the shielded oracle $\mathcal{O}$ as follows.*[3]
If the sender is corrupted

- $\mathcal{O}$ *chooses a random n-bit string I, and commits to the string I to the adversary $\mathcal{A}$ using $\langle C, R \rangle$.*
- $\mathcal{O}$ *acts as honest receiver in $2n$ sessions of $\langle C, R \rangle$ in parallel. After these sessions have completed, $\mathcal{O}$ extracts each instance of $\langle C, R \rangle$, obtaining the shares $(s_{i,b}$ for $i \in [n]$) and $b \in \{0, 1\}$. (If a commitment cannot be extracted, the corresponding share is set to $\bot$.)*
- $\mathcal{O}$ *computes $m_i := s_{i,0} \oplus s_{i,1}$ for all $i \in [n]$. (Indices i where one or both of the $s_{i,b}$ is $\bot$ are ignored.) Let $m \in \{0, 1\}$ be the most frequently occurring m_i. (If there are multiple m_i occurring with the highest frequency, m chooses $m = 0$). $\mathcal{O}$ relays (Commit, m) to $\mathcal{F}_{com}$.*
- *When $\mathcal{A}$ sends shares $s'_{1,0}, s'_{1,1}, \ldots, s'_{n,0}, s'_{n,1}$ in the unveil phase of Π, $\mathcal{O}$ acts as an honest receiver, unveiling I.*
- *Finally, if $\mathcal{A}$'s unveil is accepting, $\mathcal{O}$ instructs $\mathcal{F}_{com}$ to unveil the message.*

If the receiver is corrupted

- $\mathcal{O}$ *acts as the sender in an execution of Π, engaging in a commit session of $\langle C, R \rangle$ with the adversary. If the adversary's commitment is accepting, $\mathcal{O}$ extracts this instance of $\langle C, R \rangle$ obtaining a string I (If parts of this string cannot be extracted they are set to $\bot$).*
- $\mathcal{O}$ *picks n random bits $s_{i,0}$, and lets $s_{i,1} = s_{i,0}$ for all $i \in [n]$, as if it were honestly committing to $m = 0$. Next, it runs $2n$ instances of Π in parallel, committing to the $s_{i,b}$.*
- *In the unveil phase, when $\mathcal{O}$ learns the message m, it computes "fake" shares $t_{i,b}$ as follows: $t_I = s_I$ and $t_{\neg I} = s_{\neg I} \oplus (m, \ldots, m)$ ($\oplus$ is interpreted element-wise.). $\mathcal{O}$ sends these shares $t_{i,b}$ to the adversary.*
- $\mathcal{O}$ *acts as the honest sender in the unveil phase of Π. If $\mathcal{A}$'s unveil of I is accepting, then $\mathcal{O}$ honestly executes the unveil phase for all bit shares t_I. (Otherwise, $\mathcal{O}$ outputs nothing and ignores all further inputs.)*

---

[3] For ease of notation, we drop the prefixes `output-to-fnct` and `output-to-adv` in the messages output by $\mathcal{O}$.

If no parties are corrupted, $\mathcal{O}$ *simulates an honest execution of protocol* Π *on input* 0, *forwarding all messages to the adversary. Since* $\mathcal{O}$ *knows the index string* I *(because* $\mathcal{O}$ *has created it itself) it can create fake shares just like in the case of a corrupted receiver.*

If both parties are corrupted, $\mathcal{O}$ *just executes the dummy adversary* $\mathcal{D}$ *internally. (Note that* $\mathcal{Z}$ *only interacts with* $\mathcal{D}$ *in the real experiment if both parties are corrupted).*

This concludes the description of the shielded oracle $\mathcal{O}$. Observe that $\mathcal{O}$ can be implemented in super-polynomial time. Also note that in the case of *both or no* party being corrupted, $\mathcal{O}$ can be implemented in polynomial time.

Before we can state our theorem, we need another assumption about the commitment scheme $\langle C, R \rangle$.

Definition 20 (pCCA-secure commitment schemes). *Let* $\langle C, R \rangle$ *be a tag-based commitment scheme. A pCCA-decommitment oracle* $\mathcal{E}$ *interacts with an adversary* $\mathcal{A}$ *in polynomial many parallel sessions of* $\langle C, R \rangle$ *as an honest receiver with tags chosen by the adversary. After all sessions have been completed successfully,* $\mathcal{E}$ *simultaneously reveals all committed values to* $\mathcal{A}$ *(note that when a session has multiple compatible committed values,* $\mathcal{E}$ *reveals only one of them. Hence, there might exist many decommitment oracles).*

Consider the probabilistic experiment $\text{IND}_b(\langle C, R \rangle, \mathcal{A}^{\mathcal{E}}, 1^n, z)$ *with* $b \in \{0, 1\}$:

On input 1^n *and auxiliary input* z, *the adversary* $\mathcal{A}$ *adaptively chooses a pair of challenge values* $v_0, v_1 \in \{0, 1\}$ *together with a tag and sends them to the challenger. The challenger commits to* v_b *using* $\langle C, R \rangle$ *with that tag. The output of the experiment is the output of* $\mathcal{A}^{\mathcal{E}}$. *If any of the tags used by* $\mathcal{A}$ *for queries to the pCCA-decommitment oracle equals the tag of the challenge, the output of the experiment is replaced by* $\perp$.

$\langle C, R \rangle$ *is said to be parallel-CCA-secure if there exists an* $\mathcal{E}$ *s.t. for all PPT adversaries* $\mathcal{A}$ *it holds that:*[4]

$$\text{IND}_0(\langle C, R \rangle, \mathcal{A}^{\mathcal{E}}, 1^n, z) \overset{c}{\equiv} \text{IND}_1(\langle C, R \rangle, \mathcal{A}^{\mathcal{E}}, 1^n, z)$$

Note that previous protocols proven secure in the Angel-based framework required (adaptive) CCA-secure commitments schemes [CLP10, Goy+15, Kiy14]. For our notion it suffices to assume parallel-CCA-secure (i.e. non-adaptive) commitment schemes as a building block.

Theorem 21. *Assume that* $\langle C, R \rangle$ *is parallel-CCA-secure and immediately committing. Then* $\Pi \geq_{\mathcal{F}_{com}^{\mathcal{O}}} \mathcal{F}_{com}^{\mathcal{O}}$, *where* Π *is as defined in Construction 1 and* $\mathcal{O}$ *is the shielded oracle as defined in Construction 2.*

Proof. By Proposition 6 it suffices to find a simulator for the dummy adversary. By construction of $\mathcal{O}$ the simulator in the ideal experiment can be chosen to be identical to the dummy adversary.

---

[4] In our special case the decommitment oracle $\mathcal{E}$ is unique since we assume an immediately committing commitment scheme.

The main idea of the proof is to consider a sequence of hybrid experiments for a PPT environment $\mathcal{Z}$ that may externally invoke polynomially many $\mathcal{F}_{com}^{\mathcal{O}}$-sessions and iteratively replace those sessions by the real protocol Π in a specific order utilizing the fact that the super-polynomial computations of $\mathcal{O}$ are hidden away and thus the replacements are unnoticeable by $\mathcal{Z}$, or otherwise we would obtain a PPT adversary against the hiding property of $\langle C, R \rangle$.

Step 1: Let $\mathcal{Z}$ be a PPT environment that may externally invoke polynomial many $\mathcal{F}_{com}^{\mathcal{O}}$-sessions. We denote the output of this experiment by the random variable $\text{Exec}(\mathcal{F}_{com}^{\mathcal{O}}, \mathcal{Z})$. Let $\text{Exec}(\Pi, \mathcal{Z})$ be the output of $\mathcal{Z}$ if all instances of $\mathcal{F}_{com}^{\mathcal{O}}$ sessions are replaced by the instances of the protocol Π. We show that for all environments $\mathcal{Z}$ it holds that

$$\text{Exec}(\mathcal{F}_{com}^{\mathcal{O}}, \mathcal{Z}) \overset{c}{\equiv} \text{Exec}(\Pi, \mathcal{Z}) \tag{12}$$

Let $\mathcal{Z}$ be an environment. By a standard averaging argument we can fix some random coins r for $\mathcal{Z}$. Thus we can assume henceforth that $\mathcal{Z}$ is deterministic.

We call instances of $\mathcal{F}_{com}^{\mathcal{O}}$ (or Π) where the sender or receiver is corrupted *sender sessions* or *receiver sessions*, respectively. Since in the cases where both or no party is corrupted, the $\mathcal{O}$-adjoinded functionalities in this case can be treated as part of the environment. We therefore only need to consider $\mathcal{F}^{\mathcal{O}}$-augmented environments that only invoke either sender sessions or receiver sessions.

We say a *discrepancy* occurred, if in any ideal sender session of $\mathcal{F}_{com}^{\mathcal{O}}$ $\mathcal{O}$ extracts a value m, but later $\mathcal{Z}$ correctly unveils a value $m' \neq m$. First notice that unless a discrepancy happens, the output of an ideal sender session is identically distributed to the output of the real protocol Π.

We will now distinguish two cases.

1. The probability that $\mathcal{Z}$ causes a discrepancy is negligible.
2. The probability that $\mathcal{Z}$ causes a discrepancy is non-negligible.

Case 1: We replace all sender sessions with instances of Π, incurring only a negligible statistical distance. We are left with a hybrid experiment in which only the receiver sessions are still ideal. We will now iteratively replace ideal receiver sessions with the real protocol, beginning with the *last* session that is started.

Assume that there are at most q receiver sessions. Define hybrids $H_0, \ldots, H_q$ as follows. Hybrid H_i is the experiment where the first i receiver sessions are ideal and the remaining $q - i$ receiver sessions are replaced by instances of Π (in which the receiver is corrupted). Clearly, H_q is identical to the experiment where all receiver sessions are ideal, whereas H_0 is the experiment where all receiver sessions are real. The experiment H_i outputs whatever $\mathcal{Z}$ outputs. Let $P_i = \Pr[H_i = 1]$ denote the probability that $\mathcal{Z}$ outputs 1 in the hybrid game H_i. Assume now that $\epsilon := |P_0 - P_q|$ is non-negligible, i.e., $\mathcal{Z}$ has non-negligible advantage ϵ in distinguishing H_0 from H_q. We will now construct an adversary $\mathcal{A}_\Pi$ that breaks the hiding property of Π with advantage ϵ/q.

By the averaging principle, there must exist an index $i^* \in [q]$ such that $|P_{i^*-1} - P_{i^*}| \geq \epsilon/q$. By a standard coin-fixing argument, we can fix the coins selected by the $\mathcal{O}$-instances inside the first $i^* - 1$ (ideal) receiver sessions. Fixing these coins maintains $\mathcal{Z}$'s distinguishing advantage. Since we fixed the coins of $\mathcal{Z}$ before, the experiment is now deterministic until the start of receiver session i^*. Since $\mathcal{Z}$ is fully deterministic up until this point, the first message of $\mathcal{Z}$ in session i^*, which is a commitment on the bit string I, is also computed deterministically.

We can now construct the non-uniform adversary $\mathcal{A}$ against the hiding property of $\langle C, R \rangle$. (We note that we do not construct an adversary $\mathcal{A}$ for the standard hiding game but for a multi-instance variant.) As a non-uniform advice, $\mathcal{A}$ receives a complete trace of all messages sent until this point. This includes all bit strings $I_1, \ldots, I_{i^*}$ to which $\mathcal{Z}$ committed to in all receiver sessions $1, \ldots, i^*$ (it also includes $\mathcal{Z}$'s input). Note that all messages come from a deterministic process, and the corresponding I_i are uniquely determined by the first messages of each session i since $\langle C, R \rangle$ is immediately committing.

$\mathcal{A}$ now proceeds as follows. $\mathcal{A}$ internally simulates $\mathcal{Z}$ and all sessions invoked by $\mathcal{Z}$. This simulation can be done in *polynomial time*, since all sender sessions and the subsequent receiver sessions $i^* + 1$ through q have been replaced by instances of Π, and $\mathcal{A}$ knows the index strings I_i that are used in the (ideal) receiver sessions 1 through i^*.

Let m^* be the message that $\mathcal{Z}$ chooses as input for the sender in session i^*. $\mathcal{A}$ reads $I \stackrel{\text{def}}{=} I_{i^*}$ from its non-uniform advice and samples a tuple s_I of n random strings. It then computes $s_{\neg I} = s_I \oplus (m^*, \ldots, m^*)$ and $s'_{\neg I} = s_I$ for all $i \in [n]$. $\mathcal{A}$ sends the messages $(s_{\neg I}, s'_{\neg I})$ to the hiding experiment. It now forwards all the messages between the hiding experiment and $\mathcal{Z}$ and simultaneously commits honestly on all values s_I to $\mathcal{Z}$. When $\mathcal{Z}$ requires that the commitments for all s_I be opened, $\mathcal{A}$ honestly unveils these. When $\mathcal{Z}$ terminates, $\mathcal{A}$ outputs whatever $\mathcal{Z}$ output in the experiment. This concludes the description of $\mathcal{A}$.

We will now analyze $\mathcal{A}$'s advantage. If the challenger of the hiding game picks the messages $s'_{\neg I}$, $\mathcal{Z}$ obtains a commitment on the all-zero string in $\mathcal{A}$'s simulation. Therefore, in this case the view of $\mathcal{Z}$ is distributed identically to the view inside the hybrid H_{i^*}. If the challenger of the hiding game picks the messages $s_{\neg I}$, $\mathcal{Z}$ obtains a commitment to the message m which is identical to the view of $\mathcal{Z}$ inside the hybrid H_{i^*-1}. It follows

$$\mathsf{Adv}(\mathcal{A}) = |\Pr[\mathsf{H}_{i^*} = 1] - \Pr[\mathsf{H}_{i^*-1} = 1]| = |P_{i^*} - P_{i^*-1}| \geq \epsilon/q, \qquad (13)$$

i.e. $\mathcal{A}$ breaks the hiding property of protocol $\langle C, R \rangle$ with advantage ϵ/q, which concludes case 1. (Note that in this case $\mathcal{A}$ does not need the pCCA oracle.)

Case 2: We now turn to case 2. A first observation is that we only need to consider augmented environments that invoke exactly *one* external session where the sender is corrupted. This is because if a (general) environment $\mathcal{Z}$ causes a discrepancy with non- negligible probability, then there exists a session j^* in which a discrepancy happens *for the first time*. An environment $\mathcal{Z}'$ that invokes only one session where the sender is corrupted can then simulate $\mathcal{Z}$, guess j^*

and simulate all the other sessions where the sender is corrupted with the real protocol. It holds that $\mathcal{Z}'$ also causes a discrepancy with non-negligible probability.

So we henceforth assume that $\mathcal{Z}$ invokes at most q sessions and only one session where the sender is corrupted. In what follows, we will replace all ideal sessions where the receiver is corrupted with real protocols using the same strategy as in case 1. Define the hybrids $H_0, \ldots, H_q$ as in case 1 except that now $\mathcal{Z}$ can additionally invoke exactly one sender session in all these hybrids. Clearly, H_q is identical to the experiment where all sessions are ideal, whereas H_0 is the experiment where all receiver sessions are real. Let $P_i = \Pr[H_i = 1]$ again.

Assume now that $\mathcal{Z}$ can distinguish between H_0 and H_q with non-negligible advantage ϵ. Then there exists an index $i^* \in [q]$ such that $|P_{i^*-1} - P_{i^*}| \geq \epsilon/q$. We can now fix the coins that are used in the first $i^* - 1$ ideal sessions until the point where session i^* starts, while maintaining $\mathcal{Z}$'s distinguishing advantage.

We will construct a non-uniform adversary $\mathcal{A}'$ that breaks the parallel-cca-security of $\langle C, R \rangle$ with advantage ϵ/q. As in case 1, $\mathcal{A}'$ receives as a non-uniform advice a trace of a run of $\mathcal{Z}$ which also includes all index sets I_i to which $\mathcal{Z}$ committed in all sessions until session i^* and possibly the shares to which $\mathcal{Z}$ committed in the only sender-session (again, it also includes $\mathcal{Z}'$s input).

$\mathcal{A}'$ now proceeds the same way as in case 1. It internally runs $\mathcal{Z}$ and simulates either hybrid H_{i^*-1} or H_{i^*} for $\mathcal{Z}$ by embedding the challenge of the hiding game into the simulated session i^*. The adversary $\mathcal{A}'$ simulates all ideal receiver sessions for $i \leq i^*$ with the help of its advice while all subsequent *receiver* sessions for $i > i^*$ have already been replaced by Π. If $\mathcal{Z}$ has already started to commit to the shares in the only sender session then (by definition) these shares are also part of $\mathcal{A}'$'s advice and $\mathcal{A}'$ can simulate the sender session. (Note that $\langle C, R \rangle$ is immediately committing, hence the first message of (the parallel executions of) $\langle C, R \rangle$ uniquely determines the shares). If $\mathcal{Z}$ has not yet started to commit to the shares in the sender session then $\mathcal{A}'$ can use its parallel-cca oracle to extract them by forwarding the corresponding messages between the oracle and $\mathcal{Z}$. After the experiment terminates, $\mathcal{A}'$ outputs whatever $\mathcal{Z}$ outputs.

The analysis of $\mathcal{A}'$ is the same as in case 1 and we end up with the conclusion that $\mathcal{A}'$ breaks the parallel-cca-security of protocol $\langle C, R \rangle$ with advantage ϵ/q.

Hence, it remains to consider environments that invoke exactly one sender-session (all receiver sessions are real and hence can be treated as part of the environment). Assume that such an environment $\mathcal{Z}$ causes a discrepancy with non-negligible probability ϵ'.

We will now construct a non-uniform adversary $\mathcal{A}''$ that breaks the hiding property of the commitment scheme $\langle C, R \rangle$. $\mathcal{A}''$ takes part in a partial one-way hiding experiment where the challenger picks a random string $I = b_1 \cdots b_n$ and commits to this string using the commitment scheme $\langle C, R \rangle$. $\mathcal{A}''$ then sends a vector $(a_1, \ldots, a_n)$ to the experiment where $a_l \in \{0, 1, \bot\}$. Let $M = \{l \mid a_l \neq \bot\}$. $\mathcal{A}''$ wins if $\mathrm{card}(M) \geq n/2$ and $a_l = b_l$ for all $l \in M$. It holds that since $\langle C, R \rangle$ is hiding, $\mathcal{A}''$ can win this experiment only with negligible probability.

$\mathcal{A}''$ receives as non-uniform advice the input of $\mathcal{Z}$. $\mathcal{A}''$ now proceeds as follows: $\mathcal{A}''$ forwards the commitment it receives in the experiment to $\mathcal{Z}$ as in the commit phase of the one sender session that $\mathcal{Z}$ can invoke. When $\mathcal{Z}$ sends the commitments on the shares $s_{l,b}$, $\mathcal{A}''$ forwards them to its parallel-CCA-oracle, thus learning the values $s_{l,b}$ that $\mathcal{Z}$ committed to. $\mathcal{A}$ can now simulate the oracle $\mathcal{O}$ and reconstruct the message m defined by these shares (by defining m to be the most frequent value that occurs in $\{s_{i,0} \oplus s_{i,1}\}_{i\in[n]}$ just like $\mathcal{O}$). When $\mathcal{Z}$ sends the shares $s'_{l,b}$ in the unveil phase of the sender session, $\mathcal{A}''$ compares them to the originally extracted shares $s_{l,b}$ and defines the vector $(a_1, \ldots, a_n)$ as

$$a_l := \begin{cases} b_l & \text{if } \exists\, b_l \in \{0,1\} : s_{l,b_l} = s'_{l,b_l} \wedge s_{l,\neg b_l} \neq s'_{l,\neg b_l} \quad (\star) \\ \bot & \text{else (if no such } b_i \text{ exists)} \end{cases} \qquad (14)$$

and sends $(a_1, \ldots, a_n)$ to the experiment.

We will now analyze $\mathcal{A}'''$s success probability. Let M be the set of indices l for that condition $(\star)$ holds. If $\mathcal{Z}$ causes a discrepancy, it holds that all tuples of shares $(s'_{l,0}, s'_{l,1})$ define the same but different message $m' \neq m$ than the majority of the original shares $(s_{l,0}, s_{l,1})$, i.e. card$(M) \geq n/2$. Moreover, for each $l \in M$ b_l equals the lth bit of I. Hence, by construction, $\mathcal{A}''$ wins with non- negligible probability if $\mathcal{Z}$ causes a discrepancy with non-negligible probability.

Step 2: We will now prove that for every $\mathcal{F}^{\mathcal{O}}$-augmented environment

$$\text{Exec}\big(\Pi, \mathcal{D}, \mathcal{Z}[\mathcal{F}^{\mathcal{O}}_{\text{com}}]\big) \stackrel{c}{\equiv} \text{Exec}\big(\mathcal{F}^{\mathcal{O}}_{\text{com}}, \mathcal{D}, \mathcal{Z}[\mathcal{F}^{\mathcal{O}}_{\text{com}}]\big).$$

If the *sender is corrupted* then nothing needs to be shown, as in this case the real and ideal experiment are statistically close. This follows from the fact that by step 1, case 2, an $\mathcal{F}^{\mathcal{O}}_{\text{com}}$-augmented environment can cause a discrepancy only with negligible probability.

If the *receiver is corrupted* then by step 1 the real and ideal experiment are both indistinguishable to an experiment where all instances of $\mathcal{F}^{\mathcal{O}}_{\text{com}}$ invoked by the environment have been replaced by the real protocol. Hence the outputs of the real and ideal experiment are indistinguishable.

If *no party is corrupted* then one can first replace all sender sessions and receiver sessions with the real protocol using step 1, obtaining a polynomial time environment. Then one can prove indistinguishability by using a very similar reduction to the hiding property as in step 1, case 1.

If *both parties are corrupted* then the real and ideal experiment are identically distributed. $\square$

The premise of Theorem 21 can be further relaxed by using only a *weakly* pCCA oracle instead of a standard pCCA oracle. A weakly pCCA oracle returns $\bot$ everywhere in case that at least one commitment is not accepting. Weakly pCCA suffices because a shielded oracle in a sender session (acting as the honest receiver) aborts if at least one commitment is not accepting in the commit phase.

The underlying commitment scheme $\langle C, R \rangle$ can be instantiated with the 8-round construction in [Goy+14]. It is straightforward to see that this scheme is pCCA secure by using the extractor in its security proof. The Zero-Knowledge Argument of Knowledge inside [Goy+14] is instantiated with the Feige-Shamir protocol [FS90] and—deviating from the original work —the basic commitment scheme is instantiated by the Blum commitment [Blu81] because we require an immediately committing protocol. Since this scheme is constant-round, we obtain the following result:

Corollary 22. *Assume the existence of one-way permutations. Then there is a constant-round protocol Π_{com} and a shielded oracle $\mathcal{O}$ such that $\Pi_{com} \geq_{\mathcal{F}_{com}^{\mathcal{O}}} \mathcal{F}_{com}^{\mathcal{O}}$.*

The above construction is non-black-box since [Goy+14] (instantiated this way) is non-black-box. However, recall that the only non-black-box part of [Goy+14] is a ZK proof for proving knowledge of committed values and that these values satisfy linear relations. As already pointed out in [Goy+14], this can both be done making only black-box use of a homomorphic commitment scheme. Instantiating [Goy+14] with a perfectly binding homomorphic commitment scheme thus yields a fully black-box construction. Since we need an immediately committing scheme in the plain model for our protocol we let the sender (and not a trusted setup) generate the commitment key of the homomorphic commitment. This construction can be used as a building block in [Goy+14] if the homomorphic commitment scheme is "verifiable". A verifiable homomorphic commitment scheme allows one to (non-interactively) verify that a commitment key is well-formed. For instance, the ElGamal commitment scheme [ElG84] (which is based on the DDH assumption) is a verifiable perfectly binding homomorphic commitment scheme [AIR01]. The Linear Encryption scheme [BBS04] (which is based on the DLin assumption) can also be viewed as a commitment scheme with these properties.

Corollary 23. *Assume the existence of verifiable perfectly binding homomorphic commitment schemes. Then there exists a constant-round black-box protocol Π_{com}^{BB} and a shielded oracle $\mathcal{O}$ such that $\Pi_{com}^{BB} \geq_{\mathcal{F}_{com}^{\mathcal{O}}} \mathcal{F}_{com}^{\mathcal{O}}$.*

5   A Modular Composition Theorem for Π

We show that we can plug the protocol Π from Construction 1 into a large class of UC-secure protocols in the $\mathcal{F}_{com}$-hybrid model in such a way that the composite protocol is secure in our framework. We first define Commit-Compute protocols and parallel-CCA-UC-emulation.

Definition 24 (Commit-Compute protocols). *Let $\rho^{\mathcal{F}_{com}}$ be a protocol in the $\mathcal{F}_{com}$-hybrid model. We call $\rho^{\mathcal{F}_{com}}$ a commit-compute protocol or CC protocol if it can be broken down into two phases: An initial commit phase, where the only communication allowed is sending messages to instances of $\mathcal{F}_{com}$. After the commit phase is over, a compute phase begins where sending messages to instances of $\mathcal{F}_{com}$ except for unveil-messages is prohibited, but all other communication is allowed.*

Definition 25 (pCCA-UC-emulation). *We write $\rho \geq_{\mathcal{E}\text{-}pCCA} \phi$ if a protocol ρ UC-emulates a protocol ϕ in the presence of (non-uniform) environments that may interact with a pCCA-decommitment oracle $\mathcal{E}$ as defined in Definition 20 for tags that are not extensions of the session identifier of the challenge protocol.*

In the following, let Π be the protocol as in Construction 1 with an immediately committing and parallel-CCA secure commitment scheme $\langle C, R \rangle$. Let $\mathcal{E}$ be the (uniquely defined) pCCA-decommitment oracle of $\langle C, R \rangle$.

We are now ready to state the theorem:

Theorem 26. *Let $\rho^{\mathcal{F}_{com}}$ be a CC protocol and $\mathcal{G}$ a functionality. If $\rho^{\mathcal{F}_{com}} \geq_{\mathcal{E}\text{-}pCCA} \mathcal{G}$ then there exists a shielded oracle $\mathcal{O}'$ such that*

$$\rho^{\Pi} \underset{\mathcal{G}^{\mathcal{O}'}}{\geq} \mathcal{G}^{\mathcal{O}'}$$

Proof. Since $\rho^{\mathcal{F}_{com}} \geq_{\mathcal{E}\text{-}pCCA} \mathcal{G}$ there exists a dummy adversary simulator $\mathcal{S}_D$. Let $\mathcal{O}$ be the shielded oracle from Construction 2, s.t. $\Pi \geq_{\mathcal{F}_{com}^{\mathcal{O}}} \mathcal{F}_{com}^{\mathcal{O}}$. We define the shielded oracle $\mathcal{O}'$ as follows. $\mathcal{O}'$ internally simulates multiple instances of $\mathcal{O}$ (one for each instance of $\mathcal{F}_{com}$ in ρ) and $\mathcal{S}_D$, and forwards messages as follows.

- Messages from the adversary addressed to an instance of $\mathcal{F}_{com}$ are forwarded to the corresponding internal instance of $\mathcal{O}$.
- Messages from an internal instance of $\mathcal{O}$ to an instance of $\mathcal{F}_{com}$ are forwarded to the dummy adversary simulator $\mathcal{S}_D$.
- Messages between $\mathcal{S}_D$ and the functionality $\mathcal{G}$ are forwarded.
- Messages from the dummy adversary simulator $\mathcal{S}_D$ addressed as coming from an instance of $\mathcal{F}_{com}$ are forwarded to the respective instance of $\mathcal{O}$.
- Messages from the dummy adversary simulator $\mathcal{S}_D$ not addressed as coming from an instance of $\mathcal{F}_{com}$ are output to the adversary (without forwarding them to an internal instance of $\mathcal{O}$).

We claim that for this oracle $\rho^{\Pi} \geq_{\mathcal{G}^{\mathcal{O}'}} \mathcal{G}^{\mathcal{O}'}$ holds. By Proposition 6 it is sufficient to find a simulator for the dummy adversary. The simulator will be the dummy adversary in the ideal world.

Recall that we call instances of $\mathcal{F}_{com}^{\mathcal{O}}$ (or Π) where the sender or receiver is corrupted *sender sessions* or *receiver sessions*, respectively.

We denote by $\rho^{\Pi_S, \mathcal{F}_{com}^{\mathcal{O}}}$ the protocol $\rho^{\mathcal{F}_{com}}$ where all ideal sender sessions have been replaced by the real protocol. Let $\text{Exec}(\rho^{\Pi_S, \mathcal{F}_{com}^{\mathcal{O}}}, \mathcal{Z})$ denote an execution of an environment $\mathcal{Z}$ with (polynomially many) instances of $\rho^{\Pi_S, \mathcal{F}_{com}^{\mathcal{O}}}$. Furthermore, denote by $\text{Exec}(\mathcal{G}^{\mathcal{O}'}, \mathcal{Z})$ an execution of an environment $\mathcal{Z}$ where all instances of $\rho^{\Pi_S, \mathcal{F}_{com}^{\mathcal{O}}}$ have been replaced by instances of $\mathcal{G}^{\mathcal{O}'}$.

Let $\mathcal{Z}$ be an environment in the experiment $\text{Exec}(\rho^{\Pi_S, \mathcal{F}_{com}^{\mathcal{O}}}, \mathcal{Z})$. By a standard averaging argument we can fix some random coins r for $\mathcal{Z}$. Thus we can assume henceforth that $\mathcal{Z}$ is deterministic.

In the following hybrid argument, we will have to globally order the main sessions by the *ending* of their commit-phase and (adaptively) invoke instances of $\rho^{\Pi_S, \mathcal{F}^O_{com}}$, $\rho^{\mathcal{F}^O_{com}}$ or $\mathcal{G}^{O'}$ based on this order. Since the message scheduling may be random, however, this order is not determined a-priori.

In the following, we will therefore have the experiment in the hybrids implement the commit-phases of all invoked protocols "obliviously", i.e., interact with the environment by running the programs of the shielded oracles and store the inputs of the honest parties without following their instructions in the commit-phases. Note that the only communication that is *visible* to the environment in the commit-phase is its interaction with the shielded oracles or the receiver in an instance of Π_S. The latter interaction is identical to an interaction with the shielded oracle in a sender session. Each time the adversary commits to a value, this value is extracted (by a super-polynomial computation) and stored. Note that the inputs of the honest parties have no effect on the messages the shielded oracles output to the adversary in the commit phase.

Once the commit phases of an instance of $\rho^{\Pi_S, \mathcal{F}^O_{com}}$ has ended, the experiment in the hybrids will invoke an instance of $\rho^{\Pi_S, \mathcal{F}^O_{com}}$, $\rho^{\mathcal{F}^O_{com}}$ or $\mathcal{G}^{O'}$ depending on the position within the global order of sessions. The experiment will then invoke the honest parties with their respective inputs and follow their instructions (it will also invoke the simulator $\mathcal{S}_D$ with the extracted values if this session is $\mathcal{G}^{O'}$). Messages from $\mathcal{F}^O_{com}$ or $\mathcal{S}_D$ to instances of $\mathcal{O}$ (which are "ok" messages) are suppressed. This way, the emulation is consistent with the messages in the commit phase and distributed identically as if one of the protocols $\mathcal{G}^{O'}$, $\rho^{\Pi_S, \mathcal{F}^O_{com}}$, or $\rho^{\mathcal{F}^O_{com}}$ was executed from the beginning.

Step 1. We show that

$$\text{Exec}(\rho^{\Pi_S, \mathcal{F}^O_{com}}, \mathcal{Z}) \stackrel{c}{\equiv} \text{Exec}(\mathcal{G}^{O'}, \mathcal{Z}) \tag{15}$$

Let $q(n)$ be an upper bound on the number of instances of $\rho^{\Pi_S, \mathcal{F}^O_{com}}$ that $\mathcal{Z}$ invokes. Consider the $2q(n) + 1$ hybrids $H_{00}, H_{01}, H_{10}, H_{11}, H_{20}, \ldots, H_{q(n)0}$ which are constructed as follows:

Definition of Hybrid H_{ij}: Execute the commit phases of each session "without running the code of the parties" by invoking instances of $\mathcal{O}$. Follow the instruction of each instance of $\mathcal{O}$. Parties are only there as placeholders for the environment in the commit phase. Their instructions will be execute after the commit phase of the respective session is over. Note that this can be done since the actions of the parties in the commit phase have no effect on the view of the environment in this phase. Messages output from an instance of $\mathcal{O}$ are stored as well. After the commit phase of a session is over do the following:

1. If this is the kth session in which the commit phase has ended and $k \leq i$ then invoke an instance of the dummy adversary simulator and the functionality $\mathcal{G}$. Hand the dummy parties their respective inputs and the dummy adversary simulator the messages output by the instances of $\mathcal{O}$. Follow the instructions of the dummy adversary simulator and $\mathcal{G}$. Ignore messages of the dummy adversary simulator to the environment if these messages are coming from an

instance of $\mathcal{F}_{\text{com}}$ in the commit phase (i.e. an "ok" message). In the unveil phase, messages from the dummy adversary simulator mimicking an interaction with $\mathcal{F}_{\text{com}}$ (which are messages of the form $(\texttt{unveil}, b)$) are forwarded to the respective instance of $\mathcal{O}$ Messages from the dummy adversary simulator not mimicking an interaction with an instance of $\mathcal{F}_{\text{com}}$ are output (without forwarding them to an internal instance of $\mathcal{O}$).

2. If $k = i + 1$ and $j = 0$ or $k > i + 1$ then run the protocol parties of $\rho^{\mathcal{F}_{\text{com}}}$ with their inputs and follow their instructions. For all subsessions where the sender is corrupted invoke instances Π_{S} and execute the commit phase of Π_{S} using the same randomness for the receiver as the respective oracle (do not pass the messages to the environment). For all subsessions where the receiver or both or no party has been corrupted invoke instances of $\mathcal{F}_{\text{com}}$ and adjoin the respective oracle. Send the outputs of the instances of $\mathcal{O}$ to the respective instances of $\mathcal{F}_{\text{com}}$. Ignore "ok" messages from the instances of $\mathcal{F}_{\text{com}}$.

3. If $k = i + 1$ and $j = 1$ then run the parties of $\rho^{\mathcal{F}_{\text{com}}}$ with their inputs in the commit phase and follow their instructions. For all subsessions invoke an instance of $\mathcal{F}_{\text{com}}$ and adjoin the respective oracle. Send the extracted committed values of the $\mathcal{O}$-instances in sender sessions to the respective $\mathcal{F}_{\text{com}}$-instance. Ignore "ok" messages from the instances of $\mathcal{F}_{\text{com}}$.

Observe that $\mathsf{H}_{00} = \text{Exec}(\rho^{\Pi_{\text{S}}, \mathcal{F}^{\mathcal{O}}_{\text{com}}}, \mathcal{Z})$ and $\mathsf{H}_{q(n)0} = \text{Exec}(\mathcal{G}^{\mathcal{O}'}, \mathcal{Z})$.

Let P_{ij} denote the probability that $\mathcal{Z}$ outputs 1 in hybrid H_{ij}. Assume $|P_{00} - P_{q(n)0}|$ is non-negligible. Then there exists an index i^* such that either $|P_{i^*1} - P_{(i^*+1)0}|$ or $|P_{i^*0} - P_{i^*1}|$ is also non-negligible.

Case 1: $|P_{i^*1} - P_{(i^*+1)0}|$ is non-negligible. In this case, these neighboring hybrids are equal except that in the $(i^* + 1)$th session $\rho^{\mathcal{F}^{\mathcal{O}}_{\text{com}}}$ is replaced by $\mathcal{G}^{\mathcal{O}'}$.

We fix the coins that are used in the experiment in all sessions until the point where the (i^*+1)th commit phase has ended, while maintaining $\mathcal{Z}$'s distinguishing advantage.

We can now construct an environment $\mathcal{Z}'$ that distinguishes $\rho^{\mathcal{F}_{\text{com}}}$ from $\mathcal{G}$. As a non-uniform advice, $\mathcal{Z}'$ receives a complete trace of all messages sent until this point, including all shares s_i and strings I that $\mathcal{Z}$ committed to until the point where the $(i^* + 1)$th commit phase has ended. $\mathcal{Z}'$ internally simulates the execution experiment with $\mathcal{Z}$ using its advice. Messages to the $(i^* + 1)$th session are sent to the challenge protocol. $\mathcal{Z}'$ may (tentatively) also invoke instances of $\mathcal{F}^{\mathcal{O}}_{\text{com}}$ in order to simulate the instances of $\mathcal{F}^{\mathcal{O}}_{\text{com}}$ that are invoked after the point where the $(i^* + 1)$th commit phase has ended.

Observe that the real execution corresponds to hybrid H_{i^*1} and the ideal execution to hybrid $\mathsf{H}_{(i^*+1)0}$. By construction, $\mathcal{Z}'$ distinguishes $\rho^{\mathcal{F}_{\text{com}}}$ from $\mathcal{G}$. Since $\mathcal{F}^{\mathcal{O}}_{\text{com}}$ is polynomially simulatable, $\mathcal{Z}'$ can be replaced by a polynomial time environment that also distinguishes $\rho^{\mathcal{F}_{\text{com}}}$ from $\mathcal{G}$, using Theorem 12. This is a contradiction (to the definition of the dummy adversary simulator).

Case 2: $|P_{i^*0} - P_{i^*1}|$ is non-negligible. In this case, these neighboring hybrids are equal except that in the $(i^* + 1)$th session $\rho^{\Pi_{\text{S}}, \mathcal{F}^{\mathcal{O}}_{\text{com}}}$ is replaced by $\rho^{\mathcal{F}^{\mathcal{O}}_{\text{com}}}$.

Since $\mathcal{Z}$ distinguishes these hybrids it holds that with non-negligible probability $\mathcal{Z}$ causes a *discrepancy* in hybrid H_{i*1} as otherwise these hybrids would be statistically close. Let $\widetilde{\mathcal{Z}}$ be the environment that internally runs $\mathcal{Z}$ and outputs 1 as soon as a discrepancy occurs.[5] By construction, $\widetilde{\mathcal{Z}}$ outputs 1 with non-negligible probability in H_{i*1}. We will now consider $i^* + 1$ new hybrids $\mathsf{h}_0, \ldots, \mathsf{h}_{i*}$.

Definition of Hybrid h_j: Execute the commit phases of each session "without running the code of the parties" as described in the description of the hybrids H_{ij}. After the commit phase of a session is over do the following (for a fixed $j \in \{0, \ldots, i^*\}$):

1. If $k \le i^* - j$ then invoke an instance of the dummy adversary simulator and the functionality $\mathcal{G}$. Hand the dummy parties their respective inputs and the dummy adversary simulator the messages output by the instances of $\mathcal{O}$. Follow the instructions of the dummy adversary simulator and $\mathcal{G}$. Ignore messages of the dummy adversary simulator to the environment if these messages are coming from an instance of $\mathcal{F}_{\mathrm{com}}$ in the commit phase (i.e. an "ok" message). In the unveil phase, messages from the dummy adversary simulator mimicking an interaction with $\mathcal{F}_{\mathrm{com}}$ (which are messages of the form $(\mathtt{unveil}, b)$) are forwarded to the respective instance of $\mathcal{O}$ (with the same SID). Messages from the dummy adversary simulator not mimicking an interaction with an instance of $\mathcal{F}_{\mathrm{com}}$ are output (without forwarding them to an internal instance of $\mathcal{O}$)
2. If this is the kth session in which the commit phase has ended and $i^* - j + 1 \le k \le i^* + 1$ then run the protocol parties of $\rho^{\mathcal{F}_{\mathrm{com}}}$ with their inputs in the commit phase and follow their instructions. For all subsessions where the receiver or both or no party is corrupted invoke instances of $\mathcal{F}_{\mathrm{com}}$ and adjoin the respective oracle Send the outputs of the instances of $\mathcal{O}$ to the respective instances of $\mathcal{F}_{\mathrm{com}}$. Ignore "ok" messages from the instances of $\mathcal{F}_{\mathrm{com}}$.
3. If $k \ge i^* + 2$ then run the protocol parties of $\rho^{\mathcal{F}_{\mathrm{com}}}$ with their inputs in the commit phase and follow their instructions. For all subsessions invoke an instance of $\mathcal{F}_{\mathrm{com}}$ and adjoin the respective oracle. Send the extracted committed values of the $\mathcal{O}$-instances in sender sessions to the respective $\mathcal{F}_{\mathrm{com}}$-instance. Ignore "ok" messages from the instances of $\mathcal{F}_{\mathrm{com}}$.

---

[5] To make the environment able to learn the committed value in a $\mathcal{F}_{\mathrm{com}}^{\mathcal{O}}$-hybrid protocol, we redefine the shielded oracle $\mathcal{O}$ for the case of a corrupted sender as follows: After the unveil phase is over, the oracle first outputs the extracted committed value to the simulator and after receiving a notification message from the simulator it sends an unveil message to the functionality. Denote this modified oracle by $\widetilde{\mathcal{O}}$. Furthermore define $\widetilde{\Pi}$ to be identical to Π, except that before outputting the committed value, the receiver sends the committed value to the sender. The sender then sends a notification message to the receiver, who then outputs the committed value. It follows from the exact same arguments as in the proof of Theorem 21 that $\widetilde{\Pi} \ge_{\mathcal{F}_{\mathrm{com}}^{\widetilde{\mathcal{O}}}} \mathcal{F}_{\mathrm{com}}^{\widetilde{\mathcal{O}}}$ and that $\widetilde{\mathcal{O}}$ adjoined to $\mathcal{F}_{\mathrm{com}}$ is polynomially simulatable. Using these modified versions in the above proof one obtains $\rho^{\widetilde{\Pi}} \ge_{\mathcal{G}^{\mathcal{O}'}} \mathcal{G}^{\mathcal{O}'}$. Since Π unconditionally emulates $\widetilde{\Pi}$ it holds that $\rho^{\Pi} \ge_{\mathcal{G}^{\mathcal{O}'}} \rho^{\widetilde{\Pi}}$, hence $\rho^{\Pi} \ge_{\mathcal{G}^{\mathcal{O}'}} \mathcal{G}^{\mathcal{O}'}$.

Observe that $h_0 = H_{i*_1}$. Let j^* be the *largest index* such that $\widetilde{\mathcal{Z}}$ causes a discrepancy in hybrid h_{j*} with non-negligible probability. (j^* is well-defined, since there is an index for which this property holds, namely 0). Furthermore, $j^* \leq i^* - 1$. This follows from the following argument. Observe that the last hybrid h_{i*} only contains instances of $\rho^{\mathcal{F}_{\text{com}}}$ (since all instance of $\mathcal{G}$ have been replaced). Because Π emulates $\mathcal{F}^{\mathcal{O}}_{\text{com}}$ and due to the composition theorem $\text{Exec}(\rho^\Pi, \mathcal{Z})$ is indistinguishable from h_{i*}. Since no discrepancy occurs in $\text{Exec}(\rho^\Pi, \mathcal{Z})$ it follows that a discrepancy can occur in h_{i*} only with negligible probability.

By construction, $\widetilde{\mathcal{Z}}$ distinguishes the hybrids h_{j*} and h_{j*+1} (in the first hybrid $\widetilde{\mathcal{Z}}$ outputs 1 with non-negligible probability and in the second hybrid only with negligible probability).

We will now modify these hybrids. For $k \in \{j^*, j^* + 1\}$ define the hybrid hyb_{k-j*} to be identical to h_k except for the following: At the beginning, the experiment randomly selects one sender session in one of the commit phases $1, \ldots, i^* + 1$. In all commit phases that end *after* the $(i^* - j^*)$th commit phase the real protocol Π_S is invoked instead of $\mathcal{F}^{\mathcal{O}s}$ in all sender sessions that have not been selected at the beginning. The one sender session that has been selected at the beginning always remains ideal.

It holds that $\widetilde{\mathcal{Z}}$ also distinguishes hyb_0 from hyb_1. This is because $\widetilde{\mathcal{Z}}$ still causes a discrepancy in hyb_0 with non-negligible probability because with high probability $(1/poly)$ the first session in which $\widetilde{\mathcal{Z}}$ causes a discrepancy is selected. Furthermore, $\widetilde{\mathcal{Z}}$ causes a discrepancy in hyb_1 only with negligible probability.

We fix the coins that are used in the experiment in all sessions until the point where the $(i^* - j^*)$th commit phase has ended, while maintaining $\widetilde{\mathcal{Z}}$'s distinguishing advantage.

We can now construct an environment $\mathcal{Z}''$ that distinguishes $\rho^{\mathcal{F}_{\text{com}}}$ from $\mathcal{G}$. As a non-uniform advice, $\mathcal{Z}''$ receives a complete trace of all messages sent until this point, including all shares s_l and index sets I that $\widetilde{\mathcal{Z}}$ committed to until the point where the $(i^* - j^*)$th commit phase has ended. $\mathcal{Z}''$ proceeds as follows: It internally simulates the execution experiment with $\widetilde{\mathcal{Z}}$ using its advice, randomly picking a sender session at the beginning. Messages to the $(i^* - j^*)$th session are sent to the challenge protocol. $\mathcal{Z}''$ can simulate the only instance of $\mathcal{F}^{\mathcal{O}s}$ that may occur in a commit phase with its pCCA-oracle $\mathcal{E}$. $\mathcal{Z}''$ may (tentatively) also invoke ideal receiver sessions in order to simulate ideal receiver sessions that are invoked after the point where the $(i^* - j^*)$th commit phase has ended.

Observe that the real execution corresponds to hybrid hyb_1 and the ideal execution to hybrid hyb_0. By construction, $\mathcal{Z}''$ distinguishes $\rho^{\mathcal{F}_{\text{com}}}$ from $\mathcal{G}$. With the same argument as in the proof of Theorem 21, step 1, case 2, one can replace all ideal receiver sessions that $\mathcal{Z}''$ invokes with instances of the real protocol. By construction, an environment $\mathcal{Z}''$ was found that can query the pCCA-oracle $\mathcal{E}$ and distinguish $\rho^{\mathcal{F}_{\text{com}}}$ and $\mathcal{D}$ from $\mathcal{G}$ and $\mathcal{S}_{\mathcal{D}}$. We have thus reached a contradiction.

Step 2. We show that $\rho^\Pi \geq_{\mathcal{G}^{\mathcal{O}'}} \mathcal{G}^{\mathcal{O}'}$, completing the proof.

Let $\mathcal{Z}$ be a $\mathcal{G}^{\mathcal{O}'}$-augmented environments. By step 1, we can replace all instances of $\mathcal{G}^{\mathcal{O}'}$ with instances of $\rho^{\Pi_S, \mathcal{F}_{\text{com}}^{\mathcal{O}}}$. Since Π emulates $\mathcal{F}^{\mathcal{O}}_{\text{com}}$, it follows from the composition theorem that we can replace (the challenge protocol) ρ^{Π} also with $\rho^{\Pi_S, \mathcal{F}^{\mathcal{O}}_{com}}$. Again by step 1, we can replace all instances of $\rho^{\Pi_S, \mathcal{F}^{\mathcal{O}}_{com}}$ back with instances of $\mathcal{G}^{\mathcal{O}'}$. The theorem follows.   □

If the following property holds for the commitment scheme $\langle C, R \rangle$, the premise $\rho^{\mathcal{F}_{com}} \geq_{\mathcal{E}\text{-pCCA}} \mathcal{G}$ is automatically fulfilled.

Definition 27 (r-non-adaptive robustness). *Let $\langle C, R \rangle$ be a tag-based commitment scheme and $\mathcal{E}$ a pCCA-decommitment oracle for it as in Definition 20. For $r \in \mathbb{N}$, we say that $\langle C, R \rangle$ is r-non-adaptively-robust w.r.t. $\mathcal{E}$ if for every PPT adversary $\mathcal{A}$, there exists a PPT simulator $\mathcal{S}$, such that for every PPT r-round interactive Turing machine $\mathcal{B}$, the following two ensembles are computationally indistinguishable:*

- $\{\langle \mathcal{B}(y), \mathcal{A}^{\mathcal{E}}(z)\rangle(1^n)\}_{n\in\mathbb{N}, y\in\{0,1\}^*, z\in\{0,1\}^*}$
- $\{\langle \mathcal{B}(y), \mathcal{S}(z)\rangle(1^n)\}_{n\in\mathbb{N}, y\in\{0,1\}^*, z\in\{0,1\}^*}$

The above definition is a weakening of the (adaptive) robustness property put forward by [CLP10].

Corollary 28. *If additionally the commitment scheme $\langle C, R \rangle$ in Π is r-non-adaptively-robust, then for every r-round CC protocol $\rho^{\mathcal{F}_{com}}$ it holds that if $\rho^{\mathcal{F}_{com}} \geq_{UC} \mathcal{G}$ then there exists a shielded oracle $\mathcal{O}'$ such that*

$$\rho^{\Pi} \underset{\mathcal{G}^{\mathcal{O}'}}{\geq} \mathcal{G}^{\mathcal{O}'}$$

Up to now we could instantiate $\langle C, R \rangle$ with a modified version of [Goy+14] as described above of Corollary 22. To additionally make this scheme r-non-adaptively-robust w.r.t. $\mathcal{E}$ one can add "redundant slots" using the idea of [LP09] (the scheme needs to have at least $r + 1$ slots to be r-non-adaptively-robust).

In the following lemma we show that every UC-secure protocol $\rho^{\mathcal{F}_{com}}$ can be transformed into a UC-secure CC protocol.

Lemma 29 (CC compiler). *Let $\rho^{\mathcal{F}_{com}}$ be a protocol in the $\mathcal{F}_{com}$-hybrid model. Then there exists a CC protocol $\mathrm{Comp}(\rho)^{\mathcal{F}_{com}}$ such that $\mathrm{Comp}(\rho)^{\mathcal{F}_{com}} \geq_{UC} \rho^{\mathcal{F}_{com}}$. Furthermore, if $\rho^{\mathcal{F}_{com}}$ is constant-round then so is $\mathrm{Comp}(\rho)^{\mathcal{F}_{com}}$.*

Proof (Idea of proof). Replace each instance of $\mathcal{F}_{com}$ with a *randomized commitment* where the sender commits to a bit b by sending a random value a to $\mathcal{F}_{com}$ and $a \oplus b$ to the receiver. Note that since the protocol is PPT the number of commitments of each party is polynomially bounded. Put all randomized calls to $\mathcal{F}_{com}$ in a single commit phase.   □

Let Π_r be the constant-round protocol as in Construction 1 where $\langle C, R \rangle$ is instantiated with the immediately committing, parallel-CCA secure and r-non-adaptively-robust modified version of [Goy+14] as described above. Furthermore, let Π_r^{BB} be the same as Π_r, except that [Goy+14] is instantiated with a verifiable perfectly binding homomorphic commitment scheme, thus making the construction fully black-box. Applying Corollary 28 and Lemma 29 one obtains the following:

Corollary 30. *Assume the existence of one-way permutations. Let $\rho^{\mathcal{F}_{com}}$ be a constant-round protocol and $\mathcal{G}$ a functionality. If $\rho^{\mathcal{F}_{com}} \geq_{UC} \mathcal{G}$ then there exists a shielded oracle $\mathcal{O}'$ such that for sufficiently large r it holds that*

$$\text{Comp}(\rho)^{\Pi_r} \underset{\mathcal{G}^{\mathcal{O}'}}{\geq} \mathcal{G}^{\mathcal{O}'}$$

Furthermore, assuming the existence of verifiable perfectly binding homomorphic commitment schemes, the same property holds for Π_r^{BB}.

6   Constant-Round (Black-Box) General MPC

We can now apply Corollary 30 to obtain a constant-round general MPC protocol based on standard polynomial-time hardness assumptions that is secure in our framework. [HV15] showed that for every well-formed functionality $\mathcal{F}$ there exists a constant-round protocol $\rho^{\mathcal{F}_{com}}$ that UC-emulates $\mathcal{F}$, assuming two-round semi-honest oblivious transfer. Plugging Π_r (for a sufficiently large r) into this protocol yields a constant-round general MPC protocol based on standard assumptions (e.g. enhanced trapdoor permutations). Furthermore, since the construction in [HV15] is black-box, plugging Π_r^{BB} into [HV15] yields a fully black-box construction of a constant-round general MPC protocol based on polynomial-time hardness assumptions that is secure in our framework.

Theorem 31 (Constant-round general MPC in the plain model)

(a) *Assume the existence of enhanced trapdoor permutations. Then for every well-formed functionality $\mathcal{F}$, there exists a constant-round protocol $\pi_{\mathcal{F}}$ (in the plain model) and a shielded oracle $\mathcal{O}$ such that*

$$\pi_{\mathcal{F}} \underset{\mathcal{F}^{\mathcal{O}}}{\geq} \mathcal{F}^{\mathcal{O}} \qquad (16)$$

(b) *Assume the existence of verifiable perfectly binding homomorphic commitment schemes and two-round semi-honest oblivious transfer.*
    Then for every well-formed functionality $\mathcal{F}$, there exists a constant-round protocol $\pi_{\mathcal{F}}^{BB}$ (in the plain model) and a shielded oracle $\mathcal{O}$ such that

$$\pi_{\mathcal{F}}^{BB} \underset{\mathcal{F}^{\mathcal{O}}}{\geq} \mathcal{F}^{\mathcal{O}} \qquad (17)$$

$\pi_{\mathcal{F}}^{BB}$ *uses the underlying homomorphic commitment scheme and oblivious transfer only in a black-box way.*

7   Conclusion

Shielded super-polynomial resources allow for general concurrent composition in the plain model while being compatible with UC security. As an application a secure constant-round (black-box) general MPC protocol was modularly designed and future work will be needed to make this proof of concept a general principle.

References

[AIR01] Aiello, B., Ishai, Y., Reingold, O.: Priced oblivious transfer: how to sell digital goods. In: Pfitzmann, B. (ed.) EUROCRYPT 2001. LNCS, vol. 2045, pp. 119–135. Springer, Heidelberg (2001). doi:10.1007/3-540-44987-6_8

[Bar+04] Barak, B., et al.: Universally composable protocols with relaxed set-up assumptions. In: 45th Annual IEEE Symposium on Foundations of Computer Science, FOCS 2004, pp. 186–195. IEEE (2004)

[BBS04] Boneh, D., Boyen, X., Shacham, H.: Short group signatures. In: Franklin, M. (ed.) CRYPTO 2004. LNCS, vol. 3152, pp. 41–55. Springer, Heidelberg (2004). doi:10.1007/978-3-540-28628-8_3

[Blu81] Blum, M.: Coin flipping by telephone. In: Advances in Cryptology, CRYPTO 1981: IEEE Workshop on Communications Security. University of California, Santa Barbara, Department of Electrical and Computer Engineering, pp. 11–15 (1981)

[BS05] Barak, B., Sahai, A.: How to play almost any mental game over the net - concurrent composition via super-polynomial simulation. In: 46th Annual IEEE Symposium on Foundations of Computer Science, FOCS 2005, pp. 543–552. IEEE (2005)

[Can+02] Canetti, R., et al.: Universally composable two-party and multiparty secure computation. In: Proceedings of the 34th Annual ACM Symposium on Theory of Computing, STOC 2002, pp. 494–503. ACM (2002)

[Can+07] Canetti, R., Dodis, Y., Pass, R., Walfish, S.: Universally composable security with global setup. In: Vadhan, S.P. (ed.) TCC 2007. LNCS, vol. 4392, pp. 61–85. Springer, Heidelberg (2007). doi:10.1007/978-3-540-70936-7_4

[Can01] Canetti, R.: Universally composable security: a new paradigm for cryptographic protocols. In: 42th Annual IEEE Symposium on Foundations of Computer Science. FOCS 2001, pp. 136–145. IEEE (2001)

[CF01] Canetti, R., Fischlin, M.: Universally composable commitments. In: Kilian, J. (ed.) CRYPTO 2001. LNCS, vol. 2139, pp. 19–40. Springer, Heidelberg (2001). doi:10.1007/3-540-44647-8_2

[CGJ15] Canetti, R., Goyal, V., Jain, A.: Concurrent secure computation with optimal query complexity. In: Gennaro, R., Robshaw, M. (eds.) CRYPTO 2015. LNCS, vol. 9216, pp. 43–62. Springer, Heidelberg (2015). doi:10.1007/978-3-662-48000-7_3

[CKL03] Canetti, R., Kushilevitz, E., Lindell, Y.: On the limitations of universally composable two-party computation without set-up assumptions. In: Biham, E. (ed.) EUROCRYPT 2003. LNCS, vol. 2656, pp. 68–86. Springer, Heidelberg (2003). doi:10.1007/3-540-39200-9_5

[CLP10] Canetti, R., Lin, H., Pass, R.: Adaptive hardness and composable security in the plain model from standard assumptions. In: 51st Annual IEEE Symposium on Foundations of Computer Science, FOCS 2010, pp. 541–550. IEEE (2010)

[CLP13] Canetti, R., Lin, H., Pass, R.: From unprovability to environmentally friendly protocols. In: 54th Annual IEEE Symposium on Foundations of Computer Science, FOCS 2013, pp. 70–79. IEEE (2013)

[CPS07] Canetti, R., Pass, R., Shelat, A.: Cryptography from sunspots: how to use an imperfect reference string. In: 48th Annual IEEE Symposium on Foundations of Computer Science. FOCS 2007, pp. 249–259. IEEE (2007)

[Dac+13] Dachman-Soled, D., Malkin, T., Raykova, M., Venkitasubramaniam, M.: Adaptive and concurrent secure computation from new adaptive, non-malleable commitments. In: Sako, K., Sarkar, P. (eds.) ASIACRYPT 2013. LNCS, vol. 8269, pp. 316–336. Springer, Heidelberg (2013). doi:10.1007/978-3-642-42033-7_17

[DDN00] Dolev, D., Dwork, C., Naor, M.: Nonmalleable cryptography. SIAM J. Comput. **30**(2), 391–437 (2000)

[DS13] Damgård, I., Scafuro, A.: Unconditionally secure and universally composable commitments from physical assumptions. In: Sako, K., Sarkar, P. (eds.) ASIACRYPT 2013. LNCS, vol. 8270, pp. 100–119. Springer, Heidelberg (2013). doi:10.1007/978-3-642-42045-0_6

[ElG84] ElGamal, T.: A public key cryptosystem and a signature scheme based on discrete logarithms. In: Blakley, G.R., Chaum, D. (eds.) CRYPTO 1984. LNCS, vol. 196, pp. 10–18. Springer, Heidelberg (1985). doi:10.1007/3-540-39568-7_2

[FS90] Feige, U., Shamir, A.: Witness indistinguishable and witness hiding protocols. In: Proceedings of the 22nd Annual ACM Symposium on Theory of Computing, STOC 1990, pp. 416–426. ACM (1990)

[Gar+12] Garg, S., Goyal, V., Jain, A., Sahai, A.: Concurrently secure computation in constant rounds. In: Pointcheval, D., Johansson, T. (eds.) EUROCRYPT 2012. LNCS, vol. 7237, pp. 99–116. Springer, Heidelberg (2012). doi:10.1007/978-3-642-29011-4_8

[GGJ13] Goyal, V., Gupta, D., Jain, A.: What information is leaked under concurrent composition? In: Canetti, R., Garay, J.A. (eds.) CRYPTO 2013. LNCS, vol. 8043, pp. 220–238. Springer, Heidelberg (2013). doi:10.1007/978-3-642-40084-1_13

[GJ13] Goyal, V., Jain, A.: On concurrently secure computation in the multiple ideal query model. In: Johansson, T., Nguyen, P.Q. (eds.) EUROCRYPT 2013. LNCS, vol. 7881, pp. 684–701. Springer, Heidelberg (2013). doi:10.1007/978-3-642-38348-9_40

[GJO10] Goyal, V., Jain, A., Ostrovsky, R.: Password-authenticated session-key generation on the internet in the plain model. In: Rabin, T. (ed.) CRYPTO 2010. LNCS, vol. 6223, pp. 277–294. Springer, Heidelberg (2010). doi:10.1007/978-3-642-14623-7_15

[Goy+14] Goyal, V., et al.: An algebraic approach to non-malleability. In: 55th Annual IEEE Symposium on Foundations of Computer Science, FOCS 2014, pp. 41–50. IEEE (2014)

[Goy+15] Goyal, V., Lin, H., Pandey, O., Pass, R., Sahai, A.: Round-efficient concurrently composable secure computation via a robust extraction lemma. In: Dodis, Y., Nielsen, J.B. (eds.) TCC 2015. LNCS, vol. 9014, pp. 260–289. Springer, Heidelberg (2015). doi:10.1007/978-3-662-46494-6_12

[HV15] Hazay, C., Venkitasubramaniam, M.: On black-box complexity of universally composable security in the CRS model. In: Iwata, T., Cheon, J.H. (eds.) ASIACRYPT 2015. LNCS, vol. 9453, pp. 183–209. Springer, Heidelberg (2015). doi:10.1007/978-3-662-48800-3_8

[HV16] Hazay, C., Venkitasubramaniam, M.: Composable adaptive secure protocols without setup under polytime assumptions. In: Hirt, M., Smith, A. (eds.) TCC 2016. LNCS, vol. 9985, pp. 400–432. Springer, Heidelberg (2016). doi:10.1007/978-3-662-53641-4_16

[Kat07] Katz, J.: Universally composable multi-party computation using tamper-proof hardware. In: Naor, M. (ed.) EUROCRYPT 2007. LNCS, vol. 4515, pp. 115–128. Springer, Heidelberg (2007). doi:10.1007/978-3-540-72540-4_7

[Kiy14] Kiyoshima, S.: Round-efficient black-box construction of composable multi-party computation. In: Garay, J.A., Gennaro, R. (eds.) CRYPTO 2014. LNCS, vol. 8617, pp. 351–368. Springer, Heidelberg (2014). doi:10.1007/978-3-662-44381-1_20

[KL11] Kidron, D., Lindell, Y.: Impossibility results for universal composability in public-key models and with fixed inputs. J. Cryptol. 24(3), 517–544 (2011). Cryptology ePrint Archive (IACR): Report 2007/478. Version 2010–06-06

[KLP07] Kalai, Y.T., Lindell, Y., Prabhakaran, M.: Concurrent composition of secure protocols in the timing model. J. Cryptol. 20(4), 431–492 (2007)

[KMO14] Kiyoshima, S., Manabe, Y., Okamoto, T.: Constant-round black-box construction of composable multi-party computation protocol. In: Lindell, Y. (ed.) TCC 2014. LNCS, vol. 8349, pp. 343–367. Springer, Heidelberg (2014). doi:10.1007/978-3-642-54242-8_15

[Lin03] Lindell, Y.: General composition and universal composability in secure multi-party computation. In: 44th Annual IEEE Symposium on Foundations of Computer Science, FOCS 2003, pp. 394–403. IEEE (2003)

[LP09] Lin, H., Pass, R.: Non-malleability amplification. In: Proceedings of the 41st Annual ACM Symposium on Theory of Computing, STOC 2009, pp. 189–198. ACM (2009)

[LP12] Lin, H., Pass, R.: Black-box constructions of composable protocols without set-up. In: Safavi-Naini, R., Canetti, R. (eds.) CRYPTO 2012. LNCS, vol. 7417, pp. 461–478. Springer, Heidelberg (2012). doi:10.1007/978-3-642-32009-5_27

[LPV09] Lin, H., Pass, R., Venkitasubramaniam, M.: A unified framework for concurrent security: universal composability from stand-alone non-malleability. In: Proceedings of the 41st Annual ACM Symposium on Theory of Computing, STOC 2009, pp. 179–188. ACM (2009)

[LPV12] Pass, R., Lin, H., Venkitasubramaniam, M.: A unified framework for uc from only OT. In: Wang, X., Sako, K. (eds.) ASIACRYPT 2012. LNCS, vol. 7658, pp. 699–717. Springer, Heidelberg (2012). doi:10.1007/978-3-642-34961-4_42

[MMY06] Malkin, T., Moriarty, R., Yakovenko, N.: Generalized environmental security from number theoretic assumptions. In: Halevi, S., Rabin, T. (eds.) TCC 2006. LNCS, vol. 3876, pp. 343–359. Springer, Heidelberg (2006). doi:10.1007/11681878_18

[MPR06] Micali, S., Pass, R., Rosen, A.: Input-indistinguishable computation. In: 47th Annual IEEE Symposium on Foundations of Computer Science, FOCS 2006, pp. 367–378. IEEE (2006)

[Pas03] Pass, R.: Simulation in quasi-polynomial time, and its application to protocol composition. In: Biham, E. (ed.) EUROCRYPT 2003. LNCS, vol. 2656, pp. 160–176. Springer, Heidelberg (2003). doi:10.1007/3-540-39200-9_10

[PR05] Pass, R., Rosen, A.: Concurrent non-malleable commitments. In: 46th Annual IEEE Symposium on Foundations of Computer Science, FOCS 2005, pp. 563–572. IEEE (2005)

[PR08] Prabhakaran, M., Rosulek, M.: Cryptographic complexity of multi-party computation problems: classifications and separations. In: Wagner, D. (ed.) CRYPTO 2008. LNCS, vol. 5157, pp. 262–279. Springer, Heidelberg (2008). doi:10.1007/978-3-540-85174-5_15

[PS04] Prabhakaran, M., Sahai, A.: New notions of security: achieving universal composability without trusted setup. In: Proceedings of the 36th Annual ACM Symposium on Theory of Computing, STOC 2004, pp. 242–251. ACM (2004)

[Ven14] Venkitasubramaniam, M.: On adaptively secure protocols. In: Abdalla, M., Prisco, R. (eds.) SCN 2014. LNCS, vol. 8642, pp. 455–475. Springer, Heidelberg (2014). doi:10.1007/978-3-319-10879-7_26

Unconditional UC-Secure Computation with (Stronger-Malicious) PUFs

Saikrishna Badrinarayanan[1], Dakshita Khurana[1(✉)], Rafail Ostrovsky[1], and Ivan Visconti[2]

[1] Department of Computer Science, UCLA, Los Angeles, USA
{saikrishna,dakshita,rafail}@cs.ucla.edu
[2] DIEM, University of Salerno, Salerno, Italy
visconti@unisa.it

Abstract. Brzuska et. al. (Crypto 2011) proved that unconditional UC-secure computation is possible if parties have access to honestly generated physically unclonable functions (PUFs). Dachman-Soled et. al. (Crypto 2014) then showed how to obtain unconditional UC secure computation based on malicious PUFs, assuming such PUFs are stateless. They also showed that unconditional oblivious transfer is impossible against an adversary that creates malicious stateful PUFs.

- In this work, we go beyond this seemingly tight result, by allowing any adversary to create stateful PUFs with a-priori bounded state. This relaxes the restriction on the power of the adversary (limited to stateless PUFs in previous feasibility results), therefore achieving improved security guarantees. This is also motivated by practical scenarios, where the size of a physical object may be used to compute an upper bound on the size of its memory.
- As a second contribution, we introduce a new model where any adversary is allowed to generate a malicious PUF that may encapsulate other (honestly generated) PUFs within it, such that the outer PUF has oracle access to all the inner PUFs. This is again a natural scenario, and in fact, similar adversaries have been studied in the tamper-proof hardware-token model (e.g., Chandran et. al. (Eurocrypt 2008)), but no such notion has ever been considered with respect to PUFs. All previous constructions of UC secure protocols suffer from explicit attacks in this stronger model.

In a direct improvement over previous results, we construct *UC protocols with unconditional security* in both these models.

The full version of the paper can be found at http://eprint.iacr.org/2016/636.

R. Ostrovsky—Research supported in part by NSF grant 1619348, US-Israel BSF grant 2012366, by DARPA Safeware program, OKAWA Foundation Research Award, IBM Faculty Research Award, Xerox Faculty Research Award, B. John Garrick Foundation Award, Teradata Research Award, and Lockheed-Martin Corporation Research Award. The views expressed are those of the authors and do not reflect position of the Department of Defense or the U.S. Government.

I. Visconti—Work done in part while visiting UCLA. Research supported in part by "GNCS - INdAM" and EU COST Action IC1306.

J.-S. Coron and J.B. Nielsen (Eds.): EUROCRYPT 2017, Part I, LNCS 10210, pp. 382–411, 2017.
DOI: 10.1007/978-3-319-56620-7_14

1   Introduction

In recent years, there has been a rich line of work studying how to enhance the computational capabilities of probabilistic polynomial-time players by making assumptions on hardware [33]. Two types of hardware assumptions in particular have had tremendous impact on recent research: tamper-proof hardware tokens and physically unclonable functions (PUFs).

The tamper-proof hardware token model introduced by Katz [24] relies on the simple and well accepted assumption that it is possible to physically protect a computing machine so that it can only be accessed as a black box, via oracle calls (as an example, think of smart cards). Immediately after its introduction, this model has been studied and its power is now understood in large part. Tamper-proof hardware tokens allow to obtain strong security notions and very efficient constructions, in some cases without requiring computational assumptions. In particular, the even more challenging case of stateless tokens started by [6] has been investigated further in [1, 10, 11, 14, 15, 19, 22, 23, 26].

1.1   Physically Unclonable Functions

Physically Unclonable Functions (PUFs) were introduced by Pappu et al. [28, 29] but their actual potential has been understood only in recent years[1]. Increasing excitement over such *physical random oracles* generated various different (and sometimes incompatible) interpretations about the actual features and formalizations of PUFs.

Very roughly, a PUF is an object that can be queried by translating an input into a specific physical stimulation, and then by translating the physical effects of the stimulation to an output through a measurement. The primary appealing properties of PUFs include: (1) constructing two PUFs with similar input-output behavior is believed to be impossible (i.e. unclonability), and (2) the output of a PUF on a given input is seemingly unpredictable, i.e., one cannot "learn" the behavior of an honestly-generated PUF on any specific input without actually querying the PUF on that input.

There is a lot of ongoing exciting research on concrete constructions of PUFs, based on various technologies. As such, a PUF can only be described in an abstract way with the attempt to establish some target properties for PUF designers.

However, while formally modeling a PUF, one might (incorrectly) assume that a PUF guarantees some properties that unfortunately exceed the state of affairs in real-world scenarios. For example, assuming that the output of a genuine PUF is purely random is clearly excessive, while relying on min-entropy is certainly a safer and more conservative assumption. Various papers have proposed different models and even attempts to unify them. The interested reader

---

[1] PUFs are used in several applications like secure storage, RFID systems, anti-counterfeiting mechanisms, identification and authentication protocols [13, 16, 25, 31, 32, 35].

can refer to [2] for detailed discussions about PUF models and their connections to properties of actual PUFs. We stress that in this work we will consider the use of PUFs in the UC model of [5]. Informally, this means that we want to study protocols that can securely compose with other protocols that may be executing concurrently.

1.2 UC Security Based on Physically Unclonable Functions

Starting with the work of Brzuska et al. [4], a series of papers have explored UC-secure computation based on physically unclonable functions. The goal of this line of cryptographic research has been to build protocols secure in progressively stronger models.

The Trusted PUFs of Brzuska et al. [4]. Brzuska et al. [4] began the first general attempts to add PUFs to the simulation paradigm of secure computation. They allowed any player (malicious or honest) to create only well-formed PUFs. As already mentioned, the output of a well-formed PUF on any arbitrary input is typically assumed to have sufficient min-entropy. Furthermore, on being queried with the same input, a well-formed PUF can be assumed to always produce identical (or sufficiently close) outputs. Applying error-tolerant fuzzy extractors [9] to the output ensures that each invocation of the PUF generates a (non-programmable) random string that can be reproduced by querying the PUF again with the same input. Brzuska et al. demonstrated how to obtain *unconditional* UC secure computation for any functionality in this model.

The Malicious PUFs of Ostrovsky et al. [27]. Ostrovsky et al. [27] then showed that the constructions of [4] become insecure in case the adversary can produce a *malicious* PUF that deviates from the behavior of an honest PUF. For instance, a malicious PUF could produce outputs according to a pseudo-random function rather than relying on physical phenomena, or it could just refuse to answer to a query. They also showed that it is possible to UC-securely compute any functionality using (potentially malicious) PUFs if one is willing to additionally make computational assumptions. They left open the problem of achieving *unconditional* UC-secure computation for any functionality using malicious PUFs.

Damgård and Scafuro [8] showed that *unconditional* UC secure commitments can be obtained even in the presence of malicious PUFs[2].

The Fully Malicious but Stateless PUFs of Dachman-Soled et al. [7]. More recently, it was shown by Dachman-Soled et al. [7] that unconditional UC security for general functionalities is impossible if the adversary is allowed to create malicious PUFs that can maintain state. They also gave a complementary feasibility result in an intermediate model where PUFs are allowed to be malicious, but are required to be stateless.

---

[2] This can be extended to other functionalities but not to all functionalities.

We note that the impossibility result of [7] crucially relies on (malicious) PUFs being able to maintain a priori unbounded state.

Thus, the impossibility seems interesting theoretically, but its impact to practical scenarios is unclear. In the real world, this result implies that unconditional UC secure computation of all functionalities is impossible in a model where an honest player is unable to distinguish maliciously created PUFs *with gigantic memory*, from honest (and therefore completely stateless) PUFs. One could argue that this allows the power of the adversary to go beyond the reach of current technology. On the other hand, the protocol of [7] breaks down completely if the adversary can generate a maliciously created PUF with even one bit of memory, and pass it off as a stateless (honest) PUF. This gap forms the starting point for our work.

1.3 Our Contributions

The current state-of-the-art leaves open the following question:

Can we achieve UC-secure computation with malicious PUFs that are allowed to have a priori bounded state?

In the main contribution of this work we answer this question in the affirmative. We show that not only it is possible to obtain UC-secure computation for any functionality as proven in [27] with computational assumptions, but we prove that this can be done with unconditional security, without relying on any computational assumptions. This brings us to our first main result, which we now state informally.

Informal Theorem 1. *For any two party functionality $\mathcal{F}$, there exists a protocol π that unconditionally and UC-securely realizes $\mathcal{F}$ in the malicious bounded-stateful PUF model.*

As our second contribution, we introduce a new adversarial model for PUF-based protocols. Here, in addition to allowing the adversary to generate malicious stateless PUFs, we also allow him to encapsulate other (honestly generated) PUFs inside his own (malicious, stateless) PUF, even without the knowledge of the functionality of the inner PUFs. This allows the outer malicious PUF to make black-box (or oracle) calls to the inner PUFs that it encapsulates. In particular, the outer malicious PUF could answer honest queries by first making oracle calls to its inner PUFs, and generating its own output as a function of the output of the inner PUFs on these queries. An honest party interacting with such a malicious PUF need not be able to tell whether the PUF is malicious and possibly encapsulates other PUFs in it, or it is honest.

In this new adversarial model[3], we require all PUFs to be stateless. We will refer to this as the malicious encapsulated PUF model. It is interesting to

---

[3] A concurrent and independent work [30] considers an adversary that can encapsulate PUFs but does not propose UC-secure definitions/constructions.

note that all previously known protocols (even for limited functionalities such as commitments) suffer explicit attacks in this stronger malicious encapsulated (stateless) PUF model.

As our other main result, we develop techniques to obtain unconditional UC-secure computation in the malicious encapsulated PUF model.

Informal Theorem 2. *For any two party functionality $\mathcal{F}$, there exists a protocol π that unconditionally and UC-securely realizes $\mathcal{F}$ in the malicious encapsulated (stateless) PUF model.*

Table 1 compares our results with prior work. Our feasibility result in the malicious bound-stateful PUF model and our feasibility result in the malicious encapsulated-stateless PUF model directly improve the works of [4,7]. Indeed each of our two results strengthen the power of the adversaries of [4,7] in one meaningful and natural direction still achieving the same unconditional results of [4,7]. A natural question is whether our techniques defeating malicious bounded-stateful PUFs can be composed with our techniques defeating malicious encapsulated-stateless PUFs to obtain unconditional UC-security for any functionality against adversaries that can construct malicious bounded-stateful encapsulated PUFs. While we do not see a priori any conceptual obstacle in obtaining such even stronger feasibility result, the resulting construction would be extremely complex and heavily tedious to analyze. Therefore we defer such a stronger claim to future work hoping that follow up research will achieve a more direct and elegant construction.

Table 1. The symbol $\checkmark$ (resp. $\times$) indicates that the construction satisfies (resp. does not satisfy) the corresponding security guarantee.

Reference	Unconditional UC for any functionality	UC with stateless mal. PUFs	UC with bounded stateful mal. PUFs	UC with encapsulated stateless mal. PUFs
[4]	$\checkmark$	$\times$	$\times$	$\times$
[27]	$\times$	$\checkmark$	$\checkmark$	$\times$
[7]	$\checkmark$	$\checkmark$	$\times$	$\times$
This work	$\checkmark$	$\checkmark$	$\checkmark$	$\times$
This work	$\checkmark$	$\checkmark$	$\times$	$\checkmark$

1.4 Our Techniques

The starting point for our constructions is the UC-secure OT protocol of [7], which itself builds upon the works of [4,27]. We begin by giving a simplified description of the construction in [7].

Suppose a sender S with inputs (m_0, m_1) and a receiver R with input bit b want to run a UC secure OT protocol in the malicious stateless PUF model. Then, S generates a PUF and sends it to the receiver. The receiver queries the

PUF on a random challenge string c, records the output r and then returns the PUF to $\mathcal{S}$. Then, the sender sends two random strings (x_0, x_1) to the receiver. In turn, the receiver picks x_b, and sends $v = c \oplus x_b$ to the sender. The sender uses $\mathsf{PUF}(v \oplus x_0)$ to mask his input m_0 and $\mathsf{PUF}(v \oplus x_1)$, to mask his input m_1; and sends both masked values to the receiver. Here $\mathsf{PUF}(\cdot)$ denotes the output of the PUF on the given input. Since $\mathcal{R}$ had to return the PUF before (x_0, x_1) were revealed, with overwhelming probability, $\mathcal{R}$ only knows $r = \mathsf{PUF}(v \oplus x_b)$, and can output one and only one of the masked sender inputs.

Enhancing [7] in the Stateless PUF Model. Though this was a simplified overview of the protocol in [7], it helps us to explain a subtle assumption required in their simulation strategy against a malicious sender. In particular, the simulator against a malicious sender must return the PUF to the sender before the sender picks random messages (x_0, x_1). However, it is evident that in order to extract both messages (m_0, m_1), the simulator must know $(x_0 \oplus x_1)$, and in particular know the response of the PUF on challenges $(c, c \oplus x_0 \oplus x_1)$ for some known string c.

But the simulator only learns (x_0, x_1) after sending the PUF back to $\mathcal{S}$. Thus, in order to successfully extract the input of $\mathcal{S}$, the simulator should have the ability to make these queries *even after* the PUF has been returned to the malicious sender. This means that the PUF is supposed to remain accessible and untouched even when it is again in the hands of its malicious creator. We believe this is a very strong assumption that clearly deviates from real scenarios where the state of a PUF can easily be changed (e.g., by damaging it).

Our protocol in Fig. 1 gets rid of this strong assumption on the simulator, and we give a new sender simulation strategy that does not need to query the PUF when it is back in the hands of the malicious sender $\mathcal{S}$. This is also a first step in obtaining security against bounded-stateful PUFs. In the protocol of [7], if the PUF created by a malicious $\mathcal{S}$ is stateful, $\mathcal{S}$ on receiving the PUF can *first* change the state of the PUF (say, to output $\bot$ everywhere), and then output values (x_0, x_1). In this case, no simulation strategy will be able to extract the inputs of the sender.

We change the protocol in [7], by having $\mathcal{S}$ commit to the random values (x_0, x_1) at the beginning of the protocol, using a UC-secure commitment scheme. These values are decommitted only after $\mathcal{R}$ returns the PUF back to $\mathcal{S}$, so the scheme still remains UC-secure against a malicious receiver. Moreover, now the simulator against a malicious sender can use the straight-line extractor guaranteed by the UC-secure commitment scheme, to extract values (x_0, x_1), and query the PUF on challenges of the form $(c, c \oplus x_0 \oplus x_1)$ for some string c. It then sets $v = c \oplus x_0$ and sends it to $\mathcal{S}$. Now, the sender masks are $\mathsf{PUF}(v \oplus x_0)$ and $\mathsf{PUF}(v \oplus x_1)$, which is nothing but $\mathsf{PUF}(c)$ and $\mathsf{PUF}(c \oplus x_0 \oplus x_1)$, which was already known to the sender simulator before returning the PUF to $\mathcal{S}$. This simulation strategy works (with the simulator requiring only black-box access to the malicious PUF's code) even if the PUF is later broken or its state is reset in any way. This protocol is described formally and proven secure in Sect. 3.

Inputs: Sender $\mathcal{S}$ has private inputs $(m_0, m_1) \in \{0,1\}^{2n}$ and Receiver $\mathcal{R}$ has private input $b \in \{0,1\}$.

1. **Sender Message:** $\mathcal{S}$ does the following:
   - Generate a PUF $PUF_s : \{0,1\}^n \rightarrow \{0,1\}^n$.
   - Choose a pair of random strings $(x_0, x_1) \xleftarrow{\$} \{0,1\}^{2n}$.
   - Send PUF_s and $(t_0, t_1) = UC\text{-}Com.Commit(x_0, x_1)$ to $\mathcal{R}$.
2. **Receiver Message:** $\mathcal{R}$ does the following:
   - Choose a pair of random strings $(c_0, c_1) \xleftarrow{\$} \{0,1\}^{2n}$.
   - Compute $r_0 = PUF_s(c_0), r_1 = PUF_s(c_1)$.
   - Set $c = c_p$ and $r = r_p$ for $p \xleftarrow{\$} \{0,1\}$.
   - Store the pair (c, r) and send PUF_s to $\mathcal{S}$.
3. **Sender Message:**
   - $\mathcal{S}$ sends $(x_0, x_1) = UC\text{-}Com.Decommit(t_0, t_1)$ to $\mathcal{R}$.
4. **Receiver Message:** $\mathcal{R}$ does the following:
   - Abort if the decommitment does not verify correctly.
   - Compute and send $val = c \oplus x_b$ to $\mathcal{S}$.
5. **Sender Message:**
   - $\mathcal{S}$ computes $S_0 = m_0 \oplus PUF_s(val \oplus x_0)$, $S_1 = m_1 \oplus PUF_s(val \oplus x_1)$ and sends (S_0, S_1) to $\mathcal{R}$.

Outputs: $\mathcal{S}$ has no output. $\mathcal{R}$ outputs m_b which is computed as $(S_b \oplus r)$.

Fig. 1. Protocol Π^1 for 2-choose-1 OT in the malicious stateless PUF model.

UC Security with Bounded Stateful PUFs. A malicious PUF is allowed to maintain *state*, and can generate outputs (including $\perp$) as a function of not only the current query but also the previous queries that it received as input. This allows for some attacks on the protocol we just described, but they can be prevented by carefully interspersing coin-tossing with the protocol. Please see Sect. 4 for more details.

A stateful PUF created by the sender can also record information about the queries made by the receiver, and replay this information to a malicious sender when he inputs a secret challenge. Indeed, for PUFs with unbounded state, it is this ability to record queries that makes oblivious transfer impossible. However, we only consider PUFs that have a-priori bounded state. In this case, it is possible to design a protocol, parameterized by an upper bound on the size of the state of the PUF, that in effect exhausts the possible state space of such a malicious PUF. Our protocol then carefully uses this additional entropy to mask the inputs of the honest party.

More specifically, we repeat the OT protocol described before (with an additional coin-tossing phase) K times in parallel, using the same (possibly malicious, stateful) PUF, for sufficiently large $K > \ell$ (where ℓ denotes the upper bound on the state of the PUF). At this point, what we require essentially boils down to a *one-sided* malicious oblivious transfer extractor. This is a gadget that would yield a single OT from K leaky OTs, such that the single OT remains secure even when a malicious sender can ask for ℓ bits of universal leakage across all these OTs. This setting is incomparable to previously studied OT extractors [17,21] because: (a) we require a protocol that is secure against malicious (not just semi-honest) adversaries, and (b) the system has only one-sided leakage, i.e., a corrupt sender can request ℓ bits of leakage, but a corrupt receiver does not obtain any leakage at all.

For simplicity, we consider the setting of one-sided receiver leakage (instead of sender leakage). It is possible to consider this because OT is reversible. To protect against a malicious receiver that may obtain ℓ bits of universal leakage, the sender picks different random inputs for each OT execution, and then uses a strong randomness extractor to extract min-entropy and mask his inputs. We show that this in fact suffices to statistically hide the input messages of the sender. Please see Sect. 5 for a more detailed overview and construction.

UC Security with Encapsulated PUFs. We demonstrate the feasibility of UC secure computation, in a model where a party may (maliciously) encapsulate one or more PUFs that it obtained from honest parties, inside a malicious stateless PUF of its choice. We stress that our protocol itself does not require honest parties to encapsulate PUFs within each other.

To describe our techniques, we begin by revisiting the protocol in Fig. 1, that we described at the beginning of this overview. Suppose parties could maliciously encapsulate some honest PUFs inside a malicious PUF. Then a malicious receiver in this protocol, when it is supposed to return the sender's PUF PUF_s, could instead return a *different* malicious PUF $\widehat{\mathsf{PUF}}_s$. In this case, the receiver would easily learn both inputs of the sender. But as correctly pointed out in prior work [7,8], the sender can deflect such attacks by probing and recording the output of PUF_s on some random input(s) (known as Test Queries) before sending it to the receiver. Later the sender can check whether $\widehat{\mathsf{PUF}}_s$ correctly answers to all Test Queries.

However, a malicious receiver may create $\widehat{\mathsf{PUF}}_s$ that encapsulates PUF_s, such that $\widehat{\mathsf{PUF}}_s$ is programmed to send most outer queries to PUF_s and echo its output externally; in order to pass the sender's test. However, $\widehat{\mathsf{PUF}}_s$ may have its own malicious procedure to evaluate some of the other external queries. In particular, the "unpredictability" of $\widehat{\mathsf{PUF}}_s$ may break down completely on these queries.

It turns out that the security of the sender in the basic OT protocol of Fig. 1 hinges on the unpredictability of the output of PUF_s (in this situation, $\widehat{\mathsf{PUF}}_s$) on a "special challenge query" only, which we will denote by s. It is completely feasible for a receiver to create a malicious encapsulating PUF $\widehat{\mathsf{PUF}}_s$ that passes the sender tests, and yet its output on this special query s is completely known to the receiver, therefore breaking sender security.

We overcome this issue by ensuring that s is chosen using a coin toss, and is completely unknown to the receiver until after he has sent $\widehat{\mathsf{PUF}}_s$ (possibly a malicious encapsulating PUF) back to the sender. Intuitively, this means that $\widehat{\mathsf{PUF}}_s$ will either not pass the sender tests, or will be highly likely to deflect this the query s to the inner PUF and echo its output (thereby ensuring that the output of the PUF on input s is unpredictable for the receiver). An additional subtlety that arises is that the receiver might use an incorrect s in the protocol (instead of using the output of the coin toss): the receiver is forced to use the correct s via a special cut-and-choose mechanism. For a more detailed overview and construction, please see to Sect. 6.

UC-Secure Commitments Against Encapsulation Attacks. Finally, UC-secure commitments against encapsulation attacks play a crucial role in our UC-secure OT protocol in the encapsulation model. But, we note that the basic commitment protocol of [8] is insecure in this stronger model, and therefore we modify the protocol of [8] to achieve UC-security in this scenario. In a nutshell, this is done by having the receiver send an additional PUF at the end of the protocol, and forcing any malicious committer to query this additional PUF on the committer's input bit. We then show that even an encapsulating (malicious) committer will have to carry out this step honestly in order to complete the commit phase. Then, a simulator can extract the adversary's committed value by observing the queries of the malicious committer to this additional PUF. We illustrate in detail, how prior constructions of UC-secure commitments fail in the PUF encapsulation model in Sect. 7. Our UC-secure commitment protocol in the encapsulated malicious (stateless) PUF model is also described in Sect. 7.

1.5   Organization

The rest of this paper is organized as follows. In Sect. 2, we discuss PUFs and other preliminaries relevant to our protocols. In Sect. 3, we describe an improved version of the protocol in [7], in the stateless PUF model. In Sects. 4 and 5, we boost this protocol to obtain security in the bounded stateful PUF model. In Sects. 6 and 7, we discuss protocols that are secure in the PUF encapsulation model. In Appendix A, we discuss the formal modelling of our PUFs. The complete models and proofs that could not be included in this version owing to space restrictions, can be found in the full version of the paper.

2   Preliminaries

2.1   Physically Unclonable Functions

A PUF is a noisy physical source of randomness. The randomness property comes from an uncontrollable manufacturing process. A PUF is evaluated with a physical stimulus, called the *challenge*, and its physical output, called the *response*, is measured. Since the processes involved are physical, the function implemented by a PUF can not (necessarily) be modeled as a mathematical function, neither can be considered computable in PPT. Moreover, the output of a PUF is noisy, namely, querying a PUF twice with the same challenge, could yield distinct responses within a small Hamming distance to each other.

Moreover, the response need not be random-looking; rather, it is a string drawn from a distribution with high min-entropy. Prior work has shown that, using fuzzy extractors, one can eliminate the noisiness of the PUF and make its output uniformly random. For simplicity, we assume this in the body of the paper and give a detailed description in the full version.

A PUF-family is a pair of (not necessarily efficient) algorithms Sample and Eval. Algorithm Sample abstracts the PUF fabrication process and works as follows. On input the security parameter, it outputs a PUF-index id from the PUF-family satisfying the security properties (that we define soon) according to the security parameter. Algorithm Eval abstracts the PUF-evaluation process. On input a challenge q, it evaluates the PUF on q and outputs the response a of length rg, denoting the range. Without loss of generality, we assume that the challenge space of a PUF is a full set of strings of a certain length.

Security of PUFs. Following [4], we consider only the two main security properties of PUFs: *unclonability* and *unpredictability*. Informally, unpredictability means that the output of the PUF is statistically indistinguishable from a uniform random string. Formally, unpredictability is modeled via an entropy condition on the PUF distribution. Namely, given that a PUF has been measured on a polynomial number of challenges, the response of the PUF evaluated on a new challenge still has a significant amount of entropy. For simplicity, a PUF is unpredictable if its output on any given input appears uniformly random.

Informally, unclonability states that in a protocol consisting of several parties, only the party in whose possession the PUF is, can evaluate the PUF. When a party sends a PUF to a different party, it can no longer evaluate the PUF till the time it gets the PUF back. Thus a party not in possession of a PUF cannot predict the output of the PUF on an input for which it did not query the PUF, unless it maliciously created the PUF. A formal definition of unclonability is given in the full version of this paper.

A PUF can be modeled as an ideal functionality $\mathcal{F}_{\mathsf{PUF}}$, which mimics the behavior of the PUF in the real world. We formally define ideal functionalities corresponding to honestly generated and various kinds of maliciously generated PUFs in Appendix A. We summarize these here: the model for honestly generated PUFs and for malicious stateless/stateful PUFs has been explored in prior work [7,27], and we introduce the model for encapsulated PUFs.

- **An honestly generated PUF** can be created according to a sampling algorithm Samp, and evaluated honestly using an evaluation algorithm Eval. The output of an honestly generated PUF is *unpredictable* even to the party that created it, i.e., even the creator cannot predict the output of an honestly generated PUF on any given input without querying the PUF on that input.
- **A malicious stateless PUF**, on the other hand, can be created by the adversary substituting an Eval$_{mal}$ procedure of his choice for the honest Eval procedure. Whenever a (honest) party in possession of this PUF evaluates the PUF, it runs the stateless procedure Eval$_{mal}(c)$ instead of Eval(c) (and cannot distinguish Eval$_{mal}(c)$ from Eval(c) unless they are distinguishable with

black-box access to the PUF). The output of such a PUF cannot depend on previous queries, moreover no adversary that creates the PUF but does not possess it, can learn previous queries made to the PUF when it was not in its possession. We adapt the definitions from [27], where $Eval_{mal}$ is a polynomial-time algorithm with oracle access to Eval. This is done to model the fact that the $Eval_{mal}$ algorithm can access an (honest) source of randomness Eval, and can arbitrarily modify its output using any polynomial-time strategy.

- **A malicious stateful PUF** can be created by the adversary substituting a stateful $Eval_{mal}$ procedure of his choice for the honest Eval procedure. Whenever a party in possession of this PUF evaluates the PUF, it runs the stateful procedure $Eval_{mal}(c)$ instead of $Eval(c)$. Thus, the output of a stateful malicious PUF can possibly depend on previous queries, moreover an adversary that created a PUF can learn previous queries made to the PUF by querying it, say, on a secret input. $Eval_{mal}$ is a polynomial-time stateful Turing Machine with oracle access to Eval. Again, this is done to model the fact that the $Eval_{mal}$ algorithm can access an (honest) source of randomness, Eval, and arbitrarily modify its output using any polynomial-time strategy. Malicious stateful PUFs can further be of two types:
  - *Bounded Stateful*. Such a PUF can maintain a-priori bounded memory/state (which it may rewrite, as long as the total memory is bounded).
  - *Unbounded Stateful*. Such a PUF can maintain unbounded memory/state.
- **A malicious encapsulating PUF** can possibly encapsulate other (honestly generated) PUFs inside it[4], without knowing the functionality of these inner PUFs. Such a PUF PUF_{mal} can make black-box calls to the inner PUFs, and generate its outputs as a function of the output of the inner (honest) PUFs. This is modeled by having the adversary substitute an $Eval_{mal}$ procedure of his choice for the honest Eval procedure in the PUF_{mal} that it creates, where as usual $Eval_{mal}$ is a polynomial-time Turing Machine with oracle access to Eval. Similar to the two previous bullets, this is done to model the fact that the $Eval_{mal}$ algorithm can access an (honest) source of randomness, Eval, and arbitrarily modify its output using any polynomial-time strategy.

  In addition, $Eval_{mal}$ can also make oracle calls to polynomially many other (honestly generated) procedures $Eval_1, Eval_2, \ldots Eval_M$ that are contained in PUFs $PUF_1, PUF_2, \ldots PUF_M$, for any a-priori unbounded $M = poly(n)$. These correspond to honestly generated PUFs that the adversary may be encapsulating within its own malicious PUF. Thus on some input c, the $Eval_{mal}$ procedure may make oracle calls to $Eval_1, Eval_2, \ldots Eval_M$ on polynomially many inputs, and compute its output as a function of the outputs of the $Eval, Eval_1, Eval_2, \ldots Eval_M$ procedures. Of course, we ensure that the adversary's $Eval_{mal}$ procedure can make calls to some honestly generated procedure $Eval_i$ *only if* the adversary owns the PUF PUF_i implementing the $Eval_i$ procedure when creating the encapsulating malicious PUF. Furthermore, when the

---

[4] Since the adversary knows the code of maliciously generated PUFs, this model automatically captures real-world scenarios where an adversary may be encapsulating other malicious PUFs inside its own.

adversary passes such a PUF to an honest party, the adversary "loses ownership" of PUF_i and is no longer allowed to access the $Eval_i$ procedure, this is similar to the unclonability requirement. This is modeled by assigning an owner to each PUF, and on passing an outer (encapsulating) PUF to an honest party, the adversary must automatically pass all the inner (encapsulated) honest PUFs. Whenever an honest party is in possession of such an adversarial PUF PUF_{mal} and evaluates it, it receives the output of $Eval_{mal}$. When the adversary is allowed to construct encapsulating PUFS, we restrict all PUFs to be stateless. Therefore the model with encapsulating PUFs is incomparable with the model with bounded-stateful malicious PUFs. Further details on the modeling of malicious stateless PUFs that may encapsulate other stateless PUFs, are provided in Appendix A.

To simplify notation, we write $PUF \leftarrow Sample(1^K)$, $r = PUF(c)$ and assume that PUF is a deterministic function with random output.

2.2   UC Secure Computation

The UC framework, introduced by [5] is a strong framework which gives security guarantees even when protocols may be arbitrarily composed.

Commitments. A UC-secure commitment scheme UC-Com consists of the usual commitment and decommitment algorithms, along with (straight-line) procedures allowing the simulator to extract the committed value of the adversary and to equivocate a value that the simulator committed to. We denote these by (UC-Com.Commit, UC-Com.Decommit, UC-Com.Extract, UC-Com.Equivocate). Damgård and Scafuro [8] realized unconditional UC secure commitments using stateless PUFs, in the malicious stateful PUF model.

OT. Ideal 2-choose-1 oblivious transfer (OT) is a two-party functionality that takes two inputs m_0, m_1 from a sender and a bit b from a receiver. It outputs m_b to the receiver and $\perp$ to the sender. We use $\mathcal{F}_{ot}$ to denote this functionality. Given UC oblivious transfer, it is possible to obtain UC secure two-party computation of any functionality.

Formal definitions of these functionalities and background on prior results are provided in the full version of this paper.

3   Unconditional UC Security with (Malicious) Stateless PUFs

As a warm up, we start by considering malicious stateless PUFs as in [7] and we strengthen their protocol in order to achieve security even when the simulator does not have access to a malicious PUF that is in possession of the adversary that created it.

Construction. Let n denote the security parameter. The protocol Π^1 in Fig. 2 UC-securely and unconditionally realizes 2-choose-1 OT in the malicious stateless PUF model, between a sender S and receiver R, with the following restrictions:

1. The random variables (x_0, x_1) are chosen by S independently of PUF$_s$.[5]
2. A (malicious) R returns to S the same PUF, PUF$_s$ that it received[6].

We enforce these restrictions in this section only for simplicity and modularity purposes. We remove them in Sects. 4 and 6 respectively.

Inputs: Sender S has private inputs $(m_0, m_1) \in \{0,1\}^{2n}$ and Receiver R has private input $b \in \{0,1\}$.

1. **Sender Message:** S does the following:
   - Generate a PUF PUF$_s : \{0,1\}^n \rightarrow \{0,1\}^n$.
   - Choose a pair of random strings $(x_0, x_1) \xleftarrow{\$} \{0,1\}^{2n}$.
   - Send PUF$_s$ and $(t_0, t_1) = $ UC-Com.Commit(x_0, x_1) to R.
2. **Receiver Message:** R does the following:
   - Choose a pair of random strings $(c_0, c_1) \xleftarrow{\$} \{0,1\}^{2n}$.
   - Compute $r_0 = $ PUF$_s(c_0), r_1 = $ PUF$_s(c_1)$.
   - Set $c = c_p$ and $r = r_p$ for $p \xleftarrow{\$} \{0,1\}$.
   - Store the pair (c, r) and send PUF$_s$ to S.
3. **Sender Message:**
   - S sends $(x_0, x_1) = $ UC-Com.Decommit(t_0, t_1) to R.
4. **Receiver Message:** R does the following:
   - Abort if the decommitment does not verify correctly.
   - Compute and send val $= c \oplus x_b$ to S.
5. **Sender Message:**
   - S computes $S_0 = m_0 \oplus $ PUF$_s($val$ \oplus x_0)$, $S_1 = m_1 \oplus $ PUF$_s($val$ \oplus x_1)$ and sends (S_0, S_1) to R.

Outputs: S has no output. R outputs m_b which is computed as $(S_b \oplus r)$.

Fig. 2. Protocol Π^1 for 2-choose-1 OT in the malicious stateless PUF model.

Our protocol makes black-box use of a UC-commitment scheme, denoted by the algorithms UC-Com.Commit and UC-Com.Decommit. We use UC-Com. Commit(a, b) to denote a commitment to the concatenation of strings a and b.

---

[5] This is fixed later by using coin-tossing to generate (x_0, x_1), see Sect. 4.

[6] In Sect. 6, we consider an even stronger model where R may encapsulate PUF$_s$ within a possibly malicious $\widehat{\text{PUF}}_s$. $\widehat{\text{PUF}}_s$ externally forwards some queries to PUF$_s$ and forwards the outputs to the evaluator, while possibly replacing some or all of these outputs with other arbitrary values. We note that this covers the case where the receiver generates $\widehat{\text{PUF}}_s$ malicious and independently of PUF$_s$.

UC-secure commitments can be unconditionally realized in the malicious state-less PUF model [8]. Formally, we prove the following theorem:

Theorem 1. *The protocol Π^1 in Fig. 2 unconditionally UC-securely realizes $\mathcal{F}_{ot}$ in the malicious stateless PUF model.*

This protocol is essentially the protocol of Dachman-Soled et al. [7], modified to enable correct extraction of the sender's input. The protocol as specified in [7], even though private, does not allow for straight-line extraction of the sender's input messages, unless one is willing to make the strong assumption that the simulator can make queries to a (malicious) PUF that an adversary created, even when this malicious PUF is in the adversary's possession (i.e., the adversary is forced not to update nor to damage/destroy the PUF).

Our main modification is to have the sender commit to his values (x_0, x_1) using a UC-secure commitment scheme. In this case, it is possible for the simulator to extract (x_0, x_1) in a straight-line manner from the commitment, and therefore extract the sender's input while it remains hidden from a real receiver. The rest of the proof follows in the same manner as [7]; recall that we already gave an overview in Sect. 1.4. The formal proofs of correctness and security can be found in the full version of this paper.

4 UC-Security with (Bounded-Stateful Malicious) PUFs

Overview. A malicious *stateful* PUF can generate outputs as a function of its previous input queries. For the (previous) protocol in Fig. 2, note that in Step 2, $\mathsf{Sim}_\mathcal{S}$ makes two queries (c_1, c_2) to the PUF such that $(c_1 \oplus c_2) = (x_1 \oplus x_2)$, where (x_1, x_2) are the sender's random messages. On the other hand, an honest receiver makes two queries (c_1, c_2) to the PUF such that $(c_1 \oplus c_2) = \mathsf{rv}$, for an independent random variable rv.

Therefore, when combined with the sender's view, the joint distribution of the evaluation queries made to the PUF by $\mathsf{Sim}_\mathcal{S}$, differs from the joint distribution of the evaluation queries made to the PUF by an honest receiver. Thus, a malicious sender can distinguish the two worlds by having a malicious stateful PUF compute a reply to c_2 depending on the value of the previous challenge c_1. We will call these attacks of Type I. In this section, we will describe a protocol secure against all possible attacks where a stateful PUF computes responses to future queries as a function of prior queries.

A stateful PUF created by the sender can also record information about the queries made by the receiver, and replay this information to a malicious sender when he inputs a secret challenge. For PUFs with bounded state, we view these as 'leakage' attacks, by considering all information recorded and replayed by a PUF as leakage. We will call these attacks of Type II. We describe a protocol secure against general bounded stateful PUFs (i.e., secure against attacks of both Type I and Type II) in Sect. 5.

Repeat the following protocol K times in parallel for fresh private inputs (m_0^i, m_1^i) of the sender and b_i of the receiver for $i \in [K]$.

Inputs: Sender S has private inputs $(m_0, m_1) = (m_0^i, m_1^i) \in \{0,1\}^{2n}$ and Receiver R has private input $b = b^i \in \{0,1\}$.

1. **Sender Message:** S does the following.
   - Generate a PUF $\mathsf{PUF}_s : \{0,1\}^n \to \{0,1\}^n$.(Use the same PUF for all the K parallel sessions).
   - Choose a pair of random strings $(x_0, x_1) \xleftarrow{\$} \{0,1\}^{2n}$.
   - Send PUF_s and $(t_0, t_1) = \mathsf{UC\text{-}Com.Commit}(x_0, x_1)$ to R.
2. **Receiver Message:** R does the following.
   - Choose a pair of random strings $(c_0, c_1) \xleftarrow{\$} \{0,1\}^{2n}$.
   - Compute $r_0 = \mathsf{PUF}_s(c_0), r_1 = \mathsf{PUF}_s(c_1)$.
   - Set $c = c_p$ and $r = r_p$ for $p \xleftarrow{\$} \{0,1\}$ and store the pair (c, r).
   - Pick and send $\underline{(\hat{x}_0, \hat{x}_1) \xleftarrow{\$} \{0,1\}^{2n}}$ along with PUF_s, to S.
3. **Sender Message:**
   S sends $(x_0, x_1) = \mathsf{UC\text{-}Com.Decommit}(t_0, t_1)$ to R.
4. **Receiver Message:** If $\mathsf{UC\text{-}Com.Decommit}(t_0, t_1)$ does not verify, abort. Else, compute and send $\mathsf{val} = c \oplus x_b \underline{\oplus \hat{x}_b}$ to S.
5. **Sender Message:** S does the following.
   - Compute
     $S_0 = m_0 \oplus \mathsf{PUF}_s(\mathsf{val} \oplus x_0 \underline{\oplus \hat{x}_0})$ and $S_1 = m_1 \oplus \mathsf{PUF}_s(\mathsf{val} \oplus x_1 \underline{\oplus \hat{x}_1})$.
   - Send (S_0, S_1) to R.

Outputs: S has no output. R outputs m_b which is computed as $(S_b \oplus r)$.

Fig. 3. Protocol Π_K for K 2-choose-1 OTs (with at most ℓ-bounded leakage) in the malicious stateful PUF model. The changes from the protocol in Fig. 2 are underlined.

Our Strategy. Let ℓ denote a polynomial upper bound on the size of the memory of any malicious PUF created by the sender S. Our strategy to obtain secure oblivious transfer from any PUF with ℓ-bounded state is as follows: We use (the same) PUF_s created by the sender, to execute $K = \Theta(\ell)$ oblivious transfers in parallel. In our new protocol in Fig. 3, we carefully intersperse an additional round of coin tossing with our basic protocol from Fig. 2, to obtain security against attacks of Type I.

Specifically, we modify the protocol of Fig. 2 as follows: instead of having S generate the random strings (x_0, x_1), we set the protocol up so that *both* S and the receiver R generate XOR shares of (x_0, x_1). Furthermore, R generates his shares only after obtaining the PUF and a commitment to sender shares from S. In such a case, the PUF created by S must necessarily be independent of the receiver shares and consequently, also independent of (x_0, x_1).

Recall from Sect. 3, that the simulator against a malicious sender succeeds if it can obtain the output of the PUF to queries of the form $(c, c \oplus x_0 \oplus x_1)$ for a random c, whereas an honest receiver can only make queries of the form (c_1, c_2) for randomly chosen (c_1, c_2). Since (x_0, x_1) appear to be distributed uniformly at random to the PUF, the distributions of $(c, c \oplus x_0 \oplus x_1)$ and (c_1, c_2) are also statistically indistinguishable to the PUF[7]. Therefore, the sender simulator succeeds whenever the honest receiver does not abort and this suffices to prove security against a malicious sender.

Finally, we note that the simulation strategy against a malicious receiver remains similar to one of Sect. 3, even if the receiver has the ability to create PUFs with unbounded state.

Construction. The protocol Π_K in Fig. 3 allows us to use an ℓ-bounded stateful PUF to obtain K secure (but one-sided leaky) oblivious transfers, such that a malicious sender can obtain at most ℓ bits of additional universal leakage on the joint distribution of the receiver's choice input bits $(b_1, b_2, \ldots b_K)$. Our protocol makes black-box use of a UC-commitment scheme, denoted by the algorithms UC-Com.Commit and UC-Com.Decommit[8]. UC-secure commitments can be unconditionally realized in the malicious *stateful* PUF model [8].

Theorem 2. *The protocol Π_K unconditionally UC-securely realizes K instances of $OT(\mathcal{F}_{ot}^{[\otimes K]})$ in an ℓ-bounded-stateful PUF model, except that a malicious sender can obtain at most ℓ bits of additional universal leakage on joint distribution of the receiver's choice bits over all $\mathcal{F}_{ot}^{[\otimes K]}$.*

Correctness is immediate from inspection, and the complete proof of security is in the full version of the paper.

---

[7] We assume the simulator can control which simulator queries the adversary's PUF records (but an honest party cannot). Indeed, without our assumption, if a stateful PUF recorded every simulator query, a malicious sender on getting back PUF_s may observe the correlation between queries (c, c') recorded by the PUF when the simulator queried it, versus two random queries when an actual honest party queried it. Ours is a natural assumption and obtaining secure OT remains extremely nontrivial even with this assumption. We note that this requirement can be removed using standard secret sharing along with cut-and-choose, but at the cost of a more complicated protocol with a worse OT production rate. This protocol is described in the full version of this paper.

[8] The UC framework (and its variants) seemingly fail to capture the possibility of transfer of physical devices like PUFs across different protocols, to the best of our knowledge. Within our OT protocol, we invoke the ideal functionality for UC-secure commitments. Thus, we would like to ensure that our UC-secure commitment scheme composes with the rest of the protocol even if PUFs created in the commitment scheme are used elsewhere in the OT protocol and vice versa. In our protocol, the only situation where such an issue might arise, is if one of the parties in the main OT protocol, later maliciously passes a PUF that it received from the honest party during a commitment phase. This is avoided by requiring all parties to return the PUFs to their original creator at the end of the decommitment phase. Note that this does not violate security even if the PUFs are malicious and stateful. The creating party, like in previous works [7,8] can probe a random point before sending the PUF, and then check this point again on receiving the PUF, to ensure that they received the correct PUF. Generic results attempting to model UC security in presence of physical devices that can be transferred across different protocol executions have been presented in [3,20].

5   One-Sided Correlation Extractors with Malicious Security

From Sect. 4, in the ℓ-bounded stateful PUF model, we obtain K leaky oblivious transfers, such that the sender can obtain ℓ bits of universal leakage on the joint distribution of the receiver's choice bits over all K oblivious transfers.

Because OT is reversible [37], it suffices to consider a reversed version of the above setting, i.e., where the receiver can obtain ℓ bits of additional universal leakage on the joint distribution of all the sender's messages over all K oblivious transfers. More formally, the leakage model we consider is as follows:

One-Sided Leakage Model for Correlation Extractors. Here, we begin by describing our leakage model for OT correlations formally, and then we define one-sided correlation extractors for OT. Our leakage model is as follows:

1. **K-OT Correlation Generation Phase:** For $i \in [K]$, the sender $\mathcal{S}$ obtains $(x_0^i, x_1^i) \in \{0,1\}^2$ and the receiver $\mathcal{R}$ gets $(b_i, x_{b_i}^i)$.
2. **Corruption and Leakage Phase:** A malicious adversary corrupts the receiver and sends a leakage function $L : \{0,1\}^K \rightarrow \{0,1\}^{t_R}$. It receives $L(\{(x_0^i, x_1^i)\}_{i \in [K]})$.

Let (X, Y) be a random OT correlation (i.e., $X = (x_0, x_1), Y = (r, x_r)$, where (x_0, x_1, r) are sampled uniformly at random.) We denote a t_R-leaky version of $(X, Y)^K$ described above as $((X, Y)^K)^{[t_R]}$.

Definition 1 ((n, p, t_R, ϵ) One-Sided Malicious OT-Extractor). *An (n, p, t_R, ϵ) one-sided malicious OT-extractor is an interactive protocol between 2 parties S and R with access to $((X, Y)^n)^{[t_R]}$ described above. The protocol implements p independent copies of secure oblivious transfer instances with error ϵ.*

In other words, we want the output oblivious transfer instances to satisfy the standard ϵ-correctness and ϵ-privacy requirements for OT. In more detail, the correctness requirement is that the receiver output is correct in all p instances of OT with probability at least $(1 - \epsilon)$. The privacy requirement is that in every instance of the output OT protocol, a corrupt sender cannot output the receiver's choice bit, and a corrupt receiver cannot output the 'other message' of the sender with probability more than $\frac{1}{2} + \epsilon$.

Theorem 3 (Extracting a Single OT). *There exists a $(2\ell + 2n + 1, 1, \ell, 2^{-n})$ one-sided OT extractor according to Definition 1.*

Theorem 4 (High Production Rate). *There exists a $(2\ell + 2n, \frac{n}{\log^2 n}, \ell, \frac{1}{n \log n})$ one-sided OT extractor according to Definition 1.*

We prove these theorems by giving a construction and proof of security of such extractors in the following sections. We will make use of strong seeded extractors in our construction, and we define such extractors below.

Definition 2 (Strong seeded extractors). *A function* Ext : $\{0,1\}^n \times \{0,1\}^d \rightarrow \{0,1\}^m$ *is called a strong seeded extractor for entropy* k *if for any* (n,k)-*source* X *and an independent random variable* Y *that is uniform over* $\{0,1\}^d$, *it holds that* $(\mathsf{Ext}(X,Y),Y) \approx (U_m,Y)$.

Here, U_m is a random variable that is uniformly distributed over m bit strings and is independent of Y, namely (U_m,Y) is a product distribution. In particular, it is known [12,18,34] how to construct strong seeded extractors for any entropy $k = \Omega(1)$ with seed length $d = O(\log n)$ and $m = 0.99\,k$ output bits.

Construction. In Fig. 4, we give the basic construction of an OT extractor that securely obtains a single oblivious transfer from $K = (2\ell + 2n)$ OTs, when a receiver can obtain at most ℓ bits of universal leakage from the joint distribution of sender inputs over all the OTs.

Let $\mathcal{E} : \{0,1\}^K \times \{0,1\}^n \rightarrow \{0,1\}$ be a strong randomness $(K, 2^{-n})$-extractor for seed length $d = O(n)$.

Inputs: Sender $\mathcal{S}$ has private inputs $(x_0, x_1) \in \{0,1\}^{2n}$ and receiver $\mathcal{R}$ has private input $\mathsf{b} \in \{0,1\}$.

Given: $K = 2\ell + 2n$ OTs, such that a malicious receiver can obtain additional ℓ bits of leakage on the joint distribution of all sender inputs.

1. **Invoking OT Correlations:**
    - For $i \in [K]$, $\mathcal{S}$ picks inputs $m_0^i, m_1^i \xleftarrow{\$} \{0,1\}$.
    - For $i \in [K]$, $\mathcal{S}$ invokes the ith OT on input m_0^i, m_1^i.
    - For $i \in [K]$, $\mathcal{R}$ invokes the ith OT on input (the same) choice bit b.
2. **Sender Message:**
    - $\mathcal{S}$ picks random seed $s \xleftarrow{\$} \{0,1\}^d$ for the strong seeded extractor $\mathcal{E}$, and computes $M_0 = \mathcal{E}.\mathsf{Ext}(m_0^1||m_0^2||m_0^3 \ldots m_0^K, s)$ and $M_1 = \mathcal{E}.\mathsf{Ext}(m_1^1||m_1^2||m_1^3 \ldots m_1^K, s)$, where $||$ denotes the concatenation operator.
    - $\mathcal{S}$ sends $y_0 = M_0 \oplus x_0, y_1 = M_1 \oplus x_1$ to $\mathcal{R}$, along with seed s.
3. **Output:** $\mathcal{R}$ computes $x_{\mathsf{b}} = y_{\mathsf{b}} \oplus \mathcal{E}.\mathsf{Ext}(m_{\mathsf{b}}^1||m_{\mathsf{b}}^2||m_{\mathsf{b}}^3 \ldots ||m_{\mathsf{b}}^K, s)$.

Fig. 4. $(2\ell + 2n, 1, \ell, 2^{-n})$ one-sided malicious correlation extractor.

Correctness is immediate from inspection. Intuitively, the protocol is secure against ℓ bits of universal (joint) leakage because setting $K = 2\ell + 2n$ still leaves n bits of high entropy even when the receiver can obtain $2\ell + n$ bits of leakage. Moreover, with ℓ bits of additional universal leakage over all pairs of sender inputs $(m_0^1, m_1^1, m_0^2, m_1^2, \ldots m_0^K, m_1^K)$, the strong seeded extractor extracts an output that is statistically close to uniform, and this suffices to mask the sender input.

The formal proof of security can be found in the full version of this paper.

High Production Rate: It is possible to obtain an improved production rate at the cost of higher simulation error. This follows using techniques developed in prior work [17,36], and the details can be found in the full version.

6   UC Secure Computation in the Malicious Encapsulation Model

Let us consider the stateless protocol described in Sect. 3. In this protocol, the receiver must query PUF_s that he obtained from the sender on a random challenge c, before returning PUF_s to the sender. A malicious receiver cannot have queried PUF_s on both c and $(c \oplus x_0 \oplus x_1)$, because $(x_0 \oplus x_1)$ is chosen by the sender, independently and uniformly at random, and is revealed *only after* the receiver has returned PUF_s. If a malicious receiver was restricted to honestly returning the PUF generated by the sender, by unpredictability of PUF_s, the output of PUF_s on $(c \oplus x_0 \oplus x_1)$ would be a completely unpredictable uniform random variable from the point of view of the receiver, and this sufficed to prove sender security.

However, if a malicious receiver had no such restriction, it could possibly generate a malicious PUF $\widehat{PUF}$ of his own and give it to the sender, in place of the sender's PUF that it was actually supposed to return. The output of $\widehat{PUF}$ would no longer remain unpredictable to the receiver and this would lead to a total break of security. As already pointed out in [7], this can be fixed by having the sender make "test queries" to the PUF he generates, before sending the PUF to the receiver. Indeed, when $\widehat{PUF}$ is generated by the receiver independently of PUF_s, the response of $\widehat{PUF}$ on the sender's random test query will not match the response of PUF_s and the sender will catch such a cheating receiver with overwhelming probability.

However there could be a different attack: a malicious receiver can construct $\widehat{PUF}$ encapsulating PUF_s, such that $\widehat{PUF}$ redirects all test queries to PUF_s (and outputs the value output by PUF_s on the evaluation query), whereas it maliciously answers all protocol queries. In order to rule this out, we ensure that the protocol queries (i.e., the input c that the receiver must query PUF_s with) are generated uniformly at random, by using coin-tossing, combined with cut-and-choose tests to ensure that they are properly used. This is done carefully to ensure that the test queries and protocol queries are identically distributed in the view of $\widehat{PUF}$ (and are revealed only after the receiver has sent $\widehat{PUF}$ to the sender).

This ensures that if a maliciously generated $\widehat{PUF}$ correctly answers all test queries, then with overwhelming probability it must necessarily have answered at least one evaluation query correctly according to the output of PUF_s. At this point, an OT combiner is used to obtain one secure instance of OT.

Let the security parameter be n. The protocol in Fig. 5 UC-securely realizes 2-choose-1 OT in a stronger model, where a malicious party is allowed to create malicious PUFs that encapsulate other honest PUFs (see Sect. 2.1).

Inputs: Sender $\mathcal{S}$ has private inputs $(m_0, m_1) \in \{0,1\}^{2n}$ and Receiver $\mathcal{R}$ has private input $b \in \{0,1\}$.

1. **Coin Flip I:** For $i \in [n]$, $\mathcal{R}$ picks $c_1^i \xleftarrow{\$} \{0,1\}^n$, sends $d^i = $ UC-Com.Commit(c_1^i) to $\mathcal{S}$. $\mathcal{S}$ chooses $c_2^i \xleftarrow{\$} \{0,1\}^n$, sends c_2^i to $\mathcal{R}$. $\mathcal{R}$ computes $c^i = c_1^i \oplus c_2^i$.

2. **Sender Message:** $\mathcal{S}$ generates $\mathsf{PUF}_s : \{0,1\}^n \to \{0,1\}^n$, and does:
   - **Test Queries:** For each $i \in [n]$, choose $\mathsf{TQ}_i \xleftarrow{\$} \{0,1\}^n$ and compute $\mathsf{TR}_i = \mathsf{PUF}_s(\mathsf{TQ}_i)$. Store the pair $(\mathsf{TQ}_i, \mathsf{TR}_i)$.
   - For each $i \in [n]$, choose a pair of random strings $(x_0^i, x_1^i) \xleftarrow{\$} \{0,1\}^{2n}$. Compute $(t_0^i, t_1^i) = $ UC-Com.Commit(x_0^i, x_1^i). Send (t_0^i, t_1^i) and PUF_s to $\mathcal{R}$.

3. **Receiver Message:** For each $i \in [n]$, choose a random string $(c_0^i) \xleftarrow{\$} \{0,1\}^n$ and obtain $r^i = \mathsf{PUF}_s(c^i), r_0^i = \mathsf{PUF}_s(c_0^i)$. Abort if PUF_s aborts, else send PUF_s to $\mathcal{S}$. For $i \in [n]$, pick and send $(\hat{x}_0^i, \hat{x}_1^i) \xleftarrow{\$} \{0,1\}^{2n}$.

4. **Sender Message:** $\mathcal{S}$ does the following.
   - **Verification of TQ:** For each $i \in [n]$, if $\mathsf{TR}_i \neq \mathsf{PUF}_s(\mathsf{TQ}_i)$, abort.
   - For each $i \in [n]$, send $(x_0^i, x_1^i) = $ UC-Com.Decommit(t_0^i, t_1^i) to $\mathcal{R}$.

5. **Receiver Message:** Abort if UC-Com.Decommit(t_0^i, t_1^i) does not verify for any $i \in [n]$. Else pick $b_i \xleftarrow{\$} \{0,1\}$, compute and send $\mathsf{val}^i = c^i \oplus x_{b_i}^i \oplus \hat{x}_{b_i}^i$ to $\mathcal{S}$.

6. **Cut-and-choose:**
   - **Coin Flip II:** $\mathcal{S}$ picks $r_\mathcal{S} \xleftarrow{\$} \{0,1\}^{2K}$, sends $t_\mathcal{S} = $ UC-Com.Commit($r_\mathcal{S}$). $\mathcal{R}$ picks and sends $r_\mathcal{R} \xleftarrow{\$} \{0,1\}^{2K}$. $\mathcal{S}$ sends $r_\mathcal{S} = $ UC-Com.Decommit($t_\mathcal{S}$), and $(\mathcal{S}, \mathcal{R})$ use $(r_\mathcal{S} \oplus r_\mathcal{R})$ to pick a subset I of indices $i \in [n]$, of size $\frac{K}{2}$.
   - For $i \in [I]$, $\mathcal{R}$ sends $c_1^i = $ UC-Com.Decommit(d^i).
   - **Verification:** $\mathcal{S}$ computes $c^i = c_1^i \oplus c_2^i$ and checks if either $\mathsf{val}^i = c^i \oplus x_0^i \oplus \hat{x}_0^i$ OR $\mathsf{val}^i = c^i \oplus x_1^i \oplus \hat{x}_1^i$. If not, $\mathcal{S}$ aborts.

7. **Receiver Message:** For each $i \in [n] \setminus I$, $\mathcal{R}$ sends $bc_i = b_i \oplus b$ to $\mathcal{S}$.

8. **Sender Message:** $\mathcal{S}$ computes $\underline{S_0 = m_0 \bigoplus_{i \in n \setminus I} \mathsf{PUF}_s(\mathsf{val}^i \oplus x_{bc_i}^i \oplus \hat{x}_{bc_i}^i)}$, $\underline{S_1 = m_1 \bigoplus_{i \in n \setminus I} \mathsf{PUF}_s(\mathsf{val}^i \oplus x_{1-bc_i}^i \oplus \hat{x}_{1-bc_i}^i)}$. $\mathcal{S}$ sends (S_0, S_1) to $\mathcal{R}$.

Outputs: $\mathcal{S}$ has no output. $\mathcal{R}$ outputs $m_b := (S_b \oplus r^1 \oplus \ldots \oplus r^n)$.

Fig. 5. OT in the malicious stateless PUF model with encapsulation. We underline all differences from the protocol in the stateless malicious PUF model.

We emphasize that our protocol does not require that honest parties must have the capability to encapsulate PUFs, yet it is secure even when adversarial parties can create encapsulated PUFs. The protocol uses a UC-commitment scheme, secure in the malicious stateless encapsulated PUF model. We use Com to denote the ideal functionality for such a scheme. We construct such a scheme in Sect. 7.

Though the commitment scheme we construct is UC-secure, it is not immediately clear that it composes with the rest of the OT protocol for the same reasons as were described in Sect. 4. Namely, the UC framework seemingly does not capture the possibility of transfer of PUFs across sub-protocols, thus we would like to ensure that our UC-commitment scheme composes with the rest of the protocol even if PUFs created for the commitment scheme are used elsewhere.

Like in Sect. 4, this can be resolved by requiring both parties to return PUFs back to the respective creators at the end of the decommitment phase, and the creators performing simple verification checks to ensure that the correct PUF was returned. If any party fails to return the PUF, the other party aborts the protocol. Therefore, parties cannot pass off PUFs used by some party in a previous sub-protocol as a new PUF in a different sub-protocol.

Correctness

Claim. For all $(m_0, m_1) \in \{0, 1\}^2$ and $b \in \{0, 1\}$, the output of $\mathcal{R}$ equals m_b.

Proof. If $b = 0$, $\mathsf{bc}^i = \mathsf{b}^i$ for all i, and the receiver computes:

$$m'_0 = S_0 \bigoplus_{i \in n \setminus I} r^i = S_0 \bigoplus_{i \in n \setminus I} \mathsf{PUF}_s(c^i) = S_0 \bigoplus_{i \in n \setminus I} \mathsf{PUF}_s(\mathsf{val}^i \oplus x^i_{\mathsf{b}_i})$$
$$= m_0 \bigoplus_{i \in n \setminus I} \mathsf{PUF}_s(\mathsf{val}^i \oplus x^i_{\mathsf{bc}_i}) \bigoplus_{i \in n \setminus I} \mathsf{PUF}_s(\mathsf{val}^i \oplus x^i_{\mathsf{b}_i}) = m_0.$$

If $b = 1$, $1 - \mathsf{bc}^i = \mathsf{b}^i$ for all i, and the receiver computes:

$$m'_1 = S_1 \bigoplus_{i \in n \setminus I} r^i = S_1 \bigoplus_{i \in n \setminus I} \mathsf{PUF}_s(c^i) = S_1 \bigoplus_{i \in n \setminus I} \mathsf{PUF}_s(\mathsf{val}^i \oplus x^i_{\mathsf{b}_i})$$
$$= m_1 \bigoplus_{i \in n \setminus I} \mathsf{PUF}_s(\mathsf{val}^i \oplus x^i_{1-\mathsf{bc}_i}) \bigoplus_{i \in n \setminus I} \mathsf{PUF}_s(\mathsf{val}^i \oplus x^i_{\mathsf{b}_i}) = m_1.$$

The formal proof of security can be found in the full version of the paper.

7   UC Commitments in the Malicious Encapsulation Model

In this section we construct unconditional UC commitments using stateless PUFs. The model we consider is incomparable with respect to the one of [8] since in our model an adversary can encapsulate honest PUFs (see Sect. 2.1) when creating malicious *stateless encapsulated* PUFs. Note that the protocol does not require any honest party to have the ability to encapsulate PUFs, but is secure against parties that do have this ability.

We note that it suffices to construct an extractable commitment scheme that is secure against encapsulation. Indeed, given such a scheme, Damgård and Scafuro [8] show that it is possible to compile the extractable commitment scheme using an additional ideal commitment scheme, to obtain a UC commitment scheme that is secure in the malicious stateless PUF model. Since the compiler of [8] does not require any additional PUFs at all, if the extractable commitment and the ideal commitment are secure against encapsulation attacks, then so is the resulting UC commitment.

Extractable Commitments. We describe how to construct an extractable bit commitment scheme ExtCom = (ExtCom.Commit, ExtCom.Decommit, ExtCom.Extract) that is secure in the malicious stateless PUFs model with encapsulation. We start with the extractable commitment scheme of [8] that is secure against malicious PUFs in the non-encapsulated setting. They crucially rely on the fact that the initial PUF (let's call it PUF_r) sent by the receiver can not be replaced by the committer (as that would be caught using a previously computed test query). To perform extraction, the simulator against a malicious committer observes the queries made by the committer to PUF_r and extracts the committer's bit. However, in the encapsulated setting, the malicious committer could encapsulate the receiver's PUF inside another PUF (let's call it $\widehat{\mathsf{PUF}_r}$) that, for all but one query, answers with the output of PUF_r. For the value that the committer is actually required to query on, $\widehat{\mathsf{PUF}_r}$ responds with a maliciously chosen value. Observe that in the protocol description, this query is chosen only by the committer and hence this is an actual attack. Therefore, except with negligible probability, all the receiver's test queries will be answered by $\widehat{\mathsf{PUF}_r}$ with the output of the receiver's original PUF PUF_r. On the other hand, since the target query is no longer forwarded by $\widehat{\mathsf{PUF}_r}$ to the receiver's original PUF, the simulator does not get access to the target query and hence can not extract the committer's bit.

To overcome this issue, we develop a new technique that forces the malicious committer to reveal the target query to the simulator (but not to the honest receiver). After the committer returns $\widehat{\mathsf{PUF}_r}$, the receiver creates a new PUF (let's call it PUF_R). Now, using the commitment, the receiver queries PUF_R on two values, one of which is guaranteed to be the output of $\widehat{\mathsf{PUF}_r}$ on the target query. The receiver stores these two outputs and sends PUF_R to the committer. The malicious committer now has to query PUF_R with $\widehat{\mathsf{PUF}_r}$'s output on his target query and commit to the value that is given in output by PUF_R (using an ideal commitment scheme). In the decommitment phase, using the previously stored values and the committer's input bit, the receiver can verify that the committer indeed queried PUF_R on the correct value. Observe that since the receiver has precomputed the desired output, the malicious committer will not be able to produce an honest decommitment if he tampers with PUF_R and produces a different output. Therefore, the malicious committer *must* indeed query PUF_R and this can be observed by the simulator and used to extract the committer's bit. Our scheme is described in Fig. 6. We show that this scheme is correct, statistically hiding, and extractable; and give further details in the full version.

Inputs: Committer $\mathcal{C}$ has private input $b \in \{0, 1\}$ and receiver $\mathcal{R}$ has no input.

Commitment Phase:

1. **Receiver Message:** $\mathcal{R}$ does the following:
   - Generate a PUF $\mathsf{PUF}_r : \{0,1\}^{3n} \to \{0,1\}^{3n}$.
   - **Test Queries :** For each $i \in [n]$, choose $\mathsf{TQ}_i \xleftarrow{\$} \{0,1\}^{3n}$, and compute $\mathsf{TR}_i = \mathsf{PUF}_r(\mathsf{TQ}_i)$. Store the pair $(\mathsf{TQ}_i, \mathsf{TR}_i)$. Send PUF_r to $\mathcal{C}$.
2. **Committer Message:** $\mathcal{C}$ does the following:
   - Generate a PUF $\mathsf{PUF}_s : \{0,1\}^n \to \{0,1\}^{3n}$.
   - For each $i \in [n]$, choose $s_i \in \{0,1\}^n$. Compute $\sigma_{s_i} = \mathsf{PUF}_s(s_i)$ and $\sigma_{r_i} = \mathsf{PUF}_r(\sigma_{s_i})$. Send $\mathsf{PUF}_s, \mathsf{PUF}_r$ to $\mathcal{R}$.
3. **Receiver Message:** $\mathcal{R}$ does the following:
   - **Verification :** For each $i \in [n]$, if $\mathsf{TR}_i \neq \mathsf{PUF}_r(\mathsf{TQ}_i)$, abort.
   - For each $i \in [n]$, choose a random string $r_i \xleftarrow{\$} \{0,1\}^{3n}$ and send r_i to $\mathcal{C}$.
4. **Committer Message:** $\mathcal{C}$ does the following: If $b = 0$, set $c_i = \sigma_{s_i}$ for $i \in [n]$, else set $c_i = (\sigma_{s_i} \oplus r_i)$ for $i \in [n]$. Send c_i to $\mathcal{R}$.
5. **Receiver Message:** $\mathcal{R}$ does the following:
   - Generate a PUF $\mathsf{PUF}_R : \{0,1\}^{3n} \to \{0,1\}^{3n}$.
   - For $i \in [n]$, set $y_i = \mathsf{PUF}_R(\mathsf{PUF}_r(c_i))$, $z_i = \mathsf{PUF}_R(\mathsf{PUF}_r(c_i \oplus r_i))$, send PUF_R to $\mathcal{C}$.
6. **Committer Message:** For each $i \in [n]$, $\mathcal{C}$ computes $x_i = \mathsf{PUF}_R(\sigma_{r_i})$. $\mathcal{C}$ computes and sends $t_i = \mathsf{IdealCom.Commit}(x_i)$ to $\mathcal{R}$.

Decommitment Phase:

1. **Committer Message:** $\mathcal{C}$ does the following:
   - Send b to $\mathcal{R}$ and for each $i \in [n]$, send s_i, $\mathsf{IdealCom.Decommit}(x_i)$ to $\mathcal{R}$.
2. **Receiver $\mathcal{R}$ does the following:**
   - For any $i \in [n]$, if $\mathsf{IdealCom.Decommit}(x_i)$ does not verify, output $\perp$.
   - If $b = 0$, $c_i = \mathsf{PUF}_s(s_i)$ and $x_i = y_i$ for all $i \in [n]$, output 0, else $\perp$.
   - If $b = 1$, $c_i = (\mathsf{PUF}_s(s_i) \oplus r_i)$ and $x_i = z_i$ for all $i \in [n]$, output 1, else $\perp$.

Fig. 6. Protocol for Extractable Commitment in the malicious stateless PUF model with encapsulation.

Acknowledgements. We thank the anonymous reviewers for valuable comments, and in particular for suggesting some important updates to our functionality for encapsulated PUFs.

A   Formal Models for PUFs

While we discuss the physical behaviour of PUFs, and their various properties in detail in the full version of the paper, here, we describe the formal modelling of various honest, malicious and encapsulating PUFs.

We model honest PUFs similar to prior work. The ideal functionality for honest PUFs is described in Fig. 7. We assume that in situations where P_i is required to send a message of the form $(\ldots, P_i, \ldots)$, the ideal functionality checks

---

$\mathcal{F}_{\mathsf{HPUF}}$ uses PUF family $\mathcal{P} = (\mathsf{Sample}, \mathsf{Eval})$ with parameters $(rg, d_{\mathsf{noise}}, d_{\mathsf{min}}, m)$. It runs on input the security parameter 1^K, with parties $\mathbb{P} = \{P_1, \ldots, P_n\}$ and adversary $\mathcal{S}$.

- When a party $\hat{P} \in \mathbb{P} \cup \{\mathcal{S}\}$ writes $(\mathsf{init}_{\mathsf{PUF}}, \mathsf{sid}, \hat{P})$ on the input tape of $\mathcal{F}_{\mathsf{HPUF}}$, $\mathcal{F}_{\mathsf{HPUF}}$ checks whether $\mathcal{L}$ already contains a tuple $(\mathsf{sid}, *, *, *, *)$:
  - If this is the case, then turn into the waiting state.
  - Else, draw $\mathsf{id} \leftarrow \mathsf{Sample}_{\mathsf{mode}}(1^K)$ from the PUF family. Put $(\mathsf{sid}, \mathsf{id}, \hat{P}, \mathsf{notrans})$ in $\mathcal{L}$ and write $(\mathsf{initialized}_{\mathsf{PUF}}, \mathsf{sid})$ on the input tape of $\hat{P}$.
- When party P_i writes $(\mathsf{eval}_{\mathsf{PUF}}, \mathsf{sid}, P_i, q)$ on $\mathcal{F}_{\mathsf{HPUF}}$'s input tape, $\mathcal{F}_{\mathsf{HPUF}}$ checks if there exists a tuple $(\mathsf{sid}, \mathsf{id}, P_i, \mathsf{notrans})$ in $\mathcal{L}$.
  - If not, then turn into waiting state.
  - Else, run $a \leftarrow \mathsf{Eval}_{\mathsf{mode}}(1^K, \mathsf{id}, q)$. Write $(\mathsf{response}_{\mathsf{PUF}}, \mathsf{sid}, q, a)$ on P_i's input tape.
- When a party P_i sends $(\mathsf{handover}_{\mathsf{PUF}}, \mathsf{sid}, P_i, P_j)$ to $\mathcal{F}_{\mathsf{HPUF}}$, check if there exists a tuple $(\mathsf{sid}, *, P_i, \mathsf{notrans})$ in $\mathcal{L}$.
  - If not, then turn into waiting state.
  - Else, modify the tuple $(\mathsf{sid}, \mathsf{id}, P_i, \mathsf{notrans})$ to the updated tuple $(\mathsf{sid}, \mathsf{id}, \perp, \mathsf{trans}(P_j))$. Write $(\mathsf{invoke}_{\mathsf{PUF}}, \mathsf{sid}, P_i, P_j)$ on P_i's input tape.
- When the adversary sends $(\mathsf{eval}_{\mathsf{PUF}}, \mathsf{sid}, P_i, q)$ to $\mathcal{F}_{\mathsf{HPUF}}$, check if $\mathcal{L}$ contains a tuple $(\mathsf{sid}, \mathsf{id}, \perp, \mathsf{trans}(*))$.
  - If not, then turn into waiting state.
  - Else, run $a \leftarrow \mathsf{Eval}_{\mathsf{mode}}(1^K, \mathsf{id}, q)$ and return $(\mathsf{response}_{\mathsf{PUF}}, \mathsf{sid}, q, a)$ to P_i.
- When the adversary sends $(\mathsf{ready}_{\mathsf{PUF}}, \mathsf{sid}, P_i)$ to $\mathcal{F}_{\mathsf{HPUF}}$, check if $\mathcal{L}$ contains the tuple $(\mathsf{sid}, \mathsf{id}, \mathsf{mode}, \perp, \mathsf{trans}(P_j))$.
  - If not found, turn into the waiting state.
  - Else, change the tuple $(\mathsf{sid}, \mathsf{id}, \mathsf{mode}, \perp, \mathsf{trans}(P_j))$ to $(\mathsf{sid}, \mathsf{id}, P_i, \mathsf{notrans})$ and write $(\mathsf{handover}_{\mathsf{PUF}}, \mathsf{sid}, P_i)$ on P_j's input tape and store the tuple $(\mathsf{received}_{\mathsf{PUF}}, \mathsf{sid}, P_i)$.
- When the adversary sends $(\mathsf{received}_{\mathsf{PUF}}, \mathsf{sid}, P_i)$ to $\mathcal{F}_{\mathsf{HPUF}}$, check if the tuple $(\mathsf{received}_{\mathsf{PUF}}, \mathsf{sid}, P_i)$ has been stored. If not, return to the waiting state. Else, write this tuple to the input tape of P_i.

Fig. 7. The ideal functionality $\mathcal{F}_{\mathsf{HPUF}}$ for honest PUFs.

that the message is indeed coming from party P_i, if not the ideal functionality $\mathcal{F}_{\mathsf{HPUF}}$ turns into waiting state.

Modeling Malicious PUFs. We model malicious PUFs as in [27]. Their ideal functionality is parameterized by two PUF families in order to handle honestly and maliciously generated PUFs: The honestly generated family is a pair $(\mathsf{Sample}_{\mathsf{normal}}, \mathsf{Eval}_{\mathsf{normal}})$ and the malicious one is $(\mathsf{Sample}_{\mathsf{mal}}, \mathsf{Eval}_{\mathsf{mal}})$. Whenever a party P_i initializes a PUF, then it specifies if it is an honest or a malicious PUF by sending mode $\in \{\mathsf{nor}, \mathsf{mal}\}$ to the functionality $\mathcal{F}_{\mathsf{PUF}}$. The ideal functionality then initialises the appropriate PUF family and it also stores a tag nor or mal representing this family. Whenever the PUF is evaluated, the ideal functionality uses the evaluation algorithm that corresponds to the tag.

The handover procedure is identical to the original formulation of Brzuska et al., where each PUF has a status flag $\in \{\mathsf{trans}(\mathcal{R}), \mathsf{notrans}\}$ that indicates if a PUF is in transit or not. A PUF that is in transit can be queried by the adversary. Thus, whenever a party P_i sends a PUF to P_j, then the status flag is changed from $\mathsf{notrans}$ to trans and the attacker can evaluate the PUF. At some point, the attacker sends $\mathsf{ready_{PUF}}$ to the ideal functionality to indicate that it is not querying the PUF anymore. The ideal functionality then hands the PUF over to P_j and changes the status flag back to $\mathsf{notrans}$. The party P_j may evaluate the PUF. Finally, when the attacker sends the message $\mathsf{received_{PUF}}$ to the ideal functionality, then $\mathcal{F}_{\mathsf{PUF}}$ sends $\mathsf{received_{PUF}}$ to P_i in order to notify P_i that the handover is over. The ideal functionality for malicious PUFs is shown in Fig. 8. We refer the reader to [27] for more details on the different properties of malicious PUFs.

We additionally allow malicious PUFs to maintain $\mathsf{poly}(n)$ a-prior bounded memory. This is done by allowing $\mathsf{Eval_{mal}}$ to be a stateful procedure.

$\mathcal{F}_{\mathsf{MPUF}}$ uses PUF families $\mathcal{P}_1 = (\mathsf{Sample_{normal}}, \mathsf{Eval_{normal}})$ with parameters $(rg, d_{\mathsf{noise}}, d_{\mathsf{min}}, m)$, and $\mathcal{P}_2 = (\mathsf{Sample_{mal}}, \mathsf{Eval_{mal}})$. It runs on input the security parameter 1^K, with parties $\mathbb{P} = \{P_1, \ldots, P_n\}$ and adversary $\mathcal{S}$.

- When a party $\hat{P} \in \mathbb{P} \cup \{\mathcal{S}\}$ writes $(\mathsf{init_{PUF}}, \mathsf{sid}, \mathsf{mode}, \hat{P})$ on the input tape of $\mathcal{F}_{\mathsf{MPUF}}$, where $\mathsf{mode} \in \{\mathsf{normal}, \mathsf{mal}\}$, then $\mathcal{F}_{\mathsf{MPUF}}$ checks whether $\mathcal{L}$ already contains a tuple $(\mathsf{sid}, *, *, *, *)$: If this is the case, then turn into the waiting state. Else, draw $\mathsf{id} \leftarrow \mathsf{Sample_{mode}}(1^K)$ from the PUF family. Put $(\mathsf{sid}, \mathsf{id}, \mathsf{mode}, \hat{P}, \mathsf{notrans})$ in $\mathcal{L}$ and write $(\mathsf{initialized_{PUF}}, \mathsf{sid})$ on the input tape of $\hat{P}$.
- When party $P_i \in \mathbb{P}$ writes $(\mathsf{eval_{PUF}}, \mathsf{sid}, P_i, q)$ on $\mathcal{F}_{\mathsf{MPUF}}$'s input tape, check if there exists a tuple $(\mathsf{sid}, \mathsf{id}, \mathsf{mode}, P_i, \mathsf{notrans})$ in $\mathcal{L}$. If not, then turn into waiting state. Else, run $a \leftarrow \mathsf{Eval_{mode}}(1^K, \mathsf{id}, q)$. Write $(\mathsf{response_{PUF}}, \mathsf{sid}, q, a)$ on P_i's input tape.
- When a party P_i sends $(\mathsf{handover_{PUF}}, \mathsf{sid}, P_i, P_j)$ to $\mathcal{F}_{\mathsf{PUF}}$, check if there exists a tuple $(\mathsf{sid}, *, *, P_i, \mathsf{notrans})$ in $\mathcal{L}$. If not, then turn into waiting state. Else, modify the tuple $(\mathsf{sid}, \mathsf{id}, \mathsf{mode}, P_i, \mathsf{notrans})$ to the updated tuple $(\mathsf{sid}, \mathsf{id}, \mathsf{mode}, \bot, \mathsf{trans}(P_j))$. Write $(\mathsf{invoke_{PUF}}, \mathsf{sid}, P_i, P_j)$ on P_i's input tape to indicate that a handover occurred between P_i and P_j.
- When the adversary sends $(\mathsf{eval_{PUF}}, \mathsf{sid}, P_i, q)$ to $\mathcal{F}_{\mathsf{MPUF}}$, check if $\mathcal{L}$ contains a tuple $(\mathsf{sid}, \mathsf{id}, \mathsf{mode}, \bot, \mathsf{trans}(*))$ or $(\mathsf{sid}, \mathsf{id}, \mathsf{mode}, P_i, \mathsf{notrans})$. If not, then turn into waiting state. Else, run $a \leftarrow \mathsf{Eval_{mode}}(1^K, \mathsf{id}, q)$ and return $(\mathsf{response_{PUF}}, \mathsf{sid}, q, a)$ to P_i.
- When the adversary sends $(\mathsf{ready_{PUF}}, \mathsf{sid}, P_i)$ to $\mathcal{F}_{\mathsf{MPUF}}$, check if $\mathcal{L}$ contains the tuple $(\mathsf{sid}, \mathsf{id}, \mathsf{mode}, \bot, \mathsf{trans}(P_j))$. If not found, turn into the waiting state. Else, change the tuple $(\mathsf{sid}, \mathsf{id}, \mathsf{mode}, \bot, \mathsf{trans}(P_j))$ to $(\mathsf{sid}, \mathsf{id}, \mathsf{mode}, P_j, \mathsf{notrans})$ and write $(\mathsf{handover_{PUF}}, \mathsf{sid}, P_i)$ on P_j's input tape and store the tuple $(\mathsf{received_{PUF}}, \mathsf{sid}, P_i)$.
- When the adversary sends $(\mathsf{received_{PUF}}, \mathsf{sid}, P_i)$ to $\mathcal{F}_{\mathsf{MPUF}}$, check if the tuple $(\mathsf{received_{PUF}}, \mathsf{sid}, P_i)$ has been stored. If not, return to the waiting state. Else, write this tuple to the input tape of P_i.

Fig. 8. The ideal functionality $\mathcal{F}_{\mathsf{MPUF}}$ for malicious PUFs.

$\mathcal{F}_{\text{E-PUF}}$ uses PUF families $\mathcal{P}_1 = (\text{Sample}_{\text{normal}}, \text{Eval}_{\text{normal}})$ with parameters $(rg, d_{\text{noise}}, d_{\text{min}}, m)$, and $\mathcal{P}_2 = (\text{Sample}_{\text{mal}}, \text{Eval}_{\text{mal}})$. It runs on input the security parameter 1^K, with parties $\mathbb{P} = \{P_1, \ldots, P_n\}$ and adversary $\mathcal{S}$ corrupting some parties.

- When a party $P_i \in \mathbb{P} \cup \{\mathcal{S}\}$ writes $(\text{init}_{\text{PUF}}, \text{sid}, \text{mode}, P_i)$ on the input tape of $\mathcal{F}_{\text{E-PUF}}$, where $\text{mode} \in \{\text{normal}, \text{mal}\}$, then $\mathcal{F}_{\text{E-PUF}}$ checks whether $\mathcal{L}$ already contains a tuple $(\text{sid}, \text{id}, *, *, *, *)$ for some id. If it does, turn to waiting state. Else, draw $\text{id} \leftarrow \text{Sample}_{\text{mode}}(1^K)$ from the PUF family. Put $(\text{sid}, \text{id}, \text{mode}, P_i, \text{notrans})$ in $\mathcal{L}$ and write $(\text{initialized}_{\text{PUF}}, \text{sid})$ on the input tape of P_i. If any of the checks failed, turn to waiting state.
- When the adversary P_i writes $\text{reassign}(\text{sid}, \text{sid}', P_i)$ on the input tape of $\mathcal{F}_{\text{E-PUF}}$, check if there exists a tuple $(\text{sid}, \text{id}, \text{mode}, P_i, \text{notrans})$, and check that $\mathcal{L}$ does not already contains a tuple $(\text{sid}, \text{id}, *, *, *, *)$ for some id. If either of the conditions are not met, turn to waiting state. Else, replace the first tuple with $(\text{sid}', \text{id}, \text{mode}, P_i, \text{notrans})$.
- When the adversary P_i writes $(\text{encap}_{\text{PUF}}, \text{sid}, \text{sid}', P_i)$ on the input tape of $\mathcal{F}_{\text{E-PUF}}$, check if there exist tuples $(\text{sid}, *, *, P_i, \text{notrans})$ and $(\text{sid}', *, *, P_i, \text{notrans})$. If such tuples exist, set $\text{owner}(\text{sid}) = \text{sid}'$ [a].
- When party P_i sends $(\text{handover}_{\text{PUF}}, \text{sid}, P_i, P_j)$ to $\mathcal{F}_{\text{E-PUF}}$, check if there exists a tuple $(\text{sid}, *, *, P_i, \text{notrans})$ in $\mathcal{L}$. If not, then turn into waiting state. Else, modify the tuple $(\text{sid}, \text{id}, \text{mode}, P_i, \text{notrans})$ to $(\text{sid}, \text{id}, \text{mode}, \bot, \text{trans}(P_j))$. Write $(\text{invoke}_{\text{PUF}}, \text{sid}, P_i, P_j)$ on P_i's input tape [b].
- When a party $P_i \in \mathbb{P} \cup \{\mathcal{S}\}$ writes $(\text{eval}_{\text{PUF}}, \text{sid}, P_i, q)$ on $\mathcal{F}_{\text{E-PUF}}$'s input tape, check if there exists a tuple $(\text{sid}, \text{id}, \text{mode}, P_i, \text{notrans})$ or $(\text{sid}, \text{id}, \text{mode}, \bot, \text{trans}(*))$ in $\mathcal{L}$. If not, then turn into waiting state. Else, run $a \leftarrow \text{Eval}_{\text{mode}}(1^K, \text{id}, q)$. Write $(\text{response}_{\text{PUF}}, \text{sid}, q, a)$ on P_i's input tape.
- The Eval_{mal} procedure can either makes calls to $\text{Eval}_{\text{normal}}$, or can write $(\text{eval}_{\text{PUF}}, \text{sid}*, \text{sid}, q*)$ on $\mathcal{F}_{\text{E-PUF}}$'s input tape. If Eval_{mal} writes $(\text{eval}_{\text{PUF}}, \text{sid}*, \text{sid}, q*)$ on $\mathcal{F}_{\mathcal{F}_{\text{E-PUF}}}$'s input tape, check if $\text{owner}(\text{sid}*) = \text{sid}$. If not, turn to waiting state. Else, like the previous bullet, check if there exists a tuple $(\text{sid}*, \text{id}, \text{mode}, P_i, \text{notrans})$ or $(\text{sid}*, \text{id}, \text{mode}, \bot, \text{trans}(*))$ in $\mathcal{L}$. If not, then turn into waiting state. Else, run $a \leftarrow \text{Eval}_{\text{mode}}(1^K, \text{id}, q)$ and return $(\text{response}_{\text{PUF}}, \text{sid}*, q, a)$ as output to sid.
- When the adversary sends $(\text{ready}_{\text{PUF}}, \text{sid}, P_i)$ to $\mathcal{F}_{\text{E-PUF}}$, check if $\mathcal{L}$ contains $(\text{sid}, \text{id}, \text{mode}, \bot, \text{trans}(P_j))$. If not, turn into waiting state. Else, change $(\text{sid}, \text{id}, \text{mode}, \bot, \text{trans}(P_j))$ to $(\text{sid}, \text{id}, \text{mode}, P_j, \text{notrans})$, write $(\text{handover}_{\text{PUF}}, \text{sid}, P_i)$ on P_j's input tape and store $(\text{received}_{\text{PUF}}, \text{sid}, P_i)$.
- When the adversary sends $(\text{received}_{\text{PUF}}, \text{sid}, P_i)$ to $\mathcal{F}_{\text{E-PUF}}$, check if $(\text{received}_{\text{PUF}}, \text{sid}, P_i)$ has been stored. If not, return to waiting state. Else, write this tuple to the input tape of P_i.

---

[a] Intuitively, when a (malicious) party encapsulates a PUF, this sets the outer PUF as owner of the inner PUF. Even the adversary can access the inner PUF via evaluation queries to outer PUF. This step permits multiple iterative encapsulations.
[b] Handover does not change the owner (outer PUF) of an (inner) encapsulated PUF.

Fig. 9. The ideal functionality $\mathcal{F}_{\text{E-PUF}}$ for malicious PUFs that may *encapsulate* PUFs.

Modeling Encapsulating PUFs. We model malicious PUFs that can encapsulate functionalities as in [6,27]. This functionality formalizes the intuition that an honest user can create a PUF implementing a random function, but an adversary given the PUF can only observe its input/output characteristics.

$\mathcal{F}_{\text{E-PUF}}$ models the PUF (sent by party P_i to party P_j) encapsulating some functionality M_{ij}. The changes from the previous definition [27] that we make is that M_{ij} is now an oracle machine (instead of a functionality) which can make evaluation calls to other PUFs itself. The ideal functionality for malicious PUFs that could possibly encapsulate honest PUFs, is described in Fig. 9. $\mathcal{F}_{\text{E-PUF}}$ models the following sequence of events: (1) a party P_i samples a random PUF from the challenge space, (2) P_i then gives this PUF to another party P_j (the receiver) who can use the PUF as a black-box implementing M_{ij}, (3) On giving M_{ij}, P_i loses oracle access to all PUFs of which it was previously the owner but which M_{ij} has oracle access to. Figure 9 has the formal description of $\mathcal{F}_{\text{E-PUF}}$ based on such an algorithm M_{ij}.

We assume that every PUF has a *single* calling procedure known as its *owner*. This owner can either be a party, or another PUF (in the case of adversarially generated PUFs). This models (refer to the first bullet in Fig. 9) the fact that an adversary that receives a PUF implementing M_{xy} can either keep the PUF to make calls later or incorporate the functionality of this PUF in a black-box manner into another (maliciously created) PUF, but cannot do both. The evaluation procedure for a malicious encapsulating outer PUF, carefully checks that the outer PUF has ownership of inner PUFs (refer the second bullet in Fig. 9), before allowing the malicious outer evaluation procedure oracle access to any inner PUF. The handover operation (described in the third bullet in Fig. 9) is similarly carefully modified to ensure that the party that receives an encapsulated PUF can only access the inner PUF via evaluation queries to the outer PUF. Each PUF is uniquely identified by an identifier known as id.

Finally, we note that our model may also allow an adversary to "dismount" a PUF, i.e., separate out its inner component PUFs. For simplicity, we choose to not formalize this requirement. Our protocols trivially remain secure in this model since we never require the honest parties to hand over any "encap"-PUFs back to the adversary, where an "encap"-PUF is a malicious PUF that may be encapsulating honest PUFs.

References

1. Agrawal, S., Ananth, P., Goyal, V., Prabhakaran, M., Rosen, A.: Lower bounds in the hardware token model. In: Lindell, Y. (ed.) TCC 2014. LNCS, vol. 8349, pp. 663–687. Springer, Heidelberg (2014). doi:10.1007/978-3-642-54242-8_28
2. Armknecht, F., Moriyama, D., Sadeghi, A.-R., Yung, M.: Towards a unified security model for physically unclonable functions. In: Sako, K. (ed.) CT-RSA 2016. LNCS, vol. 9610, pp. 271–287. Springer, Cham (2016). doi:10.1007/978-3-319-29485-8_16
3. Boureanu, I., Ohkubo, M., Vaudenay, S.: The limits of composable crypto with transferable setup devices. In: Proceedings of the 10th ACM Symposium on Information, Computer and Communications Security, ASIA CCS 2015, Singapore, 14–17 April 2015, pp. 381–392. ACM (2015)

4. Brzuska, C., Fischlin, M., Schröder, H., Katzenbeisser, S.: Physically unclone-able functions in the universal composition framework. In: Rogaway, P. (ed.) CRYPTO 2011. LNCS, vol. 6841, pp. 51–70. Springer, Heidelberg (2011). doi:10.1007/978-3-642-22792-9_4

5. Canetti, R.: Universally composable security: a new paradigm for cryptographic protocols. In: Foundations of Computer Science (FOCS 2001), pp. 136–145 (2001)

6. Chandran, N., Goyal, V., Sahai, A.: New constructions for UC secure computation using tamper-proof hardware. In: Smart, N. (ed.) EUROCRYPT 2008. LNCS, vol. 4965, pp. 545–562. Springer, Heidelberg (2008). doi:10.1007/978-3-540-78967-3_31

7. Dachman-Soled, D., Fleischhacker, N., Katz, J., Lysyanskaya, A., Schröder, D.: Feasibility and infeasibility of secure computation with malicious PUFs. In: Garay, J.A., Gennaro, R. (eds.) CRYPTO 2014. LNCS, vol. 8617, pp. 405–420. Springer, Heidelberg (2014). doi:10.1007/978-3-662-44381-1_23

8. Damgård, I., Scafuro, A.: Unconditionally secure and universally composable commitments from physical assumptions. In: Sako, K., Sarkar, P. (eds.) ASIACRYPT 2013. LNCS, vol. 8270, pp. 100–119. Springer, Heidelberg (2013). doi:10.1007/978-3-642-42045-0_6

9. Dodis, Y., Ostrovsky, R., Reyzin, L., Smith, A.: Fuzzy extractors: how to generate strong keys from biometrics and other noisy data. SIAM J. Comput. 38(1), 97–139 (2008)

10. Döttling, N., Kraschewski, D., Müller-Quade, J., Nilges, T.: General statistically secure computation with bounded-resettable hardware tokens. In: Dodis, Y., Nielsen, J.B. (eds.) TCC 2015. LNCS, vol. 9014, pp. 319–344. Springer, Heidelberg (2015). doi:10.1007/978-3-662-46494-6_14

11. Döttling, N., Mie, T., Müller-Quade, J., Nilges, T.: Implementing resettable UC-functionalities with untrusted tamper-proof hardware-tokens. In: Sahai, A. (ed.) TCC 2013. LNCS, vol. 7785, pp. 642–661. Springer, Heidelberg (2013). doi:10.1007/978-3-642-36594-2_36

12. Dvir, Z., Kopparty, S., Saraf, S., Sudan, M.: Extensions to the method of multiplicities, with applications to Kakeya sets and mergers. SIAM J. Comput. 42(6), 2305–2328 (2013)

13. Eichhorn, I., Koeberl, P., van der Leest, V.: Logically reconfigurable PUFs: memory-based secure key storage. In: Proceedings of the Sixth ACM Workshop on Scalable Trusted Computing, STC 2011, pp. 59–64. ACM, New York (2011)

14. Goyal, V., Ishai, Y., Sahai, A., Venkatesan, R., Wadia, A.: Founding cryptography on tamper-proof hardware tokens. In: Micciancio, D. (ed.) TCC 2010. LNCS, vol. 5978, pp. 308–326. Springer, Heidelberg (2010). doi:10.1007/978-3-642-11799-2_19

15. Goyal, V., Maji, H.K.: Stateless cryptographic protocols. In: Ostrovsky, R. (ed.) IEEE 52nd Annual Symposium on Foundations of Computer Science, FOCS 2011, Palm Springs, CA, USA, 22–25 October 2011, pp. 678–687. IEEE Computer Society (2011)

16. Guajardo, J., Kumar, S.S., Schrijen, G.-J., Tuyls, P.: FPGA intrinsic PUFs and their use for IP protection. In: Paillier, P., Verbauwhede, I. (eds.) CHES 2007. LNCS, vol. 4727, pp. 63–80. Springer, Heidelberg (2007). doi:10.1007/978-3-540-74735-2_5

17. Gupta, D., Ishai, Y., Maji, H.K., Sahai, A.: Secure computation from leaky correlated randomness. In: Gennaro, R., Robshaw, M. (eds.) CRYPTO 2015. LNCS, vol. 9216, pp. 701–720. Springer, Heidelberg (2015). doi:10.1007/978-3-662-48000-7_34

18. Guruswami, V., Umans, C., Vadhan, S.P.: Unbalanced expanders and randomness extractors from Parvaresh-Vardy codes. J. ACM 56(4) (2009)

19. Hazay, C., Lindell, Y.: Constructions of truly practical secure protocols using stan-dardsmartcards. In: Proceedings of the 2008 ACM Conference on Computer and Communications Security, CCS 2008, Alexandria, Virginia, USA, 27–31 October 2008, pp. 491–500 (2008)
20. Hazay, C., Polychroniadou, A., Venkitasubramaniam, M.: Composable security in the tamper-proof hardware model under minimal complexity. In: Hirt, M., Smith, A. (eds.) TCC 2016. LNCS, vol. 9985, pp. 367–399. Springer, Heidelberg (2016). doi:10.1007/978-3-662-53641-4_15
21. Ishai, Y., Kushilevitz, E., Ostrovsky, R., Sahai, A.: Extracting correlations. In: 50th Annual IEEE Symposium on Foundations of Computer Science, FOCS 2009, Atlanta, Georgia, USA, 25–27 October 2009, pp. 261–270. IEEE Computer Society (2009)
22. Järvinen, K., Kolesnikov, V., Sadeghi, A., Schneider, T.: Efficient secure two-party computation with untrusted hardware tokens (full version). In: Sadeghi, A.R., Naccache, D. (eds.) Towards Hardware-Intrinsic Security - Foundations and Practice, pp. 367–386. Springer, Heidelberg (2010)
23. Järvinen, K., Kolesnikov, V., Sadeghi, A.-R., Schneider, T.: Embedded SFE: offloading server and network using hardware tokens. In: Sion, R. (ed.) FC 2010. LNCS, vol. 6052, pp. 207–221. Springer, Heidelberg (2010). doi:10.1007/978-3-642-14577-3_17
24. Katz, J.: Universally composable multi-party computation using tamper-proof hardware. In: Naor, M. (ed.) EUROCRYPT 2007. LNCS, vol. 4515, pp. 115–128. Springer, Heidelberg (2007). doi:10.1007/978-3-540-72540-4_7
25. Koçabas, Ü., Sadeghi, A.R., Wachsmann, C., Schulz, S.: Poster: practical embedded remote attestation using physically unclonable functions. In: ACM Conference on Computer and Communications Security, pp. 797–800 (2011)
26. Kolesnikov, V.: Truly efficient string oblivious transfer using resettable tamper-proof tokens. In: Micciancio, D. (ed.) TCC 2010. LNCS, vol. 5978, pp. 327–342. Springer, Heidelberg (2010). doi:10.1007/978-3-642-11799-2_20
27. Ostrovsky, R., Scafuro, A., Visconti, I., Wadia, A.: Universally composable secure computation with (malicious) physically uncloneable functions. In: Johansson, T., Nguyen, P.Q. (eds.) EUROCRYPT 2013. LNCS, vol. 7881, pp. 702–718. Springer, Heidelberg (2013). doi:10.1007/978-3-642-38348-9_41
28. Pappu, R.S., Recht, B., Taylor, J., Gershenfeld, N.: Physical one-way functions. Science **297**, 2026–2030 (2002)
29. Pappu, R.S.: Physical one-way functions. Ph.D. thesis. MIT (2001)
30. Rührmair, U.: On the security of PUF protocols under bad PUFs and PUFs-inside-PUFs attacks. Cryptology ePrint Archive, Report 2016/322 (2016). http://eprint.iacr.org/
31. Sadeghi, A.R., Visconti, I., Wachsmann, C.: Enhancing RFID security and privacy by physically unclonable functions. In: Sadeghi, A.R., Naccache, D. (eds.) Towards Hardware-Intrinsic Security. Information Security and Cryptography, pp. 281–305. Springer, Heidelberg (2010)
32. Sadeghi, A.R., Visconti, I., Wachsmann, C.: PUF-enhanced RFID security and privacy. In: Workshop on Secure Component and System Identification (SECSI) (2010)
33. Standaert, F.-X., Malkin, T.G., Yung, M.: Does physical security of crypto-graphic devices need a formal study? (Invited talk). In: Safavi-Naini, R. (ed.) ICITS 2008. LNCS, vol. 5155, p. 70. Springer, Heidelberg (2008). doi:10.1007/978-3-540-85093-9_7

34. Ta-Shma, A., Umans, C.: Better condensers and new extractors from Parvaresh-Vardy codes. In: Proceedings of the 27th Conference on Computational Complexity, CCC 2012, Porto, Portugal, 26–29 June 2012, pp. 309–315. IEEE (2012)
35. Tuyls, P., Batina, L.: RFID-tags for anti-counterfeiting. In: Pointcheval, D. (ed.) CT-RSA 2006. LNCS, vol. 3860, pp. 115–131. Springer, Heidelberg (2006). doi:10. 1007/11605805_8
36. Vadhan, S.P.: Constructing locally computable extractors and cryptosystems in the bounded-storage model. J. Cryptol. **17**(1), 43–77 (2004)
37. Wolf, S., Wullschleger, J.: Oblivious transfer is symmetric. In: Vaudenay, S. (ed.) EUROCRYPT 2006. LNCS, vol. 4004, pp. 222–232. Springer, Heidelberg (2006). doi:10.1007/11761679_14

Lattice Attacks and Constructions III

Lattice Attacks and Constructions III

Private Puncturable PRFs from Standard Lattice Assumptions

Dan Boneh[1], Sam Kim[1($\boxtimes$)], and Hart Montgomery[2]

[1] Stanford University, Stanford, USA
skim13@cs.stanford.edu
[2] Fujitsu Laboratories of America, Sunnyvale, USA

Abstract. A puncturable pseudorandom function (PRF) has a master key k that enables one to evaluate the PRF at all points of the domain, and has a punctured key k_x that enables one to evaluate the PRF at all points but one. The punctured key k_x reveals no information about the value of the PRF at the punctured point x. Punctured PRFs play an important role in cryptography, especially in applications of indistinguishability obfuscation. However, in previous constructions, the punctured key k_x completely reveals the punctured point x: given k_x it is easy to determine x. A *private* puncturable PRF is one where k_x reveals nothing about x. This concept was defined by Boneh, Lewi, and Wu, who showed the usefulness of private puncturing, and gave constructions based on multilinear maps. The question is whether private puncturing can be built from a standard (weaker) cryptographic assumption.

We construct the first privately puncturable PRF from standard lattice assumptions, namely learning with errors (LWE) and 1 dimensional short integer solutions (1D-SIS), which have connections to worst-case hardness of general lattice problems. Our starting point is the (non-private) PRF of Brakerski and Vaikuntanathan. We introduce a number of new techniques to enhance this PRF, from which we obtain a privately puncturable PRF. In addition, we also study the simulation based definition of private constrained PRFs for general circuits, and show that the definition is not satisfiable.

1 Introduction

A pseudorandom function (PRF) [GGM86] is a function $F\colon \mathcal{K}\times\mathcal{X}\to\mathcal{Y}$ that can be computed by a deterministic polynomial time algorithm: on input $(k,x)\in\mathcal{K}\times\mathcal{X}$ the algorithm outputs $F(k,x)\in\mathcal{Y}$. The PRF F is said to be a constrained PRF [BW13,KPTZ13,BGI14] if one can derive constrained keys from the master PRF key k. Each constrained key k_g is associated with a predicate $g\colon\mathcal{X}\to\{0,1\}$, and this k_g enables one to evaluate $F(k,x)$ for all $x\in\mathcal{X}$ for which $g(x)=1$, but at no other points of $\mathcal{X}$. A constrained PRF is secure if given constrained keys for predicates $g_1,\dots,g_Q$, of the adversary's choosing, the adversary cannot distinguish the PRF from a random function at points not covered by the given keys, namely at points x where $g_1(x)=\cdots=g_Q(x)=0$. We review the precise definition in Sect. 3.

© International Association for Cryptologic Research 2017
J.-S. Coron and J.B. Nielsen (Eds.): EUROCRYPT 2017, Part I, LNCS 10210, pp. 415–445, 2017.
DOI: 10.1007/978-3-319-56620-7_15

The simplest constraint, called a *puncturing constraint*, is a constraint that enables one to evaluate the PRF at all points except one. For $x \in \mathcal{X}$ we denote by k_x a *punctured key* that lets one evaluate the PRF at all points in $\mathcal{X}$, except for the punctured point x. Given the key k_x, the adversary should be unable to distinguish $F(k, x)$ from a random element in $\mathcal{Y}$. Puncturable PRFs have found numerous applications in cryptography [BW13,KPTZ13,BGI14], most notably in applications of indistinguishability obfuscation ($i\mathcal{O}$) [SW14]. Note that two punctured keys, punctured at two different points, enable the evaluation of the PRF at all points in the domain $\mathcal{X}$, and are therefore equivalent to the master PRF key k. For this reason, for puncturing constraints, we are primarily interested in settings where the adversary is limited to obtaining at most a single punctured key, punctured at a point of its choice. At the punctured point, the adversary should be unable to distinguish the value of the PRF from random.

PRFs supporting puncturing constraints can be easily constructed from the tree-based PRF of [GGM86], as discussed in [BW13,KPTZ13,BGI14]. Notice, however, that a punctured key k_x completely reveals what the point x is. An adversary that is given k_x can easily tell where the key was punctured.

Private Puncturing. Can we construct a PRF that can be punctured *privately*? The adversary should learn nothing about x from the punctured key k_x. More generally, Boneh, Lewi, and Wu [BLW17] define private constrained PRFs, where a constrained key k_g reveals nothing about the predicate g. They present applications of private constraint PRFs to constructing software watermarking [CHN+16], deniable encryption [CDNO97], searchable encryption, and more. They also construct private constrained PRFs from powerful tools, such as multilinear maps and $i\mathcal{O}$.

Several of the applications for private constraints in [BLW17] require only private puncturing. Here we describe one such application, namely the connection to distributed point functions (DPF) [GI14,BGI15] and 2-server private information retrieval (PIR) [CKGS98]. In a DPF, the key generation algorithm is given a point $x^* \in \mathcal{X}$ and outputs two keys k_0 and k_1. The two keys are equivalent, except at the point x^*. More precisely, $F(k_0, x) = F(k_1, x)$ for all key $x \neq x^*$ and $F(k_0, x^*) \neq F(k_1, x^*)$. A DPF is secure if given one of k_0 or k_1, the adversary learns nothing about x^*. In [GI14] the authors show that DPFs give a simple and efficient 2-server PIR scheme. They give an elegant DPF construction from one-way functions.

A privately puncturable PRF is also a DPF: set k_0 to be the master PRF key k, and set the key k_1 to be the punctured key k_{x^*}, punctured at x^*. The privacy property ensures that this is a secure DPF. However, there is an important difference between a DPF and a privately puncturable PRF. DPF key generation takes the punctured point x^* as input, and generates the two keys k_0, k_1. In contrast, private puncturing works differently: one first generates the master key k, and at some time later asks for a punctured key k_{x^*} at some point x^*. That is, the punctured point is chosen *adaptively* after the master key is generated. This adaptive capability gives rise to a 2-server PIR scheme that has a surprising property: one of the servers can be offline. In particular, one of the servers does its PIR computation *before* the PIR query is chosen, sends the result to the client, and goes offline. Later, when the client chooses the PIR query, it only talks to the second server.

Our Contribution. We construct the first *privately* puncturable PRF from the learning with errors problem (LWE) [Reg09] and the one-dimensional short integer solution problem (1D-SIS) [Ajt96,BV15], which are both related to worst-case hardness of general lattice problems. We give a brief overview of the construction here, and give a detailed overview in Sect. 2.

Our starting point is the elegant LWE-based PRF of Brakerski and Vaikuntanathan [BV15], which is a constrained PRF for general circuits, but is only secure if at most one constrained key is published (publishing two constrained keys reveals the master key). This PRF is not private because the constraint is part of the constrained key and is available in the clear. As a first attempt, we try to make this PRF private by embedding in the constrained key, an FHE encryption of the constraint, along with an encryption of the FHE decryption key (a similar structure is used in the predicate encryption scheme of [GVW15b]). Now the constraint is hidden, but PRF evaluation requires an FHE decryption, which is a problem. We fix this in a number of steps, as described in the next section. To prove security, we introduce an additional randomizing component as part of the FHE plaintext to embed an LWE instance in the challenge PRF evaluation.

We prove security of our private puncturable PRF in the selective setting, where the adversary commits ahead of time to the punctured point x where it will be challenged. To obtain adaptive security, where the punctured point is chosen adaptively, we use standard complexity leveraging [BB04].

In addition to our punctured PRF construction, we show in Sect. 6 that, for general function constraints, a simulation based definition of privacy is impossible. This complements [BLW17] who show that a game-based definition of privacy is achievable assuming the existence of $i\mathcal{O}$. To prove the impossibility, we show that even for a single key, a simulation-secure privately constrained PRF for general functions, implies a simulation secure functional encryption for general functions, which was previously shown to be impossible [BSW11,AGVW13].

Finally, our work raises a number of interesting open problems. First, our techniques work well to enable private puncturing, but do not seem to generalize to arbitrary circuit constraints. It would be a significant achievement if one could use LWE/SIS to construct a *private* constrained PRF for arbitrary circuits, even in the single-key case. Also, can we construct an LWE/SIS-based *adaptively* secure private puncturable PRF, without relying on complexity leveraging? We discuss these questions in more detail in Sect. 7.

1.1   Related Work

PRFs from LWE. The first PRF construction from the learning with errors assumption was given by Banerjee, Peikert, and Rosen in [BPR12]. Subsequent

PRF constructions from LWE gave the first key-homomorphic PRFs [BLMR13, BP14]. The constructions of [BV15,BFP+15] generalized the previous works to the setting of constrained PRFs.

Constrained PRFs. The notion of constrained PRFs was first introduced in three independent works [BW13,KPTZ13,BGI14] and since then, there have been a number of constructions from different assumptions. We briefly survey the state of the art. The standard GGM tree [GGM86] gives PRFs for simple constraints such as prefix-fixing or puncturing [BW13,KPTZ13,BGI14]. Bilinear maps give left/right constraints but in the random oracle model [BW13]. LWE gives general circuit constraints, but only when a single constrained key is released [BV15]. Multilinear maps and indistinguishability obfuscation provide general circuit constraints, and even for constraints represented as Turing machines with unbounded inputs [BW13,BZ14,BFP+15,CRV14,AFP16, DKW16], as well as constrained verifiable random functions [Fuc14]. Several works explore how to achieve adaptive security [FKPR14,BV15,HKW15, HKKW14].

  Private constrained PRFs were introduced by Boneh, Lewi, and Wu [BLW17]. They construct a privately constrained PRF for puncturing and bit-fixing constraints from multilinear maps, and for circuit constraints using indistinguishability obfuscation.

ABE and PE from LWE. The techniques used in this work build upon a series of works in the area of *attribute-based encryption* [SW05] and *predicate encryption* [BW07,KSW08] from LWE. These include constructions of [ABB10,GVW15a,BGG+14,GV15,BV16,BCTW16], and predicate encryption constructions of [AFV11,GMW15,GVW15b].[1]

Concurrent Work. In an independent and concurrent work, Canetti and Chen [CC17] construct a single-key privately constrained PRF for general NC^1 circuits from LWE. Their techniques are very different from the ones used in this work as their construction relies on instances of the graph-induced multilinear maps construction by Gentry, Gorbunov, and Halevi [GGH15] that can be reduced to LWE. They also analyze their construction with respect to a simulation-based definition. We note that the simulation-based definition that we consider in Sect. 6 is stronger than their definition and therefore, the impossibility that we show does not apply to their construction.

2   Overview of the Main Construction

In this section, we provide a general overview of our main construction. The complete construction and proof of security are provided in Sect. 5.1.

  Recall that the LWE assumption states that for a uniform vector $s \in \mathbb{Z}_q^n$ and a matrix $A \in \mathbb{Z}_q^{n \times m}$ for an appropriately chosen n, m, q, it holds that

---

[1] We note that LWE based predicate encryption constructions satisfy a weaker security property often referred to as *weak attribute-hiding* than as is defined in [BW07, KSW08].

$(\mathbf{A}, \mathbf{s}^T\mathbf{A} + \mathbf{e}^T)$ is indistinguishable from uniform where $\mathbf{e}$ is sampled from an appropriate low-norm error distribution. We present the outline ignoring the precise generation or evolution of $\mathbf{e}$ and just refer to it as noise.

Embedding Circuits into Matrices. Our starting point is the single-key constrained PRF of [BV15], which builds upon the ABE construction of [BGG+14] and the PRF of [BP14]. At a high level, the ABE of [BGG+14] encodes an attribute vector $\mathbf{x} \in \{0,1\}^\ell$ as a vector

$$\mathbf{s}^T(\mathbf{A}_1 + x_1 \cdot \mathbf{G} \mid \cdots \mid \mathbf{A}_\ell + x_\ell \cdot \mathbf{G}) + \mathsf{noise} \ \in \mathbb{Z}_q^{m\ell}, \qquad (2.1)$$

for public matrices $\mathbf{A}_1, \ldots, \mathbf{A}_\ell$ in $\mathbb{Z}_q^{n\times m}$, a secret random vector $\mathbf{s}$ in $\mathbb{Z}_q^n$, and a specific fixed "gadget matrix" $\mathbf{G} \in \mathbb{Z}_q^{n\times m}$. This encoding allows for fully homomorphic operations on the attributes, while keeping the noise small. In particular, given $\mathbf{x}$ and a poly-size circuit $f : \{0,1\}^\ell \to \{0,1\}$, one can compute from (2.1), the vector

$$\mathbf{s}^T(\mathbf{A}_f + f(\mathbf{x}) \cdot \mathbf{G}) + \mathsf{noise} \ \in \mathbb{Z}_q^m \qquad (2.2)$$

where the matrix $\mathbf{A}_f$ depends only on the function f, and not on the underlying attribute $\mathbf{x}$. This implies a homomorphic operation on the matrices $\mathbf{A}_1, \ldots, \mathbf{A}_\ell$ defined as $\mathsf{Eval}_{\mathsf{pk}}(f, \mathbf{A}_1, \ldots, \mathbf{A}_\ell) \to \mathbf{A}_f$.

This homomorphic property leads to the following puncturable PRF. Let $\mathsf{eq}(\mathbf{x}^*, \mathbf{x})$ be the equality check circuit (represented as NAND gates) defined as follows:

$$\mathsf{eq}(\mathbf{x}^*, \mathbf{x}) = \begin{cases} 1 \text{ if } \mathbf{x}^* = \mathbf{x}, \\ 0 \text{ otherwise.} \end{cases}$$

For $\mathbf{x} = (x_1, \ldots, x_\ell) \in \{0,1\}^\ell$ define the PRF as:

$$\mathsf{PRF}_{\mathbf{s}}(\mathbf{x}) := \lfloor \mathbf{s}^T \cdot \mathbf{A}_{\mathsf{eq}} \rceil_p \ \in \mathbb{Z}_p^m \quad \text{where} \quad \mathbf{A}_{\mathsf{eq}} := \mathsf{Eval}_{\mathsf{pk}}(\mathsf{eq}, \mathbf{B}_1, \ldots, \mathbf{B}_\ell, \mathbf{A}_{x_1}, \ldots, \mathbf{A}_{x_\ell}).$$

Here $\mathbf{s} \in \mathbb{Z}_q^n$ is the master secret key, and the matrices $\mathbf{A}_0, \mathbf{A}_1, \mathbf{B}_1, \ldots, \mathbf{B}_\ell$ are random public matrices in $\mathbb{Z}_q^{n\times m}$ chosen at setup. Note that $\mathbf{A}_{\mathsf{eq}}$ is a function of $\mathbf{x}$. The operation $\lfloor \cdot \rceil_p$ is component-wise rounding that maps an element in $\mathbb{Z}_q$ to an element in $\mathbb{Z}_p$ for an appropriately chosen p, where $p < q$.

Next, define the punctured key at the point $\mathbf{x}^* = (x_1^*, \ldots, x_\ell^*) \in \{0,1\}^\ell$ as:

$$k_{\mathbf{x}^*} = \big(\mathbf{x}^*, \ \mathbf{s}^T \cdot (\mathbf{A}_0 + 0 \cdot \mathbf{G} \mid \mathbf{A}_1 + 1 \cdot \mathbf{G} \quad \mid \quad \mathbf{B}_1 + x_1^* \cdot \mathbf{G} \mid \cdots \mid \mathbf{B}_\ell + x_\ell^* \cdot \mathbf{G}) + \mathsf{noise}\big). \qquad (2.3)$$

To use this key to evaluate the PRF at a point $\mathbf{x} \in \{0,1\}^\ell$, the user homomorphically evaluates the equality check circuit $\mathsf{eq}(\mathbf{x}^*, \mathbf{x})$, as in (2.2), to obtain the vector $\mathbf{s}^T(\mathbf{A}_{\mathsf{eq}} + \mathsf{eq}(\mathbf{x}^*, \mathbf{x}) \cdot \mathbf{G}) + \mathsf{noise}$. Rounding this vector gives the correct PRF value whenever $\mathsf{eq}(\mathbf{x}^*, \mathbf{x}) = 0$, namely $\mathbf{x} \neq \mathbf{x}^*$, as required. A security argument as in [BV15] proves that with some minor modifications, this PRF is a secure (non-private) puncturable PRF, assuming that the LWE problem is hard.

FHE to Hide Puncture Point. The reason why the construction above is not private is because to operate on the ABE encodings, one needs the description of the attributes. Therefore, the punctured key must include the point $\mathbf{x}^*$ in the clear, for the evaluator to run the equality check circuit on the punctured key (2.3).

Our plan to get around this limitation is to first encrypt the attributes $(x_1^*, \ldots, x_\ell^*)$ using a fully homomorphic encryption (FHE) scheme before embedding it as the attributes. In particular, we define our punctured key to be

$$k_{\mathbf{x}^*} = (\mathsf{ct}, \quad \mathbf{s}^T \cdot (\mathbf{A}_0 + 0 \cdot \mathbf{G} \mid \mathbf{A}_1 + 1 \cdot \mathbf{G} \quad \mid \quad \mathbf{B}_1 + \mathsf{ct}_1 \cdot \mathbf{G} \mid \cdots \mid \mathbf{B}_z + \mathsf{ct}_z \cdot \mathbf{G}$$
$$\mid \mathbf{C}_1 + \mathsf{sk}_1 \cdot \mathbf{G} \mid \cdots \mid \mathbf{C}_t + \mathsf{sk}_t \cdot \mathbf{G}) + \mathsf{noise}),$$

where $\mathsf{ct} \in \mathbb{Z}_q^z$ is an FHE encryption of the punctured point $\mathbf{x}^*$, and $\mathsf{sk} \in \mathbb{Z}_q^t$ is the FHE secret key. While it is not clear how to use this key to evaluate the PRF, at least the punctured point $\mathbf{x}^*$ is not exposed in the clear. One can show that the components of $k_{\mathbf{x}^*}$ that embed the secret key sk do not leak information about sk.

Now, given $\mathbf{x} \in \{0, 1\}^\ell$, one can now run the equality check operation inside the FHE ciphertext, which gives the *encrypted* result of the equality check circuit. The question is how the evaluator can extract this result from the ciphertext. To do this, we take advantage of another property of the underlying ABE: to homomorphically multiply two attributes, one requires knowledge of just one of the attributes, not both. This means that even without the knowledge of the FHE secret key sk, the evaluator can compute the inner product of sk and ct. Recall that for lattice-based FHE schemes (e.g., [GSW13]), the decryption operation is the rounding of the inner product of the ciphertext with the FHE secret key. This technique was also used in the lattice-based predicate encryption scheme of [GVW15b].

Rounding Away FHE Noise. The problem with the approach above is that we cannot compute the full FHE decryption. We can only compute the first decryption step, the inner product. The second step, rounding, cannot be done while keeping the FHE decryption key secret. Computing just the inner product produces the FHE plaintext, but offset by some small additive error term $e \in \mathbb{Z}_q$. More specifically, the homomorphic evaluation of $\mathsf{eq}(\mathbf{x}^*, \mathbf{x})$ followed by the inner product with sk, results in the vector

$$\mathbf{s}^T \left(\mathbf{A}_{\mathsf{fhe,eq}} + \left(\frac{q}{2} \cdot \mathsf{eq}(\mathbf{x}^*, \mathbf{x}) + e \right) \cdot \mathbf{G} \right) + \mathsf{noise} \quad \in \mathbb{Z}_q^m,$$

where $\mathbf{A}_{\mathsf{fhe,eq}}$ is the result of homomorphically computing the FHE equality test circuit, along with the inner product with the secret key, on the public matrices. Here $e \in \mathbb{Z}_q$ is some offset term. Even when $\mathsf{eq}(\mathbf{x}^*, \mathbf{x}) = 0$, the rounding of this vector will not produce the correct evaluation due to this offset term e. Moreover, the term e contains information about the original plaintext and therefore, to ensure private puncturing, we must somehow allow for correct computation without revealing the actual value of e. Resolving this issue seems difficult. It is precisely the reason why the predicate encryption scheme of [GVW15b] cannot be converted to a *fully-attribute hiding* predicate encryption scheme (and therefore a full-fledged functional encryption scheme). However, in our context, the problem of *noisy decryption* has an elegant solution.

The idea is to "shorten" the vector $(\mathbf{s}^T \cdot e \cdot \mathbf{G})$ so that it is absorbed into noise, and disappears as we round to obtain the PRF value at x. Towards this goal, we sample the secret vector $\mathbf{s}$ from the LWE noise distribution, which does not change the hardness of LWE [ACPS09]. Next, we observe that although the gadget matrix $\mathbf{G}$ is not a short matrix as a whole, it does contain a number of short column vectors. For instance, a subset of the columns vectors of the gadget matrix consist of elementary basis vectors $\mathbf{u}_i \in \mathbb{Z}_q^n$ with the ith entry set to 1 and the rest set to 0. More precisely, for $1 \le i \le n$, let the vector $\mathbf{v}_i \in \mathbb{Z}_q^m$ be an m dimensional basis vectors with its $i \cdot \lfloor \log q - 1 \rfloor$th entry set to 1 and the rest set to 0. Then, $\mathbf{G} \cdot \mathbf{v}_i = \mathbf{u}_i$.

With this observation, we can simply define the PRF with respect to these short column positions in the gadget matrix. For instance, consider defining the PRF with respect to the first column position as follows

$$\mathsf{PRF}_\mathbf{s}(\mathbf{x}) := \lfloor \mathbf{s}^T \cdot \mathbf{A}_{\mathsf{fhe,eq}} \cdot \mathbf{v}_1 \rceil_p \ \in \mathbb{Z}_p.$$

Since we are simply taking the first component of a pseudorandom vector, this does not change the pseudorandomness property of the PRF (to adversaries without a constrained key). However, for the evaluation with the punctured key, this allows the FHE error term to be "merged" with noise

$$\left(\mathbf{s}^T \left(\mathbf{A}_{\mathsf{fhe,eq}} + \left(\frac{q}{2} \cdot \mathsf{eq}(\mathbf{x}^*, \mathbf{x}) + e \right) \cdot \mathbf{G} \right) + \mathsf{noise} \right) \mathbf{v}_1$$

$$= \mathbf{s}^T \mathbf{A}_{\mathsf{fhe,eq}} \mathbf{v}_1 + \mathbf{s}^T \left(\frac{q}{2} \cdot \mathsf{eq}(\mathbf{x}^*, \mathbf{x}) + e \right) \mathbf{u}_1 + \mathsf{noise}'$$

$$= \mathbf{s}^T \mathbf{A}_{\mathsf{fhe,eq}} \mathbf{v}_1 + \left(\frac{q}{2} \cdot \mathsf{eq}(\mathbf{x}^*, \mathbf{x}) + e \right) \langle \mathbf{s}, \mathbf{u}_1 \rangle + \mathsf{noise}'$$

$$= \mathbf{s}^T \mathbf{A}_{\mathsf{fhe,eq}} \mathbf{v}_1 + \frac{q}{2} \cdot \mathsf{eq}(\mathbf{x}^*, \mathbf{x}) \langle \mathbf{s}, \mathbf{u}_1 \rangle + \underbrace{e \cdot \langle \mathbf{s}, \mathbf{u}_1 \rangle + \mathsf{noise}'}_{\text{short}}.$$

When $\mathsf{eq}(\mathbf{x}^*, \mathbf{x}) = 0$, then the rounding of the vector above results in the correct PRF evaluation since the final noise $e \cdot \langle \mathbf{s}, \mathbf{u}_1 \rangle + \mathsf{noise}'$ is small and will disappear with the rounding.

Pseudorandomness at Punctured Point. The remaining problem is to make the PRF evaluation at the punctured point look random to an adversary who only holds a punctured key. Note that if the adversary evaluates the PRF at the punctured point x^* using its punctured key, the result is the correct PRF output, but offset by the term $(\frac{q}{2} + e) \cdot s_1 + \mathsf{noise}'$, which is clearly distinguishable from random. To fix this, we make the following modifications. First, we include a uniformly generated vector $\mathbf{w} = (w_1, \ldots, w_n) \in \mathbb{Z}_q^n$ as part of the public parameters. Then, we modify the FHE homomorphic operation such that after evaluating the equality check circuit, we multiply the resulting message with one of the w_i's such that decryption outputs $w_i \cdot \mathsf{eq}(\mathbf{x}^*, \mathbf{x}) + e$, instead of $\frac{q}{2} \cdot \mathsf{eq}(\mathbf{x}^*, \mathbf{x}) + e$. Then, we define the PRF evaluation as the vector

$$\mathsf{PRF}_\mathbf{s}(\mathbf{x}) = \left\lfloor \sum_i \mathbf{s}^T \cdot \mathbf{A}_{\mathsf{fhe,eq},i} \cdot \mathbf{v}_i \right\rceil_p.$$

where $\mathbf{v}_1, \ldots, \mathbf{v}_n \in \mathbb{Z}_q^m$ are elementary basis vectors such that $\mathbf{G} \cdot \mathbf{v}_i = \mathbf{u}_i \in \mathbb{Z}_q^n$. Here, the matrix $\mathbf{A}_{\mathsf{fhe,eq},i}$ represents the matrix encoding the equality check circuit operation, followed by scalar multiplication by w_i. Now, evaluating the PRF with the punctured key at the punctured point results in the vector

$$\sum_i \left(\mathbf{s}^T \left(\mathbf{A}_{\mathsf{fhe,eq},i} + (w_i \cdot \mathsf{eq}(\mathbf{x}^*, \mathbf{x}) + e) \cdot \mathbf{G} \right) + \mathsf{noise} \right) \mathbf{v}_i$$

$$= \sum_i \mathbf{s}^T \mathbf{A}_{\mathsf{fhe,eq},i} \cdot \mathbf{v}_i + \sum_i \mathbf{s}^T \left(w_i \cdot \mathsf{eq}(\mathbf{x}^*, \mathbf{x}) + e_i \right) \mathbf{u}_i + \mathsf{noise}'$$

$$= \sum_i \mathbf{s}^T \mathbf{A}_{\mathsf{fhe,eq},i} \cdot \mathbf{v}_i + \sum_i (\mathsf{eq}(\mathbf{x}^*, \mathbf{x}) + e_i) \langle \mathbf{s}, w_i \cdot \mathbf{u}_i \rangle \cdot + \mathsf{noise}'.$$

$$= \sum_i \mathbf{s}^T \mathbf{A}_{\mathsf{fhe,eq},i} \cdot \mathbf{v}_i + \mathsf{eq}(\mathbf{x}^*, \mathbf{x}) \langle \mathbf{s}, \mathbf{w} \rangle + \mathsf{noise}''.$$

We note that when $\mathsf{eq}(\mathbf{x}^*, \mathbf{x}) = 1$, then the offset term is a noisy inner product on the secret vector $\mathbf{s}$. This allows us to embed an LWE sample in the offset term and show that the evaluation indeed looks uniformly random to an adversary with a punctured key.

3   Preliminaries

Basic Notations. For an integer n, we write $[n]$ to denote the set $\{1, \ldots, n\}$. For a finite set S, we write $x \xleftarrow{\$} S$ to denote sampling x uniformly at random from S. We use bold lowercase letters (*e.g.*, $\mathbf{v}, \mathbf{w}$) to denote column vectors and bold uppercase letters (*e.g.*, $\mathbf{A}, \mathbf{B}$) to denote matrices. For a vector or matrix $\mathbf{s}, \mathbf{A}$, we use $\mathbf{s}^T, \mathbf{B}^T$ to denote their transpose. We write λ for the security parameter. We say that a function $\epsilon(\lambda)$ is negligible in λ, if $\epsilon(\lambda) = o(1/\lambda^c)$ for every $c \in \mathbb{N}$, and we write $\mathsf{negl}(\lambda)$ to denote a negligible function in λ. We say that an event occurs with *negligible probability* if the probability of the event is $\mathsf{negl}(\lambda)$, and an event occurs with *overwhelming probability* if its complement occurs with negligible probability.

Rounding. For an integer $p \leq q$, we define the modular "rounding" function

$$\lfloor \cdot \rceil_p \colon \mathbb{Z}_q \to \mathbb{Z}_p \text{ that maps } x \to \lfloor (p/q) \cdot x \rceil$$

and extend it coordinate-wise to matrices and vectors over $\mathbb{Z}_q$. Here, the operation $\lfloor \cdot \rceil$ is the rounding operation over $\mathbb{R}$.

Norm for Vectors and Matrices. Throughout this work, we will always use the infinity norm for vectors and matrices. This means that for a vector $\mathbf{x}$, the norm $\|\mathbf{x}\|$ is the maximal absolute value of an element in $\mathbf{x}$. Similarly, for a matrix $\mathbf{A}$, $\|\mathbf{A}\|$ is the maximal absolute value of any of its entries. If $\mathbf{x}$ is n-dimensional and $\mathbf{A}$ is $n \times m$, then $\|\mathbf{x}^T \mathbf{A}\| \leq n \cdot \|\mathbf{x}\| \cdot \|\mathbf{A}\|$.

3.1   Private Constrained PRFs

We first review the definition of a pseudorandom function (PRF) [GGM86].

Definition 1 (Pseudorandom Function [GGM86]). *Fix a security parameter* λ. *A keyed function* $F \colon \mathcal{K} \times \mathcal{X} \to \mathcal{Y}$ *with keyspace* $\mathcal{K}$, *domain* $\mathcal{X}$, *and range* $\mathcal{Y}$ *is pseudorandom if for all efficient algorithms* $\mathcal{A}$,

$$\left| \Pr\left[k \xleftarrow{\$} \mathcal{K} : \mathcal{A}^{F(k,\cdot)}(1^\lambda) = 1 \right] \right| - \Pr\left[f \xleftarrow{\$} \mathsf{Funcs}(\mathcal{X}, \mathcal{Y}) : \mathcal{A}^{f(\cdot)}(1^\lambda) = 1 \right] = \mathsf{negl}(\lambda).$$

Sometimes, a PRF is defined more naturally with respect to a pair of algorithms $\Pi_{\mathsf{PRF}} = (\mathsf{PRF.Setup}, \mathsf{PRF.Eval})$ where $\mathsf{PRF.Setup}$ is a randomized algorithm that samples the PRF key k in $\mathcal{K}$ and $\mathsf{PRF.Eval}$ computes the keyed function $F(k, \cdot)$.

In a constrained PRF [BW13,KPTZ13,BGI14], an authority with a master secret key msk for the PRF can create a *restricted key* sk_f associated with some function f that allows one to evaluate the PRF only at inputs $x \in \mathcal{X}$ for which $f(x) = 0$.[2]

Definition 2 (Constrained PRF [BW13,KPTZ13,BGI14]). *A constrained PRF consists of a tuple of algorithms* $\Pi_{\mathsf{pPRF}} = (\mathsf{cPRF.Setup}, \mathsf{cPRF.Constrain}, \mathsf{cPRF.ConstrainEval}, \mathsf{cPRF.Eval})$ *over domain* $\mathcal{X}$, *range* $\mathcal{Y}$, *and circuit class* $\mathcal{C}$ *is defined as follows:*

- $\mathsf{cPRF.Setup}(1^\lambda) \to \mathsf{msk}$: *On input the security parameter* λ, *the setup algorithm outputs the master secret key* msk.
- $\mathsf{cPRF.Constrain}(\mathsf{msk}, f) \to \mathsf{sk}_f$: *On input the master secret key* msk, *and a circuit* $f \in \mathcal{C}$, *the constrain algorithm outputs a constrained key* sk_f.
- $\mathsf{cPRF.ConstrainEval}(\mathsf{sk}, x) \to y$: *On input a constrained key* sk, *and an input* $x \in \mathcal{X}$, *the puncture evaluation algorithm evaluates the PRF value* $y \in \mathcal{Y}$.
- $\mathsf{cPRF.Eval}(\mathsf{msk}, x) \to y$: *On input the master secret key* msk *and an input* $x \in \mathcal{X}$, *the evaluation algorithm evaluates the PRF value* $y \in \mathcal{Y}$.

Algorithms $\mathsf{cPRF.Setup}$ and $\mathsf{cPRF.Constrain}$ are randomized, while algorithms $\mathsf{cPRF.ConstrainEval}$ and $\mathsf{cPRF.Eval}$ are always deterministic.

Correctness. A constrained PRF is correct if for all $\lambda \in \mathbb{N}$, $\mathsf{msk} \leftarrow \mathsf{cPRF.Setup}$ (1^λ), for every circuit $C \in \mathcal{C}$, and input $x \in \mathcal{X}$ for which $f(x) = 0$, we have that

$$\mathsf{cPRF.ConstrainEval}(\mathsf{cPRF.Constrain}(\mathsf{msk}, f), x) = \mathsf{cPRF.Eval}(\mathsf{msk}, x)$$

with overwhelming probability.

Security. We require two security properties for constrained PRFs: pseudorandomness and privacy. The first property states that given constrained PRF keys, an adversary cannot distinguish the PRF evaluation at the points where it is not allowed to compute, from a randomly sampled point from the range.

---

[2] We adopt the convention that $f(x) = 0$ signifies the ability to evaluate the PRF. This is opposite of the standard convention, and is done purely for convenience in the technical section.

Definition 3 (Pseudorandomness). *Fix a security parameter* λ. *A constrained PRF scheme* $\Pi_{\mathsf{cPRF}} = (\mathsf{cPRF.Setup}, \mathsf{cPRF.Constrain}, \mathsf{cPRF.ConstrainEval}, \mathsf{cPRF.Eval})$ *is pseudorandom if for all PPT adversary* $\mathcal{A} = (\mathcal{A}_1, \mathcal{A}_2)$, *there is a negligible function* $\mathsf{negl}(\lambda)$ *such that*

$$\mathsf{Adv}^{\mathsf{rand}}_{\Pi_{\mathsf{cPRF}}, \mathcal{A}}(\lambda) = \left| \Pr[\mathsf{Expt}^{(0)}_{\Pi_{\mathsf{cPRF}}, \mathcal{A}}(\lambda) = 1] - \Pr[\mathsf{Expt}^{(1)}_{\Pi_{\mathsf{cPRF}}, \mathcal{A}}(\lambda) = 1] \right| \leq \mathsf{negl}(\lambda)$$

where for each $b \in \{0, 1\}$ *and* $\lambda \in \mathbb{Z}$, *the experiment* $\mathsf{Expt}^{(b)}_{\Pi_{\mathsf{cPRF}}, \mathcal{A}}(\lambda)$ *is defined as follows:*

1. $\mathsf{msk} \leftarrow \mathsf{cPRF.Setup}(1^\lambda)$
2. $(x^*, \mathsf{state}_1) \leftarrow \mathcal{A}_1^{\mathsf{cPRF.Constrain}(\mathsf{msk}, \cdot), \mathsf{cPRF.Eval}(\mathsf{msk}, \cdot)}(1^\lambda)$
3. $y_0 \leftarrow \mathsf{cPRF.Eval}(\mathsf{msk}, x^*)$
4. $y_1 \xleftarrow{\$} \mathcal{Y}$
5. $b' \leftarrow \mathcal{A}_2^{\mathsf{cPRF.Constrain}(\mathsf{msk}, \cdot), \mathsf{cPRF.Eval}(\mathsf{msk}, \cdot)}(y_b, \mathsf{state}_1)$
6. *Output* b'

To prevent the adversary from trivially winning the game, we require that for any query f *that* $\mathcal{A}$ *makes to the* $\mathsf{cPRF.Constrain}(\mathsf{msk}, \cdot)$ *oracle, it holds that* $f(x^*) = 1$, *and for any query* x *that* $\mathcal{A}$ *makes to the* $\mathsf{cPRF.Eval}(\mathsf{msk}, \cdot)$ *oracle, it holds that* $x \neq x^*$.

The security games as defined above is the *fully adaptive* game. One can also define a *selective* variant of the games above where the adversary commits to the challenge point before the game starts. We do so in Definition 6 below.

Next, we require that a constrained key sk_f not leak information about the constraint function f as in the setting of private constrained PRFs of [BLW17].

Definition 4 (Privacy). *Fix a security parameter* $\lambda \in \mathbb{N}$. *A constrained PRF scheme* $\Pi_{\mathsf{cPRF}} = (\mathsf{cPRF.Setup}, \mathsf{cPRF.Constrain}, \mathsf{cPRF.ConstrainEval}, \mathsf{cPRF.Eval})$ *is private if for all PPT adversary* $\mathcal{A}$, *there is a negligible function* $\mathsf{negl}(\lambda)$ *such that*

$$\mathsf{Adv}^{\mathsf{priv}}_{\Pi_{\mathsf{cPRF}}, \mathcal{A}}(\lambda) = \left| \Pr[\mathsf{Expt}^{(0)}_{\Pi_{\mathsf{cPRF}}, \mathcal{A}}(\lambda) = 1] - \Pr[\mathsf{Expt}^{(1)}_{\Pi_{\mathsf{cPRF}}, \mathcal{A}}(\lambda) = 1] \right| \leq \mathsf{negl}(\lambda)$$

where the experiments $\mathsf{Expt}^{(b)}_{\Pi_{\mathsf{cPRF}}, \mathcal{A}}$ *are defined as follows:*

1. $\mathsf{msk} \leftarrow \mathsf{cPRF.Setup}(1^\lambda)$.
2. $b' \leftarrow \mathcal{A}^{\mathsf{cPRF.Constrain}_b(\mathsf{msk}, \cdot, \cdot), \mathsf{cPRF.Eval}(\mathsf{msk}, \cdot)}(1^\lambda)$.
3. *Output* b'

where the oracle $\mathsf{cPRF.Constrain}_b(\cdot, \cdot, \cdot)$ *is defined as follows*

- $\mathsf{cPRF.Constrain}_b(\mathsf{msk}, f_0, f_1)$: *On input the master secret key* msk, *and a pair of constraint functions* f_0, f_1, *outputs* $\mathsf{sk}_{f,b} \leftarrow \mathsf{cPRF.Constrain}(f_b)$.

In the experiment above, we require an extra admissibility condition on the adversary to prevent it from trivially distinguishing the two experiments. For a circuit $f \in C$, define the set $S(f) \subseteq X$ where $\{x \in X : f(x) = 0\}$. Let d be the number of queries that A makes to cPRF.Constrain$_b$(msk, $\cdot$, $\cdot$) *and let $(f_0^{(i)}, f_1^{(i)})$ for $i \in [d]$ denote the ith pair of circuits that the adversary submits to the constrain oracle. Then we require that*

1. *For every query x that A makes to the evaluation oracle, $f_0^{(i)}(x) = f_1^{(i)}(x)$.*
2. *For every pair of distinct indices $i, j \in [d]$,*

$$S\left(f_0^{(i)}\right) \cap S\left(f_0^{(j)}\right) = S\left(f_1^{(i)}\right) \cap S\left(f_1^{(j)}\right).$$

Justification for the second admissibility condition is discussed in [BLW17, Remark 2.11].

3.2   Private Puncturable PRFs

A puncturable PRF is a special case of constrained PRFs where one can only request constained keys for point functions. That is, each constraining circuit C_{x^*} is associated with a point $x^* \in \{0,1\}^n$, and $C_{x^*}(x) = 0$ if and only if $x \neq x^*$. Concretely, a puncturable PRF is specified by a tuple of algorithms $\Pi_{\mathsf{pPRF}} = (\mathsf{pPRF.Setup}, \mathsf{pPRF.Puncture}, \mathsf{pPRF.PunctureEval}, \mathsf{pPRF.Eval})$ with identical syntax as regular constrained PRFs, with the exception that the algorithm pPRF.Puncture takes in a point x to be punctured rather than a circuit f.

In the context of private puncturing, we require without loss of generality, that algorithm pPRF.Puncture be deterministic (see [BLW17, Remark 2.14]). If it were randomized, it could be de-randomized by generating its random bits using a PRF keyed by a part of msk, and given the point x as input.

We define a slightly weaker variant of correctness than as is defined above for constrained PRF called *computational functionality preserving* as in the setting of [BV15]. In words, this property states that it is computationally hard to find a point $x \neq x^*$ such that the result of the puncture evaluation differs from the actual PRF evaluation. This is essentially a relaxation of the *statistical* notion of correctness to the *computational* notion of correctness.

Definition 5 (Computational Functionality Preserving). *Fix a security parameter λ and let $\Pi_{\mathsf{pPRF}} = (\mathsf{pPRF.Setup}, \mathsf{pPRF.Puncture}, \mathsf{pPRF.PunctureEval}, \mathsf{pPRF.Eval})$ be a private-puncturable PRF scheme. For every adversary $A = (A_1, A_2)$, consider the following experiment where we choose* msk $\leftarrow$ pPRF.Setup(1^λ), (state, x^*) $\leftarrow A_1(1^\lambda)$, *and* sk$_{x^*}$ $\leftarrow$ pPRF.Puncture(msk, x^*). *Then, the private-puncturable PRF scheme Π_{pPRF} is computational functionality preserving if*

$$\Pr\left[x \leftarrow A_2^{\mathsf{pPRF.Eval}(\mathsf{msk}, \cdot)}(\mathsf{state}, \mathsf{sk}_{x^*}) : \begin{array}{c} x \neq x^* \wedge \\ \mathsf{pPRF.Eval}(\mathsf{msk}, x) \neq \\ \mathsf{pPRF.PunctureEval}(\mathsf{sk}_{x^*}, x) \end{array}\right] \leq \mathsf{negl}(\lambda)$$

for some negligible function negl.

We next specialize the security definitions to the settings of puncturing constraints. For puncturable PRFs, the adversary in the pseudorandomness game is limited to making at most one key query to pPRF.Puncture. If it made two key queries, for two distinct punctures, it would be able to evaluate the PRF on all points in the domain, and then cannot win the game. Therefore, we need only consider two types of adversaries in the pseudorandomness game:

- an adversary that makes evaluation queries, but no key queries during the game, and
- an adversary that makes exactly one key query.

The first adversary plays the regular PRF security game. A simple reduction shows that selective security against an adversary of the second type implies security against an adversary of the first type. Therefore, when defining (selective) security, it suffices to only consider (selective) adversaries of the second type.

One technicality in defining pseudorandomness for puncturable PRFs that satisfy a computational notion of correctness is that the adversary must also be given access to an evaluation oracle. This is because given only a punctured key, the adversary cannot efficiently detect whether a point in the domain evaluates to the correct PRF evaluation with the punctured key without the evaluation oracle. Therefore, we define the following pseudorandomness definition.

Definition 6. *Fix a security parameter* λ. *A puncturable PRF scheme* Π_{pPRF} $=$ (pPRF.Setup, pPRF.Puncture, pPRF.PunctureEval, pPRF.Eval) *is selectively-pseudorandom if for every PPT adversary* $\mathcal{A} = (\mathcal{A}_1, \mathcal{A}_2)$, *there exists a negligible function* negl *such that for* msk $\leftarrow$ pPRF.Setup(1^λ), $(x^*, \text{state}) \leftarrow \mathcal{A}_1(1^\lambda)$, sk$_{x^*} \leftarrow$ pPRF.Puncture(msk, x^*), $u \overset{\$}{\leftarrow} \mathcal{Y}$, *we have that*

$$\Big| \Pr[\mathcal{A}_2^{\text{pPRF.Eval(msk,·)}}(\text{state}, \text{sk}_{x^*}, \text{pPRF.Eval(msk}, x^*)) = 1]$$

$$- \Pr[\mathcal{A}_2^{\text{pPRF.Eval(msk,·)}}(\text{state}, \text{sk}_{x^*}, u) = 1] \Big| \leq \text{negl}(\lambda) \quad (3.1)$$

To prevent the adversary from trivially breaking the game, we require that the adversary $\mathcal{A}$ *cannot query the evaluation oracle on* x^*.

We next define the notion of privacy for puncturable PRFs. Again, since we rely on the computational notion of correctness, we provide the adversary access to an honest evaluation oracle (except for at the challenge points). As in the pseudorandomness game, we only consider selective adversaries that make a *single* key query, although that results in a slightly weaker notion of privacy than in Definition 4.[3]

---

[3] We note that the admissibility condition in Definition 4 allows an adversary to make two constrained key queries (see [BLW17] Remark 2.14). However, applications of privately puncturable PRFs require pseudorandomness property to be satisfied, which can only be achieved in the single-key setting. Therefore, the restriction of privacy to the single-key setting does not affect the applications of privately puncturable PRFs.

Definition 7. *Fix a security parameter* λ. *A puncturable PRF scheme* Π_{pPRF} = ($\mathsf{pPRF.Setup}, \mathsf{pPRF.Puncture}, \mathsf{pPRF.PunctureEval}, \mathsf{pPRF.Eval}$) *is selectively-private if for every PPT adversary* $\mathcal{A} = (\mathcal{A}_1, \mathcal{A}_2)$, *there exists a negligible function* negl *such that for* $\mathsf{msk} \leftarrow \mathsf{pPRF.Setup}(1^\lambda)$, $(x^*, \mathsf{state}) \leftarrow \mathcal{A}_1(1^\lambda)$, $\mathsf{sk}_{x^*} \leftarrow \mathsf{pPRF.Puncture}(\mathsf{msk}, x^*)$, $\mathsf{sk}_0 \leftarrow \mathsf{pPRF.Puncture}(\mathsf{msk}, \mathbf{0})$, *we have that*

$$\left| \Pr[\mathcal{A}_2^{\mathsf{pPRF.Eval}(\mathsf{msk}, \cdot)}(\mathsf{state}, \mathsf{sk}_{x^*}) = 1] - \Pr[\mathcal{A}_2^{\mathsf{pPRF.Eval}(\mathsf{msk}, \cdot)}(\mathsf{state}, \mathsf{sk}_0) = 1] \right| \leq \mathsf{negl}(\lambda).$$

To prevent the adversary from trivially winning the game, we require that the adversary $\mathcal{A}$ *cannot query the evaluation oracle on* x^* *or* $\mathbf{0}$.

Remarks. We note that a *selectively*-secure privately constrained PRF can be shown to be fully secure generically through complexity leveraging. In particular, the selectivity of the definition does not hurt the applicability of privacy as it can be shown to be adaptively secure generically. Achieving adaptive security for any kind of constrained PRFs without complexity leveraging (with polynomial loss in the reduction) remains a challenging problem. For puncturable PRFs, for instance, the only known adaptively secure constructions rely on the power of indistinguishability obfuscation [HKW15, HKKW14].[4]

We also note that since constrained PRF is a symmetric-key notion, the setup algorithm just returns the master secret key msk. However, one can also consider dividing the setup into distinct parameter generation algorithm and seed generation algorithm where the parameters can be generated once and can be reused with multiple seeds for the PRF. In fact, for our construction in Sect. 5.1, a large part of the master secret key component can be fixed once and made public as parameters for the scheme. However, we maintain our current definition for simplicity.

3.3 Fully-Homomorphic Encryption

Following the presentation of [GVW15b], we give a minimal definition of fully homomorphic encryption (FHE) which is sufficient for this work. Technically, in this work, we use a leveled homomorphic encryption scheme (LHE); however, we will still refer to it simply as FHE. A leveled homomorphic encryption scheme is a tuple of polynomial-time algorithms Π_{HE} = ($\mathsf{HE.KeyGen}, \mathsf{HE.Enc}, \mathsf{HE.Eval}, \mathsf{HE.Dec}$) defined as follows:

- $\mathsf{HE.KeyGen}(1^\lambda, 1^d, 1^k) \rightarrow \mathsf{sk}$: On input the security parameter λ, a depth bound d, and a message length k, the key generation algorithm outputs a secret key sk.
- $\mathsf{HE.Enc}(\mathsf{sk}, \mu) \rightarrow \mathsf{ct}$: On input a secret key sk and a message $\mu \in \{0,1\}^k$, the encryption algorithm outputs a ciphertext ct.

---

[4] There are other adaptively secure constrained PRF constructions for prefix fixing and bit-fixing constraints as in [FKPR14, Hof14]; however, they too either require superpolynomial loss in the security parameter or rely on random oracles. The construction of [BV15] achieves adaptive security for the challenge point, but is selective with respect to the constraint.

- HE.Eval$(C, \text{ct}) \rightarrow \text{ct}'$: On input a circuit $C\colon \{0,1\}^k \rightarrow \{0,1\}$ of depth d and a ciphertext ct, the homomorphic evaluation algorithm outputs ciphertext ct$'$.
- HE.Dec$(\text{sk}, \text{ct}') \rightarrow \mu'$: On input a secret key sk and a ciphertext ct$'$, the decryption algorithm outputs a message $\mu' \in \{0,1\}$.

Correctness. We require that for all λ, d, k, sk $\leftarrow$ HE.KeyGen$(1^\lambda, 1^d, 1^k)$, $\mu \in \{0,1\}^k$, and boolean circuits $C\colon \{0,1\}^k \rightarrow \{0,1\}$ of depth at most d, we have that

$$\Pr\left[\text{HE.Dec}(\text{sk}, \text{HE.Eval}(C, \text{HE.Enc}(\text{sk}, \mu))) = C(\mu)\right] = 1$$

where the probability is taken over HE.Enc and HE.KeyGen.

Security. For security, we require standard semantic security. For any PPT adversary $\mathcal{A} = (\mathcal{A}_1, \mathcal{A}_2)$, and for all $d, k = \text{poly}(\lambda)$, there exists a negligible function negl such that

$$\Pr\left[b = b' : \begin{array}{l} \text{sk} \leftarrow \text{HE.KeyGen}(1^\lambda, 1^d, 1^k); \\ \mu \leftarrow \mathcal{A}(1^\lambda, 1^d, 1^k); \\ b \xleftarrow{\$} \{0,1\}; \\ \text{ct}_0 \leftarrow \text{HE.Enc}(\text{sk}, 0^{|\mu|}); \\ \text{ct}_1 \leftarrow \text{HE.Enc}(\text{sk}, \mu); \\ b' \leftarrow \mathcal{A}(\text{ct}_b) \end{array} \right] - \frac{1}{2} \le \text{negl}(\lambda)$$

4 LWE, SIS, Lattice FHE, and Matrix Embeddings

In this section, we present a brief background on the average case lattice problems of the Learning with Errors problem (LWE) as well as the one-dimensional Short Integer Solutions problem (1D-SIS). We also discuss the instantiations of FHE from LWE and summarize the circuit matrix embedding technique of the lattice ABE constructions.

Gaussian Distributions. We let $D_{\mathbb{Z}^m, \sigma}$ to be the discrete Gaussian distribution over $\mathbb{Z}^m$ with parameter σ. For simplicity, we truncate the distribution, which means that we replace the output by $\mathbf{0}$ whenever the norm $\|\cdot\|$ exceeds $\sqrt{m} \cdot \sigma$.

The LWE Problem. Let n, m, q be positive integers and χ be some noise distribution over $\mathbb{Z}_q$. In the LWE(n, m, q, χ) problem, the adversary's goal is to distinguish between the two distributions:

$$(\mathbf{A}, \mathbf{s}^T \mathbf{A} + \mathbf{e}^T) \quad \text{and} \quad (\mathbf{A}, \mathbf{u}^T)$$

where $\mathbf{A} \xleftarrow{\$} \mathbb{Z}_q^{n \times m}$, $\mathbf{s} \xleftarrow{\$} \chi^n$, $\mathbf{e} \leftarrow \chi^m$, and $\mathbf{u} \xleftarrow{\$} \mathbb{Z}_q^m$ are uniformly sampled.

Connection to Worst-Case. Let $B = B(n) \in \mathbb{N}$. A family of distributions $\chi = \{\chi_n\}_{n \in \mathbb{N}}$ is called B-bounded if

$$\Pr[\chi \in \{-B, -B+1 \ldots, B-1, B\}] = 1.$$

For certain B-bounded error distributions χ, including the discrete Gaussian distributions[5], the $\mathsf{LWE}(n, m, q, \chi)$ problem is as hard as approximating certain worst-case lattice problems such as GapSVP and SIVP on n-dimensional lattices to within $\tilde{O}(n \cdot q/B)$ factor [Reg09, Pei09, ACPS09, MM11, MP12, BLP+13].

The Gadget Matrix. Let $\tilde{N} = n \cdot \lceil \log q \rceil$ and define the "gadget matrix" $\mathbf{G} = \mathbf{g} \otimes \mathbf{I}_n \in \mathbb{Z}_q^{n \times \tilde{N}}$ where $\mathbf{g} = (1, 2, 4, \ldots, 2^{\lceil \log q \rceil - 1})$. We define the inverse function $\mathbf{G}^{-1} : \mathbb{Z}_q^{n \times m} \to \{0, 1\}^{\tilde{N} \times m}$ which expands each entry $a \in \mathbb{Z}_q$ of the input matrix into a column of size $\lceil \log q \rceil$ consisting of the bits of the binary representation of a. To simplify the notation, we always assume that $\mathbf{G}$ has width m, which we do so without loss of generality as we can always extend the width of $\mathbf{G}$ by adding zero columns. We have the property that for any matrix $\mathbf{A} \in \mathbb{Z}_q^{n \times m}$, it holds that $\mathbf{G} \cdot \mathbf{G}^{-1}(\mathbf{A}) = \mathbf{A}$.

The 1D-SIS Problem. Following the technique of [BV15], we use a variant of the Short Integer Solution (SIS) problem of [Ajt96] called 1D-SIS problem to show correctness and security for our scheme. Let m, β be positive integers and let q be a product of n prime moduli $p_1 < p_2 < \ldots < p_n$, $q = \prod_{i \in [n]} p_i$. Then, in the 1D-SIS$_{m,q,\beta}$, the adversary is given a uniformly random vector $\mathbf{v} \in \mathbb{Z}_q^m$ and its goal is to find $\mathbf{z} \in \mathbb{Z}^m$ such that $\|\mathbf{z}\| \leq \beta$ and $\langle \mathbf{v}, \mathbf{z} \rangle = 0 \bmod q$. For $m = O(n \log q)$, $p_1 \geq \beta \cdot \omega(\sqrt{mn \log n})$, the 1D-SIS-R$_{m,q,p,\beta}$ problem is as hard as approximating certain worst-case lattice problems such as GapSVP and SIVP to within $\beta \cdot \tilde{O}(\sqrt{mn})$ factor [Reg04, BV15].

For this work, we will use another variant called 1D-SIS-R that we define as follows. Let m, β be positive integers. We let $q = p \cdot \prod_{i \in [n]} p_i$, where all $p_1 < p_2 < \cdots < p_n$ are all co-prime and co-prime with p as well. In the 1D-SIS-R$_{m,q,p,\beta}$ problem, the adversary is given a uniformly random vector $\mathbf{v} \in \mathbb{Z}_q^m$ and its goal is to find a vector $\mathbf{z} \in \mathbb{Z}^m$ such that $\|\mathbf{z}\| \leq \beta$ and $\langle \mathbf{v}, \mathbf{z} \rangle \in [-\beta, \beta] + (q/p)(\mathbb{Z} + 1/2)$.[6] [BKM17, Appendix B] shows that 1D-SIS-R$_{m,q,p,\beta}$ is as hard as 1D-SIS$_{m,q,\beta}$ and therefore, is as hard as certain worst-case lattice problems.

4.1 FHE from LWE

There are a number of FHE constructions from LWE [BV14a, BGV12, GSW13, BV14b, ASP14, CM15, MW16, BP16, PS16]. For this work, we use the fact that these constructions can support not just binary, but field operations.

Specifically, given an encryption of a message $\mathbf{x} \in \{0, 1\}^\ell$, a circuit $C \colon \{0, 1\}^\ell \to \{0, 1\}$, and any field element $w \in \mathbb{Z}_q$, one can homomorphically compute the function

$$f_{C,w}(\mathbf{x}) = w \cdot C(\mathbf{x}) \in \mathbb{Z}_q$$

on the ciphertext. Here, we take advantage of the fact that the FHE homomorphic operations can support scalar multiplication by a field element without

---

[5] By discrete Gaussian, we always mean the truncated discrete Gaussian.

[6] The term $(q/p)(\mathbb{Z} + 1/2)$ is a slight abuse of notation when q/p is not even. In this case, we mean $(q/p) \cdot (\mathbb{Z}) + \lfloor (q/p) \cdot (1/2) \rfloor$.

increasing the noise too much. Looking ahead, we will encrypt the punctured point $\mathbf{x}^*$, and homomorphically compute the equality predicate $\mathsf{eq}_\mathbf{x}(\mathbf{x}^*) = \begin{cases} 1 & \mathbf{x} = \mathbf{x}^* \\ 0 & \text{otherwise} \end{cases}$ on the ciphertext such that it decrypts to a random element $w_\gamma \in \mathbb{Z}_q$ only if the evaluation of the PRF a point $\mathbf{x}$ equals to the punctured point. This is simply evaluating the equality check circuit on the FHE ciphertext and scaling the result by w_γ.

We formally summarize the properties of FHE constructions from LWE below.[7]

Theorem 1 (FHE from LWE). *Fix a security parameter λ and depth bound $d = d(\lambda)$. Let n, m, q, χ be LWE parameters where χ is a B-bounded error distribution and $q > B \cdot m^{O(d)}$. Then, there is an FHE scheme $\Pi_{\mathsf{HE}} = (\mathsf{HE.KeyGen}, \mathsf{HE.Enc}, \mathsf{HE.Eval}, \mathsf{HE.Dec})$ for circuits of depth bound d, with the following properties:*

- *$\mathsf{HE.KeyGen}$ outputs a secret key $\mathsf{sk} \in \mathbb{Z}_q^n$*
- *$\mathsf{HE.Enc}$ takes in a message $\mathbf{m} \in \{0,1\}^k$ and outputs a ciphertext $\mathsf{ct} \in \{0,1\}^z$ where $z = \mathsf{poly}(\lambda, d, \log q, k)$.*
- *$\mathsf{HE.Eval}$ takes in a circuit $f_{C,w}$ and a ciphertext ct and outputs ciphertexts $\mathsf{ct}' \in \{0,1\}^n$.*
- *For any boolean circuit C of depth d and scalar element $w \in \mathbb{Z}_q$, $\mathsf{HE.Eval}(f_{C,w}, \cdot)$ is computed by a boolean circuit of depth $\mathsf{poly}(d, \log z)$.*
- *$\mathsf{HE.Dec}$ on input sk and ct, when $C(\mathbf{m}) = 1$ we have that*

$$\sum_{i=1}^t \mathsf{sk}[i] \cdot \mathsf{ct}[i] \in [w - E, w + E].$$

When $C(\mathbf{m}) = 0$ we have

$$\sum_{i=1}^t \mathsf{sk}[i] \cdot \mathsf{ct}[i] \in [-E, E]$$

for some bound $E = B \cdot m^{O(d)}$.
- *Security relies on $\mathsf{LWE}(n, m, q, \chi)$.*

We note that in the predicate encryption construction of [GVW15b], the result of [BV14b] is used, which applies the *sequential* homomorphic multiplication of ciphertexts (through branching programs) to take advantage of the asymmetric noise growth of FHE. This allows the final noise from the FHE homomorphic operations to be bounded by $\mathsf{poly}(\lambda)$, but the depth of the FHE evaluation grows polynomially in the bit length of the FHE modulus. In our construction, this optimization is not needed because we will only be concerned with the equality check circuit which is already only logarithmic in the depth of the input length. Therefore, one can perform regular FHE homomorphic operations with depth logarithmic in the bit length of the FHE modulus.

---

[7] We slightly abuse the FHE syntax in Sect. 3.3.

4.2   Matrix Embeddings

In the ABE construction of [BGG+14], Boneh *et al.* introduced a method to embed circuits into LWE matrices and since then, the technique saw a number of applications in lattice-based constructions [BV15, GVW15b, GV15, BV16, BCTW16].

We provide an overview of this technique since our proof of security will rely on the specifics of this matrix encodings. Our, description will be informal, but we formally describe the properties that we need for the proofs below. We refer the readers to [BGG+14, GVW15b] for the formal treatment.

In the setting of [BGG+14], for the set of public matrix $\mathbf{A}_1, \ldots \mathbf{A}_\ell$, we encode a vector of field elements $\mathbf{x} \in \mathbb{Z}_q^t$ as an LWE sample as

$$\mathbf{a}_{x_i} = \mathbf{s}^T(\mathbf{A}_i + x_i \cdot \mathbf{G}) + \mathbf{e}_i$$

for $i = 1, \ldots, \ell$ where $\mathbf{s}$ and $\mathbf{e}_i$'s are sampled according to the standard LWE distribution. Then, given two encodings, $\mathbf{a}_{x_i}, \mathbf{a}_{x_j}$, we can add and multiply them as follows:

$$
\begin{aligned}
\mathbf{a}_{x_i + x_j} &= \mathbf{a}_{x_i} + \mathbf{a}_{x_j} \\
&= \mathbf{s}^T(\mathbf{A}_i + x_i \cdot \mathbf{G}) + \mathbf{e}_i + \mathbf{s}^T(\mathbf{A}_j + x_j \cdot \mathbf{G}) + \mathbf{e}_j \\
&= \mathbf{s}^T([\mathbf{A}_i + \mathbf{A}_j] + [x_i + x_j] \cdot \mathbf{G}) + [\mathbf{e}_i + \mathbf{e}_j] \\
&= \mathbf{s}^T(\mathbf{A}_{+,i,j} + [x_i + x_j] \cdot \mathbf{G}) + \mathbf{e}_{+,i,j}
\end{aligned}
$$

$$
\begin{aligned}
\mathbf{a}_{x_i \times x_j} &= \mathbf{a}_{x_i} \cdot x_j - \mathbf{a}_{x_j} \mathbf{G}^{-1}(\mathbf{A}_i) \\
&= \mathbf{s}^T(x_j \mathbf{A}_i + x_i x_j \cdot \mathbf{G}) + x_j \mathbf{e}_j - \mathbf{s}^T(\mathbf{A}_j \mathbf{G}^{-1}(\mathbf{A}_i) + x_j \mathbf{A}_i) + \mathbf{e}_j \mathbf{G}^{-1}(\mathbf{A}_i) \\
&= \mathbf{s}^T([-\mathbf{A}_j \mathbf{G}^{-1}(\mathbf{A}_i)] + [x_i \cdot x_j] \cdot \mathbf{G}) + [x_j \mathbf{e}_i + \mathbf{e}_j \mathbf{G}^{-1}(\mathbf{A}_i)] \\
&= \mathbf{s}^T(\mathbf{A}_{\times,i,j} + [x_i \cdot x_j] \cdot \mathbf{G}) + \mathbf{e}_{\times,i,j}
\end{aligned}
$$

Correspondingly, we can define operations on the matrices

- $\mathbf{A}_{+,i,j} = \mathbf{A}_i + \mathbf{A}_j$
- $\mathbf{A}_{\times,i,j} = -\mathbf{A}_j \mathbf{G}^{-1}(\mathbf{A}_i)$

Using these operations, one can compute an arithmetic circuit F on the encodings gate-by-gate. In particular, restricting $\mathbf{x}$ to be a binary string, we can compute the NAND operation as

$$\mathbf{a}_{\neg(x_i \wedge x_j)} = \mathbf{a}_1 - \mathbf{a}_{x_i \times x_j}$$

$$\mathbf{A}_{\neg(x_i \wedge x_j)} = \mathbf{A}^* - \mathbf{A}_{\times,i,j}$$

where $\mathbf{a}_1 = \mathbf{s}^T(\mathbf{A}^* + \mathbf{G}) + \mathbf{e}^*$ is a fixed encoding of 1.

We note that in the description above, to compute a single multiplication on the encodings $\mathbf{a}_{x_i}, \mathbf{a}_{x_j}$, one must know one of x_i or x_j, but it is not required to know both. This means that computing operations such as inner products on two vector attributes can be done without the knowledge of one of the vectors. In particular, given the encodings of $(\mathbf{x}, \mathbf{w}) \in \{0,1\}^z \times \mathbb{Z}_q^t$, and a pair $(C, \mathbf{x})$ where $C : \{0,1\}^z \to \{0,1\}^t$ and $\mathbf{x} \in \{0,1\}^z$, one can derive an encoding of $(\mathsf{IP} \circ C)(\mathbf{x}, \mathbf{w}) = \langle C(\mathbf{x}), \mathbf{w} \rangle$.

Theorem 2 [BGG+14, GVW15b]. *Fix a security parameter λ, and lattice parameters n, m, q, χ where χ is a B-bounded error distribution. Let C be a depth-d Boolean circuit on z input bits. Let $\mathbf{A}_1, \ldots, \mathbf{A}_z, \tilde{\mathbf{A}}_1, \ldots, \tilde{\mathbf{A}}_t \in \mathbb{Z}_q^{n \times m}$, $(x_1, \mathbf{b}_1), \ldots, (x_z, \mathbf{b}_z) \in \{0,1\} \times \mathbb{Z}_q^m$, and $(w_1, \tilde{\mathbf{b}}_1), \ldots, (w_t, \tilde{\mathbf{b}}_t) \in \mathbb{Z}_q \times \mathbb{Z}_q^m$ such that*

$$\left\| \mathbf{b}_i^T - \mathbf{s}^T (\mathbf{A}_i + x_i \cdot \mathbf{G}) \right\| \leq B \quad for\ i = 1, \ldots, z$$

$$\left\| \tilde{\mathbf{b}}_j^T - \mathbf{s}^T (\tilde{\mathbf{A}}_j + w_j \cdot \mathbf{G}) \right\| \leq B \quad for\ j = 1, \ldots, t$$

for some $\mathbf{s} \in \mathbb{Z}_q^n$. There exists the following pair of algorithms

- $\mathsf{Eval}_{\mathsf{pk}}((\mathsf{IP} \circ C), \mathbf{A}_1, \ldots, \mathbf{A}_z, \tilde{\mathbf{A}}_1, \ldots, \tilde{\mathbf{A}}_t) \to \mathbf{A}_{(\mathsf{IP} \circ C)}$: *On input a circuit $(\mathsf{IP} \circ C)$ for $C : \{0,1\}^z \to \{0,1\}^t$ and $z + t$ matrices $\mathbf{A}_1, \ldots, \mathbf{A}_z, \tilde{\mathbf{A}}_1, \ldots, \tilde{\mathbf{A}}_t$, outputs a matrix $\mathbf{A}_{(\mathsf{IP} \circ C)}$.*
- $\mathsf{Eval}_{\mathsf{ct}}((\mathsf{IP} \circ C), \mathbf{b}_1, \ldots, \mathbf{b}_z, \tilde{\mathbf{b}}_1, \ldots, \tilde{\mathbf{b}}_t, \mathbf{x}) \to \mathbf{b}_{(\mathsf{IP} \circ C)}$: *On input a circuit $(\mathsf{IP} \circ C)$ for $C : \{0,1\}^z \to \{0,1\}^t$, $z + t$ vectors $\mathbf{b}_1, \ldots, \mathbf{b}_z, \tilde{\mathbf{b}}_1, \ldots, \tilde{\mathbf{b}}_t$, and length z string $\mathbf{x}$, outputs a vector $\mathbf{b}_{(\mathsf{IP} \circ C)}$*

such that for $\mathbf{A}_{(\mathsf{IP} \circ C)} \leftarrow \mathsf{Eval}_{\mathsf{pk}}((\mathsf{IP} \circ C), \mathbf{A}_1, \ldots, \mathbf{A}_z, \tilde{\mathbf{A}}_1, \ldots, \tilde{\mathbf{A}}_t)$, and $\mathbf{b}_{(\mathsf{IP} \circ C)} \leftarrow \mathsf{Eval}_{\mathsf{ct}}((\mathsf{IP} \circ C), \mathbf{b}_1, \ldots, \mathbf{b}_z, \tilde{\mathbf{b}}_1, \ldots, \tilde{\mathbf{b}}_t, \mathbf{x})$, we have that

$$\left\| \mathbf{b}_{(\mathsf{IP} \circ C)}^T - \mathbf{s}^T (\mathbf{A}_{(\mathsf{IP} \circ C)} + \langle C(\mathbf{x}), \mathbf{w} \rangle \cdot \mathbf{G}) \right\| \leq B \cdot m^{O(d)}.$$

Moreover, $\mathbf{b}_{(\mathsf{IP} \circ C)}$ is a "low-norm" linear function of $\mathbf{b}_1, \ldots, \mathbf{b}_z, \tilde{\mathbf{b}}_1, \ldots, \tilde{\mathbf{b}}_z$. That is, there are matrices $\mathbf{R}_1, \ldots, \mathbf{R}_z, \tilde{\mathbf{R}}_1, \ldots, \tilde{\mathbf{R}}_t$ such that $\mathbf{b}_{(\mathsf{IP} \circ C)}^T = \sum_{i=1}^z \mathbf{b}_i^T \mathbf{R}_i + \sum_{j=1}^t \tilde{\mathbf{b}}_j^T \tilde{\mathbf{R}}_j$ and $\|\mathbf{R}_i\|, \|\tilde{\mathbf{R}}_j\| \leq m^{O(d)}$.

5 Main Construction

In this section, we present our private puncturable PRF. We first give a formal description of the construction followed by a sample instantiation of the parameters used in the construction. Then, we show correctness and security. We conclude the section with some extensions.

5.1 Construction

Our construction uses a number of parameters and indices, which we list here for reference:

- (n, m, q, χ) - LWE parameters
- ℓ - length of the PRF input

- p - rounding modulus
- z - size of FHE ciphertext (indexed by i)
- t - size FHE secret key (indexed by j)
- d' - depth of the equality check circuit
- d - depth of the circuit that computes the FHE homomorphic operation of equality check
- γ - index for the randomizers $w_1, \ldots, w_n$

For $\gamma \in [n]$ we use $\mathbf{u}_\gamma$ to denote the n dimensional basis vector in $\mathbb{Z}_q^n$ with γth entry set to 1 and the rest set to 0. Also, for $\gamma \in [n]$, we denote by $\mathbf{v}_\gamma$ the m dimensional basis vector in $\mathbb{Z}_q^m$ with the $\gamma \cdot (\lceil \log q \rceil - 1)$th component set to 1 and the rest set to 0. By construction of $\mathbf{G}$ we have that $\mathbf{G} \cdot \mathbf{v}_\gamma = \mathbf{u}_\gamma$.

For the cleanest way to describe the construction, we slightly abuse notation and define the setup algorithm pPRF.Setup to also publish a set of public parameters pp along with the master secret key msk. One can view pp as a fixed set of parameters for the whole system that is available to each algorithms pPRF.Puncture, pPRF.PunctureEval, pPRF.Eval, or it can be viewed as a component included in both the master secret key msk and the punctured key $\mathsf{sk}_{\mathbf{x}^*}$.

Fix a security parameter λ. We construct a privately puncturable PRF $\Pi_{\mathsf{pPRF}} = (\mathsf{pPRF.Setup}, \mathsf{pPRF.Puncture}, \mathsf{pPRF.PunctureEval}, \mathsf{pPRF.Eval})$ with domain $\{0,1\}^\ell$ and range $\mathbb{Z}_p$ as follows:

- pPRF.Setup(1^λ): On input the security parameter λ, the setup algorithm generates a set of uniformly random matrices in $\mathbb{Z}_q^{n \times m}$:
  - $\mathbf{A}_0, \mathbf{A}_1$ that will encode the input to the PRF
  - $\mathbf{B}_1, \ldots, \mathbf{B}_z$ that will encode the FHE ciphertext
  - $\mathbf{C}_1, \ldots, \mathbf{C}_t$ that will encode the FHE secret key
  Then, it generates a secret vector $\mathbf{s}$ from the error distribution $\mathbf{s} \leftarrow \chi^n$, and also samples a uniformly random vector $\mathbf{w} \in \mathbb{Z}_q^n$. It sets

$$\mathsf{pp} = (\{\mathbf{A}_b\}_{b \in \{0,1\}}, \{\mathbf{B}_i\}_{i \in [z]}, \{\mathbf{C}_j\}_{j \in [t]}, \mathbf{w}) \quad \text{and} \quad \mathsf{msk} = \mathbf{s}$$

- pPRF.Eval(msk, $\mathbf{x}$): On input the master secret key $\mathsf{msk} = \mathbf{s}$ and the PRF input $\mathbf{x}$, the evaluation algorithm first computes

$$\tilde{\mathbf{B}}_\gamma \leftarrow \mathsf{Eval}_{\mathsf{pk}}(C_\gamma, \mathbf{B}_1, \ldots, \mathbf{B}_z, \mathbf{A}_{x_1}, \ldots, \mathbf{A}_{x_\ell}, \mathbf{C}_1, \ldots, \mathbf{C}_t)$$

where for $\gamma \in [n]$ the circuit C_γ is defined as $C_\gamma(\cdot) = \mathsf{IP} \circ \mathsf{HE.Eval}(\mathsf{eq}_{w_\gamma}, \cdot)$ and the equality check function eq_{w_γ} is defined as:

$$\mathsf{eq}_{w_\gamma}(\mathbf{x}^*, \mathbf{x}) = \begin{cases} w_\gamma & \text{if } \mathbf{x} = \mathbf{x}^* \\ 0 & \text{otherwise} \end{cases}.$$

The algorithm outputs the following as the PRF value:

$$\left\lfloor \sum_{\gamma \in [n]} \langle \mathbf{s}^T \tilde{\mathbf{B}}_\gamma, \mathbf{v}_\gamma \rangle \right\rceil_p \in \mathbb{Z}_p.$$

- pPRF.Puncture(msk, $\mathbf{x}^*$): given msk and the point to be punctured $\mathbf{x}^* = (x_1^*, \ldots, x_\ell^*) \in \{0,1\}^\ell$ as input, the puncturing algorithm generates an FHE key he.sk $\leftarrow$ HE.KeyGen$(1^\lambda, 1^{d'}, 1^\ell)$ and encrypts $\mathbf{x}^*$ as

$$\text{he.ct} \leftarrow \text{HE.Enc}\big(\text{he.sk}, (x_1^*, \ldots, x_\ell^*)\big) \in \mathbb{Z}_q^z.$$

Then, it samples an error vector $\mathbf{e} \leftarrow \chi^{2+z+t}$ from the error distribution and computes

$$\begin{aligned}
\mathbf{a}_b &= \mathbf{s}^T(\mathbf{A}_b + b \cdot \mathbf{G}) + \mathbf{e}_{1,b}^T & \forall b \in \{0,1\} \\
\mathbf{b}_i &= \mathbf{s}^T(\mathbf{B}_i + \text{he.ct}_i \cdot \mathbf{G}) + \mathbf{e}_{2,i}^T & \forall i \in [z] \\
\mathbf{c}_j &= \mathbf{s}^T(\mathbf{C}_j + \text{he.sk}_j \cdot \mathbf{G}) + \mathbf{e}_{3,j}^T & \forall j \in [t].
\end{aligned}$$

It outputs the punctured key $\text{sk}_{\mathbf{x}^*} = (\{\mathbf{a}_b\}_{b \in \{0,1\}}, \{\mathbf{b}_i\}_{i \in [z]}, \{\mathbf{c}_j\}_{j \in [t]}, \text{he.ct})$.
- pPRF.PunctureEval($\text{sk}_{\mathbf{x}^*}, \mathbf{x}$): On input a punctured key $\text{sk}_{\mathbf{x}^*} = (\{\mathbf{a}_b\}_{b \in \{0,1\}}, \{\mathbf{b}_i\}_{i \in [z]}, \{\mathbf{c}_j\}_{j \in [t]}, \text{he.ct})$ and $\mathbf{x} \in \{0,1\}^\ell$, the puncture evaluation algorithm runs

$$\tilde{\mathbf{b}}_\gamma \leftarrow \text{Eval}_{\text{ct}}\big(C_\gamma, \ \mathbf{b}_1, \ldots, \mathbf{b}_z, \ \mathbf{a}_{x_1}, \ldots, \mathbf{a}_{x_\ell}, \ \mathbf{c}_1, \ldots, \mathbf{c}_t, \ (\text{he.ct}, \mathbf{x})\big)$$

for $\gamma = 1, \ldots, n$. Here C_γ is the circuit defined as in algorithm pPRF.Eval. The puncture evaluation algorithm then outputs the PRF value:

$$\left\lfloor \sum_{\gamma \in [n]} \langle \tilde{\mathbf{b}}_\gamma, \mathbf{v}_\gamma \rangle \right\rceil_p \in \mathbb{Z}_p.$$

As discussed in Sect. 3.2, we can de-randomize algorithm pPRF.Puncture so that it always returns the same output when run on the same input.

5.2 Parameters

The parameters can be instantiated such that breaking correctness or security translates to solving worst-case lattice problems to $2^{\tilde{O}(n^{1/c})}$ for some constant c. We set the parameters to account for the noise of both (a) the FHE decryption and (b) the homomorphic computation on the ABE encodings. The former will be bounded largely by $B \cdot m^{O(d')}$ and the latter by $B \cdot m^{O(d)}$. Here, d' is the depth of the equality check circuit and d is the depth of the *FHE operation* of the equality check circuit. We want to set the modulus of the encodings q to be big enough to account for these bounds. Furthermore, for the 1D-SIS-R assumption, we need q to be the product of coprime moduli $p_1, \ldots, p_\lambda$ such that the smallest of these primes exceeds these bounds.

Sample Instantiations. We first set the PRF input length $\ell = \text{poly}(\lambda)$. The depth of the equality check circuit is then $d' = O(\log \ell)$. We set $n = \lambda^{2c}$. We define q to be the product of λ coprime moduli $p, p_1, \ldots, p_\lambda$ where we set $p = \text{poly}(\lambda)$ and for each $i \in [\lambda]$, $p_i = 2^{O(n^{1/2c})}$ such that $p_1 < \ldots < p_\lambda$. The noise distribution χ is set to be the discrete Gaussian distribution $D_{\mathbb{Z}, \sqrt{n}}$. Then the FHE ciphertext size z and the secret key size t is determined by q. Set $m = \Theta(n \log q)$. The depth of the FHE equality check circuit is $d = \text{poly}(d', \log z)$.

5.3   Correctness and Security

We now state the correctness and security theorems of the construction in Sect. 5.1. The proofs of these theorems can be found in the full version [BKM17].

Theorem 3. *The puncturable PRF from Sect. 5.1 with parameters instantiated as in Sect. 5.2 satisfies the correctness property of Definition 5 assuming the hardness of $LWE_{n,m,q,\chi}$ and $1D\text{-}SIS\text{-}R_{q,p,\beta,m'}$ for $\beta = B \cdot m^{\tilde{O}(d)}$ and $m' = m \cdot (2 + z + t) + 1$.*

Theorem 4. *The puncturable PRF from Sect. 5.1 with parameters instantiated as in Sect. 5.2 is* selectively-pseudorandom *as defined in Definition 6 assuming the hardness of $LWE_{n,m',q,\chi}$ and $1D\text{-}SIS\text{-}R_{q,p,\beta,m'}$ for $\beta = B \cdot m^{\tilde{O}(d)}$ and $m' = m \cdot (2 + z + t) + 1$.*

Theorem 5. *Let Π_{HE} be a secure leveled homomorphic encryption scheme with parameters instantiated as in Sect. 5.2. The puncturable PRF from Sect. 5.1 with parameters instantiated as in Sect. 5.2 satisfies the security property of a private puncturable PRF as defined in Definition 6 assuming the hardness of $LWE_{n,m',q,\chi}$ and $1D\text{-}SIS\text{-}R_{q,p,\beta,m'}$ for $\beta = B \cdot m^{\tilde{O}(d)}$ and $m' = m \cdot (2 + z + t) + 1$.*

5.4   Extentions

We conclude this section with some high-level discussion on extending our scheme and how it relates to other lattice based PRF constructions.

Puncturing at Multiple Points. A private puncturable PRF can be combined to support a single-key private k-puncturable PRF generically where a constrained key can be punctured at k distinct points in the domain. One way of doing this is to simply define the PRF to be the xor of k independent instances of a 1-puncturable PRF. More precisely, let $\mathsf{msk}_i \leftarrow \mathsf{pPRF.Setup}(1^\lambda)$ for $i = 1, \ldots, k$. Then define the master secret key of the k-puncturable PRF to be the collection of these master secret keys $\mathsf{msk} = (\mathsf{msk}_1, \ldots, \mathsf{msk}_k)$. We define the evaluation of the PRF to be $F(\mathsf{msk}, \mathbf{x}) = F(\mathsf{msk}_1, \mathbf{x}) \oplus \ldots \oplus F(\mathsf{msk}_k, \mathbf{x})$. Then, to generate a punctured key at $S = \{\mathbf{x}_1, \ldots, \mathbf{x}_k\}$, we puncture each msk_i at point $\mathbf{x}_i$, to get punctured key $\mathsf{sk}_{\mathbf{x}_i} \leftarrow \mathsf{pPRF.Puncture}(\mathsf{msk}_i, \mathbf{x}_i)$, and then set $\mathsf{sk}_S = (\mathsf{sk}_{\mathbf{x}_1}, \ldots, \mathsf{sk}_{\mathbf{x}_k})$. It is easy to see that one can evaluate the PRF with the punctured key only at a point $\mathbf{x}$ in the domain $\mathbf{x} \notin S$. It is also straightforward to show that pseudorandomness and privacy follow from the security of the underlying 1-puncturable PRF.

Short Constrained Keys. In [BV15], Brakerski and Vaikuntanathan provide a way to achieve succinct constrained keys for their single-key constrained PRF, which also extends to our construction in Sect. 5.1. We provide a high level overview of this method.

In the constrained PRF construction of [BV15], a constrained key consists of the description of the constraint circuit along with the ABE encodings of the constraint circuit. To get succinct constrained keys, one can encrypt the bit encodings for each possible bits using an encryption scheme and publish it as part of the public parameters (just like in a garbling scheme). Then, as the constrained key, one can provide the decryption keys corresponding to the bit description of the constraint circuit. Now, using the attribute-based encryption construction of [BGG+14], which has short decryption keys, one can provide the ABE secret key that allows the decryption of the bits of the constraint circuit.

One difference with our construction is that we encode *field elements* in our ABE encodings for the FHE key. However, the FHE key stays the same for any punctured point. Therefore, we can garble just the bit positions corresponding to the encryption of the point to be punctured and publish the rest of the components in the clear. This allows the size of the public parameters to absorb the size of the constrained key.

Key Homomomorphism. Our PRF construction has a similar structure as the other lattice-based PRF constructions and therefore, the master secret key (LWE vector) for which the PRF is defined can be added homomorphically from the PRF evaluations. However, we note that in our construction, the PRF key (secret vector) is from a short noise distribution χ. Although there are applications of key-homomorphic PRFs with short keys, for most applications of key-homomorphic PRFs, one requires a perfect secret sharing of the PRF key, which requires it to come from a uniform distribution over a finite group. We leave it as an open problem to extend the construction to the setting of key-homomorphic PRFs with uniform keys.

6   Impossibility of Simulation Based Privacy

In this section, we show that a simulation based privacy notion for constrained PRFs for general circuit constraints is impossible. More precisely, we show that even for the single-key setting where the adversary is given one single constrained key, a natural extension of the indistinguishability privacy definition (Definition 4) to a simulation based privacy definition cannot be satisfied. We do this rather indirectly by showing that a constrained PRF (for general circuits) satisfying the simulation based privacy definition implies a simulation secure functional encryption [SS10, BSW11, O'N10], which was shown to be impossible in [BSW11, AGVW13].

6.1   Definition

We begin with the definition of a simulation based privacy notion for constrained PRFs. The simulation based privacy requires that any adversary given a constrained key sk_f does not learn any more information about the constraint other than what can be implied by comparing the output of the real evaluation cPRF.Eval(msk, ·) and cPRF.ConstrainEval(sk_f, ·). The correctness and pseudorandomness properties stay the same as how it is defined in Sect. 3.

Definition 8 (Sim-Privacy) *Fix a security parameter $\lambda \in \mathbb{N}$. A constrained PRF scheme $\Pi_{\mathsf{cPRF}} = (\mathsf{cPRF.Setup}, \mathsf{cPRF.Constrain}, \mathsf{cPRF.ConstrainEval}, \mathsf{cPRF.Eval})$ is simulation-private for single-key if there exists a PPT simulator $\mathcal{S} = (\mathcal{S}_{\mathsf{Eval}}, \mathcal{S}_{\mathsf{Constrain}})$ such that for all PPT adversary $\mathcal{A}$, there exists a negligible function $\mathsf{negl}(\lambda)$ such that*

$$\mathsf{Adv}^{\mathsf{priv}}_{\Pi_{\mathsf{cPRF}},\mathcal{A}}(\lambda) = \left| \Pr[\mathsf{Expt}^{\mathsf{REAL}}_{\Pi_{\mathsf{cPRF}},\mathcal{A}}(\lambda) = 1] - \Pr[\mathsf{Expt}^{\mathsf{RAND}}_{\Pi_{\mathsf{cPRF}},\mathcal{A}}(\lambda) = 1] \right| \leq \mathsf{negl}(\lambda)$$

where the experiments $\mathsf{Expt}^{\mathsf{REAL}}_{\Pi_{\mathsf{cPRF}},\mathcal{A}}(\lambda)$ and $\mathsf{Expt}^{\mathsf{RAND}}_{\Pi_{\mathsf{cPRF}},\mathcal{A}}(\lambda)$ are defined as follows:

$\mathsf{Expt}^{\mathsf{REAL}}_{\Pi_{\mathsf{cPRF}},\mathcal{A}}(\lambda)$:	$\mathsf{Expt}^{\mathsf{RAND}}_{\Pi_{\mathsf{cPRF}},\mathcal{A}}(\lambda)$:
1. $\mathsf{msk} \leftarrow \mathsf{cPRF.Setup}(1^\lambda)$.	1. $(f^*, \mathsf{state}_1) \leftarrow \mathcal{A}^{\mathcal{S}_{\mathsf{Eval}}(\cdot)}(1^\lambda)$.
2. $(f^*, \mathsf{state}) \leftarrow \mathcal{A}^{\mathsf{cPRF.Eval}(\mathsf{msk},\cdot)}(1^\lambda)$.	2. $\mathsf{sk}_{f^*} \leftarrow \mathcal{S}_{\mathsf{Constrain}}()$.
3. $\mathsf{sk}_{f^*} \leftarrow \mathsf{cPRF.Constrain}(\mathsf{msk}, f^*)$.	3. $b \leftarrow \mathcal{A}(\mathsf{sk}_{f^*}, \mathsf{state}_1)$.
4. $b \leftarrow \mathcal{A}(\mathsf{sk}_{f^*}, \mathsf{state})$.	4. Output b.
5. Output b	

Here, the algorithms $\mathcal{S}_{\mathsf{Eval}}$ and $\mathcal{S}_{\mathsf{Constrain}}$ share common state and the algorithm $\mathcal{S}_{\mathsf{Constrain}}$ is given the size $|f|$ and oracle access to the following set of mappings

$$\mathcal{C}_{\mathsf{constrain}} = \left\{ i \mapsto f^*(x^{(i)}) : i \in [Q] \right\}$$

where Q represents the number of times $\mathcal{A}$ queries the evaluation oracles $\mathcal{S}_{\mathsf{Eval}}$.

In words, the security definition above requires that an adversary cannot distinguish whether it is interacting with a real constrained PRF or it is interacting with a simulator that is not actually given the constraint f^* except for output of f^* applied to each of the adversary's queries to the evaluation oracle.

6.2 Functional Encryption

In this subsection, we define a simulation secure functional encryption for circuits. For simplicity, we consider functions with just binary outputs.

A (secret-key) functional encryption (FE) scheme is a tuple of algorithms $\Pi_{\mathsf{FE}} = (\mathsf{FE.Setup}, \mathsf{FE.KeyGen}, \mathsf{FE.Encrypt}, \mathsf{FE.Decrypt})$ defined over a message space $\mathcal{X}$, and a class of functions $\mathcal{F}_\lambda = \{f : \mathcal{X} \to \{0,1\}\}$ with the following properties:

- $\mathsf{FE.Setup}(1^\lambda) \to \mathsf{msk}$: On input the security parameter λ, the setup algorithm outputs the master secret key msk.
- $\mathsf{FE.KeyGen}(\mathsf{msk}, f) \to \mathsf{sk}_f$: On input the master secret key msk and a circuit f, the key generation algorithm outputs a secret key sk_f.
- $\mathsf{FE.Encrypt}(\mathsf{msk}, \mathbf{x}) \to \mathsf{ct}$: On input the master secret key msk, and a message $\mathbf{x}$, the encryption algorithm outputs a ciphertext ct.
- $\mathsf{FE.Decrypt}(\mathsf{ct}, \mathsf{sk}_f) \to \{0,1\}$: On input a ciphertext ct and a secret key sk_f, the decryption algorithm outputs a bit $y \in \{0, 1\}$.

Correctness. A functional encryption scheme $\Pi_{FE} = (FE.Setup, FE.KeyGen, FE.Encrypt, FE.Decrypt)$ is correct if for all $\lambda \in \mathbb{N}$, $msk \leftarrow FE.Setup(1^\lambda)$, $f \in \mathcal{F}$, and $sk_f \leftarrow FE.KeyGen(msk, f)$, we have that

$$\Pr[FE.Decrypt(sk_f, FE.Encrypt(msk, \mathbf{x})) = f(\mathbf{x})] = 1 - negl(\lambda).$$

Security. For security, we require that any adversary given a secret key does not learn any more information about an encrypted message other than what can be deduced from an honest decryption.

Definition 9. *Fix a security parameter* $\lambda \in \mathbb{N}$. *A functional encryption scheme* $\Pi_{FE} = (FE.Setup, FE.KeyGen, FE.Encrypt, FE.Decrypt)$ *is* simulation secure for single-key *if there exists a PPT simulator* $S = (S_{Encrypt}, S_{KeyGen})$ *such that for all PPT adversary* $\mathcal{A}$, *there exists a negligible function* $negl(\lambda)$ *such that*

$$Adv^{FE}_{\Pi_{FE}, \mathcal{A}}(\lambda) = \left| \Pr[Expt^{REAL}_{\Pi_{FE}, \mathcal{A}}(\lambda) = 1] - \Pr[Expt^{RAND}_{\Pi_{FE}, \mathcal{A}}(\lambda) = 1] \right| \leq negl(\lambda)$$

where the experiments $Expt^{REAL}_{\Pi_{FE}, \mathcal{A}}(\lambda)$ *and* $Expt^{RAND}_{\Pi_{FE}, \mathcal{A}}(\lambda)$ *are defined as follows:*

$Expt^{REAL}_{\Pi_{FE}, \mathcal{A}}(\lambda)$:	$Expt^{RAND}_{\Pi_{FE}, \mathcal{A}}(\lambda)$:
1. $msk \leftarrow FE.Setup(1^\lambda)$.	*1.* $(f^*, state) \leftarrow \mathcal{A}^{S_{Encrypt}()}(1^\lambda)$
2. $(f^*, state) \leftarrow \mathcal{A}^{FE.Encrypt(msk, \cdot)}(1^\lambda)$.	*2.* $sk_{f^*} \leftarrow S_{KeyGen}(f^*)$.
3. $sk_{f^*} \leftarrow FE.KeyGen(msk, f^*)$.	*3.* $b \leftarrow \mathcal{A}(sk_{f^*}, state)$.
4. $b \leftarrow \mathcal{A}(sk_{f^*}, state)$.	*4.* Output b.
5. Output b	

Here, the algorithms $S_{Encrypt}$ *and* S_{KeyGen} *share common state and the simulator* S_{KeyGen} *is given oracle access to the set of mappings* $\mathcal{C}_{msg} = \{i \mapsto f^*(\mathbf{x}^{(i)}) : i \in [Q]\}$ *where* Q *represents the number of queries that* $\mathcal{A}$ *makes to* $S_{Encrypt}$.

It was shown in [BSW11, AGVW13] that a functional encryption scheme satisfying the security definition above is impossible to achieve.

6.3   FE from Constrained PRFs

In this subsection, we present our construction of functional encryption. Fix a security parameter λ. Let $\Pi_{cPRF} = (cPRF.Setup, cPRF.Constrain, cPRF.ConstrainEval, cPRF.Eval)$ be a constrained PRF with domain $\{0,1\}^{\lambda+\ell}$ and range $\{0,1\}^\lambda$ where ℓ is the size of the message in the functional encryption scheme. We also use an additional regular PRF, which we denote by $F_{\mathbf{k}} : \{0,1\}^\lambda \to \{0,1\}^\ell$. We construct $\Pi_{FE} = (FE.Setup, FE.KeyGen, FE.Encrypt, FE.Decrypt)$ as follows:

- FE.Setup(1^λ): On input the security parameter λ, the setup algorithm first samples a regular PRF key $\mathbf{k} \xleftarrow{\$} \{0,1\}^\lambda$. Then, it runs cprf.msk $\leftarrow$ cPRF.Setup(1^λ) and sets msk = (cprf.msk, $\mathbf{k}$).

– FE.KeyGen(msk, f): On input the master secret key msk and a circuit f, the key generation algorithm generates a constrained PRF key $\mathsf{sk}_{C_{f,\mathbf{k}}} \leftarrow$ cPRF.Constrain(cprf.msk, $C_{f,\mathbf{k}}$) where the circuit $C_{f,\mathbf{k}}$ is defined as follows:

$$C_{f,\mathbf{k}}(\mathbf{r}, \mathbf{y}) = \begin{cases} 0 \text{ if } f(F_{\mathbf{k}}(\mathbf{r}) \oplus \mathbf{y}) = 0 \\ 1 \text{ otherwise} \end{cases}.$$

It outputs $\mathsf{sk}_f = \mathsf{sk}_{C_{f,\mathbf{k}}}$.

– FE.Encrypt(msk, $\mathbf{x}$): On input the master secret key msk, and a message $\mathbf{x} \in \{0,1\}^\ell$, the encryption algorithm first samples encryption randomness $\mathbf{r} \xleftarrow{\$} \{0,1\}^\lambda$ and computes $\mathbf{y} = F_{\mathbf{k}}(\mathbf{r}) \oplus \mathbf{x}$. Then, it returns

$$\mathsf{ct} = (\mathbf{r}, \mathbf{y}, \mathsf{cPRF.Eval}(\mathsf{cprf.msk}, \mathbf{r}\|\mathbf{y}))$$

– FE.Decrypt(sk_f, ct): On input a secret key $\mathsf{sk}_f = \mathsf{sk}_{C_{f,\mathbf{k}}}$ and ct $= (\mathbf{r}, \mathbf{y}, \tilde{\mathsf{ct}})$, the decryption algorithm returns 0 if cPRF.ConstrainEval($\mathsf{sk}_{C_{f,\mathbf{k}}}, \mathbf{r}\|\mathbf{y}$) $= \tilde{\mathsf{ct}}$ and 1 otherwise.

Correctness. To show correctness, we note that the decryption algorithm simply evaluates the PRF using the constrained key cPRF.Constrain($\mathsf{sk}_{C_{f,\mathbf{k}}}, \mathbf{r}\|\mathbf{y}$) and returns 0 if the result equals $\tilde{\mathsf{ct}}$ and 1 otherwise. Since $\tilde{\mathsf{ct}}$ is precisely the PRF evaluation using the master secret key cPRF.Eval(cprf.msk, $\mathbf{r}\|\mathbf{y}$), the two evaluations coincide if $C_{f,\mathbf{k}}(\mathbf{r}, \mathbf{y}) = 0$. Also, if Π_{cPRF} satisfies the standard notion of pseudorandomness as in Definition 3, the PRF evaluation using the master secret key and the PRF evaluation using the constrained key differs with overwhelming probability if the constraint is not satisfied $C_{f,\mathbf{k}}(\mathbf{r}, \mathbf{y}) = 1$.

6.4 Security

In this section, we prove security of construction above.

Theorem 6. *Let $\Pi_{\mathsf{cPRF}} = $ (cPRF.Setup, cPRF.Constrain, cPRF.ConstrainEval, cPRF.Eval) be a constrained PRF scheme satisfying the security properties of Definition 8. Also, let $F_{\mathbf{k}}$ be a secure PRF. Then, the functional encryption scheme Π_{FE} constructed above satisfies the simulation based security notion of Definition 9.*

Proof. We proceed through a series of hybrid experiments where the first hybrid H_0 represents the real experiment $\mathsf{Expt}^{\mathsf{REAL}}_{\Pi_{\mathsf{FE}},\mathcal{A}}$ and the final hybrid H_3 represents the ideal simulation $\mathsf{Expt}^{\mathsf{RAND}}_{\Pi_{\mathsf{FE}},\mathcal{A}}$.

– **Hybrid H_0:** This is the *real* experiment. The challenger runs the real setup algorithm to generate msk. Then the adversary makes a number of encryption queries and a key generation query which the challenger answers using its msk.

- **Hybrid H_1:** In this experiment, the challenger runs the simulator for the constrained PRF to answer the adversary's queries. More precisely, given a constrained PRF simulator $\mathcal{S} = (\mathcal{S}_{\mathsf{Eval}}, \mathcal{S}_{\mathsf{Constrain}})$, the challenger first samples a PRF key $\mathbf{k} \overset{\$}{\leftarrow} \{0,1\}^\lambda$ as msk. Then for each encryption query $\mathbf{x}$ that the adversary makes, the challenger samples $\mathbf{r} \overset{\$}{\leftarrow} \{0,1\}^\lambda$, computes $\mathbf{y} \leftarrow F_{\mathbf{k}}(\mathbf{r}) \oplus \mathbf{x}$ and invokes the simulator to generate $\tilde{\mathsf{ct}} \leftarrow \mathcal{S}_{\mathsf{Eval}}(\mathbf{r}\|\mathbf{y})$. It provides $(\mathbf{r}, \mathbf{y}, \tilde{\mathsf{ct}})$ to the adversary as the encryption of $\mathbf{x}$. To answer the single key generation query on f^* from the adversary, the challenger invokes the simulator $\mathcal{S}_{\mathsf{Constrain}}()$ to generate the key. For the set of mappings $\mathcal{C}_{\mathsf{constrain}}$ that are to be provided to $\mathcal{S}_{\mathsf{Constrain}}$, the challenger computes $f^*(\mathbf{x}^{(i)})$ itself and feeds it to the simulator. By the assumption on the simulator $\mathcal{S} = (\mathcal{S}_{\mathsf{Eval}}, \mathcal{S}_{\mathsf{Constrain}})$, we have that the hybrids H_0 and H_1 are indistinguishable to the adversary. We note that in H_1, the challenger does not actually use the PRF key $\mathbf{k}$ to generate the secret keys.

- **Hybrid H_2:** In this experiment, the challenger replaces $F_{\mathbf{k}}(\cdot)$ with a random function. Namely, to answer an encryption query $\mathbf{x}$ by the adversary, the challenger ignores the message $\mathbf{x}$ and samples $\tilde{\mathbf{y}}$ randomly $\tilde{\mathbf{y}} \overset{\$}{\leftarrow} \{0,1\}^\ell$. It then invokes $\tilde{\mathsf{ct}} \leftarrow \mathcal{S}_{\mathsf{Eval}}(\mathbf{r}\|\tilde{\mathbf{y}})$ and sets $(\mathbf{r}, \tilde{\mathbf{y}}, \tilde{\mathsf{ct}})$ as the encryption of $\mathbf{x}$. The rest of the experiment remains unchanged from H_1.

  Note that in both hybrid experiments H_1 and H_2, the challenger does not use the PRF key $\mathbf{k}$ other than in evaluating the PRF $F_{\mathbf{k}}(\cdot)$ to encrypt. Therefore, by the PRF security of $F_{\mathbf{k}}(\cdot)$, the two experiments are indistinguishable to the adversary. We note that in H_2, the challenger does not use any information about the message $\mathbf{x}_i$ other than providing the simulator $\mathcal{S}_{\mathsf{Constrain}}$ with the values $f^*(\mathbf{x}^{(i)})$.

- **Hybrid H_3:** This experiment represents the *ideal* experiment where the challenger corresponds to the simulator for the functional encryption game. The simulator runs in exactly the same way as in the previous hybrid H_2. Namely, for each encryption query $\mathbf{x}$ that the adversary makes, it samples $\mathbf{r} \overset{\$}{\leftarrow} \{0,1\}^\lambda$, $\mathbf{y} \overset{\$}{\leftarrow} \{0,1\}^\ell$ and invokes $\tilde{\mathsf{ct}} \leftarrow \mathcal{S}_{\mathsf{Eval}}(\mathbf{r}\|\tilde{\mathbf{y}})$. It sets $(\mathbf{r}, \tilde{\mathbf{y}}, \tilde{\mathsf{ct}})$ as the encryption of $\mathbf{x}$. Note that to generate the ciphertext, it does not use any information about $\mathbf{x}$. For the single key query, the simulator invokes $\mathcal{S}_{\mathsf{Constrain}}()$. For the set of mappings $\mathcal{C}_{\mathsf{constrain}}$ that are to be provided to $\mathcal{S}_{\mathsf{Constrain}}$, it uses its own oracle $\mathcal{C}_{\mathsf{msg}}$ to provide the values $f^*(\mathbf{x}^{(i)})$.

  It is easy to see that the distribution of the experiments H_2 and H_3 are identical.

We have shown that the experiment H_0, which corresponds to $\mathsf{Expt}_{\Pi_{\mathsf{FE}}, \mathcal{A}}^{\mathsf{REAL}}$ and the experiment H_3, which corresponds to $\mathsf{Expt}_{\Pi_{\mathsf{FE}}, \mathcal{A}}^{\mathsf{RAND}}$ are indistinguishable. This concludes the proof.

7   Conclusion and Open Problems

We constructed a privately puncturable PRF from worst-case lattice problems. Prior constructions of privately puncturable PRFs required heavy tools such as multilinear maps or $i\mathcal{O}$. This work provides the first privately puncturable PRF from a standard assumption. We also showed that for general functions, a natural simulation-based privacy definition for constrained PRFs is impossible to achieve.

Our PRF builds on the construction of [BV15], which supports circuit constraints. However, our construction does not extend to more general constraints, and it will be interesting to provide a private constrained PRF for a larger class of circuit constraints. For private puncturing, it will be interesting to give more constructions based on assumptions other than LWE.

Our construction satisfies the selective security game of private puncturable PRFs, and we rely on complexity leveraging for adaptive security. Recently, [HKW15] gave a way to achieve adaptively secure puncturable PRFs without complexity leveraging. Can we extend the result to *private* puncturable PRFs?

Finally, private constrained PRFs have a number of interesting applications, as explored here and in [BLW17]. It would be interesting to find further applications and relations to other cryptographic primitives.

Acknowledgements. We thank the anonymous Eurocrypt reviewers for their insightful comments which helped improve the paper. We also thank David Wu for his helpful comments on the definition of privately constrained PRFs. This work is supported by NSF, DARPA, the Simons foundation, and a grant from ONR. Opinions, findings and conclusions or recommendations expressed in this material are those of the author(s) and do not necessarily reflect the views of DARPA. Part of the work was also done while the second author was visiting Fujitsu Laboratories of America as an intern.

References

[ABB10] Agrawal, S., Boneh, D., Boyen, X.: Efficient lattice (H)IBE in the standard model. In: Gilbert, H. (ed.) EUROCRYPT 2010. LNCS, vol. 6110, pp. 553–572. Springer, Heidelberg (2010). doi:10.1007/978-3-642-13190-5_28

[ACPS09] Applebaum, B., Cash, D., Peikert, C., Sahai, A.: Fast cryptographic primitives and circular-secure encryption based on hard learning problems. In: Halevi, S. (ed.) CRYPTO 2009. LNCS, vol. 5677, pp. 595–618. Springer, Heidelberg (2009). doi:10.1007/978-3-642-03356-8_35

[AFP16] Abusalah, H., Fuchsbauer, G., Pietrzak, K.: Constrained PRFs for unbounded inputs. In: Sako, K. (ed.) CT-RSA 2016. LNCS, vol. 9610, pp. 413–428. Springer, Cham (2016). doi:10.1007/978-3-319-29485-8_24

[AFV11] Agrawal, S., Freeman, D.M., Vaikuntanathan, V.: Functional encryption for inner product predicates from learning with errors. In: Lee, D.H., Wang, X. (eds.) ASIACRYPT 2011. LNCS, vol. 7073, pp. 21–40. Springer, Heidelberg (2011). doi:10.1007/978-3-642-25385-0_2

[AGVW13] Agrawal, S., Gorbunov, S., Vaikuntanathan, V., Wee, H.: Functional encryption: new perspectives and lower bounds. In: Canetti, R., Garay, J.A. (eds.) CRYPTO 2013. LNCS, vol. 8043, pp. 500–518. Springer, Heidelberg (2013). doi:10.1007/978-3-642-40084-1_28

[Ajt96] Ajtai, M.: Generating hard instances of lattice problems. In: STOC (1996)

[ASP14] Alperin-Sheriff, J., Peikert, C.: Faster bootstrapping with polynomial error. In: Garay, J.A., Gennaro, R. (eds.) CRYPTO 2014. LNCS, vol. 8616, pp. 297–314. Springer, Heidelberg (2014). doi:10.1007/978-3-662-44371-2_17

[BB04] Boneh, D., Boyen, X.: Efficient selective-ID secure identity-based encryption without random oracles. In: Cachin, C., Camenisch, J.L. (eds.) EUROCRYPT 2004. LNCS, vol. 3027, pp. 223–238. Springer, Heidelberg (2004). doi:10.1007/978-3-540-24676-3_14

[BCTW16] Brakerski, Z., Cash, D., Tsabary, R., Wee, H.: Targeted homomorphic attribute-based encryption. In: Hirt, M., Smith, A. (eds.) TCC 2016. LNCS, vol. 9986, pp. 330–360. Springer, Heidelberg (2016). doi:10.1007/978-3-662-53644-5_13

[BFP+15] Banerjee, A., Fuchsbauer, G., Peikert, C., Pietrzak, K., Stevens, S.: Key-homomorphic constrained pseudorandom functions. In: Dodis, Y., Nielsen, J.B. (eds.) TCC 2015. LNCS, vol. 9015, pp. 31–60. Springer, Heidelberg (2015). doi:10.1007/978-3-662-46497-7_2

[BGG+14] Boneh, D., Gentry, C., Gorbunov, S., Halevi, S., Nikolaenko, V., Segev, G., Vaikuntanathan, V., Vinayagamurthy, D.: Fully key-homomorphic encryption, arithmetic circuit abe and compact garbled circuits. In: Nguyen, P.Q., Oswald, E. (eds.) EUROCRYPT 2014. LNCS, vol. 8441, pp. 533–556. Springer, Heidelberg (2014). doi:10.1007/978-3-642-55220-5_30

[BGI14] Boyle, E., Goldwasser, S., Ivan, I.: Functional signatures and pseudorandom functions. In: Krawczyk, H. (ed.) PKC 2014. LNCS, vol. 8383, pp. 501–519. Springer, Heidelberg (2014). doi:10.1007/978-3-642-54631-0_29

[BGI15] Boyle, E., Gilboa, N., Ishai, Y.: Function secret sharing. In: Oswald, E., Fischlin, M. (eds.) EUROCRYPT 2015. LNCS, vol. 9057, pp. 337–367. Springer, Heidelberg (2015). doi:10.1007/978-3-662-46803-6_12

[BGV12] Brakerski, Z., Gentry, C., Vaikuntanathan, V.: (leveled) fully homomorphic encryption without bootstrapping. In: ITCS (2012)

[BKM17] Boneh, D., Kim, S., Montgomery, H.: Private puncturable PRFs from standard lattice assumptions. Cryptology ePrint Archive, Report 2017/100 (2017). http://eprint.iacr.org/2017/100

[BLMR13] Boneh, D., Lewi, K., Montgomery, H., Raghunathan, A.: Key homomorphic PRFs and their applications. In: Canetti, R., Garay, J.A. (eds.) CRYPTO 2013. LNCS, vol. 8042, pp. 410–428. Springer, Heidelberg (2013). doi:10.1007/978-3-642-40041-4_23

[BLP+13] Brakerski, Z., Langlois, A., Peikert, C., Regev, O., Stehlé, D.: Classical hardness of learning with errors. In: STOC (2013)

[BLW17] Boneh, D., Lewi, K., Wu, D.J.: Constraining pseudorandom functions privately. In: Fehr, S. (ed.) PKC 2017. LNCS, vol. 10175, pp. 494–524. Springer, Heidelberg (2017). doi:10.1007/978-3-662-54388-7_17

[BP14] Banerjee, A., Peikert, C.: New and improved key-homomorphic pseudorandom functions. In: Garay, J.A., Gennaro, R. (eds.) CRYPTO 2014. LNCS, vol. 8616, pp. 353–370. Springer, Heidelberg (2014). doi:10.1007/978-3-662-44371-2_20

[BP16] Brakerski, Z., Perlman, R.: Lattice-based fully dynamic multi-key FHE with short ciphertexts. In: Robshaw, M., Katz, J. (eds.) CRYPTO 2016. LNCS, vol. 9814, pp. 190–213. Springer, Heidelberg (2016). doi:10.1007/978-3-662-53018-4_8

[BPR12]   Banerjee, A., Peikert, C., Rosen, A.: Pseudorandom functions and lattices. In: Pointcheval, D., Johansson, T. (eds.) EUROCRYPT 2012. LNCS, vol. 7237, pp. 719–737. Springer, Heidelberg (2012). doi:10.1007/978-3-642-29011-4_42

[BSW11]   Boneh, D., Sahai, A., Waters, B.: Functional encryption: definitions and challenges. In: Ishai, Y. (ed.) TCC 2011. LNCS, vol. 6597, pp. 253–273. Springer, Heidelberg (2011). doi:10.1007/978-3-642-19571-6_16

[BV14a]   Brakerski, Z., Vaikuntanathan, V.: Efficient fully homomorphic encryption from (standard) LWE. SIAM J. Comput. 43(2), 831–871 (2014)

[BV14b]   Brakerski, Z., Vaikuntanathan, V.: Lattice-based FHE as secure as PKE. In: ITCS (2014)

[BV15]   Brakerski, Z., Vaikuntanathan, V.: Constrained key-homomorphic PRFs from standard lattice assumptions. In: Dodis, Y., Nielsen, J.B. (eds.) TCC 2015. LNCS, vol. 9015, pp. 1–30. Springer, Heidelberg (2015). doi:10.1007/978-3-662-46497-7_1

[BV16]   Brakerski, Z., Vaikuntanathan, V.: Circuit-ABE from LWE: unbounded attributes and semi-adaptive security. In: Robshaw, M., Katz, J. (eds.) CRYPTO 2016. LNCS, vol. 9816, pp. 363–384. Springer, Heidelberg (2016). doi:10.1007/978-3-662-53015-3_13

[BW07]   Boneh, D., Waters, B.: Conjunctive, subset, and range queries on encrypted data. In: Vadhan, S.P. (ed.) TCC 2007. LNCS, vol. 4392, pp. 535–554. Springer, Heidelberg (2007). doi:10.1007/978-3-540-70936-7_29

[BW13]   Boneh, D., Waters, B.: Constrained pseudorandom functions and their applications. In: Sako, K., Sarkar, P. (eds.) ASIACRYPT 2013. LNCS, vol. 8270, pp. 280–300. Springer, Heidelberg (2013). doi:10.1007/978-3-642-42045-0_15

[BZ14]   Boneh, D., Zhandry, M.: Multiparty key exchange, efficient traitor tracing, and more from indistinguishability obfuscation. In: Garay, J.A., Gennaro, R. (eds.) CRYPTO 2014. LNCS, vol. 8616, pp. 480–499. Springer, Heidelberg (2014). doi:10.1007/978-3-662-44371-2_27

[CC17]   Canetti, R., Chen, Y.: Constraint-hiding constrained PRFs for NC1 from LWE. In: EUROCRYPT (2017)

[CDNO97] Canetti, R., Dwork, C., Naor, M., Ostrovsky, R.: Deniable encryption. In: Kaliski, B.S. (ed.) CRYPTO 1997. LNCS, vol. 1294, pp. 90–104. Springer, Heidelberg (1997). doi:10.1007/BFb0052229

[CHN+16] Cohen, A., Holmgren, J., Nishimaki, R., Vaikuntanathan, V., Wichs, D.: Watermarking cryptographic capabilities. In: STOC (2016)

[CKGS98] Chor, B., Kushilevitz, E., Goldreich, O., Sudan, M.: Private information retrieval. J. ACM 45(6), 965–981 (1998)

[CM15]   Clear, M., McGoldrick, C.: Multi-identity and multi-key leveled FHE from learning with errors. In: Gennaro, R., Robshaw, M. (eds.) CRYPTO 2015. LNCS, vol. 9216, pp. 630–656. Springer, Heidelberg (2015). doi:10.1007/978-3-662-48000-7_31

[CRV14]   Chandran, N., Raghuraman, S., Vinayagamurthy, D.: Constrained pseudorandom functions: verifiable and delegatable. IACR Cryptology ePrint Archive, 2014:522 (2014)

[DKW16]   Deshpande, A., Koppula, V., Waters, B.: Constrained pseudorandom functions for unconstrained inputs. In: Fischlin, M., Coron, J.-S. (eds.) EUROCRYPT 2016. LNCS, vol. 9666, pp. 124–153. Springer, Heidelberg (2016). doi:10.1007/978-3-662-49896-5_5

[FKPR14] Fuchsbauer, G., Konstantinov, M., Pietrzak, K., Rao, V.: Adaptive security of constrained PRFs. In: Sarkar, P., Iwata, T. (eds.) ASIACRYPT 2014. LNCS, vol. 8874, pp. 82–101. Springer, Heidelberg (2014). doi:10.1007/978-3-662-45608-8_5

[Fuc14] Fuchsbauer, G.: Constrained verifiable random functions. In: Abdalla, M., Prisco, R. (eds.) SCN 2014. LNCS, vol. 8642, pp. 95–114. Springer, Cham (2014). doi:10.1007/978-3-319-10879-7_7

[GGH15] Gentry, C., Gorbunov, S., Halevi, S.: Graph-induced multilinear maps from lattices. In: Dodis, Y., Nielsen, J.B. (eds.) TCC 2015. LNCS, vol. 9015, pp. 498–527. Springer, Heidelberg (2015). doi:10.1007/978-3-662-46497-7_20

[GGM86] Goldreich, O., Goldwasser, S., Micali, S.: How to construct random functions. J. ACM (JACM) 33(4), 792–807 (1986)

[GI14] Gilboa, N., Ishai, Y.: Distributed point functions and their applications. In: Nguyen, P.Q., Oswald, E. (eds.) EUROCRYPT 2014. LNCS, vol. 8441, pp. 640–658. Springer, Heidelberg (2014). doi:10.1007/978-3-642-55220-5_35

[GMW15] Gay, R., Méaux, P., Wee, H.: Predicate encryption for multi-dimensional range queries from lattices. In: Katz, J. (ed.) PKC 2015. LNCS, vol. 9020, pp. 752–776. Springer, Heidelberg (2015). doi:10.1007/978-3-662-46447-2_34

[GSW13] Gentry, C., Sahai, A., Waters, B.: Homomorphic encryption from learning with errors: conceptually-simpler, asymptotically-faster, attribute-based. In: Canetti, R., Garay, J.A. (eds.) CRYPTO 2013. LNCS, vol. 8042, pp. 75–92. Springer, Heidelberg (2013). doi:10.1007/978-3-642-40041-4_5

[GV15] Gorbunov, S., Vinayagamurthy, D.: Riding on asymmetry: efficient ABE for branching programs. In: Iwata, T., Cheon, J.H. (eds.) ASIACRYPT 2015. LNCS, vol. 9452, pp. 550–574. Springer, Heidelberg (2015). doi:10.1007/978-3-662-48797-6_23

[GVW15a] Gorbunov, S., Vaikuntanathan, V., Wee, H.: Attribute-based encryption for circuits. J. ACM (JACM) 62(6), 45 (2015)

[GVW15b] Gorbunov, S., Vaikuntanathan, V., Wee, H.: Predicate encryption for circuits from LWE. In: Gennaro, R., Robshaw, M. (eds.) CRYPTO 2015. LNCS, vol. 9216, pp. 503–523. Springer, Heidelberg (2015). doi:10.1007/978-3-662-48000-7_25

[HKKW14] Hofheinz, D., Kamath, A., Koppula, V., Waters, B.: Adaptively secure constrained pseudorandom functions. IACR Cryptology ePrint Archive, 2014:720 (2014)

[HKW15] Hohenberger, S., Koppula, V., Waters, B.: Adaptively secure puncturable pseudorandom functions in the standard model. In: Iwata, T., Cheon, J.H. (eds.) ASIACRYPT 2015. LNCS, vol. 9452, pp. 79–102. Springer, Heidelberg (2015). doi:10.1007/978-3-662-48797-6_4

[Hof14] Hofheinz, D.: Fully secure constrained pseudorandom functions using random oracles. IACR Cryptology ePrint Archive, 2014:372 (2014)

[KPTZ13] Kiayias, A., Papadopoulos, S., Triandopoulos, N., Zacharias, T.: Delegatable pseudorandom functions and applications. In: CCS (2013)

[KSW08] Katz, J., Sahai, A., Waters, B.: Predicate encryption supporting disjunctions, polynomial equations, and inner products. In: Smart, N. (ed.) EUROCRYPT 2008. LNCS, vol. 4965, pp. 146–162. Springer, Heidelberg (2008). doi:10.1007/978-3-540-78967-3_9

[MM11] Micciancio, D., Mol, P.: Pseudorandom knapsacks and the sample complexity of LWE search-to-decision reductions. In: Rogaway, P. (ed.) CRYPTO 2011. LNCS, vol. 6841, pp. 465–484. Springer, Heidelberg (2011). doi:10.1007/978-3-642-22792-9_26

[MP12] Micciancio, D., Peikert, C.: Trapdoors for lattices: simpler, tighter, faster, smaller. In: Pointcheval, D., Johansson, T. (eds.) EUROCRYPT 2012. LNCS, vol. 7237, pp. 700–718. Springer, Heidelberg (2012). doi:10.1007/978-3-642-29011-4_41

[MW16] Mukherjee, P., Wichs, D.: Two round multiparty computation via multi-key FHE. In: Fischlin, M., Coron, J.-S. (eds.) EUROCRYPT 2016. LNCS, vol. 9666, pp. 735–763. Springer, Heidelberg (2016). doi:10.1007/978-3-662-49896-5_26

[O'N10] O'Neill, A.: Definitional issues in functional encryption. IACR Cryptology ePrint Archive, 2010:556 (2010)

[Pei09] Peikert, C.: Public-key cryptosystems from the worst-case shortest vector problem. In: STOC (2009)

[PS16] Peikert, C., Shiehian, S.: Multi-key FHE from LWE, revisited. In: Hirt, M., Smith, A. (eds.) TCC 2016. LNCS, vol. 9986, pp. 217–238. Springer, Heidelberg (2016). doi:10.1007/978-3-662-53644-5_9

[Reg04] Regev, O.: Lattices in computer science-average case hardness. Lecture Notes for Class (scribe: Elad Verbin) (2004)

[Reg09] Regev, O.: On lattices, learning with errors, random linear codes, and cryptography. J. ACM (JACM) 56(6), 34 (2009)

[SS10] Sahai, A., Seyalioglu, H.: Worry-free encryption: functional encryption with public keys. In: CCS (2010)

[SW05] Sahai, A., Waters, B.: Fuzzy identity-based encryption. In: Cramer, R. (ed.) EUROCRYPT 2005. LNCS, vol. 3494, pp. 457–473. Springer, Heidelberg (2005). doi:10.1007/11426639_27

[SW14] Sahai, A., Waters, B.: How to use indistinguishability obfuscation: deniable encryption, and more. In: STOC (2014)

Constraint-Hiding Constrained PRFs for NC¹ from LWE

Ran Canetti[1,2] and Yilei Chen[1(✉)]

[1] Boston University, Boston, USA
{canetti,chenyl}@bu.edu
[2] Tel Aviv University, Tel Aviv, Israel

Abstract. Constraint-hiding constrained PRFs (CHCPRFs), initially studied by Boneh, Lewi and Wu (PKC 2017), are constrained PRFs where the constrained key hides the description of the constraint. Envisioned with powerful applications such as searchable encryption, private-detectable watermarking and symmetric deniable encryption, the only known candidates of CHCPRFs are based on indistinguishability obfuscation or multilinear maps with strong security properties.

In this paper we construct CHCPRFs for all NC¹ circuits from the Learning with Errors assumption. The construction draws heavily from the graph-induced multilinear maps by Gentry, Gorbunov and Halevi (TCC 2015), as well as the existing lattice-based PRFs. In fact, our construction can be viewed as an instance of the GGH15 approach where security can be reduced to LWE.

We also show how to build from CHCPRFs reusable garbled circuits (RGC), or equivalently private-key function-hiding functional encryptions with 1-key security. This provides a different approach of constructing RGC from that of Goldwasser et al. (STOC 2013).

1 Introduction

Constrained PRFs [15,16,39] are pseudorandom functions with a special mode that outputs a constrained key defined by a predicate C. The constrained key CK_C preserves the functionality over the inputs x s.t. $C(x) = 1$, while leaving the function values on inputs x s.t. $C(x) = 0$ pseudorandom. In the standard formulation of constrained PRFs, the constrained key is not required to hide the predicate C. In fact, many constructions of constrained PRFs do reveal the constraint. A quintessential example is GGM's puncturable PRF [33] where CK explicitly reveals the punctured points.

The notion of *constraint-hiding constrained PRF* (CHCPRF), proposed by Boneh, Lewi and Wu [12], makes the additional guarantee that the constraining predicate C remains hidden, even given the constrained key. Such an additional property allows the primitive to provide fairly natural constructions of searchable encryption, watermarking, deniable encryption, and others. However, they only propose candidates of CHCPRFs based on strong assumptions, like indistinguishability obfuscation (iO) or heuristic assumptions on candidate multilinear maps (multilinear-DDH or subgroup elimination).

© International Association for Cryptologic Research 2017
J.-S. Coron and J.B. Nielsen (Eds.): EUROCRYPT 2017, Part I, LNCS 10210, pp. 446–476, 2017.
DOI: 10.1007/978-3-319-56620-7_16

This Work. We further investigate the notion of CHCPRF, propose constructions based on standard cryptographic assumptions, and demonstrate more applications.

We first propose an alternative, simulation-based definition for CHCPRF. While for the cases addressed in our constructions the new style is (almost) equivalent to the indistinguishability-based one from [12], the new formulation provides a different viewpoint on the primitive.

Our main result is a construction of CHCPRF for all NC^1 circuit constraints based on the Learning with Errors (LWE) assumption [48]:

Theorem 1. *Assuming the intractability of LWE, there are CHCPRFs with 1-key simulation-based security, for all constraints recognizable by NC^1 circuits.*

The construction combines the graph-induced multilinear maps by Gentry, Gorbunov and Halevi [31], their candidate obfuscator, and the lattice-based PRFs of [5,6,11,19]. At the heart of our technical contribution is identifying a restricted (yet still powerful) variant of the GGH15 maps, whose security can be reduced to LWE. This involves formulating new "LWE-hard" secret distributions that handle the permutation matrices underlying Barrington's construction.

In addition, we construct function-hiding private-key functional encryptions (equivalently, reusable garbled circuits [34]) from CHCPRFs. This gives a construction of reusable garbled circuits from LWE that is very different from that of [34]:

Theorem 2. *For a circuit class $\mathcal{C}$, assuming 1-key simulation-based CHCPRFs for constraints in $\mathcal{C}$, and CPA secure private-key encryption whose decryption circuit is in $\mathcal{C}$, there exist 1-key secure reusable garbled circuits for $\mathcal{C}$.*

1.1 CHCPRFs, Functional Encryption and Obfuscation

We propose a simulation-based definitional approach for CHCPRF, and compare this approach to the indistinguishability-based approach of Boneh et al. [12].

Defining CHCPRFs. A constrained PRF consists of three algorithms: Master secret key generation, constrained key generation, and function evaluation. We first note that in order to have hope to hide the constraint, the function evaluation algorithm should return a random-looking value v even if evaluated on a constrained input x, as opposed to returning $\perp$ as in the standard formulation. Furthermore, we require that the value of the original function on x remains pseudorandom even given the constrained key and the value v.

The definition of CHCPRF is aimed at capturing three requirements: (1) the constrained keys preserve functionality on inputs that do not match the constraint; (2) the function values at constrained points remain pseudorandom given the constrained key; (3) the constrained key does not reveal any information on the constraining function.

Boneh et al. [12] give a number of indistinguishability-based definitions that vary in strength, depending on the level of adaptivity of the adversary in choosing the constraints and evaluation points, as well as on the number of constrained keys that the adversary is allowed to see. We take an alternative approach and give a simulation-based definition. We also compare the definitions, and show equivalence and derivations in a number of cases.

Here is a sketch of the non-adaptive single-key variant of our simulation-based definition. The definition captures all three requirements via a single interaction: We require that, for any polytime adversary, there exists a polytime simulator such that the adversary can distinguish between the outcomes of the following two experiments only with negligible probability:

- In the real experiment, the system first generates a master secret key K. The adversary can then query a constraint circuit C and many inputs $x^{(1)}, \ldots, x^{(t)}$. In return, it obtains $\mathsf{CK}_C, x^{(1)}, \ldots, x^{(t)}, y^{(1)}, \ldots, y^{(t)}$, where CK_C is a key constrained by C, and $y^{(i)}$ is the result of evaluating the original, unconstrained function with key K at point $x^{(i)}$. (This is so regardless of whether $x^{(i)}$ meets the constraint or not.)
- In the ideal experiment, the simulator samples a master secret key K^S. Once received a constraint query, the simulator obtains only the description length of C and creates a simulated constrained key CK^S. Once received input queries $x^{(1)}, \ldots, x^{(t)}$, the simulator also t indicator bits $d^{(1)}, \ldots, d^{(t)}$ where the $d^{(i)}$ denotes whether $x^{(i)}$ is in the constraint, and generates simulated values $y^{(1)S}, \ldots, y^{(t)S}$. If $d^{(i)} = 0$, then the simulated $y^{(i)S}$ is uniformly random. The output of the experiment is $\mathsf{CK}^S, x^{(1)}, \ldots, x^{(t)}, y^{(1)S}, \ldots, y^{(t)S}$.

Secret-Key Functional Encryption from Simulation-Based CHCPRFs. We sketch our construction of functional encryption from CHCPRFs. Functional encryption [13] allows the evaluator, given a functional decryption key, to learn the value of the function applied to encrypted data without learning anything else. With CHCPRFs in hand, it is rather simple to construct a private-key functional encryption scheme that is both function-private and input-private. Our functional encryption scheme proceeds as follows:

- Key generation: The master key for the scheme is a key K for a CHCPRF, and a key SK for a CPA-secure symmetric encryption scheme $(\mathsf{Enc}, \mathsf{Dec})$.
- Encrypt a message m: $\mathsf{CT} = (c, t)$, where $c = \mathsf{Enc}_{\mathsf{SK}}(m)$, and $t = \mathsf{CHCPRF}_K(c)$.
- Functional decryption key: The functional decryption key for a binary function f is a constrained-key $\mathsf{CK}_{\hat{f}}$ for the function $\hat{f}(c) = f(\mathsf{Dec}_{\mathsf{SK}}(c))$. That is, $\hat{f}$ has SK hardwired; it decrypts its input c and applies f to the plaintext.
- Functional decryption: Given ciphertext $\mathsf{CT} = (c, t)$ and the constrained decryption key $\mathsf{CK}_{\hat{f}}$, output 1 if $t = \mathsf{CHCPRF}_{\mathsf{CK}_{\hat{f}}}(c)$, 0 otherwise.

Correctness of decryption follows from the correctness and constrainability of the CHCPRF, and secrecy follows from the constraint-hiding property.

This construction is conceptually different from the previous construction [34]. In particular, the size of ciphertext (for 1-bit output) is the size of a symmetric encryption ciphertext plus the security parameter, independent of the depth of the circuit.

Two-Key CHCPRFs Imply Obfuscation. It is natural to consider an extension of the CHCPRF definition to the case where the adversary may obtain multiple constrained keys derived from the same master key. Indeed in [12] some applications of this extended notion are presented.

We observe that this extended notion in fact implies full fledged program obfuscation: To obfuscate a circuit C, choose a key K for a CHCPRF, and output two constrained keys: The constrained key $\mathsf{CK}[C]$, and the constrained key $\mathsf{CK}[I]$, where I is the circuit that always outputs 1. To evaluate $C(x)$ check whether $\mathsf{CHCPRF}_{\mathsf{CK}[C]}(x) = \mathsf{CHCPRF}_{\mathsf{CK}[I]}(x)$.

Again, correctness of evaluation follows from the correctness and constrainability of the CHCPRF. The level of security for the obfuscation depends on the definition of CHCPRF in use. Specifically, the natural extension of the above simulation-based definition to the two-key setting implies that the above simple obfuscation method is VBB (which in turn means that the known impossibility results for VBB obfuscation carry over to this variant of two-key CHCPRF). The indistinguishability-based definition of [12] implies that the above obfuscation method is IO.

1.2   Overview of Our Construction

Our construction of CHCPRFs draws heavily from the multilinear maps by Gentry et al. [31], and the lattice-based PRFs of Banerjee et al. [5,6,11,19]. We thus start with a brief review of the relevant parts of these works.

Recap GGH15. The GGH15 multilinear encoding is depicted by a DAG that defines the rule of homomorphic operations and zero-testing. For our purpose it is sufficient to consider the following special functionality (which corresponds to a graph of ℓ nodes and two parallel edges from node i to node $i+1$, see Fig. 1a): We would like to encode $2\ell + 1$ secrets $s_1^0, s_1^1, \ldots, s_\ell^0, s_\ell^1, s_T$ over some finite group G, in such a way that an evaluator who receives the encodings can test, for any given $x \in \{0,1\}^\ell$, whether $s_T = \prod_{i=1}^\ell s_i^{x_i}$, and at the same time the encodings hide "everything else" about the secrets. (Indeed, "everything else" might have different meanings in different contexts.)

To do that, GGH15 take the group G to actually be a ring R_q, where R denotes the base ring (typical choices include $R = \mathbb{Z}^{n \times n}$ or $R = \mathbb{Z}[x]/(\Phi_n(x))$, where n is a parameter related to the lattice dimension, and Φ_n is the nth cyclotomic polynomial), and q is the modulus. The encoder then samples $\ell + 1$ hard Ajtai-type matrices $\{\mathbf{A}_1, \mathbf{A}_2, \ldots, \mathbf{A}_\ell, \mathbf{A}_{\ell+1} \leftarrow R_q^{1 \times m}\}$ with trapdoors [2, 3,44], and associates each matrix with the corresponding node of the graph. These matrices and their trapdoors are treated as (universal) public and secret parameters, respectively. We refer to the indices $1 \ldots \ell + 1$ as *levels*.

The 2ℓ secrets are associated with the 2ℓ edges of the graph in the natural way. Encoding a secret s_i^b is done in two steps: First create an LWE sample for the secret s_i^b under the matrix $\mathbf{A}_{i+1}$, namely $\mathbf{Y}_i^b = s_i^b \mathbf{A}_{i+1} + \mathbf{E}_i^b$. Next, sample a preimage $\mathbf{D}_i^b$ of $\mathbf{Y}_i^b$ under the matrix $\mathbf{A}_i$, using the trapdoor of $\mathbf{A}_i$. That is, $\mathbf{A}_i \mathbf{D}_i^b = \mathbf{Y}_i^b$ and $\mathbf{D}_i^b$ is sampled from discrete Gaussian distribution of

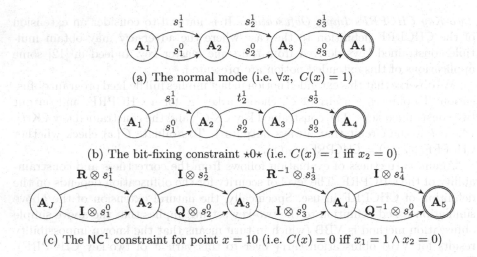

(a) The normal mode (i.e. $\forall x,\ C(x) = 1$)

(b) The bit-fixing constraint $\star 0\star$ (i.e. $C(x) = 1$ iff $x_2 = 0$)

(c) The NC^1 constraint for point $x = 10$ (i.e. $C(x) = 0$ iff $x_1 = 1 \land x_2 = 0$)

Fig. 1. Examples of the GGH15-based PRFs

small width. The encoder then lets $\mathbf{D}_i^b$ be the encoding of s_i^b. The encoding T of s_T, where $s_T = \prod_{i=1}^{\ell} s_i^{x_i}$ for some $x \in \{0,1\}^{\ell}$, is defined as $T = F(x)$, where $F(x) = \mathbf{A}_1 \prod_{i=1}^{\ell} \mathbf{D}_i^{x_i}$. Finally, the values $\mathbf{A}_1, \mathbf{D}_1^0, \mathbf{D}_1^1, \dots, \mathbf{D}_\ell^0, \mathbf{D}_\ell^1, T$ are given to the evaluator. To test a given $x' \in \{0,1\}^{\ell}$, the evaluator computes $F(x')$ and checks whether $F(x') - T$ is a matrix with small entries.

To see why this works out functionality-wise consider the following equation:

$$F(x) = \mathbf{A}_1 \prod_{i=1}^{\ell} \mathbf{D}_i^{x_i} = \prod_{i=1}^{\ell} s_i^{x_i} \mathbf{A}_{\ell+1} + \underbrace{\sum_{i=1}^{\ell} \left(\prod_{j=1}^{i-1} s_j^{x_j} \cdot \mathbf{E}_i^{x_i} \cdot \prod_{k=i+1}^{\ell} \mathbf{D}_i^{x_i} \right)}_{\mathbf{E}_x} \pmod{q}.$$

$$(1)$$

Indeed, if the secrets s_i^b are set with small norm, then the entire $\mathbf{E}_x$ term can be viewed as a small error term, so the dominant factor, $\prod_{i=1}^{\ell} s_i^{x_i} \mathbf{A}_{\ell+1}$, will be purely determined by the multiplicative relationship of the secrets. As for security, observe that the encoding $\mathbf{D}_i^b$ of each secret s_i^b amounts to an LWE encoding of s_i^b, and furthermore the encoding of $s_x = \prod_{i=1}^{\ell} s_i^{x_i}$ is also in the form of an LWE instance $\mathbf{A}_{\ell+1}, \prod_{i=1}^{\ell} s_i^{x_i} \mathbf{A}_{\ell+1} + \mathbf{E}_x \pmod{q}$. Of course, being in the form of LWE does not amount to a clear security property that is based on LWE. We discuss this point further below.

The Power and Danger in the GGH15 Approach. The GGH15 encoding embeds the plaintext s into the *secret* term of the LWE instance, unlike in other LWE-based systems (e.g. Regev [48] or dual-Regev [32]) where the plaintext is associated with the error term or the $\mathbf{A}$ matrix. While the graph structure and trapdoor sampling mechanism enables homomorphic evaluations on the LWE secrets, analyzing the security becomes tricky. Unlike the traditional case where the LWE secrets s are independent and random, here the LWE secrets, representing plaintexts, are taken from distributions that are potentially structured or correlated with each other.

Such dependencies make it hard to prove security of the trapdoor sampling: Recall that the encoding $\mathbf{D}_i$ of some secret $\hat{s}_i$ (possibly obtained from an evaluation over correlated secrets) is the preimage of $\mathbf{Y}_i := \hat{s}_i \mathbf{A}_{i+1} + \mathbf{E}$ sampled by the trapdoor of $\mathbf{A}_i$. For instance, in the extreme case where $\hat{s}_i = 0$, then the public encoding $\mathbf{D}_i$ becomes a "weak trapdoor" of $\mathbf{A}_i$, which endangers the secrets encoded on the edges heading to $\mathbf{A}_i$ [31].

Consequently, to safely use the GGH15 encoding, one has to consider the joint distribution of all the LWE secrets s_i^b, and demonstrate that the trapdoor sampling algorithm remains secure even with respect to these secrets. We demonstrate how to do that in a specific setting, by showing that there exists a "simulated" way to sample the encodings without knowing the secrets or trapdoors, and the resulting sample is indistinguishable from the real one.

LWE-Based PRFs. The example of the "subset product" type encoding may remind the readers of the lattices-based pseudorandom functions [5,6,11,19]. Indeed, recall the basic construction of Banerjee et al. [6, Sect. 5.1]. For modulus $2 \leq p < q$ chosen such that q/p is exponential in the input length ℓ. The secret keys of the PRF are exactly 2ℓ LWE secrets $s_1^0, s_1^1, \ldots, s_\ell^0, s_\ell^1$ and a uniform matrix $\mathbf{A}$ over R_q. To evaluate, compute $F(x) = \left\lfloor \prod_{i=1}^{\ell} s_i^{x_i} \mathbf{A} \right\rfloor_p$ where $\lfloor v \rfloor_p$ means multiplying v by p/q and rounding to the nearest integer. Rounding plays a crucial role in the security proof, since it allows to add fresh small noise terms without changing the functionality whp, hence one can inductively obtain fresh LWE instances on any level.

Our Construction for Bit-Fixing Constraints. A bit-fixing constraint is specified by a string $\mathbf{c} \in \{0, 1, \star\}^\ell$, where 0 and 1 are the matching bits and $\star$ denotes the wildcards. The constrain predicate C outputs 1 if the input matches $\mathbf{c}$.

The combination of GGH15 and lattice-based PRFs inspires us to construct CHCPRFs for bit-fixing constraints. In fact, after rounding $F(x)$ in Eq. (1), the functionality of $\lfloor F(x) \rfloor$ is equivalent to (up to the rounding error) both the BPR PRF [6, Sect. 5.1] and a variant of the PRF in [11, Sect. 5.1]. If we take the 2ℓ LWE secrets $s_1^0, s_1^1, \ldots, s_\ell^0, s_\ell^1$ as master secret key, the encodings $\mathbf{A}_1, \mathbf{D}_1^0, \mathbf{D}_1^1, \ldots, \mathbf{D}_\ell^0, \mathbf{D}_\ell^1$ as the evaluation key in the normal mode. An intuitive constraining algorithm is simply replacing the LWE secret of the constrained bit with an independent random element t, and reproduce its encoding $\mathbf{D}_t$. As an example, Fig. 1a and b illustrate the normal mode and constrained mode of a bit-fixing PRF.

We show that the key and the outputs from both the normal mode and the constrained mode (both modes use trapdoor sampling) are indistinguishable from an oblivious sampling procedure without using the trapdoors. The proof proceeds level-by-level (from level ℓ to level 1). Within each level i, there are two steps. The first step uses the computational hardness of LWE: observe that the LWE samples associated on $\mathbf{A}_{i+1}$ are with independent secrets, and $\mathbf{A}_{i+1}$ is trapdoor-free in that hybrid distribution by induction, so the LWE samples are indistinguishable from uniformly random. The second step uses a statistical sampling lemma by Gentry, Peikert and Vaikuntanathan [32], which says the preimage of uniform outputs can be sampled without using the trapdoor of $\mathbf{A}_i$. The proof strategy is first illustrated by Brakerski et al. where they construct an evasive conjunction obfuscator from GGH15 [20].

We note that this construction and analysis imply that a variant of the PRF from [11] also satisfies 1-key bit-fixing constraint hiding. Although the PRF from [11] does not involve the trapdoor sampling procedure and is much simpler as a bit-fixing CHCPRF, understanding the GGH15-based version is beneficial for understanding the CHCPRF for NC^1 coming next.

Embedding a General Constraint in the PRF Keys. We move on towards embedding a general constraint in the key. Consider in particular the task of puncturing the key at a single point without revealing the point, which is essential to the applications like watermarking and deniable encryption mentioned in [12]. Indeed, even that simple function seems to require some new idea.

To preserve the graph structure while handling general constraints, Barrington's Theorem [8] comes into the picture. Recall that Barrington's Theorem converts any depth-d Boolean circuits into an oblivious branching program of length $z \leq 4^d$ composed of permutation matrices $\{\mathbf{B}_i^b\}_{b\in\{0,1\},i\in[z]}$ of dimension w (by default $w = 5$). Evaluation is done via multiplying the matrices selected by input bits, with the final output $\mathbf{I}^{w\times w}$ or a w-cycle $\mathbf{P}$ recognizing 1 or 0 respectively.

To embed permutation matrices in the construction, we set the secret term for the normal mode as $\mathbf{S}_i^b = \mathbf{I}^{w\times w} \otimes s_i^b = \begin{bmatrix} s_i^b & & 0 \\ & \ddots & \\ 0 & & s_i^b \end{bmatrix}$ (where $\otimes$ is the tensor product operator); in the constrained mode as $\mathbf{S}_i^b = \mathbf{B}_i^b \otimes s_i^b$. This provides the functionality of constraining all NC^1 circuits. See Fig. 1c for an example of 2-bit point constraint $x_1 x_2 \in \{0,1\}^2$, where x_1 controls the 1st and 3rd branches, x_2 controls the 2nd and 4th branches, $\mathbf{Q}$ and $\mathbf{R}$ represent different w-cycles.

We then analyze whether the permutation matrix structures are hidden in the constrained key, and whether the constrained outputs are pseudorandom. The first observation is that the tensor product of a permutation matrix $\mathbf{B}$ and any hard LWE secret distribution s forms a hard LWE distribution, i.e. $\mathbf{A}, (\mathbf{B} \otimes s) \cdot \mathbf{A} + \mathbf{E}$ is indistinguishable from uniformly random. This means both the secret and the permutation matrices are hidden in the constrained key.

Still, the rounded constrained output $\left\lfloor (\mathbf{P} \otimes \prod_{i=1}^{\ell} s_i^{x_i}) \cdot \mathbf{A}_{z+1} \right\rceil$ is a fixed permutation of the original value. So the adversary can left-multiply $\mathbf{P}^{-1}$ to obtain the original output. To randomize the constrained outputs, we adapt the "bookend" idea from the GGH15 candidate obfuscator. That is, we multiply the output on the left by a small random vector $\mathbf{J} \in R^{1\times w}$. By a careful reduction to standard LWE, one can show that $\mathbf{A}, \mathbf{JA} + \mathbf{E}, \mathbf{J}\left(\mathbf{P} \otimes \mathbf{1}_R\right)\mathbf{A} + \mathbf{E}'$ is indistinguishable from uniformly random.

With these two additional hard LWE distributions in the toolbox, we can base NC^1 CHCPRF on LWE via the same two-step proof strategy (i.e. LWE+GPV in each level) used in the bit-fixing construction.

1.3 More on Related Work

More Background on Multilinear Maps and the Implication of this Work. The notion of cryptographic multilinear maps was introduced by Boneh and Silverberg [14]. Currently there are three main candidates [26,30,31], with a number of variants. However, what security properties hold for the candidates remains unclear. In particular, none of the candidates is known to satisfy the multilinear DDH or subgroup elimination assumptions that are sufficient for the CHCPRFs by Boneh et al. [12] (see [24,25,30,37] for the attacks on these assumptions).

Note that even our result does not imply that GGH15 satisfies the traditional assumptions like multilinear DDH, but at least it demonstrates a safe setting. To what extent can the safe setting be generalized remains an open problem. Indeed, a central task in the study of the existing candidate multilinear maps is to identify settings where they can be used based on standard cryptographic assumptions [36].

Relations to the GGH15 Candidate Program Obfuscator. Our construction for NC^1 constraints is strongly reminiscent of the candidate obfuscator from GGH15 [31, Sect. 5.2]. In particular, the "secrets" in the CHCPRF corresponds to the "multiplicative bundling scalars" from the GGH15 obfuscator. Under the restriction of releasing only 1 branch (either the functional branch or the dummy branch), our result implies that the "scalars" and permutation matrices can be hidden (without using additional safeguards such as the Kilian-type randomization and padded randomness on the diagonal).

In contrast, the recent cryptanalysis of the GGH15 obfuscator [23] shows that when releasing both the functional key and the dummy key, one can extract the bundling scalars even if the obfuscator is equipped with all the safeguards.

It might be instructive to see where our reduction to LWE fail if one attempts to apply our proof technique to the two-key setting. The point is that in this case, the adversary obtains LWE samples $\mathbf{Y}, \mathbf{Y}'$ with correlated secrets; Therefore it is not clear how to simulate the Gaussian samples of $\mathbf{D}$ conditioned on $\mathbf{AD} = \mathbf{Y}$ or of $\mathbf{D}'$ conditioned on $\mathbf{A}'\mathbf{D}' = \mathbf{Y}'$, without knowing the trapdoors of $\mathbf{A}$ and $\mathbf{A}'$.

1.4 Concurrent Work

In an independent work, Boneh, Kim and Montgomery [10] build CHCPRF from LWE, for the special case of input puncturing constraints. Their construction is very different from ours. In particular, their starting point is the (non-hiding) constrained PRF by Brakerski and Vaikuntanathan [19].

While they analyze their construction with respect to the indistinguishability-based definition, they also consider a simulation-based definition that is significantly stronger than the one here. They show that it is impossible to realize that definition for general functions. To do that, they use the same construction of functional encryption from CHCPRFs as the one presented here.

2   Preliminaries

Notations and Terminology. Let $\mathbb{R}, \mathbb{Z}, \mathbb{N}$ be the set of real numbers, integers and positive integers. The notation R is often used to denote some base ring. The concrete choices of R are $\mathbb{Z}^{n \times n}$ (the integer matrices) and $\mathbb{Z}[x]/(x^n + 1)$ (where n is a power of 2). We denote $R/(qR)$ by R_q. The rounding operation $\lfloor a \rceil_p : \mathbb{Z}_q \to \mathbb{Z}_p$ is defined as multiplying a by p/q and rounding the result to the nearest integer.

For $n \in \mathbb{N}$, $[n] := \{1, \ldots, n\}$. A vector in $\mathbb{R}^n$ is represented in column form, and written as a bold lower-case letter, e.g. $\mathbf{v}$. For a vector $\mathbf{v}$, the ith component of $\mathbf{v}$ will be denoted by v_i. A matrix is written as a bold capital letter, e.g. $\mathbf{A}$. The ith column vector of $\mathbf{A}$ is denoted $\mathbf{a}_i$.

The length of a vector is the ℓ_p-norm $\|\mathbf{v}\|_p = (\sum v_i^p)^{1/p}$. The length of a matrix is the norm of its longest column: $\|\mathbf{A}\|_p = \max_i \|\mathbf{a}_i\|_p$. By default we use ℓ_2-norm unless explicitly mentioned. When a vector or matrix is called "small" (or "short"), we refer to its norm (resp. length). The thresholds of "small" will be precisely parameterized in the article and are not necessary negligible functions.

2.1   Matrix Branching Programs

Definition 1 (Matrix branching programs). *A width-w, length-z matrix branching program over ℓ-bit inputs consists of an index-to-input map, a sequence of pairs of matrices $\mathbf{B}_i^b$, and a non-identity matrix $\mathbf{P}$ representing 0: $\mathsf{BP} = \{\iota : [z] \to [\ell], \{\mathbf{B}_i^b \in \{0,1\}^{w \times w}\}_{i \in [z], b \in \{0,1\}}, \mathbf{P} \in \{0,1\}^{w \times w} \setminus \{\mathbf{I}\}\}$. The program computes the function $f_{\mathsf{BP}} : \{0,1\}^\ell \to \{0,1\}$, defined as*

$$f_{\mathsf{BP}}(x) = \begin{cases} 1 & \text{if } \prod_{i \in [z]} \mathbf{B}_i^{x_{\iota(i)}} = \mathbf{I} \\ 0 & \text{if } \prod_{i \in [z]} \mathbf{B}_i^{x_{\iota(i)}} = \mathbf{P} \\ \perp & \text{elsewhere} \end{cases}$$

A set of branching programs $\{\mathsf{BP}\}$ is called **oblivious** if all the programs in the set have the same index-to-input map ι.

Theorem 3 (Barrington's theorem [8]). *For $d \in \mathbb{N}$, and for any set of depth-d fan-in-2 Boolean circuits $\{C\}$, there is an oblivious set of width-5 length-4^d branching programs $\{\mathsf{BP}\}$ with a index-to-input map ι, where each BP is composed of permutation matrices $\{\mathbf{B}_i^b \in \{0,1\}^{5 \times 5}\}_{i \in [z], b \in \{0,1\}}$, a 5-cycle $\mathbf{P}$, and ι.*

2.2 Lattices

An n-dimensional lattice Λ is a discrete additive subgroup of $\mathbb{R}^n$. Given n linearly independent basis vectors $\mathbf{B} = \{\mathbf{b}_1, \ldots, \mathbf{b}_n \in \mathbb{R}^n\}$, the lattice generated by $\mathbf{B}$ is $\Lambda(\mathbf{B}) = \Lambda(\mathbf{b}_1, \ldots, \mathbf{b}_n) = \{\sum_{i=1}^{n} x_i \cdot \mathbf{b}_i, x_i \in \mathbb{Z}\}$. We have the quotient group $\mathbb{R}^n / \Lambda$ of cosets $\mathbf{c} + \Lambda = \{\mathbf{c} + \mathbf{v}, \mathbf{v} \in \Lambda\}$, $\mathbf{c} \in \mathbb{R}^n$. Let $\tilde{\mathbf{B}}$ denote the Gram-Schmidt orthogonalization of $\mathbf{B}$.

Gaussian on Lattices. For any $\sigma > 0$, define the Gaussian function on $\mathbb{R}^n$ centered at $\mathbf{c}$ with parameter σ:

$$\forall \mathbf{x} \in \mathbb{R}^n, \; \rho_{\sigma, \mathbf{c}}(\mathbf{x}) = e^{-\pi \|\mathbf{x} - \mathbf{c}\|^2 / \sigma^2}$$

For any $\mathbf{c} \in \mathbb{R}^n$, $\sigma > 0$, and n-dimensional lattice Λ, define the discrete Gaussian distribution over Λ as:

$$\forall \mathbf{x} \in \Lambda, \; D_{\Lambda + \mathbf{c}, \sigma}(\mathbf{x}) = \frac{\rho_{\sigma, \mathbf{c}}(\mathbf{x})}{\rho_{\sigma, \mathbf{c}}(\Lambda)}$$

Lemma 1 [45,47]. *Let $\mathbf{B}$ be a basis of an m-dimensional lattice Λ, and let $\sigma \geq \|\tilde{\mathbf{B}}\| \cdot \omega(\log n)$, then $\Pr_{\mathbf{x} \leftarrow D_{\Lambda, \sigma}}[\|\mathbf{x}\| \geq \sigma \cdot \sqrt{m} \vee \mathbf{x} = \mathbf{0}] \leq \mathrm{negl}(n)$.*

Gentry, Peikert and Vaikuntanathan [32] show how to sample statistically close to discrete Gaussian distribution in polynomial time for sufficiently large σ (the algorithm is first proposed by Klein [40]). The sampler is upgraded in [18] so that the output is distributed exactly as a discrete Gaussian.

Lemma 2 [18,32]. *There is a p.p.t. algorithm that, given a basis $\mathbf{B}$ of an n-dimensional lattice $\Lambda(\mathbf{B})$, $\mathbf{c} \in \mathbb{R}^n$, $\sigma \geq \|\tilde{\mathbf{B}}\| \cdot \sqrt{\ln(2n + 4)/\pi}$, outputs a sample from $D_{\Lambda + \mathbf{c}, \sigma}$.*

We then present the trapdoor sampling algorithm and the corollary of GPV lemma in the general ring R.

Lemma 3 [2,3,44]. *There is a p.p.t. algorithm $\mathsf{TrapSam}(R, 1^n, 1^m, q)$ that, given the base ring R, modulus $q \geq 2$, lattice dimension n, and width parameter m (under the condition that $m = \Omega(\log q)$ if $R = \mathbb{Z}^{n \times n}$, $m = \Omega(n \log q)$ if $R = \mathbb{Z}[x]/(x^n + 1)$), outputs $\mathbf{A} \leftarrow U(R_q^{1 \times m})$ with a trapdoor τ.*

Lemma 4 [32]. *There is a p.p.t. algorithm $\mathsf{PreimgSam}(\mathbf{A}, \tau, \mathbf{y}, \sigma)$ that with all but negligible probability over $(\mathbf{A}, \tau) \leftarrow \mathsf{TrapSam}(R, 1^n, 1^m, q)$, for sufficiently large $\sigma = \Omega(\sqrt{n \log q})$, the following distributions are statistically close:*

$$\{\mathbf{A}, \mathbf{x}, \mathbf{y} : \mathbf{y} \leftarrow U(R_q), \mathbf{x} \leftarrow \mathsf{PreimgSam}(\mathbf{A}, \tau, \mathbf{y}, \sigma)\} \approx_s \{\mathbf{A}, \mathbf{x}, \mathbf{y} : \mathbf{x} \leftarrow \gamma_\sigma, \mathbf{y} = \mathbf{A}\mathbf{x}\}$$

where γ_σ represents $D_{\mathbb{Z}^{1 \times n}_{nm}, \sigma}$ if $R = \mathbb{Z}^{n \times n}$; represents $D_{R^m, \sigma}$ if $R = \mathbb{Z}[x]/(x^n + 1)$.

When the image is a matrix $\mathbf{Y} = [\mathbf{y}_1 \| \ldots \| \mathbf{y}_\ell]$, we abuse the notation for the preimage sampling algorithm, use $\mathbf{D} \leftarrow \mathsf{PreimgSam}(\mathbf{A}, \tau, \mathbf{Y}, \sigma)$ to represent the concatenation of ℓ samples from $\mathbf{d}_i \leftarrow \mathsf{PreimgSam}(\mathbf{A}, \tau, \mathbf{y}_i, \sigma)_{i \in [\ell]}$.

2.3   General Learning with Errors Problems

The learning with errors (LWE) problem, formalized by Regev [48], states that solving noisy linear equations, in certain rings and for certain error distributions, is as hard as solving some worst-case lattice problems. The two typical forms used in cryptographic applications are (standard) LWE and RingLWE. The latter is introduced by Lyubashevsky, Peikert and Regev [42].

We formulate them as the General learning with errors problems similar to those of [17], with more flexibility in the secret distribution and the base ring.

Definition 2 (General learning with errors problem). *The (decisional) general learning with errors problem (GLWE) is parameterized by the base ring R, dimension parameters k, ℓ, m for samples, dimension parameter n for lattices, modulus q, the secret distribution η over $R^{k \times \ell}$, and the error distribution χ over $R^{\ell \times m}$. The $\mathsf{GLWE}_{R,k,\ell,m,n,q,\eta,\chi}$ problem is to distinguish the following two distributions: (1) LWE samples $s \leftarrow \eta$, $\mathbf{A} \leftarrow U(R_q^{\ell \times m})$, $\mathbf{E} \leftarrow \chi^{k \times m}$, output $(\mathbf{A}, s\mathbf{A} + \mathbf{E}) \in (R_q^{\ell \times m} \times R_q^{k \times m})$; (2) uniform distributions $U(R_q^{\ell \times m} \times R_q^{k \times m})$.*

We define $\mathsf{GLWE}_{R,k,\ell,m,n,q,\eta,\chi}$-hardness for secret distributions. The subscripts are dropped if they are clear from the context.

Definition 3. *A secret distribution η is called $\mathsf{GLWE}_{R,k,\ell,m,n,q,\chi}$-hard if no p.p.t. adversary distinguishes the two distributions in the $\mathsf{GLWE}_{R,k,\ell,m,n,q,\eta,\chi}$ problem with $1/2$ plus non-negligible probability.*

Here are the connections of decisional LWE/RingLWE to the worst-case lattice problems, in the language of GLWE-hardness. For the LWE problem we present the version where the secret is a square matrix.

Lemma 5 (LWE [18,46,48]). *Let n be an integer, $R = \mathbb{Z}^{n \times n}$. q be an integer modulus, $0 < \sigma < q$ such that $\sigma > 2\sqrt{n}$. If there exists an efficient (possibly quantum) algorithm that breaks $\mathsf{GLWE}_{R,1,1,m,n,q,U(R_q),D_{\mathbb{Z},\sigma}^{n \times n}}$, then there exists an efficient (possibly quantum) algorithm for approximating SIVP and GapSVP in the ℓ_2 norm, in the worst case, to within $\tilde{O}(nq/\sigma)$ factors.*

Lemma 6 (RingLWE [29,41,42]). *Let n be a power of 2, $R = \mathbb{Z}[x]/(x^n + 1)$. Let q be a prime integer s.t. $q \equiv 1 \pmod{n}$. $0 < \sigma < q$, $\sigma > \omega(\sqrt{\log(n)})$, $\sigma' > n^{3/4} m^{1/4} \sigma$. If there exists an efficient (possibly quantum) algorithm that breaks $\mathsf{GLWE}_{R,1,1,m,n,q,U(R_q),D_{R,\sigma'}}$, then there exists an polynomial time quantum algorithm for solving SVP for ideal-lattices over R, in the worst case, to within $\tilde{O}(\sqrt{n}q/\sigma)$ factors.*

For proper choices of parameters, error distributions of small norm can be used as hard secret distribution (usually called Hermit-normal-form LWE).

Lemma 7 (HNF-LWE [4,18]). *For R, m, n, q, σ chosen as was in Lemma 5, $\mathsf{GLWE}_{R,1,1,m',n,q,D_{\mathbb{Z},\sigma}^{n \times n},D_{\mathbb{Z},\sigma}^{n \times n}}$ is as hard as $\mathsf{GLWE}_{R,1,1,m,n,q,U(R_q),D_{\mathbb{Z},\sigma}^{n \times n}}$ for $m' \leq m - (16n + 4 \log \log q)$.*

Lemma 8 (HNF-RingLWE [43]). *For $R, m, n, q, \sigma, \sigma'$ chosen as in Lemma 6, $\mathsf{GLWE}_{R,1,1,m-1,n,q,D_{R,\sigma'},D_{R,\sigma'}}$ is as hard as $\mathsf{GLWE}_{R,1,1,m,n,q,U(R_q),D_{R,\sigma'}}$.*

Pseudorandom Functions Based on GLWE. We adapt theorems from the PRF construction of Boneh, Lewi, Montgomery, and Raghunathan [11, Theorems 4.3, 5.1]. The result was originally stated for LWE. We observe that it holds for general rings under proper choices of parameter. A proof sketch is described in [21].

Lemma 9 (Adapted from [11]). *Let $\ell \in \mathbb{N}$ be the bit-length of the input. $m, n, q, p \in \mathbb{N}$, $\sigma, B \in \mathbb{R}$ s.t. $0 < \sigma < q$, $B \geq \sigma\sqrt{m}$, $q/p > B^\ell$. $\eta = U(R_q)$, γ_σ is a distribution over $R^{m \times m}$ parameterized by σ, χ_σ is a distribution over $R^{1 \times m}$ parameterized by σ. $\|\gamma_\sigma\|, \|\chi_\sigma\| \leq \sigma\sqrt{m}$.*

Consider the function $f : \{0,1\}^\ell \to R_p^{1 \times m}$, $f_{\mathbf{U}}(x) = \left\lfloor \mathbf{U} \prod_{i=1}^{\ell} \mathbf{D}_i^{x_i} \right\rfloor_p$, where $\mathbf{U} \leftarrow U(R_q^{1 \times m})$ is the private parameter, $\{\mathbf{D}_i^b \leftarrow \gamma_\sigma\}_{b \in \{0,1\}, i \in [\ell]}$ is the public parameter.

If there is an efficient algorithm that given input $\mathbf{A} \leftarrow U(R_q^{1 \times m})$, outputs $\mathbf{U} \in R_q^{1 \times m}, \mathbf{D} \in R^{m \times m}$ that are statistically close to $U(R_q^{1 \times m}) \times \gamma_\sigma$ and $\mathbf{UD} = \mathbf{A}$; then f is a PRF assuming the hardness of $\mathsf{GLWE}_{R,1,1,m,n,q,\eta,\chi_\sigma}$.

3 GLWE-Hard Distributions: Extension Package

We prove GLWE-hardness for the following "structural" secret distributions. They are used in the analysis of Construction 11.

Lemma 10. *Fix a permutation matrix $\mathbf{B} \in \{0,1\}^{w \times w}$. If a secret distribution η over R is $\mathsf{GLWE}_{R,1,1,w^2m,n,q,\eta,\chi}$-hard, then the secret distribution $\mathbf{B} \otimes \eta$ is $\mathsf{GLWE}_{R^{w \times w},1,1,m,n,q,\mathbf{B} \otimes \eta,\chi^{w \times w}}$-hard.*

Proof. For a permutation matrix $\mathbf{B} \in \{0,1\}^{w \times w}$, suppose there is a p.p.t. distinguisher between samples from

$$(\mathbf{B}, \mathbf{A}, (\mathbf{B} \otimes s)\mathbf{A} + \mathbf{E}), \text{ where } \mathbf{A} \leftarrow U(R_q^{w \times wm}), s \leftarrow \eta, \mathbf{E} \leftarrow \chi^{w \times wm}$$

and samples from the uniform distribution $(\mathbf{B}, U(R_q^{w \times wm}), U(R_q^{w \times wm}))$, then we build an attacker for $\mathsf{GLWE}_{R,1,1,w^2m,n,q,\eta,\chi}$.

The attacker is given an $\mathsf{GLWE}_{R,1,1,w^2m,n,q,\eta,\chi}$ instance

$$(\mathbf{A}', \mathbf{Y}') = (\mathbf{A}_1||...||\mathbf{A}_w, \mathbf{Y}_1||...||\mathbf{Y}_w), \text{ where } \mathbf{A}_i, \mathbf{Y}_i \in R^{1 \times wm}, \ i \in [w].$$

It then rearranges the blocks as $(\mathbf{U}, \mathbf{V}) \in R^{w \times wm} \times R^{w \times wm}$, where the ith (blocked) row of $\mathbf{U}$ is $\mathbf{A}_i$, the ith (blocked) row of $\mathbf{V}$ is $\mathbf{Y}_i$. The attacker then sends $(\mathbf{B}, \mathbf{U}, (\mathbf{B} \otimes 1_R) \mathbf{V})$ to the distinguisher. Observe that $(\mathbf{B}, \mathbf{U}, (\mathbf{B} \otimes 1_R) \mathbf{V})$ is from the $\mathbf{B} \otimes \eta$ secret distribution if $(\mathbf{A}', \mathbf{Y}')$ is from the η secret distribution, or from the uniform distribution if $(\mathbf{A}', \mathbf{Y}')$ is from the uniform distribution. Hence the attacker wins with the same probability as the distinguisher. $\square$

Lemma 11. *Let $w \in [2, \infty) \cap \mathbb{Z}$. Fix a permutation matrix $\mathbf{C} \in \{0,1\}^{w \times w}$ that represents a w-cycle. If a secret distribution η over R is $\mathsf{GLWE}_{R,1,1,wm,n,q,\eta,\chi}$-hard, then $(\eta^{1 \times w}, \eta^{1 \times w} \times (\mathbf{C} \otimes 1_R))$ is $\mathsf{GLWE}_{R,2,w,m,n,q,(\eta^{1 \times w},\eta^{1 \times w} \times (\mathbf{C} \otimes 1_R)),\chi}$-hard.*

Proof. Let $\mathbf{H} = [h_1, h_2, \ldots, h_w]$ where $\{h_i \leftarrow \eta\}_{i \in [w]}$. Let

$$\mathcal{H} := \{(\mathbf{A}_j, \mathbf{Y}_{i,j} = h_i \mathbf{A}_j + \mathbf{E}_{i,j}) | \mathbf{A}_j \leftarrow U(R_q^{1 \times m}), h_i \leftarrow \eta, \mathbf{E}_{i,j} \leftarrow \chi, i, j \in [w]\}$$

be the rearranging of w independent GLWE samples from $\mathsf{GLWE}_{R,1,1,wm,n,q,\eta,\chi}$. $\mathcal{H}$ is indistinguishable from the uniform distribution $\mathcal{U} := \{(\mathbf{A}_j, \mathbf{Y}_{i,j}) | \mathbf{A}_j \leftarrow U(R_q^{1 \times m}), \mathbf{Y}_{i,j} \leftarrow U(R_q^{1 \times m}), i, j \in [w]\}$ due to standard GLWE.

We show that if there is an attacker D' that distinguishes

$$(\mathbf{A}, \mathbf{HA} + \mathbf{E}, \mathbf{H}(\mathbf{C} \otimes 1_R)\mathbf{A} + \mathbf{E}'),$$

where $\mathbf{E}, \mathbf{E}' \leftarrow \chi^{1 \times m}$ from

$$U(R_q^{w \times m} \times R_q^{1 \times m} \times R_q^{1 \times m}),$$

then there is a distinguisher D for (a subset of) $\mathcal{H}$ and $\mathcal{U}$.

To do so, we simulate the $(\eta^{1 \times w}, \eta^{1 \times w} \times (\mathbf{C} \otimes 1_R))$ samples from $\mathcal{H}$ or $\mathcal{U}$ by setting $\mathbf{A} \in R^{w \times m}$ where the jth row of $\mathbf{A}$ is $\mathbf{A}_j$, $\mathbf{Y} := \sum_{j \in [w]} \mathbf{Y}_{j,j}$, and $\mathbf{Z} := \sum_{j \in [w]} \mathbf{Y}_{\zeta(j),j}$, where $\zeta(j) : [w] \to [w]$ outputs the row number of the 1-entry in the jth column of $\mathbf{C}$. Note that being a w-cycle indicates that the 1-entries in $\mathbf{C}$ disjoint with the 1-entries in $\mathbf{I}^{w \times w}$. Observe that the sample $(\mathbf{A}, \mathbf{Y}, \mathbf{Z})$ is from the secret distribution $(\eta^{1 \times w}, \eta^{1 \times w} \times (\mathbf{C} \otimes 1_R))$ if transformed from $\mathcal{H}$, or from the uniform distribution if transformed from $\mathcal{U}$. Hence the distinguisher D' wins with the same probability as the attacker D.   □

4   Constraint-Hiding Constrained PRFs

This section provides the definitions of constraint-hiding constrained PRFs. We first recall the indistinguishability-based definition from [12], then give our simulation-based definition, and discuss the relations among these two definitions and program obfuscation.

4.1   The Indistinguishability-Based Definition

We first recall the indistinguishability-based definition for CHCPRF from [12].

Definition 4 (Indistinguishability-based CHCPRF [12]). *Consider a family of functions* $\mathcal{F} = \{\mathcal{F}_\lambda\}_{\lambda \in \mathbb{N}}$ *where* $\mathcal{F}_\lambda = \{F_k : D_\lambda \to R_\lambda\}_{\lambda \in \mathbb{N}}$, *along with a triple of efficient functions* (Gen, Constrain, Eval). *For a constraint family* $\mathcal{C} = \{C_\lambda : D_\lambda \to \{0,1\}\}_{\lambda \in \mathbb{N}}$; *the key generation algorithm* $\mathsf{Gen}(1^\lambda)$ *generates the master secret key* MSK, *the constraining algorithm* $\mathsf{Constrain}(1^\lambda, \mathsf{MSK}, C)$ *takes the master secret key* MSK, *a constraint* C, *outputs the constrained key* CK; *the evaluation algorithm* $\mathsf{Eval}(k, x)$ *takes a key* k, *an input* x, *outputs* $F_k(x)$.

We say that $\mathcal{F}$ *is an* **indistinguishability-based CHCPRF** *for* $\mathcal{C}$ *if it satisfies the following properties:*

Functionality Preservation Over Unconstrained Inputs. *For input* $x \in D_\lambda$ *s.t.* $C(x) = 1$, $\Pr[\mathsf{Eval}(\mathsf{MSK}, x) = \mathsf{Eval}(\mathsf{CK}, x)] \geq 1 - \mathsf{negl}(\lambda)$, *where the probability is taken over the randomness in algorithms* Gen *and* Constrain.

Pseudorandomness for Constrained Inputs. *Consider the following experiment between a challenger and an adversary. The adversary can ask 3 types of oracle queries: constrained key oracle, evaluation oracle, and challenge oracle. For* $b \in \{0, 1\}$, *the challenger responds to each oracle query in the following manner:*

- *Constrained key oracle. Given a circuit* $C \in \mathcal{C}$, *the challenger outputs a constrained key* $\mathsf{CK} \leftarrow \mathsf{Constrain}(1^\lambda, \mathsf{MSK}, C)$.
- *Evaluation oracle. Given an input* $x \in D_\lambda$, *the challenger outputs* $y \leftarrow \mathsf{Eval}(\mathsf{MSK}, x)$.
- *Challenge oracle. Given an input* $x_c \in D_\lambda$, *the challenger outputs* $y \leftarrow \mathsf{Eval}(\mathsf{MSK}, x_c)$ *if* $b = 1$; *outputs* $y \leftarrow U(R_\lambda)$ *if* $b = 0$.

The queries from the adversary satisfy the conditions that $C(x_c) = 0$, *and* x_c *is not sent among evaluation queries. At the end of the experiment, the adversary chooses* b' *and wins if* $b' = b$. *The scheme satisfies the pseudorandomness property if the winning probability of any p.p.t. adversary is bounded by* $1/2 + \mathsf{negl}(\lambda)$.

Indistinguishability-Based Constraint-Hiding. *Consider the following experiment between a challenger and an adversary. The adversary can ask 2 types of oracle queries: constrained key oracle or evaluation oracle. For* $b \in \{0, 1\}$, *the challenger responds to each oracle query in the following manner:*

- *Constrained key oracle. Given a pair of circuits* $C_0, C_1 \in \mathcal{C}$, *the challenger outputs a constrained key for* C_b: $\mathsf{CK} \leftarrow \mathsf{Constrain}(1^\lambda, \mathsf{MSK}, C_b)$.
- *Evaluation oracle. Given an input* $x \in D_\lambda$, *the challenger outputs* $y \leftarrow \mathsf{Eval}(\mathsf{MSK}, x)$.

For a circuit $C \in \mathcal{C}$, *denote* $S(C) := \{x \in D_\lambda : C(x) = 1\}$. *Suppose the adversary asks* h *pairs of circuit constraints* $\{C_0^{(g)}, C_1^{(g)}\}_{g \in [h]}$, *the queries are* **admissible** *if (1)* $\forall i \neq j \in [h]$, $S(C_0^{(i)}) \cap S(C_0^{(j)}) = S(C_1^{(i)}) \cap S(C_1^{(j)})$; *(2) for all input evaluation queries* x, *for all* $g \in [h]$, $C_0^{(g)}(x) = C_1^{(g)}(x)$.

At the end of the experiment, the adversary chooses b' *and wins if* $b' = b$. *The scheme satisfies the constraint-hiding property if the winning probability of any p.p.t. adversary is bounded by* $1/2 + \mathsf{negl}(\lambda)$.

4.2 The Simulation-Based Definition

Next we give the simulation-based definition. We first present a definition that is central to the discussions and constructions in the paper, then mention its variants.

Definition 5 (Simulation-based CHCPRF). *Consider a family of functions* $\mathcal{F} = \{\mathcal{F}_\lambda\}_{\lambda \in \mathbb{N}}$ *with the same syntax as in Definition 4. We say that* $\mathcal{F}$ *is* **simulation-based CHCPRF for family** $\mathcal{C}$ **of circuits** *if for any polytime stateful algorithm* Adv, *there is a polytime stateful algorithm* Sim *such that:*

$$\{Experiment\ REAL_{\mathsf{Adv}}(1^\lambda)\}_{\lambda \in \mathbb{N}} \approx_c \{Experiment\ IDEAL_{\mathsf{Adv},\mathsf{Sim}}(1^\lambda)\}_{\lambda \in \mathbb{N}}$$

The ideal and real experiments are defined as follows for adversaries Adv *and* Sim. *Both algorithms are stateful.*

Experiment $REAL_{\mathsf{Adv}}(1^\lambda)$	*Experiment* $IDEAL_{\mathsf{Adv},\mathsf{Sim}}(1^\lambda)$
MSK $\leftarrow$ Gen(1^λ),	Sim $\leftarrow 1^\lambda$
Repeat :	*Repeat* :
Adv $\rightarrow (x, d_x)$; $y =$ Eval(MSK, x)	Adv $\rightarrow (x, d_x)$; $y =$ Sim(x, d_x)
Adv $\leftarrow y$	if $d_x = 0$ then $y = U(R)$; Adv $\leftarrow y$
Adv $\rightarrow C$;	Adv $\rightarrow C$;
if $d_x \neq C(x)$ *for some* x *then Output* $\perp$	if $d_x \neq C(x)$ *for some* x *then Output* $\perp$
else Adv $\leftarrow$ Constrain(MSK, C)	*else* Adv $\leftarrow$ Sim$(1^{|C|})$
Repeat :	*Repeat* :
Adv $\rightarrow x$; $y =$ Eval(MSK, x)	Adv $\rightarrow x$; $y =$ Sim$(x, C(x))$
Adv $\leftarrow y$	if $C(x) = 0$ then $y = U(R)$; Adv $\leftarrow y$
Adv $\rightarrow b$; *Output* b	Adv $\rightarrow b$; *Output* b

That is, in the experiments the adversary can ask a single constraint query and polynomially many input queries, in any order. For input queries x *made before the circuit query,* Adv *is expected to provide a bit* b_x *indicating whether* $C(x) = 1$. *In the real experiment* Adv *obtains the unconstrained function value at* x. *In the ideal experiment* Sim *learns the indicator bit* d_x; *if* $d_x = 1$ *then* Adv *gets a value generated by* Sim, *and if* $d_x = 0$ *then* Adv *obtains a random value from the range* R *of the function. Once* Adv *makes the constraint query* $C \in \mathcal{C}_\lambda$, *both experiments verify the consistency of the indicator bits* d_x *for all the inputs* x *queried by* Adv *so far. If any inconsistency is found then the experiment halts. Next, in the real experiment* Adv *obtains the constrained key generated by the constraining algorithm; in the ideal experiment* Adv *obtains a key generated by* Sim, *whereas* Sim *is given only the size of* C. *The handling of input queries made by* Adv *after the circuit query is similar to the ones before, with the exception that the indicator bit* d_x *is no longer needed and* Sim *obtains the value of* $C(x)$ *instead. The output of the experiment is the final output bit of* Adv.

Remark 1. One may also consider a stronger definition than Definition 5 where the adversary is not required to provide the indicator bits d_x in the queries prior to providing the constraint. However we note that this stronger definition is unachievable if the number of input queries before the constraint query is unbounded, due to an "incompressibility" argument similar to the one from [1].

Remark 2. The simulation-based definition can also be generalized to the setting where the adversary queries multiple constrained keys. That is, once received each constrained key query, the simulator has to simulate a constrained key, given only the size of the constraining circuit. We further discuss this strong variant shortly.

4.3 Relations Among the Definitions

We discuss the relation among the definitions of CHCPRF and program obfuscation.

Multiple-Key CHCPRFs Implies Obfuscation. We show that the simulation-based CHCPRF for 2 keys implies virtual black-box obfuscation (VBB), which is impossible to obtain for general functionalities [7,35]. For the indistinguishability-based definition proposed in [12], achieving 2-key security implies indistinguishability obfuscation [7].

Recall the definitions for VBB obfuscation (we present the strongest variant in [7]) and indistinguishability obfuscation.

Definition 6 (Obfuscation [7,35]). *A probabilistic algorithm O is an obfuscator for a class of circuit C if the following conditions hold:*

- *(Preservation of the function) For all inputs x, $\Pr[C(x) = O(C(x))] > 1 - \mathsf{negl}(\lambda)$.*
- *(Polynomially slowdown) There is a polynomial p s.t. $|O(C)| < p(|C|)$.*
- *(Strong virtual black-box obfuscation) For any p.p.t. adversary Adv, there is a p.p.t. simulator Sim s.t. for all C, $\{\mathsf{Adv}(1^\lambda, O(C))\} \approx_c \{\mathsf{Sim}^C(1^\lambda, |C|)\}$.*
- *(Indistinguishability obfuscation) For functionally equivalent circuits $C_0, C_1, O(C_0) \approx_c O(C_1)$.*

Construction 4 (Obfuscator from 2-key CHCPRFs). *Given a CHCPRF, we construct an obfuscator for C by create a constrained key $\mathsf{CK}[C]$, and a constrained key $\mathsf{CK}[I]$ where I is the circuit that always outputs 1. To evaluate $C(x)$, output 1 if $\mathsf{CHCPRF}_{\mathsf{CK}[C]}(x) = \mathsf{CHCPRF}_{\mathsf{CK}[I]}(x)$, 0 otherwise.*

Theorem 5. *If 2-key simulation-secure CHCPRF exists for circuit class C, then strong VBB obfuscation exists for circuit class C.*

Proof. The simulator for the VBB obfuscator (does not have to make oracle queries to C) runs the simulator for CHCPRF, produce simulated constraint keys for $\mathsf{CK}^S[C]$, $\mathsf{CK}^S[I]$, which are indistinguishable from the real constrained keys $\mathsf{CK}[C]$, $\mathsf{CK}[I]$ that are used to construct the obfuscator. $\square$

Corollary 1 [7,35]. *There are circuit classes for which 2-key simulation-secure CHCPRF does not exist.*

Theorem 6. *If 2-key indistinguishability-based CHCPRF exists for circuit class C, then indistinguishability obfuscation exists for circuit class C.*

Proof. For a circuit C, the obfuscator outputs $\mathsf{CK}[C]$, $\mathsf{CK}[I]$. For functionally equivalent circuits C_0 and C_1, $S(C_0) \cap S(I) = S(C_1) \cap S(I)$. By indistinguishability constraint-hiding, $(\mathsf{CK}[C_0], \mathsf{CK}[I]) \approx_c (\mathsf{CK}[C_1], \mathsf{CK}[I])$.   □

Simulation and Indistinguishability-Based Definitions for CHCPRF. Next we discuss the relation of the simulation and indistinguishability-based definitions for CHCPRF, under 1-key security. The two definitions are equivalent in the 1-key setting, for the corresponding order of queries and adaptivity. Below we state the theorems for the non-adaptive version of the definitions, then discuss their generalizations to the adaptive setting.

We first show that the simulation based definition implies the indistinguishability based definition.

Theorem 7. *If a CHCPRF satisfies the non-adaptive simulation-based definition, then it satisfies the non-adaptive indistinguishability-based definition.*

The proof of this theorem is via a standard hybrid argument and we describe the proof in [21]. The implication holds for the adaptive setting. In particular, the standard simulation definition from Definition 5 implies Definition 4 where the predicates on the input queries are committed; for the stronger simulation definition discussed in Remark 1, it implies the fully adaptive variant of Definition 4.

In the 1-key setting, the indistinguishability definition implies the simulation based definition.

Theorem 8. *If a CHCPRF satisfies 1-key non-adaptive indistinguishability-based definition, it satisfies the 1-key non-adaptive simulation-based definition.*

Proof. For a CHCPRF $\mathcal{F}$ that satisfies Definition 4 for one constrained key query, we construct a simulator as per Definition 5. The simulator picks an all-1 circuit $C^S = I$ such that $I(x) = 1, \forall x \in D_\lambda$, and use the indistinguishability-secure constraining algorithm to derive a constrained key CK^S for C^S. Once the simulator obtains the inputs and the indicators $\{x^{(k)}, d^{(k)}\}_{k \in [t]}$, if $d^{(k)} = 1$, outputs $\mathsf{Eval}(\mathsf{CK}^S, x^{(k)})$; if $d^{(k)} = 0$, outputs $y \leftarrow U(R_\lambda)$.

We first prove constraint-hiding. Suppose there is an adversary A' that distinguishes the simulated distribution from the real distribution, we build an adversary A that breaks the indistinguishability definition for $\mathcal{F}$. A sends constrained circuit queries $C_0 = C$ and $C_1 = I$, obtains $\mathsf{CK}[C_b]$. Then A sends input queries. For $x^{(k)}$ s.t. $C(x^{(k)}) = I(x^{(k)}) = 1$, the output is $\mathsf{Eval}(\mathsf{CK}[C_b], x^{(k)})$; for $x^{(k)}$ s.t. $C(x^{(k)}) \neq I(x^{(k)})$, it is an inadmissible query so A samples an uniform random output on its own. Then A forwards $\mathsf{CK}[C_b]$, inputs and outputs to A'. The choice of A' for the real or the simulated distribution corresponds to $b = 0$ or 1, hence the advantage of A is equivalent to A'.

The proof for pseudorandomness of constrained outputs is analogous.   □

The theorem extends to the setting where the input queries can be made after the constraint query.

5 The Constructions

In Sects. 5.1 and 5.2 we present the bit-fixing and NC^1 CHCPRFs.

5.1 Bit-Fixing CHCPRFs

Definition 7 (Bit-fixing constraint [15]). *A bit-fixing constraint is specified by a string* $\mathbf{c} \in \{0, 1, \star\}^\ell$, *where 0 and 1 are the fixing bits and* $\star$ *denotes the wildcards.* $C(x) = 1$ *if the input matches* $\mathbf{c}$, *namely* $((x_1 = c_1) \vee (c_1 = \star)) \wedge \ldots \wedge ((x_\ell = c_\ell) \vee (c_\ell = \star))$.

We start with a brief overview of the construction and then give the details. For a PRF with ℓ-bit input, the key-generation algorithm samples 2ℓ secrets from GLWE-hard distributions with small Euclidean norm $\{s_i^b \leftarrow \eta\}_{b \in \{0,1\}, i \in [\ell]}$, places them in a chain of length ℓ and width 2, and uses the GGH15 methodology to encode the chain. The evaluation key consists of the resulting $\mathbf{A}_1$ matrix and the $\mathbf{D}$ matrices $\{\mathbf{D}_i^b\}_{b \in \{0,1\}, i \in [\ell]}$.

The evaluation algorithm selects the path according to the input, computes the product of $\mathbf{D}$ matrices along the path $\prod_{i=1}^\ell \mathbf{D}_i^{x_i}$, then multiplies $\mathbf{A}_1$ on the left. The unrounded version of the output $\mathbf{A}_1 \prod_{i=1}^\ell \mathbf{D}_i^{x_i}$ is close to $\prod_{i=1}^\ell s_i^{x_i} \mathbf{A}_{\ell+1}$, where "close" hides the cumulated error terms. Finally, the resulting subset product is rounded by p where $2 \leq p < q$, $q/p > B$ with B being the maximum error bound. Rounding is required for correctness and security.

Construction 9 (Bit-fixing CHCPRFs). *We construct a function family* $\mathcal{F} = \{f : \{0,1\}^\ell \to R_p^{1 \times m}\}$ *equipped with algorithms* (Gen, Constrain, Eval) *and a set of vectors* $\mathcal{C} = \{\mathbf{c} \in \{0, 1, \star\}^\ell\}$:

- Gen(1^λ) *takes the security parameter* λ, *samples parameters* q, p, σ, m, $\mathbf{A}_{\ell+1} \leftarrow U(R_q^m)$, $\{(\mathbf{A}_i, \tau_i) \leftarrow \mathsf{TrapSam}(R, 1^n, 1^m, q)\}_{i \in [\ell]}$. *Then, sample* 2ℓ *independent small secrets from GLWE-hard distributions* $\{s_i^b \leftarrow \eta\}_{b \in \{0,1\}, i \in [\ell]}$. *Next, encode the secrets as follows: first compute* $\{\mathbf{Y}_i^b = s_i^b \mathbf{A}_{i+1} + \mathbf{E}_i^b, \mathbf{E}_i^b \leftarrow \chi^m\}_{i \in [\ell], b \in \{0,1\}}$, *then sample* $\{\mathbf{D}_i^b \leftarrow \mathsf{PreimgSam}(\mathbf{A}_i, \tau_i, \mathbf{Y}_i^b, \sigma)\}_{i \in [\ell], b \in \{0,1\}}$. *Set* MSK $:= (\{\mathbf{A}_i\}_{i \in [1,\ell+1]}, \{\tau_i\}_{i \in [\ell]}, \{s_i^b, \mathbf{D}_i^b\}_{i \in [\ell], b \in \{0,1\}})$.
- Constrain(MSK, $\mathbf{c}$) *takes* MSK *and the bit-matching vector* $\mathbf{c}$, *for* $i \in [\ell]$, *if* $c_i \neq \star$ *(i.e. specified as 0 or 1), replaces the original* $s_i^{1-c_i}$ *by a fresh* $t_i^{1-c_i} \leftarrow \eta$, *then updates the encodings on these secrets:* $\mathbf{Y}_i^{1-c_i} = t_i^{1-c_i} \mathbf{A}_{i+1} + \mathbf{E}_i'^{1-c_i}, \mathbf{E}_i'^{1-c_i} \leftarrow \chi^m$, *samples* $\mathbf{D}_i^{1-c_i} \leftarrow \mathsf{PreimgSam}(\mathbf{A}_i, \tau_i, \mathbf{Y}_i^{1-c_i}, \sigma)$. *Set* CK $:= (\mathbf{A}_1, \{\mathbf{D}_i^b\}_{i \in [\ell], b \in \{0,1\}})$.
- Eval(k, x) *takes the key* $k = (\mathbf{A}_1, \{\mathbf{D}_i^b\}_{i \in [\ell], b \in \{0,1\}})$ *and the input* x, *outputs* $\left\lfloor \mathbf{A}_1 \prod_{i=1}^\ell \mathbf{D}_i^{x_i} \right\rceil_p$.

Remark 3. We occasionally call $i \in \{1, 2, \ldots, \ell\}$ "levels", from low to high.

Setting of Parameters. Parameters shall be set to ensure both correctness (i.e. the preservation of functionality over unconstrained inputs) and security. Note that the approximation factors of the underlying worst-case (general or ideal) lattices problems are inherently exponential in ℓ.

Specifically, for $R = \mathbb{Z}^{n \times n}$, set $\eta = \chi = D_{\mathbb{Z},\sigma}^{n \times n}$, $\gamma = D_{\mathbb{Z}^{nm},\sigma}^{1 \times nm}$. The parameters are set to satisfy $m \geq 2 \log q$ due to Lemma 3; $q/p > (\sigma \cdot m)^\ell$ due to Lemma 1 for the correctness of rounding; $0 < \sigma < q$, $\sigma = 2\sqrt{n \log q}$, $nq/\sigma < 2^{\lambda^{1-\epsilon}}$ due to Lemmas 4, 5, 7, and 9. An example setting of parameters: $p = 2$, $\epsilon = 1/2$, $q = (32\ell n^2 \log n)^\ell$, $\lambda = n = (\log q)^2$.

For $R = \mathbb{Z}[x]/(x^n + 1)$, n being a power of 2, set $\eta = \chi = D_{R,\sigma}$, $\gamma = D_{R^m,\sigma}^{1 \times m}$. The parameters are set to satisfy $m \geq 2 \cdot n \log q$ due to Lemma 3; $q/p > (\sigma \cdot n^{3/4} m^{5/4})^\ell$ due to Lemma 1 for the correctness of rounding; $0 < \sigma < q$, $\sigma = 2\sqrt{n \log q}$, $nq/\sigma < 2^{\lambda^{1-\epsilon}}$ due to Lemmas 4, 6, 8, and 9. An example setting of parameters against the state-of-art ideal SVP algorithms [9,27,28]: $p = 2$, $\epsilon = 0.5001$, $q = (70\ell n^3 \log n)^\ell$, $\lambda = n = (\log q)^{2.1}$.

Theorem 10. *Assuming* $\mathsf{GLWE}_{R,1,1,m,n,q,\eta,\chi}$, *Construction 9 is a simulation-secure bit-fixing CHCPRF.*

Functionality Preservation on the Unconstrained Inputs. The constraining algorithm does not change any secrets on the unconstrained paths. So the functionality is perfectly preserved.

Security Proof Overview. The aim is to capture two properties: (1) pseudorandomness on the constrained inputs (2) the constrained key is indistinguishable from an obliviously sampled one.

We construct a simulator as follows: the simulator samples a key composed of $\mathbf{A}$ matrices from uniform distribution and $\mathbf{D}$ matrices from discrete-Gaussian distribution of small width. For the input-output pairs queried by the adversary, if the functionality is preserved on that point, then the simulator, knowing the input x, simply outputs the honest evaluation on the simulated key. If the input is constrained, it means at some level i, the secret $t_i^{x_i}$ in the constrained key is sampled independently from the original secret key $s_i^{x_i}$. Therefore the LWE instance $s_i^{x_i} \mathbf{A}_{i+1} + \mathbf{E}_i^{x_i}$, in the expression of the constrained output, provides an fresh random mask $\mathbf{U}$. The reduction moves from level $\ell + 1$ to level 1. At level 1, by the result of [11], the rounded output on x is pseudorandom if $C(x) = 0$.

Note that the evaluation algorithm only needs $\mathbf{A}_1$ but not the rest of the $\mathbf{A}$ matrices. However, in the analysis we assume all the $\mathbf{A}$ matrices are public.

Proof. The simulator samples all the $\{\mathbf{A}_j\}_{j \in [1,\ell+1]}$ matrices from random and $\{\mathbf{D}_i^b\}_{b \in \{0,1\}, i \in [\ell]}$ from γ, outputs the constrained key $(\mathbf{A}_1, \{\mathbf{D}_i^b\}_{i \in [\ell], b \in \{0,1\}})$. To respond the input queries, the simulator picks $\{y^{(k)}\}_{k \in [t]}$ according to $\{d^{(k)}\}_{k \in [t]}$: if $d^{(k)} = 1$ (i.e. the functionality is preserved on the constraint key at $x^{(k)}$), then outputs $y^{(k)} = \left\lfloor \mathbf{A}_1 \prod_{i=1}^{\ell} \mathbf{D}_i^{x_i^{(k)}} \right\rceil_p$ (the honest evaluation on the simulated key); otherwise $y^{(k)} \leftarrow U(R_p^{1 \times m})$.

The proof consists of two parts. The first part (Lemma 12) shows that the real distribution is indistinguishable from a semi-simulated one, where all the $\mathbf{D}$ matrices on the constrained key are sampled obliviously without knowing the constraint and the trapdoors of $\mathbf{A}$ matrices, and all the outputs are derived from the simulated constrained key. The second part (Lemma 13) argues that the outputs are pseudorandom if they are in the constrained area.

In the first part, we define intermediate hybrid distributions $\{\mathsf{H}_v\}_{v\in[0,\ell]}$. H_ℓ corresponds to the real constrained key and outputs, H_0 corresponds to the simulated constrained key and the semi-simulated outputs. The intermediate simulator in H_v knows the partial constraint from level 1 to v, and the level $w^{(k)} \in \{v, \ldots, \ell\}$ where the input $x^{(k)}$ starts to deviate from the constraint vector.

Descriptions of H_v, $v \in [0,\ell]$: The simulator in H_v

1. Samples $\{(\mathbf{A}_j, \tau_j) \leftarrow \mathsf{TrapSam}(R, 1^n, 1^m, q)\}_{j\in[v]}$ with trapdoors, $\{\mathbf{A}_{j'} \leftarrow U(R_q^m)\}_{j'\in[v+1,\ell+1]}$ from uniform;
2. Samples the GLWE secrets $\{s_i^b \leftarrow \eta\}_{b\in\{0,1\}, i\in[v]}$ below level v; then, with part of the constraint vector $\mathbf{c}_{[v]}$ in hand, for $i \in [v]$, if $c_i \neq \star$, samples $t_i^{1-c_i} \leftarrow \eta$;
3. For $b \in \{0,1\}, i \in [v]$, if t_i^b is sampled in the previous step, samples $\mathbf{Y}_i^b := t_i^b \mathbf{A}_{i+1} + \mathbf{E}'^b_i$; otherwise, $\mathbf{Y}_i^b := s_i^b \mathbf{A}_{i+1} + \mathbf{E}_i^b$;
4. Samples $\{\mathbf{D}_i^b \leftarrow \mathsf{PreimgSam}(\mathbf{A}_i, \tau_i, \mathbf{Y}_i^b, \sigma)\}_{b\in\{0,1\}, i\in[v]}$ as the constrained-key below level v. Samples the rest of the $\mathbf{D}$ matrices obliviously $\{\mathbf{D}_i^b \leftarrow \gamma\}_{b\in\{0,1\}, i\in[v+1,\ell]}$.
5. To simulate the outputs, the simulator maintains a list $\mathcal{U}$ of $\mathbf{U}$ matrices (to be specified) initiated empty. For $k \in [t]$, if the constraint is known to deviate in the path of $x_{[v+1,\ell]}^{(k)}$ from level $w^{(k)} \in [v+1, \ell]$, then compute

$$y^{(k)} \text{ as } \left\lfloor \prod_{i=1}^{v} s_i^{x_i^{(k)}} \mathbf{U}^{x_{[v+1,w]}^{(k)}} \prod_{j=w^{(k)}}^{\ell} \mathbf{D}_j^{x_j^{(k)}} \right\rceil_p \text{ — here } \mathbf{U}^{x_{[v+1,w]}^{(k)}} \text{ is indexed by}$$

$x_{[v+1,w]}^{(k)}$; if it is not in the list $\mathcal{U}$, sample $\mathbf{U}^{x_{[v+1,w]}^{(k)}} \leftarrow U(R_q^m)$, include it in $\mathcal{U}$; otherwise, reuse the one in $\mathcal{U}$. If $x^{(k)}$ has not deviated above level v, then

$$y^{(k)} = \left\lfloor \prod_{i=1}^{v} s_i^{x_i^{(k)}} \mathbf{A}_{v+1} \prod_{j=v+1}^{\ell} \mathbf{D}_j^{x_j^{(k)}} \right\rceil_p.$$

Lemma 12. $\mathsf{H}_v \approx_c \mathsf{H}_{v-1}$, for $v \in \{\ell, ..., 1\}$.

Proof. The difference of H_v and H_{v-1} lies in the sampling of $\mathbf{D}_v^0$, $\mathbf{D}_v^1$ and the outputs $\{y^{(k)}\}$. We first analyze the difference of the outputs between H_v and H_{v-1} by classifying the input queries into 3 cases:

1. For input $x^{(k)}$ that matches the partial constraint vector $\mathbf{c}_{[v,\ell]}$, observe that

$$\left\lfloor \prod_{i=1}^{v-1} s_i^{x_i^{(k)}} \mathbf{A}_v \prod_{j=v}^{\ell} \mathbf{D}_j^{x_j^{(k)}} \right\rceil_p = \left\lfloor \prod_{i=1}^{v-1} s_i^{x_i^{(k)}} (s_v^{x_v^{(k)}} \mathbf{A}_{v+1} + \mathbf{E}_v^{x_v^{(k)}}) \prod_{j=v+1}^{\ell} \mathbf{D}_j^{x_j^{(k)}} \right\rceil_p$$

$$\approx_s \left\lfloor \prod_{i=1}^{v} s_i^{x_i^{(k)}} \mathbf{A}_{v+1} \prod_{j=v+1}^{\ell} \mathbf{D}_j^{x_j^{(k)}} \right\rceil_p, \text{ where } \approx_s \text{ is due to the small norm of}$$

$\prod_{i=1}^{v-1} s_i^{x_i^{(k)}} \mathbf{E}_v^{x_v^{(k)}} \prod_{j=v+1}^{\ell} \mathbf{D}_j^{x_j^{(k)}}$. Hence the output is statistically close in H_{v-1} and H_v.

2. For the input $x^{(k)}$ that is preserving above level v but deviated at level v, the fresh LWE secret $t_v^{x_v^{(k)}}$ sampled in the constrained key is independent from the original key $s_v^{x_v^{(k)}}$. So $s_v^{x_v^{(k)}} \mathbf{A}_{v+1} + \mathbf{E}_v^{x_v^{(k)}}$ and $t_v^{x_v^{(k)}} \mathbf{A}_{v+1} + \mathbf{E}'^{x_v^{(k)}}_v$ are treated as independent LWE instances w.r.t. $\mathbf{A}_{v+1}$.

3. For $x^{(k)}$ that has deviated above level v, the output can be written as

$$
y^{(k)} = \left[\prod_{i=1}^{v} s_i^{x_i^{(k)}} \mathbf{U}^{x_{[v+1,w]}^{(k)}} \prod_{j=w^{(k)}}^{\ell} \mathbf{D}_j^{x_j^{(k)}} \right]_p
$$

$$
\approx_s \left[\prod_{i=1}^{v-1} s_i^{x_i^{(k)}} (s_v^{x_v^{(k)}} \mathbf{U}^{x_{[v+1,w]}^{(k)}} + \mathbf{E}') \prod_{j=w^{(k)}}^{\ell} \mathbf{D}_j^{x_j^{(k)}} \right]_p, \tag{2}
$$

where $\mathbf{U}^{x_{[v+1,w]}^{(k)}}$ is uniform by induction.

To summarize, there are less than $3(|\mathcal{U}|+1)$ matrices that are GLWE samples in H_v while uniform in H_{v-1}. The GLWE samples involves 3 independent secrets: s_v^0, s_v^1 and $t_v^{1-c_v}$ if $c_v \neq \star$. $t_v^{1-c_v}$ is only masked by $\mathbf{A}_{v+1}$; $\{s_v^b\}_{b \in \{0,1\}}$ are masked by $\mathbf{A}_{v+1}$ (in the constrained key and the outputs of cases (1) and (2)) and the uniform matrices in the list $\mathcal{U}$ (the outputs of case (3)); all the samples are associated with independently sampled noises.

If there is an attacker A' that distinguishes H_v and H_{v-1} with non-negligible probability ζ, we can build an attacker A who distinguishes (a subset among the $3(|\mathcal{U}|+1)$) GLWE samples

$$
\{[\mathbf{A}_{v+1}, \mathbf{U}^1, \mathbf{U}^2, \dots, \mathbf{U}^{|\mathcal{U}|}], [s_v^0, s_v^1, t_v^{1-c_i}]^T \cdot [\mathbf{A}_{v+1}, \mathbf{U}^1, \mathbf{U}^2, \dots, \mathbf{U}^{|\mathcal{U}|}] + \tilde{\mathbf{E}}\},
$$

where $\tilde{\mathbf{E}} \leftarrow \chi^{3 \times (|\mathcal{U}|+1)m}$ from

$$
\{U(R_q^{(|\mathcal{U}|+1)m} \times R_q^{3 \times (|\mathcal{U}|+1)m})\}
$$

To do so, once A obtains the samples, it places the samples under mask $\mathbf{A}_{v+1}$ in the constrained key and the outputs of cases (1) and (2); places the samples under masks $\mathbf{U}^1, \dots, \mathbf{U}^{|\mathcal{U}|}$ in the outputs of cases (3). Then samples $\{\mathbf{A}_j\}_{j \in [v]}$ with trapdoors, GLWE secrets $\{s_i^b \leftarrow \eta\}_{b \in \{0,1\}, i \in [v]}$. Then samples $\{\mathbf{D}_i^b \leftarrow \mathsf{PreimgSam}(\mathbf{A}_i, \tau_i, \mathbf{Y}_i^b, \sigma)\}_{b \in \{0,1\}, i \in [v]}$ as the constrained-key below level v. Samples the rest of the $\mathbf{D}$ matrices obliviously $\{\mathbf{D}_i^b \leftarrow \gamma\}_{b \in \{0,1\}, i \in [v+1, \ell]}$.

With these matrices the attacker A is able to simulate the outputs, send the outputs and constrained key to A'. If the samples are from GLWE, then it corresponds to H_v; if the samples are uniform, then the matrices $\{\mathbf{D}_v^b\}_{b \in \{0,1\}}$ sampled via $\{\mathbf{D}_v^b \leftarrow \mathsf{PreimgSam}(\mathbf{A}_v, \tau_v, \mathbf{Y}_v^b, \sigma)\}_{b \in \{0,1\}}$ are statistically close to the obliviously sampled ones due to Lemma 4, so it is statistically close to H_{v-1}. Hence A breaks GLWE with probability more than $\zeta/(3(t+1))$, which contradicts to Lemma 7. □

Lemma 13. *If* $C(x^{(k)}) = 0$*, then the output* $y^{(k)}$ *in* H_0 *is pseudorandom.*

Proof. A constrained output $y^{(k)}$ can be expressed as $\left\lfloor \mathbf{U}^{x^{(k)}_{[1,w]}} \prod_{j=w^{(k)}}^{\ell} \mathbf{D}_j^{x^{(k)}_j} \right\rfloor_p$,

where the secret $\mathbf{U}^{(k)}_{[1,w^{(k)}]}$ is uniform; the public $\mathbf{D}$ matrices are sampled from discrete-Gaussian distribution γ. By Lemma 9 $y^{(k)}$ is pseudorandom.   □

The proof completes by combining Lemmas 12 and 13.   □

5.2 Constraint-Hiding for NC^1 Circuits

Next we present the CHCPRF for NC^1 circuit constraints. For circuits of depth d, use Barrington's Theorem [8] to convert them into a set of oblivious branching program $\{BP\}$ with the same index-to-input map $\iota : [z] \to [\ell]$, the same w-cycle $\mathbf{P}$ that represents the 0 output (by default $w = 5$). Let $\{\mathbf{B}_i^b \in \{0,1\}^{w \times w}\}_{i \in [z], b \in \{0,1\}}$ be the permutation matrices in each BP.

The master secret key for the CHCPRF consists of $2z$ secrets from GLWE-hard distributions η over R with small Euclidean norm $\{s_i^b \leftarrow \eta\}_{b \in \{0,1\}, i \in [z]}$, together with a vector $\mathbf{J} \in R^{1 \times w}$. To generate an evaluation key, in the normal setting, let $\mathbf{S}_i^b := \mathbf{I}^{w \times w} \otimes s_i^b \in \{0,1\}^{w \times w} \otimes_R R$; in the constrained setting for a constraint recognized by BP, let $\mathbf{S}_i^b := \mathbf{B}_i^b \otimes s_i^b \in \{0,1\}^{w \times w} \otimes_R R$. For both settings, places $\{\mathbf{S}_i^b\}_{b \in \{0,1\}, i \in [z]}$ in a chain of length z and width 2, places $\mathbf{J}$ on the left end of the chain, and uses the GGH15 methodology to encode the chain. The encoding of $\mathbf{J}$ is merged into $\mathbf{A}_1$ and denote the resultant matrix as $\mathbf{A}_J$. The evaluation key consists of $\mathbf{A}_J$ and the $\mathbf{D}$ matrices $\{\mathbf{D}_i^b\}_{b \in \{0,1\}, i \in [z]}$.

To evaluate on x, output $\left\lfloor \mathbf{A}_J \prod_{i=1}^{z} \mathbf{D}_i^{x_{\iota(i)}} \right\rfloor_p$. To elaborate the functionality, for x s.t. $C(x) = 1$, $\left\lfloor \mathbf{A}_J \prod_{i=1}^{z} \mathbf{D}_i^{x_{\iota(i)}} \right\rfloor_p \approx_s \left\lfloor \mathbf{J}(\mathbf{I}^{w \times w} \otimes \prod_{i=1}^{z} s_i^{x_{\iota(i)}}) \mathbf{A}_{z+1} \right\rfloor_p$; for x s.t. $C(x) = 0$, $\left\lfloor \mathbf{A}_J \prod_{i=1}^{z} \mathbf{D}_i^{x_{\iota(i)}} \right\rfloor_p \approx_s \left\lfloor \mathbf{J}(\mathbf{P} \otimes \prod_{i=1}^{z} s_i^{x_{\iota(i)}}) \mathbf{A}_{z+1} \right\rfloor_p$. As a reminder, the permutation matrix $\mathbf{P}$ that represent the w-cycle is not a secret to the construction, so the use of the left-bookend $\mathbf{J}$ is essential for security.

Construction 11 (CHCPRFs for NC^1 circuits). *We construct a function family $\mathcal{F} = \{f : \{0,1\}^{\ell} \to R_p^{1 \times wm}\}$ equipped with 3 algorithms (Gen, Constrain, Eval), associated with a set of oblivious branching programs $\{BP\}$ of length z obtained by applying Lemma 3 on all the NC^1 circuits.*

- *Gen(1^{λ}) samples parameters q, p, σ, m, z (the length of branching programs), $\{(\mathbf{A}_i, \tau_i) \leftarrow \mathsf{TrapSam}(R^{w \times w}, 1^n, 1^m, q)\}_{i \in [z]}, \mathbf{A}_{z+1} \leftarrow U(R_q^{w \times wm})$. Samples $2z$ independent small secrets from GLWE-hard distributions $\{s_i^b \leftarrow \eta\}_{b \in \{0,1\}, i \in [z]}$, sets the secret matrices to be $\mathbf{S}_i^b = \mathbf{I}^{w \times w} \otimes s_i^b$. Next, encode the secrets as follows: first compute $\{\mathbf{Y}_i^b = \mathbf{S}_i^b \mathbf{A}_{i+1} + \mathbf{E}_i^b, \mathbf{E}_i^b \leftarrow \chi^{w \times wm}\}_{i \in [z], b \in \{0,1\}}$; then, sample $\{\mathbf{D}_i^b \leftarrow \mathsf{PreimgSam}(\mathbf{A}_i, \tau_i, \mathbf{Y}_i^b, \sigma)\}_{i \in [z], b \in \{0,1\}}$. Additionally, sample a small secret $\mathbf{J} \leftarrow \eta^{1 \times w}$ as the left-bookend. Compute $\mathbf{A}_J := \mathbf{J}\mathbf{A}_1 + \mathbf{E}_J$ where $\mathbf{E}_J \leftarrow \chi^{1 \times wm}$.*
Set $\mathsf{MSK} := (\{\mathbf{A}_i\}_{i \in [1,z+1]}, \{\tau_i\}_{i \in [z]}, \mathbf{A}_J, \{s_i^b, \mathbf{D}_i^b\}_{i \in [z], b \in \{0,1\}})$.

– Constrain(MSK, BP) *takes* MSK, *and a matrix branching program* BP $= \{\mathbf{B}_i^b \in R^{w \times w}\}_{i \in [z], b \in \{0,1\}}$. *For* $i \in [z]$, $b \in \{0,1\}$, *compute* $\mathbf{Y}_i^b = (\mathbf{B}_i^b \otimes s_i^b) \mathbf{A}_{i+1} + \mathbf{E}'^b_i$, $\mathbf{E}'^b_i \leftarrow \chi^{w \times wm}$, *samples* $\mathbf{D}_i^b \leftarrow$ PreimgSam$(\mathbf{A}_i, \tau_i, \mathbf{Y}_i^b, \sigma)$. *Set the constrained key* CK $:= (\mathbf{A}_J, \{\mathbf{D}_i^b\}_{i \in [z], b \in \{0,1\}})$.

– Eval(k, x) *takes the input* x *and the key* $k = (\mathbf{A}_J, \{\mathbf{D}_i^b\}_{i \in [z], b \in \{0,1\}})$, *outputs* $\left\lfloor \mathbf{A}_J \prod_{i=1}^z \mathbf{D}_i^{x_{\iota(i)}} \right\rceil_p$.

Setting of Parameters. Settings of the distributions and their dimensions: For $R = \mathbb{Z}^{n \times n}$, set $\eta = \chi = D_{\mathbb{Z}, \sigma}^{n \times n}$, $\gamma = D_{\mathbb{Z}^{nwm}, \sigma}^{1 \times nwm}$. For $R = \mathbb{Z}[x]/(x^n + 1)$, n being a power of 2, set $\eta = \chi = D_{R, \sigma}$, $\gamma = D_{R^{wm}, \sigma}^{1 \times wm}$.

The restriction on the parameters are analogous to the settings in the bit-fixing construction.

Theorem 12. *Assuming* GLWE$_{R,1,1,w^2 m, n, q, \eta, \chi}$, *Construction 11 is a simulation-secure CHCPRF for* NC^1 *constraints.*

Proof Overview. The simulation algorithm and the overall proof strategy is similar to the one for the bit-fixing constraints. Namely, we close the trapdoors for $\mathbf{A}$ matrices from level z to level 1. Within each level v, there are several GLWE instance associated with $\mathbf{A}_{v+1}$ whose trapdoor is closed in the previous hybrid. The additional complexity comes from dealing with secrets with permutation matrix structures. They are handled by the new GLWE packages from Sect. 3.

Proof. The simulator samples $\{\mathbf{A}_i \leftarrow U(R_q^{w \times wm})\}_{i \in [1, z+1]}$, and $\{\mathbf{D}_i^b \leftarrow \gamma\}_{b \in \{0,1\}, i \in [z]}$. It also samples $\mathbf{J} \leftarrow \eta^{1 \times w}$, computes $\mathbf{A}_J := \mathbf{J}\mathbf{A}_1 + \mathbf{E}_J$ where $\mathbf{E}_J \leftarrow \chi^{1 \times wm}$. Outputs the constrained key $(\mathbf{A}_J, \{\mathbf{D}_i^b\}_{i \in [z], b \in \{0,1\}})$. The simulator responds the input queries by picking $\{y^{(k)}\}_{k \in [t]}$ according to $\{d^{(k)}\}_{k \in [t]}$: if $d^{(k)} = 1$, then outputs $y^{(k)} = \left\lfloor \mathbf{A}_J \prod_{i=1}^z \mathbf{D}_i^{x_i^{(k)}} \right\rceil_p$; otherwise $y^{(k)} \leftarrow U(R_p^{1 \times wm})$.

The proof consists of two parts. The first part (Lemma 14) shows that the real distribution is indistinguishable from a semi-simulated one, where all the $\mathbf{D}$ matrices on the constrained key are sampled without knowing the constraint and trapdoors of $\mathbf{A}$ matrices, and all the outputs are expressed by these obliviously sampled $\mathbf{A}$ and $\mathbf{D}$ matrices. The second part (Lemma 15) argues that the outputs are pseudorandom if they are in the constrained area.

In the first part, we define intermediate hybrid distributions $\{H_v\}_{v \in [0, z]}$. H_z corresponds to the real constrained key and output distributions, H_0 corresponds to the simulated constrained key and the semi-simulated outputs. The simulators in $H_z, H_{z-1}, \ldots, H_1$ know the full description of the constraint BP $= \{\mathbf{B}_i^b\}_{i \in [z], b \in \{0,1\}}$; the simulator in H_0 only knows the indicators $\{d^{(k)}\}_{k \in [t]}$.

Descriptions of H_v, $v \in [0, z]$: The simulator in H_v

1. Samples $\{(\mathbf{A}_j, \tau_j) \leftarrow$ TrapSam$(R^{w \times w}, 1^n, 1^m, q)\}_{j \in [v]}$ with trapdoors; samples $\{\mathbf{A}_{j'} \leftarrow U(R_q^{w \times wm})\}_{j' \in [v+1, z+1]}$ uniformly random;

2. Samples the GLWE secrets $\{s_i^b \leftarrow \eta\}_{b\in\{0,1\}, i\in[v]}$ below level v; and a bookend vector $\mathbf{J} \leftarrow \eta^{1\times w}$;

3. Samples $\mathbf{Y}_i^b := \left(\mathbf{B}_i^b \otimes s_i^b\right)\mathbf{A}_{i+1} + \mathbf{E}'_i^b$; computes $\mathbf{A}_J := \mathbf{J}\mathbf{A}_1 + \mathbf{E}_J$;

4. Simulates $\{\mathbf{D}_i^b \leftarrow \mathsf{PreimgSam}(\mathbf{A}_i, \tau_i, \mathbf{Y}_i^b, \sigma)\}_{b\in\{0,1\}, i\in[v]}$ as the constrained-key below level v. Samples the rest of the $\mathbf{D}$ matrices obliviously $\{\mathbf{D}_i^b \leftarrow \gamma\}_{b\in\{0,1\}, i\in[v+1,z]}$.

5. Simulates the outputs. For $k \in [t]$, computes $y^{(k)}$ as

$$y^{(k)} = \left\lfloor \mathbf{J} \times \left(\left(\prod_{j=v+1}^{z} \mathbf{B}_j^{x_{\iota(j)}^{(k)}}\right)^{-1} \otimes \prod_{i=1}^{v} s_i^{x_{\iota(i)}^{(k)}}\right) \times \mathbf{A}_{v+1} \prod_{j=v+1}^{z} \mathbf{D}_j^{x_{\iota(j)}^{(k)}} \right\rceil_p \tag{3}$$

Lemma 14. $\mathsf{H}_v \approx_c \mathsf{H}_{v-1}$, *for* $v \in [z]$.

Proof. The difference of H_v and H_{v-1} lies in the sampling of $\mathbf{D}_v^0, \mathbf{D}_v^1$ and the outputs $\{y^{(k)}\}$. We first examine the outputs. For $k \in [t]$, we express the output $y^{(k)}$, starting from the expression in H_v to the one in H_{v-1}:

$$y^{(k)} = \left\lfloor \mathbf{J} \times \left(\left(\prod_{j=v+1}^{z} \mathbf{B}_j^{x_{\iota(j)}^{(k)}}\right)^{-1} \otimes \prod_{i=1}^{v} s_i^{x_{\iota(i)}^{(k)}}\right) \times \mathbf{A}_{v+1} \prod_{j=v+1}^{z} \mathbf{D}_j^{x_{\iota(j)}^{(k)}} \right\rceil_p$$

$$= \left\lfloor \mathbf{J} \times \left(\left(\prod_{j=v+1}^{z} \mathbf{B}_j^{x_{\iota(j)}^{(k)}}\right)^{-1} \otimes \prod_{i=1}^{v-1} s_i^{x_{\iota(i)}^{(k)}}\right) \times \left(\mathbf{I}^{w\times w} \otimes s_v^{x_{\iota(v)}^{(k)}}\right)\mathbf{A}_{v+1} \prod_{j=v+1}^{z} \mathbf{D}_j^{x_{\iota(j)}^{(k)}} \right\rceil_p$$

$$= \left\lfloor \mathbf{J} \times \left(\left(\prod_{j=v+1}^{z} \mathbf{B}_j^{x_{\iota(j)}^{(k)}}\right)^{-1} \otimes \prod_{i=1}^{v-1} s_i^{x_{\iota(i)}^{(k)}}\right) \right.$$
$$\left. \times \left(\mathbf{B}_v^{x_{\iota(v)}^{(k)}} \otimes 1_R\right)^{-1} \times \left(\mathbf{B}_v^{x_{\iota(v)}^{(k)}} \otimes s_v^{x_{\iota(v)}^{(k)}}\right)\mathbf{A}_{v+1} \prod_{j=v+1}^{z} \mathbf{D}_j^{x_{\iota(j)}^{(k)}} \right\rceil_p$$

$$\approx_s \left\lfloor \mathbf{J} \times \left(\left(\prod_{j=v}^{z} \mathbf{B}_j^{x_{\iota(j)}^{(k)}}\right)^{-1} \otimes \prod_{i=1}^{v-1} s_i^{x_{\iota(i)}^{(k)}}\right) \right.$$
$$\left. \times \left[\left(\mathbf{B}_v^{x_{\iota(v)}^{(k)}} \otimes s_v^{x_{\iota(v)}^{(k)}}\right)\mathbf{A}_{v+1} + \mathbf{E}'\right] \prod_{j=v+1}^{z} \mathbf{D}_j^{x_{\iota(j)}^{(k)}} \right\rceil_p$$

$$= \left\lfloor \mathbf{J} \times \left(\left(\prod_{j=v}^{z} \mathbf{B}_j^{x_{\iota(j)}^{(k)}}\right)^{-1} \otimes \prod_{i=1}^{v-1} s_i^{x_{\iota(i)}^{(k)}}\right) \times \mathbf{Y}_v^{x_{\iota(v)}^{(k)}} \prod_{j=v+1}^{z} \mathbf{D}_j^{x_{\iota(j)}^{(k)}} \right\rceil_p$$

$$= \left\lfloor \mathbf{J} \times \left(\left(\prod_{j=v}^{z} \mathbf{B}_j^{x_{\iota(j)}^{(k)}}\right)^{-1} \otimes \prod_{i=1}^{v-1} s_i^{x_{\iota(i)}^{(k)}}\right) \times \mathbf{A}_v \prod_{j=v}^{z} \mathbf{D}_j^{x_{\iota(j)}^{(k)}} \right\rceil_p \tag{4}$$

where $\mathbf{Y}_v^{x_{\iota(v)}^{(k)}} = \mathbf{A}_v \mathbf{D}_v^{x_{\iota(v)}^{(k)}}$. The correctness of this equation is a routine check. The implication is that the difference of H_v and H_{v-1} fully lies in the sampling of $\mathbf{Y}_v^0, \mathbf{Y}_v^1$ (being GLWE samples in H_v or uniform in H_{v-1}) and their preimages $\mathbf{D}_v^0, \mathbf{D}_v^1$ sampled by the trapdoor of $\mathbf{A}_v$.

Formally, suppose there is an attacker A' that distinguishes H_v and H_{v-1} with non-negligible probability ζ, we can build an attacker A who distinguishes:

$$\mathbf{A}_{v+1}, \{\mathbf{Y}_v^b = \left(\mathbf{B}_v^b \otimes s_v^b\right)\mathbf{A}_{v+1} + \mathbf{E}_v^b\}_{b \in \{0,1\}}$$

from

$$\{U(R_q^{w \times wm} \times R_q^{w \times wm} \times R_q^{w \times wm})\}$$

To do so, once A obtains the samples, it samples $\{\mathbf{A}_j\}_{j \in [v]}$ with trapdoors, and produce the preimages $\{\mathbf{D}_v^b \leftarrow \mathsf{PreimgSam}(\mathbf{A}_v, \tau_v, \mathbf{Y}_v^b, \sigma)\}_{b \in \{0,1\}}$. Then places $\mathbf{A}_{v+1}$, $\mathbf{Y}_v^0, \mathbf{Y}_v^1, \mathbf{D}_v^0, \mathbf{D}_v^1$ in the constrained key and the outputs. It further samples GLWE secrets $\{s_i^b \leftarrow \eta\}_{b \in \{0,1\}, i \in [v]}$, $\{\mathbf{D}_i^b \leftarrow \mathsf{PreimgSam}(\mathbf{A}_i, \tau_i, \mathbf{Y}_i^b, \sigma)\}_{b \in \{0,1\}, i \in [v]}$ as the constrained-key below level v. Samples the rest of the $\mathbf{D}$ matrices obliviously $\{\mathbf{D}_i^b \leftarrow \gamma\}_{b \in \{0,1\}, i \in [v+1,z]}$.

With these matrices the attacker A is able to simulate the rest of the outputs, send the outputs and constrained key to A'. If the samples are from GLWE, then it corresponds to H_v; if the samples are uniform, then the matrices $\{\mathbf{D}_v^b\}_{b \in \{0,1\}}$ sampled via $\{\mathbf{D}_v^b \leftarrow \mathsf{PreimgSam}(\mathbf{A}_v, \tau_v, \mathbf{Y}_v^b, \sigma)\}_{b \in \{0,1\}}$ are statistically close to the obliviously sampled ones due to Lemma 4, so it is statistically close to H_{v-1}. Hence A breaks GLWE with probability more than $\zeta/2$, which contradicts to Lemma 10.   □

Lemma 15. If $C(x^{(k)}) = 0$, then the output $y^{(k)}$ in H_0 is pseudorandom.

Proof. Following Eq. 3, a constrained output $y^{(k)}$ in H_0 can be expressed as:

$$y^{(k)} = \left\lfloor \mathbf{J} \times (\mathbf{P}^{-1} \otimes 1_R) \times \mathbf{A}_1 \prod_{j=1}^{z} \mathbf{D}_j^{x_{\iota(j)}^{(k)}} \right\rceil_p \approx_s \left\lfloor (\mathbf{J} \times (\mathbf{P}^{-1} \otimes 1_R) \times \mathbf{A}_1 + \mathbf{E}) \prod_{j=1}^{z} \mathbf{D}_j^{x_{\iota(j)}^{(k)}} \right\rceil_p \tag{5}$$

For $\mathbf{JA}_1 + \mathbf{E}_J$ as part of the constrained key, $\mathbf{J} \times (\mathbf{P}^{-1} \otimes 1_R) \times \mathbf{A}_1 + \mathbf{E}$ as part of the constrained output $y^{(k)}$, $(\mathbf{JA}_1 + \mathbf{E}_J, \mathbf{J} \times (\mathbf{P}^{-1} \otimes 1_R) \times \mathbf{A}_1 + \mathbf{E})$ is indistinguishable from $U(R_q^{1 \times wm}, R_q^{1 \times wm})$ due to Lemma 11. This means each constrained output $y^{(k)}$ is indistinguishable from $\left\lfloor \mathbf{U} \prod_{j=1}^{z} \mathbf{D}_j^{x_{\iota(j)}^{(k)}} \right\rceil_p$ where $\mathbf{U} \leftarrow U(R_q^{1 \times wm})$. Hence $y^{(k)}$ is pseudorandom if $C(x^{(k)}) = 0$ due to Lemma 9.   □

The proof completes by combining the Lemmas 14 and 15.   □

6   Private-Key Functional Encryption from CHCPRF

We construct private-key function-hiding functional encryptions for NC^1 circuits from (1) CHCPRFs for NC^1; (2) semantic secure private-key encryption schemes with decryption in NC^1. The scheme satisfies 1-key simulation-based security.

6.1   The Definition of Functional Encryption

Definition 8 (Function-hiding private-key functional encryption [34]).
A functional encryption scheme for a class of functions $\mathcal{C}_\mu = \{C : \{0,1\}^\mu \rightarrow \{0,1\}\}$ is a tuple of p.p.t. algorithms (Setup, FSKGen, Enc, Dec) *such that:*

- Setup(1^λ) *takes as input the security parameter 1^λ, outputs the master secret key* MSK.
- FSKGen$($MSK$, C)$ *takes* MSK *and a function $C \in \mathcal{C}_\mu$, outputs a functional decryption key* FSK$_C$.
- Enc$($MSK$, m)$ *takes* MSK *and a message $m \in \{0,1\}^\mu$, outputs a ciphertext* CT$_m$.
- Dec$($FSK$_C,$ CT$_m)$ *takes as input a ciphertext* CT$_m$ *and a functional decryption key* FSK$_C$, *outputs (in the clear) the result $C(m)$ of applying the function on the message.*

We require that:

Correctness. *For every message $m \in \{0,1\}^\mu$ and function $C \in \mathcal{C}_\mu$ we have:*

$$\Pr\left[b = C(m) \,\middle|\, \begin{array}{rl} \text{MSK} \leftarrow & \text{Setup}(1^\lambda) \\ \text{FSK}_C \leftarrow & \text{FSKGen}(\text{MSK}, C) \\ \text{CT}_m \leftarrow & \text{Enc}(\text{MSK}, m) \\ b \leftarrow & \text{Dec}(\text{FSK}_C, \text{CT}_m) \end{array} \right] = 1 - \text{negl}(\lambda)$$

Security. *We require that for all polytime, stateful algorithm* Adv, *there is a polytime, stateful algorithm* Sim *such that:*

$$\{Experiment\ REAL_{\mathsf{Adv}}(1^\lambda)\}_{\lambda \in \mathbb{N}} \approx_c \{Experiment\ IDEAL_{\mathsf{Adv},\mathsf{Sim}}(1^\lambda)\}_{\lambda \in \mathbb{N}}$$

The real and ideal experiments of stateful algorithms Adv, Sim *are as follow:*

Experiment $REAL_{\mathsf{Adv}}(1^\lambda)$	*Experiment* $IDEAL_{\mathsf{Adv},\mathsf{Sim}}(1^\lambda)$
MSK $\leftarrow$ Gen(1^λ),	Sim $\leftarrow 1^\lambda$
Repeat :	*Repeat :*
Adv $\rightarrow (m, d_m)$; Adv $\leftarrow$ Enc$($MSK$, m)$;	Adv $\rightarrow (m, d_m)$; Adv $\leftarrow$ Sim$(1^{\|m\|}, d_m)$;
Adv $\rightarrow C$;	Adv $\rightarrow C$;
if $d_m \neq C(m)$ for some m then Output $\bot$	*if $d_m \neq C(m)$ for some m then Output $\bot$*
else Adv $\leftarrow$ FSK$_C =$ FSKGen$($MSK$, C)$;	*else* Adv $\leftarrow$ FSK$_S =$ Sim$(1^{\|C\|})$;
Repeat :	*Repeat :*
Adv $\rightarrow m$; Adv $\leftarrow$ Enc$($MSK$, m)$	Adv $\rightarrow m$; Adv $\leftarrow$ Sim$(1^{\|m\|}, C(m))$
Adv $\rightarrow b$; *Output b*	Adv $\rightarrow b$; *Output b*

That is, in the experiments Adv *can ask for a single functional decryption key and polynomially many input queries, in any order. For encryption queries m made before the decryption key query,* Adv *is expected to provide a bit d_x indicating whether $C(m) = 1$. In the real experiment* Adv *obtains the encryption of m.*

In the ideal experiment Adv *obtains a value generated by* Sim, *whereas* Sim *is given only* $1^{|m|}$ *and* d_m. *Once* Adv *makes the functional key query for circuit* $C \in \mathcal{C}_\lambda$, *both experiments verify the consistency of the indicator bits* d_m *for all the encryption queries* m *made by* Adv *so far. If any inconsistency is found then the experiment halts. Next, in the real experiment* Adv *obtains the constrained key generated by the constraining algorithm; in the ideal experiment* Adv *obtains a key generated by* Sim, *whereas* Sim *is given only the size of* C. *The handling of encryption queries made by* Adv *after the circuit query is similar to the ones before, with the exception that the indicator bit* d_m *is no longer needed and* Sim *obtains the value of* $C(m)$ *instead. The output of the experiment is the final output bit of* Adv.

6.2   The Construction

Theorem 13. *If there are 1-key secure constraint-hiding constraint PRFs for constraint class* $\mathcal{C}$, *and symmetric-key encryption schemes with decryption in the class* $\mathcal{C}$, *then there are 1-key secure private-key function-hiding functional encryptions for function class* $\mathcal{C}$.

Corollary 2. *Assuming the intractability of GLWE, there are 1-key secure private-key function-hiding functional encryptions for* NC^1.

Construction 14. *Given a CHCPRF* (F.Gen, F.Constrain, F.Eval), *a semantic secure symmetric-key encryption scheme* (Sym.Gen, Sym.Enc, Sym.Dec), *we build a private-key functional encryption* FE *as follows:*

- FE.Setup(1^λ) *takes as input the security parameter* 1^λ, *runs* Sym.Gen(1^λ) $\rightarrow$ Sym.SK, F.Gen(1^λ) $\rightarrow$ F.MSK, *outputs the master secret key* FE.MSK = (Sym.SK, F.MSK).
- FE.Enc(FE.MSK, m) *parses* FE.MSK = (Sym.SK, F.MSK), *computes* Sym.CT = Sym.Enc(m), Tag = F.Eval(F.MSK, Sym.CT). *Outputs* FE.CT = (Sym.CT, Tag).
- FE.FSKGen(FE.MSK, C) *parses* FE.MSK = (Sym.SK, F.MSK), *outputs the functional decryption key* FE.FSK$_C$ = F.Constrain(F.MSK, F[Sym.SK, C]), *where the functionality of* F[Sym.SK, C]($\cdot$) *is:*
  - *On input* x, *computes* Sym.Dec(Sym.SK, x) $\rightarrow m \in \{0, 1\}^\mu \cap \bot$;
  - *if* $m = \bot$, *return 0; else, return* $C(m)$.
- FE.Dec(FE.FSK$_C$, FE.CT) *parses* FE.FSK$_C$ = F.CK$_F$, FE.CT = (Sym.CT, Tag), *computes* T = F.Eval(F.CK$_F$, Sym.CT). *Outputs 1 if* T = Tag, *0 if not.*

Correctness. Correctness follows the correctness of Sym and F.

Proof. We build the FE simulator FE.Sim from the symmetric-key encryption simulator Sym.Sim and CHCPRF simulator F.Sim:

1. Generates the simulated master secret-keys Sym.SK^S and F.MSK^S
2. Given a function-decryption key query (for function C), FE.Sim runs CK^S $\leftarrow$ F.Sim$_1$(1^λ, $1^{|F[\mathsf{Sym.SK}, C]|}$, F.MSK^S), outputs CK^S as FE.FSK^S.

3. Given a ciphertext query and the output bit $C(m)$, FE.Sim runs Sym.CT^S ← Sym.Sim($1^\lambda, 1^{|m|}$, Sym.SK^S) and Tag^S ← FSim$_2$(F.MSK^S, CK^S, Sym.CT^S, $C(m)$), outputs (Sym.CT^S, Tag^S) as FE.CT^S.

To show that the simulated outputs are indistinguishable from the real outputs, consider an intermediate simulator FE.Sim' which is the same to FE.Sim, except that it uses the real Sym ciphertexts in the ciphertext queries. Observe that the secret-key of Sym is not exposed in FE.Sim' or FE.Sim, the output distributions of FE.Sim' and FE.Sim are indistinguishable following the security of Sym.

Next, assume there is a distinguisher D for the outputs of the real FE scheme and FE.Sim', we build an attacker A for the CHCPRF F. A samples a secret key for Sym, sends a constrained circuit query, obtains the real CK if it is the real distribution, or the simulated CK^S if it is the simulated distribution; then creates symmetric-key ciphertexts, sends as the input queries to the CHCPRF. It obtains the real outputs if it is the real case, or the simulated outputs if it is the simulated case. A treats the outputs as tags. A forwards the ciphertexts, tags and FSK to D. D's success probability transfers to the one for A. □

Acknowledgments. We thank Leonid Reyzin for helpful discussions and the formalization of LWE-hardness. The first author is a member of the Check Point Institute for Information Security. This work is supported by the NSF MACS project, NSF grants CNS1012798, CNS1012910, and ISF grant 1523/14.

References

1. Agrawal, S., Gorbunov, S., Vaikuntanathan, V., Wee, H.: Functional encryption: new perspectives and lower bounds. In: Canetti, R., Garay, J.A. (eds.) CRYPTO 2013. LNCS, vol. 8043, pp. 500–518. Springer, Heidelberg (2013). doi:10.1007/978-3-642-40084-1_28
2. Ajtai, M.: Generating hard instances of the short basis problem. In: Wiedermann, J., Emde Boas, P., Nielsen, M. (eds.) ICALP 1999. LNCS, vol. 1644, pp. 1–9. Springer, Heidelberg (1999). doi:10.1007/3-540-48523-6_1
3. Alwen, J., Peikert, C.: Generating shorter bases for hard random lattices. Theory Comput. Syst. 48(3), 535–553 (2011)
4. Applebaum, B., Cash, D., Peikert, C., Sahai, A.: Fast cryptographic primitives and circular-secure encryption based on hard learning problems. In: Halevi, S. (ed.) CRYPTO 2009. LNCS, vol. 5677, pp. 595–618. Springer, Heidelberg (2009). doi:10.1007/978-3-642-03356-8_35
5. Banerjee, A., Peikert, C.: New and improved key-homomorphic pseudorandom functions. In: Garay, J.A., Gennaro, R. (eds.) CRYPTO 2014. LNCS, vol. 8616, pp. 353–370. Springer, Heidelberg (2014). doi:10.1007/978-3-662-44371-2_20
6. Banerjee, A., Peikert, C., Rosen, A.: Pseudorandom functions and lattices. In: Pointcheval, D., Johansson, T. (eds.) EUROCRYPT 2012. LNCS, vol. 7237, pp. 719–737. Springer, Heidelberg (2012). doi:10.1007/978-3-642-29011-4_42
7. Barak, B., Goldreich, O., Impagliazzo, R., Rudich, S., Sahai, A., Vadhan, S.P., Yang, K.: On the (im)possibility of obfuscating programs. J. ACM 59(2), 6 (2012)
8. Barrington, D.A.M.: Bounded-width polynomial-size branching programs recognize exactly those languages in NC^1. In: Hartmanis, J. (ed.) STOC, pp. 1–5. ACM, New York (1986)

9. Biasse, J.-F., Song, F.: Efficient quantum algorithms for computing class groups and solving the principal ideal problem in arbitrary degree number fields. In: Proceedings of the Twenty-Seventh Annual ACM-SIAM Symposium on Discrete Algorithms, pp. 893–902. SIAM (2016)

10. Boneh, D., Kim, S., Montgomery, H.: Private puncturable PRFs from standard lattice assumptions. In: Coron, J.-S., Nielsen, J.B. (eds.) EUROCRYPT 2017. LNCS, vol. 10210, pp. 415–445. Springer, Cham (2017). Cryptology ePrint Archive, Report 2017/100 (2017)

11. Boneh, D., Lewi, K., Montgomery, H.W., Raghunathan, A.: Key homomorphic PRFs and their applications. In: Canetti and Garay [22], pp. 410–428

12. Boneh, D., Lewi, K., Wu, D.J.: Constraining pseudorandom functions privately. In: Fehr, S. (ed.) PKC 2017. LNCS, vol. 10175, pp. 494–524. Springer, Heidelberg (2017). doi:10.1007/978-3-662-54388-7_17. Cryptology ePrint Archive, Report 2015/1167

13. Boneh, D., Sahai, A., Waters, B.: Functional encryption: definitions and challenges. In: Ishai, Y. (ed.) TCC 2011. LNCS, vol. 6597, pp. 253–273. Springer, Heidelberg (2011). doi:10.1007/978-3-642-19571-6_16

14. Boneh, D., Silverberg, A.: Applications of multilinear forms to cryptography. Contemp. Math. 324(1), 71–90 (2003)

15. Boneh, D., Waters, B.: Constrained pseudorandom functions and their applications. In: Sako, K., Sarkar, P. (eds.) ASIACRYPT 2013. LNCS, vol. 8270, pp. 280–300. Springer, Heidelberg (2013). doi:10.1007/978-3-642-42045-0_15

16. Boyle, E., Goldwasser, S., Ivan, I.: Functional signatures and pseudorandom functions. In: Krawczyk, H. (ed.) PKC 2014. LNCS, vol. 8383, pp. 501–519. Springer, Heidelberg (2014). doi:10.1007/978-3-642-54631-0_29

17. Brakerski, Z., Gentry, C., Vaikuntanathan, V.: (Leveled) fully homomorphic encryption without bootstrapping. In: ITCS, pp. 309–325. ACM (2012)

18. Brakerski, Z., Langlois, A., Peikert, C., Regev, O., Stehlé, D.: Classical hardness of learning with errors. In: Proceedings of the Forty-Fifth Annual ACM Symposium on Theory of Computing, pp. 575–584. ACM (2013)

19. Brakerski, Z., Vaikuntanathan, V.: Constrained key-homomorphic PRFs from standard lattice assumptions. In: Dodis, Y., Nielsen, J.B. (eds.) TCC 2015. LNCS, vol. 9015, pp. 1–30. Springer, Heidelberg (2015). doi:10.1007/978-3-662-46497-7_1

20. Brakerski, Z., Vaikuntanathan, V., Wee, H., Wichs, D.: Obfuscating conjunctions under entropic ring LWE. In: ITCS, pp. 147–156. ACM (2016)

21. Canetti, R., Chen, Y.: Constraint-hiding constrained PRFs for NC^1 from LWE. In: Coron, J.-S., Nielsen, J.B. (eds.) EUROCRYPT 2017. LNCS, vol. 10210, pp. 446–476. Springer, Cham (2017). Cryptology ePrint Archive (2017)

22. Canetti, R., Garay, J.A. (eds.): CRYPTO 2013. LNCS, vol. 8042. Springer, Heidelberg (2013). doi:10.1007/978-3-642-40041-4

23. Chen, Y., Gentry, C., Halevi, S.: Cryptanalyses of candidate branching program obfuscators. In: Coron, J.-S., Nielsen, J.B. (eds.) EUROCRYPT 2017. LNCS, vol. 10210, pp. 278–307. Springer, Cham (2017). Cryptology ePrint Archive, Report 2016/998 (2017)

24. Cheon, J.H., Han, K., Lee, C., Ryu, H., Stehlé, D.: Cryptanalysis of the multilinear map over the integers. In: Oswald, E., Fischlin, M. (eds.) EUROCRYPT 2015. LNCS, vol. 9056, pp. 3–12. Springer, Heidelberg (2015). doi:10.1007/978-3-662-46800-5_1

25. Coron, J.-S., Lee, M.S., Lepoint, T., Tibouchi, M.: Cryptanalysis of GGH15 multilinear maps. In: Robshaw, M., Katz, J. (eds.) CRYPTO 2016. LNCS, vol. 9815, pp. 607–628. Springer, Heidelberg (2016). doi:10.1007/978-3-662-53008-5_21

26. Coron, J.-S., Lepoint, T., Tibouchi, M.: Practical multilinear maps over the integers. In: Canetti and Garay [22], pp. 476–493
27. Cramer, R., Ducas, L., Peikert, C., Regev, O.: Recovering short generators of principal ideals in cyclotomic rings. In: Fischlin, M., Coron, J.-S. (eds.) EUROCRYPT 2016. LNCS, vol. 9666, pp. 559–585. Springer, Heidelberg (2016). doi:10.1007/978-3-662-49896-5_20
28. Cramer, R., Ducas, L., Wesolowski, B.: Short stickelberger class relations and application to ideal-SVP. In: Coron, J.-S., Nielsen, J.B. (eds.) EUROCRYPT 2017. LNCS, vol. 10210, pp. 324–348. Springer, Cham (2017). Cryptology ePrint Archive, Report 2016/885 (2017)
29. Ducas, L., Durmus, A.: Ring-LWE in polynomial rings. In: Fischlin, M., Buchmann, J., Manulis, M. (eds.) PKC 2012. LNCS, vol. 7293, pp. 34–51. Springer, Heidelberg (2012). doi:10.1007/978-3-642-30057-8_3
30. Garg, S., Gentry, C., Halevi, S.: Candidate multilinear maps from ideal lattices. In: Johansson and Nguyen [38], pp. 1–17
31. Gentry, C., Gorbunov, S., Halevi, S.: Graph-induced multilinear maps from lattices. In: Dodis, Y., Nielsen, J.B. (eds.) TCC 2015. LNCS, vol. 9015, pp. 498–527. Springer, Heidelberg (2015). doi:10.1007/978-3-662-46497-7_20
32. Gentry, C., Peikert, C., Vaikuntanathan, V.: Trapdoors for hard lattices and new cryptographic constructions. In: STOC, pp. 197–206 (2008)
33. Goldreich, O., Goldwasser, S., Micali, S.: How to construct random functions. J. ACM **33**(4), 792–807 (1986)
34. Goldwasser, S., Kalai, Y., Popa, R.A., Vaikuntanathan, V., Zeldovich, N.: Reusable garbled circuits and succinct functional encryption. In: Proceedings of the Forty-Fifth Annual ACM Symposium on Theory of Computing, pp. 555–564. ACM (2013)
35. Hada, S.: Zero-knowledge and code obfuscation. In: Okamoto, T. (ed.) ASIACRYPT 2000. LNCS, vol. 1976, pp. 443–457. Springer, Heidelberg (2000). doi:10.1007/3-540-44448-3_34
36. Halevi, S.: Graded encoding, variations on a scheme. Cryptology ePrint Archive, Report 2015/866 (2015)
37. Hu, Y., Jia, H.: Cryptanalysis of GGH map. In: Fischlin, M., Coron, J.-S. (eds.) EUROCRYPT 2016. LNCS, vol. 9665, pp. 537–565. Springer, Heidelberg (2016). doi:10.1007/978-3-662-49890-3_21
38. Johansson, T., Nguyen, P.Q. (eds.): EUROCRYPT 2013. LNCS, vol. 7881. Springer, Heidelberg (2013). doi:10.1007/978-3-642-38348-9
39. Kiayias, A., Papadopoulos, S., Triandopoulos, N., Zacharias, T.: Delegatable pseudorandom functions and applications. In: ACM Conference on Computer and Communications Security, pp. 669–684. ACM (2013)
40. Klein, P.N.: Finding the closest lattice vector when it's unusually close. In: SODA, pp. 937–941. ACM/SIAM (2000)
41. Langlois, A., Stehlé, D.: Worst-case to average-case reductions for module lattices. Des. Codes Cryptogr. **75**(3), 565–599 (2015)
42. Lyubashevsky, V., Peikert, C., Regev, O.: On ideal lattices and learning with errors over rings. J. ACM **60**(6), 43 (2013)
43. Lyubashevsky, V., Peikert, C., Regev, O.: A toolkit for ring-LWE cryptography. In: Johansson and Nguyen [38], pp. 35–54
44. Micciancio, D., Peikert, C.: Trapdoors for lattices: simpler, tighter, faster, smaller. In: Pointcheval, D., Johansson, T. (eds.) EUROCRYPT 2012. LNCS, vol. 7237, pp. 700–718. Springer, Heidelberg (2012). doi:10.1007/978-3-642-29011-4_41
45. Micciancio, D., Regev, O.: Worst-case to average-case reductions based on Gaussian measure. SIAM J. Comput. **37**(1), 267–302 (2007)

46. Peikert, C.: Public-key cryptosystems from the worst-case shortest vector problem: extended abstract. In: Mitzenmacher, M. (ed.) STOC 2009, pp. 333–342. ACM, New York (2009)
47. Peikert, C., Rosen, A.: Efficient collision-resistant hashing from worst-case assumptions on cyclic lattices. In: Halevi, S., Rabin, T. (eds.) TCC 2006. LNCS, vol. 3876, pp. 145–166. Springer, Heidelberg (2006). doi:10.1007/11681878_8
48. Regev, O.: On lattices, learning with errors, random linear codes, and cryptography. J. ACM 56(6), 34:1–34:40 (2009)

Zero Knowledge I

Amortized Complexity of Zero-Knowledge Proofs Revisited: Achieving Linear Soundness Slack

Ronald Cramer[1,2]($\boxtimes$), Ivan Damgård[3], Chaoping Xing[4], and Chen Yuan[4]

[1] CWI, Amsterdam, Netherlands
cramer@cwi.nl
[2] Mathematical Institute, Leiden University, Leiden, Netherlands
[3] Department of Computer Science, Aarhus University, Aarhus, Denmark
[4] School of Physical and Mathematical Sciences, Nanyang Technological University, Singapore, Singapore

Abstract. We propose a new zero-knowledge protocol for proving knowledge of short preimages under additively homomorphic functions that map integer vectors to an Abelian group. The protocol achieves amortized efficiency in that it only needs to send $O(n)$ function values to prove knowledge of n preimages. Furthermore we significantly improve previous bounds on how short a secret we can extract from a dishonest prover, namely our bound is a factor $O(k)$ larger than the size of secret used by the honest prover, where k is the statistical security parameter. In the best previous result, the factor was $O(k^{\log k} n)$.

Our protocol can be applied to give proofs of knowledge for plaintexts in (Ring-)LWE-based cryptosystems, knowledge of preimages of homomorphic hash functions as well as knowledge of committed values in some integer commitment schemes.

1 Introduction

Proofs of Knowledge. In a zero-knowledge protocol, a prover demonstrates that some claim is true (and in some cases that he knows a proof) while giving the verifier no other knowledge beyond the fact that the claim is true. Zero-knowledge protocols are essential tools in cryptographic protocol design. For instance, one needs zero-knowledge proofs of knowledge in multiparty computation to have a player demonstrate that he knows the input he is providing.

In this work, we will consider the problem of proving knowledge of a preimage under a one-way functions $f : \mathbb{Z}^r \mapsto G$ where G is an Abelian group (written

I. Damgård—Supported by The Danish National Research Foundation and The National Science Foundation of China (under the grant 61061130540) for the Sino-Danish Center for the Theory of Interactive Computation, within which part of this work was performed; and by the Advanced ERC grant MPCPRO.
C. Xing—Supported by the Singapore Ministry of Education under Tier 1 grants RG20/13 and RG25/16.

© International Association for Cryptologic Research 2017
J.-S. Coron and J.B. Nielsen (Eds.): EUROCRYPT 2017, Part I, LNCS 10210, pp. 479–500, 2017.
DOI: 10.1007/978-3-319-56620-7_17

additively in the following), and where furthermore the function is additively homormorphic, i.e., $f(a) + f(b) = f(a+b)$. We will call such functions $ivOWF$'s (for homomorphic One-Way Functions over Integer Vectors). This problem was considered in several earlier works, in particular recently in [BDLN16], from where we have borrowed most of the notation and basic definitions we use in the following.

ivOWF turns out to be a very general notion. Examples of ivOWFs include:

- The encryption function of several (Ring-)LWE-based cryptosystems (such as the one introduced in [BGV12] and used in the so-called SPDZ protocol [DPSZ12]).
- The encryption function of any semi-homomorphic cryptosystem as defined in [BDOZ11].
- The commitment function in commitment schemes for committing to integer values (see, e.g., [DF02]).
- Hash functions based on lattice problems such as [GGH96, LMPR08], where it is hard to find a short preimage.

We will look at the scenario where a prover $\mathcal{P}$ and a verifier $\mathcal{V}$ are given $y \in G$ and $\mathcal{P}$ holds a short preimage x of y, i.e., such that $||x|| \leq \beta$ for some β. $\mathcal{P}$ wants to prove in zero-knowledge that he knows such an x. When f is an encryption function and y is a ciphertext, this can be used to demonstrate that the ciphertext decrypts and $\mathcal{P}$ knows the plaintext. When f is a commitment function this can be used to show that one has committed to a number in a certain interval.

A well-known, simple but inefficient solution is the following protocol π:

(1) $\mathcal{P}$ chooses r at random such that $||r|| \leq \tau \cdot \beta$ for some sufficiently large τ, the choice of which we return to below.
(2) $\mathcal{P}$ then sends $a = f(r)$ to $\mathcal{V}$.
(3) $\mathcal{V}$ sends a random challenge bit b.
(4) $\mathcal{P}$ responds with $z = r + b \cdot x$.
(5) $\mathcal{V}$ checks that $f(z) = a + b \cdot y$ and that $||z|| \leq \tau \cdot \beta$.

If τ is sufficiently large, the distribution of z will be statistically independent of x, and the protocol will be honest verifier statistical zero-knowledge[1]. On the other hand, we can extract a preimage of y from a cheating prover who can produce correct answers z_0, z_1 to $b = 0, b = 1$, namely $f(z_1 - z_0) = y$. Clearly, we have $||z_1 - z_0|| \leq 2 \cdot \tau \cdot \beta$. We will refer to the factor 2τ as the *soundness slack* of the protocol, because it measures the discrepancy between the interval used by the honest prover and what we can force a dishonest prover to do. The value of the soundness slack is important: if f is, e.g., an encryption function, then a large soundness slack will force us to use larger parameters for the underlying

---

[1] We will only be interested in honest verifier zero-knowledge here. In applications one would get security for malicious verifiers by generating the challenge in a trusted way, e.g., using a maliciously secure coin-flip protocol.

cryptosystem to ensure that the ciphertext decrypts even if the input is in the larger interval, and this will cost us in efficiency.

The naive protocol above requires an exponentially large slack to get zero-knowledge, but using Lyubachevsky's rejection sampling technique, the soundness slack can made polynomial or even constant (at least in the random oracle model, at the cost that even the honest prover may sometimes fail to execute the protocol).

The obvious problem with the naive solution is that one needs to repeat the protocol k times where k is the statistical security parameter, to get soundness error probability 2^{-k}. This means that one needs to generate $\Omega(k)$ auxiliary f-values. We will refer to this as the *overhead* of the protocol and use it as a measure of efficiency.

One wants, of course as small overhead and soundness slack as possible, but as long as we only want to give a proof for a single f-value, we do not know how to reduce the overhead dramatically in general. But if instead we want to give a proof for k or more f-values, then we know how to reduce the *amortised* overhead: Cramer and Damgård ([CD09], see also full version in [CDK14]) show how to get amortised overhead $O(1)$, but unfortunately the soundness slack is $2^{\Omega(k)}$, even if rejection sampling is used[2]. In [DKL+13] two protocols were suggested, where one is only covertly secure. The other one can achieve polynomial soundness slack with overhead $\Omega(k)$ and works only in the random oracle model[3]. This was improved in [BDLN16]: a protocol was obtained (without random oracles) that has $O(1)$ overhead and quasi polynomial soundness slack (proportional to $n \cdot (2k+1)^{\log(k)/2}$).

1.1 Contributions & Techniques

In this paper, we improve significantly the result from [BDLN16] and [DKL+13]: we obtain $O(1)$ overhead and soundness slack $O(k)$. All results hold in the standard model (no random oracles are needed). As with any other protocol with amortised efficiency, one needs to amortise over at least some number of instances before the amortisation "kicks in", i.e., n needs to be large enough in order to achieve the amortized efficiency. Our most basic construction needs n to be $\Theta(k^2)$, and we later improve this to $\Theta(k^{3/2})$, still with the same overhead and soundness slack.

---

[2] In [CD09], the main result was first shown for functions dealing with *finite* rings and groups, and then generalised to the integers. The result is optimal for the finite case, while the integer case leaves room for improvement.

[3] The protocol in [DKL+13] is actually stated as a proof of plaintext knowledge for random ciphertexts, but generalizes to a protocol for ivOWFs. It actually offers a tradeoff between soundness slack s and overhead in the sense that the overhead is $M \cdot \log(k)$, where M has to be chosen such that the error probability $(1/s)^M$ is negligible. Thus to get exponentially small error probability in k as we do here, one can choose s to be $\mathsf{poly}(k)$ and hence M will be $\Omega(k/\log k)$.

Our protocol uses a high-level strategy similar to [BDLN16]:

(1) Do a cut-and-choose style protocol for the inputs $y_1, \ldots, y_n$. This is a relatively simple but imperfect proof of knowledge: It only guarantees that the prover knows almost all preimages.
(2) Let the verifier assign each y_i to one of several *buckets*.
(3) For each bucket, add all elements that landed in the bucket and do an imperfect proof of knowledge as in the first step, but now with all the bucket sums as input.

The reason why one might hope this would work is as follows: as mentioned, the first step will ensure that we can extract *almost* all of the required n preimages, in fact we can extract all but k preimages (we assume throughout that $n \gg k$). In the second step, since we only have k elements left that were "bad" in the sense that we could not yet extract a preimage, then if we have many more than k buckets and distribute them in buckets according to a carefully designed strategy, we may hope that with overwhelming probability, all the bad elements will be alone in one of those buckets for which we can extract a preimage of the bucket sum. This seems plausible because we can extract almost all such preimages. If indeed this happens, we can extract all remaining preimages by linearity of f: each bad element can be written as a sum of elements for which the extractor already knows a preimage.

Furthermore, the overall cost of doing the protocol would be $O(n)$, and the soundness slack will be limited by the maximal number of items in a bucket. In fact, if each bucket contains $O(k)$ elements, then the soundness slack is $O(k)$ as well. Our main technical contribution is a construction of a strategy for assignment to buckets with properties as we just outlined. We explain more about the intuition below.

In comparison, the protocol from [BDLN16] also plays a "balls and buckets" game. The difference is that they use only $O(k)$ buckets, but repeat the game $\Omega(\log k)$ times. This means that their extraction takes place in $\Omega(\log k)$ stages, which leads to the larger soundness slack. Also, they use a randomised strategy for assignment to buckets. While this makes the protocol and analysis somewhat more complicated, the randomization seems critical to make the proof go through: it makes essential use of the fact that the adversary does not know how elements are distributed in buckets until after the "bad" elements from Step 1 have been fixed. It is therefore somewhat surprising that the problem can be solved with a deterministic strategy, as we do here.

We also show a probabilistic strategy which is inferior to our deterministic one in that it requires k^3 input instances to work. On the other hand, it differs from the deterministic strategy by being more flexible: if the number of instances is less than k^3, then the protocol will not remove all bad elements, but it will reduce the number of bad elements significantly. We can therefore combine the deterministic and probabilistic methods to get a protocol that works already for $k^{3/2}$ input instances, still with the same overhead and soundness slack.

Our protocol is honest verifier zero-knowledge and is sound in the sense of a standard proof of knowledge, i.e., we extract the prover's witness by rewinding.

Nevertheless, the protocol can be readily used as a tool in a bigger protocol that is intended to be UC secure against malicious adversaries. Such a construction is already known from [DPSZ12].

We now explain how we arrive at our construction of the strategy for assigning elements to buckets: We define the buckets via a bipartite graph. Consider a finite, undirected, bipartite graph $G = (L, R, E)$ without multi-edges, where L denotes the set of vertices "on the left," R those "on the right" and E the set of edges. Write $n = |L|$ and $m = |R|$. Each vertex $w \in R$ on the right gives a "bucket of vertices" $N(\{w\}) \subset L$ on the left, where $N(\{w\})$ denotes the neighborhood of w.

We say that the bipartite graph G has the (f_1, f_2)-*strong unique neighbour property* if the following holds. For each set $N_1 \subset L$ with $|N_1| = f_1$, for each set $N_2 \subset R$ with $|N_2| = f_2$, and for each $i \in N_1$, there is $w \in R \setminus N_2$ such that $N_1 \cap N(\{w\}) = \{i\}$. Note that this property is anti-monotonous in the sense that if it holds for parameters (f_1, f_2) it also holds for parameters (f_1', f_2') with $f_1' \leq f_1$ and $f_2' \leq f_2$.

With f_1 corresponding to the failures in step 1 and f_2 corresponding to those in step 3, it should be clear that this property on (an infinite family of bipartite graphs) G, together with the conditions that $n = \text{poly}(k)$, $m = O(n)$, $f_1 = O(k)$, $f_2 = O(k)$ and the condition that the right-degrees in G are all in $O(k)$, is sufficient to pull off our claimed result. Of course, in addition, this requires *efficient* construction of G. We propose two approaches satisfying each of these requirements. The first one, based on a construction from universal hash functions, achieves $n = O(k^2)$. A second approach, based on certain excellent (nonconstant-degree) expander graphs achieves $n = O(k^3)$, but also achieves a weaker (but still useful) "neighbour property" even if n is much smaller than k^3.

Notation

Throughout this work we will format vectors such as b in lower-case bold face letters, whereas matrices such as B will be in upper case. We refer to the ith position of vector b as $b[i]$, let $[r] := \{1, \ldots, r\}$ and define for $b \in \mathbb{Z}^r$ that $\|b\| = \max_{i \in [r]}\{|b[i]|\}$. To sample a variable g uniformly at random from a set G we use $g \overset{\$}{\leftarrow} G$. Throughout this work we will let λ be a computational and k be a statistical security parameter. Moreover, we use the standard definition for polynomial and negligible functions and denote those as $\text{poly}(\cdot), \text{negl}(\cdot)$.

2   Homomorphic OWFs and Zero-Knowledge Proofs

We first define a primitive called *homomorphic one-way functions over integer vectors*. It is an extension of the standard definition of a OWF found in [KL14].

Let $\lambda \in \mathbb{N}$ be the security parameter, we consider a probabilistic polynomial time algorithm *Gen* which on input 1^λ outputs: an Abelian group G, natural numbers β, r, and a function $f : \mathbb{Z}^r \to G$. Let $\mathcal{A}$ be any algorithm. Consider the following game:

Invert$_{\mathcal{A},Gen}(\lambda)$:

(1) Run $Gen(1^\lambda)$ to get G, β, r and f.
(2) Choose $\boldsymbol{x} \in \mathbb{Z}^r, \|\boldsymbol{x}\| \leq \beta$ and compute $y = f(\boldsymbol{x})$.
(3) On input $(1^\lambda, y, G, \beta, r, f)$ the algorithm $\mathcal{A}$ computes an $\boldsymbol{x}'$.
(4) Output 1 iff $f(\boldsymbol{x}') = y, \|\boldsymbol{x}'\| \leq \beta$, and 0 otherwise.

Definition 1 (Homomorphic OWF over Integer Vectors (ivOWF)).
The algorithm Gen producing functions of form $f : \mathbb{Z}^r \to G$ is called a homomorphic one-way function generator over the integers if the following conditions hold:

(1) There exists a polynomial-time algorithm $eval_f$ such that $eval_f(\boldsymbol{x}) = f(\boldsymbol{x})$ for all $\boldsymbol{x} \in \mathbb{Z}^r$.
(2) For all $\boldsymbol{x}, \boldsymbol{x}' \in \mathbb{Z}^r$ it holds that $f(\boldsymbol{x}) + f(\boldsymbol{x}') = f(\boldsymbol{x} + \boldsymbol{x}')$.
(3) For every probabilistic polynomial-time algorithm $\mathcal{A}$ there exists a negligible function $\mathsf{negl}(\lambda)$ such that

$$\Pr[\mathsf{Invert}_{\mathcal{A},Gen}(\lambda) = 1] \leq \mathsf{negl}(\lambda)$$

In the following, we will abuse terminology slightly by referring to a fixed function $f : \mathbb{Z}^r \to G$ as an ivOWF. As mentioned in the introduction, this abstraction captures, among other primitives, lattice-based encryption schemes such as [BGV12, GSW13, BV14] where the one-way property is implied by IND-CPA and β is as large as the plaintext space. Moreover it also captures hash functions such as [GGH96, LMPR08], where it is hard to find a preimage for all *sufficiently short* vectors that have norm smaller than β.

2.1 Proving Knowledge of Preimage

We consider two parties, the prover $\mathcal{P}$ and the verifier $\mathcal{V}$. $\mathcal{P}$ holds values $\boldsymbol{x}_1, \ldots, \boldsymbol{x}_n \in \mathbb{Z}^r$, both parties have values $y_1, \ldots, y_n \in G$ and $\mathcal{P}$ wants to prove to $\mathcal{V}$ that $y_i = f(\boldsymbol{x}_i)$ and that $\boldsymbol{x}_i$ is short, while giving away no extra knowledge on the $\boldsymbol{x}_i$. More formally, the relation that we want to give a zero-knowledge proof of knowledge for is

$$R_{\mathrm{KSP}} = \left\{ (G, \beta, v, w) \ \middle| \ \begin{array}{l} v = (y_1, \ldots, y_n) \wedge w = (\boldsymbol{x}_1, \ldots, \boldsymbol{x}_n) \wedge \\[2mm] [y_i = f(\boldsymbol{x}_i) \wedge \|\boldsymbol{x}_i\| \leq \beta]_{i \in [n]} \end{array} \right\}$$

However, like all other protocols for this type of relation, we will have to live with a *soundness slack* τ as explained in the introduction. What this means more precisely is that there must exist a knowledge extractor with properties exactly as in the standard definition of knowledge soundness, but the extracted values only have to satisfy $[y_i = f(\boldsymbol{x}_i) \wedge \|\boldsymbol{x}_i\| \leq \tau \cdot \beta]_{i \in [n]}$.

3   Proofs of Preimage

3.1   Imperfect Proof of Knowledge

The first tool we need for our protocol is a subprotocol which we borrow from [BDLN16], a so-called *imperfect proof of knowledge*. This protocol is proof of knowledge for the above relation with a certain soundness slack, however, the knowledge extractor is only required to extract almost all preimages. We note that to show knowledge soundness later for our full protocol, Goldreich and Bellare [BG93] have shown that it is sufficient to consider deterministic provers, therefore we only need to consider deterministic provers in the following.

The idea for the protocol is that the prover constructs $T = 3n$ auxiliary values of form $z_i = f(r_i)$ where r_i is random and short. The verifier asks the prover to open half the values (chosen at random) and aborts if the preimages received are not correct and short. One can show that this means the prover must know correct preimages of almost all the unopened values. The prover must now reveal, for each y_i in the input, a short preimage of the sum $y_i + z_j$ for some unopened z_j. By the homomorphic property of f this clearly means we can extract from the prover also a short preimage of most of the y_i's.

The reason one needs to have more than $2n$ auxiliary values is that the protocol makes use of Lyubashevsky's rejection sampling technique [Lyu08, Lyu09], where the prover is allowed to refuse to use some of the auxiliary values. This allows for a small soundness slack while still maintaining the zero-knowledge property. For technical reasons the use of rejection sampling means that the prover should not send the auxiliary values z_i in the clear at first but should commit to them, otherwise we cannot show zero-knowledge.

The following theorem is proved in [BDLN16] (their Theorem 1):

Theorem 1. *Let f be an ivOWF, k be a statistical security parameter, Assume we are given C_{aux}, a perfectly binding/computationally hiding commitment scheme over G, $\tau = 100 \cdot r$ and $T = 3 \cdot n, n \geq \max\{10, k\}$. Then there exists a protocol $\mathcal{P}_{\text{IMPERFECTPROOF}}$ with the following properties:*

Efficiency: *The protocol requires communication of at most $T = 3n$ f-images and preimages.*

Completeness: *If $\mathcal{P}, \mathcal{V}$ are honest and run on an instance of R_{KSP}, then the protocol succeeds with probability at least $1 - \mathsf{negl}(k)$.*

Soundness: *For every deterministic prover $\hat{\mathcal{P}}$ that succeeds to run the protocol with probability $p > 2^{-k+1}$ one can extract at least $n - k$ values x'_i such that $f(x'_i) = y_i$ and $\|x'_i\| \leq 2 \cdot \tau \cdot \beta$, in expected time $O(\mathsf{poly}(s) \cdot k^2/p)$ where s is the size of the input to the protocol.*

Zero-Knowledge: *The protocol is computational honest-verifier zero-knowledge.*

In the following we will use $\mathcal{P}_{\text{IMPERFECTPROOF}}(v, w, T, \tau, \beta)$ to denote an invocation of the protocol from this theorem with inputs $v = (y_1, \ldots, y_n), w = (x_1, \ldots, x_n)$ and parameters τ, β.

3.2   The Full Proof of Knowledge

The above imperfect protocol will be used as a building block. After executing it with the (x_i, y_i) as input, we may assume that a preimage of most of the y_i's (in fact, all but k) can be extracted from the prover.

The strategy for the last part of the protocol is as follows: each y_i is assigned to one of several *buckets*. Then, for each bucket, we add all elements that landed in the bucket and have the prover demonstrate that he knows a preimage of the sum. The observation (made in [BDLN16]) is that we can now extract a preimage of every bad elements that is alone in a bucket. The question, however, is how we distribute items in buckets to maximize our chance of extracting all the missing preimages, and how many buckets we should use. One solution to this was given in [BDLN16], but it requires repeating the experiment $\log k$ times before all bad elements have been handled with good probability.

Here we propose a new strategy that achieves much better results: we need just one repetition of the game and each bucket will contain only $O(k)$ items which gives us the soundness slack of $O(k)$.

Before we can describe the protocol, we need to define a combinatorial object we use in the protocol, namely a good set system:

Definition 2. *A set system S with parameters n, m is a collection of m index sets $B_1, \ldots, B_m$, where each $B_j \subset [n]$, and $[n] = \{1, \ldots, n\}$. Both n and m depend on a security parameter k. The set system is* good *if the maximal size of a set B_j is $O(k)$, m is $O(n)$ and if for every set $N_1 \subset [n]$ of size k, every set $N_2 \subset [m]$ of size k and every $i \in N_1$, there exists $j \in [m] - N_2$ such that $B_j \cap N_1 = \{i\}$.*

The idea in the definition is that the buckets are defined by the sets $\{B_j\}$. Then, if the set system is good, and if we can extract preimage sums over all bucket except k, then we will be in business.

Procedure $\mathcal{P}_{\text{COMPLETEPROOF}}$

Let f be an ivOWF. $\mathcal{P}$ inputs w to the procedure and $\mathcal{V}$ inputs v. We assume that good set system $S = \{B_1, \ldots, B_m\}$ is given with parameters n, m.

proof(v, w, β) :
  (1) Let $v = (y_1, \ldots, y_n)$, $w = (x_1, \ldots, x_n)$. Run $\mathcal{P}_{\text{IMPERFECTPROOF}}(v, w, 3n, 100r, \beta)$. If $\mathcal{V}$ in $\mathcal{P}_{\text{IMPERFECTPROOF}}$ aborts then abort, otherwise continue.
  (2) For $j = 1, \ldots, m$, both players compute $\gamma_j = \sum_{i \in B_j} v_i$ and $\mathcal{P}$ also computes $\delta_j = \sum_{i \in B_j} x_i$. Let h be the maximal size of a bucket set B_j, and set $\gamma = (\gamma_1, \ldots, \gamma_m)$, $\delta = (\delta_1, \ldots, \delta_m)$.
  (3) Run $\mathcal{P}_{\text{IMPERFECTPROOF}}(\gamma, \delta, 3m, 100r, h\beta)$. If $\mathcal{V}$ in $\mathcal{P}_{\text{IMPERFECTPROOF}}$ aborts then abort, otherwise accept.

Fig. 1. A protocol to prove the relation R_{KSP}

Theorem 2. *Let f be an ivOWF, k be a statistical security parameter, and β be a given upper bound on the size of the honest prover's secrets. If $\mathcal{P}_{\text{COMPLETEPROOF}}$ (Fig. 1) is executed using a good set system $\mathcal{S}$, then it is an interactive honest-verifier zero-knowledge proof of the relation R_{KSP} with knowledge error 2^{-k+1}. More specifically, it has the following properties:*

Efficiency: *The protocol has overhead $O(1)$.*

Correctness: *If $\mathcal{P}, \mathcal{V}$ are honest then the protocol succeeds with probability at least $1 - 2^{-O(k)}$.*

Soundness: *For every deterministic prover $\hat{\mathcal{P}}$ that succeeds to run the protocol with probability $p > 2^{-k+1}$ one can extract n values x_i' such that $f(x_i') = y_i$ and $\|x_i'\| \leq O(k \cdot r \cdot \beta)$ except with negligible probability, in expected time $\text{poly}(s, k)/p$, where s is the size of the input to the protocol.*

Zero-Knowledge: *The protocol is computational honest-verifier zero-knowledge.*

Proof. Efficiency is immediate from Theorem 1 and the fact that we use a good set system, so that m is $O(n)$. Note also that the verifier can specify the set system for the prover using $O(m \cdot k \cdot \log n)$ bits. This will be dominated by the communication of m preimages if a preimage is larger than $k \log n$ bits, which will be the case for any realistic setting.

Correctness is immediate from correctness of $\mathcal{P}_{\text{IMPERFECTPROOF}}$.

The extractor required for knowlege soundness will simply run the extractor for $\mathcal{P}_{\text{IMPERFECTPROOF}}$ twice, corresponding to the 2 invocations of $\mathcal{P}_{\text{IMPERFECTPROOF}}$. Let N_1 be the set of k preimages we fail to extract in the first invocation, and let N_2 be the set of bucket sums we fail to extract in the second invocation. The properties of a good set system distribution now guarantee that no matter what set N_2 turns out to be, we can find, for each $i \in N_1$, a set B_j where we know a preimage of the sum over the bucket ($j \in [m] - N_2$), and furthermore $B_j \cap N_1 = \{i\}$. Concretely, we know δ_j such that $f(\delta_j) = \sum_{l \in B_j} y_l$ and we know preimages of all summands except for y_i. By the homomorphic property of f we can solve for a preimages of y_i, and the size of the preimage found follows immediately from Theorem 1 and the fact that buckets have size $O(k)$.

Honest-verifier zero-knowledge follows immediately from Theorem 1. We do the simulation by first invoking the simulator $\mathcal{P}_{\text{IMPERFECTPROOF}}$ with the input parameters for the first step. We then sample according to $\mathcal{D}$, compute the inout parameters for the second invocation and run the simulator for $\mathcal{P}_{\text{IMPERFECTPROOF}}$ again.   □

To make this theorem be useful, we need of course that good set systems exist. This is taken care of in the following theorem which we prove in the next section.

Theorem 3. *Good set systems exist with parameters $n = m \in O(k^2)$ and can be constructed in time polynomial in k.*

This theorem implies that we need to have at least $\Omega(k^2)$ instances to amortise over to get an efficient protocol. Of course, for the applicability of the protocol it is better if one could make do with less. We now sketch how to get the same overhead and soundness slack using only $O(k^{3/2})$ inputs.

This is based on a weaker, but more flexible notion of set system, namely an (k, d, s)-good set system:

Definition 3. *A set system S with parameters n, m is a collection of m index sets $B_1, \ldots, B_m$ with each $B_j \subseteq [n]$. Both parameters n, m depend on a security parameter k. We say a set system is (k, d, s)-good for N_1 if m is $O(n)$, the maximal size of a set B_j is d and if $N_1 \subseteq [n]$ of size k satisfies the following: for every set $N_2 \subseteq [m]$ of size k, there exists a subset $T \subseteq N_1$ of size at least $k - s$ such that for every $i \in T$, there exists $j \in [m] - N_2$ satisfying $B_j \cap N_1 = \{i\}$.*

As before, the idea is that the system can be used to design a protocol based on a balls-and-buckets game similar to the above, where the B_j's define the buckets, and N_1, N_2 correspond to the subset of instances we fail to extract via the weak zero-knowledge protocol. The final requirement now says that if the system is good for N_1, then we can extract witnesses for $k - s$ of the remaining bad items in N_1 using the witnesses we have for the bucket sums.

While it seem like bad news that we will not be able to kill all the bad items in N_1, the point is that this relaxed requirement enables us to construct such set systems with different parameters, in particular with much smaller n, m compared to k that we can get for a regular set system. In particular we have the following theorem which is proved in the next section.

Theorem 4. *For any constant $0 < c < 1$, there is a probabilistic polynomial time algorithm for constructing set systems where $m = n = O(k^{1+2c})$, such that for any fixed $N_1 \subseteq [n]$ of size k, the resulting system is $(k, k^c, 5k^{1-c})$-good for N_1 except with probability exponentially small in k.*

In our protocol, we set $c = 0.25$, so we get that we can construct a set system $S_1 = \{A_1, \ldots, A_m\}$ with $m = n = O(k^{1.5})$, such that for any fixed N_1, it will be $(k, k^{0.25}, 5k^{0.75})$-good for N_1, except with exponentially small probability. Note that this property does not guarantee that the system will be good for every N_1 simultaneously.

On the other hand, this property *is* guaranteed by the good set systems from Theorem 3. It is easy to see that these are *simultaneously* $(r, 2r, 0)$-good for all N_1 of size k. We are going to set $r = 5k^{0.75}$. So we obtain a $(5k^{0.75}, 10k^{0.75}, 0)$-good set system $S_2 = \{B_1, \ldots, B_m\}$ with $m = n = O(k^{1.5})$.

Here follows an informal sketch of the protocol we can now construct for an input consisting of $n = O(k^{1.5})$ f-images $\boldsymbol{y} = (y_1, \ldots, y_n)$:

(1) Both players compute bucket sums $\boldsymbol{\delta} = (\delta_1, \ldots, \delta_m)$ of the y_i's according to the set system S_2.
(2) Run the imperfect zero-knowledge proof for both $\boldsymbol{y}$ and $\boldsymbol{\delta}$. Note that at this point we cannot hope to extract all witnesses. This would require that only

$5k^{0.75}$ witnesses were left unknown by the imperfect proofs. But this is not the case. Therefore we extend the protocol to reduce this number:

(3) The verifier constructs a set system S_1 according to Theorem 4 with parameters as defined above. Both players compute bucket sums $\boldsymbol{u} = (u_1, \ldots, u_m)$ of the y_i's according to the set system S_1. Moreover, the players compute bucket sums $\boldsymbol{\omega} = (\omega_1, \ldots, \omega_m)$ of the δ_i's according to the system S_1.

(4) Run the imperfect zero-knowledge proof for $\boldsymbol{u}$ and $\boldsymbol{\omega}$.

We sketch the argument that this is sound as a proof of knowledge: after we run the extractor for the first two imperfect proofs, we know witnesses for all y_i except for a set N_1 and for all δ_i except for a set N_1'. Now, we know that except with negligible probability the set system S_1 will be good for both N_1 and N_1' (by a union bound). And we can run the knowledge extractor for the last two imperfect proofs so we will get witnesses for all u_i except a set N_2 and for all ω_i except a set N_2'. All these sets have size k.

Now, by Definition 3, and because we can assume that S_1 is $(k, k^{0.25}, 5k^{0.75})$-good for both N_1 and N_1', we can use the homomorphic property of f and the known witnesses for $\boldsymbol{y}, \boldsymbol{u}$ in the usual way to reduce the set of unknown witnesses for $\boldsymbol{y}$ (in N_1) to a set M_1 of size $5k^{0.75}$. Like wise, we can reduce the set of unknown witnesses (in N_1') for $\boldsymbol{\delta}$ to a set M_2 of size $5k^{0.75}$.

Finally, we are in a position to use that S_2 is a $(5k^{0.75}, 10k^{0.75}, 0)$-good set system, where M_1, M_2 are the set of unknown witnesses. This will allow us to extract all witnesses. Note that the set M_1 is not fixed when S_2 is constructed but this is fine since S_2 is simultaneously good for all sets of size $5k^{0.75}$.

We leave it to the reader to verify that this protocol has overhead $O(1)$ and soundness slack $O(k)$.

4 Proof of Theorem 3 and Theorem 4

4.1 Definitions and Conventions

Let $G = (L, R, E)$ be a finite, undirected bipartite graph. For simplicity we also assume G has no multi-edges.[4] Here, L denotes the set of vertices "on the left," R the set of vertices "on the right" and E the set of edges. A vertex v is said to be *adjacent* to a vertex w if $(v, w) \in E$. An edge $e \in E$ is *incident* to a vertex v if there is a vertex w such that $e = (v, w)$. Suppose $S \subset L$ and $T \subset R$. The *neighborhood* of S, denoted $N(S)$, consists of all vertices adjacent to some vertex in S. Note that

$$N(S) \subset R$$

since G is bipartite. If $S = \emptyset$ then $N(S) = \emptyset$. The neighborhood $N(T) \subset L$ of $T \subset R$ is defined similarly.

---

[4] We do not necessarily require that each of L, R is nonempty. But, of course, if at least one of them is, then also $E = \emptyset$.

The *unique neighbor set* $U(S) \subset R$ of the set $S \subset L$ consists of all $w \in R$ such that

$$|N(\{w\}) \cap S| = 1,$$

i.e., it consists of all vertices "on the right" whose respective neighborhoods have "a single vertex" intersection with S "on the left." We make extensive use of the following refinement that "prescribes" that intersection. For $v \in S$, the set $U(S, v)$ consists of all $w \in R$ such that

$$N(\{w\}) \cap S = \{v\}.$$

Note that

$$U(S) \subset N(S),$$

and that

$$U(S, v) \subset N(\{v\}).$$

Also note that, if $v, v' \in S$ and if $v \neq v'$, then

$$U(S, v) \cap U(S, v') = \emptyset.$$

The corresponding notions for $T \subset R$ may be defined similarly, but we will not need any of these.

Let $d, d', f_1, f_1', f_2', f_2, f, f'$ be nonnegative integers.

We say that the graph G is *d-left-bounded* if, for each $v \in L$, it holds that $|N(\{v\})| \leq d$. In other words, each of "the degrees on the left" is at most d. If there is equality for each vertex, i.e., each of the degrees on the left equals d, we say that the graph G is *d-left-regular*. Similarly for *d'-right-bounded*. The graph G is (d, d')-*bi-bounded* if it is d-left-bounded and d'-right-bounded. Finally, the graph G is *d-biregular* if it is d-left-regular and d-right-regular.

Definition 4 (Unique Neighbor Property). *The set S has the* unique neighbor property *if it holds that $U(S) \neq \emptyset$.*

Definition 5 (Strong Unique Neighbor Property of a Set). *The set S has the* strong unique neighbor property *if, for each $v \in S$, we have $U(S, v) \neq \emptyset$.*

Definition 6 (f-Strong Unique Neighbor Property of a Set). *The set S has the f-strong* unique neighbor property *if, for each $v \in S$, we have $|U(S, v)| > f$.*

Remark 1. The latter is equivalent to the requirement that, for an arbitrary selection of f vertices from R, the set S has the strong unique neighbor property in the bipartite subgraph G' obtained from G by removing this selection of f vertices from R and by removing their incident edges from E.

Remark 2. Unlike the unique neighbor property, the (f-)strong unique neighbor property is *anti-monotonous* in the following sense. If S has the (f-)strong unique neighbor property and if $S' \subset S$ (and if $f' \leq f$), than S' has the (f'-)strong unique neighbor property. This follows trivially by exploiting that fact that, by definition, "intersection with S can be prescribed."

Definition 7 ((f_1, f_2)-Strong Unique Neighbor Property of a Graph G).
The bipartite graph $G = (L, R, E)$ has the (f_1, f_2)-strong unique neighbor property if each set $S \subset L$ with $|S| = f_1$ has the f_2-strong unique neighbor property.

By an earlier remark, it follows that this property is anti-monotonous in the sense that the (f_1, f_2)-strong unique neighbor property implies the (f_1', f_2')-strong unique neighbor property if $f_1' \leq f_1$ and $f_2' \leq f_2$.

The unique neighbor property has been widely considered before and it has many known applications. There are also several applications of an *approximate* version of the strong unique neighbor property, namely where the property is only guaranteed to hold for a given fraction of each set S.

The following lemma collects some immediate, useful consequences of the definitions.

Lemma 1. *Let $G = (L, R, E)$ be a d'-right-bounded bipartite graph. Suppose there are nonnegative integers f_1, f_2 and a cover of L consisting of sets $S \subset L$ such that $|S| = f_1$ such that S has the f_2-strong unique neighbor property. Then each of the following holds.*

(1) $|R| \geq N(S) \geq f_1(f_2 + 1)$, for each S in the cover.
(2) For each $v \in L$, it holds that $|N(\{v\})| \geq f_2 + 1$.
(3) $d' \geq (f_2 + 1)\frac{|L|}{|R|}$ if $R \neq \emptyset$.

PROOF. Fix an arbitrary $v \in L$. Let $S \subset L$ be such that $v \in S$, $|S| = f_1$ and S has the f_2-strong unique neighbor property. Such S exists by the cover condition. Since we have $U(S, v) \subset N(\{v\})$ in general and since we have $|U(S, v)| \geq f_2 + 1$ by the choice of S, the second claim follows. As to the third claim, we have

$$d'|R| \geq |E| \geq (f_2 + 1)|L|,$$

where the inequality on the left follows by the definition of d'-right-boundedness and where the inequality on the right follows from the second claim. As to the first claim, since the sets $U(S, v) \subset R$ with $v \in S$ are pairwise disjoint in general and since each of them satisfies $|U(S, v)| \geq f_2 + 1$ by the choice of S, we have that

$$|R| \geq |N(S)| \geq f_1(f_2 + 1).$$

$\triangle$

Of course, the lemma holds if the graph has the (f_1, f_2)-unique neighbor property. But its actual formulation under the weaker cover condition is convenient for a purpose later on.

4.2 Details of the Proof

We show the following theorem, which immediately implies Theorem 3 by the correspondence between bi-partite graphs and the balls-and-buckets game explained in the introduction.

Theorem 5. *There is an effective construction that, for each $k > 1$, gives a bipartite graph $G = (L, R, E)$ such that*

(1) $|L| = |R| = ck^2$ where $4 < c < 16$,
(2) G is d'-right-bounded with $d' = k$
(3) G has the (f_1, f_2)-strong unique neighbor property with $f_1 = f_2 = k$.

Moreover, under our conditions that $f_1, f_2 \in \Omega(k)$ and that $|R| = O(|L|)$, each of the achieved parameters for $|L|$ and d' is asymptotically optimal.

To prove this theorem, we now show the claimed construction and provide its analysis. The optimality claim is an immediate consequence of Lemma 1; by substitution of the conditions (dictated by our application to Sigma-protocols), we get $|L| \in \Omega(k^2)$ and we get $d' \in \Omega(k)$.

Now let $\mathcal{H}$ be a ρ-universal family of hash functions $h : X \to Y$. Thus, for each $x, x' \in X$ with $x \neq x'$, the collision probability that $h(x) = h(x')$ is at most ρ if $h \in \mathcal{H}$ is selected uniformly random.[5]

We define a bipartite graph $G = (X, \mathcal{H} \times Y, E)$ as follows. For a pair

$$(x, (h, y)) \in X \times (\mathcal{H} \times Y),$$

we declare

$$(x, (h, y)) \in E \text{ if and only if } h(x) = y.$$

We also define

$$d' = \max_{(h,y) \in \mathcal{H} \times Y} |\{h^{-1}(y)\}|,$$

the maximum preimage size. Thus, the graph G is d'-right-bounded. Note that each of the degrees on the left equals $|\mathcal{H}|$. Thus, the graph G is $|\mathcal{H}|$-left-regular.

Before proceeding, we first argue why we may exclude the case $\rho = 0$. This case arises if and only if each of the functions is injective. Now, even if some $h \in \mathcal{H}$ is injective, this implies that $|Y| \geq |X|$. So, under our condition that $|R| = O(|L|)$, it should be the case that $|\mathcal{H}|$ is constant. But this leads to a contradiction. Namely, since G is $|\mathcal{H}|$-left-regular, it follows that G is left-bounded by a constant. But, by Lemma 1, each of the left-degrees is greater than f_2 and $f_2 \in \Omega(k)$ by our condition. So we assume $\rho \neq 0$.

Lemma 2. *Let $S \subset X$ be nonempty. Then, for each $x \in S$, it holds that*

$$(1 - \rho(|S| - 1)) |\mathcal{H}| \leq |U(S, x)| \leq |\mathcal{H}|$$

PROOF. The inequality on the RHS follows from the facts that $U(S, x) \subset N(\{x\})$ in general and that, by $|\mathcal{H}|$-left-regularity of G, we have $|N(\{x\})| = |\mathcal{H}|$. As to the inequality on the LHS, fix S. In the case that $|S| = 1$, we have $U(S, x) = N(\{x\})$ and, once again by $|\mathcal{H}|$-left-regularity, we have $|N(\{x\})| = |\mathcal{H}|$. So the inequality follows. Now assume $|S| > 1$ and fix $x \in S$. Consider the neighborhood of x, i.e., the set

$$N(\{x\}) = \{(h, h(x)) : h \in \mathcal{H}\} \subset \mathcal{H} \times Y.$$

---

[5] Note that $\rho = 0$ only if each $h \in \mathcal{H}$ is injective.

It is clear at once that

$$|U(S,x)| = |\{h \in \mathcal{H} : \text{ for each } x' \in S \setminus \{x\}, \text{ it holds that } h(x) \neq h(x')\}|$$

Fixing $x' \in S \setminus \{x\}$ for now, there are at most $\rho|\mathcal{H}|$ hash functions h such that $h(x) = h(x')$, by definition of collision probability. Hence, the number of hash functions h such that $h(x) = h(x')$ for *some* $x' \in S \setminus \{x\}$ is at most $\rho|\mathcal{H}|(|S|-1)$. In conclusion, the number of hash functions h such that $h(x) \neq h(x')$ for each $x' \in S \setminus \{x\}$ is at least $(1 - \rho(|S| - 1))|\mathcal{H}|$ and the claim follows.   △

Note that the lemma only gives a nontrivial result if $|S| < 1 + 1/\rho$.

Let p be a prime number with $p \geq 2k + 1$. By Bertrand's Postulate, there exists such prime p with $p < 4k$. Now consider the family with

$$\mathcal{H} = \mathbb{F}_p, X = \mathbb{F}_p^2, Y = \mathbb{F}_p$$

such that, for $h \in \mathbb{F}_p$, the corresponding hash function is defined as

$$h : \mathbb{F}_p^2 \to \mathbb{F}_p$$

$$(x_0, x_1) \mapsto x_0 h + x_1.$$

One verifies directly that for this family we can take

$$\rho = 1/p \text{ and } d' = p.$$

Setting $|S| = k$, it follows by Lemma 2 that, for each $x \in S$, we have

$$|U(S,x)| \geq (1 - (k-1)/p)p = p - k + 1.$$

Therefore, $|U(S,x)| > k$ if the prime p satisfies $p \geq 2k + 1$. This concludes the proof of Theorem 5.

4.3   Alternative Approaches and Generalization

An alternative constructive approach can be based on graphs G with "excellent expansion," a basic concept from the theory of expander graphs. We say that a d-left-bounded graph G *expands excellently* on a set $S \subset L$ if the neighborhood $N(S) \subset R$ of S satisfies

$$|N(S)| \geq (1 - \epsilon)d|S|$$

where ϵ is a nonnegative real number with

$$\epsilon < 1/2.$$

Excellent expansion is well-known to imply the unique neighbor property. We adapt the arguments so as to imply the (f_1, f_2)-strong unique neighbor property instead, in certain parameter regimes. Then we discuss elementary construction of suitable expander graphs. We elaborate below.

The following lemma is well-known.

Lemma 3. *Suppose G is d-left-bounded. If $N(S) \geq (1 - \epsilon)d|S|$, then*

$$|U(S)| \geq (1 - 2\epsilon)d|S|.$$

PROOF. Since G is d-left-bounded, there are at most $d|S|$ edges "emanating" from S and "arriving" at $N(S)$. Write m_1 for the number of vertices $w \in N(S)$ with $|S \cap N(\{w\})| = 1$. Then we have the obvious bound

$$m_1 + 2(|N(S)| - m_1) \leq d|S|.$$

Therefore,

$$m_1 \geq 2|N(S)| - d|S|.$$

Since $|N(S)| \geq (1 - \epsilon)d|S|$, it follows that

$$m_1 \geq (1 - 2\epsilon)d|S|,$$

as desired. $\triangle$

Using a "greedy argument" the f-strong unique neighbor property for a set is implied by a large unique neighbor set, as follows. Let δ be a real number with $0 < \delta \leq 1$.

Lemma 4. *Suppose that G is d-left-bounded ($d > 0$) and that $S \subset L$ is non-empty. Write $|U(S)| \geq (1 - \delta)d|S|$, where δ is a real number with $0 \leq \delta \leq 1$. If*

$$\delta|S| < 1 - \frac{f}{d},$$

the set S has the f-strong unique neighbor property.

PROOF. If $|S| = 1$, say $S = \{v\}$, then it follows at once that $|U(S, v)| = N(\{v\})| > f$ and the claim follows. So now assume $|S| > 1$. Using a pigeon-hole argument, we see that, if

$$\frac{(1 - \delta)d|S| - f}{|S| - 1} > d, \quad (*)$$

then the set S has the f-strong unique neighbor property. Indeed, consider the subgraph G' obtained by removing some f vertices from R and by removing their incident edges from E. Towards a contradiction, suppose S does not have the strong unique neighbor property in G'. Say it fails on some $v \in S$. Then the inequality implies that there is some $v' \in S \setminus \{v\}$ with degree greater than d, which contradicts the fact that, just as the graph G, its subgraph G' is d-left-bounded. The proof is finalized by observing that the inequality $(*)$ is equivalent to the inequality $\delta|S| < 1 - f/d$. $\triangle$

By combining Lemmas 3 and 4 we get the following sufficient condition for the f-strong unique neighbor property of a set $S \subset L$.

Corollary 1. *Suppose G is d-left-bounded $(d > 0)$ and suppose $S \subset L$ is non-empty. If, for some nonnegative real number ϵ and for some nonnegative integer f, it holds that*

(1) $N(S) \geq (1 - \epsilon)d|S|$ and
(2) $2\epsilon|S| < 1 - \frac{f}{d}$,

then S has the f-strong unique neighbor property.

Remark 3. In order to satisfy the conditions, it is necessary that $\epsilon < 1/2$ i.e., expansion is excellent.

We now discuss constructions based on this excellent expansion approach. Recall that, under the constraints that $f_1, f_2 \in \Omega(k)$ and that $|R| = O(|L|)$, we wish to minimize $|L|$ (the size of the set of left-vertices) and d' (the right-degree). From the conditions in Corollary 1, we then have that $1/\epsilon \in \Omega(k)$ and that $d \in \Omega(k)$.

Observe that the construction in Theorem 5 gives excellent expansion for all sets of size k. Namely, by Lemma 1, the size of the neighborhood of a set of size k equals $(p - k + 1)k$, where $p = c'k$ for some constant $c' > 2$. Therefore, in this case, $\epsilon = (1 - 1/k) \cdot 1/c' < 1/2$ but $1/\epsilon = c'k/(k - 1) \in O(1)$. In conclusion, the result of Theorem 5 cannot also be obtained by application of Corollary 1, except for less favorable parameter settings. Namely, it would require setting p super-linear in k, thus rendering $|L|$ super-quadratic. Furthermore, since $d \in \Omega(k)$, excellent constant left-degree expander graphs [CRVW02] do not apply here. A (well-known) variation on the greedy counting arguments above shows that a combination of excellent expansion and constant left-degree *does* imply an *approximate* version of the f-strong unique neighbor property, i.e., it holds for a certain fraction of each S. But this notion is not sufficient for our present purposes.

To illustrate this approach based on excellent expansion, we show a construction from *random* permutations instead. This is in contrast with the deterministic approach in Theorem 5 where permutations had to be excluded. We use a classical result by Bassalygo [Bas81] who showed a *Monte Carlo* construction of bipartite graphs with excellent expansion. Basically, a (d, d)-bi-bounded bipartite graph with $|L| = |R|$ is constructed by "taking the union" of d random perfect bipartite matchings (or, equivalently, permutations). In general, the probability of success of this procedure is high but not exponentially close to 1. Therefore, it is not sufficient for our purposes. However, choosing convenient parameters in the procedure, one can show that each individual set S of size k has the required expansion with probability of success exponentially (in k) close to 1. It is not hard to see that this weaker "probabilistic, set-wise" property is sufficient for our purposes as well. The downside, in addition to being Monte Carlo, is that $|L|$ here is *cubic* instead of quadratic. All in all, this leads to the following theorem.

Theorem 6. *There is an efficient construction that, for each $k \geq 1$, gives a bipartite graph $G = (L, R, E)$ such that*

(1) $|L| \in O(k^3)$ and $|R| = |L|$,
(2) G is $O(k)$-right-bounded,
(3) for each fixed set $S \subset L$ with $|S| = k$, it holds that S has the k-strong unique neighbor property, except with exponentially small (in k) probability.

Remark 4. Lemma 1 implies that such a probabilistic approach obeys the same lower bounds that $|L| \in \Omega(k^2)$ and $d' \in \Omega(k)$ as in the deterministic case, conditioned on $f_1, f_2 \in \Omega(k)$ and $|R| = O(|L|)$. In a nutshell, there is a small cover of L by sets S of size f_1 such that, by a union-bound argument, each set S in this cover has the f_2-strong unique neighbor property, with probability still extremely close to 1.

We will prove Theorem 6 by combining Corollary 1 with Proposition 1 below. Suppose $|L| = |R| = n$. Write $L = \{v_1, \ldots, v_n\}$ and $R = \{w_1, \ldots, w_n\}$. For a permutation π on $\{1, \ldots, n\}$, define $E(\pi) \subset L \times R$ as the set of edges

$$\{(v_1, w_{\pi(1)}), \ldots, (v_n, w_{\pi(n)})\}.$$

Suppose $1 \leq d \leq n$. For a d-vector $\Pi = (\pi_1, \ldots, \pi_d)$ of (not-necessarily distinct) permutations on $\{1, \ldots, n\}$, define the set

$$E(\Pi) = \bigcup_{j=1}^{d} E(\pi_j) \subset L \times R$$

and define the bipartite graph

$$G(\Pi) = (L, R, E(\Pi)).$$

Note that G is a (d, d)-bi-bounded (undirected) bipartite graph (without multi-edges). We have the following proposition.

Proposition 1. *Let $G = (L, R, E)$ be a random (d, d)-bi-bounded bipartite graph with $|L| = |R| = n$ as described above. Let α be a real number with $0 < \alpha < 1$. Then, for any **fixed** set $S \subset L$ with $|S| = \alpha n$, it holds that*

$$N(S) \geq (d - 2)|S|,$$

except with probability

$$p'_S \leq \left(\frac{d^2 \alpha e}{2(1 - \alpha)} \right)^{2\alpha n},$$

where e denotes Euler's constant.

PROOF. Choose the d permutations $\pi_1, \ldots, \pi_d$ sequentially. For convenience, write $S = \{1, \ldots, s\}$. For $i = 1, \ldots, s$ and $j = 1, \ldots, d$, consider the random variables

$$X_i^j,$$

the image of $i \in S$ under the permutation π_j. We now think of these as "ordered" $X_1^1, \ldots, X_s^1, X_1^2, \ldots, X_s^2, \ldots$, "increasing" from left to right.

For given X_i^j, condition on all "prior" random variables in the ordering. The probability that X_i^j is a *repeat*, i.e., it lands in what is $N(S)$-so-far is at most

$$\frac{d|S|}{n-i+1} \le \frac{d|S|}{n-|S|}.$$

Here the denominator on the LHS is due to the fact that when choosing the image of i, the $i-1$ distinct images of $1,\ldots,i-1$ are already taken. Hence, the probability p'_S that the event $|N(S)| \le (d-2)|S|$ occurs is at most the probability of the event that there are $2|S|$ repeats. By the union bound, the latter probability is clearly at most

$$\binom{d|S|}{2|S|} \left(\frac{d|S|}{n-|S|}\right)^{2|S|}$$

Therefore,[6]

$$p'_S \le \binom{d|S|}{2|S|} \left(\frac{d|S|}{n-|S|}\right)^{2|S|} \le \left(\frac{de}{2}\right)^{2|S|} \left(\frac{d|S|}{n-|S|}\right)^{2|S|} = \left(\frac{d^2\alpha e}{2(1-\alpha)}\right)^{2\alpha n}.$$

$$\triangle$$

The proposition and its proof are adapted from the classical expander graph construction due to Bassalygo [Bas81]. Our exposition follows (part of) the proof of Theorem 4.4 in [Vad12]. The reason we do not apply the Bassalygo result directly is that the success probability of the construction of an excellent expander is high (i.e., constant) but still much too small for our purposes. Fortunately, we can do with the slightly weaker requirement on G that, for any *fixed* set S of precisely the dictated size, the probability that the set S does not expand excellently is negligibly small. As this saves two applications of the union bound, one to quantify over all sets S of the dictated size and one to quantify over the subsets of size smaller than the dictated size, we get exponentially small failure probability instead of constant.

Now let c_1, c_2 be arbitrary positive integers. Set

(1) $f_1 = c_1 k$, $f_2 = c_2 k$.
(2) $d = c_3 k$ with $c_3 = c_1 + c_2 + 1$.
(3) $\alpha = \frac{1}{d^2 e+1}$.
(4) $n = m = \frac{c_1}{\alpha} k = (d^2 e + 1)c_1 k = (c_3^2 e k^2 + 1)c_1 k = c_1 c_3^2 e k^3 + c_1 k$.

Then, for each fixed set $S \subset L$ with $|S| = f_1$, it holds that S has the f_2-strong unique neighbor property, except with exponentially small (in k) probability

$$p' \le \left(\frac{1}{2}\right)^{2c_1 k}$$

---

[6] Note that $\left(\frac{r}{s}\right)^s \le \binom{r}{s} \le \left(\frac{re}{s}\right)^s$.

Namely, for each set S of size $K = \alpha n = c_1 k = f_1$, it holds that $N(S) \geq (d-2)|S|$. Note that $\epsilon = 2/d$ here. This means that the second condition for the f_2-strong unique neighbour property of sets of this size is $f_1 + f_2 < d$. This is satisfied by definition. Efficiency of the construction is obvious. This concludes the proof of Theorem 6.

4.4 A Generalized Construction and Proof of Theorem 4

We now generalize the construction from Theorem 6 to get one where the number of nodes can be much smaller compared to the size of the special set S at the price that the unique neighbour property holds only in a weaker sense.

Recall that $U(S, v)$ is the set of all $w \in R$ such that

$$N(\{w\}) \cap S = \{v\}.$$

The set $U(S)$ is the union of $U(S, v)$ for all $v \in S$.

Definition 8 ((s, r)-Approximate Unique Neighbour Property of a Set). *Given a bipartite graph $G = (L, R, E)$, the set $S \subseteq L$ has (s, r)-approximate unique neighbour property if there exists a subset $S_1 \subseteq S$ of size s such that for any set $T \subseteq R$ of size r, we have*

$$|U(S, v) - T| > 0 \quad \forall v \in S_1.$$

We may ask whether such set exists. Our following lemma answers this question.

Lemma 5. *Suppose $G = (L, R, E)$ is d-left-bounded. If $U(S) \geq (1-\epsilon)d|S|$, then the set S has (s, r)-approximate unique neighbour property for $s = (1-\epsilon)|S| - \frac{r}{d}$.*

Proof. Let $T \subseteq R$ of size r. Let $T_1 = U(S) - T$ and $S_1 \subseteq S$ be the set such that for any $v \in S_1$, $N(v) \cap T_1 \neq \phi$. Since $\deg(v) \leq d$, there are at least $\frac{|T_1|}{d}$ vertices contained in S_1. We are done.

Combining Lemmas 5 and 3, we get a sufficient condition for the unique neighbour property of a set $S \subseteq L$.

Corollary 2. *Suppose G is d-left-bounded. If $N(S) \geq (1 - \epsilon)d|S|$, then the set $S \subseteq L$ has (s, r)-approximate unique neighbour property for $s = (1 - 2\epsilon)|S| - \frac{r}{d}$.*

Now, if in Proposition 1, we set $|S| = k$, $n = O(k^{1+2c})$, $\epsilon = \frac{2}{d}$ and $d = k^c$ for a constant $0 < c < 1$, we can proceed in a way similar to the proof of Theorem 6, and get the following theorem, which immediately implies Theorem 4.

Theorem 7. *There is an efficient construction that, for each $k \geq 1$ and for a constant $0 < c < 1$, gives a bipartite graph $G = (L, R, E)$ such that*

(1) $|L| = |R| = O(k^{1+2c})$,
(2) G is $O(k^c)$-right-bounded,

(3) for each fixed set $S \subseteq L$ with $|S| = k$, it holds that S has the $(k - 5k^{1-c}, k)$-approximate unique neighbour property, except with exponentially small (in k) probability.

Acknowledgements. We are grateful for feedback and suggestions we received after we circulated a preprint of this work on July 6, 2016. Omer Reingold [Rei16] suggested the approach from Theorem 5 as an improvement to our "strong unique neighbor" Theorem 6, which was also stated in this preprint. This suggestion not only gave a deterministic construction instead of our Monte Carlo one but it also improved the left-vertex-set parameter from cubic to quadratic. We thank him for allowing us to incorporate his suggestion. Independently from Omer Reingold, Gilles Zémor [Z16] suggested an alternative improvement to our "strong unique neighbor" Theorem 6 which removed the probabilism as well but left the parameters essentially unchanged. His suggestion was based on combining our excellent expansion approach with an argument involving the girth of certain graphs and an application of Turán's Theorem. Furthermore, we thank Gilles Zémor for several helpful discussions and pointers to the literature (also at an earlier stage). Finally, thanks to Amin Shokrollahi and Salil Vadhan for answering questions about the literature.

References

[Bas81]   Bassalygo, L.: Asymptotically optimal switching circuits. Probl. Inf. Transm. **17**(3), 81–88 (1981)

[BDLN16] Baum, C., Damgård, I., Larsen, K.G., Nielsen, M.: How to prove knowledge of small secrets. In: Robshaw, M., Katz, J. (eds.) CRYPTO 2016. LNCS, vol. 9816, pp. 478–498. Springer, Heidelberg (2016). doi:10.1007/978-3-662-53015-3_17. Cryptology ePrint Archive, Report 2016/538 (2016)

[BDOZ11] Bendlin, R., Damgård, I., Orlandi, C., Zakarias, S.: Semi-homomorphic encryption and multiparty computation. In: Paterson, K.G. (ed.) EURO-CRYPT 2011. LNCS, vol. 6632, pp. 169–188. Springer, Heidelberg (2011). doi:10.1007/978-3-642-20465-4_11

[BG93]   Bellare, M., Goldreich, O.: On defining proofs of knowledge. In: Brickell, E.F. (ed.) CRYPTO 1992. LNCS, vol. 740, pp. 390–420. Springer, Heidelberg (1993). doi:10.1007/3-540-48071-4_28

[BGV12]   Brakerski, Z., Gentry, C., Vaikuntanathan, V.: (Leveled) fully homomorphic encryption without bootstrapping. In: Proceedings of the 3rd Innovations in Theoretical Computer Science Conference, ITCS 2012, pp. 309–325. ACM, New York (2012)

[BV14]   Brakerski, Z., Vaikuntanathan, V.: Efficient fully homomorphic encryption from (standard) LWE. SIAM J. Comput. **43**(2), 831–871 (2014)

[CD09]   Cramer, R., Damgård, I.: On the amortized complexity of zero-knowledge protocols. In: Halevi, S. (ed.) CRYPTO 2009. LNCS, vol. 5677, pp. 177–191. Springer, Heidelberg (2009). doi:10.1007/978-3-642-03356-8_11

[CDK14]   Cramer, R., Damgård, I., Keller, M.: On the amortized complexity of zero-knowledge protocols. J. Cryptol. **27**(2), 284–316 (2014)

[CRVW02] Capalbo, M., Reingold, O., Vadhan, S., Wigderson, A.: Randomness conductors and constant-degree lossless expanders. In: STOC, pp. 659–668 (2002)

[DF02]   Damgård, I., Fujisaki, E.: A statistically-hiding integer commitment scheme based on groups with hidden order. In: Zheng, Y. (ed.) ASIACRYPT 2002. LNCS, vol. 2501, pp. 125–142. Springer, Heidelberg (2002). doi:10.1007/3-540-36178-2_8

[DKL+13]   Damgård, I., Keller, M., Larraia, E., Pastro, V., Scholl, P., Smart, N.P.: Practical covertly secure MPC for dishonest majority – or: breaking the SPDZ limits. In: Crampton, J., Jajodia, S., Mayes, K. (eds.) ESORICS 2013. LNCS, vol. 8134, pp. 1–18. Springer, Heidelberg (2013). doi:10.1007/978-3-642-40203-6_1

[DPSZ12]   Damgård, I., Pastro, V., Smart, N., Zakarias, S.: Multiparty computation from somewhat homomorphic encryption. In: Safavi-Naini, R., Canetti, R. (eds.) CRYPTO 2012. LNCS, vol. 7417, pp. 643–662. Springer, Heidelberg (2012). doi:10.1007/978-3-642-32009-5_38

[GGH96]   Goldreich, O., Goldwasser, S., Halevi, S.: Collision-free hashing from lattice problems. In: Electronic Colloquium on Computational Complexity (ECCC), vol. 3, pp. 236–241 (1996)

[GSW13]   Gentry, C., Sahai, A., Waters, B.: Homomorphic encryption from learning with errors: conceptually-simpler, asymptotically-faster, attribute-based. In: Canetti, R., Garay, J.A. (eds.) CRYPTO 2013. LNCS, vol. 8042, pp. 75–92. Springer, Heidelberg (2013). doi:10.1007/978-3-642-40041-4_5

[KL14]   Katz, J., Lindell, Y.: Introduction to Modern Cryptography. CRC Press, Boca Raton (2014)

[LMPR08]   Lyubashevsky, V., Micciancio, D., Peikert, C., Rosen, A.: SWIFFT: a modest proposal for FFT hashing. In: Nyberg, K. (ed.) FSE 2008. LNCS, vol. 5086, pp. 54–72. Springer, Heidelberg (2008). doi:10.1007/978-3-540-71039-4_4

[Lyu08]   Lyubashevsky, V.: Lattice-based identification schemes secure under active attacks. In: Cramer, R. (ed.) PKC 2008. LNCS, vol. 4939, pp. 162–179. Springer, Heidelberg (2008). doi:10.1007/978-3-540-78440-1_10

[Lyu09]   Lyubashevsky, V.: Fiat-shamir with aborts: applications to lattice and factoring-based signatures. In: Matsui, M. (ed.) ASIACRYPT 2009. LNCS, vol. 5912, pp. 598–616. Springer, Heidelberg (2009). doi:10.1007/978-3-642-10366-7_35

[Rei16]   Reingold, O.: Private communication to the authors, July 2016

[Vad12]   Vadhan, S.: Pseudorandomness. Now publishers (2012)

[Z16]   Zémor, G.: Private communication to the authors, July 2016

Sublinear Zero-Knowledge Arguments
for RAM Programs

Payman Mohassel[1], Mike Rosulek[2(✉)], and Alessandra Scafuro[3]

[1] Visa Research, Palo Alto, USA
pmohasse@visa.com
[2] Oregon State University, Corvallis, USA
rosulekm@eecs.oregonstate.edu
[3] North Carolina State University, Raleigh, USA
ascafur@ncsu.edu

Abstract. We describe a new succinct zero-knowledge argument protocol with the following properties. The prover commits to a large data-set M, and can thereafter prove many statements of the form $\exists w : \mathcal{R}_i(M, w) = 1$, where $\mathcal{R}_i$ is a public function. The protocol is *succinct* in the sense that the cost for the verifier (in computation & communication) does not depend on $|M|$, not even in any initialization phase In each proof, the computation/communication cost for *both* the prover and the verifier is proportional only to the running time of an oblivious RAM program implementing $\mathcal{R}_i$ (in particular, this can be sublinear in $|M|$). The only costs that scale with $|M|$ are the computational costs of the prover in a one-time initial commitment to M.

Known sublinear zero-knowledge proofs either require an initialization phase where the work of the verifier is proportional to $|M|$ and are therefore sublinear only in an amortized sense, or require that the computational cost for the prover is proportional to $|M|$ upon *each proof*.

Our protocol uses efficient crypto primitives in a black-box way and is UC-secure in the *global*, non-programmable random oracle, hence it does not rely on any trusted setup assumption.

1 Introduction

A zero-knowledge proof (or argument) allows a prover to convince a verifier that a statement $\exists w : R(w) = 1$ is true, without revealing anything about the witness w. In this work we study the problem of zero-knowledge proofs concerning large datasets. For example, suppose Alice holds a large collection of files, and wants to prove that there is a file in her collection whose SHA3-hash equals some public value.

Most techniques for zero-knowledge proofs are a poor fit for proving things about large data, since they scale at least linearly with the size of the witness. For realistically large data, it is necessary to adopt methods that have sublinear

M. Rosulek—Partially supported by NSF awards 1149647 & 1617197.

© International Association for Cryptologic Research 2017
J.-S. Coron and J.B. Nielsen (Eds.): EUROCRYPT 2017, Part I, LNCS 10210, pp. 501–531, 2017.
DOI: 10.1007/978-3-319-56620-7_18

cost. There are several existing techniques for zero-knowledge proofs/arguments that have sublinear cost:

PCP Techniques: Kilian [27] and Micali [30] were the first to describe proof systems in which the verifier's cost is sublinear. The technique makes use of probabilistically checkable proofs (PCPs), which are proofs that can be verified by inspecting only a small (logarithmic) number of positions. Followup work has focused on improving the performance of the underlying PCP systems [4,6,9]. Besides the fact that constructing a PCP proof is still quite an inefficient procedure, the main drawback of the PCP approach is that if the prover wants to prove many statements about a single dataset M, he/she must expend effort proportional to $|M|$, *for each proof.*

SNARKs: Succinct non-interactive arguments of knowledge (SNARKs) [8,10,11,17] are the most succinct style of proof to-date. In the most efficient SNARKs, the verifier only processes a constant number of group elements. Born as a theoretically intriguing object that pushed the limit of proof length to the extreme, SNARKs have won the attention of the practical community [7,8,13,33] after an open-source library (libsnark [1]) was created, proving the concrete efficiency of such approach and resulting in its use in real-world applications such as Zerocash [5]. However, similar to the PCP approach, the main drawback of SNARKs is that each proof requires work for the prover that is proportional to the size of the dataset. Moreover, while SNARKs do require a trusted CRS, they are not directly compatible with the UC-framework due to their use of non black-box knowledge extraction (A recent work [28] put forward "snark-lifting" techniques to upgrade SNARKS into UC-secure NIZK. This transformation however results in zero-knowledge proofs whose sizes are linear in the witness instead of constant as in regular SNARKs).

Oblivious RAM: A recent trend in secure computation is to represent computations as RAM programs rather than boolean circuits [2,22,24]. This leads to protocols whose cost depends on the running time of the RAM program (which can be sublinear in the data size). Looking more closely, however, the RAM program must be an *oblivious* RAM. An inherent feature of oblivious RAM programs is that there must be an initialization phase in which every bit of memory is touched. In existing protocols, this initialization phase incurs linear cost for *all parties.* Therefore, RAM-based protocols are sublinear only in an *amortized* sense, as they incur an expensive setup phase with cost proportional to the data size.

Our Results. We construct a zero-knowledge argument based on RAM programs, with the following properties:

- A prover can commit to a large (private) dataset M, and then prove many statements of the form $\exists w_i : \mathcal{R}_i(M, w_i) = 1$, for public $\mathcal{R}_i$.
- The phase in which the prover commits to M has $|M|$ computation cost for the prover. This is the only phase in which the prover's effort is linear in M, but this effort can be reused for many proofs. Unlike prior ZK proofs

based on RAM programs [24], the cost to the verifier (in communication &
computation) is constant in this initial phase. Unlike other approaches based
on PCPs & SNARKs, the expensive step for the prover can be reused for
many proofs about the same data.

- The communication/computation cost for both parties in each proof is proportional to the running time of a (oblivious) RAM program implementing
  $\mathcal{R}_i$. In particular, if $\mathcal{R}_i$ is sublinear in $|M|$, then the verifier's cost is sublinear.
  In succinct proofs based on PCP/SNARKs, on the other hand, computation
  cost for the prover is always proportional to $|M|$.

- The protocol is proven UC-secure based only on a *global*, non-programmable
  random oracle. In particular, there are no trusted setup assumptions.

On Non-standard Assumptions. Our protocol uses a non-programmable random
oracle. We point out that if one wishes to achieve UC security in a succinct
protocol, then some non-standard-model assumption is required. In particular,
the simulator must be able to extract the dataset M of a corrupt prover during
the commitment phase. In the standard model, this would require the prover to
send at least $|M|$ bits of data in the protocol.[1]

A global (in the sense of [12]), non-programmable random oracle is arguably
the mildest non-standard-model assumption. We point out that SNARKs
also use non-standard-model assumptions, such as the knowledge of exponent
assumptions (KEA), which are incompatible with the UC framework [28].

2   Our Techniques

Our goal is to construct ZK proofs where the overhead of the verifier does not
depend on $|M|$, not even in the initialization phase. Moreover we insist the
computational overhead for P when computing a proof is proportional only to
the running time of the RAM program representing $\mathcal{R}(M, w)$, and not on $|M|$.
The latter requirement immediately rules out any circuit-based approach, such
as PCP proof, or SNARKs where the relation $\mathcal{R}(M, w)$ is unrolled into a boolean
circuit of size at least $|M|$.

Towards achieving complexity that is proportional only to the running time
of $\mathcal{R}$, the starting point is to represent $\mathcal{R}$ as a (oblivious) RAM program. An
oblivious RAM [31] is a RAM program whose *access pattern* (i.e., the set $\mathcal{I}$ of
memory addresses accessed, along with whether the accesses are reads or writes)
leaks nothing about the private intermediate values of the computation. The
transformation from an arbitrary RAM computation to an oblivious one incurs
a small polylogarithmic overhead in running time and in the size of the memory.
However, once the memory is in an ORAM-suitable format, it can be persistently
reused for many different ORAM computations.

Hu et al. [24] provide a ZK proof of knowledge protocol for RAM programs
that is sublinear in the amortized sense: the protocol has an initial setup phase

---

[1] Work that pre-dates the UC security model avoids this problem by using a simulator
that rewinds the prover — a technique that is not possible in the UC model.

in which both parties expend effort proportional to $|M|$. After this initialization phase, each proof of the form "$\exists w : \mathcal{R}(M,w) = 1$" has cost (for both parties) proportional only to the running time of $\mathcal{R}(M,w)$. There are other works [2,15,16,18,29] that can be used to construct malicious-secure two-party computation of general functionalities based on RAM programs. Compared to [24], these other techniques are overkill for the special case of ZK functionalities. All of these techniques result in sublinear performance only in the amortized sense described above.

Our goal is to achieve similar functionality as [24] without expensive effort by the verifier in the initialization phase. Looking more closely at the initialization phase of [24], the two parties engage in a secure two-party protocol where they jointly compute a shared representation of each block of M (specifically, a garbled sharing, where the verifiers has values l_0, l_1 for each bit, while the prover learns l_b if the corresponding bit of M is b).

Towards removing the verifier's initial overhead, a natural approach is to remove the participation of V in the setup phase, and have P commit succinctly to the memory using a **Merkle Tree**. Then later in the proof phase, P can prove that the RAM program accepts when executed on the values stored within the Merkle Tree.

Technical Challenge (Extraction): Unfortunately, this natural approach leads to challenges in the UC model. Consider a malicious prover who convinces a verifier of some statement. For UC security, there must exist a simulator that can extract the (large) witness M. But since the main feature of this proof is that the total communication is much shorter than $|M|$, it is information-theoretically impossible for the simulator to extract M in the standard model.

Instead, we must settle for a non-standard-model assumption. We use the global random oracle (gRO) of [12], which equips the UC model with a global, non-programmable random oracle. Global here means that the same random oracle is used by all the protocol executions that are run in the world, and this framework was introduced precisely to model the real world practice of instantiating the random oracle with a single, publicly known, hash function.

A non-programmable random oracle allows the simulator to observe the queries made by an adversary. Suppose such an oracle is used as the hash function for the Merkle tree. Then the simulator can use its ability to observe an adversary's oracle queries to reconstruct the entire contents of the Merkle tree from just the root alone.

Now that the Merkle tree is constructed with a random oracle as its hash function, authenticating a value to the Merkle tree is a computation that involves the random oracle. Hence, we cannot use a standard ZK proof to prove a statement that mentions the logic of authenticating values in the Merkle tree. Any Merkle-tree authentication has to take place "in the open." Consequently, the leaves of the Merkle tree need to be revealed "in the open" for each authentication. Therefore, the leaves of the Merkle tree must not contain actual blocks of the witness, but commitments to those blocks (more specifically, UC commitments so the simulator can further extract the RAM program's memory).

Another challenge for extraction comes from the fact that the Merkle tree contains only an ORAM-ready encoding of the logical data M. A simulator can extract the contents of the Merkle tree, but must provide the corresponding *logical* data M to the ideal functionality. We therefore require an ORAM scheme with the following nonstandard *extractability* property: Namely, there should be a way to extract, from any (possibly malicious) ORAM-encoded initial memory, corresponding logical data M that "explains" the ORAM-encoded memory. We formally define this property and show that a simple modification of the Path ORAM construction [36] achieves it.

Consistency Across Multiple ORAM Executions. An oblivious RAM program necessarily performs both physical reads and writes, even if the underlying logical RAM operations are read-only. This means that each proof about the contents of M *modifies* M. Now that the verifier has no influence on the Merkle-tree commitment of M, we need a mechanism to ensure that the Merkle-tree commitment to M remains consistent across many executions of ORAM programs.

Additionally, an ORAM also requires a persistent client state, shared between different program executions. However, in our setting it suffices to simply consider a distinguished block of memory — say, $M[0]$ — as the storage for the ORAM client state.

To manage the modifications made by RAM program executions, we have the prover present commitments to both the initial value and final value of each memory block accessed by the program. The prover (A) proves that the values inside these commitments are consistent with the execution of the program; (B) authenticates the commitments of initial values to the current Merkle tree; (C) updates the Merkle tree to contain the commitments to the updated values. In this way, the verifier can be convinced that RAM program accepts, and that the Merkle tree always encodes the most up-to-date version of the memory M.

In more detail, the protocol proceeds as follows. In the **initialization phase**, the prover processes M to make it an ORAM memory. She commits individually to each block of M and places these commitments in a Merkle tree. She sends the root of the Merkle tree to the verifier.

Then (repeatedly) to prove $\mathcal{R}(M) = 1$ for an oblivious RAM program $\mathcal{R}$, the parties do the following:

1. The prover runs $\mathcal{R}$ in her head. Let I be the set of blocks that were accessed in this execution. Let $M[I]$ denote the initial values in M at those positions, and let $M'[I]$ denote the values in those positions after $\mathcal{R}$ has terminated.
2. The prover sends I to the verifier, which leaks no information if the RAM program is oblivious.
3. The prover sends the commitments to the $M[I]$ blocks which are stored in the Merkle tree. She authenticates each of them to the root of the Merkle tree.
4. The prover generates commitments to the blocks of $M'[I]$ and sends them to the verifier. She gives authenticated updates to the Merkle tree to replace the previous $M[I]$ commitments with these new ones.

5. The prover then proves in zero-knowledge that the access pattern I, the values inside the commitments to $M[I]$, and the values inside the commitments to $M'[I]$ are *consistent with an accepting execution* of $\mathcal{R}$ (i.e., $\mathcal{R}$ indeed generates access pattern I and accepts when $M[I]$ contains the values within the commitments that the prover has shown/authenticated). Importantly, the witness to this proof consists of only the openings of the commitments to $M[I]$ and $M'[I]$ and not the entire contents of M. We can instantiate this proof using any traditional (linear-time) ZK proof protocol.

Note that the cost to the prover is linear in $|M|$ in the initialization phase, although the communication cost is constant. The cost to both parties for each proof depends only on the running time of $\mathcal{R}$. Also, all Merkle-tree authentications are "in the open," so the approach is compatible with a random-oracle-based Merkle tree.

Note that ORAM computations inherently make read/write access to their memory, even if their logical computation is a read-only computation. Hence our protocol has no choice but to deal with reads and writes by the program $\mathcal{R}$. As a side effect, our protocol can be used without modification to provably perform read/write computations on a dataset.

Technical Challenge (Black-Box Use of Commitments): Since our construction will already use the global random oracle model, we would like to avoid any further setup assumptions. This means that the UC commitments in our scheme will use the random oracle.

At the same time, the last step of our outline requires a zero-knowledge proof about the contents of a commitment scheme. We therefore need a method to prove statements about the contents of commitments in a way that *treats the commitment scheme in a black-box way*.

Towards this, we borrow well-known techniques from previous work on black-box (succinct) zero-knowledge protocol [23, 25, 32]. Abstractly, suppose we want to commit to a value m and prove that the committed value satisfies $f(m) = 1$. A black-box commitment to m will consist of UC-secure commitments to the components of $(e_1, \ldots, e_n)$, where $(e_1, \ldots, e_n) \leftarrow \mathsf{Code}(m)$ is an encoding of m in an error correcting code. The prover uses $(e_1, \ldots, e_n)$ as a witness in a standard ZK proof that $f(\mathsf{Decode}(e_1, \ldots, e_n)) = 1$. The statement being proven does not mention commitments at all. However, we show how to modify the ZK proof so that it reveals to the verifier a random subset of the e_i components as a side-effect. The verifier can then ask the prover to open the corresponding e_i-commitments to prove that they match.

Suppose the error-correcting encoding has high minimum distance. Then in order to cheat successfully, the prover must provide a witness to the ZK proof with many e_i values that don't match the corresponding commitments. But, conditioned on the fact that enough e_i's are revealed, this would lead to a high chance of getting caught. Hence the prover is bound to use a witness in the ZK proof that coincides with the contents of the commitment.

We note that each black-box commitment is used in at most two proofs — one when that block of memory is written and another when that block of memory is read. This fact allows us to choose a coding scheme for which no information about m is revealed by seeing two random subsets of e_i's.

In summary it suffices to construct a modified ZK proof protocol that reveals a random subset of the e_i witness components. We show two instantiations:

- In the "MPC in the head" approach of [25], the prover commits to views of an imagined MPC interaction, and opens some subset of them. For example, the computation of $f(\mathsf{Decode}(e_1, \ldots, e_n))$ may be expressed as a virtual 3-party computation where each simulated party has an additive share of the e_i's. The prover commits to views of these parties and the verifier asks for some of them to be opened, and checks for consistency.

  We modify the protocol so that the prover commits not only to each virtual party's view, but also commits individually to each virtual party's *share of each* e_i. A random subset of these can also be opened (for *all* virtual parties), and the verifier can check them for consistency. Intuitively, the e_i's that are fully revealed are *bound* to the ZK proof. That is, the prover cannot deny that these e_i values were the ones actually used in the computation $f(\mathsf{Decode}(e_1, \ldots, e_n))$.

- The ZK protocol of Jawurek et al. [26] is based on garbled circuits. In fact, their protocol is presented as a 2PC protocol for the special class of functions that take input from just one party (and gives a single bit of output). This special class captures zero-knowledge, since we can express a ZK proof as an evaluation of the function $f_x(w) = R(x, w)$ for an NP-relation R, public x, and private input w from the prover. In other words, ZK is a 2PC in which the verifier has no input.

  We show that their protocol extends in a very natural way to the case of 2PC for functions of the form $f(x, y) = (y, g(x, y))$ — i.e., functions where both parties have input but one party's input is made public. Then in addition to proving that $f(\mathsf{Decode}(e_1, \ldots, e_n)) = 1$, we can let the verifier have input that chooses a public, random subset of e_i's to reveal. As above, the prover cannot deny that these are the e_i values that were actually used in the computation of $f(\mathsf{Decode}(e_1, \ldots, e_n)) = 1$.

Technical Challenge (Non-interactive UC-Commitments in the gRO*):* In the above outline, we assume that the commitment scheme used in the construction is instantiated with a UC-secure commitment scheme in the gRO model. For our application we crucially need a UC-commitment with **non-interactive commitment phase**, meaning that a committer can compute a commitment without having to interact with the verifier. To see why this is crucial, recall that in the Setup phase the prover needs to commit to each block of the memory M using a UC-commitment. If the commitment procedure was interactive, then the verifier (who is the receiver of the commitment) will need to participate. This would lead to a linear (in $|M|$) effort required for the verifier.

Unfortunately, known UC-commitments in the gRO model [12] are interactive[2]. Therefore, as an additional contribution, in this work we design a new commitment scheme that is UC-secure in the gRO and has non-interactive commitment and decommitment. Our new commitment scheme is described in Fig. 5.

Optimal Complexity by Combining ORAM and PCP. It is possible to achieve optimal complexity (i.e., polylog$|M|$ for V and $O(T)$ for P, where T is the program's running time) by combining ORAM and PCP-based ZK proofs as follows. Upon each proof, P runs the ORAM in his head and succinctly commits to the ORAM states (using Merkle Tree, for example). Then P proves that the committed ORAM states are correct and consistent with the *committed* memory M, using PCP-based ZK. The use of PCP guarantees that V only reads a few positions of the proof, while the use of ORAM bounds the work of P to $O(T)$. Unfortunately, this approach requires a non-black-box use of the hash function, and as such it is not compatible with the use of random oracles, and does not yield itself to efficient implementation.

Note that plugging in the black-box succinct ZK proof developed in [23] would not give the desired complexity. Very roughly, this is because proving consistency of T committed positions using [23]'s techniques, requires to open at least T paths.

3   Preliminaries

3.1   The gRO model

This *global* random oracle model was introduced by Canetti et al. in [12] to model the fact that in real world random oracles are typically replaced with a single, publicly known, hash function (e.g., SHA-2) which is globally used by all protocols running in the world. The main advantage of adopting gRO, besides being consistent with the real world practice of using a global hash function, is that we are not assuming any trusted setup assumption. In order to be global, the gRO must be *non programmable*. This means that the power of the simulator lies exclusively in his ability to observe the queries made by an adversary to gRO. Therefore, when modeling a functionality in the gRO model, [12] provides a mechanism that allows the simulator for a session sid to obtain all queries to gRO that start with sid.

The global random oracle functionality $\mathcal{G}_{gRO}$ of [12] is depicted Fig. 1. $\mathcal{G}_{gRO}$ has the property that "leaks" to an adversary (the simulator) all the illegitimate queries[3]. The reader is referred to [12] for further details on the gRO model.

---

[2] Several techniques exist that construct equivocal commitments from an extractable commitment, and could be potentially adopted in the gRO model. Unfortunately, all such techniques require interaction.

[3] In each session sid, an illegitimate query to $\mathcal{G}_{gRO}$ is a query that was made with an index $sid' \neq sid$.

3.2 Ideal Functionalities

We require a commitment functionality $\mathcal{F}_{tcom}$ for the gRO model; we defer details to the full version.

The main difference with the usual commitment functionality is that in $\mathcal{F}_{tcom}$, the simulator of session sid, requests the set $\mathcal{Q}_{sid}$ of queries starting with prefix sid submitted to $\mathcal{G}_{gRO}$.

Our final protocol realizes the zero-knowledge functionality described in Fig. 2. It captures proving recurring statements about a large memory M where M can be updated throughout the process. This functionality consists of two phases: in the **Setup** phase, the prover sends a dataset M, for a session sid. This is a one-time phase, and all subsequent proofs will be computed by $\mathcal{F}_{zk}$ over the committed dataset M. In the **Proof** phase, P simply sends the relation $\mathcal{R}_l$ that he wishes to run over the data M, and possibly a witness w. A relation can be seen as a RAM program that takes in input (M, w). The evaluation of the RAM program can cause M to be updated.

Our main protocol can be seen as a way to reduce $\mathcal{F}_{zk}$ (succinct ZK of RAM execution) to a series of smaller zero-knowledge proofs about circuits. The functionality $\mathcal{F}_{check}^{C_1, C_2}$ (Fig. 3) captures a variant of ZK proofs for boolean circuits that we require. In particular, while in standard ZK only the prover has input (the witness), in this generalization the verifier also has input, but its input will be revealed to the prover by the end of the proof. Later we show how to instantiate this functionality using either the garbled-circuit-based protocol of [26] or the MPC-in-the-head approach of [19,25].

3.3 Encoding Scheme

A pair of polynomial time algorithms (Code, Decode) is an encoding scheme with parameters (d, t, κ) if it satisfies the following properties.

- The output of Code is a vector of length κ.
- Completeness. For all messages m, $m = \mathsf{Decode}(\mathsf{Code}(m))$.
- Minimum distance: For any $m \neq m'$, the two codewords $\mathsf{Code}(m)$ and $\mathsf{Code}(m')$ are different in at least d indices.
- Error correction: For any m, and any codeword C that is different from $\mathsf{Code}(m)$ in at most $d/2$ positions, $m \leftarrow \mathsf{Decode}(C)$.
- t-Hiding. For any m, any subset of $2t$ indices of $\mathsf{Code}(m)$ information-theoretically hide m.

Let $s \in \mathbb{N}$ denote the statistical security parameter. We observe that we can use Reed-Solomon codes to obtain an encoding satisfying the above properties with $\kappa = 4s, d = 2s$, and $t = s$. To encode a message m from a finite field $\mathbb{F}$, we generate a random polynomial P of degree $2s$ over $\mathbb{F}$ such that $P(0) = m$. The codeword is the evaluation of P at $\kappa = 4s$ different points i.e. $C = (P(1), \ldots, P(4s))$. To decode a message, we use the well-known decoding algorithm of Berlekamp and Welch for Reed-Solomon codes.

Functionality $\mathcal{G}_{gRO}$

Parameters: output length $\ell(\lambda)$ and a list $\bar{\mathcal{F}}$ of ideal functionality programs.

1. Upon receiving a query x, from some party $P = (\text{pid}, \text{sid})$ or from the adversary S do:
   - If there is a pair (x, v) for some $v \in \{0, 1\}^{\ell(\lambda)}$ in the (initially empty) list $\mathcal{Q}$ of past queries, return v to P. Else, choose uniformly $v \in \{0, 1\}^{\ell(\lambda)}$ and store the pair (x, v) in $\mathcal{Q}$. Return v to P.
   - Parse x as (s, x'). If $\text{sid} \neq s$ then add (s, x', v) to the (initially empty) list of **illegitimate** queries for SID s, that we denote by $\mathcal{Q}_{|s}$.
2. Upon receiving a request from an instance of an ideal functionality in the list $\bar{\mathcal{F}}$, with SID s, return to this instance the list $\mathcal{Q}_{|s}$ of illegitimate queries for SID s.

Fig. 1. $\mathcal{G}_{gRO}$

Functionality $\mathcal{F}_{zk}$
Parties: P and V and adversary Sim.

- **Setup.** On input $(\text{sid}, \text{INIT}, M)$ from P, if no previous init command has been given, then record M. Else, do nothing.
- l-th **Proof.** On input $(\text{PROVE}, \text{sid}|l, \mathcal{R}_l, w)$ from P: evaluate $(M', b) = \mathcal{R}_l(M, w)$. If $b = 1$ send $(\text{ACCEPT}, \text{sid}|l)$ to V. Set $M = M'$.
- When asked by the adversary, obtain from $\mathcal{G}_{gRO}$ the list $\mathcal{Q}_{sid}$ of illegitimate queries that pertain to SID sid, and send $\mathcal{Q}_{sid}$ to S.

Fig. 2. $\mathcal{F}_{zk}$

Functionality $\mathcal{F}_{check}^{C_1, C_2}$
Parameterized by: Two check circuits C_1, C_2 that each take one input.
Parties: P and V and adversary Sim.

- **Challenge.** On input $(\text{CHALLENGE}, sid, r)$ from V, if there is no previous sid and $C_2(r) = 1$, record r. Else, do nothing.
- **Proof.** On input $(\text{PROVE}, sid, \mathbf{W})$ from P, if $C_1(\mathbf{W}) = 1$, the $\mathcal{F}_{check}$ outputs $(\text{ACCEPT}, sid, (\{\mathbf{W}[i]\}_{i, r_i=1})$ to party V and (r, sid) to P.

Fig. 3. $\mathcal{F}_{check}^{C_1, C_2}$

Hiding follows from the security of Shamir's Secret Sharing: any $t = 2s$ points on a polynomial of degree $2s$ do not leak any information about the secret $P(0)$. Minimum distance $d = 2s$ follows from the observation that if two encodings agree in more than $2s$ points, then they must in fact be the same polynomial and hence encode the same value. Error correction follows from the Berlekamp-Welch decoding algorithm, which can efficiently correct errors up to half the minimum distance.

3.4 Oblivious RAM Programs

Oblivious RAM (ORAM) programs were first introduced by Goldreich and Ostrovsky [21]. ORAM provides a wrapper that encodes a logical dataset as a physical dataset, and translates each logical memory access into a series of physical memory accesses so that the physical memory access pattern leaks nothing about the underlying logical access pattern.

Syntactically, let Π be a RAM program that operates on memory M and also takes an additional auxiliary input w. We write $(M', z) \leftarrow \Pi(M, w)$ to denote that when Π runs on memory M and input w, it modifies the memory to result in M' and outputs z.

We use M to represent the logical memory of a RAM program and $\widehat{M}$ to indicate the physical memory array in Oblivious RAM program. We consider all memory to be split into **blocks**, where $M[i]$ denotes the ith block of M.

An Oblivious RAM (wrapper) consists of algorithms (RamInit, RamEval) with the following meaning:

- RamInit takes a security parameter and logical memory M as input, and outputs a physical memory $\widehat{M}$ and state st.
- RamEval takes a (plain) RAM program Π, auxiliary input w, and state st as input, and outputs an updated memory $\widehat{M}'$, updated state st, and RAM output z.

In general these algorithms are randomized. When we wish to explicitly refer to specific randomness used in these algorithms, we write it as an additional explicit argument ω. When we omit this extra argument, it means the randomness is chosen uniformly.

Definition 1. *Let* (RamInit, RamEval) *be an ORAM scheme. For all M and sequences of RAM programs $\Pi_1, \ldots, \Pi_n$ and auxiliary inputs $w_1, \ldots, w_n$, and all random tapes $\omega_0, \ldots, \omega_n$, define the following values:*

- *RealOutput$(M, \Pi_1, \ldots, \Pi_n, w_1, \ldots, w_n)$:*
  Set $M_0 = M$. Then for $i \in [n]$, do $(M_i, z_i) = \Pi_i(M_{i-1}, w_i)$. Return $(z_1, \ldots, z_n)$.
- *OblivOutput$(M, \Pi_1, \ldots, \Pi_n, w_1, \ldots, w_n, \omega_0, \ldots, \omega_n)$:*
  Set $(\widehat{M}_0, st_0) = $ RamInit$(1^k, M; \omega_0)$. Then for $i \in [n]$, do $(\widehat{M}_i, st_i, z_i') = $ RamEval$(\Pi_i, \widehat{M}_{i-1}, st_{i-1}, w_i; \omega_i)$. Return $(z_1', \ldots, z_n')$.

*The ORAM scheme is **correct** if RealOutput$(M, \Pi_1, \ldots, \Pi_n, w_1, \ldots, w_n)$ and OblivOutput$(M, \Pi_1, \ldots, \Pi_n, w_1, \ldots, w_n, \omega_0, \ldots, \omega_n)$ agree with overwhelming probability over choice of random ω_i.*

*The ORAM scheme is **sound** if for all $\omega_0, \ldots, \omega_n$, the vectors RealOutput $(M, \Pi_1, \ldots, \Pi_n, w_1, \ldots, w_n)$ and OblivOutput$(M, \Pi_1, \ldots, \Pi_n, w_1, \ldots, w_n, \omega_0, \ldots, \omega_n)$ disagree only in positions where the latter vector contains $\perp$.*

In our protocol, we allow the adversary to choose the randomness to the ORAM construction. The soundness property guarantees that the adversary cannot use this ability to falsify the output of the RAM program. At worst, the adversary can influence the probability that the RAM program aborts.

In our protocol, the simulator for a corrupt prover can extract only the ORAM-initialized memory $\widehat{M}$. However, the simulator must give the logical memory M to the ideal functionality. For this reason, we require an ORAM construction that is **extractable** in the following sense:

Definition 2. *An ORAM scheme* (RamInit, RamEval) *is **extractable** if there is a function* RamExtract *with the following property. For all (possibly maliciously generated)* $(\widehat{M}, st)$, *all M and sequences of RAM programs* $\Pi_1, \ldots, \Pi_n$ *and auxiliary inputs* $w_1, \ldots, w_n$ *define the following:*

- *Set* $M_0 \leftarrow$ RamExtract$(\widehat{M}, st)$. *Then for $i \in [n]$, do* $(M_i, z_i) = \Pi_i(M_{i-1}, w_i)$.
  Return $(z_1, \ldots, z_n)$.
- *Set* $(\widehat{M}_0, st_0) = (\widehat{M}, st)$. *Then for* $i \in [n]$, *do* $(\widehat{M}_i, st_i, z_i') =$
  RamEval$(\Pi_i, \widehat{M}_{i-1}, st_{i-1}, w_i)$. *Return* $(z_1', \ldots, z_n')$.

Then with overwhelming probability $z_i' \in \{z_i, \bot\}$ for each i.

In other words, RamExtract produces a plain RAM memory that "explains" the effect of $(\widehat{M}, st)$. The only exception is that a malicious $\widehat{M}, st$ could cause the ORAM construction to abort more frequently than a plain RAM program.

Let AccessPattern$(\Pi, \widehat{M}, w, st; \omega)$ denote the **access pattern** describing the accesses to physical memory made by RamEval$(\Pi, \widehat{M}, w, st; \omega)$. The access pattern is a sequence of tuples of the form (READ, id) or (WRITE, id), where id is a block index in $\widehat{M}$.

Definition 3. *We say that a scheme* (RamInit, RamEval) *is **secure** if there exists an efficent S such that, for all M, Π, and w, the following two distributions are indistinguishable:*

- *Run* $S(1^k, |M|, \Pi, |w|)$.
- *Run* $(\widehat{M}, st) \leftarrow$ RamInit$(1^k, M)$, *then return* AccessPattern$(\Pi, \widehat{M}, w, st)$.

In other words, the access pattern leaks no information about M or w.

Note that the output of AccessPattern contains only the memory *locations* and not the *contents* of memory. Hence, we do not require the ORAM construction to encrypt/decrypt memory contents — they will be protected via other mechanisms in our protocol.

Our definitions of *soundness* and *extractability* are non-standard. We discuss how to modify existing ORAM constructions to achieve these definitions in the full version.

3.5 Trapdoor Commitment

We construct UC commitments from trapdoor commitments with the following properties: (a) the trapdoor is used only to compute the decommitment, (b) knowledge of the trapdoor allows to equivocate any previously computed commitment (as long as the state z is known). Such a commitment scheme can be based on Pedersen's perfectly hiding commitment scheme [35]. Details and formal definitions for this instantiation are given in the full version.

4 Succinct Zero-Knowledge Proof for RAM Programs

4.1 Protocol Description

Overview. The protocol consists of two phases: a (one-time) setup phase, and a proof phase.

In the setup phase the prover commits to the ORAM memory $\widehat{M}$ in a black-box friendly manner. That is, for each memory location $\widehat{M}[i]$, P first computes an encoding of $\widehat{M}[i]$ resulting in shares $(x_{i,1}, \ldots, x_{i,\kappa})$, then it commits to each share $x_{i,j}$ independently, obtaining commitments $N_i = (cx_{i,1}, \ldots, cx_{i,\kappa})$. Committing to each share independently will allows the prover to later selectively open a subset of t shares. N_i is then placed in the i-th leaf of the Merkle Tree. Similarly, P will also commit to the ORAM state st used to computed $\widehat{M}$, by committing to its shares $(s_1, \ldots, s_\kappa)$. At the end of the setup phase, the verifier receives the root of the Merkle Tree, and the commitments to the encoding of the initial ORAM state.

In the l-th proof phase, the prover first runs the ORAM program corresponding to relation $\mathcal{R}_l$ in her head. From this, she will obtain the access pattern $\mathcal{I}$, the updated contents of memory, and the final ORAM state st'.

P will then commit to this information, using again the black-box friendly commitment outlined above. The verifier at this point receives the set of positions $\mathcal{I}$ as well as commitments to all the encodings. Then, to prove consistency of such computation in a black-box manner, P will invoke the $\mathcal{F}_{check}$ functionality (Fig. 3) that does the following:

1. Decode the shares received in input and reconstruct initial ORAM state st, initial memory blocks $\{\widehat{M}[i]\}$ read by the ORAM computatation, the final ORAM state st' and the updated value $\{\widehat{M}[i]\}$ of any memory blocks accessed during the ORAM computation.
2. Run the ORAM evaluation on input st and the given initial memory block. Check that the program indeed generates access pattern $\mathcal{I}$, updates the memory to the values provided, and outputs the updated state provided.
3. If the check above is successful, then output a subset of t shares from each encoding received in input.

This invocation of $\mathcal{F}_{check}$ is described in greater detail below. It checks only that the encodings provided by P lead to an accepting computation. As it is, this does not prove anything about whether this computation is consistent with the initial memory committed in the setup phase, and with the previous proofs. To glue such encodings to the values that P has committed outside the functionality, we have P open also to a subset of t commitments. In this way, the verifier can be convinced that the values that made the $\mathcal{F}_{check}$ functionality accept are consistent with the ones committed by P.

Notation. We use upper case letters to denote vectors, while we use lower case letters to denote a string. For example, notation $Z = (z_1, \ldots, z_n)$ means that vector Z has components $z_1, \ldots, z_n$. Notation $Z[i]$ denotes the ith component of vector Z and is equivalent to value z_i. We use bold upper case to denote a collection of vectors. For example, $\mathbf{S} = \{S_1, S_2, \ldots\}$.

Moreover, in the protocol, we shall use notation X_i to denote the value of memory block i *before* the proof is computed, while we use notation Y_i to denote the value of memory block i *after* the proof. Similarly we used notation S, S' to denote the encoding of a pre-proof and post-proof ORAM state, respectively.

Let $\mathsf{UCCom} = (\mathsf{Gen}, \mathsf{Com}, \mathsf{Dec}, \mathsf{Ver})$ be a UC-secure commitment scheme that has *non-interactive* commitment and the decommitment phase. In Sect. 5 we give an instantiation of such a scheme in the gRO model. Let $(\mathsf{Code}, \mathsf{Decode})$ be an encoding scheme with parameters (d, t, κ). Let $(\mathsf{RamInit}, \mathsf{RamEval})$ be a secure ORAM scheme. Our (stateful) ZK protocol $\Pi = (\Pi.\mathsf{Setup}, \Pi.\mathsf{Proof})$ is described in Figs. 4 and 5.

Setup $\Pi.\mathsf{Setup}(M, 1^\lambda)$
Let $m = |M|$.

- Run the initialization of the UC commitment scheme to obtain parameters pk.
- Initialize ORAM. Run $(\widetilde{M}, st) \leftarrow \mathsf{RamInit}(M, 1^\lambda)$.
- Encode memory $\widetilde{M}$. For $i \in [m]$:
    (encoding) $X_i = (x_{i,1}, \ldots, x_{i,\kappa}) \xleftarrow{\$} \mathsf{Code}(\widetilde{M}[i])$.
    (commitment) $CX_i = (cx_{i,1}, \ldots, cx_{i,\kappa}) \xleftarrow{\$} \mathsf{Com}(pk, x_{i,1}), \ldots, \mathsf{Com}(pk, x_{i,\kappa})$.
    (decommitment) $DX_i = (dx_{i,1}, \ldots, dx_{i,\kappa})$
- Encode state st.
    (encoding) $S = (s_1, \ldots, s_\kappa) \xleftarrow{\$} \mathsf{Code}(st)$.
    (commitment) $CS = (cs_1, \ldots, cs_\kappa) \xleftarrow{\$} \mathsf{Com}(pk, s_1), \ldots, \mathsf{Com}(pk, s_\kappa)$.
    (decommitment) $DS = (ds_1, \ldots, ds_\kappa)$
- Build the tree. Let $d = \log m$.
    Leaves. For $i = 1, \ldots, m$. Set leaf $N_i = \mathsf{gRO}(sid, i, \mathsf{P}, CX_i)$.
    Internal nodes. For $v \in \{0,1\}^{\leq d-1}$, set $N_v = \mathsf{gRO}(sid, N_{v0}|N_{v1})$.
    Root. Let $h = N_\epsilon$.
- Publish values.
    ◇ Commitment to the ORAM state $CS = (cs_1, \ldots, cs_\kappa)$.
    ◇ Root of the Merkle Tree h.

Fig. 4. Setup phase

$l + 1$-th **Zero-knowledge Proof.** $\Pi.\mathsf{Proof}(state)$

Public Inputs: Relation $\mathcal{R}_l = (\Pi_l, x)$. Root of the Merkle tree h, commitment to ORAM sate: $CS = (cs_1, \ldots, cs_\kappa)$.
Private Inputs for P: Witness w. Memory $\widetilde{M} = \widetilde{M}^l$, ORAM state $st = st^l$. Encodings, commitments and decommitment information for memory blocks and ORAM state. That is: (memory) X_i, CX_i, DX_i, for $i \in [m]$; (state) S, CS,DS.

1. Program evaluation and commitments to the updated memory/state.
   - **Program Evaluation.** P runs $(\mathcal{I}, \widetilde{M}', st') \xleftarrow{\$} \mathsf{RamEval}(\Pi_l, \widetilde{M}, x, w, st)$.
     Let $\mathsf{access}(\mathcal{I}) = \mathsf{read}(\mathcal{I}) \cup \mathsf{write}(\mathcal{I})$. Let $\widetilde{M}'$ denotes the final version of the memory $\widetilde{M}$.
   - **Commitment to updated memory blocks.** For $i \in \mathsf{access}(\mathcal{I})$.
     (Encoding) $Y_i = (y_{i,1}, \ldots, y_{i,\kappa}) \xleftarrow{\$} \mathsf{Code}(\widetilde{M}'[i])$.
     (Commitment) $CY_i = (cy_{i,1}, \ldots, cy_{i,\kappa}) \xleftarrow{\$} \mathsf{Com}(pk, y_{i,1}), \ldots, \mathsf{Com}(pk, y_{i,\kappa})$.
     (Decommitments) $DY_i = (dy_{i,1}, \ldots, dy_{i,\kappa}) \xleftarrow{\$} \mathsf{Dec}(cy_{i,1}), \ldots, \mathsf{Dec}(cy_{i,\kappa})$
   - **Commitment to updated state.**
     (Encoding) $S' = (s'_i, \ldots, s'_\kappa) \leftarrow \mathsf{Code}(st')$.
     (Commitment) $CS' = (cs'_1, \ldots, cs'_\kappa) \xleftarrow{\$} \mathsf{Com}(pk, s'_1), \ldots, \mathsf{Com}(pk, s'_\kappa)$.
     (Decommitments) $DS' = (ds'_1, \ldots, ds'_\kappa) \xleftarrow{\$} \mathsf{Dec}(cs'_1), \ldots, \mathsf{Dec}(cs'_\kappa)$.
   - **Update root.** Recompute the Merkle Tree root h'. Compute circuits $C_{1,\mathcal{I}}, C_{2,\mathcal{I}}$
   - Set $\mathbf{W} = (w\ S, S', \{X_i\}_{i \in \mathsf{read}(\mathcal{I})}, \{Y_i\}_{i \in \mathsf{access}(\mathcal{I})})$.
   - **Send to V.** Access pattern $\mathcal{I}$; commitments CY_i ($\forall i \in \mathsf{access}(\mathcal{I})$); CS', new root h' to V.
2. $\mathcal{F}_{check}^C$
   - V sends $(\mathsf{CHALLENGE}, sid_\ell, r)$ to $\mathcal{F}_{check}^{C_1, C_2}$, for $r \xleftarrow{\$} \{0,1\}^\kappa$. P sends $(\mathsf{PROVE}, \mathsf{sid}, \mathbf{W})$ to $\mathcal{F}_{check}^{C_1, C_2}$
   - V obtains partial encodings $Y_j[\gamma]$, $X_i[\gamma]$, $S[\gamma], S'[\gamma]$ for γ s.t. $r_\gamma = 1$. P receives r.
3. **Verification.**
   - P sends: (1) decommitments $DX_i[\gamma]$ and authentication path π_i from h to N_i, $\forall i \in \mathsf{read}(\mathcal{I})$. (2)decommitments $DY_j[\gamma]$ for $j \in \mathsf{access}(\mathcal{I})$, (3) decommitments $DS[\gamma]$, $DS'[\gamma]$.
   - V checks validity[a] of path π_i, $\forall i \in [\mathsf{read}(\mathcal{I})]$, and the validity of decommitments by running procedure $\mathsf{Ver}(pk, \cdot)$.

---

[a] An authentication path π_i for a node N_i is a chain of hash values. In order to check validity of a path, V checks that node $N_i = \mathsf{gRO}(sid, i, \mathsf{P}, CX_i)$ and that for each internal node v, $N_v = \mathsf{gRO}(sid, N_{v0} \| N_{v1})$.

Fig. 5. Proof phase

The $\mathcal{F}_{check}^{C_1, C_2}$ Circuits

$ORAM$ $Components$: Let $\mathcal{I}$ be an ORAM memory access sequence. We define $\mathsf{read}(\mathcal{I}) = \{i \mid (\mathsf{READ}, i) \in \mathcal{I}\}, \mathsf{write}(\mathcal{I}) = \{i \mid (\mathsf{WRITE}, i) \in \mathcal{I}\}$, and $\mathsf{access}(\mathcal{I}) = \mathsf{read}(\mathcal{I}) \cup \mathsf{write}(\mathcal{I})$; i.e., the indices of blocks that are read/write/accessed in $\mathcal{I}$. If $S = \{s_1, \ldots, s_n\}$ is a set of memory-block indices, then we define $M[S] = (M[s_1], \ldots, M[s_n])$.

Next, we describe the exact check circuits C_1 and C_2 we need for our main protocol. The check circuit $C_{2,\mathcal{I}}(r)$ is straightforward. Given bit string r, it returns 1 if $r_\gamma = 1$ in at most t locations.

Given an ORAM access pattern $\mathcal{I}$, we let the witness W consist of the auxiliary input w and a collection of *encodings* of: the initial ORAM state S, the final ORAM state S', the input memory blocks $\mathbf{X} = (X_1, \ldots, X_{|\mathsf{read}(\mathcal{I})|})$, and the output/resulting memory blocks $\mathbf{Y} = (Y_1, \ldots, Y_{|\mathsf{access}(\mathcal{I})|})$. The check circuit $C_{1,\mathcal{I}}(\mathrm{W})$ is defined as follows:

$\underline{C_{1,\mathcal{I}}(w, S, S', \mathbf{X}, \mathbf{Y}):}$

$st := \mathsf{Decode}(S)$

simulate $\mathsf{RamEval}(\Pi, \widehat{M}, st, w)$ in the following way:

  whenever a block i of $\widehat{M}$ is accessed:

  if $i \notin \mathsf{access}(\mathcal{I})$ then return 0

  else if the access is a READ: take $\mathsf{Decode}(X_i)$ as the result of the access

  else if the access is (WRITE, v), set $\widehat{M}'[i] = v$

if the above simulation of $\mathsf{RamEval}$ does not return 1, then return 0

if the above simulation does not result in access pattern $\mathcal{I}$, then return 0

if the above simulation results in ORAM state $st' \neq \mathsf{Decode}(S')$ then return 0

for $i \in \mathsf{access}(\mathcal{I})$: if $\widehat{M}'[i] \neq \mathsf{Decode}(Y_i)$ then return 0

return 1

4.2 Instantiation $\mathcal{F}_{check}$

Instantiating $\mathcal{F}_{check}^C$ using JKO Protocol. JKO refers to a zero-knowledge protocol of Jawurek et al. [26]. The protocol is based on garbled circuits and is quite efficient, requiring only a single garbled circuit to be sent.

We first give an overview of the JKO protocol. Abstractly, suppose the prover would like to prove knowledge of a witness w such that $R(w) = 1$, where R is a public function/circuit.

1. The verifier generates a garbled circuit implementing R. The parties then perform instances of oblivious transfer, where the verifier acts as receiver. The verifier sends the garbled inputs for the garbled circuit, and the prover picks up a garbled input encoding the witness w.
2. The verifier sends the garbled circuit and the prover evaluates it, resulting in a garbled output. Since R has a single output bit, this is a single wire label (the wire label encoding output "true", if the prover is honest). The prover commits to this garbled output.
3. The verifier opens the garbled circuit so the prover can check that it was garbled correctly. In the JKO protocol, this is done using *committed OT* in step (1). The verifier "opens" its inputs to these OTs, revealing the entire set of garbled inputs. This is enough for the prover to verify the correctness of the garbled circuit.

4. If the prover is satisfied that the circuit was garbled correctly, then she opens her commitment to the garbled output.
5. The verifier accepts the proof if the prover's commitment is opened to the "true" output wire label of the garbled circuit.

The protocol is zero-knowledge because a simulator can extract the entire set of garbled inputs from the OTs in step (1). Then the simulator can compute the "true" output wire label and commit to it instep (2).

The protocol is sound due to the *authenticity* property of the garbled circuit. Namely, given a garbled input encoding w and the garbled circuit, it should be hard to guess an output wire label other than the one encoding truth value $R(w)$. (See [3] for the formal definition) This authenticity property holds in step (2) when the prover must commit to the output wire label. After step (3), the prover can compute any garbled output for the garbled circuit, but the prover has already committed to the garbled output at that point.

Importantly, the prover is the only party with private input to the garbled circuit. But the prover plays the role of garbled circuit *evaluator*. Hence, the protocol does not use the traditional *privacy* security property of garbled circuits. This is also the reason that the same garbled circuit can be both evaluated and checked. Doing this in a more general 2PC is problematic since opening/checking a circuit would reveal the secrets of the garbled circuit's generator. In this case, that party is the verifier and has no secrets to reveal.

Modifications. With some minor modifications, the JKO protocol can be used to efficiently instantiate the $\mathcal{F}_{check}^{C_1,C_2}$ functionality. The main differences are:

- The computation gives more than a single bit output.
- The computation takes input from the verifier (r) as well as the prover. We are able to handle this in the JKO protocol paradigm because r is eventually made public to the prover.

The modified JKO protocol proceeds as follows.

1. The verifier generates a garbled circuit computing the function $\widetilde{C}(\mathbf{W}) = $ [if $C_2(r)$ then $C_1(\mathbf{W}, r)$ else $\bot$]. The parties perform a committed OT for each input bit, in which the prover obtains garbled input encoding $\mathbf{W}$.
2. The verifier sends the garbled circuit and the prover evaluates it, resulting in a garbled encoding of the (many-bit) output $z = \widetilde{C}(\mathbf{W})$. The prover commits to the garbled output.
3. The verifier opens the committed OTs, revealing all garbled inputs. The verifier also sends r at this point. The prover can check whether the garbled circuit was generated correctly.
4. The prover, if satisfied, opens the commitment to the garbled output and sends the plain output $z = \widetilde{C}(\mathbf{W})$. The prover outputs (r, sid).
5. The verifier outputs (z, sid) if the commitment is opened to the valid garbled encoding of z.

Lemma 4. *The modified JKO protocol above is a UC-secure realization of* $\mathcal{F}_{check}^{C_1,C_2}$, *in the committed-OT + commitment hybrid model, if the underlying garbling scheme satisfies the authenticity property.*

Instantiating $\mathcal{F}_{check}^{C_1,C_2}$ **using IKOS.** IKOS refers to the general approach introduced in [25] for obtaining ZK proofs in the commitment-hybrid model for arbitrary NP statements, given any generic MPC protocol. Recently, Giacomelli et. al [19] explored and implemented a concrete instantiation of the IKOS approach based on the GMW protocol [20] among three parties. Their optimized construction is only slightly less efficient than the JKO protocol [26] but instead has the advantage of being a public-coin Σ protocol that can be efficiently made a non-interactive Zero-knowledge proof using the Fiat-Shamir transform.

We first recall the IKOS approach and show how we can modify it to realize the $\mathcal{F}_{check}^{C_1,C_2}$ functionality for any circuits C_1, C_2. As mentioned above, the main ingredient is a Σ protocol with special soundness and honest-verifier Zero-knowledge property:

The prover has an input W and wants to prove that $C_1(W) = 1$ where C_1 can be any public circuit. Let Π be a t-private n-party MPC protocol with perfect correctness. The protocol proceeds as follows.

- Prover generates n random values $\mathbf{W}_i$ such that $\mathbf{W} = \bigoplus_{i=1}^n \mathbf{W}_i$.
- Prover runs (on its own) the n-party MPC Π for computing $C_1(\bigoplus_i \mathbf{W}_i)$ where party P_i's input is $\mathbf{W}_i$, and obtains the view $v_i = \mathsf{View}_{P_i}(\mathbf{W})$ for all $i \in [n]$.
- Prover commits to $v_1, \dots, v_n$.
- Verifier chooses a random subset $E \subset [n]$ where $|E| = t$, and sends E to prover.
- Prover opens the commitment to v_e for all $e \in E$.
- Verifier checks that:
  - For all $e \in E$, v_e yields the output 1 for P_e.
  - For all $e, e' \in E$, the view of P_e and $P_{e'}$ (v_e and v'_e) are consistent.
  - If any of the checks fail it rejects. Else it accepts.

The above protocol has a soundness probability that is a function of n and t. But this probability can be easily amplified by repeating the protocol multiple times in parallel for different runs of Π and using different random challenges E each time. This parallel version remains a Σ protocol as desired.

We need to enhance the above protocol to also take a random string r satisfying $C_2(r)$ for a circuit C_2 as Verifier's input and reveal those locations in the witness $W[i]$ where $r_i = 1$. The above Σ protocol can be easily extended to handle this case. We simply have the verifier send r along with E to the Prover. Prover checks that $C_2(r) = 1$ and if the case, it opens commitments $\mathbf{W}[i]$ for all i where $r_i = 1$. This is in addition to the views it opens to achieve soundness.

1. Prover generates n random values $\mathbf{W}_i$ such that $\mathbf{W} = \bigoplus_{i=1}^n \mathbf{W}_i$.
2. Prover runs (on its own) the n-party MPC Π for computing $C_1(\bigoplus_i \mathbf{W}_i)$ where party P_i's input is $\mathbf{W}_i$, and obtains the view $v_i = \mathsf{View}_{P_i}(\mathbf{W})$ for all $i \in [n]$.

3. Prover commits to $\mathbf{W}_1[j], \ldots, \mathbf{W}_n[j]$ for all $j \in [\|\mathbf{W}\|]$ and $v_1, \ldots, v_n$.
4. Verifier chooses a random subset $E \subset [n]$ where $|E| = t$, and sends E and its input r to the prover.
5. Prover aborts if $C_2(r) \neq 1$. Else it opens commitment to $\mathbf{W}_i[j]$ for all $i \in [n]$ and all j where $r_j = 1$.
6. Prover also opens the commitment to v_e for all $e \in E$ and to $\mathbf{W}_e[j]$ for all $j \in [\|\mathbf{W}\|]$.
7. Verifier checks that:
   (a) For all $e \in E$, the opened $\mathbf{W}_e$ and v_e are consistent, i.e. $\mathbf{W}_e$ is correctly embedded in v_e.
   (b) For all $e \in E$, v_e yields the output 1 for P_e.
   (c) For all $e, e' \in E$, the view of P_e and $P_{e'}$ (v_e and v'_e) are consistent.
   (d) If any of the checks fail it rejects. Else it accepts.

The above protocol is a public-coin, honest-verifier protocol. We can transform it into a zero-knowledge protocol by letting the verifier commit to his random challenge before the prover sends the first message.

Lemma 5. *The modified IKOS protocol above is a secure realization of the $\mathcal{F}_{check}^{C_1, C_2}$ functionality, when the commitments are instantiated with UC commitments.*

5 A New UC-Commitment in the gRO Model

In [12], Canetti et al. show a UC commitment scheme that is secure in the gRO model. Such a commitment scheme is based on trapdoor commitments (e.g., Pedersen's Commitment). The main idea is to have the receiver choose parameters (pk, sk) of a trapdoor commitment, have the sender commit using pk, and later, in the decommitment phase, before revealing the opening, have the receiver reveal the trapdoor sk (this is done in such a way that despite revealing sk, binding is still preserved). This trick allows to achieve equivocability without programming the RO. On the other hand, this trick has the fundamental drawback of requiring that each commitment is computed under a fresh public key pk. (To see why, note that if more than one commitment is computed under the same public key, then binding holds only if all such commitments are opened at the same time.). This is highly problematic in our setting, where the prover commits to each element of the memory, as the verifier would need to provide as many public keys as the size of the memory.

Therefore, we design a new commitment scheme in the gRO model that satisfies the crucial property that the receiver can send one public key pk at the beginning, and the sender can re-use it for all subsequent commitments.

The idea behind our new scheme is fairly simple. The receiver R will pick two public keys $(\mathsf{pk}^0, \mathsf{pk}^1)$ for a trapdoor commitment scheme. Additionally, R computes a non-interactive witness indistinguishable proof of knowledge (NIWI) π, proving knowledge of one of the secret keys sk^b. R then sets the parameters of the commitment as $pk = (\mathsf{pk}^0, \mathsf{pk}^1, \pi)$. NIWI proofs of knowledge can be constructed from any Σ-protocol in the gRO model using the transformation of [14,34]. A self-contained description of this technique is deferred to the full version. For concrete efficiency, one can instantiate the trapdoor commitment with Pedersen's commitment. In this case the public keys are of the form $\mathsf{pk}^0 = g_0, h^{\mathsf{trap}_0}$ and $\mathsf{pk}^1 = g_1, h^{\mathsf{trap}_1}$, and proving knowledge of the secret key sk^b corresponds to simply prove knowledge of the exponent trap_b. The parameters pk so generated are used for all subsequent commitments.

To commit to a message m, S first splits m as m^0, m^1 s.t. $m = m^0 \oplus m^1$. Then S computes commitments C^0 and C^1 to m^0 and m^1 as follows.

First, commit to m^b, i.e., $c^b_{\mathsf{msg}} = \mathsf{TCom}(\mathsf{pk}^b, m^b)$ using the trapdoor commitment scheme. Then, S queries gRO with the *opening* of c^b_{msg}, and receives an answer a^b_C. At this point, S commits to the answer a^b_C using again TCom, resulting in commitment c^b_{ro}. The commitment C^b will then consist of the pair $C^b = (c^b_{\mathsf{msg}}, c^b_{\mathsf{ro}})$. Intuitively, the commitment is extractable in the gRO model since S is forced to commit to the *answer* of gRO, and hence the extractor can simply extract the decommitments by observing the queries to gRO, and checking that there exists at least a query q that corresponds to a valid opening of c^b_{msg}.

In the decommitment phase S simply opens the two commitments, and R checks that c^b_{ro} is indeed the commitment of the answer of gRO, on input the decommitment of c^b_{msg}. Note that the receiver R does not reveal any trapdoor (as she already proved knowledge of one of them), and therefore the same pk can be used again for a new commitment. To equivocate, the simulator simply extracts the trapdoor sk^b from NIWI proof π (recall that π is straight-line extractable in the gRO model), and uses it to equivocate commitments $c^b_{\mathsf{msg}}, c^b_{\mathsf{ro}}$.

We describe the protocol in more details below. Further details proving knowledge of a Pedersen commitment trapdoor are given in the full version.

Protocol UCCom. A New UC Commitment in the gRO Model. Let sid denote the session identifier.

Setup Phase $\langle \mathsf{Gen}(C(1^\lambda), R(1^\lambda)) \rangle$.

- R computes $(\mathsf{pk}^0, \mathsf{sk}^0) \leftarrow \mathsf{TCGen}(1^\lambda)$, and $(\mathsf{pk}^1, \mathsf{sk}^1) \leftarrow \mathsf{TCGen}(1^\lambda)$. R computes a NIWI proof of knowledge π for proving knowledge of sk^d for a random bit d. R sends $pk = (\mathsf{pk}^0, \mathsf{pk}^1, \pi)$ to C.
- If π is accepting, C records parameters $\mathsf{pk}^0, \mathsf{pk}^1$.

i-th Commitment Phase $\mathsf{Com}(\mathsf{sid}, i, m)$: C randomly picks m^0, m^1 such that $m = m^0 \oplus m^1$. Then for each m^b:

- Commit to m^b: $(c^b_{\mathsf{msg}}, d^b_{\mathsf{msg}}) \leftarrow \mathsf{TCom}(\mathsf{pk}, m^b)$.
- Query gRO on input $(\mathsf{sid}, i, \mathsf{S}\|m^b\|d^b_{\mathsf{msg}}\|s^b)$, where $s^b \xleftarrow{\$} \{0,1\}^\lambda$. Let a_C be the answer of gRO.
- Commit to a^b_C: $(c^b_{\mathsf{ro}}, d^b_{\mathsf{ro}}) \leftarrow \mathsf{TCom}(\mathsf{pk}, a_C)$. Set $C^b = (c^b_{\mathsf{msg}}, c^b_{\mathsf{ro}})$.

Send $C = [C^0, C^1]$ to R.

i-th **Decommitment Phase:** Dec(*state*)

- S sends $D = [m^b, d^b_{msg}, d^b_{ro}, a^b_C, s^b]$ to R for each $b \in \{0,1\}$.
- Ver(pk, D). The receiver R accepts m as the decommitted value iff *all* of the following verifications succeed: (a) (b) TRec($c^b_{ro}, a^b_C, d^b_{ro}$) = 1, (c) a^b_C = gRO(sid, C$\|m^b\|d^b_{msg}\|s^b$), (d) TRec($c^b_{msg}, m^b, d^b_{msg}$) = 1.

Theorem 6. *Assume that* (TCGen, TVer, TCom, TRec, TEquiv) *is a Trapdoor commitment scheme, that on-line extractable NIWI proof of knowledge exist in the* gRO *model, then* UCCom *is UC-secure commitment scheme in the* gRO *model.*

Proof (Sketch).

Case R^* is Corrupted. We show that there exists a simulator, that for convenience we call SimCom, that is able to equivocate any commitment. The strategy of SimCom is to first extract the trapdoor of sk^b for some bit b from the NIWI π, then use the trapdoor sk^b to appropriately equivocate the commitment C^b. The key point is that, because $m = m^0 \oplus m^1$, equivocating one share m^b will be sufficient to open to any message m. The completed description of the simulator SimCom is provided below.

Simulator SimCom

To generate a simulated commitment under parameters *pk* and sid:

- Parse *pk* as pk^0, pk^1, π. Extract sk$_b$ from π (for some $b \in \{0,1\}$) running the extractor associated to the NIWI protocol, and by observing queries to gRO for session sid. If the extractor fails, output Abort and halt.
- Compute $c^{\bar{b}}_{msg}, d^{\bar{b}}_{msg} = \mathsf{TCom}(\mathsf{pk}^{\bar{b}}, m^{\bar{b}})$, where $m^{\bar{b}}$ is a random string.
- Query gRO and obtain: $a^{\bar{b}}_C = \mathsf{gRO}(\mathsf{sid}, \mathsf{C}\|m^{\bar{b}}\|d^{\bar{b}}_{msg}\|s^{\bar{b}})$.
- Compute $c^{\bar{b}}_{ro}, d^b_{ro} = \mathsf{TCom}(\mathsf{pk}^{\bar{b}}, a^{\bar{b}}_C)$.
- Compute c^b_{msg}, c^b_{ro} as commitments to 0.

To equivocate the simulated commitment to a value m:

- Compute $m^b = m \oplus m^{\bar{b}}$. Compute $d^b_{msg} = \mathsf{TEquiv}(\mathsf{sk}^b, c_{msg}, m^b)$.
- Query gRO and obtain: $a^b_C = \mathsf{gRO}(\mathsf{sid}, \mathsf{C}\|mb\|d^b_{msg}\|s^b)$. Compute $d^b_{ro} = \mathsf{TEquiv}(\mathsf{sk}^b, c^b_{ro}, a^b_C)$.
- Output $(d^e_{msg}, d^e_{ro}, s^e)$ for $e = 0, 1$.

Indistinguishability. The difference between the transcript generated by SimCom and an honest S is in the fact that SimCom equivocates the commitments using the trapdoor extracted form *pk*, and that SimCom will abort if such trapdoor is not extracted. Indistinguishability then follows from the extractability property of π (which holds unconditionally in the gRO model) and due to the trapdoor property of the underlying trapdoor commitment scheme.

Case S^* is Corrupted. We show that there exists a simulator, that we denote by SimExt, that is able to extract the messages m^0, m^1 already in the commitment phase, by just observing the queries made to $\mathcal{G}_{gRO}$ (with SID sid).

The extraction procedure follows identically the extraction procedure of the simulator shown in [12]. We describe SimExt in details below.
SimExt(sid, $pk, C = [C^0, C^b]$).

- Parse $pk = \mathsf{pk}^0, \mathsf{pk}^1, \pi$. If π is not accepting halt. Else, parse $C^b = c^b_{\mathsf{msg}}, c^b_{\mathsf{ro}}$ for $b = 0, 1$. Let $\mathcal{Q}_{\mathsf{sid}}$ be the list of queries made to gRO by any party.
- For $b = 0, 1$. If there exists a query q of the form $q = \mathsf{sid}\|\text{'C'}\|m^b\|d^b_{\mathsf{msg}}\|s^b$ such that $\mathsf{TRec}(c^b_{\mathsf{msg}}, m^b, d^b_{\mathsf{msg}}) = 1$, the record m^b, otherwise set $m^b = \perp$. Set $m = m^0 \oplus m^1$.
- Send $(\mathsf{commit}, \mathsf{sid}, \text{'C'}, \text{'R'}, m')$ to $\mathcal{F}_{\mathsf{tcom}}$.
- *Decommitment phase:* If the openings is not accepting, halt. Else, let m^* be the valid messages obtained from the decommitment. If $m^* = m$, it sends the message $(\mathsf{decommit}, \mathsf{sid}, \text{'C'}, \text{'R'})$ to the trusted party. Otherwise, if $m^* \neq m$, then output **Abort** and halt.

Indistinguishability. The indistinguishability of the output of SimExt follows from the witness indistinguishability property of the proof system, and the biding property of the trapdoor commitment.

Due to the WI of π, any S^* cannot extract secret key sk_b used by R. Thus, if SimExt fails in extracting the correct opening, it must be that S^* is breaking the binding of commitment scheme. In such a case we can build an adversary $\mathcal{A}$ that can use S^* and the queries made by S^* to gRO to extract two openings for commitment $c^{\bar{b}}_{\mathsf{msg}}, c^{\bar{b}}_{\mathsf{ro}}$.

6   Security Proof

Theorem 7. *If* UCCom $=$ (Gen, Com, Dec, Ver) *is a UC-secure commitment scheme, with non-interactive commitment and decommitment phase,* (Code, Decode) *is an encoding scheme with parameters* (d, t, κ), (RamInit, RamEval, S_{oram}) *is a secure ORAM scheme, then protocol* $\Pi = (\Pi.\mathsf{Setup}, \Pi.\mathsf{Proof})$ *(Figs. 4 and 5), securely realizes* $\mathcal{F}_{zk}$ *functionality (Fig. 2).*

Proof. The proof follows from Lemma 9 and Lemma 8.

6.1   Case P is Corrupted

Lemma 8. *If* UCCom *is UC-secure in the* gRO *model,* (Code, Decode) *is an encoding scheme with parameters* (d, t, κ), (RamInit, RamEval, S_{oram}) *is a secure ORAM scheme. Then, protocol* $\Pi = (\Pi.\mathsf{Setup}, \Pi.\mathsf{Proof})$ *in Fig. 5 and Fig. 4 securely realizes* $\mathcal{F}_{zk}$ *in the* $\mathcal{F}^{C_1, C_2}_{check}$ *(resp.,* $\mathcal{F}^{C}_{check}$) *hybrid model, in presence of malicious PPT prover* P^*.

Proof. The proof consists in two step. We first describe a simulator Sim for the malicious P^*. Then, we prove that the output of the simulator is indistinguishable from the output of the real execution.

Simulator *Intuition*. At high level, the simulator Sim proceeds in two steps. In the setup phase, Sim extracts the value committed in the nodes of the Merkle Tree. Recall that a leaf N_i of the tree is just a concatenation of commitments of shares of the memory block $\widehat{M}[i]$ (indeed, $N_i = CX_i = (cx_{i,1}, \ldots, cx_{i,\kappa})$). Sim is able to extract all commitments in CX_i by observing the queries made to gRO that are consistent with the published root h. Moreover, given such commitments, Sim is able to further extract the shares by exploiting the extractability property of UCCom (which, in turns, uses the observability of gRO.) Therefore, by the end of the setup phase, Sim has extracted shares for each block $i \in [m]$, and reconstructed "its view" of the memory, that we denote by $\widehat{M}^\star$, as well as the initial ORAM state st. Given $\widehat{M}^\star$, Sim will then be able to determine the memory $M^\star$, by running extractor $\mathsf{RamExtract}(\widehat{M}, st)$, and sends it to the ideal functionality $\mathcal{F}_{zk}$.

In the proof phase, the goal of the simulator Sim is to continuously monitor that each computation (each proof) is *consistent* with the memory $M^\star$ initially sent to $\mathcal{F}_{zk}$. Intuitively, the computation is consistent if the memory values input by $P^\star$ in each successful execution of $\mathcal{F}_{check}$ (which are represented in encoded form $X_i = [x_{i,1}, \ldots, x_{i,\kappa}]$), are "consistent" with the memory $M^\star$ that Sim has computed by extracting from the commitments; or more precisely, with the *encoding* of the block memory extracted so far.

Upon the first proof, the simulator will check that the shares of $M[i]$ submitted to $\mathcal{F}_{check}$ agree with the shares for block $M^\star[i]$ extracted in the setup phase. Here agree means that they decode to the same values. (Note that we do not require that all shares agree with the ones that were extracted by Sim, but we required that enough shares agree so that they decode to the same value).

After the first proof, $P^\star$ will also send commitments to the *updated* version of the blocks j touched during the computation. (Precisely, the shares of each block). As in Setup phase, Sim will extract these *new blocks* and update his view of $M^\star$ accordingly. In the next proof then, Sim will check consistency just as in the first proof, but consistency is checked against the newly extracted blocks.

In each proof, when checking consistency, two things can go wrong. Case 1. (Binding/extraction failure) When decommitting to the partial encodings (Step 3 of Fig. 5), $P^\star$ correctly opens values that are different from the ones previously extracted by Sim. If this happens, that $P^\star$ either has broken the extractability property of UCCom or has found a collision in the output of gRO. Thus, due to the security of UCCom, this events happens with negligible probability.

Case 2. (Encoding failure) Assume that the t shares extracted by Sim correspond to the t shares decommitment by $P^\star$, but that among the $\kappa - t$ shares that were not open, there are at least d shares that are different. This means that the values decoded by $\mathcal{F}_{check}$ are inconsistent with the values that are decoded from the extracted shares, which means that the computation in the protocol is taking a path that is inconsistent with the path dictated by the $M^\star$ initially submitted by Sim.

We argue that this events also happen with negligible probability. Indeed, due to the security of $\mathcal{F}_{check}$ we know that the position γ that $P^\star$ will need to decommit are unpredictable to P^*. Thus, the probability that P^* is able to open t consistent shares, while committing to d bad shares is bounded by: $(1 - \frac{d}{\kappa})^t$ which is negligible.

The Algorithm Sim. We now provide a more precise description of the simulator Sim. *Notation.* We use notation X^* to denote the fact that this is the "guess" that Sim has on the value X after extracting from the commitment of CX. During the proof phase, Sim will keep checking if this guess is consistent with the actual values that P^* is giving in input to $\mathcal{F}_{check}$.

Let SimExt be the extractor associated to UCCom and outlined in Sect. 5

Setup Phase. Run SimExt for the generation algorithm Gen. Upon receiving commitments: $CS = (cs_1, \ldots, cs_\kappa)$ and root h from P.

1. **(Extract Commitments at the Leaves of Merkle Tree)** For each query made by P^* to gRO $(sid, i\|l, P, C)$, set $CX_i^*[l] = C$ iff $sid' = sid$ and the outputs of gRO along the paths to i are consistent with the root h. This is done by obtaining the list of queries $\mathcal{Q}_{|sid}$ from $\mathcal{G}_{gRO}$. At the end of this phase, Sim has collected commitments $CX_i^*[i]$ that need to be extracted.
2. **(Extract Shares.)** Invoke extractor SimExt on input $(sid, pk, CX_i^*[l])$ for all $i \in [m]$ and $l \in [\kappa]$. Let $X_i^* = (x_{i,1}^*, \ldots, x_{i,\kappa}^*)$ denote the openings extracted by SimExt. Similarly, invoke SimExt on input $(sid, pk, CS[l])$ with for $l \in [\kappa]$ and obtain shares $s_1^*, \ldots, s_\kappa^*$ for the initial state. Note that the extracted values could be $\perp$. Record all such values.
3. **(Decode memory blocks $\widehat{M}^*[i]$)** For each $i \in m$, run $b_i = \mathsf{Decode}(x_{i,1}^*, \ldots, x_{i,\kappa}^*)$. If Decode aborts, then mark $b_i = \perp$. Set block memory: $\widehat{M}^*[i] = b_i$. Similarly, set $st = \mathsf{Decode}(s_1, \ldots, s_\kappa)$.
4. Determine the real memory M^* as follows: $M^* = \mathsf{RamExtract}(\widehat{M}, st)$. Send $(sid, \mathsf{INIT}, M^*)$ to $\mathcal{F}_{zk}$.

l **-proof.** Input to this phase: (Public Input) Statement $\mathcal{R}_l$, x. **Private input for** Sim. For each memory block i, Sim has recorded the most updated shares extracted: $X_i^* = [x_{i,1}^*, \ldots, x_{i,\kappa}^*]$. The first time X_i^* are simply the ones extracted in the setup phase. In the l sub-sequent proof, X_i^* is set to the values extracted from the transcript of the $l - 1$ proof. Similarly, Sim has recorded the extracted encodings of the ORAM state $S^* = [s_1^*, \ldots, s_\kappa^*]$.

1. Upon receiving commitments CY_i ($\forall i \in \mathsf{access}(\mathcal{I})$); CS', and new root h'. Run SimExt on inputs (sid, pk, CY_i) and obtain encoding Y_i^*, and on input (sid, CS') to obtain the encoding of the ORAM state S'^*.
2. Invoke $\mathsf{Sim}_{\mathcal{F}_{check}}$. If $\mathsf{Sim}_{\mathcal{F}_{check}}$ aborts, then abort and output $\mathcal{F}_{check}$ **failure!!**. If $\mathsf{Sim}_{\mathcal{F}_{check}}$ halts, then halt.
   Else, obtain P^*'s inputs to $\mathcal{F}_{check}$: $W = (w\ S, S', \mathbf{X}, \mathbf{Y})$. Recall that $\mathbf{X} = \{X_1, \ldots, X_{|\mathsf{read}(\mathcal{I})|}\}$ and $\mathbf{Y} = \{Y_1, \ldots, Y_{|\mathsf{access}(\mathcal{I})|}\}$, where X_i, Y_j are encodings of blocks in position i and j. Sim records the above values as comparison values for later.

3. Upon receiving decommitments $DX_i[\gamma]$, and authentication paths π_i for $i \in$ read$(\mathcal{I})$; $DY_j[\gamma]$ for $j \in$ access$(\mathcal{I})$ and $DS[\gamma]$, $DS'[\gamma]$. Let $X_i[\gamma], X_j[\gamma], S[\gamma], S'[\gamma]$ the value obtained from the decommitment.

   Perform the verification step as an honest verifier V (Step. 3 of Fig. 5). If any check fails, alt and output the transcript obtained so far. Else, perform the following consistency checks.

   (a) **Check consistency of the commitments stored in the Merkle tree.** If there is exists an i s.t., the commitment CX_i^* extracted in Π.Setup phase, is different from the commitment CX_i opened in the proof phase (with accepting authentication path π_i), then abort and output Collision Failure!!!.

   (b) **Check binding/extraction.** Check that, for all $i \in$ read$(\mathcal{I})$, all γ s.t. $r_\gamma = 1$ $X_i[\gamma] = X_i^*[\gamma]$, and for all $j \in$ access$(\mathcal{T})$ $Y_i[\gamma] = Y_i^*[\gamma]$, and $S[\gamma] = S^*[\gamma]$, $S'[\gamma] = S'[\gamma]$. If not, abort and output Binding Failure!!.

   (c) **Check correct decoding.** Check that, for all $i \in$ read$(\mathcal{I})$, Decode$(X_i^*) =$ Decode(X_i); that for all $j \in$ access$(\mathcal{I})$, Decode$(Y_i^*) =$ Decode(Y_i), and that Decode$(S^*) =$ Decode(S), Decode$(S') \neq$ Decode(S'^*). If any of this check fails, abort and output Decoding Failure!!.

4. Send (PROVE, $sid, \mathcal{R}_l, w$) to $\mathcal{F}_{zk}$.

5. **Update extracted memory and extracted state.** For each $i \in$ access$(\mathcal{I})$: Set $X_i^* = Y_i^*$, and $S^* = S'^*$.[4]

Indistinguishability Proof. The proof is by hybrids arguments. As outlined at the beginning of the section, the crux of the proof is to show that the memory M^* extracted by Sim in Π.Setup, is consistent with all the proofs subsequently provided by P^*. In other words, upon each proof, the updates performed to the memory in the real transcript are consistent with the updates that $\mathcal{F}_{zk}$ performs on the memory M^* sent by Sim in the ideal world.

Recall that, for each proof, Sim continuosly check that the memory blocks used in $\mathcal{F}_{check}$ are consistent with the memory blocks committed (and extracted by Sim). If this consistency is not verified, then Sim will declare failure and abort.

Intuitively, proving that the simulation is succesfull corresponds to prove that the probability that Sim declares failure is negligible. Assuming secure implementation of $\mathcal{F}_{check}$, the above follows directly from the (on-line) extractability of UCCom, the collision resistance of gRO and the d-*distance* property of the encoding scheme. We now proceed with the description of the hybrid arguments.

H_0 (**Real world**). This is the real world experiment. Here Sim runs just like an honest verifier. It outputs the transcript obtained from the executions.

H_1 (**Extracting witness from $\mathcal{F}_{check}$**). In this hybrid experiment Sim deviates from the algorithm of the verifier V in the following way. In the proof phase, Sim

---

[4] Recall that we use notation X_i to denote the in "initial version" of block i (before the proof was computed) while we use notation Y_i to denote the version of block i after the proof. Similarly, S denotes the initial ORAM state (pre-proof), while S' denotes the ORAM state obtained after the proof has been computed.

obtain the witness W used by P^* in $\mathcal{F}_{check}$, and it aborts if it fails in obtaining such inputs. Due to the security of $\mathcal{F}_{check}$ H_1 and H_0 are computationally indistinguishable.

H_2 **(Extracting the leaves of the Merkle Tree).** In this hybrid Sim uses to observability of gRO to obtain commitments $CX_i^\star[l]$ for $i \in [m], l \in [\kappa]$, and it aborts if an (accepting) path π_i, revealed by P^* in the proof phase, lead to a commitment $CX_i[l] \neq CX_i^\star[l]$. This corresponds to the event Collision Failure!!!. Due to the collision resistance property of gRO, probability of event Collision Failure!!! is negligible, and hence, the transcript generated in H_1 and H_2 are statistically close.

H_3 **(Extracting openings from commitments).** In this hybrid Sim invokes SimExt to extract the opening from all commitments. The difference between H_3 and H_2 is that in H_3 Sim aborts every time event Binding Failure!! occurs, which is negligible under the assumption that UCCom is an extractable commitment.

H_4 **(Decoding from extracted shares).** In this hybrid Sim determines each memory block $\widehat{M}^\star[i]$ by running Decode algorithm on the *extracted* shares $X_i^\star$. That is, $\widehat{M}^\star[i]=$Decode$(X_i^\star)$.

Moreover, it checks that all the extracted encodings (i.e., Y_i, S, S') decodes to the same values used in $\mathcal{F}_{check}$ (Step (c) in the algorithm Sim). In this hybrid Sim aborts everytime events Decoding Failure!! happen.

Hence, to prove that experiment H_3 and H_4 are statistically indistinguishable, it is sufficient to prove that: $Pr[Event\ \text{DecodingFailure!!}] = negl(\kappa)$. As we argued in the high-level overview, event Decoding Failure!!happens with probability $(1 - \frac{d}{\kappa})^t$, which is negligible in κ for $t = 1/2\kappa$.

H_5 **(Submit to $\mathcal{F}_{zk}$ the extracted memory $\widehat{M}^\star$) Ideal World.** In this hybrid Sim plays in the ideal world, using the memory $\widehat{M}^\star$ extracted in the Setup phase.

We have proved that the value extracted by Sim are consistent with the values sent in input to $\mathcal{F}_{check}$. (Indeed, we have proved that all the failure events happen with negligible probability). Due to the security of $\mathcal{F}_{check}$ it follows that each proof l, is the a correct computation given the input blocks and the input ORAM state[5]. Due to the above arguments we know that the value sent to $\mathcal{F}_{check}$ are consistent with the memory blocks and ORAM state extracted so far. Putting the two things together, we have that any accepting proof is computed on values that are consistent with the committed values (extracted by Sim), which in turn are generated from the first version of the memory $\widehat{M}^\star$ extracted by Sim. This experiment corresponds to the description of the simulator Sim, proving the lemma.

---

[5] Here we are also using crucially the fact that choosing a bad ORAM state does not effect the correctness of the ORAM computation, but it can only effect the security guarantees for the prover.

6.2 Case V is corrupted

Lemma 9. *If* UCCom *is an equivocal commitment scheme in the* gRO *model,* (Code, Decode) *is an encoding scheme with parameters* $(d, 2k, \kappa)$, *(RamInit, RamEval, S_{oram}) is a secure ORAM scheme. Then, protocol* $\Pi = (\Pi.\text{Setup}, \Pi.\text{Proof})$ *in Fig. 5 and Fig. 4 securely realizes* $\mathcal{F}_{zk}$ *in the* $\mathcal{F}_{check}^{C_1, C_2}$ *(resp.,* $\mathcal{F}_{check}^{C}$*) hybrid model, in presence of malicious PPT verifier* V^*.

Proof Intuition. At high-level, assuming the $\mathcal{F}_{check}$ is securely implemented, the transcript of the verifier simply consists of a set of commitments, and partial encodings for each block memory touched in the computation and the ORAM state. Due to the hiding (in fact equivocability) properties of the commitments as well as the $2k$ hiding property of the encodings, it follows that by looking at $< 2t$ shares, V^* cannot distinguish the correct values of the memory/state from commitments to 0. Moreover, due to the security of ORAM, the access pattern $\mathcal{I}$ disclosed upon each proof, does not reveal any additional information about the memory/ORAM state.

Following this intuition, the simulator for V^* follows a simple procedure. It computes all commitments so that they are equivocal (i.e., it runs procedure SimCom guaranteed by the security property of commitment scheme UCCom). Upon each proof, Sim will run S_{oram} to obtain the access pattern $\mathcal{I}$, and the simulator $\text{Sim}_{\mathcal{F}_{check}}$ to compute the transcript of $\mathcal{F}_{check}$, and to obtain the partial encodings that V^* is expected to see. Finally, Sim will simply equivocate the commitments that must be opened, so that they actually open to the correct partial encodings. The precise description of the simulator Sim is provided below.

The Algorithm Sim.

Setup Phase. Compute all commitments using algorithm SimCom(sid, pk, com, $\cdot$). Compute Merkle tree correctly.

l-**proof.** Upon receiving (PROVE, sid, $\mathcal{R}_l$, 1) from $\mathcal{F}_{zk}$.

1. Run ORAM simulator $S(1^\lambda, |\widehat{M}|)$ and obtain $\mathcal{I}$.
2. Run SimCom to obtain commitments CY_i for all $i \in \text{access}(\mathcal{I})$ and commitments CS'. Update the root of the Merkle Tree accordingly.
3. Run $\text{Sim}_{\mathcal{F}_{check}}$ to obtain the transcript for $\mathcal{F}_{check}$ and obtain the partial encodings: $X_i[\gamma], Y_j[\gamma]$ for $i \in \text{read}(\mathcal{I})$ and $j \in \text{access}(\mathcal{I})$; $S[\gamma], S'[\gamma]$, where γ is such that $r_\gamma = 1$, where r is the verifier's input to $\mathcal{F}_{check}$.
4. Equivocate commitments.
   - For each $i \in \text{read}(\mathcal{I})$, compute $DX_i[\gamma] \leftarrow \text{SimCom}(sid, pk, equiv, CX_i[\gamma], X_i[\gamma])$ Moreover, retrieve path π in the tree.
   - For each $j \in \text{access}(\mathcal{I})$, compute $DY_i[\gamma] \leftarrow \text{SimCom}(sid, pk, equiv, CY_i[\gamma], Y_i[\gamma])$.
   - Compute $DS[\gamma] \leftarrow \text{SimCom}(sid, pk, equiv, DS[\gamma], DS[\gamma])$, $DS'[\gamma] \leftarrow \text{SimCom}(sid, pk, equiv, DS'[\gamma], DS'[\gamma])$.
5. Send decommitments to V^*.

Indistinguishability Proof. The proof is by hybrid arguments. We will move from an experiment where Sim computes the transcript for V^* using real input M and following the algorithm run by P (hybrid H_0), to an hybrid where Sim has not input at all (hybrid H_3).

H_0. This is the real world experiment. Sim gets in input M and simply follows the algorithm of P (Figs. 4 and 5).

H_1 **(Compute Equivocal Commitments using SimCom).** In this hybrid Sim computes commitments using procedure $\mathsf{SimCom}(sid, pk, com, \cdot)$, which requires no inputs, and decommit using $\mathsf{SimCom}(pk, equiv, \cdots)$, using the correct encodings computed from $\widehat{M}$. The difference between H_1 and H_2 is only in the way commitments are computed. Due to the equivocability property of the commitment scheme (in the gRO) model, it follows that H_1 and H_2 are statistically indistinguishable. Note that at this point Sim still uses real values for $\widehat{M}$ to compute the shares that will be later committed, and some of which will be opened.

H_2 (Run $\mathsf{Sim}_{\mathcal{F}_{check}}$). In this hybrid argument Sim computes the transcript of $\mathcal{F}_{check}$ by running simulator $\mathsf{Sim}_{\mathcal{F}_{check}}$, and decommit to the share given in output by $\mathcal{F}_{check}$. Note that $\mathcal{F}_{check}$ will output t encodings for each block memory and state. Note also that, if a block memory was accessed in a previous execution, then t shares of the encodings have been already revealed. For example the encoding of the final state $S'[1], \ldots, S'[\kappa]$, which is the output state in an execution ℓ, will the the encoding used as initial state in proof $\ell + 1$. This means that for each encoding, the adversary R^* collects $2k$ partials encodings. Due to the security of $\Pi_{\mathcal{F}_{check}}$, and to the $2k$ hiding property of the encoding scheme, hybrids H_2 and H_1 are computationally indistinguishable.

H_3 **(Use ORAM simulator S_{oram}).** In this hybrid Sim will replace executions of RamInit and RamEval with S_{oram}. This is possible because the actual values computed by RamInit and RamEval are not used anywhere at this point. Due to the statistical security of (RamInit, RamEval, S_{oram}) hybrids H_3 and H_4 are statistically instinguishable. Note that in this experiment the actual memory M is not used anywhere. This experiment corresponds to the description of the simulator Sim, proving the lemma.

References

1. libSNARK: a C++ library for zkSNARK proofs. https://github.com/scipr-lab/libsnark
2. Afshar, A., Hu, Z., Mohassel, P., Rosulek, M.: How to efficiently evaluate RAM programs with malicious security. In: Oswald, E., Fischlin, M. (eds.) EUROCRYPT 2015. LNCS, vol. 9056, pp. 702–729. Springer, Heidelberg (2015). doi:10.1007/978-3-662-46800-5_27

3. Bellare, M., Hoang, V.T., Rogaway, P.: Foundations of garbled circuits. In: Yu, T., Danezis, G., Gligor, V.D., (eds.) ACM CCS 2012, pp. 784–796. ACM Press, October 2012
4. Ben-Sasson, E., Chiesa, A., Gabizon, A., Virza, M.: Quasi-linear size zero knowledge from linear-algebraic PCPs. In: Kushilevitz, E., Malkin, T. (eds.) TCC 2016. LNCS, vol. 9563, pp. 33–64. Springer, Heidelberg (2016). doi:10.1007/978-3-662-49099-0_2
5. Ben-Sasson, E., Chiesa, A., Garman, C., Green, M., Miers, I., Tromer, E., Virza, M., Zerocash: Decentralized anonymous payments from bitcoin. In: 2014 IEEE Symposium on Security and Privacy, pp. 459–474. IEEE Computer Society Press, May 2014
6. Ben-Sasson, E., Chiesa, A., Genkin, D., Tromer, E.: On the concrete efficiency of probabilistically-checkable proofs. In: Boneh, D., Roughgarden, T., Feigenbaum, J. (eds.) 45th ACM STOC, pp. 585–594. ACM Press, June 2013
7. Ben-Sasson, E., Chiesa, A., Genkin, D., Tromer, E., Virza, M.: SNARKs for C: verifying program executions succinctly and in zero knowledge. In: Canetti, R., Garay, J.A. (eds.) CRYPTO 2013. LNCS, vol. 8043, pp. 90–108. Springer, Heidelberg (2013). doi:10.1007/978-3-642-40084-1_6
8. Ben-Sasson, E., Chiesa, A., Tromer, E., Virza, M.: Scalable zero knowledge via cycles of elliptic curves. In: Garay, J.A., Gennaro, R. (eds.) CRYPTO 2014. LNCS, vol. 8617, pp. 276–294. Springer, Heidelberg (2014). doi:10.1007/978-3-662-44381-1_16
9. Ben-Sasson, E., Kaplan, Y., Kopparty, S., Meir, O., Stichtenoth, H.: Constant rate PCPs for circuit-SAT with sublinear query complexity. In 54th FOCS, pp. 320–329. IEEE Computer Society Press, October 2013
10. Bitansky, N., Canetti, R., Chiesa, A., Tromer, E.: Recursive composition and bootstrapping for SNARKS and proof-carrying data. In: Boneh, D., Roughgarden, T., Feigenbaum, J. (eds.) 45th ACM STOC, pp. 111–120. ACM Press, June 2013
11. Bitansky, N., Chiesa, A., Ishai, Y., Paneth, O., Ostrovsky, R.: Succinct non-interactive arguments via linear interactive proofs. In: Sahai, A. (ed.) TCC 2013. LNCS, vol. 7785, pp. 315–333. Springer, Heidelberg (2013). doi:10.1007/978-3-642-36594-2_18
12. Canetti, R., Jain, A., Scafuro, A.: Practical UC security with a global random oracle. In: Ahn, G.-J., Yung, M., Li, N. (eds.) ACM CCS 2014, pp. 597–608. ACM Press, November (2014)
13. Costello, C., Fournet, C., Howell, J., Kohlweiss, M., Kreuter, B., Naehrig, M., Parno, B., Zahur, S.: Geppetto: versatile verifiable computation. In: 2015 IEEE Symposium on Security and Privacy, pp. 253–270. IEEE Computer Society Press, May 2015
14. Fischlin, M.: Communication-efficient non-interactive proofs of knowledge with online extractors. In: Shoup, V. (ed.) CRYPTO 2005. LNCS, vol. 3621, pp. 152–168. Springer, Heidelberg (2005). doi:10.1007/11535218_10
15. Garg, S., Lu, S., Ostrovsky, R.: Black-box garbled RAM. In: Guruswami, V. (ed.) 56th FOCS, pp. 210–229. IEEE Computer Society Press, October 2015
16. Garg, S., Lu, S., Ostrovsky, R., Scafuro, A.: Garbled RAM from one-way functions. In: Servedio, R.A., Rubinfeld, R., (eds.) 47th ACM STOC, pp. 449–458. ACM Press, June 2015
17. Gennaro, R., Gentry, C., Parno, B., Raykova, M.: Quadratic span programs and succinct NIZKs without PCPs. In: Johansson, T., Nguyen, P.Q. (eds.) EUROCRYPT 2013. LNCS, vol. 7881, pp. 626–645. Springer, Heidelberg (2013). doi:10.1007/978-3-642-38348-9_37

18. Gentry, C., Halevi, S., Lu, S., Ostrovsky, R., Raykova, M., Wichs, D.: Garbled RAM revisited. In: Nguyen, P.Q., Oswald, E. (eds.) EUROCRYPT 2014. LNCS, vol. 8441, pp. 405–422. Springer, Heidelberg (2014). doi:10.1007/978-3-642-55220-5_23

19. Giacomelli, I., Madsen, J., Orlandi, C.: ZKBoo: faster zero-knowledge for Boolean circuits. In: 25th USENIX Security Symposium (USENIX Security 2016) (2016)

20. Goldreich, O., Micali, S., Wigderson, A.: How to play any mental game or a completeness theorem for protocols with honest majority. In: Aho, A. (ed.) 19th ACM STOC, pp. 218–229. ACM Press, May 1987

21. Goldreich, O., Ostrovsky, R.: Software protection and simulation on oblivious RAMs. J. ACM (JACM) **43**, 431–473 (1996)

22. Gordon, S.D., Katz, J., Kolesnikov, V., Krell, F., Malkin, T., Raykova, M., Vahlis, Y.: Secure two-party computation in sublinear (amortized) time. In: Yu, T., Danezis, G., Gligor, V.D. (eds.) ACM CCS 2012, pp. 513–524. ACM Press, October 2012

23. Goyal, V., Ostrovsky, R., Scafuro, A., Visconti, I.: Black-box non-black-box zero knowledge. In: Shmoys, D.B. (ed.) 46th ACM STOC, pp. 515–524. ACM Press, May/June 2014

24. Hu, Z., Mohassel, P., Rosulek, M.: Efficient zero-knowledge proofs of non-algebraic statements with sublinear amortized cost. In: Gennaro, R., Robshaw, M. (eds.) CRYPTO 2015. LNCS, vol. 9216, pp. 150–169. Springer, Heidelberg (2015). doi:10.1007/978-3-662-48000-7_8

25. Ishai, Y., Kushilevitz, E., Ostrovsky, R., Sahai, A.: Zero-knowledge from secure multiparty computation. In: Johnson, D.S., Feige, U. (eds.) 39th ACM STOC, pp. 21–30. ACM Press, June 2007

26. Jawurek, M., Kerschbaum, F., Orlandi, C.: Zero-knowledge using garbled circuits: how to prove non-algebraic statements efficiently. In: Sadeghi, A.-R., Gligor, V.D., Yung, M. (eds.) ACM CCS 2013, pp. 955–966. ACM Press, November 2013

27. Kilian, J.: A note on efficient zero-knowledge proofs and arguments (extended abstract). In: 24th ACM STOC, pp. 723–732. ACM Press, May 1992

28. Kosba, A.E., Zhao, Z., Miller, A., Qian, Y., Chan, T.H., Papamanthou, C., Pass, R., Shelat, A., Shi, E.: How to use SNARKS in universally composable protocols. IACR Cryptology ePrint Archive 2015:1093 (2015)

29. Lu, S., Ostrovsky, R.: Distributed oblivious RAM for secure two-party computation. In: Sahai, A. (ed.) TCC 2013. LNCS, vol. 7785, pp. 377–396. Springer, Heidelberg (2013). doi:10.1007/978-3-642-36594-2_22

30. Micali, S.: CS proofs (extended abstracts). In: 35th FOCS, pp. 436–453. IEEE Computer Society Press, November 1994

31. Ostrovsky, R.: Efficient computation on oblivious RAMs. In: 22nd ACM STOC, pp. 514–523. ACM Press, May 1990

32. Ostrovsky, R., Scafuro, A., Venkitasubramanian, M.: Resettably Sound zero-knowledge arguments from OWFs - the (semi) black-box way. In: Dodis, Y., Nielsen, J.B. (eds.) TCC 2015. LNCS, vol. 9014, pp. 345–374. Springer, Heidelberg (2015). doi:10.1007/978-3-662-46494-6_15

33. Parno, B., Howell, J., Gentry, C., Raykova, M., Pinocchio: nearly practical verifiable computation. In: 2013 IEEE Symposium on Security and Privacy, pp. 238–252. IEEE Computer Society Press, May 2013

34. Pass, R.: On deniability in the common reference string and random oracle model. In: Boneh, D. (ed.) CRYPTO 2003. LNCS, vol. 2729, pp. 316–337. Springer, Heidelberg (2003). doi:10.1007/978-3-540-45146-4_19

35. Pedersen, T.P.: Non-interactive and information-theoretic secure verifiable secret sharing. In: Feigenbaum, J. (ed.) CRYPTO 1991. LNCS, vol. 576, pp. 129–140. Springer, Heidelberg (1992). doi:10.1007/3-540-46766-1_9
36. Stefanov, E., Dijk, M., Shi, E., Fletcher, C.W., Ren, L., Yu, X., Devadas, S.: Path ORAM: an extremely simple oblivious RAM protocol. In: Sadeghi, A.-R., Gligor, V.D., Yung, M. (eds.) ACM CCS 2013, pp. 299–310. ACM Press, November 2013

Side-Channel Attacks and Countermeasures

Parallel Implementations of Masking Schemes and the Bounded Moment Leakage Model

Gilles Barthe[1], François Dupressoir[2], Sebastian Faust[3], Benjamin Grégoire[4], François-Xavier Standaert[5(✉)], and Pierre-Yves Strub[6]

[1] IMDEA Software Institute, Pozuelo de Alarcón, Spain
[2] University of Surrey, Guildford, UK
[3] Ruhr Universität Bochum, Bochum, Germany
[4] Inria Sophia-Antipolis – Méditerranée, Valbonne, France
[5] Université Catholique de Louvain, Louvain-la-Neuve, Belgium
fstandae@uclouvain.be
[6] Ecole Polytechnique, Palaiseau, France

Abstract. In this paper, we provide a necessary clarification of the good security properties that can be obtained from parallel implementations of masking schemes. For this purpose, we first argue that (i) the probing model is not straightforward to interpret, since it more naturally captures the intuitions of serial implementations, and (ii) the noisy leakage model is not always convenient, e.g. when combined with formal methods for the verification of cryptographic implementations. Therefore we introduce a new model, the bounded moment model, that formalizes a weaker notion of security order frequently used in the side-channel literature. Interestingly, we prove that probing security for a serial implementation implies bounded moment security for its parallel counterpart. This result therefore enables an accurate understanding of the links between formal security analyses of masking schemes and experimental security evaluations based on the estimation of statistical moments. Besides its consolidating nature, our work also brings useful technical contributions. First, we describe and analyze refreshing and multiplication algorithms that are well suited for parallel implementations and improve security against multivariate side-channel attacks. Second, we show that simple refreshing algorithms (with linear complexity) that are not secure in the continuous probing model are secure in the continuous bounded moment model. Eventually, we discuss the independent leakage assumption required for masking to deliver its security promises, and its specificities related to the serial or parallel nature of an implementation.

1 Introduction

The masking countermeasure is currently the most investigated solution to improve security against power-analysis attacks [26]. It has been analyzed theoretically in the so-called probing and noisy leakage models [42,53], and based on a large number of case studies, with various statistical tools (e.g. [16,60] for

© International Association for Cryptologic Research 2017
J.-S. Coron and J.B. Nielsen (Eds.): EUROCRYPT 2017, Part I, LNCS 10210, pp. 535–566, 2017.
DOI: 10.1007/978-3-319-56620-7_19

non-profiled and profiled attacks, respectively). Very briefly summarized, state-of-the-art masking schemes are currently divided in two main trends: on the one hand, software-oriented masking, following the initial work of Prouff and Rivain [56]; on the other hand hardware-oriented masking (or *threshold implementations*) following the inital work of Nikova, Rijmen and Schläffer [50].

At CRYPTO 2015, Reparaz et al. highlighted interesting connections between the circuit constructions in these two lines of works [55]. Looking at these links, a concrete difference remains between software- and hardware-oriented masking schemes. Namely, the (analyses of the) first ones usually assume a serial manipulation of the shares while the (implementations of the) second ones encourage their parallel manipulation.[1] Unfortunately, the probing leakage model, that has led to an accurate understanding of the security guarantees of software-oriented masking schemes [31], is not directly interpretable in the parallel setting. Intuitively, this is because the parallel manipulation of the shares reveals information on all of them, e.g. via their sum, but observing sums of wires is not permitted in the probing model. As will be clear in the following, this does not limit the concrete relevance of the probing model. Yet, it reveals a gap between the level of theoretical understanding of serial and parallel masked implementations.

1.1 Our Contribution

Starting from the observation that parallelism is a key difference between software and hardware-oriented masking, we introduce a new model – the bounded moment model – that allows rigorous reasoning and efficient analyses of parallel masked implementations. In summary, the bounded moment model can be seen as the formal counterpart to the notion of security against higher-order attacks [54,63], just as the noisy leakage model [53] is the formal counterpart to information theoretic leakage metrics such as introduced in [59]. It allows us to extend the consolidating work of [55] and to obtain the following results:

- First, we exhibit a natural connection between the probing model and the bounded moment model. More precisely, we prove that security in the probing model for a serial implementation implies security in the bounded moment model for the corresponding parallel implementation.
- Next, we propose regular refreshing and multiplication algorithms suitable for parallel implementations. Thanks to parallelism, these algorithms can be implemented in linear time, with the same memory requirements as a serial implementation (since masking requires to store all the shares anyway). Note that the refreshing algorithm is particularly appealing for combination with key-homomorphic primitives (e.g. inner product based [36]), since it allows them to be masked with linear (time and randomness) complexity. As for the multiplication algorithm, its linear execution time also provides improved security against multivariate (aka horizontal) side-channel attacks [17].

---

[1] This division between hardware and software is admittedly oversimplifying in view of the improved capabilities of modern microprocessors to take advantage of parallelism. So the following results in fact also apply to parallel software computing.

– Third, we exhibit the concrete separation between the probing model and the bounded moment model. For this purpose, we provide simple examples from the literature on leakage squeezing and low-entropy masking schemes showing that (for linear leakage funtions) it is possible to have a larger security order in the bounded moment model than in the probing model [25, 41]. More importantly, we show that our simple refreshing algorithm is insecure in the probing model against adversaries taking advantage of continuous leakage, while it remains secure against such (practically relevant) adversaries in the bounded moment model. This brings a theoretical foundation to the useful observation that simple refreshing schemes that are sometimes considered in practice (e.g. adding shares that sum to zero) do not lead to devastating attacks when used to refresh an immutable secret state (e.g. a block cipher key), despite their lack of security in the continuous probing model. Note that the latter result is also of interest for serial implementations.

– Finally, we illustrate our results with selected case studies, and take advantage of them to discuss the assumption of independent leakages in side-channel attacks (together with its underlying physical intuitions).

1.2 Related Work

Serial Masking and Formal Methods. The conceptual simplicity of the probing model makes it an attractive target for automated verification. Recognizing the close similarities between information-flow policies and security in the probing model, Moss, Oswald, Page and Turnstall [49] build a masking compiler that takes as input an unprotected program and outputs an equivalent program that resists first-order DPA. Their compiler performs a type-based analysis of the input program and iteratively transforms the program when encountering a typing error. Aiming for increased generality, Bayrak, Regazzoni, Novo and Ienne [18] propose a SMT-based method for analyzing statistical independence between secret inputs and intermediate computations, still in the context of first-order DPA. In a series of papers starting with [38], Eldib, Wang and Schaumont develop more powerful SMT-based methods for synthesizing masked implementations or analyzing the security of existing masked implementations. Their approach is based on a logical characterization of security at arbitrary orders in the probing model. In order to avoid the "state explosion" problem, which results from looking at higher-orders and from the logical encoding of security in the probing model, they exploit elaborate methods that support incremental verification, even for relatively small orders. A follow-up by Eldib and Wang [37] extends this idea to synthesize masked implementations fully automatically. Leveraging the connection between probabilistic information flow policies and relational program logics, Barthe, Belaïd, Dupressoir, Fouque, Grégoire and Strub [13] introduce another approach based on a domain-specific logic for proving security in the probing model. Like Eldib, Wang and Schaumont, their method applies to higher orders. Interestingly, it achieves practicality at orders up to four for multiplications and S-boxes. In a complementary line of work, Belaïd, Benhamouda, Passelegue, Prouff, Thillard and Vergnaud [19] develop an automated tool for finding probing attacks on implementations and use it to discover optimal (in randomness complexity) implementations of multiplication at order 2, 3, and 4 (with 2, 4, and 5 random bits). They also propose a multiplication for arbitrary orders, requiring $\frac{d^2}{4} + d$ bits of randomness to achieve security at order d.

All these works focus on the usual definition of security in the probing model. In contrast, Barthe, Belaïd, Dupressoir, Fouque and Grégoire introduce a stronger notion of security, called strong non-interference (or SNI), which enables compositional verification of higher-order masking schemes [14], and leads to much improved capabilities to analyze large circuits (i.e. full algorithms, typically). Similar to several other security notions for the probing model, strong non-interference is qualitative, in the sense that a program is either secure or insecure. Leaving the realm of qualitative notions, Eldib, Wang, Taha, and Schaumont [39] consider a quantitative relaxation of the usual definition of (probing) security, and adapt their tools to measure the quantitative masking strength of an implementation. Their definition is specialized to first-order moments, but the connections with the bounded moment model are evident, and it would be interesting to explore generalizations of their work to our new model.

Threshold and Parallel Implementations. The inital motivation of Nikova, Rijmen and Schläffer was the observation that secure implementations of masking in hardware are challenging, due to the risk of glitches recombining the shares [45]. Their main idea to prevent this issue is to add a condition of non-completeness to the masked computations (i.e. ensure that any combinatorial circuit never takes all shares as input). Many different works have confirmed the practical relevance of this additional requirement, making it the de facto standard for hardware masking (see [20–22,48,52] for a few examples). Our following results are particularly relevant to threshold implementations since (i) in view of their hardware specialization, they encourage a parallel manipulation of the shares, (ii) most of their security evaluations so far were based on the estimation of statistical moments that we formalize with the bounded moment model, and (iii) their higher-order implementations suggested in [55] and recently analyzed in [28] exploit the simple refreshing scheme that we study in Sect. 8.2.

Noisy Leakage Model. Note that the noisy leakage model in [53] also provides a natural way to capture parallel implementations (and in fact a more general one: see Fig. 7 in conclusions). Yet, this model is not always convenient when exploiting the aforementioned formal methods. Indeed, these tools benefit greatly from the simplicity of the probing model in order to analyze complex implementations, and hardly allow the manipulation of noisy leakages. In this respect, the bounded moment model can be seen as a useful intermediate (i.e. bounded moment security can be efficiently verified with formal methods, although its verification naturally remains slower than probing security).

Eventually, we believe it is fundamentally interesting to clarify the connections between the mainstream (probing and) noisy leakage model(s) and concrete evaluation strategies based on the estimation of statistical moments. In this respect, it is the fact that bounded moment security requires a weaker independence condition than probing security that enables us to prove the simple refreshing of Sect. 8.2, which is particularly useful in practice, especially compared to previous solutions for efficient refreshing algorithms such as [9]. Here as well, directly dealing with noisy leakages would be more complex.

2 Background

In this section, we introduce our leakage setting for serial and parallel implementations. Note that for readability, we keep the description of our serial and parallel computing models informal, and defer their definition to Sect. 5.

2.1 Serial Implementations

We start from the description of leakage traces in [32], where y is an n-bit sensitive value manipulated by a leaking device. Typically, it could be the output of an S-box computation such that $y = S(x \oplus k)$ with n-bit plaintext and key words x and k. Let $y_1, y_2, \ldots, y_d$ be the d shares representing y in a Boolean masking scheme (i.e. $y = y_1 \oplus y_2 \oplus \ldots \oplus y_d$). In a side-channel attack, the adversary is provided with some information (or leakage) on each share. Concretely, the type of information provided highly depends on the type of implementation considered. For example, in a serial implementation, we typically have that each share is manipulated during a different "cycle" c so that the number of cycles in the implementation equals the number of shares, as in Fig. 1(a). The leakage in each cycle then takes the form of a random variable L_c that is the output of a leakage function L_c, which takes y_c and a noise variable R_c as arguments:

$$L_c = \mathsf{L}_c(y_c, R_c), \quad \text{with } 1 \leq c \leq d. \tag{1}$$

That is, each subtrace L_c is a vector, the elements of which represent time samples. When accessing a single sample τ, we use the notation $L_c^\tau = \mathsf{L}_c^\tau(y_c, R_c)$. From this general setup, a number of assumptions are frequently used in the literature on side-channel cryptanalysis. We consider the following two (also considered in [32]). First, we assume that the leakage vectors L_c are independent random variables. This a strict requirement for masking proofs to hold and will be specifically discussed in Sect. 9. Second, and for convenience only, we assume that the leakage functions are made of a deterministic part $G_c(y_c)$ and additive noise R_c so that $L_c = \mathsf{L}_c(y_c, R_c) \approx G_c(y_c) + R_c$. Note that the + symbol here denotes the addition in $\mathbb{R}$ (while $\oplus$ denotes a bitwise XOR).

2.2 Parallel Implementations

We now generalize the previous serial implementation to the parallel setting. In this case, the main difference is that several shares can be manipulated in the same cycle. For example, the right part of Fig. 1 shows the leakage corresponding

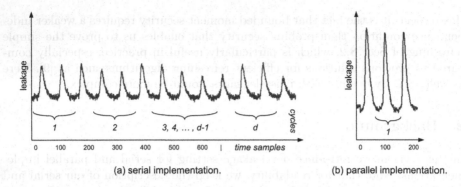

(a) serial implementation.   (b) parallel implementation.

Fig. 1. Leakage trace of d-shared secret.

to a fully parallel implementation where all the shares are manipulated in a single cycle. As a result, we have a single leakage vector $\boldsymbol{L}_1 = \mathsf{L}(y_1, y_2 \ldots, y_d, \boldsymbol{R}_1)$. More generally, we will consider N-cycle parallel implementations such that for each cycle c $(1 \leq c \leq N)$, we define the set of shares that are manipulated during the cycle as $\mathcal{Y}_c$, and the number of shares in a set $\mathcal{Y}_c$ as n_c. This means that a masked implementation requires at least that the union of these sets equals $\{y_1, y_2, \ldots, y_d\}$, i.e. all the shares need to be manipulated at least once. This model of computation is a generalization of the previous one since the serial implementation in the left part of the figure is simply captured with the case $N = d$ and, for every c, $n_c = 1$. As previously mentioned, the highly parallel implementation in the right part of the figure is captured with the case $N = 1$ and $n_1 = d$. For simplicity, we refer to this case as the parallel implementation case in the following. Any intermediate solution mixing serial and parallel computing (e.g. 2 shares per cycle, 3 shares per cycle, ...) can be captured by our model. Concretely, the impact of parallel computation is reflected both by a reduced number of cycles and by an increased instantaneous power consumption, illustrated with the higher amplitude of the curves in Fig. 1(b). A simple abstraction to reflect this larger power consumption is the following linear model:

$$\boldsymbol{L}_c = \alpha_c^1 \cdot \mathsf{G}_c^1 \left(\mathcal{Y}_c(1)\right) + \alpha_c^2 \cdot \mathsf{G}_c^2 \left(\mathcal{Y}_c(2)\right) + \ldots + \alpha_c^{nc} \cdot \mathsf{G}_c^{nc} \left(\mathcal{Y}_c(n_c)\right) + \boldsymbol{R}_c. \quad (2)$$

with all α_c^j's $\in \mathbb{R}$. Contrary to the additive noise assumption that is only used for convenience and not needed for masking proofs, this linear model is a critical ingredient of our analysis of parallel implementations, since it is needed to maintain the independent leakage assumption. As for other physical issues that could break this assumption, we assume Eq. (2) holds in the next sections and discuss its possible limitations in Sect. 9. Yet, we already note that a general contradiction of this hypothesis would imply that any (e.g. threshold) implementation manipulating its shares in parallel should be insecure.

3   Security Models

3.1   Probing Security and Noisy Leakage

We first recall two important models for analyzing masking countermeasures.

First, the conceptually simple *t-probing* and *ε-probing* (or random probing) models were introduced in [42]. In the former, the adversary obtains t intermediate values of the computation (e.g. can probe t wires if we compute in binary fields). In the latter, he rather obtains each of these intermediate values with probability ϵ, and gets $\perp$ with probability $1 - \epsilon$ (where $\perp$ means no knowledge). Using a Chernoff-bound, it is easy to show that security in the t-probing model reduces to security in the ϵ-probing model for certain values of ϵ.

Second, the noisy leakage model describes many realistic side-channel attacks where an adversary obtains each intermediate value perturbed with a "δ-noisy" leakage function [53]. A leakage function L is called δ-noisy if for a uniformly random variable Y we have $\mathrm{SD}(Y; Y | L_Y) \le \delta$, with SD the statistical distance. It was shown in [32] that an equivalent condition is that the leakage is not too informative, where informativity is measured with the standard notion of mutual information $\mathrm{MI}(Y; L_Y)$. In contrast with the ϵ-probing model, the adversary obtains noisy leakage for each intermediate variable. For example, in the context of masking, he obtains $\mathsf{L}(Y_i, \boldsymbol{R}_i)$ for all the shares Y_i, which is reflective of actual implementations where the adversary can potentially observe the leakage of all these shares, since they are all present in leakage traces (as in Fig. 1).

Recently, Duc et al. showed that security against probing attacks implies security against noisy leakages [31]. This result leads to the natural strategy of proving security in the (simpler) probing model while stating security levels based on the concrete information leakage evaluations (as discussed in [32]).

3.2   The Bounded Moment Model

Motivation. In practice, the probing model is perfectly suited to proving the security of the serial implementations from Sect. 2.1. This is because it ensures that an adversary needs to observe d shares with his probes to recover secret information. Since in a serial implementation, every share is manipulated in a different clock cycle, it leads to a simple analogy between the number of probes and the number of cycles exploited in the leakage traces. By contrast, this simple analogy no longer holds for parallel implementations, where all the shares manipulated during a given cycle can leak concurrently. Typically, assuming that an adversary can only observe a single share with each probe is counter-intuitive in this case. For example, it would be natural to allow that he can observe the output of Eq. (2) with one probe, which corresponds to a single cycle in Fig. 1(b) and already contains information about all the shares (if $n_c = d$).

As mentioned in introduction, the noisy leakage model provides a natural solution to deal with the leakages of parallel implementations. Indeed, nothing prevents the output of Eq. (2) from leaking only a limited amount of information if a large enough noise is considered. Yet, directly dealing with noisy leakages is sometimes inconvenient for the analysis of masked implementations, e.g. when it comes to verification with the formal methods listed in Sect. 1.2. In view of their increasing popularity in embedded security evaluation, this creates a strong incentive to come up with an alternative model allowing both the construction of proofs for parallel implementations and their efficient evaluation with formal methods. Interestingly, we will show in Sect. 5 that security in this alternative model is implied by probing security. It confirms the relevance of the aforementioned strategy of first proving security in the probing model, and then stating security levels based on concrete information leakage evaluations.

Definition. Intuitively, the main limitation of the noisy leakage model in the context of formal methods is that it involves the (expensive) manipulation of complete leakage distributions. In this respect, one natural simplification that fits to a definition of "order" used in the practical side-channel literature is to relate security to the smallest key-dependent statistical moment in the leakage distributions. Concretely, the rationale behind this definition is that the security of a masked implementation comes from the need to estimate higher-order statistical moments, a task that becomes exponentially difficult in the number of shares if their leakages are independent and sufficiently noisy (see the discussion in [32]). Interestingly, such a definition directly captures the parallel implementation setting, as can easily be illustrated with an example. Say we have a single-bit sensitive value Y that is split in $d = 2$ shares, and that an adversary is able to observe a leakage function where the deterministic part is the Hamming weight function and the noise is normally distributed. Then, the (bivariate) leakage distribution for a serial implementation, where the adversary can observe the leakage of the two shares separately, is shown in the upper part of Fig. 2. And the (univariate) leakage distribution for a parallel implementation, where the adversary can only observe the sum of the leakages of the two shares, is shown in the lower part of the figure. In both cases, the first-order moment (i.e. the mean) of the leakage distributions is independent of Y.

In order to define our security model, we therefore need the following definition.

Definition 1 (Mixed moment at orders $o_1, o_2, \ldots, o_r$). *Let $\{X_i\}_{i=1}^r$ be a set of r random variables. The* mixed moment at orders $o_1, o_2, \ldots, o_r$ *of $\{X_i\}_{i=1}^r$ is:*

$$\mathsf{E}(X_1^{o_1} \times X_2^{o_2} \times \ldots \times X_r^{o_r}),$$

where E denotes the expectation operator and $\times$ denotes the multiplication in $\mathbb{R}$. For simplicity, we denote the integer $o = \sum_i o_i$ as the order of this mixed moment. We further say that a mixed moment at order o is m-variate (or has dimension m) if there are exactly m non-zero coefficients o_i.

This directly leads to our defintion of security in the bounded moment model.

Definition 2 (Security in the bounded moment model). *Let $\{L_c\}_{c=1}^N$ be the leakage vectors corresponding to an N-cycle cryptographic implementation*

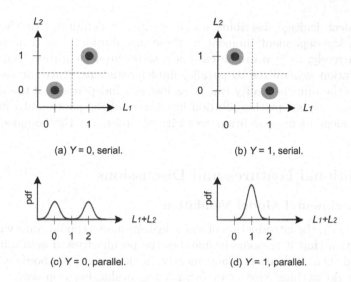

Fig. 2. Leakage distributions of a single-bit 2-shared secret.

manipulating a secret variable Y. This implementation is secure at order o *if all the mixed moments of order up to o of $\{L_c\}_{c=1}^{N}$ are independent of Y.*[2]

Say for example that we have a sensitive value Y that is split in $d = 3$ shares, for which we leak the same noisy Hamming weights as in Fig. 2. In the case of a (fully) parallel implementation, we have only one leakage sample L_1 and security at order 2 requires that $\mathsf{E}(L_1)$ and $\mathsf{E}(L_1^2)$ are independent of Y. In the case of a serial implementation, we have three samples L_1, L_2, L_3 and must show that $\mathsf{E}(L_1)$, $\mathsf{E}(L_2)$, $\mathsf{E}(L_3)$, $\mathsf{E}(L_1^2)$, $\mathsf{E}(L_2^2)$, $\mathsf{E}(L_3^2)$, $\mathsf{E}(L_1 \times L_2)$, $\mathsf{E}(L_1 \times L_3)$ and $\mathsf{E}(L_2 \times L_3)$ are independent of Y. Note that the only difference between this example and concrete implementations is that in the latter case, each cycle would correspond to a leakage vector $\boldsymbol{L_c}$ rather than a single (univariate) sample L_c.

Note also that this definition allows us to clarify a long standing discussion within the cryptographic hardware community about the right definition of security order. That is, the first definitions for secure masking (namely "perfect masking at order o" in [24] and "masking at order o" in [30]) were specialized to serial implementations, and required that any tuple of o intermediate variables is independent of any sensitive variable in an implementation. For clarity, we will now call this (strong) independence condition "security at order o in the probing model". However, due to its specialization to serial implementation, this definition also leaves a confusion about whether its generalization to parallel implementations should relate to the smallest dimensionality of a

---

[2] This definition justifies why we use raw moments rather than central or standardized ones. Indeed, to establish security at order o, we require moments of orders less than o to be independent of Y. Thus centralization (i.e. removing the mean) or normalization by the standard deviation only add terms known to be independent of Y.

key-dependent leakage distribution (i.e. m in our definition) or the smallest order of a key-dependent moment in these distributions (i.e. o in our definition). Concretely, $m \geq o$ in the case of a serial implementation, but only the second solution generalizes to parallel implementations, since for such implementations the dimensionality can be as low as 1 independent of the number of shares. Hence, we adopt this solution in the rest of the paper and will call this (weaker) independence condition "security at order o in the bounded moment model".

4   Additional Features and Discussions

4.1   Experimental Model Validation

Quite naturally, the introduction of a new leakage model should come with empirical validation that it reasonably matches the peculiarities of actual implementations and their evaluation. Conveniently, in the case of the bounded moment model, we do nothing else than formalizing evaluation approaches that are already deployed in the literature. This is witnessed by attacks based on the estimation of statistical moments, e.g. exploiting the popular difference-of-means and correlation distinguishers [33,47,57]. Such tools have been applied to various protected implementations, including threshold ones [21,22,48,52] and other masking schemes or designs running in recent high-frequency devices [11,12,44]. In all these cases, security at order o was claimed if the lowest key-dependent statistical moment of the leakage distribution was found to be of order $o + 1$.

4.2   Dimensionality Reduction

One important property of Definition 2 is that it captures security based on the statistical order of the key-dependent moments of a leakage distribution. This means that the dimensionality of the leakage vectors does not affect the security order in the bounded moment model. Therefore, it also implies that such a security definition is not affected by linear dimensionality reductions. This simple observation is formalized by the following definition and lemma.

Definition 3 (Linear dimensionality reduction). *Let $\boldsymbol{L} = [L_1, L_2, \ldots, L_M]$ denote an M-sample leakage vector and $\{\boldsymbol{\alpha}_i\}_{i=1}^m$ denote M-element vectors in $\mathbb{R}$. We say that $\boldsymbol{L}' = [L_1', L_2', \ldots, L_m']$ is a linearly reduced leakage vector if each of its (projected) samples L_i' corresponds to a scalar product $\langle \boldsymbol{L}; \boldsymbol{\alpha}_i \rangle$.*

Lemma 1. *Let $\{\boldsymbol{L}_c\}_{c=1}^N$ be the leakage vectors corresponding to an N-cycle cryptographic implementation manipulating a secret variable Y. If this implementation is secure at order o in the bounded moment model, then any implementation with linearly reduced leakages of $\{\boldsymbol{L}_c\}_{c=1}^N$ is secure at order o.*

Proof. Since the samples of $\boldsymbol{L}'$ are linear combinations of the samples of $\boldsymbol{L}$, we need the expectation of any polynomial of degree up to o of the samples of $\boldsymbol{L}'$ to be independent of Y. This directly derives from Definition 2 which guarantees that the expectation of any monomial of degree up to o is independent of Y. $\square$

Typical examples of linear dimensionality reductions are PCA [10] and LDA [58]. Note that while linearly combining leakage samples does not affect bounded moment security, it can be used to reduce the noise of the samples implied in a higher-order moment computation, and therefore can impact security in the noisy leakage model. This is in fact exactly the goal of the bounded moment model. Namely, it aims at simplifying security evaluations by splitting the tasks of evaluating the leakages' deterministic part (captured by their moments) and probabilistic part (aka noise). Concrete security against side-channel attacks is ensured by two ingredients: a high security order and sufficient noise.

4.3   Abstract Implementation Settings

In the following, we exploit our model in order to study the impact of parallelism in general terms. For this purpose, we follow the usual description of masked implementations as a sequence of leaking operations. Furthermore, and in order to first abstract away physical specificities, we consider so-called (noiseless) "abstract implementations", simplifying Eq. (2) into:

$$L_c = \alpha_1 \cdot G_1\left(\mathcal{Y}_c(1)\right) + \alpha_2 \cdot G_2\left(\mathcal{Y}_c(2)\right) + \ldots + \alpha_{n_c} \cdot G_{n_c}\left(\mathcal{Y}_c(n_c)\right). \tag{3}$$

Such simplifications allow analyzing masked implementations independent of their concrete instantiation in order to detect algorithmic flaws. Note that having $R_c \neq 0$ cannot change conclusions regarding the security order of an implementation (in the bounded moment model), which is the only metric we consider in this paper. Indeed, this order only depends on the smallest key-dependent moment of the leakage distribution, which is independent of the additive noise. By contrast, the variance of R_c affects the concrete information leakage of an implementation. We recall that algorithmically sound masked implementations do not mandatorily lead to physically secure implementations (e.g. because of the independence issues discussed in Sect. 9 or a too low noise). Yet, and as mentioned in Sect. 3.2, testing the security of abstract implementations of masking schemes (in the probing or bounded moment models) is a useful preliminary, before performing expensive evaluations of concrete implementations.

5   Serial Security Implies Parallel Security

We now provide our first result in the bounded moment model. Namely, we establish an intuitive reduction between security of parallel implementations in the bounded moment model and security of serial implementations in the probing model. For this purpose, we also formalize our serial and parallel computation models. One useful and practical consequence of the reduction is that one can adapt existing tools for proving security in the bounded moment model, either by implementing a program transformation that turns parallel implementations into serial ones, or by adapting these tools to parallel implementations.

Intuition. In order to provide some intuition for the reduction, recall that the leakage samples of an abstract parallel implementation are of the form:

$$L_c = \mathcal{Z}_c(1) + \mathcal{Z}_c(2) + \ldots + \mathcal{Z}_c(n_c),$$

with $\mathcal{Z}_c(i) = \alpha_i \cdot \mathsf{G}_i(\mathcal{Z}_c(i))$, and that the bounded moments are of the form:

$$\mathsf{E}(L_1^{o_1} \times L_2^{o_2} \times \ldots \times L_r^{o_r}).$$

Therefore, by linearity of the expectation, mixed moments at order d are independent of secrets provided all quantities of the form:

$$\mathsf{E}\left((\mathcal{Z}_1(1))^{o_{1,1}} \times \ldots \times (\mathcal{Z}_1(n_1))^{o_{1,n_1}} \times (\mathcal{Z}_r(1))^{o_{r,1}} \times \ldots \times (\mathcal{Z}_r(n_r))^{o_{r,n_r}}\right), \quad (4)$$

are independent of secrets, for all $o_{1,1}, \ldots o_{r,n_r}$ whose sum is bounded by o. Note that there are at most o pairs (i,j) such that $o_{i,j} \neq 0$. Let $(i_1, n_1) \ldots (i_k, n_k)$ with $k \leq o$ be an enumeration of these pairs. Therefore, in order to establish that Eq. (4) is independent of the secrets, it is sufficient to show that the tuple $\langle \mathcal{Z}_{i_1}(n_1), \ldots, \mathcal{Z}_{i_k}(n_k) \rangle$ is independent of these secrets. This in fact corresponds exactly to proving security in the probing model at order o.

Formalization. The theoretical setting for formalizing the reduction is a simple parallel programming language in which programs are sequences of basic instructions (note that adding for loops poses no further difficulty). A basic instruction is either a parallel assignment:

$$\langle a_1, \ldots, a_n \rangle := \langle e_1, \ldots, e_n \rangle,$$

where $e_1, \ldots, e_n$ are expressions built from variables, constants, and operators, or a parallel sampling:

$$\langle a_1, \ldots, a_n \rangle \leftarrow \langle \mu_1, \ldots, \mu_n \rangle,$$

where $\mu_1, \ldots, \mu_n$ are distributions. Despite its simplicity, this formalism is sufficient to analyse notions used for reasoning about threshold implementations, for instance non-completeness. More importantly, one can also define the notion of leakage associated to the execution of a program. Formally, an execution of a program c of length ℓ is a sequence of states $s_0 \ldots s_\ell$, where s_0 is the initial state and the state s_{i+1} is obtained from the state s_i as follows:

- If the ith-instruction is a parallel assignment, $\langle a_1, \ldots, a_n \rangle := \langle e_1, \ldots, e_n \rangle$ by evaluating the expressions $e_1 \ldots e_n$ in state s_i, leading to values $v_1 \ldots v_n$, and updating state s_i by assigning values $v_1 \ldots v_n$ to variables $a_1 \ldots a_n$;
- if the ith-instruction is a parallel sampling, $\langle a_1, \ldots, a_n \rangle \leftarrow \langle \mu_1, \ldots, \mu_n \rangle$ by sampling values $v_1 \ldots v_n$ from distributions $\mu_1 \ldots \mu_n$, and updating the state s_i by assigning the values $v_1 \ldots v_n$ to the variables $a_1 \ldots a_n$.

By assigning to each execution a probability (formally, this is the product of the probabilities of each random sampling), one obtains for every program c

of length ℓ a sequence of distributions over states $\sigma_0\sigma_1\ldots\sigma_\ell$, where σ_0 is the distribution $\mathbb{K}_{s_0}$. The leakage of a program is then a sequence $L_1\ldots L_\ell$, defined by computing for each i the sum of the values held by the variables assigned by the ith instruction, that is $a_1+\ldots+a_n$ for parallel assignments (or samplings). The mixed moments at order o then simply follow Definition 1. As for the serial programming language, instructions are either assignments $a := e$ or sampling $a \leftarrow \mu$. The semantics of a program are defined similarly to the parallel case. Order o security of a serial program in the probing model amounts to show that each o-tuple of intermediate values is independent of the secret.

Without loss of generality, we can assume that parallel programs are written in static single assignment form, meaning that variables: *(i)* appear on the left hand side of an assignment or a sampling only once in the text of a program; *(ii)* are defined before use (i.e. they occur on the left of an assignment or a sampling before they are used on the right of an assignment); *(iii)* do not occur simultaneously on the left and right hand sides of an assignment. Under such assumption, any serialization that transforms parallel assignments or parallel samplings into sequences of assignments or samplings preserve the semantics of programs. For instance, the left to right serialization transforms the parallel instructions $\langle a_1,\ldots,a_n\rangle := \langle e_1,\ldots,e_n\rangle$ and $\langle a_1,\ldots,a_n\rangle \leftarrow \langle\mu_1,\ldots,\mu_n\rangle$ into $a_1 := e_1;\ldots;a_n := e_n$ and $a_1 \leftarrow \mu_1;\ldots;a_n \leftarrow \mu_n$ respectively.

Reduction Theorem. We can now state the reduction formally:

Theorem 1. *A parallel implementation is secure at order o in the bounded moment model if its serialization is secure at order o in the probing model.*

Proof. Assume that a parallel implementation is insecure in the bounded moment model but its serialization is secure in the probing model. Therefore, there exists a mixed moment:

$$\mathsf{E}(L_1^{o_1} \times L_2^{o_2} \times \ldots \times L_r^{o_r}),$$

that is dependent of the secrets. By definition of leakage vector, and properties of expectation, there exist program variables $a_1,\ldots,a_k$, with $k \leq o$, and $o'_1,\ldots,o'_k$ with $\sum_i o_i \leq o$ such that:

$$\mathsf{E}(a_1^{o'_1} \times a_2^{o'_2} \times \ldots \times a_k^{o_k})$$

is dependent of secrets, contradicting the fact (due to security of serialization in the probing model) that the tuple $\langle a_1,\ldots,a_k\rangle$ is independent of secrets. $\square$

Note that concretely, this theorem suggests the possibility of efficient "combined security evaluations", starting with the use of the formal verification tools to test probing security, and following with additional tests in the (weaker) bounded moment model in case of negative results (see the examples in Sect. 8).

Interestingly, it also backs up a result already used in [21] (Theorem 1), where the parallel nature of the implementations was not specifically discussed but typically corresponds to the experimental case study in this paper.

6   Parallel Algorithms

In this section, we describe regular and parallelizable algorithms for secure (additively) masked computations. For this purpose, we denote a vector of d shares as $\boldsymbol{a} = [a_1, a_2, \ldots, a_d]$, the rotation of this vector by q positions as $\mathsf{rot}(\boldsymbol{a}, q)$, and the bitwise addition (XOR) and multiplication (AND) operations between two vectors as $\boldsymbol{a} \oplus \boldsymbol{b}$ and $\boldsymbol{a} \cdot \boldsymbol{b}$. For concreteness, our analyses focus on computations in $\mathsf{GF}(2)$, but their generalization to larger fields is straightforward.

6.1   Parallel Refreshing

As a starting point, we exhibit a very simple refreshing algorithm that has constant time in the parallel implementation setting and only requires d bits of fresh uniform randomness. This refreshing is given in Algorithm 1 and an example of abstract

---
Algorithm 1. Parallel refreshing algorithm.

Input: Shares $\boldsymbol{a}$ satisfying $\bigoplus_i a_i = a$, uniformly random vector $\boldsymbol{r}$.
Output: Refreshed shares $\boldsymbol{b}$ satisfying $\bigoplus_i b_i = a$.
$\boldsymbol{b} = \boldsymbol{a} \oplus \boldsymbol{r} \oplus \mathsf{rot}(\boldsymbol{r}, 1)$;
   return $\boldsymbol{b}$.

---

cycles	1	2	3	4	5
shares	a_1	r_1	r_3	$x_1{=}a_1{\oplus}r_1$	$b_1{=}x_1{\oplus}r_3$
	a_2	r_2	r_1	$x_2{=}a_2{\oplus}r_2$	$b_2{=}x_2{\oplus}r_1$
	a_3	r_3	r_2	$x_3{=}a_3{\oplus}r_3$	$b_3{=}x_3{\oplus}r_2$
leakage	$\sum a_i$	$\sum r_i$	$\sum r_i$	$\sum x_i$	$\sum b_i$
	gray	green	green	blue	blue

Fig. 3. Abstract implementation of a 3-share refreshing. (Color figure online)

6.2   Parallel Multiplication

Next, we consider the more challenging case of parallel multiplication with the similar goal of producing a simple and systematic way to manipulate the shares and fresh randomness used in the masked computations. For this purpose, our starting observation is that existing secure (serial) multiplications such as [42] (that we will mimic) essentially work in two steps: first a product phase that computes a d^2-element matrix containing the pairwise multiplications of all the shares, second a compressing phase that reduces this d^2-element matrix to a d-element one (using fresh randomness). As a result, and given the share vectors $\boldsymbol{a}$

cycles	1	2	3	4	5 (=1.3)	6 (=1.4)	7 (=2.3)
shares	a_1	a_3	b_1	b_3	$c_1=a_1.b_1$	$d_1=a_1.b_3$	$e_1=a_3.b_1$
	a_2	a_1	b_2	b_1	$c_2=a_2.b_2$	$d_2=a_2.b_1$	$e_2=a_1.b_2$
	a_3	a_2	b_3	b_2	$c_3=a_3.b_3$	$d_3=a_3.b_2$	$e_3=a_2.b_3$
leakage	$\sum a_i$	$\sum a_i$	$\sum b_i$	$\sum b_i$	$\sum c_i$	$\sum d_i$	$\sum e_i$
	gray	gray	gray	gray	red	red	red

	8	9	10 (=5⊕8)	11 (=10⊕6)	12 (=11⊕7)	13 (=12⊕9)
	r_1	r_3	$f_1=c_1\oplus r_1$	$g_1=f_1\oplus d_1$	$h_1=g_1\oplus e_1$	$x_1=h_1\oplus r_3$
	r_2	r_1	$f_2=c_2\oplus r_2$	$g_2=f_2\oplus d_2$	$h_2=g_2\oplus e_2$	$x_2=h_2\oplus r_1$
	r_3	r_2	$f_3=c_3\oplus r_3$	$g_3=f_3\oplus d_3$	$h_3=g_3\oplus e_3$	$x_3=h_3\oplus r_2$
	$\sum r_i$	$\sum r_i$	$\sum f_i$	$\sum g_i$	$\sum h_i$	$\sum x_i$
	green	green	blue	orange	orange	blue

Fig. 4. Abstract implementation of a 3-share multiplication. (Color figure online)

and b of two sensitive values a and b, it is at least possible to perform each pair of cross products $a_i \cdot b_j$'s and $a_j \cdot b_i$'s with XOR and rotation operations, and without refreshing. By contrast, the direct products $a_i \cdot b_j$ have to be separated by fresh randomness (since otherwise it could lead to the manipulation of sensitive values during the compression phase, e.g. $(a_i \cdot b_i) \oplus (a_i \cdot b_j) = a_i \cdot (b_i \oplus b_j)$. A similar reasoning holds with the uniform randomness used between the XORs of the compression phase. Namely, every fresh vector can be used twice (in its original form and rotated by one) without leaking additional information.

This rationale suggests a simple multiplication algorithm that has linear time complexity in the parallel implementation setting and requires $\lceil \frac{d-1}{4} \rceil$ random vectors of d bits (it can be viewed as an adaptation of the algorithms in [19]). We first highlight it based on its abstract implementation in Fig. 4, which starts with the loading and rotation of the input shares (gray cycles), then performs the product phase (red cycles) and finally compresses its output by combining the addition of fresh randomness (blue cycles) and accumulation (orange cycles). In general, such an implementation runs in $< 5d$ cycles for d shares, with slight variations depending on the value of d. For $d \bmod 4 = 3$ (as in Fig. 4) it is "complete" (i.e. ends with two accumulation cycles and one refreshing). But for $d \bmod 4 = 0$ it ends with a single accumulation cycle, for $d \bmod 4 = 1$ it ends with two accumulation cycles and for $d \bmod 4 = 2$ it ends with an accumulation cycle and a refrehing. An accurate description is given in [15], Algorithm 3.

Impact for Multivariate (Aka Horizontal) Attacks. In simplified terms, the security proofs for masked implementations in [31,32] state that the data complexity of a side-channel attack can be bounded by $\frac{1}{\mathrm{MI}(Y_i, L_{Y_i})^d}$, with d the number of shares and $MI(Y_i, L_{Y_i})$ the information leakage of each share Y_i (assumed identical $\forall i$'s for simplicity – we take the worst case otherwise), if $MI(Y_i, L_{Y_i}) \leq \frac{1}{d}$ (where the d factor is due to the computation of the partial products in the multiplication algorithm of [42]). In a recent work, Batistello et al. [17] showed that the manipulation of the shares in masked implementations can be exploited concretely thanks to efficient multivariate/horizontal attacks (either via combination of shares' tuples corresponding to the same sensitive variable, or via averaging of shares appearing multiple times). Interestingly, while multivariate/horizontal attacks are also possible in our parallel case, the number of leakage samples that parallel implementations provide to the side-channel adversary is reduced (roughly by a factor d), which also mitigates the impact of such attacks.

7   Case Studies

By Theorem 1, security in the bounded moment model of a parallel implementation can be established from security of its serialization in the probing model. Therefore, it is possible to use existing formal methods to test the security of parallel implementations, by first pre-processing them into a serial ones, and feeding the resulting serial programs into a verification tool. In this section, we report on the successful automated analysis of several parallel implementations, including the parallel refreshing and multiplication presented in the previous section, and serial composition of parallel S-boxes. Note that, due to the algorithmic complexity of the verification task, we only establish security at small orders. However, we also note that, although our main design constraint was for our algorithms to be easily implementable in parallel, the use of automated tools – as opposed to manual analyses – to verify their security has yielded algorithms that match or improve on the state-of-the-art in their randomness requirements at these orders. All experiments reported in this section are based on the current version of the tool of [13]. This version supports automated verification of two properties: the usual notion of probing security, and a strictly stronger notion, recently introduced in [14] under the name *strong non-interference* (SNI), which is better suited to the compositional verification of large circuits.

7.1   Parallel Refreshing

We first consider the parallel refreshing algorithm from the previous section.

Theorem 2 (Security of Algorithm 1). *The refreshing in Algorithm 1 is secure at order $d - 1$ in the bounded moment model for all $d \leq 7$.*

By Theorem 1, it is sufficient to prove $(d-1)$-probing security to get security at order $d - 1$ in the bounded moment model. We do so using the tool by Barthe et al. [13] for each order $d \leq 7$. Table 1 shows the verification time for each proof.

In addition, we consider the problem of how to construct a SNI mask refreshing gadget that behaves as well with respect to our parallel computation model.

Table 1. Probing and bounded moment security of Algorithm 1.

d	$(d-1)$-b.m	Time (s)
3	✓	1
4	✓	1
5	✓	2
6	✓	20
7	✓	420

We rely on the current version of the tool from Barthe et al. [13], which supports the verification of strong non-interference properties. Table 2 reports the verification results for some number of mask refreshing algorithms, constructed simply by iterating Algorithm 1 (denoted R_d). We denote with R_d^n the algorithm that iterates R_d n times. Table 2 also shows the randomness requirements both for our algorithm and for the only other known SNI mask refreshing gadget, based on Ishai, Sahai and Wagner's multiplication algorithm [42].

Table 2. SNI secure variants of Algorithm 1.

Alg.	d	$(d-1)$-SNI	# rand. bits		Time (s)
			Our alg.	[42]	
R_d	3	✓	3	3	1
R_d	4	✓	4	6	1
R_d	5	✗	5	10	1
R_d^2	5	✓	10	10	1
R_d^2	6	✓	12	15	1
R_d^2	7	✓	14	21	1
R_d^2	8	✗	16	28	1
R_d^3	8	✓	24	28	4
R_d^3	9	✓	27	36	36
R_d^3	10	✓	30	45	288
R_d^4	11	✓	40	55	3045

These experiments show that, for small masking orders, there exist regular mask refreshing gadgets that are easily parallelizable, suitable for the construction of secure circuits by composition, and that have small randomness requirements. This fact is particularly useful when viewed through the lens of Theorem 1. Indeed, SNI gadgets are instrumental in easily proving probing security for large circuits [14], which Theorem 1 then lifts to the bounded moment model and parallel implementations. We conjecture that iterating the simple mask refreshing gadget from Algorithm 1 $\lceil (d-1)/3 \rceil$ times always yields a $(d-1)$-SNI mask refreshing algorithm over d shares. The resulting algorithm is easily parallelizable and requires $\lceil (d-1)/3 \rceil \cdot d$ bits of randomness (marginally improving on the $d \cdot (d-1)/2$ bits of randomness from the ISW-based mask refreshing). We leave a proof of strong non-interference for all d's as future work.

7.2   Parallel Multiplication

We now consider the parallel multiplication algorithm from the previous section (specified in Algorithm 3 in [15]), and prove its security for small orders.

Theorem 3 (Security of Algorithm 3 in [15]). *The multiplication in Algorithm 3 in [15] is secure at order $d-1$ in the bounded moment model for all $d \leq 7$.*

By Theorem 1, it is sufficient to prove $(d-1)$-probing security to get security at order $d-1$ in the bounded moment model. We do so using the tool by Barthe et al. [13] for each $d \leq 7$. Table 3 shows the verification time for each instance.

Table 3. Probing and bounded moment security of Algorithm 3 in [15].

d	$(d-1)$-b.m	# rand. bits		time (s)
		our alg	[19]	
3	✓	3	2	1
4	✓	4	4	1
5	✓	5	5	2
6	✓	12	11	17
7	✓	14	15	480

We also show a comparison of the randomness requirement of our algorithm and those of Belaï et al. [19]. Note that we sometimes need one additional random bit compared to the algorithm of Belaïd et al. [19]. This is due to our parallelization constraint: instead of sampling uniform sharings of 0, we only allow ourselves to sample uniformly random vectors and to rotate them.

As before, we now investigate some combinations of Algorithms 1 and 3 in [15] in the hope of identifying regular and easily parallelizable SNI multiplication algorithms. The results of the experiments are shown in Table 4, where $\odot_d$ is Algorithm 3 in [15], specialized to d shares. In addition to showing whether or not the algorithm considered is SNI, the table shows verification times and compares the randomness requirements of our algorithm with that of the multiplication algorithm by Ishai, Sahai and Wagner, which is the best known SNI multiplication algorithm in terms of randomness. As with the original tool by Barthe et al. [13], the verification task is constrained to security orders $o \leq 8$ for circuits involving single multiplications due to the exponential nature of the problem it tackles.

We conjecture that the combination of our multiplication algorithm with a single refreshing is SNI for any d. This is intuitively justified by the fact our

Table 4. SNI security for variants of Algorithm 3 in [15].

Algorithm	d	$(d-1)$-SNI	# rand. bits		Time (s)
			Our alg.	[42]	
$\odot_d$	3	✓	3	3	1
$\odot_d$	$d \geq 4$	✗	$d(d-1)/4$	$d(d-1)/2$	-
$R_d \circ \odot_d$	4	✓	8	6	1
$R_d \circ \odot_d$	5	✓	10	10	1
$R_d \circ \odot_d$	6	✓	18	15	39
$R_d \circ \odot_d$	7	✓	21	21	2647
$R_d \circ \odot_d$	8	✓	24	28	166535

multiplication algorithm includes a number of "half refreshings", which must be combined with a final refreshing for the d's such that it ends with an accumulation step. We leave the proof of this conjecture as an open problem.

7.3   S-Boxes and Feistel Networks

In order to better investigate the effects on the security of larger circuits of reducing the randomness requirements of the multiplication and refreshing algorithms, we now consider small S-boxes, shown in Fig. 5, and their iterations.

Figure 5(a) describes a simple 3-bit S-box similar to the "Class 13" S-box of Ullrich et al. [62]. Figure 5(b) describes a 4-bit S-box constructed by applying a Feistel construction to a 2-bit function. Table 5 shows verification results for iterations of these circuits for several small orders, exhibiting some interesting compositional properties for these orders. sbox$_3$ denotes the circuit from Fig. 5(a), sbox$_4$ denotes the circuit from Fig. 5(b), and sboxr$_4$ denotes the circuit

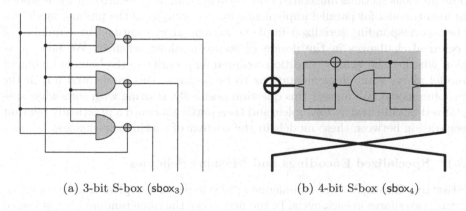

(a) 3-bit S-box (sbox$_3$)   (b) 4-bit S-box (sbox$_4$)

Fig. 5. Examples of elementary circuits.

Table 5. Probing and bounded moment security of small S-boxes.

$d = 3$				$d = 4$		
Algorithm	2-b.m.	time (s)		Algorithm	3-b.m.	time (s)
sbox_3	✓	1		sbox_3	✓	13
sbox_3^2	✓	1		sbox_3^2	✓	322
sbox_4	✓	1		sbox_4	✓	2
sbox_4^2	✓	1		sbox_4^2	✓	67
sbox_4^3	✓	714				
sboxr_4^3	✓	1				
sboxr_4^4	✓	3				
sboxr_4^5	✓	7				
sboxr_4^6	✓	12				

from Fig. 5(b), modified so that the upper output of its inner transformation is refreshed. As before, integer exponents denote sequential iteration.

We note that, although there is no evidence that iterating sbox_4 longer yields insecure circuits, obtaining convincing security results for more than 3 iterations using automated tools seems unfeasible without relying on compositional principles. In particular, inserting a single mask refreshing operation per Feistel round greatly speeds up the verification of large iterations of the 4-bit S-box from Fig. 5(b). This highlights possible interactions between tools oriented towards the verification of small optimized circuits for particular values of d [13,18,38] and tools geared towards the more efficient but less precise verification of large circuits [14]. The ability to make our algorithms SNI allows us to directly take advantage of this "randomness complexity vs. verification time" tradeoff.

8   Separation Results

The previous sections illustrated that the reduction from security in the bounded moment model for parallel implementations to security in the probing model for their corresponding serialized implementations gives solutions to a number of technical challenges in the design of secure masking schemes. We now question whether the weaker condition required for security in the bounded moment model allows some implementations to be secure in this model and not in the probing model. We answer this question positively, starting with somewhat specialized but illustrative examples, and then putting forward a practically relevant separation between these models in the context of continuous leakages.

8.1   Specialized Encodings and Masking Schemes

Starting Example. Let us imagine a 2-cycle parallel implementation manipulating two shares in each cycle. In the first cycle, the same random bit r is loaded twice, giving rise to a state (r, r). In the second cycle, a shared sensitive value a is loaded twice, giving rise to a state $(a \oplus r, \bar{a} \oplus r)$. Clearly, in the probing model two probes (on r and $a \oplus r$) are sufficient to learn a. But for an adversary observing the abstract leakages of this parallel implementations (i.e. the arithmetic sum for each cycle), and for a particular type of leakage function such that $\alpha_i^j = 1$ and $G_i^j = \mathrm{Id}$ in Eq. (2), the first cycle will only reveal $r + r$ while the second cycle will reveal a constant 1. So no combinations of these leakages can be used to recover a. An even simpler example would be the parallel manipulation of a and $\bar{a}$ which trivially does not leak any information if their values are just summed. Such implementations are known under the name "dual-rail pre-charged" implementations in the literature [61]. Their main problem is that they require much stronger physical assumptions than masked implementations. That is, the leakages on the shares a and $\bar{a}$ do not only need to be independent but identical, which turns our to be much harder to achieve in practice [27].

Leakage Squeezing and Low Entropy Masking Schemes. Interestingly, the literature provides additional examples of countermeasures where the security order is larger in the bounded moment model than in the probing model. In particular, leakage squeezing and low entropy masking schemes exploit special types of encodings such that the lowest key-dependent statistical moment of their leakage distributions is larger than the number of shares, if the leakage function's deterministic part is linear [25,41], i.e. if $G_i^j = \mathrm{Id}$ in Eq. (2). Note that this requirement should not be confused with the global linearity requirement of Eq. (2). That is, what masking generally requires to be secure is that the different shares are combined linearly (i.e. that Eq. (2) is a first-degree polynomial of the $G_i^j(\mathcal{Y}_i(j))$'s). Leakage squeezing and low entropy masking schemes additionally require that the (local) G_i^j functions are linear.

The previous examples show that in theory, there exist leakage functions such that the security order in the bounded moment model is higher than the security order in the probing model, which is sufficient to prove separation. Yet, as previously mentioned, in practice the identical (resp. linear) leakage assumption required for dual-rail pre-charged implementations (resp. leakage squeezing and low entropy masking schemes) is extremely hard to fulfill (resp. has not been thoroughly studied yet). So this is not a general separation for any implementation. We next present such a more general separation.

8.2 The Continuous Leakage Separation

A Continuous Probing Attack Against the Refreshing of Algorithm 1. Up to this point of the paper, our analyses have considered "one-shot" attacks and security. Yet, in practice, the most realistic leakage models consider adversaries who can continuously observe several executions of the target algorithms. Indeed, this typically corresponds to the standard DPA setting where sensitive information is extracted by combining observations from many successive runs [43]. Such a setting is reflected in the continuous t-probing model of Ishai, Sahai and Wagner [42], where the adversary can learn t intermediate values produced during the computation of each execution of the algorithm. It implies that over time the adversary may learn much more information than just the t values

– and in particular more than d, the number of shares. To be concrete, in a continuous attack that runs for q executions the adversary can learn up to tq intermediate values, evenly distributed between the executions of the algorithm.

Designing strong mask refreshing schemes that achieve security in the continuous t-probing model is a non-trivial task. In this section, we show that Algorithm 1 can be broken for any number of shares d, if the refreshing is repeated consecutively for d times and in each execution the adversary can learn up to 3 intermediate values. To explain the attack, we first generalize this algorithm to d executions, with $a_1^{(0)}, \ldots, a_d^{(0)}$ the initial encoding of some secret bit a, as given in Algorithm 3 in [15]. The lemma below gives the attack. Similar attacks are used in [35] for the inner product masking in the bounded leakage model.

Algorithm 2. d-times execution of the parallel refreshing algorithm.

Input: Shares $a^{(0)}$ satisfying $\bigoplus_i a_i^{(0)} = a$ and
  d random vectors $r^{(i)}$.
Output: Refreshed shares $a^{(d)}$ satisfying $\bigoplus_i a_i^{(d)} = a$.
  for $i = 1$ to d **do**
    $a^{(i)} = a^{(i-1)} \oplus r^{(i)} \oplus \text{rot}(r^{(i)}, 1)$;
  end for
  return $a^{(d)}$.

Lemma 2. *Let a be a uniformly chosen secret bit, $d \in \mathbb{N}$ a number of shares and consider Algorithm 2. In each iteration of the* for *loop there exists a set of 3 probes such that after d iterations the secret a can be learned.*

Proof. We show that, if the adversary can probe 3 intermediate values in each iteration of the parallel refreshing for d iterations, then he can recover the secret bit a. The proof is by induction, where we show that, after learning the values of his 3 probes in the ith iteration, the adversary knows the sum of the first i shares of a, that is $A_1^i := \bigoplus_{j=1}^{i} a_j^{(i)}$. Since $A_1^d := \bigoplus_{j=1}^{d} a_j^{(d)} = a$, after d iterations, the adversary thus knows the value of a. In the first iteration, a single probe on share $a_1^{(1)}$ is sufficient to learn $A_1^1 := a_1^{(1)}$. We now prove the inductive step. Let $1 < \ell \le d$. Suppose after the $(\ell-1)$th execution, we know: $A_1^{\ell-1} := \bigoplus_{j=1}^{\ell-1} a_j^{(\ell-1)}$. In the ℓth iteration, the adversary probes $r_d^{(\ell)}, r_{\ell-1}^{(\ell)}$ and $a_\ell^{(\ell)}$, allowing him to compute A_1^ℓ using the following equalities:

$$A_1^\ell = \bigoplus_{j=1}^{\ell} a_j^{(\ell)} = a_\ell^{(\ell)} \oplus \bigoplus_{j=1}^{\ell-1} a_j^{(\ell)} = a_\ell^{(\ell)} \oplus \bigoplus_{j=1}^{\ell-1} a_j^{(\ell-1)} \oplus r_j^{(\ell)} \oplus r_{j-1}^{(\ell)}$$

$$= a_\ell^{(\ell)} \oplus r_d^{(\ell)} \oplus r_{\ell-1}^{(\ell)} \oplus \bigoplus_{j=1}^{\ell-1} a_j^{(\ell-1)} = a_\ell^{(\ell)} \oplus r_d^{(\ell)} \oplus r_{\ell-1}^{(\ell)} \oplus A_1^{\ell-1},$$

where we use the convention that for any j we have $r_0^{(j))} = r_d^{(j))}$. Since all values after the last equality either are known from the previous round or have been learned in the current round the above concludes the proof.   □

Continuous Security of Algorithm 1 in the Bounded Moment Model. The previous attack crucially relies on the fact that the adversary can move his probes adaptively between different iterations, i.e. in the ith execution he must learn different values than in the $(i-1)$th execution. This implies that in practice he would need to exploit jointly $\approx 3d$ *different* time samples from the power trace. We now show that such an attack is not possible in the (continuous) bounded moment model. The only difference between the continuous bounded moment model and the one-shot bounded moment model is that the first offers more choice for combining leakages as there are q-times more cycles. More precisely, the natural extension of bounded moment security towards a continuous setting requires that the expectation of any oth-degree polynomial of leakage samples among the q leakage vectors that can be observed by the adversary is independent of any sensitive variable $Y \in \{0, 1\}$ that is produced during the q executions of the implementation. Thanks to Lemma 1, we know that a sufficient condition for this condition to hold is that the expectation of all the monomials is independent of Y. So concretely, we only need that for any tuple of o possible clock cycles $c_1, c_2 \ldots, c_o \in [1, qN]$, we have:

$$\Pr[Y = 0] = \Pr[Y = 0 | \mathbb{E}[L_{c_1} \times L_{c_2} \times \ldots \times L_{c_o}]],$$
$$\Pr[Y = 1] = \Pr[Y = 1 | \mathbb{E}[L_{c_1} \times L_{c_2} \times \ldots \times L_{c_o}]].$$

In the one-shot bounded moment model $c_1, c_2 \ldots, c_o$ would only run in $[1, N]$. Our following separation result additionally needs a specialization to stateless primitives. By stateless, we mean primitives such as block ciphers that only need to maintain a constant secret key in memory from one execution to the other.

Theorem 4. *The implementation of a stateless primitive where the secret key is refreshed using Algorithm 1 is secure at order o in the continuous bounded moment model if it is secure at order o in the one-shot probing model.*

Proof (sketch). We consider an algorithm for which a single execution takes N cycles which is repeated q times. We can view the q-times execution of the algorithm as a computation running for qN cycles. Since we are only interested in protecting stateless primitives, individual executions are only connected via their refreshed key. Hence, the q-times execution of the N-cycle implementation can be viewed as a circuit consisting of q refreshings of the secret key using Algorithm 1, where each refreshed key is used as input for the stateless masked implementation. If we show that this "inflated" circuit is secure against an adversary placing up to o probes in these qN cycles (in total and not per execution as in the continuous probing model), the result follows by Theorem 1.

For this purpose, we first observe that o probes just in the part belonging to the q-times refreshing do not allow the adversary to learn the masked secret key.

This follows from the fact that probing o values in a one-shot execution of the refreshing (Algorithm 1) does not allow the adversary to learn this masked secret key. More precisely, any such probes in the refreshing can be directly translated into probes on the initial encoding (and giving the appropriate randomness of the refreshing to the adversary for free). This means that any probe in the refreshing part allows to learn at most a single share of the masked secret key going into the stateless masked implementation. Moreover, we know by assumption that a single-shot execution of the implementation is o-probing secure. This implies that even after o probes inside the masked implementation there still must exist one share of the masked state of which these probes are independent. More generally, placing $o - i$ probes in the masked implementation must imply that these probes are independent of at least $i + 1$ shares of the masked state, since otherwise the remaining i probes can be placed at the unknown input shares to get a correlation with the masked secret key. As a result, we can also reveal all of the shares of the input encoding except for these $i + 1$ shares that are independent. Therefore, by simply adding up the probes, we get that even placing o probes inside of the inflated circuit maintains security.   □

Note that the above argument with the inflated circuit and the special use of the refreshing fails to work when we consider stateful primitives. In such a setting, the refreshing may interact with other parts of the circuit. Hence, we would need a stronger (composable) refreshing to achieve security in this case, in order to deal with the fact that Algorithm 1 could then appear at arbitrary positions in the computation. As already mentioned, the security condition of the bounded moment model is significantly weaker than in the probing model, which is what allows us to reach this positive result. Intuitively, security in the probing model requires that, given a certain number of probes, no information is leaked. By contrast, security in the bounded moment model only requires that this information is hard to exploit, which is captured by the fact that the lowest informative statistical moment in the leakage distribution observed by the adversary is bounded. This model nicely captures the reality of concrete side-channel attacks, where all the points of a leakage traces (as in Fig. 1) are available to this adversary, and we want to ensure that he will at least have to estimate a higher-order moment of this leakage distribution in order to extract sensitive information (a task that is exponentially hard in o if the distribution is sufficiently noisy). We believe this last result is particularly relevant for cryptographic engineers, since it clarifies a long standing gap between the theory and practice of masking schemes regarding the need of complex refreshing schemes. Namely, we are able to show that simple refreshing schemes such as in Sect. 6.1 indeed bring sufficient security against concrete higher-order side-channel attacks.

Note also that it is an interesting open problem to investigate the security of our simple refreshing scheme in the continuous noisy leakage model. Intuitively, extending the attack of Lemma 2 to this setting seems difficult. Take the second step for example: we have learned A_1^1 and want to learn A_1^2 with three noisy probes. If the noise is such that we do not learn A_1^2 exactly, then observing again three probes with an independent noise will not help much (since we cannot easily combine the information on the fresh A_1^2, and would need to collect information on all d shares to accumulate information on the constant secret). As for Theorem 1 in Sect. 5, we can anyway note that the bounded moment model allows obtaining much easier connections to the (more theoretical) probing model than the (more general but more involved) noisy leakage model.

9   Independence Issues

Before concluding, we discuss one important advantage of threshold implementation for hardware (parallel) implementations, namely their better resistance against glitches. We take advantage of and generalize this discussion to clarify the different independence issues that can affect leaking implementations, and detail how they can be addressed in order to obtain actual implementations that deliver the security levels guaranteed by masking security proofs.

Implementation Defaults. As a starting point, we reproduce a standard example of threshold implementation, in Fig. 6(a), which corresponds to the secure execution of a small Boolean function $f(x)$, where both the function and the inputs/outputs are shared in three pieces. In this figure, the (light and dark) gray rectangles correspond to registers, and the blue circles correspond to combinatorial circuits. From this example, we can list three different types of non-independence issues that can occur in practice:

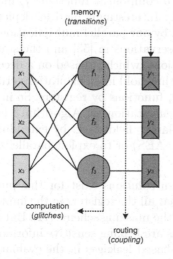 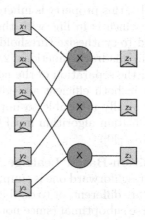

(a) Threshold implementation.   (b) 3-bit partial product.

Fig. 6. Independence issues and threshold implementations. (Color figure online)

1. *Computational re-combining (or glitches).* In this first case, transient intermediate computations are such that the combinatorial part of the circuit re-combines the shares. This effect has been frequently observed in the literature under the name "glitches", and has been exploited to break (i.e. reduce the security order) of many hardware implementations (e.g. [46]).
2. *Memory re-combining (or transitions).* In this second case, non independence comes from register re-use and the fact that actual leakage may be proportional to the transition between the register states. For example, this would happen in Fig. 6(a), if registers x_1 and y_1 (which depends on x_2, x_3) are the same. This effect has been frequently observed in the literature too, under the name "distance-based" or "transition-based" leakages, and has been exploited to break software implementations (e.g. [11,29]).
3. *Routing re-combining (or coupling).* In this final case, the re-combining is based on the physical proximity of the wires. The leakage function would then be proportional to some function of these wires. Such effects, known under the name "coupling", could break the additive model of Eq. (2) in case of complex (e.g. quadratic) function. To the best of our knowledge, they have not yet been exploited in a concrete (published) attack.

Glitches, Threshold Implementations and Non-completeness. One important contribution of threshold implementations is to introduce a sound algorithmic way to deal with glitches. For this purpose, they require their implementations to satisfy the "non-completeness" property, which requires (at order o) that any combination of up to o component functions f_i must be independent of at least one input share [21]. Interestingly, and as depicted in Fig. 6(b), this property is inherently satisfied by our parallel multiplication algorithm, which is in line with the previous observations in [55] and the standard method to synthesize threshold implementations, which is based on a decomposition in quadratic functions [23]. Note that threshold implementations crucially rely on the separation of the non-complete f_i functions by registers. So in order to obtain both efficient and glitch-free implementations of Algorithm 3 in [15], it is typically advisable to implement it in larger fields (e.g. by extending our multiplication algorithm in $GF(2^8)$ as for the AES) or to exploit parallelism via bitslicing [40].

Transition-Based Leakage. Various design solutions exist for this purpose. The straighforward one is simply to ensure that all the registers in the implementation are different, or to double the order of the masking scheme [11]. But this is of course suboptimal (since not all transitions are leaking sensitive information). So a better solution is to include transition-based leakages in the evaluation of masked implementations, a task which also benefits from the tools in [13].

Couplings. This last effect being essentially physical, there are no algorithmic/software methods to prevent it. Couplings are especially critical in the context of parallel implementation since the non-linearity they imply may break the the independent leakage assumption. (By contrast, in serial implementations this assumption is rather fulfilled by manipulating the shares at different cycles). So the fact that routing-based recombinations do not occur in parallel masked implementations is essentially an assumption that all designers have to make. In this respect, we note that experimental results of attacks against threshold implementations where several shares are manipulated in parallel (e.g. the ones listed in Sect. 4.1) suggest that this assumption is indeed well respected for current technologies. Yet, we also note that the risk of couplings increases with technology scaling [51]. Hence, in the latter case it is anyway a good design strategy to manipulate shares in larger fields, or to ensure a sufficient physical distance between them if masking is implemented in a bitslice fashion.

10   Open Problems

These results lead to two important tracks for further research.

First, the bounded moment model that we introduce can be seen as an intermediate path between the conceptually simple probing model and the practically relevant noisy leakage model. As discussed in Sect. 8 (and illustrated in Fig. 7), the bounded moment leakage model is strictly weaker than the probing model. Hence, it would be interesting to investigate whether bounded moment security implies noisy leakage security for certain classes of leakage functions. Clearly, this cannot hold in general since there exist different distributions with identical moments. Yet, and in view of the efficiency gains provided by moment-based security evaluations, it is an interesting open problem to identify the contexts in which this approach is sufficient, i.e. to find out when a leakage distribution is well enough represented by its moments. Building on and formalizing the results in [34] is an interesting direction for this purpose.

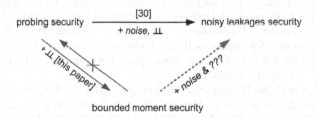

Fig. 7. Reductions between leakage security models.

Second, whenever discovering a bias in a masked implementation, our tools not only output the computation leading to this bias, but also its (possibly small) amplitude. Hence, the bounded moment model has great potential to extend the quantitative analysis in [39] (so far limited to first-order leakages) to the higher-order case. Relying on the fact that the biases may be quantitatively hard to exploit could lead to further reductions of the randomness requirements in masked implementations, e.g. by combining the evaluation of these biases with tools to analyze non-independent leakages introduced in [32] (Sect. 4.2).

Acknowledgements. Sebastian Faust is funded by the Emmy Noether Program FA 1320/1-1 of the German Research Foundation (DFG). François-Xavier Standaert is a research associate of the Belgian Fund for Scientific Research (F.R.S.-FNRS). This work has been funded in parts by projects S2013/ICE-2731 N-GREENS Software-CM, ONR Grants N000141210914 and N000141512750, FP7 Marie Curie Actions-COFUND 291803 and ERC project 280141.

References

1. Francillon, A., Rohatgi, P. (eds.): CARDIS 2013. LNCS, vol. 8419. Springer, Heidelberg (2014)
2. Prouff, E., Schaumont, P. (eds.): CHES 2012. LNCS, vol. 7428. Springer, Heidelberg (2012). doi:10.1007/978-3-642-33027-8
3. Güneysu, T., Handschuh, H. (eds.): CHES 2015. LNCS, vol. 9293. Springer, Heidelberg (2015). doi:10.1007/978-3-662-48324-4
4. Gierlichs, B., Poschmann, A.Y. (eds.): CHES 2016. LNCS, vol. 9813. Springer, Heidelberg (2016). doi:10.1007/978-3-662-53140-2
5. Wiener, M. (ed.): CRYPTO 1999. LNCS, vol. 1666. Springer, Heidelberg (1999). doi:10.1007/3-540-48405-1
6. Oswald, E., Fischlin, M. (eds.): EUROCRYPT 2015. LNCS, vol. 9056. Springer, Heidelberg (2015). doi:10.1007/978-3-662-46800-5
7. Fischlin, M., Coron, J.-S. (eds.): EUROCRYPT 2016. LNCS, vol. 9666. Springer, Heidelberg (2016)
8. Joye, M., Moradi, A. (eds.): CARDIS 2014. LNCS, vol. 8968. Springer, Heidelberg (2015). doi:10.1007/978-3-319-16763-3
9. Andrychowicz, M., Dziembowski, S., Faust, S.: Circuit compilers with $O(1/\backslash n)$ leakage rate. In: EUROCRYPT 2016, Part II [7], pp. 586–615 (2016)
10. Archambeau, C., Peeters, E., Standaert, F.-X., Quisquater, J.-J.: Template attacks in principal subspaces. In: Goubin, L., Matsui, M. (eds.) CHES 2006. LNCS, vol. 4249, pp. 1–14. Springer, Heidelberg (2006). doi:10.1007/11894063_1
11. Balasch, J., Gierlichs, B., Grosso, V., Reparaz, O., Standaert, F.-X.: On the cost of lazy engineering for masked software implementations. In: Joye, M., Moradi, A. (eds.) CARDIS 2014. LNCS, vol. 8968, pp. 64–81. Springer, Heidelberg (2015). doi:10.1007/978-3-319-16763-3_5
12. Balasch, J., Gierlichs, B., Reparaz, O., Verbauwhede, I.: DPA, bitslicing and masking at 1 GHz. In: Güneysu, T., Handschuh, H. (eds.) CHES 2015. LNCS, vol. 9293, pp. 599–619. Springer, Heidelberg (2015). doi:10.1007/978-3-662-48324-4_30
13. Barthe, G., Belaïd, S., Dupressoir, F., Fouque, P.-A., Grégoire, B., Strub, P.-Y.: Verified proofs of higher-order masking. In: Oswald, E., Fischlin, M. (eds.) EURO-CRYPT 2015. LNCS, vol. 9056, pp. 457–485. Springer, Heidelberg (2015). doi:10.1007/978-3-662-46800-5_18
14. Barthe, G., Belaid, S., Dupressoir, F., Fouque, P.-A., Gregoire, B., Strub, P.-Y., Zucchini, R.: Strong non-interference and type-directed higher-order masking. In: Proceedings of ACM CCS (2016, to appear)
15. Barthe, G., Dupressoir, F., Faust, S., Grégoire, B., Standaert, F., Strub, P.: Parallel implementations of masking schemes and the bounded moment leakage model. IACR Cryptol. ePrint Arch. **2016**, 912 (2016)
16. Batina, L., Gierlichs, B., Prouff, E., Rivain, M., Standaert, F., Veyrat-Charvillon, N.: Mutual information analysis: a comprehensive study. J. Cryptol. **24**(2), 269–291 (2011)

17. Battistello, A., Coron, J.-S., Prouff, E., Zeitoun, R.: Horizontal side-channel attacks and countermeasures on the ISW masking scheme. In: Gierlichs, B., Poschmann, A.Y. (eds.) CHES 2016. LNCS, vol. 9813, pp. 23–39. Springer, Heidelberg (2016). doi:10.1007/978-3-662-53140-2_2

18. Bayrak, A.G., Regazzoni, F., Novo, D., Ienne, P.: Sleuth: automated verification of software power analysis countermeasures. In: Bertoni, G., Coron, J.-S. (eds.) CHES 2013. LNCS, vol. 8086, pp. 293–310. Springer, Heidelberg (2013). doi:10.1007/978-3-642-40349-1_17

19. Belaïd, S., Benhamouda, F., Passelègue, A., Prouff, E., Thillard, A., Vergnaud, D.: Randomness complexity of private circuits for multiplication. In: Fischlin, M., Coron, J.-S. (eds.) EUROCRYPT 2016. LNCS, vol. 9666, pp. 616–648. Springer, Heidelberg (2016). doi:10.1007/978-3-662-49896-5_22

20. Bilgin, B., Daemen, J., Nikov, V., Nikova, S., Rijmen, V., Assche, G.: Efficient and first-order DPA resistant implementations of KECCAK. In: Francillon, A., Rohatgi, P. (eds.) CARDIS 2013. LNCS, vol. 8419, pp. 187–199. Springer, Heidelberg (2014). doi:10.1007/978-3-319-08302-5_13

21. Bilgin, B., Gierlichs, B., Nikova, S., Nikov, V., Rijmen, V.: Higher-order threshold implementations. In: Sarkar, P., Iwata, T. (eds.) ASIACRYPT 2014. LNCS, vol. 8874, pp. 326–343. Springer, Heidelberg (2014). doi:10.1007/978-3-662-45608-8_18

22. Bilgin, B., Gierlichs, B., Nikova, S., Nikov, V., Rijmen, V.: A more efficient AES threshold implementation. In: Pointcheval, D., Vergnaud, D. (eds.) AFRICACRYPT 2014. LNCS, vol. 8469, pp. 267–284. Springer, Heidelberg (2014). doi:10.1007/978-3-319-06734-6_17

23. Bilgin, B., Nikova, S., Nikov, V., Rijmen, V., Stütz, G.: Threshold implementations of all 3 × 3 and 4 × 4 S-boxes. In: Prouff, E., Schaumont, P. (eds.) CHES 2012. LNCS, vol. 7428, pp. 76–91. Springer, Heidelberg (2012). doi:10.1007/978-3-642-33027-8_5

24. Blömer, J., Guajardo, J., Krummel, V.: Provably secure masking of AES. In: Handschuh, H., Hasan, M.A. (eds.) SAC 2004. LNCS, vol. 3357, pp. 69–83. Springer, Heidelberg (2004). doi:10.1007/978-3-540-30564-4_5

25. Carlet, C., Danger, J., Guilley, S., Maghrebi, H., Prouff, E.: Achieving side-channel high-order correlation immunity with leakage squeezing. J. Cryptogr. Eng. 4(2), 107–121 (2014)

26. Chari, S., Jutla, C.S., Rao, J.R., Rohatgi, P.: Towards sound approaches to counteract power-analysis attacks. In: Wiener, M. (ed.) CRYPTO 1999. LNCS, vol. 1666, pp. 398–412. Springer, Heidelberg (1999). doi:10.1007/3-540-48405-1_26

27. Chen, C., Eisenbarth, T., Shahverdi, A., Ye, X.: Balanced encoding to mitigate power analysis: a case study. In: Joye, M., Moradi, A. (eds.) CARDIS 2014. LNCS, vol. 8968, pp. 49–63. Springer, Heidelberg (2015). doi:10.1007/978-3-319-16763-3_4

28. De Cnudde, T., Reparaz, O., Bilgin, B., Nikova, S., Nikov, V., Rijmen, V.: Masking AES with d + 1 shares in hardware. In: Gierlichs, B., Poschmann, A.Y. (eds.) CHES 2016. LNCS, vol. 9813, pp. 194–212. Springer, Heidelberg (2016). doi:10.1007/978-3-662-53140-2_10

29. Coron, J.-S., Giraud, C., Prouff, E., Renner, S., Rivain, M., Vadnala, P.K.: Conversion of security proofs from one leakage model to another: a new issue. In: Schindler, W., Huss, S.A. (eds.) COSADE 2012. LNCS, vol. 7275, pp. 69–81. Springer, Heidelberg (2012). doi:10.1007/978-3-642-29912-4_6

30. Coron, J.-S., Prouff, E., Rivain, M.: Side channel cryptanalysis of a higher order masking scheme. In: Paillier, P., Verbauwhede, I. (eds.) CHES 2007. LNCS, vol. 4727, pp. 28–44. Springer, Heidelberg (2007). doi:10.1007/978-3-540-74735-2_3

31. Duc, A., Dziembowski, S., Faust, S.: Unifying leakage models: from probing attacks to noisy leakage. In: Nguyen, P.Q., Oswald, E. (eds.) EUROCRYPT 2014. LNCS, vol. 8441, pp. 423–440. Springer, Heidelberg (2014). doi:10.1007/978-3-642-55220-5_24

32. Duc, A., Faust, S., Standaert, F.-X.: Making masking security proofs concrete. In: Oswald, E., Fischlin, M. (eds.) EUROCRYPT 2015. LNCS, vol. 9056, pp. 401–429. Springer, Heidelberg (2015). doi:10.1007/978-3-662-46800-5_16

33. Durvaux, F., Standaert, F.-X.: From improved leakage detection to the detection of points of interests in leakage traces. In: Fischlin, M., Coron, J.-S. (eds.) EURO-CRYPT 2016. LNCS, vol. 9665, pp. 240–262. Springer, Heidelberg (2016). doi:10.1007/978-3-662-49890-3_10

34. Durvaux, F., Standaert, F.-X., Pozo, S.M.: Towards easy leakage certification. In: Gierlichs, B., Poschmann, A.Y. (eds.) CHES 2016. LNCS, vol. 9813, pp. 40–60. Springer, Heidelberg (2016). doi:10.1007/978-3-662-53140-2_3

35. Dziembowski, S., Faust, S.: Leakage-resilient cryptography from the inner-product extractor. In: Lee, D.H., Wang, X. (eds.) ASIACRYPT 2011. LNCS, vol. 7073, pp. 702–721. Springer, Heidelberg (2011). doi:10.1007/978-3-642-25385-0_38

36. Dziembowski, S., Faust, S., Herold, G., Journault, A., Masny, D., Standaert, F.-X.: Towards sound fresh re-keying with hard (physical) learning problems. In: Robshaw, M., Katz, J. (eds.) CRYPTO 2016. LNCS, vol. 9815, pp. 272–301. Springer, Heidelberg (2016). doi:10.1007/978-3-662-53008-5_10

37. Eldib, H., Wang, C.: Synthesis of masking countermeasures against side channel attacks. In: Biere, A., Bloem, R. (eds.) CAV 2014. LNCS, vol. 8559, pp. 114–130. Springer, Heidelberg (2014). doi:10.1007/978-3-319-08867-9_8

38. Eldib, H., Wang, C., Schaumont, P.: Formal verification of software countermeasures against side-channel attacks. ACM Trans. Softw. Eng. Methodol. **24**(2), 11:1–11:24 (2014)

39. Eldib, H., Wang, C., Taha, M.M.I., Schaumont, P.: Quantitative masking strength: quantifying the power side-channel resistance of software code. IEEE Trans. CAD Integr. Circuits Syst. **34**(10), 1558–1568 (2015)

40. Grosso, V., Leurent, G., Standaert, F.-X., Varıcı, K.: LS-designs: bitslice encryption for efficient masked software implementations. In: Cid, C., Rechberger, C. (eds.) FSE 2014. LNCS, vol. 8540, pp. 18–37. Springer, Heidelberg (2015). doi:10.1007/978-3-662-46706-0_2

41. Grosso, V., Standaert, F.-X., Prouff, E.: Low entropy masking schemes, revisited. In: Francillon, A., Rohatgi, P. (eds.) CARDIS 2013. LNCS, vol. 8419, pp. 33–43. Springer, Heidelberg (2014). doi:10.1007/978-3-319-08302-5_3

42. Ishai, Y., Sahai, A., Wagner, D.: Private circuits: securing hardware against probing attacks. In: Boneh, D. (ed.) CRYPTO 2003. LNCS, vol. 2729, pp. 463–481. Springer, Heidelberg (2003). doi:10.1007/978-3-540-45146-4_27

43. Kocher, P., Jaffe, J., Jun, B.: Differential power analysis. In: Wiener, M. (ed.) CRYPTO 1999. LNCS, vol. 1666, pp. 388–397. Springer, Heidelberg (1999). doi:10.1007/3-540-48405-1_25

44. Longo, J., Mulder, E., Page, D., Tunstall, M.: SoC it to EM: electromagnetic side-channel attacks on a complex system-on-chip. In: Güneysu, T., Handschuh, H. (eds.) CHES 2015. LNCS, vol. 9293, pp. 620–640. Springer, Heidelberg (2015). doi:10.1007/978-3-662-48324-4_31

45. Mangard, S., Popp, T., Gammel, B.M.: Side-channel leakage of masked CMOS gates. In: Menezes, A. (ed.) CT-RSA 2005. LNCS, vol. 3376, pp. 351–365. Springer, Heidelberg (2005). doi:10.1007/978-3-540-30574-3_24
46. Mangard, S., Pramstaller, N., Oswald, E.: Successfully attacking masked AES hardware implementations. In: Rao, J.R., Sunar, B. (eds.) CHES 2005. LNCS, vol. 3659, pp. 157–171. Springer, Heidelberg (2005). doi:10.1007/11545262_12
47. Moradi, A.: Statistical tools flavor side-channel collision attacks. In: Pointcheval, D., Johansson, T. (eds.) EUROCRYPT 2012. LNCS, vol. 7237, pp. 428–445. Springer, Heidelberg (2012). doi:10.1007/978-3-642-29011-4_26
48. Moradi, A., Poschmann, A., Ling, S., Paar, C., Wang, H.: Pushing the limits: a very compact and a threshold implementation of AES. In: Paterson, K.G. (ed.) EUROCRYPT 2011. LNCS, vol. 6632, pp. 69–88. Springer, Heidelberg (2011). doi:10.1007/978-3-642-20465-4_6
49. Moss, A., Oswald, E., Page, D., Tunstall, M.: Compiler assisted masking. In: Prouff, E., Schaumont, P. (eds.) CHES 2012. LNCS, vol. 7428, pp. 58–75. Springer, Heidelberg (2012). doi:10.1007/978-3-642-33027-8_4
50. Nikova, S., Rijmen, V., Schläffer, M.: Secure hardware implementation of nonlinear functions in the presence of glitches. J. Cryptol. 24(2), 292–321 (2011)
51. Paul, C.R.: Introduction to Electromagnetic Compatibility. Wiley & Sons, Hoboken (2006)
52. Poschmann, A., Moradi, A., Khoo, K., Lim, C., Wang, H., Ling, S.: Side-channel resistant crypto for less than 2,300 GE. J. Cryptol. 24(2), 322–345 (2011)
53. Prouff, E., Rivain, M.: Masking against side-channel attacks: a formal security proof. In: Johansson, T., Nguyen, P.Q. (eds.) EUROCRYPT 2013. LNCS, vol. 7881, pp. 142–159. Springer, Heidelberg (2013). doi:10.1007/978-3-642-38348-9_9
54. Prouff, E., Rivain, M., Bevan, R.: Statistical analysis of second order differential power analysis. IEEE Trans. Comput. 58(6), 799–811 (2009)
55. Reparaz, O., Bilgin, B., Nikova, S., Gierlichs, B., Verbauwhede, I.: Consolidating masking schemes. In: Gennaro, R., Robshaw, M. (eds.) CRYPTO 2015. LNCS, vol. 9215, pp. 764–783. Springer, Heidelberg (2015). doi:10.1007/978-3-662-47989-6_37
56. Rivain, M., Prouff, E.: Provably secure higher-order masking of AES. In: Mangard, S., Standaert, F.-X. (eds.) CHES 2010. LNCS, vol. 6225, pp. 413–427. Springer, Heidelberg (2010). doi:10.1007/978-3-642-15031-9_28
57. Schneider, T., Moradi, A.: Leakage assessment methodology. In: Güneysu, T., Handschuh, H. (eds.) CHES 2015. LNCS, vol. 9293, pp. 495–513. Springer, Heidelberg (2015). doi:10.1007/978-3-662-48324-4_25
58. Standaert, F.-X., Archambeau, C.: Using subspace-based template attacks to compare and combine power and electromagnetic information leakages. In: Oswald, E., Rohatgi, P. (eds.) CHES 2008. LNCS, vol. 5154, pp. 411–425. Springer, Heidelberg (2008). doi:10.1007/978-3-540-85053-3_26
59. Standaert, F.-X., Malkin, T.G., Yung, M.: A unified framework for the analysis of side-channel key recovery attacks. In: Joux, A. (ed.) EUROCRYPT 2009. LNCS, vol. 5479, pp. 443–461. Springer, Heidelberg (2009). doi:10.1007/978-3-642-01001-9_26
60. Standaert, F.-X., Veyrat-Charvillon, N., Oswald, E., Gierlichs, B., Medwed, M., Kasper, M., Mangard, S.: The world is not enough: another look on second-order DPA. In: Abe, M. (ed.) ASIACRYPT 2010. LNCS, vol. 6477, pp. 112–129. Springer, Heidelberg (2010). doi:10.1007/978-3-642-17373-8_7

61. Tiri, K., Verbauwhede, I.: Securing encryption algorithms against DPA at the logic level: next generation smart card technology. In: Walter, C.D., Koç, Ç.K., Paar, C. (eds.) CHES 2003. LNCS, vol. 2779, pp. 125–136. Springer, Heidelberg (2003). doi:10.1007/978-3-540-45238-6_11
62. Ullrich, M., de Cannière, C., Indesteege, S., Küçük, Ö., Mouha, N., Preneel, B.: Finding optimal bitsliced implementations of 4 × 4-bit S-boxes. In: Symmetric Key Encryption Workshop 2011 (2011)
63. Waddle, J., Wagner, D.: Towards efficient second-order power analysis. In: Joye, M., Quisquater, J.-J. (eds.) CHES 2004. LNCS, vol. 3156, pp. 1–15. Springer, Heidelberg (2004). doi:10.1007/978-3-540-28632-5_1

How Fast Can Higher-Order
Masking Be in Software?

Dahmun Goudarzi[1,2] and Matthieu Rivain[1(✉)]

[1] CryptoExperts, Paris, France
{dahmun.goudarzi,matthieu.rivain}@cryptoexperts.com
[2] ENS, CNRS, Inria and PSL Research University, Paris, France

Abstract. Higher-order masking is widely accepted as a sound counter-measure to protect implementations of blockciphers against side-channel attacks. The main issue while designing such a countermeasure is to deal with the nonlinear parts of the cipher *i.e.* the so-called s-boxes. The prevailing approach to tackle this issue consists in applying the Ishai-Sahai-Wagner (ISW) scheme from CRYPTO 2003 to some polynomial representation of the s-box. Several efficient constructions have been proposed that follow this approach, but higher-order masking is still considered as a costly (impractical) countermeasure. In this paper, we investigate efficient higher-order masking techniques by conducting a case study on ARM architectures (the most widespread architecture in embedded systems). We follow a bottom-up approach by first investigating the implementation of the base field multiplication at the assembly level. Then we describe optimized low-level implementations of the ISW scheme and its variant (CPRR) due to Coron *et al.* (FSE 2013) [14]. Finally we present improved state-of-the-art polynomial decomposition methods for s-boxes with custom parameters and various implementation-level optimizations. We also investigate an alternative to these methods which is based on bitslicing at the s-box level. We describe new masked bitslice implementations of the AES and PRESENT ciphers. These implementations happen to be significantly faster than (optimized) state-of-the-art polynomial methods. In particular, our bitslice AES masked at order 10 runs in 0.48 megacycles, which makes 8 ms in presence of a 60 MHz clock frequency.

1 Introduction

Since their introduction in the late 1990's, side-channel attacks have been considered as a serious threat against cryptographic implementations. Among the existing protection strategies, one of the most widely used relies on applying *secret sharing* at the implementation level, which is known as (*higher-order*) *masking*. This strategy achieves provable security in the so-called *probing security model* [24] and *noisy leakage model* [17,32], which makes it a prevailing way to get secure implementations against side-channel attacks.

© International Association for Cryptologic Research 2017
J.-S. Coron and J.B. Nielsen (Eds.): EUROCRYPT 2017, Part I, LNCS 10210, pp. 567–597, 2017.
DOI: 10.1007/978-3-319-56620-7_20

Higher-Order Masking. Higher-order masking consists in sharing each internal variable x of a cryptographic computation into d random variables $x_1, x_2, \ldots, x_d$, called *the shares* and satisfying $x_1 + x_2 + \cdots + x_d = x$ for some group operation $+$, such that any set of $d - 1$ shares is randomly distributed and independent of x. In this paper, we will consider the prevailing *Boolean masking* which is based on the bitwise addition of the shares. It has been formally demonstrated that in the noisy leakage model, where the attacker gets noisy information on each share, the complexity of recovering information on x grows exponentially with the number of shares [12,32]. This number d, called *the masking order*, is hence a sound security parameter for the resistance of a masked implementation.

When dth-order masking is involved to protect a blockcipher, a so-called dth-order masking scheme must be designed to enable computation on masked data. To be sound, a dth order masking scheme must satisfy the two following properties: *(i) completeness*, at the end of the encryption/decryption, the sum of the d shares must give the expected result; *(ii) probing security*, every tuple of $d - 1$ or less intermediate variables must be independent of any sensitive variable.

Most blockcipher structures are composed of one or several linear transformation(s), and a non-linear function, called the *s-box* (where the linearity is considered w.r.t. the bitwise addition). Computing a linear transformation $x \mapsto \ell(x)$ in the masking world can be done in $O(d)$ complexity by applying ℓ to each share independently. This clearly maintains the probing security and the completeness holds by linearity since we have $\ell(x_1) + \ell(x_2) + \cdots + \ell(x_d) = \ell(x)$. On the other hand, the non-linear operations (such as s-boxes) are more tricky to compute on the shares while ensuring completeness and probing security.

Masked S-boxes. In [24], Ishai, Sahai, and Wagner tackled this issue by introducing the first generic higher-order masking scheme for the multiplication over $\mathbb{F}_2$ in complexity $O(d^2)$. The here-called ISW scheme was later used by Rivain and Prouff to design an efficient masked implementation of AES [34]. Several works then followed to improve this approach and to extend it to other SPN blockciphers [10,14,15,26]. The principle of these methods consists in representing an n-bit s-box as a polynomial $\sum_i a_i x^i$ in $\mathbb{F}_{2^n}[x]/(x^{2^n} - x)$, whose evaluation is then expressed as a sequence of linear functions (*e.g.* squaring, additions, multiplications by constant coefficients) and *nonlinear multiplications* over $\mathbb{F}_{2^n}$. The former are simply masked in complexity $O(d)$ (thanks to their linearity), whereas the latter are secured using ISW in complexity $O(d^2)$. The total complexity is hence mainly impacted by the number of nonlinear multiplications involved in the underlying polynomial evaluation. This observation led to a series of publications aiming at conceiving polynomial evaluation methods with the least possible nonlinear multiplications [10,15,35]. The so-called CRV method, due to Coron et al. [15], is currently the best known generic method with respect to this criteria.

Recently, an alternative to previous ISW-based polynomial methods was proposed by Carlet, Prouff, Rivain and Roche in [11]. They introduce a so-called *algebraic decomposition method* that can express an s-box in terms of polynomials of low algebraic degree. They also show that a variant of ISW due to Coron

Prouff, Rivain and Roche [14] can efficiently be used to secure the computation of any quadratic function. By combining the here-called CPRR scheme together with their algebraic decomposition method, Carlet *et al.* obtain an efficient alternative to existing ISW-based masking schemes. In particular, their technique is argued to beat the CRV method based on the assumption that an efficiency gap exists between an ISW multiplication and a CPRR evaluation. However, no optimized implementation is provided to back up this assumption.

Despite these advances, higher-order masking still implies strong performance overheads on protected implementations, and it is often believed to be impractical beyond small orders. On the other hand, most published works on the subject focus on theoretical aspects without investigating optimized low-level implementations. This raises the following question: *how fast can higher-order masking be in software?*

Our Contribution. In this paper, we investigate this question and present a case study on ARM (v7) architectures, which are today the most widespread in embedded systems (privileged targets of side-channel attacks). We provide an extensive and fair comparison between the different methods of the state of the art and a benchmarking on optimized implementations of higher-order masked blockciphers. For such purpose, we follow a bottom-up approach and start by investigating the efficient implementation of the base-field multiplication, which is the core elementary operation of the ISW-based masking schemes. We propose several implementations strategies leading to different time-memory trade-offs. We then investigate the two main building blocks of existing masking schemes, namely the ISW and CPRR schemes. We optimize the implementation of these schemes and we describe parallelized versions that achieve significant gains in performances. From these results, we propose fine-tuned variants of the CRV and algebraic decomposition methods, which allows us to compare them in a practical and optimized implementation context. We also investigate efficient polynomial methods for the specific s-boxes of two important blockciphers, namely AES and PRESENT.

As an additional contribution, we put forward an alternative strategy to polynomial methods which consists in applying bitslicing at the s-box level. More precisely, the s-box computations within a blockcipher round are bitsliced so that the core nonlinear operation is not a field multiplication anymore (nor a quadratic polynomial) but a bitwise logical AND between two m-bit registers (where m is the number of s-box computations). This allows us to translate compact hardware implementations of the AES and PRESENT s-boxes into efficient masked implementations in software. This approach has been previously used to design blockciphers well suited for masking [21] but, to the best of our knowledge, has never been used to derive efficient higher-order masked implementations of existing standard blockciphers such as AES or PRESENT. We further provide implementation results for full blockciphers and discuss the security aspects of our implementations.

Our results clearly demonstrate the superiority of the bitslicing approach (at least on 32-bit ARM architectures). Our masked bitslice implementations of AES and PRESENT are significantly faster than state-of-the-art polynomial methods with fine-tuned low-level implementations. In particular, an encryption masked at the order 10 only takes a few milliseconds with a 60 MHz clock frequency (specifically 8 ms for AES and 5 ms for PRESENT).

Other Related Works. Our work focuses on the optimized implementation of *polynomial methods* for efficient higher-order masking of s-boxes and block-ciphers, as well as on the bitslice alternative. All these schemes are based on Boolean masking with the ISW construction (or the CPRR variant) for the core non-linear operation (which is either the field multiplication or the bitwise logical AND). Further masking techniques exist with additional features that should be adverted here.

Genelle, Prouff and Quisquater suggest mixing Boolean masking and *multiplicative masking* [19]. This approach is especially effective for blockciphers with inversion-based s-boxes such as AES. Prouff and Roche turn classic constructions from multi-party computation into a higher-order masking scheme resilient to glitches [33]. A software implementation study comparing these two schemes and classical polynomial methods for AES has been published in [23]. Compared to this previous work, our approach is to go deeper in the optimization (at the assembly level) and we further investigate generic methods (*i.e.* methods that apply to any s-box and not only to AES). Another worth-mentioning line of works is the field of *threshold implementations* [29,30] in which the principle of threshold cryptography is applied to get secure hardware masking in the presence of glitches (see for instance [6,28,31]). Most of threshold implementations target first-order security but recent works discuss the extension to higher orders [5]. It should be noted that in the context of hardware implementations, the occurrence of glitches prevents the straight use of classic ISW-based Boolean masking schemes (as considered in the present work). Threshold implementations and the Prouff-Roche scheme are therefore the main solutions for (higher-order) masking in hardware. On the other hand, these schemes are not competitive for the software context (due to limited masking orders and/or to an increased complexity) and they are consequently out of the scope of our study.

Finally, we would like to mention that subsequently to the first version of this work, and motivated by the high performances of our bitslice implementations of AES and PRESENT, we have extended the bitslice higher-order masking approach to any s-box by proposing a generic decomposition method in [20]. New blockcipher designs with efficient masked bitslice implementation have also been recently proposed in [25].

Paper Organization. The next section provides some preliminaries about ARM architectures (Sect. 2). We then investigate the base field multiplication (Sect. 3) and the ISW and CPRR schemes (Sect. 4). Afterward, we study polynomial methods for s-boxes (Sect. 5) and we introduce our masked bitslice implementations of the AES and PRESENT s-boxes (Sect. 6). Eventually, we describe our implementations of the full ciphers (Sect. 7). The security aspects of our implementations are further discussed in the full version of the paper.

Source Code and Performances. For the sake of illustration, the performances of our implementations are mostly displaid on graphics in the present version. Exact performance figures (in terms of clock cycles, code size and RNG consumption) are provided in the full version of the paper (available on IACR ePrint). The source code of our implementations is also available on GitHub.

2   Preliminaries on ARM Architectures

Most ARM cores are RISC processors composed of sixteen 32-bit registers, labeled R0, R1, ..., R15. Registers R0 to R12 are known as *variable registers* and are available for computation.[1] The three last registers are usually reserved for special purposes: R13 is used as the stack pointer (SP), R14 is the link register (LR) storing the return address during a function call, and R15 is the program counter (PC). The link register R14 can also be used as additional variable register by saving the return address on the stack (at the cost of push/pop instructions). The gain of having a bigger register pool must be balanced with the saving overhead, but this trick enables some improvements in many cases.

In ARM v7, most of the instructions can be split into the following three classes: *data instructions, memory instructions*, and *branching instructions*. The data instructions are the arithmetic and bitwise operations, each taking one clock cycle (except for the multiplication which takes two clock cycles). The memory instructions are the load and store (from and to the RAM) which require 3 clock cycles, or their variants for multiple loads or stores ($n + 2$ clock cycles). The last class of instructions is the class of branching instructions used for loops, conditional statements and function calls. These instructions take 3 or 4 clock cycles.

One important specificity of the ARM assembly is the *barrel shifter* allowing any data instruction to shift one of its operands at no extra cost in terms of clock cycles. Four kinds of shifting are supported: the logical shift left (LSL), the logical shift right (LSR), the arithmetic shift right (ASR), and the rotate-right (ROR). All these shifting operations are parameterized by a shift length in $[\![1, 32]\!]$ (except for the logical shift left LSL which lies in $[\![0, 31]\!]$). The latter can also be relative by using a register but in that case the instruction takes an additional clock cycle.

Eventually, we assume that our target architecture includes a fast True Random Number Generator (TRNG), that frequently fills a register with a fresh 32-bit random strings (*e.g.* every 10 clock cycles). The TRNG register can then be read at the cost of a single load instruction.[2]

---

[1] Note that some conventions exist for the first four registers R0–R3, also called *argument registers*, and serving to store the arguments and the result of a function at call and return respectively.

[2] This is provided that the TRNG address is already in a register. Otherwise one must first load the TRNG address, before reading the random value. Our code ensures a gap of at least 10 clock cycles between two readings of the TRNG.

3   Base Field Multiplication

In this section, we focus on the efficient implementation of the multiplication over $\mathbb{F}_{2^n}$ where n is small (typically $n \in [\![4, 10]\!]$). The fastest method consists in using a precomputed table mapping the 2^{2n} possible pairs of operands (a, b) to the output product $a \cdot b$.

In the context of embedded systems, one is usually constrained on the code size and spending several kilobytes for (one table in) a cryptographic library might be prohibitive. That is why we investigate hereafter several alternative solutions with different time-memory trade-offs. Specifically, we look at the classical binary algorithm and exp-log multiplication methods. We also describe a tabulated version of Karatsuba multiplication, and another table-based method: the *half-table multiplication*. The obtained implementations are compared in terms of clock cycles, register usage, and code size (where the latter is mainly impacted by precomputed tables).

In the rest of this section, the two multiplication operands in $\mathbb{F}_{2^n}$ will be denoted a and b. These elements can be seen as polynomials $a(x) = \sum_{i=0}^{n-1} a_i x^i$ and $b(x) = \sum_{i=0}^{n-1} b_i x^i$ over $\mathbb{F}_2[x]/p(x)$ where the a_i's and the b_i's are binary coefficients and where p is a degree-n irreducible polynomial over $\mathbb{F}_2[x]$. In our implementations, these polynomials are simply represented as n-bit strings $a = (a_{n-1}, \ldots, a_0)_2$ or equivalently $a = \sum_{i=0}^{n-1} a_i \, 2^i$ (and similarly for b).

3.1   Binary Multiplication

The binary multiplication algorithm is the most basic way to perform a multiplication on a binary field. It consists in evaluating the following formula:

$$a(x) \cdot b(x) = \left(\cdots \left(\left(b_{n-1} a(x) x + b_{n-2} a(x) \right) x + b_{n-3} a(x) \right) \cdots \right) x + b_0 a(x), \quad (1)$$

by iterating over the bits of b. A formal description is given in Algorithm 1.

Algorithm 1. Binary multiplication algorithm

Input: $a(x), b(x) \in \mathbb{F}_2[x]/p(x)$
Output: $a(x) \cdot b(x) \in \mathbb{F}_2[x]/p(x)$
1. $r(x) \leftarrow 0$
2. **for** $i = n - 1$ **down to** 0 **do**
3. $r(x) \leftarrow x \cdot r(x) \bmod p(x)$
4.   **if** $b_i = 1$ **then** $r(x) \leftarrow r(x) + a(x)$
5. **end for**
6. **return** $r(x) \bmod p(x)$

The reduction modulo $p(x)$ can be done either inside the loop (at Step 3 in each iteration) or at the end of the loop (at Step 6). If the reduction is done inside the loop, the degree of $x \cdot r(x)$ is at most n in each iteration. So we have

$$x \cdot r(x) \bmod p(x) = \begin{cases} x \cdot r(x) - p(x) & \text{if } r_{n-1} = 1 \\ x \cdot r(x) & \text{otherwise} \end{cases} \quad (2)$$

The reduction then consists in subtracting $p(x)$ to $x \cdot r(x)$ if and only if $r_{n-1} = 1$ and doing nothing otherwise. In practice, the multiplication by x simply consists in left-shifting the bits of r and the subtraction of p is a simple XOR. The tricky part is to conditionally perform the latter XOR with respect to the bit r_{n-1} as we aim to a branch-free code. This is achieved using the *arithmetic right shift*[3] instruction (sometimes called signed shift) to compute $(r \ll 1) \oplus (r_{n-1} \times p)$ by putting r_{n-1} at the sign bit position, which can be done in 3 ARM instructions (3 clock cycles) as follows:

```
LSL $tmp, $res, #(32-n)         ;; tmp = r_{n-1}
AND $tmp, $mod, $tmp, ASR #32   ;; tmp = p & (tmp ASR 32)
EOR $res, $tmp, $res, LSL #1    ;; r = (r_{n-1} * p)^(r << 1)
```

Step 4 consists in conditionally adding a to r whenever b_i equals 1. Namely, we have to compute $r \oplus (b_i \times a)$. In order to multiply a by b_i, we use the rotation instruction to put b_i in the sign bit and the arithmetic shift instruction to fill a register with b_i. The latter register is then used to mask a with a bitwise AND instruction. The overall Step 4 is performed in 3 ARM instructions (3 clock cycles) as follows:

```
ROR $opB, #31                   ;; b_i = sign(opB)
AND $tmp, $opA, #opB, ASR #32   ;; tmp = a & (tmp ASR 32)
EOR $res, $tmp                  ;; r = r^(a * b_i)
```

Variant. If the reduction is done at the end of the loop, Step 3 then becomes a simple left shift, which can be done together with Step 4 in 3 instructions (3 clock cycles) as follows:

```
ROR $opB, #31                        ;; b_i = sign(opB)
AND $tmp, $opA, $opB, ASR #32        ;; tmp = a & (tmp ASR 32)
EOR $res, $tmp, $res, LSL #1         ;; r = (a * b_i)^(r << 1)
```

The reduction must then be done at the end of the loop (Step 6), where we have $r(x) = a(x) \cdot b(x)$ which can be of degree up to $2n - 2$. Let r_h and r_ℓ be the polynomials of degree at most $n - 2$ and $n - 1$ such that $r(x) = r_h(x) \cdot x^n + r_\ell(x)$. Since we have $r(x) \bmod p(x) = (r_h(x) \cdot x^n \bmod p(x)) + r_\ell(x)$, we only need to reduce the high-degree part $r_h(x) \cdot x^n$. This can be done by tabulating the function mapping the $n - 1$ coefficients of $r_h(x)$ to the $n - 2$ coefficients of $r_h(x) \cdot x^n \bmod p(x)$. The overall final reduction then simply consists in computing $T[r \gg n] \oplus (r \wedge (2^n - 1))$, where T is the corresponding precomputed table.

---

[3] This instruction performs a logical right-shift but instead of filling the vacant bits with 0, it fills these bits with the leftmost bit operand (*i.e.* the sign bit).

3.2 Exp-Log Multiplication

Let $g \in \mathbb{F}_{2^n}$ be a generator of the multiplicative group $\mathbb{F}_{2^n}^*$. We shall denote by $\exp_g$ the exponential function defined over $[\![0, 2^n - 1]\!]$ as $\exp_g(\ell) = g^\ell$, and by $\log_g$ the discrete logarithm function defined over $\mathbb{F}_{2^n}^*$ as $\log_g = \exp_g^{-1}$. Assume that these functions can be tabulated (which is usually the case for small values of n). The multiplication between field elements a and b can then be efficiently computed as

$$a \cdot b = \begin{cases} \exp_g(\log_g(a) + \log_g(b) \bmod 2^n - 1) & \text{if } a \neq 0 \text{ and } b \neq 0 \\ 0 & \text{otherwise} \end{cases} \quad (3)$$

Le us denote $t = \log_g(a) + \log_g(b)$. We have $t \in [\![0, 2^{n+1} - 2]\!]$ giving

$$t \bmod 2^n - 1 = \begin{cases} t - 2^n + 1 & \text{if } t_n = 1 \\ t & \text{otherwise} \end{cases} \quad (4)$$

where t_n is the most significant bit in the binary expansion $t = \sum_{i=0}^n t_i 2^i$, which can be rewritten as $t \bmod 2^n - 1 = (t + t_n) \wedge (2^n - 1)$. This equation can be evaluated with 2 ARM instructions[4] (2 clock cycles) as follows:

```
ADD $tmp , $tmp , LSR #n        ;;tmp = tmp + tmp>>n
AND $tmp , #(2^n-1)             ;;tmp = tmp & (2^n-1)
```

Variant. Here again, a time-memory trade-off is possible: the $\exp_g$ table can be doubled in order to handle a $(n + 1)$-bit input and to perform the reduction. This simply amounts to consider that $\exp_g$ is defined over $[\![0, 2^{n+1} - 2]\!]$ rather than over $[\![0, 2^n - 1]\!]$.

Zero-Testing. The most tricky part of the exp-log multiplication is to manage the case where a or b equals 0 while avoiding any conditional branch. Once again we can use the arithmetic right-shift instruction to propagate the sign bit and use it as a mask. The test of zero can then be done with 4 ARM instructions (4 clock cycles) as follows:

```
RSB $tmp , $opA , #0             ;; tmp = 0 - a
AND $tmp , $opB , $tmp , ASR #32 ;; tmp = b & (tmp ASR 32)
RSB $tmp , #0                    ;; tmp = 0 - tmp
AND $res , $tmp , ASR #32        ;; r = r & (tmp ASR 32)
```

---

[4] Note that for $n > 8$, the constant $2^n - 1$ does not lie in the range of constants enabled by ARM (*i.e.* rotated 8-bit values). In that case, one can use the BIC instruction to perform a logical AND where the second argument is complemented. The constant to be used is then 2^n which well belongs to ARM constants whatever the value of n.

3.3 Karatsuba Multiplication

The Karatsuba method is based on the following equation:

$$a \cdot b = (a_h + a_\ell)(b_h + b_\ell)\, x^{\frac{n}{2}} + a_h\, b_h\, (x^n + x^{\frac{n}{2}}) + a_\ell\, b_\ell\, (x^{\frac{n}{2}} + 1) \bmod p(x) \quad (5)$$

where a_h, a_ℓ, b_h, b_ℓ are the $\frac{n}{2}$-degree polynomials such that $a(x) = a_h\, x^{\frac{n}{2}} + a_\ell$ and $b(x) = b_h\, x^{\frac{n}{2}} + b_\ell$. The above equation can be efficiently evaluated by tabulating the following functions:

$$(a_h + a_\ell, b_h + b_\ell) \mapsto (a_h + a_\ell)(b_h + b_\ell)\, x^{\frac{n}{2}} \bmod p(x),$$
$$(a_h, b_h) \mapsto a_h\, b_h\, (x^n + x^{\frac{n}{2}}) \bmod p(x),$$
$$(a_\ell, b_\ell) \mapsto a_\ell\, b_\ell\, (x^{\frac{n}{2}} + 1) \bmod p(x).$$

We hence obtain a way to compute the multiplication with 3 look-ups and a few XORs based on 3 tables of 2^n elements.

In practice, the most tricky part is to get the three pairs $(a_h \| b_h)$, $(a_\ell \| b_\ell)$ and $(a_h + a_\ell \| b_h + b_\ell)$ to index the table with the least instructions possible. The last pair is a simple addition of the two first ones. The computation of the two first pairs from the operands $a \equiv (a_h \| a_\ell)$ and $b \equiv (b_h \| b_\ell)$ can then be seen as the transposition of a 2×2 matrix. This can be done with 4 ARM instructions (4 clock cycles) as follows:

```
EOR  $tmp0 ,  $opA ,  $opB , LSR #(n/2)      ;; tmp0 = [a_h|a_l^b_h]
EOR  $tmp1 ,  $opB ,  $tmp0 , LSL #(n/2)     ;; tmp1 = [a_h|a_l|b_l]
BIC  $tmp1 ,  #(2^n*(2^(n/2)-1))             ;; tmp1 = [a_l|b_l]
EOR  $tmp0 ,  $tmp1 , LSR #(n/2)             ;; tmp0 = [a_h|b_h]
```

3.4 Half-Table Multiplication

The half-table multiplication can be seen as a trade-off between the Karatsuba method and the full-table method. While Karatsuba involves 3 look-ups in three 2^n-sized tables and the full-table method involves 1 look-up in a 2^{2n}-sized table, the half-table method involves 2 look-ups in two $2^{\frac{3n}{2}}$-sized tables. It is based on the following equation:

$$a \cdot b = b_h\, x^{\frac{n}{2}} (a_h\, x^{\frac{n}{2}} + a_\ell) + b_\ell\, (a_h\, x^{\frac{n}{2}} + a_\ell) \bmod p(x), \quad (6)$$

which can be efficiently evaluated by tabulating the functions:

$$(a_h, a_\ell, b_h) \mapsto b_h\, x^{\frac{n}{2}} (a_h\, x^{\frac{n}{2}} + a_\ell) \bmod p(x),$$
$$(a_h, a_\ell, b_\ell) \mapsto b_\ell\, (a_h\, x^{\frac{n}{2}} + a_\ell) \bmod p(x).$$

Once again, the barrel shifter is useful to get the input triplets efficiently. Each look-up can be done with two ARM instructions (for a total of 8 clock cycles) as follows:

```
EOR   $tmp,$opB,$opA,LSL#n            ;;tmp=[a_h|a_l|b_h|b_l]
LDRB  $res,[$tab1,$tmp,LSR#(n/2)]     ;;res=T1[a_h|a_l|b_h]
EOR   $tmp,$opA,$opB,LSL#(32-n/2)     ;;tmp=[b_l|0..|a_h|a_l]
LDRB  $tmp,[$tab2,$tmp,ROR#(32-n/2)]  ;;tmp=T2[a_h|a_l|b_l]
```

3.5 Performances

The obtained performances are summarized in Table 1 in terms of clock cycles, register usage, and code size. For clock cycles, the number in brackets indicates instructions that need to be done only once when multiple calls to the multiplication are performed (as in the secure multiplication procedure described in the next section). These are initialization instructions such as loading a table address in a register. For $n > 8$, elements take two bytes to be stored (assuming $n \leq 16$) which implies an overhead in clock cycles and a doubling of the table size. For most methods, the clock cycles and register usage are constant w.r.t. $n \geq 8$, whereas the code size depends on n. For the sake of illustration, we therefore additionally display the code size (and corresponding LUT sizes) in Fig. 1 for several values of n.

Table 1. Multiplication performances.

	bin mult v1	bin mult v2	exp-log v1	exp-log v2	kara.	half-tab	full-tab
clock cycles ($n \leq 8$)	$10n+3$ (+3)	$7n+3$ (+3)	18 (+2)	16 (+2)	19 (+2)	10 (+3)	4 (+3)
clock cycles ($n > 8$)	$10n+4$ (+3)	$7n+15$ (+3)	35 (+2)	31 (+2)	38 (+2)	n/a	n/a
registers	5	5	5 (+1)	5 (+1)	6 (+1)	5 (+1)	5
code size ($n \leq 8$)	52	$2^{n-1}+48$	$2^{n+1}+48$	$3 \cdot 2^n+40$	$3 \cdot 2^n+42$	$2^{\frac{3n}{2}+1}+24$	$2^{2n}+12$

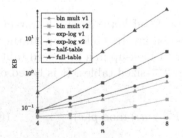

n	4	6	8	10
Binary v1	0	0	0	0
Binary v2	8 B	32 B	128 B	1 KB
Exp-log v1	32 B	128 B	0.5 KB	4 KB
Exp-log v2	48 B	192 B	0.75 KB	6 KB
Karatsuba	48 B	192 B	0.75 KB	6 KB
Half-table	0.13 KB	1 KB	8 KB	128 KB
Full-table	0.25 KB	4 KB	64 KB	2048 KB

Fig. 1. Full code size (left graph) and LUT size (right table) w.r.t. n.

We observe that all the methods provide different time-memory trade-offs except for Karatsuba which is beaten by the exp-log method (v1) both in terms of clock cycles and code size. The latter method shall then always be preferred to the former (at least on our architecture). As expected, the full-table method is by far the fastest way to compute a field multiplication, followed by the half-table method. However, depending on the value of n, these methods might be too consuming in terms of code size due to their large precomputed tables. On the other hand, the binary multiplication (even the improved version) has very poor performances in terms of clock cycles and it should only be used for extreme cases where the code size is very constrained. We consider that the exp-log method v2 (i.e. with doubled exp-table) is a good compromise between code size an speed whenever the full-table and half-table methods are not affordable (which might be the case for e.g. $n \geq 8$). In the following, we shall therefore focus our study on secure implementations using the exp-log (v2), half-table or full-table method for the base field multiplication.

4 Secure Multiplications and Quadratic Evaluations

We have seen several approaches to efficiently implement the base-field multiplication. We now investigate the secure multiplication in the masking world where the two operands $a, b \in \mathbb{F}_{2^n}$ are represented as random d-sharings $(a_1, a_2, \ldots, a_d)$ and $(b_1, b_2, \ldots, b_d)$. We also address the secure evaluation of a function f of algebraic degree 2 over $\mathbb{F}_{2^n}$ (called *quadratic function* in the following). Specifically, we focus on the scheme proposed by Ishai, Sahai, and Wagner (ISW scheme) for the secure multiplication [24], and its extension by Coron, Prouff, Rivain and Roche (CPRR scheme) to secure any quadratic function [11,14].

4.1 Algorithms

ISW Multiplication. From two d-sharings $(a_1, a_2, \ldots, a_d)$ and $(b_1, b_2, \ldots, b_d)$, the ISW scheme computes an output d-sharing $(c_1, c_2, \ldots, c_d)$ as follows:

1. for every $1 \leq i < j \leq d$, sample a random value $r_{i,j}$ over $\mathbb{F}_{2^n}$;
2. for every $1 \leq i < j \leq d$, compute $r_{j,i} = (r_{i,j} + a_i \cdot b_j) + a_j \cdot b_i$;
3. for every $1 \leq i \leq d$, compute $c_i = a_i \cdot b_i + \sum_{j \neq i} r_{i,j}$.

One can check that the output $(c_1, c_2, \ldots, c_d)$ is well a d-sharing of the product $c = a \cdot b$. We indeed have $\sum_i c_i = \sum_{i,j} a_i \cdot b_j = (\sum_i a_i)(\sum_j b_j)$ since every random value $r_{i,j}$ appears exactly twice in the sum and hence vanishes.

Mask Refreshing. The ISW multiplication was originally proved probing secure at the order $t = \lfloor (d-1)/2 \rfloor$ (and not $d-1$ as one would expect with masking order d). The security proof was later made tight under the condition that the input d-sharings are based on independent randomness [34]. In some situations, this independence property is not satisfied. For instance, one might have to multiply two values a and b where $a = \ell(b)$ for some linear operation ℓ. In that case, the shares of a are usually derived as $a_i = \ell(b_i)$, which clearly breaches the required independence of input shares. To deal with this issue, one must refresh the sharing of a. However, one must be careful doing so since a bad refreshing procedure might introduce a flaw [14]. A sound method for mask-refreshing consists in applying an ISW multiplication between the sharing of a and the tuple $(1, 0, 0, \ldots, 0)$ [2,17]. This gives the following procedure:

1. for every $1 \leq i < j \leq d$, randomly sample $r_{i,j}$ over $\mathbb{F}_{2^n}$ and set $r_{j,i} = r_{i,j}$;
2. for every $1 \leq i \leq d$, compute $a_i' = a_i + \sum_{j \neq i} r_{i,j}$.

It is not hard to see that the output sharing $(a_1', a_2', \ldots, a_d')$ well encodes a. One might think that such a refreshing implies a strong overhead in performances (almost as performing two multiplications) but this is still better than doubling the number of shares (which roughly quadruples the multiplication time). Moreover, we show hereafter that the implementation of such a refreshing procedure can be very efficient in practice compared to the ISW multiplication.

CPRR Evaluation. The CPRR scheme was initially proposed in [14] as a variant of ISW to securely compute multiplications of the form $x \mapsto x \cdot \ell(x)$ where ℓ is linear, without requiring refreshing. It was then shown in [11] that this scheme (in a slightly modified version) could actually be used to securely evaluate any quadratic function f over $\mathbb{F}_{2^n}$. The method is based on the following equation

$$f(x_1 + x_2 + \cdots + x_d) = \sum_{1 \leq i < j \leq d} f(x_i + x_j + s_{i,j}) + f(x_j + s_{i,j}) + f(x_i + s_{i,j}) + f(s_{i,j})$$

$$+ \sum_{i=1}^{d} f(x_i) + (d + 1 \bmod 2) \cdot f(0) \tag{7}$$

which holds for every $(x_i)_i \in (\mathbb{F}_{2^n})^d$, every $(s_{i,j})_{1 \leq i < j \leq d} \in (\mathbb{F}_{2^n})^{d(d-1)/2}$, and every quadratic function f over $\mathbb{F}_{2^n}$.

From a d-sharing $(x_1, x_2, \ldots, x_d)$, the CPRR scheme computes an output d-sharing $(y_1, y_2, \ldots, y_d)$ as follows:

1. for every $1 \leq i < j \leq d$, sample two random values $r_{i,j}$ and $s_{i,j}$ over $\mathbb{F}_{2^n}$,
2. for every $1 \leq i < j \leq d$, compute $r_{j,i} = r_{i,j} + f(x_i + s_{i,j}) + f(x_j + s_{i,j}) + f((x_i + s_{i,j}) + x_j) + f(s_{i,j})$,
3. for every $1 \leq i \leq d$, compute $y_i = f(x_i) + \sum_{j \neq i} r_{i,j}$,
4. if d is even, set $y_1 = y_1 + f(0)$.

According to (7), we then have $\sum_{i=1}^{d} y_i = f\left(\sum_{i=1}^{d} x_i\right)$, which shows that the output sharing $(y_1, y_2, \ldots, y_d)$ well encodes $y = f(x)$.

In [11,14] it is argued that in the gap where the field multiplication cannot be fully tabulated (2^{2n} elements is too much) while a function $f : \mathbb{F}_{2^n} \to \mathbb{F}_{2^n}$ can be tabulated (2^n elements fit), the CPRR scheme is (likely to be) more efficient than the ISW scheme. This is because it essentially replaces (costly) field multiplications by simple look-ups. We present in the next section the results of our study for our optimized ARM implementations.

4.2 Implementations and Performances

For both schemes we use the approach suggested in [13] that directly accumulates each intermediate result $r_{i,j}$ in the output share c_i so that the memory cost is

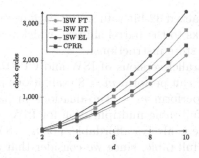

Fig. 2. Timings of ISW and CPRR schemes.

$O(d)$ instead of $O(d^2)$ when the $r_{i,j}$'s are stored. Detailed algorithms can be found in the appendix. The ARM implementation of these algorithms is rather straightforward and it does not make use of any particular trick.

As argued in Sect. 3.5, we consider three variants for the base field multiplication in the ISW scheme, namely the full-table method, the half-table method and the exp-log method (with doubled exp table). The obtained ISW variants are labeled ISW-FT, ISW-HT and ISW-EL in the following. The obtained performances are illustrated in Fig. 2 with respect to d. Note that we did not consider ISW-FT for $n > 8$ since the precomputed tables are too huge.

These results show that CPRR indeed outperforms ISW whenever the field multiplication cannot be fully tabulated. Even the half-table method (which is more consuming in code-size) is slower than CPRR. For $n \leq 8$, a CPRR evaluation asymptotically costs 1.16 ISW-FT, 0.88 ISW-HT, and 0.75 ISW-EL.

4.3 Parallelization

Both ISW and CPRR schemes work on n-bit variables, each of them occupying a full 32-bit register. Since in most practical scenarios, we have $n \in [\![4, 8]\!]$, this situation is clearly suboptimal in terms of register usage, and presumably suboptimal in terms of timings. A natural idea to improve this situation is to use parallelization. A register can simultaneously store $m := \lfloor 32/n \rfloor$ values, we can hence try to perform m ISW/CPRR computations in parallel (which would in turn enable to perform m s-box computations in parallel). Specifically, each input shares is replaced by m input shares packed into a 32-bit value. The ISW (resp. CPRR) algorithm load packed values, and perform the computation on each unpacked n-bit chunk one-by-one. Using such a strategy allows us to save multiple load and store instructions, which are among the most expensive instructions of ARM assembly (3 clock cycles). Specifically, we can replace m load instructions by a single one for the shares a_i, b_j in ISW (resp. x_i, x_j in CPRR) and the random values $r_{i,j}$, $s_{i,j}$ (read from the TRNG), we can replace m store instructions by a single one for the output shares, and we can replace m XOR instructions by a single one for some of the addition involved in ISW (resp. CPRR). On the other hand, we get an overhead for the extraction of the

n-bit chunks from the packed 32-bit values. But each of these extractions takes a single clock cycle (thanks to the barrel shifter), which is rather small compared to the gain in load and store instructions.

We implemented parallel versions of ISW and CPRR for $n = 4$ and $n = 8$. For the former case, we can perform $m = 8$ evaluations in parallel, whereas for the later case we can perform $m = 4$ evaluations in parallel. For $n = 4$, we only implemented the full-table multiplication for ISW, since we consider that a 256-byte table in code is always affordable. For $n = 8$ on the other hand, we did not implement the full-table, since we consider that a 64-KB table in code would be to much in most practical scenarios. Figures 3 and 4 give the obtained performances in terms of clock cycles.

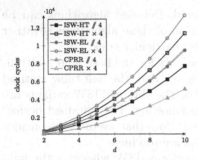
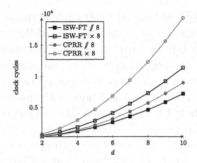

Fig. 3. Timings of (parallel) ISW and CPRR schemes for $n = 8$.

Fig. 4. Timings of (parallel) ISW and CPRR schemes for $n = 4$.

These results show the important gain obtained by using parallelism. For ISW, we get an asymptotic gain around 30% for 4 parallel evaluations ($n = 8$) compared to 4 serial evaluations, and we get a 58% asymptotic gain for 8 parallel evaluations ($n = 4$) compared to 8 serial evaluations. For CPRR, the gain is around 50% (timings are divided by 2) in both cases ($n = 8$ and $n = 4$). We also observe that the efficiency order keeps unchanged with parallelism, that is: ISW-FT > CPRR > ISW-HT > ISW-EL.

Remark 1. Note that using parallelization in our implementations does not compromise the probing security. Indeed, we pack several bytes/nibbles within one word of the cipher state but we never pack (part of) different shares of the same variable together. The probing security proofs hence apply similarly to the parallel implementations.[5]

---

[5] Putting several shares of the same variable in a single register would induce a security flaw in the probing model where full registers can be probed. For this reason, we avoid doing so and we stress that parallelization does not result in such an undesired result. However, it should be noted that in some other relevant security models, such as the single-bit probing model or the *bounded moment leakage model* [3], this would not be an issue anyway.

4.4 Mask-Refreshing Implementation

The ISW-based mask refreshing is pretty similar to an ISW multiplication, but it is actually much faster since it involves no field multiplications and fewer additions (most terms being multiplied by 0). It simply consists in processing:

$$\text{for } i = 1 \mathbin{..} d: \quad \text{for } j = i+1 \mathbin{..} d: r \leftarrow \$; a_i \leftarrow a_i + r; a_j \leftarrow a_j + r;$$

A straightforward implementation of this process is almost 3 times faster than the fastest ISW multiplication, namely the full-table one (see Fig. 5).

We can actually do much better. Compared to a standard ISW implementation, the registers of the field multiplication are all available and can hence be used in order to save several loads and stores. Indeed, the straightforward implementation performs $d - i + 1$ loads and stores for every $i \in [\![1, d]\!]$, specifically 1 load-store for a_i and $d - i$ for the a_j's. Since we have some registers left, we can actually pool the a_j's loads and stores for several a_i's. To do so, we load several shares $a_i, a_{i+1}, \ldots, a_{i+k}$ with the LDM instruction (which has a cost of $k + 2$ instead of $3k$) and process the refreshing between them. Then, for every $j \in [\![i + k + 1, d]\!]$, we load a_j, performs the refreshing between a_j and each of the $a_i, a_{i+1}, \ldots, a_{i+k}$, and store a_j back. Afterwards, the shares $a_i, a_{i+1}, \ldots, a_{i+k}$ are stored back with the STM instruction (which has a cost of $k + 2$ instead of $3k$). This allows us to load (and store) the a_j only once for the k shares instead of k times, and to take advantage of the LDM and STM instructions. In practice, we could deal with up to $k = 8$ shares at the same time, meaning that for $d \leq 8$ all the shares could be loaded and stored an single time using LDM and STM instructions.

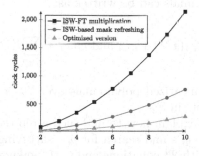

Fig. 5. Timings of mask refreshing.

The performances of our implementations of the ISW-based mask refreshing are plotted in Fig. 5. Our optimized refreshing is up to 3 times faster than the straightforward implementation and roughly 10 times faster that the full-table-based ISW multiplication.

5 Polynomial Methods for S-boxes

This section addresses the efficient implementation of polynomial methods for s-boxes based on ISW and CPRR schemes. We first investigate the two best known generic methods, namely the *CRV method* [15], and the *algebraic decomposition method* [11], for which we propose some improvements. We then look at specific methods for the AES and PRESENT s-boxes, and finally provide extensive comparison of our implementation results.

5.1 CRV Method

The CRV method was proposed by Coron, Roy and Vivek in [15]. Before recalling its principle, let us introduce the notion of *cyclotomic class*. For a given integer n, the cyclotomic class of $\alpha \in [\![0, 2^n - 2]\!]$ is defined as $C_\alpha = \{\alpha \cdot 2^i \bmod 2^n - 1; i \in \mathbb{N}\}$. We have the following properties: (i) cyclotomic classes are equivalence classes partitioning $[\![0, 2^n - 2]\!]$, and (ii) a cyclotomic class has at most n elements. In the following, we denote by x^L the set of monomials $\{x^\alpha; \alpha \in L\}$ for some set $L \subseteq [\![0, 2^n - 1]\!]$.

The CRV method consists in representing an s-box $S(x)$ over $\mathbb{F}_{2^n}[x]/(x^{2^n} - x)$ as

$$S(x) = \sum_{i=1}^{t-1} p_i(x) \cdot q_i(x) + p_t(x), \tag{8}$$

where $p_i(x)$ and $q_i(x)$ are polynomials with monomials in x^L for some set $L = C_{\alpha_1=0} \cup C_{\alpha_2=1} \cup C_{\alpha_3} \cup \ldots \cup C_{\alpha_\ell}$ such that for every $i \geq 3$, $\alpha_i = \alpha_j + \alpha_k \bmod 2^n - 1$ for some $j, k < i$ (or more generally $\alpha_i = 2^w \cdot \alpha_j + \alpha_k \bmod 2^n - 1$ with $k \in [\![0, n-1]\!]$). Such polynomials can be written as:

$$p_i(x) = \sum_{j=2}^{\ell} l_{i,j}(x^{\alpha_j}) + c_{i,0} \text{ and } q_i(x) = \sum_{j=2}^{\ell} l'_{i,j}(x^{\alpha_j}) + c'_{i,0}, \tag{9}$$

where the $l_{i,j}, l'_{i,j}$ are linearized polynomials over $\mathbb{F}_{2^n}[x]/(x^{2^n} - x)$ and where the $c_{i,0}, c'_{i,0}$ are constants in $\mathbb{F}_{2^n}$.

In [15], the authors explain how to find such a representation. In a nutshell, one randomly picks the q_i's and search for p_i's satisfying (8). This amounts to solve a linear system with 2^n equations and $t \cdot |L|$ unknowns (the coefficients of the p_i's). Note that when the choice of the classes and the q_i's leads to a solvable system, then it can be used with any s-box (since the s-box is the target vector of the linear system). We then have two necessary (non sufficient) conditions for such a system to be solvable: (1) the set L of cyclotomic classes is such that $t \cdot |L| \geq 2^n$, (2) all the monomials can be reached by multiplying two monomials from x^L, that is $\{x^i \cdot x^j \bmod (x^{2^n} - x); i, j \in L\} = x^{[\![0, 2^n - 1]\!]}$. For the sake of efficiency, the authors of [15] impose an additional constraint for the choice of the classes: (3) every class (but $C_0 = \{0\}$) have the maximal cardinality of n. Under this additional constraint, condition (1) amounts to the following inequality:

$t \cdot \left(1 + n \cdot (\ell - 1)\right) \geq 2^n$. Minimizing the number of nonlinear multiplications while satisfying this constraint leads to parameters $t \approx \sqrt{2^n/n}$ and $\ell \approx \sqrt{2^n/n}$.

Based on the above representation, the s-box can be evaluated using $(\ell - 2) + (t - 1)$ nonlinear multiplications (plus some linear operations). In a first phase, one generates the monomials corresponding to the cyclotomic classes in L. Each x^{α_i} can be obtained by multiplying two previous x^{α_j} and x^{α_k} (where x^{α_j} might be squared w times if necessary). In the masking world, each of these multiplications is performed with a call to ISW. The polynomials $p_i(x)$ and $q_i(x)$ can then be computed according to (9). In practice the linearized polynomials are tabulated so that at masked computation, applying a $l_{i,j}$ simply consists in performing a look-up on each share of the corresponding x^{α_j}. In the second phase, one simply evaluates (8), which takes $t - 1$ nonlinear multiplications plus some additions. We recall that in the masking world, linear operation such as additions or linearized polynomial evaluations can be applied on each share independently yielding a $O(d)$ complexity, whereas nonlinear multiplications are computed by calling ISW with a $O(d^2)$ complexity. The performances of the CRV method is hence dominated by the $\ell + t - 3$ calls to ISW.

Mask Refreshing. As explained in Sect. 4.1, one must be careful while composing ISW multiplications with linear operations. In the case of the CRV method, ISW multiplications are involved on sharings of values $q_i(x)$ and $p_i(x)$ which are linearly computed from the sharings of the x^{α_j} (see (9)). This contradicts the independence requirement for the input sharings of an ISW multiplication, and this might presumably induce a flaw as the one described in [14]. In order to avoid such a flaw in our masked implementation of CRV, we systematically refreshed one of the input sharings, namely the sharing of $q_i(x)$. As shown in Sect. 4.4, the overhead implied by such a refreshing is manageable.

Improving CRV with CPRR. As suggested in [11], CRV can be improved by using CPRR evaluations instead of ISW multiplications in the first phase of CRV, whenever CPRR is faster than ISW (*i.e.* when full-table multiplication cannot be afforded). Instead of multiplying two previously computed powers x^{α_j} and x^{α_k}, the new power x^{α_i} is derived by applying the quadratic function $x \mapsto x^{2^w+1}$ for some $w \in [\![1, n-1]\!]$. In the masking world, securely evaluating such a function can be done with a call to CPRR. The new chain of cyclotomic classes $C_{\alpha_1=0} \cup C_{\alpha_2=1} \cup C_{\alpha_3} \cup \ldots \cup C_{\alpha_\ell}$ must then satisfy $\alpha_i = (2^w + 1)\alpha_j$ for some $j < i$ and $w \in [\![1, n-1]\!]$.

We have implemented the search of such chains of cyclotomic classes satisfying conditions (1), (2) and (3). We could validate that for every $n \in [\![4, 10]\!]$ and for the parameters (ℓ, t) given in [15], we always find such a chain leading to a solvable system. For the sake of code compactness, we also tried to minimize the number of CPRR exponents $2^w + 1$ used in these chains (since in practice each function $x \mapsto x^{2^w+1}$ is tabulated). For $n \in \{4, 6, 7\}$ a single CPRR exponent (either 3 or 5) is sufficient to get a *satisfying chain* (*i.e.* a chain of cyclotomic class fulfilling the above conditions and leading to a solvable system). For the other values of n, we could prove that a single CPRR exponent does not suffice to get a satisfying chain. We could then find satisfying chains for $n = 5$ and $n = 8$ using 2 CPRR exponents (specifically 3 and 5). For $n > 8$, we tried all the pairs and triplets of possible CPRR exponents without success, we could only find a satisfying chain using the 4 CPRR exponents 3, 5, 9 and 17.

Optimizing CRV Parameters. We can still improve CRV by optimizing the parameters (ℓ, t) depending on the ratio $\theta = \frac{C_{\text{CPRR}}}{C_{\text{ISW}}}$, where C_{CPRR} and C_{ISW} denote the costs of ISW and CPRR respectively. The cost of the CRV method satisfies

$$C_{\text{CRV}} = (\ell - 2) C_{\text{CPRR}} + (t - 1) C_{\text{ISW}} = ((\ell - 2) \cdot \theta + t - 1)) C_{\text{ISW}}$$
$$\geq \left((\ell - 2) \cdot \theta + \left\lceil \frac{2^n}{(\ell - 1) \cdot n + 1} \right\rceil - 1 \right) C_{\text{ISW}}$$

where the inequality holds from conditions (1) and (3) above. This lower bound ensures that the system contains enough unknowns to be solvable. In practice, it was observed in [15] that this is a sufficient condition most of the time to get a solvable system (and our experiments corroborate this fact). Our optimized version of CRV hence consists in using the parameter ℓ minimizing the above lower bound and the corresponding $t = \left\lceil \frac{2^n}{(\ell-1) \cdot n+1} \right\rceil$ as parameters for given bit-length n and cost ratio θ.

It can be checked (see full version) that a ratio slightly lower than 1 implies a change of optimal parameters for all values of n except 4 and 9. In other words, as soon as CPRR is slightly faster than ISW, using a higher ℓ (*i.e.* more cyclotomic classes) and therefore a lower t is a sound trade. For our implementations of ISW and CPRR (see Sect. 4), we obtained a ratio θ greater than 1 only when ISW is based on the full-table multiplication. In that case, no gain can be obtain from using CPRR in the first phase of CRV, and one should use the original CRV parameters. On the other hand, we obtained θ-ratios of 0.88 and 0.75 for half-table-based ISW and exp-log-based ISW respectively. For the parallel versions, these ratios become 0.69 (half-table ISW) and 0.58 (exp-log ISW). For such ratios, the optimal parameter ℓ is greater than in the original CRV method (see full version for details).

For $n \in \{6, 8, 10\}$, we checked whether we could find satisfying CPRR-based chains of cyclotomic classes, for the obtained optimal parameters. For $n = 6$, the optimal parameters are $(\ell, t) = (5, 3)$ (giving 3 CPRR plus 2 ISW) which are actually the original CRV parameters. We could find a satisfying chain for these parameters. For $n = 8$, the optimal parameters are $(\ell, t) = (9, 4)$ (giving 7 CPRR plus 3 ISW). For these parameters we could not find any satisfying chain. We therefore used the second best set of parameters that is $(\ell, t) = (8, 5)$ (giving 6 CPRR plus 4 ISW) for which we could find a satisfying chain. For $n = 10$, the optimal parameters are $(\ell, t) = (14, 8)$ (giving 12 CPRR plus 7 ISW). For these parameters we could neither find any satisfying chain. So once again, we used the second best set of parameters, that is $(\ell, t) = (13, 9)$ (giving 11 CPRR plus 8 ISW) and for which we could find a satisfying chain. All the obtained satisfying CPRR-based chains of cyclotomic classes are provided in the full version of the paper.

Table 2. Performances of CRV original version and improved version (with and without optimized parameters).

	Original CRV [15]			CRV with CPRR [11]				Optimized CRV with CPPR			
	# ISW	# CPRR	Clock cycles	# ISW	# CPRR	Clock cycles	Ratio	# ISW	# CPRR	Clock cycles	Ratio
$n = 6$ (HT)	5	0	$142.5\,d^2 + O(d)$	2	3	$132\,d^2 + O(d)$	93%	2	3	$132\,d^2 + O(d)$	93%
$n = 6$ (EL)	5	0	$167.5\,d^2 + O(d)$	2	3	$142\,d^2 + O(d)$	85%	2	3	$142\,d^2 + O(d)$	85%
$n = 8$ (HT)	10	0	$285\,d^2 + O(d)$	5	5	$267.5\,d^2 + O(d)$	94%	4	6	$264\,d^2 + O(d)$	93%
$n = 8$ (EL)	10	0	$335\,d^2 + O(d)$	5	5	$292.5\,d^2 + O(d)$	87%	4	6	$284\,d^2 + O(d)$	85%
$n = 10$ (EL)	19	0	$997.5\,d^2 + O(d)$	10	9	$858\,d^2 + O(d)$	86%	8	11	$827\,d^2 + O(d)$	83%
$n = 8$ (HT) $/\!/4$	10	0	$775\,d^2 + O(d)$	5	5	$657.5\,d^2 + O(d)$	85%	4	6	$634\,d^2 + O(d)$	82%
$n = 8$ (EL) $/\!/4$	10	0	$935\,d^2 + O(d)$	5	5	$737.5\,d^2 + O(d)$	79%	4	6	$698\,d^2 + O(d)$	75%

Table 2 compares the performances of the original CRV method and the improved versions for our implementation of ISW (half-table and exp-log variants) and CPRR.[6] For the improved methods, we give the ratio of asymptotic performances with respect to the original version. This ratio ranks between 79% and 94% for the improved version and between 75% and 93% for the improved version with optimized parameters.

5.2 Algebraic Decomposition Method

The algebraic decomposition method was recently proposed by Carlet, Prouff, Rivain and Roche in [11]. It consists in using a basis of polynomials $(g_1, g_2, \ldots, g_r)$ that are constructed by composing polynomials f_i as follows

$$
\begin{cases}
g_1(x) = f_1(x) \\
g_i(x) = f_i\big(g_{i-1}(x)\big)
\end{cases}
\tag{10}
$$

The f_i's are of given algebraic degree s. In our context, we consider the algebraic decomposition method for $s = 2$, where the f_i's are (algebraically) quadratic polynomials. The method then consists in representing an s-box $S(x)$ over $\mathbb{F}_{2^n}[x]/(x^{2^n} - x)$ as

$$
S(x) = \sum_{i=1}^{t} p_i\big(q_i(x)\big) + \sum_{i=1}^{r} \ell_i\big(g_i(x)\big) + \ell_0(x),
\tag{11}
$$

with

$$
q_i(x) = \sum_{j=1}^{r} \ell_{i,j}\big(g_j(x)\big) + \ell_{i,0}(x),
\tag{12}
$$

where the p_i's are quadratic polynomials over $\mathbb{F}_{2^n}[x]/(x^{2^n} - x)$, and where the ℓ_i's and the $\ell_{i,j}$'s are linearized polynomials over $\mathbb{F}_{2^n}[x]/(x^{2^n} - x)$.

---

[6] We only count the calls to ISW and CPRR since other operations are similar in the three variants and have linear complexity in d.

As explain in [11], such a representation can be obtained by randomly picking some f_i's and some $\ell_{i,j}$'s (which fixes the q_i's) and then search for p_i's and ℓ_i's satisfying (11). As for the CRV method, this amounts to solve a linear system with 2^n equations where the unknowns are the coefficients of the p_i's and the ℓ_i's. Without loss of generality, we can assume that only ℓ_0 has a constant terms. In that case, each p_i is composed of $\frac{1}{2}n(n+1)$ monomials, and each ℓ_i is composed of n monomials (plus a constant term for ℓ_0). This makes a total of $\frac{1}{2}n(n+1)\cdot t + n\cdot r + 1$ unknown coefficients. In order to get a solvable system we hence have the following condition: (1) $\frac{1}{2}n(n+1)\cdot t + n\cdot r + 1 \geq 2^n$. A second condition is (2) $2^{r+1} \geq n$, otherwise there exists some s-box with algebraic degree greater than 2^{r+1} that cannot be achieved with the above decomposition $i.e.$ the obtained system is not solvable for every target S.

Based on the above representation, the s-box can be evaluated using $r + t$ evaluations of quadratic polynomials (the f_i's and the q_i's). In the masking world, this is done thanks to CPRR evaluations. The rest of the computation are additions and (tabulated) linearized polynomials which are applied to each share independently with a complexity linear in d. The cost of the algebraic decomposition method is then dominated by the $r + t$ calls to CPRR.

We implemented the search of sound algebraic decompositions for $n \in [\![4, 10]\!]$. Once again, we looked for *full rank* systems $i.e.$ systems that would work with any target s-box. For each value of n, we set r to the smallest integer satisfying condition (2) $i.e.$ $r \geq \log_2(n) - 1$, and then we looked for a t starting from the lower bound $t \geq \frac{2(2^n - rn - 1)}{n(n+1)}$ (obtained from condition (1)) and incrementing until a solvable system can be found. We then increment r and reiterate the process with t starting from the lower bound, and so on. For $n \leq 8$, we found the same parameters as those reported in [11]. For $n = 9$ and $n = 10$ (these cases were not considered in [11]), the best parameters we obtained were $(r, t) = (3, 14)$ and $(r, t) = (4, 22)$ respectively.

Saving Linear Terms. In our experiments, we realized that the linear terms $\ell_i(g_i(x))$ could always be avoided in (11). Namely, for the best known parameters (r, t) for every $n \in [\![4, 10]\!]$, we could always find a decomposition $S(x) = \sum_{i=1}^{t} p_i(q_i(x))$ hence saving $r + 1$ linearized polynomials. This is not surprising if we compare the number of degrees of freedom brought by the p_i's in the linear system ($i.e.$ $\frac{1}{2}n(n+1)\cdot t$) to those brought by the ℓ_i's ($i.e.$ $n\cdot r$). More details are given in the full version of the paper.

5.3   Specific Methods for AES and PRESENT

Rivain-Prouff (RP) Method for AES. Many works have proposed masking schemes for the AES s-box and most of them are based on its peculiar algebraic structure. It is the composition of the *inverse function* $x \mapsto x^{254}$ over $\mathbb{F}_{2^8}$ and an affine function: $S(x) = \mathrm{Aff}(x^{254})$. The affine function being straightforward to mask with linear complexity, the main issue is to design an efficient masking scheme for the inverse function.

In [34], Rivain and Prouff introduced the approach of using an efficient addition chain for the inverse function that can be implemented with a minimal number of ISW multiplications. They show that the exponentiation to the 254 can be performed with 4 nonlinear multiplications plus some (linear) squarings, resulting in a scheme with 4 ISW multiplications. In [14], Coron *et al.* propose a variant where two of these multiplications are replaced CPRR evaluations (of the functions $x \mapsto x^3$ and $x \mapsto x^5$).[7] This was further improved by Grosso *et al.* in [22] who proposed the following addition chain leading to 3 CPRR evaluations and one ISW multiplications: $x^{254} = (x^2 \cdot ((x^5)^5)^5)^2$. This addition chain has the advantage of requiring a single function $x \mapsto x^5$ for the CPRR evaluation (hence a single LUT for masked implementation). Moreover it can be easily checked by exhaustive search that no addition chain exists that trades the last ISW multiplication for a CPRR evaluation. We therefore chose to use the Grosso *et al.* addition chain for our implementation of the RP method.

Kim-Hong-Lim (KHL) Method for AES. This method was proposed in [26] as an improvement of the RP scheme. The main idea is to use the tower field representation of the AES s-box [36] in order to descend from $\mathbb{F}_{2^8}$ to $\mathbb{F}_{2^4}$ where the multiplications can be fully tabulated. Let δ denote the isomorphism mapping $\mathbb{F}_{2^8}$ to $(\mathbb{F}_{2^4})^2$ with $\mathbb{F}_{2^8} \equiv \mathbb{F}_{2^4}[x]/p(x)$, and let $\gamma \in \mathbb{F}_{2^8}$ and $\lambda \in \mathbb{F}_{2^4}$ such that $p(x) = x^2 + x + \lambda$ and $p(\gamma) = 0$. The tower field method for the AES s-box works as follows:

1	$a_h\gamma + a_l = \delta(x),\ a_h, a_l \in \mathbb{F}_{2^4}$	4	$a'_h = d'\, a_j \in \mathbb{F}_{2^4}$
2	$d = \lambda a_h^2 + a_l \cdot (a_h + a_l) \in \mathbb{F}_{2^4}$	5	$a'_l = d'(a_h + a_l) \in \mathbb{F}_{2^4}$
3.	$d' = d^{14} \in \mathbb{F}_{2^4}$	6.	$S(x) = \text{Aff}(\delta^{-1}(a'_h\gamma + a'_l)) \in \mathbb{F}_{2^8}$

At the third step, the exponentiation to the 14 can be performed as $d^{14} = (d^3)^4 \cdot d^2$ leading to one CPRR evaluation (for $d \mapsto d^3$) and one ISW multiplication (plus some linear squarings).[8] This gives a total of 4 ISW multiplications and one CPRR evaluation for the masked AES implementation.

$F \circ G$ Method for PRESENT. As a 4-bit s-box, the PRESENT s-box can be efficiently secured with the CRV method using only 2 (full table) ISW multiplications. The algebraic decomposition method would give a less efficient implementation with 3 CPRR evaluations. Another possible approach is to use the fact that the PRESENT s-box can be expressed as the composition of two quadratic functions $S(x) = F \circ G(x)$. This representation was put forward by Poschmann *et al.* in [31] to design an efficient *threshold implementation* of PRESENT. In our

---

[7] The original version of the RP scheme [34] actually involved a weak mask refreshing procedure which was exploited in [14] to exhibit a flaw in the s-box processing. The CPRR variant of ISW was originally meant to patch this flaw but the authors observed that using their scheme can also improve the performances. The security of the obtained variant of the RP scheme was recently verified up to masking order 4 using program verification techniques [2].

[8] The authors of [26] suggest to perform $d^3 = d^2 \cdot d$ with a full tabulated multiplication but this would actually imply a flaw as described in [14]. That is why we use a CPRR evaluation for this multiplication.

context, this representation can be used to get a masked s-box evaluation based on 2 CPRR evaluations. Note that this method is asymptotically slower than CRV with 2 full-table ISW multiplications. However, due to additional linear operations in CRV, $F \circ G$ might actually be better for small values of d.

5.4   Implementations and Performances

We have implemented the CRV method and the algebraic decomposition method for the two most representative values of $n = 4$ and $n = 8$. For $n = 4$, we used the full-table multiplication for ISW (256-byte table), and for $n = 8$ we used the half-table multiplication (8-KB table) and the exp-log multiplication (0.75-KB table). Based on our analysis of Sect. 5.1, we used the original CRV method for $n = 4$ (*i.e.* $(\ell, t) = (3, 2)$ with 2 ISW multiplications), and we used the improved CRV method with optimized parameters for $n = 8$ (*i.e.* $(\ell, t) = (8, 5)$ with 6 CPRR evaluations and 4 ISW multiplications). We further implemented parallel versions of these methods, which mainly consisted in replacing calls to ISW and CPRR by calls to their parallel versions (see Sect. 4.3), and replacing linear operations by their parallel counterparts.

We also implemented the specific methods described in Sect. 5.3 for the AES and PRESENT s-boxes, as well as their parallel counterparts. Specifically, we implemented the $F \circ G$ method for PRESENT and the RP and KHL methods for AES. The RP method was implemented with both the half-table and the exp-log methods for the ISW multiplication. For the KHL method, the ISW multiplications and the CPRR evaluation are performed on 4-bit values. It was then possible to perform 8 evaluations in parallel. Specifically, we first apply the isomorphism δ on 8 s-box inputs to obtain 8 pairs (a_h, a_l). The a_h values are grouped in one register and the a_l values are then grouped in a second register. The KHL method can then be processed in a 8-parallel version relying on the parallel ISW and CPRR procedures for $n = 4$.

Our implementation results (in terms of clock cycles) are depicted in Figs. 6 and 7 for $n = 4$ (with the $F \circ G$ method as a particular case), in Figs. 8 and 9 for $n = 8$, and in Figs. 10 and 11 for the AES s-box.

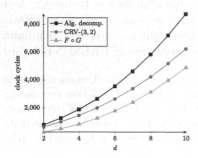
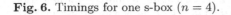
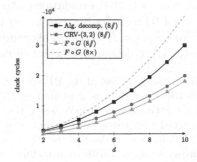

Fig. 6. Timings for one s-box $(n = 4)$.   **Fig. 7.** Timings for 8 s-boxes $(n = 4)$.

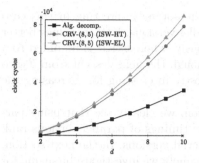

Fig. 8. Timings for one s-box $(n = 8)$.

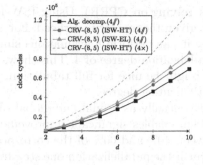

Fig. 9. Timings for 4 s-boxes $(n = 8)$.

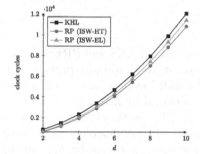

Fig. 10. Timings for one AES s-box.

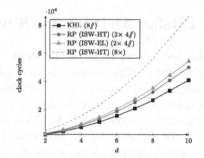

Fig. 11. Timings for 8 AES s-boxes.

We observe that the CRV method is clearly better than the algebraic decomposition method for $n = 4$ in both the serial and parallel case. This is not surprising since the former involves 2 calls to ISW-FT against 3 calls to CPRR for the latter. For $n = 8$, CRV is only slightly better than the algebraic decomposition, which is due to the use of CPRR and optimized parameters, as explained in Sect. 5.1. On the other hand, the parallel implementation of the algebraic decomposition method becomes better than CRV which is due to the efficiency of the CPRR parallelization.

Regarding the specific case of PRESENT, we see that the $F \circ G$ method is actually better than CRV for $d \in [\![2, 10]\!]$ thought it is asymptotically slower. It can be checked (see full version) that CRV becomes faster only after $d \geq 38$. In parallel, $F \circ G$ is also faster than CRV until $d \geq 11$. This shows that the $F \circ G$ method offers a valuable alternative to the CRV method for PRESENT in practice. Note that many 4-bit s-boxes have a similar decomposition (see [6] for an extensive analysis), so this method could be applied to further blockciphers.

For the AES, we observe that the RP method is better than KHL, which means that the gain obtained by using full-table multiplications does not compensate the overhead implied by the additional multiplication required in KHL compared to RP. We also see that the two versions of RP are very closed, which is not surprising since the difference regards a single multiplication (the other ones relying on CPRR). Using ISW-HT might not be interesting in this context given the memory overhead. For the parallel versions, KHL becomes better since it can perform 8 evaluations simultaneously, whereas RP is bounded to a parallelization degree of 4. This shows that though the field descent from $\mathbb{F}_{2^8}$ to $\mathbb{F}_{2^4}$ might be nice for full tabulation, it is mostly interesting for increasing the parallelization degree.

Eventually as a final and global observation, we clearly see that using parallelism enables significant improvements. The timings of parallel versions rank between 40% and 60% of the corresponding serial versions. In the next section, we push the parallelization one step further, namely we investigate bitslicing for higher-order masking implementations.

6   Bitslice Methods for S-boxes

In this section, we focus on the secure implementation of AES and PRESENT s-boxes using bitslice. Bitslice is an implementation strategy initially proposed by Biham in [4]. It consists in performing several parallel evaluations of a Boolean circuit in software where the logic gates can be replaced by instructions working on registers of several bits. As nicely explained in [27], *"in the bitslice implementation one software logical instruction corresponds to simultaneous execution of m hardware logical gates, where m is a register size [...] Hence bitslice can be efficient when the entire hardware complexity of a target cipher is small and an underlying processor has many long registers."*

In the context of higher-order masking, bitslice can be used at the s-box level to perform several secure s-box computations in parallel. One then need a compact Boolean representation of the s-box, and more importantly a representation with the least possible nonlinear gates. These nonlinear gates can then be securely evaluated in parallel using the ISW scheme as detailed hereafter. Such an approach was applied in [21] to design blockciphers with efficient masked computations. To the best of our knowledge, it has never been applied to get fast implementations of classical blockciphers such as AES or PRESENT. Also note that a bitsliced implementation of AES masked at first and second orders was described in [1] and used as a case study for practical side-channel attacks on a ARM Cortex-A8 processor running at 1 GHz.

6.1   ISW Logical AND

The ISW scheme can be easily adapted to secure a bitwise logical AND between two m-bit registers. From two d-sharings $(a_1, a_2, \ldots, a_d)$ and $(b_1, b_2, \ldots, b_d)$ of two m-bit strings $a, b \in \{0, 1\}^m$, the ISW scheme computes an output d-sharing $(c_1, c_2, \ldots, c_d)$ of $c = a \wedge b$ as follows:

1. for every $1 \leq i < j \leq d$, sample an m-bit random value $r_{i,j}$,
2. for every $1 \leq i < j \leq d$, compute $r_{j,i} = (r_{i,j} \oplus a_i \wedge b_j) \oplus a_j \wedge b_i$,
3. for every $1 \leq i \leq d$, compute $c_i = a_i \wedge b_i \oplus \bigoplus_{j \neq i} r_{i,j}$.

On the ARM architecture considered in this paper, registers are of size $m = 32$ bits. We can hence perform 32 secure logical AND in parallel. Moreover a logical AND is a single instruction of 1 clock cycle in ARM so we expect the above ISW logical AND to be faster than the ISW field multiplications. The detailed performances of our ISW-AND implementation are provided in the full version. We observe that the ISW-AND is indeed faster than the fastest ISW field multiplication (i.e. ISW-FT). Moreover it does not require any precomputed table and is hence lighter in code than the ISW field multiplications (except for the binary multiplication which is very slow).

6.2 Secure Bitslice AES S-box

For the AES s-box, we based our work on the compact representation proposed by Boyar et al. in [8]. Their circuit is obtained by applying logic minimization techniques to the tower-field representation of Canright [9]. It involves 115 logic gates including 32 logical AND. The circuit is composed of three parts: the *top linear transformation* involving 23 XOR gates and mapping the 8 s-box input bits $x_0, x_1, \ldots, x_7$ to 23 new bits $x_7, y_1, y_2, \ldots, y_{21}$; the *middle non-linear transformation* involving 30 XOR gates and 32 AND gates and mapping the previous 23 bits to 18 new bits $z_0, z_1, \ldots, z_{17}$; and the *bottom linear transformation* involving 26 XOR gates and 4 XNOR gates and mapping the 18 previous bits to the 8 s-box output bits $s_0, s_1, \ldots, s_7$. In particular, this circuit improves the usual count of 34 AND gates involved in previous tower-field representations of the AES s-box.

Using this circuit, we can perform the 16 s-box computations of an AES round in parallel. That is, instead of having 8 input bits mapped to 8 output bits, we have 8 (shared) input 16-bit words $X_0, X_1, \ldots, X_7$ mapped to 8 (shared) output 16-bit words $S_1, S_2, \ldots, S_8$. Each word X_i (resp. S_i) contains the ith bits input bit (resp. output bit) of the 16 s-boxes. Each XOR gate and AND gate of the original circuit is then replaced by the corresponding (shared) bitwise instruction between two 16-bit words.

Parallelizing AND Gates. For our masked bitslice implementation, a sound complexity unit is one call to the ISW-AND since this is the only nonlinear operation, i.e. the only operation with quadratic complexity in d (compared to other operations that are linear in d). In a straightforward bitslice implementation of the considered circuit, we would then have a complexity of 32 ISW-AND. This is suboptimal since each of these ISW-AND is applied to 16-bit words whereas it can operates on 32-bit words. Our main optimization is hence to group together pairs of ISW-AND in order to replace them by a single ISW-AND with fully filled input registers. This optimization hence requires to be able to group AND gates by pair that can be computed in parallel. To do so, we reordered the gates in the middle non-linear transformation of the Boyar et al. circuit, while keeping the computation consistent. We were able to fully parallelize the AND gates, hence dropping our bitslice complexity from 32 down to 16 ISW-AND. We thus get a parallel computation of the 16 AES s-boxes of one round with a complexity of 16 ISW-AND, that is one single ISW-AND per s-box. Since an ISW-AND is (significantly) faster than any ISW multiplication, our masked bitslice implementation breaks through the barrier of one ISW field multiplication per s-box. Our reordered version of the Boyar *et al.* circuit is provided in the full version of the paper.

Mask Refreshing. As for the CRV method, our bitslice AES s-box makes calls to ISW with input sharings that might be linearly related. In order to avoid any flaw, we systematically refreshed one of the input sharings in our masked implementation. Here again, the implied overhead is mitigated (between 5% and 10%).

6.3   Secure Bitslice PRESENT S-box

For our masked bitslice implementation of the PRESENT s-box, we used the compact representation given by Courtois *et al.* in [16], which was obtained from Boyar *et al.* 's logic minimization techniques improved by involving OR gates. This circuit is composed of 4 nonlinear gates (2 AND and 2 OR) and 9 linear gates (8 XOR and 1 XNOR).

PRESENT has 16 parallel s-box computations per round, as AES. We hence get a bitslice implementation with 16-bit words that we want to group for the calls to ISW-AND. However for the chosen circuit, we could not fully parallelize the nonlinear gates because of the dependency between three of them. We could however group the two OR gates after a slight reordering of the operations. We hence obtain a masked bitslice implementation computing the 16 PRESENT s-boxes in parallel with 3 calls to ISW-AND. Our reordered version of the circuit is depicted in the full version of the paper. For the sake of security, we also refresh one of the two input sharings in the 3 calls to ISW-AND. As for the bitslice AES s-box, the implied overhead is manageable.

6.4   Implementation and Performances

Figures 12 and 13 plot the performances obtained for our masked bitslice implementations of the AES and PRESENT s-boxes. For comparison, we also recall the performances of the fastest polynomial methods for AES and PRESENT (*i.e.* parallel versions of KHL and $F \circ G$) as well as the fastest generic methods for $n = 8$ and $n = 4$ (*i.e.* parallel versions of the algebraic decomposition method for $n = 8$ and CRV for $n = 4$).

These results clearly demonstrate the superiority of the bitslicing approach. Our masked bitslice implementations of the AES and PRESENT s-boxes are significantly faster than state-of-the art polynomial methods finely tuned at the assembly level.

7   Cipher Implementations

This section finally describes masked implementations of the full PRESENT and AES blockciphers. These blockciphers are so-called *substitution-permutation*

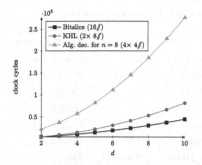

Fig. 12. Timings for 16 AES s-boxes.

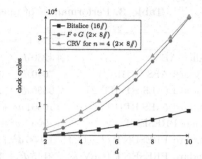

Fig. 13. Timings for 16 PRESENT s-boxes.

networks, where each round is composed of a key addition layer, a nonlinear layer and a linear diffusion layer. For both blockciphers, the nonlinear layer consists in the parallel application of 16 s-boxes. The AES works on a 128-bit state (which divides into sixteen 8-bit s-box inputs) whereas PRESENT works on a 64-bit state (which divides into sixteen 4-bit s-box inputs). For detailed specifications of these blockciphers, the reader is referred to [7,18]. For both blockciphers, we follow two implementation strategies: the standard one (with parallel polynomial methods for s-boxes) and the bitslice one (with bitslice s-box masking).

For the sake of efficiency, we assume that the key is already expanded, and for the sake of security we assume that each round key is stored in (non-volatile) memory under a shared form. In other words, we do not perform a masked key schedule. Our implementations start by masking the input plaintext with $d - 1$ random m-bit strings (where m is the blockcipher bit-size) and store the d resulting shares in memory. These d shares then compose the sharing of the blockcipher state that is updated by the masked computation of each round. When all the rounds have been processed, the output ciphertext is recovered by adding all the output shares of the state. For the bitslice implementations, the translation from standard to bitslice representation is performed before the initial masking so that it is done only once. Similarly, the translation back from the bitslice to the standard representation is performed a single time after unmasking.

The secure s-box implementations are done as described in previous sections. It hence remains to deal with the key addition and the linear layers. These steps are applied to each share of the state independently. The key-addition step simply consists in adding each share of the round key to one share of the state. The linear layer implementations are described in the full version of the paper.

7.1 Performances

In our standard implementation of AES, we used the parallel versions of KHL and RP (with ISW-EL) for the s-box. For the standard implementation of PRESENT, we used the parallel versions of the $F \circ G$ method and of the CRV method.

Table 3. Performances of masked blockciphers implementation.

	Clock cycles	Code (KB)	Random (bytes)
Bitslice AES	$3280\,d^2 + 14075\,d + 12192$	7.5	$640d(d+1)$
Standard AES (KHL $/\!/$)	$7640\,d^2 + 6229\,d + 6311$	4.8	$560d(d+1)$
Standard (AES RP-HT $/\!/$)	$9580\,d^2 + 5129\,d + 7621$	12.4	$400d(d+1)$
Standard (AES RP-EL $/\!/$)	$10301\,d^2 + 6561\,d + 7633$	4.1	$400d(d+1)$
Bitslice PRESENT	$1906.5\,d^2 + 10972.5\,d + 7712$	2.2	$372d(d+1)$
Standard PRESENT ($F \circ G$ $/\!/$)	$11656\,d^2 + 341\,d + 9081$	1.9	$496d(d+1)$
Standard PRESENT (CRV $/\!/$)	$9145\,d^2 + 45911\,d + 11098$	2.6	$248d(d+1)$

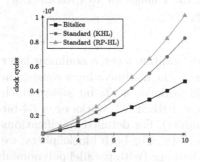

Fig. 14. Timings of masked AES.   **Fig. 15.** Timings of masked PRE-SENT.

The obtained performances are summarized in Table 3. The timings are further plotted in Figs. 14 and 15 for illustration.

These results clearly confirm the superiority of the bitslice implementations in our context. The bitslice AES implementation asymptotically takes 38% of the timings of the standard AES implementation using the best parallel polynomial method for the s-box (namely KHL). This ratio reaches 18% for PRESENT (compared to the $F \circ G$ method). It is also interesting to observe that PRESENT is slower than AES for standard masked implementations whereas it is faster for masked bitslice implementations. In the latter case, a PRESENT computation asymptotically amounts to 0.58 AES computation. This ratio directly results from the number of calls to ISW-AND which is $10 \times 16 = 160$ for AES (16 per round) and $31 \times 3 = 93$ for PRESENT (3 per round).

Table 4. Timings for masked bitslice AES and PRESENT with a 60 MHz clock.

	$d = 2$	$d = 3$	$d = 4$	$d = 5$	$d = 10$
Bitslice AES	0.89 ms	1.39 ms	1.99 ms	2.7 ms	8.01 ms
Bitslice PRESENT	0.62 ms	0.96 ms	1.35 ms	1.82 ms	5.13 ms

In order to illustrate the obtained performances in practice, Table 4 gives the corresponding timings in milliseconds for a clock frequency of 60 MHz. For a masking order of 10, our bitslice implementations only take a few milliseconds.

References

1. Balasch, J., Gierlichs, B., Reparaz, O., Verbauwhede, I.: DPA, bitslicing and masking at 1 GHz. In: Güneysu, T., Handschuh, H. (eds.) CHES 2015. LNCS, vol. 9293, pp. 599–619. Springer, Heidelberg (2015). doi:10.1007/978-3-662-48324-4_30
2. Barthe, G., Belaïd, S., Dupressoir, F., Fouque, P.-A., Grégoire, B., Strub, P.-Y.: Verified proofs of higher-order masking. In: Oswald, E., Fischlin, M. (eds.) EUROCRYPT 2015. LNCS, vol. 9056, pp. 457–485. Springer, Heidelberg (2015). doi:10.1007/978-3-662-46800-5_18
3. Barthe, G., Dupressoir, F., Faust, S., Grégoire, B., Standaert, F.-X., Strub, P.-Y.: Parallel implementations of masking schemes and the bounded moment leakage model. Cryptology ePrint Archive, Report 2016/912 (2016). http://eprint.iacr.org/2016/912
4. Biham, E.: A fast new DES implementation in software. In: Biham, E. (ed.) FSE 1997. LNCS, vol. 1267, pp. 260–272. Springer, Heidelberg (1997). doi:10.1007/BFb0052352
5. Bilgin, B., Gierlichs, B., Nikova, S., Nikov, V., Rijmen, V.: Higher-order threshold implementations. In: Sarkar, P., Iwata, T. (eds.) ASIACRYPT 2014. LNCS, vol. 8874, pp. 326–343. Springer, Heidelberg (2014). doi:10.1007/978-3-662-45608-8_18
6. Bilgin, B., Nikova, S., Nikov, V., Rijmen, V., Stütz, G.: Threshold implementations of all 3×3 and 4×4 S-boxes. In: Prouff, E., Schaumont, P. (eds.) CHES 2012. LNCS, vol. 7428, pp. 76–91. Springer, Heidelberg (2012). doi:10.1007/978-3-642-33027-8_5
7. Bogdanov, A., Knudsen, L.R., Leander, G., Paar, C., Poschmann, A., Robshaw, M.J.B., Seurin, Y., Vikkelsoe, C.: PRESENT: an ultra-lightweight block cipher. In: Paillier, P., Verbauwhede, I. (eds.) CHES 2007. LNCS, vol. 4727, pp. 450–466. Springer, Heidelberg (2007). doi:10.1007/978-3-540-74735-2_31
8. Boyar, J., Matthews, P., Peralta, R.: Logic minimization techniques with applications to cryptology. J. Cryptol. 26(2), 280–312 (2013)
9. Canright, D.: A very compact S-box for AES. In: Rao, J.R., Sunar, B. (eds.) CHES 2005. LNCS, vol. 3659, pp. 441–455. Springer, Heidelberg (2005). doi:10.1007/11545262_32
10. Carlet, C., Goubin, L., Prouff, E., Quisquater, M., Rivain, M.: Higher-order masking schemes for S-boxes. In: Canteaut, A. (ed.) FSE 2012. LNCS, vol. 7549, pp. 366–384. Springer, Heidelberg (2012). doi:10.1007/978-3-642-34047-5_21
11. Carlet, C., Prouff, E., Rivain, M., Roche, T.: Algebraic decomposition for probing security. In: Gennaro, R., Robshaw, M. (eds.) CRYPTO 2015. LNCS, vol. 9215, pp. 742–763. Springer, Heidelberg (2015). doi:10.1007/978-3-662-47989-6_36
12. Chari, S., Jutla, C.S., Rao, J.R., Rohatgi, P.: Towards sound approaches to counteract power-analysis attacks. In: Wiener, M. (ed.) CRYPTO 1999. LNCS, vol. 1666, pp. 398–412. Springer, Heidelberg (1999). doi:10.1007/3-540-48405-1_26
13. Coron, J.-S.: Higher order masking of look-up tables. In: Nguyen, P.Q., Oswald, E. (eds.) EUROCRYPT 2014. LNCS, vol. 8441, pp. 441–458. Springer, Heidelberg (2014). doi:10.1007/978-3-642-55220-5_25
14. Coron, J.-S., Prouff, E., Rivain, M., Roche, T.: Higher-order side channel security and mask refreshing. In: Moriai, S. (ed.) FSE 2013. LNCS, vol. 8424, pp. 410–424. Springer, Heidelberg (2014). doi:10.1007/978-3-662-43933-3_21

15. Coron, J.-S., Roy, A., Vivek, S.: Fast evaluation of polynomials over binary finite fields and application to side-channel countermeasures. In: Batina, L., Robshaw, M. (eds.) CHES 2014. LNCS, vol. 8731, pp. 170–187. Springer, Heidelberg (2014). doi:10.1007/978-3-662-44709-3_10
16. Courtois, N.T., Hulme, D., Mourouzis, T.: Solving circuit optimisation problems in cryptography and cryptanalysis. Cryptology ePrint Archive, Report 2011/475 (2011). http://eprint.iacr.org/2011/475
17. Duc, A., Dziembowski, S., Faust, S.: Unifying leakage models: from probing attacks to noisy leakage. In: Nguyen, P.Q., Oswald, E. (eds.) EUROCRYPT 2014. LNCS, vol. 8441, pp. 423–440. Springer, Heidelberg (2014). doi:10.1007/978-3-642-55220-5_24
18. FIPS PUB 197: Advanced Encryption Standard, November 2001
19. Genelle, L., Prouff, E., Quisquater, M.: Thwarting higher-order side channel analysis with additive and multiplicative maskings. In: Preneel, B., Takagi, T. (eds.) CHES 2011. LNCS, vol. 6917, pp. 240–255. Springer, Heidelberg (2011). doi:10.1007/978-3-642-23951-9_16
20. Goudarzi, D., Rivain, M.: On the multiplicative complexity of Boolean functions and bitsliced higher-order masking. In: Gierlichs, B., Poschmann, A.Y. (eds.) CHES 2016. LNCS, vol. 9813, pp. 457–478. Springer, Heidelberg (2016). doi:10.1007/978-3-662-53140-2_22
21. Grosso, V., Leurent, G., Standaert, F.-X., Varıcı, K.: LS-designs: bitslice encryption for efficient masked software implementations. In: Cid, C., Rechberger, C. (eds.) FSE 2014. LNCS, vol. 8540, pp. 18–37. Springer, Heidelberg (2015). doi:10.1007/978-3-662-46706-0_2
22. Grosso, V., Prouff, E., Standaert, F.-X.: Efficient masked S-boxes processing – a step forward –. In: Pointcheval, D., Vergnaud, D. (eds.) AFRICACRYPT 2014. LNCS, vol. 8469, pp. 251–266. Springer, Heidelberg (2014). doi:10.1007/978-3-319-06734-6_16
23. Grosso, V., Standaert, F.-X., Faust, S.: Masking vs. multiparty computation: how large is the gap for AES? In: Bertoni, G., Coron, J.-S. (eds.) CHES 2013. LNCS, vol. 8086, pp. 400–416. Springer, Heidelberg (2013). doi:10.1007/978-3-642-40349-1_23
24. Ishai, Y., Sahai, A., Wagner, D.: Private circuits: securing hardware against probing attacks. In: Boneh, D. (ed.) CRYPTO 2003. LNCS, vol. 2729, pp. 463–481. Springer, Heidelberg (2003). doi:10.1007/978-3-540-45146-4_27
25. Journault, A., Standaert, F., Varici, K.: Improving the security and efficiency of block ciphers based on LS-designs. Des. Codes Cryptogr. 82(1–2), 495–509 (2017)
26. Kim, H.S., Hong, S., Lim, J.: A fast and provably secure higher-order masking of AES S-box. In: Preneel, B., Takagi, T. (eds.) CHES 2011. LNCS, vol. 6917, pp. 95–107. Springer, Heidelberg (2011). doi:10.1007/978-3-642-23951-9_7
27. Matsui, M., Nakajima, J.: On the power of bitslice implementation on Intel Core2 processor. In: Paillier, P., Verbauwhede, I. (eds.) CHES 2007. LNCS, vol. 4727, pp. 121–134. Springer, Heidelberg (2007). doi:10.1007/978-3-540-74735-2_9
28. Moradi, A., Poschmann, A., Ling, S., Paar, C., Wang, H.: Pushing the limits: a very compact and a threshold implementation of AES. In: Paterson, K.G. (ed.) EUROCRYPT 2011. LNCS, vol. 6632, pp. 69–88. Springer, Heidelberg (2011). doi:10.1007/978-3-642-20465-4_6
29. Nikova, S., Rijmen, V., Schläffer, M.: Secure hardware implementation of nonlinear functions in the presence of glitches. In: Lee, P.J., Cheon, J.H. (eds.) ICISC 2008. LNCS, vol. 5461, pp. 218–234. Springer, Heidelberg (2009). doi:10.1007/978-3-642-00730-9_14
30. Nikova, S., Rijmen, V., Schläffer, M.: Secure hardware implementation of nonlinear functions in the presence of glitches. J. Cryptol. 24(2), 292–321 (2011)

31. Poschmann, A., Moradi, A., Khoo, K., Lim, C.-W., Wang, H., Ling, S.: Side-channel resistant crypto for less than 2,300 GE. J. Cryptol. **24**(2), 322–345 (2011)
32. Prouff, E., Rivain, M.: Masking against side-channel attacks: a formal security proof. In: Johansson, T., Nguyen, P.Q. (eds.) EUROCRYPT 2013. LNCS, vol. 7881, pp. 142–159. Springer, Heidelberg (2013). doi:10.1007/978-3-642-38348-9_9
33. Prouff, E., Roche, T.: Higher-order glitches free implementation of the AES using secure multi-party computation protocols. In: Preneel, B., Takagi, T. (eds.) CHES 2011. LNCS, vol. 6917, pp. 63–78. Springer, Heidelberg (2011). doi:10.1007/978-3-642-23951-9_5
34. Rivain, M., Prouff, E.: Provably secure higher-order masking of AES. In: Mangard, S., Standaert, F.-X. (eds.) CHES 2010. LNCS, vol. 6225, pp. 413–427. Springer, Heidelberg (2010). doi:10.1007/978-3-642-15031-9_28
35. Roy, A., Vivek, S.: Analysis and improvement of the generic higher-order masking scheme of FSE 2012. In: Bertoni, G., Coron, J.-S. (eds.) CHES 2013. LNCS, vol. 8086, pp. 417–434. Springer, Heidelberg (2013). doi:10.1007/978-3-642-40349-1_24
36. Satoh, A., Morioka, S., Takano, K., Munetoh, S.: A compact Rijndael hardware architecture with S-box optimization. In: Boyd, C. (ed.) ASIACRYPT 2001. LNCS, vol. 2248, pp. 239–254. Springer, Heidelberg (2001). doi:10.1007/3-540-45682-1_15

Functional Encryption I

Multi-input Inner-Product Functional Encryption from Pairings

Michel Abdalla[1], Romain Gay[1(✉)], Mariana Raykova[2], and Hoeteck Wee[1]

[1] ENS and PSL Research University, Paris, France
{michel.abdalla,romain.gay,hoeteck.wee}@ens.fr
[2] Yale University, New Haven, USA
mariana.raykova@yale.edu

Abstract. We present a multi-input functional encryption scheme (MIFE) for the inner product functionality based on the k-Lin assumption in prime-order bilinear groups. Our construction works for any polynomial number of encryption slots and achieves adaptive security against unbounded collusion, while relying on standard polynomial hardness assumptions. Prior to this work, we did not even have a candidate for 3-slot MIFE for inner products in the generic bilinear group model. Our work is also the first MIFE scheme for a non-trivial functionality based on standard cryptographic assumptions, as well as the first to achieve polynomial security loss for a super-constant number of slots under falsifiable assumptions. Prior works required stronger non-standard assumptions such as indistinguishability obfuscation or multi-linear maps.

1 Introduction

In a functional encryption (FE) scheme [11,25], an authority can generate restricted decryption keys that allow users to learn specific functions of the encrypted messages and nothing else. That is, each FE decryption key sk_f is associated with a function f and decrypting a ciphertext $\mathsf{Enc}(x)$ with sk_f results in $f(x)$. Multi-input functional encryption (MIFE) introduced by Goldwasser et al. [19] is a generalization of functional encryption to the setting of multi-input functions. A MIFE scheme has several encryption slots and each decryption key sk_f for a multi-input function f decrypts jointly ciphertexts $\mathsf{Enc}(x_1), \ldots, \mathsf{Enc}(x_n)$ for all slots to obtain $f(x_1, \ldots, x_n)$ without revealing anything more about the encrypted messages. The MIFE functionality provides the capability to encrypt independently messages for different slots. This facilitates scenarios where information, which will be processed jointly during decryption, becomes available

M. Abdalla—CNRS. Supported in part by SAFEcrypto (H2020 ICT-644729).
M. Raykova—Supported by NSF grants CNS-1633282, 1562888, 1565208, and DARPA SafeWare W911NF-15-C-0236, W911NF-16-1-0389.
H. Wee—CNRS and Columbia University. Supported in part by the ERC Project aSCEND (H2020 639554) and NSF Award CNS-1445424.

© International Association for Cryptologic Research 2017
J.-S. Coron and J.B. Nielsen (Eds.): EUROCRYPT 2017, Part I, LNCS 10210, pp. 601–626, 2017.
DOI: 10.1007/978-3-319-56620-7_21

at different points of time or is provided by different parties. MIFE has many applications related to computation and data-mining over encrypted data coming from multiple sources, which include examples such as executing search queries over encrypted data, processing encrypted streaming data, non-interactive differentially private data releases, multi-client delegation of computation, order-revealing encryption [10,19]. The security requirement for FE and MIFE is that the decryption keys are resilient to collusion attacks, namely any group of users holding different decryption keys learns nothing about the underlying messages beyond what each of them could individually learn.

We now have several constructions of MIFE schemes, which can be broadly classified as follows: (i) feasibility results for general circuits [5,6,12,19], and (ii) constructions for specific functionalities, notably comparison, which corresponds to order-revealing encryption [10]. Unfortunately, all of these constructions rely on indistinguishability obfuscation, single-input FE for circuits, or multi-linear maps [15,16], which we do not know how to instantiate under standard and well-understood cryptographic assumptions.[1]

1.1   Our Contributions

In this work, we present a multi-input functional encryption scheme (MIFE) for the inner product functionality based on the k-Lin assumption in prime-order *bilinear* groups. This is the first MIFE scheme for a non-trivial functionality based on standard cryptographic assumptions with polynomial security loss, and for any polynomial number of slots and secure against unbounded collusions.

Concretely, the functionality we consider is that of "bounded-norm" multi-input inner product: each function is specified by a collection of n vectors $\mathbf{y}_1, \ldots, \mathbf{y}_n$, takes as input n vectors $\mathbf{x}_1, \ldots, \mathbf{x}_n$, and outputs

$$f_{\mathbf{y}_1,\ldots,\mathbf{y}_n}(\mathbf{x}_1, \ldots, \mathbf{x}_n) = \sum_{i=1}^{n} \langle \mathbf{x}_i, \mathbf{y}_i \rangle.$$

We require that the $\mathbf{x}_1, \ldots, \mathbf{x}_n, \mathbf{y}_1, \ldots, \mathbf{y}_n$ have bounded norm, and inner product is computed over the integers. The functionality is a natural generalization of single-input inner product functionality introduced by Abdalla et. al [1], and studied in [1,2,4,7,13], and captures several useful computations arising in the context of data-mining. A summary of our results and prior works on single-input inner product is shown in Fig. 1.

Prior Approaches. Prior constructions of MIFE schemes in [10] requires (at least) nm-linear maps for n slots with m-bit inputs as they encode each input bit for each slot into a fresh level of a multi-linear map. In addition, there is typically a security loss that is exponential in n due to the combinatorial explosion arising from combining different ciphertexts across the slots. In the case of inner product,

---

[1] In this paper, we refer only to unbounded collusions (i.e. the adversary can request for any number of secret keys). See [12,20,21,24] for results on bounded collusions.

Reference	# inputs	setting	security	assumption	pairing
ABDP15 [1]	1	public-key	many-SEL-IND	DDH	no
ALS15 [4], ABDP16 [2]	1	public-key	many-AD-IND	DDH, k-Lin	no
BSW11 [11]	1	any	many-SEL-SIM	impossible	
[28]	1	public-key	one-SEL-SIM	k-Lin	no
LL16 [23]	2	private-key	many-SEL-IND	SXDH + T3DH	yes
KLMMRW16 [22]	2	private-key	single-key many-AD-IND[6]	function-private FE	yes
easy	multi	public-key	many-AD-IND	k-Lin	no
this work	multi	private-key	many-AD-IND	k-Lin	yes

Fig. 1. Summary of constructions from cyclic or bilinear groups. We have 8 security notions xx-yy-zzz where xx $\in$ {one, many} refers to the number of challenge ciphertexts; yy $\in$ {SEL, AD} refers to encryption queries are selectively or adaptively chosen; zzz $\in$ {IND, SIM} refers to indistinguishability vs simulation-based security.

one can hope to reduce the multi-linearity to n by exploiting linearity as in the single-input FE; indeed, this was achieved in two independent works [22, 23][2] showing how to realize a two-slot MIFE for inner product over bilinear groups. We stress that our result is substantially stronger: we show how to realize n-slot MIFE for inner product for any polynomial n over bilinear groups under standard assumptions, while in addition avoiding the exponential security loss. In particular, we deviate from the prior approaches of encoding each slot into a fresh level of a multi-linear map. We stress that prior to this work, we do not even have a candidate for 3-slot MIFE for inner product in the generic bilinear group model.

A Public-Key Scheme. Our first observation is that we can build a public-key MIFE for inner product by running n independent copies of a single-input FE for inner product. Combined with existing instantiations of the latter in [1], this immediately yields a public-key MIFE for inner product under the standard DDH in cyclic groups.

In a bit more detail, we recall the DDH-based public-key single-input FE scheme from [1]:[3]

$$\mathsf{mpk} := [\mathbf{w}], \quad \mathsf{ct}_{\mathbf{x}} = ([s], [\mathbf{x} + \mathbf{w}s]), \quad \mathsf{sk}_{\mathbf{y}} := \langle \mathbf{w}, \mathbf{y} \rangle$$

Decryption computes $[\langle \mathbf{x}, \mathbf{y} \rangle] = [\mathbf{x} + \mathbf{w}s]^{\top}\mathbf{y} \cdot [s]^{-\langle \mathbf{w}, \mathbf{y} \rangle}$ and then recovers $\langle \mathbf{x}, \mathbf{y} \rangle$ by computing the discrete log.

---

[2] This work is independent of both works.

[3] Here, we use the implicit representation notation for group elements, using $[s]$ to denote g^s and $[\mathbf{w}]$ to denote $g^{\mathbf{w}}$, etc.

Our public-key MIFE scheme is as follows:

$$\mathsf{mpk} := ([\mathbf{w}_1], \ldots, [\mathbf{w}_n]),$$
$$\mathsf{ct}_{\mathbf{x}_i} := ([s_i], [\mathbf{x}_i + \mathbf{w}_i s_i]),$$
$$\mathsf{sk}_{\mathbf{y}_1, \ldots, \mathbf{y}_n} := (\langle \mathbf{w}_1, \mathbf{y}_1 \rangle, \ldots, \langle \mathbf{w}_n, \mathbf{y}_n \rangle)$$

We note that the encryption of $\mathbf{x}_i$ uses fresh randomness s_i; to decrypt, we need to know each $\langle \mathbf{w}_i, \mathbf{y}_i \rangle$, and not just $\langle \mathbf{w}_1, \mathbf{y}_1 \rangle + \cdots + \langle \mathbf{w}_n, \mathbf{y}_n \rangle$. In particular, an adversary can easily recover each $[\langle \mathbf{x}_i, \mathbf{y}_i \rangle]$, whereas the ideal functionality should only leak the sum $\sum_{i=1}^n \langle \mathbf{x}_i, \mathbf{y}_i \rangle$. In the *public-key* setting, it is easy to see that $\langle \mathbf{x}_i, \mathbf{y}_i \rangle$ is in fact inherent leakage from the ideal functionality. Concretely, an adversary can always pad an encryption of $\mathbf{x}_i$ in the i'th slot with encryptions of $\mathbf{0}$'s in the remaining $n - 1$ slots and then decrypt.

Our Main Scheme. The bulk of this work lies in constructing a multi-input FE for inner product in the *private-key* setting, where we can no longer afford to leak $\langle \mathbf{x}_i, \mathbf{y}_i \rangle$. We modify the previous scheme by introducing additional rerandomization into each slot with the use of bilinear groups as follows:

$$\mathsf{msk} := ([\mathbf{w}_1]_1, [v_1]_1, [z_1]_1, \ldots, [\mathbf{w}_n]_1, [v_n]_1, [z_n]_1),$$

$$\mathsf{ct}_{\mathbf{x}_i} := ([s_i]_1, [\mathbf{x}_i + \mathbf{w}_i s_i]_1, [z_i + v_i s_i]_1),$$

$$\mathsf{sk}_{\mathbf{y}_1, \ldots, \mathbf{y}_n} := ([\langle \mathbf{w}_1, \mathbf{y}_1 \rangle + v_1 r]_2, \ldots, [\langle \mathbf{w}_n, \mathbf{y}_n \rangle + v_n r]_2,$$
$$[r]_2, [(z_1 + \cdots + z_n)r]_T)$$

The ciphertext $\mathsf{ct}_{\mathbf{x}_i}$ can be viewed as encrypting $\mathbf{x}_i \| z_i$ using the single-input FE, where $z_1, \ldots, z_n$ are part of msk. In addition, we provide a single-input FE key for $\mathbf{y}_i \| r$ in the secret key, where a fresh r is sampled for each key. Decryption proceeds as follows: first compute

$$[\langle \mathbf{x}_i, \mathbf{y}_i \rangle + z_i r]_T = e([\mathbf{x}_i + \mathbf{w}_i s_i]_1^\top, [\mathbf{y}_i]_2)$$
$$\cdot e([z_i + v_i s_i]_1^\top, [r]_2) \cdot e([s_i], [\langle \mathbf{w}_i, \mathbf{y}_i \rangle + v_i r]_2)^{-1}$$

and then

$$[\sum_{i=1}^n \langle \mathbf{x}_i, \mathbf{y}_i \rangle]_T = [(z_1 + \cdots + z_n)r]_T^{-1} \cdot \prod_{i=1}^n [\langle \mathbf{x}_i, \mathbf{y}_i \rangle + z_i r]_T.$$

The intuition underlying security is that by the DDH assumption $[z_i r]_T$ is pseudorandom and helps mask the leakage about $\langle \mathbf{x}_i, \mathbf{y}_i \rangle$ in $[\langle \mathbf{x}_i, \mathbf{y}_i \rangle + z_i r]_T$; in particular,

$$[\langle \mathbf{x}_1, \mathbf{y}_1 \rangle + z_1 r]_T, \ldots, [\langle \mathbf{x}_n, \mathbf{y}_n \rangle + z_n r]_T, [(z_1 + \cdots + z_n)r]_T$$

constitutes a computational secret-sharing of $[\langle \mathbf{x}_1, \mathbf{y}_1 \rangle + \cdots + \langle \mathbf{x}_n, \mathbf{y}_n \rangle]_T$, even upon reusing $z_1, \ldots, z_n$ as long as we pick a fresh r. In addition, sharing the same exponent r across n elements in the secret key helps prevent mix-and-match attacks across secret keys.

Our main technical result is that a variant of the private-key MIFE scheme we just described selective indistinguishability-based security under the k-Lin assumption in bilinear groups; a straight-forward extension of an impossibility in [3,11] rules out simulation-based security. Our final scheme as described in Fig. 6 remains quite simple and achieves good concrete efficiency. We focus on selective security in this overview, and explain at the end the additional ideas needed to achieve adaptive security.

Overview of Security Proof. There are two main challenges in the security proof: (i) avoiding leakage beyond the ideal functionality, (ii) avoiding super-polynomial hardness assumptions. Our proof proceeds in two steps: first, we establish security with a single challenge ciphertext per slot, and from which we bootstrap to achieve security with multiple challenge ciphertexts per slot. We will address the first challenge in the first step and the second challenge in the second. For notation simplicity, we focus on the setting with $n = 2$ slots and a single key query $\mathbf{y}_1 \| \mathbf{y}_2$.

Step 1. To prove indistinguishability-based security, we want to switch encryptions $\mathbf{x}_1^0, \mathbf{x}_2^0$ to encryptions of $\mathbf{x}_1^1, \mathbf{x}_2^1$. Here, the leakage from the ideal functionality imposes the restriction that

$$\langle \mathbf{x}_1^0, \mathbf{y}_1 \rangle + \langle \mathbf{x}_2^0, \mathbf{y}_2 \rangle = \langle \mathbf{x}_1^1, \mathbf{y}_1 \rangle + \langle \mathbf{x}_2^1, \mathbf{y}_2 \rangle$$

and this is the only restriction we can work with. The natural proof strategy is to introduce an intermediate hybrid that generates encryptions of $\mathbf{x}_1^1, \mathbf{x}_2^0$. However, to move from encryptions $\mathbf{x}_1^0, \mathbf{x}_2^0$ to this hybrid, we would require that $\langle \mathbf{x}_1^0 \| \mathbf{x}_2^0, \mathbf{y}_1 \| \mathbf{y}_2 \rangle = \langle \mathbf{x}_1^1 \| \mathbf{x}_2^0, \mathbf{y}_1 \| \mathbf{y}_2 \rangle$, which implies the extraneous restriction $\langle \mathbf{x}_1^0, \mathbf{y}_1 \rangle = \langle \mathbf{x}_1^1, \mathbf{y}_1 \rangle$. (Indeed, the single-input inner product scheme in [7] imposes extraneous restrictions to overcome similar difficulties in the function-hiding setting.)

To overcome this challenge, we rely on a single-input FE that achieves simulation-based security, which allows us to avoid the intermediate hybrid. See Theorem 1 and Remark 4 for further details.

Step 2. Next, we consider the more general setting with Q_1 challenge ciphertexts in the first slot and Q_2 in the second, but still a single key query. We achieve security loss $O(Q_1 + Q_2)$ for two slots, and more generally, $O(Q_1 + \cdots + Q_n)$ —as opposed to $Q_1 Q_2 \cdots Q_n$ corresponding to all possible combinations of the challenge ciphertexts— for n slots.

Our first observation is that we can bound the leakage from the ideal functionality by $O(Q_1 + Q_2)$ relations (the trivial bound being $Q_1 \cdot Q_2$). Denote the j'th ciphertext query in the i'th slot by $\mathbf{x}_i^{j,b}$, where b is the challenge bit. By decrypting the encryptions of $\mathbf{x}_1^{2,b}, \mathbf{x}_2^{1,b}$ and $\mathbf{x}_1^{1,b}, \mathbf{x}_2^{1,b}$ and substracting the two, the adversary learns $\langle \mathbf{x}_1^{2,b} - \mathbf{x}_1^{1,b}, \mathbf{y}_1 \rangle$ and more generally, $\langle \mathbf{x}_i^{j,b} - \mathbf{x}_i^{1,b}, \mathbf{y}_i \rangle$. Indeed, these are essentially the only constraints we need to work with, namely:

$$\langle \mathbf{x}_1^{1,0}, \mathbf{y}_1 \rangle + \langle \mathbf{x}_2^{1,0}, \mathbf{y}_2 \rangle = \langle \mathbf{x}_1^{1,1}, \mathbf{y}_1 \rangle + \langle \mathbf{x}_2^{1,1}, \mathbf{y}_2 \rangle$$
$$\langle \mathbf{x}_i^{j,0} - \mathbf{x}_i^{1,0}, \mathbf{y}_i \rangle = \langle \mathbf{x}_i^{j,1} - \mathbf{x}_i^{1,1}, \mathbf{y}_i \rangle, j = 2, \ldots, Q_i, i = 1, 2$$

Next, we need to translate the bound on the constraints to a $O(Q_1 + Q_2)$ bound on the security loss in the security reduction. We will switch from encryptions of $\mathbf{x}_i^{j,0}$ to those of $\mathbf{x}_i^{j,1}$ as follows: we write $\mathbf{x}_i^{j,0} = \mathbf{x}_i^{1,0} + (\mathbf{x}_i^{j,0} - \mathbf{x}_i^{1,0})$.

We can switch the first terms in the sums from $\mathbf{x}_i^{1,0}$ to $\mathbf{x}_i^{1,1}$ using security for a single challenge ciphertext, and then switch $\mathbf{x}_i^{j,0} - \mathbf{x}_i^{1,0}$ to $\mathbf{x}_i^{j,1} - \mathbf{x}_i^{1,1}$ by relying on security of the underlying single-input FE and the fact that $\langle \mathbf{x}_i^{j,0} - \mathbf{x}_i^{1,0}, \mathbf{y}_i \rangle = \langle \mathbf{x}_i^{j,1} - \mathbf{x}_i^{1,1}, \mathbf{y}_i \rangle$. Here, we will require that the underlying single-input FE satisfies a malleability property, namely given Δ, we can maul an encryption of $\mathbf{x}$ into that of $\mathbf{x} + \Delta$. Note that this does not violate security because given $\langle \mathbf{x}, \mathbf{y} \rangle, \mathbf{y}, \Delta$, we can efficiently compute $\langle \mathbf{x} + \Delta, \mathbf{y} \rangle$. See Theorem 2 for further details.

Extension to Adaptive Security. The previous argument for selective security requires to embed the challenge into the setup parameters. To circumvent this issue, we use a two-step strategy for the adaptive security proof of MIFE. The first step uses an adaptive argument (this is essentially the argument used for the selective case, but applied to parameters that are picked at setup time), while the second step uses a selective argument, with *perfect security*. Thus, we can use complexity leveraging without incurring an exponential security loss, since the exponential term is multiplied by a zero term. The idea of using complexity leveraging to deduce adaptive security from selective security when the security is perfect, already appears in [27, Remark 1].

Theoretical Perspective. The focus of this work is on obtaining constructions for a specific class of functions with good concrete efficiency. Nonetheless, we believe that our results do shed some new insights into general feasibility results for MIFE:

- First, our results are indicative of further qualitative differences between MIFE in the public-key and the private-key settings. Indeed, we already know that the security guarantees are quite different due to additional inherent leakages in the public-key setting. In the case of order-revealing encryption [10], the differences are sufficient to enable positive results in the private-key setting, while completely ruling out any construction in the public-key setting. Our results hint at a different distinction, where the private-key setting seems to require qualitative stronger assumptions than in the public-key setting, namely the use of pairings.
- Next, our results provide the first evidence supporting the intuition that MIFE requires qualitatively stronger assumptions than FE, but not too much stronger. Concretely, for the inner product FE, we have existing positive results under the DDH assumption in pairing-free groups. Prior to this work, it was not clear if we could extend the positive results to MIFE for n-ary inner

product under the same assumptions, or if n-ary inner product would already require the same complex assumptions as MIFE for circuits. Our results suggest a rather different picture, namely that going from single-input to multi-input should require no more than an extra level of multi-linearity, even for restricted functionalities. The situation is somewhat different for general circuits, where we now know that going from single-input to multi-input incurs no more than a quantitative loss in the underlying assumptions [5,12].

- Finally, we presented the first MIFE for a non-trivial functionality that polynomial security loss for a super-constant number of slots under falsifiable assumptions. Recall that indistinguishability obfuscation and generic multilinear maps are not falsifiable, whereas the constructions based on single-input FE in [5,8,12] incur a security loss which is exponential in the number of slots. Indeed, there is a reason why prior works relied on non-falsifiable assumptions or super-polynomial security loss. Suppose an adversary makes Q_0 key queries, and $Q_1, \ldots, Q_n$ ciphertext queries for the n slots. By combining the ciphertexts and keys in different ways, the adversary can learn $Q_0 Q_1 \cdots Q_n$ different decryptions. When n is super-constant, the winning condition in the security game may not be efficiently checkable in polynomial-time, hence the need for either a non-falsifiable assumption or a super-polynomial security loss. To overcome this difficulty, we show that for inner product, we can exploit linearity to succinctly characterize the $Q_0 Q_1 \cdots Q_n$ constraints by roughly $Q_0 \cdot (Q_1 + \cdots Q_n)$ constraints.

1.2 Discussion

Beyond Inner Product? Our constructions and techniques may seem a-priori largely tailored to the inner product functionality and properties of bilinear groups. We clarify here that our high-level approach (which builds upon [9,27]) may be applicable beyond inner product, namely:

i. start with a multi-input FE that is only secure for a single ciphertext per slot and one secret key, building upon a single-input FE whose security is simulation-based for a single ciphertext (in our case, this corresponds to introducing the additional $z_1, \ldots, z_n$ to hide the intermediate computation $\langle \mathbf{x}_i, \mathbf{y}_i \rangle$);

ii. achieve security for a single ciphertext per slot and multiple secret keys, by injecting additional randomness to the secret keys to prevent mix-and-match attacks (for this, we replaced $z_1, \ldots, z_n$ with $z_1 r, \ldots, z_n r, r$ in the exponent);

iii. "bootstrap" to multiple ciphertexts per slot, where we also showed how to avoid incurring an exponential security loss.

In particular, using simulation-based security for i. helped us avoid additional leakage beyond what is allowed by the ideal functionality.

Additional Related Work. Goldwasser et al. [19] showed that both two-input public-key MIFE as well as n-input private-key MIFE for circuits already implies indistinguishability obfuscation for circuits.

There have also been several works that proposed constructions for private-key multi-input functional encryption. The work of Boneh et al. [10] constructs a single-key MIFE in the private key setting, which is based on multilinear maps and is proven secure in the idealized generic multilinear map model. Two other papers explore the question how to construct multi-input functional encryption starting from the single input variant. In their work [5] Ananth and Jain demonstrate how to obtain selectively secure MIFE in the private key setting starting from any general-purpose public key functional encryption. In an independent work, Brakerski et al. [12] reduce the construction of private key MIFE to general-purpose private key (single input) functional encryption. The resulting scheme achieves selective security when the starting private key FE is selectively secure. Additionally in the case when the MIFE takes any constant number of inputs, adaptive security for the private key FE suffices to obtain adaptive security for the MIFE construction as well. The constructions in that work provide also function hiding properties for the MIFE encryption scheme.

While this line of work reduces MIFE to single-input FE for general-purpose constructions, the only known instantiations of construction for public and private key functional encryption with unbounded number of keys require either indistinguishability obfuscation [16] or multilinear maps with non-standard assumptions [17]. We stress that the transformations from single-input to MIFE in [5,12] are not applicable in the case of inner product since these transformations require that the single-input FE for complex functionalities related to computing a PRF, which is not captured by the simple inner functionality.

Open Problems. One natural open problem is to eliminate the use of pairings in MIFE for inner product; we think such a result would be quite surprising though. Another open problem is to achieve function privacy, as considered in the setting of single-input inner product functional encryption in [7,13]. Note that these latter results require pairings. Our first guess is that it would be possible to achieve private-key, function-hiding MIFE for inner product under the k-Lin assumption in bilinear groups.

2　Preliminaries

Notation. We denote by $s \leftarrow_R S$ the fact that s is picked uniformly at random from a finite set S. By PPT, we denote a probabilistic polynomial-time algorithm. Throughout, we use 1^λ as the security parameter. We use lower case boldface to denote (column) vectors and upper case boldface to denote matrices.

Cryptographic Assumptions. We follow the notation and algebraic framework for Diffie-Hellman-like assumptions in [14]. We fix a pairing group $\mathcal{PG} := (\mathbb{G}_1, \mathbb{G}_2, \mathbb{G}_T)$ with $e : \mathbb{G}_1 \times \mathbb{G}_2 \to \mathbb{G}_T$ of prime order q, where q is a prime of $\Theta(\lambda)$ bits. We use the implicit representation notation for group elements: for fixed generators g_1 and g_2 of $\mathbb{G}_1$ and $\mathbb{G}_2$, respectively, and for a matrix $\mathbf{M}$ over $\mathbb{Z}_q$, we define $[\mathbf{M}]_1 := g_1^{\mathbf{M}}$ and $[\mathbf{M}]_2 := g_2^{\mathbf{M}}$, where exponentiation is carried out component-wise.

We recall the definitions of the Matrix Decision Diffie-Hellman (MDDH) Assumption [14].

Definition 1 (Matrix Distribution). *Let $k, \ell \in \mathbb{N}$, with $\ell > k$. We call $\mathcal{D}_{\ell,k}$ a matrix distribution if it outputs matrices in $\mathbb{Z}_q^{\ell \times k}$ of full rank k in polynomial time. We write $\mathcal{D}_k := \mathcal{D}_{k+1,k}$.*

Without loss of generality, we assume the first k rows of $\mathbf{A} \leftarrow_R \mathcal{D}_{\ell,k}$ form an invertible matrix. The $\mathcal{D}_{\ell,k}$-Matrix Diffie-Hellman problem is to distinguish the two distributions $([\mathbf{A}], [\mathbf{Aw}])$ and $([\mathbf{A}], [\mathbf{u}])$ where $\mathbf{A} \leftarrow_R \mathcal{D}_{\ell,k}$, $\mathbf{w} \leftarrow_R \mathbb{Z}_q^k$ and $\mathbf{u} \leftarrow_R \mathbb{Z}_q^\ell$.

Definition 2 ($\mathcal{D}_k$-Matrix Diffie-Hellman Assumption $\mathcal{D}_k$-MDDH). *Let $\mathcal{D}_k$ be a matrix distribution. We say that the $\mathcal{D}_k$-Matrix Diffie-Hellman ($\mathcal{D}_k$-MDDH) Assumption holds relative to $\mathcal{PG}$ in $\mathbb{G}_s$ for $s \in \{1,2\}$, if for all PPT adversaries $\mathcal{A}$, there exists a negligible function Adv such that:*

$$\mathsf{Adv}_{\mathbb{G}_s,\mathcal{A}}^{\mathcal{D}_k\text{-MDDH}}(\lambda) := |\Pr[\mathcal{A}(\mathcal{PG}, [\mathbf{A}]_s, [\mathbf{Aw}]_s) = 1] - \Pr[\mathcal{A}(\mathcal{PG}, [\mathbf{A}]_s, [\mathbf{u}]_s) = 1]|$$
$$= \mathsf{negl}(\lambda),$$

where the probability is taken over $\mathbf{A} \leftarrow_R \mathcal{D}_k, \mathbf{w} \leftarrow_R \mathbb{Z}_q^k, \mathbf{u} \leftarrow_R \mathbb{Z}_q^{k+1}$.

For each $k \geq 1$, [14] specifies distributions $\mathcal{L}_k, \mathcal{SC}_k, \mathcal{C}_k$ (and others) over $\mathbb{Z}_q^{(k+1) \times k}$ such that the corresponding $\mathcal{D}_k$-MDDH assumptions are generically secure in bilinear groups and form a hierarchy of increasingly weaker assumptions. $\mathcal{L}_k$-MDDH is the well known k-Linear Assumption k-Lin with 1-Lin = DDH. In this work we are mostly interested in the uniform matrix distribution $\mathcal{U}_{\ell,k}$.

Definition 3 (Uniform Distribution). *Let $\ell, k \in \mathbb{N}$, with $\ell > k$. We denote by $\mathcal{U}_{\ell,k}$ the uniform distribution over all full-rank $\ell \times k$ matrices over $\mathbb{Z}_q$. Let $\mathcal{U}_k := \mathcal{U}_{k+1,k}$.*

Let $Q \geq 1$. For $\mathbf{W} \leftarrow_R \mathbb{Z}_q^{k \times Q}, \mathbf{U} \leftarrow_R \mathbb{Z}_q^{(k+1) \times Q}$, we consider the Q-fold $\mathcal{U}_{\ell,k}$-MDDH Assumption which consists in distinguishing the distributions $([\mathbf{A}], [\mathbf{AW}])$ from $([\mathbf{A}], [\mathbf{U}])$. That is, a challenge for the Q-fold $\mathcal{U}_{\ell,k}$-MDDH Assumption consists of Q independent challenges of the $\mathcal{U}_{\ell,k}$-MDDH Assumption (with the same $\mathbf{A}$ but different randomness $\mathbf{w}$). We recall in Lemma 1 the random self reducibility of the Q-fold $\mathcal{U}_{\ell,k}$-MDDH assumption, namely, the fact that it reduces to the 1-fold $\mathcal{U}_k$ assumption.

Lemma 1 ($\mathcal{U}_k$-MDDH $\Rightarrow$ Q-fold $\mathcal{U}_{\ell,k}$-MDDH [14,18]). *Let $\ell, k \in \mathbb{N}^*$, with $\ell > k$, and $s \in \{1,2\}$. For any PPT adversary $\mathcal{A}$, there exists a PPT adversary $\mathcal{B}$ such that*

$$\mathsf{Adv}_{\mathbb{G}_s,\mathcal{A}}^{Q\text{-}\mathcal{U}_{\ell,k}\text{-MDDH}}(\lambda) \leq \mathsf{Adv}_{\mathbb{G}_s,\mathcal{B}}^{\mathcal{U}_k\text{-MDDH}}(\lambda) + \frac{1}{q-1},$$

where $\mathsf{Adv}_{\mathbb{G}_s,\mathcal{A}}^{Q\text{-}\mathcal{U}_{\ell,k}\text{-MDDH}}(\lambda) := |\Pr[\mathcal{A}(\mathcal{PG}, [\mathbf{A}]_s, [\mathbf{AW}]_s) = 1] - \Pr[\mathcal{A}(\mathcal{PG}, [\mathbf{A}], [\mathbf{U}]) = 1]|$ and the probability is taken over $\mathbf{A} \leftarrow_R \mathcal{U}_{\ell,k}, \mathbf{W} \leftarrow_R \mathbb{Z}_q^{k \times Q}, \mathbf{U} \leftarrow_R \mathbb{Z}_q^{(k+1) \times Q}$.

Among all possible matrix distributions $\mathcal{D}_k$, the uniform matrix distribution $\mathcal{U}_k$ is the hardest possible instance, so in particular k-Lin $\Rightarrow \mathcal{U}_k$-MDDH, as stated in Lemma 2.

Lemma 2 ($\mathcal{D}_k$-MDDH $\Rightarrow \mathcal{U}_k$-MDDH, [14]). *Let $\mathcal{D}_k$ be a matrix distribution. For any PPT adversary $\mathcal{A}$, there exists a PPT adversary $\mathcal{B}$ such that* $\mathsf{Adv}^{\mathcal{U}_k\text{-MDDH}}_{\mathbb{G}_s,\mathcal{B}}(\lambda) \leq \mathsf{Adv}^{\mathcal{D}_k\text{-MDDH}}_{\mathbb{G}_s,\mathcal{A}}(\lambda)$.

3   Definitions for Multi-input Functional Encryption

We recall the definitions for multi-input functional encryption from [19]. We focus here on the private-key setting, which allows us to simplify the definitions.

Definition 4 (Multi-input Function Encryption). *Let $\{\mathcal{F}_n\}_{n\in\mathbb{N}}$ be an ensemble where each $\mathcal{F}_n$ is a family of n-ary functions. A function $f \in \mathcal{F}_n$ is defined as follows $f : \mathcal{X}_1 \times \ldots \times \mathcal{X}_n \to \mathcal{Y}$. A multi-input functional encryption scheme $\mathcal{MIFE}$ for $\mathcal{F}$ consists of the following algorithms:*

- Setup$(1^\lambda, \mathcal{F}_n)$: *on input the security parameter λ and a description of $\mathcal{F}_n \in \mathcal{F}$, outputs a master public key mpk[4] and a master secret key msk. All of the remaining algorithms get mpk as part of its input.*
- Enc(msk, i, x_i): *on input the master secret key msk, $i \in [n]$, and a message $x_i \in \mathcal{X}_i$, outputs a ciphertext ct. We assume that each ciphertext has an associated index i, which denotes what slot this ciphertext can be used for. If $n = 1$, we omit the input i.*
- KeyGen(msk, f): *on input the master secret key msk and a function $f \in \mathcal{F}_n$, outputs a decryption key sk_f.*
- Dec$(\mathsf{sk}_f, f, \mathsf{ct}_1, \ldots, \mathsf{ct}_n)$: *on input a decryption key sk_f for function f and n ciphertexts, outputs a string $y \in \mathcal{Y}$.*

The scheme $\mathcal{MIFE}$ is correct if for all $f \in \mathcal{F}$ and all $x_i \in \mathcal{X}_i$ for $1 \leq i \leq n$, we have

$$\Pr \left[\begin{array}{r} (\mathsf{mpk}, \mathsf{msk}) \leftarrow \mathsf{Setup}(1^\lambda, n); \\ \mathsf{sk}_f \leftarrow \mathsf{KeyGen}(\mathsf{msk}, f); \\ \mathsf{Dec}(\mathsf{sk}_f, f, \mathsf{Enc}(\mathsf{msk}, 1, x_1), \ldots, \mathsf{Enc}(\mathsf{msk}, n, x_n)) = f(x_1, \ldots, x_n) \end{array} \right]$$
$$= 1,$$

where the probability is taken over the coins of Setup, KeyGen and Enc.

3.1   Security Notions

Following [3], we may consider 8 security notions xx-yy-zzz where xx $\in$ {one, many} refers to the number of challenge ciphertexts; yy $\in$ {SEL, AD} refers

---

[4] We note that in the private key setting of MIFE, we can make mpk part of msk, but we allow for a separate master public key for better clarity in our proofs. In constructions where we do not need mpk we omit it.

to encryption queries are selectively or adaptively chosen; zzz $\in$ {IND, SIM} refers to indistinguishability vs simulation-based security. We have the following trivial relations: many $\Rightarrow$ one, AD $\Rightarrow$ SEL, and the following standard relations: SIM $\Rightarrow$ IND, and one-yy-IND $\Rightarrow$ many-yy-IND, the latter in the public-key setting. Here, we focus on {one,many}-SEL-IND and one-SEL-SIM, which are the notions most relevant to our positive results.

Definition 5 (xx-SEL-IND-secure MIFE). *For every multi-input functional encryption $\mathcal{MIFE} := (\mathsf{Setup}, \mathsf{Enc}, \mathsf{KeyGen}, \mathsf{Dec})$ for $\mathcal{F}$, every security parameter λ, every stateful adversary $\mathcal{A}$, and every $xx \in \{one, many\}$, the advantage of $\mathcal{A}$ is defined as*

$$\mathsf{Adv}^{\mathcal{MIFE},SEL-IND}(\lambda, \mathcal{A}) = \left| \Pr\left[\mathbf{SEL} - \mathbf{IND}_0^{\mathcal{MIFE}}(1^\lambda, \mathcal{A}) = 1\right] \right.$$
$$\left. - \Pr\left[\mathbf{SEL} - \mathbf{IND}^{\mathcal{MIFE}}(1^\lambda, \mathcal{A}) = 1\right] \right|$$

where the experiments are defined as follows:

Experiment $\mathbf{xx\text{-}SEL\text{-}IND}_\beta^{\mathcal{MIFE}}(1^\lambda, \mathcal{A})$:	*Experiment* $\mathbf{xx\text{-}SEL\text{-}IND}^{\mathcal{MIFE}}(1^\lambda, \mathcal{A})$:
	$\beta \leftarrow_{\mathrm{R}} \{0,1\}$
$\{x_i^b\}_{i \in [n], j \in [Q_i], b \in \{0,1\}} \leftarrow \mathcal{A}(1^\lambda, \mathcal{F}_n)$	$\{x_i^b\}_{i \in [n], j \in [Q_i], b \in \{0,1\}} \leftarrow \mathcal{A}(1^\lambda, \mathcal{F}_n)$
$(\mathsf{mpk}, \mathsf{msk}) \leftarrow \mathsf{Setup}(1^\lambda, \mathcal{F}_n)$	$(\mathsf{mpk}, \mathsf{msk}) \leftarrow \mathsf{Setup}(1^\lambda, \mathcal{F}_n)$
$\mathsf{ct}_i^j \leftarrow \mathsf{Enc}(\mathsf{msk}, i, x_i^{j,\beta}) \; \forall i \in [n], j \in [Q_i]$	$\mathsf{ct}_i^j \leftarrow \mathsf{Enc}(\mathsf{msk}, i, x_i^{j,\beta}) \; \forall i \in [n], j \in [Q_i]$
$\beta' \leftarrow \mathcal{A}^{\mathsf{KeyGen}(\mathsf{msk},\cdot)} \left(\mathsf{mpk}, (\mathsf{ct}_i^j)_{i \in [n], j \in [Q_i]}\right)$	$\beta' \leftarrow \mathcal{A}^{\mathsf{KeyGen}(\mathsf{msk},\cdot)} \left(\mathsf{mpk}, (\mathsf{ct}_i^j)_{i \in [n], j \in [Q_i]}\right)$
Output: β'	*Output: 1 if $\beta' = \beta$, 0 otherwise.*

where $\mathcal{A}$ only makes queries f to $\mathsf{KeyGen}(\mathsf{msk}, \cdot)$ satisfying

$$f(x_1^{j_1,0}, \ldots, x_n^{j_1,0}) = f(x_1^{j_1,1}, \ldots, x_n^{j_1,1})$$

for all $j_1, \ldots, j_1 \in [Q_1] \times \cdots \times [Q_n]$. For $xx = one$, we require additionally that the adversary $\mathcal{A}$ only sends one challenge per slot, i.e. for all $i \in [n]$, $Q_i = 1$.

The private key multi-input functional encryption $\mathcal{MIFE}$ is xx-SEL-IND-secure if for every PPT adversary $\mathcal{A}$, there exists a negligible function negl such that for all $\lambda \in \mathbb{N}$: $\mathsf{Adv}_\mathcal{A}^{\mathcal{MIFE},xx\text{-}SEL\text{-}IND}(\lambda) = \mathsf{negl}(\lambda)$.

Remark 1 (winning condition). Note that the winning condition is in general not efficiently checkable because of the combinatorial explosion in the restriction on the queries.

Next, we present the simulation-based security definition for MIFE, in the setting with a single challenge ciphertext per slot.

Definition 6 (one-SEL-SIM-secure FE). *A single-input functional encryption $\mathcal{FE}$ for function $\mathcal{F}$ is one-SEL-SIM-secure if there exists a PPT simulator[5] $(\widetilde{\mathsf{Setup}}, \widetilde{\mathsf{Encrypt}}, \widetilde{\mathsf{KeyGen}})$ such that for every PPT adversary $\mathcal{A}$ and every $\lambda \in \mathbb{N}$, the following two distributions are computationally indistinguishable:*

---

[5] That is, $\widetilde{\mathsf{Setup}}, \widetilde{\mathsf{Enc}}, \widetilde{\mathsf{KeyGen}}$ correspond respectively to the simulated $\mathsf{Setup}, \mathsf{Enc}, \mathsf{KeyGen}$.

$Experiment$ $\mathbf{REAL}^{\mathcal{FE}}(1^\lambda, \mathcal{A})$:	$Experiment$ $\mathbf{IDEAL}^{\mathcal{FE}}(1^\lambda, \mathcal{A})$:
$x \leftarrow \mathcal{A}(1^\lambda, \mathcal{F})$	$x \leftarrow \mathcal{A}(1^\lambda, \mathcal{F})$
$(\mathsf{mpk}, \mathsf{msk}) \leftarrow \mathsf{Setup}(1^\lambda, \mathcal{F})$	$(\widetilde{\mathsf{mpk}, \mathsf{msk}}) \leftarrow \widetilde{\mathsf{Setup}}(1^\lambda, \mathcal{F})$
$\mathsf{ct} \leftarrow \mathsf{Enc}(\mathsf{msk}, x)$	$\mathsf{ct} \leftarrow \widetilde{\mathsf{Encrypt}}(\widetilde{\mathsf{msk}})$
$\alpha \leftarrow \mathcal{A}^{\mathsf{KeyGen}(\mathsf{msk}, \cdot)}(\mathsf{mpk}, \mathsf{ct})$	$\alpha \leftarrow \mathcal{A}^{\mathcal{O}(\cdot)}(\widetilde{\mathsf{mpk}}, \mathsf{ct})$
$\textbf{\textit{Output:}}$ α	$\textbf{\textit{Output:}}$ α

The oracle $\mathcal{O}(\cdot)$ in the above ideal experiment has access to an oracle that provides the value $\langle \mathbf{x}, \mathbf{y} \rangle$, for each $\mathbf{y} \in \mathbb{Z}_p^m$ queried to $\mathcal{O}(\cdot)$. Then, $\mathcal{O}(\cdot)$ returns $\widetilde{\mathsf{KeyGen}}(\widetilde{\mathsf{msk}}, \mathbf{y}, \langle \mathbf{x}, \mathbf{y} \rangle)$.
Namely, for every stateful adversary $\mathcal{A}$, we define

$$\mathsf{Adv}^{\mathcal{FE}, one\text{-}SEL\text{-}SIM}(\lambda, \mathcal{A}) =$$

$$\left| \Pr\left[\mathbf{REAL}^{\mathcal{FE}}(1^\lambda, \mathcal{A}) = 1\right] - \Pr\left[\widetilde{\mathbf{IDEAL}}^{\mathcal{FE}}(1^\lambda, \mathcal{A}) = 1\right] \right|,$$

and we require that for every PPT $\mathcal{A}$, there exists a negligible function negl such that for all $\lambda \in \mathbb{N}$, $\mathsf{Adv}^{\mathcal{FE}, one\text{-}SEL\text{-}SIM}(\lambda, \mathcal{A}) = \mathsf{negl}(\lambda)$.

Zero vs Multiple Queries in Private-Key Setting. It is convenient in our proof of security to assume that $Q_1, \ldots, Q_n \geq 1$, that is, there is at least one ciphertext for each encryption slot, which is where the technical bulk of the work lies as we would need to reason about leakage from the ideal functionality. In the setting where some $Q_i = 0$, the ideal functionality leaks nothing, and here, we can easily achieve semantic security for all of the messages being encrypted in the private key MIFE setting, via the following simple generic transformation.

Lemma 3. *Let $(\mathsf{Setup}, \mathsf{Enc}, \mathsf{KeyGen}, \mathsf{Dec})$ be a private key MIFE construction for n-input functions in the class $\mathcal{F}_n$, which satisfies any xx-yy-zzz MIFE security definition when the adversary receives at least one ciphertext for each encryption slot. Let $(\mathsf{Gen_{SE}}, \mathsf{Enc_{SE}}, \mathsf{Dec_{SE}})$ be symmetric key encryption. The private key MIFE scheme $(\mathsf{Setup}', \mathsf{Enc}', \mathsf{KeyGen}', \mathsf{Dec}')$ described in Fig. 2 satisfies xx-yy-zzz security without any restrictions on the ciphertext challenge sets.*

Proof (Sketch). We consider two cases:

- Case 1: there exists some $i \in [n]$ for which $Q_i = 0$. Here, k_i and thus K is perfectly hidden from the adversary. Then, security follows readily from semantic security of $(\mathsf{Gen_{SE}}, \mathsf{Enc_{SE}}, \mathsf{Dec_{SE}})$.
- Case 2: for all i, $Q_i \geq 1$. Here, security follows immediately from that of $(\mathsf{Setup}, \mathsf{Enc}, \mathsf{KeyGen}, \mathsf{Dec})$. □

$\mathsf{Setup}'(1^\lambda, \mathcal{F}_n)$:
  $\mathsf{msk} \leftarrow \mathsf{Setup}(1^\lambda, \mathcal{F}_n)$
  $\mathsf{K} \leftarrow \mathsf{Gen}(1^\lambda)$
  $k_1, \ldots, k_{n-1} \leftarrow_R \{0,1\}^\lambda,\ k_n = \left(\bigoplus_{i \in [n-1]} k_i \right) \oplus \mathsf{K}$
  return $\mathsf{msk}' \leftarrow \left(\mathsf{msk}, \mathsf{K}, \{k_i\}_{i \in [n]} \right)$

$\mathsf{Enc}'(\mathsf{msk}, i, \mathbf{x}_i)$:
  parse $\mathsf{msk}' = (\mathsf{msk}, \mathsf{K}, \{k_i\}_{i \in [n]})$
  $\mathsf{ct} \leftarrow \mathsf{Enc}(\mathsf{msk}, i, \mathbf{x}_i)$
  $\mathsf{ct}' \leftarrow \mathsf{Enc}_{\mathsf{SE}}(\mathsf{K}, \mathsf{ct})$
  return (k_i, ct')

$\mathsf{KeyGen}'(\mathsf{msk}, f)$:
  return $\mathsf{KeyGen}(\mathsf{msk}, f)$

$\mathsf{Dec}'(\mathsf{sk}_f, f, \mathsf{ct}'_1, \ldots, \mathsf{ct}'_n)$:
  parse $\left\{ \mathsf{ct}'_i = (k_i, \mathsf{ct}_i) \right\}_{i \in [n]}$
  $\mathsf{K} \leftarrow \bigoplus_{i \in [n]} k_i$
  $\left\{ \mathsf{ct}_i \leftarrow \mathsf{Dec}_{\mathsf{SE}}(\mathsf{K}, \mathsf{ct}'_i) \right\}_{i \in [n]}$
  return $\mathsf{Dec}(\mathsf{sk}_f, f, \mathsf{ct}_1, \ldots, \mathsf{ct}_n)$.

Fig. 2. Compiler from private-key MIFE with xx-yy-zzz security when $|Q_i| > 0$ for all i to private-key MIFE with xx-yy-zzz security

3.2 Inner Product Functionality

Multi-input Inner Product. We construct a multi-input functional encryption that supports the class of multi-input bounded-norm inner product functions, which is defined as $\mathcal{F}_n^{m,B} = \{f_{\mathbf{y}_1,\ldots,\mathbf{y}_n} : (\mathbb{Z}^m)^n \to \mathbb{Z}\}$ where

$$f_{\mathbf{y}_1,\ldots,\mathbf{y}_n}(\mathbf{x}_1, \ldots, \mathbf{x}_n) = \sum_{i=1}^n \langle \mathbf{x}_i, \mathbf{y}_i \rangle.$$

We require that the norm of the inner product of any two vector components from function and input $\langle \mathbf{x}, \mathbf{y} \rangle$ is bounded by B. This bound will determine the parameters of the bilinear map groups that we will be using in our constructions; in particular, we will choose a target group that has order $q \gg n \cdot B$. To simplify naming conventions, we will omit "bounded-norm" for the rest of the paper, but we will always refer to a multi-input inner-product functionality with this property.

Remark on Leakage. Let $(\mathbf{x}_i^{j,0}, \mathbf{x}_i^{j,1})_{i \in [n], j \in [Q_i]}$ be the ciphertext queries, and $\mathbf{y}_1 \| \cdots \| \mathbf{y}_n$ be a secret key query. For all slots $i \in [n]$, all $j \in [Q_i]$, and all bits $b \in \{0,1\}$, the adversary can learn $\langle \mathbf{x}_i^{j,b} - \mathbf{x}_i^{j,b}, \mathbf{y}_i \rangle$ via the ideal functionality. In the IND security game, this means the adversary is restricted to queries satisfying $\langle \mathbf{x}_i^{j,0} - \mathbf{x}_i^{1,0}, \mathbf{y}_i \rangle = \langle \mathbf{x}_i^{j,1} - \mathbf{x}_i^{1,1}, \mathbf{y}_i \rangle$. In the hybrid, we want to avoid additional constraints such as

$$\langle \mathbf{x}_i^{j,0} - \mathbf{x}_i^{1,0}, \mathbf{y}_i \rangle = \langle \mathbf{x}_i^{j,0} - \mathbf{x}_i^{1,1}, \mathbf{y}_i \rangle = \langle \mathbf{x}_i^{j,1} - \mathbf{x}_i^{1,0}, \mathbf{y}_i \rangle = \langle \mathbf{x}_i^{j,1} - \mathbf{x}_i^{1,1}, \mathbf{y}_i \rangle$$

4   Private-Key MIFE for Inner Product

In this section, we present a private-key MIFE for inner product that achieves many-SEL-IND security. We use a pairing group $(\mathbb{G}_1, \mathbb{G}_2, \mathbb{G}_T)$ with $e : \mathbb{G}_1 \times \mathbb{G}_2 \to \mathbb{G}_T$ of prime order q, where q is a prime of $\Theta(\lambda)$ bits. Our construction relies on the k-Lin Assumption in $\mathbb{G}_1$ and in $\mathbb{G}_2$ and is shown in Fig. 6.

We present our construction in two steps: first, in Sect. 4.1, we show how to construct a selectively-secure MIFE scheme starting from a single-input one-SEL-SIM scheme that satisfies some additional structural properties. Then, we show how to instantiate the underlying single-input scheme (cf. Fig. 7) and we present a self-contained description of the scheme in Fig. 6. We refer the reader to Sect. 1.1 for an overview of the construction.

4.1   Selectively-Secure, Multi-input Scheme from Single-Input Scheme

Main Construction. We build a private key multi-input FE (Setup′, Enc′, KeyGen′, Dec′) for the class $\mathcal{F}_n^{m,B}$, starting from a private key one-SEL-SIM secure, single-input FE (Setup, Enc, KeyGen, Dec) for the class $\mathcal{F}_1^{m+k,B}$. We present our construction in Fig. 3.

Correctness. Correctness follows readily from the correctness of the underlying scheme and the equation:

$$\langle \mathbf{x}_1 \| \cdots \| \mathbf{x}_n, \mathbf{y}_1 \| \cdots \| \mathbf{y}_n \rangle = (\sum_{i=1}^{n} \langle \mathbf{x}_i \| \mathbf{z}_i, \mathbf{y}_i \| \mathbf{r} \rangle) - \langle \mathbf{z}_1 + \cdots + \mathbf{z}_n, \mathbf{r} \rangle$$

Finally, we use the fact that $\langle \mathbf{x}_1 \| \cdots \| \mathbf{x}_n, \mathbf{y}_1 \| \cdots \| \mathbf{y}_n \rangle \mod q = \langle \mathbf{x}_1 \| \cdots \| \mathbf{x}_n, \mathbf{y}_1 \| \cdots \| \mathbf{y}_n \rangle$, since for all slots $i \in [n]$, we have $\langle \mathbf{x}_i, \mathbf{y}_i \rangle \le B$, and $q > Bn$.

Additional Requirements. The construction and the analysis requires that (Setup, Enc, KeyGen, Dec) satisfies the following structural properties:

- The scheme can be instantiated over $\mathbb{G}_1$, where the ciphertext is a vector $[\mathbf{c}]_1$ over $\mathbb{G}_1$ and the secret key is a vector $\mathbf{d}_i$ over $\mathbb{Z}_q$.
- Enc is linearly homomorphic and public-key. More specifically, we only require that, given $\mathsf{mpk}, \mathsf{Enc}(\mathsf{msk}, \mathbf{x}), \mathbf{x}'$, we can generate a fresh random encryption of $\mathbf{x} + \mathbf{x}'$, i.e. $\mathsf{Enc}(\mathsf{msk}, \mathbf{x} + \mathbf{x}')$.
- For correctness, Dec should be linear in its inputs $(\mathbf{d}, \mathbf{y})$ and $\mathbf{c}$, so that $\mathsf{Dec}([\mathbf{d}]_2, [\mathbf{y}]_2, [\mathbf{c}]_1) = [\mathsf{Dec}(\mathbf{d}, \mathbf{y}, \mathbf{c})]_T \in \mathbb{G}_T$ can be computed using a pairing.

$\underline{\mathsf{Setup}'(1^\lambda, \mathcal{F}_n^{m,B}):}$

  $(\mathsf{mpk}_i, \mathsf{msk}_i) \leftarrow \mathsf{Setup}(1^\lambda, \mathcal{F}_1^{m+k,B}), i = 1, \ldots, n$

  $\mathbf{z}_i \leftarrow_{\mathrm{R}} \mathbb{Z}_q^k, i = 1, \ldots, n$

  $(\mathsf{mpk}, \mathsf{msk}) := \left(\left\{ \mathsf{mpk}_i \right\}_{i \in [n]}, \left\{ \mathsf{msk}_i, \mathbf{z}_i \right\}_{i \in [n]} \right)$

  return $(\mathsf{mpk}, \mathsf{msk})$

$\underline{\mathsf{Enc}'(\mathsf{msk}, i, \mathbf{x}_i):}$

  return $\mathsf{Enc}(\mathsf{msk}_i, \mathbf{x}_i \| \mathbf{z}_i)$

$\underline{\mathsf{KeyGen}'(\mathsf{msk}, \mathbf{y}_1 \| \cdots \| \mathbf{y}_n):}$

  $\mathbf{r} \leftarrow_{\mathrm{R}} \mathbb{Z}_q^k$

  $\mathbf{d}_i \leftarrow \mathsf{KeyGen}(\mathsf{msk}_i, \mathbf{y}_i \| \mathbf{r}), i = 1, \ldots, n$

  $z := \langle \mathbf{z}_1 + \cdots + \mathbf{z}_n, \mathbf{r} \rangle$

  $\mathsf{sk}_{\mathbf{y}_1 \| \cdots \| \mathbf{y}_n} := \left(\left\{ [\mathbf{d}_i]_2 \right\}_{i \in [n]}, [\mathbf{r}]_2, [z]_T \right)$

  return $\mathsf{sk}_{\mathbf{y}_1 \| \cdots \| \mathbf{y}_n}$

$\underline{\mathsf{Dec}'((\{ [\mathbf{d}_i]_2 \}_{i \in [n]}, [\mathbf{r}]_2, [z]_T), \mathbf{y}_1 \| \cdots \| \mathbf{y}_n, \mathsf{ct}_1, \ldots, \mathsf{ct}_n):}$

  $[a_i]_T \leftarrow \mathsf{Dec}([\mathbf{d}_i]_2, [\mathbf{y}_i \| \mathbf{r}_i]_2, \mathsf{ct}_i), i = 1, \ldots, n$

  return the discrete log of $\left(\prod_{i=1}^n [a_i]_T \right) / [z]_T$

Fig. 3. Multi-input functional encryption scheme $(\mathsf{Setup}', \mathsf{Enc}', \mathsf{KeyGen}', \mathsf{Dec}')$ for the class $\mathcal{F}_n^{m,B}$. $(\mathsf{Setup}, \mathsf{Enc}, \mathsf{KeyGen}, \mathsf{Dec})$ refers to the single-input functional encryption scheme for the class $\mathcal{F}_1^{m+k,B}$.

- For an efficient MIFE decryption, Dec must work without any restriction on the norm of the output as long as the output is in the exponent.
- Let $(\widetilde{\mathsf{Setup}}, \widetilde{\mathsf{Enc}}, \widetilde{\mathsf{KeyGen}})$ be the stateful simulator for the one-SEL-SIM security of the single-input inner-product FE scheme. We require that $\widetilde{\mathsf{KeyGen}}(\widetilde{\mathsf{msk}}, \cdot, \cdot)$ is linear in its inputs $(\mathbf{y}, a)$, so that we can compute $\widetilde{\mathsf{KeyGen}}(\widetilde{\mathsf{msk}}, [\mathbf{y}]_2, [a]_2) = [\widetilde{\mathsf{KeyGen}}(\widetilde{\mathsf{msk}}, \mathbf{y}, a)]_2$. This property is used in the proof of Lemma 5.

Remark 2 (notation). We use subscripts and superscripts for indexing over multiple copies, and never for indexing over positions or exponentiation. Concretely, the j'th ciphertext query in slot i is $\mathbf{x}_i^j$.

Security. Theorem 1 and Theorem 2 below, together with the fact that one-SEL-SIM security implies one-SEL-IND security, which itself implies many-SEL-IND security for a public-key FE, such as $(\mathsf{Setup}, \mathsf{Enc}, \mathsf{KeyGen})$ used in the construction presented in Fig. 3, implies the many-SEL-IND security of the MIFE $(\mathsf{Setup}', \mathsf{Enc}', \mathsf{KeyGen}')$.

Theorem 1 (one-SEL-IND Security of $\mathcal{MIFE}$). *Suppose the single-input FE $(\mathsf{Setup}, \mathsf{Enc}, \mathsf{KeyGen}, \mathsf{Dec})$ is one-SEL-SIM secure, and that the $\mathcal{D}_k$-MDDH assumption holds in $\mathbb{G}_2$. Then, the multi-input FE $(\mathsf{Setup}', \mathsf{Enc}', \mathsf{KeyGen}', \mathsf{Dec}')$ is one-SEL-IND-secure.*

$\underline{\text{Game}_0(1^\lambda, \mathcal{A}):}$
$\beta \leftarrow_R \{0,1\}, \mathbf{z}_i \leftarrow_R \mathbb{Z}_q^k$
$\{\mathbf{x}_i^b\}_{i\in[n],b\in\{0,1\}} \leftarrow \mathcal{A}(1^\lambda, \mathcal{F}_n)$
$(\text{mpk}_i, \text{msk}_i) \leftarrow \text{Setup}(1^\lambda, \mathcal{F}_n)$
$\text{mpk} := \{\text{mpk}_i\}_{i\in[n]}; \text{msk} := \{\text{msk}_i, \mathbf{z}_i\}_{i\in[n]}$
$\text{ct}_i := \text{Enc}(\text{msk}_i, \mathbf{x}_i^\beta \| \mathbf{z}_i)$
$\beta' \leftarrow \mathcal{A}^{\text{KeyGen}'(\text{msk},\cdot)}(\text{mpk}, (\text{ct}_i)_{i\in[n]})$
Output: 1 if $\beta' = \beta$, 0 otherwise.

$\underline{\text{KeyGen}'(\text{msk}, \mathbf{y}_1 \| \cdots \| \mathbf{y}_n):}$
$\mathbf{r} \leftarrow_R \mathbb{Z}_q^k$
$\mathbf{d}_i \leftarrow \text{KeyGen}(\text{msk}_i, \mathbf{y}_i \| \mathbf{r})$
$z := \langle \mathbf{z}_1 + \cdots + \mathbf{z}_n, \mathbf{r} \rangle$
$\text{sk}_{\mathbf{y}_1\|\cdots\|\mathbf{y}_n} := \left(\{ [\mathbf{d}_i]_2 \}_{i\in[n]}, [\mathbf{r}]_2, [z]_T \right)$
Return $\text{sk}_{\mathbf{y}_1\|\cdots\|\mathbf{y}_n}$

$\underline{\text{Game}_1(1^\lambda, \mathcal{A}):}$
$\beta \leftarrow_R \{0,1\}, \mathbf{z}_i \leftarrow_R \mathbb{Z}_q^k$
$\{\mathbf{x}_i^b\}_{i\in[n],b\in\{0,1\}} \leftarrow \mathcal{A}(1^\lambda, \mathcal{F}_n)$
$\left(\widetilde{\text{mpk}}_i, \widetilde{\text{msk}}_i \right) \leftarrow \widetilde{\text{Setup}}(1^\lambda, \mathcal{F}_1^{m+k,B})$
$\text{mpk} := \{\widetilde{\text{mpk}}_i\}_{i\in[n]}; \text{msk} := \{\widetilde{\text{msk}}_i, \mathbf{z}_i\}_{i\in[n]}$
$\text{ct}_i := \widetilde{\text{Enc}}(\widetilde{\text{msk}}_i)$
$\beta' \leftarrow \mathcal{A}^{\text{KeyGen}'(\text{msk},\cdot)}(\text{mpk}, (\text{ct}_i)_{i\in[n]})$
Output: 1 if $\beta' = \beta$, 0 otherwise.

$\underline{\text{KeyGen}'(\text{msk}, \mathbf{y}_1 \| \cdots \| \mathbf{y}_n):}$
$\mathbf{r} \leftarrow_R \mathbb{Z}_q^k$
$\mathbf{d}_i \leftarrow \widetilde{\text{KeyGen}}\left(\widetilde{\text{msk}}_i, \mathbf{y}_i \| \mathbf{r}, \langle \mathbf{x}_i^\beta \| \mathbf{z}_i, \mathbf{y}_i \| \mathbf{r} \rangle \right)$
$z := \langle \mathbf{z}_1 + \cdots + \mathbf{z}_n, \mathbf{r} \rangle$
$\text{sk}_{\mathbf{y}_1\|\cdots\|\mathbf{y}_n} := \left(\{ [\mathbf{d}_i]_2 \}_{i\in[n]}, [\mathbf{r}]_2, [z]_T \right)$
Return $\text{sk}_{\mathbf{y}_1\|\cdots\|\mathbf{y}_n}$

$\underline{\text{Game}_2(1^\lambda, \mathcal{A}):}$
$\beta \leftarrow_R \{0,1\}$
$\{\mathbf{x}_i^b\}_{i\in[n],b\in\{0,1\}} \leftarrow \mathcal{A}(1^\lambda, \mathcal{F}_n)$
$\left(\widetilde{\text{mpk}}_i, \widetilde{\text{msk}}_i \right) \leftarrow \widetilde{\text{Setup}}(1^\lambda, \mathcal{F}_1^{m+k,B})$
$\text{mpk} := \{\widetilde{\text{mpk}}_i\}_{i\in[n]}; \text{msk} := \{\widetilde{\text{msk}}_i\}_{i\in[n]}$
$\text{ct}_i := \widetilde{\text{Enc}}(\widetilde{\text{msk}}_i)$
$\beta' \leftarrow \mathcal{A}^{\text{KeyGen}'(\text{msk},\cdot)}(\text{mpk}, (\text{ct}_i)_{i\in[n]})$
Output: 1 if $\beta' = \beta$, 0 otherwise.

$\underline{\text{KeyGen}'(\text{msk}, \mathbf{y}_1 \| \cdots \| \mathbf{y}_n):}$
$\mathbf{r} \leftarrow_R \mathbb{Z}_q^k; \tilde{z}_1, \ldots, \tilde{z}_n \leftarrow_R \mathbb{Z}_q$
$\mathbf{d}_i \leftarrow \widetilde{\text{KeyGen}}\left(\widetilde{\text{msk}}_i, \mathbf{y}_i \| \mathbf{r}, \langle \mathbf{x}_i^\beta, \mathbf{y}_i \rangle + \tilde{z}_i \right)$
$z := \tilde{z}_1 + \cdots + \tilde{z}_n$
$\text{sk}_{\mathbf{y}_1\|\cdots\|\mathbf{y}_n} := \left(\{ [\mathbf{d}_i]_2 \}_{i\in[n]}, [\mathbf{r}]_2, [z]_T \right)$
Return $\text{sk}_{\mathbf{y}_1\|\cdots\|\mathbf{y}_n}$

$\underline{\text{Game}_3(1^\lambda, \mathcal{A}):}$
$\beta \leftarrow_R \{0,1\}$
$\{\mathbf{x}_i^b\}_{i\in[n],b\in\{0,1\}} \leftarrow \mathcal{A}(1^\lambda, \mathcal{F}_n)$
$\left(\widetilde{\text{mpk}}_i, \widetilde{\text{msk}}_i \right) \leftarrow \widetilde{\text{Setup}}(1^\lambda, \mathcal{F}_1^{m+k,B})$
$\text{mpk} := \{\widetilde{\text{mpk}}_i\}_{i\in[n]}; \text{msk} := \{\widetilde{\text{msk}}_i\}_{i\in[n]}$
$\text{ct}_i := \widetilde{\text{Enc}}(\widetilde{\text{msk}}_i)$
$\beta' \leftarrow \mathcal{A}^{\text{KeyGen}'(\text{msk},\cdot)}(\text{mpk}, (\text{ct}_i)_{i\in[n]})$
Output: 1 if $\beta' = \beta$, 0 otherwise.

$\underline{\text{KeyGen}'(\text{msk}, \mathbf{y}_1 \| \cdots \| \mathbf{y}_n):}$
$\mathbf{r} \leftarrow_R \mathbb{Z}_q^k; \tilde{z}_1, \ldots, \tilde{z}_n \leftarrow_R \mathbb{Z}_q$
$\mathbf{d}_i \leftarrow \widetilde{\text{KeyGen}}\left(\widetilde{\text{msk}}_i, \mathbf{y}_i \| \mathbf{r}, \tilde{z}_i \right)$
$z := \tilde{z}_1 + \cdots + \tilde{z}_n - \sum_i \langle \mathbf{x}_i^\beta, \mathbf{y}_i \rangle$
$\text{sk}_{\mathbf{y}_1\|\cdots\|\mathbf{y}_n} := \left(\{ [\mathbf{d}_i]_2 \}_{i\in[n]}, [\mathbf{r}]_2, [z]_T \right)$
Return $\text{sk}_{\mathbf{y}_1\|\cdots\|\mathbf{y}_n}$

Fig. 4. Game_i for $i \in \{0, \ldots, 3\}$ for the proof of Theorem 1.

That is, we show that our multi-input FE is selectively secure when there is only a single challenge ciphertext.

Proof (of Theorem 1). We proceed via a series of Game_i for $i \in \{0, \ldots, 3\}$, described in Fig. 4. Let $\mathcal{A}$ be a PPT adversary, and $\lambda \in \mathbb{N}$ be the security parameter.

Game$_0$: is the experiment **one-SEL-IND**$^{\mathcal{MIFE}}$ (see Definition 5).

Game$_1$: we replace (Setup, KeyGen, Enc) by the efficient simulator $(\widetilde{\text{Setup}}, \widetilde{\text{KeyGen}}, \widetilde{\text{Enc}})$, using the one-SEL-SIM security of $\mathcal{FE}$, via a hybrid argument across all slots $i \in [n]$ (cf Lemma 4).

Lemma 4 (Game$_0$ to Game$_1$). *There exists a PPT adversary $\mathcal{B}_1$ such that*

$$\text{Adv}_0(\mathcal{A}) - \text{Adv}_1(\mathcal{A}) \leq n \cdot \text{Adv}^{\mathcal{FE}, one\text{-}SEL\text{-}SIM}(1^\lambda, \mathcal{B}_1).$$

Game$_{0.\ell}(1^\lambda, \mathcal{A})$:

  $\{x_i^b\}_{i \in [n], b \in \{0,1\}} \leftarrow \mathcal{A}(1^\lambda, \mathcal{F}_1^{m+k,B})$

  $\beta \leftarrow_R \{0,1\}$

  $\left(\widetilde{\text{mpk}_i}, \widetilde{\text{msk}_i}\right) \leftarrow \widetilde{\text{Setup}}(1^\lambda, \mathcal{F}_1^{m+k,B}), i = 1, \ldots, \ell$

  $(\text{mpk}_i, \text{msk}_i) \leftarrow \text{Setup}(1^\lambda, \mathcal{F}_1^{m+k,B}), i = \ell+1, \ldots, n$

  $z_i \leftarrow_R \mathbb{Z}_q^k, i = 1, \ldots, n$

  $\text{mpk} := \{\widetilde{\text{mpk}_i}\}_{i=1,\ldots,\ell} \cup \{\text{mpk}_i\}_{i=\ell+1,\ldots,n}$

  $\text{msk} := \{\widetilde{\text{msk}_i}, z_i\}_{i=1,\ldots,\ell} \cup \{\text{msk}_i, z_i\}_{i=\ell+1,\ldots,n}$

  $\text{ct}_i := \widetilde{\text{Enc}}(\widetilde{\text{msk}_i})$, for all $i = 1, \ldots, \ell$

  $\text{ct}_i := \text{Enc}(\text{msk}_i, x_i^\beta \| z_i)$, for all $i = \ell+1, \ldots, n$

  $\beta' \leftarrow \mathcal{A}^{\text{KeyGen}'(\text{msk}, \cdot)}(\text{mpk}, \{\text{ct}_i\}_{i \in [n]})$

  Output :1 if $\beta' = \beta$, 0 otherwise.

KeyGen$'$(msk, $y_1 \| \cdots \| y_n$):

  $r \leftarrow_R \mathbb{Z}_q^k$

  $d_i \leftarrow \widetilde{\text{KeyGen}}\left(\widetilde{\text{msk}_i}, y_i \| r, \langle x_i^\beta \| z_i, y_i \| r \rangle\right)$, for all $i = 1, \ldots, \ell$

  $d_i \leftarrow \text{KeyGen}(\text{msk}_i, y_i \| r)$, for all $i = \ell+1, \ldots, n$

  $z := \langle z_1 + \cdots + z_n, r \rangle$

  $\text{sk}_{y_1 \| \cdots \| y_n} := \left(\{[d_i]_2\}_{i \in [n]}, [r]_2, [z]_T\right)$

  return $\text{sk}_{y_1 \| \cdots \| y_n}$

Fig. 5. Description of (Setup$'$, Enc$'$, KeyGen$'$) defining game $0.\ell$ for the proof of Lemma 4.

Proof. In Game$_1$, we replace (Setup, Enc, KeyGen) by $(\widetilde{\text{Setup}}, \widetilde{\text{Enc}}, \widetilde{\text{KeyGen}})$, which is a PPT simulator whose existence is ensured by the one-SEL-SIM security of (Setup, KeyGen, Enc) (see Definition 6). A complete description of Games$_0$ and Game$_1$ is given in Fig. 4.

We use a hybrid argument, which involves hybrid Game$_{0.\ell}$ for $\ell \in \{0, \ldots, n\}$, defined in Fig. 5, and we use $\text{Adv}_{0.\ell}(\lambda, \mathcal{A})$ to denote $\Pr[\text{Game}_{0.\ell}(\lambda, \mathcal{A}) = 1]$, where the probability is taken over the random coins of $\mathcal{A}$ and Game$_{0.\ell}$. Notice that Game$_0$ and Game$_1$ are identical to Game$_{0.0}$ and Game$_{0.n}$, respectively. For any $\ell \in [n]$, we build a PPT adversary $\mathcal{B}_{0.\ell}$ such that

$$\text{Adv}_{0.\ell-1}(\mathcal{A}) - \text{Adv}_{0.\ell}(\mathcal{A}) \leq \text{Adv}^{\mathcal{FE}, one\text{-}SEL\text{-}SIM}(1^\lambda, \mathcal{B}_{0.\ell}).$$

- **Simulation of mpk** : First, $\mathcal{B}_{0.\ell}$ receives the challenge $\{x_i^b\}_{i\in[n], b\in\{0,1\}}$ from $\mathcal{A}$. Then, it picks $\beta \leftarrow_R \{0,1\}$, $z_i \leftarrow_R \mathbb{Z}_q^k$ for all $i \in [n]$, and sends $x_\ell^\beta \| z_\ell$ to the experiment it is interacting with, which is either $\mathbf{REAL}^{\mathcal{FE}}$ or $\widetilde{\mathbf{IDEAL}}^{\mathcal{FE}}$. Then, $\mathcal{B}_{0.\ell}$ receives mpk'_ℓ, and a ciphertext ct, which are either of the form $\mathsf{mpk}'_\ell := \mathsf{mpk}_\ell$, where $(\mathsf{msk}_\ell, \mathsf{mpk}_\ell) \leftarrow \mathsf{Setup}(1^\lambda, \mathcal{F}_1^{m+k,B})$, and $\mathsf{ct} := \mathsf{Enc}(\mathsf{msk}_\ell, x_\ell^\beta \| z_\ell)$ if $\mathcal{B}_{3.\ell}$ is interacting with the experiment $\mathbf{REAL}^{\mathcal{FE}}$; or of the form $\mathsf{mpk}'_\ell := \widetilde{\mathsf{mpk}}_\ell$, where $(\widetilde{\mathsf{msk}}_\ell, \widetilde{\mathsf{mpk}}_\ell) \leftarrow \widetilde{\mathsf{Setup}}(1^\lambda, \mathcal{F}_1^{m+k,B})$, $\mathsf{ct} := \widetilde{\mathsf{Enc}}(\widetilde{\mathsf{msk}}_\ell)$ if $\mathcal{B}_{3.\ell}$ is interacting with the experiment $\widetilde{\mathbf{IDEAL}}^{\mathcal{FE}}$. It samples $(\widetilde{\mathsf{mpk}}_i, \widetilde{\mathsf{msk}}_i) \leftarrow \widetilde{\mathsf{Setup}}(1^\lambda, \mathcal{F}_1^{m+k,B})$ for $i = 1, \ldots, \ell-1$, $(\mathsf{mpk}_i, \mathsf{msk}_i) \leftarrow \mathsf{Setup}(1^\lambda, \mathcal{F}_1^{m+k,B})$ for $i = \ell+1, \ldots, n$, and returns $\mathsf{mpk} := (\widetilde{\mathsf{mpk}}_1, \ldots, \widetilde{\mathsf{mpk}}_{\ell-1}, \mathsf{mpk}'_\ell, \mathsf{mpk}_{\ell+1}, \ldots, \mathsf{mpk}_n)$ to $\mathcal{A}$.
- **Simulation of ct_i** : $\mathcal{B}_{0.\ell}$ computes $\mathsf{ct}_i := \mathsf{Enc}(\mathsf{msk}_i, x_i^\beta \| z_i)$ for all $i < \ell$ (note that $\mathcal{B}_{0.\ell}$ can do so since it knows msk_i, x_i^β, and z_i), and computes $\mathsf{ct}_i := \widetilde{\mathsf{Enc}}(\widetilde{\mathsf{msk}}_i)$ for all $i > \ell$ (again, $\mathcal{B}_{0.\ell}$ can do so since it knows $\widetilde{\mathsf{msk}}_i$). Finally, $\mathcal{B}_{0.\ell}$ sets $\mathsf{ct}_\ell := \mathsf{ct}$ and returns $\{\mathsf{ct}_i\}_{i\in[n]}$ to $\mathcal{A}$.
- **Simulation of $\mathsf{KeyGen}'(\mathsf{msk}, \cdot)$** : For each query $y_1\|\ldots\|y_n$ that $\mathcal{A}$ makes to $\mathsf{KeyGen}'(\mathsf{msk}, \cdot)$, $\mathcal{B}_{0.\ell}$ picks $r \leftarrow_R \mathbb{Z}_q^k$, and computes $d_i \leftarrow \widetilde{\mathsf{KeyGen}}(\widetilde{\mathsf{msk}}_i, y_i\|r, \langle x_i^\beta\|z_i, y_i\|r\rangle)$ for $i = 1, \ldots, \ell-1$, $d_i \leftarrow \mathsf{KeyGen}(\mathsf{msk}_i, y_i\|r)$ for $i = \ell+1, \ldots, n$. Then it computes d_ℓ by querying the oracle it has access to, which is $\mathsf{KeyGen}(\mathsf{msk}, \cdot)$ in the experiment $\mathbf{REAL}^{\mathcal{FE}}$, or $\mathcal{O}(\cdot)$ in the experiment $\widetilde{\mathbf{IDEAL}}^{\mathcal{FE}}$, on input $y_\ell\|r$. Then, it computes $z := \langle z_1 + \cdots + z_n, r\rangle$ and it returns $\mathsf{sk}_{y_1\|\cdots\|y_n} := (\{[d_i]_2\}_{i\in[n]}, [r]_2, [z]_T)$.

Finally, $\mathcal{B}_{0.\ell}$ outputs 1 if $\mathcal{A}$ outputs 1, 0 otherwise. It is clear that when $\mathcal{B}_{0.\ell}$ interacts with the experiment $\mathbf{REAL}^{\mathcal{FE}}$, it simulates the Game 0, whereas it simulates the Game 1 when it interacts with $\widetilde{\mathbf{IDEAL}}^{\mathcal{FE}}$. Therefore,

$$\mathsf{Adv}^{\mathcal{FE}, one\text{-}SEL\text{-}SIM}(\lambda, 1^\lambda, \mathcal{B}_{0.\ell})$$
$$= \left|\Pr\left[\mathbf{REAL}^{\mathcal{FE}}(1^\lambda, \mathcal{B}_{0.\ell}) = 1\right] - \Pr\left[\widetilde{\mathbf{IDEAL}}^{\mathcal{FE}}(1^\lambda, \mathcal{B}_{0.\ell}) = 1\right]\right|$$
$$= |\mathsf{Adv}_{0.\ell-1}(\mathcal{A}) - \mathsf{Adv}_{0.\ell}(\mathcal{A})|$$

Summing up for all $\ell \in [n]$, we obtain the lemma. $\square$

Game$_2$: we replace the values $\langle z_i, r\rangle$ used by $\mathsf{KeyGen}'(\mathsf{msk}, \cdot)$ to $\tilde{z}_i \leftarrow_R \mathbb{Z}_q$, for all slots $i \in [n]$, using the $\mathcal{D}_k$-MDDH assumption in $\mathbb{G}_2$ (cf Lemma 5).

Lemma 5 (Game$_1$ to Game$_2$). *There exists a PPT adversary $\mathcal{B}_2$ such that:*

$$\mathsf{Adv}_1(\mathcal{A}) - \mathsf{Adv}_2(\mathcal{A}) \leq \mathsf{Adv}_{\mathbb{G}_2, \mathcal{B}_2}^{\mathcal{U}_k\text{-}\mathrm{MDDH}}(\lambda) + \frac{1}{q-1}.$$

Proof. Here, we switch $\{[r]_2, [\langle z_i, r\rangle]_2\}_{i\in[n]}$ used by $\mathsf{KeyGen}(\mathsf{msk}, \cdot)$ to $\{[r]_2, [\tilde{z}_i]_2\}_{i\in[n]}$, where for all $i \in [n]$, $z_i \leftarrow_R \mathbb{Z}_q^k$, $\tilde{z}_1, \ldots, \tilde{z}_n \leftarrow_R \mathbb{Z}_p$ and $r \leftarrow_R \mathbb{Z}_q^k$.

This is justified by the fact that $[\mathbf{r}^\top \| \langle \mathbf{z}_1, \mathbf{r} \rangle \| \cdots \| \langle \mathbf{z}_n, \mathbf{r} \rangle]_2 \in \mathbb{G}_2^{1 \times (k+n)}$ is identically distributed to $[\mathbf{r}^\top \mathbf{U}^\top]_2$ where $\mathbf{U} \leftarrow_R \mathcal{U}_{k+n,k}$ (wlog. we assume that the upper k rows of $\mathbf{U}$ are full rank), which is indistinguishable from a uniformly random vector over $\mathbb{G}_2^{1 \times (k+n)}$, that is, of the form: $[\mathbf{r} \| \tilde{z}_1 \| \cdots \| \tilde{z}_n]_2$, according to the $\mathcal{U}_{k+n,k}$-MDDH assumption. To do the switch simultaneously for all calls to KeyGen, that is, to switch $\{[\mathbf{r}^j]_2, [\langle \mathbf{z}_i, \mathbf{r}^j \rangle]_2\}_{i \in [n], j \in [Q_0]}$ to $\{[\mathbf{r}^j]_2, [\tilde{z}_i^j]_2\}_{i \in [n], j \in [Q_0]}$, where Q_0 denotes the number of calls to KeyGen($\mathsf{msk}, \cdot$), and for all $i \in [n]$, $\mathbf{z}_i \leftarrow_R \mathbb{Z}_q^k$, $\tilde{z}_1^j, \ldots, \tilde{z}_n^j \leftarrow_R \mathbb{Z}_p$ and for all $j \in [Q_0]$, $\mathbf{r}^j \leftarrow_R \mathbb{Z}_q^k$, we use the Q_0-fold $\mathcal{U}_{k+n,k}$-MDDH assumption. Namely, we build a PPT adversary $\mathcal{B}_2'$ such that $\mathsf{Adv}_1(\mathcal{A}) - \mathsf{Adv}_2(\mathcal{A}) \leq \mathsf{Adv}_{\mathbb{G}_2, \mathcal{B}_2'}^{n\text{-fold } \mathcal{U}_{Q_0,k}\text{-MDDH}}(\lambda)$. This, together with Lemma 1 ($\mathcal{U}_k$-MDDH $\Rightarrow$ n-fold $\mathcal{U}_{Q_0,k}$-MDDH), implies the lemma.

- **Simulation of mpk :** Upon receiving an Q_0-fold $\mathcal{U}_{k+n,k}$-MDDH challenge

$$\left(\mathcal{PG}, [\mathbf{U}]_2 \in \mathbb{G}_2^{(k+n) \times k}, [\mathbf{h}^1 \| \cdots \| \mathbf{h}^{Q_0}]_2 \in \mathbb{G}_2^{(k+n) \times Q_0} \right),$$

and the challenge $\{\mathbf{x}_i^b\}_{i \in [n], b \in \{0,1\}}$ from $\mathcal{A}$, $\mathcal{B}_1'$ picks $\beta \leftarrow_R \{0,1\}$, samples $(\widetilde{\mathsf{mpk}_i}, \widetilde{\mathsf{msk}_i}) \leftarrow \widetilde{\mathsf{Setup}}(1^\lambda, \mathcal{F}_1^{m+k, B})$ for $i \in [n]$, and returns $\mathsf{mpk} :=$ $(\widetilde{\mathsf{mpk}_1}, \ldots, \widetilde{\mathsf{mpk}_n})$ to $\mathcal{A}$.
- **Simulation of ct_i :** $\mathcal{B}_2'$ computes $\mathsf{ct}_i := \widetilde{\mathsf{Enc}}(\widetilde{\mathsf{msk}_i})$ for all $i \in [n]$, which it can do since it knows $\widetilde{\mathsf{msk}_i}$, and returns $\{\mathsf{ct}_i\}_{i \in [n]}$ to $\mathcal{A}$.
- **Simulation of KeyGen'(msk , $\cdot$) :** On the j'th query $\mathbf{y}_1 \| \cdots \| \mathbf{y}_n$ of $\mathcal{A}$ to KeyGen', $\mathcal{B}_2'$ sets $[\mathbf{r}^j]_2 := [\overline{\mathbf{h}^j}]_2$, where $\overline{\mathbf{h}^j} \in \mathbb{Z}_q^k$ denotes the k-upper components of $\mathbf{h}^j \in \mathbb{Z}_q^{k+n}$, and for each $i \in [n]$, computes $[\mathbf{d}_i]_2 :=$ $[\widetilde{\mathsf{KeyGen}}(\widetilde{\mathsf{msk}_i}, \mathbf{y}_i \| \mathbf{r}^j, \langle \mathbf{x}_i^\beta, \mathbf{y}_i \rangle + \mathbf{h}_{k+i}^j)]_2$, where $\mathbf{h}_{k+i}^j$ denotes the $k + i$'th coordinate of the vector $\mathbf{h}^j \in \mathbb{Z}_p^{k+n}$. Here we rely on the fact that $\widetilde{\mathsf{KeyGen}}(\widetilde{\mathsf{msk}}, \cdot, \cdot)$ is linear in its inputs $(\mathbf{y}, a)$, so that we can compute $\widetilde{\mathsf{KeyGen}}(\widetilde{\mathsf{msk}}, [\mathbf{y}]_2, [a]_2) =$ $[\widetilde{\mathsf{KeyGen}}(\widetilde{\mathsf{msk}}, \mathbf{y}, a)]_2$. Note that when $[\mathbf{h}^1 \| \cdots \| \mathbf{h}^{Q_0}]_2$ is a real MDDH challenge, $\mathcal{B}_2'$ simulate Game$_1$, whereas it simulates Game$_2$ when $[\mathbf{h}^1 \| \cdots \| \mathbf{h}^{Q_0}]_2$ is uniformly random over $\mathbb{G}_1^{(k+n) \times Q_0}$. $\square$

Game$_3$: here the values $\mathbf{d}_i$ for $i \in [n]$, and z, computed by KeyGen'($\mathsf{msk}, \cdot$), are of the form: $\mathbf{d}_i \leftarrow \widetilde{\mathsf{KeyGen}} \left(\widetilde{\mathsf{msk}_i}, \mathbf{y}_i \| \mathbf{r}, \boxed{\tilde{z}_i} \right)$, and $z := \tilde{z}_1 + \cdots + \tilde{z}_n - \boxed{\sum_i \langle \mathbf{x}_i^\beta, \mathbf{y}_i \rangle}$. In Lemma 6, we prove that Game$_3$ and Game$_2$ are perfectly indistinguishable, using a statistical argument that crucially relies on the fact that Game$_3$ and Game$_2$ are selective. In Lemma 7, we prove that no adversary can win Game$_3$, using the restriction on the queries to KeyGen'($\mathsf{msk}, \cdot$) and the challenge $\{\mathbf{x}_i^b\}_{i \in [n]}$ imposed by the ideal functionality.

Lemma 6 (Game$_2$ to Game$_3$). $\mathsf{Adv}_2(\mathcal{A}) = \mathsf{Adv}_3(\mathcal{A})$.

Proof. Here, we use the fact that for all $\mathbf{y}_1\|\cdots\|\mathbf{y}_n \in (\mathbb{Z}_q^m)^n$, for all $\{\mathbf{x}_i^b \in \mathbb{Z}_q^m\}_{i\in[n],b\in\{0,1\}}$, all $\beta \in \{0,1\}$, the following are identically distributed: $\{\tilde{z}_i\}_{i\in[n]}$ and $\{\tilde{z}_i - \langle\mathbf{x}_i^\beta,\mathbf{y}_i\rangle\}_{i\in[n]}$, where $\tilde{z}_i \leftarrow_R \mathbb{Z}_q$ for all $i \in [n]$.

For each query $\mathbf{y}_1\|\cdots\|\mathbf{y}_n$, $\mathsf{KeyGen}'(\mathsf{msk},\mathbf{y}_1\|\cdots\|\mathbf{y}_n)$ picks values $\tilde{z}_i \leftarrow_R \mathbb{Z}_q$ for $i \in [n]$ that are *independent* of $\mathbf{y}_1\|\cdots\|\mathbf{y}_n$ and the challenge $\{\mathbf{x}_i^b \in \mathbb{Z}_q^m\}_{i\in[n],b\in\{0,1\}}$ (note that here we crucially rely on the fact the Game_2 and Game_3 are *selective*), therefore, using the previous fact, we can switch $\tilde{z}_i$ to $\tilde{z}_i - \langle\mathbf{x}_i^\beta,\mathbf{y}_i\rangle$ without changing the distribution of the game. This way, $\mathsf{KeyGen}'(\mathsf{msk}, \mathbf{y}_1\|\cdots\|\mathbf{y}_n)$ computes $\mathbf{d}_i \leftarrow \widetilde{\mathsf{KeyGen}}(\widetilde{\mathsf{msk}}_i, \mathbf{y}_i\|\mathbf{r}, \tilde{z}_i)$ for all $i \in [n]$, and $z := \tilde{z}_1 + \ldots + \tilde{z}_n - \sum_{i=1}^n\langle\mathbf{x}_i^\beta,\mathbf{y}_i\rangle$, as in Game_3. $\square$

Lemma 7 (Game_3). $\mathsf{Adv}_3(\mathcal{A}) = 0$.

Proof. We use the fact that for all $i \in [n]$, the query $(i, \mathbf{x}_i^0, \mathbf{x}_i^1)$ to Enc' (recall that there can be at most one query per slot $i \in [n]$), and for all queries $\mathbf{y}_1\|\cdots\|\mathbf{y}_n$ to KeyGen', by definition of the security game, we have:

$$\sum_{i=1}^n\langle\mathbf{x}_i^0,\mathbf{y}_i\rangle = \sum_{i=1}^n\langle\mathbf{x}_i^1,\mathbf{y}_i\rangle.$$

Therefore, for each call to $\mathsf{KeyGen}(\mathsf{msk},\cdot)$, the value z, which is of the form $z := \sum_i\tilde{z}_i - \sum_i\langle\mathbf{x}_i^\beta,\mathbf{y}_i\rangle$, is independent of β. Since the challenge ciphertext and the public key are also independent of β, we have $\mathsf{Adv}_3(\mathcal{A}) = 0$. $\square$

Summing up, we proved that for all security parameter $\lambda \in \mathbb{N}$ and all PPT adversaries $\mathcal{A}$, the following holds.

- In Lemma 4, we show that there exists a PPT adversary $\mathcal{B}_1$ such that $\mathsf{Adv}_0(\mathcal{A}) - \mathsf{Adv}_1(\mathcal{A}) \le n \cdot \mathsf{Adv}^{\mathcal{FE},one\text{-}SEL\text{-}SIM}(1^\lambda, \mathcal{B}_1)$.
- In Lemma 5, we show that there exists a PPT adversary $\mathcal{B}_2$ such that $\mathsf{Adv}_1(\mathcal{A}) - \mathsf{Adv}_2(\mathcal{A}) \le \mathsf{Adv}_{\mathbb{G}_2,\mathcal{B}_2}^{\mathcal{U}_k\text{-}MDDH}(\lambda) + \frac{1}{q-1}$.
- In Lemma 6, we show that $\mathsf{Adv}_2(\mathcal{A}) = \mathsf{Adv}_3(\mathcal{A})$.
- In Lemma 7, we show that $\mathsf{Adv}_3(\mathcal{A}) = 0$.

Putting everything together, we obtain:

$$\mathsf{Adv}_0(\mathcal{A}) \le n \cdot \mathsf{Adv}^{\mathcal{FE},one\text{-}SEL\text{-}SIM}(1^\lambda, \mathcal{B}_0) + \mathsf{Adv}_{\mathbb{G}_2,\mathcal{B}_2}^{\mathcal{U}_k\text{-}MDDH}(\lambda) + \frac{1}{q-1}.$$

By Definition 6, $\mathsf{Adv}_0(\mathcal{A}) = \mathsf{Adv}^{\mathcal{MIFE},one\text{-}SEL\text{-}IND}(1^\lambda, \mathcal{A})$. Therefore, by the one-SEL-SIM security of $(\mathsf{Setup}, \mathsf{Enc}, \mathsf{KeyGen})$ and the $\mathcal{D}_k$-MDDH assumption in $\mathbb{G}_2$, $\mathsf{Adv}^{\mathcal{MIFE},one\text{-}SEL\text{-}IND}(1^\lambda, \mathcal{A})$ is a negligible function of λ. $\square$

Remark 3 (decryption capabilities). As a sanity check, we note that the simulated secret keys will correctly decrypt a simulated ciphertext. However, unlike schemes proven secure via the standard dual system encryption methodology [26], a simulated secret key will incorrectly decrypt a normal ciphertext. This is not a problem because we are in the private-key setting, so a distinguisher will not be able to generate normal ciphertexts by itself.

Remark 4 (why a naive argument is inadequate). We cannot afford to do a naive hybrid argument across the n slots for the challenge ciphertext as it would introduce extraneous restrictions on the adversary's queries. Concretely, suppose we want to use a hybrid argument to switch from encryptions of $\mathbf{x}_1^0, \mathbf{x}_2^0$ in game 0 to those of $\mathbf{x}_1^1, \mathbf{x}_2^1$ in game 2 with an intermediate hybrid that uses encryptions of $\mathbf{x}_1^1, \mathbf{x}_2^0$ in Game_1. To move from game 0 to game 1, the adversary's query $\mathbf{y}_1 \| \mathbf{y}_2$ must satisfy $\langle \mathbf{x}_1^0 \| \mathbf{x}_2^0, \mathbf{y}_1 \| \mathbf{y}_2 \rangle = \langle \mathbf{x}_1^1 \| \mathbf{x}_2^0, \mathbf{y}_1 \| \mathbf{y}_2 \rangle$, which implies the extraneous restriction $\langle \mathbf{x}_1^0, \mathbf{y}_1 \rangle = \langle \mathbf{x}_2^1, \mathbf{y}_1 \rangle$.

As described in the proof above, we overcome the limitation by using simulation-based security. Note that what essentially happens in the first slot in our proof is as follows (for $k = 1$, that is, DDH): we switch from $\text{Enc}(\text{msk}_1, \mathbf{x}_1^0 \| z_1)$ to $\text{Enc}(\text{msk}_1, \mathbf{x}_1^1 \| z_1)$ while giving out a secret key which contains $\text{KeyGen}(\text{msk}_1, \mathbf{y}_1 \| r^1), [r^1]_2$. Observe that

$$\langle \mathbf{x}_1^0 \| z_1, \mathbf{y}_1 \| r^1 \rangle = \langle \mathbf{x}_1^0, \mathbf{y}_1 \rangle + z_1 r^1, \quad \langle \mathbf{x}_1^1 \| z_1, \mathbf{y}_1 \| r^1 \rangle = \langle \mathbf{x}_1^1, \mathbf{y}_1 \rangle + z_1 r^1$$

may not be equal, since we want to avoid the extraneous restriction $\langle \mathbf{x}_1^0, \mathbf{y}_1 \rangle = \langle \mathbf{x}_2^1, \mathbf{y}_1 \rangle$. This means that one-SEL-IND security does not provide any guarantee that the ciphertexts are indistinguishable. However, one-SEL-SIM security does provide such a guarantee, because

$$([\langle \mathbf{x}_1^0, \mathbf{y}_1 \rangle + z_1 r^1]_2, [r^1]_2) \approx_c ([\langle \mathbf{x}_1^1, \mathbf{y}_1 \rangle + z_1 r^1]_2, [r^1]_2)$$

via the DDH assumption in $\mathbb{G}_2$. Since the outcomes of the decryption are computationally indistinguishable, the output of the simulated ciphertext would also be computationally indistinguishable.

Theorem 2 (many-SEL-IND Security of $\mathcal{MIFE}$). *Suppose the single-input FE* (Setup, Enc, KeyGen, Dec) *is many-SEL-IND-secure and the multi-input FE* (Setup', Enc', KeyGen', Dec') *is one-SEL-IND-secure. Then, the multi-input FE* (Setup', Enc', KeyGen', Dec') *is many-SEL-IND-secure.*

That is, we show that our multi-input FE is selectively secure in the setting with multiple challenge ciphertexts (and since our multi-input FE is a private key scheme, one-SEL-IND security does not immediately imply many-SEL-IND security).

Proof Overview.

- We first switch encryptions of $\mathbf{x}_1^{1,0}, \dots, \mathbf{x}_n^{1,0}$ to those of $\mathbf{x}_1^{1,1}, \dots, \mathbf{x}_n^{1,1}$ in a "single shot", and for the remaining ciphertexts, we switch from an encryption of $\mathbf{x}_i^{j,0} = (\mathbf{x}_i^{j,0} - \mathbf{x}_i^{1,0}) + \mathbf{x}_i^{1,0}$ to that of $(\mathbf{x}_i^{j,0} - \mathbf{x}_i^{1,0}) + \mathbf{x}_i^{1,1}$. This basically follows from the setting where there is only a single ciphertext in each slot.
- Then, we apply a hybrid argument across the slots to switch from encryptions of $(\mathbf{x}_i^{2,0} - \mathbf{x}_i^{1,0}) + \mathbf{x}_i^{1,1}, \dots, (\mathbf{x}_i^{Q_i,0} - \mathbf{x}_i^{1,0}) + \mathbf{x}_i^{1,1}$ to those of $(\mathbf{x}_i^{2,1} - \mathbf{x}_i^{1,1}) + \mathbf{x}_i^{1,1}, \dots, (\mathbf{x}_i^{Q_i,1} - \mathbf{x}_i^{1,1}) + \mathbf{x}_i^{1,1}$.

As described earlier, to carry out the latter hybrid argument, the queries must satisfy the constraint

$$\langle (x_i^{j,0} - x_i^{1,0}) + x_i^{1,1}, y_i \rangle = \langle (x_i^{j,1} - x_i^{1,1}) + x_i^{1,1}, y_i \rangle$$
$$\Longleftrightarrow \langle x_i^{j,0} - x_i^{1,0}, y_i \rangle = \langle x_i^{j,1} - x_i^{1,1}, y_i \rangle$$

where the latter is already imposed by the ideal functionality.

We defer to the full version of this paper for the complete proof.

5   Achieving Adaptive Security

In this section, we show that the multi-input FE in Fig. 7 is many-AD-IND secure. Roughly speaking, xx-AD-IND security, where xx $\in$ {many, one}, is defined as xx-SEL-IND security (see Definition 5), except that the adversary does not have to commit to its challenge beforehand, and queries secret keys adaptively. See the full version of this paper for the formal definition of xx-AD-IND security.

Theorem 3. *Suppose the $\mathcal{D}_k$-MDDH assumption holds in $\mathbb{G}_1$ and $\mathbb{G}_2$. Then, the multi-input FE in Fig. 6 is many-AD-IND-secure.*

Setup($\mathbb{G}, \mathcal{F}_n^{m,B}$):
For $i \in [n]$, $A_i \leftarrow_R \mathcal{D}_k$, $W_i \leftarrow_R \mathbb{Z}_q^{m \times (k+1)}$, $V_i \leftarrow_R \mathbb{Z}_q^{k \times (k+1)}$, $z_i \leftarrow_R \mathbb{Z}_q^k$
mpk := $\{ [A_i]_1, [W_i A_i]_1 \}_{i \in [n]}$, msk := $\{ W_i, V_i, z_i \}_{i \in [n]}$
return (mpk, msk)

Enc(msk, i, $x_i \in \mathbb{Z}_q^m$):
return $([c_i]_1, [c_i']_1, [c_i'']_1) := ([A_i s_i]_1, [x_i + W_i A_i s_i]_1, [z_i + V_i A_i s_i]_1)$

KeyGen(msk, $y_1 \| \cdots \| y_n \in (\mathbb{Z}_q^m)^n$):
For $i \in [n]$: $d_i := W_i^\top y_i + V_i^\top r$, $r \leftarrow_R \mathbb{Z}_q^k$, $z := \langle z_1 + \cdots + z_n, r \rangle$
return $\left(\{ [d_i]_2 \}_{i \in [n]}, [r]_2, [z]_T \right)$

Dec$\left((\{ [d_i]_2 \}_{i \in [n]}, [r]_2, [z]_T), y_1 \| \cdots \| y_n, \{ [c_i]_1, [c_i']_1, [c_i'']_1 \}_{i \in [n]} \right)$:

out := $\left(\sum_i e([c_i']_1, [y_i]_2) \cdot e([c_i'']_1, [r]_2)/e([c_i]_1, [d_i]_2) \right)/[z]_T$
return discrete log of out

Fig. 6. Our private-key MIFE scheme for the class $\mathcal{F}_n^{m,B}$ (self-contained description). The scheme is many-AD-IND-secure under the $\mathcal{D}_k$-MDDH assumption in $\mathbb{G}_1$ and $\mathbb{G}_2$. We use $e([X]_1, [Y]_2)$ to denote $[X^\top Y]_T$.

Proof Overview. The security proof proceeds in three steps:

- First, we show that the MIFE in Fig. 6 is one-AD-IND secure, that is, it is adaptively secure when there is only a single challenge ciphertext. To achieve *adaptive* security, we borrow the techniques used in the selective security proof, using *complexity leveraging* to obtain adaptive security. Note that in our case, we can afford the exponential security loss from complexity leveraging, since this is used in the proof in combination with perfect indistinguishability, therefore, the exponential term is multiplied by a zero term.
- Then, we show that the generic construction of MIFE in Fig. 3 is many-AD-IND secure, if the underlying single-input FE is many-AD-IND secure, and the MIFE is one-AD-IND secure.
- Finally, we show that the single-input scheme in Fig. 7 is many-AD-IND.

Putting everything together, we obtain many-AD-IND security of the MIFE in Fig. 6. We defer to the full version of this paper for a complete proof, and for the definition of one-AD-IND and many-AD-IND security.

A   One-SEL-SIM, Many-AD-IND Secure Scheme for Single-Input Inner Products

In Fig. 7, we describe the scheme for Single-Input Inner Products from [28], which is essentially the same as those in [2,4], extended explicitly to the $\mathcal{D}_k$-MDDH assumption. In the full version of this paper, we recall the proof of one-SEL-SIM-security from [28] and we prove its many-AD-IND security. Moreover, note that the scheme is public key, linearly homomorphic, and satisfies additional requirements for the construction in Fig. 3.

Setup$(\mathbb{G}, \mathcal{F}_1^{m,B})$:	Enc$(\mathsf{msk}, \mathbf{x} \in \mathbb{Z}_q^m)$:
$\mathbf{A} \leftarrow_R \mathcal{D}_k,\ \mathbf{W} \leftarrow_R \mathbb{Z}_q^{m \times (k+1)}$	$\mathbf{r} \leftarrow_R \mathbb{Z}_p^k$;
$\mathsf{mpk} := ([\mathbf{A}], [\mathbf{WA}]), \mathsf{msk} := (\mathbf{W}, \mathbf{A})$;	return $([\mathbf{c}], [\mathbf{c}']) := ([\mathbf{Ar}], [\mathbf{x} + \mathbf{WAr}])$
return $(\mathsf{mpk}, \mathsf{msk})$	
KeyGen$(\mathsf{msk}, \mathbf{y} \in \mathbb{Z}_q^m)$:	Dec$(\mathsf{sk}_\mathbf{y}, \mathbf{y}, ([\mathbf{c}], [\mathbf{c}']))$:
return $\mathsf{sk}_\mathbf{y} := \mathbf{W}^\top \mathbf{y} \in \mathbb{Z}_q^{k+1}$	return discrete log of $[\mathbf{c}'^\top \mathbf{y} - \mathbf{c}^\top \mathsf{sk}_\mathbf{y}]$

Fig. 7. A one-SEL-SIM scheme for single-input inner product $\mathcal{F}_1^{m,B}$ [28].

Theorem 4 (one-SEL-SIM, many-AD-IND Security of $\mathcal{FE}$). *If the $\mathcal{D}_k$-MDDH assumption holds in $\mathbb{G}$, then the single-input FE in Fig. 7 is one-SEL-SIM secure (see Definition 6), and many-AD-IND secure.*

We defer to the full version of this paper for the complete proof. We provide the description of the simulator for the proof of one-SEL-SIM security from [28], in Fig. 8.

$$\overline{\mathsf{Setup}}(\mathbb{G}):$$

$\mathbf{A} \leftarrow_R \mathbb{Z}_q^{(k+1)\times k}, \widetilde{\mathbf{W}} \leftarrow_R \mathbb{Z}_q^{m\times(k+1)}, \mathbf{c} \leftarrow_R \mathbb{Z}_q^{k+1} \setminus \mathsf{Span}(\mathbf{A});$

compute $\mathbf{a}^\perp \in \mathbb{Z}_q^{k+1} \setminus \{\mathbf{0}\}$ s.t. $\mathbf{A}^\top \mathbf{a}^\perp = \mathbf{0}$

$\widetilde{\mathsf{mpk}} := ([\mathbf{A}], [\widetilde{\mathbf{W}}\mathbf{A}]), \widetilde{\mathsf{msk}} := (\mathbf{a}^\perp, \widetilde{\mathbf{W}}, \mathbf{c});$

return $(\widetilde{\mathsf{mpk}}, \widetilde{\mathsf{msk}})$

$$\overline{\mathsf{KeyGen}}(\widetilde{\mathsf{msk}}, \mathbf{y} \in \mathbb{Z}_q^m, a \in \mathbb{Z}_q):$$

return $\mathsf{sk}_\mathbf{y} := \widetilde{\mathbf{W}}^\top \mathbf{y} - \frac{a}{\langle \mathbf{c}, \mathbf{a}^\perp \rangle}\mathbf{a}^\perp \in \mathbb{Z}_q^{k+1}$

$$\overline{\mathsf{Enc}}(\widetilde{\mathsf{msk}}):$$

return $([\mathbf{c}], [\widetilde{\mathbf{W}}\mathbf{c}])$

Fig. 8. Simulator $(\widetilde{\mathsf{Setup}}, \widetilde{\mathsf{KeyGen}}, \widetilde{\mathsf{Enc}})$ from [28] for the one-SEL-SIM security of the single-input scheme for inner product $\mathcal{F}_1^{m,B}$ in Fig. 7

References

1. Abdalla, M., Bourse, F., Caro, A., Pointcheval, D.: Simple functional encryption schemes for inner products. In: Katz, J. (ed.) PKC 2015. LNCS, vol. 9020, pp. 733–751. Springer, Heidelberg (2015). doi:10.1007/978-3-662-46447-2_33

2. Abdalla, M., Bourse, F., De Caro, A., Pointcheval, D.: Better security for functional encryption for inner product evaluations. Cryptology ePrint Archive, Report 2016/011 (2016). http://eprint.iacr.org/2016/011

3. Agrawal, S., Gorbunov, S., Vaikuntanathan, V., Wee, H.: Functional encryption: new perspectives and lower bounds. In: Canetti, R., Garay, J.A. (eds.) CRYPTO 2013. LNCS, vol. 8043, pp. 500–518. Springer, Heidelberg (2013). doi:10.1007/978-3-642-40084-1_28

4. Agrawal, S., Libert, B., Stehlé, D.: Fully secure functional encryption for inner products, from standard assumptions. In: Robshaw, M., Katz, J. (eds.) CRYPTO 2016. LNCS, vol. 9816, pp. 333–362. Springer, Heidelberg (2016). doi:10.1007/978-3-662-53015-3_12

5. Ananth, P., Jain, A.: Indistinguishability obfuscation from compact functional encryption. In: Gennaro, R., Robshaw, M. (eds.) CRYPTO 2015. LNCS, vol. 9215, pp. 308–326. Springer, Heidelberg (2015). doi:10.1007/978-3-662-47989-6_15

6. Badrinarayanan, S., Gupta, D., Jain, A., Sahai, A.: Multi-input functional encryption for unbounded arity functions. In: Iwata, T., Cheon, J.H. (eds.) ASIACRYPT 2015. LNCS, vol. 9452, pp. 27–51. Springer, Heidelberg (2015). doi:10.1007/978-3-662-48797-6_2

7. Bishop, A., Jain, A., Kowalczyk, L.: Function-hiding inner product encryption. In: Iwata, T., Cheon, J.H. (eds.) ASIACRYPT 2015. LNCS, vol. 9452, pp. 470–491. Springer, Heidelberg (2015). doi:10.1007/978-3-662-48797-6_20

8. Bitansky, N., Vaikuntanathan, V.: Indistinguishability obfuscation from functional encryption. In: 56th FOCS, pp. 171–190. IEEE Computer Society Press, October 2015

9. Blazy, O., Kiltz, E., Pan, J.: (Hierarchical) identity-based encryption from affine message authentication. In: Garay, J.A., Gennaro, R. (eds.) CRYPTO 2014. LNCS, vol. 8616, pp. 408–425. Springer, Heidelberg (2014). doi:10.1007/978-3-662-44371-2_23

10. Boneh, D., Lewi, K., Raykova, M., Sahai, A., Zhandry, M., Zimmerman, J.: Semantically secure order-revealing encryption: multi-input functional encryption without obfuscation. In: Oswald, E., Fischlin, M. (eds.) EUROCRYPT 2015. LNCS, vol. 9057, pp. 563–594. Springer, Heidelberg (2015). doi:10.1007/978-3-662-46803-6_19
11. Boneh, D., Sahai, A., Waters, B.: Functional encryption: definitions and challenges. In: Ishai, Y. (ed.) TCC 2011. LNCS, vol. 6597, pp. 253–273. Springer, Heidelberg (2011). doi:10.1007/978-3-642-19571-6_16
12. Brakerski, Z., Komargodski, I., Segev, G.: Multi-input functional encryption in the private-key setting: stronger security from weaker assumptions. In: Fischlin, M., Coron, J.-S. (eds.) EUROCRYPT 2016. LNCS, vol. 9666, pp. 852–880. Springer, Heidelberg (2016). doi:10.1007/978-3-662-49896-5_30
13. Datta, P., Dutta, R., Mukhopadhyay, S.: Functional encryption for inner product with full function privacy. In: Cheng, C.-M., Chung, K.-M., Persiano, G., Yang, B.-Y. (eds.) PKC 2016. LNCS, vol. 9614, pp. 164–195. Springer, Heidelberg (2016). doi:10.1007/978-3-662-49384-7_7
14. Escala, A., Herold, G., Kiltz, E., Ràfols, C., Villar, J.: An algebraic framework for Diffie-Hellman assumptions. In: Canetti, R., Garay, J.A. (eds.) CRYPTO 2013. LNCS, vol. 8043, pp. 129–147. Springer, Heidelberg (2013). doi:10.1007/978-3-642-40084-1_8
15. Garg, S., Gentry, C., Halevi, S.: Candidate multilinear maps from ideal lattices. In: Johansson, T., Nguyen, P.Q. (eds.) EUROCRYPT 2013. LNCS, vol. 7881, pp. 1–17. Springer, Heidelberg (2013). doi:10.1007/978-3-642-38348-9_1
16. Garg, S., Gentry, C., Halevi, S., Raykova, M., Sahai, A., Waters, B.: Candidate indistinguishability obfuscation and functional encryption for all circuits. In: 54th FOCS, pp. 40–49. IEEE Computer Society Press, October 2013
17. Garg, S., Gentry, C., Halevi, S., Zhandry, M.: Functional encryption without obfuscation. In: Kushilevitz, E., Malkin, T. (eds.) TCC 2016. LNCS, vol. 9563, pp. 480–511. Springer, Heidelberg (2016). doi:10.1007/978-3-662-49099-0_18
18. Gay, R., Hofheinz, D., Kiltz, E., Wee, H.: Tightly CCA-secure encryption without pairings. In: Fischlin, M., Coron, J.-S. (eds.) EUROCRYPT 2016. LNCS, vol. 9665, pp. 1–27. Springer, Heidelberg (2016). doi:10.1007/978-3-662-49890-3_1
19. Goldwasser, S., Gordon, S.D., Goyal, V., Jain, A., Katz, J., Liu, F.-H., Sahai, A., Shi, E., Zhou, H.-S.: Multi-input functional encryption. In: Nguyen, P.Q., Oswald, E. (eds.) EUROCRYPT 2014. LNCS, vol. 8441, pp. 578–602. Springer, Heidelberg (2014). doi:10.1007/978-3-642-55220-5_32
20. Goldwasser, S., Kalai, Y.T., Popa, R.A., Vaikuntanathan, V., Zeldovich, N.: Reusable garbled circuits and succinct functional encryption. In: Boneh, D., Roughgarden, T., Feigenbaum, J. (eds.) 45th ACM STOC, pp. 555–564. ACM Press, June 2013
21. Gorbunov, S., Vaikuntanathan, V., Wee, H.: Functional encryption with bounded collusions via multi-party computation. In: Safavi-Naini, R., Canetti, R. (eds.) CRYPTO 2012. LNCS, vol. 7417, pp. 162–179. Springer, Heidelberg (2012). doi:10.1007/978-3-642-32009-5_11
22. Kim, S., Lewi, K., Mandal, A., Montgomery, H., Roy, A., Wu, D.J.: Function-hiding inner product encryption is practical. Cryptology ePrint Archive, Report 2016/440 (2016). http://eprint.iacr.org/2016/440
23. Lee, K., Lee, D.H.: Two-input functional encryption for inner products from bilinear maps. Cryptology ePrint Archive, Report 2016/432 (2016). http://eprint.iacr.org/2016/432
24. Sahai, A., Seyalioglu, H.: Worry-free encryption: functional encryption with public keys. In: ACM CCS 2010, pp. 463–472. ACM Press, October 2010

25. Sahai, A., Waters, B.: Fuzzy identity-based encryption. In: Cramer, R. (ed.) EURO-CRYPT 2005. LNCS, vol. 3494, pp. 457–473. Springer, Heidelberg (2005). doi:10.1007/11426639_27

26. Waters, B.: Dual system encryption: realizing fully secure IBE and HIBE under simple assumptions. In: Halevi, S. (ed.) CRYPTO 2009. LNCS, vol. 5677, pp. 619–636. Springer, Heidelberg (2009). doi:10.1007/978-3-642-03356-8_36

27. Wee, H.: Dual system encryption via predicate encodings. In: Lindell, Y. (ed.) TCC 2014. LNCS, vol. 8349, pp. 616–637. Springer, Heidelberg (2014). doi:10.1007/978-3-642-54242-8_26

28. Wee, H.: New techniques for attribute-hiding in prime-order bilinear groups (2016, in preparation)

Simplifying Design and Analysis of Complex Predicate Encryption Schemes

Shashank Agrawal[1](✉) and Melissa Chase[2]

[1] Visa Research, Palo Alto, USA
shaagraw@visa.com
[2] Microsoft Research, Redmond, USA
melissac@microsoft.com

Abstract. Wee (TCC'14) and Attrapadung (Eurocrypt'14) introduced predicate and pair encodings, respectively, as a simple way to construct and analyze attribute-based encryption schemes, or more generally predicate encryption. However, many schemes do not satisfy the simple information theoretic property proposed in those works, and thus require much more complicated analysis. In this paper, we propose a new simple property for pair encodings called *symbolic* security. Proofs that pair encodings satisfy this property are concise and easy to verify. We show that this property is inherently tied to the security of predicate encryption schemes by arguing that any scheme which is not trivially broken must satisfy it. Then we use this property to discuss several ways to convert between pair encodings to obtain encryption schemes with different properties like small ciphertexts or keys. Finally, we show that any pair encoding satisfying our new property can be used to construct a fully secure predicate encryption scheme. The resulting schemes are secure under a new q-type assumption which we show follows from several of the assumptions used to construct such schemes in previous work.

1 Introduction

Traditional public key encryption allows an encryptor to use a public key to encrypt a message so that the owner of the corresponding secret key can decrypt. In 2005, Sahai and Waters [35] introduced the concept of attribute-based encryption, in which who can decrypt is determined by some more complex attributes of the decryptor and the message. Of course this is only meaningful if there is some party that can determine the attributes of the decryption, thus the basic model assumes a trusted party who publishes parameters used in encryption, and who issues decryption keys to users based on their attributes; given such a key, a user should be able to decrypt any ciphertext which is compatible with his attributes. The initial result considered a simple threshold functionality: every ciphertext was encrypted with a set of attributes, and a user could decrypt if

S. Agrawal—Part of this work was done when the author was a graduate student at the University of Illinois, Urbana-Champaign, supported by NSF CNS 12-28856 and Andrew & Shana Laursen fellowship, and then an intern at Microsoft Research.

© International Association for Cryptologic Research 2017
J.-S. Coron and J.B. Nielsen (Eds.): EUROCRYPT 2017, Part I, LNCS 10210, pp. 627–656, 2017.
DOI: 10.1007/978-3-319-56620-7_22

they possessed sufficiently many of those attributes. This was then generalized to key-policy ABE [22], in which the user's key specifies a policy determining what attributes must be present in the ciphertext in order for that user to be able to decrypt, and ciphertext-policy ABE [10], which is the natural opposite in that the user's key corresponds to a list of attributes and ciphertexts are encrypted with a policy which determines which attributes the user must have to decrypt.

Since then the field of ABE has grown dramatically. There has been work which extends the type of policies that can be considered, for example to non-monotone formulas [32], or even regular languages [38]. There has also been work which improves the efficiency of ABE in various dimensions, for example considering schemes with very short (e.g. constant size) ciphertexts or keys [7,41], or schemes with very short parameters (again constant-size) which still support attributes from an unbounded space [29,31,33]. There has been work on distributing the job of the authority across multiple entities [14,28], on updating ciphertexts [34], or hiding the key and/or ciphertext attributes [11,12,25,36], and many other interesting directions.[1]

One weakness in much of the early work is that the schemes presented were only shown to satisfy a weak notion of security called *selective security*. Selective security essentially only guarantees security for an adversary who chooses which type of ciphertext to attack (i.e. the attributes/policy for the ciphertext) without seeing the system parameters, any ciphertexts, or any decryption keys. Thus it was a major breakthrough when Waters introduced the dual-system encryption technique [37], paving the way for schemes which satisfied the natural definition, in which the adversary may choose what type of ciphertext to attack adaptively based on any of the other information it sees while interacting with the system. Since then there has been a lot of work focused on obtaining the results above under this more natural security definition, which is usually referred to as *full security*.

One of the main downsides of this process, however, is that while most of the original constructions were simple and intuitive, many of these new constructions are significantly more complex. Also many of the first fully secure schemes relied on composite-order pairing groups, which while conceptually simpler are not really usable in practice [23]. The effort to move these results to be based on standard prime-order pairing groups has added even more complexity [18,24,27]. As a result, the intuition for the resulting constructions is often difficult to follow, and the security analysis for these schemes is much more involved, so much so that even *verifying* the security proof is often very time consuming.

Two recent works by Wee and Attrapadung [2,40] set out to simplify the process of designing and analyzing fully secure ABE schemes. They proposed a

---

[1] There has also been a very interesting line of work which uses indistinguishability obfuscation or multi-linear maps to construct ABE for circuits [19,20], and a lot of progress on building ABE schemes from lattices [13,21], although achieving the natural full security notion there still requires complexity leveraging. Here, we focus on pairing based constructions as to date they provide the best efficiency and security guarantees.

simple building block, called a predicate/pair encoding, which essentially considers what happens in the exponent of a single key and a single ciphertext. They proposed an information theoretic security property, which considers the distributions of these values, again only considering a single key and ciphertext, and showed that from any pair encoding scheme which satisfies this property one can construct a fully secure ABE scheme. The initial works proposed only composite-order group schemes; later works [1,4,15] have updated these results to prime-order groups.

These results led to very simple, intuitive, and easy to analyze constructions for several basic types of ABE schemes, that worked in efficient prime order groups, and were based on simple assumptions like DLIN or SXDH. However, there are many types of ABE schemes for which we do not know how to construct this type of pair encoding. And in fact there are many types of ABE which we do not know how to construct under simple assumptions using any approach, like ABE with short ciphertexts, or with large universe, or where an attribute can be used any number of times in a policy, etc.

To address this problem, Attrapadung [2] also proposed a different security notion for pair encodings, and showed that under this notion one could construct pair encodings for many more types of ABEs, and that this notion was sufficient to produce secure constructions under more complex q-type assumptions. However, proving that a pair encoding scheme satisfies the new security notion is again a challenging task. This property involves elements in bilinear groups rather than just the exponent, and it is no longer information-theoretic, so that it must be proved via reduction to a different q-type assumption for every encoding. These reductions are very complex, and again verifying the security becomes a matter of studying several pages of proof (9 pages for predicate encryption for regular languages, for instance), providing relatively little intuition for why the scheme is secure.

1.1 Our Contributions

Our goal in this work is to simplify the process of designing and analyzing ABE schemes for those types of ABEs which we only know how to construct from q-type assumptions. Towards this, we introduce a very different kind of security property for pair encodings that completely does away with any kind of distributions, and show that it is a very powerful and natural property through a series of results. We believe it provides a new perspective for looking at the security of predicate encryption schemes.

A pair encoding scheme, as defined by Attrapadung [2], gives a way to encode the two inputs x and y to a predicate into polynomials of a simple structure. These polynomials have three types of variables: common variables shared by the encodings of x and y, and variables specific to the encoding of x and to that of y.

A New Property for Pair Encodings. We present a new security property for pair encodings that essentially requires one to describe a mapping from the variables in the encoding to matrices and vectors. Once a mapping is specified, verifying that the property holds is just a matter of checking if the polynomials in the encoding evaluate to 0 when the variables are substituted.[2] Thus verification is much easier compared to any property known before, since they all require checking whether certain distributions are (pefectly, statistically or computationally) indistinguishable. We call our new property the *symbolic property* (Sym-Prop) since verification only involves symbolic manipulation.

We show how to convert *any* pair encoding that satisfies Sym-Prop into a *fully* secure encryption scheme whose security is based on a *fixed* q-type assumption that we call q-ratio. We use the generic transformation from Agrawal and Chase [1], henceforth called Gen-Trans, for this purpose. Gen-Trans takes an encoding scheme satisfying a certain information-theoretic property and produces an encryption scheme in dual system groups [16], which can then be instantiated in composite-order groups under subgroup decision assumptions or prime-order groups under the k-linear assumption.

We show that the security of Gen-Trans can also be argued when the pair encoding satisfies a very different security property, the symbolic property. The main novelty in our proof, and the crucial difference from AC16, is in how the form of master secret key is changed: while AC16 uses an information-theoretic property, we use Sym-Prop in conjunction with a new assumption called q-ratio$_{dsg}$ on dual system groups.[3] At a very high level, the terms that cannot be generated from q-ratio$_{dsg}$ are exactly the ones that go to zero due to Sym-Prop. Thus we are able to embed q-ratio$_{dsg}$ successfully into the reduction. Interestingly, however, as we will discuss below, Sym-Prop is not just an artifact of our proof strategy but seems to be inherently linked to the fundamental security of the resulting predicate encryption schemes.

An added advantage of borrowing AC16's transformation is that when a pair encoding is *used* in a way that can be shown to be information-theoretically secure, then the encryption scheme obtained through Gen-Trans is fully secure under a standard assumption. We show a useful application of this feature below.

We also show that the q-ratio assumption is in fact implied by several other q-type assumptions used to construct ABE schemes, in particular those used in the Lewko-Waters ABE [30] and Attrapadung's fully secure predicate encryption for regular languages [2]. This assumption is also simpler to describe than either [30] or [2] and we believe that this approach better captures the intuition for why these schemes are secure.

Analysis of Pair Encodings. We show that Sym-Prop holds for several pair encoding schemes, both new and old: multi-use CP-ABE, short ciphertext CP-ABE, large universe KP-ABE, short ciphertext KP-ABE, and predicate encryption for regular languages.

First, we present a new pair encoding $\Pi_{\text{re-use}}$ for CP-ABE that allows an attribute to be used *any* number of times in a policy. An interesting feature of

---

[2] The trivial case is ruled out because we also require that the vectors corresponding to two special variables, in the encoding of x and y respectively, are not orthogonal.

[3] q-ratio$_{dsg}$ is very similar to q-ratio. We show that Chen and Wee's instantiations of dual system groups satisfy q-ratio$_{dsg}$ if the underlying bilinear maps satisfy q-ratio.

$\Pi_{\text{re-use}}$ is that if no attribute is used more than once, then it collapses to the one-use scheme of [2], which is information-theoretically secure. So if we get an encryption scheme ES when Gen-Trans is applied on $\Pi_{\text{re-use}}$, then ES is fully secure under a *standard* assumption as long as it is used to encrypt policies where attributes are not repeated. If a policy with multiple use of attributes needs to be encrypted, then ES still fully hides the payload but under a q-type assumption. As far as we know, no multi-use scheme with this feature was known before. For instance, the Lewko-Waters' scheme [30] uses an assumption whose size scales with that of the access policy in the challenge ciphertext. So even if no attribute is used more than once, security still relies on a q-type assumption.[4]

For short ciphertext CP-ABE, we show that the pair encoding of Agrawal and Chase [1] satisfies Sym-Prop. This means that the encryption scheme that comes out after applying Gen-Trans is fully secure, not just selectively secure as they proved it (since we use the same transformation as them), under a q-type assumption. Note that it was not known earlier whether there exists a fully-secure CP-ABE scheme with constant-size ciphertexts under any kind of assumption on bi-linear maps. In fact, we can *generically* build an encryption scheme with constant-size ciphertexts for any predicate P from *any* pair encoding for P that satisfies Sym-Prop as discussed in more detail below.

The last three encodings we analyze are borrowed from the work of Attrapadung [2] with slight simplification. Previously, we only knew how to analyze them using the much more complex computational security property in [2]. Our analysis of these schemes is considerably simpler: for comparison, the proof of computational security for the regular languages pair encoding required 9 full pages, while our proof of symbolic security only takes 2.5 llncs pages. Our proofs can be seen as extracting, abstracting and somewhat simplifying the key ideas behind Attrapadung's security analysis, so that they can be very easily verified, and more easily applied to future schemes.

Symbolic Property Inherent in a Secure Scheme. While there are several security properties for encoding schemes that allow one to check if they can be used to build some type of encryption scheme, is there a property that an encoding scheme should *not* satisfy? A natural one that comes to mind is that correctness holds for an x and y that make a predicate *false*. In other words, there exists a way to combine the polynomials in the encoding to recover the blinding factor for the message even when the predicate is false. We call a pair encoding scheme that satisfies this property *trivially broken*.

Building an encryption scheme from a pair encoding scheme seems to require at least that the pair encoding *not* be trivially broken, but there is no general result that shows some type of security for a scheme that only provides such a minimal guarantee. In Sect. 4, we give the first result of this kind: *Any pair encoding scheme that is not trivially broken satisfies our symbolic property.*

---

[4] There are other ABE schemes that get much more than attribute re-use, like large universe or short keys, based on q-type assumptions [2], but proving them secure under a standard assumption when re-use does not happen would be even more difficult.

This result has several interesting broad implications. Suppose we have an encoding Π that we do not know to be secure. We apply Gen-Trans on it to get an encryption scheme ES. For this scheme to not be completely broken, there should not be a way to trivially combine some ciphertext and key to recover the message when the predicate is false. Now an interesting fact about our generic transformation Gen-Trans is that it preserves the structure of pair encodings, so that if there is way to combine the polynomials to recover the blinding factor, then the ciphertext and key coming out of Gen-Trans can be combined to recover the message. Therefore, if ES is not completely broken, Π is not broken either. This further implies that Π satisfies Sym-Prop and ES is fully secure under q-ratio. Thus we arrive at a very interesting conclusion: *Either* ES *is broken in an obvious way or it is fully secure under* q-ratio. Hence, Sym-Prop seems to be inherently linked to the fundamental security of encryption schemes, and is not just an artifact of our proof strategy.

We can take this line of argument even further. Suppose there is a generic transformation that preserves the structure of pair encodings in the sense described above. And suppose that when an encoding scheme satisfying a certain property X is given as input, it generates an encryption scheme that is not obviously broken, for example a *selectively* secure scheme. Then every encoding that satisfies X will also satisfy our symbolic property, and hence will lead to a *fully* secure encryption scheme through Gen-Trans! In this paper, we do not formalize the exact requirements a generic transformation should satisfy for such a general result to hold, leaving it as an interesting exercise for future work.

We conclude with an alternate way of proving symbolic security in case finding a mapping from an encoding's variables to matrices/vectors seems difficult: show that for all x and y for which the predicate is false, the blinding factor cannot be recovered from the encoding's polynomials.

New Generic Conversions. Thanks to the simplicity of our new symbolic property, we are able to show several useful transformations of pair encodings that preserve security. Specifically,

1. *Dual conversion.* Any secure pair encoding for a predicate can be transformed into a secure encoding scheme for the dual predicate (where the role of key and ciphertext are switched).
2. *Compact ciphertexts.* Any secure pair encoding can be converted into one that has a constant number of variables and polynomials in the ciphertext encoding. Thus, after applying Gen-Trans to the latter encoding, one gets encryption schemes with constant-size ciphertexts.
3. *Compact keys.* Analogous to above, any secure pair encoding can be converted into one that has a constant number of variables and polynomials in the key encoding, leading to encryption schemes with constant-size keys.[5]

This demonstrates the power and versatility of the new symbolic property. In contrast, only the first type of transformation is known for the security properties of Attrapadung [2,8], and none is known for Wee [40] or Chen et al. [15].

---

[5] This transformation and the one above requires some bound on the number of variables and polynomials in the respective encoding.

More New Schemes. Apart from the new scheme for unbounded attribute-reuse and showing that the constant-size ciphertext CP-ABE of [1] is fully secure, our generic conversions for pair encodings help us arrive at schemes that were not known before:

- As mentioned before, we show that the regular language pair encoding from [2] satisfies our symbolic property. Here keys are associated with regular languages, expressed as deterministic finite automata (DFA), and ciphertexts are associated with strings of any length from an alphabet set. One can first apply the dual conversion transformation to get an encoding scheme where ciphertexts and keys are associated with DFAs and strings, respectively. Then applying our compact ciphertext transformation to this encoding, and using the resulting pair encoding in Gen-Trans, one gets an encryption scheme for regular languages with constant sized ciphertexts (but with an upper bound on the size of DFAs).
- Similarly, applying our compact ciphertext/key transformation to Attrapadung's pair encodings for doubly spatial encryption (DSE) yields new encoding schemes, that then lead to encryption schemes with constant size ciphertext and keys, respectively. The only previous work on short ciphertext DSE [5] relied on a more complex series of transformations in which one type of predicate family (e.g. CP-ABE) is embedded inside another (e.g. DSE), and resulted in more expensive encodings.

1.2   Overview of Symbolic Security

This section provides a high-level *informal* treatment of pair encodings and the symbolic property with the goal of building some intuition about these concepts. Please refer to Sect. 3 for a formal presentation.

Pair Encodings. The pair encoding framework focuses on the exponent space of an encryption scheme. Suppose there is a predicate P that takes two inputs x and y. We want to encode x into a ciphertext and y into a key. An encryption scheme for P generally has terms like $g^{b_1}, g^{b_2}, \ldots$ and a special one of the form $e(g,g)^{\alpha}$ in the public parameters ($b_1, b_2, \ldots$ and α are chosen randomly). α plays the role of the master secret key. To encrypt a message m along with attribute x, some random numbers $s_0, s_1, s_2, \ldots$ are chosen and new terms are created by raising g, or some *common* term like g^{b_j}, to some s_i, and then taking a linear combination of these terms, where the terms and combination used depend on x. So, if we look at the exponent of any group element output by the encryption algorithm, it is usually a polynomial of the form $s_1 + \lambda_1 s_2 b_3 + \ldots$ where λ_1 is a constant that depends on x. Finally, m is hidden inside the ciphertext by blinding it with a re-randomization of $e(g,g)^{\alpha}$, say $e(g,g)^{\alpha s_0}$.

Similarly, the exponents of group elements in any key are of the form $r_1 + \mu r_2 b_1 + \ldots$, where $r_1, r_2, \ldots$ is fresh randomness chosen for this key. We could also have expressions that contain α because key generation involves the master secret key. Thus there are three different types of variables involved in a pair encoding:

the common variables $b_1, b_2, \ldots$, the ciphertext encoding variables $s_0, s_1, s_2, \ldots$, and the key encoding variables $\alpha, r_1, r_2, \ldots$.

Overall, it can be seen that if we focus on the exponent space of an encryption scheme, we need to deal with polynomials of a special form only. If $P(x, y) = 1$, then it should be possible to combine the ciphertext and key polynomials so that αs_0 can be recovered, and then used to unblind the message. The pair encoding framework just abstracts out such similarities between predicate encryption schemes in a formal way.

Security Properties and Transformation. Many security properties have been proposed in the literature for pair encodings, and a more restricted structure called predicate encodings [1,2,15,40]. The main contribution of these papers is to give a *generic* transformation from *any* pair encoding that satisfies their respective property into a fully secure predicate encryption scheme in composite or prime order groups (or a higher level abstraction called dual-system groups [16]). Proving that a pair encoding scheme satisfies a certain property is *significantly* easier, especially if the property is information-theoretic, than directly proving security of an encryption scheme. This is not surprising because there are no bi-linear maps, hardness assumptions, or sophisticated dual-encryption techniques involved in this process. Furthermore, verifying security of any number of encryption schemes designed through the pair encoding framework reduces to checking that the respective pair encodings are secure—a much easier task—and that the generic transformation is correct—a one-time effort. Needless to say, this saves a huge amount of work.

A Concrete Example: Unbounded Attribute Re-use. Suppose we want to design an ABE scheme that puts *no* restriction on the number of times an attribute can be used in an access policy. We know that a linear secret sharing scheme is the standard way to present a policy. It consists of a matrix $\mathbf{A}$ of size $m \times k$ and a mapping π from its rows to the universe of attributes. A value γ can be secret-shared through $\mathbf{A}$ by creating m shares, one for each row. If a user has a set of attributes S, then she gets shares for all the rows that map to some attribute in S through π. If S satisfies $(\mathbf{A}, \pi)$, then those shares can be combined to recover γ; otherwise, γ is information-theoretically hidden. In nearly all fully secure ABE schemes, the mapping π is assumed to be injective or one-to-one (this is called the one-use restriction), but we want to build an ABE scheme that supports any π whatsoever. In particular, the size of public parameters should not affect how many times an attribute can be used in a policy. (Any such scheme will likely rely on a q-type assumption [30].[6])

---

[6] In a recent work, Kowalczyk and Lewko [26] proposed a new technique to boost the entropy of a small set of (unpublished) semi-functional parameters. Using this idea, they propose a new KP-ABE scheme where the number of group elements in the public parameters grows only logarithmically in the bound on the number of attribute-uses in a policy, but note that the number of times an attribute can be reused is still affected. Furthermore, the size of ciphertexts scales with the maximum number of times an attribute can be re-used.

For a row i of $\mathbf{A}$, suppose $\rho(i)$ denotes which occurrence of $\pi(i)$ this is. (If an attribute y is attached to the second and fifth rows, then $\rho(2) = 1$ and $\rho(5) = 2$.) We now present a new pair encoding $\Pi_{\text{re-use}}$ for unbounded re-use by adapting the one-use scheme of [2]. (Some minor elements of the encoding have been suppressed for simplicity; see the full version for a full description.)

$$\text{EncCt}((\mathbf{A}, \pi)) \rightarrow s_0, s_1, \ldots, s_d, \quad \{\mathbf{a}_i(s_0 b', \hat{s}_2, \ldots, \hat{s}_k)^\mathsf{T} + s_{\rho(i)} b_{\pi(i)}\}_{i=1,\ldots,m}$$
$$\text{EncKey}(S) \rightarrow r, \quad \alpha + rb', \quad \{rb_y\}_{y \in S}$$

Here $\mathbf{a}_i$ is the ith row of $\mathbf{A}$ and d is the maximum number of times any attribute appears in it. A nice feature of $\Pi_{\text{re-use}}$ is that if no attribute is used more than once (i.e. $d = 1$), then the scheme collapses to that of [2], and one can show that α is information-theoretically hidden, or that $\Pi_{\text{re-use}}$ is *perfectly* secure.

If attributes are used multiple times, so that the ciphertext encoding has several variables $s_1, \ldots, s_d$, then α might be revealed to an unbounded adversary. Thus we need to find out if $\Pi_{\text{re-use}}$ satisfies a different type of property for which a generic transformation is known. One possibility is the computational *double selective master-key hiding* property due to Attrapadung, but then the advantages of an abstraction like pair encoding are more or less lost: we will have to work at the level of bi-linear maps instead of simple polynomials, and find a suitable q-type assumption(s) under which the property can be shown to hold.

The Symbolic Property. Our new symbolic property (Sym-Prop) can be very useful in such cases. It provides a new, clean way of reasoning about security of pair encodings: instead of arguing that one distribution is indistinguishable from another, whether information-theoretically or computationally, one needs to discover a mapping from the variables involved in an encoding to matrices and vectors, such that when the latter is substituted for the former in any cipher text/key encoding polynomial, the zero vector is obtained. Indeed, one needs to invest some effort in order to find the right matrices and vectors that will make the polynomials go to zero, but once such a discovery is made, verifying the property is just a matter of doing some simple linear algebra.

Recall that a pair encoding scheme for a predicate P that takes two inputs x and y, consists of three different types of variables: common variables $b_1, b_2, \ldots$, ciphertext encoding variables $s_0, s_1, s_2, \ldots$, and key encoding variables $\alpha, r_1, r_2, \ldots$. Sym-Prop is defined w.r.t. three (deterministic) algorithms, EncB, EncS and EncR. Among them, EncB generates matrices for the common variables; EncS and EncR generate vectors for ciphertext encoding and key encoding variables, respectively. The inputs to these three algorithms depend on what type of symbolic property we want to prove. For the selective version, the three algorithms get x as input, while EncR also gets y; and for the co-selective version, they all get y as input, while EncS also gets x. This is in line with the selective and co-selective security notions for encryption schemes. In the former, all key queries come after the challenge ciphertext, while in the latter, they come beforehand. A pair encoding scheme satisfies Sym-Prop if it satisfies both the selective and co-selective variants.

The trivial case where all the matrices and vectors output by the three algorithms are simply zero is ruled out because we also require that the vectors corresponding to two special variables, s_0 in the encoding of x and α in the encoding of y, are not orthogonal.

Proving the Symbolic Property for $\Pi_{\text{re-use}}$. To prove Sym-Prop for the multi-use encoding scheme $\Pi_{\text{re-use}}$ defined above, we need to define the outputs of the three algorithms EncB, EncS and EncR (in other words, a mapping from the variables in $\Pi_{\text{re-use}}$ to vectors and matrices) in both the selective and co-selective settings. Towards this, we make use of a simple combinatorial fact that is often used in arguing security of ABE schemes. If a set of attributes S does not satisfy an access policy $(\mathbf{A}, \pi)$, then there exists a vector $\mathbf{w} = (w_1, \ldots, w_k)$ s.t. $w_1 = 1$ and $\mathbf{a}_i$ is orthogonal to $\mathbf{w}$ for all i such that $\pi(i) \in S$. Note that $\mathbf{w}$ can be computed only by an algorithm that knows both $(\mathbf{A}, \pi)$ and S.

We also need some simple notation to describe the mapping. Let $\mathbf{E}_{i,j}$ be an $k \times d$ matrix with 1 at the (i, j)-th position and 0 everywhere else. Also, let $\mathbf{e}_i$ be the ith d-length unit vector and $\overline{\mathbf{e}}_j$ be the jth k-length unit vector. Here is the mapping for the selective version:

$$b_y : -\sum_{\ell=1}^{d} \sum_{j=1}^{k} a_{\sigma(y,\ell),j} \mathbf{E}_{j,\ell}, \qquad b' : \mathbf{E}_{1,1},$$

$$s_0 : \mathbf{e}_1, \qquad s_\ell : \mathbf{e}_\ell, \qquad \hat{s}_j : \overline{\mathbf{e}}_j, \qquad \alpha : \mathbf{e}_1, \qquad r : -\sum_{j=1}^{k} w_j \overline{\mathbf{e}}_j,$$

where $\sigma(y, \ell)$ is the index of the row in $\mathbf{A}$ which has the ℓ-th occurrence of y. Further, if $\mathbf{E}_{i,j}$, $\mathbf{e}_i$ and $\overline{\mathbf{e}}_j$ carry the meaning as above, except that their dimensions are $1 \times T$, T and 1 respectively, then the mapping for the co-selective version is:

$$b_y : \mathbf{0} \text{ for } y \in S \text{ and } -\mathbf{E}_{1,y} \text{ otherwise}, \qquad b' : \mathbf{E}_{1,1},$$

$$s_0 : w_1 \mathbf{e}_1, \qquad s_\ell : \sum_{i:\rho(i)=\ell} \mathbf{a}_i \mathbf{w}^\mathsf{T} \mathbf{e}_{\pi(i)}, \qquad \hat{s}_j : w_j \overline{\mathbf{e}}_1, \qquad \alpha : \mathbf{e}_1, \qquad r : -\overline{\mathbf{e}}_1.$$

We encourage the reader to verify that the polynomials in $\Pi_{\text{re-use}}$ (except the simples ones $s_0, s_1, \ldots, s_d, r$) go to zero when the two mappings described above are applied. (Vectors output by EncS (resp. EncR) are multiplied to the right (resp. left) of matrices output by EncB.) All it takes are simple observations like $\mathbf{E}_{i,j} \cdot \mathbf{e}_{j'}^\mathsf{T}$ gives a non-zero vector if and only if $j = j'$, and that $\mathbf{w}$ is orthogonal to every row in $\mathbf{A}$ that maps to an attribute in S. (See the full version for a formal proof.) One can consider the two mappings to be a short *certificate* of the security of $\Pi_{\text{re-use}}$.

How to Find a Mapping? Indeed, as pointed out earlier, finding an appropriate mapping is not a trivial task. Nevertheless, Sym-Prop is still the *right* property for arguing security of pair encodings for the following reasons:

- If finding the right mapping is difficult for Sym-Prop, then finding a proof for the computational property of Attrapadung [2] is several times more difficult. A typical proof of the symbolic property is 1–2 pages while computational property proofs could go up to 10 pages (see the encoding for regular languages, for instance). A central issue with computational properties is finding an appropriate q-type assumption under which it holds, which may be very difficult for a complex predicate. Our approach can be seen as extracting out the *real* challenging part of designing Attrapadung's computational proofs.
- Verification of Sym-Prop involves doing simple linear algebra, arguably a much simpler task than checking indistinguishability of distributions, and certainly a much simpler task than verifying a long computational reduction.
- The *certificate* for the symbolic security of $\Pi_{\text{re-use}}$ bears many similarities with those of other encodings that we will describe later in the paper. Thus proving Sym-Prop for a new encoding scheme is not as difficult as it might seem at first. Furthermore, modifying a short proof of the symbolic property is much easier than a long proof of a computational property.
- Recall our result that if an encoding scheme is not trivially broken then it satisfies Sym-Prop. This gives an alternate way of showing that Sym-Prop holds, by proving that the scheme is not broken.

1.3 Outline of the Paper

In Sect. 2 we define relevant notation and review the standard definition of predicate encryption. In Sect. 3 we define pair encoding schemes and our new symbolic property formally. Section 5 first reviews the notion of dual system groups, then shows how to build encryption schemes from any pair encoding by using them. This conversion is a two-step process: first we *augment* an encoding so that it satisfies a few extra properties (Sect. 5.1); next we apply the transformation from Agrawal and Chase [1] (Sect. 5.4). A proof of security of the resulting encryption scheme is provided in Sect. 7.

Section 6 gives generic transformations that can be used to reduce the number of variables and/or polynomials in an encoding, which can then be used to get encryption schemes with constant-size ciphertexts/keys. We also provide a transformation from any encoding for a predicate to an encoding for the dual predicate. However, due to space constraints, most of the details are available in the full version only. The full version also provides several examples to illustrate how symbolic property can substantially simplifying the analysis of encoding schemes.

2 Preliminaries

We use λ to denote the security parameter. A negligible function is denoted by negl. We use bold letters to denote matrices and vectors, with the former in uppercase and the latter in lowercase. The operator $\cdot$ applied to two vectors computes their entry-wise product and $\langle , \rangle$ gives the inner-product. For a vector

$\mathbf{u}$, we use u_i to denote its ith element, and for a matrix $\mathbf{M}$, $M_{i,j}$ denotes the element in the ith row and jth column. When we write $g^{\mathbf{u}}$ for a vector $u = (u_1, \ldots, u_n)$, we mean the vector $(g^{u_1}, \ldots, g^{u_n})$. $g^{\mathbf{M}}$ for a matrix $\mathbf{M}$ should be interpreted in a similar way. The default interpretation of a vector should be as a row vector.

For two matrices $\mathbf{U}$ and $\mathbf{V}$ of dimension $n \times m_1$ and $n \times m_2$ respectively, let $\mathbf{U} \circ \mathbf{V}$ denote the column-wise *join* of $\mathbf{U}$ and $\mathbf{V}$ of dimension $n \times (m_1 + m_2)$, i.e., $\mathbf{U} \circ \mathbf{V}$ has the matrix $\mathbf{U}$ as the first m_1 columns and $\mathbf{V}$ as the remaining m_2 columns. We also refer to this operation as *appending* $\mathbf{V}$ to $\mathbf{U}$. (The notation easily extends to vectors because we represent them as row matrices.) If we want to join matrices row-wise instead, we could take their transpose, apply a column-wise join, and then take the transpose of the resultant matrix.

We use $x \leftarrow_R S$, for a set S, to denote that x has been drawn uniformly at random from it. The set of integers $a, a+1, \ldots, b$ is compactly represented as $[a, b]$. If $a = 1$, then we just use $[b]$, and if $a = 0$, then $[b]^+$.

Let $\mathbb{Z}_N$ denote the set of integers $\{0, 1, 2, \ldots, N\}$. Let $\mathcal{G}_N(m)$ denote the set of all vectors of length m with every element in $\mathbb{Z}_N$. Similarly, let $\mathcal{G}_N(m_1, m_2)$ denote the set of all matrices of size $m_1 \times m_2$ that have all the elements in $\mathbb{Z}_N$.

Bilinear Pairings. We use the standard definition of pairing friendly groups from literature. A mapping e from a pair of groups $(\mathcal{G}, \mathcal{H})$ to a target group $\mathcal{G}_T$ is bilinear if there is linearity in both the first and second inputs, i.e. $e(g^a, h^b) = e(g, h)^{ab}$ for every $g \in \mathcal{G}, h \in \mathcal{H}$ and $a, b \in \mathbb{Z}$. We require e to be non-degenerate and efficiently computable. The identity element of a group G is denoted by 1_G.

Let GroupGen be an algorithm that on input the security parameter λ outputs $(N, \mathcal{G}, \mathcal{H}, \mathcal{G}_T, g, h, e)$ where $N = \Theta(\lambda)$; $\mathcal{G}$, $\mathcal{H}$ and $\mathcal{G}_T$ are (multiplicative) cyclic groups of order N; g, h are generators of $\mathcal{G}, \mathcal{H}$, respectively; and $e : \mathcal{G} \times \mathcal{H} \to \mathcal{G}_T$ is a bilinear map. In this paper our focus will be on prime-order groups because they perform much better in practice.

Predicate Family. We borrow the notation of predicate family from Attrapadung [2]. It is given by $P = \{P_\kappa\}_{\kappa \in \mathbb{N}^c}$ for some constant c, where P_κ maps an $x \in \mathcal{X}_\kappa$ and a $y \in \mathcal{Y}_\kappa$ to either 0 or 1. The first entry of κ is a number $N \in \mathbb{N}$ that is supposed to specify the size of a domain; rest of the entries are collectively referred to as par, i.e. $\kappa = (N, \text{par})$.

2.1 Predicate Encryption

An encryption scheme for a predicate family $P = \{P_\kappa\}_{\kappa \in \mathbb{N}^c}$ over a message space $\mathcal{M} = \{\mathcal{M}_\lambda\}_{\lambda \in \mathbb{N}}$ consists of a tuple of four PPT algorithms (Setup, Encrypt, KeyGen, Decrypt) that satisfy a correctness condition. These algorithms behave as follows.

- Setup(1^λ, par). On input 1^λ and par, Setup outputs a master public key MPK and a master secret key MSK. The output of Setup is assumed to also define a natural number N, and κ is set to (N, par).

- Encrypt(MPK, x, m). On input MPK, $x \in \mathcal{X}_\kappa$ and $m \in \mathcal{M}_\lambda$, Encrypt outputs a ciphertext CT.
- KeyGen(MSK, y). On input MSK and $y \in \mathcal{Y}_\kappa$, KeyGen outputs a secret key SK.
- Decrypt(MPK, SK, CT). On input MPK, a secret key SK and a ciphertext CT, Decrypt outputs a message $m' \in \mathcal{M}_\lambda$ or $\perp$.

Correctness. For all par, $m \in \mathcal{M}_\lambda$, $x \in \mathcal{X}_\kappa$ and $y \in \mathcal{Y}_\kappa$ such that $P_\kappa(x, y) = 1$,

$$\Pr[(\text{MPK}, \text{MSK}) \leftarrow \text{Setup}(1^\lambda);$$
$$\text{Decrypt}(\text{MPK}, \text{KeyGen}(\text{MSK}, y), \text{Encrypt}(\text{MPK}, x)) \neq P_\kappa(x, y)] \leq \text{negl}(\lambda),$$

where the probability is over the random coin tosses of Setup, Encrypt and KeyGen (Decrypt can be assumed to be deterministic without loss of generality).

Security. Consider the following game IND-CPA$^b_\mathcal{A}$ (λ, par) between a challenger Chal and an adversary $\mathcal{A}$ for $b \in \{0, 1\}$ when both are given inputs 1^λ and par:

1. *Setup Phase*: Chal runs Setup(1^λ, par) to obtain MPK and MSK. It gives MPK to $\mathcal{A}$.
2. *Query Phase*: $\mathcal{A}$ requests a key by sending $y \in \mathcal{Y}_\kappa$ to Chal, and obtains SK $\leftarrow$ KeyGen(MSK, y) in response. This step can be repeated any number of times.
3. *Challenge Phase*: $\mathcal{A}$ sends two messages $m_0, m_1 \in \mathcal{M}_\lambda$ and an $x^\star \in \mathcal{X}_\kappa$ to Chal, and gets CT $\leftarrow$ Encrypt(MPK, x, m_b) as the challenge ciphertext.
4. *Query Phase*: This is identical to step 2.
5. *Output*. $\mathcal{A}$ outputs a bit.

The output of the experiment is the bit that $\mathcal{A}$ outputs at the end. It is required that for all y queried in steps 2 and 4, $P_\kappa(x^\star, y) = 0$.

Definition 2.1. An encryption scheme is *adaptively* or *fully* secure if for all par and PPT adversary $\mathcal{A}$,

$$|\Pr[\text{IND-CPA}^0_\mathcal{A}(\lambda, \text{par}) = 1] - \Pr[\text{IND-CPA}^1_\mathcal{A}(\lambda, \text{par}) = 1]| \leq \text{negl}(\lambda), \qquad (1)$$

where the probabilities are taken over the coin tosses of $\mathcal{A}$ and Chal. It is *semi-adaptively* secure if (1) is satisfied with respect to a modified version of IND-CPA where the second step is omitted [17,39]. Further, it is *co-selectively* secure if (1) holds when the fourth step is removed from the IND-CPA game [6].

3 Pair Encoding Schemes

The notion of pair encoding schemes (PES) was introduced by Attrapadung [2], and later refined independently by Agrawal and Chase [1] and Attrapadung [4] himself in an identical way. As observed in the latter works, *all* pair encodings proposed originally in [2] satisfy the additional constraints in the refined versions.

We present here a more structured definition of pair encoding schemes so that the reader can easily see the different components involved. In the full version we describe the original formulation as well, and argue why our definition does not lose any generality.

3.1   Definition

A PES for a predicate family $P_\kappa : \mathcal{X}_\kappa \times \mathcal{Y}_\kappa \to \{0,1\}$ indexed by $\kappa = (N, \mathsf{par})$, where par specifies some parameters, is given by four *deterministic* polynomial-time algorithms as described below.

- Param(par) $\to n$. When given par as input, Param outputs $n \in \mathbb{N}$ that specifies the number of *common* variables, which we denote by $\mathbf{b} := (b_1, \dots, b_n)$.

- EncCt$(x, N) \to (w_1, w_2, \mathbf{c}(\mathbf{s}, \hat{\mathbf{s}}, \mathbf{b}))$. On input $N \in \mathbb{N}$ and $x \in \mathcal{X}_{(N,\mathsf{par})}$, EncCt outputs a vector of polynomials $\mathbf{c} = (c_1, \dots, c_{w_3})$ in *non-lone* variables $\mathbf{s} = (s_0, s_1, \dots, s_{w_1})$ and *lone* variables $\hat{\mathbf{s}} = (\hat{s}_1, \dots, \hat{s}_{w_2})$. (The variables $\hat{s}_1, \dots, \hat{s}_{w_2}$ never appear in the form $\hat{s}_z b_j$, and are hence called lone.) For $\ell \in [w_3]$, where $\eta_{\ell,z}, \eta_{\ell,i,j} \in \mathbb{Z}_N$, the ℓth polynomial is given by

$$\sum_{z \in [w_2]} \eta_{\ell,z} \hat{s}_z \ + \sum_{\substack{i \in [w_1]^+, \\ j \in [n]}} \eta_{\ell,i,j} s_i b_j.$$

- EncKey$(y, N) \to (m_1, m_2, \mathbf{k}(\mathbf{r}, \hat{\mathbf{r}}, \mathbf{b}))$. On input $N \in \mathbb{N}$ and $y \in \mathcal{Y}_{(N,\mathsf{par})}$, EncKey outputs a vector of polynomials $\mathbf{k} = (k_1, \dots, k_{m_3})$ in non-lone variables $\mathbf{r} = (r_1, \dots, r_{m_1})$ and lone variables $\hat{\mathbf{r}} = (\alpha, \hat{r}_1, \dots, \hat{r}_{m_2})$. For $t \in [m_3]$, where $\phi_t, \phi_{t,z'}, \phi_{t,i',j} \in \mathbb{Z}_N$ the tth polynomial is given by

$$\phi_t \alpha \ + \sum_{z' \in [m_2]} \phi_{t,z'} \hat{r}_{z'} \ + \sum_{\substack{i' \in [m_1], \\ j \in [n]}} \phi_{t,i',j} r_{i'} b_j.$$

- Pair$(x, y, N) \to (\mathbf{E}, \overline{\mathbf{E}})$. On input N, and both x and y, Pair outputs two matrices $\mathbf{E}$ and $\overline{\mathbf{E}}$ of size $(w_1 + 1) \times m_3$ and $w_3 \times m_1$, respectively.

Observe that the output of EncKey is analogous to that of EncCt, except in how the special variables α and s_0 are treated in the respective case. While α is lone variable, i.e. it never appears in conjunction with a common variable, s_0 is not. See the full version for several concrete examples of pair encodings and the different types of variables involved.

Correctness. A PES is correct if for every $\kappa = (N, \mathsf{par})$, $x \in \mathcal{X}_\kappa$ and $y \in \mathcal{Y}_\kappa$ such that $P_\kappa(x, y) = 1$, the following holds symbolically

$$\mathbf{sEk}^\mathsf{T} + \mathbf{c}\overline{\mathbf{E}}\mathbf{r}^\mathsf{T} \ = \sum_{\substack{i \in [w_1]^+, \\ t \in [m_3]}} s_i E_{i,t} k_t \ + \sum_{\substack{\ell \in [w_3], \\ i' \in [m_1]}} c_\ell \overline{E}_{\ell,i'} r_{i'} \ = \alpha s_0.$$

The matrix $\mathbf{E}$ takes a linear combination of the products of non-lone variables output by EncCt and polynomials output by EncKey. (Its rows are numbered from 0 to w_1.) Analogously, $\overline{\mathbf{E}}$ takes a linear combination of the products of polynomials output by EncCt and non-lone variables output by EncKey. Below we use ct-enc and key-enc as a shorthand for polynomials and variables output by EncCt (ciphertext-encoding) and EncKey (key-encoding), respectively.

3.2 Symbolic Property

We introduce a new symbolic property for pair encoding schemes that significantly simplifies their analysis for even complex predicates. We get the best of two worlds: not only is our symbolic property very clean to describe (like information-theoretic properties), it can also capture all the predicates that have been previously captured by any computational property. Further, the property does not involve dealing with any kind of distribution.

We now formally define the property. We use $a : b$ below to denote that a variable a is substituted by a matrix/vector b.

Definition 3.1 (Symbolic property). *A pair encoding scheme* $\Gamma = ($Param, EncCt, EncKey, Pair$)$ *for a predicate family* $P_\kappa : \mathcal{X}_\kappa \times \mathcal{Y}_\kappa \rightarrow \{0,1\}$ *satisfies* (d_1, d_2)-*selective symbolic property*[7] *for positive integers* d_1 *and* d_2 *if there exist three deterministic polynomial-time algorithms* EncB, EncS, EncR *such that for all* $\kappa = (N, \mathsf{par})$, $x \in \mathcal{X}_\kappa$, $y \in \mathcal{Y}_\kappa$ *with* $P_\kappa(x, y) = 0$,

- EncB$(x) \rightarrow \mathbf{B}_1, \ldots, \mathbf{B}_n \in \mathcal{G}_N(d_1, d_2)$;
- EncS$(x) \rightarrow \mathbf{s}_0, \ldots, \mathbf{s}_{w_1} \in \mathcal{G}_N(d_2)$, $\quad \hat{\mathbf{s}}_1, \ldots, \hat{\mathbf{s}}_{w_2} \in \mathcal{G}_N(d_1)$;
- EncR$(x, y) \rightarrow \mathbf{r}_1, \ldots, \mathbf{r}_{m_1} \in \mathcal{G}_N(d_1)$, $\quad \mathbf{a}, \hat{\mathbf{r}}_1, \ldots, \hat{\mathbf{r}}_{m_2} \in \mathcal{G}_N(d_2)$;

such that $\langle \mathbf{s}_0, \mathbf{a} \rangle \neq 0$, *and if we substitute*

$$\hat{s}_z : \hat{\mathbf{s}}_z^\mathsf{T} \qquad s_i b_j : \mathbf{B}_j \mathbf{s}_i^\mathsf{T} \qquad \alpha : \mathbf{a} \qquad \hat{r}_{z'} : \hat{\mathbf{r}}_{z'} \qquad r_{i'} b_j : \mathbf{r}_{i'} \mathbf{B}_j$$

for $z \in [w_2]$, $i \in [w_1]^+$, $j \in [n]$, $z' \in [m_2]$ *and* $i' \in [m_1]$ *in all the polynomials output by* EncCt *and* EncKey *on input* x *and* y, *respectively, they evaluate to* $\mathbf{0}$.

Similarly we say a pair encoding scheme satisfies (d_1, d_2)-*co-selective symbolic security property if there exist* EncB, EncR, EncS *that satisfy the above properties but where* EncB *and* EncR *depend only on* y, *and* EncS *depends on both* x *and* y. *Finally, a scheme satisfies* (d_1, d_2) *symbolic property if it satisfies both* (d_1', d_2')-*selective and* (d_1'', d_2'')-*co-selective properties for some* $d_1', d_1'' \leq d_1$ *and* $d_2', d_2'' \leq d_2$.

We use Sym-Prop as a shorthand for symbolic property. It is easy to see that if a scheme satisfies (d_1, d_2)-selective Sym-Prop then it also satisfies (d_1', d_2') for any $d_1' \geq d_1$ and $d_2' \geq d_2$. Just append $d_1' - d_1$ rows of zeroes and $d_2' - d_2$ columns of zeroes to the $\mathbf{B}_j$ matrices, $d_2' - d_2$ zeroes to the $\mathbf{s}_i$ vectors, $d_1' - d_1$ zeroes to the $\hat{\mathbf{s}}_z$ vectors, $d_1' - d_1$ zeroes to the $\mathbf{r}_{i'}$ vectors, and $d_2' - d_2$ zeroes to the $\hat{\mathbf{r}}_{z'}$ vectors. A similar claim can also be made about co-selective Sym-Prop. Thus if a PES satisfies (d_1, d_2)-Sym-Prop then it also satisfies selective and co-selective properties with the same parameters, as well as (d_1', d_2')-Sym-Prop for any $d_1' \geq d_1$ and $d_2' \geq d_2$.

Lastly, if a PES Γ satisfies Sym-Prop for a predicate family P_κ, we say that Γ is *symbolically secure* for P_κ, or simply that Γ is symbolically secure if the predicate family is clear from context.

---

[7] d_1, d_2 could depend on κ but we leave this implicit for simplicity of presentation.

4   Obtaining Symbolic Security Generically

In this section, we prove an interesting and useful result. If a pair encoding scheme in not *trivially* broken in the sense that for any x, y that do not satisfy the predicate, there does not exist a way to directly recover αs_0 from the encoding polynomials (note that for correctness we require exactly this, but when the predicate is true), then the scheme satisfies the symbolic property.

Definition 4.1 (Trivially broken scheme). *A pair encoding scheme* $\Gamma =$ (Param, EncCt, EncKey, Pair) *for a predicate family* $P_\kappa : \mathcal{X}_\kappa \times \mathcal{Y}_\kappa \to \{0, 1\}$ *is trivially broken if for a* $\kappa = (N, \mathsf{par})$, $x \in \mathcal{X}_\kappa$, $y \in \mathcal{Y}_\kappa$ *that satisfy* $P_\kappa(x, y) = 0$, *there exists a matrix* $\mathbf{E}$ *such that* $(\mathbf{s}, \mathbf{c})\mathbf{E}(\mathbf{r}, \mathbf{k})^\mathsf{T} = \alpha s_0$, *where* $\mathbf{c}$ *is the vector of polynomials output by* EncCt(x, N) *in variables* $\mathbf{s} = (s_0, \ldots)$, $\hat{\mathbf{s}}$, $\mathbf{b}$, *and* $\mathbf{k}$ *is the vector of polynomials output by* EncKey(y, N) *in variables* $\mathbf{r}$, $\hat{\mathbf{r}} = (\alpha, \ldots)$, $\mathbf{b}$.

Theorem 4.2. *If a pair encoding scheme is not trivially broken then it satisfies the symbolic property.*

Proof. If a scheme Γ is not trivially broken, then for all x and y for which the predicate evaluates to false, the ct-enc non-lone variables $\mathbf{s} = (s_0, \ldots, s_{w_1})$ and polynomials $\mathbf{c} = (c_1, \ldots, c_{w_3})$ cannot be paired with the key-enc non-lone variables $\mathbf{r} = (r_1, \ldots, r_{m_1})$ and polynomials $\mathbf{k} = (k_1, \ldots, k_{m_3})$ to recover αs_0. We know that the former have monomials of the form $s_0, \ldots, s_{w_1}, \hat{s}_1, \ldots, \hat{s}_{w_2}$, $s_0 b_1, \ldots, s_0 b_n, \ldots, s_{w_1} b_1, \ldots, s_{w_1} b_n$, so a total of $w_2 + (n+1)(w_1 + 1)$. Similarly, the total number of distinct monomials in the latter is $m_2 + 1 + (n + 1)m_1$ (because α is a lone variable as opposed to s_0). Let us denote the two quantities above by var_c and var_k respectively.

Define a matrix $\mathbf{\Delta}$ over $\mathbb{Z}_N$ with $(w_1 + w_3 + 1)(m_1 + m_3)$ rows and $\mathsf{var}_c\mathsf{var}_k$ columns. A row is associated with the product of a ct-enc non-lone variable or polynomial with a key-enc non-lone variable or polynomial. Each column represents a unique monomial that can be obtained by multiplying a ct-enc monomial with a key-enc monomial, with the first column representing αs_0. The (i, j)th entry in this matrix is the coefficient of the monomial associated with the jth column in the product polynomial attached with the ith row. Since Γ is not broken, we know that the rows in $\mathbf{\Delta}$ cannot be linearly combined to get the vector $(1, 0, \ldots, 0)$.

Note that it is enough to work with any subset of rows because they cannot be combined to get $(1, 0, \ldots, 0)$ either. Thus, for the rest of the proof, we consider only those rows of $\mathbf{\Delta}$ that multiply a ct-enc non-lone variable with a key-enc polynomial and vice versa (and only those columns which have monomials that can be obtained from multiplying such polynomials). Let n_1 denote the number of rows now.

Since rows in $\mathbf{\Delta}$ cannot be linearly combined to get $(1, 0, \ldots, 0)$, the first column of $\mathbf{\Delta}$, say col, can be written as a linear combination of the other columns. Because if not, one can show that there exists a vector $\mathbf{v} = (v_1, \ldots, v_{n_1})$ that is orthogonal to all the columns except the first one[8]. We can then combine the rows of $\boldsymbol{\Delta}$ using $v_1 / \langle \text{col}, \mathbf{v} \rangle, \ldots, v_{n_1} / \langle \text{col}, \mathbf{v} \rangle$ to get $(1, 0, \ldots, 0)$—a contradiction.

Let $\mathcal{Q}$ denote the set of monomials associated with the columns of $\boldsymbol{\Delta}$. These columns can be linearly combined to get the zero vector, without zeroing out col, which corresponds to αs_0. Let λ_q be the factor that multiplies the column associated with the monomial $q \in \mathcal{Q}$ in one such linear combination. Note that $\lambda_{\alpha s_0} \neq 0$.

Our first goal is to show that Γ satisfies the selective symbolic property. So we need to define matrices and vectors for various variables in the encoding such that all the polynomials evaluate to the zero vector. Towards this, pick any non-lone key-enc variable $r_{i'}$ for $i' \in [m_1]$ and consider the sub-matrix $\boldsymbol{\Delta}'$ of $\boldsymbol{\Delta}$ that consists of rows which are attached with the product of $r_{i'}$ with a ct-enc polynomial and columns which are associated with the product of $r_{i'}$ and a ct-enc monomial. (Note that it does not matter which non-lone key-enc variable we consider; the sub-matrix obtained in each case will be exactly the same.) Recall that a ct-enc polynomial c_ℓ is given by

$$\sum_{z \in [w_2]} \eta_{\ell,z} \hat{s}_z \quad + \quad \sum_{i \in [w_1]^+, j \in [n]} \eta_{\ell,i,j} s_i b_j$$

for $\ell \in [w_3]$. So more formally, rows in $\boldsymbol{\Delta}'$ are associated with $(c_\ell, r_{i'})$, and columns are associated with monomials $\hat{s}_z r_{i'}$, $s_i b_j r_{i'}$, where the range of i, j, z is as described above. For simplicity in the following, assume that the columns are ordered as $\hat{s}_1, \ldots, \hat{s}_{w_2}, s_0 b_1, \ldots, s_0, b_n, \ldots, s_{w_1} b_1, \ldots, s_{w_1} b_n$ and the rows are ordered as $(c_1, r_{i'}), \ldots, (c_{w_3}, r_{i'})$, so that the ℓth row of $\boldsymbol{\Delta}'$ is $(\eta_{\ell,1}, \ldots, \eta_{\ell,w_2}, \eta_{\ell,0,1}, \ldots, \eta_{\ell,0,n}, \ldots, \eta_{\ell,w_1,1}, \ldots, \eta_{\ell,w_1,n})$.

Let $\mathcal{T}$ be the kernel of $\boldsymbol{\Delta}'$, i.e. the set of all vectors $\mathbf{v}$ such that $\boldsymbol{\Delta}' \mathbf{v} = \mathbf{0}$. Let $\mathbf{v}_1, \mathbf{v}_2, \ldots, \mathbf{v}_{d_1}$ be a basis of $\mathcal{T}$ and write $\mathbf{v}_p$ as $(v_{p,1}, \ldots, v_{p,w_2}, v_{p,0,1}, \ldots, v_{p,0,n}, \ldots, v_{p,w_1,1}, \ldots, v_{p,w_1,n})$ for $p \in [d_1]$. (We discuss the special case of $\boldsymbol{\Delta}'$'s kernel being empty later on.) Therefore, we have that for any $\ell \in [w_3]$ and $p \in [d_1]$,

$$\sum_z \eta_{\ell,z} v_{p,z} \quad + \quad \sum_{i,j} \eta_{\ell,i,j} v_{p,i,j} \tag{2}$$

is equal to 0. Let $\mathbf{u}_z = (v_{1,z}, \ldots, v_{d_1,z})$ and $\mathbf{u}_{i,j} = (v_{1,i,j}, \ldots, v_{d_1,i,j})$ for $z \in [w_2]$, $i \in [w_1]^+$, $j \in [n]$.

We now define matrices $\mathbf{B}_1, \ldots, \mathbf{B}_n$ and vectors $\mathbf{s}_0, \ldots, \mathbf{s}_{w_1}, \hat{\mathbf{s}}_1, \ldots, \hat{\mathbf{s}}_{w_2}$ as follows. $\mathbf{B}_j$ has d_1 rows and $d_2 = w_1 + 1$ columns with the $(i+1)$th column being $\mathbf{u}_{i,j}^\mathsf{T}$ for $i = [w_1]^+$. Vector $\mathbf{s}_i$ is set to $\mathbf{e}_{i+1}$ for $i = [w_1]^+$, where $\mathbf{e}_i$ denotes the ith unit vector of size d_2, and $\hat{\mathbf{s}}_z$ is set to $\mathbf{u}_z$ for $z \in [w_2]$. These matrices and vectors depend only on $\mathbf{v}_1, \mathbf{v}_2, \ldots, \mathbf{v}_{d_1}$, which in turn depends on $\boldsymbol{\Delta}'$ only. The entries in

---

[8] The claim is similar to one made in the case of linear secret sharing schemes where we say that if a set of attributes does not satisfy a policy, i.e. the associated set of rows cannot be linearly combined to get a certain vector $\mathbf{v}$, then one can find a vector orthogonal to all those rows but not to $\mathbf{v}$. See, for instance, [9, Claim 2] for a formal proof.

$\mathbf{\Delta}'$ are the coefficients of the monomials obtained by multiplying $r_{i'}$ with various ct-enc polynomials. Hence, they only depend on x and, in particular, not on y. Further, it is easy to observe that all the operations involved in computing $\mathbf{B}_j$, s_i, $\hat{s}_z$ are efficient. Thus, one can define two deterministic polynomial time algorithms EncB and EncS that on input x only, output $\mathbf{B}_1, \ldots, \mathbf{B}_n$ and $s_0, \ldots, s_{w_1}$, $\hat{s}_1, \ldots, \hat{s}_{w_2}$ respectively.

We need to verify that if we substitute $\hat{s}_z$ with $\hat{s}_z^\mathsf{T}$ and $s_i b_j$ with $\mathbf{B}_j s_i^\mathsf{T}$ in any ct-enc polynomial c_ℓ, then we get an all zeroes vector. On performing such a substitution, we have

$$\sum_z \eta_{\ell,z} \mathbf{u}_z^\mathsf{T} \;+\; \sum_{i,j} \eta_{\ell,i,j}(\mathbf{u}_{0,j}^\mathsf{T}, \ldots, \mathbf{u}_{w_1,j}^\mathsf{T})\mathbf{e}_{i+1}^\mathsf{T} \;=\; \sum_z \eta_{\ell,z}\mathbf{u}_z^\mathsf{T} \;+\; \sum_{i,j}\eta_{\ell,i,j}\mathbf{u}_{i,j}^\mathsf{T}$$

The pth element in the column vector above is given by (2), which is equal to 0 for any p.

In the special case where $\mathbf{\Delta}'$'s kernel is empty, $\mathbf{B}_1, \ldots, \mathbf{B}_n$ are all set to $d_1 \times d_2$ matrices with zero entries; $\hat{s}_1, \ldots, \hat{s}_{w_2}$ are set to the zero vector of size d_1; $s_1, \ldots, s_{w_1}$ are set to the zero vector of size d_2; and s_0 is set to $(1, 0, \ldots, 0)$. It is easy to see that all ct-enc polynomials still evaluate to zero upon substitution.

We also need to make sure that with the appropriate choice of vectors for the key-enc variables, all the key-enc polynomials also evaluate to the zero vector. Recall that such polynomials are given by

$$k_t \;=\; \phi_t \alpha \;+\; \sum_{z' \in [m_2]} \phi_{t,z'}\hat{r}_{z'} \;+\; \sum_{\substack{i' \in [m_1], \\ j \in [n]}} \phi_{t,i',j}r_{i'}b_j$$

for $t \in [m_3]$. When they are multiplied with a non-lone ct-enc variable s_i, we get the monomials αs_i, $s_i\hat{r}_{z'}$, $s_i r_{i'}b_j$ for $i \in [w_1]^+$ and i', j, z' as above.

Recall that the columns of Δ can be linearly combined using $\{\lambda_q\}_{q \in Q}$ to get the zero vector. Going back to the product of $r_{i'}$ with c_ℓ, we can say that

$$\sum_z \eta_{\ell,z}\lambda_{\hat{s}_z r_{i'}} \;+\; \sum_{i,j}\eta_{\ell,i,j}\lambda_{s_i b_j r_{i'}} \;=\; 0$$

irrespective of what ℓ and i' are because only the entries in the columns associated with monomials $\hat{s}_z r_{i'}$, $s_i b_j r_{i'}$ are non-zero. Hence, the vector $\mathbf{w}_{i'}$ given by $(\lambda_{\hat{s}_1 r_{i'}}, \ldots, \lambda_{\hat{s}_{w_2} r_{i'}}, \lambda_{s_0 b_1 r_{i'}}, \ldots, \lambda_{s_0 b_n r_{i'}}, \ldots, \lambda_{s_{w_1} b_1 r_{i'}}, \ldots, \lambda_{s_{w_1} b_n r_{i'}})$ lies in the kernel of $\mathbf{\Delta}'$. (Recall that no matter what key-enc non-lone variable is chosen, one always gets the same $\mathbf{\Delta}'$.) In other words, there exists a vector $\mathbf{r}_{i'}$ of size d_1 such that $[\mathbf{v}_1^\mathsf{T}, \ldots, \mathbf{v}_{d_1}^\mathsf{T}]\mathbf{r}_{i'}^\mathsf{T} = \mathbf{w}_{i'}$. Now the transpose of $\mathbf{r}_{i'}\mathbf{B}_j$ is given by

$$\begin{bmatrix} \mathbf{u}_{0,j} \\ \vdots \\ \mathbf{u}_{w_1,j} \end{bmatrix} \mathbf{r}_{i'}^\mathsf{T} \;=\; \begin{bmatrix} v_{1,0,j} & \cdots & v_{d_1,0,j} \\ \vdots & \vdots & \vdots \\ v_{1,w_1,j} & \cdots & v_{d_1,w_1,j} \end{bmatrix} \mathbf{r}_{i'}^\mathsf{T} \;=\; \begin{bmatrix} \lambda_{s_0 b_j r_{i'}} \\ \vdots \\ \lambda_{s_{w_1} b_j r_{i'}} \end{bmatrix}$$

for every $j \in [n]$. In the special case where $\mathbf{\Delta}'$'s kernel is empty, set $\mathbf{r}_{i'}$ to be the zero vector of size d_1. The relation $\mathbf{r}_{i'}\mathbf{B}_j = (\lambda_{s_0 b_j r_{i'}}, \ldots, \lambda_{s_{w_1} b_j r_{i'}})$ for all j still holds because $\mathbf{w}_{i'}$ must be zero.

Define the remaining vectors as follows: $\mathbf{a}$ is set to be $[\lambda_{\alpha s_0}, \ldots, \lambda_{\alpha s_{w_1}}]$ and $\hat{\mathbf{r}}_{z'}$ to be $[\lambda_{s_0 \hat{r}_{z'}}, \ldots, \lambda_{s_{w_1} \hat{r}_{z'}}]$ for $z' \in [m_2]$. (Note that the first element of $\mathbf{a}$ is not zero.) When we substitute α with $\mathbf{a}$, $\hat{r}_{z'}$ with $\hat{\mathbf{r}}_{z'}$ and $r_{i'} b_j$ with $\mathbf{r}_{i'} \mathbf{B}_j$ in k_t for $t \in [m_3]$, we get

$$\phi_t [\lambda_{\alpha s_0}, \ldots, \lambda_{\alpha s_{w_1}}] \quad + \quad \sum_{z'} \phi_{t,z'} [\lambda_{s_0 \hat{r}_{z'}}, \ldots, \lambda_{s_{w_1} \hat{r}_{z'}}]$$
$$+ \quad \sum_{i',j} \phi_{t,i',j} [\lambda_{s_0 b_j r_{i'}}, \ldots, \lambda_{s_{w_1} b_j r_{i'}}].$$

The ith element of this sum is given by

$$\phi_t \lambda_{\alpha s_i} \quad + \quad \sum_{z'} \phi_{t,z'} \lambda_{s_i \hat{r}_{z'}} \quad + \quad \sum_{i',j} \phi_{t,i',j} \lambda_{s_i r_{i'} b_j}$$

for $i \in [w_1]^+$. It is easy to see that the above quantity is zero when we consider the row in Δ attached with the product $s_i k_t$.

One can define a deterministic polynomial time algorithm EncR that on input x and y, computes how the columns of Δ can be combined to get the zero vector, and then uses this information to define $\mathbf{a}$, $\hat{\mathbf{r}}_{z'}$, $\mathbf{r}_{i'}$ as shown above.

The proof for the co-selective symbolic property is analogous to the proof above, so we skip the details.   □

5   Predicate Encryption from Pair Encodings

In this section, we describe how any pair encoding scheme for a predicate can be transformed into an encryption scheme for the same predicate in dual system groups (DSG), introduced by Chen and Wee [16], and later used and improved by several works [1,4,15]. This transformation is a two-step process: first we *augment* an encoding so that it satisfies a few extra properties (Sect. 5.1)[9]; next we apply the transformation from Agrawal and Chase [1] (Sect. 5.4).

5.1   Augmenting Pair Encodings

We need the matrices and vectors involved in the symbolic property to have some extra features, so that we can prove the security of the derived predicate encryption scheme from our q-ratio assumption. Towards this, we show how any pair encoding scheme that satisfies Sym-Prop can be transformed into another scheme that satisfies a more constrained version of this property, with only a few additional variables and polynomials.

We note that, although they are presented monolithically, many of the pair encodings introduced by Attrapadung [2] can be viewed as the result of applying a very similar augmentation to simpler underlying encodings. Thus, our results also help explain the structure of those previous encodings.

---

[9] This step need not be applied if the properties are already satisfied.

Recall that the algorithms of symbolic security output $\mathbf{a}$ for α, $\mathbf{B}_1, \ldots, \mathbf{B}_n$ for common variables, $\mathbf{s}_0, \ldots, \mathbf{s}_{w_1}$ for non-lone ct-enc variables, and $\mathbf{r}_1, \ldots, \mathbf{r}_{m_1}$ for key-enc non-lone variables. Let $\mathbf{b}_j$ denote the first column of $\mathbf{B}_j$ and $s_{i,1}$ the first element of $\mathbf{s}_i$.

Definition 5.1 (Enhanced symbolic property). *A pair encoding scheme satisfies* (d_1, d_2)*-Sym-Prop* for a predicate P_κ if it satisfies selective and co-selective* (d_1, d_2)*-Sym-Prop for P_κ but under the following constraints for both*

1. $\mathbf{a}$ *is set to* $(1, 0, \ldots, 0)$.
2. *In every* ct-enc *polynomial, if $s_i b_j$ is replaced by*
   - $\mathbf{s}_i^\mathsf{T} \mathbf{b}_j$ *then we get a matrix with non-zero elements in the first row only;*
   - $s_{i,1} \mathbf{B}_j$ *then we get a matrix with non-zero elements in the first column only.*
   (The lone variables are replaced by the zero vector.)
3. *In every* key-enc *polynomial, if we replace $r_{i'} b_j$ with $\mathbf{b}_j^\mathsf{T} \mathbf{r}_{i'}$, then we get a diagonal matrix. (The lone variables, once again, are replaced by the zero vector.)*
4. *The set of vectors $\{\mathbf{s}_0, \ldots, \mathbf{s}_{w_1}\}$ is linearly independent, and so is the set $\{\mathbf{r}_1, \ldots, \mathbf{r}_{m_1}\}$.*

We convert any pair encoding that satisfies Sym-Prop into one that satisfies Sym-Prop* in three steps. First we show that with only one additional key-enc non-lone variable, an additional common variable, and an extra ct-enc polynomial, we can get an encoding scheme for which the vector $\mathbf{a}$ corresponding to α can be set to $(1, 0, \ldots, 0)$ (in proving that Sym-Prop holds). Next, with two extra common variables, and an additional variable and a polynomial each in the ciphertext and key encoding, one can satisfy the second and third properties from above. Finally, a simple observation can be used to satisfy the fourth property as well. More formally, we prove the following theorem in the full version.

Theorem 5.2 (Augmentation). *Suppose a* PES *for a predicate family* P_κ : $\mathcal{X}_\kappa \times \mathcal{Y}_\kappa \to \{0, 1\}$ *outputs n on input* par, $(w_1, w_2, \mathbf{c})$ *on input $x \in \mathcal{X}_\kappa$, $(m_1, m_2, \mathbf{k})$ on input $y \in \mathcal{Y}_\kappa$ and satisfies (d_1, d_2)-Sym-Prop, then there exists another* PES *for P_κ that outputs $n + 3$ on input* par, $(w_1 + 1, w_2, \overline{\mathbf{c}})$ *on input x and $(m_1 + 2, m_2, \overline{\mathbf{k}})$ on input y, where $|\overline{\mathbf{c}}| = |\mathbf{c}| + 2$ and $|\overline{\mathbf{k}}| = |\mathbf{k}| + 1$, and satisfies* $(\max(d_1, d_2 - 1) + M_1 + 1, d_2 + W_1 + 2)$*-Sym-Prop*, where M_1 and W_1 are bounds on the number of* key-enc *and* ct-enc *non-lone variables, respectively.*[10]

The extra constraints of Sym-Prop* give rise to some nice combinatorial facts. Please refer to the full version for details.

---

[10] As we will see later, when a pair encoding scheme is transformed into a predicate encryption scheme, the parameters of Sym-Prop* have no effect on the construction. They only affect the size of assumption on which the security of encryption scheme is based.

5.2 Dual System Groups

Dual system groups (DSG) were introduced by Chen and Wee [16] and generalized by Agrawal and Chase [1]. The latter work also shows that the two instantiations of DSG – in composite-order groups under the subgroup decision assumption and in prime-order groups under the decisional linear assumption – given by Chen and Wee satisfy the generalized definition as well. Here we give a brief informal description of dual system groups. See the full version or existing work [1] for a formal definition.

Dual system groups are parameterized by a security parameter λ and a number n. They have a SampP algorithm that on input 1^λ and 1^n, outputs public parameters PP and secret parameters SP. The parameter PP contains a triple of groups $(\mathbb{G}, \mathbb{H}, \mathbb{G}_T)$ and a non-degenerate bilinear map $e : \mathbb{G} \times \mathbb{H} \to \mathbb{G}_T$, a homomorphism μ from $\mathbb{H}$ to $\mathbb{G}_T$, along with some additional parameters used by SampG, SampH. Given PP, we know the exponent of group $\mathbb{H}$ and how to sample uniformly from it; let $N = \exp(\mathbb{H})$. It is required that N is a product of distinct primes of $\Theta(\lambda)$ bits. The secret parameters SP contain $\tilde{h} \in \mathbb{H}$ (where $\tilde{h} \neq 1_{\mathbb{H}}$) along with additional parameters used by $\overline{\mathsf{SampG}}$ and $\overline{\mathsf{SampH}}$.

A dual system group has several sampling algorithms: SampGT algorithm takes an element in the image of μ and outputs another element from $\mathbb{G}_T$. SampG and SampH take PP as input and output a vector of $n + 1$ elements from $\mathbb{G}$ and $\mathbb{H}$ respectively. $\overline{\mathsf{SampG}}$ and $\overline{\mathsf{SampH}}$ take both PP and SP as inputs and output a vector of $n + 1$ elements from $\mathbb{G}$ and $\mathbb{H}$ respectively. These two algorithms are used in security proofs only. $\overline{\mathsf{SampG}}_0$ and $\overline{\mathsf{SampH}}_0$ denote the first element of $\overline{\mathsf{SampG}}$ and $\overline{\mathsf{SampH}}$ respectively.

A dual system group is *correct* if it satisfies the following two properties for all PP.

- *Projective*: For all $h \in \mathbb{H}$ and coin tosses σ, $\mathsf{SampGT}(\mu(h); \sigma) = e(\mathsf{SampG}_0$ $(\mathrm{PP}; \sigma), h)$, where SampG_0 is an algorithm that outputs only the first element of SampG.
- *Associative*: If $(g_0, g_1, \ldots, g_n)$ and $(h_0, h_1, \ldots, h_n)$ are samples from SampG(PP) and SampH(PP) respectively, then for all $i \in [1, n]$, $e(g_0, h_i) = e(g_i, h_0)$.

Dual system groups have a number of interesting security properties as well that makes them very useful for building encryption schemes, see the full version for details. We additionally require that there exists a way to sample the set-up parameters so that one not only gets PP and SP, but also some trapdoor information td that can be used to generate samples from $\overline{\mathsf{SampG}}$ and $\overline{\mathsf{SampH}}$ given only the first element. We formalize this property and show that both instantiations of Chen and Wee [16] satisfy them in the full version. The new sampling algorithm will be denoted by SampP^* below.

5.3 New Computational Assumption

We introduce a new assumption, called $\mathsf{q\text{-}ratio}_{\mathsf{dsg}}$, on dual system groups parameterized by positive integers d_1 and d_2.

Definition 5.3 ((d_1, d_2)-q-ratio$_{dsg}$ assumption). *Consider the following distribution on a dual system group's elements:*

$$\text{dsg-par} := (\text{PP}, \text{SP}, \text{td}) \leftarrow \text{SampP}^*(1^\lambda, 1^n);$$

$$\hat{g} \leftarrow \overline{\text{SampG}}_0(\text{PP}, \text{SP}); \quad \hat{h} \leftarrow \overline{\text{SampH}}_0(\text{PP}, \text{SP})$$

$$u_0, u_1, \ldots, u_{d_2}, v_1, \ldots, v_{d_1} \leftarrow_R \mathbb{Z}_N^*;$$

$$D_{\mathbb{G}} := \{\hat{g}^{u_i}\}_{i \in [d_2]^+} \ \cup \ \left\{\hat{g}^{\frac{u_i}{u_j v_k}}\right\}_{i,j \in [d_2], i \neq j, k \in [d_1]};$$

$$D_{\mathbb{H}} := \{\hat{h}^{v_i}\}_{i \in [d_1]} \ \cup \ \left\{\hat{h}^{\frac{v_i}{v_j u_k}}\right\}_{i,j \in [d_1], i \neq j, k \in [d_2]};$$

$$T_0 := \hat{h}^{1/u_0}; \quad T_1 \leftarrow_R \mathbb{H}.$$

We say that the (d_1, d_2)-q-ratio$_{dsg}$ assumption holds if for any PPT *algorithm* $\mathcal{A}$,

$$\text{Adv}_{\mathcal{A}}^{\text{qr}_{dsg}}(\lambda) := \big|\Pr[\mathcal{A}(1^\lambda, \text{dsg-par}, D_{\mathbb{G}}, D_{\mathbb{H}}, T_0) = 1]$$
$$- \Pr[\mathcal{A}(1^\lambda, \text{dsg-par}, D_{\mathbb{G}}, D_{\mathbb{H}}, T_1) = 1]\big|$$

is negligible in λ.

Note that u_0 is present in exactly one of the terms in $D_{\mathbb{G}}$ and not at all in $D_{\mathbb{H}}$.

We also define a similar assumption on bilinear maps.

Definition 5.4 ((d_1, d_2)-q-ratio assumption). *Consider the following distribution:*

$$\text{par} := (N, \mathcal{G}, \mathcal{H}, \mathcal{G}_T, g, h, e) \leftarrow \text{GroupGen}(1^\lambda)$$

$$\hat{g} \leftarrow_R \mathcal{G}; \quad \hat{h} \leftarrow_R \mathcal{H}; \qquad u_0, u_1, \ldots, u_{d_2}, v_1, \ldots, v_{d_1} \leftarrow_R \mathbb{Z}_N^*;$$

$$D_{\mathcal{G}} := \{\hat{g}^{u_i}\}_{i \in [d_2]^+} \ \cup \ \left\{\hat{g}^{\frac{u_i}{u_j v_k}}\right\}_{i,j \in [d_2], i \neq j, k \in [d_1]};$$

$$D_{\mathcal{H}} := \{\hat{h}^{v_i}\}_{i \in [d_1]} \ \cup \ \left\{\hat{h}^{\frac{v_i}{v_j u_k}}\right\}_{i,j \in [d_1], i \neq j, k \in [d_2]};$$

$$T_0 := \hat{h}^{1/u_0}; \quad T_1 \leftarrow_R \mathcal{H}.$$

We say that the (d_1, d_2)-q-ratio assumption holds if for any PPT *algorithm* $\mathcal{A}$,

$$\text{Adv}_{\mathcal{A}}^{\text{qr}}(\lambda) := \big|\Pr[\mathcal{A}(1^\lambda, \text{par}, D_{\mathcal{G}}, D_{\mathcal{H}}, T_0) = 1]$$
$$- \Pr[\mathcal{A}(1^\lambda, \text{par}, D_{\mathcal{G}}, D_{\mathcal{H}}, T_1) = 1]\big|$$

is negligible in λ.

In this paper our focus is on constructions in prime-order groups because they are much more practical, so we will consider the q-ratio assumption on prime-order bilinear maps only. We show that this assumption is implied by the assumptions proposed by Lewko, Waters [30] and Attrapadung [2] in the full version. We also show that Chen and Wee's prime order DSG construction [16] (along with the new sampling algorithms we introduce) satisfies the q-ratio$_{dsg}$ assumption if the underlying group satisfies the q-ratio assumption. Thus we have,

Lemma 5.5. *A dual system group with a bilinear map* $e : \mathbb{G} \times \mathbb{H} \to \mathbb{G}_T$ *that satisfies the* (d_1, d_2)*-q-ratio*$_\mathsf{dsg}$ *assumption can be instantiated in a prime-order bilinear map* $e' : \mathcal{G} \times \mathcal{H} \to \mathcal{G}_T$ *that satisfies the* (d_1, d_2)*-q-ratio and k-linear assumptions. Further, an element of* $\mathbb{G}$ *and* $\mathbb{H}$ *is represented using* $k+1$ *elements of* $\mathcal{G}$ *and* $\mathcal{H}$*, respectively. (An element of* $\mathbb{G}_T$ *is represented by just one from* $\mathcal{G}_T$*).*

5.4 Encryption Scheme

In this section, we show how to obtain an encryption scheme from a pair encoding using the sampling algorithms of dual system groups. Our transformation is based on the one given by Agrawal and Chase [1], and is referred to as Gen-Trans. If a PES Γ_P is defined by the tuple of algorithms (Param, EncCt, EncKey, Pair) for a predicate family $P = \{P_\kappa\}_{\kappa \in \mathbb{N}^c}$, then the algorithms for $\Pi_P := $ Gen-Trans(Γ_P) are given as follows.

- Setup$(1^\lambda, \mathsf{par})$: First the pair encoding algorithm Param(par) is run to obtain n, and then the dual system group algorithm SampP$(1^\lambda, 1^n)$ is run to get PP, SP. A randomly chosen element from $\mathbb{H}$ is designated to be the master secret key MSK. Master public key MPK is set to be $(\text{PP}, \mu(\text{MSK}))$. Further, N and κ are set to exp$(\mathbb{H})$ and (N, par), respectively (where the exponent of $\mathbb{H}$ is a part of PP).

- Encrypt$(\text{MPK}, x, \mathsf{msg})$: On input $x \in \mathcal{X}_\kappa$ and $\mathsf{msg} \in \mathbb{G}_T$, EncCt$(x, N)$ is run to obtain w_1, w_2 and polynomials $(c_1, \ldots, c_{w_3})$. For $i' \in [w_1 + w_2]^+$, draw a sample $(g_{i',0}, \ldots, g_{i',n})$ from SampG using PP. Recall that the ℓth polynomial is given by

$$\sum_{z \in [w_2]} \eta_{\ell,z} \hat{s}_z \;+\; \sum_{i \in [w_1]^+, j \in [n]} \eta_{\ell,i,j} s_i b_j.$$

Set CT$_i$ to be $g_{i,0}$ for $i \in [w_1]^+$ and $\widetilde{\text{CT}}_\ell$ to be

$$\prod_{z \in [w_2]} y_{w_1+z,0}^{\eta_{\ell,z}} \;\cdot\; \prod_{i \in [w_1]^+, j \in [n]} g_{i,j}^{\eta_{\ell,i,j}}$$

for $\ell \in [w_3]$. Also, let CT$^\star = \mathsf{msg} \cdot $ SampGT$(\mu(\text{MSK}); \sigma)$ where σ denotes the coin tosses used in drawing the first sample from SampG. Output CT $:= (\text{CT}_0, \ldots, \text{CT}_{w_1}, \widetilde{\text{CT}}_1, \ldots, \widetilde{\text{CT}}_{w_3}, \text{CT}^\star)$.

- KeyGen$(\text{MPK}, \text{MSK}, y)$: On input $y \in \mathcal{Y}_\kappa$, EncKey(y, N) is run to obtain m_1, m_2 and polynomials $(k_1, k_2, \ldots, k_{m_3})$. For $i \in [m_1 + m_2]$, draw a sample $(h_{i,0}, \ldots, h_{i,n})$ from SampH using PP. Recall the tth polynomial is given by

$$\phi_t \alpha \;+\; \sum_{z' \in [m_2]} \phi_{t,z'} \hat{r}_{z'} \;+\; \sum_{i' \in [m_1], j \in [n]} \phi_{t,i',j} r_{i'} b_j.$$

Set SK$_{i'}$ to be $h_{i',0}$ for $i' \in [m_1]$ and $\widetilde{\text{SK}}_t$ to be

$$\text{MSK}^{\phi_t} \;\cdot\; \prod_{z' \in [m_2]} h_{m_1+z',0}^{\phi_{t,z'}} \;\cdot\; \prod_{i' \in [m_1], j \in [n]} h_{i',j}^{\phi_{t,i',j}}$$

for $t \in [m_3]$. Output SK $:= (\text{SK}_1, \ldots, \text{SK}_{m_1}, \widetilde{\text{SK}}_1, \ldots, \widetilde{\text{SK}}_{m_3})$.

– Decrypt($\mathrm{MPK}, \mathrm{SK}_y, \mathrm{CT}_x$): On input SK_y and CT_x, $\mathsf{Pair}(x, y, N)$ is run to obtain matrices $\mathbf{E}$ and $\overline{\mathbf{E}}$. Output

$$\mathrm{CT}^\star \cdot \left(\prod_{i \in [w_1]^+, t \in [m_3]} e(\mathrm{CT}_i, \widetilde{\mathrm{SK}}_t)^{E_{i,t}} \cdot \prod_{\ell \in [w_3], i' \in [m_1]} e(\widetilde{\mathrm{CT}}_\ell, \mathrm{SK}_{i'})^{\overline{E}_{\ell,i'}} \right)^{-1} \cdot$$

One can use the projective and associative property of DSG to show that the predicate encryption scheme defined above is correct (see [1] for details). We defer a proof of security for Π_P to Sect. 7, and conclude with the following remark.

Remark 5.6 (Size of ciphertexts and keys). Ciphertexts have $w_1 + w_3 + 1$ elements from $\mathbb{G}$ and an element from $\mathbb{G}_T$; keys have $m_1 + m_3$ elements from $\mathbb{H}$. So the size of these objects depends only on the number of non-lone variables and polynomials. Moreover, there is a one-to-one mapping between variables/polynomials and ciphertext/key elements. Thus if we can reduce the size of an encoding, we will immediately get an equivalent reduction in the size of ciphertexts or keys.

6   Transformations on Pair Encodings

In this section we present several useful transformations on pair encodings that preserve symbolic property. The first class of transformations help in reducing the size of ciphertexts and keys, and the second one provides a way to develop schemes for *dual* predicates (where the role of the two inputs to a predicate is reversed).

Compact Encoding Schemes. We show how pair encoding schemes can be made compact by reducing the number of ct-enc and/or key-enc polynomials and/or variables to a constant in a *generic* way. Importantly, we show that if the encoding scheme we start with satisfies the symbolic property, then so does the transformed scheme. As a result, building encryption schemes with constant-size ciphertexts or keys, for instance, becomes a very simple process.

Our first transformation converts any encoding scheme Γ' to another scheme Γ where the number of ct-enc variables is *just one*. Naturally, we need to assume a bound on the total number of ct-enc variables for this transformation to work. If $W_1 + 1$ and W_2 are bounds on the number of non-lone and lone ct-enc variables, respectively, and the number of common variables in Γ' is n, then Γ has $(W_1 + 1)$ $n + W_2$ common variables, 1 ct-enc non-lone variable and 0 lone variables. The number of lone key-enc variables and polynomials increases by a multiplicative factor of $W_1 + 1$.

Our second transformation brings down the number of ct-enc polynomials to *just one*. Once again the transformation is fully generic, as long as there is a bound W_3 on the number of polynomials. In this case, the number of common variables increases by a multiplicative factor of $W_3 + 1$, the number of non-lone key-enc variables by a multiplicative factor of W_3, and the number of key-enc polynomials by an additive factor of $m_1 W_3^2 n$.

When the two transformations above are applied one after the other, we obtain an encoding scheme with just one non-lone variable and one polynomial in the ciphertext encoding. After augmenting the scheme as per Theorem 5.2 which adds a non-lone variable and two polynomials, we can convert the resulting encoding scheme into a predicate encryption scheme by using the generic mechanism of Sect. 5.4. This encryption scheme will have exactly 5 dual system's source group elements in any ciphertext, a number which would only double if the instantiation from Lemma 5.5 is used under the SXDH (1-linear) assumption.

One can also reduce the number of key-enc variables and polynomials in a manner analogous to how the corresponding quantities are reduced in the ciphertext encoding, at the cost of increasing the number of common variables and ct-enc variables and polynomials. If there is a bound on both the number of variables and polynomials in the key encoding, then one can obtain an encoding scheme with just one of each. This will result in encryption schemes with constant-size key.

Finally, we remark that one can also mix-and-match. For instance, first the number of ct-enc variables can be reduced to one, and then we can do the same for key-enc variables, resulting in a scheme with just one variable each in the ciphertext and key encodings at the cost of more polynomials in both. (This might be interesting, for example, because it produces a pair encoding of the form used in [15].) Note that when the ciphertext variable reduction transformation is applied, no lone variables are left in the ciphertext encoding (the only remaining variable is a non-lone variable). Hence, the key variable reduction transformation does not affect the number of ct-enc variables.

See the full version for a formal treatment of the two transformations described above.

Dual Predicates. The dual predicate for a family $P'_\kappa : \mathcal{Y}_\kappa \times \mathcal{X}_\kappa \rightarrow \{0,1\}$ is given by $P_\kappa : \mathcal{X}_\kappa \times \mathcal{Y}_\kappa \rightarrow \{0,1\}$ where $P_\kappa(x,y) = P'_\kappa(y,x)$ for all κ, $x \in \mathcal{X}_\kappa$, $y \in \mathcal{Y}_\kappa$. For example, CP-ABE and KP-ABE are duals of each other. In the full version we show that Attrapadung's dual scheme conversion [3, Sect. 8.1] mechanism preserves symbolic property too.

7   Security of Predicate Encryption Scheme

In this section we show that the transformation Gen-Trans leads to a secure encryption scheme if the underlying encoding satisfies the (enhanced) symbolic property. More formally, we have:

Theorem 7.1. *If a pair encoding scheme Γ_P satisfies (d_1, d_2)-Sym-Prop* for a predicate family P_κ, then the scheme Gen-Trans(Γ_P) defined in Sect. 5.4 is a fully secure predicate encryption scheme for P_κ in dual system groups under the $(d_1, d_2 - 1)$-q-ratio$_{\mathsf{dsg}}$ assumption.*

When the above theorem is combined with Theorem 5.2 and Lemma 5.5, we get the following corollary:

Corollary 7.2. *If a pair encoding scheme satisfies* (d_1, d_2)*-*Sym-Prop *for a predicate family then there exists a fully secure predicate encryption scheme for that family in prime-order bilinear maps under the* $(\max(d_1, d_2 - 1) + M_1 + 1, d_2 + W_1 + 1)$*-q-ratio and k-linear assumptions, where* M_1 *and* W_1 *are bounds on the number of* key-enc *and* ct-enc *non-lone variables, respectively, in the encoding.*

The rest of this section is devoted to the proof of Theorem 7.1. We follow the same general outline as in other papers that use dual system groups [1,15,16]. The design of hybrids in our proof is closer to [15,16] rather than [1]. In particular, our hybrid structure is simpler because, unlike [1], we don't add noise to individual samples in every key. However, since we have adopted the generic transformation from [1], the indistinguishability between several hybrids follows from that of corresponding hybrids in [1]. (We briefly review these hybrids and the properties they follow from below—for full proofs see [1].) The main novelty in our proof, and the crucial difference from [1], is how the form of master secret key is changed: in [1] relaxed perfect security is used for this purpose, but we use the symbolic property in conjunction with the q-ratio$_{\mathsf{dsg}}$ assumption.

We first define auxiliary algorithms for encryption and key generation. Below we use $g_{i,0}$ (resp. $h_{i,0}$) to denote the first element of $\mathbf{g}_i$ (resp. $\mathbf{h}_i$). Also w and m denote $w_1 + w_2$ and $m_1 + m_2$, respectively.

- $\overline{\mathsf{Encrypt}}(\mathrm{PP}, x, \mathsf{msg}; (\mathbf{g}_0', \mathbf{g}_1', \ldots, \mathbf{g}_w'), \mathrm{MSK})$: This algorithm is same as Encrypt except that it uses $\mathbf{g}_i' \in \mathbb{G}^{n+1}$ instead of the samples $\mathbf{g}_i$ from SampG, and sets $\mathrm{CT}^\star$ to $\mathsf{msg} \cdot e(g_{0,0}', \mathrm{MSK})$.

- $\overline{\mathsf{KeyGen}}(\mathrm{PP}, \mathrm{MSK}, y; (\mathbf{h}_1', \ldots, \mathbf{h}_m'))$: This algorithm is same as KeyGen except that it uses $\mathbf{h}_i' \in \mathbb{H}^{n+1}$ instead of the samples $\mathbf{h}_i$ from SampH.

Using the algorithms described above, we define alternate forms for the ciphertext, master secret key, and secret keys.

- *Semi-functional master secret key* is defined to be $\overline{\mathrm{MSK}} := \mathrm{MSK} \cdot \tilde{h}^\mu$ where $\mu \leftarrow_R \mathbb{Z}_N$.

- *Semi-functional ciphertext* is given by $\overline{\mathsf{Encrypt}}(\mathrm{PP}, x, m; \mathbf{G} \cdot \hat{\mathbf{G}}, \mathrm{MSK})$, where $\mathbf{G} \cdot \hat{\mathbf{G}}$ is defined as follows: sample $\mathbf{g}_1, \ldots, \mathbf{g}_w$ from SampG and $\hat{\mathbf{g}}_1, \ldots, \hat{\mathbf{g}}_w$ from $\overline{\mathsf{SampG}}$ (which also requires SP); set $\mathbf{G}$ and $\mathbf{G}'$ to be the vector of vectors $(\mathbf{g}_1, \ldots, \mathbf{g}_w)$ and $(\hat{\mathbf{g}}_1, \ldots, \hat{\mathbf{g}}_w)$, respectively; and denote $(\mathbf{g}_1 \cdot \hat{\mathbf{g}}_1, \ldots, \mathbf{g}_w \cdot \hat{\mathbf{g}}_w)$ by $\mathbf{G} \cdot \hat{\mathbf{G}}$.

- *Ext-semi-functional ciphertext* is given by $\overline{\mathsf{Encrypt}}(\mathrm{PP}, x, m; \mathbf{G} \cdot \hat{\mathbf{G}} \cdot \hat{\mathbf{G}}', \mathrm{MSK})$, where $\mathbf{G}$, $\hat{\mathbf{G}}$ are as above, and $\hat{\mathbf{G}}'$ is defined to be $(\hat{\mathbf{g}}_1', \ldots, \hat{\mathbf{g}}_w')$, where $\hat{\mathbf{g}}_i' = (1, \hat{g}_{i,0}^{\gamma_1}, \ldots, \hat{g}_{i,0}^{\gamma_n})$ for $i \in [w]$ and $\gamma_1, \ldots, \gamma_n \leftarrow_R \mathbb{Z}_N$. (Here these $\gamma_1, \ldots, \gamma_n$ will be chosen once and used in both ciphertext and key components.)

- Table 1 lists the different types of keys we need and the inputs that should to be passed to $\overline{\mathsf{KeyGen}}$ (besides PP and y) in order to generate them.

Table 1. Six types of keys.

Type of key	Inputs to $\overline{\mathsf{KeyGen}}$ (besides PP and y)
Normal	MSK; $(\mathbf{h}_1, \ldots, \mathbf{h}_m)$
Pseudo-normal	MSK; $(\mathbf{h}_1 \cdot \hat{\mathbf{h}}_1, \ldots, \mathbf{h}_m \cdot \hat{\mathbf{h}}_m)$
Ext-pseudo-normal	MSK; $(\mathbf{h}_1 \cdot \hat{\mathbf{h}}_1 \cdot \hat{\mathbf{h}}_1', \ldots, \mathbf{h}_m \cdot \hat{\mathbf{h}}_m \cdot \hat{\mathbf{h}}_m')$
Ext-pseudo-semi-functional	$\overline{\text{MSK}}$; $(\mathbf{h}_1 \cdot \hat{\mathbf{h}}_1 \cdot \hat{\mathbf{h}}_1', \ldots, \mathbf{h}_m \cdot \hat{\mathbf{h}}_m \cdot \hat{\mathbf{h}}_m')$
Pseudo-semi-functional	$\overline{\text{MSK}}$; $(\mathbf{h}_1 \cdot \hat{\mathbf{h}}_1, \ldots, \mathbf{h}_m \cdot \hat{\mathbf{h}}_m)$
Semi-functional	$\overline{\text{MSK}}$; $(\mathbf{h}_1, \ldots, \mathbf{h}_m)$

Table 2. An outline of the proof structure.

Hybrid	Difference from previous	Properties required
Hyb_0	-	-
Hyb_1	ct semi-func	Left subgroup ind
$\vdots$	$\vdots$	$\vdots$
$\mathsf{Hyb}_{2,\varphi-1,5}$	$\varphi - 1$ keys semi-func	-
$\mathsf{Hyb}_{2,\varphi,1}$	φth key pseudo-norm	Right subgroup ind
$\mathsf{Hyb}_{2,\varphi,2}$	ct ext-semi-func, φth key ext-pseudo-norm	Parameter hiding
$\mathsf{Hyb}_{2,\varphi,3}$	φth key ext-pseudo-semi-func	Non-degeneracy, Sym-Prop*, q-ratio$_{\mathsf{dsg}}$ assumption
$\mathsf{Hyb}_{2,\varphi,4}$	ct semi-func, φth key pseudo-semi-func	Parameter-hiding
$\mathsf{Hyb}_{2,\varphi,5}$	φth key semi-func	Right subgroup ind
$\vdots$	$\vdots$	$\vdots$
$\mathsf{Hyb}_{2,\xi,5}$	All keys semi-func	-
Hyb_3	ct semi-func encryption of random msg	Projective, orthogonality, non-degeneracy

In the table, $\mathbf{h}_1, \ldots, \mathbf{h}_m$ are samples from SampH; $\hat{\mathbf{h}}_1, \ldots, \hat{\mathbf{h}}_m$ are samples from $\overline{\mathsf{SampH}}$ (which also requires SP); and $\hat{\mathbf{h}}_i' = (1, \hat{h}_{i,0}^{\gamma_1}, \ldots, \hat{h}_{i,0}^{\gamma_n})$ for $i \in [m]$, where $\gamma_1, \ldots, \gamma_n$ are the values described above for the ext-semi-functional ciphertext.

Let ξ denote the number of key queries made by the adversary. In Table 2, we give an outline of the proof-structure with the first column stating the various hybrids we have ($\varphi \in [\xi]$), second column describes the way in which a hybrid differs from the one in the previous row, and the third column lists the properties we need to show indistinguishability from the previous one. To prevent the table from overflowing, we use some shorthands like ct for ciphertext, func for functional, norm for normal, msg for message, and ind for indistinguishability.

Also, Hyb_0 is the game $\mathsf{IND\text{-}CPA}^b_{\mathcal{A}}(\lambda, \mathsf{par})$ which is formally defined in Sect. 2.1. See the full version for a more formal description of the hybrids.

Our main concern here is the indistinguishability of hybrids $\mathsf{Hyb}_{2,\varphi,2}$ and $\mathsf{Hyb}_{2,\varphi,3}$ when the φth key changes from ext-pseudo-normal to ext-pseudo semi-functional, while the ciphertext stays ext-semi-functional. (Indistinguishability of the rest of the hybrids follows from [1] as noted earlier.) We prove the following lemma in the full version.

Lemma 7.3. *For any* PPT *adversary* $\mathcal{A}$, *there exists a* PPT *adversary* $\mathcal{B}$ *such that the advantage of* $\mathcal{A}$ *in distinguishing* $\mathsf{Hyb}_{2,\varphi,2}$ *and* $\mathsf{Hyb}_{2,\varphi,3}$ *is at most the advantage of* $\mathcal{B}$ *in the* $\mathsf{q\text{-}ratio}_{\mathsf{dsg}}$ *assumption plus some negligible quantity in the security parameter.*

References

1. Agrawal, S., Chase, M.: A study of pair encodings: predicate encryption in prime order groups. In: Kushilevitz, E., Malkin, T. (eds.) TCC 2016. LNCS, vol. 9563, pp. 259–288. Springer, Heidelberg (2016). doi:10.1007/978-3-662-49099-0_10
2. Attrapadung, N.: Dual system encryption via doubly selective security: framework, fully secure functional encryption for regular languages, and more. In: Nguyen, P.Q., Oswald, E. (eds.) EUROCRYPT 2014. LNCS, vol. 8441, pp. 557–577. Springer, Heidelberg (2014). doi:10.1007/978-3-642-55220-5_31
3. Attrapadung, N.: Dual system encryption via doubly selective security: framework, fully-secure functional encryption for regular languages, and more. Cryptology ePrint Archive, Report 2014/428 (2014). http://eprint.iacr.org/2014/428
4. Attrapadung, N.: Dual system encryption framework in prime-order groups via computational pair encodings. In: Cheon, J.H., Takagi, T. (eds.) ASIACRYPT 2016. LNCS, vol. 10032, pp. 591–623. Springer, Heidelberg (2016). doi:10.1007/978-3-662-53890-6_20
5. Attrapadung, N., Hanaoka, G., Yamada, S.: Conversions among several classes of predicate encryption and applications to ABE with various compactness tradeoffs. In: Iwata, T., Cheon, J.H. (eds.) ASIACRYPT 2015. LNCS, vol. 9452, pp. 575–601. Springer, Heidelberg (2015). doi:10.1007/978-3-662-48797-6_24
6. Attrapadung, N., Libert, B.: Functional encryption for inner product: achieving constant-size ciphertexts with adaptive security or support for negation. In: Nguyen, P.Q., Pointcheval, D. (eds.) PKC 2010. LNCS, vol. 6056, pp. 384–402. Springer, Heidelberg (2010). doi:10.1007/978-3-642-13013-7_23
7. Attrapadung, N., Libert, B., de Panafieu, E.: Expressive key-policy attribute-based encryption with constant-size ciphertexts. In: Catalano, D., Fazio, N., Gennaro, R., Nicolosi, A. (eds.) PKC 2011. LNCS, vol. 6571, pp. 90–108. Springer, Heidelberg (2011). doi:10.1007/978-3-642-19379-8_6
8. Attrapadung, N., Yamada, S.: Duality in ABE: converting attribute based encryption for dual predicate and dual policy via computational encodings. In: Nyberg, K. (ed.) CT-RSA 2015. LNCS, vol. 9048, pp. 87–105. Springer, Cham (2015). doi:10.1007/978-3-319-16715-2_5
9. Beimel, A.: Secret-sharing schemes: a survey. In: Chee, Y.M., Guo, Z., Ling, S., Shao, F., Tang, Y., Wang, H., Xing, C. (eds.) IWCC 2011. LNCS, vol. 6639, pp. 11–46. Springer, Heidelberg (2011). doi:10.1007/978-3-642-20901-7_2

10. Bethencourt, J., Sahai, A., Waters, B.: Ciphertext-policy attribute-based encryption. In: IEEE Symposium on Security and Privacy, pp. 321–334 (2007)
11. Boneh, D., Raghunathan, A., Segev, G.: Function-private identity-based encryption: hiding the function in functional encryption. In: Canetti, R., Garay, J.A. (eds.) CRYPTO 2013. LNCS, vol. 8043, pp. 461–478. Springer, Heidelberg (2013). doi:10.1007/978-3-642-40084-1_26
12. Boneh, D., Waters, B.: Conjunctive, subset, and range queries on encrypted data. In: Vadhan, S.P. (ed.) TCC 2007. LNCS, vol. 4392, pp. 535–554. Springer, Heidelberg (2007). doi:10.1007/978-3-540-70936-7_29
13. Boyen, X.: Attribute-based functional encryption on lattices. In: Sahai, A. (ed.) TCC 2013. LNCS, vol. 7785, pp. 122–142. Springer, Heidelberg (2013). doi:10.1007/978-3-642-36594-2_8
14. Chase, M.: Multi-authority attribute based encryption. In: Vadhan, S.P. (ed.) TCC 2007. LNCS, vol. 4392, pp. 515–534. Springer, Heidelberg (2007). doi:10.1007/978-3-540-70936-7_28
15. Chen, J., Gay, R., Wee, H.: Improved dual system ABE in prime-order groups via predicate encodings. In: Oswald, E., Fischlin, M. (eds.) EUROCRYPT 2015. LNCS, vol. 9057, pp. 595–624. Springer, Heidelberg (2015). doi:10.1007/978-3-662-46803-6_20
16. Chen, J., Wee, H.: Dual system groups and its applications – compact HIBE and more. Cryptology ePrint Archive, Report 2014/265 (2014). http://eprint.iacr.org/2014/265
17. Chen, J., Wee, H.: Semi-adaptive attribute-based encryption and improved delegation for boolean formula. In: Abdalla, M., Prisco, R. (eds.) SCN 2014. LNCS, vol. 8642, pp. 277–297. Springer, Cham (2014). doi:10.1007/978-3-319-10879-7_16
18. Freeman, D.M.: Converting pairing-based cryptosystems from composite-order groups to prime-order groups. In: Gilbert, H. (ed.) EUROCRYPT 2010. LNCS, vol. 6110, pp. 44–61. Springer, Heidelberg (2010). doi:10.1007/978-3-642-13190-5_3
19. Garg, S., Gentry, C., Halevi, S., Raykova, M., Sahai, A., Waters, B.: Candidate indistinguishability obfuscation and functional encryption for all circuits. In: FOCS, pp. 40–49 (2013)
20. Garg, S., Gentry, C., Halevi, S., Zhandry, M.: Functional encryption without obfuscation. In: Kushilevitz, E., Malkin, T. (eds.) TCC 2016. LNCS, vol. 9563, pp. 480–511. Springer, Heidelberg (2016). doi:10.1007/978-3-662-49099-0_18
21. Gorbunov, S., Vaikuntanathan, V., Wee, H.: Attribute-based encryption for circuits. In: ACM STOC, pp. 545–554 (2013)
22. Goyal, V., Pandey, O., Sahai, A., Waters, B.: Attribute-based encryption for fine-grained access control of encrypted data. In: ACM CCS, pp. 89–98 (2006). Available as Cryptology ePrint Archive Report 2006/309
23. Guillevic, A.: Comparing the pairing efficiency over composite-order and prime-order elliptic curves. In: Jacobson, M., Locasto, M., Mohassel, P., Safavi-Naini, R. (eds.) ACNS 2013. LNCS, vol. 7954, pp. 357–372. Springer, Heidelberg (2013). doi:10.1007/978-3-642-38980-1_22
24. Herold, G., Hesse, J., Hofheinz, D., Ràfols, C., Rupp, A.: Polynomial spaces: a new framework for composite-to-prime-order transformations. In: Garay, J.A., Gennaro, R. (eds.) CRYPTO 2014. LNCS, vol. 8616, pp. 261–279. Springer, Heidelberg (2014). doi:10.1007/978-3-662-44371-2_15
25. Katz, J., Sahai, A., Waters, B.: Predicate encryption supporting disjunctions, polynomial equations, and inner products. In: Smart, N. (ed.) EUROCRYPT 2008. LNCS, vol. 4965, pp. 146–162. Springer, Heidelberg (2008). doi:10.1007/978-3-540-78967-3_9

26. Kowalczyk, L., Lewko, A.B.: Bilinear entropy expansion from the decisional linear assumption. In: Gennaro, R., Robshaw, M. (eds.) CRYPTO 2015. LNCS, vol. 9216, pp. 524–541. Springer, Heidelberg (2015). doi:10.1007/978-3-662-48000-7_26

27. Lewko, A.: Tools for simulating features of composite order bilinear groups in the prime order setting. In: Pointcheval, D., Johansson, T. (eds.) EUROCRYPT 2012. LNCS, vol. 7237, pp. 318–335. Springer, Heidelberg (2012). doi:10.1007/978-3-642-29011-4_20

28. Lewko, A., Waters, B.: Decentralizing attribute-based encryption. In: Paterson, K.G. (ed.) EUROCRYPT 2011. LNCS, vol. 6632, pp. 568–588. Springer, Heidelberg (2011). doi:10.1007/978-3-642-20465-4_31

29. Lewko, A., Waters, B.: Unbounded HIBE and attribute-based encryption. In: Paterson, K.G. (ed.) EUROCRYPT 2011. LNCS, vol. 6632, pp. 547–567. Springer, Heidelberg (2011). doi:10.1007/978-3-642-20465-4_30

30. Lewko, A., Waters, B.: New proof methods for attribute-based encryption: achieving full security through selective techniques. In: Safavi-Naini, R., Canetti, R. (eds.) CRYPTO 2012. LNCS, vol. 7417, pp. 180–198. Springer, Heidelberg (2012). doi:10.1007/978-3-642-32009-5_12

31. Okamoto, T., Takashima, K.: Fully secure unbounded inner-product and attribute-based encryption. In: Wang, X., Sako, K. (eds.) ASIACRYPT 2012. LNCS, vol. 7658, pp. 349–366. Springer, Heidelberg (2012). doi:10.1007/978-3-642-34961-4_22

32. Ostrovsky, R., Sahai, A., Waters, B.: Attribute-based encryption with non-monotonic access structures. In: ACM CCS, pp. 195–203 (2007)

33. Rouselakis, Y., Waters, B.: Practical constructions and new proof methods for large universe attribute-based encryption. In: ACM CCS, pp. 463–474 (2013)

34. Sahai, A., Seyalioglu, H., Waters, B.: Dynamic credentials and ciphertext delegation for attribute-based encryption. In: Safavi-Naini, R., Canetti, R. (eds.) CRYPTO 2012. LNCS, vol. 7417, pp. 199–217. Springer, Heidelberg (2012). doi:10.1007/978-3-642-32009-5_13

35. Sahai, A., Waters, B.: Fuzzy identity-based encryption. In: Cramer, R. (ed.) EUROCRYPT 2005. LNCS, vol. 3494, pp. 457–473. Springer, Heidelberg (2005). doi:10.1007/11426639_27

36. Shen, E., Shi, E., Waters, B.: Predicate privacy in encryption systems. In: Reingold, O. (ed.) TCC 2009. LNCS, vol. 5444, pp. 457–473. Springer, Heidelberg (2009). doi:10.1007/978-3-642-00457-5_27

37. Waters, B.: Dual system encryption: realizing fully secure IBE and HIBE under simple assumptions. In: Halevi, S. (ed.) CRYPTO 2009. LNCS, vol. 5677, pp. 619–636. Springer, Heidelberg (2009). doi:10.1007/978-3-642-03356-8_36

38. Waters, B.: Functional encryption for regular languages. In: Safavi-Naini, R., Canetti, R. (eds.) CRYPTO 2012. LNCS, vol. 7417, pp. 218–235. Springer, Heidelberg (2012). doi:10.1007/978-3-642-32009-5_14

39. Waters, B.: A punctured programming approach to adaptively secure functional encryption. In: Gennaro, R., Robshaw, M. (eds.) CRYPTO 2015. LNCS, vol. 9216, pp. 678–697. Springer, Heidelberg (2015). doi:10.1007/978-3-662-48000-7_33

40. Wee, H.: Dual system encryption via predicate encodings. In: Lindell, Y. (ed.) TCC 2014. LNCS, vol. 8349, pp. 616–637. Springer, Heidelberg (2014). doi:10.1007/978-3-642-54242-8_26

41. Yamada, S., Attrapadung, N., Hanaoka, G., Kunihiro, N.: A framework and compact constructions for non-monotonic attribute-based encryption. In: Krawczyk, H. (ed.) PKC 2014. LNCS, vol. 8383, pp. 275–292. Springer, Heidelberg (2014). doi:10.1007/978-3-642-54631-0_16

Elliptic Curves

Twisted μ_4-Normal Form for Elliptic Curves

David Kohel[✉]

Aix Marseille University, CNRS, Centrale Marseille, I2M, Marseille, France
David.Kohel@univ-amu.fr

Abstract. We introduce the twisted μ_4-normal form for elliptic curves, deriving in particular addition algorithms with complexity 9M + 2S and doubling algorithms with complexity 2M + 5S + 2m over a binary field. Every ordinary elliptic curve over a finite field of characteristic 2 is isomorphic to one in this family. This improvement to the addition algorithm, applicable to a larger class of curves, is comparable to the 7M + 2S achieved for the μ_4-normal form, and replaces the previously best known complexity of 13M + 3S on López-Dahab models applicable to these twisted curves. The derived doubling algorithm is essentially optimal, without any assumption of special cases. We show moreover that the Montgomery scalar multiplication with point recovery carries over to the twisted models, giving symmetric scalar multiplication adapted to protect against side channel attacks, with a cost of $4M + 4S + 1m_t + 2m_c$ per bit. In characteristic different from 2, we establish a linear isomorphism with the twisted Edwards model over the base field. This work, in complement to the introduction of μ_4-normal form, fills the lacuna in the body of work on efficient arithmetic on elliptic curves over binary fields, explained by this common framework for elliptic curves in μ_4-normal form over a field of any characteristic. The improvements are analogous to those which the Edwards and twisted Edwards models achieved for elliptic curves over finite fields of odd characteristic and extend μ_4-normal form to cover the binary NIST curves.

1 Introduction

Let E be an elliptic curve with given embedding in $\mathbb{P}^r$ and identity O. The addition morphism $\mu : E \times E \to E$ is uniquely defined by the pair (E, O) but the homogeneous polynomial maps which determine μ are not unique. Let $x = (X_0, \dots, X_r)$ and $y = (Y_0, \dots, Y_r)$ be the coordinate functions on the first and second factors, respectively. We recall that an *addition law* (cf. [13]) is a bihomogenous polynomial map $\mathfrak{s} = (p_0(x, y), \dots, p_r(x, y))$ which determines μ outside of the common zero locus $p_0(x, y) = \dots = p_r(x, y) = 0$. Such polynomial addition laws play an important role in cryptography since they provide a means of carrying out addition on E without inversion in the base field.

In this work we generalize the algorithmic analysis of the μ_4-normal form to include twists. The principal improvements are for binary curves, but we are able to establish these results for a family which has good reduction and efficient arithmetic over any field k, and in fact any ring. We adopt the notation **M** and **S**

© International Association for Cryptologic Research 2017
J.-S. Coron and J.B. Nielsen (Eds.): EUROCRYPT 2017, Part I, LNCS 10210, pp. 659–678, 2017.
DOI: 10.1007/978-3-319-56620-7_23

for the complexity of multiplication and squaring in k, and $\mathbf{m}$ for multiplication by a fixed constant that depends (polynomially) only on curve constants.

In Sect. 2 we introduce a hierarchy of curves in $\boldsymbol{\mu}_4$-normal form, according to the additional 4-level structure parametrized. In referring to these families of curves, we give special attention to the so-called split and semisplit variants, while using the generic term $\boldsymbol{\mu}_4$-normal form to refer to any of the families. In particular their isomorphisms and addition laws are developed. In the specialization to finite fields of characteristic 2, by extracting square roots, we note that any of the families can be put in split $\boldsymbol{\mu}_4$-normal form, and the distinction is only one of symmetries and optimization of the arithmetic. In Sect. 3, we generalize this hierarchy to quadratic twists, which, in order to hold in characteristic 2 are defined in terms of Artin–Schreier extensions. The next two sections deal with algorithms for these families of curves over binary fields, particularly, their addition laws in Sect. 4 and their doubling algorithms in Sect. 6. These establish the main complexity results of this work — an improvement of the best known addition algorithms on NIST curves to $9\mathbf{M} + 2\mathbf{S}$ coupled with a doubling algorithm of $2\mathbf{M} + 5\mathbf{S} + 2\mathbf{m}$. These improvements are summarized in the following table of complexities (see Sect. 8 for details).

Curve model	Doubling	Addition	%	NIST
Lambda coordinates	$3\mathbf{M} + 4\mathbf{S} + 1\mathbf{m}$	$11\mathbf{M} + 2\mathbf{S}$	100%	✓
Binary Edwards $(d_1 = d_2)$	$2\mathbf{M} + 5\mathbf{S} + 2\mathbf{m}$	$16\mathbf{M} + 1\mathbf{S} + 4\mathbf{m}$	50%	✗
López-Dahab $(a_2 = 0)$	$2\mathbf{M} + 5\mathbf{S} + 1\mathbf{m}$	$14\mathbf{M} + 3\mathbf{S}$	50%	✗
López-Dahab $(a_2 = 1)$	$2\mathbf{M} + 4\mathbf{S} + 2\mathbf{m}$	$13\mathbf{M} + 3\mathbf{S}$	50%	✓
Twisted $\boldsymbol{\mu}_4$-normal form	$2\mathbf{M} + 5\mathbf{S} + 2\mathbf{m}$	$9\mathbf{M} + 2\mathbf{S}$	100%	✓
$\boldsymbol{\mu}_4$-normal form	$2\mathbf{M} + 5\mathbf{S} + 2\mathbf{m}$	$7\mathbf{M} + 2\mathbf{S}$	50%	✗

To complete the picture, we prove in Sect. 7 that the Montgomery endomorphism and resulting complexity, as described in Kohel [10] carry over to the twisted families, which allows for an elementary and relatively efficient symmetric algorithm for scalar multiplication which is well-adapted to protecting against side-channel attacks. While the most efficient arithmetic is achieved for curves for which the curve coefficients are constructed such that the constant multiplications are negligible, these extensions to twists provide efficient algorithms for backward compatibility with binary NIST curves.

2   The $\boldsymbol{\mu}_4$-normal Form

In this section we recall the definition and construction of the family of elliptic curves in (split) $\boldsymbol{\mu}_4$-normal form. The notion of a canonical model of level n was introduced in Kohel [8] as an elliptic curve C/k in $\mathbb{P}^{n-1}$ with subgroup scheme $G \cong \boldsymbol{\mu}_n$ (a k-rational subgroup of the n-torsion subgroup $C[n]$ whose points split in $k[\zeta_n]$, where ζ_n is an n-th root of unity in $\bar{k}$) such that for $P = (x_0 : x_1 : \cdots : x_{n-1})$ a generator S of G acts by $P + S = (x_0 : \zeta_n^1 x_1 : \cdots : \zeta_n^{n-1} x_{n-1})$. If, in addition, there exists a rational n-torsion point T such that $C[n] = \langle S, T \rangle$, we say that the model is *split* and impose the condition that T acts by a cyclic coordinate permutation. Construction of the special cases $n = 4$ and $n = 5$ were treated as examples in Kohel [8], and the present work is concerned with a more in depth study of the former.

The Edwards curve $x^2 + y^2 = 1 + dx^2 y^2$ (see Edwards [6] and Berstein-Lange [2]) in $\mathbb{P}^3$ (by $(1 : x : y : xy)$ as the elliptic curve

$$X_1^2 + X_2^2 = X_0^2 + dX_3^2, \ X_0 X_3 = X_1 X_2,$$

with identity $O = (1 : 0 : 1 : 0)$. Such a model was studied by Hisil et al. [7], as extended Edwards coordinates, and admits the fastest known arithmetic on such curves. The twist by a, in extended coordinates, is the twisted Edwards curve (cf. Bernstein et al. [5] and Hisil et al. [7])

$$aX_1^2 + X_2^2 = X_0^2 + adX_3^2, \ X_0 X_3 = X_1 X_2$$

with parameters (a, ad). For the special case $(a, ad) = (-1, -16r)$, the change of variables

$$(X_0 : X_1 : X_2 : X_3) \mapsto (X_0, X_1 + X_2, 4X_3, -X_1 + X_2).$$

has image the canonical model of level 4 above. The normalization to have good reduction at 2 (by setting $d = 16r$ and the coefficient of X_3) as well as the following refined hierarchy of curves appears in Kohel [9], and the subsequent article [10] treated only the properties of this hierarchy over fields of characteristic 2.

Definition 1. *An elliptic curve in μ_4-normal form is a genus one curve in the family*

$$X_0^2 - rX_2^2 = X_1 X_3, \ X_1^2 - X_3^2 = X_0 X_2$$

with base point $O = (1 : 1 : 0 : 1)$. An elliptic curve in semisplit μ_4 -normal form is a genus one curve in the family

$$X_0^2 - X_2^2 = X_1 X_3, \ X_1^2 - X_3^2 = sX_0 X_2,$$

with identity $O = (1 : 1 : 0 : 1)$, and an elliptic curve is in split μ_4 -normal form if it takes the form

$$X_0^2 - X_2^2 = c^2 X_1 X_3, \ X_1^2 - X_3^2 = c^2 X_0 X_2.$$

with identity $O = (c : 1 : 0 : 1)$.

Setting $s = c^4$, the transformation

$$(X_0 : X_1 : X_2 : X_3) \mapsto (X_0 : cX_1 : cX_2 : X_3)$$

maps the split μ_4-normal form to semisplit μ_4-normal form with parameter s, and setting $r = 1/s^2$, the transformation

$$(X_0 : X_1 : X_2 : X_3) \mapsto (X_0 : X_1 : sX_2 : X_3)$$

maps the semisplit μ_4-normal form to μ_4-normal form with parameter r. The names for the μ_4-normal forms of a curve C/k in $\mathbb{P}^3$, recognize the existence of μ_4 as a k-rational subgroup scheme of $C[4]$, and secondly, its role as defining the embedding class of C in $\mathbb{P}^3$, namely it is cut out by the hyperplane $X_2 = 0$ in $\mathbb{P}^3$.

Lemma 2. *Let C be a curve in μ_4-normal form, semi-split μ_4-normal form, or split μ_4-normal form, with identity $(e, 1, 0, 1)$. For any extension containing a square root i of -1, the point $S = (e : i : 0 : -i)$ is a point of order 4 acting by the coordinate scaling $(x_0 : x_1 : x_2 : x_3) \mapsto (x_0 : ix_1 : -x_2 : -ix_3)$. In particular,*

$$\{(e : 1 : 0 : 1), (e : i : 0 : -i), (e : -1 : 0 : -1), (e : i : 0 : -i)\},$$

is a subgroup of $C[4] \subseteq C(\bar{k})$.

The semisplit μ_4-normal form with square parameter $s = t^2$ admits a 4-torsion point $(1 : t : 1 : 0)$ acting by scaled coordinate permutation. After a further quadratic extension $t = c^2$, the split μ_4-normal form admits the constant group scheme $\mathbb{Z}/4\mathbb{Z}$ acting by signed coordinate permutation.

Lemma 3. *Let C/k be an elliptic curve in split μ_4-normal form with identity $O = (c : 1 : 0 : 1)$. Then $T = (1 : c : 1 : 0)$ is a point in $C[4]$, and translation by T induces the signed coordinate permutation*

$$(x_0 : x_1 : x_2 : x_3) \longmapsto (x_3 : x_0 : x_1 : -x_2)$$

on C.

This gives the structure of a group $C[4] \cong \mu_4 \times \mathbb{Z}/4\mathbb{Z}$, whose generators S and T are induced by the matrix actions

$$A(S) = \begin{pmatrix} 1 & 0 & 0 & 0 \\ 0 & i & 0 & 0 \\ 0 & 0 & 1 & 0 \\ 0 & 0 & 0 & -i \end{pmatrix} \text{ and } A(T) = \begin{pmatrix} 0 & 1 & 0 & 0 \\ 0 & 0 & 1 & 0 \\ 0 & 0 & 0 & -1 \\ 1 & 0 & 0 & 0 \end{pmatrix}$$

on C such that $A(S)A(T) = iA(T)A(S)$. We can now state the structure of addition laws for the split μ_4-normal form and its relation to the torsion action described above.

Theorem 4. *Let C be an elliptic curve in split μ_4-normal form:*

$$X_0^2 - X_2^2 = c^2 X_1 X_3, \quad X_1^2 - X_3^2 = c^2 X_0 X_2, \quad O = (c : 1 : 0 : 1),$$

and set $U_{jk} = X_j Y_k$. A complete basis of addition laws of bidegree $(2,2)$ is given by:

$$\mathfrak{s}_0 = (U_{13}^2 - U_{31}^2,\ c(U_{13}U_{20} - U_{31}U_{02}),\ U_{20}^2 - U_{02}^2,\ c(U_{20}U_{31} - U_{13}U_{02})),$$
$$\mathfrak{s}_1 = (c(U_{03}U_{10} + U_{21}U_{32}),\ U_{10}^2 - U_{32}^2,\ c(U_{03}U_{32} + U_{10}U_{21}),\ U_{03}^2 - U_{21}^2),$$
$$\mathfrak{s}_2 = (U_{00}^2 - U_{22}^2,\ c(U_{00}U_{11} - U_{22}U_{33}),\ U_{11}^2 - U_{33}^2,\ c(U_{00}U_{33} - U_{11}U_{22})),$$
$$\mathfrak{s}_3 = (c(U_{01}U_{30} + U_{12}U_{23}),\ U_{01}^2 - U_{23}^2,\ c(U_{01}U_{12} + U_{23}U_{30}),\ U_{30}^2 - U_{12}^2).$$

The exceptional divisor of the addition law $\mathfrak{s}_\ell$ is $\sum_{k=0}^{3} \Delta_{kS+\ell T}$, where S and T are the 4-torsion points $(c : i : 0 : -i)$ and $(1 : c : 1 : 0)$, and the divisors $\sum_{k=0}^{3}(kS + \ell T)$ are determined by $X_{\ell+2} = 0$. In particular, any pair of the above addition laws provides a complete system of addition laws.

Proof. This appears as Theorem 44 of Kohel [8] for the μ_4-normal form, subject to the scalar renormalizations indicated above. The exceptional divisor is a sum of four curves of the form Δ_P by Theorem 10 of Kohel [8], and the points P can be determined by intersection with $H = C \times \{O\}$ using Corollary 11 of Kohel [8]. Taking the particular case $\mathfrak{s}_2$, we substitute $(Y_0, Y_1, Y_2, Y_3) = (c, 1, 0, 1)$ to obtain $(U_{00}, U_{11}, U_{22}, U_{33}) = (cX_0, X_1, 0, X_3)$, and hence

$$(U_{00}^2 - U_{22}^2,\ U_{00}U_{11} - U_{22}U_{33},\ U_{11}^2 - U_{33}^2,\ U_{00}U_{33} - U_{22}U_{11}),$$

which equals

$$(c^2 X_0^2, cX_0 X_1, X_1^2 - X_3^2, cX_0 X_3) = (c^2 X_0^2, cX_0 X_1, c^2 X_0 X_2, cX_0 X_3).$$

These coordinate functions cut out the divisor $X_0 = 0$ with support on the points $kS + 2T$, $0 \leq k < 4$, where $2T = (0 : -1 : -c : 1)$. The final statement follows since the exceptional divisors are disjoint. $\qquad\square$

The above basis of addition laws can be generated by any one of the four, by means of signed coordinate permutation on input and output determined by the action of the 4-torsion group. Denote translation by S and T by σ and τ, respectively, given by the coordinate scalings and permutations

$$\sigma(X_0 : X_1 : X_2 : X_3) = (X_0 : iX_1 : -X_2 : -iX_3),$$
$$\tau(X_0 : X_1 : X_2 : X_3) = (X_3 : X_0 : X_1 : -X_2),$$

as noted above. The set $\{\mathfrak{s}_0, \mathfrak{s}_1, \mathfrak{s}_2, \mathfrak{s}_3\}$ forms a basis of eigenvectors for the action of σ. More precisely for all (j, k, ℓ), we have

$$\mathfrak{s}_\ell = (-1)^{j+k+\ell} \sigma^{-j-k} \circ \mathfrak{s}_\ell \circ (\sigma^j \times \sigma^k).$$

Then τ, which projectively commutes with σ, acts by a scaled coordinate permutation

$$\mathfrak{s}_{\ell-j-k} = \tau^{-j-k} \circ \mathfrak{s}_\ell \circ (\tau^j \times \tau^k),$$

consistent with the action on the exceptional divisors (see Lemma 31 of Kohel [8]).

Consequently, the complexity of evaluation of any of these addition laws is computationally equivalent, since they differ only by a signed coordinate permutation on input and output.

Corollary 5. *Let C be an elliptic curve in split μ_4-normal form. There exist algorithms for addition with complexity $9M + 2m$ over any ring, $8M + 2m$ over a ring in which 2 is a unit, and $7M + 2S + 2m$ over a ring of characteristic 2.*

Proof. We determine the complexity of an algorithm for the evaluation of the addition law $\mathfrak{s}_2$:

$$(Z_0, Z_1, Z_2, Z_3) = (U_{00}^2 - U_{22}^2, c(U_{00}U_{11} - U_{22}U_{33}), U_{11}^2 - U_{33}^2, c(U_{00}U_{33} - U_{11}U_{22})),$$

recalling that each of the given addition laws in the basis has equivalent evaluation. Over a general ring, we make use of the equalities:

$$Z_0 = U_{00}^2 - U_{22}^2 = (U_{00} - U_{22})(U_{00} + U_{22}),$$
$$Z_2 = U_{11}^2 - U_{33}^2 = (U_{11} - U_{33})(U_{11} + U_{33}),$$

and

$$Z_1 + Z_3 = c(U_{00}U_{11} - U_{22}U_{33}) + c(U_{00}U_{33} - U_{22}U_{11}) = c(U_{00} - U_{22})(U_{11} + U_{33}),$$
$$Z_1 - Z_3 = c(U_{00}U_{11} - U_{22}U_{33}) - c(U_{00}U_{33} - U_{22}U_{11}) = c(U_{00} + U_{22})(U_{11} - U_{33}),$$

using $1M + 1m$ each for their evaluation.

- Evaluate $U_{jj} = X_j Y_j$, for $1 \leq j \leq 4$, with $4M$.
- Evaluate $(Z_0, Z_2) = (U_{00}^2 - U_{22}^2, U_{11}^2 - U_{33}^2)$ with $2M$.
- Evaluate $A = c(U_{00} - U_{22})(U_{11} + U_{33})$ using $1M + 1m$.
- Compute $Z_1 = c(U_{00}U_{11} - U_{22}U_{33})$ and set $Z_3 = A - Z_1$ with $2M + 1m$.

This yields the desired complexity $9M + 2m$ over any ring. If 2 is a unit (and assuming a negligible cost of multiplying by 2), we replace the last line with two steps:

- Evaluate $B = c(U_{00} + U_{22})(U_{11} - U_{33})$ using $1M + 1m$.
- Compute $(2Z_1, 2Z_3) = (A + B, A - B)$ and scale (Z_0, Z_2) by 2,

which gives a complexity of $8M + 2m$. This yields an algorithm essentially equivalent to that Hisil et al. [7] under the linear isomorphism with the -1-twist of Edwards normal form. Finally if the characteristic is 2, the result $7M + 2S + 2m$ of Kohel [10] is obtained by replacing $2M$ by $2S$ for the evaluation of (Z_0, Z_2) in the generic algorithm.   □

Before considering the twisted forms, we determine the base complexity of doubling for the split μ_4-normal form.

Corollary 6. *Let C be an elliptic curve in split μ_4-normal form. There exist algorithms for doubling with complexity $5M + 4S + 2m$ over any ring, $4M + 4S + 2m$ over a ring in which 2 is a unit, and $2M + 5S + 7m$ over a ring of characteristic 2.*

Proof. The specialization of the addition law $\mathfrak{s}_2$ to the diagonal gives the forms for doubling

$$(X_0^4 - X_2^4,\ c(X_0^2 X_1^2 - X_2^2 X_3^2),\ X_1^4 - X_3^4,\ c(X_0^2 X_3^2 - X_1^2 X_2^2)).$$

which we can evaluate as follows:

- Evaluate X_j^2, for $1 \leq j \leq 4$, with 4**S**.
- Evaluate $(Z_0, Z_2) = (X_0^4 - X_2^4,\ X_1^4 - X_3^4)$ with 2**M**.
- Evaluate $A = c(X_0^2 - X_2^2)(X_1^2 + X_3^2)$ using 1**M** + 1**m**.
- Compute $Z_1 = c(X_0^2 X_1^2 - X_2^2 X_3^2)$ and set $Z_3 = A - Z_1$ with 2**M** + 1**m**.

This gives the result of 5**M** + 4**S** + 2**m** over any ring. As above, when 2 is a unit, we replace the last line with the two steps:

- Evaluate $B = c(X_0^2 + X_2^2)(X_1^2 - X_3^2)$ using 1**M** + 1**m**.
- Compute $(2Z_1, 2Z_3) = (A + B, A - B)$ and scale (Z_0, Z_2) by 2.

This reduces the complexity by 1**M**. In characteristic 2, the general algorithm specializes to 3**M** + 6**S** + 2**m**, but Kohel [10] provides an algorithm with better complexity of 2**M** + 5**S** + 7**m** (reduced by 5**m** on the semisplit model). □

In the next section, we introduce the twists of these μ_4-normal forms, and derive efficient algorithms for their arithmetic.

3   Twisted Normal Forms

A quadratic twist of an elliptic curve is determined by a non-rational isomorphism defined over a quadratic extension $k[\alpha]/k$. In odd characteristic one can take an extension defined by $\alpha^2 = a$, but in characteristic 2, the general form of a quadratic extension is $k[\omega]/k$ where $\omega^2 - \omega = a$ for some a in k. The normal forms defined above both impose the existence of a k-rational point of order 4.

Over a finite field of characteristic 2, the existence of a 4-torsion point is a weaker constraint than for odd characteristic, since if E/k is an ordinary elliptic curve over a finite field of characteristic 2, there necessarily exists a 2-torsion point. Moreover, if E does not admit a k-rational 4-torsion point and $|k| > 2$, then its quadratic twist does.

We recall that for an elliptic curve in Weierstrass form,

$$E : Y^2 Z + (a_1 X + a_3 Z)YZ = X^3 + a_2 X^2 Z + a_4 X Z^2 + a_6 Z^3,$$

the quadratic twist by $k[\omega]/k$ is given by

$$E^t : Y^2 Z + (a_1 X + a_3 Z)YZ = X^3 + a_2 X^2 Z + a_4 X Z^2 + a_6 Z^3 + a(a_1 X + a_3 Z)^2 Z,$$

with isomorphism $\tau(X : Y : Z) = (X : -Y - \omega(a_1 X + a_3 Z) : Z)$, which satisfies $\tau^\sigma = -\tau$, where σ is the nontrivial automorphism of $k[\omega]/k$. The objective here is to describe the quadratic twists in the case of the normal forms defined above.

With a view towards cryptography, the binary NIST curves are of the form $y^2 + xy = x^3 + ax^2 + b$, with $a = 1$ and group order $2n$, whose quadratic twist is the curve with $a = 0$ which admits a point of order 4. While the latter admits an isomorphism to a curve in $\boldsymbol{\mu}_4$-normal form, to describe the others, we must represent them as quadratic twists.

The Twisted $\boldsymbol{\mu}_4$-normal Form

In what follows we let $k[\omega]/k$ be the quadratic extension given by $\omega^2 - \omega = a$, and set $\overline{\omega} = 1 - \omega$ and $\delta = \omega - \overline{\omega}$. In order to have the widest possible applicability, we describe the quadratic twists with respect to any ring or field k. The discriminant of the extension is $D = \delta^2 = 1 + 4a$. When 2 is invertible we can speak of a twist by D, but in general we refer to a as the twisting parameter. While admitting general rings, all formulas hold over a field of characteristic 2, and we investigate optimizations in this case.

Theorem 7. *Let C/k be an elliptic curve in $\boldsymbol{\mu}_4$-normal form, semisplit $\boldsymbol{\mu}_4$-normal form, or split $\boldsymbol{\mu}_4$-normal form, given respectively by*

$$X_0^2 - r\,X_2^2 = X_1X_3, \quad X_1^2 - X_3^2 = X_0X_2, \quad O = (1:1:0:1),$$
$$X_0^2 - X_2^2 = X_1X_3, \quad X_1^2 - X_3^2 = s\,X_0X_2, \quad O = (1:1:0:1),$$
$$X_0^2 - X_2^2 = c^2\,X_1X_3, \quad X_1^2 - X_3^2 = c^2\,X_0X_2, \quad O = (c:1:0:1).$$

The quadratic twist C^t of C by $k[\omega]$, where $\omega^2 - \omega = a$, is given by

$$X_0^2 - Dr\,X_2^2 = X_1X_3 - a(X_1 - X_3)^2, \quad X_1^2 - X_3^2 = X_0X_2,$$
$$X_0^2 - DX_2^2 = X_1X_3 - a(X_1 - X_3)^2, \quad X_1^2 - X_3^2 = s\,X_0X_2,$$
$$X_0^2 - DX_2^2 = c^2(X_1X_3 - a(X_1 - X_3)^2), \quad X_1^2 - X_3^2 = c^2X_0X_2,$$

with identities $O = (1:1:0:1)$, $O = (1:1:0:1)$ and $O = (c:1:0:1)$, respectively. In each case, the twisting isomorphism $\tau : C \to C^t$ is given by

$$(X_0 : X_1 : X_2 : X_3) \longmapsto (\delta X_0 : \omega X_1 - \overline{\omega}X_3 : X_2 : \omega X_3 - \overline{\omega}X_1),$$

with inverse sending $(X_0 : X_1 : X_2 : X_3)$ to $(X_0 : \omega X_1 + \overline{\omega}X_3 : \delta X_2 : \overline{\omega}X_1 + \overline{\omega}X_3)$.

Proof. Since the inverse morphism is $[-1](X_0 : X_1 : X_2 : X_3) = (X_0 : X_3 : -X_2 : X_1)$, the twisting morphism satisfies $\tau^\sigma = [-1] \circ \tau$ where σ is the nontrivial automorphism of $k[\omega]/k$. Consequently, the image C^t is a twist of C. The form of the inverse is obtained by matrix inversion. $\qquad\square$

Remark. In characteristic 2 we have $D = \delta = 1$, and the twisted split $\boldsymbol{\mu}_4$-normal form is $X_0^2 + X_2^2 = c^2(X_1X_3 + a(X_1 + X_3)^2)$, $X_1^2 + X_3^2 = c^2X_0X_2$, with associated twisting morphism

$$(X_0 : X_1 : X_2 : X_3) \longmapsto (X_0 : \overline{\omega}X_1 + \omega X_3 : X_2 : \omega X_1 + \overline{\omega}X_3).$$

Over a field of characteristic different from 2, we have an isomorphism with the twisted Edwards normal form.

Theorem 8. *Let C^t be an elliptic curve in twisted μ_4-normal form*

$$X_0^2 - DrX_2^2 = X_1X_3 - a(X_1 - X_3)^2, \quad X_1^2 - X_3^2 = X_0X_2,$$

with parameters (r, a) over a field of characteristic different from 2. Then C^t is isomorphic to the twisted Edwards curve

$$X_0^2 - 16DrX_3^2 = -DX_1^2 + X_2^2$$

with parameters $(-D, -16Dr)$, via the isomorphism $C^t \to E$:

$$(X_0 : X_1 : X_2 : X_3) \longmapsto (4X_0 : 2(X_1 - X_3) : 2(X_1 + X_3) : X_2),$$

and inverse

$$(X_0 : X_1 : X_2 : X_3) \longmapsto (X_0 : X_1 + X_2 : 4X_3 : -X_1 + X_2).$$

Proof. The linear transformation is the compositum of the above linear transformations with the morphism $(X_0 : X_1 : X_2 : X_3) \longmapsto (\delta X_0 : X_1 : \delta X_2 : X_3)$ from the Edwards curve to its twist. $\qquad\square$

For completeness we provide an isomorphic model in Weierstrass form:

Theorem 9. *Let C^t be an elliptic curve in twisted split μ_4-normal form with parameters (r, a). Then C^t is isomorphic to the elliptic curve*

$$y^2 + xy = x^3 + (a - 8Dr)x^2 + 2D^2r(8r - 3)x - D^3r(1 - 4r)$$

in Weierstrass form, where $D = 4a + 1$. The isomorphism is given by the map which sends $(X_0 : X_1 : X_2 : X_3)$ to

$$\left(D\big(U_0 - 4r(U_0 + U_2)\big) : D\big(U_1 - 2r(8U_1 + 2U_0 - U_2)\big) : U_2 - 2U_0) \right),$$

where $(U_0, U_1, U_2, U_3) = (X_1 - X_3, X_0 + X_3, X_2, X_1 + X_3)$.

Proof. A symbolic verification is carried out by the Echidna code [11] implemented in Magma [14]. $\qquad\square$

Specializing to characteristic 2, we obtain the following corollary.

Corollary 10. *Let C^t be a binary elliptic curve in twisted μ_4-normal form*

$$X_0^2 + bX_2^2 = X_1X_3 + aX_0X_2, \quad X_1^2 + X_3^2 = X_0X_2,$$

with parameters $(r, a) = (b, a)$. Then C^t is isomorphic to the elliptic curve

$$y^2 + xy = x^3 + ax^2 + b,$$

in Weierstrass form via the map $(X_0 : X_1 : X_2 : X_3) \mapsto (X_1 + X_3 : X_0 + X_1 : X_2)$. On affine points (x, y) the inverse is $(x, y) \longmapsto (x^2 : x^2 + y : 1 : x^2 + y + x)$.

Proof. By the previous theorem, since $D = 1$ in characteristic 2, the Weierstrass model simplifies to $y^2 + xy = x^3 + ax^2 + b$, and the map to

$$(X_0 : X_1 : X_2 : X_3) \longmapsto (U_0 : U_1 : U_2) = (X_1 + X_3 : X_0 + X_1 : X_2).$$

The given map on affine points is easily seen to be a birational inverse, valid for $X_2 = 1$, in view of the relation $(X_1 + X_3)^2 = X_0 X_2$, well-defined outside the identity. Consequently, it extends uniquely to an isomorphism.   □

As a consequence of this theorem, any ordinary binary curve (with $j = 1/b \neq 0$) can be put in twisted $\boldsymbol{\mu}_4$-normal form, via the map on affine points:

$$(x, y) \longmapsto (x^2 : x^2 + y : 1 : x^2 + y + x).$$

In particular all algorithms of this work (over binary fields) are applicable to the binary NIST curves, which permits backward compatibility and improved performance.

4   Addition Algorithms

We now consider the addition laws for twisted split $\boldsymbol{\mu}_4$-normal form. In the application to prime finite fields of odd characteristic p (see below for considerations in characteristic 2), under the GRH, Lagarias, Montgomery and Odlyzko [12] prove a generalization of the result of Ankeny [1], under which we can conclude that the least quadratic nonresidue $D \equiv 1 \bmod 4$ is in $O(\log^2(p))$, and the average value of D is $O(1)$. Consequently, for a curve over a finite prime field, one can find small twisting parameters for constructing the quadratic twist. With this in mind, we ignore all multiplications by constants a and $D = 4a + 1$.

Theorem 11. *Let C^t be an elliptic curve in twisted split $\boldsymbol{\mu}_4$-normal form:*

$$X_0^2 - DX_2^2 = c^2(X_1X_3 - a(X_1 - X_3)^2), \ X_1^2 - X_3^2 = c^2 X_0 X_2.$$

over a ring in which 2 is a unit. The projections $\pi_1 : C^t \to \mathbb{P}^1$, with coordinates (X, Z), given by

$$\pi_1((X_0 : X_1 : X_2 : X_3)) = \{(cX_0 : X_1 + X_3), (X_1 - X_3 : cX_2)\},$$

and $\pi_2 : C^t \to \mathbb{P}^1$, with coordinates (Y, W), given by

$$\pi_2((X_0 : X_1 : X_2 : X_3)) = \{(cX_0 : X_1 - X_3), (X_1 + X_3 : cX_2)\},$$

determine an isomorphism $\pi_1 \times \pi_2$ with its image:

$$((c^2/2)^2 X^2 - Z^2)W^2 = D((c^2/2)^2 Z^2 - X^2)Y^2$$

in $\mathbb{P}^1 \times \mathbb{P}^1$, with inverse

$$\sigma((X : Z), (Y : W)) = (2XY : c(XW + ZY) : 2ZW : c(ZY - XW)).$$

Proof. The morphisms σ and $\pi_1 \times \pi_2$ determine isomorphisms of $\mathbb{P}^1 \times \mathbb{P}^1$ with the surface $X_1^2 - X_3^2 = c^2 X_0 X_2$ in $\mathbb{P}^3$, and substitution in the first equation for C^t yields the above hypersurface in $\mathbb{P}^1 \times \mathbb{P}^1$. $\square$

The twisted split μ_4-normal form has 2-torsion subgroup generated by $Q = (-c : 1 : 0 : 1)$ and $R = (0 : -1 : c : 1)$, with $Q + R = (0 : -1 : -c : 1)$. Over any extension containing a square root ε of $-D$, the point $S = (c : -\varepsilon : 0 : \varepsilon)$ is a point of order 4 such that $2S = Q$.

Theorem 12. *Let C^t be an elliptic curve in twisted split μ_4-normal form over a ring in which 2 is a unit. The projections π_1 and π_2 determine two-dimensional spaces of bilinear addition law projections:*

$$
\begin{aligned}
\pi_1 \circ \mu(x, y) &= \begin{cases} \mathfrak{s}_0 = (U_{13} - U_{31} : U_{20} - U_{02}), \\ \mathfrak{s}_2 = (U_{00} + D U_{22} : U_{11} + U_{33} + 2a V_{13}), \end{cases} \\
\pi_2 \circ \mu(x, y) &= \begin{cases} \mathfrak{s}_1 = (U_{13} + U_{31} - 2a V_{13} : U_{02} + U_{20}), \\ \mathfrak{s}_3 = (U_{00} - D U_{22} : U_{11} - U_{33}), \end{cases}
\end{aligned}
$$

where $U_{k\ell} = X_k Y_\ell$ and $V_{k\ell} = (X_k - X_\ell)(Y_k - Y_\ell)$. The exceptional divisors of the $\mathfrak{s}_j$ are of the form $\Delta_{T_j} + \Delta_{T_j + Q}$, where $T_0 = O$, $T_1 = S + R$, $T_2 = R$, $T_3 = S$.

Proof. The existence and dimensions of the spaces of bilinear addition law projections, as well as the form of the exceptional divisors, follows from Theorem 26 and Corollary 27 of Kohel [8], observing for j in $\{0, 2\}$ that $T_j + (T_j + Q) = Q$ and for j in $\{1, 3\}$ that $T_j + (T_j + Q) = O$. The correctness of the forms can be verified symbolically, and the pairs $\{T_j, T_j + Q\}$ determined by the substitution $(Y_0, Y_1, Y_2, Y_3) = (c, 1, 0, 1)$, as in Corollary 11 of Kohel [8]. In particular, for $\mathfrak{s}_0$, we obtain the tuple $(U_{13} - U_{31}, U_{20} - U_{02}) = (X_1 - X_3, c X_2)$, which vanishes on $\{O, Q\} = \{(c : 1 : 0 : 1), (-c : 1 : 0 : 1)\}$, hence the exceptional divisor is $\Delta_O + \Delta_Q$. $\square$

Composing the addition law projections of Theorem 12 with the isomorphism of Theorem 11, and dividing by 2, we obtain for the pair $(\mathfrak{s}_0, \mathfrak{s}_1)$ the tuple (Z_0, Z_1, Z_2, Z_3) with

$$
\begin{aligned}
Z_0 &= (U_{13} - U_{31})(U_{13} + U_{31} - 2a V_{13}), & Z_1 + Z_3 &= -c(U_{02} - U_{20})(U_{13} + U_{31} + 2a V_{13}), \\
Z_2 &= -(U_{02} - U_{20})(U_{02} + U_{20}), & Z_1 - Z_3 &= -c(U_{13} - U_{31})(U_{02} + U_{20}),
\end{aligned}
$$

and for the pair $(\mathfrak{s}_2, \mathfrak{s}_3)$ the tuple (Z_0, Z_1, Z_2, Z_3) with

$$
\begin{aligned}
Z_0 &= (U_{00} + D U_{22})(U_{00} - D U_{22}), & Z_1 + Z_3 &= c\,(U_{11} + U_{33} + 2a V_{13})(U_{00} - D U_{22}), \\
Z_2 &= (U_{11} + U_{33} + 2a V_{13})(U_{11} - U_{33}), & Z_1 - Z_3 &= c(U_{00} + D U_{22})(U_{11} - U_{33}).
\end{aligned}
$$

The former have efficient evaluations over a ring in which 2 is a unit, yielding $(2Z_0, 2Z_1, 2Z_2, 2Z_3)$, and otherwise we deduce expressions for (Z_1, Z_3):

$$
\begin{aligned}
Z_1 &= c((U_{02} U_{13} - U_{02} U_{31}) - a(U_{02} - U_{20}) W_{13}), \\
Z_3 &= c((U_{02} U_{31} - U_{20} U_{13}) - a(U_{02} - U_{20}) W_{13}),
\end{aligned}
$$

with $W_{13} = 2(U_{13} + U_{31}) - V_{13}$, and

$$Z_1 = c(U_{00}U_{11} - DU_{22}U_{33}) - a(U_{00} - DU_{22})W_{13}),$$
$$Z_3 = c(U_{00}U_{33} - DU_{22}U_{11}) - a(U_{00} - DU_{22})W_{13}),$$

with $W_{13} = 2(U_{11} + U_{33}) - V_{13}$, respectively. We note that these expressions remain valid over any ring despite the fact that they were derived via the factorization through the curve in $\mathbb{P}^1 \times \mathbb{P}^1$ which is singular in characteristic 2.

Before evaluating their complexity, we explain the obvious symmetry of the above equations. Let τ be the translation-by-R automorphism of C^t sending $(X_0 : X_1 : X_2 : X_3)$ to

$$(X_2 : -X_3 - 2a(X_1 + X_3) : -DX_0 : X_1 + 2a(X_1 + X_3)),$$

and denote also τ for the induced automorphism

$$\tau((X : Z), (Y : W)) = ((Z : X), (-W : DY))$$

of its image in $\mathbb{P}^1 \times \mathbb{P}^1$. Then for each (i, j) in $(\mathbb{Z}/2\mathbb{Z})^2$, the tuple of morphisms $(\tau^i \times \tau^j, \tau^k)$ such that $k = i + j$ acts on the set of tuples $(\mathfrak{s}, \mathfrak{s}')$ of addition law projections:

$$(\tau^i \times \tau^j, \tau^k) \cdot (\mathfrak{s}, \mathfrak{s}') = \tau^k \circ (\mathfrak{s} \circ (\tau^i \times \tau^j), \mathfrak{s}' \circ (\tau^i \times \tau^j)).$$

Lemma 13. *Let C^t be an elliptic curve in split μ_4-normal form. The tuples of addition law projections $(\mathfrak{s}_0, \mathfrak{s}_1)$ and $(\mathfrak{s}_2, \mathfrak{s}_3)$ are eigenvectors for the action of $(\tau \times \tau, 1)$ and are exchanged, up to scalars, by the action of $(\tau \times 1, \tau)$ and $(1 \times \tau, \tau)$.*

Proof. Since an addition law (projection) is uniquely determined by its exceptional divisor, up to scalars, the lemma follows from the action of $(\tau^i \times \tau^j, \tau^k)$ on the exceptional divisors given by Lemma 31 of Kohel [8], and can be established directly by substitution.   □

Corollary 14. *Let C^t be an elliptic curve in twisted split μ_4-normal form. There exists an algorithm for addition with complexity $11\mathbf{M} + 2\mathbf{m}$ over any ring, and an algorithm with complexity $9\mathbf{M} + 2\mathbf{m}$ over a ring in which 2 is a unit.*

Proof. Considering the product determined by the pair $(\mathfrak{s}_2, \mathfrak{s}_3)$, the evaluation of the expressions

$$Z_0 = (U_{00} - DU_{22})(U_{00} + DU_{22}),$$
$$Z_2 = (U_{11} - U_{33})(U_{11} + U_{33} + 2aV_{13}),$$

requires $4\mathbf{M}$ for the U_{ii} plus $1\mathbf{M}$ for V_{13} if $a \neq 0$, then $2\mathbf{M}$ for the evaluation of Z_0 and Z_2. Setting $W_{13} = 2(U_{11} + U_{33}) - V_{13}$, a direct evaluation of the expressions

$$Z_1 = c((U_{00}U_{11} - DU_{22}U_{33}) - a(U_{00} - DU_{22})W_{13}),$$
$$Z_3 = c((U_{00}U_{33} - DU_{22}U_{11}) - a(U_{00} - DU_{22})W_{13}),$$

requires an additional $4\mathbf{M} + 2\mathbf{m}$, saving $1\mathbf{M}$ with the relation

$$(U_{00} - DU_{22})(U_{11} + U_{33}) = (U_{00}U_{11} - DU_{22}U_{33}) + (U_{00}U_{33} - DU_{22}U_{11}),$$

for a complexity of $11\mathbf{M} + 2\mathbf{m}$. If 2 is a unit, we may instead compute

$$Z_1 + Z_3 = c\,(U_{00} - DU_{22})(U_{11} + U_{33} + 2aV_{13}),$$
$$Z_1 - Z_3 = c\,(U_{00} + DU_{22})(U_{11} - U_{33}).$$

and return $(2Z_0, 2Z_1, 2Z_2, 2Z_3)$ using $2\mathbf{M} + 2\mathbf{m}$, for a total cost of $9\mathbf{M} + 2\mathbf{m}$. $\square$

Corollary 15. *Let C^t be an elliptic curve in twisted split $\boldsymbol{\mu_4}$-normal form. There exists an algorithm for doubling with complexity $6\mathbf{M} + 5\mathbf{S} + 2\mathbf{m}$ over any ring, and an algorithm with complexity $4\mathbf{M} + 5\mathbf{S} + 2\mathbf{m}$ over a ring in which 2 is a unit.*

Proof. The specialization to $X_i = Y_i$ gives:

$$Z_0 = (X_0^2 - DX_2^2)(X_0^2 + DX_2^2),$$
$$Z_2 = (X_1^2 - X_3^2)(X_1^2 + X_3^2 + 2a(X_1 + X_3)^2).$$

The evaluation of X_i^2 costs $4\mathbf{S}$ plus $1\mathbf{S}$ for $(X_1 + X_3)^2$ if $a \neq 0$, rather than $4\mathbf{M} + 1\mathbf{M}$. Setting $W_{13} = 2(X_1^2 + X_3^2) - (X_1 + X_3)^2\,[=(X_1 - X_3)^2]$, a direct evaluation of the expressions

$$Z_1 = c((X_0^2 X_1^2 - DX_2^2 X_3^2) - a(X_0^2 - DX_2^2)W_{13}),$$
$$Z_3 = c((X_0^2 X_3^2 - DX_2^2 X_1^2) - a(X_0^2 - DX_2^2)W_{13}),$$

requires an additional $4\mathbf{M} + 2\mathbf{m}$, as above, for a complexity of $6\mathbf{M} + 5\mathbf{S} + 2\mathbf{m}$. If 2 is a unit, we compute

$$Z_1 + Z_3 = c\,(X_0^2 - DX_2^2)(X_1^2 + X_3^2 + 2a(X_1 + X_3)^2),$$
$$Z_1 - Z_3 = c\,(X_0^2 + DX_2^2)(X_1^2 - X_3^2).$$

using $2\mathbf{M} + 2\mathbf{m}$, which gives $4\mathbf{M} + 5\mathbf{S} + 2\mathbf{m}$. $\square$

In the next section we explore efficient algorithms for evaluation of the addition laws and doubling forms in characteristic 2.

5   Binary Addition Algorithms

Suppose that k is a finite field of characteristic 2. The Artin-Schreier extension $k[\omega]/k$ over which we twist is determined by the additive properties of a, and half of all elements of k determine the same field (up to isomorphism) and hence an isomorphic twist. For instance, if $k/\mathbb{F}_2$ is an odd degree extension, we may take $a = 1$. As above, we assume that that multiplication by a is negligible in our complexity analyses.

Theorem 16. *Let C^t be an elliptic curve in twisted split $\boldsymbol{\mu}_4$-normal form:*

$$X_0^2 + X_2^2 = c^2(X_1X_3 + a(X_1 + X_3)^2), \ X_1^2 + X_3^2 = c^2X_0X_2,$$

over a field of characteristic 2. A complete system of addition laws is given by the two maps $\mathfrak{s}_0$ and $\mathfrak{s}_2$,

$$((U_{13} + U_{31})^2, c(U_{02}U_{31} + U_{20}U_{13} + aF), (U_{02} + U_{20})^2, c(U_{02}U_{13} + U_{20}U_{31} + aF)),$$
$$((U_{00} + U_{22})^2, c(U_{00}U_{11} + U_{22}U_{33} + aG), (U_{11} + U_{33})^2, c(U_{00}U_{33} + U_{11}U_{22} + aG)),$$

respectively, where $U_{jk} = X_jY_k$ and

$$F = (X_1 + X_3)(Y_1 + Y_3)(U_{02} + U_{20}) \text{ and } G = (X_1 + X_3)(Y_1 + Y_3)(U_{00} + U_{22}).$$

The respective exceptional divisors are $4\Delta_O$ and $4\Delta_S$ where $S = (1 : c : 1 : 0)$ is a 2-torsion point.

Proof. The addition laws $\mathfrak{s}_0$ and $\mathfrak{s}_2$ are the conjugate addition laws of Theorem 4 (as can be verified symbolically)[1] and, equivalently, are described by the reduction at 2 of the addition laws derived from the tuples of addition law projections $(\mathfrak{s}_0, \mathfrak{s}_1)$ and $(\mathfrak{s}_2, \mathfrak{s}_3)$ of Theorem 12. Since the points O and S are fixed rational points of the twisting morphism, the exceptional divisors are of the same form. As the exceptional divisors are disjoint, the pair of addition laws form a complete set.   □

Remark. Recall that the addition laws $\mathfrak{s}_1$ and $\mathfrak{s}_3$ on the split $\boldsymbol{\mu}_4$-normal form have exceptional divisors $4\Delta_T$ and $4\Delta_{-T}$ in characteristic 2 (since $S = O$). Consequently their conjugation by the twisting morphism yields a conjugate pair over $k[\omega]$, since the twisted curve does not admit a k-rational 4-torsion point T. There exist linear combinations of these twisted addition laws which extend the set $\{\mathfrak{s}_0, \mathfrak{s}_2\}$ to a basis over k (of the space of dimension four), but they do not have such an elegant form as $\mathfrak{s}_0$ and $\mathfrak{s}_2$.

Corollary 17. *Let C^t be an elliptic curve in twisted split $\boldsymbol{\mu}_4$-normal form over a field of characteristic 2. There exists an algorithm for addition with complexity $9\mathbf{M} + 2\mathbf{S} + 2\mathbf{m}$.*

Proof. Since the addition laws differ from the split $\boldsymbol{\mu}_4$-normal form only by the term aF (or aG), it suffices to determine the complexity of its evaluation. Having determined (U_{02}, U_{20}) (or (U_{00}, U_{22})), we require an additional $2\mathbf{M}$, which gives the complexity bound.   □

For the $\boldsymbol{\mu}_4$-normal form the addition law, after coefficient scaling, we find that the addition law with exceptional divisor $4\Delta_O$ takes the form

$$((U_{13} + U_{31})^2, U_{02}U_{31} + U_{20}U_{13} + aF, (U_{20} + U_{02})^2, U_{02}U_{13} + U_{20}U_{31} + aG),$$

and in particular does not involve multiplication by constants (other than a which we may take in $\{0,1\}$ in cryptographic applications). This gives the following complexity result.

---

[1]   As is verified by the implementation in Echidna [11] written in Magma [14].

Corollary 18. *Let C^t be an elliptic curve in twisted μ_4-normal form over a field of characteristic 2. There exists an algorithm for addition outside of the diagonal Δ_O with complexity* 9M + 2S.

6   Binary Doubling Algorithms

We recall the hypothesis that multiplication by a is negligible. In the cryptographic context (e.g. in application to the binary NIST curves), we may assume $a = 1$ (or $a = 0$ for the untwisted forms).

Corollary 19. *Let C^t be an elliptic curve in twisted split μ_4-normal form. The doubling map is uniquely determined by*

$$((X_0 + X_2)^4 : c((X_0X_3 + X_1X_2)^2 + a(X_0 + X_2)^2(X_1 + X_3)^2) :$$
$$(X_1 + X_3)^4 : c((X_0X_1 + X_2X_3)^2 + a(X_0 + X_2)^2(X_1 + X_3)^2))$$

Proof. This follows from specializing $X_j = Y_j$ in the form $\mathfrak{s}_2$ of Theorem 16. □

We note that in cryptographic applications we may assume that $a = 0$ (untwisted form), giving

$$((X_0 + X_2)^4 : c(X_0X_3 + X_2X_1)^2 : (X_1 + X_3)^4 : c(X_0X_1 + X_2X_3)^2),$$

and otherwise $a = 1$, in which case we have

$$((X_0 + X_2)^4 : c(X_0X_1 + X_2X_3)^2 : (X_1 + X_3)^4 : c(X_0X_3 + X_2X_1)^2).$$

It is clear that the evaluation of doubling on the twisted and untwisted normal forms is identical. This is true also for the case of general a, up to the computation of $(X_0 + X_2)^2(X_1 + X_3)^2$. We nevertheless give an algorithm which improves upon the number of constant multiplications reported in Kohel [10], in terms of polynomials in $u = c^{-1}$. With this notation, we note that the defining equations of the curve are:

$$X_1X_3 = u^2(X_0 + X_2)^2,$$
$$X_0X_2 = u^2(X_1 + X_3)^2.$$

These relations are important, since they permit us to replace any instances of the multiplications on the left with the squarings on the right. As a consequence, we have

$$X_0X_1 + X_2X_3 = (X_0 + X_3)(X_2 + X_1) + X_0X_2 + X_1X_3$$
$$= (X_0 + X_3)(X_2 + X_1) + u^2((X_0 + X_2)^2 + (X_1 + X_3)^2)$$
$$X_0X_3 + X_2X_1 = (X_0 + X_1)(X_2 + X_3) + X_0X_2 + X_1X_3$$
$$= (X_0 + X_1)(X_2 + X_3) + u^2((X_0 + X_2)^2 + (X_1 + X_3)^2).$$

Moreover these forms are linearly dependent with $(X_0 + X_2)(X_1 + X_3)$

$$(X_0X_1 + X_2X_3) + (X_0X_3 + X_2X_1) = (X_0 + X_2)(X_1 + X_3),$$

so that two multiplications are sufficient for the determination of these three forms. Putting this together, it suffices to evaluate the tuple

$$(u(X_0 + X_2)^4, (X_0X_1 + X_2X_3)^2, u(X_1 + X_3)^4, (X_0X_3 + X_2X_1)^2),$$

for which we obtain the following complexity for doubling.

Corollary 20. *Let C^t be a curve in twisted split $\boldsymbol{\mu}_4$-normal form. There exists an algorithm for doubling with complexity $2\mathbf{M} + 5\mathbf{S} + 3\mathbf{m}_u$.*

Using the semisplit $\boldsymbol{\mu}_4$-normal form, the complexity of $2\mathbf{M} + 5\mathbf{S} + 2\mathbf{m}_u$ of Kohel [10], saving one constant multiplication, carries over to the corresponding twisted semisplit $\boldsymbol{\mu}_4$-normal form (referred to as nonsplit). By a similar argument the same complexity, $2\mathbf{M} + 5\mathbf{S} + 2\mathbf{m}_u$, is obtained for the $\boldsymbol{\mu}_4$-normal form of this article.

7   Montgomery Endomorphisms of Kummer Products

We recall certain results of Kohel [10] concerning the Montgomery endomorphism with application to scalar multiplication on products of Kummer curves. We define the Montgomery endomorphism to be the map $\varphi : C \times C \to C \times C$ given by $(Q, R) \mapsto (2Q, Q + R)$. With a view to scalar multiplication, this induces

$$((n + 1)P, nP) \longmapsto ((2n + 2)P, (2n + 1)P),$$

and

$$(nP, (n + 1)P) \longmapsto (2nP, (2n + 1)P).$$

By exchanging the order of the coordinates on input and output, an algorithm for the Montgomery endomorphism computes $((2n + 2)P, (2n + 1)P)$ or $((2n+1)P, 2nP)$ from the input point $((n+1)P, nP)$. This allows us to construct a symmetric algorithm for the scalar multiple kP of P via a Montgomery ladder

$$((n_i + 1)P, n_iP) \longmapsto ((n_{i+1} + 1)P, n_{i+1}P) = \begin{cases} ((2n_i + 1)P, 2n_iP), \text{ or} \\ ((2n_i + 2)P, (2n_i + 1)P). \end{cases}$$

It is noted that the Montgomery endomorphism sends each of the curves

$$\Delta_P = \{(Q, Q - P) \mid Q \in C(\bar{k})\}, \text{ and } \Delta_{-P} = \{(Q, Q - P) \mid Q \in C(\bar{k})\},$$

to itself, and exchange of coordinates induces $\Delta_P \to \Delta_{-P}$.

We now assume that C is a curve in split $\boldsymbol{\mu}_4$-normal form, and define the Kummer curve $\mathscr{K}(C) = C/\{\pm 1\} \cong \mathbb{P}^1$, equipped with map

$$\pi((X_0 : X_1 : X_2 : X_3)) = \begin{cases} (cX_0 : X_1 + X_3), \\ (X_1 - X_3 : cX_2). \end{cases}$$

This determines a curve $\mathscr{K}(\Delta_P)$ as the image of Δ_P in $\mathscr{K}(C) \times \mathscr{K}(C)$.

Lemma 21. *For any point P of C, the Montgomery-oriented curve $\mathcal{K}(\Delta_P)$ equals $\mathcal{K}(\Delta_{-P})$.*

Proof. It suffices to note that $(\overline{Q}, \overline{Q-P}) \in \mathcal{K}(\Delta_P)(\bar{k})$ is also a point of $\mathcal{K}(\Delta_{-P})$:

$$(\overline{Q}, \overline{Q-P}) = (\overline{-Q}, \overline{-Q+P}) = (\overline{-Q}, \overline{-Q-(-P)}) \in \mathcal{K}(\Delta_{-P}),$$

hence $\mathcal{K}(\Delta_P) \subseteq \mathcal{K}(\Delta_{-P})$ and by symmetry $\mathcal{K}(\Delta_{-P}) \subseteq \mathcal{K}(\Delta_P)$. $\quad\blacksquare$

We conclude, moreover, that $\mathcal{K}(\Delta_P)$ is well-defined by a point on the Kummer curve.

Lemma 22. *The Montgomery-oriented curve $\mathcal{K}(\Delta_P)$ depends only on $\pi(P)$.*

Proof. The dependence only on $\pi(P)$ is a consequence of the previous lemmas, which we make explicit here. Let $P = (s_0 : s_1 : s_2 : s_3)$ and $\pi(P) = (t_0 : t_1)$. By Theorem 24 of Kohel [10], the curve $\mathcal{K}(\Delta_P)$ takes the form,

$$s_0(U_0V_1 + U_1V_0)^2 + s_2(U_0V_0 + U_1V_1)^2 = c(s_1 + s_3)U_0U_1V_0V_1,$$

but then $(s_0 : s_1 + s_3 : s_2) = (t_0^2 : ct_0t_1, t_1^2)$ in $\mathbb{P}^2$, hence

$$t_0^2(U_0V_1 + U_1V_0)^2 + t_1^2(U_0V_0 + U_1V_1)^2 = c^2t_0t_1U_0U_1V_0V_1.$$

which shows that the curve depends only on $\pi(P)$. $\quad\blacksquare$

We note similarly that the Kummer curve $\mathcal{K}(C) = \mathcal{K}(C^t)$ is independent of the quadratic twist, in the sense that any twisting isomorphism $\tau : C \to C^t$ over $\bar{k}$ induces a unique isomorphism $\mathcal{K}(C) \to \mathcal{K}(C^t)$. One can verify directly the twisting isomorphism τ of Theorem 7 induces the identity on the Kummer curves with their given projections. We thus identify $\mathcal{K}(C) = \mathcal{K}(C^t)$, and denote $\pi : C \to \mathcal{K}(C)$ and $\pi^t : C^t \to \mathcal{K}(C)$ the respective covers of the Kummer curve.

Theorem 23. *Let C be a curve in split μ_4-normal form and C^t be a quadratic twist over the field k. If $P \in C^t(\bar{k})$ and $Q \in C(\bar{k})$ such that $\pi^t(P) = \pi(Q)$, then $\mathcal{K}(\Delta_P) = \mathcal{K}(\Delta_Q)$.*

It follows that we can evaluate the Montgomery endomorphism on $\mathcal{K}(\Delta_P)$, for $P \in C^t(k)$, and $\pi(P) = (t_0 : t_1)$, using the same algorithm and with the same complexity as in Kohel [10]. We recall the complexity result here, assuming a normalisation $t_0 = 1$ or $t_1 = 1$.

Corollary 24. *The Montgomery endomorphism on $\mathcal{K}(\Delta_P)$ can be computed with $4\mathbf{M} + 5\mathbf{S} + 1\mathbf{m}_t + 1\mathbf{m}_c$ or with $4\mathbf{M} + 4\mathbf{S} + 1\mathbf{m}_t + 2\mathbf{m}_c$.*

By the same argument, the same Theorem 24 of Kohel [10] provides the necessary map for point recovery in terms of the input point $P = (s_0 : s_1 : s_2 : s_3)$ of $C^t(k)$.

Theorem 25. *Let C^t be an elliptic curve in twisted split μ_4-normal form with rational point $P = (s_0 : s_1 : s_2 : s_3)$. If P is not a 2-torsion point, the morphism $\lambda : C \to \mathcal{K}(\Delta_P)$ is an isomorphism, and defined by*

$$\pi_1 \circ \lambda(X_0 : X_1 : X_2 : X_3) = \begin{cases} (cX_0 : X_1 + X_3), \\ (X_1 + X_3 : cX_2), \end{cases}$$

$$\pi_2 \circ \lambda(X_0 : X_1 : X_2 : X_3) = \begin{cases} (s_0X_0 + s_2X_2 : s_1X_1 + s_3X_3), \\ (s_3X_1 + s_1X_3 : s_2X_0 + s_0X_2), \end{cases}$$

with inverse $\lambda^{-1}((U_0 : U_1), (V_0 : V_1))$ *equal to*

$$\begin{cases} ((s_1 + s_3)U_0^2V_0 : (s_0U_0^2 + s_2U_1^2)V_1 + cs_1U_0U_1V_0 : (s_1 + s_3)U_1^2V_0 : (s_0U_0^2 + s_2U_1^2)V_1 + cs_3U_0U_1V_0), \\ ((s_1 + s_3)U_0^2V_1 : (s_2U_0^2 + s_0U_1^2)V_0 + cs_3U_0U_1V_1 : (s_1 + s_3)U_1^2V_1 : (s_2U_0^2 + s_0U_1^2)V_0 + cs_1U_0U_1V_1). \end{cases}$$

This allows for the application of the Montgomery endomorphism to scalar multiplication on C^t. Using the best results of the present work, the complexity is comparable to a double and add algorithm with window of width 4.

8   Conclusion

Elliptic curves in the twisted μ_4-normal form of this article (including split and semisplit variants) provide models for curves which, on the one hand, are isomorphic to twisted Edwards curves with efficient arithmetic over non-binary fields, and, on the other, have good reduction and efficient arithmetic in characteristic 2.

Taking the best reported algorithms from the EFD [4], we conclude with a tabular comparison of the previously best known complexity results for doubling and addition algorithms on projective curves (see Table 1). We include the projective lambda model (a singular quartic model in $\mathbb{P}^2$), which despite the extra cost of doubling, admits a slightly better algorithm for addition than López-Dahab (see [15]). Binary Edwards curves [3], like the twisted μ_4-normal form of this work, cover all ordinary curves, but the best complexity result we give here is for $d_1 = d_2$ which has a rational 4-torsion point (corresponding to the trivial twist, for which the μ_4-normal form gives better performance). Similarly, the López-Dahab model with $a_2 = 0$ admits a rational 4-torsion point, hence covers the same classes, but the fastest arithmetic is achieved on the quadratic twists with $a_2 = 1$, which manage to save one squaring **S** for doubling relative to the present work, at the loss of generality (one must vary the weighted projective space according to the twist, $a_2 = 0$ or $a_2 = 1$) and with a large penalty for the cost of addition. The results stated here concern the twisted μ_4-normal form which minimize the constant multiplications. In the final columns, we indicate the fractions of ordinary curves covered by the model (assuming a binary field of odd degree), and whether the family includes the NIST curves.

Table 1. Table of binary doubling and addition algorithm complexities.

Curve model	Doubling	Addition	%	NIST
Lambda coordinates	3M + 4S + 1m	11M + 2S	100%	✓
Binary Edwards ($d_1 = d_2$)	2M + 5S + 2m	16M + 1S + 4m	50%	✗
López-Dahab ($a_2 = 0$)	2M + 5S + 1m	14M + 3S	50%	✗
López-Dahab ($a_2 = 1$)	2M + 4S + 2m	13M + 3S	50%	✓
Twisted μ_4-normal form	2M + 5S + 2m	9M + 2S	100%	✓
μ_4-normal form	2M + 5S + 2m	7M + 2S	50%	✗

All curves can be represented in lambda coordinates or in μ_4-normal form. However by considering the two cases $a_2 \in \{0,1\}$, as for the López-Dahab models, the twists of the μ_4-normal form with $a_2 = 0$ give the faster μ_4-normal form and only when $a_2 = 1$ does one need the twisted model with its reduced complexity.

By consideration of twists, we are able to describe a uniform family of curves which capture nearly optimal known doubling performance of binary curves (up to 1S), while vastly improving the performance of addition algorithms applicable to all binary curves. By means of a trivial encoding in twisted μ_4-normal form (see Corollary 10), this brings efficient arithmetic of these μ_4-normal forms to binary NIST curves.

References

1. Ankeny, N.C.: The least quadratic non residue. Ann. Math. Second Ser. **55**(1), 65–72 (1952)
2. Bernstein, D.J., Lange, T.: Faster addition and doubling on elliptic curves. In: Kurosawa, K. (ed.) ASIACRYPT 2007. LNCS, vol. 4833, pp. 29–50. Springer, Heidelberg (2007). doi:10.1007/978-3-540-76900-2_3
3. Bernstein, D.J., Lange, T., Rezaeian Farashahi, R.: Binary edwards curves. In: Oswald, E., Rohatgi, P. (eds.) CHES 2008. LNCS, vol. 5154, pp. 244–265. Springer, Heidelberg (2008). doi:10.1007/978-3-540-85053-3_16
4. Bernstein, D.J., Lange, T.: Explicit formulas database. http://www.hyperelliptic. org/EFD/
5. Bernstein, D.J., Birkner, P., Joye, M., Lange, T., Peters, C.: Twisted edwards curves. In: Vaudenay, S. (ed.) AFRICACRYPT 2008. LNCS, vol. 5023, pp. 389–405. Springer, Heidelberg (2008). doi:10.1007/978-3-540-68164-9_26
6. Edwards, H.: A normal form for elliptic curves. Bull. Am. Math. Soc. **44**, 393–422 (2007)
7. Hisil, H., Wong, K.K.-H., Carter, G., Dawson, E.: Twisted edwards curves revisited. In: Pieprzyk, J. (ed.) ASIACRYPT 2008. LNCS, vol. 5350, pp. 326–343. Springer, Heidelberg (2008). doi:10.1007/978-3-540-89255-7_20
8. Kohel, D.: Addition law structure of elliptic curves. J. Number Theory **131**(5), 894–919 (2011)

9. Kohel, D.: A normal form for elliptic curves in characteristic 2. talk at Arithmetic, Geometry, Cryptography and Coding Theory, Luminy, 15 March 2011. http://iml. univ-mrs.fr/kohel/pub/normal_form.pdf

10. Kohel, D.: Efficient arithmetic on elliptic curves in characteristic 2. In: Galbraith, S., Nandi, M. (eds.) INDOCRYPT 2012. LNCS, vol. 7668, pp. 378–398. Springer, Heidelberg (2012). doi:10.1007/978-3-642-34931-7_22

11. Kohel, D., et al.: Echidna algorithms, v. 5.0 (2016). http://echidna.maths.usyd. edu.au/echidna/index.html

12. Lagarias, J.C., Montgomery, H.L., Odlyzko, A.M.: A bound for the least prime ideal in the Chebotarev density theorem. Invent. Math. **54**, 271–296 (1979)

13. Lange, H., Ruppert, W.: Complete systems of addition laws on abelian varieties. Invent. Math. **79**(3), 603–610 (1985)

14. Magma Computational Algebra System (Version 2.20) (2015). http://magma. maths.usyd.edu.au/magma/handbook/

15. Oliveira, T., López, J., Aranha, D.F., Rodríguez-Henríquez, F.: Lambda coordinates for binary elliptic curves. In: Bertoni, G., Coron, J.-S. (eds.) CHES 2013. LNCS, vol. 8086, pp. 311–330. Springer, Heidelberg (2013). doi:10.1007/978-3-642-40349-1_18

Efficient Compression of SIDH Public Keys

Craig Costello[1]([⊠]), David Jao[2,3], Patrick Longa[1], Michael Naehrig[1],
Joost Renes[4], and David Urbanik[2]

[1] Microsoft Research, Redmond, WA, USA
{craigco,plonga,mnaehrig}@microsoft.com
[2] Centre for Applied Cryptographic Research, University of Waterloo,
Waterloo, ON, Canada
{djao,dburbani}@uwaterloo.ca
[3] evolutionQ, Inc., Waterloo, ON, Canada
david.jao@evolutionq.com
[4] Digital Security Group, Radboud University, Nijmegen, The Netherlands
j.renes@cs.ru.nl

Abstract. Supersingular isogeny Diffie-Hellman (SIDH) is an attractive candidate for post-quantum key exchange, in large part due to its relatively small public key sizes. A recent paper by Azarderakhsh, Jao, Kalach, Koziel and Leonardi showed that the public keys defined in Jao and De Feo's original SIDH scheme can be further compressed by around a factor of two, but reported that the performance penalty in utilizing this compression blew the overall SIDH runtime out by more than an order of magnitude. Given that the runtime of SIDH key exchange is currently its main drawback in relation to its lattice- and code-based post-quantum alternatives, an order of magnitude performance penalty for a factor of two improvement in bandwidth presents a trade-off that is unlikely to favor public-key compression in many scenarios.

In this paper, we propose a range of new algorithms and techniques that accelerate SIDH public-key compression by more than an order of magnitude, making it roughly as fast as a round of standalone SIDH key exchange, while further reducing the size of the compressed public keys by approximately 12.5%. These improvements enable the practical use of compression, achieving public keys of only 330 bytes for the concrete parameters used to target 128 bits of quantum security and further strengthens SIDH as a promising post-quantum primitive.

Keywords: Post-quantum cryptography · Diffie-Hellman key exchange · Supersingular elliptic curves · Isogenies · SIDH · Public-key compression · Pohlig-Hellman algorithm

---

D. Jao and D. Urbanik—Partially supported by NSERC, CryptoWorks21, and Public Works and Government Services Canada.

J. Renes—Partially supported by the Technology Foundation STW (project 13499 – TYPHOON & ASPASIA), from the Dutch government. Part of this work was done while Joost was an intern at Microsoft Research.

J.-S. Coron and J.B. Nielsen (Eds.): EUROCRYPT 2017, Part I, LNCS 10210, pp. 679–706, 2017.
DOI: 10.1007/978-3-319-56620-7_24

1 Introduction

In their February 2016 report on post-quantum cryptography [6], the United States National Institute of Standards and Technology (NIST) stated that *"It seems improbable that any of the currently known [public-key] algorithms can serve as a drop-in replacement for what is in use today,"* citing that one major challenge is that quantum resistant algorithms have larger key sizes than the algorithms they will replace. While this statement is certainly applicable to many of the lattice- and code-based schemes (e.g., LWE encryption [24] and the McEliece cryptosystem [19]), Jao and De Feo's 2011 supersingular isogeny Diffie-Hellman (SIDH) proposal [15] is one post-quantum candidate that could serve as a drop-in replacement to existing Internet protocols. Not only are high-security SIDH public keys smaller than their lattice- and code-based counterparts, they are even smaller than some of the traditional (i.e., finite field) Diffie-Hellman public keys.

SIDH Public-Key Compression. The public keys defined in the original SIDH papers [8,15] take the form

$$PK = (E, P, Q),$$

where $E/\mathbb{F}_{p^2} : y^2 = x^3 + ax + b$ is a supersingular elliptic curve, $p = n_A n_B \pm 1$ is a large prime, the cardinality of E is $\#E(\mathbb{F}_{p^2}) = (p \mp 1) = (n_A n_B)^2$, and depending on whether the public key corresponds to Alice or Bob, the points P and Q either both lie in $E(\mathbb{F}_{p^2})[n_A]$, or both lie in $E(\mathbb{F}_{p^2})[n_B]$. Since P and Q can both be transmitted via their x-coordinates (together with a sign bit that determines the correct y-coordinate), and the curve can be transmitted by sending the two $\mathbb{F}_{p^2}$ elements a and b, the original SIDH public keys essentially consist of four $\mathbb{F}_{p^2}$ elements, and so are around $8 \log p$ bits in size.

A recent paper by Azarderakhsh, Jao, Kalach, Koziel and Leonardi [2] showed that it is possible to compress the size of SIDH public keys to around $4 \log p$ bits as follows. Firstly, to send the supersingular curve E, they pointed out that one can send the j-invariant $j(E) \in \mathbb{F}_{p^2}$ rather than $(a, b) \in \mathbb{F}_{p^2}^2$, and showed how to recover a and b (uniquely, up to isomorphism) from $j(E)$ on the other side. Secondly, for $n \in \{n_A, n_B\}$, they showed that since $E(\mathbb{F}_{p^2})[n] \cong \mathbb{Z}_n \times \mathbb{Z}_n$, an element in $E(\mathbb{F}_{p^2})[n]$ can instead be transmitted by sending two scalars $(\alpha, \beta) \in \mathbb{Z}_n \times \mathbb{Z}_n$ that determine its representation with respect to a basis of the torsion subgroup. This requires that Alice and Bob have a way of arriving at the same basis for $E(\mathbb{F}_{p^2})[n]$. Following [2], we note that it is possible to decompose points into their $\mathbb{Z}_n \times \mathbb{Z}_n$ representation since for well-chosen SIDH parameters, $n = \ell^e$ is always smooth, which means that discrete logarithms in order n groups can be solved in polynomial time using the Pohlig-Hellman algorithm [23]. Given that such SIDH parameters have $n_A \approx n_B$ (see [15]), it follows that $n \approx \sqrt{p}$ and that sending elements of $E(\mathbb{F}_{p^2})[n]$ as two elements of $\mathbb{Z}_n$ (instead of an element in $\mathbb{F}_{p^2}$) cuts the bandwidth required to send torsion points in half.

Although passing back and forth between (a, b) and $j(E)$ to (de)compress the curve is relatively inexpensive, the compression of the points P and Q requires three computationally intensive steps:

- *Step 1 – Constructing the n-torsion basis.* During both compression and decompression, Alice and Bob must, on input of the curve E, use a deterministic method to generate the same two-dimensional basis $\{R_1, R_2\} \in E(\mathbb{F}_{p^2})[n]$. The method used in [2] involves systematically sampling candidate points $R \in E(\mathbb{F}_{p^2})$, performing cofactor multiplication by h to move into $E(\mathbb{F}_{p^2})[n]$, and then testing whether or not $[h]R$ has "full" order n (and, if not, restarting).
- *Step 2 – Pairing computations.* After computing a basis $\{R_1, R_2\}$ of the group $E(\mathbb{F}_{p^2})[n]$, the task is to decompose the point P (and identically, Q) as $P = [\alpha_P]R_1 + [\beta_P]R_2$ and determine (α_P, β_P). While this could be done by solving a two-dimensional discrete logarithm problem (DLP) directly on the curve, Azarderakhsh et al. [2] use a number of Weil pairing computations to transform these instances into one-dimensional finite field DLPs in $\mu_n \subset \mathbb{F}_{p^2}^*$.
- *Step 3 – Solving discrete logarithms in μ_n.* The last step is to repeatedly use the Pohlig-Hellman algorithm [23] to solve DLPs in μ_n, and to output the four scalars α_P, β_P, α_Q and β_Q in $\mathbb{Z}_n$.

Each one of these steps presents a significant performance drawback for SIDH public-key compression. Subsequently, Azarderakhsh et al. report that, at interesting levels of security, each party's individual compression latency is more than a factor of ten times the latency of a full round of uncompressed key exchange [2, Sect. 5].

Our Contributions. We present a range of new algorithmic improvements that decrease the total runtime of SIDH compression and decompression by an order of magnitude, bringing its performance close to that of a single round of SIDH key exchange. We believe that this makes it possible to consider public-key compression a default choice for SIDH, and it can further widen the gap between the key sizes resulting from practical SIDH key exchange implementations and their code- and lattice-based counterparts.

We provide a brief overview of our main improvements with respect to the three compression steps described above. All known implementations of SIDH (e.g., [1,7,8]) currently choose $n_A = \ell_A^{e_A} = 2^{e_A}$ and $n_B = \ell_B^{e_B} = 3^{e_B}$ for simplicity and efficiency reasons, so we focus on $\ell \in \{2, 3\}$ below; however, unless specified otherwise, we note that all of our improvements will readily apply to other values of ℓ.

- *Step 1 – Constructing the n-torsion basis.* We make use of some results arising from explicit 2- and 3-descent of elliptic curves to avoid the need for the expensive cofactor multiplication that tests the order of points. These results characterize the images of the multiplication-by-2 and multiplication-by-3 maps on E, and allow us to quickly generate points that are elements of $E(\mathbb{F}_{p^2}) \setminus [2]E(\mathbb{F}_{p^2})$ and $E(\mathbb{F}_{p^2}) \setminus [3]E(\mathbb{F}_{p^2})$. Therefore, we no longer need to check the order of (possibly multiple!) points using a full-length scalar multiplication by $n_A n_B$, but instead are *guaranteed* that one half-length cofactor multiplication produces a point of the correct order. For our purposes, producing points in $E \setminus [2]E$ is as easy as generating elliptic curve points whose x-coordinates are non-square (this is classical, e.g., [14, Chap. 1 (Sect. 4), Theorem 4.1]). On the other hand, to efficiently produce points in $E \setminus [3]E$, we make use of the analogous characteristic described in more recent work on explicit 3-descent by Schaefer and Stoll [26]. Combined with a tailored version of the Elligator 2 encoding [5] for efficiently generating points on E, this approach gives rise to highly efficient n-torsion basis generation. This is described in detail in Sect. 3.

- *Step 2 – Pairing computations.* We apply a number of optimizations from the literature on elliptic curve pairings in order to significantly speed up the runtime of all pairing computations. Rather than using the Weil pairing (as was done in [2]), we use the more efficient Tate pairing [4,10]. We organize the five pairing computations that are required during compression in such a way that only two Miller functions are necessary. Unlike all of the prior work done on optimized pairing computation, the pairings used in SIDH compression cannot take advantage of torsion subgroups that lie in subfields, which means that fast explicit formulas for point operations and Miller line computations are crucial to achieving a fast implementation. Subsequently, we derive new and fast inversion-free explicit formulas for computing pairings on supersingular curves, specific to the scenario of SIDH compression. Following the Miller loops, we compute all five final exponentiations by exploiting a fast combination of Frobenius operations together with either fast repeated cyclotomic squarings (from [31]) or our new formulas for enhanced cyclotomic cubing operations. The pairing optimizations are described in Sect. 4.
- *Step 3 – Solving discrete logarithms in μ_n.* All computations during the Pohlig-Hellman phase take place in the subgroup μ_n of the multiplicative group $G_{p+1} \subset \mathbb{F}_{p^2}^*$ of order $p + 1$, where we take advantage of the fast cyclotomic squarings and cubings mentioned above, as well as the fact that $\mathbb{F}_{p^2}$ inversions are simply conjugations, so come almost for free (see Sect. 5.1). On top of this fast arithmetic, we build an improved version of the Pohlig-Hellman algorithm that exploits windowing methods to solve the discrete logarithm instances with lower asymptotic complexity than the original algorithm. For the concrete parameters, the new algorithm is approximately $14\times$ (resp. $10\times$) faster in $\mu_{2^{372}}$ (resp. $\mu_{3^{239}}$), while having very low memory requirements (see Tables 1 and 2). This is all described in more detail in Sect. 5.
- *Improved compression.* By normalizing the representation of P and Q in $\mathbb{Z}_n^4$, we are able to further compress this part of the public key representation into $\mathbb{Z}_n^3$. Subsequently, our public keys are around $\frac{7}{2}\log p$ bits, rather than the $4\log p$ bits achieved in [2]. To the best of our knowledge, this is as far as SIDH public keys can be compressed in practice. This is explained in Sect. 6.1.
- *Decompression.* The decompression algorithm – which involves only the first of the three steps above and a double-scalar multiplication – is also accelerated in this work. In particular, on top of the faster torsion basis generation,

we show that the double-scalar multiplications can be absorbed into the shared secret computation. This makes them essentially free of cost. This is described in Sect. 6.2.

The combination of the three main improvements mentioned above, along with a number of further optimizations described in the rest of this paper, yields enhanced compression software that is an order of magnitude faster than the initial software benchmarked in [2].

The Compression Software. We wrote the new suite of algorithms in plain C and incorporated the compression software into the SIDH library recently made available by Costello, Longa and Naehrig [7]; their software uses a curve with $\log p = 751$ that currently offers around 192 bits of classical security and 128 bits of quantum security. The public keys in their uncompressed software were $6 \log p = 564$ bytes, while the compressed public keys resulting from our software are $\frac{7}{2} \log p = 330$ bytes. The software described in this paper can be found in the latest release of the SIDH library (version 2.0) at https://www.microsoft.com/en-us/research/project/sidh-library/.

Although our software is significantly faster than the previous compression benchmarks given by Azarderakhsh *et al.* [2], we believe that the most meaningful benchmarks we can present are those that compare the latency of our optimized SIDH compression to the latency of the state-of-the-art key generation and shared secret computations in [7]. This gives the reader (and the PQ audience at large) an idea of the cost of public-key compression when both the raw SIDH key exchange and the optional compression are optimized to a similar level. We emphasize that although the SIDH key exchange software from [7] targeted one isogeny class at one particular security level, and therefore so does our compression software, all of our improvements apply identically to curves used for SIDH at other security levels, especially if the chosen isogeny degrees remain (powers of) 2 and 3. Moreover, we expect that the relative cost of compressed SIDH to uncompressed SIDH will stay roughly consistent across different security levels, and that our targeted benchmarks therefore give a good gauge on the current state-of-the-art.

It is important to note that, unlike the SIDH software from [7] that uses private keys and computes shared secrets, by definition our public-key compression software only operates on public data[1]. Thus, while we call several of their constant-time functions when appropriate, none of our functions need to run in constant-time.

Remark 1 (Ephemeral SIDH). A recent paper by Galbraith, Petit, Shani and Ti [11] gives, among other results, a realistic and serious attack on instantiations of SIDH that reuse static private/public key pairs. Although direct public-key validation in the context of isogeny-based cryptography is currently non-trivial, there are methods of indirect public-key validation (see, e.g., [11,17]) that

---

[1] There is a minor caveat here in that we absorb part of the decompression into the shared secret computation, which uses the constant-time software from [7] – see Sect. 6.

mirror the same technique proposed by Peikert [22, Sect. 5–6] in the context of lattice-based cryptography, which is itself a slight modification of the well-known Fujisaki-Okamoto transform [9]. At present, the software from [7] only supports secure *ephemeral* SIDH key exchange, and does not yet include sufficient (direct or indirect) validation that allows the secure use of static keys. Thus, since our software was written around that of [7], we note that it too is only written for the target application of ephemeral SIDH key exchange. In this case attackers are not incentivized to tamper with public keys, so we can safely assume throughout this paper that all public keys are well-formed. Nevertheless, we note that the updated key exchange protocols in [9,11,17,22] still send values that can be compressed using our algorithms. On a related note, we also point out that our algorithms readily apply to the other isogeny-based cryptosystems described in [8] for which the compression techniques were detailed in [2]. In all of these other scenarios, however, the overall performance ratios and relative bandwidth savings offered by our compression algorithms are likely to differ from those we report for ephemeral SIDH.

Remark 2 (Trading speed for simplicity and space). Since the compression code in our software library only runs on public data, and therefore need not run in constant-time, we use a variable-time algorithm for field inversions (a variant of the extended binary GCD algorithm [16]) that runs faster than the typical exponentiation method (via Fermat's little theorem). Although inversions are used sparingly in our code and are not the bottleneck of the overall compression runtime, we opted to add a single variable-time algorithm in this case. However, during the design of our software library, we made several decisions in the name of simplicity that inevitably hampered the performance of the compression algorithms.

One such performance sacrifice is made during the computation of the torsion basis points in Sect. 3, where tests of quadratic and cubic residuosity are performed using field exponentiations. Here we could use significantly faster, but more complicated algorithms that take advantage of the classic quadratic and cubic reciprocity identities. Such algorithms require intermediate reductions modulo many variable integers, and a reasonably optimized generic reduction routine would increase the code complexity significantly. These tests are also used sparingly and are not the bottleneck of public-key compression, and in this case, we deemed the benefits of optimizing them to be outweighed by their complexity. A second and perhaps the most significant performance sacrifice made in our software is during the Pohlig-Hellman computations, where our windowed version of the algorithm currently fixes small window sizes in the name of choosing moderate space requirements. If larger storage is permitted, then Sutherland's analysis of an optimized version of the Pohlig-Hellman algorithm [32] shows that this phase could be sped up significantly (see Sect. 5). But again, the motivation to push the limits of the Pohlig-Hellman phase is stunted by the prior (pairing computation) phase being the bottleneck of the overall compression routine. Finally, we note that the probabilistic components of the torsion basis generation phase (see Sect. 3) lend themselves to an amended definition of the compressed public keys, where the compressor can send a few extra bits or bytes in their public key to make for a faster and deterministic decompression. For simplicity (and again due to this phase not being the bottleneck of compression), we leave this more complicated adaptation to future consideration.

2 Preliminaries

Here we restrict only to the background that is necessary to understand this paper, i.e., only what is needed to define SIDH public keys. We refer to the extended paper by De Feo, Jao and Plût [8] for a background on the SIDH key exchange computations, and for the rationale behind the parameters given below.

SIDH Public Keys. Let $p = n_A n_B \pm 1$ be a large prime and $E/\mathbb{F}_{p^2}$ be a supersingular curve of cardinality $\#E(\mathbb{F}_{p^2}) = (p \mp 1)^2 = (n_A n_B)^2$. Let $n_A = \ell_A^{e_A}$ and $n_B = \ell_B^{e_B}$. Henceforth we shall assume that $\ell_A = 2$ and $\ell_B = 3$, which is the well-justified choice made in all known implementations to date [1,7,8]; however, unless specified otherwise, we note that the optimizations in this paper will readily apply to other reasonable choices of ℓ_A and ℓ_B. When the discussion is identical irrespective of ℓ_A or ℓ_B, we will often just use ℓ; similarly, we will often just use n when the discussion applies to both n_A and n_B. In general, $E/\mathbb{F}_{p^2}$ is specified using the short Weierstrass model $E/\mathbb{F}_{p^2} : y^2 = x^3 + ax + b$, so is defined by the two $\mathbb{F}_{p^2}$ elements a and b.

During one round of SIDH key exchange, Alice computes her public key as the image E_A of her secret degree-n_A isogeny ϕ_A on a fixed public curve E_0, for example $E_0/\mathbb{F}_{p^2} : y^2 = x^3 + x$, along with the images of ϕ_A on the two public points P_B and Q_B of order n_B, i.e., the points $\phi_A(P_B)$ and $\phi_A(Q_B)$. Bob performs the analogous computation applying his secret degree-n_B isogeny ϕ_B to E_0 to produce the image curve E_B and to produce the images of the public points P_A and Q_A, both of order n_A. In both cases, the public keys are of the form $\mathsf{PK} = (E, P, Q)$, where $E/\mathbb{F}_{p^2}$ is a supersingular elliptic curve transmitted as two $\mathbb{F}_{p^2}$ elements, and P and Q are points on E that are each transmitted as one $\mathbb{F}_{p^2}$ element corresponding to the x-coordinate, along with a single bit that specifies the choice of the corresponding y-coordinate. Subsequently, typical SIDH public keys are specified by 4 $\mathbb{F}_{p^2}$ elements (and two sign bits), and are therefore around $8 \log p$ bits in length.

General SIDH Compression. We now recall the main ideas behind the SIDH public key compression recently presented by Azarderakhsh, Jao, Kalach, Koziel and Leonardi [2]. Their first idea involves transmitting the j-invariant $j(E) \in \mathbb{F}_{p^2}$ of E, rather than the two curve coefficients, and recomputing a and b from $j(E)$ on the other side. However, since $\ell_A = 2$ and therefore $4 \mid \#E$, all curves in the isogeny class can also be written in Montgomery form as $E/\mathbb{F}_{p^2} : By^2 = x^3 + Ax^2 + x$; moreover, since $j(E)$ is independent of B, the implementation described in [7] performs all computations and transmissions ignoring the Montgomery B

coefficient. Although the Weierstrass curve compression in [2] applies in general, the presence of $\ell_A = 2$ in our case allows for the much simpler method of curve compression that simply transmits the coefficient $A \in \mathbb{F}_{p^2}$ of E in Montgomery form.

The second idea from [2], which is the main focus of this paper, is to transmit each of the two points $P, Q \in E(\mathbb{F}_{p^2})[n]$ as their two-dimensional scalar decomposition with respect to a fixed basis $\{R_1, R_2\}$ of $E(\mathbb{F}_{p^2})[n]$. Both of these decompositions are in $\mathbb{Z}_n^2$, requiring $2 \log n$ bits each. But $n \approx \sqrt{p}$ (see [8]), so $2 \log n \approx \log p$ is around half the size of the $2 \log p$ bits needed to transmit a coordinate in $\mathbb{F}_{p^2}$. Of course, the curve in each public key is different, so there is no public basis that can be fixed once-and-for-all, and moreover, to transmit such a basis is as expensive as transmitting the points P and Q in the first place. The whole idea therefore firstly relies on Alice and Bob being able to, on input of a given curve E, arrive at the same basis $\{R_1, R_2\}$ for $E(\mathbb{F}_{p^2})[n]$. In [2] it is proposed to try successive points that result from the use of a deterministic pseudo-random number generator, checking the order of the points each time until two points of exact order n are found. In Sect. 3 we present alternative algorithms that deterministically compute a basis much more efficiently.

Assuming that Alice or Bob have computed the basis $\{R_1, R_2\}$ for $E(\mathbb{F}_{p^2})[n]$, the idea is to now write $P = [\alpha_P]R_1 + [\beta_P]R_2$ and $Q = [\alpha_Q]R_1 + [\beta_Q]R_2$, and to solve these equations for $(\alpha_P, \beta_P, \alpha_Q, \beta_Q) \in \mathbb{Z}_n^4$. To compute α_P and β_P, Azarderakhsh et al. [2] propose first using the Weil pairing $e \colon E(\mathbb{F}_{p^2})[n] \times E(\mathbb{F}_{p^2})[n] \to \mu_n$ to set up the two discrete logarithm instances that arise from the three pairings $e_0 = e(R_1, R_2)$, $e(R_1, P) = e_0^{\beta_P}$, and $e(R_2, Q) = e_0^{-\alpha_P}$; computing α_Q and β_Q then requires two additional pairings, since e_0 can be reused. In Sect. 4 we exploit the fact that these five Weil pairings can be replaced by the much more efficient Tate pairing, and we give an optimized algorithm that computes them all simultaneously.

To finalize compression, it remains to use the Pohlig-Hellman algorithm [23] to solve the four DLP instances in μ_n. In Sect. 5 we present an efficient version of the Pohlig-Hellman algorithm that exploits windowing methods to solve the discrete logarithm instances with lower complexity than the original algorithm. In Sect. 6 we show that one of the four scalars in $\mathbb{Z}_n$ need not be transmitted, since it is always possible to normalize the tuple $(\alpha_P, \beta_P, \alpha_Q, \beta_Q)$ by dividing three of the elements by a determinstically chosen invertible one. The public key is then transmitted as 3 scalars in $\mathbb{Z}_n$ and the curve coefficient $A \in \mathbb{F}_{p^2}$.

SIDH Decompression. The first task of the recipient of a compressed public key is to compute the basis $\{R_1, R_2\}$ in the same way as was done during compression. Once the recipient has computed the basis $\{R_1, R_2\}$, two double-scalar multiplications can be used to recover P and Q. In Sect. 6.2, we show that these double-scalar multiplications can be ommitted by absorbing these scalars into the secret SIDH scalars used for shared secret computations. This further enhances the decompression phase.

Concrete Parameters. As mentioned above, we illustrate our techniques by basing our compression software on the SIDH library recently presented in [7]. This library was built using a specific supersingular isogeny class defined by $p = n_A n_B - 1$, with $n_A = \ell_A^{e_A} = 2^{372}$ and $n_B = \ell_B^{e_B} = 3^{239}$, chosen such that all curves in this isogeny class have order $(n_A n_B)^2$. In what follows we will assume that our parameters correspond to this curve, but reiterate that these techniques will be equally as applicable for any supersingular isogeny class with $\ell_A = 2$ and $\ell_B = 3$.

3   Constructing Torsion Bases

For a given $A \in \mathbb{F}_{p^2}$ corresponding to a supersingular curve $E/\mathbb{F}_{p^2} : y^2 = x^3 + Ax^2 + x$ with $\#E(\mathbb{F}_{p^2}) = (n_A n_B)^2$, the goal of this section is to produce a basis for $E(\mathbb{F}_{p^2})[n]$ (with $n \in \{n_A, n_B\}$) as efficiently as possible. This amounts to computing two order n points R_1 and R_2 whose Weil pairing $e_n(R_1, R_2)$ has exact order n. Checking the order of the Weil pairing either comes for free during subsequent computations, or requires the amendments discussed in Remark 3 at the end of this section. Thus, for now our goal is simplified to efficiently computing points of order $n \in \{n_A, n_B\}$ in $E(\mathbb{F}_{p^2})$.

Let $\{n, n'\} = \{n_A, n_B\}$, write $n = \ell^e$ and $n' = \ell'^{e'}$, and let $\mathcal{O}$ be the identity in $E(\mathbb{F}_{p^2})$. The typical process of computing a point of exact order n is to start by computing $R \in E(\mathbb{F}_{p^2})$ and multiplying by the cofactor n' to compute the candidate output $\tilde{R} = [n']R$. Note that the order of $\tilde{R}$ divides n, but might not be n. Thus, we multiply $\tilde{R}$ by ℓ^{e-1}, and if $[\ell^{e-1}]\tilde{R} \neq \mathcal{O}$, we output $\tilde{R}$, otherwise we must pick a new R and restart.

In this section we use explicit results arising from 2- and 3-descent to show that the cofactor multiplications by n' and by ℓ^{e-1} can be ommitted by making use of elementary functions involving points of order 2 and 3 to check whether points are (respectively) in $E \setminus [2]E$ or $E \setminus [3]E$. In both cases this guarantees that the subsequent multiplication by n' produces a point of exact order n, avoiding the need to perform full cofactor multiplications to check order prior to the pairing computation, and avoiding the need to restart the process if the full cofactor multiplication process above fails to output a point of the correct order (which happens regularly in practice). This yields much faster algorithms for basis generation than those that are used in [2].

We discuss the 2^e-torsion basis generation in Sect. 3.2 and the 3^e-torsion basis generation in Sect. 3.3. We start in Sect. 3.1 by describing some arithmetic ingredients.

3.1   Square Roots, Cube Roots, and Elligator 2

In this section we briefly describe the computation of square roots and that of testing cubic residuosity in $\mathbb{F}_{p^2}$, as well as our tailoring of the Elligator 2 method [5] for efficiently producing points in $E(\mathbb{F}_{p^2})$.

Computing Square Roots and Checking Cubic Residuosity in $\mathbb{F}_{p^2}$.
Square roots in $\mathbb{F}_{p^2}$ are most efficiently computed via two square roots in the
base field $\mathbb{F}_p$. Since $p \equiv 3 \bmod 4$, write $\mathbb{F}_{p^2} = \mathbb{F}_p(i)$ with $i^2 + 1 = 0$. Following [27,
Sect. 3.3], we use the simple identity

$$\sqrt{a + b \cdot i} = \pm (\alpha + \beta \cdot i), \qquad (1)$$

where $\alpha = \sqrt{(a \pm \sqrt{a^2 + b^2})/2}$ and $\beta = b/(2\alpha)$; here $a, b, \alpha, \beta \in \mathbb{F}_p$. Both of
$(a + \sqrt{a^2 + b^2})/2$ and $(a - \sqrt{a^2 + b^2})/2$ will not necessarily be square, so we
make the correct choice by assuming that $z = (a + \sqrt{a^2 + b^2})/2$ is square and
setting $\alpha = z^{(p+1)/4}$; if $\alpha^2 = z$, we output a square root as $\pm(\alpha + \beta i)$, otherwise
we can output a square root as $\pm(\beta - \alpha i)$.

In Sect. 3.3 we will need to efficiently test whether elements $v \in \mathbb{F}_{p^2}$ are cubic
residues or not. This amounts to checking whether $v^{(p^2-1)/3} = 1$ or not, which we
do by first computing $v' = v^{p-1} = v^p/v$ via one application of Frobenius (i.e., $\mathbb{F}_{p^2}$
conjugation) and one $\mathbb{F}_{p^2}$ inversion. We then compute $v'^{(p+1)/3}$ as a sequence of
$e_A = 372$ repeated squarings followed by $e_B - 1 = 238$ repeated cubings. Both of
these squaring and cubing operations are in the order $p + 1$ cyclotomic subgroup
of $\mathbb{F}_{p^2}^*$, so can take advantage of the fast operations described in Sect. 5.1.

Elligator 2. The naïve approach to obtaining points in $E(\mathbb{F}_{p^2})$ is to sequentially test candidate x-coordinates in $\mathbb{F}_{p^2}$ until $f(x) = x^3 + Ax^2 + x$ is square.
Each of these tests requires at least one exponentiation in $\mathbb{F}_p$, and a further
one (to obtain the corresponding y) if $f(x)$ is a square. The Elligator 2 construction deterministically produces points in $E(\mathbb{F}_{p^2})$ using essentially the same
operations, so given that the naïve method can fail (and waste exponentiations),
Elligator 2 performs significantly faster on average.

The idea behind Elligator 2 is to let u be any non-square in $\mathbb{F}_{p^2}$, and for any
$r \in \mathbb{F}_{p^2}$, write

$$v = -\frac{A}{1 + ur^2} \qquad \text{and} \qquad v' = \frac{A}{1 + ur^2} - A. \qquad (2)$$

Then either v is an x-coordinate of a point in $E(\mathbb{F}_{p^2})$, or else v' is [5]; this is
because $f(v)$ and $f(v')$ differ by the non-square factor ur^2.

In our implementation we fix $u = i + 4$ as a system parameter and precompute
a public table consisting of the values $-1/(1 + ur^2) \in \mathbb{F}_{p^2}$ where r^2 ranges from
1 to 10. This table is fixed once-and-for-all and can be used (by any party) to
efficiently generate torsion bases as A varies over the isogeny class. Note that
the size of the table here is overkill, we very rarely need to use more than 3 or
4 table values to produce basis points of the correct exact order.

The key to optimizing the Elligator 2 construction (see [5, Sect. 5.5]) is to
be able to efficiently modify the square root computation in the case that $f(v)$
is non-square, to produce $\sqrt{f(v')}$. This update is less obvious for our field than
in the case of prime fields, but nevertheless achieves the same result. Referring
back to (1), we note that whether or not $a + b \cdot i$ is a square in $\mathbb{F}_{p^2}$ is determined

solely by whether or not $a^2 + b^2$ is a square in $\mathbb{F}_p$ [27, Sect. 3.3]. Thus, if this check deems that $a + bi$ is non-square, we multiply it by $ur^2 = (i+4)r^2$ to yield a square, and this is equivalent to updating $(a, b) = (r(4a - b), r(a + 4b))$, which is trivial in the implementation.

3.2 Generating a Torsion Basis for $E(\mathbb{F}_{p^2})[2^{e_A}]$

The above discussion showed how to efficiently generate candidate points R in $E(\mathbb{F}_{p^2})$. In this subsection we show how to efficiently check that R is in $E \setminus [2]E$, which guarantees that $[3^{e_B}]R$ is a point of exact order 2^{e_A}, and can subsequently be used as a basis element.

Since the supersingular curves $E/\mathbb{F}_{p^2} : y^2 = x(x^2 + Ax + 1)$ in our isogeny class have a full rational 2-torsion, we can always write them as $E/\mathbb{F}_{p^2} : y^2 = x(x - \gamma)(x - \delta)$. A classic result (cf. [14, Chap. 1 (Sect. 4), Theorem 4.1]) says that, in our case, any point $R = (x_R, y_R)$ in $E(\mathbb{F}_{p^2})$ is in $[2]E(\mathbb{F}_{p^2})$, i.e., is the result of doubling another point, if and only if x_R, $x_R - \gamma$ and $x_R - \delta$ are all squares in $\mathbb{F}_{p^2}$. This means that we do not need to find the roots δ and γ of $x^2 + Ax + 1$ to test for squareness, since we want the x_R such that at least one of x_R, $x_R - \gamma$ and $x_R - \delta$ are a non-square. We found it most efficient to simply ignore the latter two terms and reject any x_R that is square, since the first non-square x_R we find corresponds to a point R such that $[3^{e_B}]R$ has exact order 2^{e_A}, and further testing square values of x_R is both expensive and often results in the rejection of R anyway.

In light of the above, we note that for the 2-torsion basis generation, the Elligator approach is not as useful as it is in the next subsection. The reason here is that we want to only try points with a non-square x-coordinate, and there is no exploitable relationship between the squareness of v and v' in (2) (such a relation only exists between $f(v)$ and $f(v')$). Thus, the best approach here is to simply proceed by trying candidate v's as consecutive elements of a list $L = [u, 2u, 3u, \dots]$ of non-squares in $\mathbb{F}_{p^2}$ until $(v^3 + Av^2 + v)$ is square; recall from above that this check is performed efficiently using one exponentiation in $\mathbb{F}_p$.

To summarize the computation of a basis $\{R_1, R_2\}$ for $E(\mathbb{F}_{p^2})[2^{e_A}]$, we compute R_1 by letting v be the first element in L where $(v^3 + Av^2 + v)$ is square. We do not compute the square root of $(v^3 + Av^2 + v)$, but rather use e_B repeated x-only tripling operations starting from v to compute x_{R_1}. We then compute y_{R_1} as the square root of $x_{R_1}^3 + Ax_{R_1}^2 + x_{R_1}$. Note that either choice of square root is fine, so long as Alice and Bob take the same one. The point R_2 is found identically, i.e., using the second element in L that corresponds to an x-coordinate of a point on $E(\mathbb{F}_{p^2})$, followed by e_B x-only tripling operations to arrive at x_{R_2}, and the square root computation to recover y_{R_2}. Note that the points R_1 and R_2 need not be normalized before their input into the pairing computation; as we will see in Sect. 4, the doubling-only and tripling-only pairings do not ever perform additions with the original input points, so the input points are essentially forgotten after the first iteration.

3.3   Generating a Torsion Basis for $E(\mathbb{F}_{p^2})[3^{e_B}]$

The theorem used in the previous subsection was a special case of more general theory that characterizes the action of multiplication-by-m on E. We refer to Silverman's chapter [29, Chap. X] and to [26] for the deeper discussion in the general case, but in this work we make use of the explicit results derived in the case of $m = 3$ by Schaefer and Stoll [26], stating only what is needed for our purposes.

Let $P_3 = (x_{P_3}, y_{P_3})$ be any point of order 3 in $E(\mathbb{F}_{p^2})$ (recall that the entire 3-torsion is rational here), and let $g_{P_3}(x, y) = y - (\lambda x + \mu)$ be the usual tangent function to E at P_3. For any other point $R \in E(\mathbb{F}_{p^2})$, the result we use from [26] states that $R \in [3]E$ if and only if $g_{P_3}(R)$ is in $(\mathbb{F}_{p^2})^3$ (i.e., is a cube) for all of the 3-torsion points[2] P_3.

Again, since we do not want points in $[3]E$, but rather points in $E \setminus [3]E$, we do not need to test that R gives a cube for all of the $g_{P_3}(R)$, we simply want to compute an R where *any* one of the $g_{P_3}(R)$ is not a cube. In this case the test involves both coordinates of R, so we make use of Elligator 2 as it is described in Sect. 3.1 to produce candidate points $R \in E(\mathbb{F}_{p^2})$.

Unlike the previous case, where the 2-torsion point $(0, 0)$ is common to all curves in the isogeny class, in this case it is computing a 3-torsion point P_3 that is the most difficult computation. We attempted to derive an efficient algorithm that finds x_{P_3} as any root of the (quartic) 3-division polynomial $\Phi_3(A, x)$, but this solution involved several exponentiations in both $\mathbb{F}_{p^2}$ and $\mathbb{F}_p$, and was also hampered by the lack of an efficient enough analogue of (1) in the case of cube roots[3]. We found that a much faster solution was to compute the initial 3-torsion point via an x-only cofactor multiplication: we use the first step of Elligator 2 to produce an x-coordinate x_R, compute $x_{\tilde{R}} = x_{[2^{e_A}]R}$ via e_A repeated doublings, and then apply k repeated triplings until the result of a tripling is $(X : Z) \in \mathbb{P}^1$ with $Z = 0$, which corresponds to the point $\mathcal{O}$, at which point we can set out x_{P_3}, the x-coordinate of a 3-torsion point P_3, as the last input to the tripling function. Moreover, if the number of triplings required to produce $Z = 0$ was $k = e_B$, then it must be that $\tilde{R}$ is a point of exact order 3^{e_B}. If this is the case, we can use a square root to recover $y_{\tilde{R}}$ from $x_{\tilde{R}}$, and we already have one of our two basis points.

At this stage we either need to find one more point of order 3^{e_B}, or two. In either case we use the full Elligator routine to obtain candidate points R exactly as described in Sect. 3.1, use our point P_3 (together with our efficient test of cubic residuosity above) to test whether $g_{P_3}(R) = y_R - (\lambda x_R + \mu)$ is a cube, and if it is not, we output $\pm[2^{e_A}]R$ as a basis point; this is computed via a sequence of x-only doublings and one square root to recover $y_{[2^{e_A}]R}$ at the end. On the other hand, if $g_{P_3}(R)$ is a cube, then $R \in [3]E$, so we discard it and proceed to generate the next R via the tailored version of Elligator 2 above.

---

[2] The astute reader can return to Sect. 3.2 and see that this is indeed a natural analogue of [14, Chap. 1 (Sect. 4), Theorem 4.1].

[3] A common subroutine when finding roots of quartics involve solving the so-called *depressed cubic*.

We highlight the significant speed advantage that is obtained by the use of the result of Schaefer and Stoll [26]. Testing that points are in $E \setminus [3]E$ by cofactor multiplication requires e_A point doubling operations and e_B point tripling operations, while the same test using the explicit results from 3-descent require one field exponentiation that tests cubic residuosity. Moreover, this exponentiation only involves almost-for-free Frobenius operations and fast cyclotomic squaring and cubing operations (again, see Sect. 5.1).

Remark 3 (Checking the order of the Weil pairing). As mentioned at the beginning of this section, until now we have simplified the discussion to focus on generating two points R_1 and R_2 of exact order n. However, this does not mean that $\{R_1, R_2\}$ is a basis for $E(\mathbb{F}_{p^2})[n]$; this is the case if and only if the Weil pairing $e_n(R_1, R_2)$ has full order n. Although the Weil pairing will have order n with high probability for random R_1 and R_2, the probability is not so high that we do not encounter it in practice. Thus, baked into our software is a check that this is indeed the case, and if not, an appropriate backtracking mechanism that generates a new R_2. We note that, following [7, Sect. 9] and [11, Sect. 2.5], checking whether or not the Weil pairing $e_n(R_1, R_2)$ has full order is much easier than computing it, and can be done by comparing the values $[n/\ell]R_1$ and $[n/\ell]R_2$.

4   The Tate Pairing Computation

Given the basis points R_1 and R_2 resulting from the previous section, and the two points P and Q in the (otherwise uncompressed) public key, we now have four points of exact order n. As outlined in Sect. 2, the next step is to compute the following five pairings to transfer the discrete logarithms to the multiplicative group $\mu_n \subset \mathbb{F}_{p^2}^*$:

$$e_0 := e(R_1, R_2) = f_{n,R_1}(R_2)^{(p^2-1)/n}$$

$$e_1 := e(R_1, P) = f_{n,R_1}(P)^{(p^2-1)/n}$$

$$e_2 := e(R_1, Q) = f_{n,R_1}(Q)^{(p^2-1)/n}$$

$$e_3 := e(R_2, P) = f_{n,R_2}(P)^{(p^2-1)/n}$$

$$e_4 := e(R_2, Q) = f_{n,R_2}(Q)^{(p^2-1)/n}.$$

As promised in Sect. 1, the above pairings are already defined by the order n reduced Tate pairing $e: E(\mathbb{F}_{p^2})[n] \times E(\mathbb{F}_{p^2})/nE(\mathbb{F}_{p^2}) \mapsto \mu_n$, rather than the Weil pairing that was used in [2]. The rationale behind this choice is clear: the lack of special (subfield) groups inside the n-torsion means that many of the tricks used in the pairing literature cannot be exploited in the traditional sense. For example, there does not seem to be a straight-forward way to shorten the Miller loop by using efficiently computable maps arising from Frobenius (see, e.g., [3,12,13]), our denominators lie in $\mathbb{F}_{p^2}$ so cannot be eliminated [4], and, while distortion maps exist on all supersingular curves [33], finding efficiently computable and therefore useful maps seems hard for random curves in the isogeny class. The upshot is that the Miller loop is far more expensive than the final exponentiation in our case, and organizing the Tate pairings in the above manner allows us to get away with the computation of only two Miller functions, rather than the four that were needed in the case of the Weil pairing [2].

In the case of ordinary pairings over curves with a larger embedding degree[4], the elliptic curve operations during the Miller loop take place in a much smaller field than the extension field; in the Tate pairing the point operations take place in the base field, while in the loop-shortened ate pairing [13] (and its variants) they take place in a small-degree subfield. Thus, in those cases the elliptic curve arithmetic has only a minor influence on the overall runtime of the pairing.

In our scenario, however, we are stuck with elliptic curve points that have both coordinates in the full extension field. This means that the Miller line function computations are the bottleneck of the pairing computations (and, as it turns out, this is the main bottleneck of the whole compression routine). The main point of this section is to present optimized explicit formulas in this case; this is done in Sect. 4.1. In Sect. 4.2 we discuss how to compute the five pairings in parallel and detail how to compute the final exponentiations efficiently.

4.1   Optimized Miller Functions

We now present explicit formulas for the point operations and line computations in Miller's algorithm [20]. In the case of the order-2^{e_A} Tate pairings inside $E(\mathbb{F}_{p^2})[2^{e_A}]$, we only need the doubling-and-line computations, since no additions are needed. In the case of the order-3^{e_B} Tate pairings inside $E(\mathbb{F}_{p^2})[3^{e_B}]$, we investigated two options: the first option computes the pairing in the usual "double-and-add" fashion, reusing the doubling-and-line formulas with addition-and-line formulas, while the second uses a simple sequence of e_B tripling-and-parabola operations. The latter option proved to offer much better performance and is arguably more simple than the double-and-add approach[5].

We tried several coordinate systems in order to lower the overall number of field operations in both pairings, and after a close inspection of the explicit formulas in both the doubling-and-line and tripling-and-parabola operations, we opted to use the coordinate tuple $(X^2 : XZ : Z^2 : YZ)$ to represent intermediate projective points $P = (X : Y : Z) \in \mathbb{P}^2$ in $E(\mathbb{F}_{p^2})$. Note that all points in our routines for which we use this representation satisfy $XYZ \neq 0$, as their orders are strictly larger than 2. This ensures that the formulas presented below do not contain exceptional cases[6].

---

[4] This has long been the preferred choice of curve in the pairing-based cryptography literature.

[5] An earlier version of this article claimed that the performance was favourable in the former case, but further optimization in the tripling-and-parabola scenario since proved this option to be significantly faster.

[6] The input into the final iteration in the doubling-only pairing is a point of order 2, but (as is well-known in the pairing literature) this last iterate is handled differently than all of the prior ones.

Doubling-and-Line Operations. The doubling step in Miller's algorithm takes as input the tuple $(U_1, U_2, U_3, U_4) = (X^2, XZ, Z^2, YZ)$ corresponding to the point $P = (X : Y : Z) \in \mathbb{P}^2$ in $E(\mathbb{F}_{p^2})$, and outputs the tuple $(V_1, V_2, V_3, V_4) = (X_2^2 : X_2 Z_2 : Z_2^2 : Y_2 Z_2)$ corresponding to the point $[2]P = (X_2 : Y_2 : Z_2) \in \mathbb{P}^2$, as well as the 5 coefficients in the Miller line function $l/v = (l_x \cdot x + l_y \cdot y + l_0)/(v_x x + v_0)$ with divisor $2(P) - ([2]P) - (\mathcal{O})$ in $\mathbb{F}_q[x, y](E)$. The explicit formulas are given as:

$$l_x = 4U_4^3 + 2U_2 U_4(U_1 - U_3),$$
$$l_y = 4U_2 U_4^2,$$
$$l_0 = 2U_1 U_4(U_1 - U_3),$$
$$v_x = 4U_2 U_4^2,$$
$$v_0 = U_2(U_1 - U_3)^2,$$

together with

$$V_1 = (U_1 - U_3)^4,$$
$$V_2 = 4U_4^2(U_1 - U_3)^2,$$
$$V_3 = 16U_4^4,$$
$$V_4 = 2U_4(U_1 - U_3)((U_1 - U_3)^2 + 2U_2(4U_2 + A(U_1 + U_3))).$$

The above point doubling-and-line function computation can be computed in $9\mathbf{M} + 5\mathbf{S} + 7\mathbf{a} + 1\mathbf{s}$. The subsequent evaluation of the line function at the second argument of a pairing, the squaring that follows, and the absorption of the squared line function into the running paired value costs $5\mathbf{M} + 2\mathbf{S} + 1\mathbf{a} + 2\mathbf{s}$.

Tripling-and-Parabola Operations. The tripling-and-parabola operation has as input the tuple $(U_1, U_2, U_3, U_4) = (X^2, XZ, Z^2, YZ)$ corresponding to the point $P = (X : Y : Z) \in \mathbb{P}^2$ in $E(\mathbb{F}_{p^2})$, and outputs the tuple $(V_1, V_2, V_3, V_4) = (X_3^2 : X_3 Z_3 : Z_3^2 : Y_3 Z_3)$ corresponding to the point $[3]P = (X_3 : Y_3 : Z_3) \in \mathbb{P}^2$, as well as the 6 coefficients in the Miller parabola function $l/v = (l_y \cdot y + l_{x,2} \cdot x^2 + l_{x,1}x + l_{x,0})/(v_x x + v_0)$ with divisor $3(P) - ([3]P) - 2(\mathcal{O})$ in $\mathbb{F}_q[x, y](E)$. The explicit formulas are given as:

$$l_y = 8U_4^3,$$
$$l_{x,2} = U_3(3U_1^2 + 4U_1 A U_2 + 6U_1 U_3 - U_3^2),$$
$$l_{x,1} = 2U_2(3U_1^2 + 2U_1 U_3 + 3U_3^2 + 6U_1 A U_2 + 4A^2 U_2^2 + 6AU_2 U_3),$$
$$l_{x,0} = U_1(-U_1^2 + 6U_1 U_3 + 3U_3^2 + 4AU_2 U_3),$$
$$v_x = 8U_3 U_4^3 (3U_1^2 + 4U_1 A U_2 + 6U_1 U_3 - U_3^2)^4,$$
$$v_0 = -8U_2 U_4^3 (3U_1^2 + 4U_1 A U_2 + 6U_1 U_3 - U_3^2)^2 (U_1^2 - 6U_1 U_3 - 3U_3^2 - 4AU_2 U_3)^2,$$

together with

$$V_1 = 8U_4^3 U_1(-U_1^2 + 6U_1U_3 + 3U_3^2 + 4AU_2U_3)^4,$$

$$V_2 = 8U_2U_4^3(3U_1^2 + 4U_1AU_2 + 6U_1U_3 - U_3^2)^2(U_1^2 - 6U_1U_3 - 3U_3^2 - 4AU_2U_3)^2,$$

$$V_3 = 8U_3U_4^3(3U_1^2 + 4U_1AU_2 + 6U_1U_3 - U_3^2)^4,$$

$$V_4 = -8U_3(3U_1^2 + 4U_1AU_2 + 6U_1U_3 - U_3^2)U_1(-U_1^2 + 6U_1U_3 + 3U_3^2 + 4AU_2U_3)$$
$$\cdot\left(16U_1U_3A^2U_2^2 + 28U_1^2AU_2U_3 + 28U_1U_3^2AU_2 + 4U_3^3AU_2 + 4U_1^3AU_2\right.$$
$$\left. + 6U_1^2U_3^2 + 28U_1^3U_3 + U_3^4 + 28U_1U_3^3 + U_1^4\right)(U_3 + U_1 + AU_2)^2.$$

The above point tripling-and-parabola function computation can be computed in $19\mathbf{M} + 6\mathbf{S} + 15\mathbf{a} + 6\mathbf{s}$. The subsequent evaluation of the line function at the second argument of a pairing, the cubing that follows, and the absorption of the cubed line function into the running paired value costs $10\mathbf{M} + 2\mathbf{S} + 4\mathbf{a}$.

Remark 4 (No irrelevant factors). It is common in the pairing literature to abuse notation and define the order-n Tate pairing as $e_n(P, Q) = f_P(Q)^{(p^k-1)/n}$, where k is the embedding degree (in our case $k = 2$), and f_P has divisor $(f_P) = n(P) - n(\mathcal{O})$ in $\mathbb{F}_{p^k}[x, y](E)$. This is due to an early result of Barreto, Kim, Lynn and Scott [4, Theorem 1], who showed that the actual definition of the Tate pairing, i.e., $e_n(P, Q) = f_P(D_Q)^{(p^k-1)/n}$ where D_Q is a divisor equivalent to $(Q)-(\mathcal{O})$, could be relaxed in practical cases of interest by replacing the divisor D_Q with the point Q. This is due to the fact that the evaluation of f_P at $\mathcal{O}$ in such scenarios typically lies in a proper subfield of $\mathbb{F}_{p^k}^*$, so becomes an *irrelevant factor* under the exponentiation to the power of $(p^k - 1)/n$. In our case, however, this is generally not the case because the coefficients in our Miller functions lie in the full extension field $\mathbb{F}_{p^2}^*$. Subsequently, our derivation of explicit formulas replaces Q with the divisor $D_Q = (Q) - (\mathcal{O})$, and if the evaluation of the Miller functions at $\mathcal{O}$ is ill-defined, we instead evaluate them at the divisor $(Q+T)-(T)$ that is linearly equivalent to D_Q, where we fixed $T = (0,0)$ as the (universal) point of order 2. If $Q = (x_Q, y_Q)$, then $Q + T = (1/x_Q, -y_Q/x_Q^2)$, so evaluating the Miller functions at the translated point amounts to a permutation of the coefficients, and evaluating the Miller functions at $T = (0,0)$ simply leaves a quotient of the constant terms. These modifications are already reflected in the operation counts quoted above.

Remark 5. In the same vein as Remark 2, there is another possible speed improvement within the pairing computation that is not currently exploited in our library. Recall that during the generation of the torsion bases described in Sect. 3, the candidate basis point R is multiplied by the cofactor $n \in \{n_A, n_B\}$ to check whether it has the correct (full) order, and if so, R is kept and stored as one of the two basis points. Following the idea of Scott [27, Sect. 9], the intermediate multiples of R (and partial information about the corresponding Miller line functions) that are computed in this cofactor multiplication could be stored in anticipation for the subsequent pairing computation, should R indeed be one of the two basis points. Another alternative here would be to immediately compute the pairings using the first two candidate basis points and to absorb the point order checks inside the pairing computations, but given the overhead incurred if either or both of these order checks fails, this could end up being too wasteful (on average).

4.2 Parallel Pairing Computation and the Final Exponentiation

In order to solve the discrete logarithms in the subgroup μ_n of n-th roots of unity in $\mathbb{F}_{p^2}^*$, we compute the five pairings $e_0 := e(R_1, R_2)$, $e_1 := e(R_1, P)$, $e_2 := e(R_1, Q)$, $e_3 := e(R_2, P)$, and $e_4 := e(R_2, Q)$. The first argument of all these pairings is either R_1 or R_2, i.e., all are of the form $f_{n,R_i}(S)^{(p^2-1)/n}$ for $i \in \{1, 2\}$ and $S \in \{R_2, P, Q\}$. This means that the only Miller functions we need are f_{n,R_1} and f_{n,R_2}, and we get away with computing only those two functions for the five pairing values. The two functions are evaluated at the points R_2, P, Q during the Miller loop to obtain the desired combinations. It therefore makes sense to accumulate all five Miller values simultaneously.

Computing the pairings simultaneously also becomes advantageous when it is time to perform inversions. Since we cannot eliminate denominators due to the lack of a subfield group, we employ the classic way of storing numerators and denominators separately to delay all inversions until the very end of the Miller loop. At this point, we have ten values (five numerators and five denominators), all of which we invert using Montgomery's inversion sharing trick [21] at the total cost of one inversion and 30 $\mathbb{F}_{p^2}$ multiplications. The five inverted denominators are then multiplied by the corresponding numerators to give the five unreduced paired values. The reason we not only invert the denominators, but also the numerators, is because these inverses are needed in the easy part of the final exponentiation.

The final exponentiation is an exponentiation to the power $(p^2 - 1)/n = (p - 1)\frac{p+1}{n}$. The so-called *easy part*, i.e., raising to the power $p - 1$, is done by one application of the Frobenius automorphism and one inversion. The Frobenius is simply a conjugation in $\mathbb{F}_{p^2}$, and the inversion is actually a multiplication since we had already computed all required inverses as above. The so-called *hard part* of the final exponentiation has exponent $(p + 1)/n$ and needs to be done with regular exponentiation techniques.

A nice advantage that makes the hard part quite a little easier is the fact that after a field element $a = a_0 + a_1 \cdot i \in \mathbb{F}_{p^2}$ has been raised to the power $p - 1$, it has order $p + 1$, which means it satisfies $1 = a^p \cdot a = a_0^2 + a_1^2$. This equation can be used to deduce more efficient squaring and cubing formulas that speed up this final part of the pairing computation (see Sect. 5.1 for further details).

Finally, in the specific setting of SIDH, where $p = n_A n_B - 1$, we have that $(p + 1)/n_A = n_B$ and $(p + 1)/n_B = n_A$. When n_A and n_B are powers of 2 and 3, respectively, the hard part of the final exponentiation consists of only squarings or only cubings, respectively. These are done with the particularly efficient formulas described in Sect. 5.1 below.

5   Efficient Pohlig-Hellman in μ_{ℓ^e}

In this section, we describe how we optimize the Pohlig-Hellman [23] algorithm
to compute discrete logarithms in the context of public-key compression for
supersingular-isogeny-based cryptosystems, and we show that we are able to
improve on the quadratic complexity described in [23]. A similar result has
already been presented in the more general context of finite abelian p-groups
by Sutherland [32]. However, our software employs a different optimization of
Pohlig-Hellman, by choosing small memory consumption over a more efficient
computation, which affects parameter choices. We emphasize that there are different time-memory trade-offs that could be chosen, possibly speeding up the
Pohlig-Hellman computation by another factor of two (see Remark 2).

Following the previous sections, the two-dimensional discrete logarithm problems have been reduced to four discrete logarithm problems in the multiplicative
group $\mu_{\ell^e} \subset \mathbb{F}_{p^2}^*$ of ℓ^e-th roots of unity, where $\ell, e \in \mathbb{Z}$ are positive integers and
ℓ is a (small) prime. Before elaborating on the details of the windowed Pohlig-
Hellman algorithm, we note that the condition $\ell^e \mid p+1$ makes various operations
in μ_{ℓ^e} more efficient than their generic counterpart in $\mathbb{F}_{p^2}^*$.

5.1   Arithmetic in the Cyclotomic Subgroup

Efficient arithmetic in μ_{ℓ^e} can make use of the fact that μ_{ℓ^e} is a subgroup of the
multiplicative group $G_{p+1} \subset \mathbb{F}_{p^2}^*$ of order $p+1$. Various subgroup cryptosystems
based on the hardness of the discrete logarithm problem have been proposed
in the literature [18,30], which can be interpreted in the general framework
of torus-based cryptography [25]. The following observations for speeding up
divisions and squarings in G_{p+1} have been described by Stam and Lenstra [31,
Sect. 3.23 and Lemma 3.24].

Division in μ_{ℓ^e}. Let $p \equiv 3 \pmod 4$ and $\mathbb{F}_{p^2} = \mathbb{F}_p(i)$, $i^2 = -1$. For any $a =
a_0 + a_1 \cdot i \in G_{p+1}$, $a_0, a_1 \in \mathbb{F}_p$, we have that $a \cdot a^p = a^{p+1} = 1$, and therefore, the
inverse $a^{-1} = a^p = a_0 - a_1 \cdot i$. This means that inversion in μ_{ℓ^e} can be computed
almost for free by conjugation, i.e., a single negation in $\mathbb{F}_p$, and thus divisions
become as efficient as multiplications in μ_{ℓ^e}.

Squaring in μ_{ℓ^e}. The square of $a = a_0 + a_1 \cdot i$ can be computed as $a^2 =
(2a_0^2 - 1) + ((a_0 + a_1)^2 - 1) \cdot i$ by essentially two base field squarings. In the case
where such squarings are faster than multiplications, this yields a speed-up over
generic squaring in $\mathbb{F}_{p^2}$.

Cubing in μ_{ℓ^e}. As far as we know, a cubing formula in G_{p+1} has not been
considered in the literature before. We make the simple observation that a^3 can
be computed as $a^3 = (a_0 + a_1 \cdot i)^3 = a_0(4a_0^2 - 3) + a_1(4a_0^2 - 1) \cdot i$, which needs
only one squaring and two multiplications in $\mathbb{F}_p$, and is significantly faster than
a naïve computation via a squaring and a multiplication in μ_{ℓ^e}.

5.2   Pohlig-Hellman

We now discuss the Pohlig-Hellman algorithm as presented in [23] for the group μ_{ℓ^e}. Let $r, g \in \mu_{\ell^e}$ be such that $r = g^\alpha$ for some $\alpha \in \mathbb{Z}$. Given r and g, the goal is to determine the unknown scalar α. Denote α as

$$\alpha = \sum_{i=0}^{e-1} \alpha_i \ell^i \quad (\alpha_i \in \{0, \ldots, \ell - 1\}).$$

Now define $s = g^{\ell^{e-1}}$, which is an element of order ℓ, and let $r_0 = r$. Finally, define

$$g_i = g^{\ell^i} \quad (0 \le i \le e - 1)$$

and

$$r_i = \frac{r_{i-1}}{g_{i-1}^{\alpha_{i-1}}} \quad (1 \le i \le e - 1).$$

A straightforward computation then shows that for all $0 \le i \le e - 1$,

$$r_i^{\ell^{e-(i+1)}} = s^{\alpha_i}. \tag{3}$$

As proven in [23], this allows to inductively recover all α_i, by solving the discrete logarithms of Eq. (3) in the group $\langle s \rangle$ of order ℓ. This can be done by precomputing a table containing all elements of $\langle s \rangle$. Alternatively, if ℓ is not small enough, one can improve the efficiency by applying the Baby-Step Giant-Step algorithm [28], at the cost of some more precomputation. For small ℓ the computation has complexity $\mathcal{O}(e^2)$, while precomputing and storing the g_i requires $\mathcal{O}(e)$ memory.

5.3   Windowed Pohlig-Hellman

The original version of the Pohlig-Hellman algorithm reduces a single discrete logarithm in the large group μ_{ℓ^e} to e discrete logarithms in the small group μ_ℓ. In this section we consider an intermediate version, by reducing the discrete logarithm in μ_{ℓ^e} to $\frac{e}{w}$ discrete logarithms in μ_{ℓ^w}. Let r, g, α as in the previous section, and let $w \in \mathbb{Z}$ be such that $w \mid e$. Note that it is not necessary for e to be divisible by w. If it is not, we replace e by $e - (e \pmod{w})$, and compute the discrete logarithm for the final $e \pmod{w}$ bits at the end. However the assumption $w \mid e$ improves the readability of the arguments with little impact on the results, so we focus on this case here. Write

$$\alpha = \sum_{i=0}^{\frac{e}{w}-1} \alpha_i \ell^{wi} \quad (\alpha_i \in \{0, \ldots, \ell^w - 1\}),$$

define $s = g^{\ell^{e-w}}$, which is an element of order ℓ^w, and let $r_0 = r$. Let

$$g_i = g^{\ell^{wi}} \quad (0 \le i \le \frac{e}{w} - 1)$$

and

$$r_i = \frac{r_{i-1}}{g_{i-1}^{\alpha_{i-1}}} \quad (1 \le i \le \frac{e}{w} - 1). \tag{4}$$

A analogous computation to the one in [23] proves that

$$r_i^{\ell^{e-w(i+1)}} = s^{\alpha_i} \quad (0 \le i \le \frac{e}{w} - 1). \tag{5}$$

Hence we inductively obtain α_i for all $0 \le i \le \frac{e}{w} - 1$, and thereby α. To solve the discrete logarithm in the smaller subgroup μ_{ℓ^w}, we consider two strategies as follows.

Baby-Step Giant-Step in $\langle s \rangle$. As before, for small ℓ and w we can compute a table containing all ℓ^w elements of $\langle s \rangle$, making the discrete logarithms in (5) trivial to solve. As explained in [28], the Baby-Step Giant-Step algorithm allows us to make a trade-off between the size of the precomputed table and the computational cost. That is, given some $v \le w$, we can compute discrete logarithms in $\langle s \rangle$ with computational complexity $\mathcal{O}(\ell^v)$ and $\mathcal{O}(\ell^{w-v})$ memory. Note that the computational complexity grows exponentially with v, whereas the memory requirement grows exponentially with $w - v$. This means that if we want to make w larger, we need to grow v as well, as otherwise the table-size will increase. Therefore in order to obtain an efficient and compact algorithm, we must seemingly limit ourselves to small w. We overcome this limitation in the next section.

Pohlig-Hellman in $\langle s \rangle$. We observe that $\langle s \rangle$ has order ℓ^w, which is again smooth. This allows us to solve the discrete logarithms in $\langle s \rangle$ by using the original Pohlig-Hellman algorithm of Sect. 5.2. However, we can also choose to solve the discrete logarithm in $\langle s \rangle$ with a second windowed Pohlig-Hellman algorithm. Note the recursion that occurs, and we can ask what the optimal depth of this recursion is. We further investigate this question in Sect. 5.4.

5.4   The Complexity of Nested Pohlig-Hellman

We estimate the cost of an execution of the nested Pohlig-Hellman algorithm by looking at the cost of doing the computations in (4) and (5). Let F_n $(n \ge 0)$ denote the cost of an n-times nested Pohlig-Hellman algorithm, and set $F_{-1} = 0$. Let $w_0, w_1, \ldots, w_n, w_{n+1}$ be the window sizes, and set $w_0 = e$, $w_{n+1} = 1$ (note that $n = 0$ corresponds to the original Pohlig-Hellman algorithm). Again, assume that $w_n \mid w_{n-1} \mid \cdots \mid w_1 \mid e$, which is merely for ease of exposition. The first iteration has window size w_1, which means that the cost of the exponentiations in (5) is

$$\left(\sum_{i=0}^{\frac{e}{w_1}-1} w_1 i \right) \mathbf{L} = \frac{1}{2} w_1 \left(\frac{e}{w_1} - 1 \right) \frac{e}{w_1} \mathbf{L} = \frac{1}{2} e \left(\frac{e}{w_1} - 1 \right) \mathbf{L},$$

where $\mathbf{L}$ denotes the cost of an exponentiation by ℓ. The exponentiations in (4) are performed with a scalar of size $\log \alpha_i \approx w_1 \log \ell$, which on average costs $\frac{1}{2} w_1 \log \ell \, \mathbf{M} + w_1 \log \ell \, \mathbf{S}$. On average, to do all $\frac{e}{w_1}$ of them then costs

$$\frac{1}{2} e \log \ell \, \mathbf{M} + e \log \ell \, \mathbf{S}.$$

We emphasize that for small w_i and ℓ this is a somewhat crude estimation, yet it is enough to get a good feeling for how to choose our parameters (i.e., window sizes). We choose to ignore the divisions, since there are only a few (see Remark 6) and, as we showed in Sect. 5.1, they can essentially be done at the small cost of a multiplication. We also ignore the cost of the precomputation for the $g^{\ell^{w_i}}$, which is small as well (see Remark 7). To complete the algorithm, we have to finish the remaining $\frac{e}{w_1}$ $(n-1)$-times nested Pohlig-Hellman routines. In other words, we have shown that

$$F_n \approx \frac{1}{2} e \left(\frac{e}{w_1} - 1 \right) \mathbf{L} + \frac{1}{2} e \log \ell \, \mathbf{M} + e \log \ell \, \mathbf{S} + \frac{e}{w_1} F_{n-1}.$$

Now, by using analogous arguments on F_{n-1}, and induction on n, we can show that

$$F_n \approx \frac{1}{2} e \left(\frac{e}{w_1} + \ldots + \frac{w_{n-1}}{w_n} + w_n - n \right) \mathbf{L} + \frac{n+1}{2} e \log \ell \, \mathbf{M} + (n+1) e \log \ell \, \mathbf{S}.$$
$$(6)$$

To compute the optimal choice of $(w_1, \ldots, w_n)$, we compute the derivatives,

$$\frac{\partial F_n}{\partial w_i} = \frac{1}{2} e \left(\frac{1}{w_{i+1}} - \frac{w_{i-1}}{w_i^2} \right) \mathbf{L} \quad (1 \leq i \leq n)$$

and simultaneously equate them to zero to obtain the equations

$$w_i = \sqrt{w_{i-1} w_{i+1}} \quad (1 \leq i \leq n).$$

From this we can straightforwardly compute that the optimal choice is

$$(w_1, \ldots, w_n) = \left(e^{\frac{n}{n+1}}, e^{\frac{n-1}{n+1}}, \ldots, e^{\frac{2}{n+1}}, e^{\frac{1}{n+1}} \right).$$
$$(7)$$

Plugging this back into the Eq. (6), we conclude that

$$F_n \approx \frac{1}{2} e (n+1) \left(e^{\frac{1}{n+1}} - 1 \right) \mathbf{L} + \frac{n+1}{2} e \log \ell \, \mathbf{M} + (n+1) e \log \ell \, \mathbf{S}.$$

Observe that $F_0 \approx \frac{1}{2} e^2$, agreeing with the complexity described in [23]. However, as n grows, the complexity of the nested Pohlig-Hellman algorithm goes from quadratic to linear in e, giving a significant improvement.

Remark 6. We observe that for every two consecutive windows w_i and w_{i+1}, we need less than $\frac{w_i}{w_{i+1}}$ divisions for (4). Breaking the full computation down, it is easy to show that the total number of divisions is less than

$$\frac{e}{w_1} + \frac{e}{w_1}\left(\frac{w_1}{w_2} + \frac{w_1}{w_2}\left(\cdots + \frac{w_{n-2}}{w_{n-1}}\left(\frac{w_{n-1}}{w_n} + \frac{w_{n-1}}{w_n}w_n\right)\right)\right),$$

which can be rewritten as

$$e\left(\frac{1}{w_1} + \frac{1}{w_2} + \ldots + \frac{1}{w_n} + \frac{w_n}{w_{n-1}}\right).$$

Now we note that $w_{i+1} \mid w_i$, while $w_{i+1} \neq w_i$, for all $0 \leq i \leq n$. As $w_{n+1} = 1$, it follows that $w_{n+1-i} \geq 2^i$ for all $0 \leq i \leq n$. Therefore

$$e\left(\frac{1}{w_1} + \frac{1}{w_2} + \ldots + \frac{1}{w_n} + \frac{w_n}{w_{n-1}}\right) \leq e\left(\frac{1}{2^n} + \frac{1}{2^{n-1}} + \ldots + \frac{1}{2} + 1\right) < 2e.$$

Remark 7. As every table element is of the form g^{ℓ^i}, where i is an integer such that $0 \leq i \leq e - 1$, we conclude that we need at most $(e-1)\mathbf{L}$ to pre-compute all tables.

5.5   Discrete Logarithms in $\mu_{2^{372}}$

For this section we fix $\ell = 2$ and $e = 372$. In this case $\mathbf{L}$ is the cost of a squaring, i.e., $\mathbf{L} = \mathbf{S}$. To validate the approach, we present estimates for the costs of the discrete logarithm computations in $\mu_{2^{372}}$ through a Magma implementation. In this implementation we count every multiplication, squaring and division operation; on the other hand, some of these were ignored for the estimation of F_n above. The results are shown in Table 1 for $0 \leq n \leq 4$, choosing the window sizes as computed in (7). The improved efficiency as well as the significantly smaller table sizes are clear, and we observe that in the group $\mu_{2^{372}}$ it is optimal to choose $n = 3$.

Table 1. Estimations of F_n in $\mu_{2^{372}}$ via a Magma implementation. Here **m** and **s** are the cost of multiplications and squarings in $\mathbb{F}_p$, while $\mathbf{M} = 3 \cdot \mathbf{m}$ and $\mathbf{S} = 2 \cdot \mathbf{s}$ are the cost of multiplications and squarings in $\mathbb{F}_{p^2}$. The costs are averaged over 100 executions of the algorithm.

#	Windows				$\mathbb{F}_{p^2}$		$\mathbb{F}_p$		Table size
n	w_1	w_2	w_3	w_4	M	S	m	s	$\mathbb{F}_{p^2}$
0	–	–	–	–	372	69 378	1 116	138 756	375
1	19	–	–	–	375	7 445	1 125	14 891	43
2	51	7	–	–	643	4 437	1 931	8 847	25
3	84	21	5	–	716	3 826	2 150	7 652	25
4	114	35	11	3	1 065	3 917	3 197	7 835	27

Table 2. Estimations of F_n in $\mu_{3^{239}}$ via a Magma implementation. Here **m** and **s** are the cost of multiplications and squarings in $\mathbb{F}_p$, while $\mathbf{M} = 3 \cdot \mathbf{m}$, $\mathbf{S} = 2 \cdot \mathbf{s}$ and $\mathbf{C} = \mathbf{m} + 2 \cdot \mathbf{s}$ are the cost of multiplications, squarings and cubings in $\mathbb{F}_{p^2}$ respectively. The costs are averaged over 100 executions of the algorithm.

#	Windows				$\mathbb{F}_{p^2}$			$\mathbb{F}_p$		Table size
n	w_1	w_2	w_3	w_4	**M**	**S**	**C**	**m**	**s**	$\mathbb{F}_{p^2}$
0	–	–	–	–	239	78	28 680	58 077	28 837	242
1	15	–	–	–	349	341	3 646	8 340	4 328	35
2	39	6	–	–	612	660	2 192	6 220	3 512	22
3	61	15	3	–	656	836	1 676	5 320	3 349	17
4	79	26	8	3	954	1 252	1 427	5 717	3 932	16

5.6 Discrete Logarithms in $\mu_{3^{239}}$

We now fix $\ell = 3$ and $e = 239$ and present estimates for the costs of the discrete logarithm computations in $\mu_{3^{239}}$. Here $\mathbf{L}$ is now the cost of a cubing in $\mu_{3^{239}}$. As explained in Sect. 5.1, this is done at the cost of one multiplication and two squarings in $\mathbb{F}_p$. As shown in Table 2, the optimal case in $\mu_{3^{239}}$ is also $n = 3$.

6   Final Compression and Decompression

In this section we explain how to further compress a public key PK from $\mathbb{F}_{p^2} \times \mathbb{Z}_n^4$ to $\mathbb{F}_{p^2} \times \mathbb{Z}_2 \times \mathbb{Z}_n^3$. Moreover, we also show how to merge the key decompression with one of the operations of the SIDH scheme, making much of the decompression essentially free of cost. For ease of notation we follow the scheme described in [7], but everything that follows in this section generalizes naturally to the theory as originally described in [8].

6.1   Final Compression

Using the techniques explained in all previous sections, we can compress a triple $(E_A, P, Q) \in \mathbb{F}_{p^2}^3$ to a tuple $(A, \alpha_P, \beta_P, \alpha_Q, \beta_Q) \in \mathbb{F}_{p^2} \times \mathbb{Z}_n^4$ such that

$$(P, Q) = (\alpha_P R_1 + \beta_P R_2, \alpha_Q R_1 + \beta_Q R_2),$$

where $\{R_1, R_2\}$ is a basis of $E_A[n]$. As described in [7], the goal is to compute $\langle P + \ell m Q \rangle$, for $\ell \in \{2, 3\}$ and a secret key m. Again, we note that the original proposal expects to compute $\langle n_1 P + n_2 Q \rangle$, for secret key (n_1, n_2), but we emphasize that all that follows can be generalized to this case.

Now, since P is an element of order n, either $\alpha_P \in \mathbb{Z}_n^*$ or $\beta_P \in \mathbb{Z}_n^*$. It immediately follows that

$$\langle P + \ell m Q \rangle = \begin{cases} \langle \alpha_P^{-1} P + \ell m \alpha_P^{-1} Q \rangle & \text{if} \quad \alpha_P \in \mathbb{Z}_n^* \\ \langle \beta_P^{-1} P + \ell m \beta_P^{-1} Q \rangle & \text{if} \quad \beta_P \in \mathbb{Z}_n^* \end{cases}.$$

Hence, to compute $\langle P + \ell m Q \rangle$, we do not necessarily have to recompute (P, Q). Instead, we can compute

$$\left(\alpha_P^{-1} P, \alpha_P^{-1} Q\right) = \left(R_1 + \alpha_P^{-1}\beta_P R_2, \alpha_P^{-1}\alpha_Q R_1 + \alpha_P^{-1}\beta_Q R_2\right)$$

or

$$\left(\beta_P^{-1} P, \beta_P^{-1} Q\right) = \left(\beta_P^{-1}\alpha_P R_1 + R_2, \beta_P^{-1}\alpha_Q R_1 + \beta_P^{-1}\beta_Q R_2\right).$$

Note that in both cases we have normalized one of the scalars. We conclude that we can compress the public key to $\mathrm{PK} \in \mathbb{F}_{p^2} \times \mathbb{Z}_2 \times \mathbb{Z}_n^3$, where

$$\mathrm{PK} = \begin{cases} \left(A, 0, \alpha_P^{-1}\beta_P, \alpha_P^{-1}\alpha_Q, \alpha_P^{-1}\beta_Q\right) & \text{if} \quad \alpha_P \in \mathbb{Z}_n^* \\ \left(A, 1, \beta_P^{-1}\alpha_P, \beta_P^{-1}\alpha_Q, \beta_P^{-1}\beta_Q\right) & \text{if} \quad \beta_P \in \mathbb{Z}_n^* \end{cases}.$$

6.2   Decompression

Let $\left(A, b, \widetilde{\alpha}_P, \widetilde{\alpha}_Q, \widetilde{\beta}_Q\right) \in \mathbb{F}_{p^2} \times \mathbb{Z}_2 \times \mathbb{Z}_n^3$ be a compressed public key. Note that, by the construction of the compression, there exists a $\gamma \in \mathbb{Z}_n^*$ such that

$$\left(\gamma^{-1} P, \gamma^{-1} Q\right) = \begin{cases} \left(R_1 + \widetilde{\alpha}_P R_2, \widetilde{\alpha}_Q R_1 + \widetilde{\beta}_Q R_2\right) & \text{if} \quad b = 0, \\ \left(\widetilde{\alpha}_P R_1 + R_2, \widetilde{\alpha}_Q R_1 + \widetilde{\beta}_Q R_2\right) & \text{if} \quad b = 1. \end{cases} \tag{8}$$

The naïve strategy, analogous to the one originally explained in [2], would be to generate the basis $\{R_1, R_2\}$ of $E_A[n]$, use the public key to compute $\left(\gamma^{-1} P, \gamma^{-1} Q\right)$ via (8), and finally compute

$$\langle P + \ell m Q \rangle = \langle \gamma^{-1} P + \ell m \gamma^{-1} Q \rangle,$$

where $m \in \mathbb{Z}_n$ is the secret key. The cost is approximately a 1-dimensional and a 2-dimensional scalar multiplication on E_A, while the final 1-dimensional scalar multiplication is part of the SIDH scheme.

Instead, we use (8) to observe that

$$\langle P + \ell m Q \rangle = \langle \gamma^{-1} P + \ell m \gamma^{-1} Q \rangle$$
$$= \begin{cases} \langle (1 + \ell m \widetilde{\alpha}_Q) R_1 + \left(\widetilde{\alpha}_P + \ell m \widetilde{\beta}_Q\right) R_2 \rangle & \text{if} \quad b = 0, \\ \langle (\widetilde{\alpha}_P + \ell m \widetilde{\alpha}_Q) R_1 + \left(1 + \ell m \widetilde{\beta}_Q\right) R_2 \rangle & \text{if} \quad b = 1. \end{cases}$$

Thus, since $1 + \ell m \widetilde{\alpha}_Q, 1 + \ell m \widetilde{\beta}_Q \in \mathbb{Z}_n^*$ (recall that $n = \ell^e$), we conclude that

$$\langle P + \ell m Q \rangle = \begin{cases} \langle R_1 + (1 + \ell m \widetilde{\alpha}_Q)^{-1} \left(\widetilde{\alpha}_P + \ell m \widetilde{\beta}_Q\right) R_2 \rangle & \text{if} \quad b = 0, \\ \langle \left(1 + \ell m \widetilde{\beta}_Q\right)^{-1} (\widetilde{\alpha}_P + \ell m \widetilde{\alpha}_Q) R_1 + R_2 \rangle & \text{if} \quad b = 1. \end{cases}$$

Decompressing in this way costs only a handful of field operations in $\mathbb{F}_{p^2}$ in addition to a 1-dimensional scalar multiplication on E_A. Since the scalar multiplication is already part of the SIDH scheme, this makes the cost of decompression essentially the cost of generating $\{R_1, R_2\}$. This is done exactly as explained in Sect. 3.

7   Implementation Details

To evaluate the performance of the new compression and decompression, we implemented the proposed algorithms in plain C and wrapped them around the SIDH software from [7]. This library supports a supersingular isogeny class defined over $p = 2^{372} \cdot 3^{239} - 1$, which contains curves of order $(2^{372} \cdot 3^{239})^2$. These parameters target 128 bits of post-quantum security.

Table 3 summarizes the results after benchmarking the software with the clang compiler v3.8.0 on a 3.4 GHz Intel Core i7-4770 Haswell processor running Ubuntu 14.04 LTS with TurboBoost turned off. The details in the table include the size of compressed and uncompressed public keys, the performance of Alice's and Bob's key exchange computations using compression, the performance of the proposed compression and decompression routines, and the total costs of SIDH key exchange with and without the use of compression. These results are compared with those from the prior work by Azarderakhsh et al. [2], which uses a supersingular isogeny class defined over $p = 2^{387} \cdot 3^{242} - 1$.

Table 3. Comparison of SIDH key exchange and public key compression implementations targeting the 128-bit post-quantum and 192-bit classical security level. Benchmarks for our implementation were done on a 3.4 GHz Intel Core i7-4770 Haswell processor running Ubuntu 14.04 LTS with TurboBoost disabled. Results for [2], obtained on a 4.0 GHz Intel Core i7-4790K Haswell processor, were scaled from seconds to cycles using the CPU frequency; the use of TurboBoost is not specified in [2]. The performance results, expressed in millions of clock cycles, were rounded to the nearest 10^6 cycles.

Implementation		This work	Prior work [2]
Public key (bytes)	Uncompressed	564	768
	Compressed	328 (Alice)	385
		330 (Bob)	
Cycles (cc $\times 10^6$)	Alice's keygen + shared key	80	NA
	Alice's compression	109	6,081
	Alice's decompression	42	539
	Bob's keygen + shared key	92	NA
	Bob's compression	112	7,747
	Bob's decompression	34	493
	Total (no compression)	192	535
	Total (compression)	469	15,395

As can be seen in Table 3, the new algorithms for compression and decompression are significantly faster than those from [2]: compression is up to 66 times faster, while decompression is up to 15 times faster. Similarly, the full key exchange with compressed public keys can be performed about 30 times faster.

Even though part of these speedups can indeed be attributed to the efficiency of the original SIDH library, this only represents a very small fraction of the performance difference (note that the original key exchange from the SIDH library is only 2.8 times faster than the corresponding result from [2]).

Our experimental results show that the use of compressed public keys introduces a factor-2.4 slowdown to SIDH. However, the use of compact keys (in this case, of 330 bytes) can now be considered practical; e.g., one round of SIDH key exchange is computed in only 150 ms on the targeted platform.

Acknowledgements. We thank Drew Sutherland for his helpful discussions concerning the optimization of the Pohlig-Hellman algorithm, and the anonymous Eurocrypt reviewers for their useful comments.

References

1. Azarderakhsh, R., Fishbein, D., Jao, D.: Efficient implementations of a quantum-resistant key-exchange protocol on embedded systems. Technical report (2014). http://cacr.uwaterloo.ca/techreports/2014/cacr2014-20.pdf

2. Azarderakhsh, R., Jao, D., Kalach, K., Koziel, B., Leonardi, C.: Key compression for isogeny-based cryptosystems. In: Emura, K., Hanaoka, G., Zhang, R. (eds.) Proceedings of the 3rd ACM International Workshop on ASIA Public-Key Cryptography, AsiaPKC@AsiaCCS, Xi'an, China, 30 May–03 June 2016, pp. 1–10. ACM (2016)

3. Barreto, P.S.L.M., Galbraith, S.D., O'Eigeartaigh, C., Scott, M.: Efficient pairing computation on supersingular abelian varieties. Des. Codes Crypt. **42**(3), 239–271 (2007)

4. Barreto, P.S.L.M., Kim, H.Y., Lynn, B., Scott, M.: Efficient algorithms for pairing-based cryptosystems. In: Yung, M. (ed.) CRYPTO 2002. LNCS, vol. 2442, pp. 354–369. Springer, Heidelberg (2002). doi:10.1007/3-540-45708-9_23

5. Bernstein, D.J., Hamburg, M., Krasnova, A., Lange, T.: Elligator: elliptic-curve points indistinguishable from uniform random strings. In: Sadeghi, A.-R., Gligor, V.D., Yung, M. (eds.) 2013 ACM SIGSAC Conference on Computer and Communications Security, CCS 2013, Berlin, Germany, 4–8 November 2013, pp. 967–980. ACM (2013)

6. Chen, L., Jordan, S., Liu, Y.-K., Moody, D., Peralta, R., Perlner, R., Smith-Tone, D.: Report on post-quantum cryptography. NISTIR 8105, DRAFT (2016). http://csrc.nist.gov/publications/drafts/nistir-8105/nistir_8105_draft.pdf

7. Costello, C., Longa, P., Naehrig, M.: Efficient algorithms for supersingular isogeny Diffie-Hellman. In: Robshaw, M., Katz, J. (eds.) CRYPTO 2016. LNCS, vol. 9814, pp. 572–601. Springer, Heidelberg (2016). doi:10.1007/978-3-662-53018-4_21

8. De Feo, L., Jao, D., Plût, J.: Towards quantum-resistant cryptosystems from supersingular elliptic curve isogenies. J. Math. Cryptol. **8**(3), 209–247 (2014)

9. Fujisaki, E., Okamoto, T.: Secure integration of asymmetric and symmetric encryption schemes. In: Wiener, M. (ed.) CRYPTO 1999. LNCS, vol. 1666, pp. 537–554. Springer, Heidelberg (1999). doi:10.1007/3-540-48405-1_34

10. Galbraith, S.D., Harrison, K., Soldera, D.: Implementing the tate pairing. In: Fieker, C., Kohel, D.R. (eds.) ANTS 2002. LNCS, vol. 2369, pp. 324–337. Springer, Heidelberg (2002). doi:10.1007/3-540-45455-1_26

11. Galbraith, S.D., Petit, C., Shani, B., Ti, Y.B.: On the security of supersingular isogeny cryptosystems. In: Cheon, J.H., Takagi, T. (eds.) ASIACRYPT 2016. LNCS, vol. 10031, pp. 63–91. Springer, Heidelberg (2016). doi:10.1007/978-3-662-53887-6_3
12. Hess, F.: Pairing lattices. In: Galbraith, S.D., Paterson, K.G. (eds.) Pairing 2008. LNCS, vol. 5209, pp. 18–38. Springer, Heidelberg (2008). doi:10.1007/978-3-540-85538-5_2
13. Hess, F., Smart, N.P., Vercauteren, F.: The eta pairing revisited. IEEE Trans. Inf. Theory **52**(10), 4595–4602 (2006)
14. Husemöller, D.: Elliptic Curves. Graduate Texts in Mathematics, vol. 111. Springer, Heidelberg (2004)
15. Jao, D., Feo, L.: Towards quantum-resistant cryptosystems from supersingular elliptic curve isogenies. In: Yang, B.-Y. (ed.) PQCrypto 2011. LNCS, vol. 7071, pp. 19–34. Springer, Heidelberg (2011). doi:10.1007/978-3-642-25405-5_2
16. Kaliski Jr., B.S.: The Montgomery inverse and its applications. IEEE Trans. Comput. **44**(8), 1064–1065 (1995)
17. Kirkwood, D., Lackey, B.C., McVey, J., Motley, M., Solinas, J.A., Tuller, D.: Failure is not an option: standardization issues for post-quantum key agreement. In: Talk at NIST Workshop on Cybersecurity in a Post-Quantum World, April 2015. http://www.nist.gov/itl/csd/ct/post-quantum-crypto-workshop-2015.cfm
18. Lenstra, A.K., Verheul, E.R.: The XTR public key system. In: Bellare, M. (ed.) CRYPTO 2000. LNCS, vol. 1880, pp. 1–19. Springer, Heidelberg (2000). doi:10.1007/3-540-44598-6_1
19. McEliece, R.J.: A public-key cryptosystem based on algebraic coding theory. Coding Thv **4244**, 114–116 (1978)
20. Miller, V.S.: The Weil pairing, and its efficient calculation. J. Cryptol. **17**(4), 235–261 (2004)
21. Montgomery, P.L.: Speeding the Pollard and elliptic curve methods of factorization. Math. Comput. **48**(177), 243–264 (1987)
22. Peikert, C.: Lattice cryptography for the internet. In: Mosca, M. (ed.) PQCrypto 2014. LNCS, vol. 8772, pp. 197–219. Springer, Cham (2014). doi:10.1007/978-3-319-11659-4_12
23. Pohlig, S.C., Hellman, M.E.: An improved algorithm for computing logarithms over GF(p) and its cryptographic significance. IEEE Trans. Inf. Theory **24**(1), 106–110 (1978)
24. Regev, O.: On lattices, learning with errors, random linear codes, and cryptography. In: Gabow, H.N., Fagin, R. (eds.) Proceedings of the 37th Annual ACM Symposium on Theory of Computing, Baltimore, MD, USA, 22–24 May 2005, pp. 84–93. ACM (2005)
25. Rubin, K., Silverberg, A.: Torus-based cryptography. In: Boneh, D. (ed.) CRYPTO 2003. LNCS, vol. 2729, pp. 349–365. Springer, Heidelberg (2003). doi:10.1007/978-3-540-45146-4_21
26. Schaefer, E., Stoll, M.: How to do a p-descent on an elliptic curve. Trans. Am. Math. Soc. **356**(3), 1209–1231 (2004)
27. Scott, M.: Implementing cryptographic pairings. In: Takagi, T., Okamoto, E., Okamoto, T., Okamoto, T. (eds.) Pairing 2007. LNCS, vol. 4575, pp. 177–196. Springer, Heidelberg (2007)
28. Shanks, D.: Class number, a theory of factorization, and genera. In: Proceedings of the Symposium in Pure Mathematics, vol. 20, pp. 415–440 (1971)
29. Silverman, J.H.: The Arithmetic of Elliptic Curves. Graduate Texts in Mathematics, 2nd edn. Springer, Heidelberg (2009)

30. Smith, P., Skinner, C.: A public-key cryptosystem and a digital signature system based on the Lucas function analogue to discrete logarithms. In: Pieprzyk, J., Safavi-Naini, R. (eds.) ASIACRYPT 1994. LNCS, vol. 917, pp. 355–364. Springer, Heidelberg (1995). doi:10.1007/BFb0000447

31. Stam, M., Lenstra, A.K.: Efficient subgroup exponentiation in quadratic and sixth degree extensions. In: Kaliski, B.S., Koç, K., Paar, C. (eds.) CHES 2002. LNCS, vol. 2523, pp. 318–332. Springer, Heidelberg (2003). doi:10.1007/3-540-36400-5_24

32. Sutherland, A.V.: Structure computation and discrete logarithms in finite abelian p-groups. Math. Comput. **80**(273), 477–500 (2011)

33. Verheul, E.R.: Evidence that XTR is more secure than supersingular elliptic curve cryptosystems. J. Cryptol. **17**(4), 277–296 (2004)

Author Index

Printed in the United States
By Bookmasters